Evidence Based Validation of Traditional Medicines

Subhash C. Mandal • Raja Chakraborty •
Saikat Sen
Editors

Evidence Based Validation of Traditional Medicines

A Comprehensive Approach

Springer

Editors
Subhash C. Mandal
Pharmaceutical Technology
Jadavpur University
Kolkata, West Bengal, India

Raja Chakraborty
NEF College of Pharmacy
Guwahati, Assam, India

Saikat Sen
Faculty of Pharmaceutical Science
Assam down town University
Guwahati, Assam, India

ISBN 978-981-15-8126-7 ISBN 978-981-15-8127-4 (eBook)
https://doi.org/10.1007/978-981-15-8127-4

This Springer imprint is published by the registered company Springer Nature Singapore Pte Ltd.
The registered company address is: 152 Beach Road, #21-01/04 Gateway East, Singapore 189721, Singapore

Contents

About the Editors

Subhash C. Mandal is currently working as a Professor at Division of Pharmacognosy, Department of Pharmaceutical Technology, Jadavpur University, Kolkata India since 1997. His research achievements have been duly honored by many international organizations for awards. He has guided more than 30 PhD scholars and 25 M Pharm scholars and has completed 12 government-funded research projects on Indian Traditional Medicines and Drug Discovery. Prof. Mandal has more than 300 research publications (Citations: 14446, h-index: 59, i10-index: 322), 08 patents and several authored, edited and book chapters to his credit. He has visited more than 30 countries for various scientific deliberations, collaborations exchange programmes.

Raja Chakraborty is presently designated as Professor and Principal at NEF College of Pharmacy, Guwahati, Assam, India. He received his PhD in Pharmaceutical Sciences from Jawaharlal Nehru Technological University Anantapur, India in 2014. With 12 years of academic and research experience, he has published more than 60 research and review articles in reputed national and international journals. He authored two books Tribal Medicinal Plants of Tripura, India: A scientific Search, Serial Publications, New Delhi and Diabetes Mellitus in 21st Century. Springer Singapore. He acted as editor for two books and nineteen book chapters with different national and international publishers. Recently he received "Dr. APJ Abdul Kalam Award for Teaching Excellence—2018" He is guiding a number of Ph.D students and serving as Co-PI in a DBT-funded mega-project.

Saikat Sen is an educationist and researcher, presently working as a professor at the Faculty of Pharmaceutical Sciences at Assam down town University, Guwahati. He has authored, and edited, various books and contributed chapters in different peer-reviewed edited books published by international/national publishers. More than 60 research and review articles in peer-reviewed reputed national/international journals were added to his credit, presented several scientific papers, attended a number of national and international conferences, and delivered invitee lectures in international forums. He has received several awards during his 12 years of his professional carrier which include Sushruta Best Young Scientist Award, Dr. APJ Abdul Kalam Award for Young Scientists—2018. He is a member of different professional bodies around the world and involve enthusiastically in scientific research on traditional/folk medicinal systems and to develop active research culture.

Part I

Phytomolecules, Drug Discovery and Modern Technique

Global Approach for Drug Discovery and Development from Indian Traditional Medicine

1

Shanti Bhushan Mishra, Alok Mukerjee, and Shradhanjali Singh

Abstract

Traditional medicine is the fusion of therapeutic experiences of generations of general physicians, tribal and rural peoples of indigenous systems of medication. Medicinal plants play a crucial role for the investigation of new entities (biologically active compounds in the current market). Drug discovery approaches based on natural products and traditional medicines are re-emerging as choices for new drugs to facilitate discovery process and also for the development of synergistic herbal formulations. Moreover, the integration of Ayurvedic knowledge and drug discovery brings the need for a revolution in the extraction process from sequential to parallel extraction. Bioassay-guided fractionation is a supportive measure through which standardized extract or isolated bioactive compound is obtained as the new drug. This integrated approach would result in saving of cost, time, and increased success rate in drug discovery. In this chapter we exemplified various approaches of drug discovery and application of Ayurvedic opinion in preference of plant for its therapeutic (e.g., antidiabetic) activity.

Keywords

Biological activity · Diabetes · Drug development · Screening · Traditional medicines

S. B. Mishra (✉) · A. Mukerjee
Department of Pharmacognosy, United Institute of Pharmacy, Allahabad, Uttar Pradesh, India
e-mail: shantipharma15@gmail.com

S. Singh
Department of Pharmaceutical Chemistry, United Institute of Pharmacy, Allahabad, Uttar Pradesh, India

S. C. Mandal et al. (eds.), *Evidence Based Validation of Traditional Medicines*,
https://doi.org/10.1007/978-981-15-8127-4_1

Abbreviations

GMP Good Manufacturing Practice
GCP Good Clinical Practice
GLP Good Laboratory Practice
CMC Chemistry, Manufacturing and Controls

1.1 Introduction

Modern drug discovery and development endeavors typically come from the basic research and then gradually move on to definite sequential activities, which if successful ends in a new drug for the treatment of a human disease. The entire pathway is systematized by well-defined mileposts, which include identification of the lead compound, selection of the drug target, its modification to a compound suitable for toxicity testing in experimental animals, and choosing a drug molecule for clinical evaluation. Even before the beginning of human studies, a drug molecule suitable for clinical testing is assumed to satisfy specific and challenging safety criteria. It should bind selectively to the receptor on the target and prompt for the preferred functional response. There must be satisfactory bioavailability and distribution inside the body to reach the site of action, and this should produce the desired responses in in vivo models. Most importantly, a drug molecule suitable for testing in human being must pass toxicity evaluations to show that humans contributing in the phase 1 clinical trials are showing negligible risks only (Hefti 2008).

Presently traditional medicines are used in primary health-care systems in most countries equivalent to conventional medicine. Therefore, traditional medicine should be subjected to research for their efficacy and safety for greater health care. At present there is a requirement for evidence-based drug development with fluctuation of global economic scene. When developing novel drugs using traditional medicines, it is essential to consider novel standard parameters whenever possible (Zhang 2015). Quality control of traditional medicines is also prerequisite of standard clinical trials. It is essential to follow current standard quality controlling methods, viz., Good Manufacturing Practice (GMP); Chemistry, Manufacturing and Controls (CMC); Good Clinical Practice (GCP); and Good Laboratory Practice (GLP).

There are numerous examples of emergence of new drugs from the plant sources. Morphine was isolated from opium produced from latex of the poppy plant (*Papaver somniferum*) about 200 years ago. A number of drugs developed from natural/plant sources have certainly revolutionized medicine, like antibiotics (e.g., erythromycin, penicillin, tetracycline), anti-parasitics (e.g., avermectin), anti-malarials (e.g., quinine, artemisinin), hypolipidemics (e.g., lovastatin and analogs), immunosuppressants for organ transplants (e.g., rapamycins, cyclosporine), anticancer drugs (e.g., irinotecan, paclitaxel), and antidiabetic drugs (e.g., metformin) (Alamgir 2017).

There are about more than 100 plant-derived drugs and molecules/compounds that are in preclinical stage on which clinical trials are ongoing (Harvey 2008); undoubtedly, there are numerous species of plants in plant kingdom that contains the substance of medicinal value which will be discovered in the future; a lot of plants are continuously being screened for their possible pharmacological value (particularly for their hypotensive, anti-inflammatory, hypoglycemic, anti-fertility, antibiotic, anti-Parkinsonism, amebicidal, and cytotoxic properties) (Pan et al. 2013). The use of sole genuine compounds with synthetic drugs is also having lots of restrictions, and in the current years, there has been an immense resurgence of interest in the Ayurvedic and homeopathic systems of medicine, both of which rely profoundly on plant source (Kumar et al. 2017).

1.2 Drug Development from Natural Resources: Benefits and Drawbacks

Use of plant sources as preliminary point of the drug development program is related with few specific advantages:

- Typically, the assortment of a plant candidate species for research can be done on the basis of long-term use of folklore medicines by humans. This methodology is based on the finding that active compounds isolated from such plants are likely to be safer than those obtained from plant species without a history of human use. Subsequently, the synthesis of lead molecules could be reducing the pressure on natural resources. Drug development from *Cinchona officinalis*, *Rauwolfia serpentina*, *Digitalis purpurea*, etc. in the past fall under this category (Atanasov et al. 2015).
- The lead molecules isolated from natural source by using such methods can be of use with some limitations like low bioavailability, low toxicity, etc. Such type of limitations can be overcome through modification in the molecule like nanonization and by formulating their semisynthetic derivatives. For example, the bark of the willow tree (genus *Salix*) has been known from ancient times to have analgesic properties which is due to the presence of the natural product salicin and is hydrolyzed into salicylic acid (Jamshidi-Kia et al. 2018). A synthetic derivative acetylsalicylic acid (aspirin) is a widely used pain reliever. There are numerous examples of phyto-constituents which are obtained from natural sources and modified chemically, viz., morphine (*Papaver somniferum*), colchicine (*Colchicum autumnale*), penicillin G (*Penicillium citrinum*), paclitaxel (*Taxus brevifolia*), metformin (*Galega officinalis*), etc.

Although there is incredible growth in traditional system of health care globally, ITSM based on its different features of folklore medicines have also developed greatly. But there are several constrains in this development in a proper way which include:

- Rules and regulations imposed for traditional medicines are just similar to chemical-based drugs.
- Availability of raw material means dramatic depletion of wild populations of the plants; for example, the plants *Panax ginseng*, *Artemisia annua*, and *Taxus brevifolia* are now endangered due to the overexploitation.
- Once isolated from their source, compounds may work differently than expected. Moreover, the approach could be more time-consuming and more costly and may be less sustainable.

Already 29 plants and their worthy products have been banned by the government of India as they are considered as endangered species (Sen and Chakraborty 2016).

1.3 Colligative Properties of an Isolated Phyto-constituent

Not all lead molecules generated by the drug discovery persons are tested in complete regulatory packages. This is because the regulatory testing is very time-consuming and costly affair. A series of tests are initially conducted to help select certain candidate molecules within the desired pharmacological possibilities and safety profile for further regulatory testing. Drugs that do not meet the necessary requirements in these initial assays are less likely to be taken for testing in more expensive, time-consuming regulatory tests (Koehn and Carter 2005).

Koehn and Carter have figured out the following some elite characteristics of the compounds isolated from natural sources. They are as follows.

1.3.1 Molecular Structure

The automatic screening technologies and wide range of chemical libraries and archives have made it reasonably easy to identify initial lead candidates for new drug targets. The chemists have developed specific rules that lead molecules must fulfill.

- Molecular weight should be less than 500.
- Not more than five hydrogen bond donors.
- Not more than ten hydrogen bond acceptors.

1.3.2 Octanol/Water Partition Coefficient

The lipid solubility of drugs is stated as octanol/water coefficients of the uncharged molecules or log P. When the log P value is higher, drugs will be highly lipid soluble and believably accumulated in the body (Bergström and Larsson 2018).

1.3.3 Structure-Activity Relationship (SAR)

SAR gives information about the possible toxicity of a chemical based on chemical structure, when no experimental data is available. It can make predictions about a wide variety of toxicological properties of compounds such as neurotoxicity, carcinogenicity, skin sensitization, thyroid toxicity, teratogenicity, respiratory, and mutagenicity.

1.3.4 Cytotoxicity

An essential part of the drug discovery/approval process is determining the toxic effects of potential drugs. Following a toxic attack, cells may react with changes in size or morphology depending on the type of cell and compound. Some toxins can affect the cell's functionality by changing the physiology of organs such as lysosomes and endosomes or by causing a rise in number of lysosomes seen in the case of phospholipidosis. There are several types of diagnostic kits that are available in the market that can be used to measure these types of parameters.

1.3.5 Parallel Artificial Membrane Permeability Assay (PAMPA)

The potential of a molecule is orally absorbed as one of the most important aspects in deciding whether a molecule is a probable lead candidate. Parallel artificial membrane permeability assay is a decent alternative to cellular models for the initial absorption, distribution, metabolism, and excretion (ADME) and primary investigations of the research compounds. This method is used to measure the effective permeability, P (e), as a function of pH from 4 to 10. This technique provides quick response, low-cost, and automation-friendly method to measure a chemical entity's passive permeability (Kansy et al. 2004).

1.3.6 Derived Solubility

The water solubility of a drug is a crucial physical property that affects both its ADME profile and screenability in high-output systems.

1.3.7 Aqueous/Plasma Stability

The stability of lead compounds in plasma is an important parameter that can strongly affect the in vivo efficacy of a test compound. Drugs that are exposed to enzymatic processes (proteinase, esterase) in plasma may undergo intra-molecular rearrangement or bind covalently to the proteins. Thus the determination of plasma stability should be performed early in drug discovery phase. Measurement of plasma stability is performed at physiological pH level in plasma (Chung et al. 2015).

1.3.8 Protein Binding

In-depth understanding of plasma and tissue (brain, liver, etc.) protein binding is important for evaluating the distribution of drug molecules. The plasma or tissue homogenate is incubated with the test compound. The bound and unbound test compounds are separated using ultrafiltration or equilibrium dialysis, and the amount of test compound in both fractions is estimated using HPLC or LC/MS.

These unique characteristics of lead molecules of natural origin pose order of challenges for medicinal chemists as they start working upon development of analogs (Chung et al. 2015).

1.4 Benchmarks for Selecting the Plants for Research

It has been estimated that <10% of the approximately 300,000–500,000 species of plants worldwide have been studied for 1 or more bioactivities.

Success in identifying a new biologically active plant-based natural product can be influenced firstly by a clever choice of plant or secondly by how randomly the selection of plant extracts can be quickly and effectively screened. The following selection criteria are suggested for plant-related research (Dias et al. 2012). The sketch of possible approaches to the discovery of new drug leads has been mentioned in Fig. 1.1.

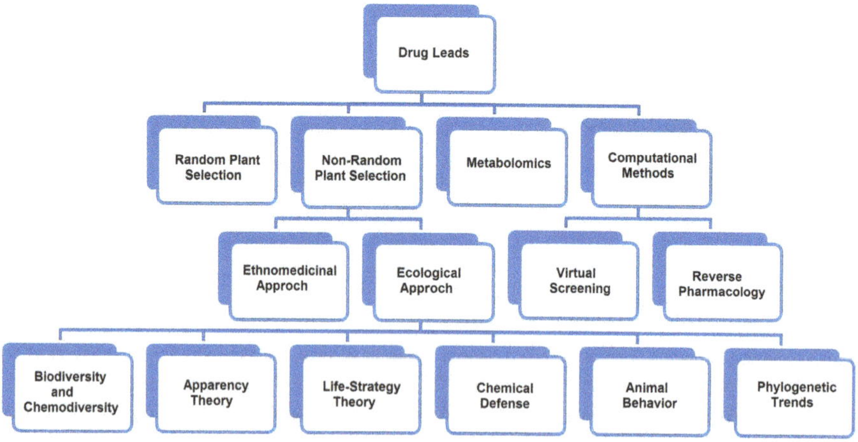

Fig. 1.1 Sketch of possible approaches to the discovery of new drug leads

1.4.1 Random Selection

In the random screening method, plant extracts, fractions, or isolated compounds are randomly selected on their convenience and availability. In the perception of plant-based drug discovery, this method can be highly beneficial when applied with samples originating from regions with high biodiversity and endemism. The random selection of test sample has the effectiveness in the identification of unpredicted biological activities that could not have been expected based on the existing information (Atanasov et al. 2015). Paclitaxel and camptothecin are the bio-actives which were isolated through this approach.

1.4.2 Selection Based on Traditional Use (Ethno-Medicinal Approach)

This is the widespread basis for selecting plants for investigation especially in societies and rural and ethnic communities where traditional medicine is whole sole part of human health care. If a traditional healer claims success in the treatment of a disease, the researcher can assume from the above selection criteria a chemical constituent with suitable pharmacological activity in the plant extract. The ethno-medicinal approach allows for better chance of finding an active compound as well as documenting and preserving local knowledge. This becomes of greater importance with the increased mobility among rural communities and the subsequent loss of local information of the use of native plant species (Lewis 2003). Regarding the ethno-medicinal approach for the selection of the plant, two important issues require attention. Firstly, the rights of the country of origin related to any drugs discovered need to be protected, as mentioned in the United Nations Convention on Biological Diversity (UNCBD) (Baker et al. 1995). Secondly, the prominence of any ethno-pharmacological field studies should be carried out before the plant selection which indicates an impact on the success of the research.

The ethno-medicinal approach has successfully been used by the researchers at Shaman Pharmaceuticals to verify the use of *Cryptolepis sanguinolenta* (Lindl.) as a treatment for type II diabetes as well as a source for the isolation of the active constituent, the alkaloid cryptolepine (Bierer et al. 1998; Luo et al. 1998). The dichloromethane and hot water extracts of the roots of *C. sanguinolenta* showed the ability to reduce the blood glucose in animal model. In vivo bioassay-guided fractionation using the same model results in the isolation of cryptolepine as an active constituent (Bierer et al. 1998; Luo et al. 1998).

1.4.3 Ecological Approach

In the ecological approach, the selection of plant candidate is dependent upon the observation of interactions between organisms and their surroundings from which bioactive natural compounds can be produced. The basic fundamental hypothesis of this approach is that secondary metabolites which possess ecological functions also exerts pharmacological activity. Different investigators have considered different phases of the ecological argument, including the relationship between biodiversity and chemodiversity (Ramesha et al. 2011), the apparency theory (de Almeida et al. 2011), the life-strategy theory (Coley et al. 2003), chemical defenses and herbivory (Albuquerque et al. 2012), animal behavior (Obbo et al. 2013), and phylogenetic trends (Zhu et al. 2012).

Another approach which is simultaneously linked with ecological approach is the zoopharmacognosy approach. In this approach, the activity of plant is sometimes evaluated through observation of animal behavior. *Khaya* species are common to Madagascar and Africa, and people use their bitter bark and seeds for treating fevers, microbial infections, and worm infestations. Baboons and chimpanzees in Western Uganda have been observed to eat the bark and seeds that are bitter in taste and have no nutritional value (Obbo et al. 2013). The petroleum extract of *Khaya anthotheca* evidenced for good activity against *Plasmodium falciparum* K1 ($IC_{50} = 0.955$ μg/mL) and *Trypanosoma brucei rhodesiense* STIB 900 ($IC_{50} = 5.72$ μg/mL). It appears that chimpanzees and baboons were using seeds and bark for self-medicating, in addition to evidence of the effectiveness of these plants used by traditional healers (Obbo et al. 2013).

1.4.4 Computational Approach: Virtual Screening and Reverse Pharmacognosy

Computational methods are supplementary knowledge-based approach that assists to select plant material with a high probability for pharmacological activity. These methods can also aid with the validation of biological activity of natural compounds and selection of test samples dependent on in silico bioactivity predictions for constituents of plant species. Virtual screening (VS) uses the availability of large compound libraries generated by combinatorial and high-throughput chemistry to select low number of potential candidates for experimental testing (López-Vallejo et al. 2011). Virtual screening can follow two general strategies: ligand-based virtual screening and structure-based virtual screening. In this method, molecular docking is broadly used to explain the mechanism of action and defend the SAR of natural products. The purpose of docking is to accurately predict the position of a ligand within the protein binding site and the ability of binding with a docking score (López-Vallejo et al. 2011).

Reverse pharmacognosy intends to find out new biological targets for natural products by either virtual screening or real screening and then to connect these findings to original or different plant sources (Do et al. 2005).

1.5 Biological Activity-Guided Fractionation for Compound Isolation

Isolation strategies for natural products are constantly evolving. Originally, all compounds that could be purified were isolated from a plant that was used traditionally to treat diseases without concern if the specific compound was responsible for activity. This ensures that the isolation of several inactive compounds and offered a way to bioassay-directed isolation, leading the disease to the responsible compound (Altemimi et al. 2017). Bioassay-guided isolations are currently the most common technique to purify the responsible compound for a certain bioactivity. Bioassay-guided fractionation starts with a crude extract from the dried plant material using either aqueous ethanol or aqueous methanol. The crude extract is then taken through a liquid-liquid extraction using solvents of increasing polarity, from hexanes to water, to produce five fractions (1–5) as shown in Fig. 1.2. Bioassay-guided fractionation of plant extracts can be achieved through chromatographic separation techniques which can lead to isolation of biological active molecules. If one of the fractions is bioactive, that fraction is further purified with either gravity column chromatography or flash column chromatography, depending on the complexity of the crude sample. The column fractions are then again screened with the bioassay to confirm the activity. This process is continued until a final pure compound is isolated responsible for the bioactivity. Final purification often requires high-performance liquid chromatography (HPLC) in order to obtain clean nuclear magnetic resonance (NMR) spectra and high-resolution mass spectrometric (HRMS) data for compound characterization. Even a small amount of an impurity can lead to mis-assignment of peaks in the spectrum. Advances in isolation and structural elucidation technologies provide a more comprehensive image of the entire plant extract.

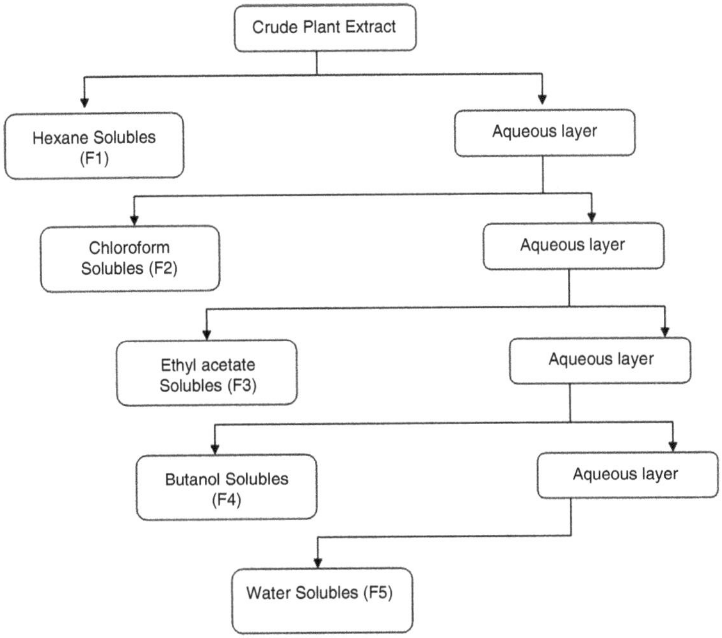

Fig. 1.2 Flow diagram of preliminary liquid-liquid extraction based on polarity

1.6 Ayurvedic Perception in Selection of Plant Candidate for its Therapeutic (e.g., Antidiabetic) Activity

On the basis of the above-said approaches of screening of medicinal plants, it is possible to apply the traditional knowledge on a variety of herbs to identify the improved lead compounds or phyto-chemicals for research and development to find out better treatment of diabetes. In Ayurvedic literature, prameha is characterized with undue urination (in both quantity and frequency) and turbidity. The nature of the turbidity may differ depending upon the body reaction with the tridoshas. Understanding of prameha is not merely related only to the patho-physiology and clinical picture of diabetes mellitus as depicted in Fig. 1.3. From the pathological and etiological state of complications, prameha is almost common in obesity and metabolic syndrome (Sharma and Chandola 2011).

The herbs from the authentic classical text *Charak Samhita*, *Sushruta Samhita*, and *Astanga Hrdaya* (Jadavaji 1992; Sharma 2001; Srikantha Murty 2000) are traditionally employed in the treatment of diabetes mentioned in Table 1.1 and scientifically explored by various investigators to evaluate the same in terms of *Rasa*, *Veerya*, *Vipaka*, *Guna*, and *Karma*. On the basis of the traditional elements for the herbs having *pramehahara/tridoshhara* effects, it can be assumed that these herbs have particular pharmacological traits in general. Going by the dominance investigation of these attributes, the following scenario appears.

Rasa: Kashaya, Tikta, and Katu
Guna: Laghu and Ruksha
Vipaka: Katu

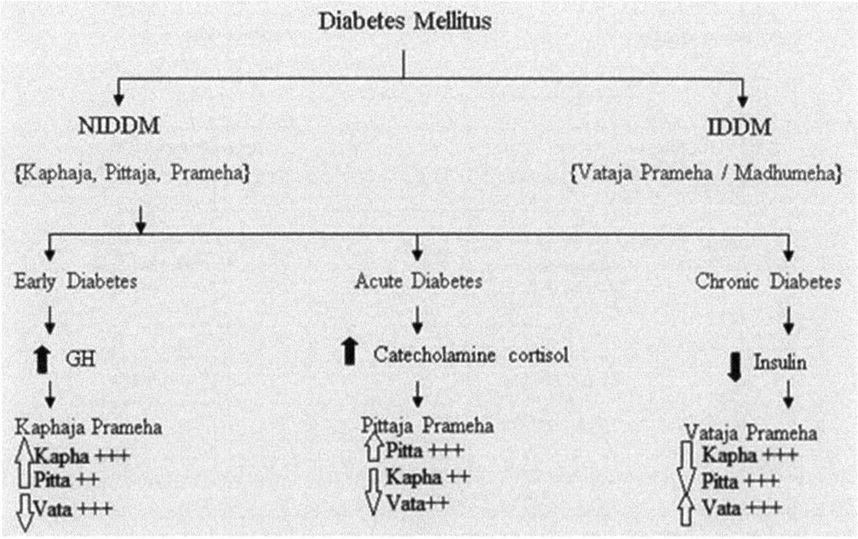

Fig. 1.3 Correlation of different types/stages of prameha with diabetes mellitus

Table 1.1 Pramehahara and madhumehahara (antidiabetic) drugs in Ayurveda

S. no.	Ayurvedic name	Botanical name	6.1.1.1.1.1.1. *Charak Samhita*	6.1.1.1.1.1.2. *Sushruta Samhita*	6.1.1.1.1.1.3. *Astanga Hrdaya*
1	Daruharidra	*Berberis aristata*	+	+	+
2	Devadaru	*Cedrus deodara*	+	+	+
3	Haritaki	*Terminalia chebula*	+	+	+
4	Vibhitaki	*Terminalia bellirica*	+	+	+
5	Amalaki	*Emblica officinalis*	+	+	+
6	Musta	*Cyperus rotundus*	+	+	+
7	Haridra	*Curcuma longa*	+	+	+
8	Katphala	*Myrica esculenta*	+	+	+
9	Lodhra	*Symplocos racemosa*	+	+	+
10	Patha	*Cyclea peltata*	+	+	+
11	Vidanga	*Embelia ribes*	+	+	+
12	Arjuna	*Terminalia arjuna*	+	+	+
13	Dhanvana	*Grewia tiliaefolia*	+	+	−
14	Tagara	*Valeriana wallichii*	+	+	+
15	Kadamba	*Anthocephalus indicus*	+	−	−
16	Shalasara	*Shorea robusta*	+	+	+
17	Yavani	*Trachyspermum ammi*	+	+	+
18	Khadira	*Acacia catechu*	+	+	+
19	Dhava	*Anogeissus latifolia*	+	+	+
20	Kustha	*Saussurea lappa*	+	+	+
21	Aguru	*Aquilaria agallocha*	+	+	−
22	Chandana	*Santalum album*	+	+	+
23	Agnimantha	*Premna integrifolia*	+	+	+
24	Murva	*Marsdenia tenacissima*	+	+	+
25	Gokshura	*Tribulus terrestris*	+	−	−

(continued)

Table 1.1 (continued)

S. no.	Ayurvedic name	Botanical name	6.1.1.1.1.1.1. *Charak Samhita*	6.1.1.1.1.1.2. *Sushruta Samhita*	6.1.1.1.1.1.3. *Astanga Hrdaya*
26	Ushira	*Vetiveria zizanioides*	+	+	+
27	Guduchi	*Tinospora cordifolia*	+	+	+
28	Chavya	*Piper retrofractum*	+	−	−
29	Chitraka	*Plumbago zeylanica*	+	+	+
30	Saptaparna	*Alstonia scholaris*	+	+	+
31	Patola	*Trichosanthes dioica*	+	+	+
32	Nimba	*Azadirachta indica*	+	+	+
33	Padmaka	*Prunus cerasoides*	+	+	−
34	Kutaja	*Holarrhena antidysenterica*	+	+	+
35	Dhataki	*Woodfordia fruticosa*	+	+	+
36	Utpala	*Nymphaea stellata*	+	+	+
37	Shirisha	*Albizia lebbeck*	+	+	+
38	Sarja	*Vateria indica*	+	+	+
39	Nagakesara	*Mesua ferrea*	+	+	+
40	Priyangu	*Callicarpa macrophylla*	+	+	+
41	Palasa	*Butea monosperma*	+	+	+
42	Aswatha	*Ficus religiosa*	+	+	+
43	Asana	*Pterocarpus marsupium*	+	+	+
44	Vetasa	*Salix caprea*	+	+	+
45	Kampillaka	*Mallotus philippinensis*	+	+	−
46	Rohitaka	*Tecoma undulata*	+	+	−
47	Kapitha	*Feronia elephantum*	+	−	−
48	Asmantaka	*Ficus rumphii*	+	−	−
49	Soma	*Ephedra gerardiana*	+	+	+

50	Ativisha	*Aconitum heterophyllum*	+	+	+
51	Vacha	*Acorus calamus*	+	−	+
52	Manjistha	*Rubia cordifolia*	+	+	+
53	Sati	*Hedychium spicatum*	+	+	+
54	Pushkaramula	*Inula racemosa*	+	+	+
55	Kramuka	*Areca catechu*	+	+	+
56	Kiratatikta	*Swertia chirayita*	+	+	+
57	Katurohini	*Picrorhiza kurroa*	+	+	+
58	Bharangi	*Clerodendrum serratum*	+	+	+
59	Pippali	*Piper longum*	+	+	+
60	Indravaruni	*Citrullus colocynthis*	+	+	+
61	Vyaghranakha	*Capparis horrida*	+	−	−
62	Tejapatra	*Cinnamomum tamala*	+	+	+
63	Maricha	*Piper nigrum*	+	+	+
64	Danti	*Baliospermum montanum*	+	+	+
65	Bhallataka	*Semecarpus anacardium*	+	+	+
66	Aragvadha	*Cassia fistula*	−	+	+
67	Madanaphala	*Randia spinosa*	−	+	+
68	Vikankata	*Flacourtia ramontchi*	−	+	+
69	Patala	*Stereospermum suaveolens*	−	+	+
70	Kuruntaka	*Barleria prionitis*	−	+	+
71	Gunja	*Abrus precatorius*	−	+	+
72	Kakajangha	*Peristrophe bicalyculata*	−	+	−
73	Karanja	*Pongamia pinnata*	−	+	+
74	Chirabilva	*Holoptelea integrifolia*	−	+	+
75	Karavellaka	*Momordica charantia*	−	+	+
76	Kadara	*Acacia suma*	−	+	+

(continued)

Table 1.1 (continued)

S. no.	Ayurvedic name	Botanical name	6.1.1.1.1.1.1. *Charak Samhita*	6.1.1.1.1.1.2. *Sushruta Samhita*	6.1.1.1.1.1.3. *Astanga Hrdaya*
77	Bhurja	*Betula utilis*	−	+	+
78	Shyonaka	*Oroxylum indicum*	−	+	+
79	Meshashringi	*Gymnema sylvestre*	−	+	+
80	Tinisa	*Ougeinia oojeinensis*	−	+	+
81	Raktachandana	*Pterocarpus santalinus*	−	+	+
82	Shinshapa	*Dalbergia sissoo*	−	+	+
83	Talamuli	*Curculigo orchioides*	−	+	+
84	Shaka	*Tectona grandis*	−	+	+
85	Aswakarna	*Dipterocarpus turbinatus*	−	−	+
86	Mushkaka	*Schrebera swietenioides*	−	+	+
87	Snuhi	*Euphorbia neriifolia*	−	+	+
88	Sunthi	*Zingiber officinale*	−	+	+
89	Paribhadra	*Erythrina variegata*	−	+	+
90	Sheivalam	*Ceratophyllum demersum*	−	+	+

Veerya: Ushna and Sheet
Dosha Karma: Kapha-Pitta-Vata Shamana (Tridoshahara)

1.7 Effects of Medicinal Plant Extract on Type II Diabetes Mellitus

Botanical agents show promise for the development of new compounds to treat type II diabetes mellitus. Till now, over 400 traditional and folklore plant treatments for diabetes have been reported, although only few of these have established a scientific and medical evaluation to assess their effectiveness. The anti-hyperglycemic effect of a number of plant extracts has been confirmed in individuals and animal models of type II diabetes (Modak et al. 2007). The WHO Expert Committee on Diabetes Mellitus has also recommended that traditional medicinal herbs should be further investigated. Several phyto-constituents including glycosides, flavonoids, alkaloids, saponins, glycolipids, dietary fibers, peptidoglycans, polysaccharides, carbohydrates, amino acids, and others obtained from various plant sources have been reported as antidiabetic agents with different mechanisms of action which are mentioned in Table 1.2, and scientifically explored plants as antidiabetics are mentioned in Table 1.3 (Mishra et al. 2010; Alam et al. 2019).

Table 1.2 Mechanism of action of phyto-chemicals of different chemical categories involved in diabetic pathway

S. no.	Constituents	Mode of activity
1	Alkaloids	Inhibit alpha-glucosidase and reduce glucose transport through the intestinal epithelium
2	Imidazoline compounds	Stimulates insulin secretion in a glucose-dependent manner
3	Polysaccharides	Increased the levels of insulin, decrease the blood glucose levels, and improve tolerance of glucose
4	Flavonoids	Suppressed the glucose level, decrease plasma triglycerides and cholesterol significantly, and increased their hepatic glucokinase activity possibly by improving the insulin release from pancreatic islets
5	Dietary fibers	Effectively adsorbed glucose, delay glucose diffusion, and reduce the alpha-amylase activity and possibly responsible for decreasing the rate of glucose absorption and concentration of postprandial serum glucose
6	Saponin, (triterpenoid +steroidal glycosides)	Stimulates the secretion of insulin and blocks the formation of glucose in the bloodstream
7	Ferulic acid	Stimulatory effects on insulin secretion (secretagogue)

Reproduced with permission from Mishra et al. 2010

Table 1.3 Scientifically explored plants investigated for their antidiabetic activity (Mishra et al. 2010; Alam et al. 2019)

S. no.	Botanical name/ family	Common name	Parts used	Chemical constituents	Antidiabetic mechanism in relation to chemical constituents
1	*Abies pindrow* (Pinaceae)	Silver fir	Entire plant	D-Pinitol, myrcene, limonene	Exert an insulin-like effect to improve glycemic control
2	*Abroma augusta* (Sterculiaceae)	Devil's cotton, Ulatkambal	Roots and leaves	Abromin, friedelin, abromasterol, taroxerylacetate, taraxeral	Reduces the absorption of glucose, thus assisting in glucose tolerance
3	*Acacia arabica* (Leguminosae)	Babool	Seed	Epicatechin, strictinin, Arabin	Inhibited the α-amylase and α-glucosidase enzymes
4	*Achyranthes aspera* (Amaranthaceae)	Apamarga	Entire plant	Oleanolic acid, betaine, triterpenoid saponins	Provide some necessary nutrients like zinc, calcium, manganese, magnesium, and copper to the β-cells and inhibit oxidative stress
5	*Agrimonia eupatoria* (Rosaceae)	Church steeples	Leaves	Essential oils, quercetin, luteolin, and tannins	Inhibited the formation of advanced glycation end products
6	*Allium sativum* (Liliaceae)	Garlic	Roots, leaves	Allin, allicin, essential oils, saponin steroids	Inhibited aldose reductase and alpha-glucosidase
7	*Allium cepa* (Liliaceae)	Onion	Bulb, leaves	Flavonoids and organo-sulfur compounds and allyl propyl disulfide	Decrease plasma glucose levels in alloxan-induced diabetic rats; alter enzyme activity of hexokinases and glucose-6-phosphate
8	*Aloe barbadensis* (Liliaceae)	Aloe	Leaves	Barbaloin, isobarbaloin, resin, pseudo-prototinosaponin, and prototinosaponins	Stimulate synthesis and release of insulin from pancreatic β-cells in vivo
9	*Anacardium occidentale* (Anacardiaceae)	Cashew, Kaju	Entire plant	Terpenoid, flavonols, coumarin, phenolic compound, essential oil	Enhance glucose metabolism and inhibited alpha-glucosidase
10	*Andrographis paniculata* (Acanthaceae)	Kalmegh	Entire plant	Diterpenoid lactone andrographolide	Delayed absorption of glucose

11	*Annona squamosa* (Annonaceae)	Sugar apple	Leaves	Isosquamosin, acetogenins-squamosin B, reticulatain-2, squamosamide	Stimulate and enhance glucose uptake
12	*Artemisia pallens* (Compositae)	Davana	Aerial parts	Davanone, artabsin, umbelliferone	Increase consumption of glucose via glucose transporter type 4
13	*Azadirachta indica* (Meliaceae)	Neem	Leaves	Nimbidin, nimbin, nimbidol, nimbosterol	Inhibited intestinal glucosidases
14	*Beta vulgaris* (Chenopodiaceae)	Chukandar	Leaves	Betaines, indicaxanthin, anthoxanthin	Reduce the level of skeletal hexokinases
15	*Bidens pilosa* (Compositae)	Spanish needle	Aerial parts	Polyyne, (2-β-D- glucopyranosyloxy-1-hydroxytrideca-5,7,9,11-tetrayne)	Protects β-cells of the pancreas and rejuvenates them
16	*Bixa orellana* (Bixaceae)	Annatto plant	Entire plant	Geranylgeranyl octadecanoate	Improving glucose uptake by adipose tissue and muscle
17	*Boerhaavia diffusa* (Nyctaginaceae)	Punarnava	Leaves and entire plant	Punarnavine and punarnavoside	Retard the intestinal absorption of glucose
18	*Brassica juncea* (Cruciferae)	Rai	Leaves and seed	Isothiocyanate glycosides sinigrin, protein, and fixed oil	Reduces plasma glucose and postprandial hyperglycemia
19	*Caesalpinia bonducella* (Leguminosae)	Kathkaranj	Seed kernels	Bonducin, caesalpin, voucapane diterpenoids, α-, β-, and γ-caesalpins	Insulin secretagogue property
20	*Camellia sinensis* (Theaceae)	Green tea	Leaves	Polyphenolic constituents (EGCG)	Enhance insulin secretion in vivo and affect gene expression
21	*Capparis decidua* (Capparidaceae)	Kair	Powder	Cappariloside A, stachydrin, capparines A and B, apigenin, kaempferol	Carbohydrate absorption and exert its postprandial hypoglycemic effect
22	*Capsicum frutescens* (Solanaceae)	Red chili	Entire plant	Capsaicin, capsicidin	Inhibited the alpha-glucosidase and alpha-amylase in vitro
23	*Carum carvi* (Umbelliferae)	Zeera, caraway	Fruits	Limonene, carvacrol, carvone, α-pinene, linalool, p-cymene	Lowers serum glucose level and inhibited the aldose reductase A

(continued)

Table 1.3 (continued)

S. no.	Botanical name/ family	Common name	Parts used	Chemical constituents	Antidiabetic mechanism in relation to chemical constituents
24	*Cassia auriculata* (Caesalpiniaceae)	Senna	Flower, roots	Goratensidine, flavonoids, and glycosides Sennosides, sennidin, gluco-aloe-emodin, auriculacacidin	Enhances the activity of hepatic hexokinase and phosphofructokinase and suppresses fructose 1,6-bisphosphatase and glucose 6-phosphatase
25	*Catharanthus roseus* (Apocynaceae)	Sadabahar	Leaves, twig, and flower	Vincristine, vinblastine, serpentine, ajmalicine, tetrahydroalstonine, yohimbine	Increase glucose metabolism and inhibited the protein tyrosine phosphatase 1B
26	*Cinnamomum zeylanicum* (Lauraceae)	Dalchini	Bark	Essential oils, flavonoids, coumarins, phlobatannins, glycosides, terpenoids, and anthraquinones	Enhance the glycogen synthase activity and activate the insulin receptor by diverse mechanisms
27	*Clausena anisata* (Rutaceae)	Maggot killer plant	Roots	Scopoletin, pimpenellin, methyl chavicol, myrcene, umbelliferone, xanthotoxin, bergapten; clausanitine	Stimulate the pancreatic beta-cells and subsequent secretion of insulin
28	*Coriandrum sativum* (Umbelliferae)	Dhania	Seed, leaves, fruits	Coriandrol, D-mannitol, β-sitosterol, flavonoid glycoside, coumarins, phthalides	Enhance glucose uptake and glucose oxidation in vivo and function as secretagogue in vitro
29	*Coscinium fenestratum* (Menispermaceae)	Yellow wine	Stem	Berberine, glycoside, and saponin	Inhibited the intestinal glucose
30	*Cryptolepis sanguinolenta* (Asclepiadaceae)	Senegal	Entire plant	Neocryptolepine, cryptolepine, quindoline	Enhance glucose uptake by 3T3-L1 cells
31	*Eclipta alba* (Compositae)	False daisy	Leaves	Ecliptin alkaloid, wedelolactone Resins	Decrease activity of fructose 1,6-bisphosphatase and glucose 6-phosphatase
32	*Enicostemma littorale* (Gentianaceae)	Chhota chirayata	Entire plant	Enicoflavin, gentiocrucine, betulin, apigenin, genkwanin, isovitexin, swertisin, saponarin	Increase antioxidant enzymes and reduce the blood glucose level

33	*Eugenia jambolana* (Myrtaceae)	Jamun	Seed, fruits leaves, kernel	Anthocyanins, raffinose, gallic acid, and cyanidin diglycoside	Decrease plasma glucose level in vivo by changing glucose metabolism
34	*Eucalyptus globulus* (Myrtaceae)	Eucalyptus	Leaves	Essential oil and cineole	Increase insulin secretion from clonal pancreatic beta line BRIN-BD 11
35	*Euphrasia officinale* (Scrophulariaceae)	Eyebright	Leaves	Iridoids, flavonoids, phenolic acids, and etheric oils	Insulin mimetic property
36	*Ficus religiosa* (Moraceae)	Pipal	Entire plant	Bergapten, bergaptol, lanosterol, β-sitosterol, stigmasterol	Modulated the enzymes of antioxidant defense system to combat oxidative stress
37	*Ficus benghalensis* (Moraceae)	Bargad	Bark	Tannin, leucocynidin, 3-*O*-beta-D-galactosyl cellobioside, leucopelargonidin	Acts as insulin secretagogue
38	*Ficus carica* (Moraceae)	Fig, Anjeer	Leaves	Quercetin-3-*O*-glucoside, ferulic acid, triterpenoids, and sesquiterpenes	Causes glucose-lowering effect in vivo by stimulating insulin secretion
39	*Gymnema sylvestre* (Asclepiadaceae)	Gurmar	Leaves	Gymnemic acid, gymnemosides, and quercetin	Stimulate secretion of insulin from existing β-cells of islets and significant enhancement in the level of insulin
40	*Gentiana olivieri* (Gentianaceae)	Asbarg	Flowers, roots	β-Myrcene, isoorientin, α-pinene, limonene, and C-glycoside	Reduces blood glucose in vivo
41	*Glycyrrhiza glabra* (Leguminosae)	Mulethi	Root/ rhizome	Triterpenoid, saponin, and glycyrrhizin	Lowers blood glucose in vivo by inducing hepatic enzymes
42	*Gynura procumbens* (Compositae)	Sambung Nyawa	Leaves	Rutin, kaempferol, quercetin, and astragalin	Promoting glucose uptake by muscles
43	*Hibiscus rosa-sinensis* (Malvaceae)	Gudhal, China rose	Entire plant	Rutin, quercetin, kaempferol, myricetin	Regulating the activities of glycogen-metabolizing enzymes

(continued)

Table 1.3 (continued)

S. no.	Botanical name/ family	Common name	Parts used	Chemical constituents	Antidiabetic mechanism in relation to chemical constituents
44	*Helicteres isora* (Sterculiaceae)	Indian screw tree	Root	Fibers, phytosterols, carotenoids, antioxidants, proteins, saponin, tannin, and lignins	Initiate insulin release and reduce plasma triglycerides
45	*Hordeum vulgare* (Gramineae)	Barley	Barley seed	Beta-glucan, vitamins, and proteins	The dietary supplement to control diabetes
46	*Hovenia dulcis* (Rhamnaceae)	Japanese raisin tree	Entire plant	Tocopherol, flavonoids, ascorbic acid, anthocyanins	Stimulate hepatic enzymes to lower plasma glucose levels
47	*Ipomoea batatas* (Convolvulaceae)	Sweet potato	Tubers	Batatinoside I, citrusin, caffeic acid, β-carotene, manganese, and vitamins	Decrease insulin resistance in vivo
48	*Juniperus* communis (Cupressaceae)	Common juniper	Fruits	Apigenin, essential oils, sitosterol, cupressuflavone	Increase peripheral glucose consumption and induce insulin secretion
49	*Luffa aegyptiaca* (Cucurbitaceae)	Sponge gourd	Seed, fruits	Elaterin 2-*O*-β-D-glucopyranoside, cucurbitacin S, gypsogenin, and sitosterol	Lowers blood glucose level
50	*Leucas lavandulifolia* (Labiatae)	Halkusha	Entire plant	Acacetin, chrysoeriol, rhamnoglucoside, lupeol, taraxerone	Lowers plasma glucose in vivo
51	*Lagerstroemia speciosa* (Lythraceae)	Crepe myrtle	Leaves	p-Coumaric acid, kaempferol, quercetin ellagitannins, corosolic acid, gallic acid, 4-hydroxybenzoic acid, 3-*O*-methyl protocatechuic acid, caffeic acid, and isoquercitrin	Stimulated glucose uptake and inhibited adipocyte differentiation that could be responsible for reducing the blood glucose level
52	*Mangifera indica* (Anacardiaceae)	Mango	Leaves, fruits	Mangiferin	Inhibited the alpha-glucosidase
53	*Musa sapientum* (Musaceae)	Banana	Flower	Vitamins, starch, and minerals	Lowers blood glucose and glycosylated hemoglobin

54	*Momordica charantia* (Cucurbitaceae)	Bitter gourd	Fruit	Momordicine alkaloid and ascorbic acid	Increase oral glucose tolerance
55	*Morus indica* (Moraceae)	Shehtoot, mulberry	Leaves, fruits	Polyphenols and flavonoids	Regulates glucose uptake and aldose reductase in vivo
56	*Murraya koenigii* (Rutaceae)	Curry leaf	Leaves	Essential oils	Increase glycogenesis and decrease glycogenolysis and gluconeogenesis
57	*Nelumbo nucifera* (Nymphaeaceae)	Lotus	Rhizome	Nuciferin and nornuciferin	Reduce sugar level in diabetic rats
58	*Ocimum sanctum* (Labiatae)	Tulsi	Leaves	Alkaloid, tannin, volatile oil, phenol, aldehyde, fixed oil, and ascorbic acid	Leaf extract showed hypoglycemic effect in vivo
59	*Olea europaea* (Oleaceae)	Olive	Leaves, fruits	Oleuropeoside	Potentiate glucose, induced insulin released, and increase peripheral uptake of glucose
60	*Punica granatum* (Punicaceae)	Pomegranate	Seed, fruits	Vit. C, protein, tannin, gallic acid, and pelletierine	Inhibition of alpha-amylase
61	*Phaseolus vulgaris* (Papilionaceae)	Red beans	Pod, seed, whole plant	Iridoid, flavonoids, lignins, and phenols	Hypolipidemic, hypoglycemic, inhibit alpha-amylase activity and antioxidant
62	*Salacia reticulata* (Celastraceae)	Marking nut tree	Stem and root	3-Oxofriedelane, 3β-hydroxyfriedelane, β-sitosterol, 28-hydroxy-3-oxofriedelane, and dulcitol	Anti-hyperglycemic
63	*Swertia chirayita* (Gentianaceae)	Chiretta	Entire plant	Amerogentin, ameroswerin, mangiferin, gentiopicrin, sweroside, swertiamarin, swerchirin	Insulin release from isolated beta-cells of the pancreas
64	*Trigonella foenum-graceum* (Leguminosae)	Methi	Seed	Protein, fat, volatile oil, fixed oil, and carbohydrate	Inhibition of alpha-amylase
65	*Tinospora cordifolia* (Menispermaceae)	Guduchi	Root, leaves	Isocolumbin, palmatine, tinosporin, tinocordiside, cordioside, and β-sitosterol	Stimulation of insulin release via modulation of β-cell and Ca^{2+} concentration

(continued)

Table 1.3 (continued)

S. no.	Botanical name/ family	Common name	Parts used	Chemical constituents	Antidiabetic mechanism in relation to chemical constituents
66	*Urtica dioica* (Urticaceae)	Nettle leaf	Leaves	Beta-sitosterol, trans-ferulic acid, dotriacontane, erucic acid, ursolic acid, scopoletin, rutin, quercetin, and p-hydroxylbenzalcohol	Decreasing blood glucose in both pancreatic and extra-pancreatic pathways and inhibitory effects on the α-amylase activity
67	*Viscum album* (Loranthaceae)	Mistletoe	Leaves, entire plant	Viscotoxins, mistletoe lectin, quercetin, naringenin, chlorogenic acid, ferulic acid, rosmarinic acid, vanillic acid, ursolic acid, betulinic acid	Stimulates insulin secretion from β-cells, attenuates lipid peroxidation, and lowers the production of free radical derivatives, therefore contributing to protection against oxidative stress
68	*Withania somnifera* (Solanaceae)	Ashwagandha	Root	Withanolides, withaferins, isopelletierine, anaferine, cuscohygrine, anahygrine	Increased serum level of insulin, decreased serum level of lipids
69	*Xanthium strumarium* (Compositae)	Common cocklebur	Fruits	Xanthinin, xanthatin, xanthinosin, caffeic acid, ferulic acid, caffeoylquinic acid, strumaroside	Stimulate insulin secretion from pancreatic beta-cells and/or sensitizing insulin receptors, inhibit amylolytic enzyme
70	*Zingiber officinale* (Zingiberaceae)	Ginger	Rhizome	Essential oils, sesquiterpene, and phenols	Stimulate insulin secretion in beta-cells

1.8 Conclusion

There is a reasonable need to renew scientific interest toward natural products for inclusion in the drug discovery program. One of the vital concerns related to plant products is the prediction of hit rate during several stages of drug development. Such a prediction is expected to be lower in case of random selection of plant species considering the overall complexity of botanical sources for new chemical entities. The best drug of the future will come from a combination of a natural product research and synthetic approaches.

Clinical experience with herbal medicine as classified in traditional medicine may simplify issues associated with deprived prognosis. New functional leads taken from traditional knowledge and experiential databases can help to reduce the time, money, and toxicity, which are the three specific barriers to drug development. Furthermore, the trend today, especially in an industrial setting, is to seek biologically active compounds from plants that will serve as lead compounds for synthetic or semisynthetic development to assure patent protection.

References

Alam F, Shafique Z, Amjad ST, BinAsad MHH (2019) Enzymes inhibitors from natural sources with anti-diabetic activity: a review. Phytother Res 33:41–54

Alamgir ANM (2017) Drugs: their natural, synthetic, and biosynthetic sources. In: Therapeutic use of medicinal plants and their extracts, volume 1. Progress in drug research, vol 73. Springer, Cham

Albuquerque UP, Ramos MA, Melo JG (2012) New strategies for drug discovery in tropical forests based on ethnobotanical and chemical ecological studies. J Ethnopharmacol 140:197–201

Altemimi A, Lakhssassi N, Baharlouei A, Watson DG, Lightfoot DA (2017) Phytochemicals: extraction, isolation, and identification of bioactive compounds from plant extracts. Plants (Basel, Switzerland) 6(4):42. https://doi.org/10.3390/plants6040042

Atanasov AG, Waltenberger B, Pferschy-Wenzig EM, Linder T, Wawrosch C, Uhrin P et al (2015) Discovery and resupply of pharmacologically active plant-derived natural products: a review. Biotechnol Adv 33(8):1582–1614. https://doi.org/10.1016/j.biotechadv.2015.08.001

Baker JT, Borris RP, Carte B, Cordell GA, Soejarto DD, Cragg GM et al (1995) Natural product drug discovery and development: new perspectives on international collaboration. J Nat Prod 58:1325–1357

Bergström C, Larsson P (2018) Computational prediction of drug solubility in water-based systems: qualitative and quantitative approaches used in the current drug discovery and development setting. Int J Pharm 540(1–2):185–193. https://doi.org/10.1016/j.ijpharm.2018.01.044

Bierer DE, Fort DM, Mendez CD, Luo J, Imbach PA, Dubenko LG et al (1998) Ethnobotanical -directed discovery of the anti-hyperglycemic properties of Cryptolepine: its isolation from *Cryptolepis sanguinolenta*, synthesis and its in vitro and in vivo activities. J Med Chem 41:894–901

Chung TDY, Terry DB, Smith LH (2015) In vitro and in vivo assessment of ADME and PK properties during lead selection and lead optimization—guidelines, benchmarks and rules of thumb. In: Sittampalam GS, Grossman A, Brimacombe K et al (eds) Assay guidance manual. Eli Lilly & Company and the National Center for Advancing Translational Sciences, Bethesda, MD. https://www.ncbi.nlm.nih.gov/books/NBK326710/

Coley PD, Heller MV, Aizprua R, Araúz B, Flores N, Correa M et al (2003) Using ecological criteria to design plant collection strategies for drug discovery. Front Ecol Environ 1:421–428 de Almeida CFCBR, de Amorim ELC, de Albuquerque UP (2011) Insights into search for new drugs from traditional knowledge: an ethnobotanical and chemical-ecological perspective. Pharm Biol 49:864–873

Dias DA, Urban S, Roessner U (2012) A historical overview of natural products in drug discovery. Metabolites 2(2):303–336. https://doi.org/10.3390/metabo2020303

Do QT, Renimel I, Andre P, Lugnier C, Muller CD, Bernard P (2005) Reverse pharmacognosy: application of Selnergy, a new tool for lead discovery. The example of ε-viniferin. Curr Drug Discov Technol 2:1–7

Harvey AL (2008) Natural products in drug discovery. Drug Discov Today 13:894–901

Hefti FF (2008) Requirements for a lead compound to become a clinical candidate. BMC Neurosci 9(Suppl 3):S7. https://doi.org/10.1186/1471-2202-9-S3-S7

Jadavaji T (ed) (1992) Charaka Samhita. Chaukhamba Surbharati Prakashan, Varanasi

Jamshidi-Kia F, Lorigooini Z, Amini-Khoei H (2018) Medicinal plants: past history and future perspective. J Herbmed Pharmacol 7(1):1–7. https://doi.org/10.15171/jhp.2018.01

Kansy M, Avdeef A, Fischer H (2004) Advances in screening for membrane permeability: high-resolution PAMPA for medicinal chemists. Drug Discov Today Technol 1(4):349–355. https://doi.org/10.1016/j.ddtec.2004.11.013

Koehn FE, Carter GT (2005) The evolving role of natural products in drug discovery. Nat Rev Drug Discov 4:206–220

Kumar S, Dobos GJ, Rampp T (2017) The significance of ayurvedic medicinal plants. J Evid Based Complement Alternat Med 22(3):494–501. https://doi.org/10.1177/2156587216671392

Lewis WH (2003) Pharmaceutical discoveries based on ethnomedicinal plants: 1985 to 2000 and beyond. Econ Bot 57:126–134

López-Vallejo F, Caulfield T, Martínez-Mayorga K, Giulianotti MA, Nefzi A, Houghten RA et al (2011) Integrating virtual screening and combinatorial chemistry for accelerated drug discovery. Comb Chem High Throughput Screen 14:475–487

Luo J, Fort DM, Carlson TJ, Noamesi BK, nii-Amon-Kotei D, King SR et al (1998) *Cryptolepis sanguinolenta*: an ethnobotanical approach to drug discovery and the isolation of a potentially useful new anti-hyperglycaemic agent. Diabet Med 15:367–374

Mishra SB, Rao CV, Ojha SK, Vijayakumar M, Verma A, Alok S (2010) An analytical review of plants for anti diabetic activity with their phyto-constituents & mechanism of action. Int J Pharm Sci Res 1(1):29–46

Modak M, Dixit P, Londhe J, Ghaskadbi S, Devasagayam TP (2007) Indian herbs and herbal drugs used for the treatment of diabetes. J Clin Biochem Nutr 40(3):163–173. https://doi.org/10.3164/jcbn.40.163

Obbo CJD, Makanga B, Mulholland DA, Coombes PH, Brun R (2013) Antiprotozoal activity of Khaya anthotheca (Welv.) C.D.C. a plant used by chimpanzees for self medication. J Ethnopharmacol 147:220–223

Pan SY, Zhou SF, Gao SH, Yu ZL, Zhang SF, Tang MK (2013) New perspectives on how to discover drugs from herbal medicines: CAM's outstanding contribution to modern therapeutics. Evid Based Complement Alternat Med 2013:627375. https://doi.org/10.1155/2013/627375

Ramesha BT, Gertsch J, Ravikanth G, Priti V, Ganeshaiah KN, Shaanker RU (2011) Biodiversity and chemodiversity: future perspectives in bioprospecting. Curr Drug Targets 12:1515–1530

Sen S, Chakraborty R (2016) Revival, modernization and integration of Indian traditional herbal medicine in clinical practice: importance, challenges and future. J Trad Complement Med 7 (2):234–244. https://doi.org/10.1016/j.jtcme.2016.05.006

Sharma PV (ed) (2001) Sushruta Samhita. Chaukhamba Vishwabharati, Varanasi

Sharma H, Chandola HM (2011) Prameha in ayurveda: correlation with obesity, metabolic syndrome, and diabetes mellitus. Part 1–etiology, classification, and pathogenesis. J Altern Complement Med 17(6):491–496. https://doi.org/10.1089/acm.2010.0396

Srikantha Murty KR (ed) (2000) Astanga Hridaya. Krishnadas Academy, Varanasi

Zhang Q (2015) Traditional and complementary medicine in primary health care. In: Medcalf A, Bhattacharya S, Momen H et al (eds) Health for all: the journey of universal health coverage. Orient Blackswan, Hyderabad. https://www.ncbi.nlm.nih.gov/books/NBK316267/

Zhu F, Ma XH, Qin C, Tao L, Liuy X, Shi Z et al (2012) Drug discovery prospect from untapped species: indicators from approved natural product drugs. PLoS One 7:e39782. https://doi.org/10.1371/journal.pone.0039782

Bioactive Natural Leads Targeting Cancer Cell Metabolism

2

Prasanta Dey, Amit Kundu, Babli Kar, Anushree Bhakta, Vineet Vishal, S. Keerthana, Anoop Kumar, Tejendra Bhakta, Suvakanta Dash, and Hyung Sik Kim

Abstract

Therapeutic interest in targeting cancer metabolism has become emerging in recent years. However, a very few numbers of molecules has been identified as a potential modulator of cancer cell metabolism and are in the investigation. Therefore discovery and development of potent and selective drugs from natural sources like plants, marine organisms, or invertebrates are very essential to check the global health disaster because of cancer. To address these emerging complications potential leads from the natural origin are a judicious choice of interest. An organized and summarized evidence concerning cancer cell metabolic proteins and natural lead molecules against cancer cell metabolism with an emphasis on molecular/cellular mechanism(s) is crucial in a single podium. This chapter precisely discusses the antimetabolic potential of natural lead molecules and their mechanisms of action (MOA) against cancer cell metabolism. This complete review on different molecular mechanisms of bioactive leads will toward cancer cell metabolism support researchers in understanding the

P. Dey · A. Kundu · H. S. Kim (✉)
School of Pharmacy, Sungkyunkwan University, Suwon, Republic of Korea
e-mail: deyprasantadey@yahoo.com; hkims@skku.edu

B. Kar
Bengal Homoeopathic Medical College and Hospital, Asansol, India

A. Bhakta
Tripura Medical College and Dr. BRAM Teaching Hospital, Agartala, Tripura, India

V. Vishal
Department of Botany, Bangabasi Evening College, Kolkata, India

S. Keerthana · A. Kumar
Department of Pharmacology and Toxicology, NIPER, Rupnagar, Punjab, India

T. Bhakta · S. Dash
Regional Institute of Pharmaceutical Science and Technology, Agartala, Tripura, India

S. C. Mandal et al. (eds.), *Evidence Based Validation of Traditional Medicines*,
https://doi.org/10.1007/978-981-15-8127-4_2

targetable metabolic protein and their interacting pathways, to discover safe and potent drugs for better therapeutic applications against cancer.

Keywords

Natural leads · Cancer cell metabolism · Cellular metabolic targets · Bioinformatic approaches

Abbreviations

2-PG	2-phosphoglycerate
3-PG	3-phosphoglycerate
6PGD	6-phosphogluconate dehydrogenase
ACC	Acetyl-CoA carboxylase
ACL	ATP citrate lysate
ACS	Acyl-CoA synthetases
ADR	Anticancer drug resistance
AH	Aconitase
AMPK	AMP-activated protein kinase
ATP	Adenosine triphosphate
DMAPP	Dimethylallyl diphosphate
DNA	Deoxyribonucleic acid
EGFR	Epidermal growth factor receptor
FAS	Fatty acid synthase
FH	Fumarate hydratase
G6PD	Glucose-6-phosphate dehydrogenase
GIST	Gastrointestinal stromal tumor
GLUTs	Glucose transporters
GPI	Glucose-6-phosphate isomerize
GSH	Glutathione
HIF-1α	Hypoxia-inducible factor-1α
HKs	Hexokinases
HMGCR	3-hydroxy-3-methylglutaryl-CoA reductase
HMGR	3-hydroxy-3-methylglutaryl-CoA reductase
HSF1	Heat shock factor
IDH	Isocitrate dehydrogenase
IPP	Isopentenyl pyrophosphate
LDH	Lactate dehydrogenase
LKB1	Liver kinase B1; also known as serine/threonine kinase 11—STK11
MALDI-TOF-MS	Matrix-assisted laser desorption/ionization-time-of-flight-mass spectrometry
MCTs	Monocarboxylate transporters

MDH1	Malate dehydrogenase
MFS	Major facilitator superfamily
mTOR	Mammalian target of rapamycin
NADPH	Nicotinamide adenine dinucleotide phosphate
OXPHOS	Oxidative phosphorylation
PC	Phosphatidylcholine
PDK	Pyruvate dehydrogenase
PEP	Phosphoenolpyruvate
PFKFB	6-phosphofructo-2-kinase/fructose-2, 6-bisphosphatases
PFKs	Phospho-fructokinases
PGAM1	Phosphoglycerate mutase 1
PGD	Phosphogluconate dehydrogenase
PGM	Phosphoglucomutase
PHGDH	Phosphoglycerate dehydrogenase
PKM2	Pyruvate kinase M2
PPP	Pentose phosphate pathway
QSAR	Quantitative structure-activity relationship
RNA	Ribonucleic acid
SCD-1	Stearoyl-CoA desaturase-1
SCDs	Stearoyl-CoA desaturases
SDH	Succinate dehydrogenase
SIRT3	Sirtuin 3
SLC	Solute carrier
SREBP1	Sterol regulatory element-binding protein 1
TAL	Transaldolase
TALDO1	Trans-aldolase
TCA	Tricarboxylic acid cycle
TKT	Human transketolase
TKT	Transketolase
TPI	Triosephosphate isomerase

Cell Line Details

NSCLC	Non-small-cell lung carcinoma
AsPC-1	Human pancreatic cell lines
BXPC-3	Human pancreatic cell lines
SK-BR-3	Breast cancer cells
MCF-7	Human breast adenocarcinoma cells
PC-3	Human prostate cancer cells
AGS	Human stomach adenocarcinoma cell line
PANC-1	Human pancreatic cancer cell line
MUM28	Uveal melanoma cells
OCM-1	Uveal melanoma cells
SW620	Human colorectal cancer cell line

SW480	Human colorectal cancer cell line
MDAH2774	Human ovarian cancer cells
PA-1	Human ovarian cancer cells
SKOV3	Human ovarian cancer cells
OVCAR3	Human ovarian cancer cells
HCC-LM3	Human hepatocellular carcinoma cells
H1650	Non-small cell lung cancer (NSCLC) cells
H460	Non-small cell lung cancer (NSCLC) cells
HCC827	Non-small cell lung cancer (NSCLC) cells
LY18	Human diffuse large B-cell lymphoma
LnCap	Human prostate cancer cell line
K-562	Human erythroleukemia cell line
Caco-2	Human colon cancer cells
A549	human lung adenocarcinoma cells
Ec109	Esophageal cancer cells
MIAPaCa-2	Human pancreatic adenocarcinoma
HT29	Human colon cancer cells
DLD-1	Human colon cancer cells
HCT116	Human colon cancer cells
SKBR3 cells	Human breast cancer cell line
TSCCa	Human tongue carcinoma cells
Tca8113	Human tongue carcinoma cells
NCI-H1299	Human lung cancer cells

2.1 Introduction

Reprogramming metabolism is one of the surviving techniques associated with a cancer cell. Several publications have been already reported various common traits that help in tumor cell growth and metastasis. Cancer cell requires a huge blast of energy to maintain their macromolecular biosynthesis, controlling the redox balance to support their tumor growth and progression, and reprogramming the cell metabolism is one of the reliable techniques.

In oncogenesis, tumor cells carry out aerobic glycolysis, and this metabolic rearrangement was first observed by Otto Warburg, and the phenomenon is known as "Warburg effect." In this case, to maintain the nutrient uptake and gene regulation, cancer cell undertaken a range of alterations in their intracellular metabolism and other metabolic interaction with the microenvironment (Pavlova and Thompson 2016). Over the past two decades, extensive research works have been carried out to explore deep insights into this metabolic reprogramming of cancer cells and how it provides survival and growth advantages to them by energy acquisition and biomass synthesis. Glucose and glutamine are the two major sources of glycolysis and the tricarboxylic acid cycle (TCA) process. These metabolic processes synthesize many unconventional nutrient sources from cells like ketone, lactate, ammonia, acetate, and other exogenous proteins which act as a source for ATP acquisition, biomass synthesis for building blocks during cancer cell reaction in the frazzled microenvironment (Fig. 2.1). Therefore, targeting cancer metabolism has become a topic of research interest to achieve therapeutic benefits against cancer. Various metabolic targets have been already under exploration for their therapeutic potential, and various small molecules or drug targeting metabolism are being tested both in preclinical and clinical trials also (Luengo et al. 2017; Martinez-Outschoorn et al. 2017; Dey et al. 2019a, b, c).

Different natural products have conventionally been used to treat various human diseases (Islam et al. 2018; Debnath et al. 2012; Karuna et al. 2018; Dey et al. 2012a, b; Dey et al. 2014; Kundu et al. 2020; Sachan et al. 2020). They hold a huge bank of bioactive compounds for the new drug discovery. Natural compounds are very useful because of their wide range of pharmacophores and a high degree of stereochemistry. Natural-derived compounds are intricately appropriate candidates for the development of novel therapeutic agents because of their structural diversity, remarkable bioactivity, and significant bioavailability. Fusion with the bioinformatics tools provides deep insights about their molecular targets and, hence, simplifying the drug discovery steps. Again combining with the synthetic chemistry will generate more potent chemical entities. Various plant-based compounds are being extensively studied in various cancer models to identify the potent molecules and their mechanism of action, of which few of them also are in clinical trials (Butler et al. 2014; Seca and Pinto 2018). Among them, some natural product-inspired anticancer compounds are also available in the market (Newman and Cragg 2016). These compounds can be used alone or can be used as combination therapy along with standard drugs to promote the synergistic effect and minimize any adverse effect. On the other hand, many natural lead molecules target multiple proteins or genes and can modulate various carcinogenic pathways, hence increase the chance of success.

In this chapter, we will systematically address important metabolic proteins or enzymes, their role and also mechanism of action of various natural lead molecules (belonging to the alkaloid, phenolic, and isoprenoid group) that modulate the cancer cell metabolism and exerts the anticancer effects along with the different in silico approaches.

2.2 Global Trends in Cancer

2.2.1 General Features

Cancer is a group of diseases characterized by uncontrolled cell growth and failure to control anti-growth signals. The immortalized cells can grow beyond normal limits, metastasize, and have the potential to invade adjacent tissues and other parts of the body via the bloodstream or lymphatic system due to malfunctioning apoptosis or programmed cell death systems. The main causes of cancer include genetic mutations, carcinogens, immune system problems, lifestyle, environmental factors such as pollutants, certain viruses, and bacterial infections.

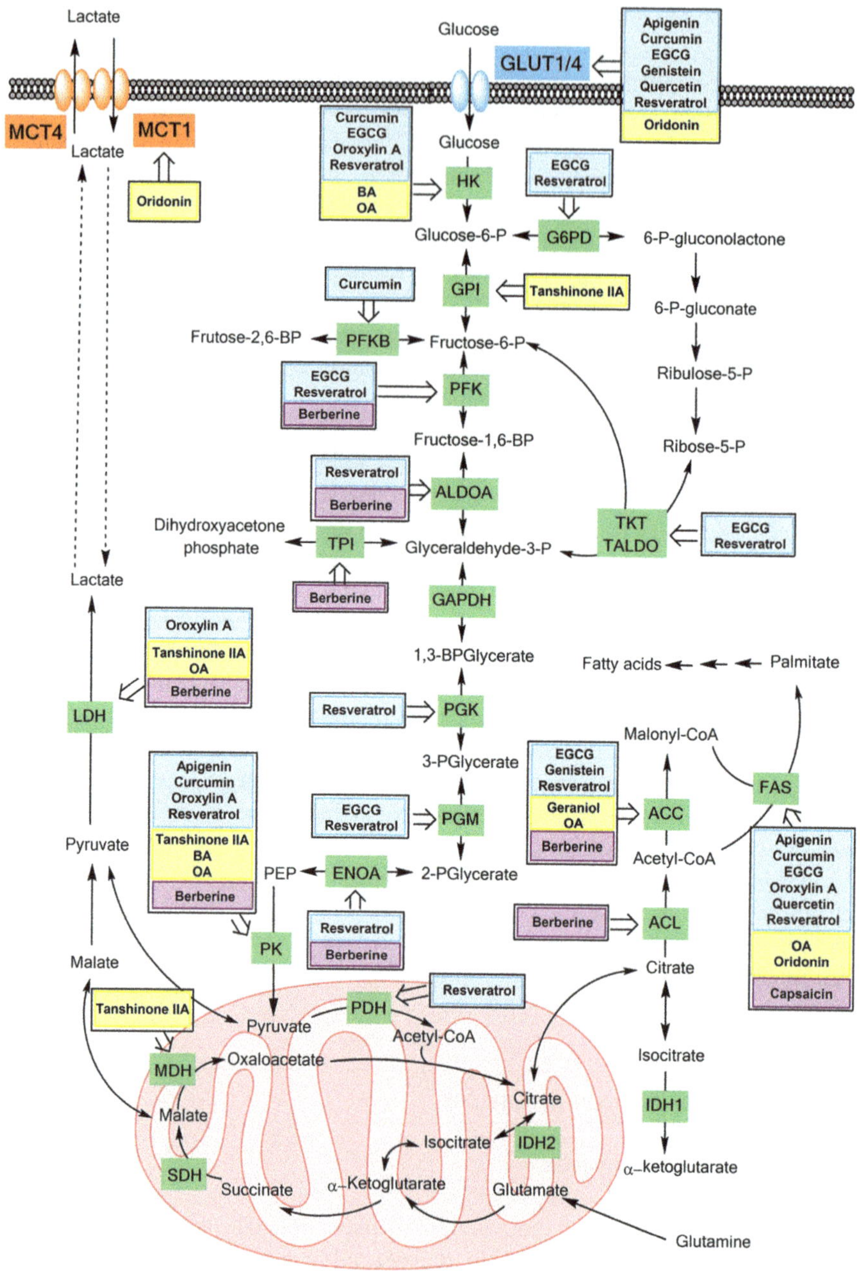

Fig. 2.1 Major metabolic signaling molecules of cancer cell metabolism pathway and the targeting natural leads. Reproduced with permission from the publisher of Guerra et al. 2018. Copyright©2018, American Chemical Society

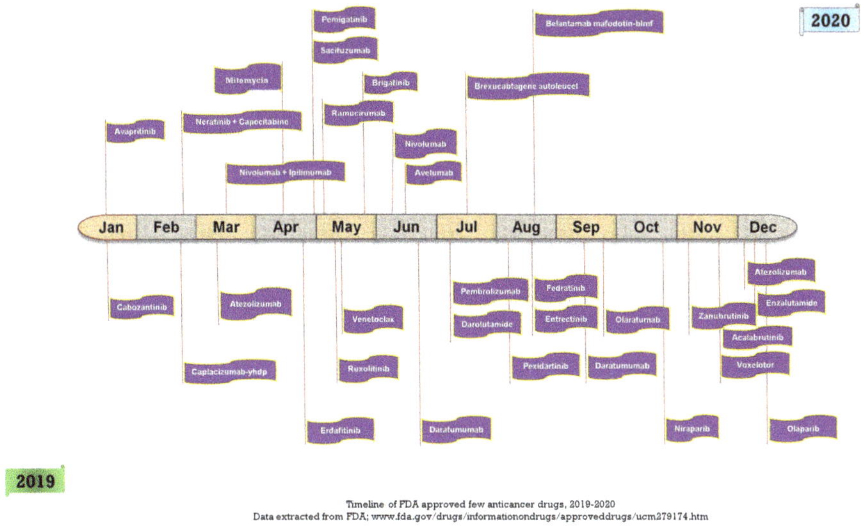

Fig. 2.2 FDA has approved certain new anticancer drugs

To combat cancer, different diagnostic approaches, including natural- and synthetic-based therapies, are being used to target cancerous cells and cancer stem cells (Block et al. 2015; Dey and Das 2013; Boohaker et al. 2012; Jeon et al. 2017; Markman et al. 2013; Thakor and Gambhir 2013). The FDA has approved certain new anticancer drugs to improve available therapeutic options (Fig. 2.2). The optimal therapeutic approach is determined by the type and stage of cancer. To date, there are ~200 different types of cancer that have been well identified, and numerous validated and effective therapies (chemotherapeutic agents, radiotherapy, surgery, hormonal drugs, complementary therapy like acupuncture, yoga, physical activity to reduce some side effects of cancer treatment, and combinations) are clinically available; however, these therapies have various limitations (Qi et al. 2010). Due to poor diagnosis and other factors, most patients are diagnosed too late to undergo surgery. Existing chemo- and radiotherapies have serious adverse effects and complications, such as diarrhea, nausea, vomiting, and alopecia, which, along with increased chemo−/radioresistance, make this a devastating and belligerent disease worldwide (Pereira et al. 2012; Qi et al. 2010). Consequently, there is a persisting requirement to develop novel, effective, and affordable anticancer drugs. One of the oldest effective strategies in drug discovery involved isolating compounds from natural sources and mimicking the active natural product. The use of medicinal plants is a common alternative for cancer treatment in many countries globally. In this regard targeting cancer energetic pathways can be a promising therapeutic approach.

2.2.2 Global Cancer Statistics

Despite technological and social development, cancer has become one of the leading causes of mortality worldwide. According to GLOBOCAN 2018, produced by the International Agency for Research on Cancer, estimations of the mortality incidence and the prevalence of major cancer types across 20 world regions have been made. The data revealed 18.1 million new cancer cases and 9.6 million cancer-associated mortalities worldwide in 2018 (Bray et al. 2018). Lung cancer is the most leading cause of cancer-associated death in both sexes. By 2030, it is projected there will be 26 million new cancer cases and 17 million cancer mortalities/year (Thun et al. 2009).

2.2.3 Chemotherapy Resistance in Cancer

Drug resistance is a common phenomenon wherein a susceptible cell becomes tolerant in the presence of a drug. Cancer is a prodigious problem for global economies and public health. Anticancer drugs, an imperative tool in medicine, are lethal to cancer cells, being designed and developed to kill cancerous cells. Resistance against chemotherapeutic drugs was first observed in the 1940s with alkylating agents, and the rapidity of the developing resistance is concerning (Cree and Charlton 2017).

Anticancer agents that kill or prevent the growth of cancer cells function by blocking DNA and RNA biosynthesis, interfering with transcription, protein synthesis, and function, and influencing hormone homeostasis. The efficacy of many, but not all, anticancer drugs is severely compromised by the emergence of anticancer drug resistance (ADR) (Housman et al. 2014). Imatinib in GIST, BRAF inhibitors in melanoma, EGFR inhibitors in NSCLC, and HER2 inhibitors in breast cancer are some examples (Cree and Charlton 2017). The six most common mechanisms of ADR are depicted in Fig. 2.3. Among them, mutations in drug targets could alter target sequences, reduction, or overexpression of targets or efflux pumps that may diminish uptake or drug efflux. Additionally, cancer cells can alter drug metabolism profiles and upregulate drug-detoxification mechanisms. Decreased susceptibility to apoptosis, changes in proliferation, and enhancement of DNA damage repair capacity can contribute to resistance development. Multiple methods of ADR confer advantages to cancer cells and allow them to grow in the presence of anticancer drugs. Thus, the number of resistant mutants increase and spread rapidly in the body.

2.3 Important Cellular Metabolic Targets

Cancer cells undertake notable metabolic reprogramming to survive and proliferate by maintaining their signaling networks and metabolic alterations (Luengo et al. 2017). These metabolic alterations actively participate in the tumor cell development and proliferation and hence positively provide growth advantages to them. These

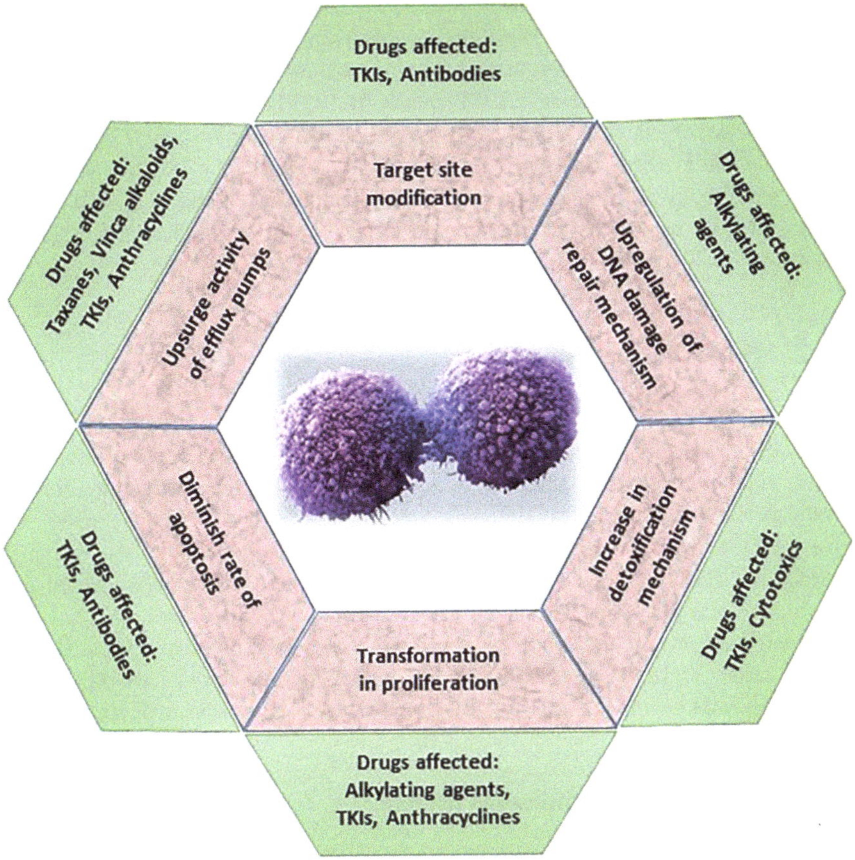

Fig. 2.3 Six most important benchmarks of anticancer drugs

changes in metabolism led us to identify and target novel therapeutic molecules in the cancer cell metabolic and bioenergetics pathway.

Glucose is the most plentiful nutrient in the blood and the main source of energy of the cell, and its metabolism is quite complex. A series of glycolytic enzymes are involved in the breakdown of glucose. Here, we will discuss those metabolic enzymes and transporters which involves in the process of cell metabolism.

2.3.1 Glucose Transporters (GLUTs)

The first rate-limiting step of glucose metabolism is the transportation of glucose across the cell membrane, and two types of hexose transporter, the family of glucose transporters (GLUTs) (Fig. 2.1) and the sodium-dependent glucose co-transporters (SGLTs), help in the glucose transportation, which is often found to be dysregulated in some of the tumor cells (Macheda et al. 2005). These GLUTs are the part of major facilitator superfamily (MFS) with alike transmembrane anatomy, but they mainly differ from each other in tissue distribution and substrate recognition. Fourteen members constitute a human GLUT family (GLUT1–14 or SLC2A1–14) (Cao et al. 2007; Macheda et al. 2005; McBrayer et al. 2012). Among all, GLUT1 is the most abundant isoform and is frequently upregulated in most of the human cancers which are directly correlated with poor patient outcomes (Wang et al. 2017). This makes GLUT1 as a possible therapeutic target of various anticancer molecules (Barron et al. 2016).

2.3.2 Hexokinases (HKs)

Tumor cells typically undergo huge glycolytic flux and lactate production as compared to normal cells which acquire most of their energy by oxidative phosphorylation in mitochondria. This facilitates tumor cells to fast energy production, biomass synthesis and accumulation, maintenance redox balance, and cancer metastasis and invasion (Vander Heiden et al. 2009). Among all rate-limiting glycolytic enzymes, hexokinases (HKs) (Fig. 2.1) participate in the first step of glycolysis which irreversibly convert phosphorylate glucose to glucose-6-phosphate, and HK2 has been extremely expressed in the cancer cells as compared to HK1 (highly expressed in the normal cells) (Pedersen et al. 2002). This makes HK2 a novel biomarker for the cancer cells and becomes a possible therapeutic target of anticancer therapy.

2.3.3 Phospho-Fructokinases (PFKs)

Phospho-fructokinases (PFKs) (Fig. 2.1) is another important glycolytic enzyme which converts fructose-6-phosphate to fructose-1, 6-bisphosphate. Various oncogenes, hypoxia-inducible factor (HIF)-1α, and Akt inhibition can trigger the upregulation of PFK1 (Scatena et al. 2008; Yalcin et al. 2009). Again the dysfunction or altered regulation of fructose-2, 6-bisphosphate, an important allosteric activator of PFK, maintains the cellular level of PFK and hence modulates the cancer metabolism (Lincet and Icard 2015). On the other hand, a bifunctional enzyme, 6-phosphofructo-2-kinase/fructose-2, 6-bisphosphatases (PFKFB), more specifically PFKFB3, which is highly expressed in various human cancers and regulate the cell cycle, also maintain the expression of fructose-2, 6-bisphosphate (Lincet and Icard 2015). Both of these PFKs can be targeted for inhibiting the cancer cell metabolism.

2.3.4 Pyruvate Kinase M2 (PKM2)

Pyruvate kinase (PK) is the last rate-limiting glycolytic enzyme that catalyzes the conversion of phosphoenolpyruvate (PEP) to pyruvate with simultaneous production of ATP (Dey et al. 2019a, b, c). Human PKs have four distinct isoenzymes: M1 (encoded by *PKM*; primarily present in muscle, heart, and brain), M2 (encoded by *PKM*; express in different types of cells and tissues), L (encoded by *PKLR*; mainly expressed in the liver), and R (encoded by *PKLR*; present in the erythrocytes) (Zhao et al. 2013). Cells prefer glycolysis, and oxidative phosphorylation depends on the relative expression of PKM1 and PKM2 (Fig. 2.1) (He et al. 2017). PKM2 is highly expressed in normal proliferating cells and rapidly proliferating cancer cells, whereas PKM1 is especially expressed in normal adult cells (Dayton et al. 2016). PKM2 performs both metabolic and nonmetabolic functions of cancer cells and promotes cellular growth (Yang and Lu 2015). Various cancer cells highly express PKM2 which is associated with the malignant type of tumor and poor patient outcomes (Dey et al. 2019a). Various cancer alterations of PKM2 expression are associated with drug resistance which makes PKM2 a potential therapeutic target (Zhao et al. 2013).

2.3.5 Lactate Dehydrogenase (LDH)

High production of lactate from pyruvate is one of the signatures and final glycolytic steps of cancer cells which is catalyzed by LDH (Fig. 2.1). In human tissue, two different subunits, H and M (encoded by LDHA and LDHB), form five active LDH isoenzymes (Miao et al. 2013). Among these, LDHA is abundantly expressed in the cancer cells, while knockdown or inhibition results increased mitochondrial respiration, reduced cell proliferation under hypoxic condition, decreased cellular metastasis and invasion potential, and repressed tumorigenicity (Fantin et al. 2006; Miao et al. 2013). High serum level of LDH has been directly correlated with the increased risk of colorectal, prostate, hematological, pulmonary, gynecological, and gastroesophageal cancer-associated death (Wulaningsih et al. 2015). Therefore, LDHA has become a predictive cancer biomarker and potential therapeutic target.

2.3.6 Monocarboxylate Transporters (MCTs)

MCTs (Fig. 2.1) are the member of the solute carrier (SLC) family protein and consist of 14 isoforms that exports intracellular lactate (Jones and Morris 2016). Among all MCT1–4 isoforms are mostly studied because of their importance in transporting various metabolites like lactate, pyruvate, etc. (Jones and Morris 2016). Most importantly a high expression of MCT4 is positively correlated with poor patient prognosis in various types of cancer (Bovenzi et al. 2015). Maintaining the metabolic phenotype in cancer cells makes it an attractive target or biomarker for cancer and to explore the MCTs in the pathway of anticancer therapy research.

2.3.7 Glucose-6-Phosphate Dehydrogenase (G6PD)

Various anabolic pathways like the pentose phosphate pathway (PPP; also known as the phosphogluconate pathway or the hexose monophosphate shunt) and serine biosynthesis pathway are also contributing cancer cell growth and proliferation. They also maintain the fatty acid synthesis, oxidative defense, and nucleotide biosynthesis (Patra and Hay 2014). G6PD (Fig. 2.1) is a cytosolic and first enzyme of PPP (an alternative pathway of glucose metabolism) that produces NADPH by reducing $NADP^+$. Many human cancers express a high level of G6PD which makes an attractive biomarker for cancer therapy. Some anticancer molecules are thought to control PPP by modulating the G6PD activity (Cho et al. 2018).

2.3.8 6-Phosphogluconate Dehydrogenase (6PGD)

6PGD (Fig. 2.1) is an oxidative carboxylase in the PPP and catalyzes the conversion of ribulose 5-phosphate from 6-phosphogluconate. This enzyme is highly upregulated in several cancers, and inhibition of this enzyme can lead to increased ROS production, suppression of RNA biosynthesis, and lipogenesis, which can be mediated by the inhibition of LKB1 (liver kinase B1; also known as serine/threonine kinase 11—STK11)-AMPK (AMP-activated protein kinase) (Lin et al. 2015b).

2.3.9 Transketolase (TKT) and Trans-aldolase (TALDO1)

Transketolase (encoded by TKT gene) and trans-aldolase (encoded by *TALDO1* gene) (Fig. 2.1) is an enzyme of the non-oxidative phase of PPP. Cancer cells utilize these two enzymes to produce nucleotides which will be further utilized to synthesize DNA and RNA (Patra and Hay 2014).

2.3.10 Phosphoglycerate Dehydrogenase (PHGDH)

Serine biosynthetic pathway is one of the metabolic reactions which produce non-essential amino acids, necessary for purine and GSH (glutathione) biosynthesis. PHGDH (Fig. 2.1) is an enzyme that catalyzes the first step of serine biosynthesis and highly overexpressed in various types of cancer cells (Tennant et al. 2010). This enzyme can be targeted to inhibit the growth and survival of the cancer cells with upregulated serine biosynthesis.

2.3.11 Glutaminase

Unlike glucose, glutamine, the non-essential amino acid, helps in energy generation as well as supplies nitrogen for the biosynthesis of nucleic acid and other amino acids and helps in the growth and metabolism of tumor cells. Glutamine helps in the synthesis of GSH and an alternative source of carbon donor in lipid biosynthesis. This enzyme glutaminase (Fig. 2.1) helps in the formation of glutamate from glutamine, and hence it can be another therapeutic target for anticancer drug discovery (Tennant et al. 2010).

2.3.12 Heat Shock Factor (HSF1)

Apart from regulating the heat shock response in the eukaryotes, HSF1 (Fig. 2.1) also has some non-heat shock function which provides advantages to cancer cell growth. HSF1 enhances the glucose consumption, LDH activity, and lactate production (Dai et al. 2007). Thus downregulation or inhibition of HSF1 hinders glycolysis (Zhao et al. 2009). It has been reported that cancer cells with high expression of HSF1 get resistant to trastuzumab, and the inhibition of HSF1 sensitizes them to trastuzumab treatment. This confirms the role of HSF1 in the trastuzumab resistance, and therefore, it can be an excellent target to overcome the trastuzumab drug resistance in cancer patients.

2.3.13 Pyruvate Dehydrogenase (PDK)

Four types of PDK (Fig. 2.1) isotypes are identified so far. Typically, HIF activation initiates the overexpression of these PDKs that negatively regulate the pyruvate dehydrogenase (PDH) which converts pyruvate into acetyl-CoA and enters into the tricarboxylic acid (TCA) cycle to produce ATP (Lu et al. 2008). Inhibition of PDKs potentiates the pyruvate to acetyl-CoA production and stimulates the TCA cycle. Dichloroacetate is one of those kinds of small molecules that has already entered into the clinical trials (Saunier et al. 2016). Generally, enzymes involved in the TCA cycle (fumarate hydratase or FH; aconitase or AH; succinate dehydrogenase or SDH; and isocitrate dehydrogenase or IDH) are either mutated or deregulated in the cancer cells, and small molecules targeting those enzymes are now under examination (Luengo et al. 2017).

2.3.14 ATP Citrate Lysate (ACL)

All rapidly proliferating cells need a huge blast of lipids and steroids to continue their phospholipid bilayers synthesis and maintain signaling cascades. In the pathways of tumor progression, various enzymes that are involved in lipid biosynthesis can be served as potential therapeutic targets (Liu et al. 2017). ACL (Fig. 2.1) which is highly expressed in various tumors are directly regulated by the PI3K/Akt pathway. This enzyme links glucose and glutamine metabolism to fatty acid synthesis by converting citrate into lipogenic precursor acetyl-CoA (Chypre et al. 2012).

2.3.15 Acetyl-CoA Carboxylase (ACC)

ACC1 (Fig. 2.1) converts acetyl-CoA into malonyl-CoA in the fatty acid synthesis pathway and has been overexpressed in various cancer which is correlated to tumor progression and poor patient outcomes (Currie et al. 2013). AMPK interacts and inhibits the sterol regulatory element-binding protein 1 (SREBP1), which, in turn, phosphorylates and inactivate ACC1 (Li and Zhang 2016).

2.3.16 Fatty Acid Synthase (FAS)

FAS (Fig. 2.1) is highly expressed in various human cancers (Liu et al. 2017) and an important enzyme in lipogenesis that catalyzes the terminal step of fatty acid synthesis and converts malonyl-CoA and acetyl-CoA substrate into palmitate (Flavin et al. 2010).

2.3.17 Others

Apart from these proteins, various other proteins are highly expressed in the human tumors and also involved in fatty acid synthesis and modification, including stearoyl-CoA desaturases (SCDs) (Igal 2016) and acyl-CoA synthetases (ACS) (Currie et al. 2013). Additionally, choline and cholesterol are synthesized by choline kinase and 3-hydroxy-3-methylglutaryl-CoA reductase (HMGCR), respectively, have a role in tumor development and progression, and can be considered as a potential target for anticancer drug development (Martinez-Outschoorn et al. 2017).

2.4 Natural Leads Targeting Cancer Cell Metabolism

2.4.1 Alkaloids

Alkaloids (Fig. 2.1) are the natural molecules found in plants, animals, bacteria, and fungus, having a basic nitrogen atom in their structure (except in peptide bond or amide) (Kushiro and Ebizuka 2010). In the ancient times, after the breakthrough discovery of the molecule morphine, in 1804, plants having alkaloids are being used in various traditional medicines by various ethnic groups of people in this world (Aniszewski 2007). Vinca alkaloids are the first alkaloids (discover in the 1950s) having anticancer potential against various types of human cancers (Cragg et al. 2009). In this section, we will cover major important anticancer natural lead molecules that target biosynthetic metabolic pathways.

2.4.1.1 Capsaicin

Capsaicin (*trans*-8-methyl-*N*-vanillyl-6-nonenamide) (Fig. 2.1) is a derivative of homovanillic acid and the key constituent in chili peppers that are consumed in various cultures as a spice (Clark and Lee 2016). Because of its vast medicinal properties, capsaicin is widely explored for its anti-obesity (Kang et al. 2007), antioxidant (Galano and Martínez 2012), analgesic (Brederson et al. 2013), and anti-inflammatory properties (Kim et al. 2003). Capsaicin possesses in vitro chemopreventive and chemotherapeutic effect (Amantini et al. 2009; Jun et al. 2007; Yang et al. 2009) and antitumorigenic activity in vivo (Bhutani et al. 2007; Lu et al. 2010; Pramanik et al. 2011; Yoshitani et al. 2001; Zhang et al. 2008) against various cancers (Bley et al. 2012). In the HepG2 cell line, capsaicin notably downregulated the FAS protein levels at a 0.05–0.5 mM concentration as compared with control groups, without affecting other lipid-related proteins (like ACL and ACC) (Impheng et al. 2014). Capsaicin also suppressed intracellular triglycerides and long-chain fatty acid levels by inhibiting the synthesis of de novo fatty acid (Impheng et al. 2014). At a concentration of 150 μM, capsaicin disrupts the functionality of mitochondrial complex I and III in human pancreatic cell lines (AsPC-1 and BXPC-3) and induces the ROS generation which finally led to apoptosis (Pramanik et al. 2011).

2.4.1.2 Tetrandrine

Tetrandrine (Fig. 2.1) is a bisbenzylisoquinoline alkaloid isolated from the root of *Stephania tetrandra* S. Moore. MALDI-TOF-MS found that tetrandrine (5 ± 0.6 μg/mL) downregulated phosphoglycerate mutase 1 (PGAM1) and transaldolase (TAL) protein level and inhibited the glycolysis and pentose phosphate pathway in HepG2 cells (Cheng et al. 2010).

2.4.1.3 Piperine

Piperine (Fig. 2.1) is isolated from *Piper nigrum* Linn. In breast cancer cells (SK-BR-3), it inhibited the mRNA expression and protein level of FAS and ultimately inhibited fatty acid synthesis (Do et al. 2013).

2.4.1.4 Berberine

Reprogramming the metabolism including OXPHOS, glycolysis and fatty acid synthesis provide growth advantages to the cancer cells. Several potential targets involved in this reprogramming mechanism were investigated by various novel molecules to find a profit in cancer treatment. Among them, berberine (Fig. 2.1), an isoquinoline quaternary alkaloid isolated from the herbal plants like *Rhizoma coptidis*, is nontoxic to humans (Jantová et al. 2003). In breast cancer cells berberine treatment (36.91 μg/mL) inhibited various glycolytic enzymes like enolase α, triosephosphate isomerase (TPI) and fructose-diphosphate aldolase A (Chou et al. 2012). In MCF-7 cells the p-ACL level was decreased, while the p-ACC level was increased after berberine (2–100 μM) treatment (Tan et al. 2015). In another study on MCF-7 cells, berberine (2–50 μM) upregulated the p-PKM2 level and downregulated PFK and LDH level and shifts the metabolism toward OXPHOS (Fan et al. 2013). PKM2 activity was inhibited after berberine treatment in HeLa and HCT116 cells (Li et al. 2017c). Other metabolic proteins like 1,2-dihydroxy-3-keto-5-methylthiopentene dioxygenase, kynurenine 3-monooxygenase, succinyl-CoA:3-

ketoacid-coenzyme A transferase 1, mitochondrial/OXCT1, and triosephosphate isomerase were also differentially expressed in MCF-7 cells after berberine treatment (Chou et al. 2012).

2.4.2 Isoprenoids

Isoprenoids are plant secondary metabolites originated from the five carbon precursor's isopentenyl pyrophosphate (IPP) and its isomer dimethylallyl diphosphate (DMAPP) (Withers and Keasling 2007). Depending on the number of five-carbon skeleton, isoprenoids can be classified as monoterpenoids (C10), sesquiterpenoids (C15), diterpenoids (C20), triterpenoids (C30), and tetraterpenoids (C40) (MA Domingues et al. 2014) which have significant pharmacological activities, among which some candidates enter the clinics. For example, artemisinin (a sesquiterpene lactone) which was isolated from *Artemisia annua*, with its semi-synthetic compounds, is potent antimalarial drug (Slezakova and Ruda-Kucerova 2017). On the other hand, a diterpenoid, namely, paclitaxel, which was isolated from *Taxus brevifolia* is being used in various chemotherapeutic regimes against lung, breast, and ovarian cancer (Bernabeu et al. 2017). Nowadays plant-derived metabolites have received growing interest because of their antimetabolic activity in the anticancer pathway. Table 2.1 provides a collected list of those plant-derived isoprenoids, while some of them are mentioned below.

2.4.2.1 Oleanolic Acid

Oleanolic acid (Fig. 2.1) is an oleanane-type pentacyclic triterpenoid isolated from *Olea europaea* (family: Oleaceae) having various pharmacological activities including anticancer activity. Treatment with oleanolic acid (50–100 μg/mL) downregulated the PKM2 protein along with the concomitant increase of PKM1 which ultimately impairs glucose uptake and lactate production in the human breast cancer (MCF-7) and prostate cancer cells (PC-3) (Liu et al. 2014b, c). In these same cell lines, oleanolic acid (10–100 μg/mL) activate AMPK which, in turn, phosphorylate and inactivate 3-hydroxy-3-methylglutaryl-CoA reductase (HMGR) and acetyl-CoA carboxylase 1 (ACC1) and downregulated fatty acid synthase (FAS) protein level which ultimately ends up into reduced fatty acid synthesis (Liu et al. 2014c). In breast cancer cells, oleanolic acid (5 μM) inhibited the glucose consumption and lactate production which was induced by high salt-mediated osmotic stress (Amara et al. 2016). Various rate-limiting glycolytic enzymes like HK, PKM2, and LDH was also downregulated after the treatment of oleanolic acid (Fig. 2.1).

2.4.2.2 Tanshinone IIA

Tanshinone IIA (Fig. 2.1) is the most important lipophilic compound found in *Salvia miltiorrhiza* Bunge roots and can inhibit tumor cell growth and proliferation by inhibiting glucose metabolism (Lin et al. 2015a). At a concentration of 5.3 μM, tanshinone IIA downregulated the glucose-6-phosphate isomerize (GPI) in human stomach adenocarcinoma AGS cell line, which ultimately inhibits the glucose

Table 2.1 Antagonistic role of isoprenoids in the bioenergetics pathways of human cancer cells

Isoprenoids	Source	Part used	Cancer cell line studied	Inhibitory concentration	Treatment time (h)	Mechanism of action	Refs.
Thymoquinone	*Nigella sativa*	Seed oil	PANC-1; MIA PaCa-2	IC_{50} values were PANC-1: 23 ± 2 μM, MIA PaCa-2: 36 ± 0.28 μM	24 h	The anticancer mechanism of thymoquinone is due to the downregulation of PKM2 protein level in PANC-1 cells which results in inhibition of PKM2 protein activity. Whereas PKM2 protein activity was increased after thymoquinone treatment in MIA PaCa-2 cells	Pandita et al. (2014)
Galbanic acid	*Ferula* genus	Roots	NIH: OVCAR-3	IC_{50} values were NIH: OVCAR-3: 37, 12.1, 10 μM	24, 48 and 72 h	Galbanic acid downregulated the mRNA expression level of both HIF1α and HIF1β in both hypoxic or normoxic conditions. In hypoxic conditions, galbanic acid downregulated GLUT1 and Eno1 expression level, whereas in the normoxic condition, it	Eskandani et al. (2015)

(continued)

Table 2.1 (continued)

Isoprenoids	Source	Part used	Cancer cell line studied	Inhibitory concentration	Treatment time (h)	Mechanism of action	Refs.
						increased the degradation of EGFR	
Andrographolide	*Andrographis paniculata* Nees	Leaves	NB4; MV-4-11	IC_{50} values were NB4: 26 μM; MV-4-11: 43 μM	72 h	In MV-4-11 cells, andrographolide downregulated the expression of ASN, ACACA, and stromal interaction molecule 1 (STIM1) protein to suppress fatty acid synthesis and lipid metabolism, respectively, which results in inhibition of cell proliferation. Andrographolide also inhibited various fatty acid contents like palmitoleic acid, oleic acid, palmitic acid, and stearic acid	Chen et al. (2017b)
Ursolic acid	Plants (thyme, lavender, marjoram, and rosemary), fruits (apple fruit peel), flowers, and berries	Whole plant or fruit peel	MCF-7; MDA-MB-231; SK-BR-3	IC_{50} values were MCF-7: 20.44 μM; MDA-MB-231: 22.9 μM; SK-BR-3: 14.58 μM	24 h	At 20 μM concentration, ursolic acid downregulated the HK2, PKM2 protein expression, and ATP, lactate production via	Lewinska et al. (2017)

						inhibiting the Akt signaling pathway which ultimately suppressed glycolysis	
US597	Ursolic acid derivative		HepG2	IC_{50} values were HepG2: 7.20 ± 2.04 μM	24 h	US597 inhibited glycolysis by downregulating the HK activity which leads to reduced ATP and lactate production	Wang et al. (2014b)
Ginsenoside 20 (S)-Rg3	*Panax ginseng*		3AO; SKOV3	Data not available	24 h	Ginsenoside 20 (S)-Rg3 suppressed glycolysis by downregulating HK2, GLUT1, PKM2, and LDH. This inhibition leads to reduced glucose consumption and lactate production. This compound inhibited the HK2 activity by downregulating the phosphorylation of STAT3	Li et al. (2015)
2-Cyano-3,12-dioxooleana-1,9-dien-28-oic imidazolide (CDDO-Im)	Oleanolic acid derivative		LiSa-2	Data not available	Data not available	Treatment with CDDO-Im inhibited the mRNA expression, protein synthesis, and gene promoter activity of FAS. As a result,	Hughes et al. (2008)

(continued)

Table 2.1 (continued)

Isoprenoids	Source	Part used	Cancer cell line studied	Inhibitory concentration	Treatment time (h)	Mechanism of action	Refs.
						fatty acid synthesis was also downregulated	
Dihydroartemisinin	Artemisinin derivative		A549, PC-3, and H1975	IC_{50} values were A549: 42.2 μM (24 h), 25.1 μM (48 h); PC-3: 35.9 μM (24 h), 25.3 μM (48 h); H1975: ~60 μM	24 and 48 h	This compound suppressed the NF-?β signaling which also inhibits the GLUT1 translocation and leads to inhibition of the Warburg effect	Jiang et al. (2016)
Artesunate	Artemisinin derivative		HCT116	IC_{50} values were HCT116: 2.2 μM	24 h	Artesunate inhibited fatty acid metabolism by suppressing fatty acid biosynthetic proteins like fatty acid synthase (FASN), Acyl-CoA synthetase 5 (ACSL5), and hydroxyacyl-coenzyme A dehydrogenase (HADH)	Chen et al. (2017a)
Acetyltanshinone IIA	Tanshinone IIA derivative		SK-BR-3; MDA-MB-453	IC_{50} values were SK-BR-3: 9.17 ± 0.42 μM; MDA-MB-453: 1.97 ± 1.16 μM	24 h	This compound inhibits the lipid biosynthesis in these cancer cells by downregulating the	Guerram et al. (2015)

						key proteins ACC, p-ACL, and FAS	
Geranylgeranoic acid	Data not available	Data not available	Huh-7	Data not available	24 h	Geranylgeranoic acid inhibits fructose 6-phosphate and spermine and increases fructose 1, 6-bisphosphate, spermidine level. This compound shifts the energetic state of these cells from aerobic glycolysis to mitochondrial respiration by upregulating synthesis of cytochrome c oxidase 2 (SCO2) and TP53- induced glycolysis and apoptosis regulator (TIGAR) level	Iwao and Shidoji (2015)
Pseudolaric acid B	*Pseudolarix kaempferi* Gordon	Root bark	A549	Data not available	24 h	This compound upregulated the HK2 and GLUT1 protein level and thus increases the glucose uptake, ATP, and lactate production	Yao et al. (2017a)
Pristimerin	*Salacia cochinchinensis*	Roots	SK-BR-3	IC_{50} values were 2.4 μM	24 h	This quinone methide triterpenoid compound	Lee et al. (2013)

(continued)

Table 2.1 (continued)

Isoprenoids	Source	Part used	Cancer cell line studied	Inhibitory concentration	Treatment time (h)	Mechanism of action	Refs.
						inhibits the mRNA and protein expression of FAS and, hence, suppresses de novo fatty acid synthesis	
Cacalol	*Cacalia delphiniifolia*		MDA-MB-231; MCF-7	Data not available	48 h	Cacalol inhibits the mRNA and protein expression of FAS	Liu et al. (2011)

metabolism, intracellular ATP, and lactate production. Downregulation of malate dehydrogenase (MDH1) and L-lactate dehydrogenase B chains (LDHB) and upregulation of PCK2 by tanshinone IIA inhibited the gluconeogenesis process. In esophageal cancer cell line, Ec109, tanshinone IIA at a concentration of 0.40–1.70 μM downregulated the PKM2 mRNA and its protein expression and hence inhibited glycolysis (Zhang et al. 2016).

2.4.2.3 Geraniol

Geraniol (Fig. 2.1) is a monoterpenoid originated in the volatile oil of various aromatic plants and possesses anticancer activity (Cho et al. 2016). Geraniol inhibited the mevalonate synthesis by downregulating HMGCR activity in human breast cancer cells (MCF-7) (Duncan et al. 2004). In human prostate cancer cells (PC3), geraniol (0.25–1 μM) activated the AMPK pathway which increases the p-ACC protein level (Kim et al. 2012). In HepG2 HCC cells, geraniol inhibits fatty acid metabolism at a concentration of 50–200 μmol/L (Crespo et al. 2012; Polo and De Bravo 2006). It has entered that geraniol inhibits the HMGCR and CTP-PC cytidyl transferase which ultimately downregulated the mevalonate and phosphatidylcholine (PC) synthesis (Crespo et al. 2012).

2.4.2.4 Betulinic Acid

Betulinic acid (Fig. 2.1) is a pentacyclic lupine-triterpenoid (found mainly in outer barks of *Betula* spp.) that showed a negative impact on the tumor metabolism (Domingues et al. 2014). PKM2 protein level was suppressed after treatment with this dietary molecule in MIAPaCa-2 (25 μM) and PANC-1 (20 μM) cell line (Pandita et al. 2014). In MCF-7 and SK-BR-3 cells, betulinic acid downregulated the HK and PKM2 protein levels and inhibited the glycolysis process (Lewinska et al. 2017). In HeLa cells, betulinic acid inhibits the de novo fatty acid synthesis by downregulating the activity of SCD-1 (Potze et al. 2016).

2.4.2.5 Oridonin

This diterpenoid oridonin (Fig. 2.1) was isolated from *Rabdosia rubescens*. In colorectal cancer (SW480) cells, oridonin (1.25–20 μM) downregulated the mRNA expression and protein level of GLUT1 and MCT1 and inhibited the glucose uptake and lactate production and hence induce autophagy (Yao et al. 2017b). In uveal melanoma cells (MUM28 and OCM-1), oridonin (5 μM) suppressed the FAS protein level (Gu et al. 2015). The mRNA expression and protein level of FAS were inhibited after oridonin (5–10 μM) treatment in colorectal cancer (SW620 and SW480) cells which suppressed cellular fatty acid level (Kwan et al. 2013).

2.4.3 Phenolic Leads

These compounds comprise a major group of secondary metabolites containing hydroxylated aromatic ring system that may be flavonoids, stilbenes, phenolic acids, xanthones, and coumarins and possess various pharmacological activities like anticancer, antibacterial, antiviral, antioxidant, and antiatherosclerotic (Han et al. 2007). The anticancer potential of different phenolic compounds boasts researchers to explore the modulation of tumor cell metabolism by this group of compounds. In this chapter, we have provided the metabolic changes in various human cancer cells governed by phenolic compounds (Table 2.2). A few phenolic compounds with antimetabolite potential are described below.

2.4.3.1 Resveratrol

Resveratrol (Fig. 2.1) is a polyphenol compound mostly found in the berries, grapes, and other food sources, having various roles in cancer cell metabolism. Treatment of resveratrol in lung cancer cells (Li et al. 2016) and hepatocellular carcinoma cells (Dai et al. 2015; Guerra et al. 2018; Iqbal and Bamezai 2012; Massimi et al. 2012) decrease the glucose uptake and lactate production. In human ovarian cancer cells (MDAH2774, PA-1, SKOV3, and OVCAR3) resveratrol treatment (50 μM) inhibits the glucose uptake by suppressing the GLUT1 membrane localization (Gwak et al. 2015; Tan et al. 2016). In HCC cells (Bel-7402 and HCC-LM3) (Dai et al. 2015), non-small cell lung cancer (NSCLC) cells (H1650, H460, and HCC827) (Li et al. 2016), and LY18 human diffuse large B-cell lymphoma (Faber et al. 2006), resveratrol inhibit the glycolysis process by inhibiting the hexokinase 2 (HK2) protein expression level. At a concentration of 10–50 μM, resveratrol inhibits the expression of glycolytic enzyme phosphofructokinase (PFK), glucose consumption, and ATP production in human breast cancer cell lines (MCF-7) (Gomez et al. 2013). In MCF-7, HeLa, and HepG2 cells, resveratrol (50 μM) inhibits PKM2 expression via downregulating mammalian target of rapamycin (mTOR) protein, which, in turn, inhibits the glycolysis and macromolecular synthesis (Iqbal and Bamezai 2012). Resveratrol also inhibits the PFK and PGAM1 expression levels in the LnCap (Narayanan et al. 2004) and LY18 lymphoma cells (Faber et al. 2006). At a concentration of 50–150 μM, resveratrol inhibits the expression of phosphogluconate dehydrogenase (PGD), G6PD, and TKT (Vanamala et al. 2011). Resveratrol decreases lactate production and increases the ATP production in colon cancer cell lines at a concentration of 10 μM for 48 h (Saunier et al. 2017). It has been also proved that resveratrol can modulate cancer cell lipid metabolism. In HCC cells, resveratrol activates AMPK expression which ultimately increases the ACC phosphorylation (Hou et al. 2008; Shin et al. 2009). Resveratrol at a concentration of 30 μM showed a protective effect against mitochondrial dysfunction, arachidonic acid, and iron-induced ROS production by AMPK-mediated inhibitory phosphorylation of GSK3β downstream of poly (ADP-ribose) polymerase-LKB1 pathway (Shin et al. 2009). Resveratrol showed antiproliferative activity toward breast cancer cells by inhibiting both FAS mRNA and protein expression (Khan et al. 2014; Pandey et al. 2011). At a concentration of 20–100 μM, resveratrol increases the arachidonic acid and its metabolite 12S-HETE concentration in breast cancer cells (MDA-MB-231 and MCF-7) (Jäger et al. 2011). In MDA-MB-231 cells, resveratrol altered lipid metabolism by increasing the ceramide level which is an important component of sphingolipids associated in cell proliferation and apoptosis. In human erythroleukemia cell line (K-562), resveratrol (50–100 μM) treatment

Table 2.2 Antagonistic role of phenolic compounds on the bioenergetics pathways of human cancer cells

Phenolic compound	Source	Part used	Cancer cell line studied	Inhibitory concentration	Treatment time (h)	Mechanism of action	Refs.
1-(+)-Acetoxypinoresinol	Extra virgin olive oil (EVOO)		SK-BR-3, MCF-7	50 μM	48 h	Inhibit the expression of the lipogenic enzyme FASN in HER2-overexpressing breast carcinoma cells	Menendez et al. (2008)
Baicalin	*Scutellaria baicalensis*	Root	AGS	NA	48 h	Under hypoxic conditions, baicalin inhibited the enzymes responsible for glycolysis such as HK2, LDH-A, and PDK1. It decreased the glucose uptake as well as the lactate production rate	Chen et al. (2015)
Bavachinin	*Psoralea corylifolia*	Seed	KB	NA	24 h	Bavachinin suppressed angiogenesis and energy metabolism of transcription of genes controlled by glut 1 and hexokinase 2 under hypoxic conditions	Nepal et al. (2012)

(continued)

Table 2.2 (continued)

Phenolic compound	Source	Part used	Cancer cell line studied	Inhibitory concentration	Treatment time (h)	Mechanism of action	Refs.
Demethoxycurcumin	*Curcuma longa*	Rhizome	LNCaP, DU145, PC-3	~20	48 h	It inhibits the expression or activity of fatty acid synthase (FASN) and acetyl-CoA carboxylase (ACC)	Hung et al. (2012)
Daidzein	*Glycine max*	NA	MCF-7, MDA-MB-231, LNCap, PC-3	LNCap: 35.2, PC-3; 100	48 h	It potentiates the glucose uptake. Increases the GLUT-1 and GLUT-4 proteins level	Uifălean et al. (2016)
Deguelin	*Mundulea sericea*	NA	H460, H1650, H1299, H520, HCC827, H1975, H358)		48 h	It decreased glucose metabolism by blocking Akt-mediated hexokinase 2 expression Inhibit lactate production	Li et al. (2017a)
Gallic acid	Gallnuts, grape seeds and skin, tea leaves	Seed, leaves	B16F10		24 h	Aldolase, pyruvate kinase, glucokinase, and α-enolase are responsible for	Liu et al. (2014a)

						glycolysis; however, GA constantly upregulated these proteins and promote cellular apoptosis in B16 melanoma cells	
Genistein	Glycine max	NA	MCF-7, MDA-MB-231 HT29 MIA PaCa-2 LNCap PC-3	IC20 0f MCF-7: 22.44 and MDA-MB-231: 11.04 IC50 LNCap: 35.2, PC-3; 100	24 h (MCF-7, MDA-MB-231) 48 h	Inhibit the GLUT1 mRNA levels and stimulate the uptake p-ACC protein (in breast and colon cancer) In prostate cancer (LNCap) increased glucose uptake rates as well as GLUT1 and GLUT4 protein levels However, in PC-3 cells glucose uptake reduced GLUT1 level decreased and GLUT4 protein levels increased	Engel et al. (2012)
Genistein derivatives Gen-27	NA	NA	MDA-MB-231, MCF-7,	MDA-MB-231: 18.48 ± 0.46, 10.10 ± 0.68, 5.34 ± 0.32	(MDA-MB-231:24,48,72), MDA-MB-	Genistein derivatives Gen-27 potentially	Tao et al. (2017)

(continued)

Table 2.2 (continued)

Phenolic compound	Source	Part used	Cancer cell line studied	Inhibitory concentration	Treatment time (h)	Mechanism of action	Refs.
			MDA-MB-468	MDA-MB-468: 19.49 ± 0.06, 11.67 ± 1.01, 3.13 ± 0.23 MCF-7: 45.43 ± 4.09,16.94 ± 0.18, 6.67 ± 0.62	468:24,48,72), MCF-7	suppressed hexokinase 2, PKM2, and LDHA protein expressions. It downregulated the lactate production as well as ATP generation	
Hesperetin	Citrus fruit	Fruit	MDA-MB-231		24 h	Hesperetin block the glucose uptake via downregulation of glucose transporter 1 (GLUT1)	Yang et al. (2013)
Hydroxytyrosol	Olive oil	NA	SW620	~20	72 h	Hydroxytyrosol inhibits the fatty acid synthase (FAS) and decreased the proliferation rate and inducing apoptosis in colon cancer cells	Notarnicola et al. (2011)
Kaempferol		NA	MCF-7 MDA-MB-	MCF-7 and MDA-MB-231: IC50 of 4 μM (1.6–9.8)	24 h	MCF-7: Kaempferol inhibits the glucose	Azevedo et al. (2015)

			231, LNCap			cellular uptake by decreasing GLUT1 MDA-MB-231: It inhibits the lipid synthesis as well as fatty acid synthase (FAS). By suppressing lactate reuptake by breast cancer cells, kaempferol may also starve cells from lactate which is responsible for cell death	
Luteolin	Extra virgin olive oil (flavonoids)	NA	MCF-7, MDA-MB-231: MIA PaCa-2, LNCap	MCF-7 and MDA-MB-231: IC50 of 4 μM (1.6–9.8)	24 h	Luteolin inhibited the oncoprotein FASN, important chemoprevention and/or management of breast cancer in which FASN highly expressed results from HER2-driven oncogenic signaling. Luteolin significantly inhibited cholesterol, phospholipid, and triglyceride	Brusselmans et al. (2005)

(continued)

Table 2.2 (continued)

Phenolic compound	Source	Part used	Cancer cell line studied	Inhibitory concentration	Treatment time (h)	Mechanism of action	Refs.
						synthesis. Luteolin also showed its potentiality in suppressing fatty acid synthesis as well as other lipid derivatives in prostate and pancreatic cell lines	
Phloretin	Flavonoid	NA	LNCap PC-3	LNCap: 25 PC-3; 39	48 h	Phloretin increased the glucose uptake via stimulation of GLUT1 and GLUT4 proteins in prostate cells (LNCap). However, when treated in PC-3 cells, it reveals the opposite effect, glucose uptake reduced via decreased sensitivity of GLUT1 proteins	Gonzalez-Menendez et al. (2014)
Osthole	*Cnidium monnieri*	Fruit	SK-OV-3	~45	48 h	Osthole inhibited the	Lin et al. (2010)

						phosphorylation of Akt and mTOR. It also suppressed the lipid synthesis as well as fatty acid synthase (FAS)	
Rosmarinic acid	*Rosmarinus officinalis*	NA	MKN45 HCT8, HCT116	240.2 μM 298.1 μmol/L 319.8 μmol/L	24 h	Rosmarinic acid interferes in the glycolytic pathway and significantly inhibits the glucose consumption as well as lactate production in both gastric and colon carcinoma	Han et al. (2015)
Scutellarein	Related flavones	NA	MDA-MB-231		24 h	Inhibition of ECAR and oxygen consumption rate (OCR)	Chen et al. (2012)
Silibinin	*Silybum marianum*	Seeds	SW480 LNCaP, 22Rv1		6 and 12 h 24 and 48 h	Silibinin generated MAP2K1/2-MAPK1/3 and suppressed PIK3CA-AKT–mTOR pathways. For this reason, endoplasmic reticulum stress was activated and glucose uptake	Raina et al. (2013)

(continued)

Table 2.2 (continued)

Phenolic compound	Source	Part used	Cancer cell line studied	Inhibitory concentration	Treatment time (h)	Mechanism of action	Refs.
						capacity reduced. Thus silibinin interferes with glucose metabolism. Silibinin significantly inhibited angiogenesis and proliferation as well as decreased FASN, HIF-1α, and ACC levels. These outcomes correlate that silibinin effectiveness against PCa through inhibiting hypoxia-induced signaling	
Wogonin	*Scutellaria baicalensis* Georgi	NA	HCT116		24 h	Wogonin markedly inhibits the expression of glycolysis-associated proteins (PDHK1, HK2, LDHA), lactate	Wang et al. (2014a)

						generation, as well as glucose uptake in a dose-dependent manner	
Xanthohumol	*Humulus lupulus*	NA	HeLa A549		48 h	Xanthohumol interacts with mitochondrial electron transfer chain complex I, potentially suppressed oxidative phosphorylation, generates reactive oxygen species, and induces apoptosis. Xanthohumol also inhibits the extracellular acidification rate	Zhang et al. (2015a)

downregulated sphingomyelin, sphingosine-1-phosphate (SIP), and upregulated transcription level of acid sphingomyelinase mRNA (ASMase) which leads to accumulation of ceramide (Mizutani et al. 2016). On the other hand, resveratrol (10 μM) treatment in human colon cancer cells (Caco-2) reduced the sphingomyelins level which results increased level of phosphatidylethanolamines and ceramides (Saunier et al. 2017).

2.4.3.2 Curcumin

Curcumin (Fig. 2.1) is a major polyphenol found in the turmeric *Curcuma longa*, which has been reported to have activity on glucose uptake, metabolism, and transport. In human A549 (15–30 μmol/L) (Liao et al. 2015) and MCF-7 (5–10 μM) cells, curcumin inhibited GLUT1 mRNA and protein expression (Vaughan et al. 2013). Interestingly curcumin treatment increased or suppressed glycolysis in the human breast cancer cells has been reported. In MCF-7 cells, curcumin (5 μM) treatment improved HK2 activity, glucose uptake, and lactate production and inhibited mitochondrial respiration (Jung et al. 2016). In MDA-MB-231 cells, curcumin (5 μM) inhibited the HK2 expression and sensitize these cells to 4-hydroxytamoxifen treatment (Geng et al. 2016). In both human breast cancer cells (SKBR-3; 5–20 μM) and HCC HepG2 cells (5–15 μg/mL), curcumin inhibited the FAS protein expression and activity (Younesian et al. 2017). Curcumin treatment (10–40 μmol/L) inhibited the expression and activity of HK2 which leads to inhibition of glucose uptake, lactate production, and ATP generation without affecting other glycolytic enzyme (PFK, LDH, and phosphoglucomutase (PGM)) activities in human colorectal cancer cells (HCT29 and HCT116) (Wang et al. 2015). Furthermore, curcumin (1–10 μM) altered AMPK activity which, in turn, downregulated the mRNA and protein expression of HK2, PFKFB3, GLUT4, and PKM2 proteins in esophageal cancer cells (Ec109) (Zhang et al. 2015b). Curcumin was also reported to inhibit the glucose uptake and lactate production by inhibiting the PKM2 expression via downregulating mTOR-HIF1α pathway in various cancer cell lines (Siddiqui et al. 2018).

2.4.3.3 Quercetin

Quercetin (Fig. 2.1) is a flavonol found in various fruits and vegetables and has been investigated its role in glucose uptake and lactate efflux in various cancer cells (Amorim et al. 2015; Moreira et al. 2013; Santos et al. 2014). Quercetin (5–100 μM) translocates GLUT1, increases its expression, but decreases its function by competitive inhibition in cholangiocarcinoma (TFK-1) and liver cancer (HepG2, Hep3B2.1-7, and HuH7) cells(Brito et al. 2016). In MCF-7 and MDA-MB-231 cells, quercetin (100 μM) competitively inhibit the GLUT1 expression (Moreira et al. 2013). Quercetin (25 μM) effectively inhibits the lipogenesis process to 38% as compared to control cells by inhibiting the enzymatic activity of FAS protein in human breast cancer (MDA-MB-231) and prostate cancer (LnCaP) cells (Brusselmans et al. 2005). In the triple-negative breast cancer model, quercetin (150–415 μM) inhibits the FAS and β-catenin levels and induced cell apoptosis (Sultan et al. 2017). In human pancreatic adenocarcinoma (MIAPaCa-2) cells, quercetin (50 μM) inhibits glycogen synthesis (Harris et al. 2012). Quercetin (25–100 μM) has been reported to inhibit the expression and activity of FAS protein level in the HepG2 cell line (Zhao et al. 2014). In the HCC cell line, quercetin inhibits glycolysis by downregulating the HK2 and Akt/mTOR pathway (Wu et al. 2019).

2.4.3.4 Apigenin

Apigenin (Fig. 2.1) is a flavone that mostly originates in the vegetables and fruits and possesses potent anticancer activity. In human pancreatic cancer cells, apigenin (6.25–50 μM) inhibited the mRNA and protein level of GLUT1 by altering the PI3K-Akt pathway (Melstrom et al. 2008). Apigenin (10 μM) inhibited the glucose flux and ATP production by downregulating GLUT1 mRNA and protein expression without affecting other glycolytic enzymes in human lung cancer cells (Lee et al. 2016b). After apigenin exposure, both GLUT1 and GLUT4 levels were upregulated in androgen-sensitive LNCaP cells (7.1 μM), whereas in androgen-insensitive PC3 cells (15.7 μM), the GLUT1 protein expression was decreased and the expression of GLUT4 increased (Gonzalez-Menendez et al. 2014). In human colon cancer (HT29, DLD-1, and HCT116) cells, apigenin (10–40 μM) binds directly with PKM2 and altered its expression and activity and deceased glucose consumption, lactate, and ATP production (Shan et al. 2017). In breast cancer cells, specifically in epidermal growth factor receptor 2 (HER2)-overexpressing SKBR3 cells, apigenin (50 μM) significantly downregulated the FAS protein expression (Menendez et al. 2008).

2.4.3.5 Oroxylin A

Oroxylin A (Fig. 2.1) is a flavone isolated from *Oroxylum indicum* and *Scutellaria baicalensis and can modulate the glucose metabolism in various cancer cells* (Lee et al. 2016a). Oroxylin A (12.5–50 μM) inhibited the mRNA and protein expression of PDK, HK, PKM2, and LDHA in HepG2 cells under hypoxic (1% oxygen) condition rather than normoxic condition via modulating the HIF1α expression (Dai et al. 2016). Glycolysis was inhibited by oroxylin A (100–200 μM) in the human lung adenocarcinoma (A549) cells by downregulating the glycolytic protein HK2 expression (Wei et al. 2013). In MDA-MB-231 cells, oroxylin A (50–200 μM) downregulated the HK2 expression by suppressing HIF1α via sirtuin 3 (SIRT3) activation and hence inhibited glucose uptake and lactate production (Wei et al. 2015). Oroxylin A at a concentration of 12.5–50 μM modulates the TP53-induced glycolysis in HepG2 cells (Dai et al. 2013). Under hypoxic condition, oroxylin A (50–150 μM) inactivates HIF1α protein and downregulates the lipid synthesis and uptake by reprogramming fatty acid metabolism whereas decreases the intracellular fatty acid level and increases fatty acid oxidation in HCT116 cells (Ni et al. 2017).

2.4.3.6 Epigallocatechin Gallate (EGCG)

Epigallocatechin gallate (Fig. 2.1) is polyphenol mostly found in green tea with an antagonistic effect on cancer cell metabolism. EGCG (50–400 μM) downregulated the mRNA level of GLUT1, AMPK activation, and vascular endothelial growth factor (VEGF) reduction in human colon cancer cells (HT-29) (Hwang et al. 2007).

In these same cell lines, suppression of glucose uptake and upsurge of glutamine consumption was reported after EGCG (70–140 μM) treatment (Sánchez-Tena et al. 2013). In human tongue carcinoma (TSCCa and Tca8113) cells, EGCG (20–80 μM) inhibited glycolysis by downregulating the HK2 (Gao et al. 2015). EGCG (80 μM) inhibits the conversion of 3-phosphoglycerate (3-PG) to 2-phosphoglycerate (2-PG) by downregulating the expression of phosphoglycerate mutase 1 (PGAM1) in the breast (MDA-MB-231) and lung (NCI-H1299) cancer cells (Li et al. 2017b). Glucose uptake and lactate production were reduced after EGCG (10–100 μM) treatment in MCF-7 cells (Moreira et al. 2013). In MCF-7 cells, EGCG (10–40 μM) altered EGFR/PI3K/Akt/Sp-1 signaling network which ultimately inhibited the epidermal growth factor (EGF)-induced FAS expression (Yeh et al. 2003). The important constituent of cellular membrane palmitate synthesis was diminished after the downregulation of acetyl-CoA by ECGC (50 μM) treatment in human pancreatic adenocarcinoma (MIA PaCa-2) cells (Lu et al. 2015).

2.5 Bioinformatics Approaches

Traditionally, pharmacology and chemical science-based drug discovery of natural product techniques face various criticism in finding new drugs due to time constrain, a huge investment of money, and toxicity occurrence. The increasing demand for time-saving strategy to discover and develop selective and safe natural lead molecules has signaled the start of a new era of bioinformatics (Li et al. 2020). In bioinformatics analysis, it is possible by designing various computer algorithm to recognize or structurally transform a natural product, to design a lead (molecule) with the required function, and to assess its in silico therapeutic effects which further paved major foundation stone in the branch of science what is now called computer-aided drug designing or virtual screening of drugs (Fig. 2.4).

In the post-genomic era, computer-aided drug designing is one of the promising rational drug designing approaches to develop anticancer leads with plant-derived natural products or synthetic compounds that target cellular metabolic signaling networks. Multiple bioinformatics methodologies can be taken into encounter (like virtual screening, molecular docking, high-throughput screening, QSAR, pharmacophore, and modeling) to minimize the cost and time of drug identification, candidate screening, and refinement but also provide insights about the characterization of side effects and predict drug resistance (Jiang and Zhou 2005; Xia 2017). Several web-based servers or standalone softwares and online public repositories of natural product libraries are available for target identification, hit to lead generation, and lead optimization that help to understand clinical and preclinical findings. All these approaches are very useful for researchers targeting the antimetabolic lead discovery, assessment, and development.

The rise of cutting-edge technology, especially the use of high-throughput next-generation data such as metagenomic, transcriptomic, proteomics, epigenetic, ribosome profiling, and pharmacogenomics data, encompasses weighty contribution to mechanism-based drug discovery and drug repurposing (Tietz and Mitchell 2016).

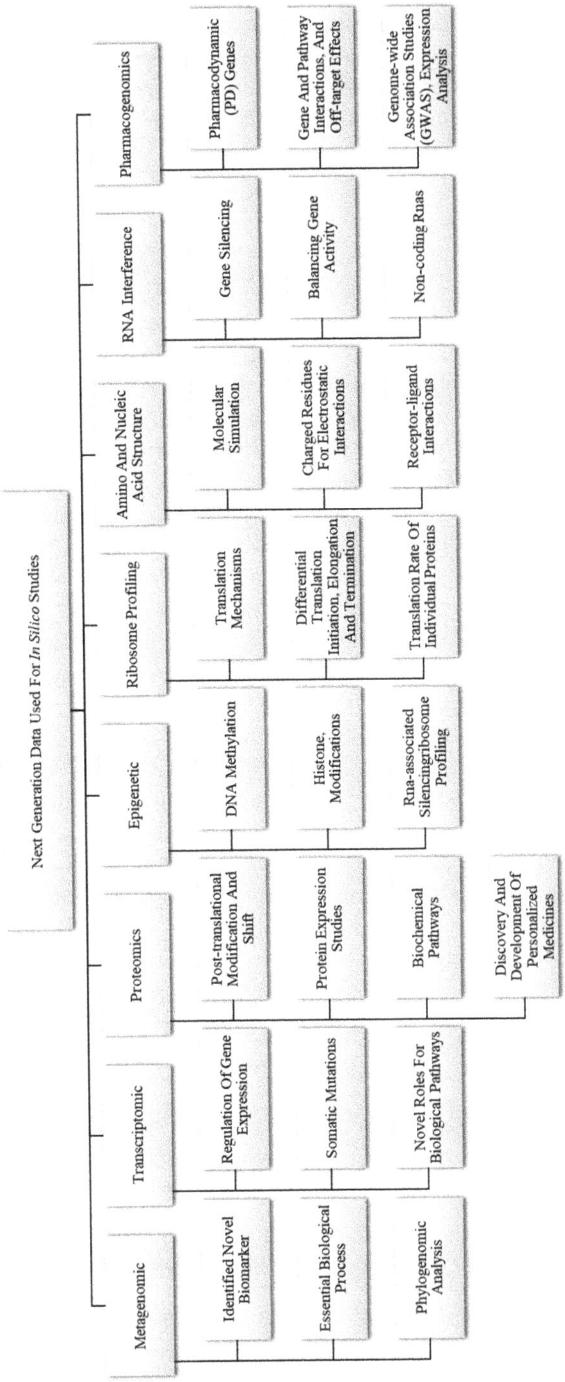

Fig. 2.4 In silico approaches for next-generation drug designing

Accumulation of amino acid and nucleotide structures, as well as the development of in silico homology modeling and molecular simulation, together with large structure databases of small molecules and metabolites, paved the way for more realistic receptor-ligand docking experiments and more informative virtual screening being uncovered biological insights whose deciphering can be the basis for scientific and economical success (Johnston et al. 2015).

2.6 Summary

Targeting cancer cell metabolism is a promising therapeutic approach that is still unexplored. In this chapter, we have highlighted many of the responsible proteins or enzymes and their modifications in the pre-carcinogenesis stage. Therefore, targeting them may be a promising strategy for chemopreventive purposes.

Identification of highly expressed or active proteins that perversely regulated the key metabolic signatures is important to develop a potential therapeutic drug. Although numerous efforts have been made for this, the effective number of lead molecules is still very low and in the preliminary stage. A promising candidate should possess toxicity against cancer cells rather than normal cells and also have favorable pharmacological profiles including stability, half-life, and bioavailability. We have reported that many natural products have been identified so far as a promising anticancer agent that targets various cancer metabolic proteins. Although these kinds of lead molecules provide many health benefits, sometimes poor pharmacokinetic properties like absorption, bioavailability is becoming a major concern that needs to explore.

Nature provides a vast source of different and diversified molecules. The rapid modernization of the screening system allows the selection and analysis of isolated natural compounds that are potent antimetabolic actions with reduced side effects. The application of synthetic chemistry and bioinformatics tools will be very much helpful to generate a more selective and less toxic lead in the distant future. Alternatively, inexpensive production of those scaffolds will make them commercially interesting.

References

Amantini C et al (2009) Triggering of transient receptor potential vanilloid type 1 (TRPV1) by capsaicin induces Fas/CD95-mediated apoptosis of urothelial cancer cells in an ATM-dependent manner. Carcinogenesis 30:1320–1329

Amara S, Zheng M, Tiriveedhi V (2016) Oleanolic acid inhibits high salt-induced exaggeration of Warburg-like metabolism in breast cancer cells. Cell Biochem Biophys 74:427–434

Amorim R, Pinheiro C, Miranda-Gonçalves V, Pereira H, Moyer MP, Preto A, Baltazar F (2015) Monocarboxylate transport inhibition potentiates the cytotoxic effect of 5-fluorouracil in colorectal cancer cells. Cancer Lett 365:68–78

Aniszewski T (2007) Alkaloids-secrets of life: alkaloid chemistry, biological significance, applications and ecological role. Elsevier, Netherland

Azevedo C, Correia-Branco A, Araújo JR, Guimarães JT, Keating E, Martel F (2015) The chemopreventive effect of the dietary compound kaempferol on the MCF-7 human breast cancer cell line is dependent on inhibition of glucose cellular uptake. Nutr Cancer 67:504–513

Barron CC, Bilan PJ, Tsakiridis T, Tsiani E (2016) Facilitative glucose transporters: implications for cancer detection, prognosis and treatment. Metabolism 65:124–139

Bernabeu E, Cagel M, Lagomarsino E, Moretton M, Chiappetta DA (2017) Paclitaxel: what has been done and the challenges remain ahead. Int J Pharm 526:474–495

Bhutani M, Pathak AK, Nair AS, Kunnumakkara AB, Guha S, Sethi G, Aggarwal BB (2007) Capsaicin is a novel blocker of constitutive and interleukin-6–inducible STAT3 activation. Clin Cancer Res 13:3024–3032

Bley K, Boorman G, Mohammad B, McKenzie D, Babbar S (2012) A comprehensive review of the carcinogenic and anticarcinogenic potential of capsaicin. Toxicol Pathol 40:847–873

Block KI et al. (2015) A broad-spectrum integrative design for cancer prevention and therapy. In: Seminars in cancer biology, vol Suppl. NIH Public Access, p S276

Boohaker RJ, Lee MW, Vishnubhotla P, Perez JM, Khaled AR (2012) The use of therapeutic peptides to target and to kill cancer cells. Curr Med Chem 19:3794–3804

Bovenzi CD et al (2015) Prognostic indications of elevated MCT4 and CD147 across cancer types: a meta-analysis. BioMed Res Int 2015:1–14

Bray F, Ferlay J, Soerjomataram I, Siegel RL, Torre LA, Jemal A (2018) Global cancer statistics 2018: GLOBOCAN estimates of incidence and mortality worldwide for 36 cancers in 185 countries. CA Cancer J Clin 68:394–424

Brederson J-D, Kym PR, Szallasi A (2013) Targeting TRP channels for pain relief. Eur J Pharmacol 716:61–76

Brito AF et al (2016) New approach for treatment of primary liver tumors: the role of quercetin. Nutr Cancer 68:250–266

Brusselmans K, Vrolix R, Verhoeven G, Swinnen JV (2005) Induction of cancer cell apoptosis by flavonoids is associated with their ability to inhibit fatty acid synthase activity. J Biol Chem 280:5636–5645

Butler MS, Robertson AA, Cooper MA (2014) Natural product and natural product derived drugs in clinical trials. Nat Prod Rep 31:1612–1661

Cao X et al (2007) Glucose uptake inhibitor sensitizes cancer cells to daunorubicin and overcomes drug resistance in hypoxia. Cancer Chemother Pharmacol 59:495–505

Chen V, Staub RE, Baggett S, Chimmani R, Tagliaferri M, Cohen I, Shtivelman E (2012) Identification and analysis of the active phytochemicals from the anti-cancer botanical extract Bezielle. PLoS One 7:e30107

Chen F et al (2015) Baicalein reverses hypoxia-induced 5-FU resistance in gastric cancer AGS cells through suppression of glycolysis and the PTEN/Akt/HIF-1α signaling pathway. Oncol Rep 33:457–463

Chen X, Wong Y, Lim T, Lim W, Lin Q, Wang J, Hua Z (2017a) Artesunate activates the intrinsic apoptosis of HCT116 cells through the suppression of fatty acid synthesis and the NF-κB pathway. Molecules 22:1272

Chen X et al (2017b) Andrographolide suppresses MV4-11 cell proliferation through the inhibition of FLT3 signaling, fatty acid synthesis and cellular iron uptake. Molecules 22:1444

Cheng Z, Wang K, Wei J, Lu X, Liu B (2010) Proteomic analysis of anti-tumor effects by tetrandrine treatment in HepG2 cells. Phytomedicine 17:1000–1005

Cho M, So I, Chun JN, Jeon J-H (2016) The antitumor effects of geraniol: Modulation of cancer hallmark pathways. Int J Oncol 48:1772–1782

Cho ES, Cha YH, Kim HS, Kim NH, Yook JI (2018) The pentose phosphate pathway as a potential target for cancer therapy. Biomol Ther 26:29

Chou H-C et al (2012) Proteomic and redox-proteomic analysis of berberine-induced cytotoxicity in breast cancer cells. J Proteomics 75:3158–3176

Chypre M, Zaidi N, Smans K (2012) ATP-citrate lyase: a mini-review. Biochem Biophys Res Commun 422:1–4

Clark R, Lee S-H (2016) Anticancer properties of capsaicin against human cancer. Anticancer Res 36:837–843
Cragg GM, Grothaus PG, Newman DJ (2009) Impact of natural products on developing new anticancer agents. Chem Rev 109:3012–3043
Cree IA, Charlton P (2017) Molecular chess? Hallmarks of anti-cancer drug resistance. BMC Cancer 17:10
Crespo R, Montero Villegas S, Abba MC, de Bravo MG, Polo MP (2012) Transcriptional and posttranscriptional inhibition of HMGCR and PC biosynthesis by geraniol in 2 Hep-G2 cell proliferation linked pathways. Biochem Cell Biol 91:131–139
Currie E, Schulze A, Zechner R, Walther TC, Farese RV Jr (2013) Cellular fatty acid metabolism and cancer. Cell Metab 18:153–161
Dai C, Whitesell L, Rogers AB, Lindquist S (2007) Heat shock factor 1 is a powerful multifaceted modifier of carcinogenesis. Cell 130:1005–1018
Dai Q et al (2013) Two p53-related metabolic regulators, TIGAR and SCO2, contribute to oroxylin A-mediated glucose metabolism in human hepatoma HepG2 cells. Int J Biochem Cell Biol 45:1468–1478
Dai W et al (2015) By reducing hexokinase 2, resveratrol induces apoptosis in HCC cells addicted to aerobic glycolysis and inhibits tumor growth in mice. Oncotarget 6:13703
Dai Q et al (2016) Oroxylin A regulates glucose metabolism in response to hypoxic stress with the involvement of hypoxia-inducible factor-1 in human hepatoma HepG2 cells. Mol Carcinog 55:1275–1289
Dayton TL, Jacks T, Vander Heiden MG (2016) PKM2, cancer metabolism, and the road ahead. EMBO Rep 17:1721–1730
Debnath P, Dey P, Chanda A, Bhakta T (2012) A survey on pineapple and its medicinal value. Scholars Acad J Pharm 1:24–29
Dey P, Bardalai D, Kumar NR, Subramani C, Mukherjee M, Bhakta T (2012a) Ethnomedicinal knowledge about various medicinal plants used by the tribes of Tripura. Res J Pharmacogn Phytochem 4:297–302
Dey P, Debnath P, Bhakta T (2012b) Evaluation of anthelmintic activity of Molineria Recurvata leaf extracts. Int Res J Pharm Appl Sci 2:16–20
Dey P, Das N (2013) Carbon nanotubes: it's role in modern health care. Int J Pharm Pharm Sci 5:9–13
Dey P, Karuna DS, Bhakta T (2014) Medicinal plants used as anti-acne agents by tribal and non-tribal people of Tripura, India. Am J Phytomed Clin Ther 2:556–570
Dey P et al (2019a) PKM2 knockdown induces autophagic cell death via AKT/mTOR pathway in human prostate cancer cells. Cell Physiol Biochem 52:1535–1552
Dey P et al (2019b) Knockdown of pyruvate kinase M2 inhibits cell proliferation, metabolism, and migration in renal cell carcinoma. Int J Mol Sci 20:5622
Dey P et al (2019c) Therapeutic value of steroidal alkaloids in cancer: current trends and future perspectives. Int J Cancer 9999:1–14
Do MT et al (2013) Antitumor efficacy of piperine in the treatment of human HER2-overexpressing breast cancer cells. Food Chem 141:2591–2599
Domingues RMA, Guerra AR, Duarte M, Freire CSR, Neto CP, Silva CMS, Silvestre AJD (2014) Bioactive triterpenic acids: from agroforestry biomass residues to promising therapeutic tools. Mini Rev Org Chem 11:382–399
Duncan RE, Lau D, El-Sohemy A, Archer MC (2004) Geraniol and β-ionone inhibit proliferation, cell cycle progression, and cyclin-dependent kinase 2 activity in MCF-7 breast cancer cells independent of effects on HMG-CoA reductase activity. Biochem Pharmacol 68:1739–1747
Engel N, Lisec J, Piechulla B, Nebe B (2012) Metabolic profiling reveals sphingosine-1-phosphate kinase 2 and lyase as key targets of (phyto-) estrogen action in the breast cancer cell line MCF-7 and not in MCF-12A. PLoS One 7:e47833

Eskandani M, Abdolalizadeh J, Hamishehkar H, Nazemiyeh H, Barar J (2015) Galbanic acid inhibits HIF-1α expression via EGFR/HIF-1α pathway in cancer cells. Fitoterapia 101:1–11

Faber AC, Dufort FJ, Blair D, Wagner D, Roberts MF, Chiles TC (2006) Inhibition of phosphatidylinositol 3-kinase-mediated glucose metabolism coincides with resveratrol-induced cell cycle arrest in human diffuse large B-cell lymphomas. Biochem Pharmacol 72:1246–1256

Fan L-X, Liu C-M, Gao A-H, Zhou Y-B, Li J (2013) Berberine combined with 2-deoxy-d-glucose synergistically enhances cancer cell proliferation inhibition via energy depletion and unfolded protein response disruption. Biochim Biophys Acta 1830:5175–5183

Fantin VR, St-Pierre J, Leder P (2006) Attenuation of LDH-A expression uncovers a link between glycolysis, mitochondrial physiology, and tumor maintenance. Cancer Cell 9:425–434

Flavin R, Peluso S, Nguyen PL, Loda M (2010) Fatty acid synthase as a potential therapeutic target in cancer. Future Oncol 6:551–562

Galano A, Martínez A (2012) Capsaicin, a tasty free radical scavenger: mechanism of action and kinetics. J Phys Chem B 116:1200–1208

Gao F, Li M, Liu W-B, Zhou Z-S, Zhang R, Li J-L, Zhou K-C (2015) Epigallocatechin gallate inhibits human tongue carcinoma cells via HK2-mediated glycolysis. Oncol Rep 33:1533–1539

Geng C et al (2016) Curcumin suppresses 4-hydroxytamoxifen resistance in breast cancer cells by targeting SLUG/hexokinase 2 pathway. Biochem Biophys Res Commun 473:147–153

Gomez LS, Zancan P, Marcondes MC, Ramos-Santos L, Meyer-Fernandes JR, Sola-Penna M, Da Silva D (2013) Resveratrol decreases breast cancer cell viability and glucose metabolism by inhibiting 6-phosphofructo-1-kinase. Biochimie 95:1336–1343

Gonzalez-Menendez P, Hevia D, Rodriguez-Garcia A, Mayo JC, Sainz RM (2014) Regulation of GLUT transporters by flavonoids in androgen-sensitive and-insensitive prostate cancer cells. Endocrinology 155:3238–3250

Gu Z et al (2015) Oridonin induces apoptosis in uveal melanoma cells by upregulation of Bim and downregulation of fatty acid synthase. Biochem Biophys Res Commun 457:187–193

Guerra AR, Duarte MF, Duarte IF (2018) Targeting tumor metabolism with plant-derived natural products: emerging trends in cancer therapy. J Agric Food Chem 66:10663–10685

Guerram M et al (2015) The potential utility of acetyltanshinone IIA in the treatment of HER2-overexpressed breast cancer: induction of cancer cell death by targeting apoptotic and metabolic signaling pathways. Oncotarget 6:21865

Gwak H, Haegeman G, Tsang BK, Song YS (2015) Cancer-specific interruption of glucose metabolism by resveratrol is mediated through inhibition of Akt/GLUT1 axis in ovarian cancer cells. Mol Carcinog 54:1529–1540

Han X, Shen T, Lou H (2007) Dietary polyphenols and their biological significance. Int J Mol Sci 8:950–988

Han S et al (2015) Anti-Warburg effect of rosmarinic acid via miR-155 in gastric cancer cells. Drug Design Dev Ther 9:2695

Harris DM, Li L, Chen M, Lagunero FT, Go VLW, Boros LG (2012) Diverse mechanisms of growth inhibition by luteolin, resveratrol, and quercetin in MIA PaCa-2 cells: a comparative glucose tracer study with the fatty acid synthase inhibitor C75. Metabolomics 8:201–210

He X et al (2017) PKM2 in carcinogenesis and oncotherapy. Oncotarget 8:110656

Hou X et al (2008) SIRT1 regulates hepatocyte lipid metabolism through activating AMP-activated protein kinase. J Biol Chem 283:20015–20026

Housman G, Byler S, Heerboth S, Lapinska K, Longacre M, Snyder N, Sarkar S (2014) Drug resistance in cancer: an overview. Cancers 6:1769–1792

Hughes DT, Martel PM, Kinlaw WB, Eisenberg BL (2008) The synthetic triterpenoid CDDO-Im inhibits fatty acid synthase expression and has antiproliferative and proapoptotic effects in human liposarcoma cells. Cancer Invest 26:118–127

Hung C-M, Su Y-H, Lin H-Y, Lin J-N, Liu L-C, Ho C-T, Way T-D (2012) Demethoxycurcumin modulates prostate cancer cell proliferation via AMPK-induced down-regulation of HSP70 and EGFR. J Agric Food Chem 60:8427–8434

Hwang J-T, Ha J, Park I-J, Lee S-K, Baik HW, Kim YM, Park OJ (2007) Apoptotic effect of EGCG in HT-29 colon cancer cells via AMPK signal pathway. Cancer Lett 247:115–121
Igal RA (2016) Stearoyl CoA desaturase-1: new insights into a central regulator of cancer metabolism. Biochim Biophys Acta 1861:1865–1880
Impheng H, Pongcharoen S, Richert L, Pekthong D, Srisawang P (2014) The selective target of capsaicin on FASN expression and de novo fatty acid synthesis mediated through ROS generation triggers apoptosis in HepG2 cells. PLoS one 9:e107842
Iqbal MA, Bamezai RN (2012) Resveratrol inhibits cancer cell metabolism by down regulating pyruvate kinase M2 via inhibition of mammalian target of rapamycin. PLoS One 7:e36764
Islam et al (2018) Phytol: a review of biomedical activities. Food Chem Toxicol 121:82–94
Iwao C, Shidoji Y (2015) Upregulation of energy metabolism-related, p53-target TIGAR and SCO2 in HuH-7 cells with p53 mutation by geranylgeranoic acid treatment. Biomed Res 36:371–381
Jäger W, Gruber A, Giessrigl B, Krupitza G, Szekeres T, Sonntag D (2011) Metabolomic analysis of resveratrol-induced effects in the human breast cancer cell lines MCF-7 and MDA-MB-231. Omics 15:9–14
Jantová S, Čipák L, Čerňáková M, Košt'álová D (2003) Effect of berberine on proliferation, cell cycle and apoptosis in HeLa and L1210 cells. J Pharm Pharmacol 55:1143–1149
Jeon M et al (2017) Site-selective rhodium (III)-catalyzed C– H amination of 7-azaindoles with anthranils: synthesis and anticancer evaluation. Adv Synth Catal 359(20):3471–3479
Jiang Z, Zhou Y (2005) Using bioinformatics for drug target identification from the genome. Am J Pharmacogenomics 5:387–396
Jiang J et al (2016) Repurposing the anti-malarial drug dihydroartemisinin suppresses metastasis of non-small-cell lung cancer via inhibiting NF-κB/GLUT1 axis. Oncotarget 7:87271
Johnston CW et al (2015) An automated genomes-to-natural products platform (GNP) for the discovery of modular natural products. Nat Commun 6:1–11
Jones R, Morris M (2016) Monocarboxylate transporters: therapeutic targets and prognostic factors in disease. Clin Pharmacol Ther 100:454–463
Jun H-S et al (2007) Capsaicin induced apoptosis of B16-F10 melanoma cells through down-regulation of Bcl-2. Food Chem Toxicol 45:708–715
Jung K-H, Lee JH, Park JW, Moon S-H, Cho YS, Choe YS, Lee K-H (2016) Effects of curcumin on cancer cell mitochondrial function and potential monitoring with 18F-FDG uptake. Oncol Rep 35:861–868
Kang J-H, Kim C-S, Han I-S, Kawada T, Yu R (2007) Capsaicin, a spicy component of hot peppers, modulates adipokine gene expression and protein release from obese-mouse adipose tissues and isolated adipocytes, and suppresses the inflammatory responses of adipose tissue macrophages. FEBS Lett 581:4389–4396
Karuna DS, Dey P, Das S, Kundu A, Bhakta T (2018) In vitro antioxidant activities of root extract of Asparagus racemosus Linn. J Trad Comp Med 8:60–65
Khan A, Aljarbou AN, Aldebasi YH, Faisal SM, Khan MA (2014) Resveratrol suppresses the proliferation of breast cancer cells by inhibiting fatty acid synthase signaling pathway. Cancer Epidemiol 38:765–772
Kim C-S, Kawada T, Kim B-S, Han I-S, Choe S-Y, Kurata T, Yu R (2003) Capsaicin exhibits anti-inflammatory property by inhibiting IkB-a degradation in LPS-stimulated peritoneal macrophages. Cell Signal 15:299–306
Kim S-H et al (2012) Geraniol induces cooperative interaction of apoptosis and autophagy to elicit cell death in PC-3 prostate cancer cells. Int J Oncol 40:1683–1690
Kushiro T, Ebizuka Y (2010) In comprehensive natural products II. In: Mander L, Liu HW (eds) Natural products structural diversity–I. Secondary metabolites: organization and biosynthesis, vol 1. Elsevier, Amsterdam
Kundu et al (2020) Protective effects of Croton hookeri on streptozotocin-induced diabetic nephropathy. Food Chem Toxicol 135:110873

Kwan H-Y, Yang Z, Fong W-F, Hu Y-M, Yu Z-L, Hsiao W-LW (2013) The anticancer effect of oridonin is mediated by fatty acid synthase suppression in human colorectal cancer cells. J Gastroenterol 48:182–192

Lee JS, Yoon IS, Lee MS, Cha EY, Thuong PT, Diep TT, Kim JR (2013) Anticancer activity of pristimerin in epidermal growth factor receptor 2-positive SKBR3 human breast cancer cells. Biol Pharm Bull 36:316–325

Lee A-Y, Kang S, Park S-J, Huang J, Im D-S (2016a) Anti-allergic effect of oroxylin a from oroxylum indicum using in vivo and in vitro experiments. Biomol Ther 24:283

Lee Y-M et al (2016b) Inhibition of glutamine utilization sensitizes lung cancer cells to apigenin-induced apoptosis resulting from metabolic and oxidative stress. Int J Oncol 48:399–408

Lewinska A, Adamczyk-Grochala J, Kwasniewicz E, Deregowska A, Wnuk M (2017) Ursolic acid-mediated changes in glycolytic pathway promote cytotoxic autophagy and apoptosis in phenotypically different breast cancer cells. Apoptosis 22:800–815

Li Z, Zhang H (2016) Reprogramming of glucose, fatty acid and amino acid metabolism for cancer progression. Cell Mol Life Sci 73:377–392

Li J, Liu T, Zhao L, Chen W, Hou H, Ye Z, Li X (2015) Ginsenoside 20 (S)-Rg3 inhibits the Warburg effect through STAT3 pathways in ovarian cancer cells. Int J Oncol 46:775–781

Li W et al (2016) Resveratrol inhibits hexokinases II mediated glycolysis in non-small cell lung cancer via targeting Akt signaling pathway. Exp Cell Res 349:320–327

Li W, Gao F, Ma X, Wang R, Dong X, Wang W (2017a) Deguelin inhibits non-small cell lung cancer via down-regulating Hexokinases II-mediated glycolysis. Oncotarget 8:32586

Li X et al (2017b) Identification of epigallocatechin-3-gallate as an inhibitor of phosphoglycerate mutase 1. Front Pharmacol 8:325

Li Z, Li H, Lu Y, Yang P, Li Z (2017c) Berberine inhibited the proliferation of Cancer cells by suppressing the activity of tumor pyruvate kinase M2. Nat Prod Commun 12:1934578X1701200909

Li K, Du Y, Li L, Wei D-Q (2020) Bioinformatics approaches for anti-cancer drug discovery. Curr Drug Targets 21:3–17

Liao H, Wang Z, Deng Z, Ren H, Li X (2015) Curcumin inhibits lung cancer invasion and metastasis by attenuating GLUT1/MT1-MMP/MMP2 pathway. Int J Clin Exp Med 8:8948

Lin VC-H et al (2010) Osthole suppresses fatty acid synthase expression in HER2-overexpressing breast cancer cells through modulating Akt/mTOR pathway. J Agric Food Chem 58:4786–4793

Lin L-L, Hsia C-R, Hsu C-L, Huang H-C, Juan H-F (2015a) Integrating transcriptomics and proteomics to show that tanshinone IIA suppresses cell growth by blocking glucose metabolism in gastric cancer cells. BMC Genomics 16:41

Lin R et al (2015b) 6-Phosphogluconate dehydrogenase links oxidative PPP, lipogenesis and tumour growth by inhibiting LKB1–AMPK signalling. Nat Cell Biol 17:1484

Lincet H, Icard P (2015) How do glycolytic enzymes favour cancer cell proliferation by nonmetabolic functions? Oncogene 34:3751

Liu W et al (2011) Cacalol, a natural sesquiterpene, induces apoptosis in breast cancer cells by modulating Akt-SREBP-FAS signaling pathway. Breast Cancer Res Treat 128:57–68

Liu C, Lin J-J, Yang Z-Y, Tsai C-C, Hsu J-L, Wu Y-J (2014a) Proteomic study reveals a co-occurrence of gallic acid-induced apoptosis and glycolysis in B16F10 melanoma cells. J Agric Food Chem 62:11672–11680

Liu J, Wu N, Ma L, Liu M, Liu G, Zhang Y, Lin X (2014b) Oleanolic acid suppresses aerobic glycolysis in cancer cells by switching pyruvate kinase type M isoforms. PLoS One 9:e91606

Liu J, Zheng L, Wu N, Ma L, Zhong J, Liu G, Lin X (2014c) Oleanolic acid induces metabolic adaptation in cancer cells by activating the AMP-activated protein kinase pathway. J Agric Food Chem 62:5528–5537

Liu Q, Luo Q, Halim A, Song G (2017) Targeting lipid metabolism of cancer cells: a promising therapeutic strategy for cancer. Cancer Lett 401:39–45

Lu C-W, Lin S-C, Chen K-F, Lai Y-Y, Tsai S-J (2008) Induction of pyruvate dehydrogenase kinase-3 by hypoxia-inducible factor-1 promotes metabolic switch and drug resistance. J Biol Chem 283:28106–28114

Lu H-F et al (2010) Antitumor activity of capsaicin on human colon cancer cells in vitro and colo 205 tumor xenografts in vivo. J Agric Food Chem 58:12999–13005

Lu Q-Y, Zhang L, Yee JK, Go V-LW, Lee W-N (2015) Metabolic consequences of LDHA inhibition by epigallocatechin gallate and oxamate in MIA PaCa-2 pancreatic cancer cells. Metabolomics 11:71–80

Luengo A, Gui DY, Vander Heiden MG (2017) Targeting metabolism for cancer therapy. Cell Chem Biol 24:1161–1180

Macheda ML, Rogers S, Best JD (2005) Molecular and cellular regulation of glucose transporter (GLUT) proteins in cancer. J Cell Physiol 202:654–662

Markman JL, Rekechenetskiy A, Holler E, Ljubimova JY (2013) Nanomedicine therapeutic approaches to overcome cancer drug resistance. Adv Drug Deliv Rev 65:1866–1879

Martinez-Outschoorn UE, Peiris-Pages M, Pestell RG, Sotgia F, Lisanti MP (2017) Cancer metabolism: a therapeutic perspective. Nat Rev Clin Oncol 14:11

Massimi M, Tomassini A, Sciubba F, Sobolev AP, Devirgiliis LC, Miccheli A (2012) Effects of resveratrol on HepG2 cells as revealed by 1H-NMR based metabolic profiling. Biochim Biophys Acta 1820:1–8

McBrayer SK, Cheng JC, Singhal S, Krett NL, Rosen ST, Shanmugam M (2012) Multiple myeloma exhibits novel dependence on GLUT4, GLUT8, and GLUT11: implications for glucose transporter-directed therapy. Blood 119:4686–4697

Melstrom LG, Salabat MR, Ding X-Z, Milam BM, Strouch M, Pelling JC, Bentrem DJ (2008) Apigenin inhibits the GLUT-1 glucose transporter and the phosphoinositide 3-kinase/Akt pathway in human pancreatic cancer cells. Pancreas 37:426–431

Menendez JA, Vazquez-Martin A, Oliveras-Ferraros C, Garcia-Villalba R, Carrasco-Pancorbo A, Fernandez-Gutierrez A, Segura-Carretero A (2008) Analyzing effects of extra-virgin olive oil polyphenols on breast cancer-associated fatty acid synthase protein expression using reverse-phase protein microarrays. Int J Mol Med 22:433–439

Miao P, Sheng S, Sun X, Liu J, Huang G (2013) Lactate dehydrogenase A in cancer: a promising target for diagnosis and therapy. IUBMB Life 65:904–910

Mizutani N et al (2016) Resveratrol-induced transcriptional up-regulation of ASMase (SMPD1) of human leukemia and cancer cells. Biochem Biophys Res Commun 470:851–856

Moreira L, Araújo I, Costa T, Correia-Branco A, Faria A, Martel F, Keating E (2013) Quercetin and epigallocatechin gallate inhibit glucose uptake and metabolism by breast cancer cells by an estrogen receptor-independent mechanism. Exp Cell Res 319:1784–1795

Narayanan NK, Narayanan BA, Nixon DW (2004) Resveratrol-induced cell growth inhibition and apoptosis is associated with modulation of phosphoglycerate mutase B in human prostate cancer cells: two-dimensional sodium dodecyl sulfate-polyacrylamide gel electrophoresis and mass spectrometry evaluation. Cancer Detect Prevent 28:443–452

Nepal M et al (2012) Anti-angiogenic and anti-tumor activity of Bavachinin by targeting hypoxia-inducible factor-1α. Eur J Pharmacol 691:28–37

Newman DJ, Cragg GM (2016) Natural products as sources of new drugs from 1981 to 2014. J Nat Prod 79:629–661

Ni T, He Z, Dai Y, Yao J, Guo Q, Wei L (2017) Oroxylin A suppresses the development and growth of colorectal cancer through reprogram of HIF1α-modulated fatty acid metabolism. Cell Death Dis 8:e2865

Notarnicola M et al (2011) Effects of olive oil polyphenols on fatty acid synthase gene expression and activity in human colorectal cancer cells. Genes Nutr 6:63

Pandey PR et al (2011) Resveratrol suppresses growth of cancer stem-like cells by inhibiting fatty acid synthase. Breast Cancer Res Treat 130:387–398

Pandita A, Kumar B, Manvati S, Vaishnavi S, Singh SK, Bamezai RN (2014) Synergistic combination of gemcitabine and dietary molecule induces apoptosis in pancreatic cancer cells and down regulates PKM2 expression. PLoS One 9:e107154

Patra KC, Hay N (2014) The pentose phosphate pathway and cancer. Trends Biochem Sci 39:347–354

Pavlova NN, Thompson CB (2016) The emerging hallmarks of cancer metabolism. Cell Metab 23:27–47

Pedersen PL, Mathupala S, Rempel A, Geschwind J, Ko YH (2002) Mitochondrial bound type II hexokinase: a key player in the growth and survival of many cancers and an ideal prospect for therapeutic intervention. Biochim Biophys Acta 1555:14–20

Pereira DM, Valentao P, Correia-da-Silva G, Teixeira N, Andrade PB (2012) Plant secondary metabolites in cancer chemotherapy: where are we? Curr Pharm Biotechnol 13:632–650

Polo MP, De Bravo MG (2006) Effect of geraniol on fatty-acid and mevalonate metabolism in the human hepatoma cell line Hep G2. Biochem Cell Biol 84:102–111

Potze L et al (2016) Betulinic acid induces a novel cell death pathway that depends on cardiolipin modification. Oncogene 35:427

Pramanik KC, Boreddy SR, Srivastava SK (2011) Role of mitochondrial electron transport chain complexes in capsaicin mediated oxidative stress leading to apoptosis in pancreatic cancer cells. PLoS One 6:e20151

Qi F et al (2010) Chinese herbal medicines as adjuvant treatment during chemoor radio-therapy for cancer. Biosci Trends 4(6):297–307

Raina K, Agarwal C, Wadhwa R, Serkova NJ, Agarwal R (2013) Energy deprivation by silibinin in colorectal cancer cells: a double-edged sword targeting both apoptotic and autophagic machineries. Autophagy 9:697–713

Sachan et al (2020) Dendropanax morbifera protects against renal fibrosis in streptozotocin-induced diabetic rats. Antioxidants 9:84

Sánchez-Tena S, Alcarraz-Vizán G, Marín S, Torres JL, Cascante M (2013) Epicatechin gallate impairs colon cancer cell metabolic productivity. J Agric Food Chem 61:4310–4317

Santos FM et al (2014) Differential sensitivities to lactate transport inhibitors of breast cancer cell lines. Endocr Relat Cancer 21(1):27–38

Saunier E, Benelli C, Bortoli S (2016) The pyruvate dehydrogenase complex in cancer: an old metabolic gatekeeper regulated by new pathways and pharmacological agents. Int J Cancer 138:809–817

Saunier E et al (2017) Resveratrol reverses the Warburg effect by targeting the pyruvate dehydrogenase complex in colon cancer cells. Sci Rep 7:6945

Scatena R, Bottoni P, Pontoglio A, Mastrototaro L, Giardina B (2008) Glycolytic enzyme inhibitors in cancer treatment. Expert Opinion Investig Drugs 17:1533–1545

Seca A, Pinto D (2018) Plant secondary metabolites as anticancer agents: successes in clinical trials and therapeutic application. Int J Mol Sci 19:263

Shan S, Shi J, Yang P, Jia B, Wu H, Zhang X, Li Z (2017) Apigenin restrains colon cancer cell proliferation via targeted blocking of pyruvate kinase M2-dependent glycolysis. J Agric Food Chem 65:8136–8144

Shin SM, Cho IJ, Kim SG (2009) Resveratrol protects mitochondria against oxidative stress through AMP-activated protein kinase-mediated glycogen synthase kinase-3β inhibition downstream of poly (ADP-ribose) polymerase-LKB1 pathway. Mol Pharmacol 76:884–895

Siddiqui FA et al (2018) Curcumin decreases Warburg effect in cancer cells by down-regulating pyruvate kinase M2 via mTOR-HIF1α inhibition. Sci Rep 8:8323

Slezakova S, Ruda-Kucerova J (2017) Anticancer activity of artemisinin and its derivatives. Anticancer Res 37:5995–6003

Sultan AS, Khalil MI, Sami BM, Alkhuriji AF, Sadek O (2017) Quercetin induces apoptosis in triple-negative breast cancer cells via inhibiting fatty acid synthase and β-catenin. Int J Clin Exp Pathol 10:156–172

Tan W et al (2015) Berberine interfered with breast cancer cells metabolism, balancing energy homeostasis. Anticancer Agents Med Chem 15:66–78

Tan L et al (2016) Resveratrol inhibits ovarian tumor growth in an in vivo mouse model. Cancer 122:722–729

Tao L et al (2017) Gen-27, a newly synthesized flavonoid, inhibits glycolysis and induces cell apoptosis via suppression of hexokinase II in human breast cancer cells. Biochem Pharmacol 125:12–25

Tennant DA, Durán RV, Gottlieb E (2010) Targeting metabolic transformation for cancer therapy. Nat Rev Cancer 10:267

Thakor AS, Gambhir SS (2013) Nanooncology: the future of cancer diagnosis and therapy. CA Cancer J Clin 63:395–418

Thun MJ, DeLancey JO, Center MM, Jemal A, Ward EM (2009) The global burden of cancer: priorities for prevention. Carcinogenesis 31:100–110

Tietz JI, Mitchell DA (2016) Using genomics for natural product structure elucidation. Curr Top Med Chem 16:1645–1694

Uifălean A, Schneider S, Gierok P, Ionescu C, Iuga C, Lalk M (2016) The impact of soy isoflavones on MCF-7 and MDA-MB-231 breast cancer cells using a global metabolomic approach. Int J Mol Sci 17:1443

Vanamala J, Radhakrishnan S, Reddivari L, Bhat VB, Ptitsyn A (2011) Resveratrol suppresses human colon cancer cell proliferation and induces apoptosis via targeting the pentose phosphate and the talin-FAK signaling pathways-A proteomic approach. Proteome Sci 9:49

Vander Heiden MG, Cantley LC, Thompson CB (2009) Understanding the Warburg effect: the metabolic requirements of cell proliferation. Science 324:1029–1033

Vaughan RA, Garcia-Smith R, Dorsey J, Griffith JK, Bisoffi M, Trujillo KA (2013) Tumor necrosis factor alpha induces Warburg-like metabolism and is reversed by anti-inflammatory curcumin in breast epithelial cells. Int J Cancer 133:2504–2510

Wang H et al (2014a) Wogonin reverses hypoxia resistance of human colon cancer HCT116 cells via downregulation of HIF-1α and glycolysis, by inhibiting PI3K/Akt signaling pathway. Mol Carcinog 53:E107–E118

Wang J et al (2014b) Synergism of ursolic acid derivative US597 with 2-deoxy-D-glucose to preferentially induce tumor cell death by dual-targeting of apoptosis and glycolysis. Sci Rep 4:5006

Wang K et al (2015) Curcumin inhibits aerobic glycolysis and induces mitochondrial-mediated apoptosis through hexokinase II in human colorectal cancer cells in vitro. Anticancer Drugs 26:15–24

Wang J et al (2017) Glucose transporter GLUT1 expression and clinical outcome in solid tumors: a systematic review and meta-analysis. Oncotarget 8:16875

Wei L, Dai Q, Zhou Y, Zou M, Li Z, Lu N, Guo Q (2013) Oroxylin A sensitizes non-small cell lung cancer cells to anoikis via glucose-deprivation-like mechanisms: c-Src and hexokinase II. Biochim Biophys Acta 1830:3835–3845

Wei L et al (2015) Oroxylin A inhibits glycolysis-dependent proliferation of human breast cancer via promoting SIRT3-mediated SOD2 transcription and HIF1α destabilization. Cell Death Dis 6:e1714

Withers ST, Keasling JD (2007) Biosynthesis and engineering of isoprenoid small molecules. Appl Microbiol Biotechnol 73:980–990

Wu H et al (2019) Quercetin inhibits the proliferation of glycolysis-addicted HCC cells by reducing hexokinase 2 and Akt-mTOR pathway. Molecules 24:1993

Wulaningsih W et al (2015) Serum lactate dehydrogenase and survival following cancer diagnosis. Br J Cancer 113:1389

Xia X (2017) Bioinformatics and drug discovery. Curr Top Med Chem 17:1709–1726

Yalcin A, Telang S, Clem B, Chesney J (2009) Regulation of glucose metabolism by 6-phosphofructo-2-kinase/fructose-2, 6-bisphosphatases in cancer. Exp Mol Pathol 86:174–179

Yang W, Lu Z (2015) Pyruvate kinase M2 at a glance. J Cell Sci 128:1655–1660

Yang KM et al (2009) Capsaicin induces apoptosis by generating reactive oxygen species and disrupting mitochondrial transmembrane potential in human colon cancer cell lines. Cell Mol Biol Lett 14:497

Yang Y, Wolfram J, Boom K, Fang X, Shen H, Ferrari M (2013) Hesperetin impairs glucose uptake and inhibits proliferation of breast cancer cells. Cell Biochem Funct 31:374–379

Yao G-D et al (2017a) Blocking the utilization of glucose induces the switch from senescence to apoptosis in pseudolaric acid B-treated human lung cancer cells in vitro. Acta Pharmacol Sin 38:1401

Yao Z et al (2017b) Oridonin induces autophagy via inhibition of glucose metabolism in p53-mutated colorectal cancer cells. Cell Death Dis 8:e2633

Yeh C, Chen W, Chiang C, Lin-Shiau S, Lin J (2003) Suppression of fatty acid synthase in MCF-7 breast cancer cells by tea and tea polyphenols: a possible mechanism for their hypolipidemic effects. Pharmacogenomics J 3:267

Yoshitani S-I, Tanaka T, Kohno H, Takashima S (2001) Chemoprevention of azoxymethane-induced rat colon carcinogenesis by dietary capsaicin and rotenone. Int J Oncol 19:929–939

Younesian O et al. (2017) Effect of curcumin on fatty acid synthase expression and enzyme activity in breast cancer cell line SKBR3. Int J Cancer Manag 10:e8173

Zhang R, Humphreys I, Sahu RP, Shi Y, Srivastava SK (2008) In vitro and in vivo induction of apoptosis by capsaicin in pancreatic cancer cells is mediated through ROS generation and mitochondrial death pathway. Apoptosis 13:1465–1478

Zhang B, Chu W, Wei P (2015a) Liu Y, Wei T. Xanthohumol induces generation of reactive oxygen species and triggers apoptosis through inhibition of mitochondrial electron transfer chain complex I Free Radical Biology and Medicine 89:486–497

Zhang F-J, Zhang H-S, Liu Y, Huang Y-H (2015b) Curcumin inhibits Ec109 cell growth via an AMPK-mediated metabolic switch. Life Sci 134:49–55

Zhang H-S, Zhang F-J, Li H, Liu Y, Du G-Y, Huang Y-H (2016) Tanshinone IIA inhibits human esophageal cancer cell growth through miR-122-mediated PKM2 down-regulation. Arch Biochem Biophys 598:50–56

Zhao Y et al (2009) Upregulation of lactate dehydrogenase A by ErbB2 through heat shock factor 1 promotes breast cancer cell glycolysis and growth. Oncogene 28:3689

Zhao Y, Butler EB, Tan M (2013) Targeting cellular metabolism to improve cancer therapeutics. Cell Death Dis 4:e532

Zhao P, Mao JM, Zhang SY, Zhou ZQ, Tan Y, Zhang Y (2014) Quercetin induces HepG2 cell apoptosis by inhibiting fatty acid biosynthesis. Oncol Lett 8:765–769

Omics Technologies and Development of Anti-diabetic Therapies from Prospective Natural Products

3

Madeleine Fleming, G. David Lin, and Rachel W. Li

Abstract

The emergence of omics technologies has revolutionised modern biology and also offers promise to addressing epidemic chronic diseases including type 2 diabetes mellitus (T2DM). Sustained peripheral insulin resistance and an irreversible systemic alteration and progressive decline of physiological functions in T2DM patients are borne of modern unhealthy lifestyle and contribute to disability and mortality worldwide. Earlier-onset and prolonged disease state underscore the need for both prophylactic and ameliorating therapeutics to impact the chronicity of T2DM and progressive comorbidities. This complexity has not been solved with prolonged use of tissue-target screens for synthetic drug discovery. The modern disease of T2DM can be impacted with recognised anti-diabetic agents derived from traditional medicines and phytochemicals. Exploit of phytochemicals employed in medicinal practices can now be validated with omics technologies. Constructive omics testing offers new and meaningful insights to improving drug discovery and implementation. This chapter reviews the application of genomics, transcriptomics, proteomics, metabolomics and network analysis to ascertain candidature of anti-diabetic agents arising from green tea, citrus, and garlic isolates including (−)-epigallocatechin gallate, tangeretin, and allicin. The restoration of the old natural products and new technologies will progress pharmacological and pathological insights into T2DM beyond current allopathic measures.

M. Fleming · R. W. Li (✉)
The Medical School and John Curtin School of Medical Research, Australian National University, Canberra, ACT, Australia
e-mail: Rachel.li@anu.edu.au

G. D. Lin
Research School of Chemistry, Australian National University, Canberra, ACT, Australia

S. C. Mandal et al. (eds.), *Evidence Based Validation of Traditional Medicines*,
https://doi.org/10.1007/978-981-15-8127-4_3

Keywords

Type 2 diabetes mellitus · Tangeretin · (−)-Epigallocatechin gallate · Allicin · Transcriptomics · Proteomics · Metabolomics

Abbreviations

^{1}H-NMR	Proton nuclear magnetic resonance
BW	Body weight
EGCG	(−)-Epigallocatechin 3-gallate
GC-MS	Gas chromatography- mass spectroscopy
GLUT 2	Glucose transporter 2
GWAS	Genome-wide association studies
HPLC	High-performance liquid chromatography
MS	Mass spectroscopy
PPAR	Peroxisome proliferator activator receptor
RNA	Ribonucleic acid
SGLT2	Sodium/glucose co-transporter 2
T2DM	Type 2 diabetes mellitus
UPLC	Ultrahigh performance liquid chromatography

3.1 Introduction

Diabetes (*passing through*) Mellitus (*sweet urine*), describes the observed excess of glucose present in copious patient urine. The disease is classified into two syndromes, differentiated by mode of insulin depletion. Type 1 diabetes is a failure to produce adequate insulin due to pancreatic beta-cell destruction or loss of function leading to peripheral excess of glucose and other metabolites. Type 2 diabetes (T2DM) is considered a failure to compensate for excessive glucose intake, whereby pancreatic beta-cell secretion is exhausted and ultimately fails. The prevalence of T2DM is rising (Krug 2016) facilitated by the modern lifestyle, yet incidence reveals remarkable heterogeneity obscuring mechanistic knowledge. Indeed, the unprecedented increase in total diabetic disease has raised informal discourse of a broader spectrum of mellitus diseases, with some categorisation into *Type 1.5* and/or *Type 3* to conjecturally support the notion of interrelated phenotypes. Diagnostically, a glycated haemoglobin level exceeding 7.5% is a well-accepted measure of unmanaged T2DM disease. Glucose inundation creates systemic toxicity and ensues metabolic and immune dysregulation, promoting chronic low-grade inflammation and chronic homeostatic compensations perpetuating the diabetic milieu.

Current anti-diabetic drugs remain expensive and are limited in outcomes. The pharmacokinetic balance of optimal specificity and minimal side effects is evidently challenging with the exceptional multiplicity of diabetic pathology encompassing all tissues. In contrast to historical incidence, rising chronicity and accompanying

Table 3.1 Representative anti-diabetic natural products

Code	Name	Structure
a	(−)-epigallocatechin 3-gallate	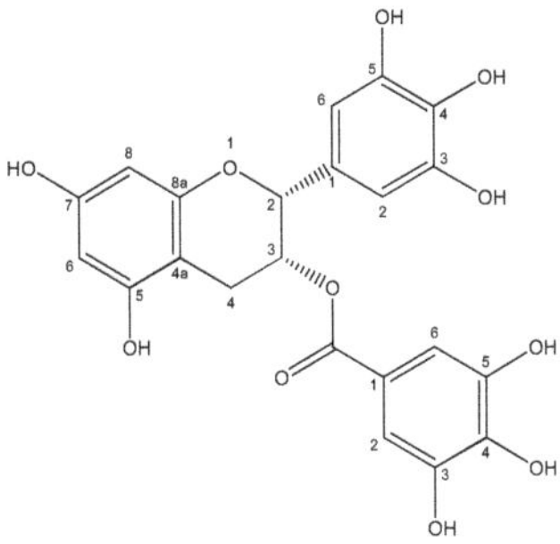
b	Allicin	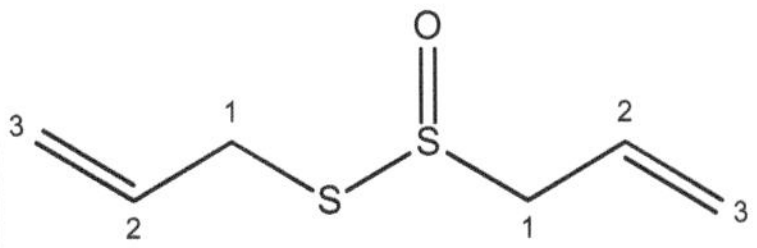
c	Tangeretin	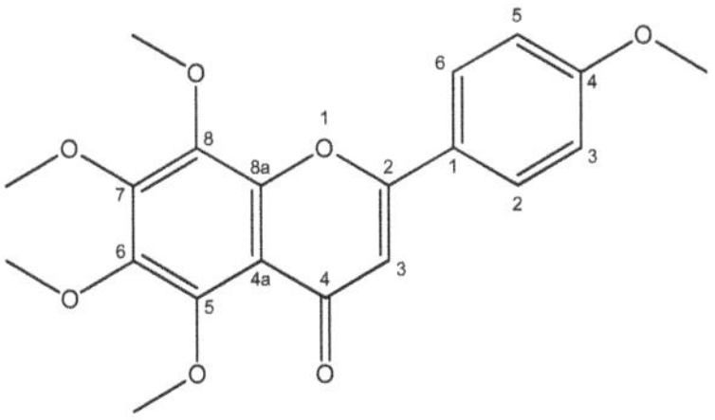

burden of disease demands capitalisation of *known* anti-inflammatory, anti-oxidant and anti-diabetic sources drawn from conserved cultural practices of traditional medicine. The abundance of natural products and their broad use in medicinal and biological applications present foundations to develop impactful anti-diabetic therapies. This chapter discusses natural products tangeretin, allicin and (−)-epigallocatechin gallate (Table 3.1) as rational candidates for derivative pursuit. The discussion features the superior means offered by 'omics' technologies to expand the evidence base obscured at population level owing to incomplete disease knowledge. Such repositioning could realise anti-diabetic potential of natural products; however, studies must confer translatability to contribute the critical data for pharmacological benefit.

3.2 Omics Technology Appraisal

The suffix '-omics' denotes the modern comprehensive technologies to assess cellular response at discrete biological demarcations of a cell population from genome to metabolome (and beyond). The superior approach of modern omics (using transcriptomic analysis for train variants in T2DM) surpasses conventional biological methods (genome wide asociation studies or expression quantitative trait locus) (Jenkinson et al. 2016). The extensive arrays based on high-throughput and multi-omics approaches provide quantitative assessment of functional discrepancies between groups, such as those present in a disease versus non-disease cohort whereas deep homogeneous sampling cannot infer effect causation with the substantial variation within and between individuals, underscoring ambiguous directionality of putative associations (Hasin et al. 2017). Computational biology has enabled processing of perturbed cell responses non-directionally to represent complex biological interactions of T2DM and prospective therapeutic interventions.

Superior interaction assessment of omics could address the articulated shortcomings of conventional drug discovery studies. High-throughput screening attests to the improved hit rate of natural substrate interactions with biological membranes (0.03%) compared to synthetics (<0.001%) (Isgut et al. 2018). Biological plausibility of natural products is further supported by conserved medicinal referral and complementary scientific findings. Previous studies have been unable to account for molecular interactions within a given product, let alone in subject response system using conventional biological techniques, diminishing outcomes and development following clinical trials.

The repositioning of traditional medical products is justified by both social and biological acceptability for semi-novel T2DM therapeutics. T2DM omics assessment complemented by animal and cell studies holds promise to construing inadvertent mechanistic insights of disease in the pursuit of drug discovery. Further analysis is essential to constructing true associations and resolving specific mechanisms of diabetic disease in pursuit of natural product transformation for successful drug development.

3.2.1 Genomics

Proceeding the fundamental *Human Genome Project*, genome-wide association studies (GWAS) have been performed on large cohorts for diabetic pathophysiology. Resultant single nucleotide polymorphisms and comparative analyses of variant profiles have marginally improved T2DM aetiology (Siddiqui and Tyagi 2015; Jain et al. 2013). True discrimination of contributing variants from coincident mutations is not possible: nor is the extent of cooperation or synergism posed by mutant loci, preventing causative interpretations between non- and diabetic cohorts.

Later iterative modifications to differential gene expression also confound direct causality thereby preventing direct inference to specific pathways or genes for therapeutic exploit. Thus the complexity of polygenic diabetes in the context of lifestyle-mediation remains to be assessed by combining genome sequencing with other omics data. Such comprehensive analyses with reference to the genome may address the articulated simplification and oversight of shared disease pathways augmented in modern T2 diabetic disease. Integrating genomics and other data sets could also combat the criticism of underwhelming and superficial genomic-drug studies conducted for T2DM (Siddiqui and Tyagi 2015).

3.2.2 Transcriptomics

Transcriptomics is the expression of total RNA population in a sample to quantify differential variation at transcriptional and post-transcriptional phases with RNA sequencing and component assays. Choreography of component RNA and associated proteins forms the dynamically attuned signalling responses constituting significant variation unexplained with GWAS alone.

Transcriptome scrutiny may support the notion of a rare variant, common disease phenotype for T2DM which may be unapparent with genomic analysis (Jenkinson et al. 2016). Alternatively RNA processing may contribute greater diversity to observed heterogeneity of T2DM than previously appreciated (Jain et al. 2013). This facet of analysis could contribute subtle new indicators of discrepancy in minority or neglected cohorts to better elucidate disease pleiotropy, for comprehensive profiling and subsequent therapeutic drug development.

3.2.3 Proteomics

Proteomics provides the quantitative assessment of cellular protein content as the functional product of gene expression. Protein composition and structures are assessed with high throughput analysis of cell modifications in response to a stimulus. Measurements of abundance, molecular weight and conformation are used to infer enrichment, reduction and post-translational modifications: reflecting stability, affinity, transport, activation or degradation. Mass spectroscopy analysis can differentiate functional effects indirectly with comparative shifts in fragment composition and prevalence.

Functional modifications can be further probed with protein imaging and expression techniques including gel electrophoresis, digestion assays and peptide sequencing to confirm aberrations. To cite a study on Traditional Chinese Medicines, their mechanisms of action including anti-diabetic properties confirmed with bioinformatics analysis of proteomic data (Yang et al. 2019).

3.2.4 Metabolomics

Metabolomics characterises the metabolite aftermath of cellular processes and considers a 'fingerprint' of molecular markers constituting amino acids, fatty acids, carbohydrates and other small molecules. The investigations of intermediates, secondary metabolites, hormones and other signalling molecule pathway up- or downregulated levels manifest altered cellular metabolism, indicative of disease or adaptation. Systematic profiling is performed with proton nuclear magnetic resonance imaging (e.g. ^{1}H NMR), providing visual comparisons of metabolic spectra. Metabolites are also quantified by various chromatography techniques coupled with mass spectroscopy for chemical composite data. Metabolomic profiling is crucial to detecting and differentiating the notorious complexity of natural products, often polyherbal or combination products requiring differentiation of active and spectator compounds for mechanistic and pharmacological clarification. For example, Sun et al. (2014) conducted a study of urinary metabolites in a T2DM rat model to posit reconstruction of associated known metabolites implicated in the disease target pathway.

3.2.5 Multi-omics and Network Analysis

Multi-omics and network analysis encompass the integration of composite omics data sub-sets for network construction and pattern recognition (Fig. 3.1) to map putative associations representing cellular interactions for biological response caused by natural products from functional protein to signal and gene expression changes. A biological cascade cannot be inferred with the multi-metric complexity and inability to resolve directionality when representing systemic and combinatorically diverse interactions of high throughput data. Such inadvertent interrelationships require correction for trans-omics prevalent objects with the advent of mediation analysis. Higher-order processing of multi-omics holds prospect to associating pharmacodynamics altered by diabetic-stasis if the inevitable 'matrix

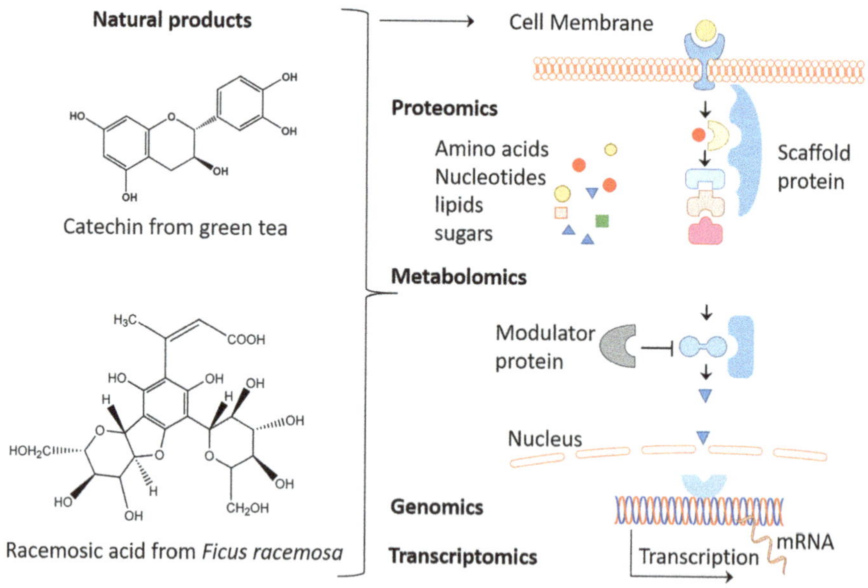

Fig. 3.1 Sketch of prospective natural products (e.g. catechin and racemosic acid) impacting on multi-omics and network analysis

effect' is addressed with multi-metric analyses (Hasin et al. 2017). Other omics iterations, immunolomics, ionomics, toxicogenomics and lipidomics, have emerged, prompting the integration of rich data for pathological and pharmacological insights.

Larger network analysis is also implicated in the paradigm shift from systems to network pharmacology. Integrative analysis can address the modern complexed diabetes and comorbid diseases such as obesity and hypertension. A study by Bezerra and colleagues analysed diabetes exclusively and in ascending comorbid combinations, finding 217 associated genes (Bezerra et al. 2016). In addition to this astounding interactivity, the notion of individualising Hemoglobin A1c thresholds has been proposed by Siddiqui and Tyagi (2015) as another measure to better stratify individual variation. Multi-metric data present prime opportunity for data scalability to deduce meaningful associations at individual and population levels, which underpins the advancement of disease knowledge and therapeutic direction.

3.3 Anti-diabetic Natural Products for Consideration

Medicinal phytochemicals have been exhaustively characterised for reverting or preventing cellular stress in the context of chronic diseases (Thomson 2010). Resultant findings of anti-oxidant, anti-inflammatory and free radical scavenging effects concur with traditional medicinal utility to promote longevity and youth (Shen et al. 2017). Beyond prophylactic means, scientific findings validate and implicate the concurrence of these practices for chronic disease impact including T2DM. The modest repositioning of phytochemicals to diabetic disease models and cell line experiments has posited compelling evidence for metabolic corrections mediated by phytochemical effects. The selected foci of this chapter, tangeretin, allicin and (−)-epigallocatechin 3-gallate, are contending sources of anti-diabetic development. Each case, discussed in current and future evaluations, builds the evidence base to demonstrate the potential of omics technologies to address the issues encountered with prior methods and the impinged translation to current populations.

3.3.1 (−)-Epigallocatechin3-Gallate

(−)-Epigallocatechin 3-gallate (Table 3.1a) or (−)-EGCG is the most abundant of the flavan-3-ols found in green tea products originating from the leaves and buds of the *Camellia sinensis* plant. The gallate ester originates from gallic acid condensation with 3*R*-hydroxy groups, responsible for biological reactivity. The established pathways and molecular targets of (−)-EGCG acting on mulptiple mediators have been comprehensively reviewed elsewhere (Lin and Li 2018), supporting traditional consumption for wellbeing and justifying it as a tangible drug candidate.

3.3.1.1 Absorption

Human uptake is limited to ingestion, with an oral bioavailability of EGCG estimated at 5% with intestinal uptake (Scholl et al. 2018) and imposes limitations reconciling animal and cell studies with excess compound exposure which may not recapitulate the preliminary interactions yielding low assimilation, due to the substantial number of hydroxyl moieties which promote hydrophilic interactions (Del Rio et al. 2013). Hydroxyl reactivity induces rapid degradation and biotransformation that appear to confer some anti-diabetic properties reflected in animal studies.

Wistar rats on a high-fat diet regime and stratified (−)-EGCG dosage (0.2, 0.4, 0.7 g/day/kg BW, respectively) showed reduction in lipid micelle uptake at the intestinal lumen, attributed to steric hindrance of the (−)-EGCG hydroxyl-motifs (Raederstorff et al. 2003). Insulin sensitisation arose from the competitive inhibition of cholesterol uptake thereby decreasing free fatty acids in circulation (Raederstorff et al. 2003). Peripheral reduction of insulin resistance was also supported by lipid-EGCG co-injection of Wistar rats (continuoue intravenous infusion of 5, 10 mg/kg for 48 h, respectively) with further metabolic improvements including reduction in oxidative stress markers (Li et al. 2011). Similar mode of impedance was demonstrated for glucose transporter uptake in human *Caco-2* intestinal cell monolayer for which several catechins (100 μM concentration, 2 min exposure) sterically hindered binding sites of GLUT2 and SGLT2 uptake of exogenous glucose, most marked for EGCG (Johnston et al. 2005). Pharmacokinetics coupled to temporal transcriptomics and proteomics at physiological relevance is necessary to establish diabetic-dependent uptake considering inflammation and compromised transport to the plasma. Conjugation to toxicogenomics screening and appreciation of hepatic processing at physiological and supra-physiological dosages to confirm bio-interactive modes of EGCG for diabetic therapy is also required.

3.3.1.2 Distribution

The scarce remainder of (−)-EGCG post-absorption is rapidly eliminated from blood plasma at an estimated half-life of 2.6 h following ingestion (Scholl et al. 2018). Recognition as a secondary plant metabolite or xenobiotic substance may prompt the observed rapid processing and clearance from circulation (Del Rio et al. 2013). Adversely, the inherent reactivity inducing cell function interference could also prompt elimination (Mihaela et al. 2009). The instability of the hydroxyl motifs and aftermath of decomposition pathway could also contribute detrimental intermediates and pro-oxidant species necessitating removal (Krupkova et al. 2016). The balance needs to be ascertained for cell responses already protracted with diabetic disease and hostile circulatory environment. The serum exclusion of the majority of the parent metabolite warrants cell screening of peripheral tissues to quantify uptake of primary and derivative metabolites which are unaccounted in plasma-based assays (Del Rio et al. 2013).

Even at low circulating levels, EGCG was documented to modify lymphocytes from T2DM patients, increasing fluidity and hyperpolarisation of cell membranes (Mihaela et al. 2009). The bi-layer interference is consistent with other cell type findings, with functional modifications yet to be determined with multi-omics assessment. Deeper analysis at diabetic conditions, controlling for inherent dysfunction of disease, offers a thorough platform to mechanistically ascertain cell responses to EGCG exposure.

Host categorisation of (−)-EGCG as a mild toxin or antigen element could also support the plasma clearance mechanism, which could offer exploit to pharmacological flushing and transient immune upregulation. The diabetic spectrum of perturbed temperature, oxygenation and pH remains to be assessed for influence on (−)-EGCG-facilitated membrane interactions and specificity. For example, (−)-EGCG exerts superior affinity than gallocatechin gallate (GCG) at pH greater than 3.5 (Muzolf-Panek et al. 2012). Consideration of local and systemic physiological parameters, and variable presence of advanced glycation end products, free fatty acids, and potentiated inflammation with upregulated stress signalling of adipokines and cytokines of T2DM is required. Comprehensive analysis with high-throughput inquisition could appreciate the deviations in cell response with (−)-EGCG exposure compared to non-diabetic controls; furthermore the establishment of a baseline spectrum may enable T2DM sub-phenotype investigations for unappreciated subtleties in disease.

3.3.1.3 Sampling Recovery

Despite attenuation in circulation, substantial inter-individual variation may underscore new avenues to therapeutic development previously neglected with plasma-based assays. Scholl and colleagues' nutrikinetics study on green tea capsule administration in a non-diabetic human population revealed sixfold variation in plasma catechin concentrations (Scholl et al. 2018). The development of diverse and non-invasive complements to plasma testing can now provide new insights that may account for the unknown sources of substantial variation, namely, lipidomics, and other facets of emerging omics assessment (Ulaszewska et al. 2019). New data sets offer complement to GWAS for population-variants exerting differential disease effects. Greater exploration of emerging omics sub-sets and integration with multi-omics appraisal could prelude probable symbiotic therapies which modulate microbiome to assist in compound targeting and delivery with definitive evidence base for (−)-EGCG therapy.

The extensive variation in plasma concentrations could also be attributed to polymorphic variants for metabolic transport and enzymes, proposed by Scholl et al. (2018). Aside from genomic discrepancies, polymorphisms can be complexed by cultural and regional exposure contributing additional variation. A Czech survey on dietary polyphenol intake compared the demographically defined exposure of consumers as a function of cultural practices including endemic low-polyphenol beer and lower consumption of polyphenol-rich food types including tea (Zloch et al. 2018). The considerable and inherent pleiotropy in a non-disease population may be further obscured by the heterogeneous diabetic spectrum and now-recognised sophisticated disease beyond insulin and glucose modulation.

The advance of omics studies must therefore correct for Caucasian biases when validating current findings, with inclusion of diverse and adequately powered population cohorts. Santangelo and colleagues demonstrate this critical gap with tissue- and time-dependent miRNA sequencing of a minority T2DM cohort, appreciating the apparent pleiotropy of T2DM pathology in broader analyses: underscoring demand for diverse omics evaluations and consulting explicit intervention data (Santangelo et al. 2016).

3.3.1.4 Standardisation

Failed conversion of clinical trial to successful therapies is further encumbered by the non-standardisation of tea extracts. Plant source material and processing unaccompanied by a chromatogram has prevented reproducibility and quality assurance especially when using commercial products (Gosslau et al. 2011). Confidential combination product compositions employed in animal studies have not published an assessment of individual isolate purity, nor contrasted with discrete effects from selected product. This is a wider argument in combination therapy screens, which require subjugated and composite analysis to truly appreciate the interaction effects between the considered active ingredient and supposedly negligible minor elements, which may actually stabilise or induce molecular dynamics (Fernandez-Ochoa et al. 2018). For example, a study of diet-induced T2DM in hamsters used a commercial (−)-EGCG capsule, standardised at 50% content, however accompanied by rice flour and gelatin, without validation of separated versus pair-wise combination effects (Li et al. 2006a). Equally, human studies reporting green tea capsule use standardise dosage of constituents, but are yet to distinguish whole product from isolate effects, preventing mechanistic insight to additive, synergistic or antagonistic properties of test products. Whilst routine exposure does not affect trial responses, dietary exposure to composite products or a range of dosage (850–2200 mg/L EGCG) cannot differentiate compound-specific effects from a whole product (Scholl et al. 2018).

Prior to product manufacture, regional variability in composition and climatic exposure affecting yield and composition of raw plant material. A metabolomics study of *Camellia sinensis* (green tea) leaves grown in three provinces of South Korea were compared for differential metabolic profiles measured by ^{1}H NMR and assessed with multi-variate analysis (Lee et al. 2010). Environmental light exposure; age of leaves; rainfall; and temperature significantly varied the catechin synthesis within tea leaves (Lee et al. 2010). Metabolomics and chemical analyses are therefore inherent to reporting (−)-EGCG product use, especially when examining traditional product cultivation, processing and crude preparation methods to modify content and structure and interaction effects. Knowledge of molecular integrity and co-stabilisation offer pharmacological insight to optimising desired isolate acquisition.

3.3.2 Allicin and Other Labile Compounds

Physical damage to the *Allium sativum* (garlic) bulb catalyses conversion of sulfoxides to rare thiosulfinates in this plant of the Liliaceae family (Perera and Li 2012). The conversion to low molecular weight sulphur-organic compounds promotes the cysteine interactions posited for medicinal applications (Kimbaris et al. 2006). The volatile sulphur metabolites are implicated for the anti-oxidant, immunomodulating, anti-inflammatory properties of garlic reviewed by (Shang et al. 2019). The instability and totality of organosulphur transformations has limited structural and composition analyses for therapeutic validation of *A. sativum's* health benefits.

Allicin (Table 3.1b) is the prevalent labile compound detected with Gas Chromatography coupled to Mass Spectrometry (GC-MS). The established conversion of alliin precursor to allicin is facilitated by the alliinlyase enzyme, which is permanently inactivated by heat and at pH $\leq$ 3.6, thereby excluding allicin conversion at stomach pH (Amagase 2006). Acidic inhibition challenges conventional attributions of allicin as the predominant bioactive—perhaps other compounds may convene the health benefits designated by its detection. The potency of garlic's many medicinal properties implores further investigation afforded with omics appraisal to contribute new ground to the evidence base for integral compounds. This is underscored with the pharmacological need to deduce synergism, antagonism or additive effects of allicin and other garlic metabolites relative to the diabetic physiological context, drawing upon traditional medicinal uses for alliaceous drug development to impact T2DM.

3.3.2.1 Variability

The sensitivity of product conversions is differentially enacted by processing method and subsequent catalysis of phytochemicals. The rarity of these compounds complicates their recovery, with substantial product loss or transient capture (Hu et al. 2002). Despite its known composition, allicin was only detected for 20 s before the reaction ceased in a bulb puncture study (Chang et al. 2016). The volatility of allicin facilitates immediate binding to fatty acids and proteins in system preventing internalisation, conjecturally acting as a spectator molecule and irrelevant as a quality assurance marker (Amagase 2006). However, other labile compound biochemical mechanisms remain incomplete, leaving allicin as the surrogate standardisation marker. The validity of this attribution remains unclear until bioactive compounds can be elucidated, stabilised and tested for their effects in physiological conditions, let alone under T2DM recapitulation.

Significant efforts to resolve and attribute other compounds of biological relevance require specific deduction of structure and function in isolation and combination assessment. Geometric isomerism was applied to differentiate sulphur mono-isotopes to improve capture and detection of volatile reactions via ultrahigh resolution mass spectrometry (Nakabayashi et al. 2016). Differential compositions may improve discrimination of unknown metabolites or better distinguish fragments and intermediates in metabolic analysis. Discovery of new isolates also requires consideration of climatic variation and plant cell response manifested in proteins and metabolites of the source material to account for diverse phenotypic influence. For example, cold stress was associated with compromised garlic bulb composition (Dufoo-Hurtado et al. 2015). Characterisations of starting plant material serves to aid exploration of garlic extract content and the reconciliation of the resultant phytochemical profile for optimisation of desired isolates for therapeutic pursuit.

Beyond standardising starting material, the various extraction processes manipulate natural product composition necessitating omics appraisal. Aged extracts, such as the Chinese 'laba' garlic preparation utilise vinegar to liberate allicin, enhancing palatability (Liu et al. 2018; Kimbaris et al. 2006). Product modification included enrichment of *S*-allylcysteine; *S*-allylmercaptocysteine content; and up- or downregulation of 20 volatile, 16 primary, and 15 phenolic compounds evident with metabolomics assessment (Liu et al. 2018). With this information, chemometric studies can then evaluate stability and optimisation protocols to reduce transformations and maximise recovery of volatile compounds and development of viable authenticity markers using HPLC (Hrbek et al. 2018). Nutritional studies have investigated garlic supplementation, employing HPLC for metabolomic analysis of plasma assays (Fernandez-Ochoa et al. 2018). This approach is arguably limited in interpretation with the incomplete profile of primary and derivative metabolites. Thus high-throughput analyses must comprehensively assess the scope of compound interactions to quantify subtle differences in product quality, exerting plausibly differential effects. This limitation has been challenged with an isotope dilution study collating UPLC-MS/MS with multiple reaction monitoring to report recovery ratios of 89–105% of four known organosulphur compounds (Lin et al. 2017). Such existing techniques to verify compositions paired with high throughput analysis in cell screening assays could decipher complex interactions and confirm the allicin-present versus allicin-independent effects of garlic preparations for optimisation and pharmacological progress.

3.3.2.2 Interaction Effects

Natural product interaction effects in T2DM cohort are still emerging and require further assessment with the imposed challenges of labile compound recovery and stabilisation. The inevitable reactivity is implicated for comprehensive anti-diabetic effects now appreciated in concomitant disease of both animal models (Qamar et al. 2015) and human cohorts (Choudhary et al. 2018). Despite compelling findings illustrating the pharmacological benefits of garlic extracts, systematic analysis has been limited. The first meta-analysis of nine studies on T2DM and garlic use, reported promising risk reduction ratios for pathological parameters, yet neither dosage nor duration of treatment was significantly associated with fasting blood glucose at a 95% confidence interval (Wang et al. 2017). Thus complex effects of disease and garlic interaction remain to be well-established in absence of definitive mechanisms and substrates for anti-diabetic therapies.

Implications for gene drug targets have emerged, with construction of an interaction network between toxicogenome and altered gene expression in T2DM *vastuslateralis* biopsy (Muhammad et al. 2017). Such analysis lays the groundwork for more integrated analyses accounting for allicin and allicin-independent garlic supplementation. A breath biomarker test revealed an unexpected increase in gamma-glutamyl-*S*-allylcysteine and *S*-allylcysteine metabolism in absence of allicin (Lawson and Hunsaker 2018). Taken together, these findings support the feasibility of integrated study designs to correctly attribute metabolite influence and associate interventions with genomic data to appraise differential outcomes for T2DM cohorts.

Beyond disease-specific interactions, the inevitable herb-drug interaction requires extensive assessment considering the conjugative potential partially demonstrated for garlic therapies. A clinical trial of metformin with garlic or placebo (300 mg tablet, thrice/day) showed enhanced anti-diabetic effects and manifested reduction in plasma lipids and blood glucose for diabetic patients, superior to Metformin therapy alone (Ashraf et al. 2011). Allicin has also been implicated in anti-coagulant enhancement, with an additive effect on warfarin in two cases, which may hold further ramifications for diabetic circulation and likelihood of vascular injury requiring assessment (Borrelli et al. 2007). In either case, omics offers deep interrogation of cell response to interpret the prevalence of these effects, with the capacity to quantify modified cell signalling and comprehensive appraisal of systemic and local effects on T2DM and the accompanying physiological milieu.

3.3.2.3 Repositioning

Expansive use of garlic, despite its obscured mechanistic properties provides other bodies of evidence supporting anti-diabetic therapies and chronic disease impact. Traditional medicinal use of garlic extracts as vitality agents and anti-ageing natural products has now been demonstrated for *Caenorhabditis elegans* life extension (Huang et al. 2015). Metabolomics appraisal of the nematode revealed 20% increase in total lifespan, beneficial gene expression for enhanced function and disease resistance (Huang et al. 2015). This study presents one case of applying omics technologies to validate traditional medicinal applications and derive new knowledge from deeper analysis. For example, the capacity for temporal omics sampling can monitor changes over time to traits like insulin resistance, improving reconciliation of true interaction effects not practicable with conventional methods (Wang et al. 2017).

Greater therapeutic exploration of garlic effects manifested in other disciplines reveals two concepts raised but inadequately assessed for therapeutic relevance, pertaining to greater questions of garlic administration, cytotoxicity and bioavailability. Persulfidation, the emergence of H_2S originating from cysteine protein residues, as a biological signalling molecule has been posited for vascular relaxation and improved circulation, yet remains untested for diabetic effects (Filipovic 2015). A second phenomenon yet to be appraised in humans is the oxidative toxicity of garlic plant matter to domestic animals. Garlic exposure corrupted canine erythrocytes, yielding methemoglobin and haemolytic anaemia (Hu et al. 2002). These effects have not been well studied for their detriment or benefit in diabetic context, and warranted pharmacokinetic studies to establish true effects for drug consideration. Broader limitations encountered in conventional compound isolation and testing imploring omics designs to appreciate mechanistic outcomes in cell response to assist in pharmacological interpretations and the validation of garlic product derivatives for T2DM therapy also require further investigation.

3.3.3 Tangeretin

Tangeretin, or 4′, 5, 6, 7, 8-pentamethoxyflavone (Table 3.1c), is one of two predominant compounds obtained from various citrus peels. The compound is water insoluble, requiring ethanol extraction confirmed in the *Citrus tanaka* (Hwang et al. 2015). As the 'poly'-methoxylated flavone, tangeretin is capable of direct gut absorption (Lin et al. 2011). Full methoxylation of both the B and A rings gives superior biological interactivity compared to incomplete methoxy-flavones. The redundancy of both rings enables these interactions, despite partial catabolism of the 4′ ring upon absorption (Xiao et al. 2011).

Tangeretin conforms an ideal structure to capitalise on metabolic lipid interactions, shown to reduce circulating apoB and other lipids exacerbated abundance with diabetic disease (Lin et al. 2011). A flavone screen of *caco-2* intestinal cells demonstrated the full methylated flavone ability to permeate the cell line, compared to unmethylated counterparts, concluding greater intestinal uptake and metabolic stability (Wen and Walle 2006). Despite favourable isomerism enabling demonstrable biological uptake, tangeretin remains modestly appraised for biological and pharmacological applications.

3.3.3.1 Preliminary Prospects

Studies specific to tangeretin are rare; instead broader inclusion of citrus composites, including tangeretin, have demonstrated a spectrum of favourable properties to ameliorate T2DM. Wounds of diabetic mice had superior collagen synthesis and tissue repair with oral administration of various citrus peels compared to controls (400 mg/kg BW) (Ahmad et al. 2013). The C2C12 muscle cell exposure to 100 μM tangeretin was comprehensively assessed for a dose-dependent increase in AMPK activity, validated in a mouse model of high-fat diet induced diabetes: observing AMPK activation in muscle with 200 mg/kg BW dosage and improved lipid, serum and anthropometric profiles compared to untreated controls (Kim et al. 2012). This finding was also supported with extrapolation to *Tanaka* citrus peel extract, verifying increased PPAR-gamma and AMPK activity in C2C12 cell line, corroborated with a high-fat diet mouse model: concluding similar benefit with 5% extract inclusion in diet (Kim et al. 2013).

Tangeretin has also shown cardioprotective effects in diabetic rats, with insulin sensitisation of cardiac muscle, reduction in plasma lipids and inhibition of NF-kappaB oxidation: anti-inflammatory effects likened to the anti-diabetic drug glibenclamide, with 100 mg/kg BW administration for 30 days (Sundaram et al. 2015). An insulin-resistant hamster model also exhibited metabolic and inflammatory improvements with reduction in pro-inflammatory cytokines and lipid parameters with polymethoxylated flavone treatment, with increased PPAR-alpha and PPAR-gamma expression (Li et al. 2006b). More recently, tangeretin exposure was shown to increase glucose absorption in 3T3-differitate adipocytes in a concentration-dependent manner (range 15–50 μM, 24 h), implicating favourable PI3K and Akt pathway modulation (Onda et al. 2013). Studies thereby demonstrate robust effects of tangeretin in glucose homeostasis pathways, with potential correction of the metabolic dysregulation potentiated in diabetes.

3.3.3.2 Elusive Recovery

Despite these promising results, tangeretin examination remains critically limited, pending mechanistic insight to deduce metabolite recovery. The first pharmacokinetics study was performed by Elhennawy and Lin (2018), to establish bioavailability and pharmacological parameters for tangeretin (50 mg/kg BW) in Sprague-Dawley rats. Despite predicted membrane permeability conferred by full methoxylation, true bioavailability of oral tangeretin suspension was <3.05% (Elhennawy and Lin 2018). Bioavailability and plasma concentrations for human ingestion have not been published to date, despite capacity to detect trace quantities to pico grams of tangeretin in animal samples (Manthey et al. 2010). The limitation in conventional focus on flavone metabolite recovery raises argument for complex analysis of the understudied transformed derivatives, via glucuronation and sulfination (Manthey et al. 2010). This observation extends to a broader criticism of the lag in tangeretin-specific studies in biological systems, despite promising upregulation of AMPK, PPARs and other metabolic pathways dysregulated in T2DM.

Gavage co-administration of nobiletin and tangeretin (50 mg/kg BW) to Sprague-Dawley rats confirmed absorption from the intestine, with plasma concentrations of tangeretin monitored for 24 h (Manthey et al. 2010). Another study reported oral bioavailability of 27.11% with peak tissue concentration of tangeretin achieved between 4 and 8 h (Hung et al. 2018). Fourty-eight hour egestion recovery in the Sprague Dawley rats totalled 0.00026% in urine, and 7.54% in faeces, indicating majority processed into other metabolites, totalling only 24% recovery of tangeretin (Hung et al. 2018). Therapeutic pursuit is thereby limited by the lack of vital biochemical data for pathway mechanisms and cellular transport processes, to be determined with proteomics and transcriptomics analysis of perturbed cell responses. In addition, the migration and degradation of parent molecule into derivatives requires broad cell screening array and metabolomic verification of migration and sequestration in aftermath of exposure. The microbiome interaction has not been studied, warranting further investigation with the marked persistence of tangeretin in the colon, peaking concentration at 12 h, compared to lesser stomach and intestine duration (Hung et al. 2018).

3.3.3.3 Isolation

Each facet of omics is implored to deduce these independent effects on tissue for local and systemic appreciation of therapeutic potential and further dictate optimisation pathway for further elaboration. Without deciphering mechanistic pathways and appreciating relevant metabolites for further analysis, interpretation is limited to whole citrus extracts which include tangeretin as an abundant, yet partial component. *Tanaka* peel, including 0.7 mg/100 g tangeretin in an insulin-resistant human cohort, reflected consensus to previous studies, with appreciable metabolic improvements, however no attribution to specific compound effects (Hwang et al.

2015). Enhanced tangeretin in the curing of 'chenpi', aged *Citrus reticulata* peel, shown to partially improve diet-induced dyslipidemia in diabetic C57/BL6 mice (Guo et al. 2016). Tangeretin has shown multi-effector promise in rectifying diabetic metabolism, warranting deeper analysis borne of omics technology to translate it promising effects to therapeutics.

3.4 Summary and Conclusion

From the discussion of allicin, EGCG and tangeretin above, it is clear that each is abundant in both cultivation and consumption with prolific use and social acceptability of perceived health benefits to consumers. Indeed, it is difficult to find human cohorts naïve to each phytochemical or its origin. The known compound compositions present a short-cut to isolate extraction and derivative generation bolstered by superior biological affinity and interaction. This presents the challenge of achieving desired cellular action yet minimising damage and other off-target effects in the balance of optimisation (Odeyemi and Bradley 2018). This compromise is notoriously complex with the intricate pathology of T2DM presenting systemic inflammation, microvascular changes and metabolic instability; with the specific tissue and organ responses including pancreas, gut lumen and nerve endings; and with the interface of converging and additive co-morbidities including obesity, CVD, hypertension, metabolic syndrome (Bezerra et al. 2016). This physiological milieu is difficult to completely corroborate with experimental variants, due to the expansive and continuing spectrum of diabetic disease.

Pharmacodynamics determination of compound utility and specific effect balance is limited in a partially representative system. The incomplete recapitulation of diabetic physiology is contrasted by incomplete profiling of traditional medicines as largely combination products. Single isolate screening cannot determine the additive, synergistic or antagonistic compilations of multi-potent preparations. Inherent variation in traditional preparations is similar to the non-standardised or unreported conditions of preparation, storage, sterility and integrity of consumer products employed in clinical studies. Unaccounted variation in starting plant material and cultivation conditions introduce further sources of variation which limit interpretation between studies and constrict broader findings with the reliance on composite, commercially confident food grade products. Rich data generation must minimise practicable noise to focus outcomes of perturbed cell response variations for conclusive interpretations.

Multi-omics discovery strategies can investigate compound interactions within and between subject, microbiome and the multiplex of environment, behaviour and cell responses. The strategies pose approach of monitoring multiple axes of variation to delineate changes observed at different levels of gene expression in both disease and resistant cohorts (Hasin et al. 2017). Comprehensive probing of overlooked conjugates with deep analysis could reveal the speculated obscurity of seemingly inactive or irrelevant components, which could actually influence toxicity, stability or other pharmacological parameters and thereby dictate drug design.

Diabetic-mediated deviations (e.g. acidosis, body temperature, advanced glycation end products, chronic inflammation) must be assessed for alterations to pharmacokinetics. Omics can evaluate these multiple phases of gene expression with temporal sampling to construct improved networks to proposed pathways of response. Tangibility is cautioned by the difficulties of establishing herb-drug interaction networks with known substrates and targets still lagging critical data of herbal natural products (Gupta et al. 2017). Omics also provides the means to address the under-representation of non-Caucasian populations in allopathic medical studies, and the greater bias of historical Western medicine prevailing deficits in T2DM disease understanding. The three natural products' premise, the potential of omics to address conventional gaps and the considerations for improved study design offer hope to champion feasible drug development with substantiated and reproducible outcomes originating from traditional medicinal practices.

References

Ahmad M, Ansari M, Alam A et al (2013) Oral dose of citrus peel extracts promotes wound repair in diabetic rats. Pak J Biol Sci 16(20):1086–1094

Amagase H (2006) Clarifying the real bioactive constituents of garlic. J Nutr 136(3):716S–725S

Ashraf R, Khan R, Ashraf I (2011) Garlic (Allium sativum) supplementation with standard antidiabetic agent provides better diabetic control in type 2 diabetes patients. Pak J Pharm Sci 24(4):565–570

Bezerra APM, Silva-Santiago SC, Da Silva JFD et al (2016) Confluence of genes related to the combined etiology DOISm (diabetes, obesity, inflammation and metabolic syndrome) in dissecting nutritional phenotypes. Lect N Bioinformat 9656:22–37. https://doi.org/10.1007/978-3-319-31744-1_3

Borrelli F, Capasso R, Izzo A (2007) Garlic (Allium sativum L.): adverse effects and drug interactions in humans. Mol Nutr Food Res 51(11):1386–1397

Chang Q, Peng Y, Shi B et al (2016) In situ identification of labile precursor compounds and their short-lived intermediates in plants using in vivo nanospray high-resolution mass spectrometry. Phytochem Anal 27(3–4):184–190. https://doi.org/10.1002/pca.2614

Choudhary PR, Jani RD, Sharma MS (2018) Effect of raw crushed garlic (Allium sativum L.) on components of metabolic syndrome. J Diet Suppl 15(4):499–506. https://doi.org/10.1080/19390211.2017.1358233

Del Rio D, Rodriguez-Mateos A, Spencer J et al (2013) Dietary (poly) phenolics in human health: structures, bioavailability, and evidence of protective effects against chronic diseases. Antioxid Redox Signal 18(14):1818–1892

Dufoo-Hurtado M, Huerta-Ocampo J, Barrera-Pacheco A et al (2015) Low temperature conditioning of garlic (Allium sativum L.) "seed" cloves induces alterations in sprouts proteome. Front Plant Sci 6:1–15. https://doi.org/10.3389/fpls.2015.00332

Elhennawy M, Lin H-S (2018) Determination of tangeretin in rat plasma: assessment of its clearance and absolute oral bioavailability. Pharmaceutics 10(1):3

Fernandez-Ochoa A, Borras-Linares I, Banos A et al (2018) A fingerprinting metabolomic approach reveals deregulation of endogenous metabolites after the intake of a bioactive garlic supplement. J Funct Foods 49:137–145. https://doi.org/10.1016/j.jff.2018.08.003

Filipovic M (2015) Persulfidation (S-sulfhydration) and H2S. In: Chemistry, biochemistry and pharmacology of hydrogen sulfide. Handbook of experimental pharmacology, vol 230. Springer, pp 29–59

Gosslau A, Li S, Ho C-T et al (2011) The importance of natural product characterization in studies of their anti-inflammatory activity. Mol Nutr Food Res 55(1):74–82

Guo J, Tao H, Cao Y et al (2016) Prevention of obesity and type 2 diabetes with aged citrus peel (Chenpi) extract. J Agric Food Chem 64(10):2053–2061

Gupta R, Chang D, Nammi S et al (2017) Interactions between antidiabetic drugs and herbs: an overview of mechanisms of action and clinical implications. Diabetol Metab Syndr 9(1):59

Hasin Y, Seldin M, Lusis A (2017) Multi-omics approaches to disease. Genome Biol 18(1):1–15

Hrbek V, Rektorisova M, Chmelarova H et al (2018) Authenticity assessment of garlic using a metabolomic approach based on high resolution mass spectrometry. J Food Composit Anal 67:19–28. https://doi.org/10.1016/j.jfca.2017.12.020

Hu Q, Yang Q, Yamato O et al (2002) Isolation and identification of organosulfur compounds oxidizing canine erythrocytes from garlic (Allium sativum). J Agric Food Chem 50 (5):1059–1062

Huang C, Hsu F, Wu Y et al (2015) Analysis of lifespan-promoting effect of garlic extract by an integrated metabolo-proteomics approach. J Nutr Biochem 26(8):808–817. https://doi.org/10.1016/j.jnutbio.2015.02.010

Hung W-L, Chang W-S, Lu W-C et al (2018) Pharmacokinetics, bioavailability, tissue distribution and excretion of tangeretin in rat. J Food Drug Anal 26(2):849–857

Hwang J, Yang H, Ha K-C et al (2015) A randomized, double-blind, placebo-controlled clinical trial to investigate the anti-diabetic effect of Citrus junos Tanaka peel. J Funct Foods 18:532–537

Isgut M, Rao M, Yang C et al (2018) Application of combination high-throughput phenotypic screening and target identification methods for the discovery of natural product-based combination drugs. Med Res Rev 38(2):504–524. https://doi.org/10.1002/med.21444

Jain P, Vig S, Datta M et al (2013) Systems biology approach reveals genome to phenome correlation in type 2 diabetes. PLoS One 8(1):e53522. https://doi.org/10.1371/journal.pone.0053522

Jenkinson C, Göring H, Arya R et al (2016) Transcriptomics in type 2 diabetes: bridging the gap between genotype and phenotype. Genom Data 8:25–36

Johnston K, Sharp P, Clifford M et al (2005) Dietary polyphenols decrease glucose uptake by human intestinal Caco-2 cells. FEBS Lett 579(7):1653–1657

Kim M, Hur H, Kwon D et al (2012) Tangeretin stimulates glucose uptake via regulation of AMPK signaling pathways in C2C12 myotubes and improves glucose tolerance in high-fat diet-induced obese mice. Mol Cell Endocrinol 358(1):127–134

Kim S, Hur H, Yang H et al (2013) Citrus junos tanaka peel extract exerts antidiabetic effects via AMPK and PPAR-both *in vitro* and *in vivo* in mice fed a high-fat diet. Evid Based Complement Alternat Med 2013:921012

Kimbaris A, Siatis N, Daferera D et al (2006) Comparison of distillation and ultrasound-assisted extraction methods for the isolation of sensitive aroma compounds from garlic (Allium sativum). Ultrason Sonochem 13(1):54–60

Krug EG (2016) Trends in diabetes: sounding the alarm. Lancet 387(10027):1485–1486. https://doi.org/10.1016/S0140-6736(16)30163-5

Krupkova O, Ferguson S, Wuertz-Kozak K (2016) Stability of (−)-epigallocatechin gallate and its activity in liquid formulations and delivery systems. J Nutr Biochem 37:1–12

Lawson L, Hunsaker S (2018) Allicin bioavailability and bioequivalence from garlic supplements and garlic foods. Nutrients 10(7):812–861

Lee JE, Lee BJ, Chung JO et al (2010) Geographical and climatic dependencies of green tea (Camellia sinensis) metabolites: a (1)H NMR-based metabolomics study. J Agric Food Chem 58 (19):10582–10589. https://doi.org/10.1021/jf102415m

Li R, Douglas T, Maiyoh G et al (2006a) Green tea leaf extract improves lipid and glucose homeostasis in a fructose-fed insulin-resistant hamster model. J Ethnopharmacol 104 (1–2):24–31

Li R, Theriault A, Au K et al (2006b) Citrus polymethoxylated flavones improve lipid and glucose homeostasis and modulate adipocytokines in fructose-induced insulin resistant hamsters. Life Sci 79(4):365–373

Li Y, Zhao S, Zhang W et al (2011) Epigallocatechin-3-O-gallate (EGCG) attenuates FFAs-induced peripheral insulin resistance through AMPK pathway and insulin signaling pathway in vivo. Diabetes Res Clin Pract 93(2):205–214

Lin GD, Li RW (2018) Natural products targeting inflammation processes and multiple mediators. In: Mandal S, Mandal V, Konishi T (eds) Natural products and drug discovery: an integrated approach. Elsevier, Oxford, pp 277–308

Lin Y, Vermeer M, Bos W et al (2011) Molecular structures of citrus flavonoids determine their effects on lipid metabolism in HepG2 cells by primarily suppressing apoB secretion. J Agric Food Chem 59(9):4496–4503

Lin Y, Huang M, Wu Y et al (2017) Reductive amination derivatization for the quantification of garlic components by isotope dilution analysis. Food Chem 230:1–5. https://doi.org/10.1016/j.foodchem.2017.02.119

Liu J, Guo W, Yang M et al (2018) Investigation of the dynamic changes in the chemical constituents of Chinese "Laba" garlic during traditional processing. RSC Adv 8 (73):41872–41883. https://doi.org/10.1039/c8ra09657k

Manthey J, Cesar T, Jackson E et al (2010) Pharmacokinetic study of nobiletin and tangeretin in rat serum by high-performance liquid chromatography– electrospray ionization– mass spectrometry. J Agric Food Chem 59(1):145–151

Mihaela I, Denisa M, Eva K et al (2009) Quercetin and epigallocatechin gallate effect on the lipid order parameter of peripheral blood mononuclear cells from diabetes patients. Rom Biotechnol Lett 14(6):4804–4811

Muhammad S, Raza W, Nguyen T et al (2017) Cellular signaling pathways in insulin resistance-systems biology analyses of microarray dataset reveals new drug target gene signatures of type 2 diabetes mellitus. Front Physiol 8:13

Muzolf-Panek M, Gliszczyńska-Świgło A, Szymusiak H et al (2012) The influence of stereochemistry on the antioxidant properties of catechin epimers. Eur Food Res Technol 235 (6):1001–1009

Nakabayashi R, Sawada Y, Aoyagi M et al (2016) Chemical assignment of structural isomers of sulfur-containing metabolites in garlic by liquid chromatography fourier transform ion cyclotron resonance mass spectrometry. J Nutr 146(2):397S–402S. https://doi.org/10.3945/jn.114.202317

Odeyemi S, Bradley G (2018) Medicinal plants used for the traditional management of diabetes in the Eastern Cape, South Africa: pharmacology and toxicology. Molecules 23(11):2759

Onda K, Horike N, Suzuki T et al (2013) Polymethoxyflavonoids tangeretin and nobiletin increase glucose uptake in murine adipocytes. Phytother Res 27(2):312–316

Perera P, Li Y (2012) Functional herbal food ingredients used in type 2 diabetes mellitus. Pharmacogn Rev 6(11):37–45

Qamar A, Siddiqui A, Kumar H (2015) Fresh garlic amelioration of high-fat-diet induced fatty liver in albino rats. J Pak Med Assoc 65(10):1102–1107

Raederstorff D, Schlachter M, Elste V et al (2003) Effect of EGCG on lipid absorption and plasma lipid levels in rats. J Nutr Biochem 14(6):326–332

Santangelo C, Zicari A, Mandosi E et al (2016) Could gestational diabetes mellitus be managed through dietary bioactive compounds? Current knowledge and future perspectives. Br J Nutr 115(7):1129–1144

Scholl C, Lepper A, Lehr T et al (2018) Population nutrikinetics of green tea extract. PLos One 13 (2):e0193074. https://doi.org/10.1371/journal.pone.0193074

Shang A, Cao S, Xu X et al (2019) Bioactive compounds and biological functions of garlic (Allium sativum L.). Foods 8(7):31. https://doi.org/10.3390/foods8070246

Shen C-Y, Jiang J-G, Yang L et al (2017) Anti-ageing active ingredients from herbs and nutraceuticals used in traditional Chinese medicine: pharmacological mechanisms and implications for drug discovery. Br J Pharmacol 174(11):1395–1425

Siddiqui K, Tyagi S (2015) Genetics, genomics and personalized medicine in type 2 diabetes: a perspective on the Arab region. Per Med 12(4):417–431. https://doi.org/10.2217/pme.15.11

Sun H, Zhang S, Zhang A et al (2014) Metabolomic analysis of diet-induced type 2 diabetes using UPLC/MS integrated with pattern recognition approach. PLoS One 9(3):e93384

Sundaram R, Shanthi P, Sachdanandam P (2015) Tangeretin, a polymethoxylated flavone, modulates lipid homeostasis and decreases oxidative stress by inhibiting NF-κB activation and proinflammatory cytokines in cardiac tissue of streptozotocin-induced diabetic rats. J Funct Foods 16:315–333

Thomson GE (2010) Further consideration of Asian medicinal plants in treating common chronic diseases in the west. J Med Plants Res 4(2):125–130

Ulaszewska M, Weinert C, Trimigno A et al (2019) Nutrimetabolomics: an integrative action for metabolomic analyses in human nutritional studies. Mol Nutr Food Res 63(1):1800384

Wang J, Zhang X, Lan H et al (2017) Effect of garlic supplement in the management of type 2 diabetes mellitus (T2DM): a meta-analysis of randomized controlled trials. Food Nutr Res 61 (1):1377571

Wen X, Walle T (2006) Methylated flavonoids have greatly improved intestinal absorption and metabolic stability. Drug Metab Dispos 34(10):1786–1792

Xiao J, Zhao Y, Wang H et al (2011) Non-covalent interaction of dietary polyphenols with total plasma proteins of type II diabetes: molecular structure/property-affinity relationships. Integr Biol 3(11):1087–1094. https://doi.org/10.1039/c1ib00076d

Yang Y-Y, Yang F, Gao J (2019) Differential proteomics for studying action mechanisms of traditional Chinese medicines. Chin Med 14(1):1–15

Zloch Z, Sedláček P, Langmajerová J et al (2018) Intake and profile of plant polyphenols in the diet of the Czech population. Pol J Food Nutr Sci 68:57–62. https://doi.org/10.1515/pjfns-2017-0007

4

Amaryllidaceae Alkaloids as Anti-inflammatory Agents Targeting Cholinergic Anti-inflammatory Pathway: Mechanisms and Prospects

Rachel W. Li, Partha Palit, Paul N. Smith, and G. David Lin

Abstract

Amaryllidaceae alkaloids are well-known for inhibiting acetylcholine esterase and for its representative galanthamine to treat Alzheimer's disease in clinical practice. Recently galanthamine is known for modulating the cholinergic anti-inflammatory pathway (CAP), in which the parasympathetic vagus nerve inhibits the release of pro-inflammatory cytokine and protects against systemic inflammation during immune challenge. This cholinergic modulation of inflammation involves with the neurotransmission of acetylcholine in the vagus nerve, acetylcholine esterase, alpha 7 subunit of nicotinic acetylcholine receptors (α7nAChR) in macrophages, and pro-inflammatory cytokines in the immune system. All of these involved factors are possible targets for the potential anti-inflammatory action of amaryllidaceae alkaloids. In addition, amaryllidaceae alkaloids inhibit p-38 MAP kinase, NFκB activation, cyclooxygenases, and nitric oxide, all of which are key inflammatory mediators. This chapter focuses on the mechanisms and prospects of four amaryllidaceae alkaloids, galanthamine, lycorine, narciclasine (lycoricidinol), and crinamine for their anti-inflammatory potential targeting CAP. It discusses their molecular and cellular action on the acetylcholine esterase, the vagus nerve, α7nAChR, pro-inflammatory cytokines, p-38 MAP kinase, NFκB, cyclooxygenases, and nitric oxide. Following the discussion, it is clear that amaryllidaceae alkaloids are not only a useful tool for studying the

R. W. Li · P. N. Smith
The Medical School, Australian National University, Canberra, ACT, Australia

P. Palit
The Medical School, Australian National University, Canberra, ACT, Australia

Department of Pharmaceutical Sciences, Assam University, Silchar, India

G. D. Lin (✉)
Research School of Chemistry, Australian National University, Canberra, ACT, Australia
e-mail: David.lin@anu.edu.au

S. C. Mandal et al. (eds.), *Evidence Based Validation of Traditional Medicines*,
https://doi.org/10.1007/978-981-15-8127-4_4

cholinergic anti-inflammatory pathway but also offer a potential as a new class of anti-inflammatory natural products, acting on multiple targets to combat inflammatory disorders and having different modes of action from the classical nonsteroidal anti-inflammatory drugs.

Keywords

Amaryllidaceae alkaloids · Cholinergic anti-inflammatory pathway · Anti-inflammatory natural products · Alpha 7 subunit of nicotinic acetylcholine receptor · The vagus nerve · NSAIDs · Galanthamine · Lycorine · Narciclasine · Crinamine

Abbreviations

AChE	Acetylcholine esterase
CAP	Cholinergic anti-inflammatory pathway
ChAT	Choline acetyltransferase
COX	Cyclooxygenase
IL	Interleukin
iNOS	Inducible nitric oxide synthase
ip	Intraperitoneally
LPS	Lipopolysaccharide
NFκB	Nuclear factor kappa-light-chain-enhancer of activated B cells
NSAIDs	Nonsteroidal anti-inflammatory drugs
p38 MAPK	p38 mitogen-activated protein kinase
PBMC	Peripheral blood mono-nuclear cell
TNF	Tumour necrosis factor
α7nAChR	Alpha 7 subunit of nicotinic acetylcholine receptor

4.1 Introduction

4.1.1 Amaryllidaceae Alkaloids

Mother Nature has created a distinguish group of natural products known as alkaloids. The NAPRALERT database contains about 27,000 known alkaloids among 150,000 natural products documented in the database (Jin 2005). The alkaloids are so diverse in structures and so important to our healthcare that the scientific community has documented 82 volumes under the lead title of *The Alkaloids* for nearly 70 years (Royal Society of Chemistry 2019). The series started in 1950 under the subtitle of Chemistry and Physiology (volumes 1–20), then in 1983 under Chemistry and Pharmacology (volumes 21–49) and from 1998 under Chemistry and Biology (volumes 50–82). The topic of Amaryllidaceae alkaloids

Table 4.1 The Amaryllidaceae family

Family	Subfamily	Genus	Species
Amarillidaceae	Agapanthoideae	*Agapanthus*	5 + unclassified
	Allioideae	*Allieae* *Gilliesieae* *Tulbaghieae*	2 10 2
	Amaryllidoideae	67 genera including *Amaryllis, Crinum, Galanthus, Haemanthus, Hippeastrum, Ismene, Lycoris, Narcissus*, and *Zephyranthes*	569 + unclassified
Total	**3**	**71**	**588 + unclassified**

was featured in volumes 2 (1952, p331), 6 (1960, p289), 11 (1968, p307), 15 (1975, p83), 30 (1987, p251), and 63 (2006, p87).

Amaryllidaceae alkaloids have been isolated almost exclusively from family Amaryllidaceae and thus bear the name. Based on the NCBI's Taxonomy Browser (National Center for Biotechnology Information 2019), family Amaryllidaceae has three subfamilies, 71 genera, and 588 species plus unclassified species; see more details summarized in Table 4.1.

The first Amaryllidaceae alkaloid lycorine was isolated in 1877 (Cook and Loudon 1952). Up to now more than 500 alkaloids are estimated to have been obtained from the plant source (Jin and Yao 2019). Based on their ring structures, these alkaloids can be categorized into 20 types, see Tables 4.2 and 4.3. Ten groups are popular and widely distributed thus classified as major types (Table 4.2) whereas other 10 groups are treated as minor types (Table 4.3).

There have been a number of classifications of Amaryllidaceae alkaloids. Norbelladine, lycorine, homolycorine, crinine, haemanthamine, narciclasine, tazettine, montanine, and galanthamine are nine major types with an unifying numbering system of the different skeleton types (Bastida et al. 2006). 15 structure types (Jin 2007) and 18 types (Jin 2013) were also grouped. Detailed structures, species, and references of 357 Amaryllidaceae alkaloids in 14 different types covering the past three decades were also published (Ding et al. 2017).

Chemistry and biological activities of Amaryllidaceae alkaloids have been investigated extensively, and one latest review series (Jin and Yao 2019) has updated all publications from 2015 to 2017. The IC_{50} or EC_{50} of Amaryllidaceae alkaloids published between 2005 and 2014 for anti-cancer, AChE inhibition, antiviral, antibacterial, antiparasitic, antimalarial, anticonvulsant, and antidepressant activities were also presented (He et al. 2015). Ding and co-workers provided a detailed account of the biological activities of many representative Amaryllidaceae alkaloids under the following areas (Ding et al. 2017):

1. Effects on the central nervous system
   (a) Anti-Alzheimer's disease and cholinesterase inhibitory activity
   (b) Anxiolytic activity
   (c) Anti-depressive and anti-epilepsy activities

Table 4.2 Ten major types of Amaryllidaceae alkaloids (based on key structure difference)

Major type 1	Lycorine	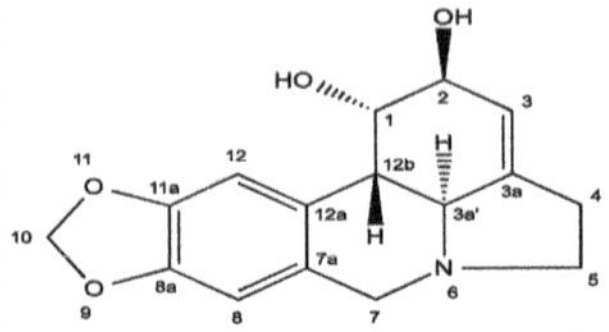
Representative alkaloid	Lycorine	
CAS RN	476-28-8	
Formula	$C_{16} H_{17} N O_4$	
Molecular weight (g/mol)	287.31	
Major type 2	Norbelladine	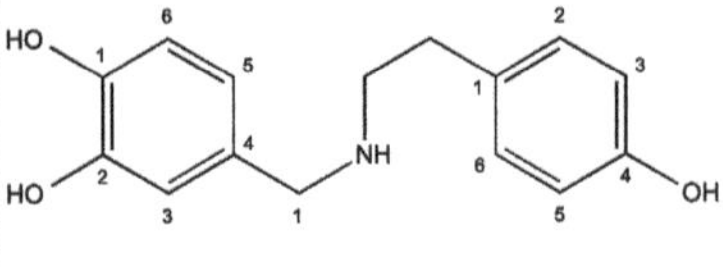
Representative alkaloid	Norbelladine	
CAS RN	6053-00-5	
Formula	$C_{15} H_{17} N O_3$	
Molecular weight (g/mol)	259.30	
Major type 3	Homolycorine	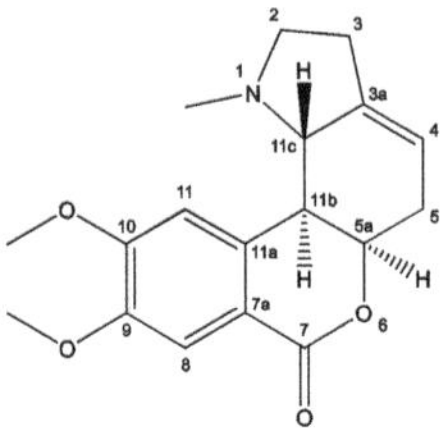
Representative alkaloid	Homolycorine	
CAS RN	477-20-5	
Formula	$C_{18} H_{21} N O_4$	
Molecular weight (g/mol)	315.36	
Major type 4	Crinine	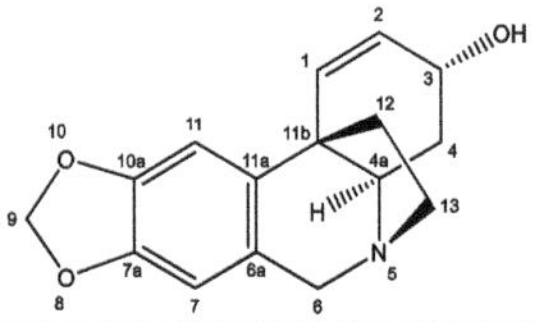
Representative alkaloid	Crinine	
CAS RN	510-67-8	
Formula	$C_{16} H_{17} N O_3$	
Molecular weight (g/mol)	271.31	
Major type 5	Haemanthamine	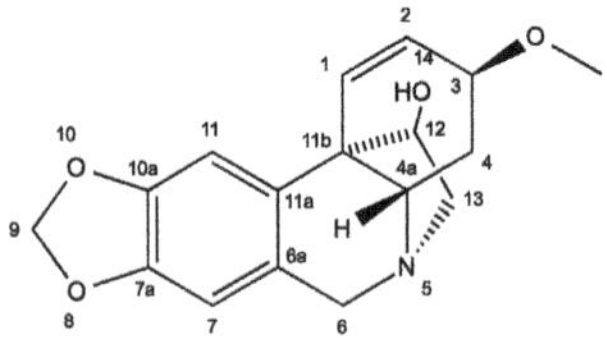
Representative alkaloid	Haemanthamine	
CAS RN	466-75-1	
Formula	$C_{17} H_{19} N O_4$	
Molecular weight (g/mol)	301.34	
Major type 6	*Galanthamine*	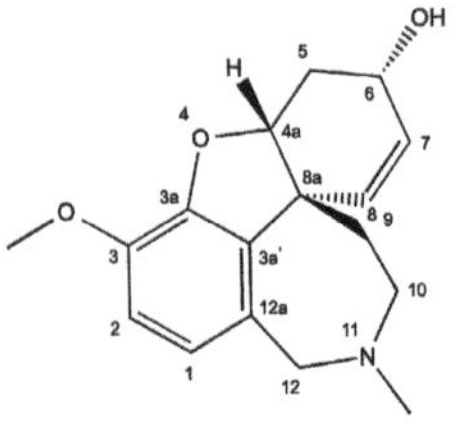
Representative alkaloid	Galanthamine	
CAS RN	357-70-0	
Formula	$C_{17} H_{21} N O_3$	
Molecular weight (g/mol)	287.35	
Major type 7	Tazettine	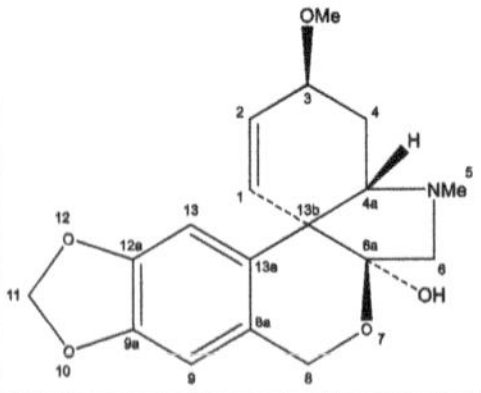
Representative alkaloid	Tazettine	
CAS RN	507-79-9	
Formula	$C_{18} H_{21} N O_5$	
Molecular weight (g/mol)	331.36	

(continued)

Table 4.2 (continued)

Major type 8	Montanine	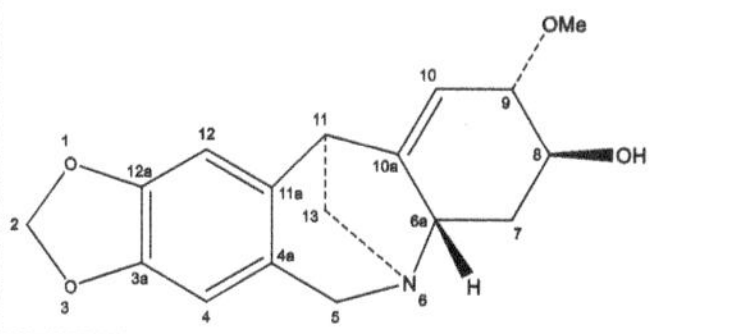
Representative alkaloid	Montanine	
CAS RN	642-52-4	
Formula	$C_{17}\ H_{19}\ N\ O_4$	
Molecular weight (g/mol)	301.34	
Major type 9	Phenanthridone	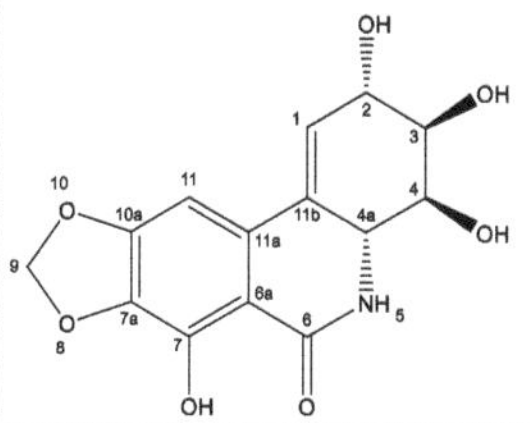
Representative alkaloid	Narciclasine	
CAS RN	29,477-83-6	
Formula	$C_{14}\ H_{13}\ N\ O_7$	
Molecular weight (g/mol)	307.26	
Major type 10	Phenanthridine	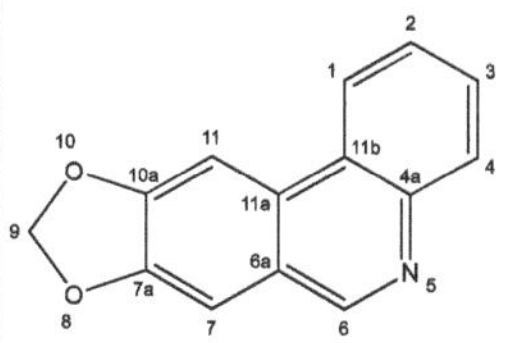
Representative alkaloid	Trisphaeridine	
CAS RN	224-11-3	
Formula	$C_{14}\ H_9\ N\ O_2$	
Molecular weight (g/mol)	223.23	

2. Antitumor activity
3. Antimicrobial properties
   (a) Antibacterial and antifungal activities
   (b) Antiviral activity
   (c) Antimalarial and antiparasitic activities
4. Anti-inflammatory activity

In the anti-Alzheimer's disease and cholinesterase inhibitory activity area, galanthamine is particularly important as it has been used clinically to treat mild to moderate severity of Alzheimer's disease since its approval by the FDA under the brand name Reminyl™ (2001) and now known as Razadyne™. Galanthamine inhibits AChE with an IC_{50} at 1 μM and increases the concentration of acetylcholine, which will be harder to be hydrolysed by acetylcholinesterase to choline and acetic acid (Fig. 4.1). Other significant pharmacological activities have also been discovered during its clinical use for nearly two decades. In a nutshell, patients taking galanthamine for a substantial time have additional benefits of reduced systematic inflammation. The recent discovery of cholinergic anti-inflammatory pathway (CAP, Fig. 4.1) has explained the anti-inflammatory action of galanthamine in part, which is summarized in Fig. 4.1 and will be discussed at various sections in this chapter.

In addition to galathamine, this chapter will showcase three more Amaryllidaceae alkaloids: lycorine, narciclasine (lycoricidinol), and crinamine. The drug-like properties, in vitro and in vivo anti-inflammatory activity of the four alkaloids, are summarized in Table 4.4 and discussed in details in Sect. 4.3.

Table 4.3 Ten minor types of Amaryllidaceae alkaloids (based on key structure difference)

Minor type i	Plicamine	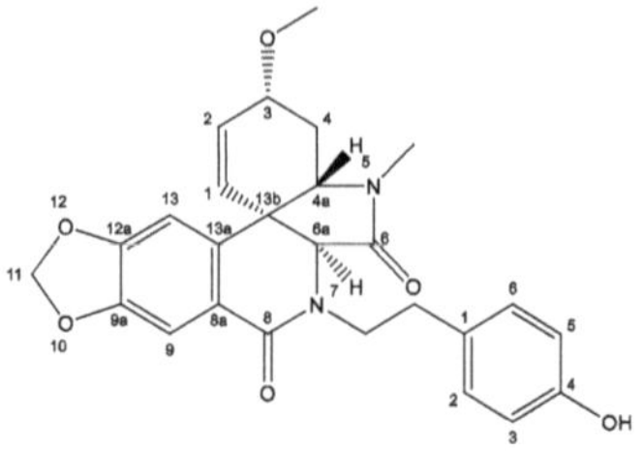
Representative alkaloid	Plicamine	
CAS RN	232,597-76-1	
Formula	$C_{26}H_{26}N_2O_6$	
Molecular weight (g/mol)	462.49	
Major type ii	Graciline	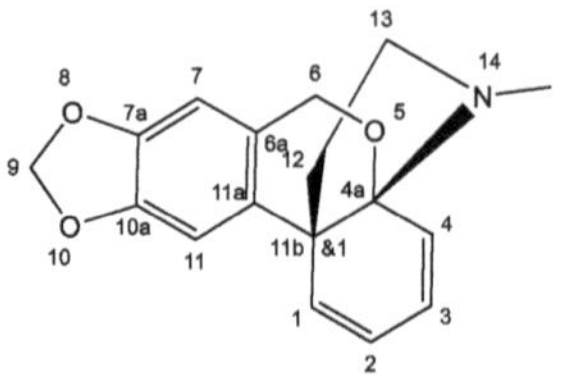
Representative alkaloid	Graciline	
CAS RN	216,444-04-1	
Formula	$C_{17}H_{17}NO_3$	
Molecular weight (g/mol)	283.32	
Major type iii	Galanthindole	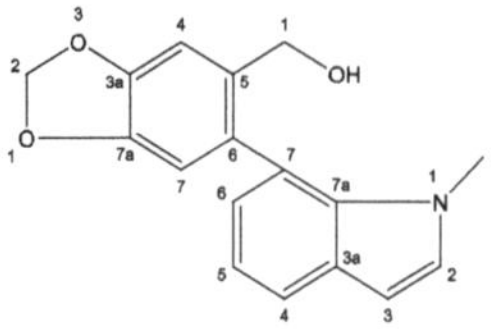
Representative alkaloid	Galanthindole	
CAS RN	666,174-48-7	
Formula	$C_{17}H_{15}NO_3$	
Molecular weight (g/mol)	281.32	
Ajor type iv	Cherylline	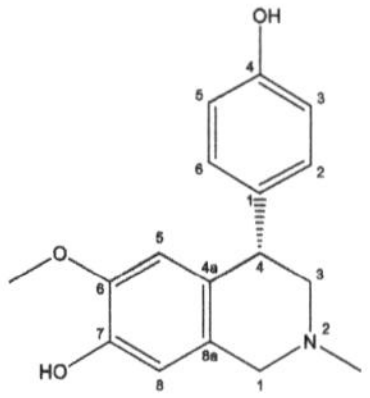
Representative alkaloid	Cherylline	
CAS RN	23,367-61-5	
Formula	$C_{17}H_{19}NO_3$	
Molecular weight (g/mol)	285.34	
Major type v	Cripowelline	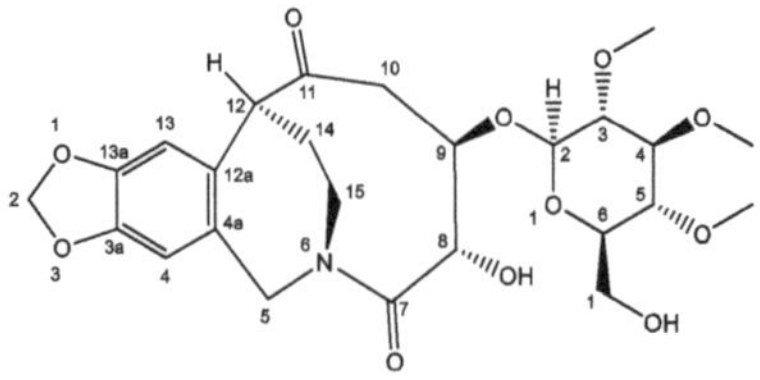
Representative alkaloid	Cripowelline B	
CAS RN	196,812-55-2	
Formula	$C_{25}H_{33}NO_{11}$	
Molecular weight (g/mol)	523.53	
Minor type vi	Augustamine	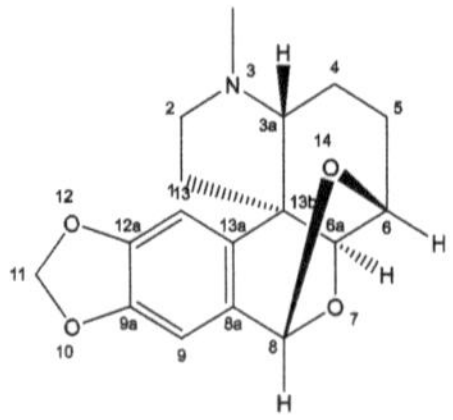
Representative alkaloid	Augustamine	
CAS RN	86,734-72-7	
Formula	$C_{17}H_{19}NO_4$	
Molecular weight (g/mol)	301.34	

(continued)

Table 4.3 (continued)

Minor type vii	Mesembrenone	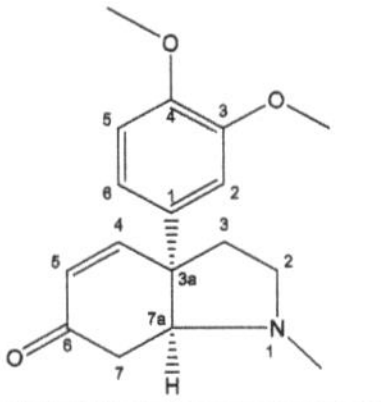
Representative alkaloid	Mesembrenone	
CAS RN	468-54-2	
Formula	$C_{17}\,H_{21}\,N\,O_3$	
Molecular weight (g/mol)	287.35	
Minor type viii	Gracilamine	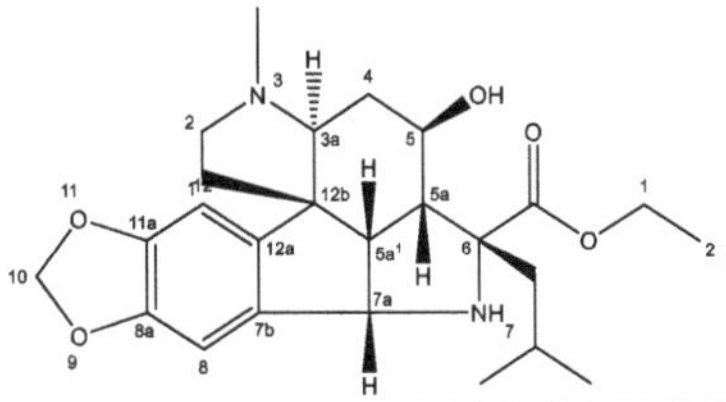
Representative alkaloid	Gracilamine	
CAS RN	879,561-55-4	
Formula	$C_{25}\,H_{34}\,N_2\,O_5$	
Molecular weight (g/mol)	442.55	
Minor type ix	Pallidiflorine	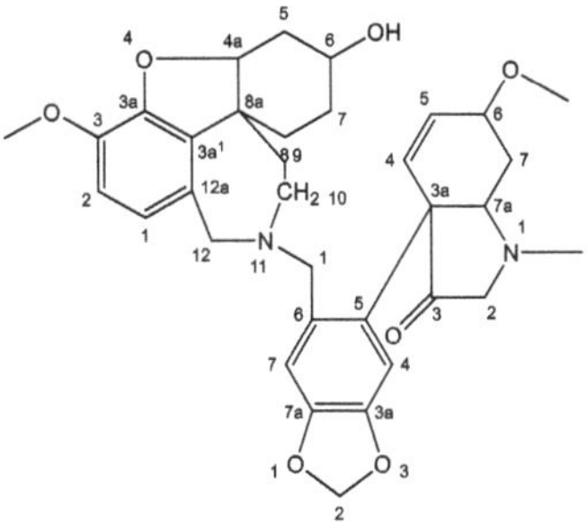
Representative alkaloid	Pallidiflorine	
CAS RN	133,740-60-0	
Formula	$C_{34}\,H_{40}\,N_2\,O_7$	
Molecular weight (g/mol)	588.69	
Minor type x	Buflavine	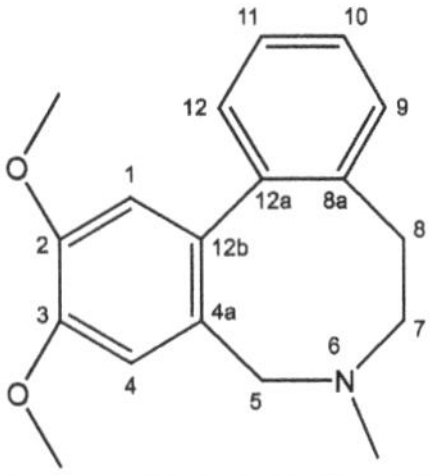
Representative alkaloid	Buflavine	
CAS RN	65,762-70-1	
Formula	$C_{18}\,H_{21}\,N\,O_2$	
Molecular weight (g/mol)	283.36	

4.1.2 Cholinergic Anti-inflammatory Pathways

The cholinergic anti-inflammatory pathway (CAP) is a novel function of the vagus nerve (Pavlov and Tracey 2005). The pathway involves the neurotransmission of acetylcholine and interaction with peripheral α7 subunit of nicotinic acetylcholine receptors (α7nAChR, or α7 subunit) expressed on macrophages (Pavlov et al. 2003). Neural signals transmitted via the vagus nerve in the CAP pathway modulate systemic and local inflammation, for example, by inhibiting the release pro-inflammatory cytokines from macrophages (Fig. 4.1).

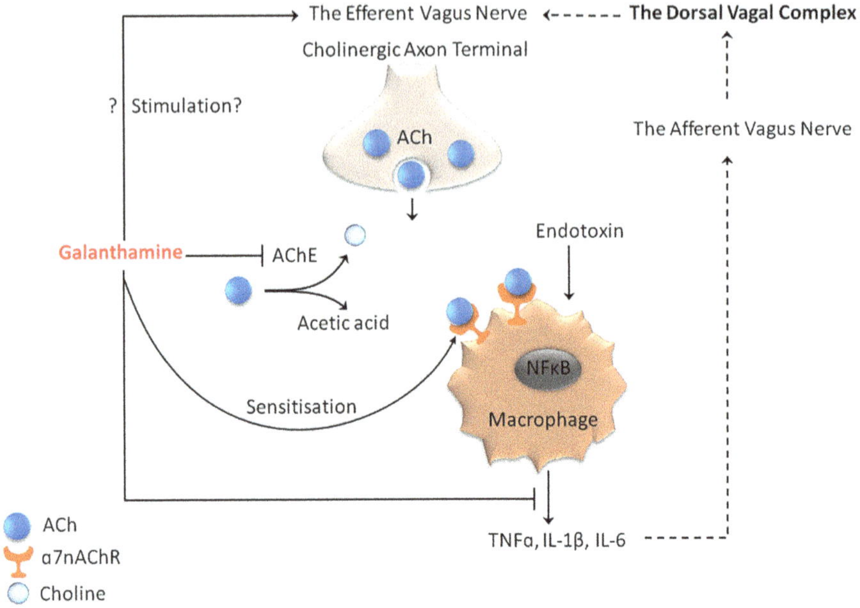

Fig. 4.1 Cholinergic anti-inflammatory pathway and targets by galanthamine. Macrophages are activated by endotoxins from foreign pathogens via NFκB and produce pro-inflammatory cytokines TNFα, IL-1β, and IL-6. Inflammatory signals interact with the dorsal vagal complex of medulla oblongata via the afferent vagus nerve and regulate inflammatory responses including the involvement of humoral pathway, the efferent vagus nerve, and neurotransmitters such as acetylcholine. Acetylcholine allosterically binds to α7nAChR on the surface of macrophages and prevents macrophages to produce pro-inflammatory cytokines. Galanthamine sensitizes the binding of acetylcholine with α7nAChR and also inhibits acetylcholine esterase to break down acetylcholine, resulting in the stronger reduction of pro-inflammatory cytokines production by macrophages. This action by galanthamine seems similar to the stimulation of the efferent vagus nerve

Macrophages can express α7nAChR, and its interaction with acetylcholine in the vagus nerve is one of mechanisms connecting the brain and immune system (Pavlov et al. 2009).

Amarillidaceae alkaloid galanthamine exhibits anti-inflammatory action by suppressing the pro-inflammatory cytokines, and its molecular mechanisms via the cholinergic anti-inflammatory pathway, deciphered in Fig. 4.1, are discussed in Sect. 4.2.

Table 4.4 Anti-inflammatory activities of selected Amaryllidaceae alkaloids

Alkaloids	Targets	Results	References
Galanthamine *Lipinski's rule of 5* MW < 500 H-bond donor 1 H-bond acceptor 3 Log*P* 1.8	C57BL/6 mice with α7nAChR deficiency and wild-type littermates were used to obtain homozygous knockout mice or wild-type mice (*n* = 7–8 per group)	Galanthamine (4 mg/kg, i.p.), injected 1 h prior to endotoxin (6 mg/kg, i.p.) suppressed significantly both serum and splenic TNF levels in wild-type mice, but did not alter serum TNF levels in α7nAChR knockout mice compared with saline-treated controls	Pavlov et al. (2009)
	(a) Rats with peritonitis induced by lipopolysaccharide (*E. coli* 055:B5, 5 mg/kg, ip), 5/group; control group (saline) (b) Rats treated as above plus sham and bilateral cervical vagotomy	(a) Galanthamine at 3 mg/kg ip significantly inhibited TNFα release by about 30% of the control group (b) Galanthamine at 3 mg/kg ip in the sham group significantly inhibited TNFα release by about 30% of the control group, but did not inhibit TNFα release in the vagotomy group	Liu et al. (2010)
	C57BL/6J mice fed with high-fat diet (HFD) for 12 weeks. At week 8 galanthamine 4 mg/kg, ip, once daily for 4 weeks (HFD-G group). In parallel, mice on the low-fat diet (LFD) were treated with saline 4 mg/kg, ip (LFD-S control)	Plasma IL-6 levels were elevated (about 50%) significantly in the HFD-S mice compared with the LFD-S mice. Galanthamine at 4 mg/kg significantly reduced the IL-6 levels in the HFD-G mice (about 50%) compared with HFD-S mice	Satapathy et al. (2011)
	5 groups of healthy rats and adjuvant arthritis (AA) rats (6/group) were used: group 1 (healthy control), group 2 (14 days' AA rat control), group 3 (galanthamine 2.5 mg/kg gavage, day 14–18), group 4 (selective α7nAChR blocker methyllycaconitine citrate (MLA) 5.6 mg/kg, ip, day14–18) and group 5 (MLA treatment 15 min before galanthamine treatment). Vehicle control injection was saline	Splenic TNF-α levels in AA rats increased 615% from the healthy rat control. In joint tissues, significant up-regulation of TNFα mRNA and increase of TNFα protein levels were observed in groups 3–5. Group 3 (galanthamine alone) significantly suppressed TNFα and mRNA levels. But groups 4 (MLA alone) and 5 (MLA + galanthamine) significantly reversed some of the suppression compared with group 3	Gowayed et al. (2019)

(continued)

Table 4.4 (continued)

Alkaloids	Targets	Results	References
Lycorine MW < 500 H-bond donor 2 H-bond acceptor 5 Log*P* 0	Sprague-Dawley rat (200–250 g) with paw oedema induced by lambda-carrageenan (type IV, 0.05 mL). Group I (saline 0.2 mL), group II (ethanol), group III (indomethacin 3 mg/kg, ip), groups IV-VI (lycorine 0.5, 1.0 and 1.5 mg/kg, ip)	ED_{50} 0.51 mg/kg based on group IV to VI results. Lycorine at 2.0 mg/kg was comparible to indomethacin 3 mg/kg (group III) with 95% inhibition of foot volume against controls (groups I and II)	Citoglu et al. (2012)
	Murine macrophage-like RAW264.7 cells, pre-treated with lycorine (5 μM) or dexamethasone (5 μM) for 3 h, then stimulated with LPS (100 ng/mL) for 10–12 h	Lycorine at 5 μM significantly inhibited the level of TNF-α, IL-6, NO and PGE_2 in the cells stimulated by LPS. Also lycorine at 5 μM significantly inhibited the level of COX-2 and iNOS protein expression in the cells with or without LPS stimulation	Kang et al. (2012)
	Mouse RAW264 macrophage cells, pre-treated with lycorine (0.1–10 μM), and then stimulated by LPS (100 μg/mL in medium)	Lycorine inhibited TNFα production with IC_{50} of 2.1 μM. Lycorine inhibited NO production with IC_{50} of 1.2 μM	Yamazaki and Kawano (2011)
	LPS-stimulated macrophage cells from the peritoneal cavities of male ddY mice stimulated with 4% thioglycolate 4 days earlier	Lycorine inhibited NO production with IC_{50} of 2.5 μM	Abdel-Halim et al. (2004)
Narciclasine (Lycoricidinol) MW < 500 H-bond donor 5 H-bond acceptor 7 Log*P* –1.2	Male Wistar Lewis rat (8 weeks old) with arthritis induced by adjuvant (0.05 mL paraffin containing 5 mg heat-killed *Mycobacterium butyricum*)	Narciclasine at 1 mg/kg (ip) for 10 day suppressed the degree of swelling and inflammation, which was comparable to hydrocortisone (20 mg/kg, ip)	Mikami et al. (1999)
	Mouse RAW264 macrophage cells, pre-treated with lycoricidinol (0.1–10 μM) and then stimulated by LPS (100 μg/mL in medium)	Lycoricidinol inhibited TNFα production with IC_{50} of 0.020 μM. Lycoricidinol inhibited NO production with IC_{50} of 0.010 μM	Yamazaki and Kawano (2011)
	Adult male Lewis rats (200–250 g) with arthritis induced by complete Freund's adjuvant (0.1 mL of 3 mg/mL), group I (healthy control), II (arthritis control), III (saline), IV (prodrug 1.75 mg/kg/day, ip), V (3.5 mg/kg/day, ip) and VI (5.0 mg/kg/day, ip) for 28 days	Narciclasine prodrug sodium narcistatin was used. It dramatically reduced joint inflammation about 70%, and bone loss about 50% at dose of 5 mg/kg/day for 28 days, compared to group II	Lubahn et al. (2012)

(continued)

Table 4.4 (continued)

Alkaloids	Targets	Results	References
Crinamine MW < 500 H-bond donor 1 H-bond acceptor 3 Log*P* 1.8	Rat liver hepatoma cancer cells	Crinamine at 25 μM after 48 h treatment produced an apoptotic index of about 90%	McNulty et al., (2007)
	LPS-stimulated macrophage cells from the peritoneal cavities of male ddY mice stimulated with 4% thioglycolate 4 days earlier	Crinamine inhibited NO production with IC_{50} of 1.8 μM	Abdel-Halim et al. (2004)

4.2 Molecular Mechanisms of Cholinergic Anti-inflammatory Pathway

4.2.1 Acetylcholine and Acetylcholine Esterase

Acetylcholine is the principal neurotransmitter in the vagus nerve signalling and can significantly attenuate the release of TNFα, IL-1β, IL-6, and IL-18, but no IL-10 in human macrophage stimulated by lipopolysaccharide (LPS) in vitro (Borovikova et al. 2000). Other cholinergic agonists including muscarine and nicotine have similar effects. Interestingly, anticholinergic drug atropine does not restore the LPS-stimulated release of TNFα in the human macrophage culture treated with acetylcholine (Borovikova et al. 2000).

Acetylcholine esterase catalyses the conversion of acetylcholine to choline and acetic acid. Galanthamine inhibits AChE and stops the breakdown of acetylcholine (Fig. 4.1). This increases acetylcholine concentration available for interaction with α7nAChR.

Through direct acetylcholine action and indirect AChE inhibition, galanthamine and other AChE inhibitors evoke cholinergic enhancement and through interaction with α7nAChR in CAP pathway suppress the inflammatory cytokines in microglia and attenuate the central and peripheral inflammation in brain for the benefit of patients with Alzheimer's disease (Tabet 2006).

Non-neuronal acetylcholine can be assembled in lymphocytes expressing choline acetyltransferase (ChAT) and is in close proximity with macrophage and interacts with α7nAChR (Fujii et al. 2008). More about α7nAChR is presented in Sect. 4.2.3.

4.2.2 The Vagus Nerve and Pro-inflammatory Cytokines

The vagus nerve is the tenth cranial nerve of the central nervous system and the longest nerve of the autonomic nervous system in the human body. It has an autonomic ganglion and innervates all the organs except the adrenal glands where epinephrine and norepinephrine are produced.

In a rat model of lethal endotoxaemia, direct stimulation of the peripheral vagus nerve inhibits the synthesis of TNFα in liver, reduces the peak serum concentration of TNFα, and prevents shock as measured by mean arterial blood pressure (Borovikova et al. 2000). The study supports that the cholinergic parasympathetic (vagus) nerve is involved with the immune regulation process or with an immunomodulatory reflux mechanism which is mediated by the vagus nerve (Pavlov and Tracey 2005). Indeed there is possible existence of "the inflammatory reflex", which controls innate immunity and inflammation in real time (Tracey 2002).

Galanthamine decreased the TNF-α level in rats with peritonitis induced by lipopolysaccharide when the vagus nerve was not cut (sham operation). With the vagotomy where the vagus nerve was cut, the TNF-α level did not decrease (Liu et al. 2010). The vagotomy studies also demonstrated that the α7nAChR-mediated cholinergic anti-inflammatory pathway was required for the anti-inflammatory effect of galanthamine (Table 4.4).

4.2.3 α7 Subunit of Nicotinic Acetylcholine Receptors

The α7 subunit of nicotinic acetylcholine receptors (α7nAChRs, or α7 subunit) is one of 16 subunits in a family of ligand-gated pentameric ion channels in the mammalian (Albuquerque et al. 2009). The α7nAChR is distributed in peripheral immune tissues with different synaptic locations (Schilstrom et al. 2007). The α7 subunit is an essential regulator of inflammation as its deficiency renders the vagus nerve ineffective as a physiological pathway to inhibit TNFα release (Wang et al. 2003).

Knockout mice lacking α7nAChR could not control the rise of TNF-α level from peripheral macrophages even after systematic treatment with galanthamine. This contrasts to the significant reduction of serum TNFα levels and the protection against lethality during murine endotoxemia wild-type mice by the peripheral adminstration of galanthamine (Pavlov et al. 2009).

The α7nAChR ligands such as choline, nicotine, epibatidine, lobeline, varenicline, and cytisine can mimic the action of acetylcholine. Galanthamine can bind allosterically with α7nAChR to increase the sensitisation of the α7 subunit receptor and regulates peripheral anti-inflammatory response. Other Amaryllidaceae alkaloids may have the similar action but need to be investigated.

In carrageenan-induced inflammatory pain and chronic constriction injury (CCI) neuropathic pain models, synthetic allosteric modulators of the α7nAChR have been tested to activate the cholinergic anti-inflammatory pathway and attenuated inflammation and pain (Freitas et al. 2013). Amaryllidaceae alkaloids may act similarly via α7AChR modulation and may emerge as broad-spectrum non-opioid analgesics like (*R*)-5-(2-azetidiylmethoxy)2-chloropyridine (Bannon et al. 1998).

4.2.4 p38 Mitogen-Activated Protein Kinase

The inhibition of p38 MAP kinase activity induced by injury or shock at the central nervous system can attenuate the peripheral inflammation through CAP-mediated vagus nerve activation. The p38 MAP kinase has been linked to the α7nAChR (Gubbins et al. 2010). The inhibition of p38 MAP kinase increases the high-frequency power spectrl compoment of heart rate variability, a parasympathetic activity parameter directly linked to the activation of the CAP pathway (Waldburger et al. 2008).

The p38 MAP kinase inhibition promotes the recovery of bone destruction in rheumatoid arthritis. This is supported by markedly suppressed paw swelling, inhibited synovial inflammation, and decreased radiographic evidence of joint destruction in rats with adjuvant arthritis when the p38 inhibitor SB203580 was administered via intrathecal (IT) catheters to selective block the spinal cord p38 MAP kinase (Boyle et al. 2006). When the same dose of SB203580 was delivered systemically, there was no effect. This means that the effect is mediated by the local concentrations of the p38 inhibitor in the neural compartment (Boyle et al. 2006).

When another potent and selective inhibitor of the p38 MAP kinase, SB 242235, was administered orally to rats at doses of 10, 5, and 2.5 mg/kg, significant inhibition of the LPS-induced production of TNFα at 80%, 68%, and 28% was observed, respectively, compared with vehicle control (Badger et al. 2000). Also there was a significant inhibition of serum IL-6 levels when rats with adjuvant induced arthritics were administered with 60 mg/kg oral dose of SB242235 from day 10 to day 20 and IL-6 was measured at day 21, but not at the lower doses than 60 mg/kg.

The Amaryllidaceae alkaloid lycorine also inhibit p38 MAP kinase expression in RAW264.7 cells challenged with LPS (Kang et al. 2012). It is postulated that the anti-inflammatory activity of the AChEIs may link with the central p38 MAP kinase inhibition and the vagus nerve mediated cholinergic outflow (Waldburger et al. 2008).

4.2.5 The Nuclear Factor NFκB, Nitric Oxide, and Cyclooxygenases

The nuclear factor kappa-light-chain-enhancer of activated B cells (NFκB) within macrophages plays a pivotal role in the expression of pro-inflammatory genes including cytokines, chemokines, and adhesion molecules during the regulation of local and systemic inflammation (Lawrence 2009). It is a prototypical pro-inflammatory signalling pathway, for example, increasing pro-inflammatory cytokine IL-1 production during inflammation, that activates the NFκB within the macrophage and triggers genes responsible for the inflammatory response (Baeuerle and Henkel 1994).

NF-κB activation was induced when the cholinergic anti-inflammatory pathway was activated using electrical stimulation of efferent vagus nerves in animal. This also results in the inhibition of TNF-α production and protection of the animal from splanchic artery occlusion shock (Altavilla et al. 2006).

Amaryllidaceae alkaloids may indirectly attenuate the NFκB activation evoked by IL-1 to alleviate the inflammatory response. Amaryllidaceae alkaloids as AChE inhibitors may also attenuate the receptor activator of NFκB ligand (RANKL)-induced osteoclastogenesis in arthritic disorders.

Nitric oxide production requires the induction of nitric oxide synthase (iNOS). The promoters of murine and human iNOS genes related to NO production comprise a unique harmony sequence for the activation of NFκB, which is essential to induce inflammatory cytokines challenged by LPS (Nunokawa et al. 1996). Amaryllidaceae alkaloids may stimulate the vagus nerve, attenuate the peripheral inflammation, inhibit inflammatory iNOS, and suppress NO production.

Nicotine exposed to the activated human astrocytes inhibited the production of critical pro-inflammatory cytokines IL-6 (60%), IL-1β (42%), and TNF-α (68%). However, nicotine's inhibitory effect on IL-6 production was reversed by the specific cyclooxygenase-2 (COX-2) inhibitor NS-398 (Revathikumar et al. 2016). It was observed that the activity of COX-2 was blocked by NS-398 at 1 and 10 μM, whereas IL-6 release in the presence of nicotine (10 μM) was not reduced when the cells were exposed to NS-398.

4.3 Anti-inflammatory Activities of Amaryllidaceae Alkaloids

4.3.1 Galanthamine

Galanthamine is a centrally acting acetylcholine esterase (AChE) inhibitor with an IC_{50} of 1 μM. It became commercially available for the treatment of poliomyelitis in Bulgaria in 1980s. In 2000 galanthamine was licensed in Iceland, Ireland, Sweden, and the UK for the treatment of Alzheimer's disease. In 2001, FDA-approved galanthamine (hydrobromide) for moderate to severe Alzheimer's disease. Since then more European countries and also Asian countries have used the drug.

Interestingly treatment of the Alzheimeric patients with galanthamine for 1 month showed marked decrease of pro-inflammatory cytokine such as IL-1β, TNF-α, and IL-6 from blood PBMC (Pollak et al. 2005). This is in contrast with untreated Alzheimeric patients who produce large amount of IL-1 in both central and peripheral systems (Dinarello 1996).

Galanthamine inhibits AChE and increases the acetylcholine neurotransmitter in the parasympathetic vagus nerve, which subsequently interacts with the α7 subunit of nicotinic acetylcholine receptors on macrophages and reduces TNFα production. Galanthamine also sensitizes the α7nAChR and attenuates the systemic inflammation (Fig. 4.1).

As summarized in Table 4.4, galanthamine at 4 mg/kg intraperitoneally (ip) injected 1 hour prior to endotoxin challenge (6 mg/kg, ip) suppressed significantly both serum and splenic TNFα levels in wild-type mice, but did not alter serum TNFα levels in α7nAChR knockout mice, compared with saline-treated controls. In rats with peritonitis induced by lipopolysaccharide (*E. coli* 055:B5, 5 mg/kg, ip), galanthamine at 3 mg/kg, ip, significantly inhibited TNFα release by about 30% of the control group (saline). Compared with the sham-operated group of the peritonitis rats, bilateral cervical vagotomy did not inhibit TNFα release significantly (Liu et al. 2010). In C57BL/6J mice fed with high fat diet (HFD) for 12 weeks, plasma IL-6 levels were elevated significantly (about 50%) compared with low-fat diet (LFD).

However, at week 8 of the 12 weeks galanthamine at 4 mg/kg (ip) daily to the HFD mice at week 8 to week 12, the IL-6 levels were reduced significantly (about 50%) at the end of 12 weeks study when compared with the LFD mice injected with saline (Satapathy et al. 2011). Splenic TNF-α levels in adjuvant arthritis rats increased 615% from the healthy rat control. In joint tissues, significant upregulation of TNFα mRNA and the increase of TNFα protein levels were observed in groups 3-5, compared with the control. Group 3 (galanthamine alone) significantly suppressed both TNFα and mRNA levels. But groups 4 (MLA alone, MLA is methyllycacronitine citrate, a selective 7αnAChR inibitor) and 5 (MLA + galanthamine) significantly reversed the suppression compared with group 3 (Gowayed et al. 2019). All these studies demonstrated that galanthamine possesses anti-inflammatory action through the cholinergic anti-inflammatory pathway. Other Amaryllidaceae alkaloids with the similar drug-like property of galanthamine based on Lipinski's Rule of 5 (a molecular weight less than 500, H-bond donors not more than 5, H-bond acceptors not more than 10 and an octanol-water partition coefficient log P not exceeding 5) may elicit similar anti-inflammatory action to galanthamine.

4.3.2 Lycorine

Lycorine is the most reported alkaloid isolated from family Amaryllidaceae, and 67 alkaloids of the lycorine-type (Table 4.2) have been isolated from 42 species (Ding et al. 2017).

Lycorine at 2.0 mg/kg has been shown to be comparable to the standard anti-inflammatory drug indomethacin 3 mg/kg, based on an ED_{50} of 0.51 mg/kg of lycorine in reducing paw oedema in a carrageenan-induced rat paw oedema model (Table 4.4).

Lycorine at 5 μM significantly inhibited the level of TNF-α, IL-6, NO, and PGE_2 in murine mancrophage-like RAW264.7 cells stimulated by LPS (Kang et al. 2012). At 5 μM, lycorine also significantly inhibited the level of COX-2 and iNOS protein expression in the cells with or without LPS stimulation (Table 4.4).

In mouse RAW264 macrophage cells, lycorine inhibited TNFα production with an IC_{50} of 2.1 μM and NO production with an IC_{50} of 1.2 μM (Yamazaki and Kawano 2011). In LPS-stimulated macrophage cells from the peritoneal cavities of male ddY mice stimulated with 4% thioglycolate 4 days earlier, lycorine inhibited NO production with an IC_{50} of 2.5 μM (Abdel-Halim et al. 2004).

4.3.3 Narciclasine (Lycoridinol)

In male Wistar Lewis rats (8 weeks old) with arthritis induced by adjuvant (Table 4.4) narciclasine at 1 mg/kg (ip) for 10 day suppressed the degree of swelling and inflammation, which was comparable to hydrocortisone at 20 mg/kg, ip (Mikami et al. 1999). In the mouse RAW264 macrophage cells pre-treated with lycoricidinol

(0.1–10 μM) and then stimulated by LPS (100 μg/mL in the medium), lycoricidinol inhibited TNFα production with an IC_{50} of 0.020 μM (Table 4.4). Lycoricidinol also inhibited NO production with and IC_{50} of 0.010 μM (Yamazaki and Kawano 2011).

Adult male Lewis rats (200–250 g) with arthritis induced by Complete Freund's Adjuvant (0.1 mL of 3 mg/mL) were divided into six groups: group I healthy control; II arthritis control; III saline; IV 1.75 mg/kg/day, ip; V 3.5 mg/kg/day, ip; and VI 5.0 mg/kg/day, ip (Table 4.4). A narciclasine prodrug sodium narcistatin dramatically reduced joint inflammation about 70% and saved bone loss about 50% at a dose of 5 mg/kg/day for 28 days, when compared to group II (Lubahn et al. 2012).

4.3.4 Crinamine

Crinamine belongs to the crinine type of Amaryllidaceae alkaloids (Table 4.2). Crinamine at 25 μM after 48 hours treatment produced an apoptotic index of about 90% in rat liver hepatoma cancer cells (Table 4.4). In the LPS-stimulated macrophage cells from the peritoneal cavities of male ddY mice stimulated with 4% thioglycolate 4 days earlier, crinamine inhibited NO production with an IC_{50} of 1.8 μM (Abdel-Halim et al. 2004).

4.4 Anti-inflammatory Potential of Amaryllidaceae Alkaloids Targeting CAP

4.4.1 The Potential as New Anti-inflammatory Agents

This section will extend the immunomodulation and anti-inflammatory action of galanthamine, lycorine, narciclasine (lycoricidinol), and crinamine (Table 4.4) to other major types (Table 4.2) and minor types (Table 4.3) of Amaryllidaceae alkaloids in search for novel anti-inflammatory agents. Would all 20 types of amaryllidaceae alkaloids have the anti-inflammatory action if they were shown to inhibit AChE or modulate α7nAChR?

From the discussion of the CAP pathway above, it is notable that the vagus nerve, AChE, and α7nAChR are emerged as potential new targets for anti-inflammatory drug design and development, in addition to classical targets of cycloxoygenases, nitric oxide, NFκB, and cytokines. It is reasonable to believe that all galanthamine-type alkaloids may hold good prospects to have anti-inflammatory activity with favourable pharmacokinetic properties for drug development.

In search of Amaryllidaceae alkaloids with one-molecule-multiple-targets including anti-inflammatory activities, some or all of the following requirements may need to be met:

1. Is a strong AChE inhibitor
2. Stimulates the efferent vagus nerve mediated by CAP
3. Binds peripheral α7nAChR allosterically or directly

4. Inhibits p38 MAP-kinase
5. Attenuates NFκB activation

Galanthamine has been demonstrated to meet all the criteria above and a clinically useful drug for Alzheimer's disease and additional benefits of reducing systematic inflammation. A multi-targeted drug can be designed by many combinations of the five targets above to screen over 500 Amaryllidaceae alkaloids for various inflammatory conditions or diseases.

4.4.2 Testing Methods

As alpha 7 nAChR is expressed locally in liver and spleen and some peripheral tissues including synovial fibroblast cells in RA diseases, in vitro screening of nACh receptors with Amaryllidaceae alkaloids via receptor binding assay may be used to develop the lead molecules as nAChR agonists or modulators from the vast and diverse Amaryllidaceae alkaloids. *In silico* screening using molecular docking of α7nAChR with a range of Amaryllidaceae alkaloids (Castillo-Ordonez et al. 2017) can be employed.

In vivo tests can use acute and chronic inflammatory models of footpad or ear swelling induced by carrageenan, lipopolysaccharide, Freund's adjuvant, or cotton pellets. Graft versus host reaction using spleen cells or marrow graft to induce nonspecific chronic inflammation, and animal model of experimental allergic encephalomyelitis, aspermatogenesis, and arthritis can also be tested.

4.5 Prospects and Conclusion

Based on the discussion in this chapter, alpha 7 nAChR is a new player in the neuronal regulation of inflammation as demonstrated in the recent discovery of the cholinergic anti-inflammatory pathway and the action of galanthamine on the pathway. Alpha 7 nAChR may be an emerging drug target for anti-inflammatory action of Amaryllidaceae alkaloids in addition to the traditional targets of COXs, NO, NFκB, TNFα, IL-1β, p38 MAP kinase, and other.

It is clear that Amaryllidaceae alkaloids are not only a useful tool for studying the cholinergic anti-inflammatory pathway but also offer a potential as a new class of anti-inflammatory natural products, acting on multiple targets to combat inflammatory disorders and having different modes of action from the classical nonsteroidal anti-inflammatory drugs. It is hoped that continuing research in this areas would lead to the development of new anti-inflammatory drugs with one-molecule-multiple-targets (Lin and Li 2018) and reduced unwanted side effects of classical NSAIDs.

Acknowledgement PP wishes to thank the Australian Government for the support of an Endeavour Fellowship.

The database access to the repository and collection of Amaryllidaceae alkaloids provided by Dr. Henry M. Fales, NHLBI Laboratory of Applied Mass Spectrometry, National Institutes of Health, Bethesda, to Professor Martin G. Banwell, Research School of Chemistry, Australian National University, Canberra is acknowledged for this work.

References

Abdel-Halim OB, Morikawa T, Ando S, Matsuda H, Yoshikawa M (2004) New crinine-type alkaloids with inhibitory effect on induction of inducible nitric oxide synthase from Crinum yemense. J Nat Prod 67:1119–1124

Albuquerque EX, Pereira EF, Alkondon M, Rogers SW (2009) Mammalian nicotinic acetylcholine receptors: from structure to function. Physiol Rev 89:73–120

Altavilla D, Guarini S, Bitto A, Mioni C, Giuliani D, Bigiani A et al (2006) Activation of the cholinergic anti-inflammatory pathway reduces NF-kappab activation, blunts TNF-alpha production, and protects against splanchnic artery occlusion shock. Shock 25:500–506

Badger AM, Griswold DE, Kapadia R, Blake S, Swift BA, Hoffman SJ et al (2000) Disease-modifying activity of SB 242235, a selective inhibitor of p38 mitogen-activated protein kinase, in rat adjuvant-induced arthritis. Arthritis Rheum 43:175–183

Baeuerle PA, Henkel T (1994) Function and activation of NF-kappa B in the immune system. Annu Rev Immunol 12:141–179

Bannon AW, Decker MW, Holladay MW, Curzon P, Donnelly-Roberts D, Puttfarcken PS et al (1998) Broad-spectrum, non-opioid analgesic activity by selective modulation of neuronal nicotinic acetylcholine receptors. Science 279(5347):77–80

Bastida J, Lavilla R, Viladomat F (2006) Chapter 3 chemical and biological aspects of narcissus alkaloids. In: Cordell G (ed) The alkaloids: chemistry and biology. Elsevier, Amsterdam, pp 87–179

Borovikova LV, Ivanova S, Zhang M, Yang H, Botchkina GI, Watkins LR et al (2000) Vagus nerve stimulation attenuates the systemic inflammatory response to endotoxin. Nature 405:458–462

Boyle DL, Jones TL, Hammaker D, Svensson CI, Rosengren S, Albani S et al (2006) Regulation of peripheral inflammation by spinal p38 MAP kinase in rats. PLoS Med 3:e338

Castillo-Ordonez WO, Tamarozzi ER, da Silva GM, Aristizabal-Pachon AF, Sakamoto-Hojo ET, Takahashi CS et al (2017) Exploration of the Acetylcholinesterase inhibitory activity of some alkaloids from Amaryllidaceae family by molecular docking in Silico. Neurochem Res 42:2826–2830

Citoglu GS, Acikara OB, Yilmaz BS, Ozbek H (2012) Evaluation of analgesic, anti-inflammatory and hepatoprotective effects of lycorine from Sternbergia fisheriana (Herbert) Rupr. Fitoterapia 83:81–87

Cook J, Loudon J (1952) Chapter XI alkaloids of the amaryllidaceae. In Manske RHF and Holmes HL (eds) The Alkaloids Chemistry and Physiology, vol 2. Academic Press, New York, pp 331–352

Dinarello CA (1996) Biologic basis for interleukin-1 in disease. Blood 87:2095–2147

Ding Y, Qu D, Zhang KM, Cang XX, Kou ZN, Xiao W et al (2017) Phytochemical and biological investigations of Amaryllidaceae alkaloids: a review. J Asian Nat Prod Res 19:53–100

Freitas K, Ghosh S, Ivy Carroll F, Lichtman AH, Imad Damaj M (2013) Effects of alpha7 positive allosteric modulators in murine inflammatory and chronic neuropathic pain models. Neuropharmacology 65:156–164

Fujii T, Takada-Takatori Y, Kawashima K (2008) Basic and clinical aspects of non-neuronal acetylcholine: expression of an independent, non-neuronal cholinergic system in lymphocytes and its clinical significance in immunotherapy. J Pharmacol Sci 106:186–192

Gowayed MA, Rothe K, Rossol M, Attia AS, Wagner U, Baerwald C et al (2019) The role of alpha7nAChR in controlling the anti-inflammatory/anti-arthritic action of galantamine. Biochem Pharmacol 170:113665

Gubbins E, Gopalakrishnan M, Li J (2010) α7 nAChR-mediated activation of MAP kinase pathways in PC12 cells. Brain Res 1328:1–11

He M, Qu C, Gao O, Hu X, Hong X (2015) Biological and pharmacological activities of amaryllidaceae alkaloids. RSC Adv 5:16562–16574

Jin Z (2005) Amaryllidaceae and Sceletium alkaloids. Nat Prod Rep 22:111–126

Jin Z (2007) Amaryllidaceae and Sceletium alkaloids. Nat Prod Rep 24:886–905

Jin Z (2013) Amaryllidaceae and Sceletium alkaloids. Nat Prod Rep 30:849–868

Jin Z, Yao G (2019) Amaryllidaceae and Sceletium alkaloids. Nat Prod Rep 36:1462–1488

Kang J, Zhang Y, Cao X, Fan J, Li G, Wang Q et al (2012) Lycorine inhibits lipopolysaccharide-induced iNOS and COX-2 up-regulation in RAW264.7 cells through suppressing P38 and STATs activation and increases the survival rate of mice after LPS challenge. Int Immunopharmacol 12:249–256

Lawrence T (2009) The nuclear factor NF-kappaB pathway in inflammation. Cold Spring Harb Perspect Biol 1:a001651

Lin GD, Li RW (2018) Natural products targeting inflammation processes and multiple mediators. In: Mandal SC, Mandal V, Konishi T (eds) Natural products & drug discovery: an integrated approach. Elsevier Ltd, pp 277–308

Liu ZH, Ma YF, Wu JS, Gan JX, Xu SW, Jiang GY (2010) Effect of cholinesterase inhibitor galanthamine on circulating tumor necrosis factor alpha in rats with lipopolysaccharide-induced peritonitis. Chin Med J 123:1727–1730

Lubahn C, Schaller JA, Shewmacker E, Wood C, Bellinger DL, Byron D et al (2012) Preclinical efficacy of sodium narcistatin to reduce inflammation and joint destruction in rats with adjuvant-induced arthritis. Rheumatol Int 32:3751–3760

McNulty J, Nair JJ, Codina C, Bastida J, Pandey S, Gerasimoff J et al (2007) Selective apoptosis-inducing activity of crinum-type Amaryllidaceae alkaloids. Phytochemistry 68:1068–1074

Mikami M, Kitahara M, Kitano M, Ariki Y, Mimaki Y, Sashida Y et al (1999) Suppressive activity of lycoricidinol (narciclasine) against cytotoxicity of neutrophil-derived calprotectin, and its suppressive effect on rat adjuvant arthritis model. Biol Pharm Bull 22:674–678

National Center for Biotechnology Information (2019) Taxonomy ID 4668, Amaryllidaceae. https://www.ncbi.nlm.nih.gov/Taxonomy/Browser/wwwtaxcgi?mode=Tree&id=4668&lvl=3&keep=1&srchmode=1&unlock. Accessed 26 Sep 2019

Nunokawa Y, Oikawa S, Tanaka S (1996) Human inducible nitric oxide synthase gene is transcriptionally regulated by nuclear factor-kappaB dependent mechanism. Biochem Biophys Res Commun 223:347–352

Pavlov VA, Tracey KJ (2005) The cholinergic anti-inflammatory pathway. Brain Behav Immun 19:493–499

Pavlov VA, Wang H, Czura CJ, Friedman SG, Tracey KJ (2003) The cholinergic anti-inflammatory pathway: a missing link in neuroimmunomodulation. Mol Med 9:125–134

Pavlov VA, Parrish WR, Rosas-Ballina M, Ochani M, Puerta M, Ochani K et al (2009) Brain acetylcholinesterase activity controls systemic cytokine levels through the cholinergic anti-inflammatory pathway. Brain Behav Immun 23:41–45

Pollak Y, Gilboa A, Ben-Menachem O, Ben-Hur T, Soreq H, Yirmiya R (2005) Acetylcholinesterase inhibitors reduce brain and blood interleukin-1beta production. Ann Neurol 57:741–745

Revathikumar P, Bergqvist F, Gopalakrishnan S, Korotkova M, Jakobsson PJ, Lampa J et al (2016) Immunomodulatory effects of nicotine on interleukin 1beta activated human astrocytes and the role of cyclooxygenase 2 in the underlying mechanism. J Neuroinflammation 13:256

Royal Society of Chemistry (2019) The alkaloids: chemistry and biology. https://www.sciencedirect.com/bookseries/the-alkaloids-chemistry-and-biology/vol/82/suppl/C. Accessed 12 Dec 2019

Satapathy SK, Ochani M, Dancho M, Hudson LK, Rosas-Ballina M, Valdes-Ferrer SI et al (2011) Galantamine alleviates inflammation and other obesity-associated complications in high-fat diet-fed mice. Mol Med 17:599–606

Schilstrom B, Ivanov VB, Wiker C, Svensson TH (2007) Galantamine enhances dopaminergic neurotransmission in vivo via allosteric potentiation of nicotinic acetylcholine receptors. Neuropsychopharmacology 32:43–53

Tabet N (2006) Acetylcholinesterase inhibitors for Alzheimer's disease: anti-inflammatories in acetylcholine clothing! Age Ageing 35:336–338

Tracey K (2002) The inflammation reflex. Nature 420:853–859

Waldburger JM, Boyle DL, Edgar M, Sorkin LS, Levine YA, Pavlov VA et al (2008) Spinal p38 MAP kinase regulates peripheral cholinergic outflow. Arthritis Rheum 58:2919–2921

Wang H, Yu M, Ochani M, Amella CA, Tanovic M, Susarla S et al (2003) Nicotinic acetylcholine receptor alpha7 subunit is an essential regulator of inflammation. Nature 421:384–388

Yamazaki Y, Kawano Y (2011) Inhibitory effects of herbal alkaloids on the tumor necrosis factor-alpha and nitric oxide production in lipopolysaccharide-stimulated RAW264 macrophages. Chem Pharm Bull 59:388–391

5

Neurodegenerative Diseases and Small Molecule Protein Chaperone Activator of Natural Origin

Naibedya Dutta, Suvranil Ghosh, and Mahadeb Pal

Abstract

Progressive neurodegenerative diseases such as Alzheimer's, Parkinson's, Huntington's disease, and amyotrophic lateral sclerosis affect millions of people around the world. In addition, these diseases cause emotional, medical, and financial burden on the caregivers and the society. Common symptoms of these patients in include depression, memory impairment, anxiety, and defective motor function. A major pathological signature of these diseases is the accumulation of aggregates of different proteins in different parts in the brain. Unfortunately, there is no cure yet; presently management options are available which provides some relief from the disease symptoms. Intense investigations are ongoing by independent research groups to develop an effective novel therapy to protect these helpless patients. Because misfolding of proteins is a common pathological marker, upregulation of protein chaperones in the affected area is being considered as a promising therapeutic approach to treat these diseases. Small molecules of natural origins of distinct mode of actions have been identified that modulate activity of different components of chaperone signaling pathway with a positive outcome in the animal models. In this book chapter, we provide an account on the association of deregulated protein chaperone pathway and the prospect of small molecule modulators of natural origin with therapeutic prospect.

Keywords

Neurodegenerative disease · Heat shock protein · Heat shock factor · Heat shock response · Small molecule · Natural product

N. Dutta · S. Ghosh · M. Pal (✉)
Division of Molecular Medicine, Bose Institute, Kolkata, India
e-mail: palmahadeb@gmail.com; mahadeb@jcbose.ac.in

S. C. Mandal et al. (eds.), *Evidence Based Validation of Traditional Medicines*,
https://doi.org/10.1007/978-981-15-8127-4_5

Abbreviations

AD	Alzheimer's disease
ALS	Amyotrophic lateral sclerosis
GFP	Green fluorescent protein
HD	Huntington's disease
HSE	Heat shock element
HSF1	Heat shock factor 1
HSPs	Heat shock proteins
HSR	Heat shock response
MJD	Machado-Joseph disease
PD	Parkinson's disease
PTM	Posttranslational modifications
SMBA	Spinal and bulbar muscular atrophy

5.1 Neurodegenerative Diseases Associated with Protein Misfolding

A number of neurodegenerative diseases such as Alzheimer's disease (AD), Parkinson's disease (PD), Huntington's disease (HD), and amyotrophic lateral sclerosis (ALS) are associated with misfolding and aggregation of distinct proteins (Ross and Poirier 2004). Some of these diseases, for example, AD and PD, are associated with aging; however many neurodegenerative diseases like ALS, HD, and PD can occur in younger ages (Longinetti et al. 2017; Jarrett and Lansbury Jr. 1993). An inefficient protein quality control mechanism is found to be common in all of these diseases and found to be partly responsible for the accumulation of protein aggregates in a particular tissue type. An accumulation of SOD1 and TDP-43 in the spinal cord is associated with patients with ALS; the deposition of β-amyloid fragments occurs in the brain of AD patients; the formation of α-synuclein plaques in specific regions of the brain including substantia nigra, hippocampus, neocortex, cerebellum, and thalamus is observed in PD patients (Ross and Poirier 2004; Dobson 2003; Jha et al. 2009). Huntington's disease occurs with the pathology of aggregates with huntingtin in the brain (cortex, striatum, and hippocampus). Ataxin 3 protein is found in aggregates that accumulate in the different regions of the brain with Machado-Joseph disease (MJD). Expansion of polyglutamine repeats as a result of the mutation in the huntingtin protein-encoding HTT gene located in chromosome 4 is a signature of Huntington's disease and different forms of familial spinocerebellar ataxia (Jimenez-Sanchez et al. 2017; Zoghbi and Orr 2000). In healthy condition α-synuclein, a water-soluble protein helps to maintain the release of dopamine from the synaptic vesicles to control the voluntary and involuntary movement of muscles, but in PD, dementia or multiple system atrophy, this protein misfolds and accumulates as a spherical filamentous structure called Lewy bodies causing premature cell death (Stefanis 2012). Similarly, it is widely believed that the production and deposition of β-amyloid peptides in brain parenchyma drives the pathogenesis of AD (Murphy and LeVine 3rd 2010).

5.2 Role of Molecular Chaperones in Neurodegenerative Condition

For normal growth proteins in a cell must fold into distinct three-dimensional conformations, and the misfolded protein must be degraded immediately. A complex network of molecular chaperones as a component of protein quality control mechanism plays a critical role in the maintenance of protein homeostasis (Krobitsch and Lindquist 2000; Kim et al. 2002). Under numerous environmental stress conditions like hyperthermia and hypoxia, genes encoding the inducible heat shock proteins are rapidly upregulated to supply the heat shock proteins (HSPs) for repairing and clearance of damaged proteins to maintain cellular homeostasis. Transcriptional activation of those genes is orchestrated by a master transcription activator, heat shock factor 1 (HSF1) functionally conserved from prokaryote to human. HSF1, a 529 amino acid long protein, carries a winged helix-turn-helix-type amino-terminal DNA binding domain, spanning from aa 16 to 120 from the N-terminal end. Two leucine zippers HR-A and HR-B (located between aa 130 and 203) composed of two heptad repeats mediate intermolecular oligomerization (Jee 2016). LZ1-3 domain that occurs within HR-A and HR-B regions was shown to be essential for oligomerization but counteracted by LZ4 region located in HR-C (aa 384–409). The regulatory domain (RD) spanning from aa 203 to 384 undergoes various posttranslational modifications and controls HSF1 functions as well as stability. The activation domain resides on the C-terminus of the LZ-4 region (Soares et al. 2019). In normal condition HSF1 exists mainly in the cytoplasm as an inactive monomer in association with HSP70 and TRiC/CCT (Shi et al. 1998; Anckar and Sistonen 2011). Upon a cellular stress, monomeric HSF1 dissociates from that complex and undergoes several posttranslational modifications (PTM) and translocates to the nucleus. The activated HSF1 in the form of a trimer binds to the heat shock element (HSE) of consensus sequence motif 5′-nGAAn-3 located in the upstream region of the HSP genes to drive their upregulation (Kroeger and Morimoto 1994). Evidence suggests that HSF1 is activated in a signal-specific manner. Once the damaged proteins are refolded to their native form or degraded for clearance from the cell via ubiquitin-mediated proteasomal degradation, HSF1 goes to an inactive monomeric state by a negative feedback mechanism involving the chaperones. Though it is considered that the sequences of HSF1 activation events are universal across different cells, some studies indicated cell-type-specific differences of HSF1 activation and the expression of its target HSPs (Batulan et al. 2003). Many studies show the co-localization of chaperons like HSP70, HSP90, and HSP40 with different protein aggregates, which confirm direct involvement of chaperones in the re-establishment of cellular homeostasis (Kuiper et al. 2017). Stability of HSF1 signaled by distinct PTM of HSF1 has been shown to play an important role in cell homeostasis. Stability of HSF1 was shown to be severely compromised in various neurodegenerative disease conditions including AD, PD, and HD resulting in depletion of inducible protein chaperone levels in the affected cells (Gomez-Pastor et al. 2018; Kim et al. 2016; Jiang et al. 2013; Abner et al. 2016).

5.3 HSF1 and Various Neurodegenerative Diseases

Numerous studies demonstrated impairment of HSF1 function in various neurodegenerative diseases. The inducible HSP levels also diminish in various neurodegenerative conditions. Notably, deletion of HSF1 was shown not to produce a particular neurodegenerative disease in a healthy cell; nevertheless upregulation of HSF1 or the inducible HSPs provided some relief in the neurodegenerative disease pathology indicating therapeutic potential in the processes. In polyQ expansion disease of spinal and bulbar muscular atrophy (SMBA), hsf−/− mouse showed more androgen receptor aggregation and severe disease pathology and behavioral phenotype than hsf+/+ mouse. Both in SMBA and HD, HSF1 protein level was reported to be abnormally low (Gomez-Pastor et al. 2018). Upregulation of HSF1 activity was shown to ameliorate HD with an extension of lifespan (Singh et al. 2018). In PD condition as well, the levels of HSF1 and the HSPs are depleted, and a correspondingly forced expression of HSF1 or HSP70 provided protection from α-synuclein aggregation-induced toxicity (Auluck et al. 2002). In ALS and AD upregulation of HSF1 led to the amelioration of the toxicity induced by aggregation of respective proteins (Lin et al. 2013; Chen et al. 2014).

5.4 Activation of Chaperone Activity by Small Molecules with Natural Source

Numerous studies since the last two decades showed ameliorative role of HSPs in various neurodegenerative diseases. A growing number of evidence suggests that forceful induction of chaperones or their master regulator HSF1 produced beneficial effects in animal models of different neurodegenerative diseases (Westerheide and Morimoto 2005; Singh et al. 2018). Number of small molecules including the those of HSP90 inhibitors were shown to activate HSR through HSF1 (Table 5.1).

5.4.1 Celastrol

Celastrol, a triterpene, isolated from the root extracts of a common Chinese medicinal plant *Tripterygium wilfordii* belongs to the Celastraceae family of plants. Celastrol was first identified as an antioxidant and anti-inflammatory drug at the beginning of the twentieth century (Allison et al. 2001). It was later identified to carry potent neuroprotective activity. In 2004, Richard Morimoto and his group showed that celastrol induces heat shock response in different types of human cells including those of neuronal origin involving HSF1 (Westerheide et al. 2004).

Table 5.1 Small molecules of natural origin that induces cellular heat shock response

Compound name	Source	Structure	Function	References
Gedunin	*Azadirachta indica*, *Carapa guianensis*, *Entandrophragma angolense*, *Xylocarpus granatum*	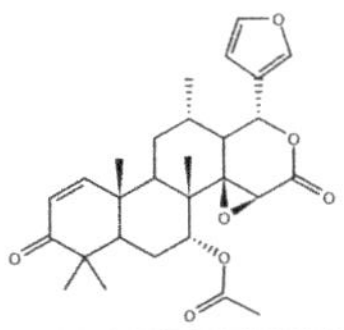	Inhibits the function of cellular HSP90	Brandt et al. (2008); Misra et al. (2011); Patwardhan et al. (2013)
Withaferin A	*Withania somnifera*, *Iochroma arborescens*	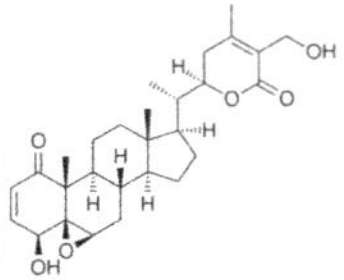	Inhibits the function of cellular HSP90	Yu et al. (2010); Zhang et al. (2012)
Azadiradione	*Azadirachta indica*	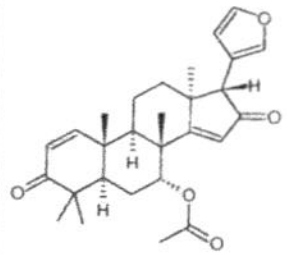	Binds directly with HSF1 and facilitates HSF1's binding with its specific recognition sequence heat shock element (HSE)	Nelson et al. (2016); Singh et al. (2018)
Celastrol	*Tripterygium wilfordii*, *Celastrus regelii*	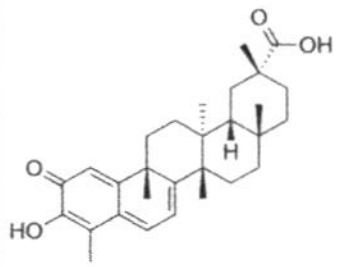	Inhibits ATPase activity of HSP90, leads to hyperphosphorylation of HSF1 and binding to the heat shock element (HSE)	Allison et al. (2001); Westerheide et al. (2004); Cleren et al. (2005); Zhang et al. (2008)
Sulforaphane	Cruciferous vegetables like broccoli (*Brassica oleracea*), cabbage, cauliflower	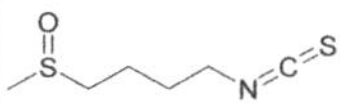	Activates heat shock response and induces proteasome activity	Gan et al. (2010); Zhang et al. (2017)

(continued)

Table 5.1 (continued)

Compound name	Source	Structure	Function	References
Curcumin	*Curcuma longa*	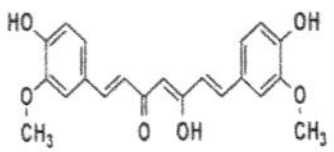	Inhibits the function of HSP90 to induce HSF1	Teiten et al. (2009); Hamaguchi et al. (2010)
Coniferyl aldehyde	*Eucommia ulmoides*	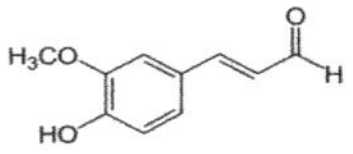	It increases the stability of HSF1 by activating the MAPKs, ERK1/2, c-Jun N-terminal kinase (JNK)1, and p38; promotes HSF1 phosphorylation at serine 326	Kim et al. (2015)

Celastrol disrupts the association between HSP90 and its co-chaperone CDC37 to induce heat shock proteins including HSP70B aiding to cellular protein homeostasis (Zhang et al. 2008). Interestingly, celastrol activity in a cell was shown to depend on its dose; at submicromolar concentration it acted as an antioxidant and inflammatory molecule, while at micromolar concentration, the activity was attributed to the ROS it produced (Chen et al. 2011). In contrast to other small molecule heat shock response modulators, celastrol has better therapeutic potential through its robustness in inducing HSF1 activity (Westerheide et al. 2004). In the mouse model of Parkinson's disease, celastrol was shown to efficiently ameliorate the loss of dopaminergic neurons induced by MPTP-treatment. Similarly, in the rat model of Huntington's disease, celastrol significantly subsidized the striatal lesion induced by 3-nitropropionic acid (Cleren et al. 2005).

5.4.2 Gedunin

Gedunin, a pentacyclic triterpenoid, is one of the prime compounds found in Indian medicinal plant neem (*Azadirachta indica*). It has been in use in numerous homeopathic medicines for a long time (Subapriya and Nagini 2005). Apart from its neuroprotective activity this small molecule also shows anticancer, antifungal, and antiparasitic activity (Misra et al. 2011; Brandt et al. 2008). In recent studies it was found that like celastrol, gedunin induces the HSF1 activity through inhibiting its negative regulator HSP90 by blocking the interaction with co-chaperone p23 although gedunin was found as a less potent activator of HSR compared to celastrol (Brandt et al. 2008; Patwardhan et al. 2013). ROS was shown to be an important mediator of gedunin activity (Braga et al. 2020).

5.4.3 Withaferin A

Withaferin A is a steroidal lactone occurring naturally in Indian plant *Withania somnifera* with anticancer and neuroprotective activity (Zhang et al. 2012). HSF1-inducing activity of withaferin was indicated through a reporter cell-based assay (Santagata et al. 2012). Like previously described molecules, withaferin A induces HSF1 activity through inhibiting the function of HSP90 (Yu et al. 2010) and ROS production (Ghosh et al. 2016).

5.4.4 Curcumin

Curcumin traditionally has been in use against wounds and inflammation for its antibacterial and anti-inflammatory property, and isolated from the Indian plant *Curcuma longa*, commonly known as turmeric (Marathe et al. 2011). Curcumin was shown to ameliorate Alzheimer's disease and was included in different clinical trials (Hamaguchi et al. 2010; Salehi et al. 2019). Curcumin activates both HSP70A and HSP70B expression through inducing HSF1 activity (Teiten et al. 2009).

5.4.5 Sulforaphane

Sulforaphane [1-isothiocyanato-4-(methyl-sulfinyl)butane] is majorly found in cruciferous vegetables like broccoli, cabbages, and cauliflower. It is a very well-known compound with anticancer activity; numerous studies with different cancer cells and animal models supported its activity in reducing human cancer risk (Zhang et al. 1994). Recently, sulforaphane appeared to carry a promising neuroprotective activity; it was found to reduce the load of β-amyloid peptide deposition in the mice model of Alzheimer's disease (Zhang et al. 2017). Sulforaphane was shown to play an important role in clearing the protein aggregates by stimulating the proteasome activity involving heat shock protein (Gan et al. 2010) and NRF2 pathway (Li et al. 2020).

5.4.6 Resveratrol

Resveratrol [3,5,4′-trihydroxy-trans-stilbene], an antioxidant is produced by several plants in response to injury by pathogens. It can be found in a major quantity of different fruits including grapes, blueberries, and peanuts. This molecule is very well-known for its anti-carcinogenic, neuroprotective, cardio-protective, and anti-aging activities (Pangeni et al. 2014). Apart from its antioxidant activity, resveratrol is also found to efficiently clear β-amyloid aggregates in animal model of Alzheimer's disease (Bastianetto et al. 2015). It activates the cellular heat shock response to protect from thermal shock heat (Putics et al. 2008). Resveratrol is an activator of SirT1 deacetylase (Borra et al. 2005).

5.4.7 Azadiradione

Azadiradione is the only molecule which stimulates HSF1 activity by facilitating the binding of HSF1 to its recognition site (HSE) present at its target promoters. Azadiradione interacts with HSF1 physically with high specificity in a test tube. Activation of azadiradione was demonstrated to be independent of HSP90 and proteasome function. In 2016, Nelson et al. have reported azadiradione as a heat shock response inducer for the first time. Notably azadiradione treatment does not result in the elevation of cellular reactive oxygen species. They have isolated azadiradione from the neem (*Azadirachta indica*) seed extract and identified its activity through a cell-based reporter assay. It was found that azadiradione significantly ameliorates protein aggregates both in the cell and *Drosophila* model of PolyQ disease (Nelson et al. 2016). Its activity against neurodegenerative diseases was further confirmed in the animal model of Huntington's disease (Singh et al.

2018). The unique activities highlighted azadiradione as a potential therapeutic agent to combat various neurodegenerative diseases.

References

Abner EL, Jicha GA, Shaw LM et al (2016) Plasma neuronal exosomal levels of Alzheimer's disease biomarkers in normal aging. Ann Clin Transl Neurol 3(5):399–403. https://doi.org/10.1002/acn3.309

Allison AC, Cacabelos R, Lombardi VR et al (2001) Celastrol, a potent antioxidant and anti-inflammatory drug, as a possible treatment for Alzheimer's disease. Prog Neuro-Psychopharmacol Biol Psychiatry 25(7):1341–1357. https://doi.org/10.1016/s0278-5846(01)00192-0

Anckar J, Sistonen L (2011) Regulation of HSF1 function in the heat stress response: implications in aging and disease. Annu Rev Biochem 80:1089–1115. https://doi.org/10.1146/annurev-biochem-060809-095203

Auluck PK, Chan HY, Trojanowski JQ et al (2002) Chaperone suppression of alpha-synuclein toxicity in a Drosophila model for Parkinson's disease. Science 295(5556):865–868. https://doi.org/10.1126/science.1067389

Bastianetto S, Menard C, Quirion R (2015) Neuroprotective action of resveratrol. Biochim Biophys Acta 1852(6):1195–1201. https://doi.org/10.1016/j.bbadis.2014.09.011

Batulan Z, Shinder GA, Minotti S et al (2003) High threshold for induction of the stress response in motor neurons is associated with failure to activate HSF1. J Neurosci 23(13):5789–5798

Borra MT, Smith BC, Denu JM (2005) Mechanism of human SIRT1 activation by resveratrol. J Biol Chem 280(17):17187–17195. https://doi.org/10.1074/jbc.M501250200

Braga TM, Rocha L, Chung TY et al (2020) Biological activities of Gedunin-A Limonoid from the meliaceae family. Molecules (Basel, Switzerland) 25(3):493. https://doi.org/10.3390/molecules25030493

Brandt GE, Schmidt MD, Prisinzano TE et al (2008) Gedunin, a novel hsp90 inhibitor: semisynthesis of derivatives and preliminary structure-activity relationships. J Med Chem 51(20):6495–6502. https://doi.org/10.1021/jm8007486

Chen G, Zhang X, Zhao M et al (2011) Celastrol targets mitochondrial respiratory chain complex I to induce reactive oxygen species-dependent cytotoxicity in tumor cells. BMC Cancer 11(1):170. https://doi.org/10.1186/1471-2407-11-170

Chen Y, Wang B, Liu D et al (2014) Hsp90 chaperone inhibitor 17-AAG attenuates Abeta-induced synaptic toxicity and memory impairment. J Neurosci 34(7):2464–2470. https://doi.org/10.1523/jneurosci.0151-13.2014

Cleren C, Calingasan NY, Chen J et al (2005) Celastrol protects against MPTP- and 3-nitropropionic acid-induced neurotoxicity. J Neurochem 94(4):995–1004. https://doi.org/10.1111/j.1471-4159.2005.03253.x

Dobson CM (2003) Protein folding and misfolding. Nature 426(6968):884–890. https://doi.org/10.1038/nature02261

Gan N, Wu YC, Brunet M et al (2010) Sulforaphane activates heat shock response and enhances proteasome activity through up-regulation of Hsp27. J Biol Chem 285(46):35528–35536. https://doi.org/10.1074/jbc.M110.152686

Ghosh K, De S, Das S et al (2016) Withaferin A induces ROS-mediated paraptosis in human breast cancer cell-lines MCF-7 and MDA-MB-231. PLoS One 11(12):e0168488. https://doi.org/10.1371/journal.pone.0168488

Gomez-Pastor R, Burchfiel ET, Thiele DJ (2018) Regulation of heat shock transcription factors and their roles in physiology and disease. Nat Rev Mol Cell Biol 19(1):4–19. https://doi.org/10.1038/nrm.2017.73

Hamaguchi T, Ono K, Yamada M (2010) REVIEW: curcumin and Alzheimer's disease. CNS Neurosci Ther 16(5):285–297. https://doi.org/10.1111/j.1755-5949.2010.00147.x
Jarrett JT, Lansbury PT Jr (1993) Seeding "one-dimensional crystallization" of amyloid: a pathogenic mechanism in Alzheimer's disease and scrapie? Cell 73(6):1055–1058. https://doi.org/10.1016/0092-8674(93)90635-4
Jee H (2016) Size dependent classification of heat shock proteins: a mini-review. J Exerc Rehabil 12 (4):255–259. https://doi.org/10.12965/jer.1632642.321
Jha SK, Dhar D, Krishnamoorthy G et al (2009) Continuous dissolution of structure during the unfolding of a small protein. Proc Natl Acad Sci U S A 106(27):11113–11118. https://doi.org/10.1073/pnas.0812564106
Jiang YQ, Wang XL, Cao XH et al (2013) Increased heat shock transcription factor 1 in the cerebellum reverses the deficiency of Purkinje cells in Alzheimer's disease. Brain Res 1519:105–111. https://doi.org/10.1016/j.brainres.2013.04.059
Jimenez-Sanchez M, Licitra F, Underwood BR et al (2017) Huntington's disease: mechanisms of pathogenesis and therapeutic strategies. Cold Spring Harb Perspect Med 7(7):a024240. https://doi.org/10.1101/cshperspect.a024240
Kim S, Nollen EA, Kitagawa K et al (2002) Polyglutamine protein aggregates are dynamic. Nat Cell Biol 4(10):826–831. https://doi.org/10.1038/ncb863
Kim SY, Lee HJ, Nam JW et al (2015) Coniferyl aldehyde reduces radiation damage through increased protein stability of heat shock transcriptional factor 1 by phosphorylation. Int J Radiat Oncol Biol Phys 91(4):807–816. https://doi.org/10.1016/j.ijrobp.2014.11.031
Kim E, Wang B, Sastry N et al (2016) NEDD4-mediated HSF1 degradation underlies alpha-synucleinopathy. Hum Mol Genet 25(2):211–222. https://doi.org/10.1093/hmg/ddv445
Krobitsch S, Lindquist S (2000) Aggregation of huntingtin in yeast varies with the length of the polyglutamine expansion and the expression of chaperone proteins. Proc Natl Acad Sci U S A 97(4):1589–1594. https://doi.org/10.1073/pnas.97.4.1589
Kroeger PE, Morimoto RI (1994) Selection of new HSF1 and HSF2 DNA-binding sites reveals difference in trimer cooperativity. Mol Cell Biol 14(11):7592–7603. https://doi.org/10.1128/mcb.14.11.7592
Kuiper EF, de Mattos EP, Jardim LB et al (2017) Chaperones in polyglutamine aggregation: beyond the Q-stretch. Front Neurosci 11:145. https://doi.org/10.3389/fnins.2017.00145
Li D, Shao R, Wang N, et al (2020) Sulforaphane activates a lysosome-dependent transcriptional program to mitigate oxidative stress. Autophagy 1–16. https://doi.org/10.1080/15548627.2020.1739442
Lin PY, Simon SM, Koh WK et al (2013) Heat shock factor 1 over-expression protects against exposure of hydrophobic residues on mutant SOD1 and early mortality in a mouse model of amyotrophic lateral sclerosis. Mol Neurodegener 8:43. https://doi.org/10.1186/1750-1326-8-43
Longinetti E, Mariosa D, Larsson H et al (2017) Physical and cognitive fitness in young adulthood and risk of amyotrophic lateral sclerosis at an early age. Eur J Neurol 24(1):137–142. https://doi.org/10.1111/ene.13165
Marathe SA, Dasgupta I, Gnanadhas DP et al (2011) Multifaceted roles of curcumin: two sides of a coin! Expert Opin Biol Ther 11(11):1485–1499. https://doi.org/10.1517/14712598.2011.623124
Misra S, Verma M, Mishra SK et al (2011) Gedunin and photogedunin of Xylocarpus granatum possess antifilarial activity against human lymphatic filarial parasite Brugia malayi in experimental rodent host. Parasitol Res 109(5):1351–1360. https://doi.org/10.1007/s00436-011-2380-x
Murphy MP, LeVine H 3rd (2010) Alzheimer's disease and the amyloid-beta peptide. J Alzheimers Dis 19(1):311–323. https://doi.org/10.3233/jad-2010-1221
Nelson VK, Ali A, Dutta N et al (2016) Azadiradione ameliorates polyglutamine expansion disease in Drosophila by potentiating DNA binding activity of heat shock factor 1. Oncotarget 7 (48):78281–78296. https://doi.org/10.18632/oncotarget.12930

Pangeni R, Sahni JK, Ali J et al (2014) Resveratrol: review on therapeutic potential and recent advances in drug delivery. Expert Opin Drug Deliv 11(8):1285–1298. https://doi.org/10.1517/17425247.2014.919253

Patwardhan CA, Fauq A, Peterson LB et al (2013) Gedunin inactivates the co-chaperone p23 protein causing cancer cell death by apoptosis. J Biol Chem 288(10):7313–7325. https://doi.org/10.1074/jbc.M112.427328

Putics A, Vegh EM, Csermely P et al (2008) Resveratrol induces the heat-shock response and protects human cells from severe heat stress. Antioxid Redox Signal 10(1):65–75. https://doi.org/10.1089/ars.2007.1866

Ross CA, Poirier MA (2004) Protein aggregation and neurodegenerative disease. Nat Med 10 (Suppl):S10–S17. https://doi.org/10.1038/nm1066

Salehi B, Stojanovic-Radic Z, Matejic J et al (2019) The therapeutic potential of curcumin: a review of clinical trials. Eur J Med Chem 163:527–545. https://doi.org/10.1016/j.ejmech.2018.12.016

Santagata S, Xu YM, Wijeratne EM et al (2012) Using the heat-shock response to discover anticancer compounds that target protein homeostasis. ACS Chem Biol 7(2):340–349. https://doi.org/10.1021/cb200353m

Shi Y, Mosser DD, Morimoto RI (1998) Molecular chaperones as HSF1-specific transcriptional repressors. Genes Dev 12(5):654–666. https://doi.org/10.1101/gad.12.5.654

Singh BK, Vatsa N, Nelson VK et al (2018) Azadiradione restores protein quality control and ameliorates the disease pathogenesis in a mouse model of Huntington's disease. Mol Neurobiol 55(8):6337–6346. https://doi.org/10.1007/s12035-017-0853-3

Soares TR, Reis SD, Pinho BR et al (2019) Targeting the proteostasis network in Huntington's disease. Ageing Res Rev 49:92–103. https://doi.org/10.1016/j.arr.2018.11.006

Stefanis L (2012) α-synuclein in Parkinson's disease. Cold Spring Harb Perspect Med 2(2): a009399. https://doi.org/10.1101/cshperspect.a009399

Subapriya R, Nagini S (2005) Medicinal properties of neem leaves: a review. Curr Med Chem Anticancer Agents 5(2):149–146

Teiten MH, Reuter S, Schmucker S et al (2009) Induction of heat shock response by curcumin in human leukemia cells. Cancer Lett 279(2):145–154. https://doi.org/10.1016/j.canlet.2009.01.031

Westerheide SD, Morimoto RI (2005) Heat shock response modulators as therapeutic tools for diseases of protein conformation. J Biol Chem 280(39):33097–33100. https://doi.org/10.1074/jbc.R500010200

Westerheide SD, Bosman JD, Mbadugha BN et al (2004) Celastrols as inducers of the heat shock response and cytoprotection. J Biol Chem 279(53):56053–56060. https://doi.org/10.1074/jbc.M409267200

Yu Y, Hamza A, Zhang T et al (2010) Withaferin A targets heat shock protein 90 in pancreatic cancer cells. Biochem Pharmacol 79(4):542–551. https://doi.org/10.1016/j.bcp.2009.09.017

Zhang Y, Kensler TW, Cho CG et al (1994) Anticarcinogenic activities of sulforaphane and structurally related synthetic norbornyl isothiocyanates. Proc Natl Acad Sci U S A 91 (8):3147–3150. https://doi.org/10.1073/pnas.91.8.3147

Zhang T, Hamza A, Cao X et al (2008) A novel Hsp90 inhibitor to disrupt Hsp90/Cdc37 complex against pancreatic cancer cells. Mol Cancer Ther 7(1):162–170. https://doi.org/10.1158/1535-7163.Mct-07-0484

Zhang X, Timmermann B, Samadi AK et al (2012) Withaferin a induces proteasome-dependent degradation of breast cancer susceptibility gene 1 and heat shock factor 1 proteins in breast cancer cells. ISRN Biochem 2012:707586. https://doi.org/10.5402/2012/707586

Zhang J, Zhang R, Zhan Z et al (2017) Beneficial effects of sulforaphane treatment in Alzheimer's disease may be mediated through reduced HDAC1/3 and increased P75NTR expression. Front Aging Neurosci 9:121–121. https://doi.org/10.3389/fnagi.2017.00121

Zoghbi HY, Orr HT (2000) Glutamine repeats and neurodegeneration. Annu Rev Neurosci 23:217–247. https://doi.org/10.1146/annurev.neuro.23.1.217

Role of Phytomolecules on the Basic Biology of Aging

6

Swapnil Pandey and Puneet Singh Chauhan

Abstract

Aging is an inevitable process influenced by genetic, lifestyle, and environmental components which increases the primary risk of multiple age-related chronic diseases. The efficacy of any herbal medication depends on delivering adequate levels of plant extract from the therapeutically active phytomolecules. Phytomolecules such as flavonoids, phenolics, and hydrophilic molecules are predominant in vegetables, nuts, teas, wines, fruits, grains, olive oil, and chocolate. Polyphenols are secondary plant metabolites and are packed with potential health benefits of plant-rich dietary polyphenols as plant-rich antioxidant diets. In this chapter, we have focused on several recently identified phytochemicals having potent antiaging properties, i.e., silymarin, 18α-glycyrrhetinic acid, piceatannol, withanolide, and other polyphenols, on the lifespan of model organisms and summarize the current understanding of phytomolecules interaction with various signaling mechanism pathways of aging context relevant to human wellness. Natural phytochemicals are widely approved for antiaging properties that have less side effects compared to synthetic and that are easier to manage for human beings.

Keywords

Phytochemicals · Antiaging · Antioxidant

S. Pandey · P. S. Chauhan (✉)
Microbial Technology Division, CSIR-National Botanical Research Institute, Lucknow, India

Academy of Scientific and Innovative Research (AcSIR), Ghaziabad, India
e-mail: puneet@nbri.res.in

S. C. Mandal et al. (eds.), *Evidence Based Validation of Traditional Medicines*,
https://doi.org/10.1007/978-981-15-8127-4_6

Abbreviations

AD	Alzheimer's diseases
C. elegans	*Caenorhabditis elegans*
DR	Dietary restriction
PD	Parkinson's diseases
ROS	Reactive oxygen species

6.1 Introduction

Caenorhabditis elegans (*C. elegans*) is a free living model organism, with well-characterized genome, short life cycle, and small body size, and conserves 65% of the genes linked with human disease (Walker et al. 2000; Hill et al. 2000). *C. elegans* is a well-established model organism for the work on aging emerged in several fields, including genetics, neurobiology, and developmental biology (Luyten et al. 2016). Aging is a natural phenomenon that deteriorates the total physiology of an organism due to widespread systematic dysfunction of almost all organs, leading to enhanced exposure to environmental problems and increased risk of destruction and disease (Fulop et al. 2010). Free radicals are an atom, molecule, or compound generated by metabolic reaction in the physical structure. Reactive oxygen species (ROS) generated by normal metabolism directly regulate the aging process. Mitochondrial dysfunction is a hallmark of aging diseases and modification in mitochondrial dynamics related to a loss of mitochondrial fitness with unhealthy aging (Sebastián et al. 2017). Phytomolecules are secondary metabolites which have been used for decades in traditional system of medicine in the form of antioxidants, drugs, flavors, dyes, scents, insecticides, pheromones, and so forth by more than 80% of the world population (Ekor 2014). Polyphenols and other food phenolics are the subject of increasing scientific interest because of their potential beneficial effects on human wellness. Phytochemicals are present in various edibles, including fruit, veggies, grains, and nuts as well as in beverages including juice, tea, chocolate, and wine (Prakash et al. 2012). Various plant-based molecules have been reported for their properties to prevent chronic diseases like prevention of malignant neoplastic disease, diabetes, obesity, cardiovascular disease, as well as neurological dysfunctions (Zhang et al. 2015; Si and Liu 2014). Phytomolecules act as a significant part in the maintenance of age-associated pathologies via a positive feedback mechanism. This chapter discusses the applications of several new phytomolecules with different pharmacological properties (Table 6.1).

6.2 Antiaging Phytochemicals

Phytomolecules are plant-based secondary metabolites which have antioxidative and anti-inflammatory characteristics that promote better wellness and prevent diseases, such as malignant neoplastic disease, diabetes, cardiovascular disease, and

Table 6.1 List of phytomolecules with different pharmacological properties

Phytomolecules/ chemical structure	Source	Different properties	References
Sesamin 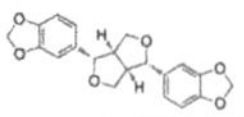	Sesame oil	Antioxidant, anticancer, anti-inflammation, antiaging	Liu et al. (2014); Nakatani et al. (2018)
Silymarin 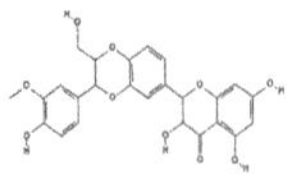	*Silybum marianum*	Antioxidant, anti-inflammatory, diabetes, sepsis, osteoporosis, arthritis, hypercholesterolemia, cancer, viral infections, Alzheimer's, Parkinson's antiaging	Kumar et al. (2015); Liu (2016); Srivastava et al. (2017); Stolf et al. (2017)
Piceatannol 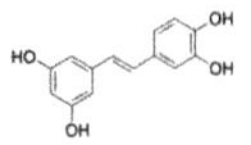	Red wine, grapes, fruits	Antioxidant, anti-inflammatory, antiatherogenic, antiaging	Wang et al. (2002); Lyons et al. (2003); Ovesná et al. (2006); Shen et al. (2017)
Ursolic acid 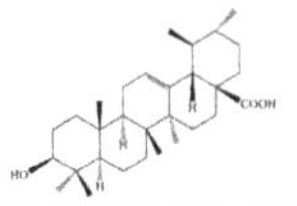	Basil, rosemary, etc.	Antioxidant, anti-inflammatory, antiaging activity	Liobikas et al. (2011); Kashyap et al. (2016); Bahrami and Bakhtiari (2016); Negi et al. (2016)
Acacetin 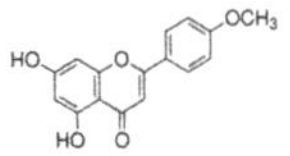	*Premna integrifolia*	Antioxidant, anti-inflammatory, antidiabetic, antibacterial, antihyperglycemic, anti-obesity	Mali (2014); Asthana et al. (2016)
Chlorogenic acid 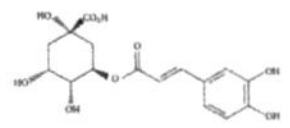	Apple peels, green coffee beans	Antibacterial, antioxidant, anticarcinogenic, anti-obesity, antidiabetic, antiaging	Sato et al. (2011); Meng et al. (2013); Jeszka-Skowron et al. (2016); Zheng et al. (2017)
Specioside 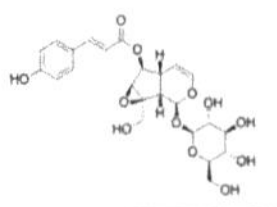	*Stereospermum suaveolens* "patala"	Analgesic, antidyspeptic, astringent, antiaging, liver stimulating, wound healing	Garg et al. (1994); Asthana et al. (2015)
Withanolide A 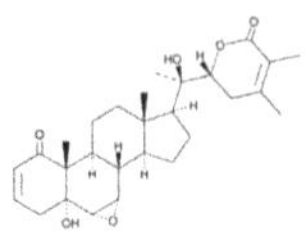	*Withania somnifera*	Antioxidant, antistress, anti-inflammatory, antiaging	Kurapati et al. (2013); Baitharu et al. (2014); Akhoon et al. (2016)

neurodegenerative disorders, in animals (Zhang et al. 2015). In summation to the health-improving and disease-preventing effects, phytochemicals also have been demonstrated to delay the aging process and prolong life in several experimental animal studies using yeast, insects, flies, fishes, and rodents (Sadowska-Bartosz and

Bartosz 2014). The presence of phytochemicals is also capable of cutting down the side effects and enhances promoting lifespan. In this chapter, the longevity benefits and putative underlying mechanisms of phytochemicals help them to survive under environmental stresses and to protect them from microbial infections and environmental pollutants as are described briefly.

6.2.1 Sesamin

A polyphenolic compound sesamin is found in sesame seeds and extracted from sesame oil which contains significant amounts of lignins. It has antioxidant, anticancer, anti-inflammatory, and diverse pharmacological functions which are reported to exert a variety of health benefits such as modulation of lipid biosynthesis and alleviation of oxidative stress (Liu et al. 2014). Sesamin is also reported as an antiaging activity to promote lifespan in *C. elegans*. It has been reported that sesamin is more resistant to oxidative stress and extended the life of a *mev-1* mutant that produces abundant superoxide anions. Interestingly, sesamin can also extend the lifespan by overexpression of dietary restriction-related signaling pathways, in RNAi-treated worms which is a key regulator of autophagy (Nakatani et al. 2018).

6.2.2 Silymarin

Silymarin is a flavanone derivative extracted from the sources of the milk thistle *Silybum marianum*. Silymarin has been demonstrated to exert antioxidant and anti-inflammatory properties. It has also been reported for the treatment of various chronic disorders, such as hepatitis, gallstones, and cirrhosis. Apart from these properties, silymarin possess diabetes, sepsis, osteoporosis, arthritis, cancer, viral infections, and Alzheimer's diseases (AD) and Parkinson's diseases (PD) activity (Srivastava et al. 2017; Stolf et al. 2017). Silymarin was reported as an antiaging agent in the *C. elegans* model system. In addition to increased lifespan, silymarin also allows increased locomotion rate, increased stimulus response, and improved stress tolerance (Kumar et al. 2015; Liu 2016).

6.2.3 Piceatannol

Piceatannol, a resveratrol metabolite, is also present in many fruits and vegetables, as a natural stilbene. This is found even in some beverages like red wine (Lyons et al. 2003; Wang et al. 2002). In addition, piceatannol is more active in antioxidant, anti-inflammatory, and antiatherogenic activities than resveratrol (Ovesná et al. 2006). Piceatannol is also reported as an antiaging potential activity in *C. elegans* in terms of reducing locomotive activity, improving pharyngeal pumping rate, and protecting the worms from oxidative and heat stress through insulin/IGF-1 signaling and sir-2.1-dependent pathways (Shen et al. 2017).

6.2.4 Ursolic Acid

Ursolic acid (3β-hydroxyurs-12-en-28-oic acid, UA) is a lipophilic pentacyclic triterpenoid, the main component of herbal plants, found in a wide variety of vegetarian and medicinal foods. It has been reported as an antioxidant, anti-inflammatory, and antiaging activity and possesses many beneficial effects, notably hepatoprotection, antitumoral, antihyperlipidemic, and anti-inflammation (Liobikas et al. 2011). UA is one of these plant-based therapeutic metabolite groups that have separate extracellular and intracellular marks that play a role in metastasis, apoptosis, and inflammatory processes. In addition, UA's synthetic derivatives were also found to be active in a series of pharmacological applications related to age-related disease prevention (Kashyap et al. 2016). Ursolic acid regulates the aging process for better health span and lifespan in *Caenorhabditis elegans* via modulating JNK pathway (Bahrami and Bakhtiari 2016; Negi et al. 2016).

6.2.5 Acacetin

Acacetin is extracted in nature from *Premna integrifolia* (*P. integrifolia*) and is an essential flavanoid. It has been known for its potential actions such as antioxidant, immunomodulatory, anti-inflammatory, antidiabetic, antibacterial, hepatoprotective, antihyperglycemic, and anti-obesity (Mali 2014). Acacetin has been reported as a stress buster, antiaging, and ROS scavenger and also able to reduced aging marker lipofuscin level in *C. elegans* (Asthana et al. 2016).

6.2.6 Chlorogenic Acid

Chlorogenic acid compounds are the most abundant polyphenols typically found in apple peels and green coffee beans and have many biological properties, including antibacterial, antioxidant, anticarcinogenic, and anti-obesity activity and are capable of lowering blood sugar rates (Sato et al. 2011; Meng et al. 2013; Jeszka-Skowron et al. 2016). Chlorogenic acid has beneficial effects on human health and ameliorates aging diseases, and it also slows age-related decrease in body activity and increases stress tolerance via insulin/IGF-1 signaling pathway in *C. elegans*. Chlorogenic acid is also able to activate the FOXO transcription factors DAF-16, HSF-1, and SKN-1 but not SIR-2.1 (Zheng et al. 2017).

6.2.7 Specioside

Specioside (6-*O*-coumaroylcatalpol) is an iridoid glucoside isolated from the plant *Stereospermum suaveolens* (Bignoniaceae), commonly known as "patala." The plant produces apigenin, lapachol, dianatin-7-glucuroniside, dinatin, β-sitosterol, saponins, palmitic, stearic, and oleic acids that have multifunctional properties, viz., analgesic, antidyspeptic, astringent, hepatic stimulant, antiaging, and wound healing (Asthana et al. 2015). The iridoids, present in a wide range of plants, are groups of natural molecules with a monoterpene cyclic ring. Iridoid glucoside has been reported to have many functions, such as antitumoral, hypotensive, sedative, and hepatoprotective (Garg et al. 1994).

6.2.8 Withanolide A

Withania somnifera, also known as ashwagandha, is an essential therapeutic medicinal plant worldwide. Withanolide A is a steroidal lactone important bioactive compound extracted from *W. somnifera* (root) (Baitharu et al. 2014). Many studies are reported for the human health benefits of *Withania somnifera* because it is one of the most important plants which has been reported that have antioxidant, antistress, anti-inflammatory, immunomodulatory, and rejuvenating properties (Alam et al. 2012). Withanolide A also has many therapeutic properties, such as neurological disorders, including Alzheimer's disease-associated amyloid pathology, neuritis regeneration, synapsis recovery, axonal outgrowth, etc. (Baitharu et al. 2014; Kurapati et al. 2013). Withanolide A is also reported as antiaging agent and promotes stress resistance via IIS pathway (Akhoon et al. 2016).

6.3 Other Phytochemicals

Many of the phytochemicals are secondary plant metabolites that are found in a wide range of fruits, vegetables, grains, nuts, tea, and wine as well as in beverages including juice and coffee. More than 1 g of phytochemicals per day is commonly ingested with diet (Septembre-Malaterre et al. 2018; Leitzmann 2016). Many natural antioxidants, nutraceutical, and functional foods have been identified as radical or reactive oxygen-free scavengers. Functional and nutraceutical foods with antioxidant activity may play a significant role in slowing aging (Liu et al. 2018; Pant and Pandey 2015). Various pathways interact with each other and modulate lifespan by controlling the cellular stress response (Pan and Finkel 2017). These nutrient-sensing pathways include IGF, TOR, AMPK, and SIRTs (Aiello et al. 2017). Numerous phytomolecules have been reported for their therapeutic effects without compromising the quality of life (Pant and Pandey 2015). The natural molecules extend lifespan and reduce age-related pathologies. Polyphenols, resveratrol, found, for example, in grape berry skins, increased lifespan in *Caenorhabditis elegans* and prolonged lifespan and reduced oxidative stress and appeared to occur via a dietary restriction (DR) mechanism and performed to validate SIR-2-dependent and prevent age-related disease (Bhullar and Hubbard 2015; Conti et al. 2016). Quercetin is the primary representative of flavonoids found in elevated levels in herbal edibles such as onions, apples, and broccoli, as well as in red wine, tea, and *Ginkgo biloba* extracts (Pant and Pandey 2015; Conti et al. 2016; Koch et al. 2014). These flavonoids extend mean lifespan by improving antioxidant potential, reducing

ROS level, and increasing *mev-1* mean lifespan and daf-16 translocation (Koch et al. 2014; Büchter et al. 2015; Kampkötter et al. 2008). Curcumin is also one of the active ingredients that has been widely used as a spice and herbal medicine found in turmeric (*Curcuma longa*). It has biological activities, such as anti-inflammatory, antioxidative, anticancer, and anti-neurodegenerative and chemopreventive properties. Curcumin 20 μM dose extends the lifespan by 39.28% as well as improves health span and enhances resistance to oxidative and thermal stresses and prevents the age-related disease in *C. elegans* (Liao et al. 2011). 4-Hydroxy-*E*-globularinin (4-HEG) is major component of *Premna integrifolia* and enhanced the mean lifespan of worms by over 18.8% at 20 μM dose and also reduced reactive oxygen species (ROS) levels and fat accumulation in the worms (Shukla et al. 2012a, b). Some other phytomolecules reported as antiaging agent are l-theanine, 10-*O*-trans-*p*-coumaroylcatalpol, ferulsinaic acid, and baicalin (Zarse et al. 2012; Shukla et al. 2012a, b; Sayed 2011).

6.4 The Mechanisms of Age-Promoting Factor of Phytomolecules

Over a few decades, aging research has seen a breakthrough, particularly in the fact that the rate of aging is regulated, at least partly, by genetic mechanisms and biochemical procedures preserved in evolution. On the other side, it is further distinguished that age-related developments developed by genetics can be strongly affected by environmental variables (Stephan et al. 2013; López-Otín et al. 2013). Aging is induced by mitochondrial dysfunction, telomeric attrition, genomic instability, epigenetic modifications, deregulated nutrient sensing, cellular senescence, and stem cell fatigue. Aging would describe procedures that offer promising outcomes in animal models or even in human studies, leading in lifespan elongation or health improvement with the positive impacts of plant-derived extracts on aging and age-related diseases. Many phytochemicals can encourage longevity and enhance health span by synthesizing to boost plant fitness by enabling it to communicate with its surroundings, including herbivorous pathogens and insects. Phytochemicals can stimulate a cellular stress response in animals that is useful for life and health through various stress response mechanisms and low-dose exposures and can then stimulate adaptive stress resistance, also called hormesis (Mattson 2008). Functional foods and nutraceutical/dietary ingredients are an enormous promise to promote health and longevity and to prevent age-related chronic diseases through bioenergetic pathways that encourage aging via well-characterized antiaging pathways (López-Lluch and Navas 2016; Aiello et al. 2017). The increase in DNA damage with the consequence of impaired DNA repair system efficiency is the primary cause of cellular senescence. However, age-related declines in AMPK (AMP-activated protein kinase) pathways have improved the lifespan of the model organism, and AMPK kinase can also control several signaling pathways engaged in senescence and aging, such as those involving p53, mTOR, and NFB (Salminen and Kaarniranta 2012). The principal applicants for targeted measures are the mTOR signaling inhibitors. This signaling pathway has been related to lifespan and health span in model organisms because both of these events have benefited from decreased mTOR signaling (Johnson et al. 2013).

6.5 Conclusion

Many Ayurvedic nutraceutical products are available in markets which provide health benefits made by phytodrug for the management of aging. Different natural compounds are an attractive research area because of their beneficial effects on human health and preventing degenerative disorders of aging for their antioxidant as well as antiaging activity. Different natural compounds are an attractive research area due to their beneficial effects on human health and the prevention for their antioxidant and antiaging activity of degenerative aging disorders. It will be interesting to see if such long-term exposure of different model organisms to natural plant molecules can guide to a variety of species that live longer than their predecessors.

Acknowledgments The Authors are grateful to the director of CSIR–NBRI, Lucknow, India, for his kind support. SP was financially supported by ICMR, India (45/11/2018-PHA/BMS/OL) through, Senior Research Fellowship grant.

References

Aiello A, Accardi G, Candore G, Gambino CM, Mirisola M, Taormina G, Caruso C (2017) Nutrient sensing pathways as therapeutic targets for healthy ageing. Expert Opin Ther Targets 21 (4):371–380

Akhoon BA, Pandey S, Tiwari S, Pandey R (2016) Withanolide A offers neuroprotection, ameliorates stress resistance and prolongs the life expectancy of *Caenorhabditis elegans*. Exp Gerontol 78:47–56

Alam N, Hossain M, Khalil MI, Moniruzzaman M, Sulaiman SA, Gan SH (2012) Recent advances in elucidating the biological properties of Withania somnifera and its potential role in health benefits. Phytochem Rev 11(1):97–112

Asthana J, Yadav AK, Pant A, Pandey S, Gupta MM, Pandey R (2015) Specioside ameliorates oxidative stress and promotes longevity in *Caenorhabditis elegans*. Comp Biochem Physiol C Toxicol Pharmacol 169:25–34

Asthana J, Yadav D, Pant A, Yadav AK, Gupta MM, Pandey R (2016) Acacetin 7-O-α-l-rhamnopyranosyl (1–2) β-D-xylopyranoside elicits life-span extension and stress resistance in *Caenorhabditis elegans*. J Gerontol Ser A Biol Med Sci 71(9):1160–1168

Bahrami SA, Bakhtiari N (2016) Ursolic acid regulates aging process through enhancing of metabolic sensor proteins level. Biomed Pharmacother 82:8–14

Baitharu I, Jain V, Deep SN, Shroff S, Sahu JK, Naik PK, Ilavazhagan G (2014) Withanolide A prevents neurodegeneration by modulating hippocampal glutathione biosynthesis during hypoxia. PLoS One 9(10):e105311

Bhullar KS, Hubbard BP (2015) Lifespan and healthspan extension by resveratrol. Biochim Biophys Acta 1852(6):1209–1218

Büchter C, Ackermann D, Honnen S, Arnold N, Havermann S, Koch K, Wätjen W (2015) Methylated derivatives of myricetin enhance life span in *Caenorhabditis elegans* dependent on the transcription factor DAF-16. Food Funct 6(10):3383–3392

Conti V, Izzo V, Corbi G, Russomanno G, Manzo V, Lise D, Filippelli A (2016) Antioxidant supplementation in the treatment of aging-associated diseases. Front Pharmacol 7:24

Ekor M (2014) The growing use of herbal medicines: issues relating to adverse reactions and challenges in monitoring safety. Front Pharmacol 4:177

Fulop T, Larbi A, Witkowski JM, McElhaney J, Loeb M, Mitnitski A, Pawelec G (2010) Aging, frailty and age-related diseases. Biogerontology 11(5):547–563

Garg HS, Bhandari SPS, Tripathi SC, Patnaik GK, Puri A, Saxena R, Saxena RP (1994) Antihepatotoxic and immunostimulant properties of iridoid glycosides of Scrophulariakoelzii. Phytother Res 8(4):224–228

Hill AA, Hunter CP, Tsung BT, Tucker-Kellogg G, Brown EL (2000) Genomic analysis of gene expression in *C. elegans*. Science 290(5492):809–812

Jeszka-Skowron M, Sentkowska A, Pyrzyńska K, De Peña MP (2016) Chlorogenic acids, caffeine content and antioxidant properties of green coffee extracts: influence of green coffee bean preparation. Eur Food Res Technol 242(8):1403–1409

Johnson SC, Rabinovitch PS, Kaeberlein M (2013) mTOR is a key modulator of ageing and age-related disease. Nature 493(7432):338–345

Kampkötter A, Timpel C, Zurawski RF, Ruhl S, Chovolou Y, Proksch P, Wätjen W (2008) Increase of stress resistance and lifespan of *Caenorhabditis elegans* by quercetin. Comp Biochem Physiol B: Biochem Mol Biol 149(2):314–323

Kashyap D, Tuli HS, Sharma AK (2016) Ursolic acid (UA): a metabolite with promising therapeutic potential. Life Sci 146:201–213

Koch K, Havermann S, Büchter C, Wätjen W (2014) *Caenorhabditis elegans* as model system in pharmacology and toxicology: effects of flavonoids on redox-sensitive signalling pathways and ageing. ScientificWorldJournal 2014:920398

Kumar J, Park KC, Awasthi A, Prasad B (2015) Silymarin extends lifespan and reduces proteotoxicity in *C. elegans* Alzheimer's model. CNS Neurol Disord Drug Targets 14 (2):295–302

Kurapati KRV, Atluri VSR, Samikkannu T, Nair MP (2013) Ashwagandha (Withania somnifera) reverses β-amyloid 1-42 induced toxicity in human neuronal cells: implications in HIV-associated neurocognitive disorders (HAND). PLoS One 8(10):e77624

Leitzmann C (2016) Characteristics and health benefits of phytochemicals. Complement Med Res 23(2):69–74

Liao VHC, Yu CW, Chu YJ, Li WH, Hsieh YC, Wang TT (2011) Curcumin-mediated lifespan extension in Caenorhabditis elegans. Mech Ageing Dev 132(10):480–487

Liobikas J, Majiene D, Trumbeckaite S, Kursvietiene L, Masteikova R, Kopustinskiene DM, Bernatoniene J (2011) Uncoupling and antioxidant effects of ursolic acid in isolated rat heart mitochondria. J Nat Prod 74(7):1640–1644

Liu L (2016) Silymarin as a promising natural agent for Alzheimer's disease. Int J Sci Res 5 (4):482–483

Liu N, Wu C, Sun L, Zheng J, Guo P (2014) Sesamin enhances cholesterol efflux in RAW264. 7 macrophages. Molecules 19(6):7516–7527

Liu Z, Ren Z, Zhang J, Chuang CC, Kandaswamy E, Zhou T, Zuo L (2018) Role of ROS and nutritional antioxidants in human diseases. Front Physiol 9:477

López-Lluch G, Navas P (2016) Calorie restriction as an intervention in ageing. J Physiol 594 (8):2043–2060

López-Otín C, Blasco MA, Partridge L, Serrano M, Kroemer G (2013) The hallmarks of aging. Cell 153(6):1194–1217

Luyten W, Antal P, Braeckman BP, Bundy J, Cirulli F, Fang-Yen C, Metspalu A (2016) Ageing with elegans: a research proposal to map healthspan pathways. Biogerontology 17(4):771–782

Lyons MM, Yu C, Toma RB, Cho SY, Reiboldt W, Lee J, Van Breemen RB (2003) Resveratrol in raw and baked blueberries and bilberries. J Agric Food Chem 51(20):5867–5870

Mali PY (2014) Beneficial effect of extracts of Premna integrifolia root on human leucocytes and erythrocytes against hydrogen peroxide induced oxidative damage. Chron Young Sci 5(1):53

Mattson MP (2008) Dietary factors, hormesis and health. Ageing Res Rev 7(1):43–48
Meng S, Cao J, Feng Q, Peng J, Hu Y (2013) Roles of chlorogenic acid on regulating glucose and lipids metabolism: a review. Evid Based Complement Alternat Med 2013:801457
Nakatani Y, Yaguchi Y, Komura T, Nakadai M, Terao K, Kage-Nakadai E, Nishikawa Y (2018) Sesamin extends lifespan through pathways related to dietary restriction in *Caenorhabditis elegans*. Eur J Nutr 57(3):1137–1146
Negi H, Shukla A, Khan F, Pandey R (2016) 3β-Hydroxy-urs-12-en-28-oic acid prolongs lifespan in *C. elegans* by modulating JNK-1. Biochem Biophys Res Commun 480(4):539–543
Ovesná Z, Kozics K, Bader Y, Saiko P, Handler N, Erker T, Szekeres T (2006) Antioxidant activity of resveratrol, piceatannol and 3, 3′, 4, 4′, 5, 5′-hexahydroxy-trans-stilbene in three leukemia cell lines. Oncol Rep 16(3):617–624
Pan H, Finkel T (2017) Key proteins and pathways that regulate lifespan. J Biol Chem 292 (16):6452–6460
Pant A, Pandey R (2015) Bioactive phytomolecules and aging in *Caenorhabditis elegans*. Healthy Aging Res 4:1–15
Prakash D, Gupta C, Sharma G (2012) Importance of phytochemicals in nutraceuticals. J Chin Med Res Dev 1:70–78
Sadowska-Bartosz I, Bartosz G (2014) Effect of antioxidants supplementation on aging and longevity. BioMed Res Int 2014:404680
Salminen A, Kaarniranta K (2012) AMP-activated protein kinase (AMPK) controls the aging process via an integrated signaling network. Ageing Res Rev 11(2):230–241
Sato Y, Itagaki S, Kurokawa T, Ogura J, Kobayashi M, Hirano T, Sugawara M, Iseki K (2011) In vitro and in vivo antioxidant properties of chlorogenic acid and caffeic acid. Int J Pharm 403 (1):136–138
Sayed AA (2011) Ferulsinaic acid attenuation of advanced glycation end products extends the lifespan of *Caenorhabditis elegans*. J Pharm Pharmacol 63(3):423–428
Sebastián D, Palacín M, Zorzano A (2017) Mitochondrial dynamics: coupling mitochondrial fitness with healthy aging. Trends Mol Med 23(3):201–215
Septembre-Malaterre A, Remize F, Poucheret P (2018) Fruits and vegetables, as a source of nutritional compounds and phytochemicals: changes in bioactive compounds during lactic fermentation. Food Res Int 104:86–99
Shen P, Yue Y, Sun Q, Kasireddy N, Kim KH, Park Y (2017) Piceatannol extends the lifespan of *Caenorhabditis elegans* via DAF-16. Biofactors 43(3):379–387
Shukla V, Phulara SC, Yadav D, Tiwari S, Kaur S, Gupta MM, Nazir A, Pandey R (2012a) Iridoid compound 10-O-trans-p-coumaroylcatalpol extends longevity and reduces alpha synuclein aggregation in *Caenorhabditis elegans*. CNS Neurol Disord Drug Targets 11(8):984–992
Shukla V, Yadav D, Phulara SC, Gupta MM, Saikia SK, Pandey R (2012b) Longevity-promoting effects of 4-hydroxy-E-globularinin in *Caenorhabditis elegans*. Free Radic Biol Med 53 (10):1848–1856
Si H, Liu D (2014) Dietary antiaging phytochemicals and mechanisms associated with prolonged survival. J Nutr Biochem 25(6):581–591
Srivastava S, Sammi SR, Laxman TS, Pant A, Nagar A, Trivedi S, Bhatta RS, Tandon S, Pandey R (2017) Silymarin promotes longevity and alleviates Parkinson's associated pathologies in *Caenorhabditis elegans*. J Funct Foods 31:32–43
Stephan J, Franke J, Ehrenhofer-Murray AE (2013) Chemical genetic screen in fission yeast reveals roles for vacuolar acidification, mitochondrial fission, and cellular GMP levels in lifespan extension. Aging Cell 12(4):574–583
Stolf AM, Cardoso CC, Acco A (2017) Effects of silymarin on diabetes mellitus complications: a review. Phytother Res 31(3):366–374
Walker DW, McColl G, Jenkins NL, Harris J, Lithgow GJ (2000) Natural selection: evolution of lifespan in *C. elegans*. Nature 405(6784):296–297

Wang Y, Catana F, Yang Y, Roderick R, Van Breemen RB (2002) An LC-MS method for analyzing total resveratrol in grape juice, cranberry juice, and in wine. J Agric Food Chem 50 (3):431–435

Zarse K, Jabin S, Ristow M (2012) L-Theanine extends lifespan of adult *Caenorhabditis elegans*. Eur J Nutr 51(6):765–768

Zhang YJ, Gan RY, Li S, Zhou Y, Li AN, Xu DP, Li HB (2015) Antioxidant phytochemicals for the prevention and treatment of chronic diseases. Molecules 20(12):21138–21156

Zheng SQ, Huang XB, Xing TK, Ding AJ, Wu GS, Luo HR (2017) Chlorogenic acid extends the lifespan of *Caenorhabditis elegans* via insulin/IGF-1 signaling pathway. J Gerontol A 72 (4):464–472

Role of Phytomedicine in Alleviating Oxidative Stress-Mediated Vascular Complications in Diabetes

7

Rakhi Chakraborty and Vivekananda Mandal

Abstract

Oxidative stress development is inevitably associated with diabetes mellitus. Hyperglycemia triggers excessive free radical generation, frequent oxidation of stable macromolecules, destabilizing conformation of antioxidative enzymes and transcription factors, modulating metabolic pathways, and promoting inflammatory reactions, and ultimately leads to endothelial dysfunctions and vasculopathy. Conventional diabetes management strategies are found inadequate to counter all these associated health issues and often exert many adverse side effects. Phytotherapy in the form of complementary and alternative medicine has raised attention in the treatment of diabetes over the years. Ethnomedicinal knowledge of several plants from traditional literatures have enlighten the path for natural regulation of ailments with safe, economic, and prolonged cure. Extensive pharmacological experiments has also been conducted to affirm the potency of plant-derived active constituents in this regard, and the scientific reports deciphering the underlying mechanism of these molecules in bringing metabolic homeostasis in diabetes is gradually increasing. However, the success rate of developing and marketing of herbal drugs is not satisfactory till date. This book chapter attempts to focus on the detailed pathophysiology of oxidative stress development in diabetes and the important contributions of phytomedicine along with pros and cons in herbal drug development in this context.

R. Chakraborty
Department of Botany, A.P.C. Roy Govt. College, Siliguri, West Bengal, India

Plant Physiology and Biochemistry Laboratory, Department of Botany, University of Gour Banga, Malda, West Bengal, India

V. Mandal (✉)
Plant Physiology and Biochemistry Laboratory, Department of Botany, University of Gour Banga, Malda, West Bengal, India
e-mail: vivekugb@gmail.com

S. C. Mandal et al. (eds.), *Evidence Based Validation of Traditional Medicines*,
https://doi.org/10.1007/978-981-15-8127-4_7

Keywords

Diabetes · Oxidative stress · Endothelial dysfunction · Phytotherapy · Herbal drug

Abbreviations

AGEs	Advanced glycation end products
AHP	American Herbal Pharmacopoeia
AYUSH	Department of Ayurveda, Yoga & Naturopathy, Unani, Siddha and Homoeopathy
BFAD	The Bureau of Food and Drugs in the Philippines
BHP	British Herbal Pharmacopoeia
CAM	Complementary and Alternative Medicines
CAT	Catalase
eNOS	Endothelial nitric oxide synthase
GPx	Glutathione peroxidase
GR	Glutathione reductase
GSH	Glutathione
HbA1c	Glycated haemoglobin
HNE	4-hydroxynonenal
HNF1A	Hepatocyte nuclear factor α
IA-2	Autoantibodies to glutamic acid decarboxylase
IAA	Insulin autoantibodies
ICA	Islet cell autoantibodies
IHP	Indian Herbal Pharmacopoeia
INGAP	Islet Neogenesis Associated Protein
IsoLGs	Isolevuglandins
JNK	c-Jun *N*-terminal kinase pathways
KHP	Korean Herbal Pharmacopoeia
LADA	Latent autoimmune diabetes in adults
LHPs	Lipid hydroperoxides
MDA	Malondialdehyde
MODY	Mmaturity-onset diabetes of the young
NAFDAC	National Agency for Food and Drug Administration and Control in Nigeria
NHDs	Nanotized herbal drugs
ONE	4-oxononenal
PDX-1	Pancreatic and duodenal homeobox factor-1
PKC	Protein kinase C
PPARGC1A	Peroxisome proliferator-activated receptor–γ coactivator-1α
RCS	Reactive chlorine species
RNS	Reactive nitrogen species
ROS	Reactive oxygen species
SOD	Superoxide dismutase

7.1 Introduction

According to the reports of the World Health Organization (2018), current global scenario of diabetes mellitus is really scary, as the increasing prevalence of diabetic patients are very rapid (approximately 9% of the global population). It was also estimated that about four million deaths in recent times were caused due to direct or indirect effects of hyperglycaemia and diabetes. Diabetes is thus considered as a chronic endocrine disorder characterized by elevated blood sugar levels and insufficiency in insulin secretion or action inside the body (Ullah et al. 2016). Typically the disease is categorized into three major types, type 1, type 2, and gestational diabetes, with some subgroups viz. LADA (latent autoimmune diabetes in adults), MODY (maturity-onset diabetes of the young), lipoatrophic diabetes (severe lipodystrophy along with other diabetic symptoms), type A insulin resistance, Leprechaunism and the Rabson-Mendenhall syndrome (occur due to genetic anomalies in insulin receptor function), etc. (American Diabetes Association 2011). Autoimmune destruction of pancreatic beta islet cells results into inadequate insulin secretion in type-1 diabetes, also known as insulin-dependent diabetes. Environmental and genetic factors and autoimmunity are mainly responsible for the development of type-1 diabetes. Type-1 diabetes often results in unpredictable or irregular blood sugar level along with ketosis and other endocrinopathies (Siddiqui et al. 2013). The situation becomes more complicated in type-2 diabetes, where body remains unresponsive to insulin in spite of the adequate secretion of the hormone, a phenomenon known as "insulin resistance". The exact reason of developing such resistance is still unknown, though functional defects of insulin receptors may be responsible. Besides this, unhealthy lifestyle, poor diet, frequent consumption of tobacco, and alcohol can increase the risk of developing type-2 diabetes (Olokoba et al. 2012). About 2–10% of expecting mothers have to face the high blood sugar levels during their pregnancy, known as gestational diabetes. Severe neonatal complications, viz. high birth weight, congenital coronary and cerebral anomalies, and skeletal muscle malformations, may occur if not properly treated at time. LADA is characterized by certain features in common with both type-1 and type-2 diabetes, viz. lower BMI; autoimmunity to cellular antibodies, islet cell autoantibodies (ICA), autoantibodies to glutamic acid decarboxylase (GAD), tyrosine phosphatase-related islet antigen 2 (IA-2), and insulin autoantibodies (IAA); intermediate level of beta cell destruction; and insulin resistance as well as faster depletion of C-peptide levels. It can occur anytime between the age of 30 and 70 years and is often misdiagnosed due to such complex signs and symptoms (O'Neal et al. 2016). MODY is considered as one of the subgroup of monogenic diabetes developed due to single gene mutations in glycolytic enzymes or transcription factors involved in carbohydrate metabolism. Two very abundant forms of MODY results from mutation in the genes encoding glucokinase (also known as MODY 2) and hepatocyte nuclear factor α or HNF1A (also known as MODY3). Post-prandial hyperglycaemia, fasting glucose tolerance, and increased ketoacidosis with other type-2 symptoms are the main diagnostic features of the patients, who are often under the age of 25 (Steenkamp et al. 2014). In all the above-mentioned cases, it was observed that prolonged hyperglycaemic condition and impaired glucose tolerance in turn results into lipotoxicity and inflammatory reactions followed by varying degree of endothelial cell degeneration, injuries, inhibition of mobilization of progenitor cells, and diffuse vascular disorders. It is also responsible for developing severe oxidative stress, atherosclerosis, renal and hepatic dysfunction, visual loss, and neurodegenerative diseases (Maiese 2015).

Fasting blood glucose level $\geq$126 mg/dL and 2 h post-prandial glucose level $\geq$200 mg/dL are considered as important markers for diabetic measurement. Presence of histocompatibility antigens (HLA-DR3 or HLA-DR4) and circulating antibodies (insulin antibodies, glutamic acid decarboxylase antibodies, protein tyrosine phosphatase antibodies) in the bloodstream, increased production of hepatic glucose and glycated haemoglobin (HbA1c; above 6.5%), decreased endogenous insulin and C-peptide levels are also diagnosed for convenience of the diabetes treatment (Woldu et al. 2014). The criteria for selection of appropriate therapeutics in diabetes management practice depend on several factors, condition of body's own immune system, severity of hyperglycaemia, liver and kidney functions, hyperlipidemia and obesity, risks of hypoglycaemia, etc. Stimulation of insulin secretion by GLP analogues (e.g. dulaglutide), exogenous insulin injection, improving incretin secretion by DPP-4 inhibitors (e.g. sitagliptin), pancreatic beta cell regeneration through islet neogenesis-associated protein (INGAP) peptide therapy, etc. are being applied for type-1 diabetes. For type-2 diabetes, the conventional medications include biguanides and thiazolidinediones for increasing insulin sensitization in liver tissues, sulphonylureas and meglitinides for promoting insulin secretion, and α-glucosidase inhibitors for delayed carbohydrate digestion, respectively (Tiwari 2015). Sometimes these medications are used in combinations according to the requirement of particular situation. Unfortunately, almost all the synthetic antidiabetic medicines exert harmful side effects, viz. hypoglycaemia, weight gain, nausea, diarrhoea, lactic acidosis, etc., thereby posing limitations in their usage.

The disease with its multifaceted health issues put forth a huge burden over the scientific community as complete cure seems to be rather impossible. Moreover, the usual diabetes management is expensive and requires long period of monitoring, thus affecting the socio-economic backbone of both the developed and developing countries (Cho et al. 2018). To manage the worsening situation carefully, scientists are now looking forward for some alternatives that could be used in combination with the synthetic medication or alone to regulate the body system in bringing glucose homeostasis more naturally and economically. Physical activities, dietary modifications, lifestyle management, etc. are prescribed to the patients as the first line of alternative treatment. However, the most stable form of such treatment is offered by complementary and alternative medicines (CAM). Herbal medicines or phytomedicines in the form of plant-based or plant-derived products are highly acclaimed for its proven efficacy towards diabetes management in several clinical and preclinical studies worldwide. Traditional or indigenous knowledge of medicinal plants of several countries have contributed significantly in the field of herbal medicine. Extensive pharmacological experiments are being carried out for isolation, purification, identification, and characterization of the bioactive principles from selected medicinal plants to decipher the molecular mechanism of their action in detail.

7.2 Pathophysiology of Oxidative Stress Development

7.2.1 Production of Free Radicals

Free radicals are short-living, highly reactive chemical species that are naturally produced inside the body and are involved in signal transduction, neurotransmission, gene transcription, and many other cellular activities. They are also significantly responsible for damaging the molecular structure of biological membranes and macromolecules by repetitive oxidation, resulting in several vascular diseases, when produced in excess (Bansal and Bilaspuri 2011). Free radicals can be grouped into three major types: reactive oxygen species (ROS), reactive nitrogen species (RNS), and reactive chlorine species (RCS). All these are produced by several exogenous and endogenous factors, viz. biochemical reactions involving oxygen, neutrophil and macrophage production, radiations, chemicals, industrial waste products, smoking, etc. Level of such reactive free radicals remains under rigid control of the protective effects of cellular antioxidative machinery that includes both enzymatic and non-enzymatic antioxidant molecules. In diseased condition like diabetes, the reaction cascade of oxidative stress is triggered by hyperglycaemia and insulin resistance, leading to the overproduction of several reactive species. Highly oxidative metabolism of glucose and free fatty acids increases the successive production of superoxide and hydroxyl radicals and hydrogen peroxide inside mitochondria (Evans et al. 2002). Other major free radicals and non-radical reactive species responsible for diabetic complications include singlet oxygen, hypochlorous acid, peroxynitrite, ozone, nitric oxide, etc. Under hyperglycaemic condition, increased levels of NADPH and $FADH_2$ cause blocking of electron transport inside Complex III, which in turn causing the electrons to flow back to coenzyme Q_{10} and ultimately to molecular oxygen, resulting in the overproduction of superoxide (Fiorentino et al. 2013). Superoxide alone or in combination with H_2O_2 plays an important role in diabetes as it regulates the spontaneous generation of nitric oxide in model organisms. Hydroxyl radical is the most reactive free radical, frequently produced by reaction of H_2O_2 with glucose (Sheikhpour 2013).

7.2.2 Effects on Antioxidant Defence System (Enzymatic and Non-enzymatic)

During normal physiological condition, production of free radicals is maintained at a steady state by scavenging activities of endogenous and exogenous antioxidants. The endogenous antioxidants include both enzymatic and non-enzymatic molecules, viz. superoxide dismutase (SOD), glutathione peroxidase (GPx), glutathione reductase (GR), catalase (CAT), glutathione (GSH), α-lipoic acid, and vitamins C and

E. Sometimes micronutrients and supplements are applied externally to reduce oxidative stress, known as exogenous antioxidants. These include flavonoids; carotenoids; vitamins C and E, and some trace metals (zinc, manganese, selenium, copper etc.) (Halliwell and Gutteridge 2007). However, in diabetic condition, detoxification capacity of the cellular antioxidative machinery becomes insufficient due to the structural modifications of the active sites of the enzymes and the other non-enzymatic antioxidants, leading to overproduction of RNS and ROS and activation of oxidative reaction cascade and several biochemical damages that occur within the body system.

Superoxide dismutase, considered as the first line of antioxidative defence against free radicals, is either deactivated by loss of essential co-factor Cu^{+2} ion, by glycosylation in red blood cells, or by increased level of malondialdehyde (MDA) in serum of diabetic patients (Taheri et al. 2012). Decreased activity of glutathione peroxidase, catalase, and glutathione was also observed in several diabetic patients. Increased glucose oxidation, protein glycation, activation of protein kinase C, and excessive consumption of NADPH through polyol pathway are thought to be the major sources of free radical-mediated oxidative damages of these antioxidants (Adel et al. 2018).

7.2.3 Effects on Metabolic Pathways and Cellular Macromolecules

7.2.3.1 Proteins

Proteins and amino acids are vulnerable to oxidative stress due to denaturation, fragmentation, and overall inactivation of primary and higher-order structural organization. In vivo studies with diabetic model rats showed significant reduction in total protein concentration. This may be due to decreased uptake of amino acids, reduced concentration and synthesis of essential amino acids, increased conversion of glycogenic amino acids to CO_2 and water, decreased de novo synthesis of mRNA and protein, etc. (Matough et al. 2012). Formation of dityrosine, an unusual amino acid, was widely designated as one of the convenient biomarkers of protein oxidation during hyperglycaemic condition. Oxidative stress-mediated cross-linking of proteins with this modified amino acid serve as a signal for rapid acceleration of in vivo proteolysis by cellular proteases (Berlett and Stadtman 1997). Excessive glycation of proteins also cause structural and functional alteration of enzymatic and non-enzymatic antioxidant molecules that affect their radical scavenging activities. The degree of such protein glycation can be estimated by using two important markers, viz. glycated haemoglobin and fructosamine levels (Maritim et al. 2003).

7.2.3.2 Lipids

It has already been established that carbohydrate and lipid metabolism are closely related with each other and has important clinical consequences. Diabetic patients often showed dyslipidemic condition where the cells become very susceptible to lipid peroxidation. Cell membrane polysaccharides are found to be extremely sensitive to free radical attack due to multiple bonds and readily form highly reactive and toxic lipid hydroperoxides (LHPs) (Matough et al. 2012). Malondialdehyde formation during lipid peroxidation is considered as a critical biomarker for diabetes-induced oxidative stress and widely used for quantitative estimation of lipid peroxidation (Esterbauer et al. 1991). Apart from MDA, other reactive aldehydes formed during peroxidation reactions are acrolein, 4-hydroxynonenal (HNE), 4-oxononenal (ONE), and isolevuglandins (IsoLGs). Peroxidative injuries are thought to be responsible for the development of many diabetic complications including atherosclerosis and neurological disorders. Studies showed that increased level of lipid peroxidation products in red blood cells and serum during diabetic condition cause concomitant decrease in antioxidant enzyme activities in erythrocytes (Singh and Shin 2009).

7.2.3.3 DNA

Hyperglycaemia-induced oxidative stress leads to inactivation of tuberin and downregulation of DNA repair enzyme OGG1 via the redox-dependent activation of Akt in the kidney cells (Simone et al. 2008). Habib and Liang (2014) also showed that overactivation of Akt was associated with decreased level of tuberin and increased 8-OHdG concentration in kidney cancer tissues from diabetic patients. Imbalanced expression in genetic variants of peroxisome proliferator–activated receptor–γ coactivator-1α (PPARGC1A) were also associated with development of oxidative stress and increased risk of DNA damage (urine 8-OHdG) during diabetes (Lai et al. 2008). The interrelationship of hyperglycaemia-induced oxidative stress and the associated cellular damages is shown in Fig. 7.1.

7.2.4 Histological Alterations

7.2.4.1 Pancreas

Structural and functional alterations of exocrine and endocrine pancreas were observed during diabetic condition. Varying degree of atrophy of the acinar cells and diffused to moderate hyperplasia of the centroacinar and the intercalated duct cells in the exocrine pancreas was observed in diabetic model organisms (Fujisawa et al. 2012). Prolonged exposure to increased glucose concentrations and proliferation of beta islet cells were extremely suppressed in humans and *Psammomys obesus*, an animal model of type-2 diabetes. This might be due to free radical-induced inflammatory reactions along with several other factors, viz. high levels of triglycerides and LDL and low levels of HDL (Donath et al. 2008). According to Deng et al. (2004), several in vitro and in vivo studies showed a significant reduction of total islet mass that could be correlated to impaired islet function. Pancreatic beta cells are also very prone to oxidative damage due to defective expression of antioxidant enzymes, viz. catalase and glutathione peroxidase. Isolated rat islet cells and HIT-T15 cells showed significantly decreased insulin gene expression when exposed to oxidative stress. It was observed in several experimental studies that downregulation of pancreatic and duodenal homeobox factor-1 (PDX-1) and

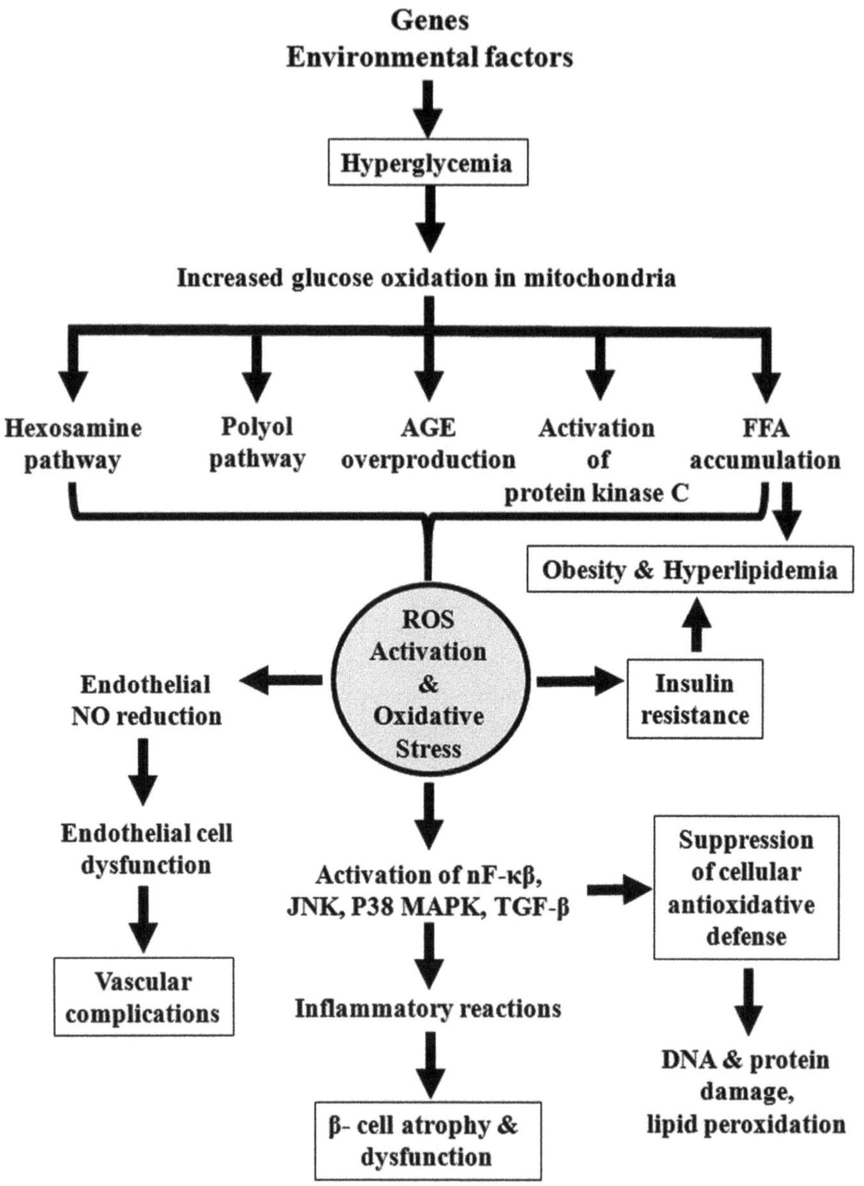

Fig. 7.1 The interrelationship of hyperglycaemia-induced oxidative stress and the associated cellular damages

subsequent activation of c-Jun *N*-terminal kinase (JNK) pathways are responsible for beta cell dysfunction in diabetic condition (Kajimoto and Kaneto 2004).

7.2.4.2 Kidney

Ultrastructural and functional alterations in kidney cells during hyperglycaemic condition create severe distresses in the pathophysiology of diabetic nephropathy. Multifocal lesions in the renal cortex cells are also an important histopathological characteristic of diabetic nephropathy (Fioretto and Mauer 2007). Vacuolization of renal tubules, abnormal thickening of the tubular membrane, and interstitial infiltration with mononuclear cells are regarded as early histological changes in diabetes which ultimately lead to interstitial fibrosis of the tubules and tubular atrophy (Pourghasem et al. 2015). In vivo experimental data on diabetic nephropathy models (ZFS_1 rats) showed an increased NO production during early stages along with arteriolar thickening, tubular dilation and atrophy, glomerular basement membrane thickening, and mesangial expansion. These major changes were also associated with increased renal expression and urinary levels of 8-hydroxydeoxyguanosine, an indicator of mitochondrial oxidative stress, as well as with increased renal peroxynitrite formation (Sharma et al. 2007).

7.2.4.3 Liver

Hyperglycaemia-induced oxidative stress leads to severe cellular damages by disrupting metabolic processes and triggering inflammatory cascade in the liver. Microvesicular steatosis (excessive accumulation of fat globules); distortion and glycogenation of nuclei; infiltration of inflammation-causing leukocytes, macrophages, and other mononucleate cells; formation of peribiliary cysts; dilatation of central and portal veins; etc. are considered as notable changes in diabetic liver (Kini et al. 2016). Presence of small inconspicuous Mallory bodies, ballooning, and necrosis of hepatocytes are also found to be associated with these changes. Significant reduction of rough endoplasmic reticulum and densities of mitochondrial cristae, decrease in glycogen content, increased nuclear pycnosis and perinuclear area along with damaged nuclear membrane are frequently observed in diabetic liver (Mohamed et al. 2016).

7.2.4.4 Heart

Immunohistochemical studies on diabetic rats showed increased protein accumulation and significantly increased staining for collagen in cardiac cells. Electron microscopic examination revealed degenerative signs of cardiac tissue including dense packing of collagen fibrils between cardiomyocytes and irregular microfibrillar organization in cytoplasm. Disintegration of mitochondrial membrane and cristae were also observed indicating progression of diabetic cardiomyopathy and interstitial fibrosis (Aragno et al. 2008). In another study with streptozotocin-induced diabetic rats, heart cells showed myonecrosis with oedema, infiltration of inflammatory cells, and separation of cardiac muscle fibres (Naderi et al. 2015). Peripheral condensation and marginalization of chromatin fibres in the nuclei and cytoplasmic vacuolations of cardiac myocytes were also observed in diabetic condition (Dallak et al. 2008). Histological alterations of the major organs are shown in Fig. 7.2.

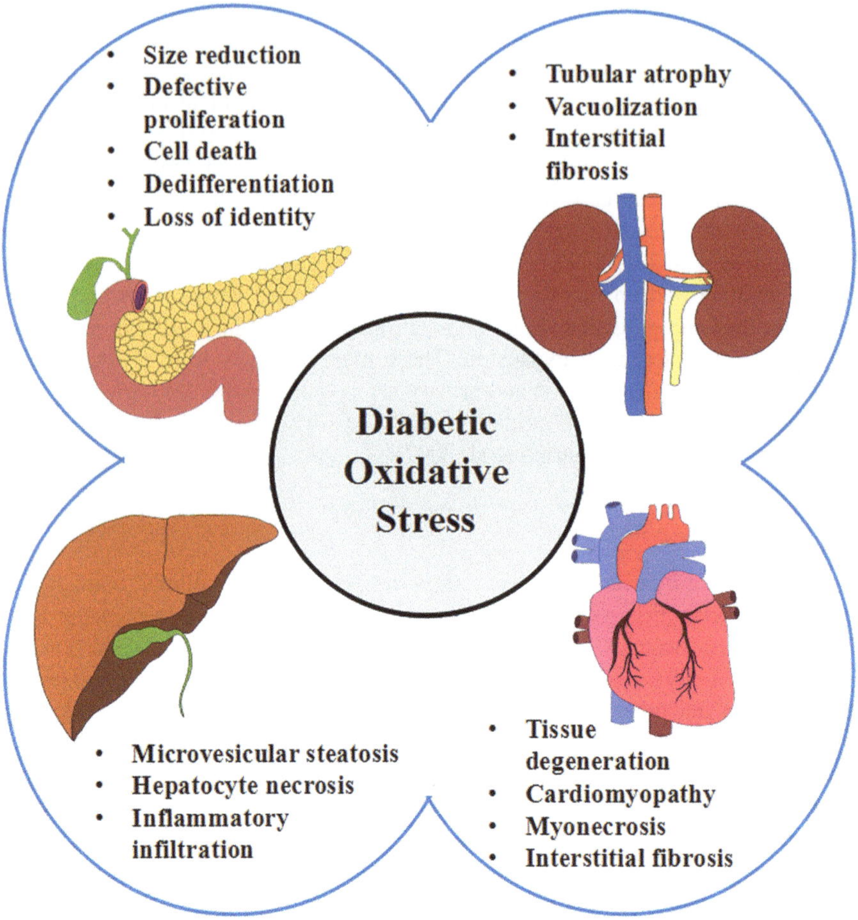

Fig. 7.2 Histological alterations of the major organs induced by oxidative stress

7.2.5 Effects on Blood Vessels (Endothelial Dysfunction and Vascular Complications)

Nitric oxide (NO) and ROS produced during diabetic condition play a pivotal role in developing endothelial dysfunction. Reduced expression of endothelial nitric oxide synthase (eNOS) and lower bioavailability of NO exert negative effect on the survival rate of endothelial cells and their function (Magenta et al. 2014). The pathological nature of endothelial dysfunction differs in type-1 and type-2 diabetes. However, in general, three major factors, viz. hyperglycaemia, excessive release of free fatty acids, and insulin resistance, are considered responsible in this regard. All of them impose oxidative stress that create an imbalance in vasoconstriction and vasodilatation and ultimately foster diabetic endothelial dysfunction. Apart from the reduced NO synthesis, increased production of advanced glycation end products

(AGEs), activation of protein kinase C (PKC), proliferation of vascular smooth muscle cells, etc. contribute for the endothelial dysfunction and vascular complications (Sena et al. 2013). According to Giugliano et al. (1996), glucose and LDL oxidation, protein glycation, and defective antioxidant response during hyperglycemic condition attribute to several micro- and macrovascular complications, viz. retinopathy, neuropathy, nephropathy, and cardiovascular diseases (Fig. 7.1).

7.3 Role of Ethnomedicinal Knowledge in Diabetes Management

Ethnomedicinal knowledge from different corners of the world is undoubtedly the prime contributor of developing the backbone of phytomedicine. The age-old practices of traditional medicinal knowledge of ethnic communities is considered the major inspirations of flourishing the natural healthcare strategies. Among 60 angiospermic families, members of Asteraceae, Fabaceae, Apiaceae, Scrophulariaceae, Liliaceae, Cucurbitaceae, Lauraceae, etc. are extensively used for diabetes treatment in indigenous culture. Phenolics, flavonoids, alkaloids, terpenoids, saponins, and polysaccharides are the major bioactive phytochemicals with hypoglycemic properties that have been isolated and purified from the leaves, roots, stems, fruits, seeds, and other parts. It was also reported that these potent metabolites could reverse or prevent diabetic complications by stimulating insulin release, increasing insulin sensitization, reducing oxidative stress, inhibiting AGEs, etc. (Singh et al. 2013). Many species are identified with significant antioxidant activities along with their antidiabetic and antihyperlipidemic properties, *Achillea santolina*, *Achyrocline satureioides*, *Azadirachta indica*, *Bauhinia forficate*, *Cuminum cyminum*, *Curcuma longa*, *Emblica officinalis*, *Gongronema latifolium*, *Hibiscus rosa-sinensis*, *Teucrium polium*, *Tinospora sinensis*, *Adhatoda vasica*, *Ocimum sanctum*, *Litsea cubeba*, *Salvia moorcroftiana*, *Gymnema sylvestre*, etc. (Jeeva and Anlin 2014; Chakraborty et al. 2016). According to Vaidya and Devasagayam (2007), the medicinal plants exert their antioxidative potential in the following major levels, viz. suppression of radical formation, scavenging of primary radicals, scavenging of secondary radicals, reconstitution of membranes, and repairing of damage. Table 7.1 listed few selected antidiabetic medicinal plants with their active phytoconstituents and possible mode of action.

7.4 Pharmacological Evidences: Clinical Trials, Case Studies, Chemical Nature of Active Constituents and Mechanism of Action

Several pharmacological experiments were conducted in vitro and in vivo to validate the previously reported ethnomedicinal knowledge of the plants. Sabu and Kuttan (2002) showed that methanolic extract of *Terminalia chebula*, *Terminalia belerica*,

Table 7.1 List of some selected antidiabetic active phytoconstituents with antioxidative properties and their possible mode of action

Name of the plants	Name of active constituents	Probable mode of action	References
Aegle marmelos	D-limonene (monoterpene)	Reduces plasma glucose and oxidative stress	Murali et al. (2013)
Trigonella foenum-graecum	Trigonelline (alkaloid)	Reinforces activities of antioxidant enzymes, improves glycemic control and liver function	Hamden et al. (2013)
Cornus officinalis	Ursolic acid (triterpenoid)	Increases SOD activities; decreases levels of MDA, TNF-α and IL-6	Qi et al. (2014)
Glycyrrhiza uralensis, G. inflata, and *G. glabra*	Glycyrrhizic acid (saponin)	Reduces oxidative stress, enhances AMPK, SIRT1 and Mn-SOD expression, exerts antioxidant activity	Hou et al. (2014)
Larrea tridentata	Nordihydroguaiaretic acid (phenolic derivative)	Alleviates diabetic renal dysfunction and oxidative stress, inhibits lipoxygenase	Anjaneyulu and Chopra (2004)
Artemisia scoparia, A. capillaris, Ceratostigma willmottianum, and *Citrus limonia*	Esculetin (lactone)	Increases activities of SOD, CAT, GPx, GST and GSH	Prabakaran and Ashokkumar (2013)
Piper longum, P. nigrum	Piperine (alkaloid)	Increases activities of SOD, CAT and GPx	Arcaro et al. (2014)
Tripterygium wilfordii	Celastrol (triterpenoid)	Reduces oxidative stress and proinflammatory cytokine activity in diabetic mice	Kim et al. (2013)
Cathanthrus roseus	Catharanthine, vindoline, vindolinene vinblastine, vincristine (alkaloid)	Free radical scavenging activities	Kar et al. 2003
Camellia sinensis, Punica granatum, Satureja khuzestanica, Bauhinia forficata	Epigallocatechin-gallate, gallocatechin, epicatechin, (+) catechin, (−) Epicatechin (flavonoid derivative)	Free radical scavenging activities	Waltner-Law et al. (2002)
Panax ginseng	Ginsenosides Rg2, panaxan A, B, C, D, E (saponins)	Regeneration of pancreatic β cells, free radical scavenging	Ma et al. (2008)
Ginkgo biloba	Kaempferol, isorhamnetin (flavonoid derivatives)	Free radical scavenging activity	Jellin et al. (1999)

(continued)

Table 7.1 (continued)

Name of the plants	Name of active constituents	Probable mode of action	References
Curcuma longa	Ferulic acid, Curcumin (phenolic derivative)	Free radical scavenging activity, insulin secretion, inhibition of lipid peroxidation, superoxide radical generation and remove insulin resistance	Ohnishi et al. (2004); Ahangarpour et al. (2019)
Morus alba, *Pisum sativum*, *Polygonum cuspidatum*	Resveratol (polyphenol)	Prevent oxidative damage by modulation of lipid metabolism and glycation reaction in diabetes	Szkudelska et al. (2009)
Smallanthus sonchifolius, *Prunus* sp., *Malus* sp.	Chlorogenic acid (polyphenol)	Prevent lipid peroxidation and formation of hydroxyl radicals during diabetic condition	Park et al. (2009)
Olea europaea	Hydroxytyrosol (phenolic compound)	Potent scavenger of superoxide anion and hydroxyl radical and protects pancreatioc beta cells from oxidative damage	Jemai et al. (2009)
Fragaria sp., *Rubus ideaus*, *Daucas carota*, *Solanum lycopersicum*, *Punica granatum*	Ellagic acid (polyphenol)	Prevention and scavenging of ROS and RNS, protection of DNA from alkylating damage	Priyadarsini et al. (2002)
Sophora flavescens	Oxymatrine (quinolizidine alkaloid)	Reduces the level of AGEs, TGF-β1 and inflammatory cytokines	Guo et al. (2014)
Peumus boldus	Boldine (Benzylisoquinoline alkaloid)	Reduces overproduction of reactive oxygen species By inhibiting angiotensin II-stimulated BMP4 expression	Lau et al. (2013)
Citrus spp.	Rutin (flavonoid glycoside)	Increases cellular antioxidant capacity and inhibits TGF-β1 mRNA expression	Tang et al. (2011)
Passiflora caerulea, *Passiflora incarnata*, *Oroxylum indicum*	Chrysin (flavone)	Suppresses TGF-β expression and inflammatory cytokines in diabetic nephropathy	Ahad et al. (2014a)
Allium sativum, *Vitis vinifera*	Quercetin (flavonol)	Increases antioxidant capacity and Smad 7 expression in diabetic kidney	Tang et al. (2011)

(continued)

Table 7.1 (continued)

Name of the plants	Name of active constituents	Probable mode of action	References
Vitis vinifera	Naringenin (flavanone)	Depletes lipid peroxidation and NO, elevates reduced glutathione	Rahigude et al. (2012)
Oroxylum indicum, *Scutellaria baicalensis*	Baicalein (flavonoid)	Mitigate renal oxidative stress and suppress the activation of NF-κB, decrease expression of iNOS and TGF-β1	Ahad et al. (2014a, b)
Catalpa sp.	Catalpol (iridoid glycoside)	Increases serum SOD and reduces ROS-mediated endothelial damage in type-2 diabetic rats	Liu (2014)
Corni Fructus	Loganin (iridoid glycoside)	Reduces oxidative stress, increase ratio of GSH/GSSG in liver and kidney	Yamabe et al. (2010)
Litchi sinensis	Oligonol (polyphenol)	Reduces renal glucose concentration and ROS, inhibits AGEs and its receptor expression	Park et al. (2014)
Salvia miltiorrhiza	Salvianolic acid A (polyphenol)	Reduces oxidative stress and expression of TGF-β1	Qiang et al. (2014)

and *Emblica officinalis* and their combination named "Triphala" showed promising hypoglycaemic activity along with antioxidant activity (inhibition of lipid peroxidation, scavenging of superoxide and hydroxyl radicals) in alloxan-induced diabetic rats. Aqueous extracts of leaves of *Murraya koenigii*, *Psidium guajava*, and *Catharanthus roseus* (500 mg/kg body weight) showed significant lowering of blood sugar and improved histological parameters in streptozotocin-induced diabetic rats, compared to the standard drug glibenclamide (Prasad et al. 2009). Ethanolic bark extract of *Sorbus decora* showed increased insulin sensitivity and decreased glycaemia in diabetic rats at a dose of 200 mg/kg body weight (Vianna et al. 2011). Oral administration of ethanolic corm extract (400 mg/kg body weight) of *Stephania hernandifolia* also showed significant antihyperlipidemic and antioxidant activity in streptozotocin-induced diabetic rats (Sharma et al. 2010). Aslan et al. (2007) showed that aqueous and ethanolic extracts of capitulums of *Helichrysum plicatum* ssp. *plicatum* exhibited marked inhibitory effects on blood glucose concentrations and lipid peroxidation in kidney and liver cells. Ethanolic extract of *Asparagus racemosus* showed promising increment in antioxidant status of alloxan-induced diabetic rats (Vadivelan et al. 2011). *Ficus racemosa* leaf extract showed hypoglycemic and antioxidant activity in diabetic model rats (Yadav et al. 2015). In vivo studies with *Saraca asoka* leaf extract also showed hypoglycemic and antihyperlipidemic activity along with antioxidant activities in streptozotocin-induced rats (Jain et al. 2013).

Different phytochemicals showed hypoglycemic activity along with antioxidative action includes alkaloids, terpenoids, flavonoids, phenolics, etc. They exert their bioactivity in various different molecular pathways. Flavonoids increase glucose transport, improve insulin sensitization and activities of antioxidative enzymes (SOD and CAT), inhibit carbohydrate-degrading enzymes, and attenuate oxidative stress-mediated inflammatory reactions and vascular complications (Vinayagam et al. 2017). Monoterpenes, sesquiterpenes and diterpenes are major constituents of plant essential oils, that exhibit antioxidative activities by preventing free radical generation, LDL oxidation and stimulating other antioxidants (vitamin E and flavonoids). Carotenoids is regarded as the effective quencher of singlet oxygen and peroxyl radicals (Grabmann 2005). Alkaloids also show antioxidative action by preventing lipid peroxidation and H_2O_2-mediated oxidative damages in diabetic condition (Tiong et al. 2013).

7.5 Herbal Medicine: Present Status and Drawbacks in Drug Development

Herbal medicine or phytomedicine has acclaimed great attention and relevance among the alternative management practices of diabetes, especially in the developing countries. In developed countries also, the use of herbal medicine is continually encouraged to avoid the adverse effect of synthetic drugs and to gently replenish body's metabolic control (Samad et al. 2009). More than 1200 antidiabetic plants are being recognized and utilized around the world for the control of diabetes and out of this only 30% of them were chemically and pharmacologically investigated.

Many countries have formed regulatory bodies to monitor these parameters, for example, Department of Ayurveda, Yoga & Naturopathy, Unani, Siddha and Homoeopathy (AYUSH) in India, the Bureau of Food and Drugs (BFAD) in the Philippines, National Agency for Food and Drug Administration and Control (NAFDAC) in Nigeria, Therapeutic Goods Administration in Australia, Food and Drug Administration in the USA, the European Medicine Agency in Europe, etc. Several pharmacopoeias have been published by these authorities, viz. American Herbal Pharmacopoeia (AHP), British Herbal Pharmacopoeia (BHP), Korean Herbal Pharmacopoeia (KHP), Indian Herbal Pharmacopoeia (IHP), etc., with useful informations in the form of monographs which are regarded as most important criteria for safety and quality control measures of herbal drugs (Sharma 2015). Table 7.2 shows some marketed herbal formulations and their proposed mode mechanisms.

According to Amjad et al. (2019), presently nanotechnology-based approach in delivering herbal products viz. nanorobots, nanopumps, smart cells, nanotized herbal drugs (NHDs), etc. in the diabetes treatment is gaining much attention. However, it

Table 7.2 Some herbal formulations along with their proposed mode of action

Herbal drugs/ formulations	Name of the plant and active constituents	Possible mode of action	References
Pancreas tonic 180 cp	*Pterocarpus marsupium, Gymnema sylvestre, Momordica charantia, Syzygium cumini, Trigonella foenum graceum, Azadirachta indica, Ficus racemosa, Aegle marmelos, Cinnamomum tamala*	Regulates glucose homeostasis	Hsia et al. (2004)
Divya Madhunashini vati (Patanjali ayurveda)	*Andrographis paniculata, Aegle marmelos, Withania somnifera, Curcuma longa,* etc.	Promotes insulin secretion, regulates carbohydrate metabolism	Patil et al. (2016)
Diasulin (ACI limited)	*Scoparia dulcis* (Scoparic acid), *Momordica charantia, Syzygium cumini, Tinospora cordifolia*, and *Trigonella foenum-graecum* (Trigonellin, Scopoletin), *Cassia auriculata, Coccinia indica, Curcuma longa, Emblica officinalis* (Emblicanin A and B), *Gymnema sylvestre*	Decreases gluconeogenesis, increases glycolysis, antihyperlipidemic and antioxidant activities	Saravanan and Pari (2005)
Diabecon (Himalaya herbal healthcare)	*Glycyrrhiza glabra, Asparagus racemosus, Phyllanthus niruri*, etc. along with minerals and vitamins	Hypoglycaemic, reduces level of glycated haemoglobin, restores normal pancreatic activity	Sridharan et al. (2011)
Epiinsulin (Swastik formulations)	Epicatechin from *Pterocarpus marsupium*	Promotes insulin release, maintains overall glucose homeostasis, inhibits Na/K ATPase activity	Rajesham et al. (2012)
Madhumardan Churna (Sri Jain Ayurvedic pharmacy)	32 plant ingredients including *Azadirachta indica, Terminalia chebula, Adhatoda vasica* etc., and minerals	Lowers blood glucose level, improves liver and kidney function	Savant et al. (2013)
BGR-34 tablet (Aimil pharmaceuticals)	*Berberis aristata, Pterocarpus marsupium, Gymnena sylvestre*	DPP-4 inhibitor	Gupta et al. (2018)
Diabeta plus (Planet ayurveda)	*Momordica charantia, Gymnema sylvestre, Pterocarpus marsupium, Ocimum tenuiflorum, Withania somnifera*, and *Salacia chinensis*	Prevents hyperglycemia, effective against diabetic neuropathy	Modak et al. (2007)

was observed that plant-based products may induce toxicity when applied through nanocarriers. Proper standardization of the herbal products in terms of safety and efficacy, therapeutic risks, and benefits should be evaluated through rigorous clinical investigations prior to administration as a drug. Pan et al. (2013) have pointed towards the major problems of herbal drug development, which should be followed strictly: maintaining ecological ethics and conservation of biodiversity of medicinal plants, prevention of overexploitation of natural resources, wide cultivation and propagation of medicinal plants for bioactive compounds, search for alternative species, rational usage of herbal medicine, etc. According to Williamson (2001), isolation and purification of bioactive principles from the phytocomplex matrix are important constraints for herbal drug development. They also suggest "herbal shotgun" approach as a preferable method of administration, because of the presence of large number of bioactive compounds in extracts or formulations rather than isolated compounds. Carmona and Pereira (2013) specified that there is still a huge gap between the present usage of medicinal plants or plant-based products and proper scientific evidence of pharmacokinetic and pharmacodynamic properties, molecular mechanism of action, adverse effects, etc.

7.6 Conclusion

Concurrence of oxidative stress and diabetes is inexorable. In fact, overproduction of free radicals and hyperglycaemia complement each other and in turn bring about destabilization of metabolic homeostasis inside the body. Several phytochemicals have shown their potent antioxidative activities by several different mechanisms and thereby protecting cellular antioxidative defence. Plant-based products are now being considered as the most potential, sustainable alternative exogenous source of antioxidants to supplement the endogenous defence system of human. Acceptance and concerns regarding the usage of herbal products are rapidly increasing across the world, suggesting the positive response and confidence of people towards the natural remedy of ailments. However, the isolation of the bioactive molecules in purest form from phytocomplex matrix, proper molecular characterization, depiction of their mode of action, effective mode of consumption, and, most importantly, toxicity symptoms (if any) are quite a big challenging task in the field of herbal medicine. Extensive and elaborative research in the laboratory as well as clinical level is thus required to strengthen the future of phytomedicine.

References

Adel AM, Waled ME, Mahmoud AL, Rehab GK (2018) Association between antioxidant enzyme activities and enterovirus-infected type 1 diabetic children. Med Princ Pract 27:86. https://doi.org/10.1159/000486718

Ahad A, Ganai AA, Mujeeb M, Siddiqui WA (2014a) Chrysin, an anti-inflammatory molecule, abrogates renal dysfunction in type 2 diabetic rats. Toxicol Appl Pharmacol 279(1):1–7

Ahad A, Mujeeb M, Ahsan H, Siddiqui WA (2014b) Prophylactic effect of baicalein against renal dysfunction in type 2 diabetic rats. Biochimie 106:101–110

Ahangarpour A, Sayahi M, Sayahi M (2019) The antidiabetic and antioxidant properties of some phenolic phytochemicals: a review study. Diabetes Metab Syndr 13(1):854–857

American Diabetes Association (2011) Standards of medical care in diabetes--2011. Diabetes Care 34(Suppl 1):S11–S61

Amjad S, Jafri A, Sharma AK, Serajuddin M (2019) A novel strategy of nanotized herbal drugs and their delivery in the treatment of diabetes: present status and future prospects. J Herb Med 17-18:100279. https://doi.org/10.1016/j.hermed.2019.100279

Anjaneyulu M, Chopra K (2004) Nordihydroguairetic acid, a lignin, prevents oxidative stress and the development of diabetic nephropathy in rats. Pharmacology 72:42–50

Aragno M, Mastrocola R, Alloatti G, Vercellinatto I, Bardini P, Geuna S, Catalano MG, Danni O, Boccuzzi G (2008) Oxidative stress triggers cardiac fibrosis in the heart of diabetic rats. Endocrinology 149(1):380–388

Arcaro CA, Gutierres VO, Assis RP, Moreira TF, Costa PI, Baviera AM, Brunetti IL (2014) Piperine, a natural bioenhancer, nullifies the antidiabetic and antioxidant activities of curcumin in streptozotocin-diabetic rats. PLoS One 9:e113993. https://doi.org/10.1371/journal.pone.0113993

Aslan M, Orhan DD, Orhan N, Sezik E, Yesilada E (2007) *In vivo* antidiabetic and antioxidant potential of *Helichrysum plicatum* ssp. plicatum capitulums in streptozotocin-induced-diabetic rats. J Ethnopharmacol 109(1):54–59

Bansal AK, Bilaspuri GS (2011) Impacts of oxidative stress and antioxidants on semen functions. Vet Med Int 2011:1. https://doi.org/10.4061/2011/686137

Berlett BS, Stadtman ER (1997) Protein oxidation in aging, disease and oxidative stress. J Biol Chem 272:20313–20316

Carmona F, Pereira AMS (2013) Herbal medicines: old and new concepts, truths and misunderstandings. Rev Bras 23(2):379–385

Chakraborty R, Roy S, Mandal V (2016) Assessment of traditional knowledge of the antidiabetic plants of Darjeeling and Sikkim Himalayas in the context of recent phytochemical and pharmacological advances. J Integr Med 14(5):336–358

Cho NH, Shaw JE, Karuranga S, Huang Y, da Rocha Fernandes JD, Ohlrogge AW, Malanda B (2018) IDF diabetes atlas: global estimates of diabetes prevalence for 2017 and projections for 2045. Diabetes Res Clin Pract 138:271–281

Dallak MM, Mikhailidis DP, Haidara MA et al (2008) Oxidative stress as a common mediator for apoptosis induced-cardiac damage in diabetic rats. Open Cardiovasc Med J 2:70–78

Deng S, Vatamaniuk M, Huang X, Doliba N, Lian M, Frank A, Velidedeoglu E, Desai NM, Koeberlein B, Wolf B, Barker CF, Naji A, Matschinsky FM, Markmann JF (2004) Structural and functional abnormalities in the islets isolated from type 2 diabetic subjects. Diabetes 53 (3):624–632

Donath MY, Schumann DM, Faulenbach M, Ellingsgaard H, Perren A, Ehses JA (2008) Islet inflammation in type 2 diabetes from metabolic stress to therapy. Diabetes Care 31(Suppl 2):161–164

Esterbauer H, Schaur RJ, Zollner H (1991) Chemistry and biochemistry of 4-hydroxynonenal, malonaldehyde and related aldehydes. Free Radic Biol Med 11(1):81–128

Evans JL, Goldfine ID, Maddux BA, Grodsky GM (2002) Oxidative stress and stress-activated signaling pathways: a unifying hypothesis of type 2 diabetes. Endocr Rev 23(5):599–622

Fiorentino TV, Prioletta A, Zuo P, Folli F (2013) Hyperglycemia-induced oxidative stress and its role in diabetes mellitus related cardiovascular diseases. Curr Pharm Des 19:5695–5703

Fioretto P, Mauer M (2007) Histopathology of diabetic nephropathy. Semin Nephrol 27 (2):195–207

Fujisawa H, Zhang Z, Sun W, Huang M, Kobayashi J, Yasuda H, Kinoshita Y, Ando R, Tamura K (2012) Histopathological changes in the pancreas from a spontaneous hyperglycemic cynomolgus monkey. J Toxicol Pathol 25(3):215–219

Giugliano D, Ceriello A, Paolisso G (1996) Oxidative stress and diabetic vascular complications. Diabetes Care 19(3):257–267
Grabmann J (2005) Terpenoids as plant antioxidants. Vitam Horm 72:505–535
Guo C, Han F, Zhang C, Xiao W, Yang Z (2014) Protective effects of oxymatrine on experimental diabetic nephropathy. Planta Med 80(4):269–276
Gupta BP, Sharma I, Kohli N, Sharma S, Rathi A, Sharma AK (2018) Preliminary clinical assessment and non- toxicity evaluation of an ayurvedic formulation BGR-34 in NIDDM. J Tradit Complement Med 8(4):506–514
Habib SL, Liang S (2014) Hyperactivation of Akt/mTOR and deficiency in tuberin increased the oxidative DNA damage in kidney cancer patients with diabetes. Oncotarget 5:2542–2550
Halliwell B, Gutteridge JMC (2007) Free radical in biology and medicine, 4th edn. Clarendon Press, Oxford
Hamden K, Bengara A, Amri Z, Elfeki A (2013) Experimental diabetes treated with trigonelline: effect on key enzymes related to diabetes and hypertension, β-cell and liver function. Mol Cell Biochem 381(1–2):85–94
Hou S, Zheng F, Li Y, Gao L, Zhang J (2014) The protective effect of glycyrrhizic acid on renal tubular epithelial cell injury induced by high glucose. Int J Mol Sci 15(9):15026–15043
Hsia SH, Bazargan M, Davidson MB (2004) Effect of pancreas tonic (an Ayurvedic herbal supplement) in type 2 diabetes mellitus. Metabolism 53(9):1166–1173
Jain A, Jasmine, Sharma S, Saini V (2013) Hypolipidemic, hypoglycemic and antioxidant potential of *Saraca asoca* ethanolic leaves extract in streptozotocin induced experimental diabetes. Int J Pharm Pharm Sci 5(1):1–7
Jeeva S, Anlin SY (2014) A review of antidiabetic potential of ethnomedicinal plants. Med Aromat Plants 3:4
Jellin JM, Batz F, Hitchens K (1999) Pharmacist's letter/prescriber's letter natural medicines comprehensive database. Therapeutic Research Faculty, Stockton, CA
Jemai H, El Feki A, Sayadi S (2009) Antidiabetic and antioxidant effects of hydroxytyrosol and oleuropein from olive leaves in alloxan-diabetic rats. J Agric Food Chem 57(19):8798–8804
Kajimoto Y, Kaneto H (2004) Role of oxidative stress in pancreatic β-cell dysfunction. Ann N Y Acad Sci 1011:168–176
Kar A, Choudhary BK, Bandyopadhyay NG (2003) Comparative evaluation of hypoglycaemic activity of some Indian medicinal plants in alloxan diabetic rats. J Ethnopharmacol 84:105–108
Kini S, Tripathi P, Amarapurkar AD (2016) Histopathology of liver in diabetes mellitus—an autopsy study. Int J Sci Stud 4(5):110–113
Kim JH, Lee JO, Lee SK et al (2013) Celastrol suppresses breast cancer MCF-7 cell viability via the AMP-activated protein kinase (AMPK)-induced p53-polo like kinase 2 (PLK-2) pathway. Cell Signal 25:805–813
Lai CQ, Tucker KL, Parnell LD, Adiconis X, García-Bailo B, Griffith J, Meydani M, Ordovás JM (2008) PPARGC1A variation associated with DNA damage, diabetes, and cardiovascular diseases: the Boston Puerto Rican Health Study. Diabetes 57:809–816
Lau YS, Tian XY, Mustafa MR, Murugan D, Liu J, Zhang Y, Lau CW, Huang Y (2013) Boldine improves endothelial function in diabetic db/db mice through inhibition of angiotensin II-mediated BMP4-oxidative stress cascade. Br J Pharmacol 170(6):1190–1198
Liu JY (2014) Catalpol protect diabetic vascular endothelial function by inhibiting NADPH oxidase. Zhongguo Zhong Yao Za Zhi 15:2936–2941
Ma SW, Benzie IF, Chu TT, Fok BS, Tomlinson B, Critchley LA (2008) Effect of *Panax ginseng* supplementation on biomarkers of glucose tolerance, antioxidant status and oxidative stress in type 2 diabetic subjects: results of a placebo-controlled human intervention trial. Diabetes Obes Metab 10:1125–1127
Magenta A, Greco S, Capogrossi MC, Gaetano C, Martelli F (2014) Nitric oxide, oxidative stress, and p66shc interplay in diabetic endothelial dysfunction. Biomed Res Int 2014:1. https://doi.org/10.1155/2014/193095

Maiese K (2015) New insights for oxidative stress and diabetes mellitus. Oxidative Med Cell Longev 2015:1. https://doi.org/10.1155/2015/875961
Maritim AC, Sanders RA, Watkins JB (2003) Diabetes, oxidative stress, and antioxidants: a review. J Biochem Mol Toxicol 17(1):24–38
Matough FA, Budin SB, Hamid ZA, Alwahaibi N, Mohamed J (2012) The role of oxidative stress and antioxidants in diabetic complications. SQU Med J 12(1):5–18
Modak M, Dixit P, Londhe J, Ghaskadbi S, Devasagayam TPA (2007) Indian herbs and herbal drugs used for the treatment of diabetes. J Clin Biochem Nutr 40:163–173
Mohamed J, Nazratun NAH, Zariyantey AH, Budin SB (2016) Mechanisms of diabetes-induced liver damage-the role of oxidative stress and inflammation. Sultan Qaboos Univ Med J 16 (2):132–141
Mohammad MD, Dimitri PM, Mohamed AH, Ismaeel MB, Olaa MT, Moshira AR, Hanaa ZY, Zeinb AA, Ibrahim MI, Samy ME, Laila AR, Noha AA (2008) Oxidative stress as a common mediator for apoptosis induced-cardiac damage in diabetic rats. Open Cardiovasc Med J 2:70–78
Murali R, Karthikeyan A, Saravanan R (2013) Protective effects of D-limonene on lipid peroxidation and antioxidant enzymes in streptozotocin-induced diabetic rats. Basic Clin Pharmacol Toxicol 112(3):175–181
Naderi R, Mohaddes G, Mohammadi M, Alihemmati A, Badalzadeh R, Ghaznavi R, Ghyasi R, Mohammadi S (2015) Preventive effects of garlic (*Allium sativum*) on oxidative stress and histopathology of cardiac tissue in streptozotocin-induced diabetic rats. Acta Physiol Hung 102 (4):380–390
O'Neal KS, Johnson JL, Panak RL (2016) Recognizing and appropriately treating latent autoimmune diabetes in adults. Pharm Ther 29(4):249–252
Ohnishi M, Matuo T, Tsuno T, Hosoda A, Nomura E, Taniguchi H, Sasaki H, Morishita H (2004) Antioxidant activity and hypoglycemic effect of ferulic acid in STZ-induced diabetic mice and KK-Ay mice. Biofactors 21:315–319
Olokoba AB, Obateru OA, Olokoba LB (2012) Type 2 diabetes mellitus: a review of current trends. Oman Med J 27(4):269–273
Pan SY, Zhou SF, Gao SH, Yu ZL, Zhang SF, Tang MK, Sun JN, Ma DL, Han YF, Fong WF, Ko KM (2013) New perspectives on how to discover drugs from herbal medicines: CAM's outstanding contribution to modern therapeutics. Evid Based Complement Alternat Med. https://doi.org/10.1155/2013/627375
Park JS, Yang JS, Hwang BY, Yoo BK, Han K (2009) Hypoglycemic effect of yacon tuber extract and its constituent, chlorogenic acid, in streptozotocin-induced diabetic rats. Biomol Ther 17 (3):256–262
Park CH, Yokozawa T, Noh JS (2014) Oligonol, a low-molecular-weight polyphenol derived from lychee fruit, attenuates diabetes-induced renal damage through the advanced glycation end product-related pathway in db/db mice. J Nutr 144(8):1150–1157
Patil KN, Patil PS, Dudhgaonkar TD, Mohite SK, Magdum CS (2016) Ayurvedic and allopathic antidiabetic formulations avilable in market. Int J Res Sci Technol 2(2):232–242
Pourghasem M, Shafi H, Babazadeh Z (2015) Histological changes of kidney in diabetic nephropathy. Caspian J Intern Med 6(3):120–127
Prabakaran D, Ashokkumar N (2013) Protective effect of esculetin on hyperglycemia-mediated oxidative damage in the hepatic and renal tissues of experimental diabetic rats. Biochimie 95 (2):366–373
Prasad SK, Kulshreshtha A, Qureshi TN (2009) Antidiabetic activity of some herbal plants in streptozotocin induced diabetic albino rats. Pak J Nutr 8:551–557
Priyadarsini KI, Khopde SM, Kumar SS, Mohan H (2002) Free radical studies of ellagic acid, a natural phenolic antioxidant. J Agric Food Chem 50(7):2200–2206
Qi MY, Yang JJ, Zhou B, Pan DY, Sun X (2014) Study on the protective effect of ursolic acid on alloxan-induced diabetic renal injury and its underlying mechanisms. Zhongguo Ying Yong Sheng Li Xue Za Zhi 30:445–448

Qiang G, Yang X, Xuan Q, Shi L et al (2014) Salvianolic acid A prevents the pathological progression of hepatic fibrosis in high-fat diet-fed and streptozotocin-induced diabetic rats. Am J Chin Med 42(5):1183–1198

Rahigude A, Bhutada P, Kaulaskar S, Aswar M, Otari K (2012) Participation of antioxidant and cholinergic system in protective effect of naringenin against type-2 diabetes-induced memory dysfunction in rats. Neuroscience 226:62–72

Rajesham VV, Ravindernath A, Bikshapathi DVRN (2012) A review on medicinal plant and herbal drug formulation used in diabetes mellitus. Indo Am J Pharm Res 2(10):1200–1212

Sabu MC, Kuttan R (2002) Anti-diabetic activity of medicinal plants and its relationship with their antioxidant property. J Ethnopharmacol 81:155–160

Samad A, Shams MS, Ullah Z, Wais M, Nazish I, Sultana Y, Aqil M (2009) Status of herbal medicines in the treatment of diabetes: a review. Curr Diabetes Rev 5:102–111

Saravanan R, Pari L (2005) Antihyperlipidemic and antiperoxidative effect of Diasulin, a polyherbal formulation in alloxan induced hyperglycemic rats. BMC Complement Altern Med 5:14. https://doi.org/10.1186/1472-6882-5-14

Savant D, Panchal P, Page A (2013) Evaluation of antidiabetic, and antioxidant activities of madhumardan churna- a polyherbal ayurvedic formulation. Int J Pharm Res Dev 5(3):191–199

Sena CM, Pereira AM, Seiça R (2013) Endothelial dysfunction—a major mediator of diabetic vascular disease. Biochim Biophys Acta 1832:2216–2231

Sharma S (2015) Current status of herbal product: regulatory overview. J Pharm Bioallied Sci 7 (4):293–296

Sharma P, Joel S, Shuping S, Betty L, Ruc T (2007) Diabetic nephropathy is associated with oxidative stress and decreased renal nitric oxide production. J Am Soc Nephrol 18:2945–2952

Sharma U, Sahu RK, Roy A, Golwala DK (2010) *In vivo* antidiabetic and antioxidant potential of *Stephania hernandifolia* in streptozotocin-induced-diabetic rats. J Young Pharm 2(3):255–260

Sheikhpour R (2013) Diabetes and oxidative stress: the mechanism and action. Iran J Diabetes Obes 5(1):40–45

Sridharan K, Mohan R, Ramaratnam S et al (2011) Ayurvedic treatments for diabetes mellitus. Cochrane Database Syst Rev (12):CD008288

Siddiqui AA, Siddiqui SA, Ahmad S, Siddiqui S, Ahsan I, Sahu K (2013) Diabetes: mechanism, pathophysiology and management—a review. Int J Drug Dev Res 5(2):1–23

Simone S, Gorin Y, Velagapudi C, Abboud HE, Habib SL (2008) Mechanism of oxidative dna damage in diabetes tuberin inactivation and downregulation of DNA repair enzyme 8-Oxo-7,8-Dihydro-2′-Deoxyguanosine-DNA glycosylase. Diabetes 57:2626–2636

Singh M, Shin S (2009) Changes in erythrocyte aggregation and deformability in diabetesmellitus: a brief review. Indian J Exp Biol 47(1):7–15

Singh R, Kaur N, Kishore L, Gupta GK (2013) Management of diabetic complications: a chemical constituents based approach. J Ethnopharmacol 150(1):51–70

Steenkamp DW, Alexanian SM, Sternthal E (2014) Approach to the patient with atypical diabetes. CMAJ 186(9):678–684

Szkudelska K, Nogowski L, Szkudelski T (2009) Resveratrol, a naturally occurring diphenolic compound, affects lipogenesis, lipolysis and the antilipolytic action of insulin in isolated rat adipocytes. J Steroid Biochem Mol Biol 113(1–2):17–24

Taheri E, Djalali M, Saedisomeolia A, Moghadam AM, Djazayeri A, Qorbani M (2012) The relationship between the activities of antioxidant enzymes in red blood cells and body mass index in Iranian type 2 diabetes and healthy subjects. J Diabetes Metab Dis 11:3

Tang DQ, Wei YQ, Yin XX, Lu Q, Hao HH, Zhai YP, Wang JY, Ren J (2011) *In vitro* suppression of quercetin on hypertrophy and extracellular matrix accumulation in rat glomerular mesangial cells cultured by high glucose. Fitoterapia 82(6):920–926

Tiong SH, Looi CY, Hazni H, Arya A, Paydar M, Wong WF, Cheah S, Mustafa MR, Awang K (2013) Antidiabetic and antioxidant properties of alkaloids from *Catharanthus roseus* (L.) G. Don. Molecules 18:9770–9784

Tiwari P (2015) Recent trends in therapeutic approaches for diabetes management: a comprehensive update. J Diabetes Res 2015:1. https://doi.org/10.1155/2015/340838
Tiwari AK, Madh J (2002) Diabetes mellitus and multiple therapeutic approaches of phytochemicals: present status and future prospects. Curr Sci 83:30–38
Ullah A, Khan A, Khan I (2016) Diabetes mellitus and oxidative stress—a concise review. Saudi Pharm J 24:547–553
Vadivelan R, Mandal D, Payyavala U, Dhanabal SP, Satishkumar MN, Antony S, Elango K (2011) Hypoglycemic, antioxidant and hypolipidemic activity of *Asparagus racemosus* on streptozotocin-induced diabetes in rats. Adv Appl Sci Res 2(3):179–185
Vaidya ADB, Devasagayam TPA (2007) Current status of herbal drugs in India: an overview. J Clin Biochem Nutr 41:1–11
Vianna R, Brault A, Martineau LC et al (2011) In vivo anti-diabetic activity of the ethanolic crude extract of *Sorbus decora* C.K.Schneid. (Rosaceae): a medicinal plant used by Canadian James Bay Cree Nations to treat symptoms related to diabetes. Evid Based Complement Alternat Med 2011:237941
Vinayagam R, Xiao J, Xu B (2017) An insight into antidiabetic properties of dietary phytochemicals. Phytochem Rev 16:535. https://doi.org/10.1007/s11101-017-9496-2
Waltner-Law ME, Wang XL, Law BK et al (2002) Epigallocatechin gallate: a constituent of green tea, represses hepatic glucose production. J Biol Chem 277:34933–34940
Williamson EM (2001) Synergy and other interactions in phytomedicines. Phytomedicine 8:401–409
Woldu MA, Lenjisa JL, Satessa GD (2014) Recent advancements in diabetes pharmacotherapy. Biochem Pharmacol 3:143
Yadav RK, Keshari AK, Singh AK, Saha S (2015) Antidiabetic and hypolipidemic effect of *Ficus racemosa* petroleum ether extract in streptozotocin induced diabetic albino rats. Int J Pharm Pharm Sci 7(4):283–287
Yamabe N, Noh JS, Park CH, Kang KS, Shibahara N, Tanaka T, Yokozawa T (2010) Evaluation of loganin, iridoid glycoside from *Corni Fructus*, on hepatic and renal glucolipotoxicity and inflammation in type 2 diabetic db/db mice. Eur J Pharmacol 648(1–3):179–187

Plant-Based β-Secretase (BACE-1) Inhibitors: A Mechanistic Approach to Encounter Alzheimer's Disorder

8

Atanu Bhattacharjee, Raja Chakraborty, and Saikat Sen

Abstract

BACE-1 (β-secretase) has emerged as a promising drug to treat Alzheimer's disease (AD). BACE-1, an aspartic protease, acts on amyloid precursor protein (APP) prompting the formation and deposition of amyloid-β peptide (Aβ). BACE-1 hindrance has direct ramifications in AD pathology without generally influencing viability. However, inhibition of BACE-1 specifically through in vivo models has introduced numerous issues to the researchers. Since its identification in 2000, inhibitors covering a wide range of auxiliary classes have been structured and created. Advancement in these fields brought about targeted drug therapy that may reduce deposition of Aβ plaque in the brain. Natural product derivatives have always been a potential source of drug discovery process, and they are considered to be safe and economical. Flavonoids are an abundance of naturally delivered BACE-1 inhibitors. Citrus bioflavonoids are distinguished as potential noncompetitive BACE-1 inhibitors. Further, xanthones displayed the hindrance of Aβ conglomeration and BACE-1 activity in vitro and in cells, notwithstanding their radical scavenging and metal chelation capacity. The manuscript has been focused to deliver a refreshed learning with array of phytoextracts and lead molecules with BACE-1 inhibitory potential.

Keywords

β-secretase · Medicinal plants · Alzheimer's disease

A. Bhattacharjee (✉) · S. Sen
Faculty of Pharmaceutical Science, Assam Down Town University, Guwahati, Assam, India
e-mail: atanu1983@gmail.com

R. Chakraborty
NEF College of Pharmacy, Guwhati, Assam, India

S. C. Mandal et al. (eds.), *Evidence Based Validation of Traditional Medicines*,
https://doi.org/10.1007/978-981-15-8127-4_8

8.1 Introduction

Alzheimer's disease (AD) is a progressive neurodegenerative disorder and irreversible age-related form of dementia that slowly erodes the brain and causes severe changes in memory, cognitive skills, personality, and behavior. The WHO perceived the effect of AD as a worldwide general wellbeing need. AD is the main cause of dementia in elderly people (>65 years), with an average duration of around 9 years between the onset of clinical side effects and death. By 2040, more than 71% of dementia cases is expected in developing nations (Zhang and Jiang 2015). Caring of AD patients requires tremendous efforts for both medical fraternity and family members (Anderson 2002). At present more than five million AD patients are under treatment in the USA, whereas in India the progression of AD is on a higher note (Gupta and Bala 2013).

Aβ plaques which decline levels of acetylcholine (Ach) in the brain are considered as a pathological hallmark of AD (Azimi et al. 2017; Berrino 2002). Aβ is formed from APP by beta- and gamma-secretases. Initial cleavage by BACE-1 generates a soluble N-terminal fragment and C-terminal fragment of peptides. The latter upon proteolysis by γ-secretase yields insoluble peptides (Skovronsky et al. 2006; Brinton and Yamazaki 1998). BACE-1 has emerged as a promising therapeutic target as it is involved in Aβ plaque formation (Barao et al. 2016). Conversely, neuro-inflammation of AD likely begins as a host resistance reaction to the harming impacts of the amyloid plaques in the cerebrum. Hence, anti-inflammatory drugs could be another potential therapeutic target to encounter progress of AD (Zhang and Jiang 2015). Medicinal plants are considered as potential sources of anti-inflammatory agents, BACE-1, and cholinesterase inhibitors (Bachurin et al. 2017; Ben Halima et al. 2016). Therefore, exploration of plant-based medicine either as primary or adjuvant therapy in early prevention of AD has become necessary as they are considered to be safe and economical by the WHO.

8.2 Pathophysiology of AD

Extracellular amyloid-beta (Aβ) plaques and neurofibrillary tangles are considered as hallmarks of AD resulting in behavioral changes and cognitive decline periodically. Hyperphosphorylation of tau protein and activation of N-methyl-d-aspartate (NMDA) receptors are other important contributing factors of AD (Hardy and Selkoe 2002; Tanzi and Bertram 2005; Crouch et al. 2008). Premature synaptotoxicity, changes in synapse articulation, neurophils count decline, and gathering of amyloid-β protein plaques and mind decay are totally connected with phases of AD movement. A few ongoing examinations have inspected the connection among Aβ and NMDA receptors. Aβ-actuated spine misfortune is related with diminishing glutamate receptors and is reliant upon the calcium subordinate phosphatase calcineurin, which has likewise been connected to long-term depression (Selkoe 2004; Armstrong 2009).

8.3 Treatments of AD

A list of existing drugs used in AD is summarized in Table 8.1.

8.4 Disease-Modifying Therapies and Drug Development

Current medications primarily aim to delay the progression of AD. However, the side effects associated with those drugs remain a big concern for medical fraternities. There is an earnest requirement for the advancement of disease-altering treatments or medicines focusing on essential pathophysiology of AD (Mei et al. 2010; Carlsson 2008).

While there is no remedy for AD or an approach to stop or moderate its progression, there are drug and nondrug alternatives that may help treat indications. Understanding accessible alternatives can help people living with the illness and their parental figures to adapt to manifestations and improve personal satisfaction. Since the last two decades, only memantine was able to clear phase III clinical trial (Vladimir and Sophia 2018). The summary of current treatment strategies for AD is mentioned below (Table 8.2).

Table 8.1 Approved drugs for the treatment of AD (adopted and modified from Miguel and Emília 2019)

Name	Structure	Chemical name	Trade name	References
Tacrine	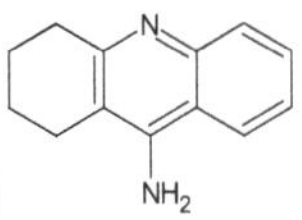	1,2,3,4-tetrahydroacridin-9-amine	Cognex	Mahesh and Tasneem (2014)
Donepezil	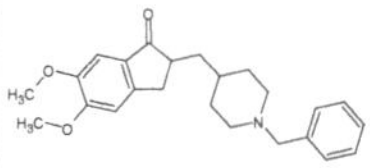	2-[(1-benzylpiperidin-4-yl) methyl]-5,6-dimethoxy-2,3-dihydroinden-1-one	Aricept	Tariot et al. (2004);
Rivastigmine	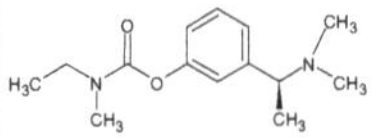	[3-[(1S)-1-(dimethylamino)ethyl] phenyl]*N*-ethyl-*N*-methylcarbamate	Exelon	Scarpini et al. (2003);
Galantamine	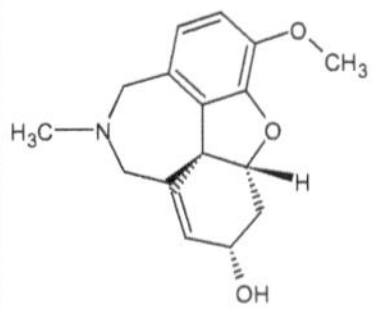	(1S,12S,14R)-9-methoxy-4-methyl-11-oxa-4-azatetracyclo heptadeca-6 (17),7,9,15-tetraen-14-ol	Razadyne	Parihar and Hemnani (2004)
Memantine	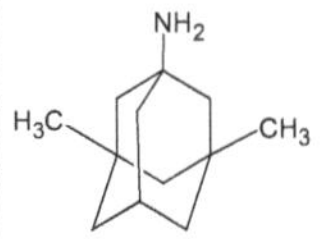	3,5-dimethyladamantan-1-amine	Axura	Shankar et al. (2007)

Table 8.2 Current treatment strategies for AD (modified from Yu et al. 2015)

Treatment strategy	Mechanism of action	Stages of AD
Modulation of neurotransmitter (existing/approved therapy)	1. **AChE inhibitor** 2. **NMDA antagonist**	Early
Disease-modifying therapies (under investigation)	1. **Amyloid cascade** • Modulation of α-, β-, γ-secretase • β-amyloid depleter • Promoting β-amyloid cleaning 2. **Tau protein modulation** • Inhibition of tau hyperphosphorylation • Degradation of tau protein • Inhibition of tau oligomerization	Mild to moderate
Immunotherapy (under investigation)	1. **Passive immunization** • Solanezumab • Crenezumab 2. **Active immunization CAD106**	Mild to moderate
Gene-based therapy (under investigation)	1. **Modulation of expression of presenilin**	Mild to moderate
Antioxidants (existing/approved therapy)	1. **Oxidative stress reduction** 2. **Free radical scavenger**	Mild to moderate
Anti-inflammatory agents (existing/approved therapy)	1. **Caspase inhibitors** 2. **Metal chelators** 3. **Statins** 4. **PPAR-γ**	Mild to moderate

8.5 The Amyloid Hypothesis

Aβ is a complex protein oligomer cluster typically found in Alzheimer's brain formed via catalysis of APP by secretase enzymes. As per the "β-amyloid theory," Aβ peptide clusters trigger neuro-inflammatory mediators resulting in inflammation of neurons (Dewachter and Van 2002). This leads to synchronization of neuron concentration in the brain resulting in dementia (Choi et al. 2008a, b; Jeon et al. 2008; Jeon et al. 2005; Atanu 2018) (Fig. 8.1). Therefore, modulating the chain of events like production of Aβ peptide fragments from APP, its deposition as extra-cellular plaques is believed to be a possible approach toward the treatment of AD.

Aβ is produced from APP by β-site APP-cleaving enzyme (BACE-1) resulting in the formation of soluble APPΆ (sAPPΆ) and a 99-amino acid fragment (C99). The C99 fragments in the presence of γ-secretase form ΆA40 and ΆA42 responsible for neuritic plaque formation (Kinghorn 2001; Balunas and Kinghorn 2005; Gurib-Fakim 2006; Huang and Mucke 2012).

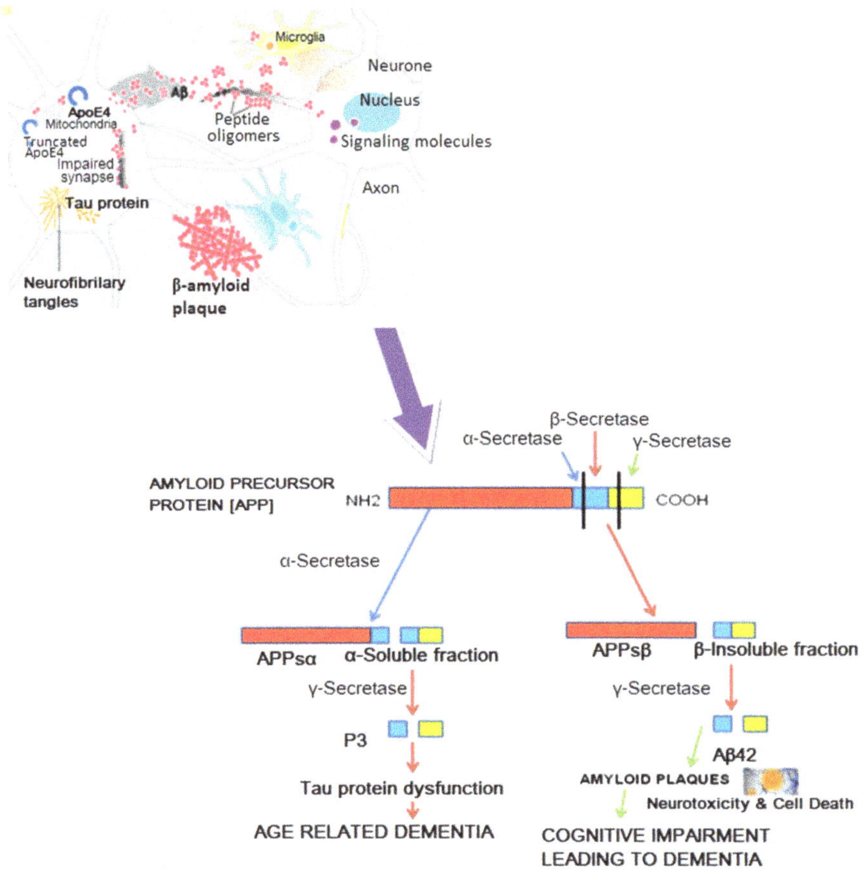

Fig. 8.1 Mechanism of toxic amyloid-β formation leading to cognitive impairment in AD (adopted and modified from Eric and Bard 2016)

8.6 β-Secretase (BACE-1)

BACE-1, a transmembrane aspartyl protease, promotes sequential proteolytic cleavage of APP and thereby formation of Aβ peptide in neurons. Hence, BACE-1 is the new intervention for AD therapy (Roberds et al. 2001). Scientific investigation on BACE-1 inhibitors is considered as a potential therapeutic approach to minimize Aβ fibril concentrations in the brain of AD patients (Guo and Hobbs 2006; Salloway et al. 2008; Vassar 2004; McArdle and Quinn 2007).

8.7 Medicinal Plants with BACE-1 Inhibitory Effect

Notwithstanding the obvious away from of medication revelation from therapeutic plants, future undertakings face numerous difficulties. Researchers should primarily focus on natural sources for the development of medication or adjuvant therapy to ease AD patient agony (Fenical and Jensen 2006). Raw crude drugs, extract, and lead molecules serve an indispensable role to treat neurological illness either as standard or adjuvant based on their ethnopharmacological aspects (Philomena 2011). Natural products not only act synergistically but also enhance the activity of other analogs and reduce the toxicity. Various alternative systems of medicine documented the usefulness of numerous plants to treat neurological disorders associated with cognitive dysfunctions (Perry et al. 1999; Vaidya 1997) (Table 8.3).

8.8 Phytoconstituetns with BACE-1 Inhibitory Activity

8.8.1 Phenyl Propanoids

Phenyl propanoids are a widely distributed plant secondary metabolite derived from the precursor amino acid phenylalanine. It has been observed that they inhibit protein aggregation in a nonstoichiometric fashion (Feng and Shoichet 2006; Coan et al. 2009). Sebesteniods A–D (Fig. 8.2 (i)), isolated from *Cordia sebestena*, showed potent BACE-1 inhibitory activity in vitro (Dai et al. 2010). These findings postulated sebestenoids as possible future drug candidate for AD (Eglen 2002; Gunasekera et al. 2010).

8.8.2 Flavonoid Derivatives

Flavonoids, found in a variety of edible plants, showed significant BACE-1inbhibitory activity in vitro along with their antioxidant activity (Youdim et al. 2004; Ross and Kasum 2002; Commenges et al. 2000). Antioxidant potentiality of flavonoids reportedly inhibits Aβ oligomerization. Myricetin (Fig. 8.2 (ii)), an isolated flavonoid, showed protective activity against Aβ fibrils in vitro (Shimmyo et al. 2008). *Morus lhou* stem bark extract and isolated compounds, viz., norartocarpetin, kuwanon C, morusin, kuwanon A, morusinol, cyclomorusin, and neocyclomorusin (Fig. 8.2 (iii–vii)), exhibited significant BACE-1 inhibitory activity in vitro (Ono et al. 2003; Jeong et al. 2009; Kang et al. 2004).

8.8.3 Napthoquinone Derivatives

Hsp90, a 1,4-napthoquinone derivative, has shown BACE-1 inhibitory activity in vitro (Muto et al. 1987). Hsp90 augmented α-secretase activity and thus increased α-APP formation (Tauraite et al. 2009; Montenegro et al. 2010). Resveratrol

Table 8.3 Potential phytoextracts and their derivatives to encounter AD (adopted and modified from Abdul et al. 2018)

Plant name	Extract	Extracted compound(s)	Structure	Mechanism of action	Pharmacological study	References
Sophora flavescens	Chloroform extract of root	Lavandulyl flavanones (Saphora flavones)	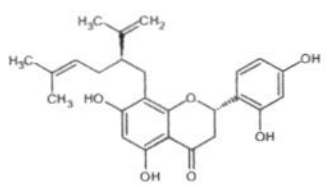	BACE-1 inhibitory activity	In vitro	Hwang et al. (2008)
Fructus gardeniae	Methanol extract of whole plant	Geniposide	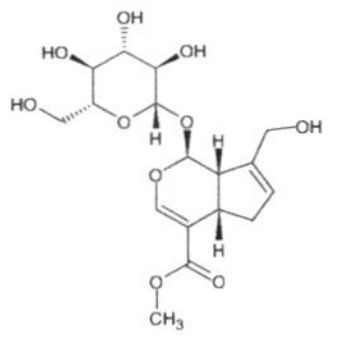	AChE inhibitor	In vitro	Nam and Lee (2013)
Paeonia lactiflora	Ethanol extract of seeds	Resveratrol oligomer	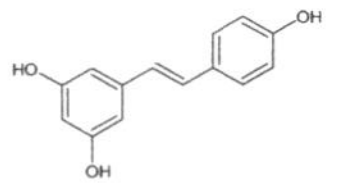	BACE-1 inhibitory activity	In vitro	Choi et al. (2011)
Psoralea corylifolia	Methanol extract of seeds	Isoflavones	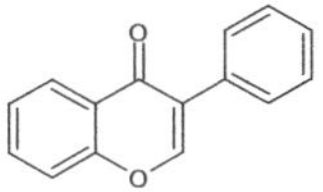	BACE-1 inhibitory activity	In vitro	Choi et al. (2008a, b)
Cordia sebestena	Ethanol extract of fruit	Sebestenoids	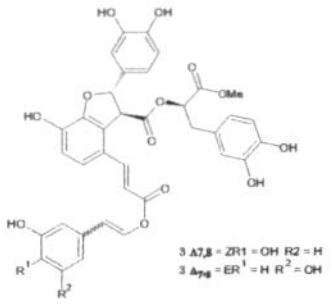	Aspartic protease inhibitor	In vitro	Dai et al. (2010)

(continued)

Table 8.3 (continued)

Plant name	Extract	Extracted compound(s)	Structure	Mechanism of action	Pharmacological study	References
Perilla frutescens var. *acuta*	Methanol extract of leaves	Luteolin and rosmarinic acid	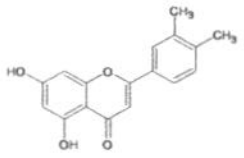 Luteolin 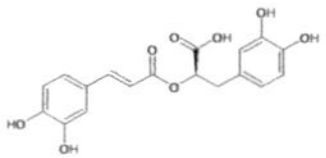 Rosmarinic acid	BACE-1 inhibitory activity	In vitro	Choi et al. (2008a, b)
Smilax china L. (Smilax Rhizoma)	Methanol extract of rhizomes	Trans−/cis-resveratrol mixture, oxyresveratrol, veraphenol, and cis-scirpusin A	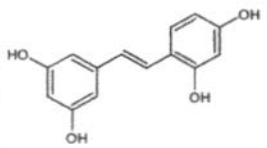 Oxyresveratrol 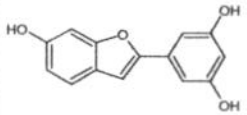 Veraphenol 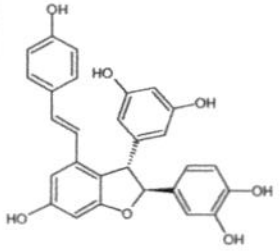 Scirpusin A	BACE-1 inhibitory activity	In vitro	Jeona et al. (2007)
Sanguisorba officinalis L. (Sanguisorbae radix)	Methanol extract of roots	1,2,3-trigalloyl-4,6-hexahydroxydiphenoyl-β-D-glucopyranoside and 1,2,3,4,6-pentagalloyl-β-D-glucopyranoside	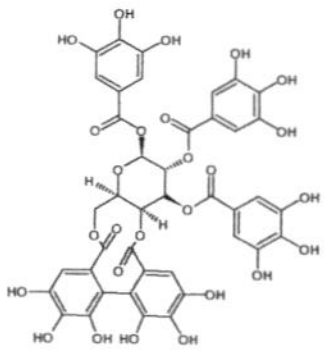	BACE-1 inhibitory activity	In vitro	Lee et al. (2005)

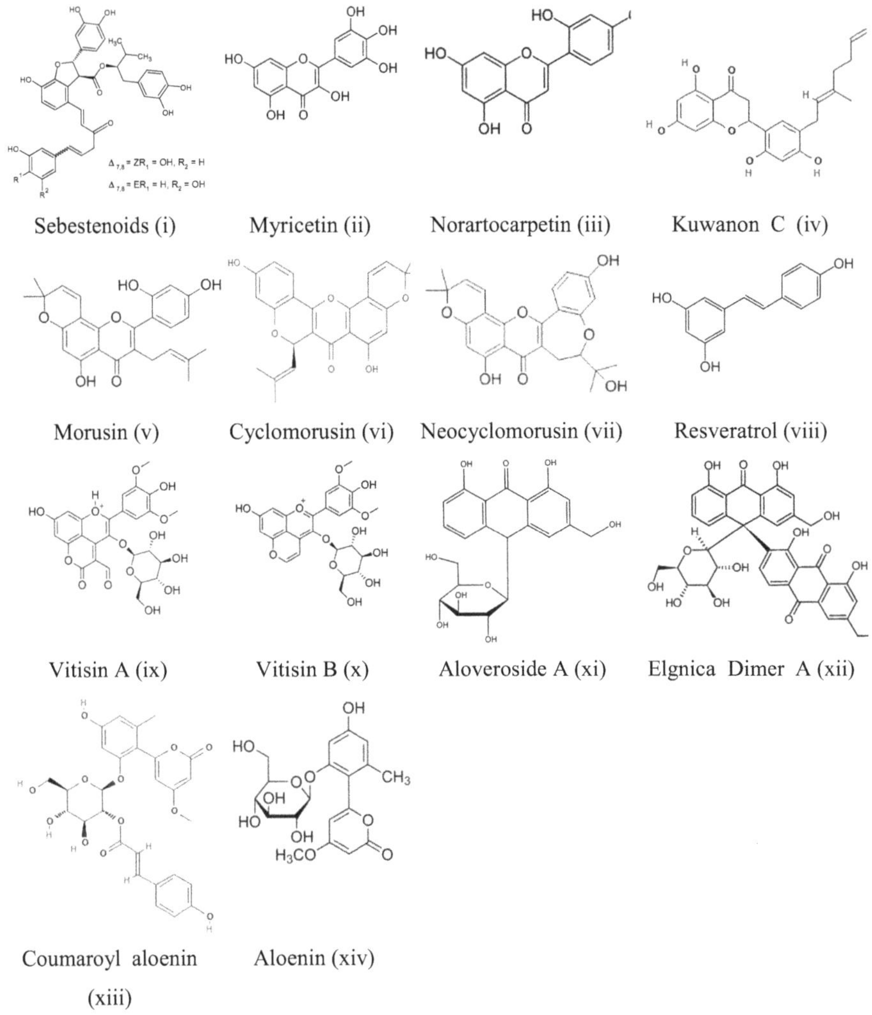

Fig. 8.2 Structure of various potential phytoconstituents to encounter AD

oligomers (Fig. 8.2 (viii)) isolated from the seed extract of *Paeonia lactiflora* was experimented successfully as a biomarker to encounter AD (Hadden et al. 2009). Vitisin A and B (Fig. 8.2 (ix–x)) are reported to decline Aβ-induced oxidative stress in PC12 cell lines and considered as promising candidates for new BACE-1 inhibitor (Ko et al. 2005).

8.8.4 Glycoside Derivatives

Nine glycosides were isolated with bioassay-guided fractionation techniques, and structural characterization ascertain them as aloveroside A, elgnica dimer A, elgnica dimer B, p-coumaroyl aloenin, and aloenin (Fig. 8.2 (xi-xiv)). BACE-1 inhibitory activity was assessed using a fluorescence resonance energy transfer (FRET) peptide cleavage assay. Among them coumaroyl aloenin showed the most potent inhibitory activity with IC50 (68.3 μg/mL) (Speranza et al. 1986; Hyun et al. 1997; Jung et al. 2010).

8.9 Conclusion

Despite recent advancement in targeted therapy, most of AD drugs fail in clinical trials due to bioavailability issues and inability to cross the BBB (Ansari and Fariba 2013). In this context BACE-1 inhibitors may be a drug of choice for the researchers. BACE-1 inhibitors prevent amyloid fibril formation and thereby may prevent malfunction of synapses. In conclusion, plant-based BACE-1 inhibitors can be considered as potent therapeutic alternative due to their less side effects and better affordability.

Acknowledgments The authors express gratitude to Assam Down Town University and NEF College of Pharmacy, Assam, for support.

References

Abdul BAA, Vidhyavathi R, Magesh J, Vijayakumar M, Muneeb MM, Faiz MMTM (2018) Synthesis and Development of BACE 1 Inhibitor for Alzheimer's Diseases from Medicinal Plants. Annual Research & Review in Biology 28(3):1–18

Anderson RN (2002) Deaths: leading causes for 2000. Natl Vital Stat Rep 50(16):1–85

Ansari N, Fariba K (2013) Molecular mechanism aspect of ER stress in Alzheimer's disease: current approaches and future strategies. Curr Drug Targets 14(1):114–122

Armstrong RA (2009) The molecular biology of senile plaques and neurofibrillary tangles in Alzheimer's disease. Folia Neuropathol 47:289–299

Atanu B (2018) Phytoextracts and their derivatives affecting neurotransmission relevant to Alzheimer's disease: an ethno-medicinal perspective. In: Akula R, Roshchina VV (eds) Neurotransmitters in plants: new achievements and perspectives. Taylor & Francis, New York, NY, pp 357–386

Azimi S, Zonouzi A, Firuzi O et al (2017) Discovery of imidazopyridines containing isoindoline-1,3-dione framework as a new class of BACE1 inhibitors: design, synthesis and SAR analysis. Eur J Med Chem 138:729–737. https://doi.org/10.1016/j.ejmech.2017.06.040

Bachurin SO, Bovina EV, Ustyugov AA (2017) Drugs in clinical trials for Alzheimer's disease: the major trends. Med Res Rev 37:1186–1225. https://doi.org/10.1002/med.21434

Balunas MJ, Kinghorn AD (2005) Drug discovery from medicinal plants. Life Sci 78:431–441

Barao S, Moechars D, Lichtenthaler SF et al (2016) BACE1 physiological functions may limit its use as therapeutic target for Alzheimer's disease. Trends Neurosci 39(3):158–169. https://doi.org/10.1016/j.tins.2016.01.003

Ben Halima S, Mishra S, Raja KMP et al (2016) Specific inhibition of β-secretase processing of the Alzheimer disease amyloid precursor protein. Cell Rep 14:2127–2141. https://doi.org/10.1016/j.celrep.2016.01.076
Berrino F (2002) Western diet and Alzheimer's disease. Epidemiol Prev 26(3):107–115
Brinton RD, Yamazaki RS (1998) Advances and challenges in the prevention and treatment of Alzeimer's disease. Pharm Res 15:386–398
Carlsson CM (2008) Lessons learned from failed and discontinued clinical trials for the treatment of Alzheimer's disease: future directions. J Alzheimers Dis 15:327–338
Choi SH, Hur JM, Yang EJ et al (2008a) Beta-secretase (BACE 1) inhibitors from Perilla frutescens var. acuta. Arch Pharm Res 31:183–187
Choi YH, Yon GH, Hong KS et al (2008b) In vitro BACE 1 inhibitory phenolic components from the seeds of Psoralea corylifolia. Planta Med 74:1405–1408
Choi CW, Choi YH, Cha MR et al (2011) In vitro BACE 1 inhibitory activity of resveratrol oligomers from the seed extract of Paeonia lactiflora. Planta Med 77:374–376
Coan KE, Maltby DA, Burlingame AL et al (2009) Promiscuous aggregate-based inhibitors promote enzyme unfolding. J Med Chem 52:2067–2075
Commenges D, Scotet V, Renaud S et al (2000) Intake of flavonoids and risk of dementia. Eur J Epidemiol 16:357–363
Crouch PJ, Harding SM, White AR et al (2008) Mechanisms of a beta mediated neurodegeneration in Alzheimer's disease. Int J Biochem Cell Biol 40:181–198
Dai J, Sorribas A, Yoshida WY, Williams PG (2010) Sebestenoids A-D, BACE 1 inhibitors from Cordia sebestena. Phytochemistry 71:2168–2173
Dewachter I, Van LF (2002) Secretases as targets for the treatment of Alzheimer's disease: the prospects. Lancet Neurol 1:409–416
Eglen RM (2002) Enzyme fragment complementation: a flexible high throughput screening assay technology. Assay Drug Dev Technol 1:97–104
Eric K, Bard DS (2016) The amyloid cascade hypothesis: are we poised for success or failure? J Neurochem 139(2):237–252
Feng BY, Shoichet BK (2006) A detergent-based assay for the detection of promiscuous inhibitors. Nat Protoc 1:550–553
Fenical WH, Jensen PJ (2006) Developing a new resource for drug discovery: marine actinomycete bacteria. Nat Chem Biol 2:666–673
Gunasekera SP, Miller MW, Kwan JC et al (2010) Molassamide, a dipeptide serine protease inhibitor from the marine cyanobacterium Dichothrix utahensis. J Nat Prod 73:459–462
Guo T, Hobbs DW (2006) Development of BACE 1 inhibitors for Alzheimer's disease. Curr Med Chem 13:1811–1829
Gupta BM, Bala A (2013) Alzheimer's disease research in India: a scientometric analysis of publications output during 2002-11. Res Neurol Int J 2013:1–11. https://doi.org/10.5171/2013.204542
Gurib-Fakim A (2006) Medicinal plants: traditions of yesterday and drugs of tomorrow. Mol Aspect Med 27:1–93
Hadden MK, Hill SA, Davenport J et al (2009) Synthesis and evaluation of Hsp90 inhibitors that contain the 1,4-naphthoquinone scaffold. Bioorg Med Chem 17:634–640
Hardy J, Selkoe DJ (2002) The amyloid hypothesis of Alzheimer's disease: progress and problems on the road to therapeutics. Science 297:353–356
Huang Y, Mucke L (2012) Alzheimer mechanisms and therapeutic strategies. Cell 148 (6):1204–1222
Hwang EM, Ryu YB, Kim HY et al (2008) BACE 1 inhibitory effects of lavandulyl flavanones from Sophora flavescens. Bioorg Med Chem 16:6669–6674
Hyun SK, Sick WW, Sung LS et al (1997) Elgonica-dimers A and B, two potent alcohol metabolism inhibitory constituents of Aloe arborescens. J Nat Prod 60:1180–1182

Jeon YH, Heo YS, Kim CM et al (2005) Phosphodiesterase: overview of protein structures, potential therapeutic applications and recent progress in drug development. Cell Mol Life Sci 62:1198–1220
Jeon M, Huong N, Sami B, Kai Z (2008) Redundancy and compensation in axon guidance: genetic analysis of the Drosophila Ptp10D/Ptp4E receptor tyrosine phosphatase subfamily. Neural Dev 3(3):1–18
Jeona SY, Kwona SH, Seong YH, Bae K, Hur JM, Lee YY, Suh DY, Song KS (2007) β-secretase (BACE1)-inhibiting stilbenoids from Smilax Rhizoma. Phytomedicine 14(6):403–408
Jeong SH, Ryu YB, Curtis-Long MJ et al (2009) Tyrosinase inhibitory polyphenols from roots of Morus ihou. J Agric Food Chem 57:1195–1203
Jung HA, Jung YJ, Hyun HK et al (2010) Selective cholinesterase inhibitory activities of a new monoterpene diglycoside and other constituents from Nelumbo nucifera stamens. Biol Pharm Bull 33:267–272
Kang HS, Kim HR, Byun DS et al (2004) Tyrosinase inhibitors isolated from the edible brown alga Ecklonia stolonifera. Arch Pharm Res 27:1226–1232
Kinghorn AD (2001) Pharmacognosy in the 21st century. J Pharm Pharmacol 53:135–148
Ko PC, Higgins JA, Kilduff PT et al (2005) Evidence for intact selective attention in Alzheimer's disease patients using a location priming task. Neuropsychology 19:381–389
Lee HJ, Seong YH, Bae KH et al (2005) β-Secretase (BACE 1) inhibitors from Sanguisorbae radix. Arch Pharm Res 28:799–803
Mahesh G, Tasneem S (2014) Biomarker controversies and diagnostic difficulties in Alzheimer's disease. Am J Phytomed Clin Ther 2(4):463–468
McArdle BM, Quinn RJ (2007) Identification of protein fold topology shared between different folds inhibited by natural products. Chem Bio Chem 8:788–798
Mei Z, Situ B, Tan X et al (2010) Cryptotanshinione upregulates alpha-secretase by activation PI3K pathway in cortical neurons. Brain Res 12:165–173
Miguel AM, Emília S (2019) BACE-1 and γ-secretase as therapeutic targets for Alzheimer's disease. Pharmaceuticals (Basel) 12(41):1–31
Montenegro RC, Araujo AJ, Molina MT et al (2010) Cytotoxic activity of naphthoquinones with special emphasis on juglone and its 5-O-methyl derivative. Chem Biol Interact 184:439–448
Muto N, Inouye K, Inaba A et al (1987) Inhibition of cytochrome P-450-linked monooxygenase systems by naphthoquinones. Biochem Biophys Res Commun 146:487–494
Nam Y, Lee D (2013) Ameliorating effect of Zhizi (Fructus Gardeniae) extract and its glycosides on scopolamine-induced memory impairment. J Tradit Chin Med 15:223–227
Ono K, Yoshiike Y, Takashima A et al (2003) Potent anti-amyloidogenic and fibril-destabilizing effects of polyphenols in vitro: implications for the prevention and therapeutics of Alzheimer's disease. J Neurochem 7:172–181
Parihar MS, Hemnani T (2004) Alzheimer's disease pathogenesis and therapeutic interventions. J Clin Neurosci 11(5):456–467
Perry E, Walker M, Grace J, Perry R (1999) Acetylcholine in mind: a neurotransmitter correlate of consciousness? Trends Neurosci 22(6):273–280
Philomena G (2011) Concerns regarding the safety and toxicity of medicinal plants—an overview. J Appl Pharm Sci 1(6):40–44
Roberds SL, Anderson J, Basi G et al (2001) BACE knockout mice are healthy despite lacking the primary beta-secretase activity in brain: implications for Alzheimer's disease therapeutics. Hum Mol Genet 10:1317–1324
Ross JA, Kasum CM (2002) Dietary flavonoids: bioavailability, metabolic effects, and safety. Annu Rev Nutr 22:19–34
Salloway S, Mintzer J, Weiner MF et al (2008) Disease-modifying therapies in Alzheimer's disease. Alzheimers Dement 4:65–79
Scarpini E, Schelterns P, Feldman H (2003) Treatment of Alzheimer's disease: current status and new perspectives. Lancet Neurol 2:539–547

Selkoe DJ (2004) Alzheimer's disease: mechanistic understanding predicts novel therapies. Ann Intern Med 140:627–638

Shankar GM, Brenda LB, Matthew T, Dominic MW, Dennis JS, Bernardo LS (2007) Natural oligomers of the Alzheimer amyloid-beta protein induce reversible synapse loss by modulating an NMDA-type glutamate receptor-dependent signaling pathway. J Neurosci 27(11):2866–2875

Shimmyo Y, Kihara T, Akaike A et al (2008) Multifunction of myricetin on Aβ: neuroprotection via a conformational change of Aβ and reduction of Aβ via the interference of secretases. J Neurosci Res 86:368–377

Skovronsky DM, Lee VMY, Trojanowsky JQ (2006) Neurodegenerative diseases: new concepts of pathogenesis and their therapeutic implications. Annu Rev Pathol 1:151–170

Speranza G, Data G, Lunazzi L et al (1986) A new Diglucosylated 6-Phenyl-2-pyrone from Kenya Aloe. J Nat Prod 45:800–805

Tanzi RE, Bertram L (2005) Twenty years of the Alzheimer' disease amyloid hypothesis: a genetic perspective. Cell 120:545–555

Tariot PN, Farlow MR, Grossberg GT et al (2004) Memantine treatment in patients with moderate to severe Alzheimer disease already receiving donepezil: a randomized controlled trial. JAMA 291:317–324

Tauraite D, Razumas V, Butkus E (2009) Lipophilic 1,4-naphthoquinone derivatives: synthesis and redox properties in solution and entrapped in the aqueous cubic liquid-crystalline phase of monoolein. Chem Phys Lipids 159:45–50

Vaidya AB (1997) The status and scope of Indian medicinal plants acting on the central nervous system. Indian J Pharmacol 29:S340:228–34

Vassar R (2004) BACE1: the beta-secretase enzyme in Alzheimer's disease. J Mol Neurosci 23 (1–2):105–114

Vladimir V, Sophia R (2018) Results of beta secretase-inhibitor clinical trials support amyloid precursor protein-independent generation of beta amyloid in sporadic Alzheimer's disease. Med Sci (Basel) 6(2):45–58

Youdim KA, Qaiser MZ, Begley DJ et al (2004) Flavonoid permeability across an in situ model of the blood-brain barrier. Free Radical Bio Med 36:592–604

Yu PN, Terry CTO, Nancy YIP (2015) Plant alkaloids as drug leads for Alzheimer's disease. Neurochem Int 32(4):1e–11e

Zhang F, Jiang L (2015) Neuroinflammation in Alzheimer's disease. Neuropsychiatr Dis Treat 11:243–256

Bioactive Compounds of Mangroves as Potent Drug and in Nanoparticle Synthesis: Play a Pivotal Role in Combating Human Pathogens

9

Sunayana Saha, Soumaryya Bhattacharyya, Mausumi Sikdar (nee Bhakta), and Rabindranath Bhattachryya

Abstract

Plants like mangroves house a large volume of bioactive compounds, as they thrive in challenging environments. Various interesting literatures are there stating successful exploration of these bioactive compounds as antimicrobials or as traditional medicines against various common diseases used by the locals. In this report, the use of mangrove plants as traditional medicines is discussed in a systematic and detailed way. A glimpse of the validation of antimicrobial potency of two common mangrove plant extracts of *Avicennia* spp. against the most predominant uropathogen of eastern India, *Escherichia coli*, is discussed in this report. Mangrove plant extract reduced silver nanoparticles can be a promising option for alternative drug and are discussed. Their efficacy as antimicrobials and their characterization were also investigated. Extracts of leaves of two selected mangrove plants, *Avicennia alba* and *Avicennia marina*, were assayed for antimicrobial efficacy. The silver nanoparticles (AgNP) synthesized using these two mangrove leaf extracts were characterized by UV-Vis spectrophotometer, Fourier transform infrared spectroscopy, zeta potential technology, transmission electron microscopy, and scanning electron microscopy. Application of AgNP as antimicrobial agents was also explored. Synergistic effect with common broad-spectrum antibiotic and cytotoxicity of the green synthesized silver nanoparticles were also evaluated in this report. The antibacterial activity of both the mangrove-synthesized AgNP and their synergistic effect with some common antibiotics exhibited significant and varying amount of antibacterial potentialities against the

S. Saha · M. Sikdar (nee Bhakta) · R. Bhattachryya (✉)
Department of Life Sciences, Presidency University, Kolkata, West Bengal, India
e-mail: sunayanaadak@gmail.com; mausumi.dbs@presiuniv.ac.in; rabindranathbpc@gmail.com

S. Bhattacharyya
Department of Botany, Ramkrishna Mission Vivekananda Centenary College, Rahara, Kolkata, India

S. C. Mandal et al. (eds.), *Evidence Based Validation of Traditional Medicines*,
https://doi.org/10.1007/978-981-15-8127-4_9

test pathogen. This study can be beneficial in the formulation of alternate drug and drug modifications for combating various human pathogens.

Keywords

Bioactivity · Antimicrobial activity · Green synthesized silver nanoparticles · Nanoparticle characterization · Cytotoxicity · Synergistic effect

9.1 Introduction

9.1.1 Urinary Tract Infection (UTI)

The ever-rising development of microbial resistance to a large number of common antibiotics is the major cause of global health hazards. Currently, infectious pathogens are mostly resistant to numerous antibiotics, and this challenges the capacity of the antibiotics to arrest infections efficiently (Tseng et al. 2008; Haider et al. 2010; Sood and Gupta 2012). UTI, either nosocomial infections or community-acquired infections, is one of the most common infections encountered. About 150 million patients per annum are evaluated to be suffering from UTI, worldwide. UTI incidences are reported to rise up to 75% in the female population by the age of 24, and 15–25% of this group are described to suffer from a recurrence of this disease (Foxman 2002; Hunstad and Justice 2010; Pouwels et al. 2012; Mukherjee et al. 2013). ARESC (Antimicrobial Resistance Epidemiological Survey on Cystitis), a global survey on antimicrobial resistance of uropathogens, validated that *E. coli* exhibited high resistance to sulfonamide, SXT (sulfamethoxazole—double-strength trimethoprim), ciprofloxacin, and fluoroquinolone in nine European countries and in Brazil (Schito et al. 2009). SXT resistance developed among uropathogens appears wide-ranging in the United States, and it appears to be probable that SXT, in due course, will need to be substituted by alternative therapies (Gupta et al. 2001). Regardless of the circumstance that UTI is the third most common infection found in our country, only a few fragmented studies on UTI in India have been reported (Akram et al. 2007; Kothari and Sagar 2008; Hussain et al. 2012; Prakash and Saxena 2013). In eastern India, UTI is a common infection reported among all ages from newborns to aged persons. However, concrete studies on UTI and the antimicrobial resistance pattern in eastern India are still continuing. There exist extensive discussions on selection of antibiotics in empirical treatment attributable to the deficiency of elaborate and authenticate guidelines. Information and awareness of the etiology and antibiotic susceptibility pattern of the causative agents of UTI is utterly crucial. In the international plus national background, the use of antibiotics in the empirical treatment of UTI pathogens has been considered as the most probable origin for the appearance and emergence of antibiotic resistance among several classes of bacterial pathogens. Even though clinically empirical, antibiotic treatment against uropathogens is conventional, and bacteria are evolving resistance to antibiotics quicker than the advances in new groups of antibiotics.

Frequently, clinical practitioners advise broad-spectrum antibiotics in its place of a precise antibiotic in the course of empirical treatment, to avoid the difficulties in treating resistance of the pathogens against the particular choice of drug. Antibiotic exploitation and patients who are noncompliant or who fail to finish the course of antibiotic treatment also effect to a surge of antibiotic-resistant bacteria (Saha et al. 2014).

Incidences of UTI occurred typically among female population aged between 2 and 13 years, and about seven million acute uncomplicated infections are reported to occur per annum among young women exclusively in the United States (Schappert 1992).

Catheter-associated UTI comprise 40% incidences among all hospital-acquired infections. This is the most common category of opportunistic nosocomial infection. UTI infections are a main pool for grouping and spreading of multidrug-resistant pathogenic strains (Kunin and Kunin 1997; Gupta and Stamm 1999).

Studies from various parts of India and several other countries have described that the most predominant uropathogen is *E. coli*, followed by *Klebsiella* spp. (denHeijer et al. 2012; Mukherjee et al. 2013; Prakash and Saxena 2013; Renuart et al. 2013). ESbL (extended spectrum β-lactamase)-producing organisms triggering a major surge in infection epidemics have been indicated from every continent globally. ESbL-producing pathogens frequently display resistance factors to numerous antibiotic groups like aminoglycoside and fluoroquinolones, which limit the range of effective antibiotics (Akram et al. 2007; Mukherjee et al. 2013). The frequency of ESbL-producing strains among clinical *Klebsiella* isolates has been gradually rising over current years and estimates for 6–17% of all nosocomial UTI incidences (Subha and Ananthan 2002; Hussain et al. 2012).

9.1.2 Mangroves and Weeds

Mangroves are the ecosystems situated between land and sea enclosing almost 18.1 million hectares of the globe (de Souza Sebastianes et al. 2013). Mangrove vegetation comprises almost 70 species from 20 different angiosperm families (Duke et al. 1998). Mangroves are considered among the most productive ecosystems on the earth as they serve as custodians of their juvenile stock and form most valuable biomass (Singh et al. 2014). Mangroves are an ecological group of halophytic plant species that grow in saline coastal sediment habitats in the tropics and subtropics and act as the salt-tolerant forest ecosystem supporting the ecology, economy, and ecosystem of the area (Singh et al. 2014). Mangroves are reported to dominate three-quarters of tropical coastlines.

India has a long coastline of approximately 7516.6 km which also includes the island territories which has the fourth largest mangrove area in the world covering an area of about 6749 km^2 (Naskar and Mandal 1999). These mangrove habitats (69–89.5°E longitude and 7–23°N latitude) is made up of three distinct zones: east coast habitats with a coastline of about 2700 km, facing the Bay of Bengal; west coast habitats with a coastline of about 3000 km, facing the Arabian Sea; and Island Territories comprising about 1816.6 km coastline (Singh et al. 2014).

Plants are known to tolerate a number of biotic and abiotic stresses (Dey and Corina Vlot 2015; Gupta et al. 2013). Stress signals modulate the secondary metabolites in plant cells (Ramakrishna and Ravishankar 2011) by changing the metabolic signaling and regulatory networks (Krasensky and Jonak 2012). Nitric oxide (NO) modulation (Liu et al. 2015) and reactive oxygen species homeostasis (Miller et al. 2010) play crucial roles in various stresses which might contribute to the secondary metabolite composition in plants. Mangroves exposed to high salt, flood, high temperature, and irradiance are always exposed to stress which might contribute to its diverse and novel secondary metabolite composition as well as increased bioactivity. Endophytic fungi associated with mangroves similarly may show unique chemical composition with enhanced bioactivity of their crude extracts or isolated compounds. Mangroves are considered as the botanical amphibians occupying a zone of desiccating heat, choking mud, and elevated salt levels (http://ngm.nationalgeographic.com/2007/02/mangroves/warne-text). Antinociceptive, anti-inflammatory, and antipyretic activities have also been exhibited by many mangrove plants (Shilpi et al. 2012).

Indian mangroves form Goa such as *Rhizophora mucronata*, *Sonneratia alba*, and *Excoecaria agallocha* were reported for possessing antibacterial activity against human pathogens such as *Staphylococcus aureus*, *Streptococcus* sp., *Salmonella typhi*, *Proteus vulgaris*, and *Proteus mirabilis* (Sahoo et al. 2012). *Avicennia marina*, *A. officinalis*, *Bruguiera sexangula*, *Excoecaria agallocha*, *Lumnitzera racemosa*, and *Rhizophora apiculata* were evaluated against antibiotic-resistant pathogenic bacteria (Abeysinghe 2010). *Suaeda maritima*, another mangrove from Bhitarkanika, India, was investigated for in vitro antioxidant and antimicrobial activities (Patra et al. 2011). Furthermore, the mangrove plant extracts were found to possess great antioxidative ability and potent antibacterial ability against seven foodborne pathogens (Suganthy et al. 2009).

Sonneratia alba (Saad et al. 2012), *Sonneratia caseolaris* (Yompakdee et al. 2012), *Lumnitzera littorea* (Saad et al. 2012), *Heritiera fomes* (Wangensteen et al. 2009), etc. are among the other mangroves reported from various parts of the world for possessing antimicrobial potential. Mangrove plant *Avicennia marina*-synthesized abietane diterpenoids with cytotoxic and antimicrobial activities (Han et al. 2008). Phytochemical evaluation of mangroves has been done in *Bruguiera gymnorrhiza* for bruguierols A–C (Han et al. 2005), *Avicennia marina* for naphthoquinone derivatives (Han et al. 2008), *Derris indica* for flavonoids (Koysomboon et al. 2006), etc., some of which showed profound bioactivity.

Another current research area involves the antimicrobial potential of some endophytic fungi occurring in mangrove plants especially from the foliage because of the greater and well-known potential of foliar extracts against human pathogens (Bhimba et al. 2012). Natural microbial and fungal consortium of mangrove has been reported for antimicrobial potential (Liu et al. 2007). *Hypocrea lixii* VB1 from the foliar extracts of *Rhizophora mucronata*, *Avicennia officinalis*, and *Avicennia marina* were reported for anticancer activity against Hep2 and MCF7 cell line in vitro as well as for having significant biocidal properties (Bhimba et al. 2012). Mangrove endophytic fungus *Nigrospora* sp. from China produced 4-deoxybostrycin, a natural anthraquinone with potent antimycobacterial activity (Wang et al. 2013). Aniquinazolines A–D, four new quinazolinone alkaloids, and 4-phenyl-3,4-dihydroquinolone derivatives were isolated from *Aspergillu nidulans* MA-143, an endophytic fungus of the mangrove plant *Rhizophora stylosa* which was tested for antibacterial and cytotoxic activities (An et al. 2013). Flavodonfuran, a new difuranylmethane derivative, was isolated from the mangrove endophytic fungus *Flavodon flavus* PSU-MA201 which was tested for antibacterial and antifungal activities (Klaiklay et al. 2013). Mangrove-derived fungus *Xylaria cubensis* PSU-MA34 yielded xylacinic acids A (1) and B (2) and other compounds investigated for cytotoxicity against KB cells and antibacterial activity against drug-resistant bacteria (Klaiklay et al. 2012). An endophytic *Streptomyces* sp. HKI0595 isolated from the mangrove plant *Kandelia candel* produced Kandenols A–E (eudesmenes) active against bacteria *Bacillus subtilis* ATCC 6633 and *Mycobacterium vaccae* IMET 10670 (Ding et al. 2012). The antibacterial compound 3-hydroxypropionic acid was isolated from mangrove endophytic fungi *Diaporthe phaseolorum* (de Souza Sebastianes et al. 2013). Another mangrove plant *Scyphiphora hydrophyllacea*-derived endophytic fungus A1 produced monoterpenes and a new fatty acid glucoside known to be active against *Staphylococcus aureus* and methicillin-resistant *S. aureus* (Zeng et al. 2012; Mei et al. 2012). The marine semi-mangrove plant *Pongamia pinnata*-derived *Nigrospora* sp. MA75 produced diverse secondary metabolites in culture media showing activity against cancer cell lines (Shang et al. 2012). Antimycobacterial activity of fusaric acid from a mangrove fungal endophyte *Fusarium* sp. was reported against *M. tuberculosis* H37Rv strain with an MIC of 10 μg/mL (Pan et al. 2011). A mangrove-derived endophytic *Streptomyces albidoflavus* produced antimycin A18 (Yan et al. 2010). *Eremophilane sesquiterpenes* from the mangrove endophytic fungus *Xylaria* sp. BL321 was reported for α-glucosidase inhibitory activity (Song et al. 2012). Nigerapyrones A–H, α-pyrone derivatives, were also isolated from marine mangrove-derived endophytic fungus *Aspergillus niger* MA-132 (Liu et al. 2011). Meroterpenoid and diphenyl ether derivatives were isolated from *Penicillium* sp. MA-37, a fungus from marine mangrove rhizospheric soil, and were evaluated for brine shrimp lethality and antibacterial activity (Zhang et al. 2012). Mangrove rhizosphere soil-derived fungus *Aspergillus effuses* H1-1 produced diketopiperazine alkaloids with potent activity against P388 cells and also exhibited selectivity against topoisomerase I (Gao et al. 2012).

Leaf extracts of *Biophytum sensitivum* in different solvents were studied for their antibacterial activity against several human pathogenic bacteria (Battu and Kumar 2012), and their activity is quite comparable with that of standard antibiotics.

The antimicrobial activities of medicinal plants are imparted by the active antioxidant components of those plants. Some reports are being published on the determination of the antioxidant activity of medicinal plants. *Aerva lanata* leaves showed high antioxidant content through various antioxidant evaluation studies like superoxide radical scavenging activity, hydroxyl radical scavenging activity, DPPH radical activity, etc. (Tsanova-Savova et al. 2018). Antioxidant content and antimicrobial properties are shown by *Asparagus racemosus*, *Ocimum sanctum*, *Cassia fistula*, *Piper betel*, *Citrus aurantifolia*, *Catharanthus roseus*, and *Polyalthia longifolia* (Kumar and Arunachalam 2009).

9.1.3 Nanotechnology and Nanoparticles

Currently, nanotechnology is projected to be the basis of several vital and innovative technologies, and exploration of this field is increasing rapidly universally. Matter at the nanometer scale (typically less than 100 nm) possesses novel and unique properties than their macro counterparts (Bar et al. 2009). Metal nanoparticles have been of immense interest due to their unique characteristics such as biotic, optical, and magnetic behaviors (Rassaei et al. 2008). This characteristic gives rise to an interdisciplinary field, nanobiotechnology, which is growing rapidly recently.

9.1.3.1 Overview

A nanoparticle is defined as particles with a dimension of less than 100 nm, and even the nanoparticles might have zero level sizes in materials like quantum dots (QD). Nanoparticles are classified based on its size and shape. Coarse particle sizes are in the range of 2500 nm and more. Fine nanoparticles range between 100 and 2500 nm, and the fine-sized nanoparticles are between 1 and 100 nm. Nanoparticles show a size-dependent nature that differs significantly from smaller fine particles to bulk particles (Buzea et al. 2007).

Nanoparticles show unique properties compared with bulk materials, and this kind of innovative characteristic features results from variation in specific features reminiscent of size, distribution, and shape. Nanoparticles have a higher surface-area-to-volume ratio, - this special features like specific surface area is pertinent to catalytic activity and other properties such as antibacterial and antifungal activity like antimicrobial activity exhibited by silver nanoparticles (AgNPs). The applications of nanoparticles are also important in the area of functional nanostructures, combined materials, biological organization, and environment.

Nanoparticles can be synthesized using both chemical agents and biological agents. But use of biological agents for preparing nanoparticles is advantageous as biological methods are free of pollution. Biological agents used as sole reducing agent for preparing nanoparticles are much cost-effective than regular chemical means of synthesis. Nanoparticles synthesized by biological means are so aptly known as green synthesized nanoparticles.

9.1.3.2 Characterization

Characterization of the nanoparticle is an important process in understanding the synthesis and applications in specific fields. Characterization is carried out by using different techniques, mainly obtained from materials science. Reports exist which defined ultraviolet- visible (UV-Vis) spectroscopy, atomic force microscopy (AFM), dynamic light scattering (DLS), electron microscopy methods such as scanning electron microscopy (SEM) and transmission electron microscopy (TEM), X-ray photoelectron spectroscopy, Fourier transform infrared spectroscopy (FTIR), X-ray diffraction (XRD) and nuclear magnetic resonance (NMR) spectroscopy, and matrix-assisted laser desorption/ionization time-of-flight mass spectrometry as the most frequently performed tests for characterization (Taylor et al. 2013).

9.1.3.3 Cytotoxicity

Most of the nanoparticles present potential dangers, both medically and environmentally (Mnyusiwalla et al. 2003), because nanoparticles have a high surface-to-volume ratio, which can make them extremely reactive or catalytic. They also possess the ability to pass through the cell membranes of microorganisms, and their interactions with biological systems are unidentified. A current study on the effects of ZnO nanoparticles on the human immune cells was observed to have varying levels of susceptibility to cytotoxicity (Hanley et al. 2009). In the present scenario, the pharmaceutical companies, looking for regulatory approval for nano-formulations of existing medicines, are depending on safety and clinical data produced during clinical studies of earlier, pre-reformulation description of the medicine. The result of the nano-reformulation of the drug may lead to change the norms in regulatory bodies, such as FDA, and they missing the analysis of specific side effects (Vines and Faunce 2009).

9.1.3.4 Application of Nanoparticles

Nanoparticles have found applications in various field ranging from electronics and communications, to optics, chemistry, energy, and, of course, biology. But, its application in medical science is increasing every day significantly from drug delivery, to drug development, imaging (McAteer et al. 2008), diagnosis (Agarwal et al. 2007), etc.

9.2 Materials and Methods

9.2.1 Plant Material and Extract Preparation

Fresh leaves of *A. alba and A. marina* were collected from Sundarban parts of West Bengal, India. The leaves were cleaned and washed in tap water and dried in shade for 3 weeks. Dried leaves were weighed and crushed in motorized grinder and stored in an airtight container and maintained at 4 °C for further use. Dried plant dust were weighed and mixed individually with excess ethanol. The mixture was kept in an incubator shaker at 35 °C in continuous shaking for 24 h. After 24 h the mixture was filtered through Whatman filter paper, and the filtrate was allowed to evaporate in a rotary evaporator under reduced pressure to concentrate the plant extract (Trifunschi et al. 2015) to a final concentration of 100 mg plant dust/mL of ethanol. The extract were stored at 4 °C to prevent any degradation.

9.2.2 Synthesis of Silver Nanoparticle

In different volumes of 1 mM aqueous AgNO3 solution, 5 mL of *A. alba* plant leaf extract and 3 mL of *A. marina* plant leaf extract were individually added to make a final volume of 10 mL. The resulting solution was heated at 70 °C for 30 min for complete formation of nanoparticles (Emeka et al. 2014). The formations of nanoparticles were confirmed by the characteristic peak of Ag in a UV-Vis spectroscopy.

9.2.3 Characterization

The reduction of Ag^+ to Ag^0 and thus formation of green synthesized silver nanoparticles were monitored and confirmed by scanning from 300 to 700 nm in UV-Vis spectrophotometer. The functional groups responsible for the bioreduction were detected by FTIR analysis. The FTIR analysis was carried out between 4000 and 400 cm^{-1} to identify the probable biomolecules of *A. alba* and *A. marina* mangrove extracts responsible for capping and stabilizing the synthesized silver nanoparticles. XRD analysis was done to study the 3D structure of the synthesized nanoparticles. The quantitative measures of the samples, like the size of the synthesized nanoparticles, were investigated by transmission electron microscopy (TEM). Scanning electron microscopy (SEM) analysis was also done to obtain surface information and shape of the green synthesized nanoparticles.

9.2.4 Test Organisms

The most common drug-resistant uropathogens of West Bengal, *Escherichia coli*, was selected as the test organisms. The clinical isolates from urinary tract-infected patients were collected from a tertiary hospital, Kolkata. Working cultures were prepared by adding fresh sterile nutrient broth on pre-cultured stabs of the test pathogens. Proper care was taken, to conserve the generation number of the pathogenic strains and diminish loss of antibiotic resistance factors.

9.2.5 Antibacterial Potentialities

The antibacterial screening of the green synthesized silver nanoparticles (AgNPs) was carried out against the epidemiologically studied most prevalent antibiotic-resistant human pathogens, *E. coli*. The pathogen was isolated from urinary tract-infected patients admitted in a tertiary care hospital and cultured overnight in sterile nutrient broth. The following day the 24 h young cultures were washed, and the cells were resuspended in freshly prepared sterile nutrient broth. The optical density of this broth was adjusted to 0.1 at 600 nm corresponding to 10^8 cfu/mL of the test organisms (Pal et al. 2007). 1 mL of the prepared broth, i.e., broth containing 10^8cfu, and 1 mL of the individual nanoparticles are added to 8 mL of fresh nutrient broth. Absorbances were noted immediately at 0 h and incubated at 37 °C for 24 h. The following day absorbance is taken at 600 nm to check the turbidity of the cells.

9.2.6 Hemolytic Assay

Prior to the antimicrobial assay, the toxicity level in RBC cells are demonstrated in this assay by study of hemocompatibility using standard protocol. In brief, blood was collected from 6-week-old male BALB/c mice in a heparinized tube, and red blood cells (RBC) were obtained by centrifugation at 1500 × g for 5 min in 4 °C. The collected RBC pellet was diluted in 20 mM PBS-buffered saline (pH 7.4) to make a 5% (v/v) solution. The RBC suspension was added to PBS-buffered saline (−ve control) and 0.1% Triton X-100 (+ve control) and incubated with different concentrations of hydrogel containing silver nanoparticles for 1 h at 37 °C. After centrifugation at 12,000 rpm at 4 °C, the supernatants were transferred to a 96-well plate. Hemolytic activity was determined by measuring the absorption at 570 nm. Control samples of 0% lysis (in PBS buffer) and 100% lysis (in 0.1% Triton X-100) were employed in the experiment. All assays were performed in triplicate. Hemolytic effect of each treatment was expressed as percent cell lysis relative to the +ve control cells (% control) using the following formula: [(Abs570 of samples)/(Abs570 of (+) ve control cells)] × 100, where absorbance is abbreviated as Abs.

9.2.7 Antibacterial Activity of AgNPs and Antibiotics

After trail with many resistant drugs, ofloxacin, a common broad-spectrum antibiotic, was chosen for this study to investigate the efficacy of green synthesized AgNP. The test organisms showed complete resistance against this drug, ofloxacin. The AgNPs (0.48 μg/mL) and the antibiotics (5 μg/mL) were mixed. A bacterial growth curve study was plotted in which four experimental setups were designed. In the first experimental setup, *E.coli* was grown in Mueller Hinton Broth (MHB), the second setup contained *E.coli* grown in MHB along with a standard antibiotic (oflaxacin), in the third setup, *E.coli* was grown in MHB with green synthesized silver nanoparticles, and in the fourth setup, *E.coli* was grown in the presence of both the oflaxacin and nanoparticles, to investigate their synergistic effect. The absorbance was measured for 600 min at 600 nm.

9.3 Results and Discussion

9.3.1 Nanoparticle Characterization

The synthesized silver nanoparticle using mangrove leaf extract was observed to be dark brown to reddish-brown color in aqueous solution as there is excitation and changes in electronic energy levels by the reduction of Ag^+ into Ag^0 (Dare et al. 2012). The reddish color formation of the nanoparticles was observed after completion of the chemical reduction of $AgNO_3$. AgNP synthesized from both the mangrove leaf extracts showed characteristic plasmon band appearing around 420 nm (Fig. 9.1), confirming the formation of nanoparticles. The FTIR analysis of the *A. alba*-synthesized nanoparticle sample shows peaks at 3796.24, 3723.82, 3443.41, 2956.97, 2879.73, 2406.98, 2314.25, 1629.85, 1383.07, 1070.55, 562.71, and 466.13 cm^{-1} (Fig. 9.2). The band at 3796.24, 3723.82, and 3443.41, cm^{-1} is assigned to the O–H stretching of H-bonded alcohols and phenols. The band at 2956.97 and 2879.73 cm^{-1} is attributed to O–H stretching of carboxylic acids. The peak value at 1629.85 cm^{-1} corresponds to N-H bonded primary amines. The peak at 2956.97 and 2879.73 cm^{-1} corresponds to C-H stretching meant for carboxylic acids. The peak at 1383.07 cm^{-1} corresponds to C-N stretching of the aromatic amine group, and the bands observed in 1070.55 cm^{-1} corresponds to C-N stretching of carboxylic acids, ethers, and esters. The peak at 562.71 cm^{-1} corresponds to C-H stretching of the alkene group. These investigations indicate the presence of some proteins and secondary metabolites of *A. alba* such as alkaloids and terpenoids having functional groups of alcohols, ketones, aldehydes, and carboxylic acids on the surface of the silver nanoparticles. These observations signify the term green synthesized silver nanoparticles.

Like *A. alba* leaf extract, the sample synthesized by *A. marina* leaf extract showed peaks at different wave numbers which corresponds to H-bonded alcohols and

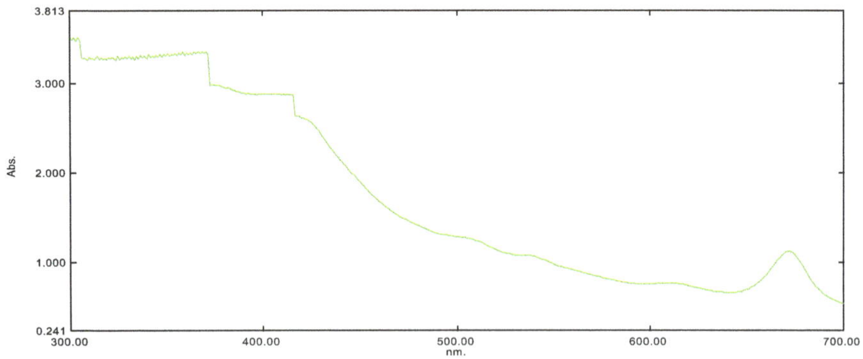

Fig. 9.1 UV-Vis spectroscopy study of green synthesized AgNP by *A. alba*. Peak around 420 nm (blue shift) confirms reduction of $AgNO_3$ to AgNP with particle size lesser than the reactants. The ethanolic extracts of *A. alba* have reacted with 1 mM $AgNO_3$ to produce silver nanoparticles. Standard used is chemically reduced 1 mM $AgNO_3$ by sodium borohydride ($NaBH_4$)

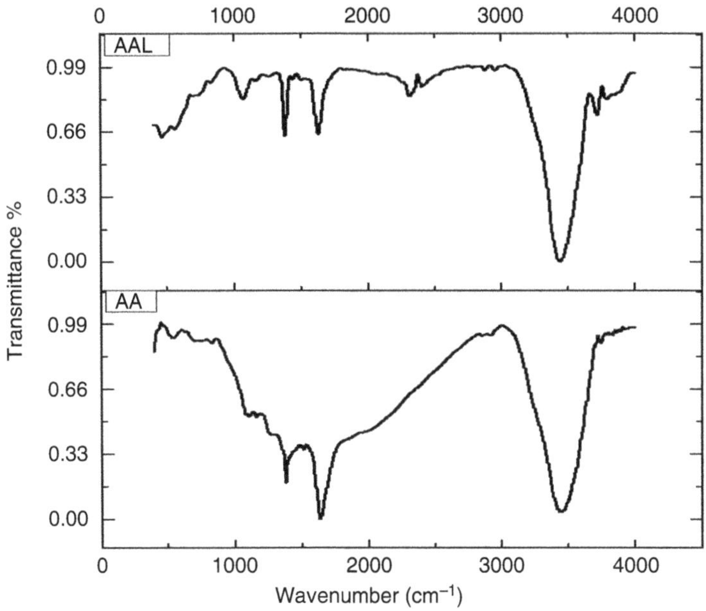

Fig. 9.2 The FTIR of AgNP synthesized from *A. alba*. Pure plant extract of *A. alba* leaves in ethanol is used as standard. The upper graph (AAL) is the graph of *A. alba*-synthesized AgNP. The graph when compared to lower graph (AA) of ethanolic extract of *A. alba* used as standard shows that there is change in band size signifying involvement of secondary metabolites of *A. alba* extract in bioreduction, capping, and stabilization of AgNP. Technical support: CECRI, Tamil Nadu

phenols, N–H bonding of primary amines, and C-N stretching of the aromatic amine group again indicating the presence of some proteins, terpenoids, and alkaloids present on the surface of the silver nanoparticles. The investigation of FTIR studies confirmed that the carbonyl group from the amino acid residues and proteins has the strong ability to bend metallic part of the AgNP, demonstrating that the proteins could possibly play an important role in capping of silver nanoparticles. The studies further indicated the role of plant secondary metabolites in avoidance of agglomeration and thereby become constant in the medium. In a nutshell, the FTIR studies also indicate that the biomolecules could possibly achieve functions of creation, stabilization, and antibacterial activities of silver nanoparticles.

The zeta potential value of nanoparticles synthesized using *A. alba* plant extract was computed to be -27.23 ± 0.32.

The profile, size, and morphology of the synthesized silver nanoparticles were revealed with the help of transmission electron microscopy. The TEM images established the green synthesis of silver nanoparticles. Nanoparticles synthesized from different mangroves leaf produced different TEM images. The nanoparticles synthesized using *A. alba* (Fig. 9.3) are homogeneous and spherical which make them conform to the shape of SPR band in the UV-Vis spectrum with an average diameter of 17.4 nm. The *A. marina*-synthesized nanoparticles show heterogeneous

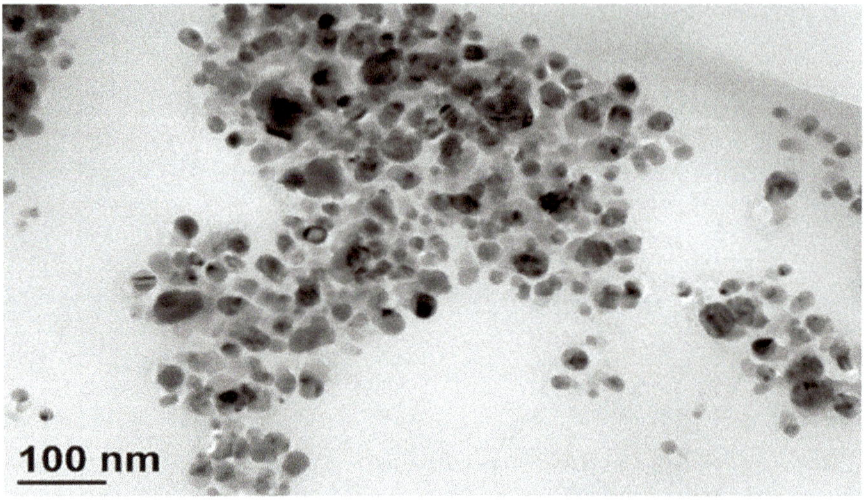

Fig. 9.3 TEM image of *A. alba*-synthesized nanoparticles. The profile, size, and morphology of the synthesized silver nanoparticles were revealed by 200 keV transmission electron microscopy. The nanoparticles synthesized by *A. alba* are homogeneous and spherical which make them conform to the shape of SPR band in the UV-Vis spectrum with an average diameter of 71.4 nm. Technical support: SINP, Kolkata

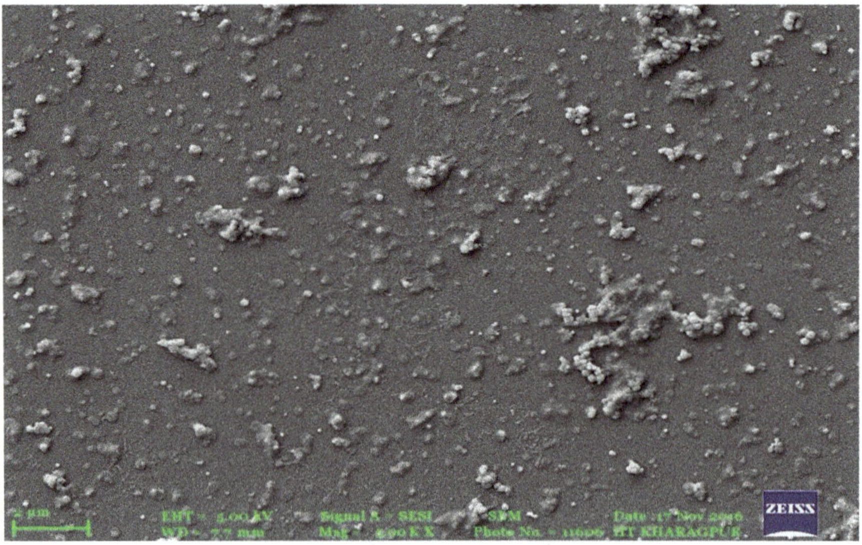

Fig. 9.4 SEM image of green synthesized AgNP. The SEM images show fairly smooth and homogenous *A. alba*-synthesized nanoparticles. Technical Support: IIT, KGP

mixture of AgNP with average particle size of 86 nm. In scanning electron microscopy, this field of AgNP synthesized by *A. alba* (Fig. 9.4) clearly reveals smooth and spherical nanoparticles.

But *A. marina*-synthesized AgNP showed spherical nanoparticles with irregular contour. The size of the nanoparticles synthesized by *A. marina* extract was seemed to be larger than that of the *A. alba*-synthesized AgNP.

9.3.2 Antibacterial Efficacy of Green Synthesized AgNP

AgNP synthesized using *A. alba* plant extracts showed more promising and significant antibacterial effectiveness than AgNP synthesized by *A. marina* plant extracts against pathogenic *E. coli.* The absorbance of the sample setup containing AgNP synthesized using *A. alba* plant extracts and AgNP synthesized using *A. marina* plant extract was found to be 0.31 and 0.67, respectively. The O.D. of the control setup was noted to be 0.94.

9.3.3 Hemolytic Assay

No significant difference in hemolysis could be detected between samples and (−) ve control after 1 h of incubation up to 1 mg/mL concentration of the nanoparticles synthesized by *A. alba*. *A. marina*-synthesized AgNP showed high cytotoxicity. The toxicity to RBC cells are regulated in a dose-dependent manner. The particle showed no cytotoxic activity at 0.48 μg/mL and below this concentration. So, the synthesized nanoparticles are completely safe against RBCs up to a concentration of 0.48 μg/mL.

9.3.4 Synergistic Activity

We have observed the complete nontoxic activity of nanoparticles synthesized by *A. alba* plant extracts at 0.48 μg/L. Therefore we have investigated a combination study of ofloxacin with different doses along with nanoparticles (0.5 μg/L).

Interestingly, it was observed that nanoparticles of very minimum dose along with minimum doses of antibiotics nullify the resistant pattern (Fig. 9.5). Between the tested nanoparticles, those synthesized using *A. alba* plant extract produced more significant results than those using *A. marina* plant extract. The killing kinetics of nanoparticle from *A. alba*, ofloxacin, and their combination were examined (Fig. 9.5) and showed an excellent synergistic effect, suggesting their combined application to control the infection by multi-antibiotic-resistant *E.coli*. Synergistic killing kinetics also performed to check the efficiency of AgNP at nontoxic dose (0.5 μg/mL) along with highly resistant antibiotics ofloxacin. From this study it is seen that the setup with no antimicrobials and in the presence of ofloxacin (4 μg/mL) reached the log phase, followed by the setup with nanoparticles only. The log phase

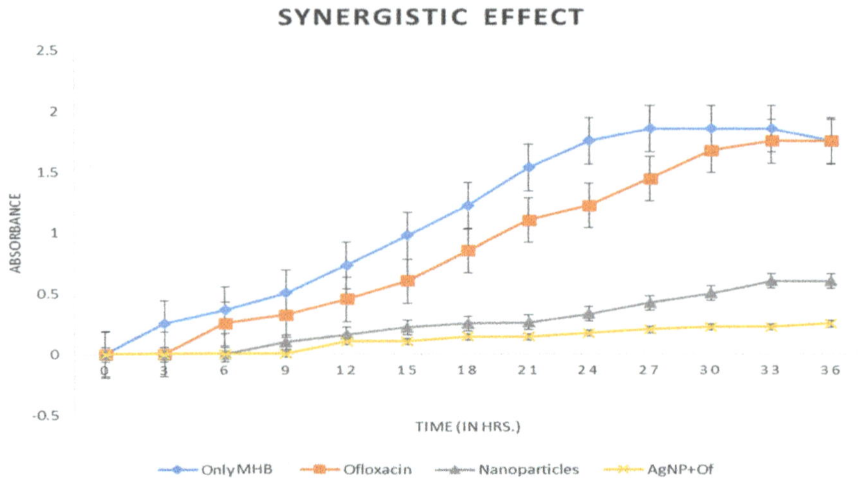

Fig. 9.5 Growth curve of *E.coli* in the presence of *A.* alba-synthesized AgNP, ofloxacin, and their combination

started after 300 min for nanoparticles only, and no growth was observed for their combination (Fig. 9.5). This study thus indicates that antibiotics and nanoparticles in synergy can inhibit resistant pathogens in a much effective way.

Three control setups were one with only MHB and *E.coli*, second with *E. coli* in the presence of ofloxacin, and third with *E. coli* in the presence of nanoparticles. The sample setup consisted of *E. coli* in the presence of ofloxacin and AgNP in synergy. The setups were incubated overnight at 37 °C, and the bacterial growth curve was measured. The absorbance was measured for 36 h at 600 nm.

The setup with no antimicrobials and in the presence of ofloxacin (4 μg/mL) reached the log phase first, followed by the setup with nanoparticles only (12 h), but no *E. coli* growth was observed for the conjugated nanoparticles indicating that the conjugated nanoparticles with antibiotics in synergy can inhibit resistant pathogens in a much more effective way.

9.4 Conclusion

From the above study, it can be concluded that as the UV-Vis spectroscopic analysis showed formation of stable nanoparticles, the reducing power of *A. alba* and *A. marina*, two common mangroves of Sundarbans, West Bengal, under the same genus, is thus established. The FTIR analysis of the AgNP suggested proteins and secondary metabolites of the mangroves which could be responsible for the bioreduction of AgNO3 and also capping and stabilizing the reduced AgNP. The release of Ag imparts the antimicrobial potency to the green synthesized silver nanoparticles (Dwivedi and Gopal 2010). On characterization by TEM and SEM, the nanoparticles showed satisfying size, morphology, and homogeneity synthesized using *A. alba* plant extract, though nanoparticles synthesized using *A. marina* failed to satisfy all criteria of being an ideal nanoparticle. The silver nanoparticles formed by the reduction of AgNP with different antibiotics showed promising results, enhancing the antibacterial activity of the antibiotics. However, further research on the plant extracts are needed to identify the active principle of the extract to pinpoint the reducing ability of the extracts and in turn can be helpful in drug designing. The information obtained from the study of synergistic effect of the AgNP and antibiotics can be employed in drug modification to augment their activity. The above mentioned study can throw light on alternative treatment and thus prove to be very helpful in clinical sciences.

Acknowledgments This work was supported by the WB-DST project [771(Sanc.)/ST/P/S&T/IG-14/2014 dt:22-12- 2015], entitled "Possible Therapeutic Application of Some Major Mangroves of Sundarban Against Some Potent Human Pathogens."

References

Abeysinghe PD (2010) Antibacterial activity of some medicinal mangroves against antibiotic resistant pathogenic bacteria. Indian J Pharm Sci 72(2):167

Agarwal A, Huang SW, O'Donnell M, Day KC, Day M, Kotov N, Ashkenazi S (2007) Targeted gold nanorod contrast agent for prostate cancer detection by photoacoustic imaging. J Appl Phys 102(6):064701

Akram M, Shahid M, Khan AU (2007) Etiology and antibiotic resistance patterns of community-acquired urinary tract infections in JNMC Hospital Aligarh, India. Ann Clin Microbiol Antimicrob 6(1):4

An CY, Li XM, Luo H, Li CS, Wang MH, Xu GM, Wang BG (2013) 4-Phenyl-3, 4-dihydroquinolone derivatives from Aspergillus nidulans MA-143, an endophytic fungus isolated from the mangrove plant Rhizophora stylosa. J Nat Prod 76(10):1896–1901

Bar H, Bhui DK, Sahoo GP, Sarkar P, De SP, Misra A (2009) Green synthesis of silver nanoparticles using latex of Jatropha curcas. Colloids Surf A Physicochem Eng Asp 339 (1–3):134–139

Battu GR, Kumar BM (2012) In-vitro antioxidant activity of leaf extract of Aerva lanata Linn. Int J Pharm Sci 2(4):74–78

Bhimba BV, Franco DAD, Mathew JM, Jose GM, Joel EL, Thangaraj M (2012) Anticancer and antimicrobial activity of mangrove derived fungi Hypocrea lixii VB1. Chin J Nat Med 10 (1):77–80

Buzea C, Pacheco II, Robbie K (2007) Nanomaterials and nanoparticles: sources and toxicity. Biointerphases 2(4):MR17–MR71

de Souza Sebastianes FL, Romao-Dumaresq AS, Lacava PT, Harakava R, Azevedo JL, de Melo IS, Pizzirani-Kleiner AA (2013) Species diversity of culturable endophytic fungi from Brazilian mangrove forests. Curr Genet 59(3):153–166

denHeijer CD, Beerepoot MA, Prins JM, Geerlings SE, Stobberingh EE (2012) Determinants of antimicrobial resistance in Escherichia coli strains isolated from faeces and urine of women with recurrent urinary tract infections. PLoS One 7(11):e49909

Dare EO, Makinde OW, Ogundele KT, et al. (2012) Zinc-salt-mediated synthesis, growth kinetic, and shaped evolution of silver nanoparticles. ISRN Dent 2012. https://doi.org/10.5402/2012/376940

Dey S, Corina Vlot A (2015) Ethylene responsive factors in the orchestration of stress responses in monocotyledonous plants. Front Plant Sci 6:640

Ding L, Maier A, Fiebig HH, Lin WH, Peschel G, Hertweck C (2012) Kandenols A–E, eudesmenes from an endophytic Streptomyces sp. of the mangrove tree Kandeliacandel. J Nat Prod 75 (12):2223–2227

Duke N, Ball M, Ellison J (1998) Factors influencing biodiversity and distributional gradients in mangroves. Glob Ecol Biogeog Lett 7(1):27–47

Dwivedi AD, Gopal K (2010) Biosynthesis of silver and gold nanoparticles using Chenopodium album leaf extract. Colloids Surf A Physicochem Eng Asp 369(1–3):27–33

Emeka EE, Ojiefoh OC, Aleruchi C, Hassan LA, Christiana OM, Rebecca M, Dare EO, Temitope AE (2014) Evaluation of antibacterial activities of silver nanoparticles green-synthesized using pineapple leaf (Ananas comosus). Micron 57:1–5

Foxman B (2002) Epidemiology of urinary tract infections: incidence, morbidity, and economic costs. Am J Med 113(1):5–13

Gao H, Liu W, Zhu T, Mo X, Mándi A, Kurtán T, Li J, Ai J, Gu Q, Li D (2012) Diketopiperazine alkaloids from a mangrove rhizosphere soil derived fungus Aspergillus effuses H1-1. Org Biomol Chem 10(47):9501–9506

Gupta K, Stamm WE (1999) Pathogenesis and management of recurrent urinary tract infections in women. World J Urol 17(6):415–420

Gupta K, Hooton TM, Roberts PL, Stamm WE (2001) Patient-initiated treatment of uncomplicated recurrent urinary tract infections in young women. Ann Intern Med 135(1):9–16

Gupta B, Sengupta A, Saha J, Gupta K (2013) Plant abiotic stress: 'Omics' approach. J Plant Biochem Physiol 1(3):e108

Haider G, Zehra N, Munir AA, Haider A (2010) Risk factors of urinary tract infection in pregnancy. J Pakistan Med Assoc 60(3):213–216

Han L, Huang X, Sattler I, Moellmann U, Fu H, Lin W, Grabley S (2005) New aromatic compounds from the marine mangrove Bruguiera gymnorrhiza. Planta Med 71(2):160–164

Han L, Huang X, Dahse HM, Moellmann U, Grabley S, Lin W, Sattler I (2008) New abietane diterpenoids from the mangrove Avicennia marina. Planta Med 74(04):432–437

Hanley C, Thurber A, Hanna C, Punnoose A, Zhang J, Wingett DG (2009) The influences of cell type and ZnO nanoparticle size on immune cell cytotoxicity and cytokine induction. Nanoscale Res Lett 4(12):1409–1420

http://ngm.nationalgeographic.com/2007/02/mangroves/warne-text

Hunstad DA, Justice SS (2010) Intracellular lifestyles and immune evasion strategies of uropathogenic Escherichia coli. Annu Rev Microbiol 64:203–221

Hussain A, Ewers C, Nandanwar N, Guenther S, Jadhav S, Wieler LH, Ahmed N (2012) Multiresistant uropathogenic Escherichia coli from a region in India where urinary tract infections are endemic: genotypic and phenotypic characteristics of sequence type 131 isolates of the CTX-M-15 extended-spectrum-β-lactamase-producing lineage. Antimicrob Agents Chemother 56(12):6358–6365

Klaiklay S, Rukachaisirikul V, Sukpondma Y, Phongpaichit S, Buatong J, Bussaban B (2012) Metabolites from the mangrove-derived fungus Xylariacubensis PSU-MA34. Arch Pharm Res 35(7):1127–1131

Klaiklay S, Rukachaisirikul V, Phongpaichit S, Buatong J, Preedanon S, Sakayaroj J (2013) Flavodonfuran: a new difuranylmethane derivative from the mangrove endophytic fungus Flavodonflavus PSU-MA201. Nat Prod Res 27(19):1722–1726

Kothari A, Sagar V (2008) Antibiotic resistance in pathogens causing community-acquired urinary tract infections in India: a multicenter study. J Infect Dev Ctries 2(05):354–358

Koysomboon S, Van Altena I, Kato S, Chantrapromma K (2006) Antimycobacterial flavonoids from Derris indica. Phytochemistry 67(10):1034–1040

Krasensky J, Jonak C (2012) Drought, salt, and temperature stress-induced metabolic rearrangements and regulatory networks. J Exp Bot 63(4):1593–1608

Kumar L, Arunachalam G (2009) Genetic variability in *Morusalba* L by biochemical and bioassay methods for increased silk productivity. J Biomed Sci and Res 1(1):11–18

Kunin CM, Kunin CM (1997) Urinary tract infections: detection, prevention, and management. Baltimore: Williams & Wilkins 419(5):226–278

Liu A, Wu X, Xu T (2007) Research advances in endophytic fungi of mangrove. Chin J Appl Ecol 18(4):912–918

Liu D, Li XM, Meng L, Li CS, Gao SS, Shang Z, Proksch P, Huang CG, Wang BG (2011) Nigerapyrones A-H, α-pyrone derivatives from the marine mangrove-derived endophytic fungus Aspergillus niger MA-132. J Nat Prod 74(8):1787–1791

Liu W, Li RJ, Han TT, Cai W, Fu ZW, Lu YT (2015) Salt stress reduces root meristem size by nitric oxide-mediated modulation of auxin accumulation and signaling in Arabidopsis. Plant Physiol 168(1):343–356

McAteer MA, Schneider JE, Ali ZA, Warrick N, Bursill CA, von zur Muhlen C, Greaves DR, Neubauer S, Channon KM, Choudhury RP (2008) Magnetic resonance imaging of endothelial adhesion molecules in mouse atherosclerosis using dual-targeted microparticles of iron oxide. Arterioscler Thromb Vasc Biol 28(1):77–83

Mei WL, Zheng B, Zhao YX, Zhong HM, Chen XLW, Zeng YB, Dong WH, Huang JL, Proksch P, Dai HF (2012) Meroterpenes from endophytic fungus A1 of mangrove plant Scyphiphora hydrophyllacea. Mar Drugs 10(9):1993–2001

Miller GAD, Suzuki N, Ciftci-Yilmaz SULTAN, Mittler RON (2010) Reactive oxygen species homeostasis and signalling during drought and salinity stresses. Plant Cell Environ 33 (4):453–467

Mnyusiwalla A, Daar AS, Singer PA (2003) 'Mind the gap': science and ethics in nanotechnology. Nanotechnology 14(3):R9–R13

Mukherjee M, Basu S, Mukherjee SK, Majumder M (2013) Multidrug-resistance and extended spectrum beta-lactamase production in uropathogenic *E. coli* which were isolated from hospitalized patients in Kolkata, India. J Clin Diagn Res 7(3):449–453

Naskar K, Mandal R (1999) Ecology and biodiversity of Indian mangroves. Daya Books, Delhi, p 1

Pal S, Tak YK, Song JM (2007) Does the antibacterial activity of silver nanoparticles depend on the shape of the nanoparticle? A study of the gram-negative bacterium Escherichia coli. Appl Environ Microbiol 73(6):1712–1720

Pan JH, Chen Y, Huang YH, Tao YW, Wang J, Li Y, Peng Y, Dong T, Lai XM, Lin YC (2011) Antimycobacterial activity of fusaric acid from a mangrove endophyte and its metal complexes. Arch Pharm Res 34(7):1177–1181

Patra JK, Dhal NK, Thatoi HN (2011) In vitro bioactivity and phytochemical screening of Suaeda maritima (Dumort): a mangrove associate from Bhitarkanika, India. Asian Pac J Trop Med 4 (9):727–734

Pouwels KB, Visser ST, Hak E (2012) Effect of pravastatin and fosinopril on recurrent urinary tract infections. J Antimicrob Chemother 68(3):708–714

Prakash D, Saxena RS (2013) Distribution and antimicrobial susceptibility pattern of bacterial pathogens causing urinary tract infection in urban community of Meerut City, India. ISRN Microbiol 2013:1–13

Ramakrishna A, Ravishankar GA (2011) Influence of abiotic stress signals on secondary metabolites in plants. Plant Signal Behav 6:1720–1731

Rassaei L, Sillanpää M, French RW, Compton RG, Marken F (2008) Arsenite determination in phosphate media at electroaggregated gold nanoparticle deposits. Electroanalysis 20 (12):1286–1292

Renuart AJ, Goldfarb DM, Mokomane M, Tawanana EO, Narasimhamurthy M, Steenhoff AP, Silverman JA (2013) Microbiology of urinary tract infections in Gaborone, Botswana. PLoS One 8(3):1–8

Saad S, Taher M, Susanti D, Qaralleh H, Awang AFIB (2012) In vitro antimicrobial activity of mangrove plant Sonneratia alba. Asian Pac J Trop Biomed 2(6):427–429

Saha S, Nayak S, Bhattacharyya I, Saha S, Mandal A, Chakraborty S, Bhattacharyya R, Chakraborty R, Franco OL, Mandal SM, Basak A (2014) Understanding the patterns of antibiotic susceptibility of bacteria causing urinary tract infection in West Bengal, India. Front Microbiol 5:463–469

Sahoo G, Mulla NSS, Ansari ZA, Mohandass C (2012) Antibacterial activity of mangrove leaf extracts against human pathogens. Indian J Pharm Sci 74(4):348–351

Schappert SM (1992) National ambulatory medical care survey: 1992 Survey. Hyattsville, MD: national center for health statistics, 1994; DHHS publication no. (PHS) 94–1250. (Advanced data for vital and health statistics. No. 253)

Schito GC, Naber KG, Botto H, Palou J, Mazzei T, Gualco L, Marchese A (2009) The ARESC study: an international survey on the antimicrobial resistance of pathogens involved in uncomplicated urinary tract infections. Int J Antimicrob Agents 34(5):407–413

Shang Z, Li XM, Li CS, Wang BG (2012) Diverse secondary metabolites produced by marine-derived fungus Nigrospora sp. MA75 on various culture media. Chem Biodivers 9 (7):1338–1348

Shilpi JA, Islam ME, Billah M, Islam KMD, Sabrin F, Uddin SJ, Nahar L, Sarker SD (2012) Antinociceptive, anti-inflammatory, and antipyretic activity of mangrove plants: a mini review. Adv Pharmacol Sci 2012(576086):1–7

Singh SK, Srivastava PK, Gupta M, Thakur JK, Mukherjee S (2014) Appraisal of land use/land cover of mangrove forest ecosystem using support vector machine. Environ Earth Sci 71 (5):2245–2255

Song Y, Wang J, Huang H, Ma L, Wang J, Gu Y, Liu L, Lin Y (2012) Four eremophilane sesquiterpenes from the mangrove endophytic fungus Xylaria sp. BL321. Mar Drugs 10 (2):340–348

Sood S, Gupta R (2012) Antibiotic resistance pattern of community acquired uropathogens at a tertiary care hospital in Jaipur, Rajasthan. Indian J Commu Med 37(1):39–44

Subha A, Ananthan S (2002) Extended spectrum beta lactamase (ESBL) mediated resistance to third generation cephalosporins among Klebsiella pneumoniae in Chennai. Indian J Med Microbiol 20(2):92–95

Suganthy N, Kesika P, Pandian SK, Devi KP (2009) Mangrove plant extracts: radical scavenging activity and the battle against food-borne pathogens. Complement Med Res 16(1):41–48

Taylor RA, Otanicar TP, Herukerrupu Y, Bremond F, Rosengarten G, Hawkes ER, Jiang X, Coulombe S (2013) Feasibility of nanofluid-based optical filters. Appl Opt 52(7):1413–1422

Trifunschi S, Munteanu MF, Agotici V, Pintea S, Gligor R (2015) Determination of flavonoid and polyphenol compounds in Viscum album and Allium Sativum extracts. Int Curr Pharm J 4 (5):382–385

Tsanova-Savova S, Denev P, Ribarova F (2018) Flavonoids in foods and their role in healthy nutrition. In Diet, Microbiome and Health 2018(7):165–198

Tseng MH, Lo WT, Lin WJ, Teng CS, Chu ML, Wang CC (2008) Changing trend in antimicrobial resistance of pediatric uropathogens in Taiwan. Pediatr Int 50(6):797–800

Vines T, Faunce T (2009) Assessing the safety and cost-effectiveness of early nanodrugs. J Law Med 16(5):822–845

Wang C, Wang J, Huang Y, Chen H, Li Y, Zhong L, Chen Y, Chen S, Wang J, Kang J, Peng Y (2013) Anti-mycobacterial activity of marine fungus-derived 4-deoxybostrycin and nigrosporin. Molecules 18(2):1728–1740

Wangensteen H, Dang HCT, Uddin SJ, Alamgir M, Malterud KE (2009) Antioxidant and antimicrobial effects of the mangrove tree Heritierafomes. Nat Prod Commun 4(3):371–376

Yan LL, Han NN, Zhang YQ, Yu LY, Chen J, Wei YZ, Li QP, Tao L, Zheng GH, Yang SE, Jiang CX (2010) Antimycin a 18 produced by an endophytic Streptomyces albidoflavus isolated from a mangrove plant. J Antibiot 63(5):259–261

Yompakdee C, Thunyaharn S, Phaechamud T (2012) Bactericidal activity of methanol extracts of crabapple mangrove tree (Sonneratia caseolaris Linn.) against multi-drug resistant pathogens. Indian J Pharm Sci 74(3):230–236

Zeng YB, Wang H, Zuo WJ, Zheng B, Yang T, Dai HF, Mei WL (2012) A fatty acid glycoside from a marine-derived fungus isolated from mangrove plant Scyphiphora hydrophyllacea. Mar Drugs 10(3):598–603

Zhang B, Zheng LP, Wang JW (2012) Nitric oxide elicitation for secondary metabolite production in cultured plant cells. Appl Microbiol Biotechnol 93(2):455–466

Medicinal Plants: A Rich Source of Bioactive Molecules Used in Drug Development

10

Laxmi Rathor

Abstract

Medicinal plants are a rich repository of various biologically active molecules that have various pharmacological effects in mammals. The majority of the world's population (around 60–80%) still rely on the traditional medicinal method to treat common illnesses. In addition, finding a plant-derived antioxidant, which scavenges reactive oxygen species (ROS), has become a central focus of drug development research. The increment in oxidative stress and impaired cellular redox homeostasis due to elevated ROS lead to age-related diseases, including type 2 diabetes, cancer, and neurodegenerative disorders. Although drugs of plant extraction origin are available to prevent these age-related disorders, living cells have various defense mechanisms to prevent the harmful effects of ROS generation. Antioxidant enzymes neutralize free radicals and help the cell to overcome stress. Phytomolecules having antioxidative properties can help in stress modulation by scavenging ROS. This book chapter summarizes how bioactive molecules can be utilized as a tool for screening plant extract/bioactive molecules of natural origin and advance our understanding of molecular mechanisms of drug action and diseases.

Keywords

Bioactive molecules · Antioxidant · Drug discovery

L. Rathor (✉)
Department of Aging and Geriatric Research, Institute on Aging, College of Medicine, University of Florida, Gainesville, FL, USA
e-mail: l.rathore.12res@gmail.com; l.rathor@ufl.edu

S. C. Mandal et al. (eds.), *Evidence Based Validation of Traditional Medicines*,
https://doi.org/10.1007/978-981-15-8127-4_10

Abbreviations

AD	Alzheimer's disease
C. elegans	*Caenorhabditis elegans*
HDL	High-density lipoprotein
LDL	Low-density lipoprotein
MAPs	Medicinal and aromatic plants
PD	Parkinson's disease
ROS	Reactive oxygen species

10.1 Introduction

The reverse pharmacological consideration of conventional herbal medicine can be a better alternative and intelligent choice to systematically develop new, safer drugs (Patwardhan and Mashelkar 2009; Yuan et al. 2016). Since ancient times, medicinal and aromatic plants (MAPs) have been used as medicine. Medicinal herbs are a rich reservoir of huge bioactive molecules that are synthesized as "sidetracks" of the primary metabolism of crops (Petrovska 2012; Yuan et al. 2016). The majority of the world's population still relies on the traditional herbal medicine system for their primary health needs (Ekor 2014; Oyebode et al. 2016). There were more than 100 herbal prescriptions at the clinical stage (Atanasov et al. 2015; Chugh et al. 2018). More understanding of the side effects of synthetic drugs and the least side effects of plant-based medicines has led to the re-emergence of the demand for herbal products, which is expected to grow continuously by 2050 (Ekor 2014; Karimi et al. 2015; Welz et al. 2018). For decades, natural plant-derived medicines have developed to treat multiple human pathologies and ailments, including AIDS (Ji et al. 2009; Yin et al. 2013). Drug discovery strategies based on medicinal plant development and traditional herbal medicines are all of a sudden on the rise from the past decade. India accounts for approximately 1.6% of the ever-growing herbal industry (Gurib-Fakim 2006; Veeresham 2012; Ekor 2014). India has been a wealthy repository of maps and medicinal herbs used for multiple medical procedures since time immemorial in the Indian medical scheme (Pandey et al. 2013; Sen and Chakraborty 2017). Approximately 25,000 efficient herbal formulations are used by Indian rural groups to treat serious ailments. More than 7800 drug-manufacturing units in Indian consume a total of 2000 tons of herbs annually for the manufacturing of herbal medicines (Pandey et al. 2013; Sen and Chakraborty 2017). A tedious and expensive method to find robust and viable candidate herbs and their compounds is large-scale screening of traditionally used herbs. Successful completion of broad screening requires a better understanding of various approaches, and key past learning with correct future strategies is necessary (Leelananda and Lindert 2016; Harvey 2008; Atanasov et al. 2015; Yuan et al. 2016). The development of drugs of natural origin is currently limited to the discovery of new active molecules, but these traditional formulations need to be systematically evaluated to develop a better alternative to synthetic drugs (Pan et al.

2013; Li and Weng 2017). Hour-needed is a useful tool which consumes less time and cost. To this end, the model organisms like *C. elegans*, *D. melanogaster*, and *mice* emerged as an important tool for biomedical and toxicological research, especially for functional characterization of novel drug targets discovered using genomics technologies and for compound screens and broad-scale target validation (National Research Council 2000; Giacomotto and Ségalat 2010; Bulla and Cheng 2013).

10.2 Aging Impact on Health

Aging is a debilitating process that is associated with physiological deterioration (Kenyon 2010). It is characterized by a gradual deterioration that results in increased vulnerability to disease and death (Kenyon 2010). The rise in age was associated with the increased incidence of degenerative diseases such as diabetes, Alzheimer's disease (AD), and Parkinson's disease (PD) which are responsible for morbidity and higher socioeconomic costs (Blagosklonny and Hall 2009; Patrícia et al. 2017). The delay of aging and the avoidance of aging-related chronic disease pathogenesis are, therefore, an important part of the strategy to promote healthy aging (Topp et al. 2004; Shilsky et al. 2017). Chronic disease associated with aging and age is highly associated with high levels of stress and metabolic cellular load. Disrupting cellular homeostasis contributes to damages to DNA, lipids, and proteins that eventually decrease an organism's health and lifespan (Pomatto and Davies 2017; Srivastava 2017). Recent developments in gerontological research have resulted in the production of plant extracts and bioactive phytomolecules as a potential candidate for the management of age-related pathologies (Pant and Pandey 2015; Pohl and KongThoo Lin 2018). With the discovery of different cellular signaling pathways controlling aging and age-related diseases in many organisms, it appears targeting nutrient-sensing pathways and energy metabolism can prolong survival in living beings and preserve the vitality of later life (Kenyon 2010; Carmona and Michan 2016; Johnson 2018).

10.3 Medicinal Plants: An Age Source that Defies Bioactive Molecules

Medicinal plants is health-promoting influence has uncovered a gold mine of natural bioactive molecules (~1.00,000) (Zhang and Reddy 2018). The world is filled with a rich abundance of strength, medicinal plants. According to the active bioactive molecule, each plant is associated with its own unique therapeutic properties (Thomford et al. 2018; Koparde et al. 2019). Natural drug substances are stated to be vital in the modern medical system and have appreciable roles. By the existence of their bioactive molecules, their therapeutic function was justified. These are extremely useful as natural drugs because they possess disease-inhibiting properties, they contain less toxic and more stable natural bioactive compounds, and they integrate chemical or biological means of alteration and extraction of natural products into potent drugs (Ji et al. 2009; Zhang and Reddy 2018; Koparde et al. 2019). Bioactive molecules formed as sidetracks of the primary metabolic process of plants are typically found in various foods including fruits, vegetables, nuts, berries, and drinks. Fruits like grapes, pears, apples, berries, and cherries contain up to 200–300 mg of polyphenols per 100 g of fresh weight (Pandey and Rizvi 2009; Pant and Pandey 2015). In different plant species, more than 8000 polyphenolic compounds have been identified. Phytomolecules are groups of phenols, terpenes, indole/glucosinolates/sulfur compounds, organosulfides, betalains, enzyme inhibitors, and other organic acids (Tsao 2010; Upadhyay and Dixit 2015).

Such bioactive molecules play a significant role in the treatment of various human diseases and drug development due to their least side effects in comparison to their chemical counterparts (Pan et al. 2013; Tungmunnithum et al. 2018). Previous research on medicinal plants centered on the beneficial effect of phytomolecules as a dietary supplement, such as a source of vitamins (Vitamin A, E) or minerals (Nasri et al. 2014). Nevertheless, with advances in modern medical science and the discovery of various side effects of chemically synthesized medicines, biologists are looking for drugs from natural sources to treat various age-related neural diseases and others (Lublin and Link 2013; Durães et al. 2018). Various phytomolecules and plant extract are found to boost age-related pathologies for this reason (Pant and Pandey 2015). The ever-increasing geriatric population is of great concern to people worldwide, as it puts an immense socioeconomic burden on the world community. Hence, this hour is required to discover and improve antiaging medication with the least side effects. Antiaging work has boomed with increasing studies and articles on antiaging plant molecules. In *C. elegans*, several plant molecules and extracts are found to cause lifespan extension (Carretero et al. 2015; Pant and Pandey 2015). The higher homology of *C. elegans* model worth noting in human and higher mammals also extends the pharmacological use of antiaging phytomolecules in human (Pant and Pandey 2015; Tissenbaum 2015). It has been found that these bioactive molecules modulate cellular signaling pathways that control age-related disorders (Rea et al. 2018). Bioactive molecules can link fruit, herbs, vegetables, and health benefits (Pant and Pandey 2015). *Ginkgo biloba*, bananas, red wine, and tea extract, for example, are rich in flavonoids such as quercetin (Pant and Pandey 2015).

10.4 Phytochemicals as a Human Reality: Potential and Challenges for the Development of Drugs

It is a gift from God to find new herbal medicines for scientific research. A modern driving force for developing new drugs, biologically active metabolites derived from such products, which were led to drug success (Lahlou 2013; Thomford et al. 2018). Supplements to dietary habits play a crucial role in the everyday life of humans and also play an important role in improving human health. Most pharmaceutical compounds contain secondary plant metabolites which are critical to drug design (Katz and Meller 2014; Panche et al. 2016). Due to active bioactive molecules, each plant is known by its own unique therapeutic properties. Natural drug products are known to be important and have essential roles in the modern medicine system. Their therapeutic function was explained by their bioactive molecules (Veeresham 2012; Koparde et al. 2019). They are tremendously valuable as natural drugs because they possess disease-inhibiting properties; consist of simple, less harmful, and more powerful bioactive compounds; and combine chemical and biological means of modifying and converting natural products into effective drugs (Dias et al. 2012; Yuan et al. 2016). However, other considerations, such as environmental changes, varied geographical distribution, labor costs, and superior crop choice, should be taken care of by green plant developers to ensure adequate supply of the source material so that ample plants can assist the pharmaceutical industry in developing high-quality bioactive medicinal drugs (Atanasov et al. 2015; Koparde et al. 2019). Herbs, vegetables, and fruits have been used in many fields including medicine, food, flavoring, drinks, dyeing, repellents, fragrances, cosmetics, and other industrial applications. Such plants have been the basis for almost all-medicinal treatment and the use of synthetic drugs since the pre-historic period (Wang 2002; Kennedy and Wightman 2011). The medicinal benefit of these plants is correlated with their phytochemical elements, which perform different physiological behaviors on the human body (Pant and Pandey 2015). Alkaloids, tannins, flavonoids, and phenolic compounds are the most common of these elements. Several herbs have been reported to exhibit antioxidant activity, and polyphenols are a major potential source of antioxidant (Tungmunnithum et al. 2018). The benefit of these plants is linked to the secondary metabolites generated by the plants, because plants developed such secondary metabolites for the benefit of the plant itself as a protection against infection and injury; however secondary metabolites have beneficial effects on the human health and the potential to cure human diseases (Takshak and Agrawal 2019). Herbs in modern medicine have provided the very important life-saving medicines. But among the approximate 4 lakh plant species, only 6% were studied for their behavior, and not more than 20% of phytochemicals were studied (Atanasov et al. 2015; Yuan et al. 2016; Koparde et al. 2019). Therefore, to achieve the dreams of herbal drug discovery, there is a need to examine the various bioactive fractions and phytoanalysis and phytopharmacological assessment of herbal products.

10.5 Phytochemicals Can Enhance the Production of Antioxidants

The importance of the antioxidant components of plant materials in preserving health and protecting against coronary heart disease and cancer also raises interest among researchers, food producers, and consumers, as the step toward functional foods with unique health effects (Abuajah et al. 2015; Wilson et al. 2017). Phytochemicals can be present in a wide variety of foods, including fruits, vegetables, cereals, nuts, and cacao/chocolate, as well as soda, tea, coffee, and wine (Upadhyay and Dixit 2015). Polyphenolic compounds commonly distributed in higher plants have possible health benefits that are thought to be derived primarily from their antioxidant activity (Tungmunnithum et al. 2018). Antioxidant protection mechanism involves the exogenous and endogenous antioxidants. The main endogenous antioxidants are superoxide dismutase, glutathione peroxidase, catalase, and glutathione (Ighodaro and Akinloye 2018). ROS may also induce enzymes such as metalloproteinase collagenase, mucopolysaccharase hyaluronidase, and serine protease (elastase), contributing to invisible skin aging. Nevertheless, numerous in vitro scientific studies have shown that phytochemicals can diminish oxidant levels and thus inhibit collagenase and hyaluronidase tyrosinase enzymes (Garg 2017; Madan and Nanda 2018). Exogenous antioxidants include vitamins, carotenoids, and polyphenols, of which the diet is the primary source. There is growing evidence that polyphenols, as antioxidants, can protect cell constituents from oxidative damage and the risk of various degenerative diseases linked to oxidative stress (Pandey and Rizvi 2009). Table 10.1 included various bioactive extract of medicinal plants used in the treatment of various diseases/disorders.

10.6 Polyphenols Uses in Human Healthspan and Lifespan

Several studies have shown that plant-derived bioactive compounds can postpone age-related declines and increase the lifespan and health of a variety of species (Argyropoulou et al. 2013; Leonov et al. 2015; Ergen et al. 2018). To prevent or slow down normal aging processes like cancer growth, one approach is to reduce stress such as shortening telomeres, non-telomeric DNA damage, and severe mitogenic signals (Shammas 2011; Bär and Blasco 2016). Several model organisms are concerned with the effect of exogenous antioxidants such as several phytomedicines on the aging process and age-related diseases (Pant and Pandey 2015).

10.6.1 Cardioprotective Effect

Polyphenols are powerful inhibitors of LDL oxidation, and this form of oxidation is seen as a key factor in the development of atherosclerosis. Antioxidant, antiplatelet, and anti-inflammatory behavior and improved HDL and endothelial function are many forms of defending polyphenols against cardiovascular diseases (Khurana et al. 2013; Cheng et al. 2017). The causal association between cardiovascular diseases (CVDs) and LDL oxidation is now well established. LDL particulate oxidation is closely associated with the risk of heart failure and myocardial infarction. Several studies have shown that many phytomolecules potentially inhibit LDL particle oxidation by chelating copper or directing free radical scavenging (Brites et al. 2017; Ahmad and Leake 2018).

Table 10.1 Bioactive extract from medicinal plants used in the treatment of a various diseases/ disorders

Natural drugs	Bioactive molecules	Therapeutics activity	References
Himalaya Boswellia	Boswellic acid 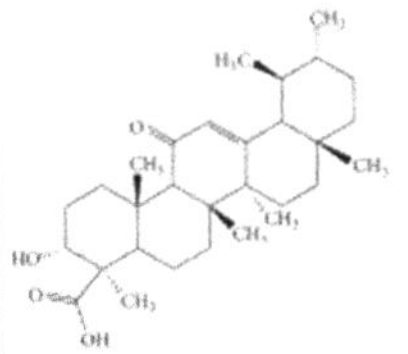	Inflammation, healthy joints, immune-modulating	Siddiqui (2011)
Zingiber officinale (ginger)	6-Gingerol 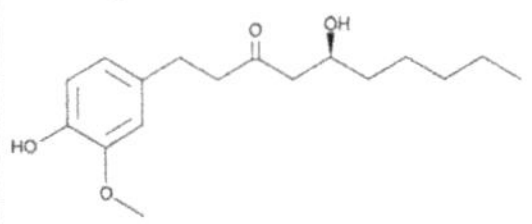	Antifungal, antibacterial, anti-inflammatory, antidiabetic, nausea	Gunathilake and Rupasinghe (2015)
Allium sativum L. (garlic)	Caffeic acid 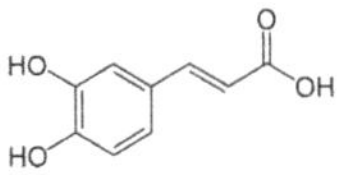 *S*-allyl cysteine	Antiaging, anticancer, hepatoprotective activity	Moutia et al. (2018)
Ocimum sanctum	Eugenol 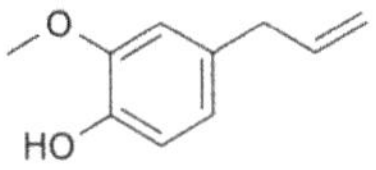	Bronchitis asthma, malaria, diarrhea, dysentery, skin diseases, arthritis, chronic fever	Garg and Sardana (2016)
Origanum vulgare (oregano)	Thymol, carvacrol 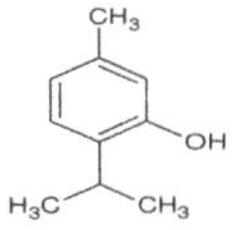 Thymol 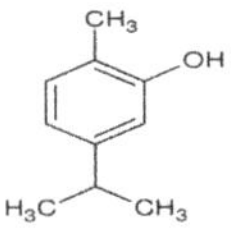 Carvacrol	Antibiotics, anti-allergies, tumors	Gutiérrez-Grijalva et al. (2018)
Bacopa monnieri	Bacosides 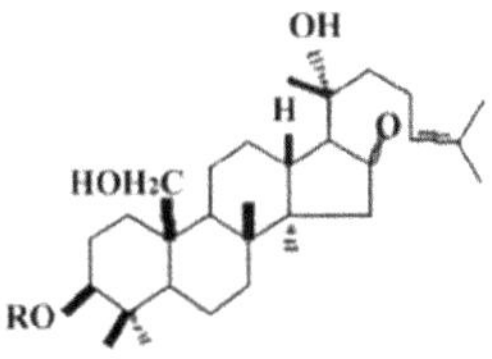	Antiaging, neuroprotection	Singh et al. (2016)
Emblica officinalis (amla)	Quercetin, 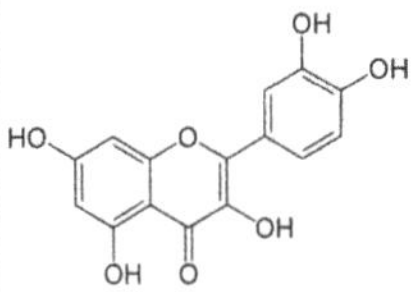 Gallic acid, 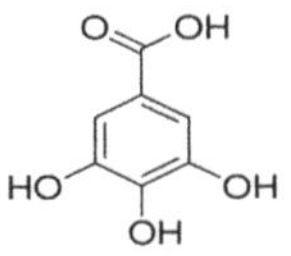 Myricetin	Cardioprotective, gastroprotective, antidiarrheal, and neuroprotective properties	Bhandari and Kamdod (2012)

(continued)

Table 10.1 (continued)

Natural drugs	Bioactive molecules	Therapeutics activity	References
Curcumin longa (turmeric)	Curcumin 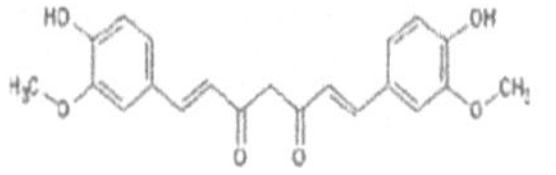	Antiaging, neuroprotection	Panda et al. (2017)
Premna integrifolia	4-hydroxy-E-globularinin 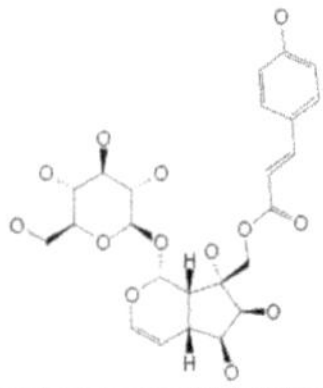	Antiaging, skin disorders, asthma, diabetes	Mali (2016); Asthana et al. (2015)
Ocimum basilicum	Beta-caryophyllene 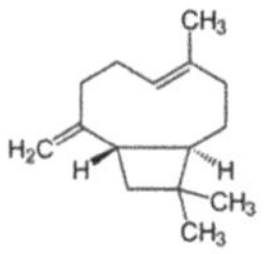	Neuroprotective activity, antiaging	Asthana et al. (2015)
Rauwolfia serpentina	Reserpine 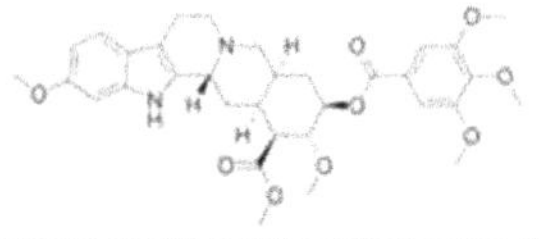	Antiaging	Kaur (2018)
Rosmarinus officinalis L. (rosemary)	Carnosol 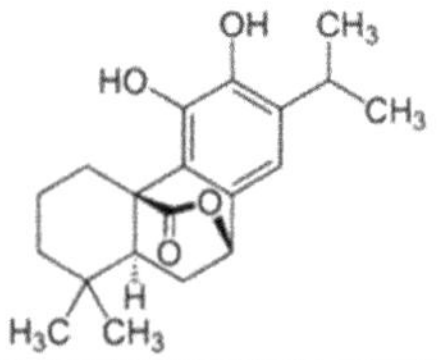	Antiaging, anticancer	Kompelly et al. (2019)
Glycyrrhiza glabra	Licorice 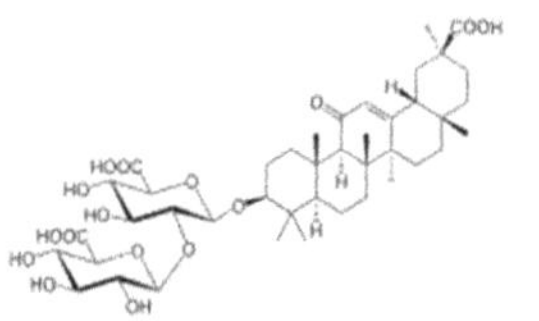	Antiaging, hepatitis, HIV, influenza, pain relief	Ishtiyaq et al. (2019)
Ginkgo biloba	Kaempferol 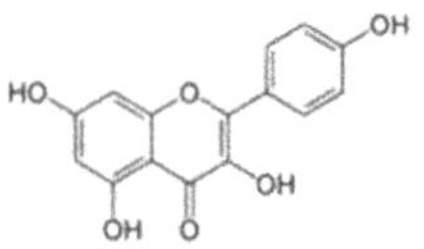	Neuroprotection, antiaging	Luo (2001); Nuhu et al. (2017)

(continued)

Table 10.1 (continued)

Natural drugs	Bioactive molecules	Therapeutics activity	References
B. diffusa	Boeravinone B 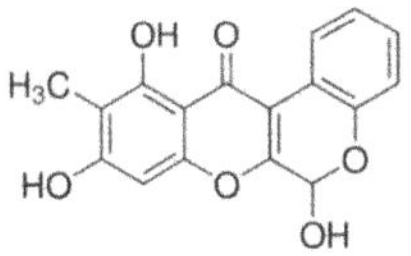	Antiaging	Rathor and Pandey (2018); Akhter et al. (2019)
Withania somnifera	Withanolide A 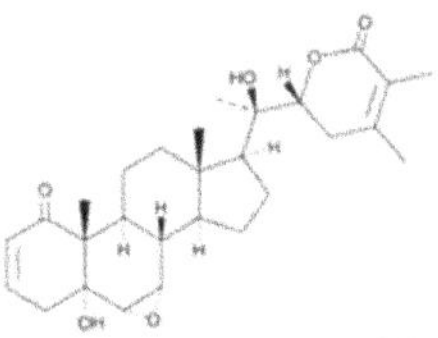	Antiaging, anticancer, neuroprotective	Verma and Kumar (2011); Mandal and Reddy (2017)
Asparagus racemosus	Shatavarin IV 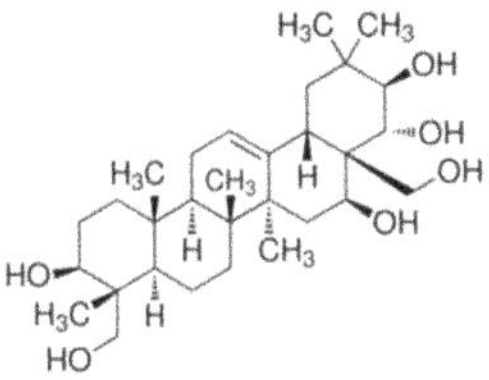	Antiaging, anti-parkinsonism	Singh and Geetanjali (2016); Selvaraj et al. (2019)
Cinnamomum cassia	Cinnacassosides	Antidiabetes, antiulcerogenic	Upadhyay (2017)

10.6.2 Anticancer Effect

The impact of bioactive compounds on human cancer cell lines is most often protective and helps to minimize the number or growth of tumors. Some polyphenols including catechins, quercetin, isoflavones, lignans, flavanones, ellagic acid, red wine polyphenols, resveratrol, and curcumin have protective effects in some models, but their mechanisms of action have been found to be unique (Pandey and Rizvi 2009; Działo et al. 2016). Multiple mechanisms of action for the chemical prevention of polyphenols have been established, including estrogenic/antiestrogenic behavior, antiproliferation and initiation of cell cycle arrest or apoptosis, and oxidation prevention (Cipolletti et al. 2018). It has also been shown that theaflavins and thearubigins, the abundant polyphenols in black tea, possess significant anticancer properties (Pandey and Rizvi 2009). Some compounds inhibit all stages of cancer growth and have been shown to be active in most cancers, including lung, hair, breast, prostate, stomach, and colorectal cancer (Wang et al. 2012).

10.6.3 Antidiabetic Effect

Impairment of the glucose metabolism causes metabolic instability with the onset of hyperglycemia and then diabetes mellitus. There are two major diabetes categories: type 1 and type 2. Numerous studies document polyphenols' antidiabetic effects. Tea catechins are tested for their antidiabetic potential (Pandey and Rizvi 2009; Umeno et al. 2016). Polyphenols can affect glycemia through a variety of mechanisms, including inhibition or absorption of glucose in the intestine by peripheral tissues. A recent study showed that quercetin can help protect changes in patients with diabetes during oxidative stress. Several research papers have shown that by working at many levels, some polyphenols act as an effective antidiabetic agent. These molecules have shown decreased blood glucose, followed by significantly increased plasma insulin and a negative correlation between blood glucose and plasma insulin (Bahadoran et al. 2013; Mukhopadhyay and Prajapati 2015).

10.6.4 Antiaging Effect

Aging is associated with gradual degeneration of organ function in organism. This is due to excess amount of cellular ROS which can cause oxidative stress. A certain amount of oxidative damage occurs even under normal conditions, but the rate of this damage increases as the efficacy of antioxidant and repair mechanisms declines throughout aging process (Nita and Grzybowski 2016; Tan et al. 2018). Fruit and vegetable extracts with high flavonoids also exhibit high total antioxidant activity such as blueberries, strawberries, and lettuce (Pandey and Rizvi 2009).

10.6.5 Neuroprotective Effects

Oxidant inflammation and brain macromolecular damage are important factors in neurodegenerative diseases. Polyphenols are highly antioxidant in nature, and their use in neurological diseases can provide protection (Tan et al. 2018). Furthermore, the antioxidant function is also associated with the induction of antioxidant expression and detoxifying enzymes, particularly in the brain (Kurutas 2015).

10.7 Future Avenues in Herbal Drug Discovery

Phytochemistry or natural product phytoanalysis is the foundation and core of both the herbal and food industries of chemical science. Scientific research on medicinal plants in many developed countries is increasingly needed for an hour at various research institutes, universities, and pharmaceutical laboratories as well as in their clinics. Nearly all countries have their own herbal pharmaceutical copeias and make adjustments to new monographs and procedures from time to time to maintain their interest of herbal products that come from common people. India's Ayurvedic pharmacopeia includes many specific quality standards and techniques of isolation, separation, and spectroscopic identification of hundreds of common herbal medicines. Both research and advances on drug discovery have immense potential to exploit chemical and natural product varieties. Newly established techniques are fast-growing, with strong results in the discovery of natural drugs.

10.8 Conclusion

Nature's exceptional phenomenon is always a golden mark of completing the discovery of herbal medicines. In recent days, the majority of the prevalent diseases and metabolic disorders have been treated with natural medicines. The chemical constituents of the plant and the pharmacological screening will provide the basis for creating a lead molecule through the discovery of herbal medicines. A increasing interest in the production of herbal medicinal products which increases minimal side effects, there are better opportunities to explore medicinal and other biological properties of previously inaccessible natural products. Life was made possible or extended many years ago only because of natural herbs as per the literature references that can be obtained. In the new era of the twenty-first century, without herbal medicines or products obtained through the discovery of natural herbal drugs, no life on earth is possible.

References

Abuajah CI, Ogbonna AC, Osuji CM (2015) Functional components and medicinal properties of food: a review. J Food Sci Technol 52(5):2522–2529

Ahmad F, Leake DS (2018) Antioxidants inhibit low density lipoprotein oxidation less at lysosomal pH: a possible explanation as to why the clinical trials of antioxidants might have failed. Chem Phys Lipids 213:13–24

Akhter F, Alvi SS, Ahmad P, Iqbal D, Alshehri BM, Khan MS (2019) Therapeutic efficacy of *Boerhaavia diffusa* (Linn.) root methanolic extract in attenuating streptozotocin-induced diabetes, diabetes-linked hyperlipidemia and oxidative-stress in rats. Biomed Res Ther 6 (7):3293–3306

Argyropoulou A, Aligiannis N, Trougakos IP, Skaltsounis AL (2013) Natural compounds with anti-ageing activity. Nat Prod Rep 30(11):1412–1437

Asthana J, Pant A, Yadav D, Lal RK, Gupta MM, Pandey R (2015) *Ocimum basilicum* (L.) and *Premna integrifolia* (L.) modulate stress response and lifespan in *Caenorhabditis elegans*. Ind Crop Prod 76:1086–1093

Atanasov AG, Waltenberger B, Pferschy-Wenzig EM, Linder T, Wawrosch C, Uhrin P, Rollinger JM (2015) Discovery and resupply of pharmacologically active plant-derived natural products: a review. Biotechnol Adv 33(8):1582–1614

Bahadoran Z, Mirmiran P, Azizi F (2013) Dietary polyphenols as potential nutraceuticals in management of diabetes: a review. J Diabetes Metab Disord 12(1):43

Bär C, Blasco MA (2016) Telomeres and telomerase as therapeutic targets to prevent and treat age-related diseases. F1000 Res 5:89

Bhandari PR, Kamdod MA (2012) Emblica officinalis (Amla): a review of potential therapeutic applications. Int J Green Pharm 6(4):257–269

Blagosklonny MV, Hall MN (2009) Growth and aging: a common molecular mechanism. Aging 1 (4):357
Brites F, Martin M, Guillas I, Kontush A (2017) Antioxidative activity of high-density lipoprotein (HDL): mechanistic insights into potential clinical benefit. BBA Clin 8:66–77
Bulla LA, Cheng TC (2013) Invertebrate models for biomedical research, vol 4. Springer Science & Business Media, Berlin
Carmona JJ, Michan S (2016) Biology of healthy aging and longevity. Rev Investig Clin 68 (1):7–16
Carretero M, Gomez-Amaro RL, Petrascheck M (2015) Pharmacological classes that extend lifespan of *Caenorhabditis elegans*. Front Genet 6:77
Cheng YC, Sheen JM, Hu WL, Hung YC (2017) Polyphenols and oxidative stress in atherosclerosis-related ischemic heart disease and stroke. Oxid Med Cell Longev 2017:8526438
Chugh NA, Bali S, Koul A (2018) Integration of botanicals in contemporary medicine: road blocks, checkpoints and go-ahead signals. Integrat Med Res 7(2):109–125
Cipolletti M, Solar Fernandez V, Montalesi E, Marino M, Fiocchetti M (2018) Beyond the antioxidant activity of dietary polyphenols in cancer: the modulation of estrogen receptors (ers) signaling. Int J Mol Sci 19(9):2624
Dias DA, Urban S, Roessner U (2012) A historical overview of natural products in drug discovery. Meta 2(2):303–336
Durães F, Pinto M, Sousa E (2018) Old drugs as new treatments for neurodegenerative diseases. Pharmaceuticals 11(2):44
Działo M, Mierziak J, Korzun U, Preisner M, Szopa J, Kulma A (2016) The potential of plant phenolics in prevention and therapy of skin disorders. Int J Mol Sci 17(2):160
Ekor M (2014) The growing use of herbal medicines: issues relating to adverse reactions and challenges in monitoring safety. Front Pharmacol 4:177
Ergen N, Coşkun SH, Orhan DD, Aslan M, Sezik E, Atalay A (2018) Evaluation of the lifespan extension effects of several Turkish medicinal plants in *Caenorhabditis elegans*. Turk J Biol 42 (2):163–173
Garg C (2017) Molecular mechanisms of skin photoaging and plant inhibitors. Int J Green Pharm 11 (02):S217–S232
Garg P, Sardana S (2016) Pharmacological and therapeutic effects of *Ocimum sanctum*. Eur J Pharm Med Res 3(8):637–640
Giacomotto J, Ségalat L (2010) High-throughput screening and small animal models, where are we? Br J Pharmacol 160(2):204–216
Gunathilake KDPP, Rupasinghe H (2015) Recent perspectives on the medicinal potential of ginger. Bot Targets Ther 5:55–63
Gurib-Fakim A (2006) Medicinal plants: traditions of yesterday and drugs of tomorrow. Mol Asp Med 27(1):1–93
Gutiérrez-Grijalva EP, Picos-Salas MA, Leyva-López N, Criollo-Mendoza MS, Vazquez-Olivo G, Heredia JB (2018) Flavonoids and phenolic acids from oregano: occurrence, biological activity and health benefits. Plants 7(1):2
Harvey AL (2008) Natural products in drug discovery. Drug Discov Today 13(19–20):894–901
Ighodaro OM, Akinloye OA (2018) First line defence antioxidants-superoxide dismutase (SOD), catalase (CAT) and glutathione peroxidase (GPX): their fundamental role in the entire antioxidant defence grid. Alex J Med 54(4):287–293
Ishtiyaq A, Alam A, Siddiqui JI, Kazmi MH (2019) Therapeutic potential of widely used unani drug asl-us-soos (Glycyrrhiza glabra Linn.): a systematic review. J Drug Deliv Ther 9(4-s):765–773
Ji HF, Li XJ, Zhang HY (2009) Natural products and drug discovery. EMBO Rep 10(3):194–200
Johnson SC (2018) Nutrient sensing, signaling and ageing: the role of IGF-1 and mTOR in ageing and age-related disease. Subcell Biochem 90:49–97
Karimi A, Majlesi M, Rafieian-Kopaei M (2015) Herbal versus synthetic drugs; beliefs and facts. J Nephropharmacol 4(1):27

Katz DL, Meller S (2014) Can we say what diet is best for health? Annu Rev Public Health 35:83–103

Kaur A (2018) Pharmacobotanical and pharmacological evaluation of ayurvedic crude drug: Rauwolfia serpentina (Apocynaceae). Int J Green Pharm 11(04):S686–S693

Kennedy DO, Wightman EL (2011) Herbal extracts and phytochemicals: plant secondary metabolites and the enhancement of human brain function. Adv Nutr 2(1):32–50

Kenyon CJ (2010) The genetics of ageing. Nature 464(7288):504–512

Khurana S, Venkataraman K, Hollingsworth A, Piche M, Tai TC (2013) Polyphenols: benefits to the cardiovascular system in health and in aging. Nutrients 5(10):3779–3827

Kompelly A, Kompelly S, Vasudha B, Narender B (2019) Rosmarinus officinalis L.: an update review of its phytochemistry and biological activity. J Drug Deliv Ther 9(1):323–330

Koparde AA, Doijad RC, Magdum CS (2019) Natural products in drug discovery. In: Pharmacognosy-medicinal plants. IntechOpen, Croatia

Kurutas EB (2015) The importance of antioxidants which play the role in cellular response against oxidative/nitrosative stress: current state. Nutr J 15(1):71

Lahlou M (2013) The success of natural products in drug discovery. Pharmacol Pharm 4:17–31

Leelananda SP, Lindert S (2016) Computational methods in drug discovery. Beilstein J Org Chem 12(1):2694–2718

Leonov A, Arlia-Ciommo A, Piano A, Svistkova V, Lutchman V, Medkour Y, Titorenko VI (2015) Longevity extension by phytochemicals. Molecules 20(4):6544–6572

Li FS, Weng JK (2017) Demystifying traditional herbal medicine with modern approach. Nat Plants 3(8):1–7

Lublin AL, Link CD (2013) Alzheimer's disease drug discovery: in vivo screening using *Caenorhabditis elegans* as a model for β-amyloid peptide-induced toxicity. Drug Discov Today Technol 10(1):e115–e119

Luo Y (2001) Ginkgo biloba neuroprotection: therapeutic implications in Alzheimer's disease. J Alzheimers Dis 3(4):401–407

Madan K, Nanda S (2018) In-vitro evaluation of antioxidant, anti-elastase, anti-collagenase, anti-hyaluronidase activities of safranal and determination of its sun protection factor in skin photoaging. Bioorg Chem 77:159–167

Mali PY (2016) Pharmacological potentials of Premna integrifolia L. Anc Sci Life 35(3):132

Mandal A, Reddy JM (2017) A review on potential therapeutic uses of *Withania somnifera*. World J Pharm Res 6:846–860

Moutia M, Habti N, Badou A (2018) In vitro and in vivo immunomodulator activities of *Allium sativum* L. Evid Based Complement Alternat Med 2018:1

Mukhopadhyay P, Prajapati AK (2015) Quercetin in anti-diabetic research and strategies for improved quercetin bioavailability using polymer-based carriers–a review. RSC Adv 5 (118):97547–97562

Nasri H, Baradaran A, Shirzad H, Rafieian-Kopaei M (2014) New concepts in nutraceuticals as alternative for pharmaceuticals. Int J Prevent Med 5(12):1487

National Research Council (2000) Using model animals to assess and understand developmental toxicity. In: Scientific frontiers in developmental toxicology and risk assessment. National Academies, Washington DC

Nita M, Grzybowski A (2016) The role of the reactive oxygen species and oxidative stress in the pathomechanism of the age-related ocular diseases and other pathologies of the anterior and posterior eye segments in adults. Oxid Med Cell Longev 2016:1

Nuhu AA, Abdulrazak S, Yashim ZI (2017) Gingko biloba leaf extract mitigates Chlorpyrifos-induced oxidative stress in albino rats. Asian J Med Pharm Res 7(4):42–47

Oyebode O, Kandala NB, Chilton PJ, &Lilford RJ (2016) Use of traditional medicine in middle-income countries: a WHO-SAGE study. Health Policy Plan, 31(8), pp. 984–991

Pan SY, Zhou SF, Gao SH, Yu ZL, Zhang SF, Tang MK, Ko KM (2013, 2013) New perspectives on how to discover drugs from herbal medicines: CAM's outstanding contribution to modern therapeutics. Evidence-Based Complementary and Alternative Medicine:1–26

Panche AN, Diwan AD, Chandra SR (2016) Flavonoids: an overview. J Nutr Sci 5:e47
Panda AK, Chakraborty D, Sarkar I, Khan T, Sa G (2017) New insights into therapeutic activity and anticancer properties of curcumin. J Exp Pharmacol 9:31
Pandey KB, Rizvi SI (2009) Plant polyphenols as dietary antioxidants in human health and disease. Oxidative Med Cell Longev 2(5):270–278
Pandey MM, Rastogi S, Rawat AKS (2013) Indian traditional ayurvedic system of medicine and nutritional supplementation. Evid Based Complement Alternat Med 2013:1
Pant A, Pandey R (2015) Bioactive phytomolecules and aging in *Caenorhabditis elegans*. Healthy Aging Res 4:1–15
Patrícia B, Pereira A, SM Group Open Access eBooks (2017) Impact and prevention of neurodegenerative diseases in society: Alzheimer and Parkinson. Neurodegenerative diseases. SM Group Open Access eBooks
Patwardhan B, Mashelkar RA (2009) Traditional medicine-inspired approaches to drug discovery: can Ayurveda show the way forward? Drug Discov Today 14(15–16):804–811
Petrovska BB (2012) Historical review of medicinal plants' usage. Pharm Rev 6(11):1
Pohl F, KongThoo Lin P (2018) The potential use of plant natural products and plant extracts with antioxidant properties for the prevention/treatment of neurodegenerative diseases: in vitro, in vivo and clinical trials. Molecules 23(12):3283
Pomatto LC, Davies KJ (2017) The role of declining adaptive homeostasis in ageing. J Physiol 595 (24):7275–7309
Rathor L, Pandey R (2018) Age-induced diminution of free radicals by Boeravinone B in *Caenorhabditis elegans*. Exp Gerontol 111:94–106
Rea IM, Gibson DS, McGilligan V, McNerlan SE, Alexander HD, Ross OA (2018) Age and age-related diseases: role of inflammation triggers and cytokines. Front Immunol 9:586
Selvaraj K, Sivakumar G, Veeraraghavan VP, Dandannavar VS, Veeraraghavan GR, Rengasamy G (2019) Asparagus Racemosus-a review. Syst Rev Phar 10(1):87–89
Sen S, Chakraborty R (2017) Revival, modernization and integration of Indian traditional herbal medicine in clinical practice: importance, challenges and future. J Tradit Complement Med 7 (2):234–244
Shammas MA (2011) Telomeres, lifestyle, cancer, and aging. Curr Opin Clin Nutr Metab Care 14 (1):28
Shilsky J, Bloom DE, Beaudreault AR, Tucker KL, Keller HH, Freund-Levi Y, Meydani SN (2017) Nutritional considerations for healthy aging and reduction in age-related chronic disease. Adv Nutr 8(1):17–26
Siddiqui MZ (2011) Boswellia serrata, a potential antiinflammatory agent: an overview. Indian J Pharm Sci 73(3):255
Singh R, Geetanjali (2016) Asparagus racemosus: a review on its phytochemical and therapeutic potential. Nat Prod Res 30(17):1896–1908
Singh A, Singh S, Prasad SM (2016) Role of medicinal plants for health perspective: special reference to antioxidant potential. J Chem Biol Ther 1(2):106
Srivastava S (2017) The mitochondrial basis of aging and age-related disorders. Genes 8(12):398
Takshak S, Agrawal SB (2019) Defense potential of secondary metabolites in medicinal plants under UV-B stress. J Photochem Photobiol B Biol 193:51–88
Tan BL, Norhaizan ME, Winnie-Pui-PuiLiew HS (2018) Antioxidant and oxidative stress: a mutual interplay in age-related diseases. Front Pharmacol 9:1162
Thomford N, Senthebane D, Rowe A, Munro D, Seele P, Maroyi A, Dzobo K (2018) Natural products for drug discovery in the 21st century: innovations for novel drug discovery. Int J Mol Sci 19(6):1578
Tissenbaum HA (2015) Using *C. elegans* for aging research. Invertebr Reprod Develop 59(Suppl 1):59–63
Topp R, Fahlman M, Boardley D (2004) Healthy aging: health promotion and disease prevention. Nurs Clin North Am 39(2):411–422
Tsao R (2010) Chemistry and biochemistry of dietary polyphenols. Nutrients 2(12):1231–1246

Tungmunnithum D, Thongboonyou A, Pholboon A, Yangsabai A (2018) Flavonoids and other phenolic compounds from medicinal plants for pharmaceutical and medical aspects: an overview. Medicines 5(3):93
Umeno A, Horie M, Murotomi K, Nakajima Y, Yoshida Y (2016) Antioxidative and antidiabetic effects of natural polyphenols and isoflavones. Molecules 21(6):708
Upadhyay RK (2017) Therapeutic and pharmaceutical potential of *Cinnamomum tamala*. Pharm Pharm Sci 6(3):18–28
Upadhyay S, Dixit M (2015) Role of polyphenols and other phytochemicals on molecular signaling. Oxid Med Cell Longev 2015:1
Veeresham C (2012) Natural products derived from plants as a source of drugs. J Adv Pharm Technol Res 3(4):200–201
Verma SK, Kumar A (2011) Therapeutic uses of Withania somnifera (ashwagandha) with a note on withanolides and its pharmacological actions. Asian J Pharm Clin Res 4(1):1–4
Wang SY (2002) Antioxidant capacity of berry crops, culinary herbs and medicinal herbs. In XXVI International Horticultural Congress: Asian Plants with Unique Horticultural Potential: Genetic Resources, Cultural, vol 620, pp. 461–473
Wang H, OoKhor T, Shu L, Su ZY, Fuentes F, Lee JH, Tony Kong AN (2012) Plants vs. cancer: a review on natural phytochemicals in preventing and treating cancers and their druggability. Anti Cancer Agents Med Chem 12(10):1281–1305
Welz AN, Emberger-Klein A, Menrad K (2018) Why people use herbal medicine: insights from a focus-group study in Germany. BMC Complement Alternat Med 18(1):92
Wilson D, Nash P, Buttar H, Griffiths K, Singh R, De Meester F, Takahashi T (2017) The role of food antioxidants, benefits of functional foods, and influence of feeding habits on the health of the older person: an overview. Antioxidants 6(4):81
Yin SY, Wei WC, Jian FY, Yang NS (2013) Therapeutic applications of herbal medicines for cancer patients. Evid Based Complement Alternat Med 2013:1
Yuan H, Ma Q, Ye L, Piao G (2016) The traditional medicine and modern medicine from natural products. Molecules 21(5):559
Zhang L, Reddy N (2018) Bioactive molecules from medicinal herbs for life threatening diseases. J Mol Sci 2(1):4

Cell-Based Assays in Natural Product-Based Drug Discovery

11

Alka Mukne, Misbah Momin, Parag Betkar, Tanvi Rane, and Saurabh Valecha

Abstract

Natural products have been crucial and bounteous sources for the discovery of lead compounds and drugs for various therapeutic indications. The intrinsic complexity of the natural products poses several obstacles in the HTS programs. Besides, drug screening in animals is a long-drawn, costly process with many hurdles and low productivity. Cell-based HTS assays on the other hand are excellent means for preliminary screening of drugs of natural origin for numerous therapeutic indications as they help in designing more relevant in vivo pharmacological models to study drugs and thus expedite drug discovery process. In this book chapter, we have discussed a plethora of cell-based HTS assays for drug discovery, employing optical and electrochemical methods of detection and for quantification of cells and information on their proliferation rate, biomarker production, and activation of various signaling pathways. Also, the potential applications of these cell-based assays are discussed briefly for screening of natural products for their anti-inflammatory, anti-diabetic, anti-obesity/hypolipidemic, anti-microbial, anti-mycobacterial, anti-cancer, and anti-viral activities.

Keywords

Cell-based assays · Natural products · Luciferase assay · Luminescence · Fluorescence readout

A. Mukne (✉) · M. Momin · P. Betkar · T. Rane · S. Valecha
Department of Pharmacognosy & Phytochemistry, Bombay College of Pharmacy, Mumbai, Maharashtra, India
e-mail: alka.mukne@gmail.com

S. C. Mandal et al. (eds.), *Evidence Based Validation of Traditional Medicines*,
https://doi.org/10.1007/978-981-15-8127-4_11

Abbreviations

2DG	2-deoxy-d-[^{14}C] glucose or 2-deoxy-d- [^{3}H] glucose
2-NBDG	2-[N-(7-nitrobenz-2-oxa-1, 3-diazol-4-yl)amino]-2-deoxy-d-glucose
3-O-MG	3-O-methyl glucose
ABC1	ATP-binding cassette transporter A1
AK	Adenylate kinase
AMPK	5′-adenosine monophosphate-activated protein kinase
BLV	Bovine leukemia virus
CAM	Calcein acetoxymethyl
CFU	Colony-forming units
CHIKV	Chikungunya virus
CPE	Cytopathic effect
FITC	Fluorescein 5(6)-isothiocyanate
GFPs	Green fluorescent proteins
GLP-1	Glucagon-like peptide-1
GLUT-4	Glucose transporter type 4
HCS	High-content screening
HEK 293 cells	Human embryonic kidney 293 cells
HTS	High-throughput screening
IL	Interleukin
LDH	Lactate dehydrogenase
MIC	Minimum inhibitory concentration
MTBH	*Mycobacterium tuberculosis* $H_{37}R_v$
Nrf2	Nuclear factor erythroid 2-related factor 2
PPAR γ	Peroxisome proliferator-activated receptor γ
Rluc	Renilla luciferase
SGLT2	Sodium/glucose cotransporter 2
SREBP1	Sterol regulatory element-binding protein 1

11.1 Introduction

Humans have been using natural products since prehistoric times as medicines to prevent and cure diseases. Natural products serve as the cornerstone of drug discovery and development; most drugs available in the market today are of natural origin. Some of the notable drugs derived from natural origin include antibiotics, paclitaxel, hypericin, curcumin, artemisinin, and so on. The present definition of a natural product can vary from "a naturally occurring compound that has been chemically modified" to "any FDA-approved, unmodified natural material or compound, semi-synthetic derivatives, or synthetic structures which were conceptually derived from a natural product." According to the Herbal Medicine Market Research Report, the global herbal medicine market is expected to grow at CAGR of 5.88% and reach around USD 1,29,690 million by 2023. Recent advances in computational techniques such as the computer-aided drug design in the discovery of bioactive natural products have aided researchers in developing the otherwise difficult to process natural products and derivatives into novel drugs.

The development of a medication in the lab to its commercialization is a detailed, prolonged, and costly procedure. It ordinarily takes 10–12 years and over $775 million for getting a new drug from bench to bedside. By and large, after screening millions of compounds, around 250 enter preclinical testing, 10 of which continue to clinical preliminaries and just about 1 gets approved by the FDA as a new drug. The emergence of combinatorial chemistry and increase in the number of identified protein targets for drug screening have triggered the development of highly selective and reliable HTS assays for comparatively faster drug discovery (Zang et al. 2012).

11.1.1 HTS Assays: Biochemical and Cell-Based Assays

HTS assays can be broadly categorized into two types: **biochemical assays** and **cell-based assays**. Biochemical assays are target-based assays in which the binding affinity of the ligand for receptor/isolated targets (such as enzymes) is determined, whereas cell-based assays are physiology-based assays performed in cell cultures. Biochemical assays, in spite of being the basis of HTS in the pharma industry, are accompanied with several limitations. Firstly, not all of the targets used for assaying can be prepared in a manner suitable to quantify the binding or affinity. Secondly, the activity shown by a drug compound in a biochemical assay may not necessarily represent its activity at the cellular level because the drug targets are generally located in the cell interior and inadequate cellular drug exposure/uptake might lead to a lower bioactivity. Such compounds are known to exhibit a "cell drop-off," i.e., they show high affinity to isolated target proteins but fail to perform intracellularly which is the major cause of failure in clinical drug development. To address this challenge, the focus has therefore shifted from target-based approach to a more appropriate in vitro physiology-based approach with cell-based assays increasingly being used in HTS screenings (Mateus et al. 2017). In addition to the conventional cell-based assays, HCS technologies are also being increasingly used, wherein cellular events can be monitored by integrating advanced microscopy with cell-based assays. HCS is any cell-based experiment, monitored through techniques such as reporter signals, phenotypic profiling, and morphological analysis to measure cellular responses. Most cell-based assays which use a small number of cells show deviation due to alterations in cell morphology or gene expression upon treatment with drug. Hence, newer techniques associated with HCS that are multidimensional/multi-parametric at the single-cell level are indispensable and help us to gain a better understanding of the interaction and effect of genes, extracellular matrix, and signal networks in the regulation of phenotypes in heterogeneous populations of cells (Nierode et al. 2016). The worldwide market for cell-based drug discovery assays is estimated to reach USD 18.9 billion by 2024 from USD 13.9 billion in 2019.

11.1.2 Cell-Based Assays

Cell-based assays are versatile tools that can differentiate between agonists and antagonists, recognize allosteric modulators, and provide immediate data on cell permeability and stabilization compounds within cells and its associated acute cytotoxicity. Cell-based assays have been used during early drug discovery since they can provide representative tissue-specific reactions, from target detection and validation to primary testing, identification, and optimization of lead compounds and screening for safety and toxicology. Cells, devices that culture cells, and systems for detecting and quantifying cells or cellular activities are the major components of cell-based HTS assay. Assays involving secondary messengers, assays for reporter genes, and assays for cell proliferation are the three main types of cell-based assays. Cells used in cell-based assays should be capable of testing, accurately portraying the system, and expressing the essential aspects and signaling intermediates. The various cell lines used for performing cell-based assays are immortalized cell lines, primary cells, human cancer cells, cancer stem cells, mesenchymal stem cells (MSCs), embryonic stem cells (ESCs), and induced pluripotent stem cells (iPSCs) (Zang et al. 2012).

Cell culture systems used in cell-based assays majorly employ multi-well plates which enable high drug screening rates. These assays can also be made automated. There are two types of cell culture models used in cell-based assays: two-dimensional (monolayered, single cells grown on a two-dimensional surface) and three-dimensional (multilayered cells grown on a three-dimensional surface). Though 3D assays provide advantages such as close cell-cell and cell-matrix interactions, owing to the significant difference in cost and ease in performing 2D cell-based assays, these are preferred as assays of choice over 3D cell-based assays in preliminary screening.

The various conventional methods for recording responses/readout include hemocytometer, coulter counter, trypan blue dye exclusion, and neutral red uptake methods which require large workforce harmful chemicals and have low throughput. Owing to these reasons, these techniques are less preferred as detection techniques in HTS vis-à-vis **electrochemical** and **optical** methods which are the most commonly used methods of quantification for cell-based assays.

Electrochemical methods are based on measurement of electrons generated and charge transferred due to redox reactions in a cell. Such reactions also cause changes in ionic composition which is a measure of cell viability in a homogenous solution. However, electrochemical methods of detection are deficient in providing cell function and cell signaling pathway data which is important to understand the mechanism of action of drug. Besides, this method can be used only for 2D cultures since direct contact of the electrode and the cells is necessary. Optical methods on the other hand can be carried out using principles of colorimetry, fluorimetry, luminescence, and radiometry as described in Fig. 11.1.

This chapter provides an update of the various cell-based assays that have been reported for screening of drugs of natural origin for various therapeutic conditions.

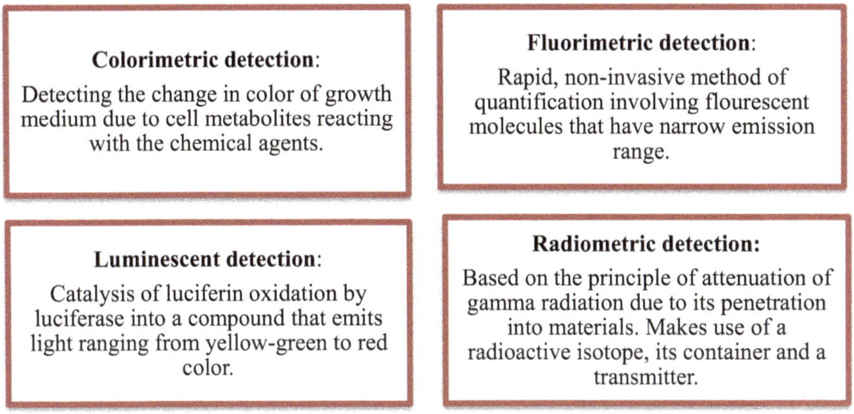

Fig. 11.1 Overview of optical detection techniques for various cell-based assays

The chapter also gives an insight into newer strategies that can be deployed for screening drugs of natural origin.

11.2 Cell-Based Assays for Screening of Natural Origin Drugs

11.2.1 Inflammatory and Autoimmune Diseases

Reports in the literature suggest that the relation between the impaired metabolism and disproportionate redox levels is the underlying cause behind the development of autoimmune diseases. Cell metabolism plays a key role in the pathogenesis and development of diseases which is indicated by the role of cross-organ control in managing multi-organ diseases. Inflammatory diseases are characterized with increase in levels of cytokines (IL-8, IL-1, IL-6, TNF-α), chemokines, platelet-activating factor (PAF), NO, inducible NO (iNO) synthase, etc. One of the tried and tested methods for screening drugs for inflammatory diseases has been to eliminate the generated ROS which is reported to be involved in the initiation and progression of inflammatory responses. Pathogenic cells exhibit imbalance in the metabolism and redox state and differ from the normal cells with generation of biomass, unusual rates of autophagy, and unmanageable cell growth. This leads to the development of complex diseases such as cancer, type I diabetes, systemic lupus erythematosus, and rheumatoid arthritis (Fan et al. 2018). Various cell-based assays reported in literature for screening drugs of natural origin for inflammatory and autoimmune disorders are discussed in this section.

Autoimmune diseases like rheumatoid arthritis exhibit indiscriminate cell proliferation. One of the most rudimentary assays reported to evaluate drugs for their activity against autoimmune or inflammatory diseases is by the **cell proliferation assay**. These assays are based on colorimetric methods and employ tetrazolium dyes, alamar blue, or resazurin sodium. However, these methods are laborious and time-consuming and require addition of chemicals at different intervals, which can hamper the normal functioning of targeted cells. Besides, such assays only provide endpoint data and have low sensitivity (Zang et al. 2012). Okamoto (2008) has reported the use of **3-(4,5-dimethylthiazol-2-yl)-2,5-diphenyltetrazolium bromide (MTT) assay** to screen drugs such as vitamin K2 [menaquinone-4 (MK-4)] and vitamin K1 against rheumatoid arthritis. It is based on the principle that MTT converts to formazan in the mitochondria of the viable cells. The researchers proposed that vitamin K exhibits activity against rheumatoid arthritis by suppressing NF-κB. NF-κB is responsible for generating many pro-inflammatory signals (TNF-α, IL-1β, IL-6, IL-8, NO, iNO synthase, COX-2), which may play a crucial role in the development of oxidative stress-induced inflammatory and degenerative/autoimmune diseases. The suppression of these pro-inflammatory signals is considered as an important strategy to counter inflammation. Inactive NF-κB is captured in the cytoplasm by the inhibitor of κB, IκBα. Cytokines such as TNF-α induce the phosphorylation of inhibitor of κB kinase (IKK), resulting in the degradation of IκBα and release of NF-κB. The free NF-κB is transported to the nucleus where it transcribes for a number of diseases such as cancer, AIDS, autoimmune diseases, and inflammatory disorders (Meijer et al. 2015).

Targeting the **AMPK-SREBP1 lipogenesis pathway** to screen drugs for rheumatoid arthritis was proposed by Fan et al. (2018). Strong lipogenic activity and high expression of SREBP1 is found in synovial fibroblasts of rheumatoid arthritis (RASFs) patients. SREBP1 causes increased sterol synthesis, and the elevated levels of lipids exacerbate rheumatoid arthritis. Besides, increased levels of inflammatory cells produce uncontrolled proliferation of cells leading to hyperplasia. In this study, the researchers have estimated the levels of triglycerides, cholesterol, and other lipid metabolites to estimate the levels of SREBP1 in the cell milieu post exposure to drugs. G-R NSCLC cell lines H1975 and H1650 with the maximum SREBP1 expression were used in the study, and levels of palmitic acid were measured. Cell viability assay was performed using MTT assay. The authors reported that berberine effectively controls SREBP1 and lipogenesis, via the ROS/AMPK pathway, while inhibiting the growth of RASFs but not that of normal cells. Berberine was found to protect the bones better than methotrexate via the inhibition of AMPK-SREBP1 lipogenesis pathway. It effectively switches on the ROS/AMPK pathway, causes mitochondrial dysfunction, and inhibits cellular lipogenesis and cell proliferation. Targeting ROS/AMPK/lipogenesis signaling pathway inhibits the progress of RASFs in vitro opening a new approach for treating proliferative diseases.

***Escherichia coli*-derived lipopolysaccharide (LPS) stimulation of cells** is the most routinely used approach to understand the underlying mechanism behind the activity of anti-inflammatory drugs. LPS is reported to be a potent activator of NF-κB. Perez del Palacio et al. (2016) devised a high-throughput 96-well method and screened 5976 noncytotoxic microbial extracts from a natural product collection (comprising of filamentous fungi, actinomycetes, and unicellular bacteria) for their potential immunomodulatory activity in LPS-stimulated murine macrophage cells (RAW 264.7) and IL-1β-stimulated intestinal epithelial cells (Caco-2). Levels of NO and IL-8 were measured. The assay developed was successful in overcoming the disadvantages of being time-consuming and expensive, posed by 24-well plate format or manual liquid addition for whole cell assay-based HTS approaches. Recent developments in spectrophotometry-based detection techniques and technological innovations have enabled the application of these assays to 96-well plate format with advantages such as decreased assay volumes, increased plate throughput, high-sensitivity readout systems, reduced consumption of cells, costly reagents, and shorter assay time. The efficiency of this assay was validated by evaluating the immunomodulatory compounds diketocoriolin B and eremoxylarin A from crude extracts.

DaSilva et al. (2019) studied the effects of urolithins (gut microbe-derived metabolites of pomegranate ellagitannin) on inflammatory biomarkers in LPS-stimulated BV-2 microglia. Urolithins decreased levels of NO, IL-6, prostaglandin E_2, and TNF-α from LPS-BV-2 microglia. This attenuation of neuroinflammation by urolithins may be responsible for pomegranate's neuroprotective effects against Alzheimer's disease. Cells were pre-treated with the test drug before incubation with LPS, and the levels of cytokines were estimated in the supernatants either directly by ELISA kits or indirectly by RT-PCR technique. The inhibition of the production of these pro-inflammatory factors can in turn provide protection from neuronal damage in neurodegenerative conditions. Usually all neurodegenerative diseases are characterized by chronic immune activation, especially of microglia of the central nervous system. Many approaches have been devised to inhibit/modulate microglial activation via the important anti-inflammatory target proteins, iNOS and IL-8. NO regulates neurotransmission, smooth muscle contractility, platelet reactivity, and cytotoxic activity of immune cells. Inappropriate release of NO has been associated with the development of a number of disease states. The inhibitors of iNOS are potential effective agents in the treatment of such conditions.

One of the more recent approaches to screen anti-inflammatory drugs is by using a type of **luminescent assay**, i.e., **luciferase reporter assay**, based on the principle of luciferase enzyme (a heterodimer comprising of α and β subunits, encoded by *luxA* and *luxB*, respectively)-mediated oxidation of luciferin to inactive oxyluciferin with emission of light. This light is then detected by illuminometer or optical microscope. Dual-reporter luciferase assays make use of both firefly luciferase (isolated from beetles, *Photinus pyralis*) and Rluc (isolated from sea pansy *Renilla reniformis*) as both differ in their substrate and cofactor requirements and can enable detection of different metabolic pathways. Firefly luciferase produces a greenish yellow light in the 550–570 nm range from luciferin in the presence of oxygen, ATP, and magnesium, whereas Rluc requires only coelenterazine (a luminescent enzyme substrate) and oxygen and produces a blue light at 480 nm. This assay can be used to detect the upregulation of the transcription factor Nrf2. Nrf2 exerts cytoprotective role in the control of oxidative stress-induced cellular damage to the body. Besides, this assay indirectly measures the levels of NF-κB which is discussed earlier in this chapter. NF-κB is also the site of action for glucocorticoid-mediated IL-8 suppression. In this assay, cells transfected with NF-κB luciferase reporter plasmid are incubated with drugs prior to stimulation with pro-inflammatory cytokine TNF-α which causes the NF-κB promoter to carry out transcription of luciferase. The amount of light emitted corresponds to the levels of NF-κB activity which is measured quantitatively in terms of relative light units by a luminometer. A high-throughput cell-based glucocorticoid activity screening platform that measures dexamethasone-mediated NF-κB repression was developed by Jiang et al. (2017). The developed assay could identify glucocorticoid receptor-modulating compounds that may be designed to improve current glucocorticoid-based therapies. Lung epithelial A549/NF-κB-luc reporter cells were seeded into plates overnight, incubated with drugs, followed by addition of either IL-1β or IL-1β + dexamethasone and luciferase agent. The repression in NF-κB was measured. About 8000 natural and synthetic compounds were screened, and several compounds that suppressed glucocorticoid receptor activity were identified including pyrromycin, camptothecin, podophyllotoxin, colchicine, vinblastine sulfate, vincristine sulfate, and sanguinarine chloride.

NF-κB detection in H293-NF-κB-RE-*luc*2P reporter cell line was used as an effective tool for screening of bioactives with anti-inflammatory activity by Meijer et al. (2015). Activity of minced broccoli seedlings, short-chain fatty acids, and lutein was evaluated to validate the method used by the researchers. Broccoli seedlings ($p < 0.01$) and lutein ($p < 0.05$) were found to significantly reduce NF-κB activity. Among the short-chain fatty acids, butyrate was found to be the most potent inhibitor of NF-κB activity, followed by propionate and acetate. Short-chain fatty acids (acetate, butyrate, and propionate) are actives produced by fermentation of dietary fiber in the colon and have been reported to possess anti-cancer and anti-inflammatory activity besides playing a vital role in the prevention of inflammation of the adipose tissue and insulin resistance.

Metabolic inflammation possesses risk for various disease conditions such as insulin resistance, type 2 diabetes, and cardiovascular diseases. Imbalance in the dietary factors is responsible for many lifestyle issues. Literature is replete with examples stating that consumption of active phytomolecules such as flavonoids, carotenoids, plant sterols, and isothiocyanates decreases adipose tissue inflammation in turn reducing the risk of insulin resistance and atherosclerosis. Qin et al. (2016) used HEK293 cells transfected with heme oxygenase-1 (HO-1) reporter plasmid DNA and Rluc plasmid to evaluate the efficacy of Sinomenine (7, 8-didehydro-4-hydroxy-3, 7-dimethoxy-17-methyl-α, 13α, 14 α-morphinan-6-one), an alkaloid originally derived from *Sinomenium acutum*. Sinomenine possesses anti-inflammatory activity and is effective with minimal side effects in patients afflicted with arthritis and mesangial proliferative nephritis. The luciferase activities, measured with a dual luciferase reporter assay system are expressed by dividing the value of reporter luciferase activity by that of Rluc activity. It was observed that sinomenine produced a dose-dependent increase in the levels of HO-1 and NAD (P)H quinone oxidoreductase 1 (NQO1) in both HEK293 cells and mouse macrophage RAW 264.7 cells. HO-1 is a crucial Nrf2 target gene, and the results obtained establish sinomenine as an activator of Nrf2 signaling pathway.

Fluorescence-based assays such as the one involving caspase activation are highly sensitive when compared to the luminescence assays. Caspase-3 belongs to the apoptotic family of caspases which cause cell death. An easy cell-based assay to screen compounds inhibiting polyglutamine-induced caspase-3 activation was developed by Piccioni et al. (2004). This assay is based on the expression of androgen receptor with an expanded polyglutamine tract in HEK 293T cells and measurement of caspase-3 activation by employing a fluorogenic substrate. Cells were incubated in assay plates before transfection with AR112 plasmid DNA. Post transfection, drugs were added and the cells harvested for analysis. To determine caspase-3 activity, lysed cells were incubated with fluorogenic substrate. Upon screening of drugs, digitoxin, nerifolin, and peruvoside, known inhibitors of Na^+ K^+- ATPase, were identified as drugs blocking polyglutamine-induced caspase-3 activation and thus providing protection against polyglutamine cytotoxicity. Spinobulbar muscular atrophy or Kennedy's disease is a neurodegenerative disorder caused by multiplication of a trinucleotide CAG repeat sequence which encodes for the polyglutamine tract in the androgen receptor. This toxic mutant protein initiates an apoptotic cascade, activating caspase-3 and leading to the generation of toxic, polyglutamine-containing fragments with consequent initiation or amplification of the disease. Caspase-3 cleaves the androgen receptor with expanded polyglutamine, resulting in cytotoxicity and elimination of these caspase-3 cleavage sites and in turn reducing the toxicity of mutant androgen receptors. Similar mutations are observed in other polyglutamine diseases like Huntington's disease, dentatorubral-pallidoluysian atrophy, and six forms of spinocerebellar ataxia. Hence, caspase assay can be used to screen drugs against the abovementioned polyglutamine-induced diseases.

One of the latest techniques to be used in cell-based assays is flow cytometry which employs sophisticated instruments with multiple lasers and detectors to measure signals and record multiple events simultaneously within single cells. However, it is tedious to scan individual cells as minute events such as subcellular localization of proteins are indiscernible. A flow cytometry method that combines accurate tracking of cell movement with a high-resolution multispectral imaging system can overcome the drawback associated with the conventional system (Nierode et al. 2016). The levels of $CD4^+T$ and $CD8^+T$ cells can be measured using **flow cytometric analysis**. T cells differentiate into $CD4^+T$ and $CD8^+T$ cells and are responsible for playing different roles in immunomodulation. The assay is performed by incubating cells with various concentrations of drug, followed by treatment with mouse CD4-FITC or CD8-FITC monoclonal antibody, and a washing step to remove unbound antibody. Yang et al. (2006) deployed this technique to evaluate the immunomodulatory effect of a polysaccharide obtained from the fresh roots of *Angelica sinensis*. The percentage of $CD4^+T$ cells within total spleen cells was greatly increased by the polysaccharide (30–100 μg/mL), whereas $CD8^+T$ cells decreased.

The Janus kinase (JAK) pathway and the T helper 17 (Th17) pathway [IL-17, IL-21, IL-22, and granulocyte-macrophage colony-stimulating factor (GM-CSF)] are the newer pathways being studied in screening of rheumatoid arthritis drugs. Anti-arthritic drugs such as tofacitinib and ruxolitinib have reported activity on the JAK pathway, whereas ustekinumab and briakinumab act against the Th17 pathway

(Koenders and van den Berg 2015). Also, proteins such as hexokinase 3, glucose-6-phosphate isomerase, fructose-bisphosphate aldolase, phosphoglycerate kinase 1, phosphoglycerate mutase 1, enolase 1-α, pyruvate kinase, LDH A, and malate dehydrogenase 1 are upregulated during the glycolysis pathway in rheumatoid arthritis. Assays can be developed to screen drugs that modulate these protein levels, thus acting as effective rheumatoid arthritis agents (Fan et al. 2018; Balakrishnan et al. 2014). Table 11.1 illustrates the various assays used to screen drugs of natural origin against inflammatory and autoimmune diseases.

11.2.2 Noncommunicable Lifestyle Diseases

Noncommunicable lifestyle diseases (NCDs) will be the prime cause of death in the world by 2030. Noncommunicable diseases include CVS diseases, hypertension, diabetes, and chronic respiratory diseases and aren't limited to high-income countries alone. Of the several lifestyle disorders mentioned, given the scope of this chapter, we are restricting our discussion to diseases with impaired glucose and lipid metabolism etiology.

11.2.2.1 Diabetes

Diabetes mellitus is a metabolic disorder in which the blood glucose levels are elevated. This mainly occurs due to decrease in insulin secretion. Diabetes mellitus is classified into three types: type 1 diabetes mellitus, type 2 diabetes mellitus, and gestational diabetes mellitus. In type 1 diabetes mellitus, there is autoimmune destruction of pancreatic β-cells mediated by T cells, autoreactive lymphocytes, insulin antibodies, macrophages, and cytokines (like IFN-γ and TNF-α) which leads to insulin insufficiency and hyperglycemia. In type 2 diabetes mellitus, there is deficiency in insulin secretion and insulin resistance. The former leads to reduction of glucose uptake in adipose tissue and skeletal muscles, as well as increase in gluconeogenesis in the liver. Due to insulin resistance, insulin responsive cells don't respond to action of insulin secreted, which leads to hyperglycemia. In type 2 diabetes mellitus, there is decrease in glycogen synthase activity, due to which there is decrease in glycogen synthesis. Abnormalities in hormone secretion like GLP-1 which helps in insulin secretion also contribute in manifestation of type 2 diabetes mellitus. The major targets of insulin are adipose tissues, skeletal muscle, and hepatic cells; hence the most commonly used cell lines to assess the glucose uptake activity of test compounds are 3T3-L1 (mouse fibroblast), L6 (rat myoblasts), C2C12 (mouse myoblasts), HepG2 (human hepatic carcinoma), human liver cells, and yeast cells.

The most basic screening assay for screening anti-diabetic drugs is **glucose uptake assay** which measures the ability of test drug to increase the glucose uptake into target tissues of insulin. To measure the glucose uptake within the cell lines, radioactive glucose analogues/radiotracers like 2DG and 3-O-MG are commonly utilized due to their structural similarities to glucose. The amount of radioactivity in the cell lysate is determined radiometrically. Radioactive waste disposal and hazards

Table 11.1 Assays employed in the search for novel drugs of natural origin for the treatment of inflammatory and autoimmune diseases

Cell-based assay	Compound/herb	Characteristic/function	References
LPS-stimulation	Narciclasine was isolated from *Lycoris radiata*	• Inhibited the mRNA expression of iNOS, IL-6, TNF-α, IL-1β, and COX-2 in RAW 264.7 cells • Inhibited NF-κB and mitogen-activated protein kinase (MAPK) pathways	Shen et al. (2019)
	Capuli cherry fruits (*Prunus serotina*)	Upregulates the activation of the AMPK/Nrf2/ARE signalling pathway	Alvarez-Suarez et al. (2017)
	Fermented preparations of *Rhizoma atractylodis macrocephalae*	Reduced the production of TNF-α, IL-1β, and NO	Bose and Kim (2013)
	Glucan, a polysaccharide extracted from a mushroom, *Lentinus edodes*	Regulate the production of NO, TNF-α, and IL-6 levels	Xu et al. (2012)
	17-*O*-acetylacuminolide, diterpenoid labdane from *Neouvaria foetida*	Reduced the production of TNF-α and IL-1β	Achoui et al. (2010)
	Dietary supplements α-tocopherol, *N*-acetylcysteine, catechin, and epigallocatechin gallate	Inhibition of TNF-α release	Singh et al. (2005)
LPS-stimulation and luciferase assay	Ethanolic extracts of *Castanea sativa* leaves, *Cinchona pubescens* bark, *Cinnamomum verum* bark, *Salix alba* bark, *Rheum palmatum* root, *Alchemilla vulgaris* plant, *Humulus lupulus* cones, *Vaccinium myrtillus* berries, *Curcuma longa* root, and *Arctostaphylos uva-ursi* leaves	• Affected inflammatory signaling pathways through toll-like receptors (TLRs), TLR2 and TLR4 in LPS-stimulated THP-1 monocytes and HeLa-TLR4 transfected reporter cells • Inhibition of the NF-κB pathway	Schink et al. (2018)
Luciferase assay	Ethanolic extract of rhizome of *Atractylodes chinensis*	Inhibition of the NF-κB pathway	Hossen et al. (2019)
	Aqueous extract from *Artemisia capillaris*	Inhibition of the NF-κB pathway	Yeo et al. (2018)
	Calycosin from *Radix astragali*	Suppression of IL-6, IL-33, COX-2	Su et al. (2016)
	Swertiamarin, sweroside, secoxyloganin, chlorogenic acid, loganic acid, and aldosecologanin from green and flower buds of *Lonicera japonica*	Inhibition of the NF-κB pathway	Jiang et al. (2014)

(continued)

Table 11.1 (continued)

Cell-based assay	Compound/herb	Characteristic/function	References
	Plumericin from stem bark of *Himatanthus sucuuba*	Inhibition of the NF-κB pathway	Fakhrudin et al. (2014)
Flow cytometry	Combination of *Caryophyllus aromaticus*, *Mentha haplocalyx*, *Magnolia biondii*, *Xanthium sibiricum*, *Asarum sieboldii*	• Cytotoxicity estimation • Activation of Nrf2	Han et al. (2019)
	Streptonigrin	Pathway-selective inhibitors of JAK-Stat and MAP kinase signaling	Krutzik et al. (2008)
LPS-stimulation/ flow cytometry	Ethanolic extract of *Anemarrhena asphodeloides* (components identified-timosaponin B and timosaponin B II)	Inhibition of production of NO, ROS, and pro-inflammatory cytokines	Ji et al. (2019)
	Isorhamnetin: -flavonoid extracted from the Chinese herb *Hippophae rhamnoides*	• Measurement of apoptosis by FITC Annexin V/propidium iodine kits • Downregulation of TNF-α, IL-6, IL-1β, and IL-12p70 and upregulation of IL-10	Shi et al. (2018)

of handling radioactive material are the major concerns and limitations of this method. An enzymatic, fluorometric resazurin assay is also used to measure the glucose (2DG and 3-O-MG) uptake within the cell which overcomes the disadvantages of radiometric method. In this assay, the intensity of fluorescence produced due to NADPH/NADH-mediated conversion of alamar blue/resazurin to reddish-pink colored resorufin in the presence of diaphorase is measured with a microplate reader (Yamamoto et al. 2015). Zou et al. (2005) for the first time used 2-NBDG as a fluorescent glucose analogue to measure the glucose uptake activity of lead compounds in HepG2 and L6 rat skeletal muscles using flow cytometry as a detection technique. Huang et al. (2018) evaluated the anti-diabetic potential of flavonoid-rich extract of *Sophora davidii* in L6 muscles using 2-NBDG as a glucose tracer. In vitro *S. davidii* extract was found to enhance GLUT4 translocation and its expression and increase uptake of glucose via the 5′-AMPK-mediated GLUT4 translocation in L6 cell line. To understand the mechanism of action of glucose uptake, GLUT4 translocation assay is routinely performed. Within the body, glucose uptake within the insulin-responsive target cells is due to the movement of insulin-responsive transporter GLUT4 on the cell surface. Hence, insulin- dependent GLUT4 translocation to the cell surface is an important molecular regulation for glucose homeostasis. Kadan et al. (2018) evaluated the GLUT4-mediated glucose uptake activity of *Teucrium polium* extracts: {water/ethanol, methanol, and hexane} in L6 muscle cell line by ELISA technique. This cell line used stably expressed myc-tagged GLUT4. Out of three extracts, the hexane extract at 32 μg/mL and methanol extract at 63 μg/mL effectively enhanced GLUT4 translocation. In contrast, water extract did not have any effect in the presence of insulin, whereas in the absence of insulin, water extract slightly enhanced GLUT4 translocation.

Radiotracers 2DG, 3-O-MG, and 2-NBDG are also used for screening of SGLT2 inhibitors. SGLT2 is a transporter protein in kidney cells responsible for reabsorption of glucose back into blood. SGLT2 inhibitors block these SGLT2 proteins which leads to the excretion of glucose from the body via urine, helping in normalizing the blood glucose levels and decreasing glucotoxicity to various organs and tissues (Huan et al. 2013). The practicability of the assay for the discovery of SGLT2 inhibitor was validated by measuring the inhibitory activity of phlorizin (a dihydrochalcone belonging to the family of bicyclic flavonoid) on the sodium-dependent uptake of 2-NBDG in stable CHO cells transfected with the plasmid vector for expression of SGLT2 receptor. Results showed that NBDG uptake decreased as the concentration of phlorizin was increased.

In type 1 diabetes mellitus, there is autoimmune destruction of β-cells. β-cell proliferation assay is an attractive and widely used assay to screen compounds that can normalize the function of β-cells or aid their differentiation and growth (Dsouza and Lakshmidevi 2015). RIN, HIT, MIN, INS-1, and β-TC cells are the most widely used cell lines in β-cell proliferation assay. Use of a new and relevant model for human β-cells, i.e., EndoC-βH1 cell line, has been reported by Tsonkova et al. (2018). In **β-cell proliferation assay**, replicating insulin-secreting β-cells are incubated with radiolabelled nucleoside analogue like methyl ^{3}H-thymidine and Bromodeoxyuridine/5-bromo-2′-deoxyuridine. These radioactive analogues help in direct measurement of proliferation. Kasabri et al. (2015) evaluated the anti-diabetic effect of *Gymnema sylvestre* aqueous extracts using β-cell MIN6 cells. Results showed that *Gymnema sylvestre* aqueous extract significantly induced β-cell MIN6 cell proliferation ($p < 0.001$) which enhanced the insulin secretion capacity and glucose uptake activity sequentially.

In addition to autoimmune destruction of β-cells, defects/decrease in insulin secretion also plays a key role in the pathogenesis of type 2 diabetes mellitus. In **insulin secretion assay**, insulin-secreting cell lines RIN, β-TC, MIN6, and INS-1 are most commonly used. The amount of insulin secreted in the presence or absence of test drug is measured using radioimmunoassay or ELISA. Kumar et al. (2017) investigated the β-cell proliferation effect of *Picrorhiza kurroa* extract in RIN5f cells. *P. kurroa* extract treatment significantly supported β-cell proliferation in a concentration-dependent manner. It was found that *P. kurroa* extract significantly increased the insulin secretion 3.1-fold at 6.25 mg/mL ($p < 0.01$) as compared to glucose control and glibenclamide control. This significantly enhanced insulin secretion was due to increase in proliferation of β-cells and enhanced insulin gene expression. Khazaei and Pazhouhi (2018) reported that there is a significant increase ($p < 0.01$) in glucose-stimulated insulin-releasing capacity of RIN-5F cells treated with *Trifolium pratense* extract. This effect of increase in insulin secretion of *T. pratense* extract is due to presence of genistein, a naturally occurring isoflavone compound with anti-diabetic effects. This anti-diabetic effect of genistein on β-cells is mediated through cyclic AMP/protein kinase A signalling and epigenetic regulation of gene expression.

Due to autoimmune destruction of β-cells, there is decrease in the number of GLP-1 receptors present on β-cells. GLP-1, an incretin-derived peptide, secreted by enteroendocrine L cells in the small intestine in response to the food intake, binds to GLP-1 receptor. GLP-1 is mainly responsible for increasing the glucose-dependent insulin release from β-cells, reducing glucose-dependent glucagon secretion, leading to decreased hepatic glucose output, inhibition of gastric acid secretion and gastric emptying, inhibition of appetite, and lowering of food intake. On binding of GLP-1 to GLP-1 receptor on β-cells, GLP-1R binds to Gs and Gq types of heterotrimeric G proteins, resulting in cAMP induction and Ca^{+2} release, followed by activation of downstream signalling such as extracellular signal-regulated kinase. Additionally, activated GLP-1R stimulates β-arrestin signalling and opening of voltage-dependent Ca^{+2} channels. To identify compounds possessing GLP-1 agonist activity, calcium mobilization assay, cAMP accumulation assay, β-arrestin recruitment assay, and receptor internalization assay are performed. CHO and INS cells stably expressing GLP-1 receptor are used in these assays. Huang et al. (2013) reported the hypoglycemic activity of *Momordica charantia* water extract and its fractions mediated by increase in secretion of GLP-1. The levels of GLP-1 were evaluated in murine enteroendocrine cell line by ELISA technique.

Decrease in glycogen synthesis is frequent in insulin resistance and type 2 diabetes. Glycogen synthase is one of the therapeutic targets for treatment of type 2 diabetes mellitus because glycogen synthase is involved in synthesis of glycogen in skeletal muscle cells. Skeletal muscle is the significant site for glucose uptake and glycogen storage in humans, both processes being regulated by insulin. In this assay, human muscle cells are incubated with test drug and [^{14}C] UDP-glucose and glycogen synthase. The amount of radiotracer incorporated into glycogen is measured radiometrically or with the help of glycogen kit assay (Nakano et al. 2017).

In addition to glycogen synthase, glucokinase is a glucose sensor, rate-controlling enzyme responsible for hepatic glucose metabolism and glycogen synthesis. Glucokinase activator activates glucokinase by binding to the allosteric site of glucokinase which increases the uptake of glucose in liver and also potentiates glucose-dependent insulin secretion, leading to decrease in blood glucose levels. Min et al. (2017) reported mangiferin (*C*-glycosyl xanthone) as a glucokinase activator by performing molecular docking studies on glucokinase-mangiferin complex and determining glucose consumption in HepG2 and C2C12 cells.

The uptake of glucose within the cells is also mediated through activation of PPAR-γ, a nuclear receptor, responsible for proliferation of adipocytes and enhanced uptake of lipogenesis, fatty acids, and glucose. PPAR-γ is expressed mainly in the adipose and β islet cells, and activation of PPAR-γ is a very critical step in glucose uptake which affects the glucose homeostasis and lipid metabolism as well as inflammation. It is also known that PPAR-γ affects glucose-stimulated insulin secretion. Therefore, assay based on measuring glucose uptake mediated by PPAR-γ can be used for screening PPAR-γ agonist which can be used as a potential anti-diabetic drug used in the treatment of type 1 or type 2 diabetes and related metabolic syndromes. HeLa cell line is extensively used in this assay. Gao et al. (2016) evaluated the anti-diabetic potential of 37 traditional Chinese medicine extracts by evaluating the activation of PPAR-γ by the PPAR-γ transactivation assay and seven medicines (root bark of *Lycium barbarum*, *Anoectochilus roxburghii*, the rhizome of *Phragmites australis*, *Pterocephalus hookeri*, *Polygonatum sibiricum*, fruit of *Gleditsia sinensis*, and *Epimedium brevicornu*) were able to significantly activate PPAR-γ ($p < 0.05$).

11.2.2.2 Obesity

Obesity is a significant risk factor and may lead to development of type 2 diabetes, hypertension, and cardiovascular disorders. In obesity, there is accumulation of lipid in insulin-responsive adipose tissues which leads to insulin resistance. Decreasing the accumulation of lipid in insulin-responsive tissues thus becomes one of the potential targets for screening of anti-diabetic and anti-obesity drugs. 3T3-L1 cell line is the most frequently used cell line in lipid accumulation assay since it exhibits appropriate features of lipid storage and glucose homeostasis. The amount of glucose uptake and lipid content within cell is the endpoint of this assay. To measure the amount of glucose uptake, cells are incubated with 2-NBDG, and the intensity of fluorescence is measured, whereas lipid content within the cell is measured using Oil Red O staining method by incubating the cells with test drug and Oil Red O stain, post which the dye retained in 3T3L1 cell is completely eluted with isopropyl alcohol, and absorbance was measured (Gulati et al. 2015). Gulati et al. (2015) evaluated the anti-diabetic activity of 7 Australian and 5 Indian Ayurvedic plant extracts using lipid accumulation assay, out of which 3 Australian plant extracts of *Acacia tetragonophylla*, *Beyeria leschenaultii*, and *Euphorbia drummondii* and the Indian plants extract of *Pterocarpus marsupium*, *Andrographis paniculata*, and *Curculigo orchioides* significantly reduced lipid accumulation in 3T3-L1 adipocytes ($p < 0.001$) and thus are promising drugs with anti-obesity activity. Kim et al. (2018) evaluated the lipid droplet accumulation inhibitory activity of quercetin and orlistat on polydimethylsiloxane-based adipocyte micro-cell pattern chips. Oil Red O staining method was used to determine the inhibitory activity of quercetin and orlistat. It was reported that orlistat exhibited lipid accumulation inhibition capacity of 19.9%, whereas quercetin exhibited lipid accumulation inhibition capacity of 24.4% in 3T3-L1 cells.

The cell-based lipid uptake assay is also used to screen for PCSK9 inhibitors. PCSK9 is a factor secreted from hepatocytes responsible for destruction of LDL receptors in hepatocytes by promoting its metabolism (Xu and Liu 2013). Xu and Liu (2013) developed a cell-based fluorescent-labelled LDL uptake model for screening PCSK9 inhibitors using HepG2 cell line to screen NINDS library. The amount of LDL uptake by HepG2 cell line was measured by fluorescence plate reader. On screening of NINDS library, it was found that colchicine inhibited PCSK9. Western blot analysis was done to confirm the PCSK9 inhibitory activity of colchicine.

In addition to lipid accumulation in obese individuals, leptin is a hormone to which obese people are resistant. Leptin is secreted by white adipocytes and is involved in whole-body weight regulation which exerts its effect by activating the leptin receptor (i.e., two main isoforms, OBRa and OBRb), and its blood level directly corresponds to the white tissue mass (Kim et al. 2014). Nowadays, preventing and overcoming leptin resistance is a major target for research of anti-obesity drugs. By increasing the number of OBR receptors, leptin signal transduction can be increased, and leptin sensitivity can be improved. HEK293, HeLa, and COS-7 cell lines are commonly used in this assay since they express OBR receptors.

As discussed earlier, obesity is a significant factor for development of atherosclerosis. Due to obesity, there is a rise in TG and LDL levels in the body which often leads to deposition of TG, LDL, etc. in the arterial walls leading to formation of atherosclerotic plaque. Initially in atherosclerosis, due to endothelial cell injury, leukocytes adhere and migrate into the endothelium with the help of various cell adhesion molecules and gap junctions. Monocytes then proliferate to macrophage cells inside the vessel which further take up oxidized LDL (Ox-LDL). These cells then interact with T cells and vascular smooth muscle cells and undergo differentiation which leads to atherosclerosis progression. Since proliferation of monocytes to macrophages is one of the most significant events of atherosclerosis, new drugs inhibiting this process can be good candidates for atherosclerosis treatment. With the help of cell differentiation assay, we can screen for drugs that inhibit the conversion of monocytes to macrophages. In cell differentiation assay, various monocytic cell lines like THP and RAW 264.7 are taken. These cells are differentiated in the presence of phorbol 12-myristate 13-acetate (PMA), Ox-LDL, MCSF, or GMCSF. Increase in upregulation of various differentiation markers like CD11, CD14, and type I SR can be monitored for evaluating differentiation. Similarly, drugs inhibiting the process of macrophage proliferation/macrophage foam cell formation are also good candidate molecules for treatment of atherosclerosis. The number of macrophages is determined after 24 h of exposure to varying concentrations of the test drug. To study the proliferation of macrophages, trypan blue dye and MTT reagent are used. Singh et al. (2015) evaluated the anti-atherosclerotic bioactivity and underlying mechanism of *Curcuma longa* oil in THP-1 macrophage cells. They found that the cells pre-treated with *C. longa* oil prevented lipid accumulation in the cells. To study the mechanism in detail, further RT-PCR was performed. Cells treated with *C. longa* oil decreased the mRNA-mediated levels of TNF-α, IL-1β, IL-6, and IFN-γ and enhanced the production of TGF-β in peritoneal macrophages. THP-1 macrophages treated with *C. longa* oil inhibited production of TNF- α and IL-1β by OxLDL and increased the levels of TGF-β, thus acting as an anti-atherosclerosis agent.

Similarly, drugs that inhibit the process of migration of leucocytes to injured endothelial cells can be screened with the help of leucocyte migration assay. To evaluate the migration of cells in response to activation due to endothelium injury, THP-1 monocyte migration can be assayed in Boyden chamber. The lower portion of the Boyden chamber is filled with medium obtained after Ox-LDL or TNF-α treatment of endothelial cells. Monocytes are incubated with/without test compound, and after incubation for around 6 h at 37 °C, cells that migrate into the lower chamber are counted by flow cytometry or various other conventional methods. Song et al. (2016) evaluated the anti-atherosclerotic effect of crocetin (obtained from

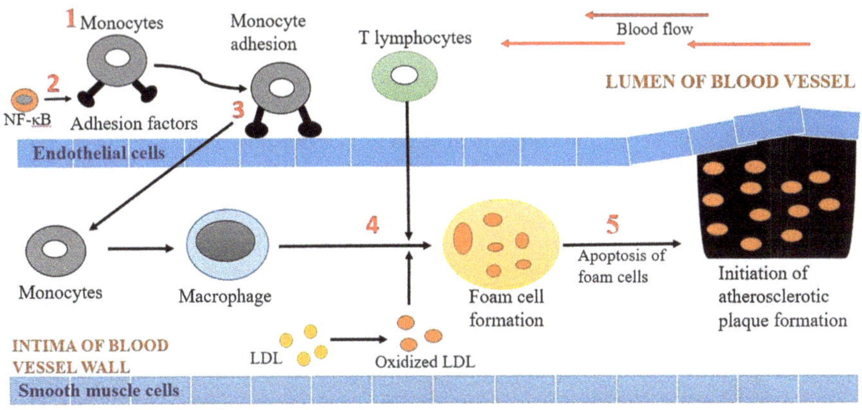

1- Leucocyte migration assay, 2- NF-κB assay, 3- Cell adhesion assay, 4-Macrophage foam cell formation assay, 5- Apoptotic assay.

Fig. 11.2 Schematic representation of cell-based assays used to screen anti-atherosclerotic agents along with targets involved in each assay

Crocus sativus) by performing functional assay. The researchers found that crocetin decreased the cell infiltration rate to the damaged site in endothelial cells. Cell adhesion assay is used to screen test drug compounds that prevent the adhesion of human monocytes to endothelial cells. Inhibition of the cellular adhesion step of monocytes to endothelial cells is due to inhibition of the expression of adhesion molecules on human monocyte cell line which further prevents progression of atherosclerosis. In this assay, endothelial cells and human monocyte cell line (THP-1) are incubated in the absence or presence of test drug compound. The total number of platelets adhered is calculated either microscopically or by protein estimation method. Hsueh et al. (2016) evaluated the anti-atherosclerotic activity of Naringin (flavanone), by performing cell adhesion assay. The researchers found that naringin prevented the expression of various adhesion molecules and chemokine which resulted in inhibiting the adhesion of THP-1 cells to endothelial cells by inhibiting the NF-κB translocation and phosphorylation of NF-κB and IκB signalling pathways. Macrophage apoptosis is one of the factors promoting atherosclerosis. Drugs which inhibit the apoptosis pathway are promising anti-atherosclerotic agents. With the help of **apoptosis assay**, we can screen new drug compounds that target macrophage necroptosis. Luo et al. (2015) examined the activity of isorhamnetin a flavonoid compound extracted from *Hippophae rhamnoides* on THP-1 macrophage cell line. To determine the anti-apoptotic effect of isorhamnetin, western blot analysis was performed. Isorhamnetin was found to enhance the upregulation of anti- apoptotic proteins and decrease the downregulation of pro-apoptotic proteins. The various cell-based assays used to screen anti-atherosclerotic agents are schematically represented in Fig. 11.2.

In addition to the abovementioned assays for atherosclerosis, a high-throughput screening assay using human HDL labelled with 1,1′-dioctadecyl-3,3,3′,3-′-tetramethylindocarbocyanine perchlorate and CHO cells has been reported. With the help of this assay, agents that inhibit the transfer of HDL into the cells were measured. Human hepatoma cells transfected with a plasmid encoding the promoter region of the human ABCA1 gene linked to a luciferase reporter along with LDL and test drug was used to study drugs which increase the expression of ABC1A1 and promote the formation of HDL particle, protecting against the development of atherosclerosis. Pyrromycin and aclarubicin (anthracycline antibiotic), daidzein, and pratensein (isoflavone compounds) were studied by this assay. It was found that both the antibiotics and isoflavone compounds promote expression of ABCA1 in cultured HepG2 cells which was determined by quantitative real-time RT-PCR (Etzion and Muslin 2009).

Various target cell-based assays like Liver X receptor agonist cell-based assay, NPY receptor cell-based assay, and β3 adrenoceptor assay are also used to screen anti-obesity agents. These assays help in target-based screening of natural drugs for treatment of obesity (Vogel 2002).

Cardiovascular disorders like ischemic heart disease arise due to atherosclerotic plaque formation in blood vessels. Ischemic heart disease is a condition which causes extensive death of cardiac cells due to activation of apoptotic pathways via regulation of *Bcl-2* and *Bax*. Therefore, compounds with anti-apoptotic and anti-necrosis activity can be positioned as good cardioprotective agents. Wang et al. (2017) evaluated the cardioprotective activity of hypaconitine and glycyrrhetinic acid on H9c2 cells under oxygen-deprived glucose condition (OGD)-induced injury. To evaluate the cardioprotective effects of hypaconitine and glycyrrhetinic acid, pathology and morphology of the nucleus and ultrastructure of H9c2 cells was determined by TEM and Hoechst 33342 staining techniques. LDH, creatine kinase-myocardial band isoenzyme, and aspartate transaminase levels were determined from the supernatant of cultured H9c2 cells by ELISA. To determine apoptosis, FITC-AV/PI double staining assay was performed which showed a decrease in the rate of apoptosis via the PI3K/Akt signalling pathway.

Guo et al. (2012) developed a cell-based assay model which is helpful to recognize cardioprotective agents. A plate-based respirometry apparatus was designed to measure mitochondrial O_2 consumption rate and glycolysis by intact cells on a 24-well plate. There is disposable cartridge with fluorescent probes sensitive to PO_2 and pH resting on the top of the plate which travels in a vertical axis. On lowering of plunger, cells are trapped in the microchamber which allows for the measurement of changes in PO_2 and pH. On continuous lowering of the plungers, cells consume all available O_2 in the microchamber, creating an ischemic-like state. In the extracellular flux apparatus, an argon gas flow was adapted in the head space of the cartridge due to which a better control over O_2 level in the media is achieved. These modifications produced a 24-well model of IR injury which can be used for HTS of ischemic drugs from natural source. Figure 11.3 represents the various cell-based assays covered in this section which are used to screen anti-obesity and hypolipidemic agents.

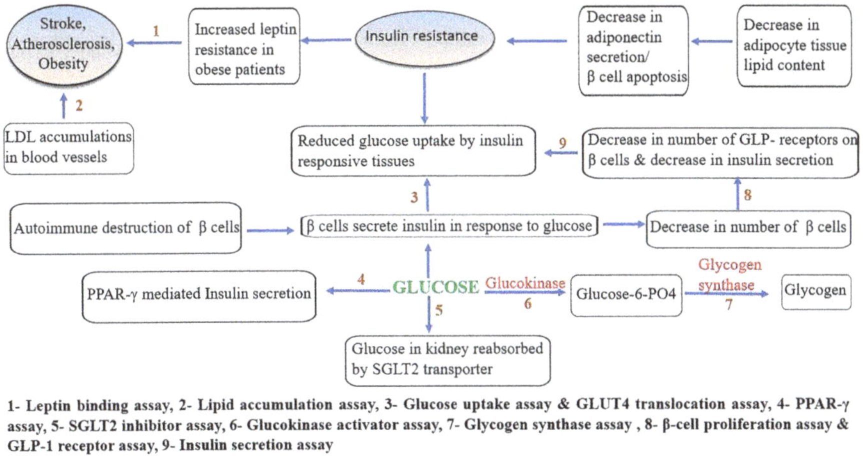

Fig. 11.3 Schematic flow chart of cell-based assays used to screen anti-diabetic and anti-obesity agents along with targets involved in each assay

11.2.3 Anti-cancer Activity

A study by the American Cancer Society in 2019 has revealed that there will be an estimated 1,762,450 newly diagnosed cancer cases and 606,880 cancer deaths in the USA alone. Cancer is a deadly disease affecting the health of humans from all strata of life. Mutations in genes can lead to altered cellular functions leading to cancer. In cancers, the cell cycle is disturbed leading to abnormal proliferation. Cell division and growth under normal conditions is governed by proto-oncogenes which convert to oncogenes during genetic mutation and threaten cell existence. Besides, the lack of tumor suppressor genes causes uncontrolled cell division. In this section, we have reviewed various cell-based screening assays for anti-cancer drugs of natural origin.

As described earlier, colorimetric techniques of detection are the most routinely used methods to determine cell viability. Alamar blue/resazurin; sulforhodamine B; tetrazolium reagents such as (3-(4,5-dimethylthiazol-2-yl)-5-(3-carboxymethoxyphenyl)-2-(4-sulfophenyl)-2H-tetrazolium) MTS, (2,3-bis-(2-methoxy-4-nitro-5-sulfophenyl)-2H-tetrazolium-5-carboxanilide) XTT, and the (water-soluble tetrazolium salts) WST; and electron acceptor reagents such as phenazine methyl sulphate/phenazine ethyl sulphate and CAM are the dyes used in cell viability assay. These assays are based on the principle of conversion of these dyes to their fluorescent/colored reduced forms in metabolically viable cells. The cytotoxicity of Thai herb *Clinacanthus nutans* was validated by both MTT and SRB assays by Vajrabhaya and Korsuwannawong (2018). An excellent correlation was observed between the two cytotoxicity assays ($p > 0.05$). Wu et al. (2017) evaluated the anti-cancer activity of Pinicolol B, a natural product from *Antrodia cinnamomea* by determining cell viability in nasopharyngeal carcinoma cell line using MTS assay.

Clonogenic cell survival assay has been studied as an effective tool for screening compounds acting against tumor cells. Clonogenic cells have the potential to proliferate indefinitely which results in formation of a clone or colony which is visible to the naked eye. In this assay, a known number of cells are plated post-treatment. The cells are allowed to grow, followed by a staining step wherein the colony-forming surviving cells are counted. This is measured in terms of plating efficiency, i.e., the percentage of cells seeded into a plate that result in formation of a colony/clone. Jerantinine B, one of the seven Aspidosperma indole alkaloids isolated from the leaf extract of *Tabernaemontana corymbosa*, was evaluated for its anti-cancer activity by clonogenic assay. This antiproliferative assay revealed that jerantinine B and jerantinine B acetate considerably inhibited growth and colony formation in HCT 116 human cancer cell lines (Qazzaz et al. 2016).

The two major forms of cell death in normal and disease pathologies include apoptosis and necrosis. Many assays for detection of apoptosis are reported, but a very few assays measure necrosis which is characterized primarily by permeabilization of plasma membrane. Necrosis can be quantified by measuring the levels of enzyme LDH. LDH is a soluble cytoplasmic enzyme which upon damage to the plasma membrane is released into the extracellular space and detected with the aid of a tetrazolium salt (Fotakis and Timbrell 2006). Using a colorimeter or fluorimeter, the released LDH can be measured which relates to cell viability. Acetone extract of roots of *Rubus fairholmianus* was evaluated on human colorectal cancer cells for its antiproliferative activity by LDH assay (PlackalAdimuriyil George et al. 2015).

Demchenko (2013) screened anti-cancer drugs using annexin V affinity assay which involves binding of the highly fluorescent annexin V to phosphatidylserine externalized from cells due to loss in cell membrane integrity. In cells which are viable, phosphatidylserine is located on the surface of the cytoplasm of the cell membrane. During the intermediate stages of apoptosis, phosphatidylserine is ferried from the inner to the outer surface of the membrane, making phosphatidylserine available for detection in the external cellular environment. Apoptosis is programmed cell death, which includes morphological changes and a set of biochemical steps, resulting in activation of caspases and fragmentation of cells into small apoptotic bodies. Activation of caspases in cancer is important as they are a family of protease enzymes playing an essential role in programmed cell death. The researchers developed a method using hapten (biotin)-labelled annexin V which could be detected fluorometrically. Annexin V is unable to bind to viable cells since the molecule cannot penetrate intact phospholipid bilayer. Besides annexin V, a DNA stain, propidium iodide can be used to differentiate viable and apoptotic cells. The cells are analyzed by flow cytometry in the annexin V and propidium iodide assays. Honokiol, an active principle isolated from the aqueous extracts of *Magnolia grandiflora*, was found to be a potent anti-tumor agent when screened using this assay in endothelial cell lines (Bai et al. 2003).

Bcl-2 family of proteins which consist of both pro-apoptotic and anti-apoptotic factors play a crucial role in development of malignancies by promoting its overexpression in pathogenesis and can be effectively tapped as an anti-cancer target. The anti-apoptotic Bcl-2 family proteins mainly exhibit their activity by sequestering pro-apoptotic BH3-only proteins (Bim, Bid, Bad). Therapeutic anti-cancer molecules are developed by mimicking BH3 molecule which antagonize the interaction between Bcl-2 proteins and BH3-only proteins. Bcl-2 inhibitors are reported to exhibit caspase-dependent cell death by liberating the pro-apoptotic BH3-only proteins which result in mitochondrial outer membrane permeabilization causing the release of cytochrome c and other pro-apoptotic factors in the region of apoptosome and activation of caspases-9/3/7 (Hassig et al. 2014). Cell-based orthogonal assays were performed to screen samples for their enhanced pro-apoptotic Bcl-2 family activity. Gossypol, a natural phenol derived from the cotton plant, was found to be one of the drugs actively targeting Bcl-2 proteins. Also, during the effector phase of apoptosis, caspases and proteases are activated and result in cleavage of intracellular substrates such as cytokeratin-18. This results in disruption and disassembly of the skeleton and cytoplasmic inclusions. These cleavage products are indicative of cell death. Anti-cancer drugs can be effectively screened by measuring their ability to activate caspase-3/7 in tumor cells or by measuring the levels of cytokeratin-18. Single multiplex assays have been used to measure cell viability and caspase-3/7 activation in order to determine the capability of extracts to produce programmed cell death. Caspase-3/7 activity can also be monitored with the aid of luminescent assays. Li et al. (2006) carried out caspase-3/7 assay to study apoptosis in HK-60 and MDA-231 cells. Activated caspases-3/7 cleave the firefly luciferase substrate Z-DEVD-aminoluciferin sodium salt to liberate aminoluciferin which reacts with luciferase resulting in generation of measurable light. This luminescence directly correlates to cell death. Potent anti-cancer drugs enhance the caspase activity which is indicated by increased luminescence. Oxadiazine nocuolin A (NoA) extracted from various cyanobacterial strains of genera *Nostoc*, *Nodularia*, and *Anabaena* was screened by this assay (Voráčová et al. 2017). NoA-induced cell death has been attributed to caspase-dependent apoptosis. Besides, NoA exhibits considerable antiproliferative activity against several human cancer lines. Another assay based on the caspase-dependent cleavage of cytokeratin-18 (CK-18) was carried out by Hägg et al. (2002) using the M30 Apoptosense ELISA kit in MDA-MB-231 cell line. Levels of caspase-cleaved CK-18 and the accumulated cleavage-products in cells were measured. A library of natural products was screened by Fayad et al. (2009) for their apoptotic activity in HCT116 colon carcinoma cells. The alkaloid thaspine obtained from *Croton lechleri*, daunorubicin obtained from *Streptomyces peucetius*, and streptonigrin obtained from *Streptomyces flocculus* were found to be potent apoptosis-inducing agents.

Some other novel targets for screening anti-cancer drugs include enzymes indoleamine 2,3-dioxygenase (IDO1) and tryptophan 2,3-dioxygenase (TDO), which catalyze the conversion of tryptophan to N-formylkynurenine (NFK). Overexpression of IDO1 has been associated with the progression of cancer, whereas preclinical studies have indicated TDO as a target for cancer immunotherapy and neurodegenerative disease. By consuming tryptophan, IDO modulates the function of effector T cells and increases autoantibody titers produced by B cells. Assays have been developed using NFK Green, a chemical probe, which reacts with

NFK to produce green fluorescence. This assay evaluated the tryptophan-metabolizing ability of cells and estimated the activity of IDO and TDO inhibitors. Natural IDO inhibitors such as curcumin, exguamine A, *Crinum latifolium* leaf extract, neem (*Azadirachta indica*), tryptanthrin, brassinin, and annulin A and B have been screened by this assay (Zulfiqar et al. 2017).

Altering the cell cycle can serve as excellent target for screening anti-tumor drugs. Cell cycle arrest assay operates through the checkpoints of the cell cycle. A group of researchers have reported that terpenoids and mixed isoprenoids suppressed the growth of tumor cells by initiating apoptosis and arrest of the cells in G1 phase in the cell cycle (Mo and Elson 2004) Ursolic acid present in plants like rosemary and thyme induces apoptosis in HCT-116 cell lines by arresting the cell cycle in G1 phase (Dar et al. 2016). Table 11.2 lists the various assays that have been reported for screening drugs of natural origin for their anti-cancer activity.

11.2.4 Anti-viral Activity

Viruses are small agents which invade body and replicate inside host cells, producing toxins that cause viral infections. Although the immune system helps to destroy viruses, anti-viral immune responses may result in tissue damage and illness. Inhibiting infectious virus growth inside cells requires the ability of the drugs to traverse across the cell membrane. Assays to screen anti-viral drugs are usually based on automated readouts, i.e., either microscopy or luminescence-based assays, which measure virus yield reduction (VYR) and CPE reduction as their endpoints and are described below. CPE assays monitor cell viability by measuring the drug-induced inhibition/reduction in viral CPE by means of employing assays such as colorimetric and luminometric cell death assays, viral plaque assays, and microscopy-based assays (Alves et al. 2018).

The simplest and most basic techniques used to screen anti-viral drugs employ colorimetric viability assays using neutral red, resazurin, MTS dyes, etc. The neutral red (3-amino-7dimethyl-2-methylphenazine hydrochloride) dye uptake cytotoxicity assay provides a quantitative estimate of the viable cell numbers in a culture by measuring the ability of a drug to inhibit the uptake of the dye. The same group evaluated the anti-reovirus (double-stranded RNA virus) activity of kuraridin isolated from the roots of *Sophora flavescens* against human type 1–3 reoviruses and Korean porcine reovirus using neutral red assay.

Another simple assay to measure cell viability was reported by Liang et al. (2013). A luminescence-based assay was performed which relies on ATP measurement to estimate the number of viable cells. The assay measures inhibition of virus-induced CPE by a test compound using luminescent cell titer kits and exhibits a luminescent signal which is directly proportional to the amount of cellular ATP present, which corresponds to the number of metabolically active cells. The researchers evaluated the anti-HIV-1 properties of *Sanguisorba officinalis* which was found to exhibit therapeutic efficacy against HIV1 (ADA and HXB2) and also

Table 11.2 Assays employed in the search for novel drugs of natural origin for their anti-cancer activity

Cell-based assay	Compound/herb	Characteristic/function	References
Anti-proliferation	Honokiol (magnifolia extract)	Inhibited angiogenesis by interfering with phosphorylation of vascular endothelial growth factor receptor (VEGFR2) in human endothelial cells. It also inhibited the growth of transformed epithelial cells in vitro	Bai et al. (2003)
	Curcumin	Inhibited COP9 signalosome-associated kinase activity resulting in downstream effects of potent angiogenic factor VEGF	
	Epicatechin gallate and genistein	Inhibited the activity of the 26S proteasome, may also regulate VEGF synthesis	
	Resveratrol	Inhibited tumor promotion by protease inhibitors activity	
Rapid colorimetric assay (MTT) assay	*Ginkgo biloba* L	MTT assay converts MTT into formazan crystals by living cells and determines the mitochondrial activity. Mitochondrial activity relates to the viability of cells	El-Menshawi et al. (2010)
The ATP bioluminescence assay	Ellagic acid extracted from *Galla chinensis*	Determines tumor cell viability by measuring the ATP (exhibited light) which is produced by adding luciferase-luciferin reagent. Inhibition in luminescence relates to no cell growth	Ceci et al. (2018)
[^{3}H] thymidine assay	Paclitaxel	The amount of thymidine incorporated into DNA co-relates to the rate of proliferation. Inhibitors of proliferation will result in reduction of thymidine signal and is measured by scintillation	Li et al. (2006)
Neutral red assay	*Glandularia selloi* (Spreng)Tronc. leaf extract	Neutral red determines cell viability by binding to anionic sites. Cell viability is expressed as concentration-dependent NR uptake into the cell	Figueiró et al. (2016)
Colony-forming assay	Chinese herbal compound Songyou Yin	Anti-tumor agents after adding to colony-forming cell line results in decrease in colony formation. Inhibits cell growth	Jia et al. (2012)

showed inhibitory activity against simian immunodeficiency virus infection by a luminescence assay.

In order to determine titers of viruses, plaque assay plays a vital role in identifying inhibitors of Enterovirus 71 (EV-A71). Plaque assay works on the principle that replication of a single infectious virus in a host cell leads to invasion of its progeny to adjacent monolayered cells under the cover medium. Yin et al. (2019) used *Homo sapiens* muscle rhabdomyosarcoma cell line to examine the anti-plaque-forming activity of nobiletin, a flavone which showed complete prevention of plaque formation at 10uM. The integrity of this monolayer was used as a measure of the compound's inhibitory activity.

An alternative to the dye-based and luminescence-based viability assays, **reporter virus/cell assays** can be used to monitor viral replication in a much efficient way. In reporter virus assays, a reporter gene, inserted into a nonessential region of viral genome, causes the expression of corresponding reporter protein when virus replicates inside cells. Hence, when viral inhibitors are added, there is a decreased expression in the reporter protein indicated by the loss-of-signal endpoint. Three types of reporter assays have been reported
for indirect measurement of anti-viral activity: the first type of assay involves viral minigenome system wherein the cells with transfected plasmids produce a virus-like RNA encoding a reporter gene, controlled by viral polymerase regulatory sequences. The second type of assay involves the engineering of a replication-incompetent virion in which a reporter gene replaces a viral gene. The third type is to generate a replication-competent reporter virus. However, the major drawback of these assays is their requirement for expression of a reporter gene, an indicator of viral polymerase activity. Hence, these assays cannot be applied to new emerging viral strains, and the need for developing a non-reporter-based screening assay arises (Wang et al. 2016).

Lucas-Hourani et al. (2013) screened 3040 molecules from a natural product library using luciferase-based reporter gene assay to identify small molecules targeting nsP2-mediated transcriptional shutoff. nsP2 protein is a nonstructural protein of CHIKV which induces a transcriptional shutoff leading to blockade of cellular anti-viral response. On evaluating the results, they found one natural compound ID1452-2, a derivative of taxoid oxetan, partially blocked nsP2 activity.

Vasou et al. (2018) proposed a class of viral interferon antagonists to counteract host IFN response as an important target for anti-viral drugs. These sites are crucial as most viruses encode these proteins and their suppression can help in exerting anti-viral activity. The researchers have developed a cell-based platform using two reporter cell lines to identify the activation of IFN induction via an eGFP gene which is controlled by IFNβ or IFN-stimulated response element (ISRE)-containing promoter, respectively. Hence, an IFN antagonist in the reporter cell line will result in IFN response blockade and subsequent inhibition in the eGFP expression. An anti-viral drug would therefore inhibit IFN antagonist function and facilitate IFN response restoring eGFP expression. The same group demonstrated the assay using human respiratory syncytial virus and cytomegalovirus, and both the screens performed robustly with Z-factor scores of >0.6.

Since the enterovirus EV71 is one of the majorly studied organisms for screening of anti-viral agents, it becomes really interesting to know how exactly it spreads infection in host cells. One of the assays targeting the first step of virus infection is the **cell-binding assay** as studied by Cao et al. (2015). Initial screening involved expression of GFP fluorescence in order to define the effective concentration range of test compound. Further, anti-EV71 activity was evaluated by infecting RD cells with EV71, followed by treatment with varying concentrations of compounds under investigation. The amount of EV71 RNA remaining in host cells after incubation was determined using qRT-PCR which showed that the reduction in the number of EV71 RNA was concentration-dependent.

Another important assay for screening of anti-virals is the syncytium induction assay. Syncytia are formed due to fusion of virus-infected host cells with neighboring cells to form a multinucleate mass leading to CPE. Therefore, this assay was employed to evaluate enzootic bovine leukosis, a neoplastic disease common in cattles, caused by BLV. This assay referred to as luminescence syncytium induction assay involves use of a reporter cell line CC81-BLU3G transfected with a reporter plasmid pBLU3-EGFP (BLV long terminal U3 region as the promoter and enhanced GFP, i.e., eGFP, as the reporter gene). Expression of the green fluorescent protein leads to direct visualization and quantification of BLV infectivity due to formation of fluorescent syncytia. This assay was found to be more sensitive to detect cell-to-cell and cell-free BLV infection. The applicability of this assay was also checked using plasma from BLV-infected cows. Chathuranga et al. (2019) demonstrated the antiviral activity of *Plantago asiatica* and *Clerodendrum trichotomum* using the syncytia assay against respiratory syncytial virus.

The major limitation of viral replication assay to screen anti-viral compounds of natural origin is its low-throughput nature. Viral replicon assay has been established to overcome this limitation. In this assay, two reporter genes Rluc and neomycin phosphotransferase are engineered into the flaviviruses like hepatitis C virus, West Nile virus, and Zika virus replicon which results in Rluc/NeoRep. BHK-21 or A549 cell lines transfected with Rluc/NeoRep are the most commonly used cell lines in this assay. In replicon-based assay, the compound to be screened is added to cells with replicon RNAs under full replication. On the other hand, in the standard viral infection assay, the compound to be screened is added either concomitantly or before viral infection of cells. Drugs which decrease the Rluc activity in the cell line are potential candidates to be developed as anti-viral agents. To measure the reduction in Rluc activity, real time RT-PCR is performed. Besides, it provides added advantages such as higher success rate in animal studies and no formation of infectious virions. Hishiki et al. (2017) used the viral replicon assay to check at which stage hirsutine (indole alkaloid from *Uncaria rhynchophylla*) target the dengue virus (DENV) life cycle. The researchers found that hirsutine did not decrease the luciferase activity in culture supernatants which helped the researchers establish that hirsutine doesn't inhibit DENV replication during viral-genome RNA translation and synthesis. Varghese et al. (2016) evaluated berberine, abamectin, and ivermectin as anti-viral drugs against chikungunya and other alphaviruses. The researchers evaluated the anti-viral activity of 3000 compounds using BHK-21 cell line transfected with a stable CHIKV replicon with a luciferase reporter, out of which abamectin ($EC_{50} = 1.5\ \mu M$) and ivermectin ($EC_{50} = 0.6\ \mu M$) which are fermentation products generated by a soil-dwelling actinomycete, *Streptomyces avermitilis*, and berberine an isoquinoline alkaloid ($EC_{50} = 1.8\ \mu M$) inhibited CHIKV replication in a dose-dependent manner and also had inhibitory effect against other alphaviruses.

11.2.5 Anti-mycobacterial Activity

Tuberculosis is an intracellular infection caused by MTB that remain housed within macrophages and is of two types – pulmonary and extrapulmonary tuberculosis. Mycobacterial infection of macrophages is indicated by initiation of a localized inflammatory response which serves as the driving factor for tissue destruction leading to cavity formation in pulmonary tuberculosis. The key players in the pathogenesis of both pulmonary and extrapulmonary tuberculosis are the matrix metalloproteinases (MMPs); pro-inflammatory TNF-α, IL-12/23p40, and IFN-γ; and anti-inflammatory (IL-10) cytokines. Cell-based assays reported in literature for screening of anti-mycobacterial and anti-microbial drugs of natural origin are as below.

Currently used cell-based assays include two basic methods for quantification: direct and indirect method. Direct method measures the number of CFUs which is simple and cost-effective. However, it is accompanied with longer testing duration and higher chances of error. Nair et al. (2017) have reported anti-tubercular activity of garlic extracts and its isolates in RAW 264.7 mouse macrophages. Initial testing for anti-TB activity of various isolates was done using Resazurin Microtiter Plate Assay (REMA). Later on, these results were validated using mouse macrophages RAW 264.7, and it was found that the anti-tubercular activity of *Allium sativum* extract against MTB-infected macrophages was appreciable and concentration-dependent.

Due to innate immune responses being suboptimal in tuberculosis, modulation of host immune responses using polysaccharides derived from medicinal plants can become an essential strategy to develop novel anti-mycobacterial agents. Gupta et al. (2016) performed CFU assay where a polysaccharide G1-4A derived from crude extract of *Tinospora cordifolia* inhibited survival of susceptible and resistant MTB strains. This study was done using in vitro and ex vivo macrophage models that were infected with MTB, followed by G1-4A treatment, and the number of colonies were counted post incubation. Induction of NO, pro-inflammatory cytokines, and surface expression of CD-86 and MHC-II happened in a TLR4-MyD88-dependent manner. Surface expression of CD-86 and MHC-II was studied by immunophenotyping, and phagocytosis assay was performed using GFP-expressing MTBH.

An indirect method involves testing of activity by measuring either fluorescence or luminescence. However, it involves high expenses due to rigid infrastructural requirements.

Of the widely accepted causes of anti-microbial resistance in patients with bacterial infections, one is a subpopulation of non-replicating persistent (NRP) bacteria. An assay was developed by Cho et al. 2007, targeting such non-replicating bacteria called luminescence-based low oxygen recovery assay (LORA) for MTB that contains a plasmid with a bacterial luciferase gene driven by an acetamidase promoter. A comparative study was also performed by exposing microplates to aerobic and anaerobic conditions and results showed that MICs of 31 selected antibiotics like rifampin, ciprofloxacin, streptomycin and so on were higher under anaerobic conditions.

The disease-causing MTB strain used for screening of anti-mycobacterial compounds could either exhibit auto-luminescence or plasmid-induced luminescence or fluorescence. A BSL-2-compatible infection model involving macrophage-passaged MTB was developed by Schaaf et al. (2016) to demonstrate that rifampicin, fluoroquinolones, and streptomycin show higher MICs against intracellular MTB. It involved THP-1 monocytes as host cells along with $H_{37}R_{V}$-derived auxotroph strain $mc^2$6206. Insertion of a plasmid pMN437 into this strain allowed direct quantification of levels of infection due to expression of green fluorescent protein. A sensitive and accurate model also offered the advantage of being easily scalable and simple to perform.

Assays for monitoring cell viability involving luminometric measurements in bacterial cells such as genetically modified strains of MTBH-expressing *lux* genes {bacterial luciferase either from *Photorhabdus luminescens* (Sharma et al. 2014) or from *Vibrio harveyi* (Larsson et al. 2014)} are reported in literature. The oxidation of reduced flavin mononucleotide (FMNH2) leads to bacterial luminescence confirming cell viability. The number of luminescence producing viable cells was determined and verified using colony-forming unit analysis (Larsson et al. 2014).

Nonpathogenic, biosafety level 2 bacteria like *M. aurum*, *M. smegmatis* $mc^2$155, and *M. neoaurum* are fast-growing and can be generally employed as surrogate models for MTB in preliminary screening of anti-mycobacterial agents since they do not require very stringent biosafety conditions. Gupta and Bhakta (2012) reported the use of an integrated host model using unlabeled and FITC-labelled *M. aurum* to infect mouse macrophage RAW 264.7 cells and measured the fluorescence intensity. This study was carried out at different pH ranges and varying time intervals in order to estimate the antibiotic susceptibility and growth kinetics of the selected host-model system. Owing to the limitation of selectivity by the use of nonpathogenic models as surrogates for MTB, Diop et al. (2018) explored utility of mammalian and protozoal hosts infected by a pathogenic bacterium. *M. marinum* is a fast-growing bacterium with lower safety risks and easy manipulation and transition into a potent zebrafish infection model. A phenotypic assay for host-pathogen was designed based on the monitoring of protozoal host amoeba *Acanthamoeba castellanii* infected with GFP-expressing strain of *M. marinum*. The difference in fluorescence readouts between the point of infection of host amoeba with GFP-expressing *M. marinum* and post its incubation period were measured. Aqueous extracts of *Combretum aculeatum* and *Guiera senegalensis* from Senegal exhibited anti-mycobacterial activities on the authenticated amoeba GFP-expressing *M. marinum* host-pathogen assay. Host model systems could also include mammalian host cells that offer better phenotypic resemblance to the disease in humans and also conserve the basic cellular functions of a housekeeping gene. In a whole-cell-based phenotypic screen developed by Trofimov et al. (2018), a microglial phagocytic cell line, BV2, and the amoeboid organisms *Dictyostelium discoideum* and *Acanthamoeba castellanii* have been employed to characterize anti-microbial activity of selected antibiotics. GFP-expressing *M. marinum* was employed to measure total fluorescence exhibited by the bacterial mass. Also, luxCDABE-expressing auto-bioluminescent *M. marinum* strain was employed for antibiotic activity testing of selected compounds. Fluorescence signal intensity was proportional to the cell count. Membrane permeability and loss of substrate attachment were indicative of cell death. The study also evaluated mode of action of the selected anti-TB agents using generation of overexpressor *M. bovis* BCG strains.

11.2.6 Anti-microbial Activity

The remarkable spread of resistance to anti-microbials used for therapy has had a great impact on public health. Natural products become excellent sources of newer drug molecules that can combat this looming healthcare emergency. Conventional methods involving luciferases, light-emitting enzymes, or even the autofluorescent proteins have been widely used and reported for the discovery of novel anti-microbial agents.

A very interesting methodology has been proposed by Lyu et al. (2018) which targets induction of host defense peptides (HDPs) that would help in controlling the disease. HDPs are a part of innate immune system that can control infection and limit it by employing various immune cells to the site of infection and further counteracting endotoxin-induced inflammation. HDP induction can be done by constructing a stable chicken HTC macrophage cell line using avian β-defensin 9 (*AvBD9*) gene promoter (readily inducible) as demonstrated by Lyu et al. (2018). Induction of luminescence was recorded as a measure of HDP induction. A potent natural lead, wortmannin, was explored as a powerful inducer of HDP synthesis. It was also found to act synergistically with butyrate in enhancing anti-bacterial activity of chicken monocytes.

An approach involving luminescent bioreporters in conjunction with HPLC was thought to be more sensitive and speedier in comparison to the traditional anti-microbial assays like turbidimetry. Such bioluminescence assays can be applied to natural product crude extracts to establish anti-microbial activity of the isolated fractions. One novel method that uses HPLC micro-fractionation of crude drug extracts coupled with a recombinant, luminescent bioreporter *Escherichia coli* K-12 strain, has been developed. After preliminary screening by TLC bioautography, the crude ethyl acetate extract of *Pycnoporus cinnabarinus* (fungal strain FBCC130) was further analyzed for anti-microbial activity using HPLC-UV of its micro-fractionations. Luminescence was measured to quantify the dose-dependent anti-microbial response (Järvinen et al. 2016).

To exemplify a novel approach highlighting the applicability of sensitive pathway-based assay, Fischer et al. (2004) focused on bacterial fatty acid synthesis (FAS) pathway (type II) which is essential for bacterial viability. FAS type II is completely different from human FAS type I and is unexploited as a target for screening of anti-microbials. Phylogenetic relevance of *Bacillus subtilis* with most of the pathogens made it a model organism of choice. The reference gene yjaX (also called fabHA) represented a central enzyme of the FAS pathway. This promoter gene was fused with firefly luciferase to construct a reporter strain that selectively indicated induction of reporter system by FAS inhibitors. This assay was successfully employed to identify a well-known FAS inhibitor, cerulenin, which inhibits enoyl-ACP reductase FabI in *Bacillus subtilis*.

Assays based on mRNA expression profiling in bacteria have found utility as targets in anti-bacterial screening and mechanism of action studies. An assay was developed and validated involving **antibiotic biosensors** for five biosynthetic pathways in bacteria: biosynthesis of DNA, RNA, protein, cell wall, and fatty acid. Cellular biosensors were based on fusion of promoters to reporter genes by genetic engineering. *yheI*, *yorB*, *yvgS*, *ypuA*, and *fabHB genes* were taken from *B. subtilis*, and each of these were fused with firefly luciferase in order to create a biosensor strain. Signal transduction was observed in the form of luminescence. An increase in luminescence indicated induction of the biomarker system. Subsequent mechanism of action studies was performed by incorporation of radiolabelled precursors using *S. aureus* revealing that the gene *fabHB* was selectively induced by fatty acid synthesis inhibitors; *yvgS* induced by RNA biosynthesis inhibitors; *ypuA* induced by cell wall biosynthesis inhibitors; *yorB* induced by DNA biosynthesis inhibitors; and finally *yheI* induced by inhibitors of protein biosynthesis. This assay led to identification of the mechanistically unexploited antibiotic, Ferrimycin A1, as a selective inhibitor of protein synthesis (Urban et al. 2007).

Antibiotic biosensor strains have been utilized in the identification of inhibitors of DNA metabolism through initiation of SOS response, an indicator of bacterial DNA damage. *RecA* promoter and GFP gene were fused to generate a plasmid-based reporter vector which was introduced into an *Escherichia coli* K-12 MG1655 strain. *RecA* senses accumulation of single-stranded DNA and interacts with *LexA* (transcriptional repressor for SOS response) to cause its cleavage and thus allow expression of SOS response to deal with DNA damage. Inhibition of bacterial cell cycle was recorded as GFP fluorescence readout (Fan et al. 2014).

One essential bacteria-specific target is the S1-dependent translation initiation pathway. Initiation is the first and the rate-limiting step of translation. Existing antibiotics have been explored to have translation as the target pathway. However, they are yet to be explored as targets for initiation. This is because initiation in bacteria can be carried out by using different mechanisms. The ribosomal protein S1 is specifically recruited to promote the binding of 30s ribosome to the translation initiation region in bacterial mRNA, i.e., 5′ UTR, encompassing the start codon. In an experiment carried out by Raneri et al. (2015) an aminoglycoside antibiotic, kasugamycin, was found to be an inhibitor of translation initiation. This was done using *E. coli* strain (AS19/DH10B) to screen inhibitors of S1-dependent pathway. eGFP was used as the reporter gene due to its ease of monitoring and amenability to high-throughput screening. pGM991 or pGM999 have been used as plasmids. Quenching of fluorescence was used as a measure of inhibitory activity.

An assay specific for gram-negative bacteria (pathway-based) has been proposed by Zhang et al. (2018). LPS is a component of the outer layer which prevents many antibiotics from travelling inside the cell, thus acting as a protective barrier for gram-negative organisms and a reason for antibiotic treatment failure. An ATP-dependent enzyme, MsbA, is responsible for translocation of LPS across inner membrane and forms an important target for screening of agents active against infections of gram-negative organisms. The utility of this assay was exploited using *Acinetobacter baumannii*. Inhibition of MsbA leads to accumulation of LPS inside the cell, thereby disrupting the essential steps that occur while LPS levels are normal. Visualization of LPS was done using a fluorescent probe, Dansyl-PMBN. Since bacterial growth was used as a readout, this pathway-based phenotypic screen was inexpensive and easy to implement. These conventional growth-based assays provide poor sensitivity to detect anti-microbial agents that are present in low concentrations, failing to screen agents that kill small colony variants or bacteria present within biofilms.

AK is an intracellular enzyme that acts as a reporter for cell lysis. The enzyme is released into the extracellular space only when a cell is lysed. AK assay helps to confirm whether the anti-microbial agent is bactericidal or bacteriostatic. Forbes et al. (2015) demonstrated selectivity of AK assay for bactericidal agents. Conversion reaction of ATP to ADP and AMP by the released AK acts as a driving force for activation of luciferases leading to luminescence as an indicator of cell lysis. Similarly, AK assay has also been used to screen bactericidal activity of known anti-microbial agents or agents that can inhibit bacterial biofilms (Jacobs et al. 2012).

Various other target-based assays that could be exploited in the discovery of anti-microbials of natural origin are assays to identify regulators of TNF-α (Stanley et al. 2014), macrophage-based assay selective for *Salmonella typhimurium* (Ellis et al. 2019), assays targeting FtsZ protein (Yuan et al. 2019), and nuclear-receptor based screening assays (Ahn et al. 2008). Table 11.3 illustrates the various assays used to screen drugs of natural origin for their anti-mycobacterial and anti-microbial activity.

11.3 Future Perspectives

Drugs from natural sources or chemical analogues of molecules of natural origin present a chemically complex and diverse array of molecules that can be the answer to many modern-day, difficult-to-treat infections and diseases. Developing these as drugs necessitates validation of their bioactivity by techniques that also ascertain their molecular-level mechanism of action.

HTS has been widely employed to achieve these goals of drug discovery in early stages so as to be able to forecast the man-drug interactions prior to conducting clinical trials. The use of sophisticated biological assays becomes essential for more accurate assessment of therapeutic activity and toxicity of compound libraries. A new approach which can perform simultaneous monitoring of cellular events with spatial and temporal resolution called the HCS approach has been proposed. HCS refers to cell-based assays that are examined through numerous techniques to quantify cellular responses. Such assays shall require urbane biochemical and molecular biology tools for quantifying and observing cell morphology, protein expression, localization, and posttranslational modifications. Methods such as immunochemistry-based assays, fluorescent reporter gene assays, assays for detecting protein interactions, and so on could be easily extended to HCS purposes. Majority of the one-dimensional cell-based assays are simple readouts of the

Table 11.3 Assays employed in the search for novel anti-mycobacterial/ anti-microbial agents which can further be developed for screening of natural product libraries

Cell line used	Protocol details	Plasmid/bacterial strain used	Quantification method	Reference
1. Determinations based on CFU enumeration				
Murine macrophages RAW 264.7	Intracellular killing of methicillin-resistant *Staphylococcus aureus* (MRSA) was studied by infecting macrophages with MRSA-MW2. Phagocytosis of the bacteria was then allowed to proceed. Macrophages were lysed and plated onto TSA plates. CFUs were enumerated to check anti-bacterial activity.	Methicillin-resistant *Staphylococcus aureus* (MRSA)	CFUs	Tharmalingam et al. (2018)
Human macrophage U937 cell line	Acetone extract of *Glycyrrhiza glabra* was tested for MIC and MBC using REMA. The results from these assays were validated using U937 cell line infected with Mtb. Number of CFUs were found to decrease with increase in time of exposure and concentration of extract. The extract showed anti-TB activity better than the standard available drugs	*Mycobacterium tuberculosis* H37Rv	CFUs	Nair et al. (2015)
Mycobacteriophage D29	Suspension of D29 by cultivation on a lawn of *M. smegmatis*. Bacterial cell cultures were infected using this suspension and incubated. Plaque-forming units were measured. The degree of killing was compared with the number of viable units of bacteria at the start of the experiment. Found correlation between PFUs and CFUs. Characterization	*M. tuberculosis* strain H37Rv *Mycobacterium smegmatis* ATCC 607	Plaque-forming units (PFUs)	Marcel et al. (2008)

(continued)

Table 11.3 (continued)

Cell line used	Protocol details	Plasmid/bacterial strain used	Quantification method	Reference
	of first line anti-TB drugs was also done using in vitro killing kinetics			
2. Determinations based on fluorescence/luminescence measurements				
THP-1 human monocytes	THP-1 monocytes were infected with Mtb $H_{37}Rv$, and the total bacterial load was measured using luminescence. Secondary assay involved infection using the Erdman strain and quantification using GFP fluorescence	H37Rv strain, Erdman strain	Luminescence/ fluorescence	Sorrentino et al. (2016)
Murine macrophage cell line RAW 264.7	Phenotypic cell-based assay that uses high-content imaging of Mtb-infected macrophages to quantify intracellular mycobacteria and estimate viability of host macrophages. These findings were confirmed on primary C57BL6 murine bone marrow macrophages and human peripheral blood-derived macrophages	GFP-expressing H37Rv Ms6 plasmid carrying a *gfp* gene (β-lactamase promoter *pBlaF*)	Fluorescence	Christophe et al. (2010)
RAW 264.7 macrophages	High-content assay involving confocal images of live bacterial samples and kinetic analysis of intracellular bacterial growth. Structure-activity relationship (SAR) studies of the synthesized derivatives were also performed	GFP-expressing *M. tuberculosis* H37Rv. H37Ra, BCG. The recombinant H37Rv with Ms6 plasmid carrying a *gfp*gene (β-lactamase promoter *pBlaF*)	Automated confocal fluorescent microscopy	Christophe et al. (2009)

population change in a biological process. However, they might miss the subtle but essential morphology or change that is present only in a particular subset of cells. Therefore, HCS can be extended to multidimensional cell-based assays in order to study the complex phenotypic population in a heterogeneous cell (Nierode et al. 2016).

References

Achoui M, Appleton D, Abdulla MA et al (2010) *In vitro* and *in vivo* anti-inflammatory activity of 17-O-acetylacuminolide through the inhibition of cytokines, NF-κB translocation and IKKβ activity. PLoS One 5(12):e15105

Ahn KC, Zhao B, Chen J et al (2008) *In vitro* biologic activities of the antimicrobials triclocarban, its analogs, and triclosan in bioassay screens: receptor-based bioassay screens. Environ Health Perspect 116(9):1203–1210

Alvarez-Suarez JM, Carrillo-Perdomo E, Aller A et al (2017) Anti-inflammatory effect of Capuli cherry against LPS-induced cytotoxic damage in RAW 264.7 macrophages. Food Chem Toxicol 102:46–52

Alves MP, Vielle NJ, Thiel V et al (2018) Research models and tools for the identification of antivirals and therapeutics against Zika virus infection. Viruses 10(11):593

Bai X, Cerimele F, Ushio-Fukai M et al (2003) Honokiol, a small molecular weight natural product, inhibits angiogenesis *in vitro* and tumor growth *in vivo*. J Biol Chem 278(37):35501–35507

Balakrishnan L, Bhattacharjee M, Ahmad S et al (2014) Differential proteomic analysis of synovial fluid from rheumatoid arthritis and osteoarthritis patients. Clin Proteomics 11(1):1

Bose S, Kim H (2013) Evaluation of *in vitro* anti-inflammatory activities and protective effect of fermented preparations of Rhizoma *Atractylodis macrocephalae* on intestinal barrier function against lipopolysaccharide insult. J Evid Based Complement Altern Med 2013:363076

Cao L, Zhu S, Wang Y et al (2015) A comprehensive procedure for antiviral inhibitor discovery using EV71 as an example. Biophys Rep 1(2):81–89

Ceci C, Lacal PM, Tentori L et al (2018) Experimental evidence of the antitumor, antimetastatic and antiangiogenic activity of ellagic acid. Nutrients 10(11):1756

Chathuranga K, Kim MS, Lee HC et al (2019) Anti-respiratory syncytial virus activity of *Plantago asiatica* and *Clerodendrum trichotomum* extracts *in vitro* and *in vivo*. Viruses 11(7):604

Cho SH, Warit S, Wan B et al (2007) Low-oxygen-recovery assay for high-throughput screening of compounds against nonreplicating *Mycobacterium tuberculosis*. Antimicrob Agents Chemother 51(4):1380–1385

Christophe T, Jackson M, Jeon HK et al (2009) High content screening identifies decaprenyl-phosphoribose 2′ epimerase as a target for intracellular antimycobacterial inhibitors. PLoS Pathog 5(10):e1000645

Christophe T, Ewann F, Jeon HK et al (2010) High-content imaging of *Mycobacterium tuberculosis*-infected macrophages: an *in vitro* model for tuberculosis drug discovery. Future Med Chem 2(8):1283–1293

Dar BA, Lone AM, Shah WA et al (2016) Synthesis and screening of ursolic acid-benzylidine derivatives as potential anti-cancer agents. Eur J Med Chem 111:26–32

DaSilva NA, Nahar PP, Ma H et al (2019) Pomegranate ellagitannin-gut microbial-derived metabolites, urolithins, inhibit neuroinflammation *in vitro*. Nutr Neurosci 22(3):185–195

Demchenko AP (2013) Beyond annexin V: fluorescence response of cellular membranes to apoptosis. Cytotechnology 65(2):157–172

Diop EA, Queiroz EF, Kicka S et al (2018) Survey on medicinal plants traditionally used in Senegal for the treatment of tuberculosis (TB) and assessment of their antimycobacterial activity. J Ethnopharmacol 216:71–78

Dsouza DORIN, Lakshmidevi N (2015) Models to study *in-vitro* antidiabetic activity of plants: a review. Int J Pharm Bio Sci 6:732–741

Ellis MJ, Tsai CN, Johnson JW et al (2019) A macrophage-based screen identifies antibacterial compounds selective for intracellular *Salmonella Typhimurium*. Nat Commun 10(1):197

El-Menshawi BS, Fayad W, Mahmoud et al (2010) Screening of natural products for therapeutic activity against solid tumors. Indian J Exp Biol 48(03):258–264

Etzion Y, Muslin AJ (2009) The application of phenotypic high-throughput screening techniques to cardiovascular research. Trends Cardiovas Med 19(6):207–212

Fakhrudin N, Waltenberger B, Cabaravdic M et al (2014) Identification of plumericin as a potent new inhibitor of the NF-κB pathway with anti-inflammatory activity *in vitro* and *in vivo*. Br J Pharmacol 171(7):1676–1686

Fan J, Jonge BLMD, Maccormack K et al (2014) A novel high-throughput cell-based assay aimed at identifying inhibitors of DNA metabolism in bacteria. Antimicrob Agents Chemother 58 (12):7264–7272

Fan XX, Leung ELH, Xie Y et al (2018) Suppression of Lipogenesis via reactive oxygen species–AMPK signaling for treating malignant and proliferative diseases. Antioxid Redox Signal 28 (5):339–357

Fayad W, Fryknäs M, Brnjic S et al (2009) Identification of a novel topoisomerase inhibitor effective in cells overexpressing drug efflux transporters. PLoS One 4(10):e7238

Figueiró LR, Comerlato LC, Da Silva MV et al (2016) Toxicity of *Glandulariaselloi* (Spreng.) Tronc. leaf extract by MTT and neutral red assays: influence of the test medium procedure. Interdiscip Toxicol 9(1):25–29

Fischer HP, Brunner NA, Wieland B et al (2004) Identification of antibiotic stress-inducible promoters: a systematic approach to novel pathway-specific reporter assays for antibacterial drug discovery. Genome Res 14(1):90–98

Forbes L, Ebsworth-Mojica K, Didone L et al (2015) A high throughput screening assay for anti-mycobacterial small molecules based on Adenylate kinase release as a reporter of cell lysis. PLoS One 10(6):e0129234

Fotakis G, Timbrell JA (2006) *In vitro* cytotoxicity assays: comparison of LDH, neutral red, MTT and protein assay in hepatoma cell lines following exposure to cadmium chloride. Toxicol Lett 160(2):171–177

Gao D, Zhang Y, Yang F et al (2016) *In vitro* screening and evaluation of 37 traditional Chinese medicines for their potential to activate peroxisome proliferator-activated receptors-γ. Pharmacogn Mag 12(46):120

Gulati V, Gulati P, Harding IH et al (2015) Exploring the anti-diabetic potential of Australian aboriginal and Indian Ayurvedic plant extracts using cell-based assays. BMC Complement Altern Med 15(1):8

Guo S, Olm-Shipman A, Walters A et al (2012) A cell-based phenotypic assay to identify cardioprotective agents. Circ Res 110(7):948–957

Gupta A, Bhakta S (2012) An integrated surrogate model for screening of drugs against *Mycobacterium tuberculosis*. J Antimicrob Chemother 67(6):1380–1391

Gupta PK, Chakraborty P, Kumar S et al (2016) G1-4A, a polysaccharide from *Tinospora cordifolia* inhibits the survival of *Mycobacterium tuberculosis* by modulating host immune responses in TLR4 dependent manner. PLoS One 11(5):e0154725

Hägg M, Bivén K, Ueno T et al (2002) A novel high-through-put assay for screening of pro-apoptotic drugs. Invest New Drug 20(3):253–259

Han JW, Kim KH, Kwun MJ et al (2019) Suppression of lung inflammation by the ethanol extract of Chung-Sang and the possible role of Nrf2. BMC Complement Altern Med 19(1):15

Hassig CA, Zeng FY, Kung P et al (2014) Ultra-high-throughput screening of natural product extracts to identify proapoptotic inhibitors of Bcl-2 family proteins. J Biomol Screen 19 (8):1201–1211

Hishiki T, Kato F, Tajima S et al (2017) Hirsutine, an indole alkaloid of *Uncaria rhynchophylla*, inhibits late step in dengue virus lifecycle. Front Microbiol 8:1674

Hossen MJ, Chou JY, Li SM et al (2019) An ethanol extract of the rhizome of *Atractylodes chinensis* exerts anti-gastritis activities and inhibits Akt/NF-κB signaling. J Ethnopharmacol 228:18–25

Hsueh TP, Sheen JM, Pang JH et al (2016) The anti-atherosclerotic effect of naringin is associated with reduced expressions of cell adhesion molecules and chemokines through NF-κB pathway. Molecules 21(2):195

Huan Y, Li L, Liu Q et al (2013) A cell-based fluorescent glucose transporter assay for SGLT2 inhibitor discovery. Acta Pharm Sin B 3(2):97–101

Huang TN, Lu KN, Pai YP et al (2013) Role of GLP-1 in the hypoglycemic effects of wild bitter gourd. Evid Based Complement Altern Med 2013:65892

Huang Y, Hao J, Tian D et al (2018) Antidiabetic activity of a flavonoid-rich extract from *Sophora davidii* (Franch.) skeels in KK-Ay mice via activation of AMP-activated protein kinase. Front Pharmacol 9:760

Jacobs AC, Didone L, Jobson J et al (2012) Adenylate kinase release as a high-throughput-screening-compatible reporter of bacterial lysis for identification of antibacterial agents. Antimicrob Agents Chemother 57(1):26–36

Järvinen P, Nybond S, Marcourt L et al (2016) Cell-based bioreporter assay coupled to HPLC micro-fractionation in the evaluation of antimicrobial properties of the basidiomycete fungus *Pycnoporus cinnabarinus*. Pharm Biol 54(6):1108–1115

Ji KY, Kim KM, Kim YH et al (2019) The enhancing immune response and anti-inflammatory effects of *Anemarrhena asphodeloides* extract in RAW 264.7 cells. Phytomedicine 59:152789

Jia QA, Ren ZG, Bu Y et al (2012) Herbal compound "Songyou Yin" renders hepatocellular carcinoma sensitive to oxaliplatin through inhibition of stemness. Evid Based Complement Altern Med 2012:908601, 12p

Jiang M, Han Y, Zhou Q, Zhao MG et al (2014) The screening research of anti-inflammatory bioactive markers from different flowering phases of FlosLoniceraeJaponicae. PLoS One 9(5): e96214

Jiang X, Dahlin A, Weiss ST et al (2017) A high-throughput chemical screen identifies novel inhibitors and enhancers of anti-inflammatory functions of the glucocorticoid receptor. Sci Rep 7(1):7405

Kadan S, Sasson Y, Abu-Reziq R et al (2018) *Teucrium polium* extracts stimulate GLUT4 translocation to the plasma membrane in L6 muscle cells. Adv Med Plant Res 6(1):1–8

Kasabri V, Afifi FU, Bustanji Y et al (2015) *In vitro* enhancement of pancreatic β cells MIN6 proliferation by insulinotropic *Gymnema sylvestre* aqueous extracts: evidence-based regenerative therapeutic capacity of a medicinal herb. Br J Med Med Res 7(3):180–194

Khazaei M, Pazhouhi M (2018) Protective effect of hydroalcoholic extracts of *Trifoliumpratense* L. on pancreatic β cell line (RIN-5F) against cytotoxicty of streptozotocin. Res Pharm Sci 13 (4):324

Kim TH, Choi DH, Vauthier V et al (2014) Anti-obesity phenotypic screening looking to increase OBR cell surface expression. J Biomol Screen 19(1):88–99

Kim G, Yeom SJ, Jang SC et al (2018) Simple analysis of lipid inhibition activity on an adipocyte micro-cell pattern chip. Biomol Ther 8(2):37

Koenders MI, van den Berg WB (2015) Novel therapeutic targets in rheumatoid arthritis. Trends Pharmacol Sci 36(4):189–195

Krutzik PO, Crane JM, Clutter MR et al (2008) High-content single-cell drug screening with phosphospecific flow cytometry. Nat Chem Biol 4(2):132

Kumar S, Patial V, Soni S et al (2017) *Picrorhiza kurroa* enhances β-cell mass proliferation and insulin secretion in streptozotocin evoked β-cell damage in rats. Front Pharmacol 8:537

Larsson MC, Lerm M, Ängeby K et al (2014) A luciferase-based assay for rapid assessment of drug activity against *Mycobacterium tuberculosis* including monitoring of macrophage viability. J Microbiol Methods 106:146–150

Li W, Lam MS, Birkeland A et al (2006) Cell-based assays for profiling activity and safety properties of cancer drugs. J Pharmacol Toxicol Methods 54(3):313–319

Liang J, Chen J, Tan Z et al (2013) Extracts of the medicinal herb *Sanguisorba officinalis* inhibit the entry of human immunodeficiency virus-1. J Food Drug Anal 21(4):S52–S58

Lucas-Hourani M, Lupan A, Desprès P et al (2013) A phenotypic assay to identify Chikungunya virus inhibitors targeting the nonstructural protein nsP2. J Biomol Screen 18(2):172–179

Luo Y, Sun G, Dong X et al (2015) Isorhamnetin attenuates atherosclerosis by inhibiting macrophage apoptosis via PI3K/AKT activation and HO-1 induction. PLoS One 10(3):e0120259

Lyu W, Deng Z, Sunkara LT et al (2018) High throughput screening for natural host defense peptide-inducing compounds as novel alternatives to antibiotics. Front Cell Infect Microbiol 8:191

Marcel N, Nahta A, Balganesh M (2008) Evaluation of killing kinetics of anti-tuberculosis drugs on *Mycobacterium tuberculosis* using a bacteriophage-based assay. Chemotherapy 54(5):404–411

Mateus A, Gordon LJ, Wayne GJ et al (2017) Prediction of intracellular exposure bridges the gap between target-and cell-based drug discovery. Proc Natl Acad Sci U S A 114(30):E6231–E6239

Meijer K, Vonk RJ, Priebe MG et al (2015) Cell-based screening assay for anti-inflammatory activity of bioactive compounds. Food Chem 166:158–164

Min Q, Cai X, Sun W et al (2017) Identification of mangiferin as a potential Glucokinase activator by structure-based virtual ligand screening. Sci Rep 7:44681

Mo H, Elson CE (2004) Studies of the isoprenoid-mediated inhibition of mevalonate synthesis applied to cancer chemotherapy and chemoprevention. Exp Biol Med 229(7):567–585

Nair SS, Pharande RR, Bannalikar AS et al (2015) *In vitro* antimycobacterial activity of acetone extract of *Glycyrrhiza glabra*. J Pharm Pharmacogn Res 3(4):80–86

Nair SS, Gaikwad SS, Kulkarni SP et al (2017) *Allium sativum* constituents exhibit anti-tubercular activity *in vitro* and in RAW 264.7 mouse macrophage cells infected with *Mycobacterium tuberculosis* $H_{37}Rv$. Pharmacogn Mag 13(Suppl 2):S209

Nakano K, Takeshita S, Kawasaki N et al (2017) AJS1669, a novel small-molecule muscle glycogen synthase activator, improves glucose metabolism and reduces body fat mass in mice. Int J Mol Med 39(4):841–850

Nierode G, Kwon PS, Dordick JS et al (2016) Cell-based assay design for high-content screening of drug candidates. J Microbiol Biotechnol 26(2):213

Okamoto H (2008) Vitamin K and rheumatoid arthritis. IUBMB Life 60(6):355–361

Perez del Palacio J, Díaz C, de la Cruz M et al (2016) High-throughput screening platform for the discovery of new immunomodulator molecules from natural product extract libraries. J Biomol Screen 21(6):567–578

Piccioni F, Roman BR, Fischbeck KH et al (2004) A screen for drugs that protect against the cytotoxicity of polyglutamine-expanded androgen receptor. Hum Mol Genet 13(4):437–446

PlackalAdimuriyil George B, Tynga IM, Abrahamse H et al (2015) *In vitro* antiproliferative effect of the acetone extract of *Rubus fairholmianus* gard. Root on human colorectal cancer cells. Biomed Res Int 2015:165037

Qazzaz ME, Raja VJ, Lim KH et al (2016) *In vitro* anticancer properties and biological evaluation of novel natural alkaloid jerantinine B. Cancer Lett 370(2):185–197

Qin T, Du R, Huang F et al (2016) Sinomenine activation of Nrf2 signaling prevents hyperactive inflammation and kidney injury in a mouse model of obstructive nephropathy. Free Radic Biol Med 92:90–99

Raneri M, Sciandrone B, Briani F (2015) A whole-cell assay for specific inhibitors of translation initiation in bacteria. J Biomol Screen 20(5):627–633

Schaaf K, Hayley V, Speer A et al (2016) A macrophage infection model to predict drug efficacy against *Mycobacterium Tuberculosis*. Assay Drug Dev Technol 14(6):345–354

Schink A, Neumann J, Leifke AL et al (2018) Screening of herbal extracts for TLR2-and TLR4-dependent anti-inflammatory effects. PLoS One 13(10):e0203907

Sharma S, Gelman E, Narayan C et al (2014) Simple and rapid method to determine antimycobacterial potency of compounds by using autoluminescent *Mycobacterium tuberculosis*. Antimicrob Agents Chemother 58(10):5801–5808

Shen CY, Xu XL, Yang LJ et al (2019) Identification of narciclasine from *Lycoris radiata* (L'Her.) herb and its inhibitory effect on LPS-induced inflammatory responses in macrophages. Food Chem Toxicol 125:605–613
Shi H, He J, Li X et al (2018) Isorhamnetin, the active constituent of a Chinese herb *Hippophae rhamnoides* L, is a potent suppressor of dendritic-cell maturation and trafficking. Int Immunopharmacol 55:216–222
Singh U, Tabibian J, Venugopal SK et al (2005) Development of an in vitro screening assay to test the antiinflammatory properties of dietary supplements and pharmacologic agents. Clin Chem 51(12):2252–2256.
Singh V, Rana M, Jain M et al (2015) Curcuma oil attenuates accelerated atherosclerosis and macrophage foam-cell formation by modulating genes involved in plaque stability, lipid homeostasis and inflammation. Br J Nutr 113(1):100–113
Song L, Kang C, Sun Y et al (2016) Crocetin inhibits lipopolysaccharide-induced inflammatory response in human umbilical vein endothelial cells. Cell Physiol Biochem 40(3–4):443–452
Sorrentino F, del Rio RG, Zheng X et al (2016) Development of an intracellular screen for new compounds able to inhibit *Mycobacterium tuberculosis* growth in human macrophages. Antimicrob Agents Chemother 60(1):640–645
Stanley SA, Barczak AK, Silvis MR et al (2014) Identification of host-targeted small molecules that restrict intracellular *Mycobacterium tuberculosis* growth. PLoS Pathog 10(2):e1003946
Su X, Huang Q, Chen J et al (2016) Calycosin suppresses expression of pro-inflammatory cytokines via the activation of p62/Nrf2-linked heme oxygenase 1 in rheumatoid arthritis synovial fibroblasts. Pharmacol Res 113:695–704
Tharmalingam N, Rajmuthiah R, Kim W et al (2018) Antibacterial properties of four novel hit compounds from a methicillin-resistant *Staphylococcus aureus–Caenorhabditis elegans* high-throughput screen. Microb Drug Resist 24(5):666–674
Trofimov V, Kicka S, Mucaria S et al (2018) Antimycobacterial drug discovery using mycobacteria-infected amoebae identifies anti-infectives and new molecular targets. Sci Rep 8 (1):3939
Tsonkova VG, Sand FW, Wolf XA et al (2018) The EndoC-βH1 cell line is a valid model of human beta cells and applicable for screenings to identify novel drug target candidates. Mol Metab 8:144–157
Urban A, Eckermann S, Fast B et al (2007) Novel whole-cell antibiotic biosensors for compound discovery. Appl Environ Microbiol 73(20):6436–6443
Vajrabhaya LO, Korsuwannawong S (2018) Cytotoxicity evaluation of a thai herb using tetrazolium (MTT) and sulforhodamine b (SRB) assays. J Anal Sci Technol 9(1):15
Varghese FS, Kaukinen P, Gläsker S et al (2016) Discovery of berberine, abamectin and ivermectin as antivirals against chikungunya and other alphaviruses. Antivir Res 126:117–124
Vasou A, Paulus C, Narloch J et al (2018) Modular cell-based platform for high throughput identification of compounds that inhibit a viral interferon antagonist of choice. Antivir Res 150:79–92
Vogel HG (ed) (2002) Drug discovery and evaluation: pharmacological assays. Springer Science & Business Media, Berlin
Voráčová K, Hájek J, Mareš J et al (2017) The cyanobacterial metabolite nocuolin A is a natural oxadiazine that triggers apoptosis in human cancer cells. PLoS One 12(3):e0172850
Wang L, Li W, Li S (2016) A high throughput assay for screening host restriction factors and antivirals targeting influenza a virus. Front Microbiol 7:858
Wang LQ, He Y, Wan HF et al (2017) Protective mechanisms of hypaconitine and glycyrrhetinic acid compatibility in oxygen and glucose deprivation injury. J Zhejiang Univ Sci B 18 (7):586–596
Wu TR, Huang TT, Martel J et al (2017) Pinicolol B from *Antrodia cinnamomea* induces apoptosis of nasopharyngeal carcinoma cells. J Ethnopharmacol 201:117–122
Xu W, Liu L (2013) An *in vitro* cell-based LDL uptake model for screening PCSK9 modulators. J Bioequiv Bioavailab 5:248–252

Xu X, Yan H, Zhang X (2012) Structure and immuno-stimulating activities of a new heteropolysaccharide from *Lentinula edodes*. J Agric Food Chem 60(46):11560–11566

Yamamoto N, Ueda-Wakagi M, Sato T et al (2015) Measurement of glucose uptake in cultured cells. Curr Protoc Pharmacol 71(1):12–14

Yang T, Jia M, Meng J et al (2006) Immunomodulatory activity of polysaccharide isolated from *Angelica sinensis*. Int J Biol Macromol 39(4–5):179–184

Yeo D, Hwang SJ, Kim WJ et al (2018) The aqueous extract from *Artemisia capillaris* inhibits acute gastric mucosal injury by inhibition of ROS and NF-kB. Biomed Pharmacother 99:681–687

Yin Y, Xu Y, Ou Z et al (2019) An antiviral drug screening system for enterovirus 71 based on an improved plaque assay: a potential high-throughput method. J Med Virol 91:1440

Yuan W, Yu Z, Song W et al (2019) Indole-core-based novel antibacterial agent targeting FtsZ. Infect Drug Resist 12:2283

Zang R, Li D, Tang IC et al (2012) Cell-based assays in high-throughput screening for drug discovery. Int J Biotechnol Wellness Ind 1(1):31–51

Zhang G, Baidin V, Pahil KS et al (2018) Cell-based screen for discovering lipopolysaccharide biogenesis inhibitors. Proc Natl Acad Sci U S A 115(26):6834–6839

Zou C, Wang Y, Shen Z (2005) 2-NBDG as a fluorescent indicator for direct glucose uptake measurement. J Biochem Biophys Methods 64(3):207–215

Zulfiqar B, Mahroo A, Nasir K et al (2017) Nanomedicine and cancer immunotherapy: focus on indoleamine 2, 3-dioxygenase inhibitors. Onco Targets Ther 10:463

Allelochemicals: An Emerging Tool for Weed Management

12

Chandan Das, Avishek Dey, and Abhijit Bandyopadhyay

Abstract

Allelopathy represents a mechanism by which plant releases biochemical compounds that influence cell division, seed germination, physiology, overall growth, development, and survival of other plants. Presently, allelochemicals find applications in crop field, especially in weed management, and in agricultural systems as growth regulator, as well as herbicides. Applying such allelochemicals is safer than synthetic harmful chemicals as these natural biodegradable phytometabolites hardly leave residual toxicity on targets. Allelopathy is mostly reported to produce inhibitory action of allelochemicals against targeted weeds and promises potent alternative to chemical herbicides. This article evaluates promising aspects of diverse allelochemicals as an upcoming tool in weed management.

Keywords

Allelopathy · Allelochemicals · Weed management · Herbicides

12.1 Introduction

The phenomenon allelopathy is an interaction between different organisms that is induced by various chemical compounds termed as allelochemicals. The word allelopathy was coined by Molisch in 1937 and is derived from the Greek word

C. Das · A. Bandyopadhyay (✉)
Plant Genetics and Molecular Biology Unit, Department of Botany, University of Burdwan, Burdwan, West Bengal, India
e-mail: cdburdwan@gmail.com; palmajua@gmail.com

A. Dey
Department of Botany, Ramananda Centenary College, Purulia, West Bengal, India
e-mail: de.avik97@gmail.com

S. C. Mandal et al. (eds.), *Evidence Based Validation of Traditional Medicines*,
https://doi.org/10.1007/978-981-15-8127-4_12

"alleon" which means of each other and "pathos" which means to suffer or injurious effect of one upon another (Rizvi et al. 1992). According to Einhellig (1995), allelopathy is also a subdiscipline of chemical ecology that is concerned with the effects of chemicals produced by plants or microorganisms on the growth, development, and distribution of different plants and microorganisms in natural environments as well as agricultural systems. However, presently the term is familiar with the stimulatory and inhibitory effects of one plant on another plant (Rice 1984). The allelochemicals are basically secondary metabolites of plants as well as microbes and control a number of processes in the environment and ecosystems (Rizvi et al. 1992; Seigler 1996).

Different parts of plants including the stems, leaves, roots, rhizomes, pollen, seeds, and flowers are the source of allelochemicals, which are released into the environment by the process of root exudation, leaching from aboveground parts and volatilization or by decomposition of different plant parts (Rice 1984). It has been reported that there are several categories of allelochemicals with potential phytotoxicity including alkaloids, benzoxazinones, cyanogenic compounds, cinnamic acid derivatives, ethylene, and flavonoids (Putnam 1988). These compounds are indispensable for the dealings between plants and the living part of the environment and also act as a defense strategy of plants against natural enemies (Kroymann 2011). Interestingly, the survival as well as reproductive ability of plants is associated with the released allelochemicals into their environment. However, the connection between ecological aspects and allelopathy is a complex and layered phenomenon.

The effect of allelopathy has been investigated in agricultural systems in terms of crop establishment and performance (Weston and Duke 2003). The application of several allelopathic crops to effectively suppress the invasive weeds in agricultural system was reported by Inderjit et al. (2007). In agroecosystems, the effect of allelopathy on weed management and plant reproduction was reported by Chon et al. (2006). Invasion of weeds in crop field is a constant problem in agricultural production. The application of chemical herbicides is the key approach to control the growth of invasive weeds in conventional agriculture system. Constant use of overdoses of synthetic chemical regulators worsened the situation with simultaneous increase of resistance towards different pests and weeds, thus endangering ecosystems. Resistance to herbicides among weeds has reached an alarming level in agricultural fields. Stephenson (2000) noticed that most of the agricultural systems collectively use three million tones of herbicides per year and simultaneously cause the selective growth of weeds. As these synthetic weedicides are not biodegradable, mounting applications of such notorious chemicals invited simultaneous rise in weedicide tolerance (thus, generating "super weeds"), which contaminates our food and devastates ecosystem. This vicious cycle rotates to demand more toxic chemicals to kill the weeds and thus generate more recalcitrant weeds. Therefore, alternative strategies to toxic herbicide application are gaining importance and have become a natural choice. In this challenging backdrop, this review attempts to explore the recent advancements in the application of allelochemicals in agroecosystem with lasting impact on weed control measures.

12.2 Production, Bioavailability, and Types of Allelochemicals in Plants

Plants synthesize different allelochemicals via secondary metabolism, and such synthesis depends on several factors including availability of precursor molecules, regulation of specific genes, as well as environmental stimuli (Croteau et al. 2000). It is reported that several environmental factors including UV radiation, unavailability of nutrients, and pathogenic stress can promote the allelochemicals accumulation in plants (Dixon and Paiva 1995). Dayan (2006) reported that the occurrence of other plants surrounding *Sorghum* found to be connected with the production of sorgoleone (an oily exudate of root hair of *Sorghum*). Cheema (1988) observed the phytotoxic effects of the soluble allelochemicals from *Sorghum* on reduction of the growth of certain weeds including *Phalaris*, *Chenopodium*, and *Convolvulus*.

Moreover, it was also studied that rice seedlings produced almost seven times higher amount of momilactone B when cultivated with *Echinochloa* (Kato-Noguchi 2011). Dudareva et al. (2013) reported that momilactone B is synthesized in rice through methylerythritol phosphate (MEP) pathway and used as weedicide for its tremendous level of allelopathic potential. Fujii (1994) noticed that the aqueous extract of *Pueraria thunbergiana* leaves shows allelopathic effect on the germination percentage of lettuce seed and growth of the seedlings, and such effect might be associated with the production of xanthoxins (Kato-Noguchi 2002). Furthermore, many weeds are now showing significant role in weed management for having allelopathic potential. *Parthenium* is a well-known weed for its massive proliferation rate and negative effects on agriculture, ecology, and the environment through the production of parthenin (Kohli and Rani 1994). The allelopathic potential of this weed is primarily due to the existence of parthenin, a sesquiterpene lactone in various parts of the plant (Kanchan and Chandra 1980). Parthenin is identified as a specific inhibitory compound and significantly affected the root and shoot growth of *Cassia*, *Ocimum*, and *Hordeum* (Khosla and Sobti 1979).

The bioavailability as well as activity of allelochemicals in the soil is dependent on several factors including adsorption, leaching, degradation, soil pH, organic matter, and the availability of water (Kobayashi 2004). It was noticed that sorgoleone showed reduced allelopathic potential due to its lipophilic nature and strong binding tendency with soil colloids (Trezzi et al. 2006). Therefore, the allelochemicals that are not adsorbed onto colloids can be absorbed by plants or leached because it was revealed in a study with rice plants that flavonoids with a high mobility in the soil were found to be less phytotoxic than those with low soil mobility (Li et al. 2013; Kong et al. 2007). According to Wang et al. (2010), allelochemicals can be grouped into ten classes according to their structures and properties including (1) water-soluble organic acids, straight-chain alcohols, aliphatic aldehydes, and ketones, (2) simple lactones, (3) long-chain fatty acids and polyacetylenes, (4) quinines (benzoquinone, anthraquinone, and complex quinines), (5) phenolics, (6) cinnamic acid and its derivatives, (7) coumarins, (8) flavonoids, (9) tannins, and (10) steroids and terpenoids (sesquiterpene lactones, diterpenes, and triterpenoids). Different types of allelochemicals from plants and their target weeds are presented in Table 12.1.

Table 12.1 Different allelochemicals from plants and their target weeds

Name of the source plants	Allelochemicals	Sensitive weeds	References
Oryza sativa	Phenolic acids, Momilactone	*Echinochloa colonum*, *Amaranthus lividus*	Rimando et al. (2001); Soltys et al. (2013)
Cucumis sativus	Benzoic and cinnamic acids	*Panicum miliaceum* and *Brassica hirta*	Yu and Matsui (1994)
Triticum aestivum	Hydroxamic acids	*Ipomoea hederacea*, *Echinochloa colonum*	Niemeyer (1988)
Avena sativa	Phenolic acids and scopoletin	*Amaranthus* sp., *Setaria* sp.	Weston (1996)
Sorghum bicolor	Sorgoleone	*Phalaris minor*, *Coronopus didymus*, *Cyperus rotundus*	Netzley and Butler (1986)
Tinospora tuberculata	Ethanolamine, hydrazinecarboxamide 3-Carene	*Echinochloa colonum*	Aslani et al. (2015)
Piper methysticum	Ferulic acid, protocatechuic acid, gallic acid, *trans*-o-coumaric acid	*Leptochloa filiformis*, wild mustard, wild spinach	Xuan et al. (2003); Soltys et al. (2013)
Medicago sativa	Vanillin, vanillic acid, sinapic acid	Paddy weeds	Xuan et al. (2003)
Brassica nigra	Glucosinolates	*Sonchus asper*, *Amaranthus hybridus*, *Cuscuta* spp., *Alopecurus myosuroides*	Turk and Tawaha (2003); Soltys et al. (2013)
Raphanus sativus	Isothiocyanates	*Matricaria inodora*, *Echinochloa crus-galli*, *Alopecurus myosuroides*, *Hirschfeldia incana*	Soltys et al. (2013)
Artemisia annua	Artemisinin	*Ipomoea lacunose*, *Portulaca oleracea*, *Lemna minor*, *Pseudokirchneriella subcapitata*	Soltys et al. (2013)
Leptospermum scoparium	Leptospermone	*Setaria glauca*, *Brassica juncea*, *Rumex crispus*	Soltys et al. (2013)
Eucalyptus sp.	Essential oils	*Cassia occidentalis*, *Lolium rigidum*	Soltys et al. (2013)
Piper sp.	Sarmentine	*Leptochloa filiformis*, *Chenopodium album*, *Leptochloa filiformis*	Soltys et al. (2013)

12.3 Mode of Action of Different Allelochemicals

Identification of the mechanisms of action of different allelochemicals became an important area of research to explore their role in agroecosystems and to find out leads in the discovery of herbicides. Allelochemicals are very diverse in nature, and therefore, it's difficult to establish a general model of action (Fig. 12.1), since it depends on the compound type and the type of receiving plants and their interaction. At internal level, allelochemicals affect a large number of physiological parameters, including action on the cellular membrane, disruption of the activity of different enzymes, as well as structural proteins and alteration of hormonal balance. In addition, allelochemicals can also inhibit or reduce cellular respiration and chlorophyll biosynthesis that cause subsequent reduction in vitality, growth, and development of the plant. Conversely, at external level, allelochemicals may be associated with release or limitation of nutrients in the soil. Despite of the discovery of thousands of allelochemicals and the determination of their chemical structure from plants, the mechanisms of actions have only been elucidated in a limited number of these substances (Vyvyan 2002). The major mechanisms of action of allelochemicals are revealed including (1) photosystem II electron transport inhibition; (2) inhibition of photosystem I; (3) interfering with reactive oxygen species; (4) alteration of the synthesis of amino acid and plant growth regulators (auxins and gibberellins); and (5) inhibition of tubulin polymerization (Duke et al. 2002).

Benzoxazolinone (BOA) is exuded from the roots of various cultivated grasses and exerts multiple physiological effects on plants, and it was noticed that the generation of reactive oxygen species and oxidative stress were the key mechanisms operating behind the actions of BOA in plants (Schulz et al. 2013). Dayan et al. (2012) reported that essential oil of citrus can disrupt microtubule polymerization and used as a desiccant herbicide. A primary action on ATP production is indicated for juglone (5-hydroxy-1,4-napthalenedione, found in bark of black walnut) and sorgoleone, since they strongly control the chloroplast oxygen evolution and significantly affect mitochondrial functions. The chloroplast block by sorgoleone was found to be associated with the inhibition of photosystem II and 4-hydroxyphenylpyruvate dioxygenase (Trezzi et al. 2006). In addition, it was studied that the secretion of glucose and phenylalanine by plants reduces the adsorption and biodegradation of *p*-coumaric acid and synergistically inhibiting *Ipomoea* growth (Gerig and Blum 1993).

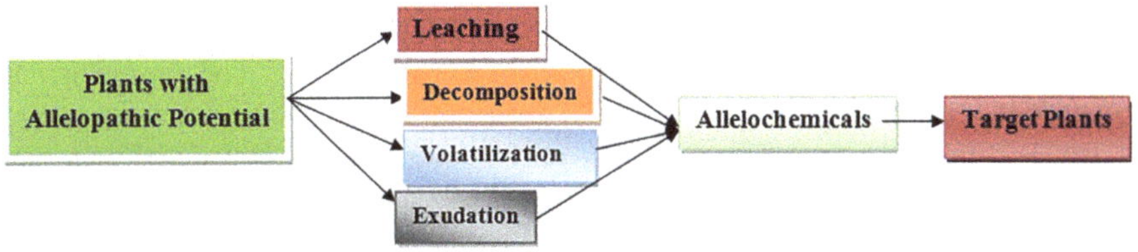

Fig. 12.1 Conceptual route for release of allelochemicals

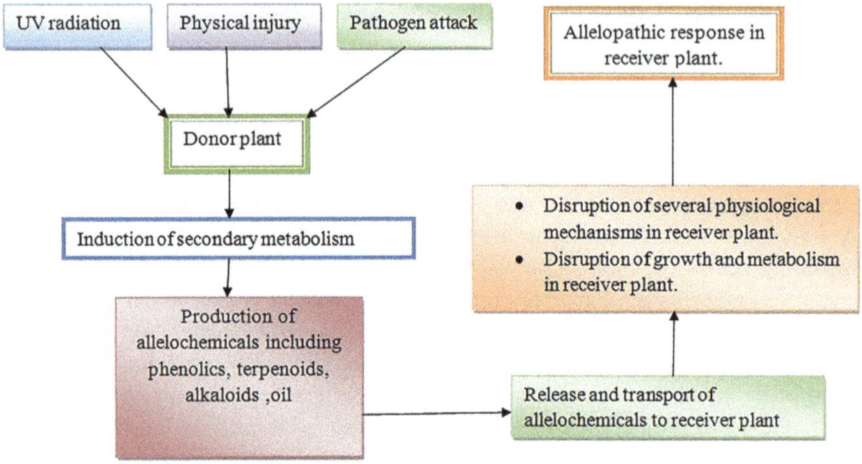

Fig. 12.2 A schematic representation showing action of allelochemicals in plants

The aqueous leaf leachate of *Shorea robusta* contains diverse type of potential allelochemicals. Routine test with *Allium cepa* bulb assay was carried out to assess antagonistic potential of allelochemicals. For this, cytotoxicity parameters like mitotic index, abnormality index, and nucleolar frequencies are generally considered (Das and Bandyopadhyay 2011). Possible action of allelochemicals in plants is presented in Fig. 12.2.

12.4 Use of Allelochemicals in Agronomy and Weed Management

The impact of allelochemicals on weeds is highly connected with the behavior of these compounds in the soil. After release from the source, the soil is the main means of transportation that makes the contact between allelochemicals and their respective target plants. Even highly volatile oily compounds, such as terpenoids, can be effective within the soil. After being released into the soil, the allelochemicals need to reach the target plant to execute their function, and this movement in the soil is associated with a series of processes including mass flow, diffusion, and root interception. It was further noted that during this movement, allelochemicals undergo retention that involves the attraction of allelochemicals to soil colloids, transformation, and degradation (Kah and Brown 2006). The properties of allelochemicals play an important role in determining their destiny in the environment. For example, the water solubility of the compounds is connected with their mobility within the soil; in addition, the vapor pressure can trigger their volatilization, and the chemical structure can modulate their attraction with the soil surface (Souza and Alves 2002).

Investigation of the allelopathic potential of different plant species is connected with the introduction of alternative methods for weed management, and this approach not only decreases the costs of herbicide application but also improves crop production. Field experiments were conducted using water extracts of *Sorghum*, sunflower, and rapeseed for controlling purple *Cyperus rotundus* in cotton. It was noticed that the density and dry weight of *Cyperus rotundus* were suppressed significantly, when different crop water extracts were used in combination with a reduced rate of glyphosate (Iqbal et al. 2009). Herbicidal property of juglone on field poppy without hindering the growth of different crops like wheat and barley was reported by Topal et al. (2007). Isothiocyanates in *Brassica* found to be strong suppressants of germination of *Sonchus asper*, *Matricaria inodora*, and *Alopecurus myosuroides* (Petersen et al. 2001). Extracts from different parts of *Brassica nigra* found to inhibit the germination and radicle length of wild oat (Turk and Tawaha 2003). The efficacy of the aqueous extracts from *Sorghum bicolor* and *Helianthus annuus* was noticed in rice on significant reduction of *Echinochloa crus-galli* biomass (Cheema and Khaliq 2000). Moreover, experiments were conducted using *Raphanus sativus* extract on germination of different weed species, and it was revealed that extracts of *R. sativus* totally inhibited germination *Sorghum halepense*, *Alhagi* sp., *Alopecurus myosuroides*, *Capsella bursa-pastoris*, *Cuscuta* sp., and *Sisymbrium polyceratium* (Uygur et al. 1990). Furthermore, the application of legumes as a source of allelochemicals has been studied, and it was noticed that *Pea* cover crop has regulated germination and growth of *Polygonum persicaria*, smooth pigweed, small flower galinsoga, and common lambs quarters (Akemo et al. 2000). Correspondingly, the aqueous extract of *Mucuna deeringiana*, *Canavalia ensiformis*, exhibited strong phytotoxic effects on the radicle growth of barnyard grass, *Amaranthus* sp., and amaranth *Amaranthus hypochondriacus* (Caamal-maldonado et al. 2001). The overall outcome of all these complex processes results in the allelopathic effect and shows weedicidal potential for agronomic application.

12.5 Nanotechnology in Weed Management

Herbicide resistance and the development of "super weed" pose serious hurdle to tackle weed eradications. Such tolerance to herbicides perhaps generates due to inadequate bioavailability of herbicides at target tissues. Promising field of nanotechnology is now being explored widely to address the weed problem. Various methods of synthesis of nanoparticles, nanoemulsions, nanocapsules with allelochemicals, and their deployment as nanoherbicides offer alternative potentials for weed management. The controlled release of nanoherbicides with their effective minimum concentration ensures least toxic impact onto target field crops along with their sustained bioavailability. In coming decades, allelochemical-based eco-friendly nanopesticides would come up in a big way for better weed management (Amna et al. 2019).

12.6 Limitations of Allelopathy in Weed Management

Despite of huge potential, there are many restrictions in using allelochemicals as a weed management system, and such restrictions are due to plant itself and the environmental condition. Many abiotic and biotic factors such as plant age, temperature, light and soil conditions, microflora, nutritional status, and herbicide treatments influence the production, release, and phytotoxic levels of allelochemicals (Duke 1985; Huang et al. 1999). Some allelochemicals are found to be very much expensive to synthesize (Duke et al. 2000). In addition, some allelochemicals are found to be toxic as well as carcinogenic to human beings, for example, AAL-toxin and fumonisin (Duke et al. 2000). Inderjit and Bhowmik (2002) reported that sorgoleone causes dermatitis to human beings. Isothiocyanates and thiocyanates have been found to be harmful when consumed by humans and animals and might be associated with thyroid, liver, and kidney diseases (Tookey et al. 1980). Furthermore, it was noted that some allelopathic agents such as monoterpenes are active only under hot and dry climate due to their specific active nature in vapor phase (Kohli and Singh 1991).

12.7 Conclusion

The application and mechanisms of allelopathic interactions among plants are upcoming areas of research in weed management. It is comprehensible from the above discussion that there is an immense prospect of allelochemicals in the area of weed management. Allelochemicals from several plants have been identified, and their activities have been established, but detailed understanding of mode of action needs to be explored. The immense diversity of allelochemicals, their easy biodegradability, and their promising weedicidal properties offer novel approach in weed management in days to come. Applications of allelochemicals have already been used to defend crops from destructive and invasive weeds. In spite of these developments, the following issues demand careful considerations during application of allelochemicals in crops fields:

1. Stringent field trials to study interaction of applied allelochemicals with various physical, chemical, biological properties of soil
2. Monitoring dynamics of allelochemicals and understanding mode of action
3. The impact assessment of allelochemicals in diverse agroecosystems

Judicious application of such wide array of allelochemicals promises to revolutionize weed management in particular and expected to provide lasting impact in global agricultural scene, in general.

References

Akemo MC, Regnier EE et al (2000) Weed suppression in spring-sown rye (*Secale cereale*)-pea (*Pisum sativum*) cover crop mixes. Weed Technol 14(3):545–549

Amna AHF, Hakeem KR et al (2019) Weed control through herbicide-loaded nanoparticles. In: Husen A, Iqbal M (eds) Nanomaterials and plant potential. Springer, Cham, pp 507–522

Aslani F, Juraimi A et al (2015) Phytotoxic interference of volatile organic compounds and water extracts of *Tinospora tuberculata* Beumee on growth of weeds in rice fields. S Afr J of Bot 100:132–140

Caamal-maldonado JA, Jiménez-osornio JJ et al (2001) The use of allelopathic legume cover and mulch species for weed control in cropping systems. Agron J 93(1):27–36

Cheema ZA (1988) Weed control in wheat through *sorghum* allelochemicals. Ph.D. Thesis, University of Agriculture, Faisalabad, Pakistan

Cheema ZA, Khaliq A (2000) Use of *sorghum* allelopathic properties to control weeds in irrigated wheat in semi arid region of Punjab. Agric Ecosyst Environ 79:105–112. https://doi.org/10.1016/S0167-8809(99)00140-1

Chon SU, Jennings JA et al (2006) Alfalfa (*Medicago sativa* L.) autotoxicity: current status. Allelopath J 18:57–80

Croteau RB, Gershenzon J, McConky ME (2000) Regulation of monoterpene accumulation in leaves of peppermint. Plant Phys 122:205–213

Das C, Bandyopadhyay A (2011) Search for allelopathic potential of leaves of *Shorea robusta* Gaertn.f. Bionature 31(1):29–35

Dayan FE (2006) Factors modulating the levels of the allelochemical sorgoleone in *Sorghum bicolor*. Planta 224:339–346

Dayan FE, Owen DK et al (2012) Rationale for a natural products approach to herbicide discovery. Pest Manag Sci 68:519–528

Dixon RA, Paiva NL (1995) Stress-induced phenylpropanoid metabolism. Plant Cell 7:1085–1097

Dudareva N, Klempien A, Muhlemann JK et al (2013) Biosynthesis, function and metabolic engineering of plant volatile organic compounds. New Phytol 198:16–32

Duke SO (1985) Biosynthesis of phenolic compounds—chemical manipulation in higher plants. In. Thompson AC (ed) The chemistry of allelopathy: biochemical interactions among plants. Am. Chem. Soc. Symp. Ser. 268. Am. Chem. Soc., Washington, DC, p 113–131

Duke SO, Dayan FE, Romagni JG et al (2000) Natural products as sources of herbicides: current status and future trends. Weed Res 40:99–111

Duke SO, Dayan FE, Rimando AM et al (2002) Chemicals from nature for weed management. Weed Sci 50:138–151

Einhellig FA (1995) Allelopathy-current status and future goals. In: Inderjit A, Dakshini KMM, Einhellig FA (eds) Allelopathy: organisms, processes, and applications. American Chemical Society Press, Washington DC, pp 1–24

Fujii Y (1994) Screening of allelopathic candidates by new specific discrimination, and assessment methods for allelopathy, and the identification of l-dopa as the allelopathic substance from the most promising velvetbean (*Mucuna pruriens*). Bull Natl Inst Agro-Environ Sci 10:115–218

Gerig TM, Blum U (1993) Modification of an inhibition curve to account for effects of a second compound. J Chem Ecol 19:2783–2790

Huang PM, Wang MC, Wang MK (1999) Catalytic transformation of phenolic compounds in the soil. In: Inderjit (ed) Principles and practices in plant ecology: allelochemical interactions. CRC Press, Boca Raton, FL, pp 287–306

Inderjit, Bhowmik PC (2002) The importance of allelochemicals in weed invasiveness and the natural suppression. In: Inderjit, Mallik AU (eds) Chemical ecology of plant: allelopathy of aquatic and terrestrial ecosystems. Birkhauser Verlag AG, Basal, pp 187–192

Inderjit, Weston LA, Duke SO (2007) Challenges, achievements and opportunities in allelopathy research. J Plant Interact 1:69–81

Iqbal J, Cheema ZA, Mushtaq M (2009) Allelopathic crop water extracts reduce the herbicide dose for weed control in cotton (*Gossypium hirsutum*). Int J Agric Biol 11(4):360–366
Kah M, Brown CD (2006) Adsorption of ionizable pesticides in soil. Rev Environ Contam Toxicol 188:149–217
Kanchan S, Chandra J (1980) Pollen allelopathy-a new phenomenon. New Phytol 84(4):739–746
Kato-Noguchi H (2002) Allelopathic substances in *Pueraria thunbergiana*. Phytochemistry 63:577–580
Kato-Noguchi H (2011) Barnyard grass-induced rice allelopathy and momilactone B. J Plant Physiol 168:1016–1020
Khosla SN, Sobti SN (1979) Parthenin-A national health hazard, its control and utility-a review. Pesticides 13:21–27
Kobayashi K (2004) Factors affecting phytotoxic activity of allelochemicals in soil. Weed Biol Manag 4:1–7
Kohli RK, Rani D (1994) *Parthenium hysterophorus* L.—a review. Res Bull. (Sci.) Pb Univ 44:105–149
Kohli RK, Singh D (1991) Allelopathic impact of volatile components from *Eucalyptus* on crop plants. Biol Plantarum (Praha) 33:475–483
Kong CH, Zhao H, Xu XH et al (2007) Activity and allelopathy of soil of flavone O-glycosides from rice. J Agric Food Chem 55:6007–6012
Kroymann J (2011) Natural diversity and adaptation in plant secondary metabolism. Curr Opin Plant Biol 14:246–251
Li XJ, Xia ZC, Kong CH et al (2013) Mobility and microbial activity of allelochemicals in soil. J Agric Food Chem 61:5072–5079
Netzley DH, Butler LG (1986) Roots of *sorghum* exude hydrophobic droplets containing biologically active components. Crop Sci 26:775–778
Niemeyer HM (1988) Hydroxamic acids (4-hydroxy-1, 4-benzoxaazin-3-ones), defence chemicals in the Gramineae. Phytochemistry 27:267–292
Petersen J, Belz R, Walker F et al (2001) Weed suppression by release of isothiocyanates from turnip-rape mulch. Agron J 93(1):37–43
Putnam AR (1988) Allelochemicals plants as herbicides. Weed Technol 2:510–518
Rice EL (1984) Allelopathy. Academic, Orlando, FL
Rimando AM, Olofsdotter M, Duke SO (2001) Searching for rice allelochemicals. Agron J 93:16–20
Rizvi SJH, Haque H, Singh VK et al (1992) A discipline called allelopathy. In: Rizvi SJH, Rizvi V (eds) Allelopathy. Springer, Dordrecht, pp 1–10
Schulz M, Marocco A, Tabaglio V et al (2013) Benzoxazinoids in rye allelopathy—from discovery to application in sustainable weed control and organic farming. J Chem Ecol 39:154–174
Seigler DS (1996) Chemistry and mechanism of allelopathic interactions. Agron J 88:876–885
Soltys D, Krasuska U, Bogatek R et al (2013) Allelochemicals as bioherbicides—present and perspectives. In: Price AJ, Jessica AK (eds) Herbicides—current research and case studies in use. InTech, Croatia, pp 517–542
Souza FAPS, Alves SM (2002) Alelopatia: princípios básicos e aspectos gerais. Embrapa Amazônia Oriental, Belém, p 206
Stephenson GR (2000) Herbicide use and world food production: Risk and Benefits. In: Abstract of 3rd international Weed Science Congress, Foz Do Iguassu, Brazil, 6–11 June 2000
Tookey HL, VanEtten CH, Daxenbichler ME (1980) In: Liener IE (ed) Toxic constituents of plant foodstuff. Academic, New York
Topal S, Kocacaliskan I, Arslan O et al (2007) Herbicidal effects of juglone as an allelochemicals. Phyton 46(2):259–269
Trezzi MM, Vidal RA, Dick DP et al (2006) Sorptive behavior of sorgoleone in ultisol in two solvent systems and determination of its lipophilicity. J Environ Sci Health Part B 41:345–356
Turk MA, Tawaha AM (2003) Allelopathic effect of black mustard (*Brassica nigra* L.) on germination and growth of wild oat (*Avena fatua* L.). Crop Prot 22(4):673–677

Uygur FN, Koseli F, Cinar A (1990) Die allelopathische Wirkung von *Raphanus sativus* L. J Plant Dis Prot 12:259–264

Vyvyan JR (2002) Allelochemicals as leads to herbicides and agrochemicals. Tetrahedron 58:1631–1646

Wang Q, Li ZH, Ruan X et al (2010) Phenolics and plant allelopathy. Molecules 15 (12):8933–8952. https://doi.org/10.3390/molecules15128933

Weston LA (1996) Utilization of allelopathy for weed management in agro-ecosystems. Agron J 88:860–866

Weston LA, Duke SO (2003) Weed and crop allelopathy. Crit Rev Plant Sci 22(3-4):367–389

Xuan T, Eiji T, Hiroyuki T et al (2003) Identification of potential allelochemicals from kava (*Piper methysticum* L.) root. Allelopath J 12:197–203

Yu JQ, Matsui Y (1994) Phytotoxic substances in the root exudates of *Cucumis sativus* L. J Chem Ecol 20:21–31

Secondary Metabolites of Higher Plants as Green Preservatives of Herbal Raw Materials and Their Active Principles During Postharvest Processing

13

Somenath Das, Vipin Kumar Singh, Neha Upadhyay, Bijendra Kumar Singh, Jitendra Prasad, Shikha Tiwari, and Nawal Kishore Dubey

Abstract

Herbal raw materials get contaminated with different microbes during collection and postharvest processing, causing both quantitative and qualitative damage leading to degradation of active principles. Currently, a wide range of plant products including essential oils have been found efficacious against fungal and mycotoxin contamination of raw materials and have been recommended as green preservatives for postharvest protection of herbal raw materials. Extracts of different higher plants and bioactive constituents, viz., carvone (from *Carum carvi*), azadirachtin (from *Azadirachta indica*), and isothiocyanate (from horse-radish and brassicaceous plants) have exhibited promising efficacy against microbial contamination. Currently, nanotechnological approach has enhanced the bio-efficacy of phytochemicals through controlled release, site specific delivery, and enhanced stability. The present article throws light on different plant products recommended in postharvest preservation of herbal raw materials and also explains the future technological advancements in their better utilization as preservative of herbal raw materials. Such documentation of pharmacological efficacy of traditionally used plant products would also be helpful in bioprospection of plant diversity against biopiracy.

Keywords

Plant products · Essential oils · Herbal drugs · Bioprospection · Microbes · Biodeterioration · Mycotoxins

S. Das · V. K. Singh · N. Upadhyay · B. K. Singh · J. Prasad · S. Tiwari · N. K. Dubey (✉)
Department of Botany, Centre of Advanced Study, Institute of Science, Banaras Hindu University, Varanasi, India
e-mail: nkdubeybhu@gmail.com

S. C. Mandal et al. (eds.), *Evidence Based Validation of Traditional Medicines*,
https://doi.org/10.1007/978-981-15-8127-4_13

Abbreviations

AFs	Aflatoxins
EOs	Essential oils
FB_1	Fumonisin B_1
FB_2	Fumonisin B_2
FB_3	Fumonisin B_3
GRAS	Generally recognized as safe
OTA	Ochratoxin A
OTB	Ochratoxin B
OTC	Ochratoxin C

13.1 Introduction

Herbal medicines have completed a long journey and are still well accepted in Asia, Africa, Americas, and other European countries. The herbal products also have deep root in formulation of different modern drugs (Silambarasan et al. 2017). About 30% of drugs are of herbal origin and approved by Food and Drug Administration (Martino et al. 2019). Recent report of World Health Organization suggests the present earning through medicinal plants equivalent to US$ 14 billion worldwide, whereas it is expected to reach up to US$ 5 trillion in the upcoming year of 2050 (Aneesh et al. 2009; Nirmal et al. 2013). The major reason behind increasing popularity is its cost effectiveness and no side effects in comparison to allopathic medicines. Herbal medicines are generally secondary metabolites with different bio-functional molecules containing a set of varied alkaloids, phenols and glycosidic metabolites of cell (Jones and Kinghorn 2006). Recently, the traditional herbal medicines have been merged with modern Ayurvedic system and given the birth of "Green medicine" with lesser side effects and long-term drug utilizing efficiency. However, biodeterioration of herbal drugs through microbial infestation and toxin contamination has shifted the global attention for their qualitative and quantitative maintenance. High moisture content, pH, relative humidity, and variable a_w level easily provide a wide and suitable platform for fungal and mycotoxin contamination in the herbal raw materials during the prolonged storage conditions (Roy 2003). Occurrence of different mold species and their toxic secondary metabolites decline the medicinal property of herbal drugs by affecting the cellular alkaloids and phenolic contents. Microbial contaminants generally include diversified fungal conidiospores, bacterial endospores, and viruses leading to severe pathogenicity. Fungal infestation of herbal drugs has attracted greater attention due to their mycotoxin contamination and toxicity as evident in form of carcinogenesis, teratogenesis, neurotoxicity, and immunotoxicity (Rizzo et al. 2004). Among different mycotoxins, aflatoxins produced by *Aspergillus flavus*, *A. parasiticus*, and *A. nomius* are common contaminants of herbal drugs and classified as class 1 human carcinogen. In addition to aflatoxins (AFs), ochratoxin A (OTA), ochratoxin B (OTB), ochratoxin C (OTC), fumonisin B_1 (FB_1), fumonisin B_2

(FB_2) fumonisin B_3 (FB_3), penicillic acids, and citreoviridin are abundantly found in different medicinal herbal preparations. Geographical variation, altitudinal differences, and climatic variability have marked impact on mycotoxin biosynthesis and transportation. Different endospore forming bacteria such as *Enterococcus faecalis*, *Acinetobacter baumannii*, *Escherichia coli* and *Pseudomonas aeruginosa* easily proliferate in postharvest stored herbal drugs in between the range of 3.8×10^3 CFU/mL and 1.6×10^6 CFU/mL (Wilson et al. 2004).

Till now, different synthetic preservatives have been utilized to control the bacterial, fungal, and mycotoxin contamination in herbal raw materials, but their significant mammalian toxicities, effect on food chain organisms, environmental persistence, and long-term bio-incompatibility have restricted their wide-scale applications (Nasreddine and Parent-Massin 2002). Several synthetic preservatives induce mycotoxin secretion by promoting reactive oxygen species generation which ultimately reduces the shelf life of stored herbal drugs (Dórea 2008). On the other hand, plant essential oils have exhibited the potential as safer alternative of synthetic antioxidants and antimicrobials due to vapor phase botanical fumigants and category of exemption under generally recognized as safe (GRAS) status. Essential oil components are more judicious for preparation of compound-based synergistic formulations with significant efficacy against pathogenic bacteria, fungi, and mycotoxins.

In spite of a lot of proved antimicrobial and antioxidant efficacy, there are various limitations for practical scale essential oil applications because of bioactive components degradation under variable pressure, heat, light, and oxygen. Moreover, water insolubility and problem of sustained release are the key parameters which could be resolved through modern nanoencapsulation technology. Nanoparticles preserve the bioactive functionality of phenolic and terpenoid components of essential oil as well as facilitate in controlled release of nanoencapsulated substances having greater surface to volume ratio. Further, essential oil and their encapsulated products also preserve the active ingredients of herbal drugs from severe biodeterioration and metabolic non-functionality.

13.2 Herbal Medicine: Global Status

Herbal medicines are in use since very ancient time of human civilization, and it is the oldest healthcare form for the humans. About 500 plants of medicinal importance are mentioned in the old literature, and almost 800 plants are mentioned of being used in native systems of medicines (Verma and Singh 2008). Europeans and other nations such as Canada, Russia, the United States, New Zealand, and Australia use herbal medicines in place of generally used allopathic medicines (Enioutina et al. 2017). Herbs are obtained from natural sources and considered as safe in comparison to allopathic medicines (Harish 2001). In the western world during the second half of twentieth century, there was decline in use of herbal medicines causing its replacement by allopathic medicines. In present time allopathic medicines have occupied most of the share of herbal medicines which were used for the sake of humankinds before the development of allopathic medicines (Bhardwaj et al. 2018).

Presently, the utilization and consumption of herbal medicines are increasing in developed as well as in developing countries. The highest consumers of herbal medicines are recorded from China and India. The practice of herbal medicines is common in almost all cultures, but it is well established in Japanese and Chinese medicine system and in their culture. In Japanese culture, commonly used Kampo medicine system is completely based on herbal medicines (Hakamatsuka 2011).

Traditional herbal drugs are obtained from herbal medicinal plants which are utilized in generally small amount or as a whole plant or its part without industrial process for treatment of disease at local level. Currently, the herbal medicines are being utilized by large number of people of the world population in the developing and developed countries (Tilburt and Kaptchuk 2008). 'WHO Traditional Medicine 2013–2023' highlighting the significance of traditional herbal drug was launched in 2013. Utilization of herbal drugs results in universal health protection and promotes the safety and quality of herbal drug (Maithani et al. 2019). The use of traditional herbal medicine has been increased in present scenario because of its natural origin, almost no side effects, and simple nature. In Europe, North America, and Asia, utilization of wild life has been raised from 8 to 15% per annum (Verma and Singh 2008). Currently, researchers have investigated different bioactive chemical compounds such as quinine, digoxin, aspirin, ephedrine, atropine, and colchicines extracted from different parts of *Ephedra* sp., *Cinchona* sp., and *Digitalis purpurea* as potent ingredients of herbal drugs (Moteriya et al. 2015). In South Africa, Francisco, and London, varied utilization of herbal drugs for the treatments of AIDS patients is evidenced, and most notably 70–90% population of Germany and Canada is described to use herbal drug at least once in their life.

Herbal drug utilization is continuously rising. About 158 million adult population of the United States used herbal drug for healthcare. Finally, at global market level, utilization of herbal medicine is US$60 billion per year (Gunjan et al. 2015). Traditional herbal drugs are utilized for the control of malaria in 60% population of Nigeria, Zambia, Ghana, and Mali. Nowadays, most of the countries have preferred the utilization of traditional herbal drugs for the healthcare, viz., 80% in Germany, 39% in Belgium, 42% in the United States, 48% in Australia, 70% in Canada, and 76% in France (Maithani et al. 2019). Herbal health product, herbal cosmetics, herbal medicines, nutraceuticals, food supplements, and herbal pharmaceuticals are non-toxic, exert less side effect, and proved better compatibility with intestinal flora; therefore plant-based products are rising continuously throughout the world (Sharma et al. 2008). Many medicinal plants are also used by primitive system, viz., Ayurveda, Unani, Allopathy, and Siddha, and documented for protection from different diseases (Rabe and Van Staden 1997). In 1983–1994, researchers observed that about 520 drugs are derived from different parts of angiospermic medicinal plants with richest source of antibacterial, anticancerous, and antispasmolytic bioactive ingredients. In 2010, China occupied first rank in the export of herbal product in the world with export value US$1329.72 million, whereas, currently India have exported herbal product with cost value of US$ 790.56 million.

The delivery of herbal materials by AYUSH will be enhanced up to 8800 crores in 2020. Presently, the United States is the largest contributor in the herbal market, while Japan is the second largest country for the export of traditional herbal medicine with transport value 505 million USD in 2016. Value of Chinese patent medicine was 225 million USD in 2016 which is identical to 13.94% decrease in transport value of herbal medicine in the last year (Lin et al. 2018). In India, marketing of aromatic and herbal drug is increasing by 8–10% per year.

13.3 Postharvest Losses of Raw Material of Herbal Drugs

Herbal drugs are prepared from different parts of medicinal plants like roots, rhizome, bulbs, bark, leaf, fruit, and seed; however, the quality of medicinal plants also depends on several factors, viz., geographical origin, growth stage at the time of collection, and their postharvest handling. Although herbal drugs possess distinct flavor, color, and aroma, their unscientific cultivation, collection methods, inappropriate harvesting, cleaning techniques, unsuitable transportation, prolonged storage, as well as warm and humid climatic conditions result into microbial contamination which further causes different health disorders (Stević et al. 2012). Mold infestation is the major threat to the stored and processed herbal raw material which further lead to its discoloration, quality deterioration, reduction in therapeutic potential, and accumulation of several mycotoxins such as aflatoxins, ochratoxins, citrinin, trichothecenes, zearalenone, and fumonisin produced by *Aspergillus flavus*, *A. ochraceous*, *Penicillium citrinum*, and *Fusarium* sp. (WHO 2007). Mycotoxins are known to induce several disorders in the liver, kidney, muscular system, skin, respiratory organs, digestive tract, and genital organs of human (Rai and Mehrotra 2005). Aflatoxins produced by different species of *Aspergillus* are of great concern in the current generation due to its mutagenic, carcinogenic, teratogenic, neurotoxic, nephrotoxic, and immunosuppressive activities (FAO 2000). Aziz et al. (1998) reported fungal and mycotoxin contamination in 84 drug samples. Spoilage of herbal medicines due to several bacterial and fungal contaminants such as *Bacillus*, *Staphylococcus*, *Klebsiella*, *Listeria*, *Micrococcus*, *Corynebacterium*, *Proteus*, *Escherichia*, *Streptococcus*, *Acinetobacter*, *Citrobacter*, *Lactobacillus*, *Serratia*, *Aspergillus*, *Penicillium*, *Candida*, and *Mucor* was also reported by Esimone et al. (2007). Microbial contamination of herbal drugs degrades its medicinal property by deteriorating its medicinally important chemical constituents (Gupta et al. 2009). Due to microbial contamination, ingredients of Triphala drug were declined, and drug potency was declined up to 50% within 2 months (Prasad et al. 2003). It has been reported that fruits of *Emblica officinalis*, *Terminalia bellirica*, and *Terminalia chebula* exhibited excessive contamination of *Aspergillus*, *Penicillium*, *Curvularia*, *Alternaria*, *Helminthosporium*, and *Rhizopus* along with different mycotoxins, viz., aflatoxins, ochratoxins, and patulin (Gautam and Bhadauria 2010). Different investigators suggested the active participation of *Aspergillus* and *Penicillium* for contamination of a range of herbal drugs leading to alarming situations of mycotoxin on their consumptions (Gautam and Bhadauria 2011). Postharvest contaminations of some herbal raw materials are presented in Table 13.1.

13.4 Higher Plant Secondary Metabolites in Postharvest Management of Herbal Raw Materials

Management of postharvest microbial and mycotoxin-mediated biodeterioration of herbal raw materials may be achieved through different physical methods as well as utilization of chemical methods like application of synthetic preservatives. Common physical methods include radiation treatment, cold shock, heat therapy, and modified packaging system. Utilization of synthetic fungitoxicant such as organochlorine, benzimidazole, organophosphate, and methyl bromide has been reported to stimulate mycotoxin secretion in herbal materials. Also, the synthetic preservatives cause generation of toxic free radicals leading to oxidation which degrade the active constituents of herbal drugs. Therefore, plant products are represented as safer and green alternatives for inhibition of microbial infestation and toxin production.

13.4.1 Inhibition of Microbial and Fungal Contamination in Herbal Materials

Since long time, phytochemicals or botanicals have been categorized under green antimicrobial for preservation of medicinal values and food materials. From a large biochemical pool of plant secondary metabolites, essential oils (EOs) extracted from aromatic plants act as a boon for pharmaceutical and agricultural industries based on non-toxicity, renewable nature, and environmental safety of medicinal formulations.

Botanical preservatives are more acceptable to the pharmaceutical and agri-industries due to its negligible mammalian toxicity and vapor phase fungitoxic activity leading to less chance of residual toxicity (Mishra et al. 2015; Das et al. 2019). Moreover, based on the volatile nature, significant antioxidant, and antimicrobial efficacy, different formulations of essential oils, viz., Protecta 1, Protecta 2, FL-USA, Bavaria Corp. Apopka, and DMC Base Natural, have been prepared and reached to the international markets (Burt 2004). Mishra et al. (2013) reported the active participation of different essential oil components such as eugenol, thymol, linalool, menthol, asarone, and caryophyllene against fungal contamination of different stored herbal raw materials. Raw materials from different medicinal plants such as *Withania somnifera*, *Boerhaavia diffusa*, *Holarrhena antidysenterica*, and *Terminalia arjuna* were significantly protected against microbial contamination by utilization of *Cymbopogon flexuosus* essential oil (Kumar et al. 2009). Leaf extract of *Adenocalymma alliaceum* has actively participated in the inhibition of fungal contamination in stored herbal materials of *Acorus calamus* and *Withania somnifera* (Shukla et al. 2008). Stević et al. (2014) suggested potential efficacy of 16 different essential oils isolated from commonly occurring aromatic plants against fungal contamination in traditionally used medicinal plants such as mint, marigold, and

Table 13.1 Microbial and toxin contamination of stored herbal raw materials

Herbal raw materials	Plant parts	Fungal/bacterial/mycotoxin contamination	References
Mentha piperita	Mint leaf	*Aspergillus* sp. (*Aspergillus flavus*, *Aspergillus niger*), *Fusarium* sp., *Alternaria* sp., *Mucor* sp., *Rhizopus* sp., etc.	Stevic et al. (2012)
Zea mays	Corn silk	*Bacillus cereus* and *Clostridium perfringens* bacterial strains and among molds *Fusarium* sp., *Penicillium* sp., *Aspergillus flavus*, and *Aspergillus niger* are dominant	Martins et al. (2001)
Emblica officinalis, *Terminalia bellirica*, and *Terminalia chebula*	Fruits	*Aspergillus* sp., *Penicillium* sp., *Helminthosporium* sp., *Rhizopus* sp., *Syncephalastrum* sp., *Alternaria* sp., and *Curvularia* sp. fungus and mycotoxins such as AFB_1, AFB_2, AFG_1, AFG_2, citrinin, and sterigmatocystin	Gautam and Bhadauria (2009), Gautam and Bhadauria (2010)
Aframomum melegueta	Seeds	*Aspergillus* sp., *Penicillium* sp., *Rhizopus* spp., *Mucor* sp., *Fusarium* sp. among which *A. flavus* and *A. niger* were dominant	Efuntoye (1996)
Azadirachta indica	Stem bark		
Xylopia aethiopica	Seeds		
Plumbago zeylanica	Seeds		
Morinda lucida	Whole stem		
Jatropha curcas	Stem bark		
Plumbago zeylanica	Roots and leaves	*Aspergillus flavus*, *Aspergillus niger*, *Penicillium italicum*, *Fusarium* sp., *Cladosporium herbarum*, *Trichoderma viride*, *Alternaria alternata*, *Absidia corymbifera*, *Rhizopodopsis* sp., *Curvularia* sp., *Aspergillus terreus*, *Umbelopsis* sp., *Mucor* sp.	Singh et al. (2008)
Evolvulus alsinoides	Whole plant		
Asparagus racemosus	Tuberous root		
Adhatoda vasica	Leaves and stem bark		
Glycyrrhiza glabra	Stem and root		
Aloysia triphylla	Whole plant	*Aspergillus flavus*, *Aspergillus ochraceus*, *A. sojae* fungi, and aflatoxin, ochratoxin, and fumonisin	Rizzo et al. (2004)
Asparagus racemosus	Tuberous root	*Aspergillus flavus*, *Aspergillus niger*, *Aspergillus terreus*, *Aspergillus fumigatus*, *Fusarium oxysporum*, *Cladosporium herbarum*, *Trichoderma viride*, *Mucor* sp., *Alternaria alternata*, *Curvularia lunata*, *Rhizopodopsis* sp., *P. italicum*, *Umbelopsis* sp., and aflatoxin B_1	Singh et al. (2010)

(continued)

Table 13.1 (continued)

Herbal raw materials	Plant parts	Fungal/bacterial/mycotoxin contamination	References
Phyllanthus emblica	Dried fruit rind	Fungi such as *Aspergillus flavus*, *Aspergillus niger*, *Mucor* sp., and *Escherichia coli* bacteria	Amina et al. (2017)
Thea sinensis (herbal black and green tea)	Whole plant	Fumonisins B_1, fumonisins B_2	Omurtag and Yazicioğilu (2004)
Red pepper, black pepper and white pepper	Fruits	Aflatoxin B_1, B_2, G_1, G_2, and ochratoxin A	Fazekas et al. (2005)
Rauvolfia serpentina	Root	*Aspergillus flavus*, *Aspergillus niger*, *Penicillium* sp., *Alternaria alternata*, *Curvularia lunata*, *Fusarium oxysporum*	Kumar et al. (2013)
Basil	Leaves	Aflatoxins B_1, B_2, G_1, and G_2	Romagnoli et al. (2007)
Coriander	Seeds		
Rosemary	Whole plant		
Cumin	Seeds		
Marjoram	Whole plant		
Fennel	Seeds		
Alpinia galanga	Rhizome	*Aspergillus terreus*, *Aspergillus niger*, *Aspergillus flavus*, *Aspergillus parasiticus*, *Penicillium verrucosum*, *Penicillium steckii*, *Rhizopus stolonifer*, *Mucor pusillus*, *Cladosporium herbarum*, *Trichoderma viride*, *Absidia corymbifera*, and *Paecilomyces liliacinus*	Mandeel (2005)
Elettaria cardamomum	Seeds		
Amomum angustifolium	Seeds		
Syzygium aromaticum	Flower buds		
Bunium persicum	Seeds	*Aspergillus* sp., *Curvularia* sp., *Alternaria* sp., *Fusarium* sp., *Cladosporium* sp., *Mucor* sp., *Rhizopus* sp., *Trichoderma* sp., *Phoma* sp., *Emericella* sp., *Paecilomyces* sp.	Gupta et al. (2013)
Zanthoxylum armatum	Dried fruits		
Glycyrrhiza glabra	Dried root		
Capsicum annum	Fruit	Aflatoxins B_1, B_2, G_1, and G_2	Jalili (2016)
Piper nigrum	Fruit		
Curcuma longa	Rhizome		
Cinnamomum verum	Bark		
Laurus nobilis	Leaf	*Aspergillus flavus*, *A. niger*, *A. fumigatus*, *A. terreus*, *A. versicolor*, *A. fumigatus*), *Penicillium chrysogenum*, *Alternaria alternata*, and *Rhizopus stolonifer*	Migahed et al. (2017)
Elettaria cardamomum	Seeds		
Foeniculum vulgare	Seeds		
Trigonella foenum graceum	Seeds		
Petroselinum sativum	Seeds		

horse tail. Mishra et al. (2012) reported the application of jamrosa essential oil and its major component linalyl acetate toward significant inhibition of fungi-mediated biodeterioration and oxidative damages in *Holarrhena antidysenterica* and *Rauvolfia serpentina*. Essential oil of *Citrus reticulata* and *Cymbopogon citratus* exhibited significant in vitro and in vivo fungitoxicity in tuberous root of *Asparagus racemosus* (Singh et al. 2010). Kumar et al. (2013) suggested the fungitoxic nature of *Ocimum sanctum* essential oil against fungal infestation-mediated deterioration in stored roots of *Rauvolfia serpentina*. Determination of per cent protection of ajmaline, the major bioactive ingredients of *Rauvolfia serpentina,* in *Ocimum sanctum* essential oil fumigated sets as determined through high-performance thin layer chromatography (HPTLC) suggested the preservative potency of *Ocimum sanctum* essential oil under in situ conditions. Dimkic et al. (2015) observed significant synergistic and additive effects of thyme and savory essential oils against the in situ fusarial infection in *Calendula officinalis*. Recently, Singh et al. (2019) reported the synergistic activity of *Salvia sclarea* essential oil and its major component linalyl acetate against major herbal drug contaminating fungi such as *A. flavus*, *A. niger*, *A. parasiticus*, and *Alternaria alternata*.

13.4.2 Inhibition of Mycotoxin Contamination in Herbal Materials

Globally, mycotoxin contamination of herbal raw materials is one of the major factors responsible for decreased medicinal value of herbal raw materials. The application of synthetic preservative to control the mycotoxin contamination is not recommended because of its persistence, bioaccumulation, and hazardous impact on human health and environment. On the other hand, the application of essential oils for controlling the mycotoxin contamination is promising due to rapid loss from the system, non-toxicity, safety to non-target species, and inclusion under generally recognized as safe (GRAS) status. Numbers of studies have reported remarkable potential of essential oils as well as plant-based products in controlling the mycotoxin contamination of stored herbal raw materials; however, more detailed studies are still warranted to recommend such plant metabolites for industrial applications. Mishra et al. (2013) reported the activity of essential oil components such as eugenol, thymol, linalool, menthol, asarone, and caryophyllene against aflatoxin contamination of different stored herbal raw materials. Effective retardation of AFB_1 contamination in herbal drugs such as *Terminalia chebula*, *Asparagus racemosus*, and *Glycyrrhiza glabra* by application of *Cinnamomum camphora* essential oil has been reported by Singh et al. (2008). Raw materials from different medicinal plants such as *Withania somnifera*, *Boerhaavia diffusa*, *Holarrhena antidysenterica*, and *Terminalia arjuna* were reported to be significantly protected against aflatoxin contamination by utilization of *Cymbopogon flexuosus* essential oil (Kumar et al. 2009). Leaf extract of *Adenocalymma alliaceum* is demonstrated to actively participate in inhibition of mycotoxin contamination of stored herbal materials of *Acorus calamus* and *Withania somnifera* (Shukla et al. 2008). Mishra et al. (2012) reported the biological activity of jamrosa essential oil and its major component linalyl acetate in significantly inhibiting the aflatoxin-mediated biodeterioration and oxidative damages in *Holarrhena antidysenterica* and *Rauvolfia serpentina.* Essential oil of *Citrus reticulata* and *Cymbopogon citratus* exhibited significant in vitro and in vivo aflatoxin inhibitory efficacy in tuberous root of *Asparagus racemosus* (Singh et al. 2010). Kumar et al. (2013) suggested the application of *Ocimum sanctum* essential oil against aflatoxin-mediated deterioration in stored roots of *Rauvolfia serpentina.* The determination of percent protection of ajmaline, the major bioactive ingredients of *Rauvolfia serpentina,* in *Ocimum sanctum* essential oil fumigated sets through high-performance thin layer chromatography (HPTLC) suggested the antifungal potency of *Ocimum sanctum* essential oil under in situ conditions. Mishra et al. (2015) have also reported the efficacy of essential oils from *Caesulia axillaris*, *Cymbopogon khasans*, and *Cymbopogon martinii* in preserving the active component andrographolide from stored raw materials of *Andrographis paniculata.* Fumigation of the raw materials with EOs caused considerable protection of active component and controlled fungal as well as aflatoxin contamination, while the samples were significantly biodeteriorated, and the active component was declined in control sets. Figure 13.1 shows microbial and toxin contamination of herbal drugs and their inhibition by bioactive plant products.

13.5 Technological Advancements in Application of Plant Products as Green Preservative in Postharvest Processing of Herbal Raw Materials

Application of plant products for preservation of herbal drugs against postharvest fungal and mycotoxin contamination is a green approach in the modern era. Most of the postharvest deteriorating fungi qualitatively and quantitatively biodeteriorate the active constituents of herbal drugs and effectively reduce their shelf life. Although numbers of plant extracts, EOs, and their components have shown effectiveness and employed as ecofriendly green pesticides against postharvest deterioration of herbal raw materials, however, volatile, unstable, and fragile nature of EOs makes them inconvenient for long-term utilization as preservative (Kedia and Dubey 2018). EOs and its bioactive components get oxidized and degraded under prolonged storage condition as well as during interaction with food components such as fat, starch, and protein, eventually reducing their antimicrobial potential (Hyldgaard et al. 2012). Such challenges of EOs and its components may be resolved through incorporation of nanotechnological approach which boost up the bio-efficacy of these natural preservatives through controlled release, site specific delivery, and enhanced stability. Also, encapsulation not only protects the degradation of volatile bioactive components from external factors such as light, temperature, and pH but also reduces alteration in organoleptic properties (Ezhilarasi et al. 2013; Sansukcharearnpon et al. 2010). Nanoencapsulation technology deals with the entrapment of bioactive molecules within polymeric substances such as chitosan, gelatin, starch, casein, cyclodextrin, cellulose, alginate, and zein that are of biodegradable, biocompatible, and user friendly nature (Ribeiro-Santos et al. 2017; Chaudhari et al. 2019).

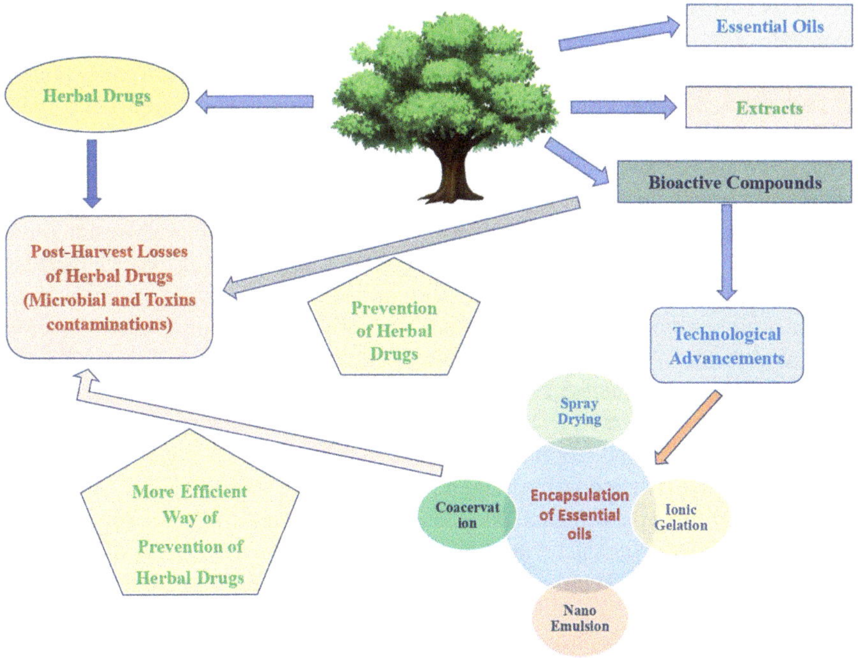

Fig. 13.1 Flowchart showing microbial and toxin contamination of herbal drugs and their inhibition by bioactive plant's products

Nanoencapsulated EOs are highly efficacious in comparison to unencapsulated form due to their increased surface area to volume ratio facilitating easy penetration into pathogen (Gupta et al. 2016) and controlled release of aroma. Singh et al. (2019) demonstrated that chitosan-encapsulated *Ocimum sanctum* EO significantly enhanced the shelf life of stored herbal drugs against fungal and aflatoxin contamination as compared to its unencapsulated form. The two fold increased antioxidant potential of encapsulated *Ocimum sanctum* was ascribed to inhibit the AFB_1 synthesis under in vitro conditions. The IC_{50} value as determined through $DPPH^{\bullet}$ assay was 2810 μL L^{-1} and 1390 μL L^{-1} for unencapsulated and encapsulated EO, respectively. Further, the study revealed decreased biosynthesis of methylglyoxal, an aflatoxin inducer, suggesting novel mechanism of antiaflatoxigenic potential of encapsulated essential oils. Increase in antioxidant activity of encapsulated essential oil of *Coriandrum sativum* is also reported by Das et al. (2019). The investigation also demonstrated boosted antimicrobial activity and in situ efficacy. The increased antiaflatoxigenic potential and fungitoxicity was the resultant of sustained release of entrapped bioactive components, nanoscale size of emulsion, and improved phenol content. The enhancement in antioxidant activity and phenolic content may also be due to association of entrapped essential oils with nanomatrix coating system. The radical scavenging potential measured through $DPPH^{\bullet}$ and $ABTS^{\bullet+}$ assay for unencapsulated and encapsulated *Coriandrum sativum* essential oil was reported as 16.04, 10.26 μL mL^{-1}, and 3.26, 2.41 μL mL^{-1}, respectively. Therefore, increased antioxidant activity and phenolic content of essential oils after encapsulation are the important features while designing the nanoformulation for the purpose of herbal raw material protection from fungal and mycotoxin contamination. Some of the important studies performed in food system using encapsulated essential oil may serve as the base for protection of herbal raw materials from fungal and aflatoxin contamination. For instance, Dwivedy et al. (2018) reported active participation of *Illicium verum* EO-loaded chitosan nanoparticle with potential fungitoxic activity and aflatoxin inhibitory action against fungi infesting stored products. Although, the authors have reported the antioxidant activity of unencapsulated essential oil but not for encapsulated one. The antioxidant potential of test essential oil was determined through $DPPH^{\bullet}$ (IC_{50} was 25.89 μL/mL) and $ABTS^{\bullet+}$ assay (2.86 μL/mL). Encapsulated cardamom essential oil exhibited strong antimicrobial activity against *Escherichia coli* and *Staphylococcus aureus* (Jamil et al. 2016). However, they did not report the antioxidant activity of essential oil and their encapsulated formulation. Chitosan-entrapped EOs of *Cinnamomum zeylanicum* (cinnamon), *Syzygium aromaticum* (clove), and *Thymus capitatus* (thyme) have been found effective against *Aspergillus parasiticus*, *Fusarium verticillioides*, and selective mycotoxins such as aflatoxins, patulin, ochratoxin, and fumonisins (Villegas-Rascón et al. 2018). Both unencapsulated and encapsulated essential oils exhibited fungistatic behavior over mycotoxin-secreting fungi. Further, they did not measure the radical scavenging potential of either form of selected essential oils. Encapsulation of clove and lemongrass EO within chitosan nanoparticle exhibited enhanced antifungal potential against *Aspergillus niger* and *Fusarium graminearum* due to biphasic and cumulative release of EOs in controlled manner (Hasheminejad et al. 2019). Although antioxidant activity plays a major role in controlling the fungal growth and subsequent mycotoxin secretion, the radical scavenging property was not determined by the workers. Thymol and carvacrol nanoformulations in zein nanoparticle were reported to show better dispersion in water with enhanced potential against different microbial pathogens and oxidative deterioration (Wu et al. 2012). The $DPPH^{\bullet}$- and ferric ion-based antioxidant activity was assessed for encapsulated thymol and carvacrol. Encapsulated thymol, for the most of the samples tested, had better radical scavenging potential as compared to unencapsulated carvacrol. The $DPPH^{\bullet}$ reduction by encapsulated bioactive compounds ranged from 37.5 to 75%. Encapsulated *Coriandrum sativum* EO resulted in enhanced antifungal potential against several toxigenic molds such as *A. flavus*, *A. niger*, *A. candidus*, *A. fumigatus*, *A. terreus*, *A. sydowii*, *P. italicum*, *F. verticillioides*, *Alternaria alternata*, *Curvularia lunata*, *Cladosporium cladosporioides*, and aflatoxin secretion (Das et al. 2019). Investigations from abovementioned study led to the conclusion that higher the antioxidant potential, higher will be protection from fungal and mycotoxin contamination. Although, lots of investigations have been made toward the application of essential oil nanoformulation against fungal and mycotoxin contamination of stored food materials, limited study has been performed to protect the herbal raw materials.

Different physical, mechanical, and chemical methods have been used to encapsulate EOs and their bioactive components, among them, liposome-based strategy, coacervation, nanoemulsion, nanoprecipitation, inclusion complexation, ionic gelation, and drying techniques, viz., lyophilization and spray drying, have been widely accepted and most commonly used for encapsulation of EOs (Bakry et al. 2016).

In coacervation technique, single or mixture of compounds is separated from solution with its subsequent deposition on active biomolecules as coacervate and generally glutaraldehyde or trans-glutamine are utilized as cross-linkers to enhance robustness of coacervate (Zuidam and Shimoni 2010). *Pimenta dioica* EO encapsulated in chitosan or chitosan/k-carrageenan through coacervation technique effectively inhibited postharvest fungal and bacterial contamination (Dima et al. 2014). In ionic gelation process, polyelectrolytes are developed by cross-linking with counter ions in order to develop nanoparticles (Giri et al. 2013). Nanoemulsion technique is based on liquid dispersion principle of two completely immiscible or partially miscible compounds where one is dispersed in form of droplet into another; suitable emulsifiers such as tween 20, tween 80, and acetone improve the stability of emulsion system (Jaiswal et al. 2015). Nanoemulsion of clove bud (*Syzygium aromaticum*) and oregano (*Origanum vulgare*) EO show effective antimicrobial potential against *Aspergillus niger* and *Penicillium* sp. (Otoni et al. 2014). To avoid the irreversible aggregation and chemical instability of nanoformulations, nano-suspensions are converted into particulate forms via spray drying technique which removes moisture, prevents from chemical and biological degradation, enables lower cost for storage and transportation, and enhances solubility of product. In spray drying technique, core and wall material are solubilized in suitable solvents followed by formation of their dispersion or emulsion through high-speed mixing leading to formation of spherical particles (Dos Santos Da Veiga et al. 2019). Nanoencapsulation of eugenol and thymol within zein/casein exhibited bactericidal potential against *E. coli* (Chen et al. 2015), while spray-dried microencapsulation of *Lippia sidoides* EO exhibited potential fungitoxicity (Fernandes et al. 2008). Hence, encapsulation technique facilitates the formulation of green preservative as effective tool in controlling herbal raw materials biodeterioration under postharvest condition and as a safer alternative of synthetic preservatives.

13.6 Conclusion and Future Perspectives

Microbial and toxin contamination in herbal raw materials cause degradation of bioactive components having important medicinal properties. Different essential oils of traditionally used plants and their bioactive components comprising of terpenoids, sesquiterpenoids, and phenolics are currently well acknowledged for effective control of microbial and mycotoxin contaminations of herbal raw materials. Nanoencapsulation of essential oils into carbohydrate and protein polymers is a modern technological approach currently practiced to enhance their efficacy and facilitate the formulation of green preservative as effective tool in controlling herbal raw materials biodeterioration under postharvest storage condition.

Acknowledgments Authors are thankful to Head and Coordinator, Centre of Advanced Study (CAS) in Botany, Banaras Hindu University for providing necessary facilities.

References

Amina RH, Eapen J, Deepa MS (2017) Microbial evaluation of commercial samples of dried Amalaki (*Phyllanthus emblica* Linn.) fruit rind procured from herbal drug market in Kerala, India. J Pharmacogn Phytochem 6(5):286–288

Aneesh TP, Hisham M, Sekhar S, Madhu M, Deepa TV (2009) International market scenario of traditional Indian herbal drugs–India declining. Int J Green Pharm 3(3)

Aziz NH, Youssef YA, El-Fouly MZ, Moussa LA (1998) Contamination of some common medicinal plant samples and spices by fungi and their mycotoxins. Bot Bull Acad Sinica 39:279–285

Bakry AM, Abbas S, Ali B, Majeed H, Abouelwafa MY, Mousa A et al (2016) Microencapsulation of oils: a comprehensive review of benefits, techniques, and applications. Compr Rev Food Sci 15:143–182

Bhardwaj S, Verma R, Gupta J (2018) Challenges and future prospects of herbal medicine. Int Res Med Health Sci 1(1):12–15

Burt S (2004) Essential oils: their antibacterial properties and potential applications in foods—a review. Int J Food Microbiol 94(3):223–253

Chaudhari AK, Dwivedy AK, Singh VK, Das S, Singh A, Dubey NK (2019) Essential oils and their bioactive compounds as green preservatives against fungal and mycotoxin contamination of food commodities with special reference to their nanoencapsulation. Environ Sci Pollut Res 26 (2):1–18

Chen H, Zhang Y, Zhong Q (2015) Physical and antimicrobial properties of spray-dried zein–casein nanocapsules with co-encapsulated eugenol and thymol. J Food Eng 144:93–102

Das S, Singh VK, Dwivedy AK, Chaudhari AK, Upadhyay N, Singh A et al (2019) Antimicrobial activity, antiaflatoxigenic potential and in situ efficacy of novel formulation comprising of *Apium graveolens* essential oil and its major component. Pestic Biochem Physiol 160:102–111

Dima C, Cotârlet M, Alexe P, Dima S (2014) Microencapsulation of essential oil of pimento (*Pimenta dioica* (L) Merr.) by chitosan/k-carrageenan complex coacervation method. Innov Food Sci Emerg Technol 22:203–211

Dimkic, I., Beric, T., Stevic, T., Pavlovic, S., Savikin, K., Fira, D. and Stankovic, S., 2015. Additive and synergistic effects of Bacillus spp. isolates and essential oils on the control of phytopathogenic and saprophytic fungi from medicinal plants and marigold seeds. Biological Control, 87, pp.6-13.

Dórea JG (2008) Persistent, bioaccumulative and toxic substances in fish: human health considerations. Sci Total Environ 400(1-3):93–114

Dos Santos Da Veiga R, Da Silva-Buzanello RA, Corso MP, Canan C (2019) Essential oils microencapsulated obtained by spray drying: a review. J Essent Oil Res 31:1–17

Dwivedy AK, Singh VK, Prakash B, Dubey NK (2018) Nanoencapsulated *Illicium verum* Hook. f. essential oil as an effective novel plant-based preservative against aflatoxin B_1 production and free radical generation. Food Chem Toxicol 111:102–113

Efuntoye MO (1996) Fungi associated with herbal drug plants during storage. Mycopathologia 136 (2):115–118

Enioutina EY, Salis ER, Job KM, Gubarev MI, Krepkova LV, Sherwin CM (2017) Herbal medicines: challenges in the modern world. Part 5. Status and current directions of complementary and alternative herbal medicine worldwide. Expert Rev Clin Pharmacol 10(3):327–338

Esimone CO, Olighe PO, Dibaa UE, Ibezium EC (2007) Gross microbial contamination of herbal medicinal products (HMPs) marketed in mid-western Nigeria. Int J Mol Sci 3(2):87–92

Ezhilarasi PN, Karthik P, Chhanwal N, Anandharamakrishnan C (2013) Nanoencapsulation techniques for food bioactive components: a review. Food Bioprocess Technol 6(3):628–647

FAO (2000) Food safety and quality as affected by animal feedstuff. Twenty second FAO regional conference for Europe, Portugal

Fazekas B, Tar A, Kovacs M (2005) Aflatoxin and ochratoxin A content of spices in Hungary. Food Addit Contam 22(9):856–863

Fernandes LP, Turatti IC, Lopes NP, Ferreira JC, Candido RC, Oliveira WP (2008) Volatile retention and antifungal properties of spray-dried microparticles of *Lippia sidoides* essential oil. Dry Technol 26(12):1534–1542

Gautam AK, Bhadauria R (2009) Mycoflora and mycotoxins in some important stored crude and powdered herbal drugs. Biol Forum Int J 1(1):1–7

Gautam AK, Bhadauria R (2010) Fungal and mycotoxin contamination of some common stored herbal fruit samples. J Indian Bot Soc 89:74–79

Gautam AK, Bhadauria R (2011) Diversity of fungi and mycotoxins associated with stored Triphala churn and its ingredients. J Biol Sci 11(3):226–235

Giri TK, Verma S, Alexander A, Ajazuddin A, Badwaik H, Tripathy M et al (2013) Crosslinked biodegradable alginate hydrogel floating beads for stomach site specific controlled delivery of metronidazole. Farmacia 61:533–550

Gunjan M, Naing TW, Saini RS, Ahmad A, Naidu JR, Kumar I (2015) Marketing trends and future prospects of herbal medicine in the treatment of various disease. World J Pharm Res 4 (9):132–155

Gupta KK, Prasad G, Chopra AK, Khanna DR (2009) Contamination of Asian herbal drugs: need for its critical evaluation. J Appl Nat Sci 1(2):291–297

Gupta D, Sharma S, Sharma YP (2013) Mycoflora and natural Aflatoxin contamination in some traditional medicinal plants from Jammu, India. J Mycol Plant Pathol 43(3):360

Gupta A, Eral HB, Hatton TA, Doyle PS (2016) Controlling and predicting droplet size of nanoemulsions: scaling relations with experimental validation. Soft Matter 12(5):1452–1458

Hakamatsuka T (2011) Revision of Kampo extra formulae that recorded in Japanese pharmacopoeia and standard approval for OTC Kampo products (in Japanese). Farumashia 47:413–417

Harish P (2001) Herbal drugs. Meeting report. Curr Sci 81(1):15–16

Hasheminejad N, Khodaiyan F, Safari M (2019) Improving the antifungal activity of clove essential oil encapsulated by chitosan nanoparticles. Food Chem 275:113–122

Hyldgaard M, Mygind T, Meyer RL (2012) Essential oils in food preservation: mode of action, synergies, and interactions with food matrix components. Front Microbiol 3:12

Jaiswal M, Dudhe R, Sharma PK (2015) Nanoemulsion: an advanced mode of drug delivery system. Biotech 5(2):123–127

Jalili M (2016) Natural occurrence of aflatoxins contamination in commercial spices in Iran. Iran J Health Saf Environ 3(2):513–517

Jamil B, Abbasi R, Abbasi S, Imran M, Khan SU, Ihsan A, Javed S, Bokhari H (2016) Encapsulation of cardamom essential oil in chitosan nano-composites: in-vitro efficacy on antibiotic-resistant bacterial pathogens and cytotoxicity studies. Front Microbiol 7:1580

Jones WP, Kinghorn AD (2006) Extraction of plant secondary metabolites. In: Natural products isolation. Humana Press, Totowa, pp 323–351

Kedia A, Dubey NK (2018) Nanoencapsulation of essential oils: a possible way for an eco-friendly strategy to control postharvest spoilage of food commodities from pests. In: Nanomaterials in plants, algae, and microorganisms. Academic, Cambridge, pp 501–522

Kumar A, Shukla R, Singh P, Dubey NK (2009) Biodeterioration of some herbal raw materials by storage fungi and aflatoxin and assessment of *Cymbopogon flexuosus* essential oil and its components as antifungal. Int Biodeter Biodegr 63(6):712–716

Kumar A, Dubey NK, Srivastava S (2013) Antifungal evaluation of *Ocimum sanctum* essential oil against fungal deterioration of raw materials of *Rauvolfia serpentina* during storage. Ind Crop Prod 45:30–35

Lin AX, Chan G, Hu Y, Ouyang D, Ung COL, Shi L et al (2018) Internationalization of traditional Chinese medicine: current international market, internationalization challenges and prospective suggestions. Chin Med 13(1):9

Maithani M, Grover H, Raturi R, Gupta V, Bansal P (2019) Ethanol content in traditionally fermented ayurvedic formulations: compromised good manufacturing practice regulations–compromised health. Am J Drug Alcohol Abuse 45(2):208–216

Martino E, Della Volpe S, Cavalloro V, Amri B, Kaab LBB, Marrubini G, Rossi D, Collina S (2019) The use of a microwave-assisted solvent extraction coupled with HPLC-UV/PAD to assess the quality of Marrubium vulgare L.(white horehound) herbal raw material. Phytochem Anal 30(4):377–384

Mandeel QA (2005) Fungal contamination of some imported spices. Mycopathologia 159 (2):291–298

Martins HM, Martins ML, Dias MI, Bernardo F (2001) Evaluation of microbiological quality of medicinal plants used in natural infusions. Int J Food Microbiol 68(1–2):149–153

Migahed FF, Abdel-Gwad MM, Mohamed SR (2017) Aflatoxigenic fungi associated with some medicinal plants. Annu Res Rev Biol 14(6):1–20

Mishra PK, Shukla R, Singh P, Prakash B, Kedia A, Dubey NK (2012) Antifungal, anti-aflatoxigenic, and antioxidant efficacy of *Jamrosa* essential oil for preservation of herbal raw materials. Annu Res Rev Biol 74:11–16

Mishra PK, Singh P, Prakash B, Kedia A, Dubey NK, Chanotiya CS (2013) Assessing essential oil components as plant-based preservatives against fungi that deteriorate herbal raw materials. Int Biodeter Biodegr 80:16–21

Mishra PK, Prakash B, Kedia A, Dwivedy AK, Dubey NK, Ahmad S (2015) Efficacy of *Caesulia axillaris*, *Cymbopogon khasans* and *Cymbopogon martinii* essential oils as plant-based preservatives against degradation of raw materials of *Andrographis paniculata* by fungal spoilage. Int Biodeter Biodegr 104:244–250

Moteriya P, Satasiya R, Chanda S (2015) Screening of phytochemical constituents in some ornamental flowers of Saurashtra region. Int J Res Phytochempharmacol 3(5):112–120

Nasreddine L, Parent-Massin D (2002) Food contamination by metals and pesticides in the European Union. Should we worry? Toxicol Lett 127(1–3):29–41

Nirmal SA, Pal SC, Otimneyin SO, Aye T, Elachouri M, Kumar Kundu S, Mandal SC (2013) Contribution of herbal products in global market. Pharma Rev:95–104

Omurtag GZ, Yazicioğilu D (2004) Determination of fumonisins B1 and B2 in herbal tea and medicinal plants in Turkey by high-performance liquid chromatography. J Food Prot 67 (8):1782–1786

Otoni CG, Pontes SF, Medeiros EA, Soares NDF (2014) Edible films from methylcellulose and nanoemulsions of clove bud (*Syzygium aromaticum*) and oregano (*Origanum vulgare*) essential oils as shelf life extenders for sliced bread. J Agric Food Chem 62(22):5214–5219

Prasad G, Metiyan P, Gupta KK, Sharma P, Prashant (2003) Microbial flora and drug efficacy related with enhancement of therapeutic spectrum of three ingredients of Triphala drug during storage. Hamdard Med 56(4):82–87

Rabe T, Van Staden J (1997) Antibacterial activity of south African plants used for medicinal purposes. J Ethnopharmacol 56(1):81–87

Rai V, Mehrotra S (2005) Toxic contaminants in herbal drugs. Environ News Arch 11(4):1–3

Ribeiro-Santos R, Andrade M, Sanches-Silva A (2017) Application of encapsulated essential oils as antimicrobial agents in food packaging. Curr Opin Food Sci 14:78–84

Rizzo I, Vedoya G, Maurutto S, Haidukowski M, Varsavsky E (2004) Assessment of toxigenic fungi on Argentinean medicinal herbs. Microbiol Res 159(2):113–120

Romagnoli B, Menna V, Gruppioni N, Bergamini C (2007) Aflatoxins in spices, aromatic herbs, herb-teas and medicinal plants marketed in Italy. Food Control 18(6):697–701

Roy AK (2003) Mycological problems of crude herbal drugs – overview and challenges. Indian Phytopathol 56:1–13

Sansukcharearnpon A, Wanichwecharungruang S, Leepipatpaiboon N, Kerdcharoen T, Arayachukeat S (2010) High loading fragrance encapsulation based on a polymer-blend: preparation and release behavior. Int J Pharm 391(1–2):267–273

Sharma A, Shanker C, Tyagi LK, Singh M, Rao CV (2008) Herbal medicine for market potential in India: an overview. Acad J Plant Sci 1(2):26–36

Shukla R, Kumar A, Prasad CS, Srivastava B, Dubey NK (2008) Antimycotic and antiaflatoxigenic potency of Adenocalymma alliaceum Miers. on fungi causing biodeterioration of food commodities and raw herbal drugs. Int Biodeterior Biodegradation 62(4):348–351

Singh P, Srivastava B, Kumar A, Dubey NK (2008) Fungal contamination of raw materials of some herbal drugs and recommendation of Cinnamomum camphora oil as herbal fungitoxicant. Microb Ecol 56(3):555–560

Singh P, Shukla R, Kumar A, Prakash B, Singh S, Dubey NK (2010) Effect *of Citrus reticulata* and *Cymbopogon citratus* essential oils on *Aspergillus flavus* growth and aflatoxin production on *Asparagus racemosus*. Mycopathologia 170(3):195–202

Singh VK, Das S, Dwivedy AK, Rathore R, Dubey NK (2019) Assessment of chemically characterized nanoencapuslated *Ocimum sanctum* essential oil against aflatoxigenic fungi contaminating herbal raw materials and its novel mode of action as methyglyoxal inhibitor. Postharvest Biol Technol 153:87–95

Stević T, Pavlović S, Stanković S, Šavikin K (2012) Pathogenic microorganisms of medicinal herbal drugs. Arch Biol Sci 64(1):49–58

Stević T, Berić T, Šavikin K, Soković M, Gođevac D, Dimkić I, Stanković S (2014) Antifungal activity of selected essential oils against fungi isolated from medicinal plant. Ind Crop Prod 55:116–122

Tilburt JC, Kaptchuk TJ (2008) Herbal medicine research and global health: an ethical analysis. Bull World Health Organ 86:594–599

Verma S, Singh SP (2008) Current and future status of herbal medicines. Veterinary World 1 (11):347

Villegas-Rascón RE, López-Meneses AK, Plascencia-Jatomea M, Cota-Arriola O, Moreno-Ibarra GM et al (2018) Control of mycotoxigenic fungi with microcapsules of essential oils encapsulated in chitosan. J Food Sci Technol 38(2):335–340

WHO (2007) WHO guidelines for assessing quality of herbal medicines with reference to contaminants and residues. WHO Library Cataloguing-in-Publication Data, Spain, pp 14–15

Wilson C, Dettenkofer M, Jonas D, Daschner FD (2004) Pathogen growth in herbal teas used in clinical settings: a possible source of nosocomial infection. Am J Infect Control 32(2):117–119

Wu Y, Luo Y, Wang Q (2012) Antioxidant and antimicrobial properties of essential oils encapsulated in zein nanoparticles prepared by liquid–liquid dispersion method. LWT Food Sci Technol 48(2):283–290

Zuidam NJ, Shimoni E (2010) Overview of microencapsulation use in food products or processes and methods to make them. In: Zuidamand NJ, Nedovic VA (eds) Encapsulation technique for active food ingredients and food processing. Springer, New York, pp 3–29

Importance of Chromatography Techniques in Phytomedicine Research 14

Aditi Gujrati, Sourabh Jain, Veenu Joshi, Shiv Shankar Shukla, Amber Vyas, and Vikas Jain

Abstract

Chromatography is a technique including thin-layer chromatography, column chromatography, high-performance liquid chromatography, and gas-liquid chromatography. These isolation techniques play a significant role in authentification, identification, isolation, and enrichment of phyto-molecules belonging to specific aromatic and medicinal plants. In thin-layer chromatography, the standard compound is used for authentification of the plant material after small-scale extraction. The large-scale extraction method was then carried accordingly with suitable solvent. Thin-layer chromatography also gives the basic idea for isolation because it requires proper combination of solvents for optimum Rf value. Column chromatography is the most basic technique for isolation of phyto-molecules. During isolation process both thin-layer chromatography and column chromatography are used simultaneously for identification of a compound in various fractions at

A. Gujrati · V. Jain (✉)
Mahakal Institute of Pharmaceutical Studies, Ujjain, Madhya Pradesh, India
e-mail: aditigujrati6107@gmail.com; vikasjain111180@gmail.com

S. Jain
Sagar Institute of Pharmaceutical Sciences, Sagar, Madhya Pradesh, India
e-mail: sourabhj47@gmail.com

V. Joshi
Center for Basic Science, Pt. Ravishankar Shukla University, Raipur, Chhattisgarh, India
e-mail: vinu.jsh@gmail.com

S. S. Shukla
Columbia Institute of Pharmacy, Tekari, Raipur, Chhattisgarh, India
e-mail: shivpharma007@gmail.com

A. Vyas
University Institute of Pharmacy, Pt. Ravishankar Shukla University, Raipur, Chhattisgarh, India
e-mail: ambervyas@gmail.com

S. C. Mandal et al. (eds.), *Evidence Based Validation of Traditional Medicines*,
https://doi.org/10.1007/978-981-15-8127-4_14

different combinations of solvent used for isolation. For increasing the polarity of a particular solvent combination, thin-layer chromatography gives the basic information. The quantitative assessment of phyto-constituents is through high-performance thin-layer chromatography (HPTLC) and high-performance liquid chromatography (HPLC) for nonvolatile components, whereas gas-liquid chromatography is used for volatile components in general context. High-performance thin-layer liquid chromatography is effective in quantification of biomarkers in extracts and fractions. High-performance liquid chromatography (HPLC) and gas chromatography (GC) are coupled with mass spectrophotometer (MS) to give a unique combination of LC-MS and GC-MS respectively, which are used for quantification of biomarkers along with their molecular weight. Recently, the work is carried out on enrichment of phyto-molecules using chromatographic techniques. In this chapter, we have focused on the recent advances in using chromatographic techniques at different levels of phytochemical investigation.

Keywords

Phyto-molecules · Thin-layer chromatography · Column chromatography · High-performance liquid chromatography · Gas chromatography · High-performance thin-layer chromatography

Abbreviations

APCI	Atmospheric pressure chemical ionization
CI	Chemical ionization
EI	Electron impact
ESI	Electrospray ionization
GC	Gas chromatography
GC-MS	Gas chromatography-mass spectroscopy
GLC	Gas-liquid chromatography
HPLC	High-performance liquid chromatography
HPTLC	High-performance thin-layer chromatography
LC-MS	Liquid chromatography-mass spectroscopy
MALDI	Matrix-assisted laser desorption/ionization
PC	Paper chromatography
TLC	Thin-layer chromatography

14.1 Introduction

The discovery of new drugs from plant sources involves a systematic scientific investigation using a multifaceted approach. It requires the knowledge of ethno-pharmacological uses, chemistry, biology, and toxicology of the plants leading to better understanding of concept correlated to drugs. It is also necessary to carry out community surveys, previous literature survey, and analysis of the outcomes of community interventions to evaluate the ethno-pharmacological uses of plants against the various diseases. Discovery of drugs from plant sources is based on the constituents and their properties, which can be used for the treatment of any disease. Selection of appropriate isolation method involves the acquisition of knowledge related to ethno-use, which can help to work on the isolation of active drug substance from the complex framework of plant materials using the concepts of chemistry for the drug development. Chromatography represents the most flexible and readily available separation technique. It refers to a physical method for the isolation and detection of components or compounds from a mixture into separate components using stationary phase (consist of solid phase or layer of liquid adsorbed onto the surface of solid support) and a mobile phase (composed of liquid or a gaseous component). Thus, phytoconstituents can be easily separated and purified by the help of various chromatographic techniques (Archana Khale and Anubha 2011; Coskun 2016). However, the quality evaluation of drugs like physical and chemical stability, bioavailability, and identity characterization requires the proper isolation and drug development procedure.

There are several techniques currently applied to separate and identify the phytoconstituents. The extraction process is followed by separation of the constituents and isolation of the compounds using different chromatographic techniques. Further, application of spectroscopic techniques leads to identification and confirmation of the pure compounds. On the basis of the nature of the functional groups of compounds, if already known, they can be extracted at high, low, or room temperatures. Moreover, the extraction process is based on mass transfer, which is a unit operation, transferring soluble matter from solid to liquid phase. When a crude particle is immersed in a solvent, it is surrounded by boundary layer of the solvent, which starts penetrating inside the particle and subsequently forms solution of the constituents within the cells. Escape of this dissolved constituent takes place through the cell wall throughout the boundary layer, and the process continues till equilibrium is set up between both the sides.

14.1.1 Extraction Methods

Extraction methods involve the separation of the soluble plant constituents from the insoluble cellular residue. In small-scale extraction processes like maceration and percolation are generally slow, time-consuming and give insufficient extraction of the crude drugs. Whereas in large scale extraction process like modified soxhlet extraction large batches of crude drugs are extracted more easily and quickly.

14.1.1.1 Maceration

Maceration is the process of extracting medicinally active components of the plant material. It involves immersing of coarse or powdered crude drugs in a stoppered vessel containing suitable solvent. The above mixture is allowed to stand for at least 3–7 days in a warm place with intermittent shaking and finally the solvent containing dissolved constituents is collected for further processing. Maceration method is modified to multiple-stage extraction to increase the yield of the constituents in the extracts. The crude plant material is taken in the extractor, which is joined with a circulatory pump and spray distributor along with number of connected tanks to collect the extraction solution. This is called as multiple-stage extraction. The solvent is added and circulated in the extractor containing plant material and is removed and stored as extracted solution in the receiver tanks. This operation is repeated thrice. When the fresh plant material is charged in the extractor, the stored solution is once again circulated and then removed as an extract. Likewise after three extractions, the plant material is removed from the extractor, again recharged with fresh drug, and the whole cycle is repeated. At the end of the process, the solvent moves out, and the crude extract is separated from the plant residue by centrifuging or pressing. Maceration is an old method in which active ingredients cannot be completely extracted (Silva et al. 2017; Raaman 2006).

14.1.1.2 Percolation

Percolation involves the constant flow of the solvent through the crude plant material, which continuously replaces the saturated solvent by fresh solvent. It involves a number of steps. In the first step of the process, powdered material is treated with sufficient amount of the solvent to make it uniformly wet and allowed to stand for 15–20 min. After which the above mixture is transferred into percolator which is generally a "V"-shaped vessel open at both ends. Again sufficient amount of the solvent is added for the saturation of the plant material, and the liquid starts to fall out from the outlet of the percolator (Fig. 14.1). Thereafter, the plant material is allowed to macerate in the vessel for 24 h, and percolation is continued gradually using sufficient solvent.

This process is based upon the flow of the solvent which is used until its point of saturation through the powdered material (Raaman 2006). This process yields products of greater concentration than the macerated products.

14.1.1.3 Modified Percolation

The conventional percolation process is modified by including evaporation, for getting more concentrated extract, especially when alcohol is used as solvent. In this method, the first portion of the percolate is collected and kept aside, and then subsequent portions are collected, concentrated, and added to the first portion. This method is known as reserve percolate method. By this method, the first portion which contains a large amount of the constituents is not subjected to heat, and also higher concentration of the extract is obtained.

In the modified process of percolation techniques, continuous or semi-extraction devices are used in some industries for handling the batches of varying size. The extraction battery which consists of a number of vessels in series is interconnected through pipelines and arranged so that the solvent can be added directly and the product can be removed from the vessel. Such type of extraction battery gives maximum efficiency of extraction with minimum use of the solvent; thus the extract obtained is more concentrated.

14.1.1.4 Continuous Extraction

Continuous extraction is used widely when the constituents and impurities both have partial solubility in the particular solvent (Silva et al. 2017). Soxhlet extraction is the

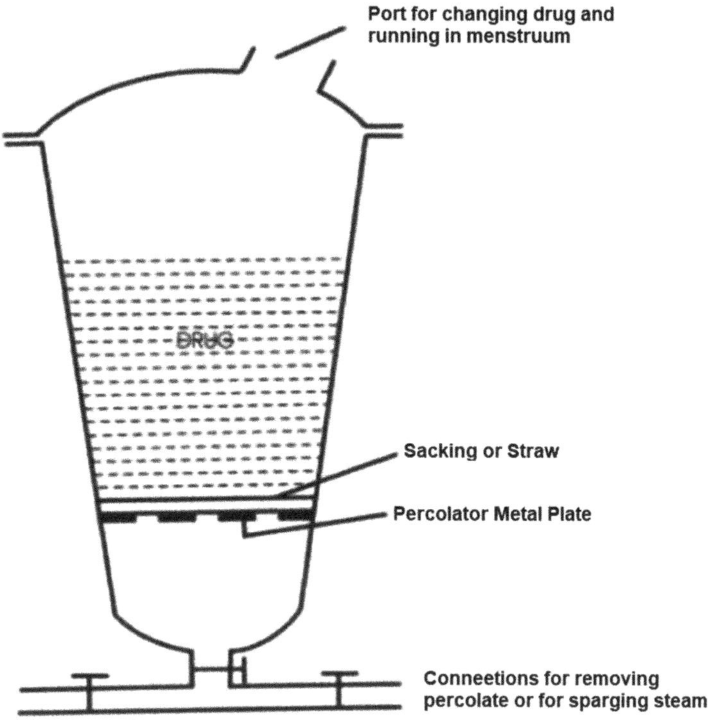

Fig. 14.1 Percolator

method of continuous extraction, in which the solvent is circulated through the extractor for a number of times. In this process, extraction is done by evaporating the solvent and collecting the vapors of the solvent into a condenser, and the condensed liquid is again returned for continuous extraction of the plant material.

Soxhlet apparatus (Fig. 14.2) designed for continuous extraction process includes the following:

- Extractor: Body of extractor is attached with siphon tube and side tube.
- Distillation flask: The extractor from the base is attached to the distillation tank.
- Condenser: Mouth of extractor is set to condenser with the help of standard joints.

In the soxhlet extraction process, first the crude powdered material is filled in the soxhlet apparatus directly or in a thimble of filter paper, and then the solvent is heated, evaporated, and siphoned back into the flask through the thimble. This sequence is repeated a number of times without changing the solvent to get sufficient extract. Fresh activated pieces of porcelain are added into the flask to avoid bumping of solvent.

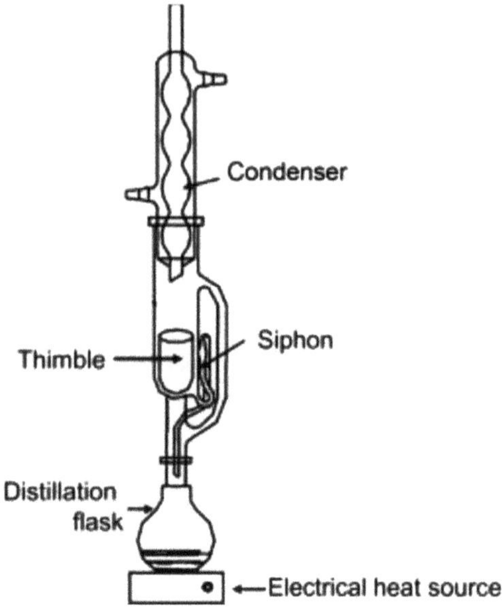

Fig. 14.2 Soxhlet extractor

14.1.1.5 Large-Scale Extraction

Large-scale extraction is meant for the extra-large batches of drug/plant material, in which various soxhlet extractor assemblies are modified (Fig. 14.3). The pilot plant extractor generally has a separate extractor and condenser unit. Separate inlet for loading the drug and outlet for drug discharge are also provided into it. The extractor body is divided into two parts: the upper one which contains plant material and lower one which is a distillation chamber. The distillation chamber is electrically heated, and the vapors of the solvent are passed to the condenser. The condensed liquid is sprayed on the bed of the powdered material with the help of a solvent distribution nozzle. Such large-scale extractors are provided with the outlet from the lower side of the extractor for removing the extract.

14.2 Method of Separation

The crude extract obtained by extraction process is subjected to separation and isolation of the pure compound, which requires specific and specialized techniques. The selection of appropriate method and/or technique depends on solubility and volatility of the compounds to be separated from the crude extract. Techniques which are used for separation and purification include the following:

(a) Paper chromatography (PC)
(b) Thin-layer chromatography (TLC)
(c) Gas-liquid chromatography (GLC)
(d) High-performance liquid chromatography (HPLC)

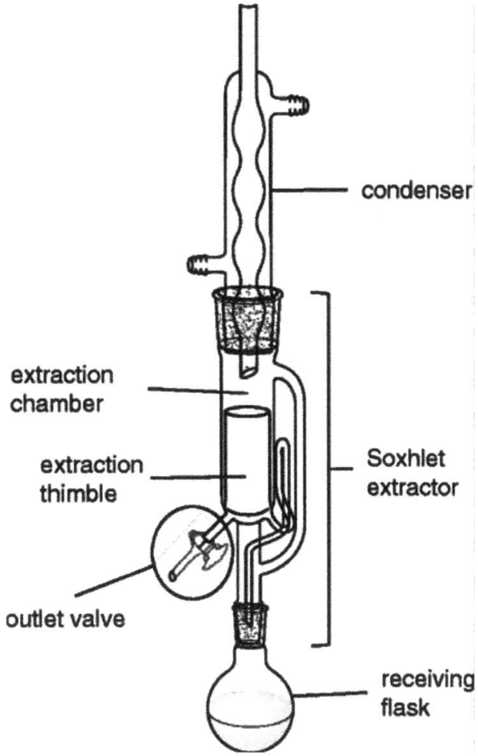

Fig. 14.3 Modified soxhlet extractor

(e) High-performance thin-layer liquid chromatography (HPTLC)

Paper chromatography is generally useful for hydrophilic components like amino acids, nucleic acid bases, carbohydrates, phenolic compounds, and organic acids. While TLC is preferred for the separation of lipophilic components like carotenoids, steroids, lipids, and chlorophyll, the third technique, i.e., GLC is mainly used for the separation of volatile components like fatty acids, mono as well as sesquiterpenes, sulfur compounds, and hydrocarbons. However, the volatility of the compounds can be enhanced by conversion to esters or ethers. In contrast, the less volatile substances can be separated by using HPLC, which has the benefit of column efficiency along with the high speed analysis. All of these methods are applied for the identification, purification, isolation, and/or separation of the compounds. For large-scale isolation column containing adsorbents (as used in PC and TLC) are used and the technique is known as column chromatography. For a large amount of crude material, these columns are coupled with automatic fraction collectors to get a high yield of purified compound (Fan et al. 2006; Hamburger and 193 Cordell 1987; Zygmunt and Namiesnik 2003). Generally, a combination of the aforementioned techniques is the best strategy to isolate a pure compound.

14.2.1 Paper Chromatography

Paper chromatography is a very simple method, in which a filter paper sheet is used as stationary phase for separation of components of a mixture. The Rf values obtained in this method are quite reproducible, which are important to authenticate the results, especially in the case of constituents like anthocyanins, which do not have defined physical properties.

Paper chromatography is of two types: partition chromatography and adsorption chromatography. In the former, compounds are partitioned between water-immiscible alcoholic solvent (e.g., n-butanol) and water. On the other hand, the latter involves the selective adsorption of compounds in the presence of aqueous solvent as in case of common purines and pyrimidines, phenolics, and plant glycosides. Generally PC is done by descending the Whatman filter paper in tanks which is suitable for two-dimensional separation of the compounds. A variety of "modified" filter papers are commercially available for separation of specific compounds. For instance, addition of alumina or salicylic acid onto cellulose paper helps in reducing its polarity, while soaking it in silicone or paraffin oil makes it suitable to carry out "reversed-phase" chromatography.

In this technique, detection of particular class of compounds is done by spraying or dipping the developed paper in specific chromatographic reagent and the compounds are eluted as colored or fluorescent spots under UV light. In few cases the paper may be heated in order to develop the colors. The Rf value can be calculated by measuring the distance from the origin to the center of the spot formed by the compound to the distance travelled by the solvent. This always appears as a fraction between the range of 0.001 and 0.99. Also, Rm value can be calculated for the comparison of structurally correlated compounds (Ye et al. 2007) by the following formula:

$$Rm = \log(1/Rf - 1)$$

14.2.2 Thin-Layer Chromatography (TLC)

TLC is a versatile and sensitive technique. Its versatility is due to the variety of adsorbents which can be spread on glass plate. Generally silica gel is used, but ciliate, polyamide, ion-exchange resin, calcium hydroxide, magnesium phosphate, aluminum oxide, sephadex, cellulose, polyvinylpyrrolidone, and a mixture of two or more of the above materials can also be used. In this technique, time taken in running the sample is less in comparison to PC due to the more compact nature of adsorbent. In ancient times, a major drawback with the TLC was the labor of spreading glass plates with adsorbent, but nowadays automatic spreading devices as well as a variety of pre-coated plates of glass, plastic, or aluminum sheets are commercially available, which has eased the routine phytochemical work. In TLC, there is a sequence of steps to be followed with all the precautions for obtaining good results. First, the glass plates are cleaned properly, and slurry of silica and/or other materials which are used as stationary phase is prepared. Some additives like calcium sulfate

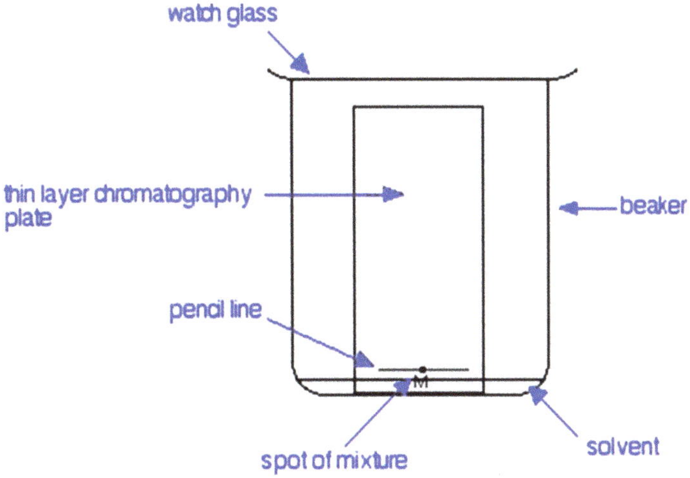

Fig. 14.4 Thin-layer chromatography

hemihydrate and inorganic salts can also be added into the slurry for effective binding and improving the adsorbents properties, respectively. Then the slurry is spread onto the glass plates, dried, and heated at 100–110 °C for 30 min for activation. Thereafter, loaded plates are put into an ascent tank saturated with the solvent system consisting of a combination of solvents in exact proportions. However, in this technique, Rf values are considerably less reproducible; thus it is essential to include one or more standard compounds for reference (Fig. 14.4).

In TLC, glass plates can be easily sprayed with concentrated H_2SO_4 for the detection of lipids and steroids. Also, the separated compounds can be recovered by scraping the adsorbent at the appropriate places on the developed glass plate, eluting the powder with a solvent like ether and centrifuging to remove the adsorbent. TLC plates can be repeatedly developed with one or several different solvent systems with intermittent drying (Archana Khale and Anubha 2011; Mohammad et al. 2010; Beckett and Stenlake 2005; Kasture et al. 2008; Chatwal and Anand 2008). Table 14.1 includes a few examples of analytes evaluated by TLC.

14.2.3 Gas-Liquid Chromatography

GLC is a well-developed analytical technique frequently used for the identification, quantification, and characterization of volatile substances. It is a highly efficient and sensitive technique, which can be used for the separation and detection of essential oils (Bombarda et al. 2008). GLC requires sophisticated and expensive apparatus (Fig. 14.5) consisting of four important components:

Table 14.1 Examples of analytes evaluated by TLC

S. No.	Analyte	Parameters of TLC	References
1	Harhra' (*Terminalia chebula* and gallic acid)	Stationary phase, i.e., silica gel Mobile phase, i.e., ethyl acetate with toluene and formic acid in the ratio of 5:5:1	Lalla et al. (2000)
2	*Azadirachta indica*, *Catharanthus roseus*, and *Momordica charantia*	Stationary phase, i.e., silica gel Mobile phase, i.e., methanol with dichloromethane in the ratio of 8:2	Habib et al. (2000)
3	*Strychnos nux-vomica*	Stationary phase, i.e., silica gel Mobile phase, i.e., diethylamine with chloroform and ethyl acetate in the ratio of 1:0.5:8.5	Jadhav et al. (2009)
4	Fruit of *Piper chaba* constituents are piperamine, piperine, methyl piperate, and piperlonnguminine	Stationary phase, i.e., silica gel Mobile phase, i.e., ethyl acetate with n-hexane in the ratio of 1:1	Richter et al. (2003)
5	Quinones	Stationary phase, i.e., silica gel 60 Mobile phase, i.e., n-hexane with dichloromethane in the ratio of 2:8	Pyka (2014)

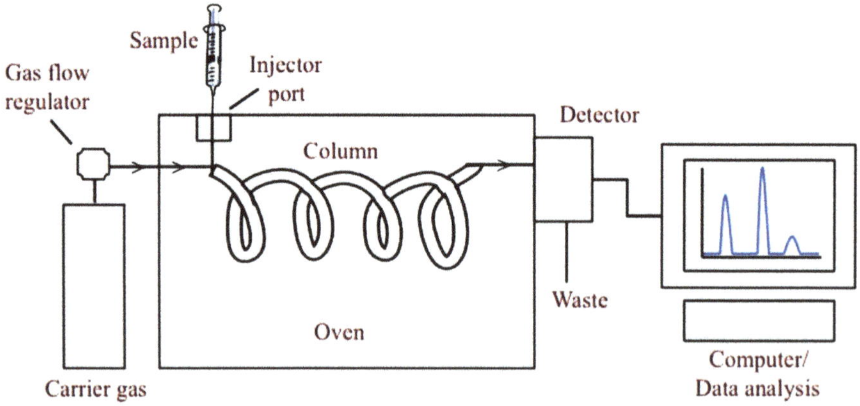

Fig. 14.5 Gas-liquid chromatography

1. Column—made up of metal in the form of a coil to conserve space, packed with a stationary phase like 5–15% silicone oil on an inert powder (Chromosorb W, celite, etc.).
2. Heater—provides heat to the column progressively from 50 to 350 °C at a standard rate and holds the temperature at the higher limit if needed. The temperature of column inlet is controlled separately so that the sample can be

vaporized rapidly as it is passed on to the column. The sample is dissolved in hexane or ether and is injected by hypodermic syringe into the inlet port through a rubber septum.

3. Gas flow—consists of an inert carrier gases, i.e., argon or nitrogen. Separation of the compounds on the column depends on passing this gas at a controlled rate.
4. Detector—based on either electron capture capacity or flame ionization, detects the compounds as they are swept through the column, and the potentiometric recorder generates peaks for different compounds with varying intensities.

The results of GLC are expressed as retention volume R", which is the volume of carrier gas required to elute a component from the column, or in terms of retention time Re, which is the time required for elution of the sample. These parameters are nearly always expressed in terms relative to a standard compound (as RR" or RRe), which may be added to the sample extract or which could take the form of the solvent used for dissolving that sample.

There are three important parameters, namely, stationary phase, the column, and the temperature of operation which varies with the volatility and polarity of the compounds to be isolated. Conversion of the compounds to derivatives can be a strategy to carry out separation at a lower temperature. GLC provides both quantitative and qualitative data on plant substances, since measurements of the area under the peaks shown on the GLC trace are directly related to the concentrations of the different components in the original mixture. There two general formulae for measuring these areas: (a) peak height $\times$ peak width at half height $=$ 94% of the peak area (this only applies to symmetrical peaks), and (b) peak area is equivalent to that of a triangle produced by drawing tangents through the points of inflection. Peak areas can be determined automatically, e.g., by the use of an electronic integrator.

14.2.4 High-Performance Liquid Chromatography

High-performance liquid chromatography is analogous to gas-liquid chromatography as it can provide qualitative and quantitative data in a single operation. Presently, HPLC is the most accepted technique for the phytochemical separation (Faghihi et al. 2001; Lu et al. 2005; Li et al. 2010).

HPLC has high rates of precision, resolution, sensitivity, and selectivity. It has the column filled with small spherical particles enclosed in stationary phase held in a narrow bore stainless steel column making it especially sensitive to impurities, so it is essential to purify and filter the plant extracts before injecting into the column (Fig. 14.6). The liquid mobile is a combination of solvents and is forced under considerable pressure. It either remains constant (isocratic separation) or may be changed continuously in its proportions by including a mixing chamber into the setup called as gradient dilution. The HPLC apparatus is more costly than GLC because of the pumping system which can withstand the high pressure. In this technique, elution process can be monitored by a detector using ultraviolet light. A computing integrator may be added to handle the whole operation, which can be

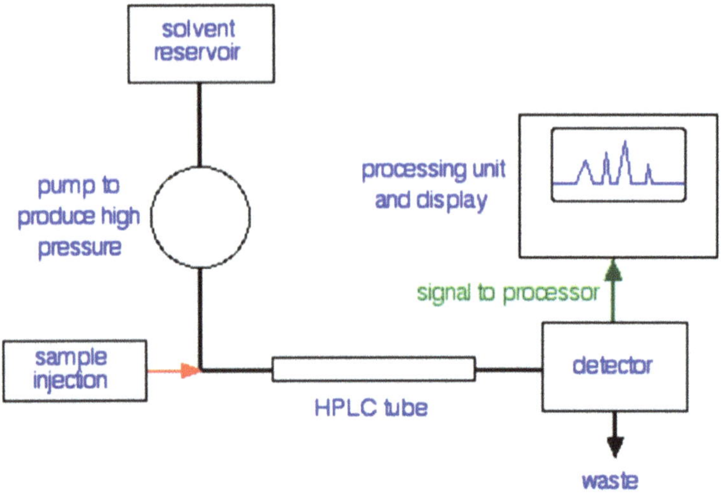

Fig. 14.6 Instrumentation of HPLC

controlled through a microprocessor. The major difference between HPLC and GLC is that the former operates at ambient temperature and the compounds are not subjected to rearrangements during separation. Moreover, the thermostatically controlled jacket of HPLC column provides effective temperature control. HPLC is used for those classes of compounds which are nonvolatile in nature, e.g., phenolics, higher terpenoids, lipids, alkaloids, and sugars. It is more advantageous for compounds which can be detected in the visible or ultraviolet regions of the spectrum. Proteins can be separated by HPLC on columns of modified silica gels, sephadex, or ion exchangers.

Preparative HPLC and analytical HPLC are used extensively in pharmaceutical industry for the isolation as well as purification of herbal compounds. The objective of the former is to purify and isolate the compounds, whereas the latter aims to get information regarding the sample. There are commonly two types of preparative HPLC, i.e., low-pressure HPLC (pressure under 5 bar) and high-pressure HPLC (pressure $>$ 20 bar). In preparative high-pressure HPLC (pressure $>$ 20 bar), larger stainless steel columns and packing materials of (particle size 1030 μm) are required. Applications of HPLC for evaluation of herbal drug are enlisted in Table 14.2.

14.2.5 High-Performance Thin-Layer Chromatography

High-performance thin-layer chromatography is commonly employed in the herbal pharmaceutical industries for identification, quality control and quality assurance of herbs and herbal products. Also the pesticide content, mycotoxins and adulterants in health food products are detected by HPTLC (Patel and Patel 2016).

Table 14.2 Applications of HPLC for evaluation of herbal drugs

Active compounds	Flow rate	Mobile phase	Column	Gradient detector	References
Atropine	1.0 mL/min	A is 0.05M KH2P04 in water of pH 3, B is acetonitrile	Zorbax Eclipse XDBC18, 3.5 μm	UV [diode array detector]	Pharmaceutical Applications with HPLC (2000)
Quercetin Kaempferol	2.0 mL/min	A is 0.5% H2.0 mL/min 3P04 in water, B is methanol	Hypersil ODS, 5 μm	Diode array detector	Pharmaceutical Applications with HPLC (2000)
Rhein Emodin	1.0 mL/min	A is 0.05M NH 40Ac in water of pH 2.5, B is acetonitrile	Hypersil ODS, 5 μm	Diode array detector	Pharmaceutical Applications with HPLC (2000)
Ephedrine Norephedrine	1.0 mL/min	A is 0.025M KH2PO4 in water of pH 3, B is acetonitrile	Zorbax SBC18, 3.5 μm	UV [diode array detector]	Pharmaceutical Applications with HPLC (2000)
Quinidine Quinine	0.8 mL/min	A is 0.05M KH2P04 in water of pH 3, B is acetonitrile	Purospher RP-18, 5 μm	UV [diode array detector]	Pharmaceutical Applications with HPLC (2000)

It is also known as planer chromatography or HPTLC. Conventional TLC is modified in order to utilize the full potential of the method. The principle of HPTLC is the same as that of TLC. The samples to be chromatographed are applied to the self-coated plates in the spot or a band without damage to the layer. The volume is governed by disposable glass capillaries or sample applicator. The sample is concentrated into a narrow band of selectable length, which needs precision and exact positioning. The growth of chromatogram takes place in the same way as that of TLC. As the solvent migrates on the plate by capillary, the samples are separated into fractions. For the evaluation of chromatogram, the peaks are scanned in a densitometer with a light beam in the visible or ultraviolet range of spectrum. The fluorescence is calculated by diffused reflectance. HPTLC systems are provided with the photo and video documentation systems. Densitometric scanning of individual or fractions can be repeated with the same or different parameters, stored on the plate operators, and evaluated or documented before and after derivatization. This technique is largely used in the study of natural product botanicals and herbal cosmetics. Some mobile phases and common derivatization used in HPTLC are tabulated in Tables 14.3 and 14.4. Applications of HPTLC in the evaluation of herbal drugs are mentioned in Table 14.5.

Table 14.3 Some mobile phase used in HPTLC for herbal compounds (Srivastava and Springer Verlag 2011; Atlas 1996; Shivatare et al. 2013)

Chemical constituents (herbal)	Mobile phase used
Hydrophilic constituents: Aanthraglycosides, arbutin, alkaloids, bitter principles, cardiac glycosides, saponin, flavonoids	Ethyl acetate with methanol and water in the ratio of 100:13.5:10
Lipophilic constituents: Eessential oils, coumarin, naphthoquinones, terpenes	Ethyl acetate with toluene in the ratio of 7:93
Flavonoids	Ethyl acetate with glacial acetic acid, formic acid, and water in the ratio of 100:11:11:26
Alkaloids	Toluene with ethyl acetate and diethylamine in the ratio of 70:20:10
Cardiac glycosides	Ethyl acetate with methanol and water in the ratio of 100:13.5:10 or 81:11:8
Terpenes	Chloroform with water and methanol in the ratio of 65:4:25
Triterpenes	Ethyl acetate with toluene and formic acid in the ratio of 50:50:15 Toluene with chloroform and ethanol in the ratio of 40:40:10
Essential oil	Ethyl acetate with toluene in the ratio of 7:93
Saponins	Chloroform with glacial acetic acid, methanol, and water in the ratio of 64:32:12:8

14.3 Recent Advances in the Chromatographic Techniques with Their Applications

The necessity to identify the components of complex mixtures stimulated the development of combination of chromatographic techniques with mass spectrometry (MS).

14.3.1 Liquid Chromatography-Mass Spectroscopy (LC-MS)

Liquid chromatography-mass spectrometry or high-performances liquid chromatography-mass spectrometry (HPLC-MS) is an analytical technique joined together for high-resolution separation with sensitive and specific mass spectrum detection. Combination of LC with MS is a significant development in the chromatography techniques. Mass spectrometry in LC-MS helps to find out the elemental composition and elucidate the structure of a sample (Pitt 2009).

Typical LC-MS system is a combination of HPLC with MS using interface, i.e., ionization source. The sample is separated using liquid chromatography, and the separated sample is sprayed into atmospheric pressure ionic source, which transformed ions into the gas phase. There is a mass analyzer which is used to sort ions depending on their mass-to-charge ratio, and a detector is used to count the ions

Table 14.4 Common derivatization agents used in HPTLC (Jork and Funk 1990; Knap et al. 2006)

S. No.	Reagent	Chemical compounds (herbal)	Color
1.	**Dragendorff reagent** This reagent forms complex reaction with nitrogen-containing compounds	Alkaloids	Red-brown zone (visible)
2.	**Natural products** PEG reagent, i.e., diphenylboric acid 2-aminoethyl ester, forms complexes via condensation reaction with 3-hydroxyflavones	Flavonoids	Intense yellow, green, and Orange Fluorescent zones at UV 366 nm
3.	**Iodine** It produces iodine reaction which results in oxidative products	Quinolone derivative, indole, thiols, and organic compounds	Dark zone at UV 254
4.	**Ethanolic KOH (10%)**	Anthrones (cascarosides, aloin)	Yellow zone (visible) Yellow fluorescence at UV 366 nm
5.	**Ninhydrin reagent**	Amines, peptides, amino acids, and amino sugars	Yellow, brown to pink, and violet (visible)
6.	**Vanillin sulfuric acid or Aanisaldehyde sulfuric acid**	Bitter principle, saponins, essential oils	Red-brown, yellow-brown, dark green zone (visible) Colored zones (visible) Blue, brown, or red zones (visible)

rising from the mass analyzer, and it may also amplify the signals generated from each ion. As a consequence, spectrum of mass is obtained, which is used to determine the mass of molecules and elemental or isotopic nature of a sample and to elucidate the chemical structure of molecules (Korfmacher 2005; Lim and Lord 2002).

Generally the LC-MS consists of an assembly of ion generation unit or ionization source, liquid chromatography, mass analyzer, and mass spectrometric data acquisition. The effluent mobile phase with separated compound from the liquid chromatography is interfaced with the ionization source of the mass spectrometer. Ionization sources include matrix-assisted laser desorption/ionization (MALDI), atmospheric pressure chemical ionization (APCI), and electrospray ionization (ESI). Apart from this chemical ionization (CI), electron impact (EI), or negative chemical ionization is also used as a source of ionization in MS (Sparkman 2006).The analyzer is a component of the mass spectrometer which takes ionized molecules and separates it on the basis of charge-to-mass ratios to the detector to be detected and later converted to a digital output (Fig. 14.7).

Table 14.5 HPTLC application in the evaluation of herbal drugs (Knap et al. 2006)

Herbal constituents	Reagent for spraying	Quantification
Panaxadiol and panaxatriol (ginseng)	10% H_2SO_4 acid in CH_3OH	UV spectroscopy absorbance at 544 nm and 522 nm, silica gel is used as an adsorbent with solvent system ether/chloroform (1:1)
Flavonol glycosides (ginkgo biloba)	8% $AlCl_3$ in Ethanol	UV spectroscopy absorbance at 370 nm, silica gel is used as an adsorbent with solvent system chloroform/benzene/ethanol/water/acetic acid (11:4:2:2:1)
Glycyrrhetinic acid (liquorice)	None	UV spectroscopy absorbance at 260 nm, silica gel is used as an adsorbent with solvent system ethyl acetate/ammonia/methanol: (10:1:3)
Aloin (aloe vera)	None	UV spectroscopy absorbance at 350 nm, silica gel is used as an adsorbent with solvent system ethyl acetate/acetic acid/water (17:2:3)
Carvone (*Cuminum cyminum*)	Anisaldehyde sulfuric acid	UV spectroscopy absorbance at 410 nm, silica gel is used as an adsorbent with solvent system acetone/chloroform (2:100)
Cholesterol (bear gall bladder)	10% H_2SO_4 acid in alcohol	UV spectroscopy absorbance at 400 nm, silica gel is used as an adsorbent with solvent system ethyl acetate/acetone/petroleum ether (2:1:11)

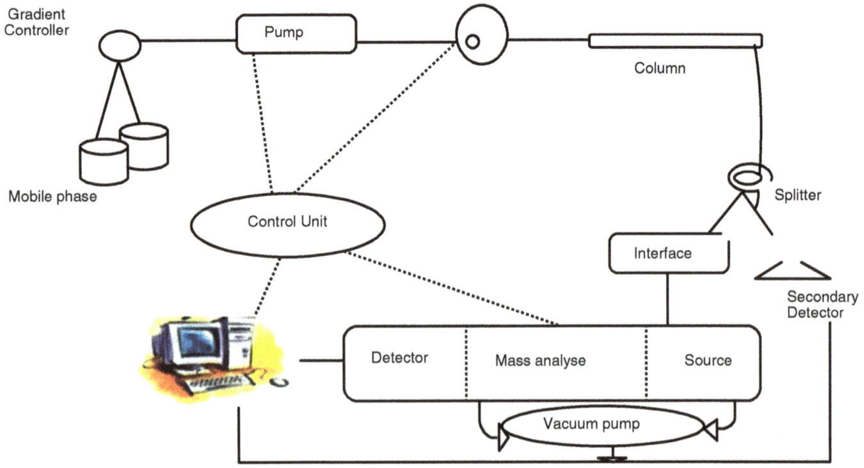

Fig. 14.7 Instrumentation of LC-MS

14.3.2 Gas Chromatography-Mass Spectroscopy (GC-MS)

The most extensively and effectively used combination technique is the GC coupled with MS. The combination of GC and MS was first reported in 1958, and it was made available commercially in 1967 (Ardrey 2003).

GC-MS is a synergistic union, extremely favorable as the compounds which can be analyzed by GC (low molecular weight, medium or low polarity in ppb-ppm

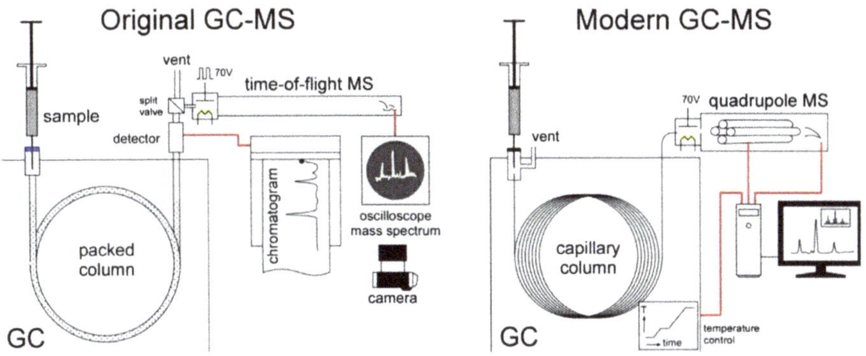

Fig. 14.8 Instrumentation of GC-MS

concentrations) are also compatible with the MS requirements. In addition, both analyses proceed in the same aggregation state, i.e., vapor phase. But both the techniques differ in the working pressures, i.e., pressure in the GC column exists low at 10.5–10.6 Tor than in the ionization chamber. To solve this problem, efficient vacuum pumps (gas jet pump and turbo molecular pump) and gas chromatography capillary columns (having internal diameter of 0.18–0.32 mm, traditionally used in GC-MS) directly inserted into the ionization chamber of a mass detector (Hubschmann 2009; McMaster 2011) are used.

There are several advantages of GC-MS including the capillary column, which provides effective separation capacity, producing clear chemical fingerprint. Another is the coupling with mass spectroscopy and corresponding mass spectral database provides the complete information of the components of the plant extract which can be used for exploring phytoconstituents and their properties especially the volatile compounds (Guetens et al. 2002; Gong et al. 2000a, b; Gong et al. 2001a, b, c; Li et al. 2003; Li and Liang 2003; Wang et al. 2003; Tsai et al. 2002; Vieira and Simon 2002; Gherman et al. 2000; Velasco-Negueruela et al. 2003). Instrumentation of original and modern GC-MS is included in Fig. 14.8.

14.4 Summary

The aim of chromatographic techniques was to separate the individual substances into a mixture depending on their color as in the case of herbal pigments. With time, its application was extended numerously. Nowadays, chromatographic techniques are widely acceptable for separation of phyto-constituents. Advanced chromatographic techniques like HPLC, HPTLC, and GC are extremely reliable for quantification and purification of carbohydrates, nucleic acids, amino acids, steroids, hydrocarbons, proteins, antibiotics and many other phyto constituents due to their high sensitivity, accuracy and reproducibility. GC is a separation technique which is used to analyze thermally stable volatile substances. For example, gas chromatography is used for the determination of fatty acids, flavored compounds, triglycerides, and several other food components, as well as aroma compounds, pesticides, polychlorinated biphenyls, and other volatile components.

High-performance thin-layer liquid chromatography is effective for quantification of biomarkers in extracts and fractions. Hyphenated techniques like LC-MS and GC-MS are highly authenticate for quantification and molecular weight determination of phyto-pharmaceuticals. The GC-MS is also used for quantification of volatile compounds too. Phytomedicine research has a very important role in contributing to the development of new and better drugs for effective phytotherapy. Fortunately, chromatography offers very powerful separation ability, such that the complex chemical components in HM extracts can be separated into many relatively simple subfractions. Further more, the recent hyphenated chromatographic techniques for separation, isolation, purification and quantification of the phytoconstituents are the stepping for global acceptance of phyto medicines. Separation, isolation, and purification of the phytoconstituents are the stepping stones to achieve this goal. Therefore, development and constant improvement of the existing chromatographic techniques is very crucial to mine the magical compounds from the huge plant repositories.

References

Archana Khale A, Anubha B (2011) An overview on thin layer chromatography. IJPSR 2 (2):256–267

Ardrey RE (2003) Liquid chromatography-mass spectrometry: an introduction, vol 4. Wiley, Hoboken, pp 182–189

Atlas WH (1996) Plant drug analysis: a thin layer chromatography. 2nd edn. Springer, p 87–92

Beckett AH, Stenlake JB (2005) Practical pharmaceutical chemistry, thin layer chromatography. 4th edn. CBS Publishers, p 115–128

Bombarda I, Dupuy N, Le Van Da JP (2008) Comparative chemometric analysis of geographic origins and compositions of lavandin var. Grosso essential oils by mid infrared spectroscopy and gas chromatography. Anal Chim Acta 613:31–39

Chatwal GR, Anand SK (2008) Instrumental methods of chemical analysis, vol. 2. 5th edn. Himalaya Publishing House, p 599–616

Coskun O (2016) Separation techniques: chromatography. North Clin Istanb 3(2):156–160

Faghihi J, Jiang X, Vierling R (2001) Reproducibility of the high- performance liquid chromatographic fingerprints obtained from two soybean cultivars and a selected progeny. J Chromatogr A915:61–74

Fan XH et al (2006) Multiple chromatographic fingerprinting and its application to the quality control of herbal medicines. Anal Chim Acta 555:217–224

Gherman C, Culeaand M, Cozar O (2000) Comparative analysis of some active principles of herb plants by GC/MS. Talanta 53:253

Gong F et al (2000a) Chin Anal Chem 28:860

Gong F, Song YQ et al (2000b) Acta Pharm Sin 35:394

Gong F et al (2001a) J Chromatogr 193:905

Gong F et al (2001b) Chem J Chin Univ 22:1481

Gong F et al (2001c) Anal Chim Acta 99:450

Guetens G et al (2002) J Chromatogr A 6:229–308

Habib MY et al (2000) J Nutr Pak 4:17

Hamburger MO, Cordell GA (1987) A direct bioautographic TLC assay for compounds possessing antibacterial activity. J Nat Prod 50, 19:–22
Hubschmann HJ (2009) Handbook of GC-MS: fundamentals and applications, 2nd edn. Wiley, Hoboken, pp 124–132
Jadhav MV et al (2009) J Int J Chem Tech Res 1:826
Jork H, Funk W (1990) LC reagents & detection methods-physical & chemical detection methods: fundamentals, vol 4. Wiley, Hoboken, pp 15–40
Kasture AV, et al (2008) A textbook of pharmaceutical analysis. Instrumental methods. Niraliprakashan, 13th edn, p 148–164
Knap H, Jork L, Liang Y, Fastre S (2006) Low-pressure gas chromatography: recent trends and developments. J Pharm Anal 23(3):212–315
Korfmacher WA (2005) Foundation review: principles and applications of LC-MS in new drug discovery. Drug Discov Today 10(20):1357–1367
Lalla JK et al (2000) J Planar Chromatogr 13:390
Li XN, Liang YZ (2003) Phytochem Anal 14:23–34
Li XN et al (2003) Phytochem Anal 14:23
Li Y, Wu T, Zhu J (2010) Combinative method using HPLC fingerprint and quantitative analyses for quality consistency evaluation of an herbal medicinal preparation produced by different manufacturers. J Pharm Biomed Anal 52:597–602
Lim CK, Lord G (2002) Current development in LC-MS for pharmaceutical analysis. Biol Pharm Bull 25(5):547–557
Lu GH, Chan K, Liang YZ (2005) Development of high- performance liquid chromatographic fingerprints for distinguishing Chinese Angelica from related umbelliferae herbs. J Chromatogr A 1073:383–392
McMaster M (2011) GC/MS: a practical user's guide, vol 4, New York, Wiley, pp 38–45
Mohammad A et al (2010) Analysis of herbal products by thin-layer chromatography: a review. Int J Pharma Biosci 1(2):1–50
Patel R, Patel K (2016) Significance of various chromatographic techniques in herbal drug analysis. J Chem Pharm Res 8(7):877–882
Pharmaceutical Applications with HPLC (2000) Agilent library
Pitt JJ (2009) Principles and applications of liquid chromatography-mass spectrometry in clinical biochemistry. Clin Biochem Rev 30(1):19–34
Pyka A (2014) Detection progress of selected drugs in TLC. Biomed Res Int 2014:732078
Raaman N (2006) Phytochemical techniques, vol. 1. New India Publishing Agency, p 19–25
Richter J et al (2003) J Pharm Anal 23(3):167
Shivatare RS, Nagore DH, Nipanikar SU (2013) HPTLC: an important tool in standardization of herbal medical product: a review. J Sci Innov Res 2(6):1086–1096
Silva GO et al (2017) Extraction methods: qualitative and quantitative techniques for screening of phytochemicals from plants. Am J Essential Oils Natural Prod 5(2):29–32
Sparkman OD (2006) Mass spectrometry. J Am Soc Mass Spectromet 17(6):873–884
Srivastava MM, Springer Verlag (2011) High performance thin-layer chromatography (HPTLC), vol. 6. Berlin, p 32–60
Tsai TR et al (2002) Identification and determination of geniposide contained in Gardenia jasminoides and in two preparations of mixed traditional Chinese medicines. J Chromatogr A 961:83–88
Velasco-Negueruela MJ et al (2003) J Chromatogr A 1011:241
Vieira RF, Simon JE (2002) Econ Bot 54:207
Wang XS et al (2003) Am J Chin Med 31:927
Ye M et al (2007) Analysis of phenolic compounds in rhubarbs using liquid chromatography coupled with electrospray ionization mass spectrometry. J Am Soc Mass Spectrom 18:82–91
Zygmunt B, Namiesnik J (2003) Preparation of samples of plant material for chromatographic analysis. J Chromatogr Sci 41:109–116

Reverse Pharmacology: A Tool for Drug Discovery from Traditional Medicine

15

V. Suba

Abstract

Reverse pharmacology approach reduces the three major bottlenecks of drug discovery. In India this approach is pioneered with the traditional basis of ayurveda. In reverse pharmacology, traditional drugs which have good efficacy and lesser side effects are chosen as a starting material for drug development. The major issue with the traditional drugs is they are polycomponent in nature with variety of biological nature. It is difficult to identify the components responsible for therapeutic action and their mechanism of action. In recent days various approaches are proposed to solve this issue. In India many hits and leads are developed from ayurvedic drugs using this approach. But still those leads remain at research level and do not reach market due to inadequate involvement from pharma industries. Various initiatives have been taken to bridge this gap. Further various strategies are suggested by Indian researchers to improve the methodology and to effectively utilize this approach for drug discovery.

Keywords

Reverse pharmacology · Traditional drugs · Drug discovery · New drug development

Abbreviations

APE	Ayurvedic pharmacoepidemiology
RP	Reverse pharmacology

V. Suba (✉)
Department of Pharmacology, National Institute of Siddha, Chennai, India
e-mail: vsubanis@gmail.com

S. C. Mandal et al. (eds.), *Evidence Based Validation of Traditional Medicines*,
https://doi.org/10.1007/978-981-15-8127-4_15

15.1 Introduction

15.1.1 Reverse Pharmacology and Its Origin

This chapter discusses the role of reverse pharmacology in the development of drug discovery from traditional formulation. The recent trends in the development of new chemical entities of pharmaceutical industries proved the fact that a lot of cost and time are incurred in the production of new drugs, but the success of the drugs developed is questionable. The drug development process takes about 8–12 years and screening of more than 25,000 molecules (O'Connell and Roblin 2006). In this classical method of drug development, the drugs are developed using reductionist approach that is testing the chemical entities against the targets. This mechanism-based models led to the development of inefficient drugs. Though millions of chemical compounds were synthesized as research drugs, they did not prove to be successful drugs for various ailments. Further the classical drug development process involves robust preclinical and clinical phases of study. Even after lot of research involved in the development of new drugs, many of the blockbuster drugs raised out of this strategy have been withdrawn from market. One of the problems with the modern medicine is it considers the human as an isolated system, whereas the traditional system of medicine considers human and health in the context of environment.

Recently large populations show interest toward the use of traditional system of medicine. The well-recognized traditional system of medicine is Chinese and Indian. India has its own system of medicine includes ayurveda, siddha, unani, yoga, and homeopathy. In India more than 500 million people use traditional system of medicine (Raut 2013). Over the last decade, there is a paradigm shift in the drug discovery starting the research with traditional drugs which are used effectively for a long time with safety. Their action targets are identified, and bioactive compounds responsible for the biological effects are identified and isolated for further development of lead and drug. This research field is called as reverse pharmacology (RP) which shifts drug discovery from clinic to laboratory reversing the classical drug development process which is from laboratory to clinic. It is a science of integrating documented experiential hits into leads by transdisciplinary exploratory studies and further developing into drug candidates by experimental research. Reverse pharmacology is also known as target-based drug discovery (Lee et al. 2012). It is an approach integrating clinical experience, traditional knowledge, and experimental observations for reversing the traditional drug development process from bedside to laboratory approach. This approach also paves the way for the finding of new use of old drugs. The leads developed through this science provide chemical scaffolds for structural modifications with drug targets.

Sir Ram Nath Chopra and Gananath Sen were the pioneers of RP in India for ayurvedic drugs. Sen and Bose 1931 took a lead from *Rauwolfia serpentina* for antihypertensive and tranquilizing effect through this approach. They also identified the side effect of this drug as depression, Parkinson-like syndrome, prolactinoma, gynecomastia, and nasal congestion. Further this research led the isolation of reserpine as bioactive compound and depletion of catecholamines as its mechanism. These findings paved the way for new drugs against Parkinson's disease, depression, prolactinoma, and nasal congestion. The need for approaching the traditional drugs different from modern drugs was understood. Then the scope of RP is extended to provide scientific evidence for safety and efficacy of traditional drugs already in practice, to find lead from natural products, and to find new clinical facts which provide insights about human biology and provide chemical scaffolds for further drug discovery (Raut et al. 2011).

15.1.2 Natural Products as Traditional Drugs and Reverse Pharmacology in Drug Discovery

Natural products and the traditional medicine system play a vital role to meet the medical and pharmaceutical needs for centuries. Indian and Chinese traditional system of medicine is the living system of medicine. India has very long history of traditional system of medicine including ayurveda, siddha, yoga, unani, and homeopathy. Natural products like plants, animals, minerals, and metals are used as medicaments in Indian traditional system of medicine. The presence of natural medicaments in traditional system provides merits over other forms of medicine. Plant-based formulations play a major role in Indian system of medicine. Approximately 90% of ayurvedic and siddha formulations are plant-based. Literature claimed around 20,000 medicinal plants in India (Pandey et al. 2008). Most of the plant-originated drugs in modern medicine are derived from traditional medicine (Li-Weber 2009). Today, approximately 80% of antimicrobial, cardiovascular, immunosuppressive, and anticancer drugs are derived from medicinal plants (Gordaliza 2009). Plants have chemically diverse phyto-constituents with varied biological actions and are exploited for the development of new lead compounds and scaffolds for drug discovery. Many lead compounds identified from natural products and traditional medicine paved the way for rational drug design. Coral, sponges, fish, and other microorganisms derived from marine sources which are used in traditional system of medicine show promising anti-inflammatory, antiviral, and anticancer activity. Venoms and toxins from snakes, spiders, scorpions, and insects which are practiced in traditional system of medicine served as lead for drugs like anticholinergics, antihypertensives, etc.

During the last two centuries, drugs were discovered incidentally from medicinal plants referred as drug serendipity. Advances in drug discovery technologies including high throughput screening, combinatorial and asymmetric synthesis, and system biology created new path in drug discovery. But drug industries faced serious challenges in terms of cost, risk, and ineffective drug molecules. There were many drugs derived from medicinal plants (aspirin, calcium antagonists, α and β blockers) before the introduction of innovative approaches, but the time lag between traditional use and modern medicine proved that many opportunities were lost. So researchers felt the need to shift the drug discovery from single to multi-targeted drugs. Drug discovery from natural products offered multi-targeted approach which is the need of the hour since now we are encountering polygenic syndromes. Natural products can provide synergistic effects simultaneously due to multicomponent nature. Current drug development approach of single target-single compound is reductionism strategy which makes use of molecular assays. (Verpoorte et al. 2005). This approach is not suitable for studies on traditional medicines. Pharma industries felt the need for the shift of drug discovery and development process from the conventional approach due to failures in the new chemical entities produced. This crisis led to the new model in the drug discovery called reverse pharmacology. This organized path of drug discovery exploiting the traditional drugs emerged as alternative drug discovery engine (Patwardhan and Vaidya 2010). These drugs are explored on fast rack using innovative approaches (Patwardhan et al. 2008). The understanding of human genome and structural characterization of human protein provide clear views by which diverse chemical structured plant products act on precise targets (Cragg and Newman 2005).

RP approach showed that standardized phytomedicine can be developed faster and more cheaply than the conventional drugs. This approach has recently emerged to reduce the three bottlenecks time, cost, and toxicity encountered in classical drug research. The research starts with the selection of traditional drugs which have clinically proven safety and efficacy. Then their targets like receptors, enzymes, and ion channels are identified on the basis of mechanism of action, and the corresponding active compounds are identified as hits and leads. They are further developed into new drug candidate through systematic and precise preclinical and clinical research. The scope is to understand the mechanism of action at different biological levels and to optimize the safety, efficacy, and quality of the leads from the natural products (Vaidya 2006).

Traditional system of medicine like ayurveda is time tested with centuries, and together with extensive scientific research, they proved to be very useful. This traditional drug discovery is not only useful to pharma industries; it also provides immense benefit to patients in terms of being closer to nature by using natural products, reducing cost and provide solution to ailments which has no answer in modern medicine. Recently, India has amended the Drug Act to include a category of phytopharmaceuticals to be developed from medicinal plants by reverse pharmacology, with evidence of quality, safety, and efficacy. These drugs will be distinct from traditional medicines like ayurvedic, unani, or siddha. India with its pluralistic healthcare system provides great opportunities for natural product drug discovery through RP using traditional knowledge and clinical observations (Patwardhan et al. 2004). The Indian Council of Scientific and Industrial Research and the Indian Council of Medical Research have successfully developed many leads through RP approach. Some of the plants subjected to RP include *Mucuna pruriens* for Parkinson's disease, *Zingiber officinale* for nausea/vomiting, *Curcuma longa* for oral cancer, *Picrorhiza kurroa* for hepatitis, *Panchvalkala* spp. for burns and wounds, and *Azadiracha indica* for malaria. Because of inadequate involvement from industries, these leads remain only at academic levels and not pursued further to druggable candidates. CSIR under the New Millennium Indian Technology Leadership Initiative (NMITLI) project brings industry and academia together to solve this issue. This project includes development of drugs for psoriasis, osteoarthritis, hepatitis, and diabetes (Padma 2005). Further ICMR established an advance center of RP which took initiatives of drug development for malaria, sarcopenia, and cognitive deficit.

15.2 Hits and Leads from Traditional Drugs for Drug Discovery

Indian and Chinese traditional system of medicine is researched through RP approach. There is large number of leads obtained from these systems of medicine. Reserpine from *Rauwolfia serpentina* for hypertension (Sen and Bose 1931) was the first lead from ayurvedic knowledge obtained through RP. This finding also documented the side effects of reserpine which was the forerunner for the discovery of newer drugs. *Artemisia annua* was used in Chinese traditional medicine for fever and malaria. Artemisinin the lead from the herb offered a new class of antimicrobial agents (Klayman 1985).

In Mucuna pruriens, the ayurvedic drug was subjected to RP for its efficacy against Parkinson's disease (Vaidya et al. 1978). This research led the development of Zandopa which may replace the synthetic L-DOPA derivatives in the market. Arogyawardhani (a herbo-mineral formulation) was practiced in ayurveda for the treatments of jaundice. Antarkar et al. (1980) reported the safety and efficacy of this drug against viral hepatitis in clinical studies. Tubocurarine, from the plant species *Strychnos guianensis* and *Chondrodendron tomentosum* was developed as a muscle relaxant from curare which was used as arrowhead poison (Verpoorte et al. 2005). Huperzine A was derived from *Huperzia serrata*, a Chinese herbal medicine for senile dementia (Ved et al. 1997).

Picrocides from *Picrorhiza kurroa* served as lead for its effectiveness against viral hepatitis (Vaidya et al. 1996). The herb *Commiphora wightii* (*guggulu*), is an ayurvedic drug. The hypolipidemic effects of the plant were studied by RP approach, and guggulsterones were identified as the lead (Satyavati et al. 1969). The antimicrobial property of berberine from *Berberis aristata* was reported (Amin et al. 1969). Further *Tinospora cordifolia* as immunomodulator (Dahanukar et al. 1988); curcumins as anti-inflammatory agents (Chandra and Gupta 1972); withanolides as immune modulators (Davis and Kuttan 1998); *Papaver somniferum* as analgesic (Martin 1983); *Physostigma venenosum* as anticholinesterase (Holmstedt 1972); Psoralens, the lead for vitiligo; *Holarrhena* alkaloids in amoebiasis; and baccosides in memory are pharmacological correlates of ayurvedic drugs.

15.3 Methodology of Reverse Pharmacology

15.3.1 Clinical and Paraclinical Models Employed in Reverse Pharmacology

The classical research methodology of traditional drugs are based on providing scientific evidence for the traditional claim of the drug and standardizing its activity or else to screen the drug against various models, isolation of the active constituents, and subjecting the isolated molecules to preclinical and clinical study. Those methodologies failed in many ways to bring the successful lead for drug discovery wherein, RP practices an organized path for drug discovery. This research starts with traditional drugs which are clinically proven to be successful for long time with lesser side effects. Then their targets are identified by the study of action mechanism; further, the bioactive compounds responsible for the pharmacological effects are selected as lead for new drug candidate.

The following are the three tracks of RP with overlapping steps with each other (Patwardhan et al. 2008).

- Experiential: This phase includes clinical observation of biodynamic effects of the traditional drug, its documentation, and record keeping.
- Exploratory: This phase cover the dose range finding studies clinically, tolerance, drug interaction studies, and in vitro and in vivo models to assess the drug targets.
- Experimental: The experimental domain involves well-planned experimental and clinical investigation at different levels of biological organization and extended clinical trials to identify and validate the pharmacological correlates of traditional drugs.

In the experiential domain, the traditional drugs are selected on the basis of clinical experiences, observation, and available data from human use. These include disease prevalence, thrust area of research where alternative medicines are optimal, availability of raw material for formulation, method of fractionation and isolation, and safety and efficacy of the formulation. These drugs are subjected to clinical trials. Extensive clinical evaluations as used in conventional clinical trial may not be feasible for all traditional drugs in many times. Hence case studies conducted in the existing clinical setup are a relevant option which provides valuable information regarding safety and efficacy of drugs within the measurable parameters. This is recommended by the National Institute for Health and National Center for Complementary and Alternative Medicine for ayurvedic practitioners. Clinical assessment including the effect of the drugs against the disease studied and various biodynamic effects are observed from clinical notes and case studies. This phase includes researchers of RP and physician in charge to provide the treatment as per the principles of traditional practice. In most cases the traditional practice includes individual variation, diet, and lifestyle modification in addition to the therapeutic intervention. In this domain traditional formulations are analyzed for efficacy and safety through precise biological and clinical end point assays, clinical interactions, and interindividuality variable measures. This domain also makes use of further information sourced from pharmacoepidemiology, observational therapeutics, and serendipitous findings. Pharmacoepidemiology is an important tool providing lot of information for experiential domain. In the Indian context, since ayurveda is largely practiced, ayurvedic pharmacoepidemiology (APE) was very relevant (Vaidya et al. 2003). APE studies the ayurvedic drugs in the context of its use, efficacy, safety, interaction, and documentation of adverse and unexpected beneficial effects.

In the exploratory studies, traditional drugs are subjected to rigorous in vitro and in vivo (preclinical and clinical components). The clinical design should involve studies of pharmacodynamics, dose finding, dose optimizing, and pharmacokinetics if assay is available. The clinical study design should be modified according to the intervention and indication of the drug. Extracts of multicomponent traditional formulations are subjected to bioguided fractionation using animal models, assay-based screening techniques, etc., to identify single compound or fewer combinations. The recent analytical techniques like HPLC and tandem mass spectrometry is commonly used for the separation. High-resolution NMR spectroscopy is a good tool for analysis and structural elucidation. Further the high throughput screening assays also find their application in an attempt to isolation and elucidation of compounds. Once the compounds are isolated, they are subjected to various synthetic approaches to convert them into suitable moiety for drug development. Several steps of chemical purification and synthesis are involved in identification and isolation of hits and further to scale-up lead. This includes porphyrin-mediated catalytic oxidation (Chorghade et al. 1996a, b), microwave-mediated methods (Collins Jr 2010; de la Hoz et al. 2005; Huber and Jones 1992; Jones and Mathews 1997; Jones and Chapman 1993; LaBeaume et al. 2010; Torregrossa et al. 2006), and glycosylation coupling chemistries (Dong et al. 2008; Kallmerten and Jones 2010; Ma et al. 2009). Then these refined molecules should be subjected to clinical studies safety and efficacy for the disease which the formulation is originally indicated for.

In these fractionation techniques, some of the compounds lacking activity in the specific assay procedures are eliminated. Hence further testing of these fractions against other assays will yield complete spectrum of bioactive compounds. Sometimes multicomponents present at different order of concentration are responsible for different level of therapeutic actions. This fact should also to be taken into account while identifying the hits and leads out of the formulations. The experimental domain has preclinical and clinical research similar to research methods used in any new drug development. In the preclinical study design, the isolated compound or combinations are subjected to conventional drug development step.

15.3.2 Traditional Drug Discovery Through Reverse Pharmacology: Essential Consideration and Requirements

In India, RP approach initiated with the discovery of lead from ayurvedic drugs. Researchers encountered many challenges while subjecting ayurvedic drugs to this approach. This traditional drug discovery mandates the inclusion of traditional knowledge and principles both in clinical and preclinical research. The attributes of ayurvedic formulations such as dosage form, dosage schedule, dosage regimen, vehicles used, diet, and lifestyle determine the therapeutic outcome. Further patient-related determinants like patient's constitution, stage of disease, causative factor, pathological factors, and clinical features also have a role in the logical outcome. Hence clinical protocol should be designed accordingly. Conventional clinical designs for efficacy and pharmacodynamics of traditional drugs are critical. According to Verpoorte et al. (2005), holistic approach based on systems biology may be more suitable. Further double-vision trials which include factors that influence the drug actions are proposed instead of double-blind trial for RP methodology (Raut et al. 2011).

The important problem of the methodology of RP includes the way of testing and analyzing the relationship between the pharmacological effects of traditional drug and its components. Many a times the multicomponent therapeutic activity of the formulation cannot be shown by single compound or combinations developed from it. According to Xiangming et al. (2014) who studied analgesic effect of *Resina draconis* through RP methodology, the study should be designed to observe the component/combination which causes the same pharmacological effects as that of the original traditional drug. While in some cases, the active principles of formulations fail to produce the desired effects when isolated individually. This may be due to the synergistic activity of several components present in a formulation.

The other important lacunae with ayurvedic and other traditional herbal products are the lack of drug standardization and quality control procedures. Since the majority of formulations contain crude extracts in mixtures of different ingredients, it is very difficult to apply modern tests to fix the standards for the same. Unlike the modern drugs, the quality of material for traditional medicines varies between, and even within, plants because of genetic differences, environmental conditions, harvesting, transport, and storage. The research on traditional drugs requires alternative steps to be included to characterize the widely diverse chemical entities. Difficulties in conducting pharmacokinetic and pharmacodynamics activities of the formulations are also major hurdles in subjecting them to RP methodology.

The scientific research carried out on ayurvedic formulations proved a fact that some of the compounds or herbal extract included in these formulations don't carry any biological effects, but instead they act as potentiator or moderator of efficacy and toxicity of the active principles. For example, milk and ghee are added to *Semecarpus anacardium* to counteract hot, acidity, and dryness of the formulation. Further in an ayurvedic practice, diverse vehicles are identified along with the drug administration for specific indication. All these facts should be taken into account while designing the study.

According to Patwardhan (2012), the basic principles of traditional system such as its knowledge, diagnosis, drug preparation, materials used, processing methods, dosage forms, diagnosis, diet, therapeutics, and personalized approach should be understood and followed while designing preclinical and clinical studies. Further epistemological difference between traditional medicine and biomedicine should be well understood.

Though there is enormous potential for this discovery approach, only few leads have found their way in the market. Most of this work remains either in the academic level or with the practitioner. This discipline requires regulatory recognition and support from pharma industries for further development. There is a need to develop this discipline in all drug research institutes including medical, pharmaceutical and life sciences colleges, and drug research centers.

15.3.3 Reverse Pharmacology and Novel Biodynamic Action

Traditional medicines always pave the way for novel biodynamic action. RP is one of the good tools to bring the biodynamic effects and mechanistic understanding of the disease. There are many such effects reported from ayurvedic drugs. According to Raut et al. (2011), the study of structure of trichomes of fruit of *Mucuna pruriens* provided them the mechanism of pruritis as mast cell degeneration. Forskolin an alkaloid from *Coleus forskohlii* (Seamon et al. 1981) and compound from *Stephania glabra* are identified to be useful in obesity and atherosclerosis (Das et al. 2009). Berberine alkaloids originally identified as antimicrobials are now found to possess cholesterol lowering property (Kong et al. 2004). Piperine was shown to improve bioavailability of many synthetic drugs, and its potential as bioenhancer was identified (Atal et al. 1985) by this approach.

15.4 Advantages of Reverse Pharmacology Approach

- RP reduces three major bottlenecks of drug discovery time, cost, and toxicity. This approach is much faster, less than 5 years of drug development process, and more efficient than the classical methods.
- The leads developed through RP methodology serve as good scaffolds for drug discovery. Since the leads are from traditional drugs already used for long years, their safety and tolerance are well-known.
- Through this approach traditional drugs are found to have new application.
- This strategy is not only useful for new drug development but also for improving current drugs. A good example is the improvement of current antimalarial medications which show a synergistic effect against quinine resistant plasmodium falciparum (Lewis 2003).
- It provides new clinical facts which provide insights about human biology. For example, involvement of metabotropic glutamate receptors and glycine reuptake inhibitors was studied in schizophrenia as a result of clinical observations and reverse pharmacology (Moghaddam 2003; Olney 2003). This led to the development of psychotomimetic drugs phencyclidine and ketamine as NMDA receptor antagonists (Anis et al. 1983; Zukin and Zukin 1979).

- It also provides clues about the plants which deserve further drug development process.
- It is a rationale strategy for the development of new drug molecule. It adopts system biology principles, so holistic and rational analysis is done in this approach.

15.5 Future Scope

The modern drug discovery process revisited traditional knowledge to improve the innovation for many reasons. Multidrug target of traditional drugs provides advantage over modern drugs due to genetic diversities of human. Ayurvedic principles of classifying human as three prakriti type encourage the researchers to correlate the genomics for individualized therapy. RP is grown in India only in the context of ayurvedic knowledge. But RP was only sporadically applied to drug development process. Hence it is important to document unknown, unintended, and desirable novel prophylactic and therapeutic effects in observational therapeutics of traditional practice. Efforts must be taken to link observational therapeutics and pharmacoepidemiology to identify clinical hits. In India, major steps have been taken already both in the private and the public sectors of pharma research and developments. India has already adopted the golden triangular research for correlating the three fields by R&D network, viz., modern medicine, Indian systems of medicine, and life and pharmaceutical sciences (Mashelkar 2003). Indian institutes are already involved in making traditional knowledge and experiential database for making this RP approach more fruitful. This approach will fast-forward the drug discovery when it is effectively linked with observational therapeutics, epidemiology, and system biology.

References

Amin AH, Subbaiah TV, Abbassi KM (1969) Berberine sulfate: antimicrobial activity, bioassay and mode of action. Can J Microbiol 15:1067–1076

Anis NA, Berry SC, Burton NR, Lodge D (1983) The dissociative anaesthetics, ketamine and phencyclidine, selectively reduce excitation of central mammalian neurones by N-methyl-aspartate. Br J Pharmacol 79:565–575

Antarkar D, Vaidya A, Doshi J, Athavale A, Vinchoo K, Natekar M, Tathed MR, Ramesh PS, Kale V (1980) A double-blind clinical trial of arogyavardhini-an Ayurvedic drug in acute viral hepatitis. Indian J Med Res 72:588–593

Atal CK, Dubey RK, Singh J (1985) Biochemical basis of enhance drug bioavailability by piperine: evidence that piperine is a potent inhibitor of drug metabolism. J Pharmacol Exp Ther 232 (1):258–262

Chandra D, Gupta SS (1972) Anti-inflammatory and antiarthritic activity of volatile oil of Curcuma longa (Haldi). Indian J Med Res 60:138–142

Chorghade MS, Dolphin DH, Dupre D, Hill DR, Lee EC, Wijesekara TP (1996a) Improved protocols for the synthesis and halogenation of sterically hindered metalloporphyrins. Synthesis 11:1320–1324

Chorghade MS, Dolphin DH, Hill DR, Hino F, Lee EC, Zhang LY, Pariza JR (1996b) Metalloporphyrins as chemical mimics of cytochrome P-450 systems. Pure Appl Chem 68 (3):753–756
Collins M Jr (2010) Future trends in microwave synthesis. Future Med Chem 2(2):151–155
Cragg GM, Newman DJ (2005) Plants as a source of anti-cancer agents. J Ethnopharmacol 100 (1-2):72–79
Dahanukar S, Thatte U, Pai N, More P, Karandikar S (1988) Immunotherapeutic modification by Tinospora cordifolia of abdominal sepsis induced by caecal ligation in rats. Indian J Gastroenterol 7:21–23
Das B, Tandon V, Lyndem LM, Grey AI, Ferro VA (2009) Phytochemicals from Flemingia vestita (Fabaceae) and Stephania glabra (Menispermaceae) alter cGMP concentration on the cestode Raillietina echinobothrida. Comp Biochem Pysiol C Toxicol Pharmacol 149:397–402
Davis L, Kuttan G (1998) Suppressive effect of cyclophosphamide-induced toxicity by Withania somnifera extract in mice. J Ethnopharmcol 62(3):209–214
de la Hoz A, Diaz-Ortiz A, Moreno A (2005) Microwaves in organic synthesis. Thermal and non-thermal microwave effects. Chem Soc Rev 34(2):164–178
Dong M, Sitkovsky M, Kallmerten A, Jones G (2008) Synthesis of 8-substituted xanthines via 5,6-diaminouracils: an efficient route to A2A adenosine receptor antagonists. Tetrahedron Lett 49(31):4633–4635
Gordaliza M (2009) Terpenyl-purines from the sea. Mar Drugs 7(4):833–849
Holmstedt B (1972) The ordeal bean of old Calabar: the pageant of Physostigma venenosum in medicine. In: Swain T (ed) Plants in the development of modern medicine. Harvard University Press, Cambridge, pp 303–360
Huber R, Jones G (1992) Acceleration of the orthoester Claisen rearrangement by clay catalyzed microwave thermolysis: expeditious route to bicyclic lactones. J Org Chem 57(21):5778–5780
Jones G, Chapman B (1993) Decarboxylation of indole-2-carboxylic acids: improved procedures. J Org Chem 58(20):5558–5559
Jones G, Mathews J (1997) Bifunctional antitumor agents. Derivatives of pyrrolo[9, 10-b] phenanthrene—a DNA intercalative delivery template. Tetrahedron 53(43):14599–14614
Kallmerten A, Jones G (2010) Microwave accelerated synthesis of PET image contrast agents for AD research. Curr Alzheimer Res 7(3):251–254
Klayman DL (1985) Qinghaosu (artemisinin): an antimalarial drug from China. Science 228:1049–1055
Kong W, Wei J, Abidi P, Lin M, Inaba S, Li C, Wang Y, Wang Z, Si S, Pan H, Wang S, Wu J, Wang Y, Li Z, Liu J, Jiang JD (2004) Berberine is a novel cholesterol lowering drug working through a unique mechanism distinct from statins. Nat Med 10(12):1344–1351
LaBeaume P, Placzek M, Daniels M, Kendrick I, Ng P, McNeel M et al (2010) Microwave-accelerated fluorodenitrations and nitrodehalogenations: expeditious routes to labeled PET ligands and fluoropharmaceuticals. Tetrahedron Lett 51(14):1906–1909
Lee JA, Uhlik MT, Moxham CM, Tomandl D, Sall DJ (2012) Modern phenotypic drug discovery is a viable, neoclassic pharma strategy. J Med Chem 55:4527–4538
Lewis WH (2003) Pharmaceutical discoveries based on ethnomedicinal plants: 1985 to 2000 and beyond. Econ Bot 57(1):126–134
Li-Weber M (2009) New therapeutic aspects of flavones: the anticancer properties of Scutellaria and its main active constituents Wogonin, Baicalein and Baicalin. Cancer Treat Rev 35:57–68
Ma D, Lin Y, Xiao Z, Kappen L, Goldberg I, Kallmerten A et al (2009) Designed DNA probes from the neocarzinostatin family: impact of glycosyl linkage stereochemistry on bulge base binding. Bioorg Med Chem 17(6):2428–2432
Martin WR (1983) Pharmacology of opioids. Pharmacol Rev 35:283–323
Mashelkar RA (2003) Chitrakoot declaration, convention of National Botanical Research Institute, published by NBRI, Lucknow
Moghaddam B (2003) Bringing order to the glutamate chaos in schizophrenia. Neuron 40:881–884

O'Connell D, Roblin D (2006) Translational research in the pharmaceutical industry: from bench to bedside. Drug Discov Today 11:833–838
Olney JW (2003) Excitotoxicity, apoptosis and neuropsychiatric disorders. Curr Opin Pharmacol 3 (1):101–109
Padma TV (2005) Ayurveda: outlook. Nature 436:486
Pandey MM, Rastogi S, Rawat AKS (2008) Indian herbal drug for general healthcare: an overview. J Alternat Med 6(1):1–12
Patwardhan B (2012) The quest for evidence-based Ayurveda: lessons learned. Curr Sci 102 (10):1406–1417
Patwardhan B, Vaidya ADB (2010) Natural products drug discovery: accelerating the clinical candidate development using reverse pharmacology approaches. Indian J Exp Biol 48:220–227
Patwardhan B, Vaidhya ADB, Chorghade M (2004) Ayurveda and natural products drug discovery. Curr Sci 86:789–799
Patwardhan B, Vaidya ADB, Chorghade M, Joshi PS (2008) Reverse pharmacology and systems approaches for drug discovery and development. Curr Bioact Comp 4(4):201–212
Raut A (2013) Scope and potential of integrative medicine in current healthcare scenario' in conference Samyukti 2013, an evidence-based approach to integrating Ayurveda and Allopathy, organized by MS Ramaiah Academy of Health and Applied Sciences, and Institute of Transdisciplinary Health Sciences and Technology, Bangalore
Raut AA, Chorghade MS, Ashok DB (2011) Reverse pharmacology. In: Kapetanovic IM (ed) Drug discovery and development—present and future, 1st edn. Janeza Trdine, Croatia, pp 88–126
Satyavati GV, Dwarkanath C, Tripathi SN (1969) Experimental studies on the hypocholesterolemic effect of Commiphora mukul Engl. (Guggul). Indian J Med Res 57:1950–1952
Seamon KB, Padgett W, Daly JW (1981) Forskolin: unique diterpene activator of adenylate cyclase in membranes and in intact cells. Proc Natl Acad Sci U S A 78(6):3363–3367
Sen G, Bose K (1931) Rauwolfia serpentina, a new Indian drug for insanity and hypertension. Indian M World 21:194–201
Torregrossa J, Bubley G, Jones G (2006) Microwave expedited synthesis of 5-aminocamptothecin analogs: inhibitors of hypoxia inducible factor HIF-1alpha. Bioorg Med Chem Lett 16 (23):6082–6085
Vaidya ADB (2006) Reverse pharmacological correlates of Ayurvedic drug actions. Ind J Pharmacol 38(5):311–315
Vaidya RA, Aloorkar SD, Sheth AR, Pandya SK (1978) Activity of bromoergocryptine, mucuna pruriens and L-dopa in the control of hyperprolactinaemia. Neurol India 26:179–182
Vaidya AB, Antarkar DS, Doshi JC, Bhatt AD, Ramesh V, Vora PV, Perissond D, Baxi AJ, Kale PM (1996) Picrorhiza kurroa (kutki) Royle ex Benth as a hepatoprotective agent-experimental and clinical studies. J Postgrad Med 42:105–108
Vaidya RA, Vaidya ADB, Patwardhan B, Tillu G, Rao Y (2003) Ayurvedic Pharmacoepidemiology: a proposed new discipline. J Assoc Phys India 51:528–531
Ved HS, Koenig ML, Dave JR (1997) Huperzine A, a potential therapeutic agent for dementia, reduces neuronal cell death caused by glutamate. Neuroreport 8(4):963–968
Verpoorte R, Choi YH, Kim HK (2005) Ethnopharmacology and system biology: a perfect holistic match. J Ethnopharmacol 100(1–2):53–56
Xiangming L, Su C, Zhou S, Lisi W, Ming G (2014) The development and application of methodology of reverse pharmacology illustrated with the research on analgesic effect of Resina Draconis. J Homeop Ayurv Med 3(3):1–7
Zukin SR, Zukin RS (1979) Specific [3H]phencyclidine binding in rat central nervous system. Proc Natl Acad Sci U S A 76(10):5372–5376

Part II

Standardization and Validation of Traditional Medicine and Medicinal Plant

Role of Modern Biological Techniques in Evidence-Based Validation of Ayurvedic Herbometallic Preparations

16

Koel Sinha, Sutapa Som Chaudhury, Bhuban Ruidas, Rabindranath Majumder, Tania Pal, Tapas K. Sur, Prasanta K. Sarkar, and Chitrangada Das Mukhopadhyay

Abstract

An alternative to synthetic drugs having severe side effects, the use of Ayurvedic preparations from natural sources like herbs, metals, and minerals may be efficient due to better activity profile to combat the harmful nature of the diseases. Ayurvedic therapy can treat those diseases better, which do not respond to the treatment by western medical practices. This preparation may have anticancer, antimicrobial, and immunomodulatory effect as evidenced earlier. Further, it may have a specific role in preventing cancer metastasis, neuro-diseases, diabetes, atherosclerosis, and many other chronic diseases. Modern biotechnology and molecular biology-based techniques have contributed to the identification of active components and mechanical effects of these preparations having ethnobiological importance. Treatment with Ayurvedic herbometallic preparations is practiced in India since 5000 BC to prevent, delay, or diminish the incidence of significant ailments. These are having a holistic approach as traditional medicine and have less toxicity and ignorable side effects. However, the experimental validity and nano-materialistic method of different Ayurvedic herbometallic preparations are not sufficiently acknowledged. An evidence-based validation of Ayurvedic herbometallic preparations, their mechanism of action in disease prevention or remedy is the main focus of this chapter. The information from the in vitro and in vivo studies on these preparations and mechanism-based analyses

K. Sinha · S. S. Chaudhury · B. Ruidas · R. Majumder · T. Pal · T. K. Sur · C. D. Mukhopadhyay (✉)
Centre for Healthcare Science and Technology, Indian Institute of Engineering Science and Technology, Howrah, West Bengal, India
e-mail: chitrangadadas@yahoo.com

P. K. Sarkar
Department of Rasashastra, J. B. Roy State Ayurvedic Medical College and Hospital, Kolkata, West Bengal, India

S. C. Mandal et al. (eds.), *Evidence Based Validation of Traditional Medicines*,
https://doi.org/10.1007/978-981-15-8127-4_16

and challenges to evaluate its efficacy can add on therapy to improve the quality of life in patients.

Keywords

Ayurveda · Herbometallic · Anticancer · Antimicrobial · Immunomodulatory

Abbreviations

AAS	Atomic absorption spectroscopy
DCFHDA	2,7-Dichlorofluorescein diacetate
FTIR	Fourier-transform infrared spectroscopy
GSH	Glutathione
ICP-AES	Inductively coupled plasma atomic emission spectroscopy
RNS	Reactive nitrogen species
ROS	Reactive oxygen species
SEM	Scanning electron microscopy
SOD	Superoxide dismutase
TEM	Transmission electron microscopy
TGA	Thermogravimetric analyses
XRD	X-ray diffraction

16.1 Introduction

Ayurveda is a native ethnic medical system popularly practiced in India since time immemorial. The foremost strength of the system is its comprehensive approach toward health and disease using naturally occurring resources derived from medicinal plants and minerals. This system of medicine also emphasizes self-discipline and modest lifestyle and exercise in daily healthy livings.

In Ayurvedic medicine, minerals and metals used are mostly water as well as fat insoluble. The ancient researchers tried to transform these metals and minerals into nanoforms that they should have excellent bioavailability and therapeutic potentials, as mentioned in *Charaka Samhita* in 1500 B.C. These modified forms of the metals and minerals used in therapeutics of Ayurveda are known as Bhasma and Sindura. Most importantly, the Bhasmas and Sinduras are fine medicinal powders containing various elements, including carbon (C), hydrogen (H), and sulfur (S). The carbon particles formed in the Bhasma at a high temperature (>600° C) may be in the form of carbon nanotubes (fullerenes), help in targeted delivery of metal and mineral preparations, and can even cross the blood-brain barrier (BBB) (Kumar et al. 2007). In recent scenario, the Bhasmas are claimed to be bioprocessed nanoparticles, prescribed with several plant drugs and are taken along with milk, butter, honey, or ghee; thus, this makes these elements easily soluble, capable of being assimilated, eliminating their harmful effects and enhancing their biocompatibility (Sarkar and

Chaudhary 2010). Lauhadi Rasayana, an essential metallic preparation, as mentioned in *Charaka Samhita*, requires iron for its preparation. This iron is heated up to red hot and quenched in some liquid media immediately until the flakes of iron get transformed into fine powder form for medicinal use (Sharma and Dash 2000).

Again, one of the most effective Ayurveda preparations is Swarna Bhasma (gold ash). It deals with globular particles of gold with an average size of 56–57 nm using modern techniques, i.e., transmission electron microscopy (TEM) analysis. It indicates the high cell penetration possibilities for potential bioactivity (Brown et al. 2007). Again, comprehensive physicochemical characterization of Yashada Bhasma using modern techniques reveals that these particles are in the oxygen-deficient state and are identifiable for being nanometer in size. Therefore, these properties might contribute to the therapeutic possibility efficiently as biomedicine (Brown et al. 2007). Furthermore, the particle size of Rasa Sindura (preparation of mercury and sulfur) is 30 nm along with increasing milling time. From this observation, it concluded that further improvisation using modern techniques in these preparations might enhance the future possibilities in an active drug preparation according to requirements (Bhowmick et al. 2009).

Interestingly, the final form of the metals and minerals (Bhasma and Sindura) is nothing, but ethno-nanomedicine. It includes not only an ancient traditional medicine system but also the potential applications based on the physicochemical properties for treatment of various ailments. Therefore, it considered the medicinal system of Ayurveda as one of the pioneers for the implementation as biomedicine based on the proper scientific knowledge and validation. Herein, we have depicted overall possibilities of Ayurveda system as biomedicine based on medicinal potential and its future aspect regarding the growing attention followed by the advancement in various scientific techniques.

16.2 Ayurveda: A System of Traditional Preparations

The Bhasma and Sindura are the most critical terms in preparation of metal and mineral-based constituents. A repeated levigation and incineration of metal or mineral with herbal extract or juice followed by sequential heating help to achieve those fractions. The compositions are prepared mostly by the following two different methods, i.e., Putapaka (incineration) and Kupipakwa (sublimation) method.

16.2.1 Putapaka Method

It is a complex process which includes four unit operations, i.e., Shodhana process, Jarana process, Bhavana process, and Marana process.

16.2.1.1 Shodhana Process

In Ayurveda formulations, Shodhana means purification. After the procurement of raw metals or minerals, these are transformed into coarse powder by hammering.

Then, prior to the Shodhana process, the final fractions are repeatedly heated, melted, and quenched immediately in particular solvents for further refinements.

16.2.1.2 Jarana Process

This is an intermediate process in between Shodhana and Marana process, followed for some specific metals having low melting point. For the process, the metal is melted, mixed with some dried plant materials, and rubbed continuously until it becomes powder form in open air.

16.2.1.3 Bhavana Process

The Shodhita or Jarita materials are then subjected to Bhavana (levigation) process. Briefly, an incineration of Jarita material and specific drugs is simply specifying the Maraka Dravyas. These are further triturated with specified liquid media to make a doughy mass.

16.2.1.4 Marana Process

Followed by the Bhavana process, the levigated doughy mass transformed into pellets called Chakrika. Next, keep those in between two earthen crucibles facing each other after the sealing of junction by mud-smeared clothes which introduce the apparatus, called Sharava Samputa. It mainly works as an electric muffle furnace or traditional Puta (heating grade) for heating. A continuation of this heating process of materials is called Putapaka in the system of Ayurveda. After the completion of the heating process for a specified time, the apparatus (Sharava Samputa) is taken out and is made open to get the incinerated powder. Finally, the powder form of incinerated metal or Bhasma is ready for collection and storage after cooling.

16.2.2 Kupipakwa Method

Herein, the finally prepared inorganic preparations are called Sindura or Kupipakwa Rasayana. It is called "Kupipakwa," because here the medicine is prepared in a specially prepared glass bottle. The Kupipakwa method is carried out by including four different steps, Shodhana process, preparation of Kajjali, Bhavana process, and Kupipaka process.

16.2.2.1 Shodhana Process

In Sindura preparation, Shodhana process includes the purification of raw materials prior to Kajjali preparation in Ayurvedic validation.

16.2.2.2 Preparation of Kajjali

After the Shodhana process, the purified materials are called mostly Parada and Gandhaka. Next, a successive trituration of purified materials completes the Kajjali process for a long time to transform the mixture in a black, lusterless, fine, impalpable powder of uniform consistence. Again in Makaradhwaja process, a mixture of Shuddha Parada and Shuddha Dhatu (metal like gold, etc.) is amalgamated prior to titration for several hours in Kajjali preparation. Finally, the Kajjali preparations are ready to get further transformation through Bhavana process.

16.2.2.3 Bhavana Process

In Bhavana process, the prepared "Kajjali" is then levigated by juice or decoction of some plant drugs for specific duration. It is then allowed to dry completely. After drying, again it is triturated to make powder.

16.2.2.4 Kupipaka Process

After Bhavana processing of prepared Kajjali, the materials are transferred and filled into a specially prepared glass bottle (seven layers of mud-smeared clothes) up to 1/3 portion, then it is placed in a Valuka Yantra (sand bath), and the Valuka Yantra is heated. Here temperature is increased gradually and provided for specific duration. The final product is sublimed to deposit in the bottleneck inner side. Therefore, it completes the Kupipakwa Rasayana prior to self-cooling and collected by breaking the bottleneck.

16.2.3 Varna (Color)

It indicates the color of the specific Bhasma. Different metals and minerals possess different colors in preparation of Bhasma. The specific color or Varna indicates the proper preparation of Bhasma, but changes in desired color suggest the inappropriate preparations.

16.2.4 Nishchandratvam

In therapeutic application, the Bhasmas are prescribed to be lusterless, i.e., Nishchandratvam. In general, a luster or Chandratva is an important characteristic in metal, and it is not desired in Bhasma preparation because of insolubility and toxic effect. Therefore, metallic characters in Bhasma preparation should be checked properly. Nishchandratvam signify the transformation of the specific metallic luster to lusterless compound after incineration. For the confirmation of lusterless, Bhasma is checked under the sunlight for further incineration process if needed.

16.2.5 Varitara

Varitara process includes the study of lightness and fineness of prepared Bhasma. It is a floating character of Bhasma based on the law of surface tension on stagnant water. In this process, the prepared Bhasma floats over the water surface without breaking the surface tension of stagnant water, and the final preparation is checked taking in between index finger and thumb followed by the sprinkling it on stagnant surface water (Kulkarni 1998). Therefore, it signifies the quality of the Bhasma preparation.

16.2.6 Unama Test

A fine-tuned Varitara process is further called Unama test. In this test, a rice grain is kept carefully on the floated layer of prepared Bhasma. Therefore, if the grain remains over the Bhasma layer without sinking, this indicates the excellent preparation, but sinking opposes its excellence (Kulkarni 1998).

16.2.7 Rekhapurnata

The Rekhapurnata test is applied to investigate the size of the Bhasma particles for easy absorption and assimilation capacity in therapeutic purposes, as size barrier in cellular system plays one of the important parts for absorption and assimilation. In case of large size, the bio-system may not allow the particles to absorb and assimilate properly, thereby causing irritation in the gastrointestinal tract (Kulkarni 1998).

16.2.8 Slakshnatvam

It is the tactile sensation produced by Bhasma by simple touch with finger tips. The properly incinerated Bhasma attains this quality. Slakshna Bhasma can be absorbed and assimilated in the body without producing any irritation to the mucous membrane of gastrointestinal tract.

16.2.9 Susukshma

Susukshma includes the further fineness of the Bhasma following both Varitara and Rekhapurnata tests.

16.2.10 Anjana Sannibha

Anjana (collyrium) is also a similar process like Slakshnatvam indicating smooth and fine character of Bhasma without any irritation whenever applied.

16.2.11 Particle Size

Particle size of Bhasma is the most important characteristic for therapeutic purposes as mentioned in Rekhapurnata. In Ayurveda system, Churna (powder) form or nanosize of the Bhasma particles are always favored which is mostly like pollen grains of *Pandanus odoratissimus* flower (*Ketaki Rajah*).

16.2.12 Gatarasatvam

Gatarasatvam indicates the specific taste of metallic preparations based on the taste perception. Some of the specified preparation also gets high attention by unique pharmaceutical procedure as every metal has its specific metallic taste.

16.2.13 Apunarbhavata

This process indicates the irreversible state of the prepared Bhasma which includes the inability to reform its original metallic form from transformed form. It also indicates the proper or improper incineration of Bhasma. A successive Apunarbhavata indicates the proper incineration of Bhasma (Kulkarni 1998).

16.2.14 Niruttha

Niruttha test is also similar to Apunarbhavata, but it emphasizes the ability of further physical condition to reform the previous stage. Briefly, a fixed weight of silver leaf and prepared Bhasma were further processed by Ayurveda system keeping in earthen pots with similar grade of heat. After cooling, the increase in silver leaf weight indicates the improper Bhasma preparations (Kulkarni 1998).

16.2.15 Inorganic Substance Processing

During Shodhana by the application of force in the form of heat, the tension in matter is increased, causing linear expansion. After heating, immediate cooling in liquid media leads to decrease in tension and increase in compression force. Repetition in heating and cooling causes disruption in compression tension equilibrium and leads to increased brittleness, reduction in hardness, and finally reduction in the particle size.

The applied force by the form of heat is initially taken on the high portion of the surface. As a result, high stress may be set up locally in the particles. The bonds at this place become weak, which may be responsible for creating flaws. The particle with the weakest flaw fractures most easily and produces largest possible pieces. The particle with the weakest flaw, fractures most easily and produces largest possible pieces. In the next step, another weak flaw fractures. In this way particle size is reduced.

It is very interesting to note that the same metal is processed (i.e., given Bhavana) with different sets of herbs, to be used for different therapeutic indications. In this context, it is all the more interesting to study as to what changes the metal goes through during the different steps of Shodhana and subsequently during the process of Bhavana and the incineration (Marana) process that it acquires a nontoxic, therapeutically efficacious form. It has applied importance also, for example, Vanga (tin, Sn) Bhasma prepared by Jarana of Apamarga (*Achyranthes aspera*) or Palasa (*Butea monosperma*) should not be used for Vrishya (spermatogenesis) purpose, because Kshara has anti-Vrishya properties. The Apamarga (*A. aspera*) or Palasa (*B. monosperma*) Kshara present in the prepared Vanga Bhasma may affect the Vrishya property of Vanga. For Vrishya purpose, Vanga Bhasma prepared by Jarana of Shuddha Haritala (orpiment) should be used.

16.3 Ingredients of Herbometallic Preparations and Their Importance in Biological System

A perfect balance between herbal and metallic components is a key requisite of Ayurvedic medicines. Gold (Au), silver (Ag), zinc (Zn), copper (Cu), iron (Fe), and tin (Sn) are frequently used in Ayurvedic drugs as "Bhasma" (the incinerated form of metal) (Sarkar, Das, Prajapati 2010). In Ayurvedic practice, the heavy metals like arsenic (As), chromium (Cr), mercury (Hg), lead (Pb), and many others are also in use but in a very low concentration and in an holistic approach (Kumar et al. 2006). These metals are then treated with herbal juice or decoction to blend properly (Pal et al. 2014).

The Ayurvedic preparations, namely, Swarna Bhasma (incinerated gold), Rajat Bhasma (silver ash), Hirak Bhasma (diamond ash), Tamra Bhasma (incinerated copper), Yashada Bhasma (zinc ash), Lauha Bhasma (incinerated iron), Vanga Bhasma (incinerated tin), and Naga Bhasma (incinerated lead), consist of one main metal component after which these are named. These key metals in the respective preparations play a central role for the cure of the ailments (Table 16.1) (Sarkar et al. 2010; Pal et al. 2014).

These Ayurvedic Bhasma preparations are suggested to consume with some specific adjuvants to enhance the drug activity. These adjuvants also complement the activity of the metal components of Bhasmas. Honey, jaggery, milk, sesame oil, and sugar are some common adjuvants prescribed with the herbometallic Ayurvedic drugs.

16.3.1 Modern Techniques to Assess the Herbometallic Preparations

For a long time, the metal-based Ayurvedic drugs were in use, but the detailed concentration-based composition was not reported in the scientific proceedings. Thus, the mechanism of drug action was unknown which has been raising debates about the use of metals or even the heavy metals of these herbometallic drugs. With the aid of modern techniques, namely, atomic absorption spectroscopy (AAS) and X-ray diffraction (XRD), analysis researchers have now shown the presence of

Table 16.1 The key metal elements of Ayurvedic herbometallic drugs, their importance, and use in respective diseases

Metals used in Ayurvedic preparation	Importance in biological system	Used in the disease
Gold	Cardiac stimulant, aphrodisiac, immunomodulator, to regain body potentiality and longevity, complexion, and to increase memory, intellect, and attentiveness	Tuberculosis, fever, dyspnea, cough, anorexia, ophthalmic disorders, and schizophrenia
Silver	Antiaging, immunomodulator, aphrodisiac, to increase potentiality and intellect	Diabetes, vitiligo, tuberculosis, anemia, dyspnea, cough, piles, etc.
Zinc	Immunomodulator, ophthalmic nourisher, to increase strength, potentiality, vitality, and intellect	Ulcer, emaciation, depression, tremor, ophthalmic disorder, diabetes, cough, dyspnea, anemia
Iron	Aphrodisiac, immunomodulatory, antiaging, emaciating agent, increases appetite and potentiality	Cachexia, obesity, bowel syndrome, hyperlipidemia, splenic disorder
Copper	Wound healer, emaciating agent, rejuvenator, and purgative	Anorexia, dyspepsia, abdominal tumor, liver disorder, ascites, piles, coughs, fever, and so on
Tin	Appetizer, rejuvenator, aphrodisiac, and immunomodulator. Increases vitality and intellect	Diabetes, hyperlipidemia, dyspnea, cough, emaciation, oligospermia
Arsenic	Antimicrobial, anticancer, immunomodulator	Sepsis, leukemia, skin diseases, psoriasis
Lead	Appetizer, aphrodisiac, immunomodulator	Rheumatoid arthritis, tetanus, cachexia, edema, ulcer, diarrhea
Mercury	Memory enhancer, antimicrobial, anticancer.	Syphilis, high fever, pneumonia, neurological disorders

metallic fraction along with their concentrations in such herbometallic preparations (Ruidas et al. 2019). XRD study revealed the crystalline nature of the Bhasmas. Also, the ratio of metallic and organic parts has been revealed via the thermogravimetric analyses (TGA) (Ruidas et al. 2019). The advanced spectroscopic techniques, namely, Fourier-transform infrared spectroscopy (FTIR) and Raman spectroscopy, have also appeared to be helpful for the detection of aromatic, aliphatic, or other organic functional groups (Balmain et al. 1999).

Pal has reported an elemental analysis of Swarna Bhasma (gold ash) via AAS study (Pal 2015). In 2016, R Sharma and colleagues reported the chemical characterization of Rajat Bhasma (silver ash) via inductively coupled plasma atomic emission spectroscopy (ICP-AES) and UV spectroscopy; also they detected the size and morphology through scanning electron microscopy (SEM) and transmission electron microscopy (TEM) (Sharma et al. 2016) The ICP-AES and AAS studies of these Bhasmas profoundly suggest that the Bhasmas are a mixture of many metal components with one key element and thus explain their synergistic action on specific target. Moreover, the results of TGA analyses suggest the higher fraction of organic material in Bhasmas is responsible to lower the toxicity of the heavy metals (Ruidas et al. 2019). Along with the progress of the new age, instrumental-based analyses of all these herbometallic compounds would bridge the lacking scientific evidences in the near future.

16.4 Safety and Efficacy of Ayurvedic Herbometallic Preparations

As discussed above, Ayurveda uses various herbometallic preparations of diverse origins which contain many metals like in gold, silver, copper, zinc, tin, lead, mercury, etc. These metals have been complexed with herbal ingredients through a scrupulous preparation protocol, to form the unique complex formulation called Bhasmas (herb-mineral ashes). It has been estimated that most of the Ayurvedic preparations (near about 40%) among 6000 reported preparations contain at least 1 metal as an important therapeutic ingredient (Saper et al. 2004). Again, metallic formulation has gained high attention due to rapid and prolonged action in small dose (Parikh et al. 2012).

The assumption exists among the public that "natural" equates with "safe" and may believe that plant drugs are not toxic in any of the concentration. However, some findings suggest that the natural products can be toxic without proper medical supervision (Barakat and Fatma 2003; Harrison's Internal Medicine n.d.; Ernst 1999; Rasheed et al. 2011). According to modern physicians, the addition of heavy metals with herbal medicine increases the toxic effects of herbal medicine at the lowest concentrations in the biological system (Desai et al. 1996). Any imbalance of essential substances in our body, whether excess or deficiency, can alter the normal activity of physiological function (Rasheed et al. 2011).

The efficacy and safety of an herbometallic preparation depend on its method of preparation, chemical nature, and therapeutic factors. The main purpose of herbometallic preparations is the transformation of starting elemental materials into an oxidized state which is the effective therapeutic ingredients of the preparation. Repeated heat treatment cycles are used to remove impurities and detoxify the harmful ingredients present in elemental materials. The effectiveness of nano-dimensional materials is greater than normal materials because their absorption, transport, and penetration into cells are relatively better (Desai et al. 1996; Rabinow 2004). The efficacy and safety of a drug depend on its pharmacokinetics parameters like digestion, absorption, metabolism, elimination, etc. These parameters are further attuned by the physicochemical properties of the herbometallic preparations. Without proper medical supervision, any drug can cause harmful effects to the body and even be life-threatening to a person. Long-term medication of any drug without proper scientific knowledge might be dangerous to health (Lucock 2004). Hence, an expert medical supervision is always welcome to overcome these backlogs.

Before the therapeutic use of modern medicine, many preclinical evaluations are done for its effectiveness, toxicity, dose-related information, etc. In the field of

Ayurvedic pharmacology, some special experiments are performed to assess the quality of metallic ingredients present in herbometallic preparation. The fineness of the metal in a Bhasma preparation can be tested by using a finger-based Rekhapoorva experiment. If the test sample settles in between the finger lines and it can be seen with a very narrow margin, then it is considered as a well-prepared Bhasma. A well-prepared Bhasma will lose its elementary nature, and monoxide metallic glaze is completely lost. The final form of metal cannot form an alloy with silver even at higher temperatures. This can be tested by the Nischandratwa test. The complete transformation of elemental metal to oxidized state is essential to its action; this is done by the Apunarbhava test. The density and surface energy of an herbometallic preparation depends on the oxidized state of metallic particles present in plasma. A properly calcined preparation tends to float on the water, rather than settling at the bottom (Nagarajan et al. 2012). Varitara test is done to determine the floating capability of Bhasma which is the indication of the density and surface energy of the preparation (Krishnamacharya et al. 2012).

16.5 Important Factors in Determining Quality and Therapeutic Potency

Ayurveda means a science of life (*Ayur* = life, *Veda* = science or knowledge). According to Vedic culture in India, Ayurveda includes a 5000-year-old system having natural healing properties. Furthermore, Ayurveda has a strong root in the preparation of the traditional Chinese medicine and Tibetan medicine (Patwardhan et al. 2005; Vetrov and Sorokina 2012). Therefore Ayurveda has shown rejuvenation not only in its native palace but also throughout the world (WHO 2001; Ragozin 2016). Despite a broad range of activity, Ayurveda failed to achieve high importance due to lack of proper scientific evidences. In recent scenario, the drug development from Ayurvedic Bhasma has gained high attention due its overwhelming response against both microbes and cancer without any toxic effect in limited dose (Ruidas et al. 2019).The proper scientific knowledge and effective validation of Ayurveda Bhasma increases the possibilities as therapeutic drug nowadays. The advancement of new techniques including the introduction of more advanced extraction method (Soxhlet method) and quantification techniques (pharmacokinetics analysis) for effective assessment of Ayurveda drugs also intensified its acceptance as potential therapeutics.

Furthermore, analysis of molecular ingredients such as organic and inorganic fractions in traditional preparations having biological and medicinal functions is widely accepted in a defined dose as most of the molecules have an important role in biological system either individually or synergistically (Jungwrith et al. 2011; Nieboer and Richardson 1980). Again, both excessive and reduced amounts of molecular ingredients in biological system may show an adverse effect, thereby needing a balance system for active functioning of cellular system (Tchounwou et al. 2012). Majority of the diseases have shown an imbalance in molecular ingredients; therefore fixation of the imbalance might be a potential approach to encounter those diseases (Dabrowiak 2017).Therefore identification of exact problem behind the disease progression and proper validation based on quality control and therapeutic potency are the most important parameters prior to the preparation of therapeutic drugs. Some of the major factors including high cost of new therapeutics, increased side effects of marketed novel drugs, and lack of effective remedial treatment for several chronic diseases, multidrug resistance of microbes, and other emerging diseases have enhanced the importance of traditional drug preparations. In ancient times, Ayurveda Bhasma has been used as therapeutic drug without proper scientific knowledge including proper medications and dose selection which also had increased the adverse side effects (Sarkar et al. 2010). In recent scenario the proper medication and dose selection including quality control, i.e., enhanced solubility, maintaining of biological pH 7.2–7.4, with minimal or without side effect, broad-spectrum activity range, low aggregation, high metabolism, etc. has gained the prime focus for Ayurveda Bhasma preparation for the enhancement of therapeutic potency. Therefore it will be worth declaring the drug prepared from Ayurveda Bhasma having high potential as therapeutic agent in the near future both quantitatively and qualitatively.

16.6 Antimicrobial Potential and Importance

Herbometallic compounds like Rasa Manikya, Hirak Bhasma, Swarna Bhasma, Parada, Yashada Bhasma, and so on possess antimicrobial potential. The key metal ingredients, namely, arsenic, diamond, gold, mercury, and zinc, respectively, of these compounds bring the microbicidal and/or microbistatic effect. Now, with the advancement in the modern scientific techniques, reports are coming very often

Table 16.2 Antimicrobial property of ancient medicines

Ayurvedic preparations	Microbes affected	References
Swarna Bhasma	*Staphylococcus aureus, Escherichia coli*	Pal (2015)
Tamra Bhasma	Enteric pathogenic bacteria e.g., *E. coli, S. aureus, Enterobacter aerogenes, S. typhi*, etc.	Tambekar and Dahikar (2011)
Lauha Bhasma		
Mandura Bhasma		
Kashis Bhasma		
Rajata Bhasma	*S. aureus, Enterococcus faecalis, E. coli*, and *Pseudomonas aeruginosa*	Sharma et al. (2016)
Rasa Manikya	Pathological MRSA/MSSA strains of *E. coli, S. aureus, Enterobacter* sp.	Ruidas et al. (2019)
Hartala Bhasma (As_2S_3)	*S. aureus, P. aeruginosa, Streptococcus pneumoniae, Klebsiella pneumoniae*	Kumar et al. (2015)
Hartalagodanti Bhasma ($CaSO_4$, $2H_2O$)		

with the emergence of scientific data proving the antimicrobial properties of these ancient medicines (Table 16.2).

Normally, in the herbometallic compounds, the metal ions are found in the oxygenated form, or bonded with carbon and nitrogen (e.g., As_2O_3, of Rasa Manikya; C≡C, C=C of Hirak Bhasma; HgO of Parada). In cellular metabolism, these oxygenated or nitrogenized metals act to release the cationic metals and eventually increase the reactive oxygen species (ROS) and/or reactive nitrogen species (RNS). This increased ROS and RNS level helps to acquire the primary defense of immunity against the pathogenic microbes. Moreover, these Ayurvedic preparations enact to regain the vitality in a sepsis condition by balancing the redox potential managing the glutathione (GSH) and superoxide dismutase (SOD) levels which reduce excessive ROS/RNS (Sharma et al. 2017). Apart from the ROS/RNS-mediated destruction of microbes, the herbometallic preparations exploit the ligand-receptor-based entry of microorganisms into the host body through metal chelation of receptor proteins. Microbes get entry inside the host body via receptor protein-mediated phagocytosis. Being the resource of cationic metals (Ca^{2+}, Fe^{2+}/Fe^{3+}, Zn^{2+}, Mn^{2+}, Mg^{2+}, Cr^{3+}, As^{3+}), herbometallic compounds help in chelating the host receptor proteins so that microbes do not get an easy access to the host cells anymore (Bharti and Singh 2009).

16.6.1 Techniques Adapted to Prove Antimicrobial Activity of Herbometallic Preparations

16.6.1.1 Agar Well Diffusion/Disc Diffusion

For testing the antimicrobial activity of antibiotics, agar well diffusion and disc diffusion technique are a routine practice. Researchers have adapted this technique for documenting the antimicrobial potential of herbometallic Ayurvedic preparations (Table 16.2).

16.6.1.2 Minimal Inhibitory Concentration/Minimal Bactericidal Concentration

MIC and MBC determine the minimal concentrations of a drug to inhibit and kill a bacterial culture in solution, respectively. Though in Ayurveda the doses are predetermined following norms, the practice of this MIC and MBC determination would help in considering the dose scientifically and thus avoiding lethal toxicity for human being (Ruidas et al. 2019). This would also lessen the side effects.

16.6.1.3 Bacterial MTT Test

Besides MBC and well diffusion techniques, another test to confirm bactericidal effect of herbometallic preparations is a modified reduction test of 3-(4,5-dimethylthiazol-2-yl)-2,5-diphenyltetrazolium bromide (MTT assay) through which the percent bacterial cell viability can be counted after the treatment with targeted herbometallic drug with respect to the control study (Wang et al. 2010).

16.6.2 Antimicrobial Mechanism Revealing Techniques

16.6.2.1 Bacterial ROS/RNS Measurement

2,7-Dichlorofluorescein diacetate (DCFHDA) is used as an indicator of ROS generation. The change in fluorescence intensity after treating with DCFHDA indicates the presence of active microbial growth. By measuring the bacterial ROS level, the mechanism of antimicrobial drug can also be revealed (Su et al. 2009). Similarly, measurement of change in SOD, glutathione, and RNS level would be complementary to find out the oxidative stress and produce evidence against the drug mechanism of the herbometallic preparations.

16.6.3 Importance of Herbometallic Preparations in the Premise of Multidrug Resistance

To date, the antibiotics are considered most reliable fighting against pathogenic bacterial strains. However, an alarming bell has already rung due to the fast emergence of multidrug-resistant (MDR) bacterial strains. Reports have confirmed that we are thus far living in the post antibiotic era (Alanis 2005). Here lies the importance of the herbometallic compounds prescribed in Ayurveda. With the proven effectiveness of metal ions against bacteria, many metal-based nanoparticles (Ag, Au, Pt, etc.) are also tested as antibacterial agent, but they come up with high toxicity and other side effects. In herbometallic compounds the metal elements are blended with the organic decoction and thus cause almost no toxicity to the host. Moreover, the nanoformulation of the metals in these compounds functions as chelating agents and stimulator of ROS to combat with the MDR strains of bacteria and also helps to regain the vitality unlike the antibiotics which affect the good gut microflora. The growing scientific evidences for these Ayurvedic drugs (Table 16.2) present a glimpse of reconsidering the cost-effective traditional drugs as antimicrobial therapeutics.

16.7 Anticancer Activity of Ayurvedic Herbometallic Preparations and Some Promising Drug Candidates

Based on the modern concepts for searching the anticancer drugs from natural resources, numerous researches have been conducted based on the information from folk and traditional medicines throughout the globe. Several anticancer drugs extracted from plant sources after purification are tested in both in vitro and in vivo models and then sent to clinical trials (Khazir et al. 2014). The substances of natural origin that exhibit antitumor or anticancer properties belong to various groups of compounds, such as alkaloids, diterpenes, lactonic sesquiterpene, peptides, cyclic depsipeptide, proteins, etc. (Subramaniam et al. 2019). Previous reports implied that on the basis of pharmacological activities and chemical structural information of natural compounds, antitumor or anticancer drugs have been synthetically developed, viz., vincristine from *Catharanthus roseus* and paclitaxel and taxanes from the yew tree or *Taxus baccata* (Mukhtar et al. 2014). Some of the important active constituents present and isolated from the natural resources having anticancer activities are classified in (Lichota and Gwozdzinski 2018) (Table 16.3).

Studies have shown that natural resource-derived compounds in combination with anticancer drugs have great potential to destroy tumor or cancer cells while not affecting normal cells such as lymphocytes and fibroblasts. The side effects of anticancer drugs may be reduced by using nanoparticle encapsulations to transport the drugs to their target sites. Drugs may also be administered in the form of liposomes, which serve as carriers for the drug. Considering the huge costs associated with the discovery and development of effective anticancer drugs, natural compounds can be an inexhaustible source in near future.

16.8 Antidiabetic and Anti-inflammatory Role of Traditional Herbometallic Preparation

Chaturmukha Rasa has four metal constituents, viz., mercury, sulfur, iron and mica based Ayurvedic preparation. Recently, Sharma and colleagues extensively studied the antidiabetic role of Chaturmukha Rasa in streptozotocin-induced diabetic rat model (Sharma et al. 2019). Diabetes mellitus, the condition of hyperglycemia, can be treated with the approach to delay the digestion of carbohydrate. Through the inhibition of salivary amylase, α-glucosidase, and sucrose, Sharma et al. have shown that Chaturmukha Rasa can lower down the carbohydrate digestion rate and thus combat hyperglycemia (Sharma et al. 2019). Earlier, in 1989 it was reported that the blood zinc level varies between diabetic and nondiabetic persons (Prasad and Sharma 1989). Also, Zn has been found in the alpha and beta cells of islet of Langerhans where it stabilizes the insulin being an integral part of insulin crystal. These findings indicate the role of Zn in diabetes mellitus. With the advancement in preclinical evaluation of the drugs, scientists now deserted the antidiabetic effect of Yashada Bhasma which contains Zn as the key metal ingredient (Rao et al. 1997). On the other hand, the anti-inflammatory role of copper and copper complexes has been known since long. Even the nonsteroidal anti-inflammatory drugs (NSAID) contain copper complexes which play the central role to heal inflammation (Bafna and Patil 2018). Bafna and Patil have reported a scientific evaluation of Tamra Bhasma in carrageenan, cotton pellet, and complete Freund's adjuvant (CFA) model (Bafna and Patil 2018). Here in this report, a significant reduction in carrageenan-induced paw edema, cotton pellet-induced granuloma, and CFA-induced arthritis has comprehensively demonstrated the anti-inflammatory action of Tamra Bhasma.

Table 16.3 Anticancer activities of promising new molecules from natural resources

Active ingredients	Herbs/natural products' name	Information of anticancer activities	References
Camptothecin (quinoline alkaloids)	*Camptotheca acuminata*	Irinotecan and topotecan have been developed as semisynthetic compounds which have clinical efficacies on second-line lung cancer	Wagner (2015)
Combretastatin (stilbene derivative)	*Combretum caffrum*	Combretastatin A-1 has been synthesized and found in drugs targeting the microtubules similar to taxanes and *vinca* alkaloids	Zweifel et al. (2011)
Podophyllotoxin (toxin lignin)	*Podophyllum peltatum*	This compound is effective in lymphomas, brain tumors, gastrointestinal cancer, colon cancer, breast cancer, small-cell lung cancer, and testicular carcinomas	Zhang et al. (2018)
Geniposide (aglycone)	*Gardenia jasminoides*	Effective for lung cancer	Habtemariam and Lentini (2018)
Artesunate	*Artemisia annua*	Semisynthetic derivative of artemisinin is effect on pancreatic tumor and brain tumor	Konstat-Korzenny et al. (2018)
Homoharringtonine	*Omacetaxine mepesuccinate*	Effective in leukemia and breast cancer	Hansz (2000)
Salvicine (diterpenoquinone)	*Salvia prionitis*	Salvicine, chemical synthesized product acts as cytotoxic on nonintercalative topoisomerase II poisons	Deng et al. (2011)
Ellipticine (alkaloid)	*Ochrosia elliptica*	Synthetic ellipticine has antitumoral activity	Stiborova et al. (2011)
Roscovitine	*Raphanus sativus*	Synthetic roscovitine has efficacies on lung cancer and breast cancer	Cicenas et al. (2015)
Maytansine	*Maytenus serrata*	Synthetic maytansine is effective on metastatic breast cancer	Lopus et al. (2010)
Thapsigargin (sesquiterpene)	*Thapsia garganica*	Synthetic thapsigargin is effective on solid tumors	Ganley et al. (2011)
Curcumin	*Curcuma longa*	Effective on neck, head, oral, colon, pancreas, bladder, prostate, and breast cancers	Mohammadi et al. (2005)
Resveratrol (phytoalexin)	*Vitis vinifera*	Antileukemic activity	Sahpazidou et al. (2014)
Carnosol (diterpene)	*Rosmarinus officinalis*	Antitumor activities	Johnson et al. (2008)
Crocetin	*Crocus sativus*	Effective on lung, liver, pancreas, colon, and breast carcinoma	Aung et al. (2007)
Silibinin	*Silybum marianum*	Effective in colon, lung, prostate, and skin cancers	Ramasamy et al. (2011)
Withanolides	*Withania somnifera*	Effective in breast cancers and brain tumors	Samadi (2015)

16.9 Traditional Ayurvedic Herbometallic Preparations in Treatment of Neurological Disorders

According to several reports, the use of traditional Indian system of Ayurvedic herbometallic preparation is very common in treatment of several neurological ailments due to its easy accessibility, low cost, and minimal side effects. Ayurvedic herbometallic formulations refer to metal-based herbal remedies prescribed on the basis of traditional Asian system of medication. From time immemorial metals play a vital role in human physiology, and the deficiency in metals led to the occurrence of various diseases. In Indian Ayurvedic history, numerous metals like gold, silver, copper, iron, lead, zinc, tin, etc. are considered as important elements of body, but very few of them has got an effective role in treating neurological diseases. Therefore any imbalance in this metal content hampers the body metabolism. In this regard, metallic formulation Bhasmas are highly effective in prevention and treatment of various diseases (Lee et al. 2018). Generally, Bhasmas are metallic Ayurvedic preparations made up of herbal juices/fruits used in India from the seventh century B.C. and are widely used against chronic ailments.

16.9.1 Efficacy of Swarna Bhasma in Neurological Disorders

Gold along with other metals are therapeutically effective Ayurvedic medicine since ancient time. In the Vedic period, gold was used to increase the strength, potency, and longevity and also to combat aging in humans. Since the eighth century AD, by proper purification and incineration, gold was utilized as Bhasma (ash) which is referred to as Swarna Bhasma (gold ash) (Mitra et al. 2002).Several reports suggest that cognitive disorders such as dementia, delirium, amnesia, etc. can be characterized by various symptoms like memory impairment, and massive cognitive decline, gait disturbances, and language disturbances can be improved by Swarna Bhasma (Warad et al. 2014). This Swarna Bhasma acts as a nootropic agent and is primarily used to improve memory, behavior, and mood. Recently, a study evidenced the antioxidant/restorative property of Swarna Bhasma in cerebrovascular diseases (Singh and Chaudhary 2012). It has been demonstrated that the different enzymatic parameters were measured to assess ischemic brain damage, and the result showed significant restoration of altered values near normal level proving the efficacy of Swarna Bhasma in treatment for cerebrovascular disease. Not only cognitive and cerebrovascular disorder, but other neurological disorders also can be effectively treated by Swarna Bhasma. Alzheimer's disease (AD), the common form of age-related dementia, can be characterized by gradual deterioration of cognitive function specially functions related to memory. The main pathogenic hallmark of this disease is the gradual aggregation of misfolded protein like amyloid beta and tau. Currently, a study with Swarna Bhasma has been performed to illustrate the decrease in amyloid-beta aggregation. The study also paved the path of hope to treat AD if proper medication with Swarna Bhasma is initiated immediately after early diagnosis of AD (Agrawal 2010). Like AD, Parkinson's disease (PD) is also a common geriatric disorder, and the initial manifestations may be tremor, rigidity, akinesia, slowness, and many others. The formulation of Swarna Bhasma plays a key role in the effective management of PD (Kaviya et al. 2016).

16.9.2 Efficacy of Raupya Bhasma in Neurological Disorders

Raupya Bhasma also called Chandi Bhasma or calcined silver ash is an Ayurvedic herbometallic formulation used in eye diseases, jaundice, anemia, hysteria, epilepsy, and neurological disorders. This type of calcined silver ash possesses medicinal properties such as a potent cognitive enhancer as well as acts as an antidepressant, antianxiety, anti-stress, and neuroprotective agent. Generally, excessive use of the brain, insomnia, anxiety, fear, working on computers, etc. cause nervous aggravations in the body which lead to reduction in mental power and strength. Therefore, Raupya Bhasma works widely in all these conditions and can be effectively used in treating mental fatigue (https://www.ayurtimes.com/rajat-bhasma-raupya-bhasma/#mental-fatigue; 2014). Despite all these properties, very few evidences have been found regarding the successful use of Raupya Bhasma in treating neurological disorders.

16.9.3 Efficacy of Naga Bhasma in Neurological Disorders

Naga Bhasma (incinerated lead) is the herbometallic preparation containing lead as the main ingredient. Manifestations of some extraordinary medical properties prove its therapeutic efficiency in treatment of various diseases. Nowadays, modern research revealed that in PD, APO-E 4/3 and 4/4 genotypes can excrete heavy metals in minimal amount. Those abundant APO-E proteins with this version of genotype in the cerebral spinal fluid surrounding the brain will have the highest affinity for becoming ill from exposure to neurotoxic heavy metals (Rajput and Patgiri 2013). This situation can be treated by Naga Bhasma due to its untoward side effects and responsive efficacy.

16.10 Effective Medications Against Other Common Lifestyle Diseases

In Ayurveda, the diseases associated with aging and cardiovascular diseases and many such diseases are correlated with obesity, food habits, and change in lifestyle. It is also considered that the normal physiology, metabolism, and the driven force of life are dependent on related state of metal ions and even a little imbalance in metal ions in body may lead to the irregular functioning of the biomolecules associated with those metal ions (Bharti and Singh 2009).Thus, it is prescribed in Ayurveda that these kinds of diseases can be treated following a common route. Virupaksha et al. have reviewed the works reflecting the success of Ayurvedic Bhasmas in treating the lifestyle diseases, viz., type II diabetes, cardiac diseases, depression, obesity, stroke, and so on (Virupaksha et al. 2011). Similarly, Dr. Bhanu Priya and colleagues reported the evaluations of herbometallic compounds to achieve a state of well-being (Priya et al. 2018).

16.11 Major Issues of Clinical Trials of Ayurvedic Medicine

The Department of AYUSH (Government of India) already has prescribed various clinical as well as preclinical trials for new herbal Ayurvedic drug formulations following proper evaluation methods to facilitate the development of regulation and registration in Ayurveda along with other traditional medicine systems. Presently, various reports evidenced the use of HMPs in lead toxicity. Thereafter, from early reports it has been evidenced that 65 cases of heavy metal intoxicity in adults and children have been found to be associated with Ayurvedic HMPs (Mishra and Gupta 2010). A report suggested that clinical trials with herbal medicine are a challenging phenomenon. For instance, the use of placebo group in evaluation of efficacy of herbal drugs leads to unethical issues as the patients are unaware of its use as an available therapy effective in treating the disastrous condition (Mishra and Gupta 2010).Recent reports suggested that there lie immense difficulties in estimation of active biomolecules in the pharmacokinetic study of drug discovery (Mishra and Gupta 2010). Another study reported that storage conditions generally alter the bioavailability of herbal medicines. This leads to loss of fungal or bacterial activity resulting into batch-to-batch variation (Mishra and Gupta 2010). On the other hand, dose selection is also a major issue. This dosage must be calculated on the basis of extractive value. Few herbal drugs have been immensely studied to prove their safety and efficacy. However, the clinical trials of *Ginkgo biloba* extract effective for treatment of CNS disorders and *Hypericum perforatum* effectively used as antidepressant evidenced the safety and efficacy of these drugs. Besides these two drugs, other herbal drugs include *Panax ginseng* (ginseng) effectively used as tonic, *Tanacetun parthenium* (feverfew) for headache and migraine, *Allium sativum* (garlic) for lowering cholesterol level, *Arnica montana* (arnica) to treat post-traumatic conditions, and *Serenoa repens* (saw palmetto) for the treatment of benign prostatic hyperplasia. Thus, all of these herbal medicines have been effectively evaluated in various clinical trials. However, well-organized and appropriate randomized clinical trials are still needed to be performed in order to prove their safety as well as efficacy.

16.12 Status of Clinical Trials of Ayurvedic Medicine

Plants cannot be patented, due to insufficient research, and validation has been performed on plants as medicinal agents. The clinical trials on ancient Ayurvedic medicine need proper validation for betterment of treatment modalities by improving dose forms and side effects of any given drug. In the USA, it almost takes 15 years with an estimated cost of $500 million for elucidating safety and efficacy of drug.

Due to the regulatory essentials regarding proof of safety, it turned out to be very uneconomical for private industry to carry out clinical trials regarding herbal Ayurvedic medicine. In this harsh condition, public funds are required in a huge amount to confirm the validity of herbal remedies, as this will help the pharmaceutical companies to earn meager incentive for development of an herbal drug against a patented drug. Owing to this fact, it is debated that there is a huge demand regarding a proper study, and unless and until it has been conducted on human subjects, no effective conclusion can be carried regarding the safety and efficacy of the drug.

16.13 Future Scope

In a developing country, most of the rural people depend on herbal products for their health and used many medicinal plants without proper supervision. Traditional medicine is based on the universal principle like cold, heat, and five sense organs (Panchami elements). So in this modern era, our challenging issue is how we can change the form of Ayurveda without distorting its principal. The common people do not know the ingredients of these Ayurvedic preparations. They know the local names of this plant, but do not know the general name and scientific name. Many plants have some ingredients that work very well at low doses, but all of these ingredients can do a lot of harmful effect in high doses and long-term usage; even life-threatening events can happen if dose and duration of an herbometallic preparation is not adjusted. Only a small fraction of medicinal plants used worldwide has been tested rigorously in randomized control trials. If clinical trials of all herbal products are possible in the future and standardization of the ingredients can be done, then the efficacy, dose-related adverse effect, etc. can be predicted. Ayurveda is not only evolving, but it is also growing and changing day by day. At the same time, there are a lot of challenges being faced with Ayurveda. More specific experiments on animal models and also clinical trials are required to understand the exact molecular mechanisms of function of different ingredients present in herbometallic preparations. As traditional medicine is the first level of contact for rural people for their health system, it is necessary for the government to take immediate steps to introduce the use of traditional medicine to supplement Primary Health Care (PHC). All medicinal plants that have been reported in the rural area should be scientifically examined and detailed about their functional contents along with its pharmacokinetics. Health education should be given to the people concerning the use of the indigenous herbal product. PHC should impart education regarding the identification of various medicinal plants and their usage for the treatment of common diseases. The government should provide financial support to promote the potential role of traditional medicine in primary health care.

16.14 Conclusion

It is observed that most of the processed inorganic materials (Bhasma) have some common pharmacological properties like Rasayana, Yogavahi, Agni Deepana, Shighravyapti, Kshipram or Shighrakari, Alpamatra, and others (Mishra and Gupta 2010). The Rasayana exhibits immunomodulatory as well as antiaging properties, whereas Yogavahi shows ability of targeted drug delivery of the Bhasma. These ancient preparations are prescribed in very minute dose of 15–250 mg/day. Besides, other properties of these herbometallic preparations include that Bhasmas are readily absorbable, adaptable, and easily assimilable in the body and will be nontoxic in nature (Rasibhavan). Therefore, these features of Bhasma are comparable with the action of nanoparticles in the body as they are easily biodegradable as well as biocompatible and nonantigenic in nature. Besides, these nanoparticles are efficiently used to provide targeted drug delivery to specific site of action in the body including the blood-brain barrier. These can also be used to extend the bioavailability time and to protect the drug from chemical as well as enzymatic decomposition. However, these results in reduced peripheral side effect by decreasing the overall dose of drugs in the body.

In recent years, there has been concerted research regarding the potentiality of Ayurvedic medicines and its efforts to understand real-life treatment paradigm. Therefore, there is an immense need to interpret logic of Ayurveda while adopting modern scientific tools in development and validation of drugs. Hence, validation of Ayurvedic medicines using the latter approach may lead to evidence-based interdisciplinary study of modern Ayurvedic medicine. Further, owing to build better treatment opportunities, we ought to step beyond the realm of only drugs and attempt validation of comprehensive specific treatment package as per classical Ayurveda.

References

Agrawal DSA (2010) Swarnabhasma (gold ash): treatment/prevention of Alzheimer's disease. Alzheimers Assoc 6:S589

Alanis AJ (2005) Resistance to antibiotics: are we in the post-antibiotic era? Arch Med Res 36:697–705

Aung HH, Wang CZ, Ni M et al (2007) Crocin from Crocus sativus possesses significant anti-proliferation effects on human colorectal cancer cells. Exp Oncol 29:175–180

Bafna PS, Patil SD (2018) Physicochemical characterisation and anti-inflammatory activity of ayurvedic herbo-metallic Tamra bhasma in acute and chronic models of inflammation. Mater Technol 33:681–688

Balmain J, Hannoyer B, Lopez E (1999) Fourier transform infrared spectroscopy (FTIR) and Xray diffraction analyses of mineral and organic matrix during heating of mother of pearl (nacre) from the shell of the mollusk Pinctada maxima. J Biomed Mater Res 48:749–754

Barakat EA, Fatma UA (2003) Herbal medicine in Jordan with special emphasis on commonly used herbs. J Ethnopharmacol 89:193–197

Bharti SK, Singh SK (2009) Metal based drugs: current use and future potential. Pharm Lett 1:39–51

Bhowmick TK, Suresh AK, Kane SG et al (2009) Physicochemical characterization of an Indian traditional medicine, jasada bhasma: detection of nanoparticles containing non-stoichiometric zinc oxide. J Nanopart Res 11(3):655–664

Brown CL, Bushell G, Whitehouse MW et al (2007) Presentation: nanogold-pharmaceutics: (i) the use of colloidal gold to treat experimentally-induced arthritis in rat models; (ii) characterization of the gold in Swarna Bhasma, a microparticulate used in traditional Indian medicine. Gold Bull 40(3):245–250

Cicenas J, Kalyan K, Sorokinas A et al (2015) Roscovitine in cancer and other diseases. Ann Transl Med 3:1–12

Dabrowiak JC (2017) Metal ion imbalance in the body. Metal Med:329–356

Deng F, Lu JJ, Liu HY et al (2011) Synthesis and antitumor activity of novel salvicine analogues. Chin Chem Lett 22:25–28

Desai MP, Labhasetwar V, Amidon GL et al (1996) Pharm Res 13:1838–1835

Ernst E (1999) Why complementary medicine? Perfusion 11:437

Ganley IG, Wong PM, Gammoh N et al (2011) Distinct autophagosomal-lysosomal fusion mechanism revealed by thapsigargin-induced autophagy arrest. Mol Cell 42:731–743

Habtemariam S, Lentini G (2018) Plant-derived anticancer agents: lessons from the pharmacology of geniposide and its aglycone, genipin. Biomedicine 6:39

Hansz J (2000) Contemporary therapy of chronic myeloid leukemia. Borgis–Post Nauk Med 2000 (4):33–39

Harrison's Internal Medicine 20th edn (n.d.)

Johnson JJ, Syed DN, Heren CR et al (2008) A dietary diterpene, displays growth inhibitory effects in human prostate cancer PC3 cells leading to G2-phase cell cycle arrest and targets the 5'-AMP-activated protein kinase (AMPK) pathway. Pharm Res 25:2125–2134

Jungwrith U, Kowol RC, Keppler BK et al (2011) Anticancer activity of metal complexes: involvement of redox processes. Antioxid Redox Signal 15(4):1085–1127

Kaviya R, Mishra PK, Soni A (2016) The role of ayurveda in the management of Parkinson's disease—a review article. Int Ayurvedic Med J

Khazir J, Mir BA, Pilcher L et al (2014) Role of plants in anticancer drug discovery. Phytochem Lett 7:173–181

Konstat-Korzenny E, Ascencio-Aragón JA, Niezen-Lugo S et al (2018) Artemisinin and its synthetic derivatives as a possible therapy for cancer. Med Sci 6:19

Krishnamacharya B, Pemiah B, Krishnaswamy S et al (2012) Elucidation of a core-shell model for lauha bhasma through physicochemical characterization. Int J Pharm Pharm Sci 4:644–649

Kulkarni DA (1998) Vagbhattachariya rasa ratna samucchaya, hindi commentary meharchand lachhmandas publications, New Delhi, p 198

Kumar A, Nair AG, Reddy AV (2006) Availability of essential elements in bhasmas: analysis of ayurvedic metallic preparations by INAA. J Radioanal Nucl Chem 270:173–180

Kumar A, Nair AGC, Reddy AVR et al (2007) Unique ayurvedic metallic-herbal preparations, chemical characterization. Biol Trace Element Res 109(3):231–254

Lee B, Kwon CY, Chang GT (2018) Oriental herbal medicine for neurological disorders in children: an overview of systematic reviews. Am J Chin Med 46:1701–1726

Lichota A, Gwozdzinski K (2018) Anticancer activity of natural compounds from plant and marine environment. Int J Mol Sci 19(11):3533–3539

Lopus M, Oroudjev E, Wilson L et al (2010) Maytansine and cellular metabolites of antibody-maytansinoid conjugates strongly suppress microtubule dynamics by binding to microtubules. Mol Cancer Ther 9:2689–2699

Lucock M (2004) Is folic acid the ultimate functional food component for disease prevention? Br Med J 328:211–214

Mishra A, Gupta R (2010) Problems associated with clinical trials of ayurvedic medicine. Rev Bras Farmacog 20:276–281

Mitra A, Chakraborty S, Auddy B (2002) Evaluation of chemical constituents and free-radical scavenging activity of Swarnabhasma (gold ash), an ayurvedic drug. J Ethnopharmacol 80:147–153

Mohammadi K, Thompson KH, Patrick BO et al (2005) Synthesis and characterization of dual function vanadyl, gallium and indium curcumin complexes for medicinal applications. J Inorg Biochem 99:2217–2225

Mukhtar E, Adhami VM, Mukhtar H (2014) Targeting microtubules by natural agents for cancer therapy. Mol Cancer Ther 13:275–284

Nagarajan S, Pemiah B, Krishnan UM et al (2012) Physico-chemical characterization of lead-based indian traditional medicine- naga bhasma. Int J Pharm Pharm Sci 4:69–74

Nieboer E, Richardson DHS (1980) The replacement of the nondescript term 'heavy metals' by a biologically and chemically significant classification of metal ions. Environ Pollut (Series B) 1:3–26

Pal SK (2015) The ayurvedic bhasma: the ancient science of nanomedicine. Recent Pat Nanomed 5:12–18

Pal D, Sahu CK, Haldar A (2014) Bhasma: the ancient Indian nanomedicine. J Adv Pharm Technol Res 5:4–12

Parikh M, Choudhury AK, Patgiri BJ et al (2012) IJPBA 3(6):1512–1517

Patwardhan B, Warude D, Pushpangadan P (2005) Ayurveda and traditional Chinese medicine: a comparative overview. Evid Based Complement Alternat Med 2(4):465–473

Prasad CM, Sharma AV (1989) Yasada bhasma: an effective hypoglycaemic drug. Anc Sci Life 9:69–70

Priya B, Aggarwal K, Barwal V (2018) Rasa dravya and their use in the management of life style diseases: an ayurveda perspective. World J Pharm Med Res 4:61–63

Rabinow BE (2004) Nanosuspensions in drug delivery. Nat Rev Drug Discov 3:785–796

Ragozin BV (2016) The history of the development of ayurvedic medicine in Russia. Anc Sci Life 35(3):143–149

Rajatbhasma (Chandibhasma)-Raupyabhasma Ayur Times https://www.ayurtimes.com/rajat-bhasma-raupya-bhasma/#mental-fatigue. 2014

Rajput D, Patgiri B (2013) Ayurvedic view on heavy metal poisoning with special reference to Nagabhasma (incinerated lead)—a literary study. Int J Pharm Biol Arch 4:41–45

Ramasamy K, Dwyer-Nield LD, Serkova NJ et al (2011) Silibinin prevents lung tumorigenesis in wild-type but not in iNOS−/− mice: potential of real-time micro-CT in lung cancer chemoprevention studies. Clin Cancer Res 17:753–761

Rao VN, Suresh P, Dixit SK (1997) Effect of yashada bhasma in streptozotocin-induced diabetes. Anc Sci Life 17:114–116

Rasheed A, Marri A, Naik MM (2011) J Pharm Res 4(6):1931–1933

Ruidas B, Som Chaudhury S, Pal K et al (2019) A novel herbometallic nanodrug has the potential for antibacterial and anticancer activity through oxidative damage. Nanomedicine 14:1173–1189

Sahpazidou D, Geromichalos GD, Stagos D et al (2014) Anticarcinogenic activity of Samadi AK (2015) potential anticancer properties and mechanisms of action of withanolides. Enzyme 37:73–94

Saper BR, Kales NS, Paquin J et al (2004) Heavy metal content of ayurvedic herbal medicine products. JAMA 292(23):2868

Sarkar PK, Chaudhary AK (2010) Ayurvedic bhasma: the most ancient application of nanomedicine. J Sci Indus Res 69:901–905

Sarkar PK, Das S, Prajapati PK (2010) Ancient concept of metal pharmacology based on ayurvedic literature. Anc Sci Life 29:1–6

Sharma RK, Dash B (2000) Acharya agnivesa charaka samhita english translation sutrasthana vol I chaukhamba sanskrit series office Varanasi 26/13

Sharma R, Bhatt A, Thakur M (2016) Physicochemical characterization and antibacterial activity of rajata bhasma and silver nanoparticle. Ayu 37:71–75

Sharma R, Amin H, Galib R (2017) Molecular targets of common ayurvedic herbal anti-oxidants. J Ayu Herb Med 3:36–40
Sharma A, Tiwari RK, Sharma V (2019) Antidiabetic activity of an ayurvedic formulation chaturmukha rasa: a mechanism based study. J Pharmacopuncture 22:115–121
Singh N, Chaudhary A (2012) Swarna bhasma and gold compounds: an innovation of pharmaceutics for illumination of therapeutics. Int J Res Ayurveda Pharm 3:5–9
Stiborova M, Poljakova J, Martinkova E et al (2011) Ellipticine cytotoxicity to cancer cell lines—a comparative study. Interdiscip Toxicol 4:98–105
Su HL, Chou CC, Hung DJ (2009) The disruption of bacterial membrane integrity through ROS generation induced by nanohybrids of silver and clay. Biomaterials 30:5979–5987
Subramaniam S, Selvaduray KR, Radhakrishnan AK (2019) Bioactive compounds: natural defense against cancer. Biomol Ther 9:758–768
Tchounwou PB, Yedjou CG, Patlolla AK (2012) Heavy metal toxicity and the environment. Exp Suppl 101:133–164
Vetrov II, Sorokina YV (2012) St. Petersburg: Dhanvantari Ayurveda Center, LLC basic principles of Ayurvedic phytotherapy
Virupaksha GK, Pallavi G, Patgiri BJ (2011) Relevance of rasa shastra in the 21st century with special reference to lifestyle disorders (LSDs). Int J Res Ayurveda Pharm 2:1628–1632
Wagner L (2015) Fifteen years of irinotecan therapy for pediatric sarcoma: where to next? Clin Sarcoma Res 5:1–7
Wang H, Cheng H, Wang F (2010) An improved 3-(4, 5-dimethylthiazol-2-yl)-2, 5-diphenyl tetrazolium bromide (MTT) reduction assay for evaluating the viability of Escherichia coli cells. J Microbiol Methods 82:330–333
Warad VB, Shastri R, Habbu P (2014) Preparation and screening of swarna prashana for nootropic activity. Int J Nutr Pharmacol Neurol Dis 4:170–178
WHO (2001) Legal status of traditional medicine, complementary/alternative medicine: a world-wide review. World Health Organization, Geneva
Zhang X, Rakesh KP, Shantharam CS et al (2018) Podophyllotoxin derivatives as an excellent anticancer aspirant for future chemotherapy: a key current imminent needs. Bioorg Med Chem 26:340–355
Zweifel M, Jayson GC, Ree NS et al (2011) Phase II trial of combretastatin A4 phosphate, carboplatin, and paclitaxel in patients with platinum-resistant ovarian cancer. Ann Oncol 22:2036–2041

HPTLC Fingerprinting Analysis of Phytoconstituents from Indigenous Medicinal Plants

17

Shradhanjali Singh, Shanti Bhushan Mishra, and Alok Mukerjee

Abstract

High-performance thin-layer chromatography (HPTLC) is a versatile and flexible technique for the analysis of herbal and medicinal plants. Standardization of plant materials was essential for their identification and assessment of quality in herbal medicines. Moreover, HPTLC fingerprinting analysis represents as an important analytical tool for semiquantitative, qualitative and quantitative estimation of phytochemicals. This incorporates the TLC fingerprinting analysis and estimation of biomarkers. Quantification of various pharmacological active constituents was estimated by fingerprinting in herbal plants which leads to conquer the therapeutic effects of phytochemicals incorporated in herbal medicines. The phytochemicals obtained by HPTLC fingerprinting from herbal plants will tell the number of constituents present. Authentication of these constituents can be done by marker compounds which will ensure the therapeutic potential of the herbal plants used in Indian traditional medicine. Due to its simplicity and reliability, this technique is popular amongst other techniques. Chromatographic fingerprinting analysis will represent the qualitative method for the purpose of authentication, quality and ratio of constituents in herbal plants.

Keywords

β-sitosterol · Herbal · HPTLC · Phytochemical · Quercetin · Rutin

S. Singh (✉)
Department of Pharmaceutical Chemistry, United Institute of Pharmacy, Allahabad, Uttar Pradesh, India
e-mail: singhshradhanjali@gmail.com

S. B. Mishra · A. Mukerjee
Department of Pharmacognosy, United Institute of Pharmacy, Allahabad, Uttar Pradesh, India
e-mail: shantipharma15@gmail.com; alokmukerjee@united.ac.in

S. C. Mandal et al. (eds.), *Evidence Based Validation of Traditional Medicines*,
https://doi.org/10.1007/978-981-15-8127-4_17

Abbreviations

HPTLC	High-performance thin-layer chromatography
ICH	International Council for Harmonisation
LOD	Limit of detection
LOQ	Limit of quantitation
mg/mL	Milligram/millilitre
ng	Nanogram
nm	Nanometre
R_f	Retardation factor
RSD	Relative standard deviation
S/N	Signal-to-noise ratio
SD	Standard deviation
v/v	Volume/volume
w/w	Weight/weight
μg/μL	Microgram/millilitre
μm	Micrometre

17.1 Introduction

Due to high variability of chemical compounds in herbal plants, fingerprinting becomes a necessary tool to ensure that proportionate amount of phytochemical constituents will be present in the herbal plants. Moreover complexity of the herbal plants represents a challenge amongst Indian traditional medicine. So to overcome this challenge, fingerprinting is a versatile tool to resolve the complex phytochemicals or constituents in a simpler way (Ekor 2014). The authenticity of these complex phytoconstituents in herbal plants can be assessed by marker compounds which can be estimated by HPTLC fingerprinting analysis. To estimate the constituents from these complex herbal plants, methods such as gas chromatography and high-performance liquid chromatography have also been employed (Geethika and Sunojkumar 2017). In this chapter, an attempt has been made to quantify a wide variety of constituents/phytochemicals in herbal plants. This includes the estimation of phytoconstituents qualitatively and quantitatively from various plant extracts and fractions. The chromatographic fingerprinting characterizes the different and similar components in the herbal plants.

17.2 Requirements for Analysis of Phytoconstituents

The protocol for analysis of phytoconstituents is required to carry out the experimentation procedure effortlessly (Chavan et al. 2011).

17.2.1 Stationary Phase

Precoated plates are available with various stationary materials of different sizes and thickness. For qualitative and quantitative analysis of phytochemicals from herbal plants, plates with 100–250 μm thickness were used. Silica gel 60F plates are versatile, and more than 80% of samples (natural component, synthetic drugs) can be analysed, hence preferred for analysis. Plates placed in the air acquire humidity, so activation of plates is necessary for proper separation of constituents. Activation of plates can be done by placing the plates in an oven at 100–120 °C for 30 min before spotting.

17.2.2 Preparation and Spotting of Sample and Standard

The interferences from the impurities can be overcome by decreasing the thickness layer of stationary phase, hence reducing S/N ratio. Other parameters which include straight baseline and LOD were also been considered and improved during the analysis. Methanol is the preferred solvent used in the preparation of samples. At a concentration range, 0.1–1 μg/μL samples and standard (marker compounds) were spotted on the plates using automatic applicator as bands since bandwise spotting leads to better separation.

17.2.3 Selection of Mobile Phase and Chamber Saturation

Polarity of mobile phase solvent is being taken under consideration since non-polar mobile phase and polar stationary phase will lead to elution of non-polar compounds first due to lesser affinity towards stationary phase. Moreover, different composition of not more than three solvents used as mobile phase should be tried on the basis of trial-and-error method for increased resolution and spot definition. Chamber saturation is extremely required for R_f reproducibility, hence should be done by lining the filter paper absorbed in the mobile phase around the chamber for 30 min prior to plate development.

17.2.4 Development of Chromatographic Plates

For the development of chromatographic plates, automated multiple development systems were used for gradient elution. The successive elution of the constituents on the basis of strength of solvents will provide a gradient pattern by which constituents will be separated in spot form.

17.2.5 Detection and Visualization

Detection of fluorescent compounds can be performed at 254 nm, 366 nm and visible light. The densitometry was done by using winCATS software, CAMAG, Switzerland. Being a non-destructive method, the property of constituents can be retained and detected easily. One percent iodine solution and derivatization is one of the method used to detect non-UV-absorbing compounds.

17.3 Quantitative Estimation

The scanner is the instrument which is used to measure the absorption and fluorescence of the constituent applied on chromatoplate. The area under the curve can be estimated by densitogram. The peaks in densitogram can be resolved by estimating optical density of the constituents. The amount of sample is compared by running the standard (marker compound) under the same condition. The data can be evaluated by measuring the optical density of sample component by concentration of the sample placed on plate.

17.4 Analytical Method Validation

HPTLC is a versatile technique for estimation of various phytochemicals from plant extract and their fractions. Method validation which includes linearity, precision, LOD, LOQ, specificity, robustness, accuracy and assay imparts that the quantification procedure is intended to be used for further analysis. To justify the above comment, some examples have been given below with quantification of phytoconstituents extracted from plant extract from the reference.

17.4.1 Estimation of Rutin from *Alhagi pseudalhagi* (M. Bieb) Desv

17.4.1.1 Preparation of Ethanolic Extract of Alhagi pseudalhagi

The freshly aerial parts of the plant as in Fig. 17.1 were dried under shade and then dried in tray drier under controlled conditions and then powdered. The powdered plant material (500 g) was extracted at 60 °C using 95% v/v ethanol by continuous extraction method (Kokate et al. 2008). The extract was collected and concentrated on rotavapor and then dried in lyophilizer under reduced pressure, and therefore 170 g of solid residue were obtained and the percentage yield was found to be 34% w/w.

Fig. 17.1 Aerial part of *A. pseudalhagi* (Bieb.) Desv

17.4.2 Preparation of Standard Solution of Rutin

Standard stock solution of rutin (1000 μg/mL) was prepared by dissolving 25 mg rutin in 25 mL of methanol. 2.0 mL aliquot of the above solution was made up to 10 mL with methanol (200 ng/μL). From the above solution, 2 μL, 3 μL, 4 μL, 5 μL and 6 μL were spotted on the HPTLC plate which corresponds to 400 ng/spot, 600 ng/spot, 800 ng/spot, 1000 ng/spot and 1200 ng/spot, respectively. The densitogram of standard solution of marker compound rutin was shown in Fig. 17.2.

17.4.2.1 Chromatographic Conditions

Analysis was performed on 5 cm × 10 cm HPTLC silica gel G60 F_{254} plates. Different compositions of mobile phase were tried using binary and ternary mixtures of solvents like toluene, methanol, chloroform, ethyl acetate, acetic acid, formic acid, etc. with chamber saturation to achieve optimum resolution. After several trials, mixture of ethyl acetate/acetic acid/formic acid/water (10:1:1:1 v/v/v/v) is chosen as the mobile phase for analysis. The details were given in Table 17.1.

The ethanolic extract and standard samples were applied on the plate as bands with the band size of 6 mm and at position of 15 mm. The plate was developed to a

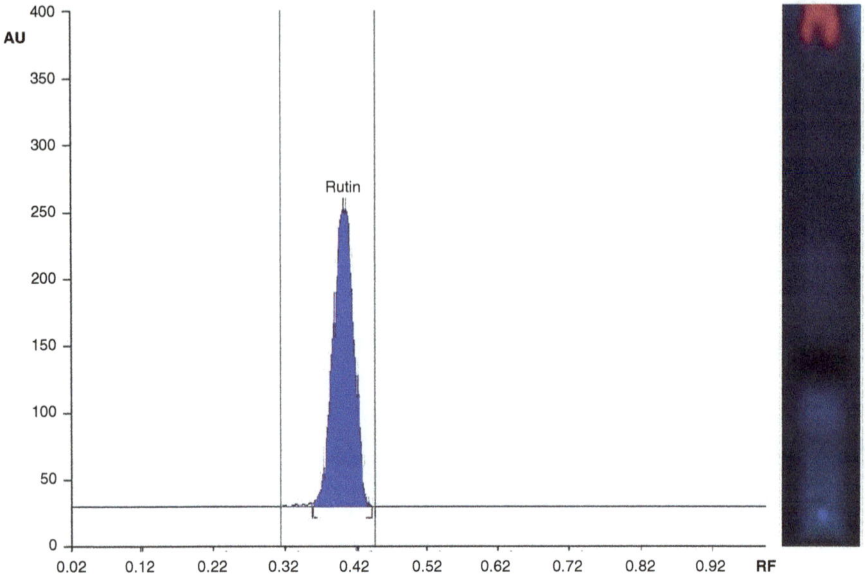

Fig. 17.2 Densitogram of rutin (marker compound) by HPTLC

Table 17.1 Selection of solvent system for method validation of rutin from *A. pseudalhagi*

Solvent system	Ratio
Chloroform/methanol/water	9:1:0.1 (v/v/v)
Toluene/ethyl acetate/formic acid	7 :2 :1 (v/v/v)
Ethyl acetate/acetic acid/formic acid/water	10:1:1:2.5 (v/v/v/v)
Ethyl acetate/acetic acid/formic acid/water	10:1:1:1 (v/v/v/v)

distance of 80 mm in the glass chamber previously saturated with ethyl acetate/acetic acid/formic acid/water (10:1:1:1 v/v) mobile phase for 30 min. The plates were completely dried in air at room temperature, were derivatized with anisaldehyde-sulphuric acid solution and were scanned. The areas for sample and standard were recorded at 366 nm by densitometry.

17.4.2.2 Validation as per ICH Guidelines

Linearity

Linearity of the method was analysed by using five different concentrations in the range of 200–1000 ng/spot for rutin. Results were expressed in terms of correlation coefficient in Fig. 17.3. The method validation parameters were shown in Table 17.2.

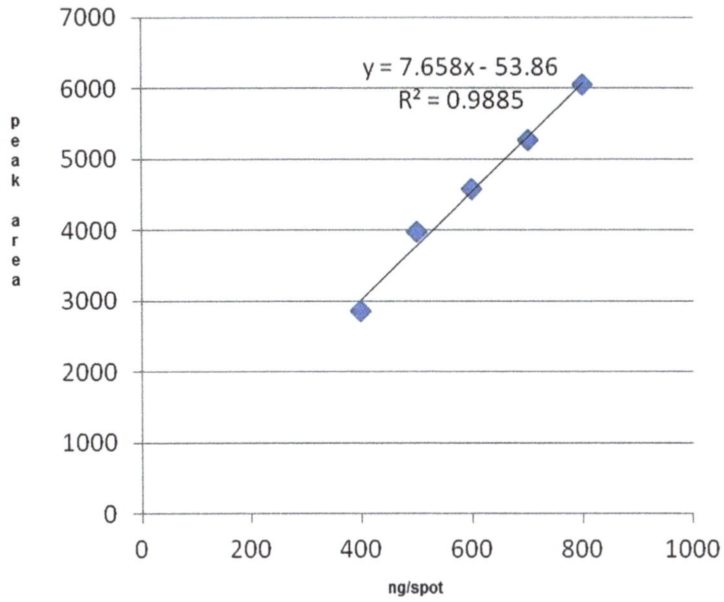

Fig. 17.3 Calibration curve of rutin at various concentrations (200–1000 ng/spot)

Table 17.2 Method validation parameters

Parameters	Rutin
Linearity range (ng/spot)[a]	200–1000
Correlation coefficient	0.9885
Slope	7.658
Intercept	53.86
Intraday precision (%RSD)[b]	1.37
Interday precision (%RSD)	1.46
LOD	5.24
LOQ	15.90
% recovery	99.5–101.12

[a]Five concentrations
[b]Nine determinations

Precision

The precision of the method was demonstrated by intraday and interday variation studies. Three different concentrations 400, 600 and 800 ng/spots of standard plant extract were analysed in a day for intraday variation and 3 consecutive days for interday variation studies. The results were determined by calculating % RSD.

Limit of Quantitation and Limit of Detection

From the calibration curve, the standard deviation (SD) of the intercept (slope) was determined. Then LOD was determined using Eq. 17.1.

$$\text{LOD} = 3.3\rho/S \quad (17.1)$$

and LOQ was determined using Eq. 17.2.

$$\text{LOQ} = 10\rho/S \quad (17.2)$$

where S is the slope of the calibration curve and ρ is the standard deviation of Y-intercept of regression line.

Accuracy

The accuracy of the method was determined by recovery experiments. The recovery studies were carried out at three levels of 80, 100 and 120%, and the percentage recovery was calculated as shown in Table 17.3.

Robustness

The robustness was determined by performing small changes in mobile phase ratio and chamber size. The effect on the R_f values and peak areas were noted.

Specificity

Specificity is the ability to access unequivocally the analyte in the presence of impurities and degradation products. It is done by comparing R_f value of samples with that of standard rutin.

Assay

A 0.025 mL aliquot of the test solution was made up to 10 mL with methanol. 3 μL was spotted on the plate for HPTLC assay. The assay was performed in triplicates and mean ± SD were calculated.

17.4.3 Quantitative Estimation of β-Sitosterol from Natural Plants

17.4.3.1 Preparation of Chloroform Fraction from Ethanolic Extract of Clerodendrum serratum

The powdered leaves as in Fig. 17.4 was considered, and its ethanolic extract was prepared by hot percolation technique using 95% ethanol at 55 °C. The percentage yield was found to be 28% w/w. 100 g of ethanolic remnant was taken with water and chloroform (1:1) in a separating funnel, and by liquid-liquid partitioning method, chloroform fraction was obtained which further concentrated on rotavapor

Table 17.3 Recovery studies

%Amount estimated	Amount present (ng/spot)	Amount recovered (ng/spot)	% Recovery ± SD
80	400	395.53	101.12 ± 1.40
100	400	398.25	99.55 ± 0.92
120	400	400.10	100.02 ± 0.21

Fig. 17.4 Leaves of *Clerodendrum serratum*

Fig. 17.5 Leaves part of *Amaranthus spinosus*

(Buchi, USA) at a temperature of <40 °C and then dried in lyophilizer (Labconco, USA) under reduced pressure. The percentage yield of fraction was 24% w/w.

17.4.4 Quantitative Estimation of Quercetin from Natural Plants

17.4.4.1 Preparation of Ethanolic Extract of Amaranthus spinosus

Powdered plant material (leaf) (2 g) as in Fig. 17.5 extracted with methanol/water (70:30 v/v) (3× 100 mL, 8 h each) was concentrated at low temperature by rotatory evaporation and was freeze-dried under high vacuum. The percentage yield was 2.3% w/w.

17.4.4.2 Preparation of Standard Solution of Quercetin

Dissolve 1 mg of pure compound (marker compound) in 1 mL of pure solvent methanol. This solution was considered as test solution.

17.4.4.3 Assessment of Quercetin from Amaranthus spinosus

10μL of the ethanolic solutions was applicated on precoated HPTLC plate; the plate was then developed in the solvent system (toluene/ethyl acetate/formic acid (5:6:2 v/v). The plate was then developed in the solvent system to a height of 8 cm, dried and scanned densitometrically at 368 nm. The percentage of quercetin was found to be 10.2% (w/w). The densitogram was shown in Fig. 17.6.

17.4.4.4 Preparation of Ethanolic Extract of Strychnos potatorum

Powdered plant material (seed) as shown in Fig. 17.7 (2 g) extracted with methanol/ water (70:30 v/v) (3× 100 mL, 8 h each) was concentrated at low temperature by rotatory evaporation and was freeze-dried under high vacuum. The percentage yield was 3.3% w/w.

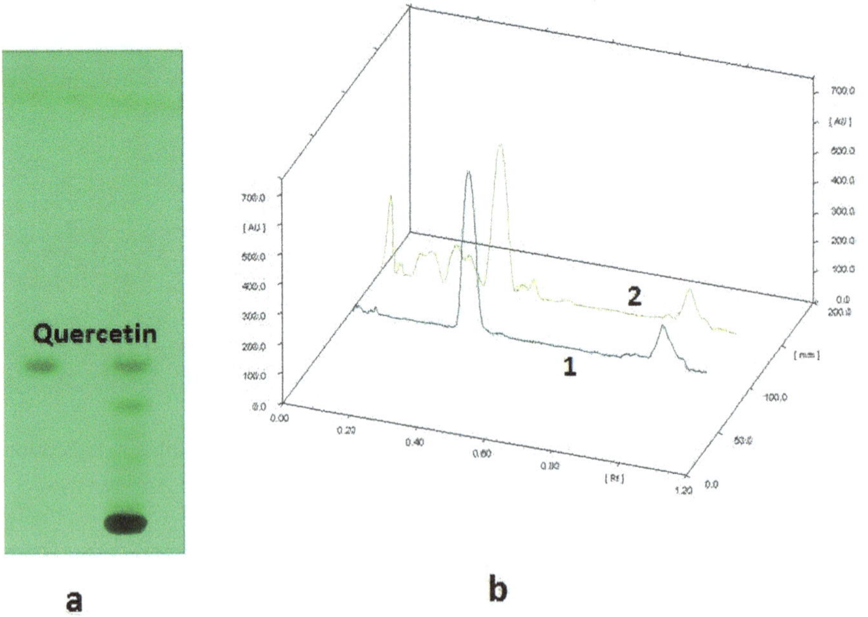

Fig. 17.6 HPTLC densitogram of (**a**) standard quercetin on HPTLC plate, (**b**) quercetin (1), ethanolic extract of *A. spinosus* (2)

Fig. 17.7 Seeds of *Strychnos potatorum*

17.4.4.5 Assessment of Quercetin from Strychnos potatorum

10μL of the standard quercetin solution was applicated on precoated HPTLC plate; the plate was then developed in the solvent system (toluene/ethyl acetate/formic acid (5:6:2). The plate was then developed in the solvent system to a height of 80 mm, dried and scanned densitometrically at 368 nm. The composition of the developing solvent was selected from the method previously developed. The percentage of quercetin was found to be 10.5% (w/w). The densitogram was shown in Fig. 17.8.

17.4.4.6 Assessment of Quercetin in Saraca asoca

Preparation of Ethanolic Extract of Saraca asoca

Powdered plant material (flower) as shown in Fig. 17.9 (2 g) extracted with methanol/:water (70:30v/v) (3× 100 mL, 8 h each) was concentrated at low temperature by rotatory evaporation and was freeze- dried under high vacuum. The percentage yield was 2.9% w/w.

Assessment of Quercetin from Saraca asoca

10μL of standard quercetin solution was applicated on precoated HPTLC plate; the plate was then developed in the solvent system (toluene/ethyl acetate/formic acid (5:6:2). The plate was then developed in the solvent system to a height of 8 cm, dried and scanned densitometrically at 368 nm. The percentage of quercetin was found to be 0.18% (w/w) Fig. 17.10.

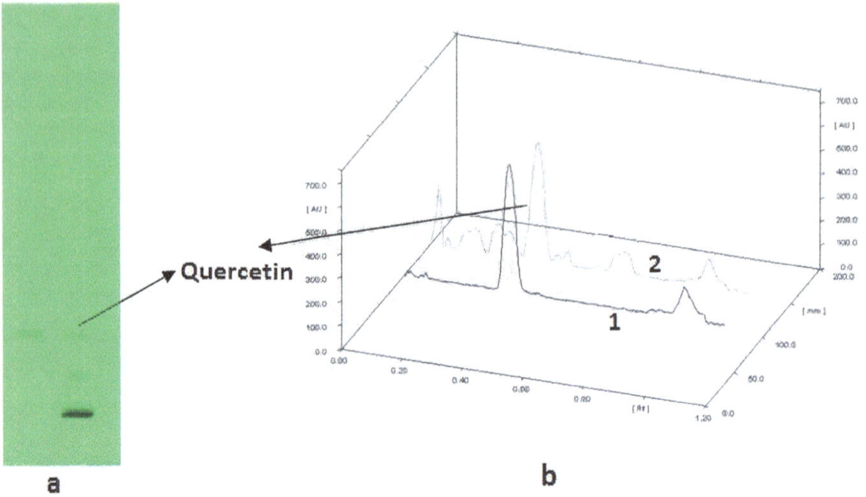

Fig. 17.8 HPTLC profile (**a**) densitogram of HPTLC plates and (**b**) densitogram of quercetin (1) and densitogram of ethanolic extract of seeds of *Strychnos potatorum* (2)

Fig. 17.9 Flower of *Saraca asoca*

17.4.5 Qualitative Assessment of Various Extracts of Traditional Plants

17.4.5.1 Preparation of 50% w/v Ethanolic Extract of Plants

Powdered plant material (2 g) extracted with methanol:water (70:30 v/v) (3× 100 mL, 8 h each), concentrate at low temperature by rotatory evaporation and was freeze dried under high vaccum.

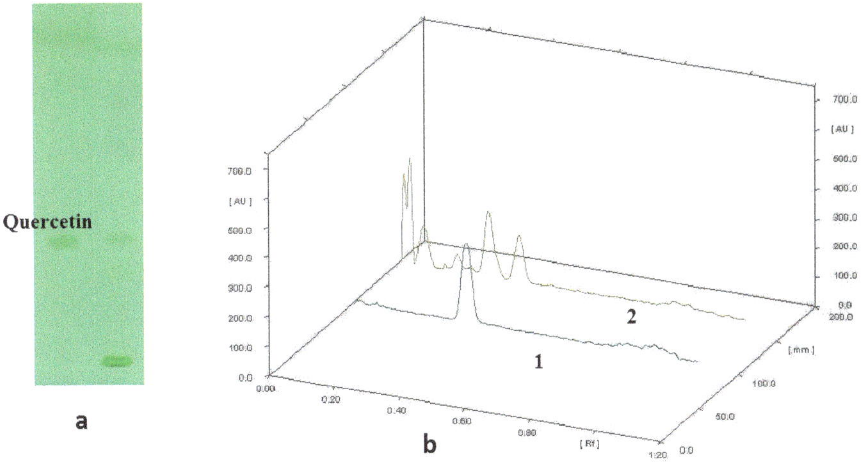

Fig. 17.10 HPTLC densitometric profile of *Saraca asoca* (**a**) HPTLC densitogram plate, (**b**) quercetin standard (1) HPTLC densitogram of methanolic extract of *Saraca asoca* (2)

17.4.5.2 Preparation of 50% w/v Aqueous Extract

Powdered plant material (2 g) extracted with water by cold percolation method. The aqueous extract was concentrated at low temperature by rotatory evaporation and was freeze dried under high vaccum. The percentage yield was 3.4% w/w.

17.4.5.3 HPTLC Fingerprinting of the Extract of Manilkara hexandra (Stem Bark)

The stem bark of the plant was shown in Fig. 17.11. The ethanolic extract of stem bark (8 μL) solution was applicated by Linomat 5 Applicator (CAMAG) on precoated HPTLC plate; the plate was then developed in the solvent system toluene/ethyl acetate/acetone/formic acid (10:5:15:1). The plate was then developed in the solvent system to a height of 8 cm, dried and scanned densitometrically at 254 nm, the peak area was recorded, and the calibration curve was prepared by plotting the peak area against concentration of the component. The HPTLC profile of *M. hexandra* revealed the presence of eight spots at R_f 0.09 (60.14%), 0.26 (22.24%), 0.55 (3.17%), 0.65 (5.07%), 0.85 (1.28%), 1.04 (1.82%), 1.15 (2.68%) and 1.35 (3.60%). The maximum percentage of area of component in the extract was found to be 60.14% w/w at R_f 0.09. The densitogram was shown in Fig. 17.12.

17.4.5.4 HPTLC Fingerprinting of Ethanolic Extract of Tecomella undulata (Stem Bark)

The stem bark of the plant was shown in Fig. 17.13. The ethanolic extract of stem bark (12 μL) was applied on precoated HPTLC plate; the plate was then developed in the solvent system toluene/ethyl acetate/acetic acid (55:45:2). The plate was then developed in the solvent system to a height of 8 cm, dried and scanned densitometrically at 254 nm, the peak area was recorded, and the calibration curve

Fig. 17.11 Stem bark of *M. hexandra*

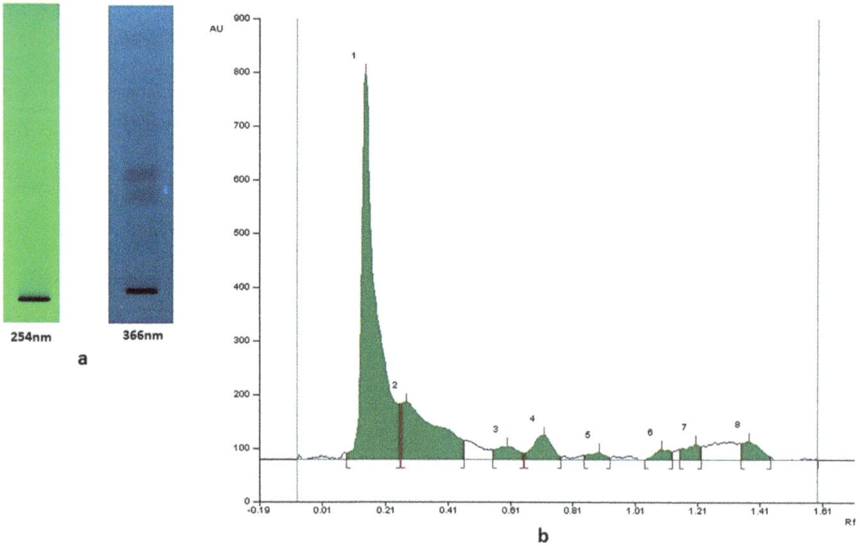

Fig. 17.12 Densitogram of ethanolic extract of *M. hexandra* (**a**) HPTLC plates at different wavelengths, (**b**) HPTLC profile of *M. hexandra*

was prepared by plotting the peak area against concentration of the component. The HPTLC profile of *T. undulata* revealed the presence of 11 spots at R_f 0.08 (55.44%), 0.25 (15.29%), 0.35 (5.09%), 0.44 (2.21%), 0.55 (2.05%), 0.72 (1.32%), 0.81 (2.45%), 0.94 (0.76%), 1.02 (2.10%), 1.10 (11.39%) and 1.26 (1.91%). The maximum percentage of area of component in the extract was found to be 55.44% w/w at R_f 0.08. The densitogram was shown in Fig. 17.14.

Fig. 17.13 Stem bark of *Tecomella undulata*

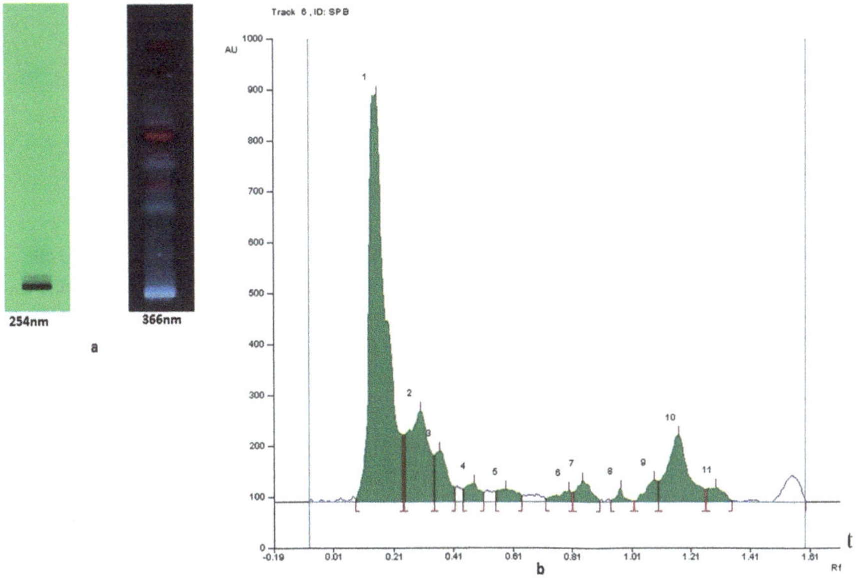

Fig. 17.14 Densitogram of ethanolic extract of *T. undulata* (**a**) HPTLC plates at different wavelengths, (**b**) HPTLC profile of *T. undulata*

17.4.5.5 HPTLC Fingerprinting Profile of the Extract of Zanthoxylum armatum (Stem Bark)

The stem bark of *Zanthoxylum armatum* was shown in Fig. 17.15. The ethanolic extract (12 μL) of stem bark was applied on precoated HPTLC plate; the plate was then developed in the solvent system toluene/ethyl acetate/formic acid (7:3:0.1). The plate was then developed in pre-saturated CAMAG twin trough chamber with solvent system to a height of 8 cm, dried and scanned densitometrically at 254 nm, the peak area was recorded, and the calibration curve was prepared by plotting the peak area against concentration of the component. The HPTLC profile of *Z. armatum* revealed the presence of 13 spots at R_f 0.05 (17.46%), 0.12 (10.08%), 0.17 (3.17%), 0.21 (2.15%), 0.25 (2.99%), 0.29 (3.16%), 0.33 (3.88%), 0.37 (6.56%), 0.42 (5.19%), 0.46 (11.11%), 0.54 (4.07%), 0.65 (25.46%) and 0.74 (4.71%). The maximum percentage of area of component in the extract was found to be 25.46% w/w at R_f 0.65. The densitogram was shown in Fig. 17.16.

17.4.5.6 HPTLC Fingerprinting Profile of Extract of Strychnos potatorum

2 g of ethanolic extract of *S. potatorum* was dissolved in 25 mL of methanol separately and used for HPTLC analysis. 10 μL of the above solution was applied, and the plates were then developed in toluene/ethyl acetate (7:3) to a distance of 8 cm. After development, the plates were dried and densitometrically scanned at 376 nm. The HPTLC profile of extract in solvent system toluene/ethyl acetate (7:3) revealed the presence of 14 spots at R_f 0.05 (8.54%), 0.26 (4.88%), 0.34 (10.26%), 0.37 (6.35%), 0.44 (18.60%), 0.48 (15.01%), 0.55 (6.07%), 0.59 (3.96%), 0.63 (8.50%), 0.71 (2.55%), 0.73 (2.80%), 0.78 (3.18%), 0.81 (3.92%) and 0.93 (5.39%); the maximum percentage were found to be of components 5 and 6 at R_f 0.44 (18.60%) and 0.48 (15.01%), respectively. The densitogram was shown in Fig. 17.17.

17.4.5.7 HPTLC Fingerprinting Profile of the Extract of Saraca asoca

10 mg of ethanolic extract was dissolved in 1.0 mL of 80% aqueous methanol and used for HPTLC analysis. 10μL of the above solution was applied, and the plates were then developed in ethyl acetate/acetic acid/formic acid/methanol (7.5:1:0.5:1) to a distance of 9 cm. Developed plate was then observed under visible light after

Fig. 17.15 Stem bark of *Zanthoxylum armatum*

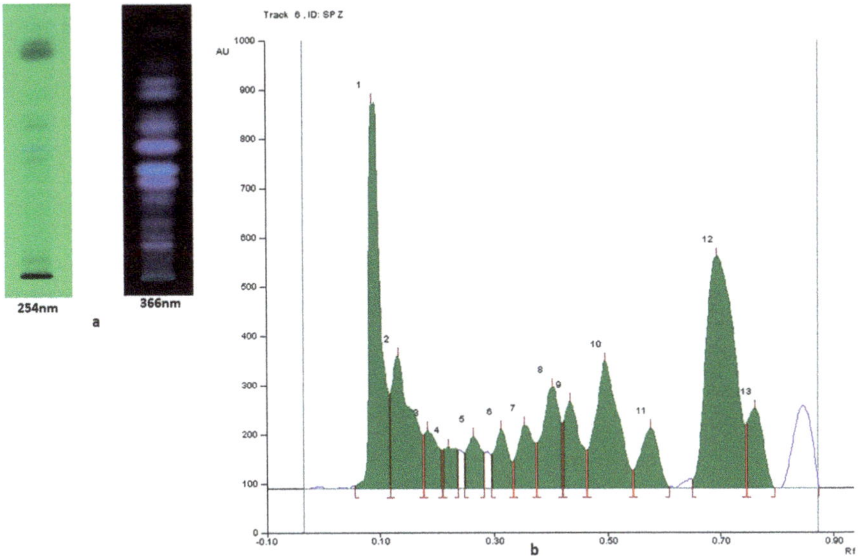

Fig. 17.16 Densitogram of ethanolic extract of *Z. armatum* (**a**) HPTLC plates at different wavelengths, (**b**) HPTLC profile of *Z. armatum*

derivatization with anisaldehyde sulphuric acid reagent and scanned densitometrically at 254 nm which showed the presence of nine spots at R_f of 0.04, 0.10, 0.15, 0.23, 0.52, 0.58, 0.70, 0.78 and 0.92. The densitogram is shown in Fig. 17.18.

17.4.5.8 HPTLC Fingerprinting of Ethanolic Extract of Amaranthus spinosus

10 μL of the ethanolic extract of *Amaranthus spinosus* was applicated on precoated HPTLC plate; the plate was then developed in the solvent system toluene/ethyl acetate (7: 3). The plate was then developed in the solvent system to a height of 8 cm, dried and scanned densitometrically at 310 nm. The HPTLC profile of *A. spinosus* revealed the presence of nine spots at R_f 0.19 (3.09%), 0.29 (1.26%), 0.40 (8.01%), 0.47 (11.76%), 0.57 (10.72%), 0.63 (4.60%), 0.66 (1.92%), 0.75 (2.55%) and 0.93 (40.23%). The maximum percentage of area of component in the extract was found to be 40.23% w/w at R_f 0.93. The densitogram was shown in Fig. 17.19.

17.4.5.9 HPTLC Fingerprinting Profile of Extract of Thespesia lampas (Flower)

10μL of the aqueous extract solution was applied, and the plates were then developed in toluene/ethyl acetate (7:3 v/v) to a distance of 9 cm. Developed plate was then observed under visible light after derivatization with anisaldehyde sulphuric acid reagent and scanned densitometrically at 310 nm which showed the presence of nine spots at R_f of 0.04, 0.10, 0.15, 0.23, 0.52, 0.58, 0.70, 0.78 and 0.92. The

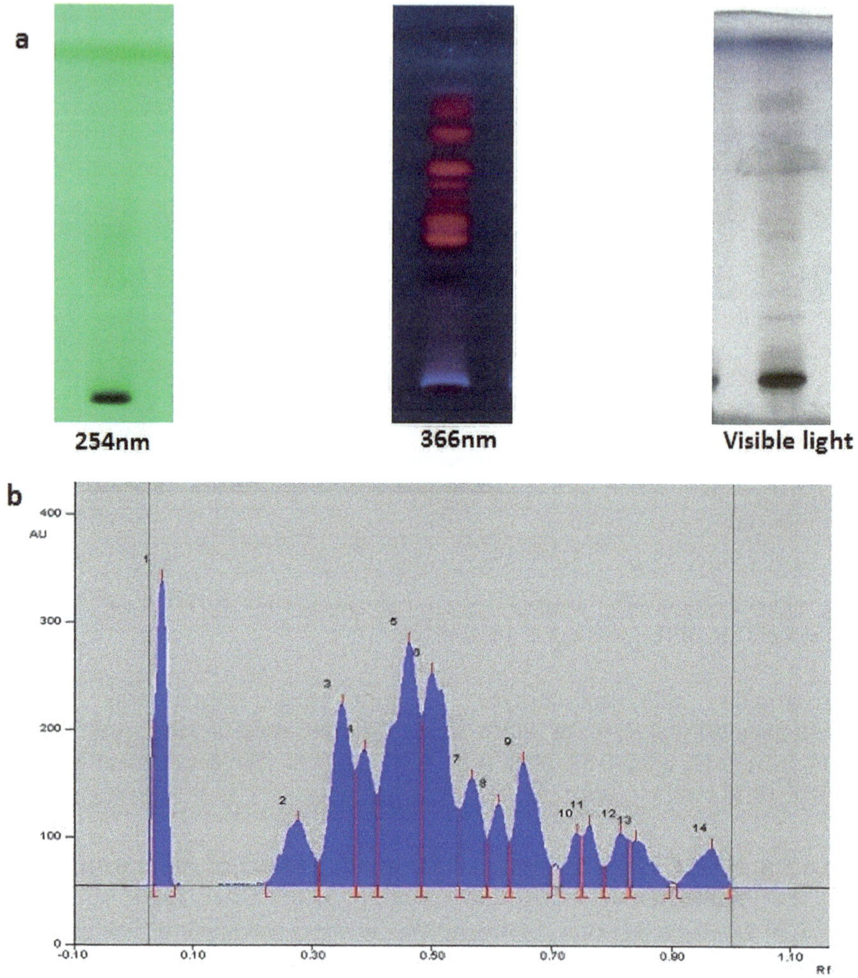

Fig. 17.17 Densitogram of ethanolic extract of *Strychnos potatorum* (**a**) HPTLC plates at various wavelengths, (**b**) HPTLC fingerprint profile of *Strychnos potatorum*

densitogram is shown in Fig. 17.11.The HPTLC profile of the plant revealed the presence of 11 spots at R_f 0.20 (4.15%), 0.33 (3.86%), 0.38 (6.05%), 0.47 (6.89%), 0.51 (2.41%), 0.57 (10.28%), 0.61 (7.67%), 0.67 (2.72%), 0.68 (1.13%), 0.75 (3.64%) and 0.92 (38.92%). The maximum percentage was found to be of component 11 at R_f 0.92 (38.92%) as shown in Fig. 17.20.

17.4.5.10 HPTLC Fingerprinting of Ethyl Acetate Fraction of Alhagi pseudalhagi

100 g of ethanolic residue was taken in separating funnel. 100 mL of ethyl acetate was considered with 100 mL distilled water and mixed for 15–20 min. The ethyl

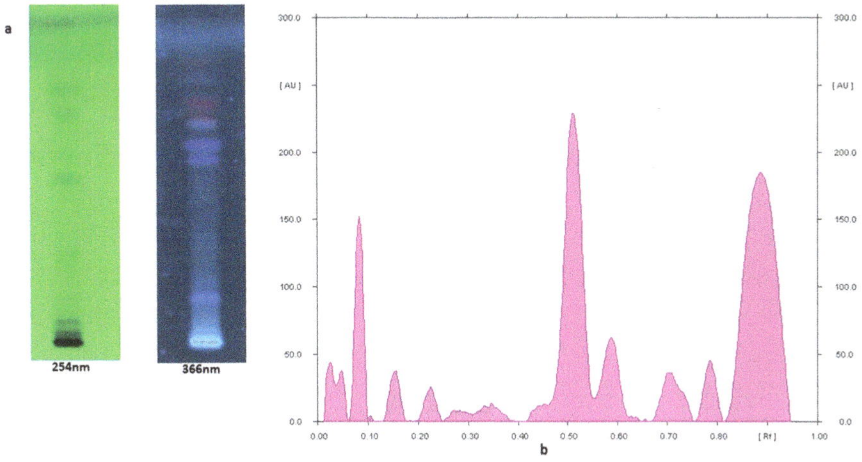

Fig. 17.18 Densitogram of ethanolic extract of *Saraca asoca* (**a**) HPTLC plates at various wavelengths, (**b**) HPTLC fingerprint profile of *Saraca asoca*

acetate section was above the aqueous layer. The ethyl acetate was subjected to aridness in an evaporating dish. 25.5 mg of dried residue of ethyl acetate fraction was dissolved in 1 mL methanol (solution 'a'); 1 mL from solution 'a' was dissolved in 1 mL of methanol (solution 'b'). From solution 'b', 10 μL of test solution was spotted on the HPTLC plates (20 × 20 cm). The plates were developed in the solvent system ethyl acetate/acetic acid/formic acid/water (10:1:1:1 v/v/v/v). The plate was then developed in the solvent system to a height of 8 cm, dried and scanned densitometrically at 366 nm.

The HPTLC profile of ethyl acetate fraction in the solvent system ethyl acetate/formic acid/acetic acid/water (10:1:1:1 v/v/v/v) revealed the presence of seven spots at R_f 0.07 (0.71%), 0.19 (1.55%), 0.36 (7.83%), 0.43 (8.19%), 0.49 (2.23%), 0.63 (3.26%) and 0.71(2.78%). The maximum percentage was found to be component 4 at R_f 0.43 (8.19%). The details were shown in Fig. 17.21.

17.5 Conclusion

With the help of special sampling techniques, quantification and separation of various constituents in herbal plant are possible by densitometry. Most of the other techniques including infrared spectroscopy and high-performance liquid chromatography were limited to examining the plant microscopically or comparing and quantifying the known separated markers. HPTLC allows for the standardization and validation of botanicals by comparing and quantifying different marker compounds simultaneously and figuring out the similarities and dissimilarities between them.

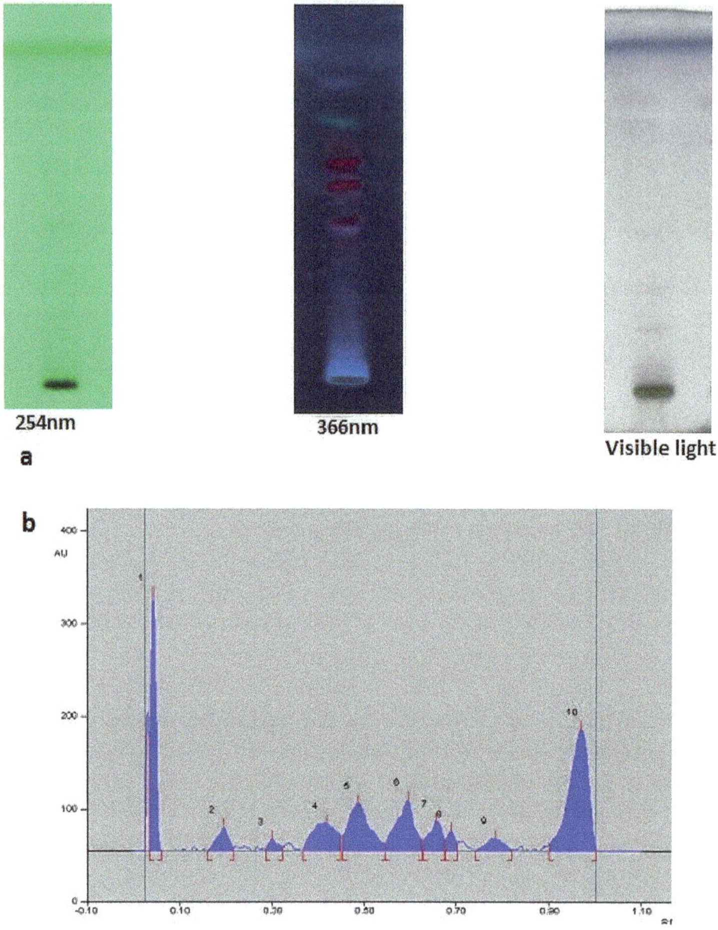

Fig. 17.19 Densitogram of ethanolic extract of *A. spinosus* (**a**) HPTLC plates at various wavelengths, (**b**) HPTLC fingerprint profile of *A. spinosus*

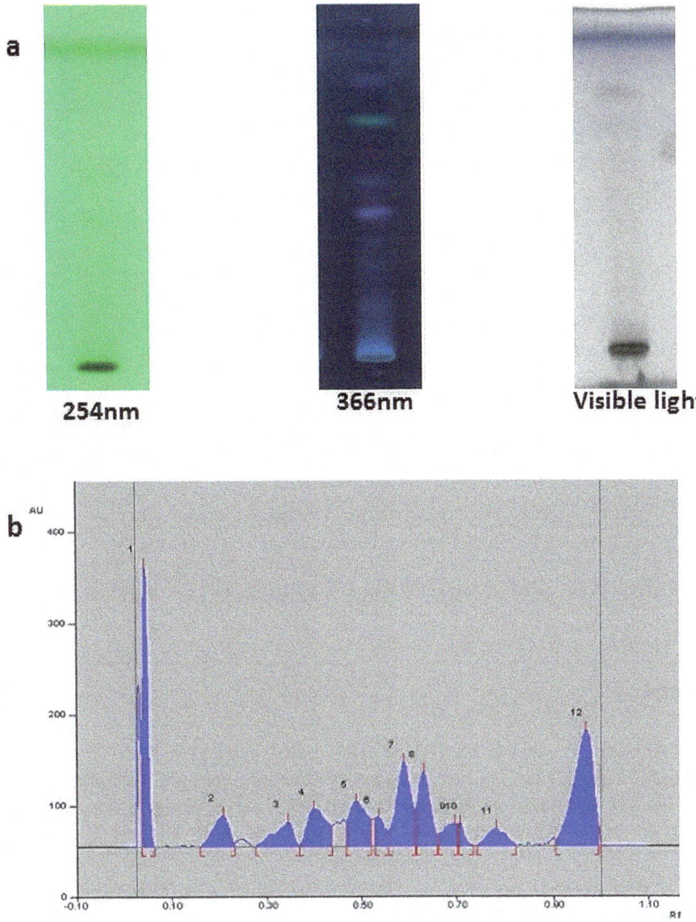

Fig. 17.20 Densitogram of aqueous extract of *T. lampas* (**a**) HPTLC plates at various wavelengths, (**b**) HPTLC fingerprint profile of *T. lampas*

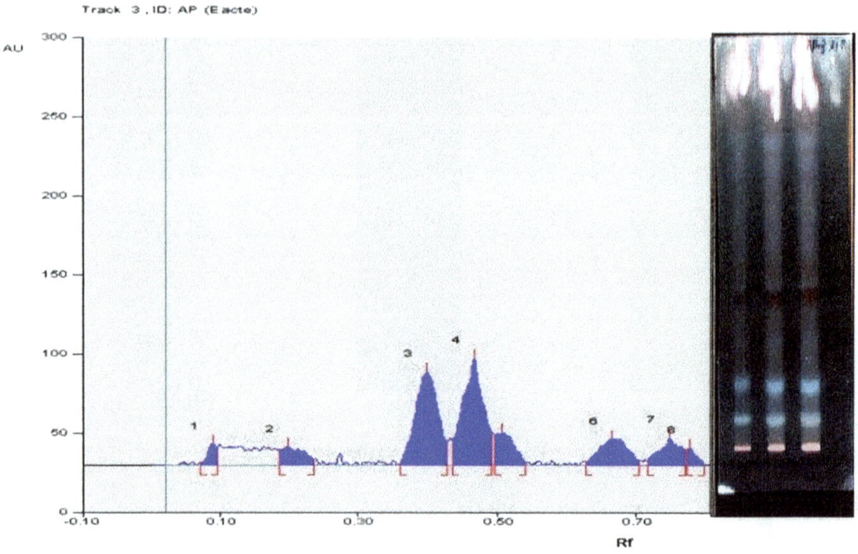

Fig. 17.21 Densitogram of ethyl acetate fraction of ethanolic extract of *Alhagi pseudalhagi*

References

Chavan MJ, Shinde DB, Wakte PS (2011) High-performance thin layer chromatography. In: Srivastava MM (ed) HPTLC in herbal drug quantification. Springer, Heidelberg, pp 118–120

Ekor M (2014) The growing use of herbal medicines: issues relating to adverse reactions and challenges in monitoring safety. Front Pharmacol 4:177. https://doi.org/10.3389/fphar.2013.00177

Geethika K, Sunojkumar P (2017) Phytochemical screening and high-performance thin-layer chromatography fingerprint profile of three species of *Leucas* (Lamiaceae). Ancient Sci Life 37(2):102–107

Kokate CK, Purohit AP, Gokhale SB (2008) Pharmacognosy. Nirali Prakashan, India

Climate Change, Geographical Location, and Other Allied Triggering Factors Modulate the Standardization and Characterization of Traditional Medicinal Plants: A Challenge and Prospect for Phyto-drug Development

18

Partha Palit and Subhash C. Mandal

Abstract

Evidenced-based traditional plants are believed to be a prospective and alternative source of safe and effective drug development both in developed and developing countries for fruitful remedy against acute and chronic diseases, as because mainline allopathic medicines demonstrate versatile untoward effect and become ineffective in the treatment of chronic therapy.

In order to fulfil this motto, large number of traditional plants needs to be cultivated in homogeneous environmental condition, favorable geographical and climatic condition for optimise the yield and increase the production rate of the plant metabolites. Satisfactory climate, good quality of soil, and advantageous geographical condition yield huge number of cultivated medicinal plants with suitable growth. By maintaining such parameters of cultivation protocol, the farmers and cultivators may be capable to give homogeneous and optimized phytoconstituents containing crude extract for herbal drug discovery. However, global changing of climate is modulating the weather condition of the same geographical condition, which in turns drastically jeopardize the secondary metabolite production rate, their composition in the organized medicinal plant extracts. Vis-a-vis, weather condition, rainfall, and humidity are versatile in different geographical locations, which could change the phytometabolites composition and yield when the same species medicinal plant is growing in different geographical locations. These variations are augmented due to difference in the

P. Palit (✉)
Drug Discovery Research Laboratory, Division of Pharmacognosy, Department of Pharmaceutical Sciences, Assam University (A Central University), Silchar, India
e-mail: itspartha_p@yahoo.com

S. C. Mandal
Department of Pharmaceutical Technology, Pharmacognosy and Phytotherapy Research Laboratory, Jadavpur University, Kolkata, India

S. C. Mandal et al. (eds.), *Evidence Based Validation of Traditional Medicines*,
https://doi.org/10.1007/978-981-15-8127-4_18

soil quality, collection time, and dispensation methods of cultivated medicinal plants for secondary metabolites extraction. These consequences are resulting in the huge discrepancy in the reproducibility of bioactive phytometabolites extractive yield and composition from the same species of medicinal plant. Therefore, patient may not obtain uniform and optimized bioactivity from the same herbal formulations from different batch. In order to attenuate this challenges and hurdles, different standardization techniques and characterization method with optimized fingerprinting technology based on marker compounds are urgently required. Present review will be discussed critically the different issues related with above-cited challenges and their probable solution to optimize the lead and traditional herbal formulation standardization techniques for establishing the smooth avenue of medicinal plant-based drug discovery program.

Keywords

Plant metabolites · Climatic factors · Collection place and time · Processing · Standardization · Annual variation

Abbreviation

ADR Adverse drug reaction
PSM Plant secondary metabolites

18.1 Introduction

There is no doubt that evidenced-based traditional folk medicinal plants are promising resource of modern ayurvedic and natural medicine development. This exercise and culture were introduced in India since the era of Charak and Sushruta's ayurvedic practice. Different tribal folk people use such medicinal and aromatic plants for mitigating and managing their own physical illness and lifestyle diseases as therapeutic and prophylactic phytomedicine (Atanasov et al. 2015), although those ethnic plants are being cultivated in their own flora and fauna-based biodiversified geographical area. That is why, those plants are very popular for phytomedicine development at local tribal zone. Apart from Indian traditional folk medicinal plants, Chinese and African people also rely on the evidenced-based traditional plants for therapeutic medicine and food development on their daily life (Kayne 2010). They utilize the ethnic plants as prophylactic along with preventive agents to combat the disease progression and its subsequent development, too (Pan 2014). Several advantages have been attributed for using the traditional medicinal plants as alternative sources toward modern natural drug discovery process over allopathic drugs in the developing and developed countries. Easy availability in copious amount, lesser side effect, lower chances of ADR, high therapeutic index with higher safety level, and synergistic action are key and advantageous factors to make the traditional plant-based drugs very impressive and acceptable.

In order to manufacture modern herbal and medicinal plant-based natural drugs, scientific cultivation, processing, and extraction is mandatory (Sachan et al. 2016). Moreover, scientific validation through preclinical biological evaluation in animal model as well as clinical trial needs to be carried out prior to marketing toward therapy against human diseases. This pharmacological effect of the herbal formulation derived from the traditional folk medicinal plants is very much dependent on the phytochemical composition and important marker metabolite's pharmacophore scaffolds. This bioactive metabolite's yield and composition are varied, if the collection time, processing, and method of extraction are not homogeneous. Furthermore, environmental weather change and soil quality due to geographical location variation also influence the biological effect of the same plant extract from batch to batch formulation due to altered metabolite's composition and percentage of yield. In this aspect, the characterization and standardization of the raw medicinal plants is very difficult to optimize the outcome biological effect (Gupta et al. 2019). Nowadays the natural drug discovery process from traditional herbs and folk plants are facing numerous hurdles. However, the development of bioprospecting and eco-friendly traditional herbal formulation in the context of modern pharmacognosy is very successful approach. Hence, we are encouraged to write this present book chapter to address the issue of climate change, geographical location variation, and other allied factors on the cultivation processing and pharmacological activity of traditional folk plants. We also highlight how could those factors influence the agricultural cultivation of same plants in respect to time and place. We try to emphasize how could the variation of biological impact of ayurvedic formulation batch to batch be solved out successfully. We have described the matter related to fix up the climate change, geographical location of the cultivating zone of traditional plants, collection time, the way of processing of phytometabolites, etc. to standardize the raw medicinal plants for obtaining the safe and effective biological effect of traditional plant extract formulation (Das et al. 2016).

18.2 Threatening Factors for Biodiversity and Ecosystem Malfunctioning on Traditional Medicinal Plant Cultivation and Bioactive Phytometabolite Production

18.2.1 Description of Different Triggering Factors Which Influence the Cultivation, Growth, and Active Secondary Metabolite Productions

18.2.1.1 Climatic Change in Local Cultivating Zone and Growth of Medicinal Plants

Climate and weather change intensely influence the cultivation and growth of active medicinal plants, because this climate change in connection to global warming (Rivaes et al. 2013) and environmental pollution is serious threat to mankind and plant kingdom for their survival and growth across the globe (Lotter and le Maitre 2014). Literature suggested that unfavorable climate may adversely affect the biosynthetic pathway for primary and secondary metabolite production in medicinal plant, for which the yield and composition of primary and secondary metabolites will be varied from time to time (Zhang et al. 2019), because plants produce active metabolites in their tissues via shikimic acid and mevalonic acid pathway. In order to biosynthesize successfully those several bioactive metabolites via biochemical reaction, plants need to be exposed into the ideal weather condition such as favorable rainfall, humidity, air temperature, atmospheric vapor pressure, precipitation, sunshine duration, etc. (Verma and Shukla 2015; He et al. 2013;Liu et al. 2011). Furthermore, the weather condition of the same place nowadays has been drastically altered in comparison to earlier time (Li 2019). This phenomenon may modify the reproducibility of percentage of total extractive matter comprising of bioactive phytoconstituents from traditional medicinal plants, because the alteration of such climatic factors completely changes the rate and nature of biochemical reaction of plant biosynthetic pathway, resulting in the lack of reproducibility of phytoconstituents composition. Thus, the ratio of the different phytopharmaceuticals derived from the traditional medicinal plant extract and their modified yield may revise the pharmaceutical property including therapeutic efficacy of the same herbal formulation batch to batch (Lotter and le Maitre 2014). This outcome is very alarming threat to the natural drug discovery process due to global climate change in intolerable manner. The variation of the phytochemical composition in the crude extract in connection to changing climatic condition need to be solved to optimize or standardize the phytoextract in proper and scientific manner; otherwise dose-dependent pharmacological effect of the same plant extract in constant dose from different batches would not be achieved. In order to fix up this issue, the traditional plants cultivating and grown up in respect to present altered climate need to be standardized with proper chromatographic fingerprint characterization and investigated their changed phytochemical composition for subsequent fixed herbal dose formulation development. The revised standardized data including marker compound's yield need to be preserved as reference results in phytochemical library for future reference. This approach could help and guide the pharmaceutical scientist to produce reproducible herbal formulation with optimized therapeutic effect and improved safety level.

18.2.1.2 Impact of Geographical Variation Location-Wise on Medicinal Plant's Cultivation and Phytoconstituents Yield

Climatic scenario is variable in connection to changing of geographical location. Therefore, understanding medicinal plant's favored territory is of great significance for their cultivation, conservation, artificial cultivation, and assessment of their biological importance, because changing of geographical location may vary the different climatic factors extremely, which could influence the phytometabolic pathway of plants as we discussed above. Consequently, the same plant species with identical taxonomical classification may change their phytometabolite composition, if they cultivate in different zones (Verma and Shukla 2015; He et al. 2013; Liu et al. 2011). Apart from climatic change, the high water and nutrient supply for rhizome and root production in the cultivating soil are also most important factors for increasing rate of cultivation and growth for medicinal plant (Bernath 1997).The soil nutrient composition and groundwater content are varied by changing of cultivating zone. Therefore, the same species plant may change their phytochemical composition, anabolism, and accumulation of secondary metabolites, if it grows in different geographical cultivating zone across the world (Radusiene et al. 2012).

It is very tedious to standardize the composition, identity, and degree of phytochemical quality of medicinal plant extract for herbal formulation development. Therefore, we may postulate that strong relationships are coexisting between the specific and favorable ecological environments and active phytometabolite production of medicinal plants. Earlier reports suggest that high water content in the ground soil may increase the total polyphenolic acid yield of the plants. Excessive temperature may reduce the proportion of bioactive flavonoids and anti-inflammatory tanshinone contents in the plant extract like *S. miltiorrhiza* (Jayabalan et al. 2007; Zhang et al. 2019). Thus, issue related to geographical location variation need to be solved out very carefully during cultivation, processing, and supply of traditional plants as raw herbs for industrial herbal formulation development. Moreover, the supplier should intimate the industrial analytical pharmaceutical chemist regarding the details of collection region for carrying out proper standardization and characterization of raw medicinal herbs. Thus, the effect of extreme climatic factor variation on active ingredients production by the medicinal plants could be rationalized for future reference data. On the other hand, farmers and analytical chemists should identify the suitable habitat for cultivation of each species of medicinal plants, so that favorable climatic factors could trigger relatively the stable growth for those cultivated plants species. These investigative studies may provide the practical guidance for proper exploration of cultivating region against the suitable medicinal plant growth.

18.2.1.3 Impact of Collection Time on Medicinal Plant Metabolite Production

Collection time is one of the key regulatory factors for plants primary and secondary metabolite production rate and their total yield content, which also affects the % of active marker compound content in the extract. The key marker phytocompound are mainly responsible for biological activity of the crude ayurvedic formulation. Its content along with other secondary metabolites composition in plants may fluctuate due to seasonal and daily variation of plucking time despite other climatic, genetic factors being constant (Gobbo-Neto and Lopes 2007; Wallaart et al. 2000). Report suggests that opium poppy (*P. somniferum*) contains in excess of 80 alkaloids in which morphine, codeine, thebaine, narcotine, and papaverine are lead marker compounds and widely used for pharmacological therapeutic efficacy. The contents of those alkaloids have varied from season to season during developmental stages. Moreover, bacopaside and bacoside a lead biomarker molecule (as anti-dementia molecules) content is maximum yield, if brahmi (*Bacopa monnieri*) leaves are plucked in August months (Munish et al. 2013).

If *Muhlenbergia glomerata* leaves are collected during the sunny daytime, then the lead bioactive coumarin content would be obtained as larger amount. Because longer sunlight in daytime favors the significant accumulation of coumarins in *M. glomerata* leaves and stem. The coumarin content accumulation is significantly affected by the photoperiod and sunlight intensity in leaves and stems (Castro et al. 2006). Similarly, *Digitalis* leaf's cardiotonic glycoside digitoxin A, B, and C would be obtained as highest quantity, if the leaves are collected in daytime in presence of sunny light, because sunlight triggers the digitoxin biosynthetic pathway production rate. So, the light intensity and photoperiod has significant role on the accumulation and biosynthetic production of primary and secondary metabolites (Kokate et al. 2008; Jacobsohn and Jacobsohn 1976).

Edible shoots of Buck's-beard [*Aruncus dioicus* (Walter) (Rosaceae)] contain the cyanogenic compound prunasin, if the leaves of the plants are plucked in vegetative stages during plant development. Interestingly shoots could not be used for cooking purposes any longer, after first mature green development (Fusani et al. 2016).

Henceforth, the optimum harvesting time of the primary and secondary metabolites, diurnal variation, optimized annual seasonal collection time, and exact time of plucking of crude drugs during the plant developmental stages need to be taken simultaneously into full consideration; otherwise the active ingredient's composition or yield would be different from reference data and may transform or degrade. It might be due to modification in the enzyme activity in metabolite biosynthetic pathway following an irreversible path, so once converted into another compound, previous compound becomes absent from the tissues (Shukla et al. 1996, 2006). Moreover, commercial supplier should provide reliable and exact data regarding the above-cited factors to the industrial chemist and lab research during the supply of any bioactive traditional plants or herbs as raw materials. If the chemist and researchers obtain satisfactory data regarding the collection profiles from supplier, then the impressive standardization may be followed for future reference, and patient may get reproducible therapeutic effect irrespective of batches of the formulation.

18.2.1.4 Impact of Soil Quality on Production of Secondary Metabolite Production

Another important abiotic factor, i.e., soil quality, also regulates the production, yield, and quantity of primary as well as secondary metabolite and their accumulation in the respective plant-organized tissues. Favorable soil quality for cultivation and growth of the plants functionally depends on the soil salinity, flooding, and low soil temperature, soil composition including mineral nutrients, heavy metals, soil pH, soil microbial actinomycete diversity, etc. Earlier reports suggest that low soil moisture and water contents reduce the bioactive anti-oxidant polyphenolic yield in tea beverages. The soil water percentage, i.e., 54 in respect to field capacity, may render the optimum level of aromatic phytoconstituents in the tea extract. Increasing level of aluminum and selenium in the soil may influence the catechin, polyphenols, and essential amino acid production positively (Ahmed et al. 2019). Earlier reports suggested that eco-friendly actinomycete bacteria-enriched soil profoundly increased the production rate of secondary metabolites from the economically important cornflakes (wheat, barley, oat, maize, and sorghum) and their photosynthesis, growth, and grain yield. This type of fertile soil improved vitamins, amino acids, and organic acids contents in grains through novel approach of organic farming (Hozzein et al. 2019). High concentration Na^+ in soil causes soil stress by increasing salinity that may uprise the primary or secondary metabolite accumulation in the plant tissue. Because report supported that *Solanum nigrum* (Solasodine), *C. roseus* (indole alkaloids), and *Achillea fragrantissima* demonstrated noteworthy rise in alkaloid content in the plants exposed to salt stress. The increase in the content of phenolic acid in the plants of *A. fragrantissima* was observed too under soil salinity-induced stress condition (Abd El-Azim and Ahmed 2009; Said-Al Ahl and Omer 2011). Mineral contents like soil phosphorous (Nell et al. 2009) and aluminum (Khalid 2006) enhance the PSM contents and their tissue accumulation like hypericum's flavonoid and total phenolic and rosmarinic acid (RA) concentrations of holy basil through modulating the enzymatic activity of plant biosynthetic pathway.

Furthermore, elevated soil pH significantly could reduce the chances of accumulation and up taking of toxic heavy metals like Ni, Cu, Pb, Cd, mercury, and arsenic in edible vegetables. This could be very detrimental for consumption and create enormous health hazards (Bai et al. 2016; Hu et al. 2017); on the other hand, soil micronutrients like Fe, Zn, Cu, Ca, Mg, Se, etc. may enhance the plant metabolite production of protein and polyphenols, whereas soil ammonia and phosphorous increase the production rate of flavonoids and alkaloids in plant tissue (Shitole and Dhumal 2012; Azizi and Dias 2004).

Soil water is key modulator for optimum biosynthesis and transportation of PSM and nutrients through the plant parts. Under drought condition or deficiency of groundwater soil, plants face the aqueous scarcity limiting the transpiration and up taking of water into plant tissue to carry out the enzymatic biosynthesis of PSM and other nutrients. It normally hampers the plant physiological process; plant growth and photosynthesis may be due to stomatal blockage, disrupted ATP synthesis, and membrane damage. Thus, it reduces the PSM and nutrient production in plants due to lowered photosynthetic rate. Scientific reports suggest that aqueous stress leads to reduce the essential oil contents of plants, total flavonoids content in *G. longituba*, hypericin, and pseudohypericin content in *H. perforatum* (Razmjoo et al. 2008; Zhang 2012). However, water scarcity under drought stress increases the amount of few PSM like artemisinin, botulinic acid, quercetin, and rutin in bioactive herbs (*Artemisia annua*, *Hypericum brasiliense*) (Zobayed et al. 2007).

The study illustrates that ideal soil quality with optimized soil pH, soil nutrient composition, soil water capacity and its content, and salinity should be considered for eco-friendly and productive harvesting of herbs, edible vegetables, and crops for obtaining optimized active constituents. Otherwise the after-formulation development and the active constituent's composition would be diversified in fixed dose from batch to batch that could lead to undesirable biological effect at reference therapeutic dose.

18.2.1.5 Impact of Processing and Extraction Method on Plant Secondary Metabolites

After collection of raw herbs, crops, and vegetables, the way how those are processed and extracted are important triggering factors for obtaining the optimum standard composition of active constituents for natural drug and nutraceuticals development. The effective health benefits and biological effects of the herbal and nutraceutical formulation also depend on the processing and adopted techniques of extraction (Jacobsohn and Jacobsohn 1976). As, for example, post harvesting of tea, way of processing like withering, oxidation, drying influence the polyphenols, catechin, and flavonoid contents of the samples. Less oxidized and withering for 2 h of post collected tea gives maximum non-hydrolysable tannins and saponins and flavonoids content (Tanaka et al. 2002). Maximum quantity of bioactive metabolites such as polyphenol, flavonoids, tannins, etc. are accumulated in coffee beans, if it is processed via dry and wet processing techniques (De Bruyn et al. 2017).

18.3 Proper Method of Standardization and Characterization of Collected Medicinal Plants

18.3.1 Different Challenges Created Due to Variation of Phytoconstituents and Its Probable Management

As we discussed above that when we concern about the natural drug development from traditional herbs and spices or aromatic plants in the form of bioactivity-guided fractions or crude extract, then serious issues like climatic change of the same place, as well as weather variation of different location may hamper the quality and reproducibility of the plant extract-based formulation at fixed dose. Because ayurvedic and herbal formulation consist of versatile phytocompounds along with lead marker in a fixed dose with different ratio and composition, which may intensely be varied due to climate change, geographical variation, and annual plucking time variation. This outcome may change the ratio of different active phytoconstituent's composition and yield of marker compounds; however total weight or volume of the crude extracts is constant in the formulated dose. As a result, the lead marker compounds yield out of total crude extracts would be either lesser or greater. The actual amount of the lead phytocompounds, for which the biological efficacy mainly exhibited, is deviated from the desirable reference dose. If the quantified dose of the lead phytocompounds is higher than the reference value, then possible toxic effect or side effect may be raised. In contrast, if the quantified dose of the lead phytocompounds in the formulated herbal dose is smaller than the reference standard value, then patient would not get the optimum pharmacological response for therapeutic management, even after taking full course of dose regimen as per prescribed value of herbal medicines. This awkward outcome could be an upsetting for natural drug discovery approaches. These challenges should be handled carefully for developing an effective, safe, and reproducible phyto-drug formulation batch to batch. Following approaches may need to be considered during or before formulation development.

1. Ideal and favorable climatic factors require to be selected for plant cultivation. The optimum rainfall, moisture content, soil water and salinity stress, soil pH, soil-friendly microbiome status, environmental temperature, soil temperature, and altitude of plantation for harvesting need to be chosen before initiation of harvesting for the suitable cultivating place, where the medicinal or crops could be grown with all desirable phytoconstituents yield.
2. The tribal folk plants should not be migrated in other places from their original cultivating region. If such practices are undertaken, then alteration of phytoconstituents composition and yield may change the biological effect of the formulated herbal dose due to climatic change related issues.
3. Agricultural farmers and suppliers should provide the relevant details of the cultivating places and collection time to the herbal pharmaceutical chemist or industrial quality control analyst to avoid the error or faults in the formulation standardization in respect to dose-oriented biological response for the fruitful treatment of target diseases. The collection time and place-related data also guide the chemists for future characterization of the same herbs from different batches as reference data.
4. If the medicinal plants collected from different harvesting region are used for formulation development, then the HPLC profiles of the same plants as per collection places need be characterized and kept in the fingerprinting library as reference standard. These approaches may help the analytical chemists to standardize the new sample of same herbal plants through finger printing matching with reference HPLC chromatogram. Thus, the interpreted results could have solid basis for accepting or rejecting the raw herbs toward the further formulation development.
5. Proper processing techniques such as drying time, drying temperature, moisture contents of the crude drugs, and suitable extraction techniques need to be followed and adopted for phytoconstituent-based herbal drug development. This tactic may give the effective and safe herbal drug formulation with reproducible results irrespective of batches.

18.4 Discussion and Conclusion

Herbal drugs and nutraceutical formulation development are significant alternative approach for prevention, management, mitigation, and cure of the functional or microorganism invading diseases, where mainline allopathic drugs are nowadays being offered a series of health hazards including ADR and side effects. Moreover, there is no permanent cure is possible via such treatment regimen. Whereas, bioactivity-guided fraction or crude drug extracts formulated from traditional herbs and medicinal aromatic plants may give much more potential and effective biological effect subjected to proper scientific standardization following the above-cited guidelines. They may provide better, effective, and safe therapeutic efficacy due to the synergistic action and complementary effect elicited by the homogeneous composition of pre-standardized active chemical constituents of herbal samples.

References

Abd El-Azim WM, Ahmed STH (2009) Effect of salinity and cutting date on growth and chemical constituents of Achillea fragrantissima Forssk, under Ras Sudr conditions. Res J Agric Biol Sci 5:1121–1129

Ahmed S et al (2019) Environmental factors variably impact tea secondary metabolites in the context of climate change: a systematic review. Front Plant Sci 10:939

Atanasov AG et al (2015) Discovery and resupply of pharmacologically active plant-derived natural products: a review. Biotechnol Adv 33(8):1582–1614

Azizi M, Dias A (2004) Nitrogen and phosphorous fertilizers affect flavonoids content of St. John's wort (*Hypericum perforatum* L.) proceedings of the fourth international Iran and Russia conference, pp 458–462

Bai J et al (2016) Spatial and temporal dynamics of heavy metal pollution and source identification in sediment cores from the short-term flooding riparian wetlands in a Chinese delta. Environ Pollut 219:379–388

Bernath J (1997) Cultivation of valerian. In: Houghton PJ (ed) Valerian—the genus Valeriana. Harwood Academic Publishers, Reading, pp 77–100

Castro EM et al (2006) Coumarin contents in young Mikania glomerata plants (Guaco) under different radiation levels and photoperiod. Acta Farm Bonaer 25:387–392

Das M, Jain V, Malhotra S (2016) Impact of climate change on medicinal and aromatic plants. Indian J Agric Sci 86:1375–1382

De Bruyn F et al (2017) Exploring the impacts of postharvest processing on the microbiota and metabolite profiles during green coffee bean production. Appl Environ Microbiol 83(1): e02398–e02316

Fusani P et al (2016) Seasonal variation in secondary metabolites of edible shoots of Buck's beard [Aruncus dioicus (Walter) Fernald (Rosaceae)]. Food Chem 202:23–30

Gobbo-Neto L, Lopes NP (2007) Medicinal plants: factors influence on the content of secondary metabolites. Quim Nova 30:374–381

Gupta A et al (2019) Medicinal plants under climate change: impacts on pharmaceutical properties of plants. In: Climate change and agricultural ecosystems. Woodhead Publishing, Sawston, pp 181–209

He CE et al (2013) Effects of nitrogen on root development and contents of bioactive compounds in Salvia miltiorrhiza Bunge. Crop Sci 53:2028–2039

Hozzein WN et al (2019) Exploring the potential of actinomycetes in improving soil fertility and grain quality of economically important cereals. Sci Total Environ 651:2787–2798

Hu W et al (2017) Heavy metals in intensive greenhouse vegetable production systems along Yellow Sea of China: levels, transfer and health risk. Chemosphere 167:82–90

Jacobsohn MK, Jacobsohn GM (1976) Annual variation in the sterol content of digitalis purpurea L. seedlings. Plant Physiol 58(4):541–543

Jayabalan R, Marimuthu S, Swaminathan K (2007) Changes in content of organic acids and tea polyphenols during kombucha tea fermentation. Food Chem 102(1):392–398

Kayne SB (ed) (2010) Traditional medicine: a global perspective. Pharmaceutical Press, London

Khalid KA (2006) Influence of water stress on growth essential oil, and chemical composition of herbs (Ocimum sp.). Int Agrophys 20:289–296

Kokate CK, Purohit AP, Gokhale SB (2008) Pharmacognosy, vol 464. Nirali Prakashan, Pune

Li J (2019) Simulating the effects of climate change across the geographical distribution of two medicinal plants in the genus Nardostachys. Peer J 7:e6730
Liu HY et al (2011) Effect of drought stress on growth and accumulation of active constituents in Salvia miltiorrhiza Bunge. Ind Crop Prod 33:84–88
Lotter D, le Maitre D (2014) Modelling the distribution of Aspalathus linearis (rooibos tea): implications of climate change for livelihoods dependent on both cultivation and harvesting from the wild. Ecol Evol 4(8):1209–1221
Munish S, Khajuria RK, Sharada M (2013) Annual variation in bacoside content of *Bacopa monnieri* (L.) Wettst plants. Int J Pharma Biosci 4(4):266–271
Nell M et al (2009) Effect of phosphorous uptake on growth and secondary metabolites of garden sage (*Salvia officinalis* L.). J Sci Food Agric 89:1090–1096
Pan SY (2014) Historical perspective of traditional indigenous medical practices: the current renaissance and conservation of herbal resources. Evid Based Complem Alternat Med 2014:525340
Radusiene J, Karpaviciene B, Stanius Z (2012) Effect of external and internal factors on secondary metabolites accumulation in St. John's wort. Bot Lithuanica 18:101–108
Razmjoo K, Heydarizadeh P, Sabzalian MR (2008) Effect of salinity and drought stresses on growth parameters and essential oil content of Matricaria chamomilla. Int J Agric Biol 10:451–454
Rivaes R et al (2013) Riparian vegetation responses to altered flow regimes driven by climate change in Mediterranean rivers. Ecohydrology 6(3):413–424
Sachan AK, Vishnoi G, Kumar R (2016) Need of standardization of herbal medicines in modern era. Int J Phytomed 8:300–307
Said-Al Ahl HAH, Omer EA (2011) Medicinal and aromatic plants production under salt stress. A review. Herba Pol 57:72–87
Shitole SM, Dhumal KN (2012) Influence of foliar applications of micronutrients on photosynthetic pigments and organic constituents of medicinal plant Cassia angustifolia Vahl. Ann Biol Res 3:520–526
Shukla S, Khanna KR, Singh SP (1996) Diurnal variation of major alkaloids in Papaver somniferum. Phyton 36:1–8
Shukla S, Singh SP, Yadav HK et al (2006) Alkaloid spectrum of different germplasm lines of opium poppy (Papaver somniferum L.). Genet Resour Crop Evol 53:533–540
Tanaka T, Mine C, Kouno I (2002) Structures of two new oxidation products of green tea polyphenols generated by model tea fermentation. Tetrahedron 58(43):8851–8856
Verma N, Shukla S (2015) Impact of various factors responsible for fluctuation in plant secondary metabolites. J Appl Res Med Aroma Plant 2:105–113
Wallaart TE et al (2000) Seasonal variation of artemisinin and its biosynthetic precursors in plants of Artemisia annua of different geographical origin: proof for the existence of chemotypes. Planta Med 66:57–62
Zhang L (2012) Growth, physiological characteristics and total flavonoid content of Glechoma longituba in response to water stress. J Med Plants Res 6:1015–1024
Zhang C et al (2019) Climatic factors control the geospatial distribution of active ingredients in Salvia miltiorrhiza Bunge in China. Sci Rep 9:904
Zobayed SMA, Afreen F, Kozai T (2007) Phytochemical and physiological changes in the leaves of St. John's wort plants under a water stress condition. Environ Exp Bot 59:109–116

Molecular Docking Studies of Plant-Derived Bioactive Compounds: A Comprehensive In Silico Standardization Approach

19

Damiki Laloo, Jun Moni Kalita, and Satyendra K. Prasad

Abstract

In the current scenario of medicine, products from natural source have been significantly contributing towards the therapeutic approach in the treatment of disease ailments ranging from simple to complicated ones. The World Health Organization has stated that approximately 80% of the global population rely on herbal medicines/products for their healthcare benefits. But in many cases, the utilization of herbal products as medicine has slowed down due to their lack of standardization involving botanical, chemical and biological (activity/toxicity) aspects. Standardization of medicinal plants and its active constituents has been the major event in the field of plant science involving ethnopharmacology, phytomedicine and pharmacognosy. The conventional or traditional standardization method of medicinal plant research is a cumbersome process, expensive, time-consuming and to a less extent outdated. Hence, a computational technique incorporating in silico molecular docking simulation study has become an essential tool for drug discovery, standardization and screening of phytochemicals. In this chapter, a comprehensive information have been discussed on the approach, significance and application of molecular docking study towards the ligand-protein interaction taking examples of pure active

D. Laloo (✉)
Department of Pharmacognosy, Girijananda Chowdhury Institute of Pharmaceutical Science, Guwahati, Assam, India
e-mail: damiki.laloo@gmail.com; dlaloo@gips-guwahati.ac.in

J. M. Kalita
Department of Pharmaceutical Chemistry, Girijananda Chowdhury Institute of Pharmaceutical Science, Guwahati, Assam, India
e-mail: pjmk84@gmail.com

S. K. Prasad
Department of Pharmaceutical Science, R.T.M. Nagpur University, Nagpur, Maharashtra, India
e-mail: skprasad.itbhu@gmail.com

S. C. Mandal et al. (eds.), *Evidence Based Validation of Traditional Medicines*,
https://doi.org/10.1007/978-981-15-8127-4_19

phytoconstituents (as ligand) against the targeted protein for therapeutic screening.

Keywords

In silico technique · Molecular docking · Standardization · Natural products · Drug discovery

Abbreviations

3D	Three-dimensional
Lys	Lysine
Ala	Alanine
Met	Methionine
Ans	Asparagine
NMR	Nuclear magnetic resonance
Arg	Arginine
PDB	Protein Data Bank
BBR	Berberine
Phe	Phenylalanine
Cys	Cysteine
Pro	Proline
Glu	Glutamine
Ser	Serine
Gly	Glycine
Thr	Threonine
His	Histidine
Trp	Tryptamine
HTS	High-throughput screening
Tyr	Tyrosine
Ile	Isoleucine
Val	Valine
kcal/mol	Kilocalorie per mole
ADME/T	Absorption, distribution, metabolism, excretion and toxicity
Leu	Leucine

19.1 Introduction

Since time immemorial, medicinal plants have been utilized as a valuable source for the production of numerous therapeutic agents, and they are still contributing significantly as a major source of medicines to counter the demand. As there are many limitations related to the natural product-based drug discovery, the pharmaceutical industry has primarily focus towards the synthetic molecule libraries and high-throughput screening (HTS) for the discovery of new leads. However, the results obtained from those trials did not produce promising results that meet with the expectations, which were evident through a consistent decline in the number of new drugs reintroduced into the market (David et al. 2015; Kingston 2011; Scannell et al. 2012). Such circumstance again encourages the researchers working in the field of natural products to develop new leads from natural origin, despite its high complexity and involvement of interdisciplinary research approaches (Heinrich 2010). The first scientific drug discovery started from the nineteenth century, when the famous German apothecary 'Friedrich Sertürner' created history by successfully isolating a sleep inducing agent and pain reliever compound 'morphine' from opium and named it 'morphium' after the Greek God of dreams, 'Morpheus' (Sertürner 1817). Later, following decades of the nineteenth century, such discovery triggered many eminent researchers of the world to examine other medicinal herbs by isolating many potent bioactive natural compounds, including alkaloids like codeine, quinine, colchicine, cocaine, nicotine, caffeine, capsaicin and atropine from their respective natural sources (Corson and Crews 2007; Kruse 2007; Zenk and Juenger 2007). Further, to facilitate the quantitative production of natural products at higher percentage yield with lower costs, research was extended by means of chemical synthesis, and the best example was the successful synthesis of the first natural compound 'salicylic acid' in 1853 (Kaiser 2008). Another revolution in the drug discovery from microbial sources was initiated in 1930s, which is attributed with the discovery of penicillin, which is till date of pharmaceutical importance (David et al. 2015). Despite in recent times with the growing dominance of combinatorial chemistry and study on HTS, the impact of the drug discovery process from natural products still remains valuable, whereby out of 1073 new chemical entities approved between the year 1981 and 2010 from the classes of small molecules, only 36% belongs to purely synthetic category, while more than half of the drugs were retrieved from natural resources (Newman and Cragg 2012). In fact, some of the natural products which are derived from plant source have become very essential for modern therapeutics especially to treat cancer; and some of the potentially active plant components of great demand include paclitaxel and its derivatives from Himalayan yew (*Taxus baccata* L.), camptothecin from *Camptotheca acuminate* Decne., and vincristine and vinblastine from *Catharanthus roseus* (L.) G. Don. (Cragg and Newman 2013; Kinghorn et al. 2011).

19.1.1 Approaches for Drug Discovery from Natural Resources

The selection criterion for starting material during drug discovery from natural origin is a very critical and important step, for example, if the extracts, its bioactive fractions, or isolated compounds under investigation are collected from their originating source or from areas with very high biodiversity and abundance, and may have better bioactivity as the chemical variety directly reflects the biodiversity of their source organisms (Henrich and Beutler 2013). Further, a knowledge-based collection of raw material can also be applied, which may help us in identifying the potential bioactive molecules from a very less number of test samples using some selected and authentic pharmacological assays. The approach is based on systematic ethnopharmacological knowledge, where the use of traditional medicinal plants plays a critical role in the selection of test material under investigation. Another approach for selection of plant material includes chemosystematic or phylogenetic characterization, by using molecular phylogenetic data and chemotaxonomic knowledge, which helps in selection of authentic plant species from identified genus and families having compounds or classes of compounds with certain therapeutic uses in a targeted manner. The ecological approach can also be one of the criteria for selecting the plant material that depends on the interactions between organisms and their original environment, which results in production of bioactive natural molecules (Atanasov et al. 2015).

In recent years, computational knowledge-based approach has gained immense value in selecting plant material for drug discovery having a very high likelihood of biological activity with very high success rates. Molecular docking simulation studies help us in understanding the binding characteristics between protein and ligand for a molecular structure, especially in case of known phytoconstituents. Phytoconstituents showing significant results in in silico model predictions can be selected as most prominent starting materials for investigation (Hein et al. 2010), thus focusing on the main phytoconstituents of herbal therapy in addition to any other natural compounds with relevant biological activity retrieved from the literature (Rollinger et al. 2009).

Virtual screening based on pharmacophore is also one of the highly demanded computational methods (success rates between 2 and 30%), which includes a 3D arrangement of physicochemical characteristics related to the formation of H-bond between the donor and acceptor, incorporation of hydrophobic areas and involvement of aromatic ring which represents the interactions between a ligand molecule with that of the targeted protein (Zhang et al. 2011; Hein et al. 2010). Molecular docking studies are widely and frequently used computational approach for elucidating the preferred mechanism of action and also to rationalize the observed structure-activity relationships (SAR) of natural compounds by accurately predicting their position within the protein binding pockets and also help to estimate the probable strength of the binding by providing a docking score (Waszkowycz et al. 2011).

19.1.2 Limitations Associated with Natural Product Drug Discovery

Correct identification and nomenclature of medicinal plants is one of the major issues related to natural product's authenticity as these are the very first step associated with the discovery of natural products. Therefore, it is essential that the medicinal plant under investigation should undergo both genetic and chemical analyses along with their morphological and histological characterization (Bucar et al. 2013). Incompetency in sharing of knowledge by the traditional healers or tribal people from various tribes and rural areas of globe regarding the use of medicinal plant is also a major issue in exploration of new medicinal plants and its therapeutic potential. Availability of starting material and the yield of the natural product derived from the starting material is also one of the areas of concern. Generally, it has been observed that the yield of the bioactive natural compounds is found to be very low and is quite insufficient for carrying out various biological activities. The problem of low availability of active biomolecule from natural products becomes more critical, if the molecule shows very high promising activity, which needs to be further evaluated in developing it as a pharmaceutical lead (David et al. 2015).

In case of highly potent herbal plant or drug, their extensive wild crafting and unsustainable harvesting techniques for collection has resulted in its depletion from the society to a very high frequency. The most common example includes 'Taxol supply crisis', where it showed remarkable clinical efficacy in ovarian cancer that resulted in a very high demand of the compound, which was obtained earlier from the barks of *Taxus brevifolia* L. This resulted in unethical and an irrelevant collection of the barks which ultimately raised concerns over the ecological impact on the methods of collection (Kingston 2011).

The quality of medicinal plants and its composition is also a very essential criterion for proper assessment of the therapeutic claims of the plant-derived drug. The main factor contributing towards the quality and composition includes correct species identity and right harvesting time along with actual climate, altitude, processing, storage conditions and type of extraction techniques (Bucar et al. 2013; Jones and Kinghorn 2012).

Another challenge in drug discovery from natural product includes the incompatibility of natural products with HTS, which are basically conducted incorporating cell-based or cell-free bioassays. This requires a comprehensive strategy of accuracy, robustness, reproducibility and reliable sample handling systems, where the test compound must not precipitate and decompose or should not interact with the components of assay, which is very rare in natural products (Coan et al. 2011). Further, it is also very difficult to identify the specific mode of action of medicinal plant extracts or bioactive fractions as they are composed of multicomponents. In addition, studies related to interaction of drug molecule to that of molecular targets are still very less explored. Rigorous clinical trial required for getting approval of phytoconstituents as drug candidate is also one of the major areas of concern due to lack of industrial support, which may be attributed to the high cost and time involved especially in case of those compounds which cannot be synthetically modified (Corson and Crews 2007).

19.1.3 Initiation in In Silico-Based Drug Discovery

It has been observed for many years that drug discovery and their development are considered to be a time-taking as well as resource-consuming method. To minimize these time and resources, efforts have been taken by researchers by application of computational technique in combination with chemical and biological space; and this has significantly attributed with streamlining drug discovery, development, design and their optimization. In scientific fields such as biomedical engineering, utilization of computer-aided drug design virtually helps to accelerate and validate the process of drug discovery. Such technique helps to facilitate identification of hit, hit-to-lead selection criterion, optimization of absorption, distribution, metabolism, excretion studies or toxicity profiling (ADME/Tox), and safety issues (Singh and Sharma 2011). The commonly and conventionally used computational approaches in in silico studies include quantitative structure-activity relationships as well as ligand-based and structure-based drug design. The applications of such structure-based design are implied quite early for optimizing the leads into drugs in the drug discovery and development process. Generally, protein structures are used for selection and target identification by assessing their 'druggability' or tractability of targets for searching the hits either by virtual screening or by fragmentation screening process. Thus, we may assume that the critical role of structural biology in drug discovery and its optimization will always maintain its priority forever (Singh and Sharma 2011; Rollinger et al. 2006).

Most of the drugs available in the market were either discovered from chance observations or from screening the libraries available for natural and/or synthetic product. Natural products obtained either in the form of pure isolated compounds or as standardized extracts have always provided unlimited opportunities for the search of new drug leads due to their diverse chemical nature. The phytoconstituents undergoing chemical modification on the basis of trial and error have typically resulted in a higher selectivity and bioavailability along with improved potency and reduced toxicity of the compounds under investigation. However, such approach is highly laborious and time-consuming; therefore scientists from the relevant sector are continuously developing new methods for enhancing the efficiency of the process related to drug discovery (Giersiefen et al. 2003). The rationale behind the protein and structure-based drug discovery chiefly depends on the functional identification of the basic chemical structure of the targeted protein, which is being followed by the hypothetical ligand prediction for the target protein from molecular modelling and subsequent chemical and biological assays of the compounds. This has resulted in a vast increase in the data collection of the Protein Data Bank (PDB, http://www.rcsb.org/pdb); however, only 1–2% 3D structure of all known proteins till date have been experimentally characterized (Berman et al. 2000; Hein et al. 2010; Atanasov et al. 2015).

However, recent advances in the field of fold recognition, sequence comparison and protein modelling algorithms has minimised the so-called sequence-structure gap up to a great extent and has also enabled the extension of experimental protein structure information to homologous proteins. The advancement in threading procedures has helped in correlation of the protein sequences available in the library of known structures and has enabled identification of the probable protein fold even when there is no clear sequence homology. The quality of homology models and their applications in discovering drug largely depends on the similarity of sequence between the known structured proteins taken as a template and the targeted proteins. Recent studies have reported that homology modelling has proven to be significantly

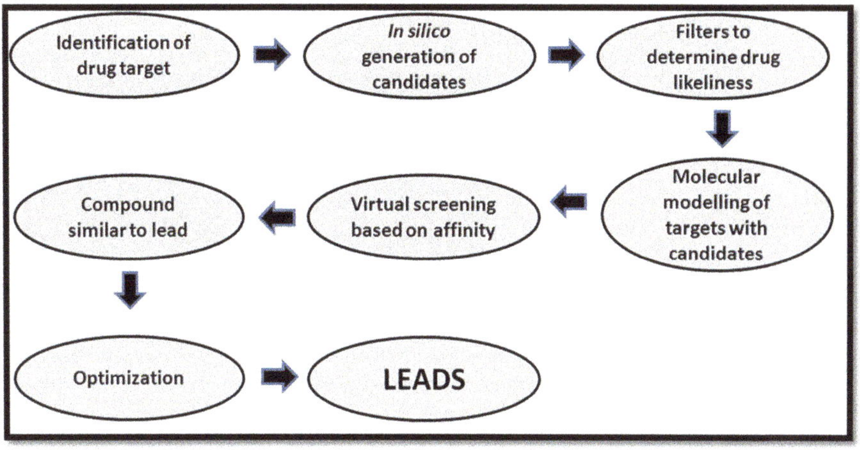

Fig. 19.1 Schematic flowchart of in silico intervention in drug discovery

advantageous in proper identification, validation and optimization of drug targets of lead compounds (Fig. 19.1) (Rollinger et al. 2006; Atanasov et al. 2015).

19.2 Molecular Docking Studies

In the last two decades, computational methods are extensively used in the study of the complex formation particularly at intermolecular level in the field of scientific research. With the well establishment of the fact that the activity of a drug depends on its binding capacity towards its specific target and along with the advancement in technology, computational methods of drug discovery have gained extensive popularity in the field of scientific research in the last few years. The drug molecules while binding to the target site exhibits close geometric and chemical complimentary properties, and the computer-based method of finding such properties of a test molecule towards the target is known as 'molecular docking'. Further, this kind of drug design methods that utilizes known structural properties of receptors in finding new drug is called as 'receptor-based drug design'. With the main target to find out newer small molecules with probably less toxic properties and greater drug likeliness, molecular docking became a vital and essential tool in the field of drug design and drug discovery system (Krovat and Langer 2005; Lengauer and Rarey 1996).

During the molecular docking program, the initial requirement is to obtain the three-dimensional structure of the target molecule. In general, the target is chemically a protein molecule; this three-dimensional structure is obtained from different biophysical-based techniques such as NMR spectroscopy and X-ray crystallography. The three-dimensional structure of both the protein and a set of ligands is used as the initial input to a docking protocol (Fig. 19.2). Search algorithm and scoring

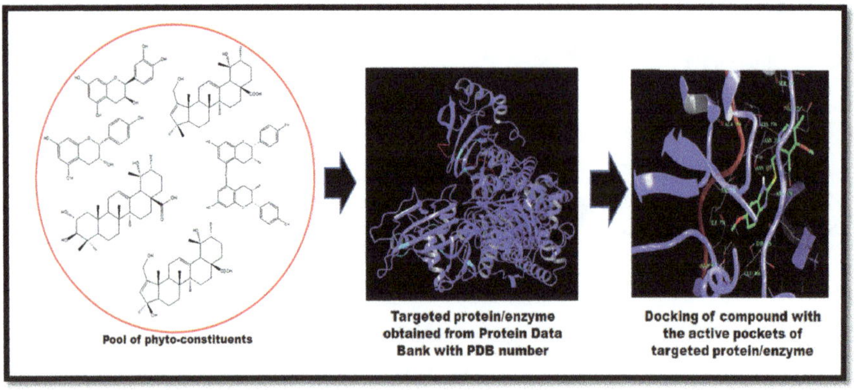

Fig. 19.2 Sample demonstration of molecular docking simulation study

function are the main key components of docking program on which its success depends upon. Both search algorithm and scoring function work parallel to find out the minimum energy binding complex.

Search algorithm: It works to find out the energetically more suitable conformation or binding pattern of the ligand-receptor complex. Different simulation techniques are applied for this purpose such as simulated annealing, molecular dynamics, stepwise bond rotation, Monte Carlo simulation, genetic algorithm, tabu search, etc. The prime goal of all these simulation techniques is to identify the most stable and suitable site by calculating the lowest binding energy conformation of the receptor-ligand complex (Meng et al. 1992).

Scoring function: It is a parameter which selects the most suitable conformation of the ligand-receptor interaction as an input, whereas, a score is generated as an output. The input is also known as 'docking pose'. Three main and important scoring functions employed in docking applications are listed below:

- Force field-based scoring
- Empirical scoring
- Knowledge-based scoring

In force field-based scoring function, the binding affinity of the ligand towards the target molecule is calculated by adding up the strength of different non-covalent interactions such as electrostatic and van der Waals interactions, H-bond energy, intramolecular strain of the ligand and the target, desolvation energy, etc. Same way the numbers of ligand and receptor atoms in contact with each other are counted in case of empirical scoring functions. PLP, ChemScore and the FlexX are some common examples of empirical scoring functions. Based on the statistical observations of intermolecular close contacts in 3D database, knowledge-based scoring functions are assigned. The potential of mean force (PMF) score is an appropriate example of a knowledge-based scoring function (Kitchen et al. 2004;

Table 19.1 Software used in molecular docking study

Software	Conformational searching method	Scoring function
AutoDoc	Genetic algorithm	Empirical
DOCK	Incremental construction	Force field
eHITS	Exhaustive systematic	Knowledge-based/eEmpirical
FLexX	Incremental construction	Empirical
FRED	Shape matching	Gaussian/empirical
Glide	Incremental construction	Empirical
GOLD	Genetic algorithm	Force field
HammerHead	Incremental construction	Empirical
ICM	Monte Carlo simulation	Force field/empirical
LigandFit	Shape matching/Monte Carlo simulation	Empirical
QXP	Monte Carlo simulation	Force field

Wei et al. 2004). The information supplied in Table 19.1 represents the best examples of different software's incorporated for molecular docking of compounds.

19.2.1 Approaches and Applications of Molecular Docking Studies

In the case of molecular docking studies, there are primarily two different types of approaches which are basically followed. They are as follows:

19.2.1.1 Molecular Simulation Approach

Here the chemical substance taken as a ligand molecule is allowed to bind with the amino acid active site of the targeted proteins. The ligand molecule in its different conformations moves within the target in search of the active sites for binding purpose. Molecular simulation then predicts the ligand conformation and orientation that is best suited to the active site. Each and every orientation/conformation the ligand binds with the receptor, the free energy is calculated. And then best on the free energy, the most suitable conformation and binding pose of the molecule is assigned (Totrov and Abagyan 2008; Wang et al. 1999).

19.2.1.2 Shape Complementary Approach

In this approach of molecular docking, shape recognition algorithms are used. These shape recognition algorithms search for the perfect shape complementary happening between the ligand taken as a small molecule and the targeted protein. Based on the shape complementary, a score is provided, and binding capability of the ligand is assigned (Shoichet et al. 2002).

19.2.2 Applications of Molecular Docking

At the molecular level, interaction between a ligand and a targeted enzyme not only signifies the process of activation or inhibition, but it also allows the researcher to understand about the mode of binding suitability of the ligand-enzyme complex. Similarly, if the macromolecule is a receptor, then it may result into agonist or antagonist. Therefore the final biological activity can only be predicted by molecular docking studies, but cannot be determined as it shows only about the interaction of the ligand and the target. It does not refer anything about the type of biological activity. Based on this criterion docking method can be applied for hit identification, for lead optimization and even for drug DNA interaction.

Hit identification: Molecular docking incorporating proper scoring function is used to find out a potent hit molecule from a set test molecules using in silico studies. If the target structure is known, it becomes very easy to find out probable potent molecule by employing molecular docking techniques (Gschwend et al. 1996).

Lead optimization: Molecular docking methods further can be used to modify the hit molecules to have a superior ligand with more affinity towards the target and probably with less toxicity. This method is used in the design of newer potent and effective ligand (Ferreira et al. 2015).

19.2.3 Significance of Molecular Docking Studies over Conventional Standardization Methods Towards Plant-Derived Molecules

Standardization of medicinal plants and its active constituents has been the major event in the field of plant science involving ethnopharmacology, phytomedicine and herbal medicine. The conventional or traditional standardization method of medicinal plant research is a cumbersome process which involves the extraction of active constituents (plant extracts), fractionation process, qualitative and quantitative identification, isolation and characterization and finally screening of the overall isolated phytoconstituents for pharmacological activity (forward pharmacology). In addition, when numerous phyto-components have been isolated from one single plant, then screening process for pharmacological activity of the most active molecule from a pool of numerous compounds is a colossal problem, and this will lead to the utilization of a huge number of animals for experimentation. In general, the entire research process to systematically evaluate the pharmacological activities of huge numbers of active ingredients is time-consuming and expensive and may not generate reproducible result outcomes. Hence, there is a need to develop a quick, effective and convenient screening technique to accurately predict the activity of a small potential group of active molecule(s) from a large pool of inactive chemical compounds which can be subjected for further in vivo or in vitro assays. Reverse pharmacology in combination with computational screening technique is one good example for fast screening of active molecule from a pool of inactive ones (Fig. 19.3). With the great advancement in science and technology, the concepts of utilizing the virtual screening technique involving in silico technique have paved

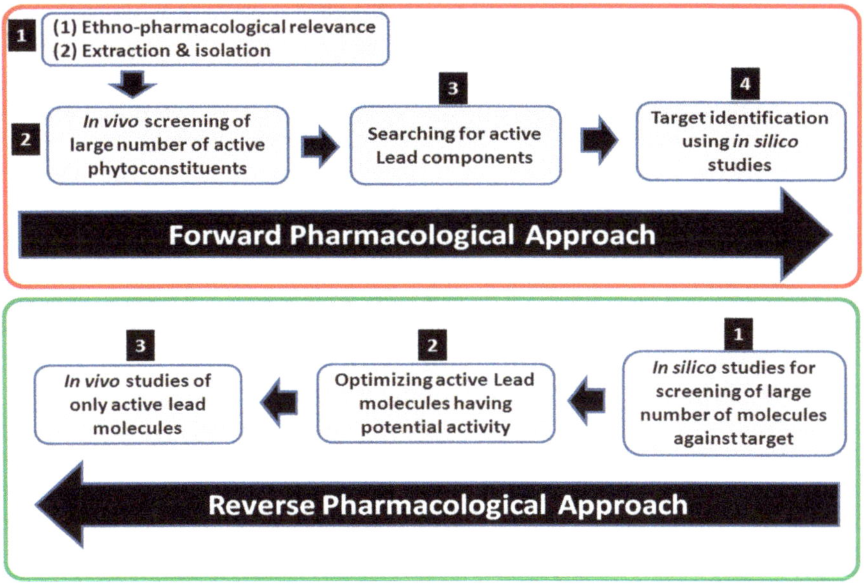

Fig. 19.3 Significance of in silico techniques in reverse pharmacological approach

ways to correct and update the limitations of the conventional screening methods for both synthetic and natural products. Molecular docking studies are virtual computational method which is widely used in the development and drug discovery system with great significance to predict the ligand-protein binding property and ADME/T properties, elucidating the molecular mechanism of action, and to rationalize the structure-activity relationships of natural products (Atanasov et al. 2016).

19.2.4 Other In Silico Virtual Screening Methods

19.2.4.1 Pharmacophore Modelling

Pharmacophore modelling has become one of the widely accepted methods in the field of drug design and discovery. Such method helps in predicting the capability of test molecules to bind with the receptors at active site. Pharmacophore is a cluster of all the features (both steric and electronic) which the molecule possessed to functionally bind with the desired target. The assembly of characters can be created either by comparing the common properties of a set of similarly active molecules or by mapping the geometric and electronic properties of a bound ligand at the active site. In both the above cases, features like the H-bond donor, H-bond acceptor, aliphatic chain, aromatic or lipophilic group, presence of acidic or basic group and presence of charge on the molecules are considered; and this logical molecular representation entity is used in the software packages to find out potential binder to a target. Commercial software packages like LigandScout (Wolber and Langer 2005) and

The Pocket v.2 (Chen and Lai 2006) use such similar algorithms in protein-ligand complex data to chart interactions between ligand and target.

Both the leading methods of virtual screening, i.e. molecular docking and pharmacophore modelling, have been extensively used in finding newer ligands. Best examples is the study in 2008 conducted by Pierce and his associates where the discovery of the four potential anticancer candidates happened by incorporating the molecular docking screening techniques from a library of databases consisting of over 700,000 molecules (Pierce et al. 2008). In an another study, Allen and his associates screened six millions of commercially available compounds using virtual screening methods, and a total of 24 molecules were selected and tested in BRD4 and EGFR biochemical assay (Allen et al. 2015).

19.2.4.2 Molecular Similarity Screening Method

Screening molecules on the basis of molecular similarity is a logical method and has been widely used in the field of drug discovery. As per this concept, two molecules may share similar physical properties and biological activities if they have got similar chemical properties. This concept of similarity between the molecules and their properties has been utilized in drug discovery to screen a large database of molecules (Armstrong et al. 2010).

Descriptors for Molecular Similarity Calculations

A significant number of descriptors are categorized and are now widely used in molecular similarity search and calculation in drug design. The descriptors can be based on molecular graph and the wave function of molecules or even based on the conformational molecular shape. Molecules are then compared in three stages, including (a) chemical space representation of molecular structures, (b) feature assigning (this step is optional) and (c) comparison of the structures. This description can be constitutional, topological, geometrical, electrostatic and even quantum chemical based on the requirement of the studies. Some common descriptors used are like molecular weight, LogP, refractivity, polarizability, hydration energy, surface area, volume, E_{HOMO}, E_{LUMO}, total energy, electronic energy, heat of formation, dipole moment, etc. (Zauhar et al. 2013; Venkatraman et al. 2009; Goldman and Wipke 2000).

19.3 Molecular Docking Study of Some Potential Active Phyto-molecules Towards Pharmacological Targeted Proteins/Enzymes

For the past decade, there has been an extensive and profound research happening on plant active constituents incorporating in silico virtual screening techniques. In silico molecular docking technique is the most valuable tool utilized by many research scientists to investigate and understand the nature of the active site of the protein and also to corroborate the binding interactions between ligands and targeted enzymes/ receptors. Molecular docking techniques predict binding properties at the molecular level, which include the energy profile (binding free energy) and by incorporating scoring function to predict the strength and stability of complexes in terms of binding affinity and binding constant (Agarwal and Mehrotra 2016). The free energy for ligand binding is taken as the reference parameter to compare the binding properties of the other unknown ligands and the possible inhibition or activation of the targeted protein/enzyme/receptor. The amount of energy liberated upon the binding affinity between a ligand to its enzyme/protein/receptor is a direct criterion to measure the stability of the complex, which designates how strongly a ligand may produce inhibitory effects towards its target (Mazumder et al. 2019). The utility of molecular docking study to screen the active components from plants against the virtual design pharmacological targeted protein/enzyme has paved ways for the scientist to predict the ligand-substrate interaction. The inclusion and application of virtual standardization process in natural product chemistry will provide precise results for the proper identification of natural lead molecules at the molecular level. The following are some of the best examples of in silico standardization techniques incorporating molecular docking simulation studies taking plant active constituent (s) as ligand and virtually docked with the targeted protein/enzyme/receptor.

19.3.1 Caffeine

Caffeine (1,3,7-trimethylpurine-2,6-dione) is a purine alkaloid which is famous for its psychoactive stimulant activity in the brain. It has a wide array of pharmacological efficacy which includes respiratory, cardiovascular, renal, smooth muscle effects and also CNS stimulant activity with significant effects on moods, memory, alertness and cognition [Institute of Medicine (US) Committee on Military Nutrition Research 2001]. In contrast to in silico study, there are many research studies published on the interaction of caffeine scaffold and its analog(s) against various targets such as monoamine oxidase (MAO), caspase-3, collagenase, elastase, tyrosinase, acetylcholinesterase (AChE), MAP kinase, etc.

19.3.1.1 Interaction with Caspase 3

Caspases are mediators which are responsible for programmed cell death (apoptosis). Caspase-3 an enzyme of the endoproteases family has a potential role in regulating the various phases of inflammation and apoptosis signalling networks (Porter and Janicke 1999). Recently, a molecular docking simulation study of caffeine molecule versus caspase-3 was reported by Erukainure and associates (2019) using the AutoDock Vina software. Results revealed that caffeine strongly binds with caspase-3 enzyme, thereby showing higher binding energy of −5.0 kcal/mol. Caffeine forms conventional H-bond with amino acid Arg-B:207, π-π T-shaped interaction with amino acid His-A:121 and alkyl interaction with Trp-B:204 and Cys-A:163. Thus, the strong interaction of caffeine with Caspase-3 signifies potential anticancer activity.

19.3.1.2 Interaction with Chk1 Receptor

Chk1 receptor is the main target in epithelial cancer where alkylxanthines actively bind with. Incorporating AutoDock 4.2.6 software, da Silva Costa and associates (2018) reported that caffeine has potential binding efficacy with such receptor via carbon-hydrogen bond associated with Chk1 amino acid residue Glu-85 and Cys-87 and π-σ bond with that of Leu-15. The binding free energy of Caffeine-Chk1 interaction was reported to be −5.14 kcal/mol. Such binding potential signifies the inhibitory action of caffeine against the epithelial cancer.

19.3.1.3 Interaction with Monoamine Oxidase

Monoamine oxidase (MAO) an enzyme found in the mitochondria is involved in the process of catabolism and breakdown of biogenic amine neurotransmitters such as serotonin, dopamine, norepinephrine and epinephrine. In humans, two isoenzymes of MAO have been identified, including MAO-A and MAO-B. Both the isoenzymes have different locations in the body and specific substrates binding to them (Markey 2007). As per the literature, structural modification at C-8 position substitution of caffeine with an array of moieties yields potential MAO inhibitors (Petzer et al. 2013). Caffeine binds with both the MAO subunits at different active sites of amino acids. There is an H-bond interaction between the carbonyl oxygen at the C-6 position of the caffeinyl ring with that of the phenolic hydrogen of Tyr-435 in MAO-B, and H-Bond interaction also takes place between the carbonyl oxygen at C-2 of the caffeinyl ring with that of the phenolic hydrogen of Tyr-444 in MAO-A. Such binding was evaluated incorporating AutoDock 4.0 software. The molecular interaction results signify that caffeine may contribute as potential MAO inhibitors.

19.3.1.4 Interaction with Human Serum Albumin (HSA)

HSA is the most abundant protein in human serum with a negatively charged surface and is basically comprised of three homologous domains—I, II and III. HSA comprises of 585 amino acids with 17 disulphide bonds, thereby forming 9 structural loops. Islam and associates (2016) as well as Wang and associates (2013) incorporate the fluorescence quenching technique and reported the positive interaction of caffeine with HSA subdomain IIA via Trp-214 amino acid residue binding site of the protein. In addition, amino acids such as Leu-198, Leu-481, Ser-202, Ser-454, Phe-211, Lys-199, Trp-214, and Val-344 also form stable hydrophobic pockets in a subsite A of the HSA protein. This probably facilitates the strong hydrophobic interaction between the pyrimidine ring of caffeine with that of the amino acid Trp-214 via aromatic group stacking. The docking study was ascertained using GOLD docking suite 5.1.

19.3.1.5 Interaction with MAPK-1 (Mitogen-Activated Protein Kinase)

MAPK is a category of protein kinases, which play an essential role in signal transduction, and it is involved in distinct biological events such as cell proliferation and differentiation, immune cell function, motility, metabolism, survival and apopto sis. Inactivation of MAPK may result in a serious defect related to cell lysis, which is particularly attributed to a deficiency in cell wall construction (Wang and Anderson

2012). A study on the interaction of caffeine with MAPK-1 was reported where caffeine induces phosphorylation reaction, thereby binding with the C-terminal extension of MAPK-1 at two serine amino acid residues Ser-423 and Ser-428 (Truman et al. 2009). The positive activation of MAPK-1 by the caffeine results in probable regulation of cellular integrity and thereby helps to correct the deficiency of cell wall construction.

19.3.1.6 Inhibitory Interaction Against Collagenase, Elastase and Tyrosinase

Collagenase, elastase and tyrosinase are intracellular enzymes, which are responsible for skin ageing, thereby resulting in the loss of skin elasticity. Inhibition of such enzyme will contribute to rejuvenating of the skin morphological characters. Caffeine has been reported to have inhibitory activity towards collagenase, elastase and tyrosinase with good binding affinity of −4.66 kcal/mol, −3.36 kcal/mol and −2.86 kcal/mol, respectively. The intermolecular interaction was ascertained with caffeine forming an H-bond with Glu-498 and His-527 amino acid residues of collagenase enzyme and moderate hydrogen bonding with Gly-216 and Ser-195 of elastase, and no sort of H-bond was observed with tyrosinase amino acid residue. However, the binding efficacy towards tyrosinase was observed via polar and hydrophobic interactions with His-61, His-85, His-244, His-259, Ans-260, His-263, Ser-282 and Phe-90 amino acid residues (Lee et al. 2019). Thus, the inhibitory effect of caffeine against the above three enzymes will result in maintaining and rejuvenating various types of skin disorders.

19.3.2 Berberine

Berberine (BBR) is a yellow coloured benzylisoquinoline alkaloid which is found in many families, including Berberidaceae, Rutaceae, Menispermaceae, Annonaceae and Ranunculaceae. It is widely distributed in plants such as *Coptis chinensis*, *Hydrastis canadensis*, *Berberis aristata*, *Coptis japonica*, *Phellodendron amurense*, etc. Many studies suggested the therapeutic value of berberine as antiulcer, antiviral, antipyretic, antibacterial, anti-inflammatory agent, antidiabetes, antiobesity, etc. (Hu et al. 2018). The following are some of the best examples of molecular interaction of berberine against various target proteins.

19.3.2.1 Inhibitory Interaction Against Hepatitis C Virus (HCV)

HCV is a pathogenic virus infecting the global population which causes serious problems to the liver, including cirrhosis and hepatocellular carcinoma (HCC). There are currently no effective vaccines, which are used to protect against HCV. Incorporating molecular docking simulation study, Hung and his associates (2018) made an attempt to screen the inhibitory binding efficacy of berberine (BBR) towards the E1/E2 terminal glycoproteins of HCV taken as the target. BBR forms polar interaction by binding to Ser-599 amino acid residue of the E2 terminal glycoproteins of HCV with the average binding energy of 6.7 kcal/mol. The strong binding and inhibitory efficacy of berberine to the viral glycoprotein signifies the potential anti-HCV activity of the compound.

19.3.2.2 Inhibitory Interaction Against Epidermal Growth Factor Receptor (EGFR)

EGFR is a membranous protein, which is a fundamental biomarker for breast cancer. This protein is also highly associated with the signalling pathway related to the dysregulation process of EGFR mitogen-activated protein kinase (MAPK). The pharmacological effects of EGFR are governed by the over-activation of MAPK and other extracellular signal-regulated kinase (ERK) which is involved in tumour cells. A molecular docking study on the inhibitory interaction of BBR against the EGFR was reported by Kaboli and his associates (2018). BBR form strong hydrogen bonding with the EGFR specifically at amino acid residue Lys-745 with the binding energy of −7.22 kcal/mol. In addition, BBR has unsaturated bonds in its structure which also provide π-π interactions with the neighbouring amino acid residues. Moreover, there is also a strong hydrophobic interaction between BBR and other surrounding amino acid residues including Leu-844, Leu-718, Pro-794, Gly-796 and Val-726.

19.3.2.3 Inhibitory Interaction Against Neuraminidase-1

Neuraminidase is a class of surface glycoprotein antigen, which is one of the significant biomarkers for the classification of subtype of influenza A virus. It is due to the activation of this enzyme that triggers the liberation of the virus by forming hydrolysing linkage with terminal sialic acid residue. In one of the study, a report on the in silico inhibitory effect of berberine (BER) and epiberberine (EBER) on neuraminidase-1 (NA-1) enzyme (Zhou et al. 2017) was analysed. Both the compounds were reported to have potential interaction against the enzyme which is comparable with zanamivir (a neurominidase-1 inhibitor). BER forms strong H-bonding (binding energy of −7.4 kcal/mol) with amino acid residue of NA-1 at Ser-179 (3.3 Å) and Asn-294 (3.0 Å) and also forms several hydrophobic contacts with amino acids Arg-152, Arg-292, Glu-277, Trp-178, Ile-222, Ser-179, Ser-246, Arg-224, Glu-227, Glu-276, Asn-294 and Asn-347, whereas at a particular binding energy of −7.8 kcal/mol, epiberberine forms H-bond with amino acid Ser-246 of NA-1. Thus, from the study it can be incurred that berberine may contribute to possible management of influenza virus.

19.3.2.4 Inhibitory Interaction Against Thrombin

During the blood coagulation process, thrombin is likely a key enzyme which is responsible for the conversion of fibrinogen to fibrin. It is a multifunctional serine protease which is produced by the cleavage of prothrombin. Inhibition of thrombin clinically leads to the regulation and management of diverse disease events such as thrombosis-fibrinolysis process, blood clotting, stroke, cancer and many neurodegenerative diseases (Wojtukiewicz et al. 2016). Berberine (BER) binds at the active sites of the thrombin amino acid residue and produces inhibitory effect. The binding involves the C-10 methoxy group of BER which is inserted into the thrombin catalytic site, thereby forming two H-bond interactions with two amino acid residues of NA-1, viz. Phe-227 and Trp-215. In addition, π-π interaction was also observed between the aromatic ring A of BER with the side chain of Trp-60D amino acid.

19.3.2.5 Inhibitory Interaction Against Urease

Urease, a urea amidohydrolase, is a nickel-dependent metalloenzyme which is widely distributed in plants, animals and microbes. In bacteria, urease facilitates the breakdown of urea to ammonia, thereby promoting the growth of various microorganisms. Virulent factors of bacterial ureases are commonly involved in disease pathogenesis of which the best example is the colonization of *Helicobacter pylori* urease (HPU) in the GIT that clinically leads to the development of peptic ulcer. Urease inhibitors are currently used to treat peptic ulcer and other urease-related diseases. In a study reported by Tan and his associates (2017), berberine and epiberberine showed potential inhibitory efficacy against *H. pylori* urease. Molecular docking assay signifies that epiberberine binds better than berberine towards the substrate by inhibiting the bacterial enzyme urease, thereby forming strong H-bond interactions with the two basic amino acid residues (Met-366 and Asn-168) of HPU. The binding free energy was reported to be −5.38 kcal/mol. Thus, berberine may be a useful drug for the treatment of peptic ulcer caused by pathogenic bacteria.

19.3.2.6 Inhibitory Interaction Against ZIKA and Dengue Virus

Zika virus (ZIKV) belonging to Flaviviridae family is a mosquito-borne pathogenic virus which produces symptoms similar to viral infections such as fever, headache, malaise, conjunctivitis, arthralgia and skin rashes. There is also a report on risk development of microcephaly with severe defects on the fetus brain of a pregnant woman. ZIKV possess a single-stranded RNA genome-bearing seven nonstructural (NS) proteins, viz. NS1, NS2A, NS2B, NS3, NS4A, NS4B and NS5. Of all these proteins, NS3 is proven to be an important drug target for drug cleavage and discovery. The protease activity of NS3 is functionally required for viral replication; and any inhibitory effect against NSE protease site may be considered as a strategic drug discovery for the treatment of ZIKV infection. In one study reported by Sahoo and his associates (2016), berberine showed potential binding efficacy against the ZIKV-NS3 terminal, thereby producing an inhibitory action. The binding affinity was reported to be −5.8 kcal/mol of which berberine forms H-bond with Ser-135 amino acid residue of ZIKV protein. In the same study, an attempt was also made to docked berberine with dengue virus (DENV), and the result was quite promising of which a potential inhibitory activity was observed. The drug berberine forms H-bonding with DENV amino acid residues His-51, Ser-135 and Asn-152, thereby producing a free binding energy of −6.6 kcal/mol. Hence, berberine may be a good antiviral agent for the treatment of dengue and zika virus.

19.3.3 Piperine

Piperine (1-piperoyl piperidine) is a yellow pungent alkaloid which is widely distributed in the plant genus *Piper* belonging to the family Piperaceae. It has a wide array of pharmacological activity which includes antitumour, antimutagenic, antidiabetic and anti-oxidant and also to treat complicated disease like Parkinson's and Alzheimer' (Cho and Yoon 2015). Some of the interesting facts about the molecular interaction of piperine with the target protein are shown below.

19.3.3.1 Interaction with Serum Albumin

The molecular docking of piperine against the human serum albumin (HSA) was reported by Yeggoni and his associates (2015) incorporating the AutoDock software (4.2.3 version). Binding of piperine-HSA complex was reported to be thermodynamically and conformationally stable, showing binding energy of −6.8 kcal/mol. Piperine develops strong hydrogen bonding specifically at Lys-212 amino acid residue of HSA. In addition, there is also a strong hydrophobic and hydrophilic interaction of HSA amino acid residues (Lys-351, Ala-350, Ala-213, Glu-354, Gly-328, Leu-347,Asp-324, Val-216 and Val-235) with piperine molecule.

In contrast with rat serum albumin (RSA), three active binding sites (site 1, site 2 and site 3) were reported. Piperine showed strong affinity towards site 1 of RSA with a binding energy of −6.7 kcal/mol. The mechanism of interaction revealed that piperine forms H-bonds with Arg-484 and Arg-485 at pH 8.4. The study also showed that there is involvement of both polar (Asn-242, Ser-454) and a polar (Phe-211, Trp-214, Met-203) amino acid residues of RSA with piperine (Zazeri et al. 2019).

19.3.3.2 Inhibitory Interaction Against Acetylcholinesterase (AChE)

AChE is a class of serine hydrolase enzyme which is found in brain cholinergic synapse and neuromuscular junction. It is mainly involved in the biological termination of impulse transmission by promoting the hydrolysis of acetylcholine (ACh) to choline and acetate. Inhibition of AChE may produce therapeutic management of various disease complications, including Alzheimer's disease and glaucoma (Colovic et al. 2013). Piperine-AChE molecular interaction was reported by Manap and associates (2019). Piperine has a strong binding interaction at both the peripheral anionic site (PAS) and catalytic action site (CAS) of the AChE enzyme with a binding energy of −10.5 kcal/mol. The mechanism of molecular binding of piperine showed conventional H-bond interaction with AChE amino acid residues Phe-295, Ser-203, Gly-122 and Gly-121. In addition, the piperidine ring of piperine also formed π-alkyl and π-σ interaction with nearby amino acids Tyr-337 and His-447 and Trp-86, respectively. Thus, piperine can be a potential drug for targeting AChE enzyme associated with the management of Alzheimer's disease.

19.3.3.3 Inhibitory Interaction with Monoamine Oxidase (MAO)

With the involvement of the active methylenedioxyphenyl (MDP) ring present in piperine structure, there exist strong inhibitory binding properties of piperine with

MAO-B by forming a water bridge with amino acid residue Tyr-188 and Cys-172. Piperine also forms H-bond interaction with the aromatic ring of Tyr-398. This selective inhibitory effect of piperine may lead to the novel discovery as potential MAO-B inhibitors for the management of Parkinson's disease (Al-Baghdadi et al. 2012).

19.3.3.4 Inhibitory Interaction with NF-kB

Nuclear factor-kappa B. (NF-kB) is a transcription factor which regulates multiple aspects of adaptive and innate immune functions and is mainly involved in inflammatory responses including cancer. As reported by Verma and associates (2017), piperine may act as anticancer molecule by strongly forming inhibition at the active sites of NF-kB, thereby producing an interaction binding energy of −24.68 kcal/mol. The molecular interaction showed that piperine forms hydrophobic bonds with Tyr-306, Lys-301 and Lys-310 and H-bonds with Arg-302 and Thr-305 of NF-kB.

19.3.3.5 Inhibitory Interaction with Kir6.2 Protein

Kir6.2 protein is a major subunit of the ATP-sensitive K^+ channel. Mutations of Kir6.2 subunits may result in the permanent blockade of the calcium channel and promote the opening of pancreatic K_{ATP} channel, thus leading to the obstruction of insulin exocytose which probably leads to type 1 diabetes. Molecular docking of piperine against Kir protein was reported incorporating AutoDock 4.2 software (Jagadeb et al. 2014). Piperine shows H-bonding with Kir6.2 protein amino acid residues Arg-301, Lys-39, Thr-302 and Phe-183 with an interaction distance of 2.01, 2.1, 1.94 and 1.66 Å, respectively. The piperine-Kir6.2 binding mode may result in the future implications for the management of type 1 diabetes mellitus.

19.3.3.6 Inhibitory Interaction with Tyrosine Kinase (EGFR Tyrosine Kinase)

Tyrosine kinase is an enzyme which on activation can cause proliferation and growth of tumour cells, induce antiapoptotic effects and promote angiogenesis with probable metastasis. Inhibition of tyrosine kinase enzyme is the key target for developing potential anticancer drugs. The inhibitory interaction of piperine against EGFR-tyrosine kinase enzyme has been studied by molecular docking study incorporating in silico techniques (Paarakh et al. 2015). The result revealed that piperine showed two H-bond interactions with EGFR tyrosine kinase amino acid residues Pro-699 and Arg-83, with a respective binding energy of −7.6 kcal/mol and docking energy of −7.06. This inhibitory binding efficacy of piperine against the tyrosine kinase may result in a new lead discovery of the natural component for cytotoxic potency.

19.3.4 Curcumin

Curcumin [1,7-bis-(4-hydroxy-3-methoxyphenyl)-1,6-heptadiene-3,5-dione] is a yellow-orange colour pigment which is distributed widely in Indian turmeric (*Curcuma longa*) belonging to the family Zingiberaceae. It is commonly known as diferuloylmethane and is one of the highly researched and the most valuable components with a wide array of pharmacological activities including anti-inflammatory, anti-HIV, anticancer, anti-angiogenic, antioxidant, antidiabetes, anti-malarial, anti-amyloid, antimicrobial, wound healing, hepatoprotective, etc. (Ji and Shen 2009).

19.3.4.1 Interaction with HIV-1 Protease and Integrase

Human immunodeficiency virus type 1 (HIV-1) is the main culprit for the cause of AIDS, and its replication cycle provides many potential targets for the action of chemotherapeutic drugs. Following the infection, there are three functional enzymes, which are involved in the propagation of the life cycle of the virus, viz. protease, integrase and reverse transcriptase. Curcumin has been reported to act against the HIV infection by inhibiting all the functional enzymes. A report on the inhibitory action of curcumin on HIV-1 protease and integrase with molecular docking simulation was ascertained by Vajragupta and associates (2005) incorporating AutoDock 3 software. Curcumin strongly binds with both integrase and protease with a binding energy of −8.79 and −9.77 kcal/mol, respectively. One terminal phenolic moiety of curcumin forms H-bond with amino acid residue of integrase at His-67, Thr-66 and Lys-155, whereas the other terminal phenolic moiety forms H-bond with Asn-120, Ser-119 and Thr-93. In addition, the *o*-methoxy group of curcumin also interacts with His-67 and Thr-66 amino acid residues of integrase, thereby engaging in 3′-processing activity. In contrast to curcumin-protease interaction, the keto-enol oxygens of curmumin selectively bind to both the chain-A and chain-B site of protease particularly involving the carboxylic acid group of Asp-25 and the carbonyl oxygen (backbone) of Gly-27, respectively. Moreover, both the terminal methoxy and phenolic groups also form H-bonding with the protease amino acid residues at Asp-29 and Asp-30, thus producing inhibitory action on the enzyme.

19.3.4.2 Inhibitory Interaction Against COX-2 Protein

Cyclooxygenase-2 is an important enzyme which is involved in inflammatory process and is also a catalytic enzyme which is responsible for the first step synthesis of prostaglandin, eicosanoids and thromboxanes. Expression of COX-2 has been widely reported to induce the growth of tumour cells by stimulating proliferation and angiogenesis process. COX-2 protein has three active sites, viz. 1CX2, 1PZZ and 1CVU. The protein-ligand docking simulation study of curcumin versus COX-2 was reported by Maldonado-Rojas and Olivero-Verbel (2011) using the AutoDock Vina software. Curcumin produces inhibitory effects at all COX-2 active sites (1CX2, 1PZZ and 1CVU) with a binding energy of −8.4, −8.7 and −9.3 kcal/mol, respectively. It was reported that curcumin binds more specifically with 1CVU active sites of the enzyme. For 1CVU-curcumin, most interactions are aromatic and hydrophobic in nature (involving COX-2 amino acid residues Val-116, Val-523, Val-349, Phe-518, Met-113, Ile-345, Leu-359, Leu-384, Trp-387, Ala-527), except for Ser-530 which forms H-bond with curcumin. This molecular inhibitory effect on COX-2 signifies a potential anti-inflammatory activity of curcumin.

19.3.4.3 Inhibitory Interaction Against PfATP6 Protein

PfATP6 is a calcium ATPase gene encoded by the malarial parasite *Plasmodium falciparum.* This gene has been reported to be the active sites and novel molecular target for the action of 'artemisinins' which are the currently prescribed antimalarial drugs. An attempt to screen the antimalarial activity of curcumin inhibitory action against PfATP6 by in silico technique was reported by Ji and Shen (2009). Result revealed that the keto-enol and phenolic groups of curcumin form H-bond with the surrounding amino acid residue of PfATP6. The keto oxygen group of curcumin forms H-bond with Gln-267, one terminal phenolic oxygen of curcumin forms H-bond with Leu-1040 and Ile-1041, whereas the other terminal phenolic oxygen forms H-bond with Ala-985 of PfATP6. Thus, from the active interaction, it can be incurred that curcumin may be considered among the potential drug for the management of malaria cases.

19.3.4.4 Inhibitory Interaction Against Cancer Proteins EGFR, GST-PI and PDGFA

Cancer is a disease which is governed by a number of virulent factors involving proteins. Epidermal growth factor receptor (EGFR) is one virulent factor which is involved in various cancer events including oral, gastric, bladder and cervical cancer. Glutathione S-transferase Pi gene (GST-Pi) is involved in prostate cancer, whereas platelet-derived growth factor alpha chain (PDGFA) is involved in malignant pleural mesothelioma and glioma cancer. Inhibition of all these virulent factors is the key target for antitumour activity. Curcumin is one potential drug which produces inhibitory effect against all these virulent factors (Mahajanakatti et al. 2014). The binding energies of curcumin against EGFR, PDGFA and GST-Pi were reported to be −7.59, −7.93 and −7.98 kcal/mol, respectively. Inhibitory interaction of curcumin was enhanced by forming H-bond with amino acid residue Lys-120 of GST-Pi, whereas H-bond formation was also observed between curcumin-PDGFA interaction which involves four amino acid residues Glu-43, Glu-125, Ile-83 and Cys-179 of PDGFA. In the case of curcumin-EGFR interaction complex, H-bond formation was observed with EGFR amino acid residues Lys 384, His 10, Ala 89, Met 87, Gly 288, Glu 306 and Asp 290. The study revealed that curcumin may serve as a potential candidate that can be used to manage various types of cancers when present treatments in oncology fail to subside cancer.

In addition to the above data, a brief information on the molecular docking commonly known as phytoconstituents are represented in Table 19.2.

19.4 Summary and Conclusion

Since time immemorial medicinal plants and their active products have been continuously serving the society with potential health benefits in treating various categories of disease ailments. Nature has given us many potential medicinal plants, which are comparatively having lesser side effects, better efficacy and ease of availability for developing new potential and therapeutic lead molecule. Worldwide, the utilization

Table 19.2 Molecular docking simulation study of some therapeutically active phytoconstituents

Sl. No	Name of compound	Targeted protein/ enzyme/receptor	Nature of interaction	Mechanism of molecular interaction	Binding energy (kcal/ mol) or docking score	Software utilized	References
1.	Allicin	Aβ-Peptide (amyloid plague)	Inhibition	Allicin form H-bond interaction with Lys-28 and hydrophobic interaction with Ile-32, Met-35, Val-36 and Val-39 of Aβ-peptide	–	–	Kumar et al. (2019)
2.	AMANTADIG (3β-[2-(1-amantadine)-1-on-thylamine]-digitoxigenin)	Na^+/K^+-ATPase	Inhibition	H-bond interaction with Thr: A-797 and van der Waals interaction with Aal:A-323, Leu:A-125, Phe:A-783, Ile:A-315 and Leu:A-311 amino acid residues of Na^+/K^+-ATPase	−11.68 kcal/ mol	AutoDock Vina	Silva et al. (2018)
3.	Digoxin	P-gp	–	Digoxin form H-bond interaction with Tyr-303, Tyr-110 and Ala981 of amino acid residue of P-gp	–	Discovery Studio	Bai et al. (2019)
4.	Ephedrine	Dipeptidyl peptidase-IV (DPP-IV) enzyme	Inhibition	Ephedrine forms salt bridge and/or H-bond interaction with Glu-205 and Glu-206; and π-π interaction with Tyr-666 amino acid residue of DPP-IV	−7.87 kcal/ mol	Phase v4.5 (Schrodinger, LLC)	Ojeda-Montes et al. (2017)
5.	Linalool	5-HT_3 receptor	Inhibition	Linalool form H-bond interaction between hydroxyl group of linalool and Thr-6 amino acid residue of 5-HT_3	–	GOLD docking program (version 3.0)	Jarvis et al. (2016)

6.	1,8-Cineole	Bcl-2 (β-cell lymphoma)	Inducing	Cineole forms polar interaction with the N-H group of Phe-112 and also exhibit hydrophobic interaction with various amino acid pockets Phe-63, Phe-112, Leu-96, Ala108, Val-92 and Val-115	−4.99 kcal/mol	AutoDock Vina	Sampath et al. (2018)
7.	Eucalyptol	Caspase-3	Inhibition	Eucalyptol forms bifurcated H-bonds with Ser-381 and Phe-381	−4.21 kcal/mol	AutoDock Vina	Bhattacharjee and Chatterjee (2013)
8.	Eugenol	Tumour necrosis factor alpha (TNF-α)	Inhibition	Eugenol forms H-bonds with Gln-102 and Glu-116 and also exhibit hydrophobic interaction with Cys-101 amino acid residue of TNF- α	−5.5 kcal/mol	AutoDock Vina	Mateen et al. (2019)
		Interleukin-6 (IL-6)	Inhibition	Eugenol forms H-bond with Thr-43, Thr-163 and Arg-104 and also exhibits two hydrophobic interactions with Arg-101 and Phe-105 amino acid residue of IL-6	−5.2 kcal/mol	AutoDock Vina	Mateen et al. (2019)
		Acetylcholinesterase (AchE)	Inhibition	(1) Eugenol forms H-bonds with Trp-84 and Glu-199 in the catalytic site of AchE (2) It also forms H-bond with His-440 in the peripheral anionic site of AchE	–	Discovery Studio	Farag et al. (2016)
		β-Glucosidase	Inhibition	Eugenol forms five H-bonds interaction with five amino acid residues at His-147, Arg-113, Asp-225, Lys-146 of β-glucosidase	−83.89 kcal/mol	Molegro Virtual Docker	da Silva et al. (2014)

(continued)

Table 19.2 (continued)

Sl. No	Name of compound	Targeted protein/ enzyme/receptor	Nature of interaction	Mechanism of molecular interaction	Binding energy (kcal/ mol) or docking score	Software utilized	References
9.	Hesperidin	NF-κB	Inhibition	Hesperidin forms H-bonds with amino acid residues of NF-κB specifically at Lys-275, Arg-305, Asp-276, Val-303 and Gln-306	−22.49 kcal/ mol	Lead IT (FlexX)	Nandeesh et al. (2018)
10.	Eriodictyol	Protein kinase A (PKA)	Inducing	(1) Hydroxyl group of eriodictyol forms H-bonds with PKA oxygen atom of amino acid group at Gln-302 and Val-313 (2) Ring C of eriodictyol forms hydrophobic interaction with Tyr-371 amino acid residue	−6.7 kcal/ mol	–	Hameed et al. (2018)
11.	Kaempferol	Protein kinase A (PKA)	Activation	(1) Benzene ring of kaempferol is stabilized by π-π and π-CH_3 interaction of PKA at Tyr-371 and Val-371 site (2) H-bond interaction of 8-OH group of kaempferol with NH moiety of Ala-326, Arg-333 and Ala-334	−5.9 kcal/ mol	–	Hameed et al. (2018)
12.	Naringenin	D_2 dopamine receptor (D_2R)	Antagonist	H-bond interaction with Asp-114 and His-393 and also exhibits hydrophobic interaction with Ile-184,	−9.3 kcal/ mol	AutoDock Vina	He et al. (2018)

				Phe-189, Val-190, Phe-198, Phe-382, Trp-386, Phe-389, Phe-390, Trp-413 amino acid residues of D_2R			
		NF-κB	Inhibition	Naringenin forms H-bond with NF-κB amino acid residue at Gln-306, Arg-305 and Val-303	−21.99 kcal/mol	Lead IT (FlexX)	Nandeesh et al. (2018)
13.	Reserpine	Nor A efflux pump	Inhibition	(1) Reserpine forms arene-arene interaction with aromatic ring of Phe-317 (2) It also forms arene-cationic interaction with Lys-125 (3) Reserpine also forms two H-bond interactions by accepting the electrons from Arg-324 and Phe-129	−8.7 kcal/mol	AutoDock Vina	Bhaskar et al. (2016)
14.	6-Gingerol	VEGFa/VGFR2	Activation	6-Gingerol forms one H-bond with amino acid residue Asp-34 of VEGFa protein and also forms two H-bond with amino acid residues at Lys-286 and Lys-287 of VGFR2	–	Glide XP	Zhong et al. (2019)
		AP-1 Activator protein in cancer	Inhibition	6-Gingerol forms two H-bonds with AP-1 amino acid residue at Arg-140 and Arg-143	−10.23 kcal/mol	Discovery Studio	Teimouri et al. (2016)
		Phosphodiesterase (PDE4B and PDE4D)	Inhibition	(1) 6-Gingerol formed H-bonds with PDE4B amino acid residues at Gln-443, His-234, Tyr-233 and	−7.05 kcal/mol for PDE4B −6.13 kcal/	AutoDock Vina	Xing et al. (2017)

(continued)

Table 19.2 (continued)

Sl. No	Name of compound	Targeted protein/ enzyme/receptor	Nature of interaction	Mechanism of molecular interaction	Binding energy (kcal/ mol) or docking score	Software utilized	References
				Asp-392. It also formed hydrophobic interactions with nearby amino acid residues Met-503, Phe-446, Tyr-403, Ile-410, Phe-414, Phe-506 and Leu-502 in PDE4B (2) 6-Gingerol formed two H-bonds with PDE4D amino acid residue at Tyr-325 and also forms hydrophobic interaction with amino acid residue Met-503, Phe-538, Tyr-495, Ile-502 and Phe-506 of PDE4D	mol for PDE4D		
		Leukotrienes LTA_4H	Inhibition	6-Gingerol forms H-bond with NH_2 terminal of LTA_4H at amino acid residue Glu-271	–	QM-polarized ligand docking	Jeong et al. (2009)
15.	Ursolic acid	β-Catenin/Wnt	Inhibition	Ursolic acid forms five H-bonds with amino acid residues Arg65, Lys66, Tyr159, Leu156 of β-catenin/ Wnt	−9.6 kcal/ mol	Discovery Studio	Teimouri et al. (2016)
16.	Hypericin	Glutathione reductase (GR)	Inhibition	(1) Hypericin forms H-bonds with amino acid residue Arg-81, Glu-381, and Ser-444	–	PatchDock	Dalmizrak et al. (2018)

				from monomer A as well as with Tyr-74, Asp-77 and Arg-81 from monomer B of GR (2) Hypericin aromatic rings produce cation-π interactions with Arg-81 from monomer A and Arg-81 from monomer B			
		P-glycoprotein (P-gp)	Activation	Hypericin forms H-bond with P-gp amino acid residue at Tyr-949 and Tyr-303 and also showed strong interaction with the P-gp amino acid residues (Met-68, Phe-953, Leu-328, Leu-971, Phe974, Ser-729, Ser-975, Phe-728, Ala-981, Val-978 and Phe 332)	–	CDOCKER	Verebova et al. (2016)
16.	Menthol	$5HT_3$ receptor	Antagonist	Menthol hydroxyl group forms strong interaction with the amino acid residue of $5HT_3$ receptor, thereby generating inhibitory effects	−6.35 kcal/mol	AutoDock Vina	Ashoor et al. (2013)
		TRP melastatin-8 receptor (TRPM8)	Activation	Menthol hydroxyl group form strong H-bond with Tyr-745 amino acid residue of TRPM8 receptor	−10.87 kcal/mol	Open Eye Scientific Software package FRED	Pedretti et al. (2009)
17.	Capsaicin	Sigma 1 glutathione S-transferase (Sigma 1 GST)	Activation	Capsaicin forms H-bonds with amino acid residues Asn-140 and Asn-146	−6.4 kcal/mol	Schrodinger Maestro 9.3	Gawande et al. (2014)

(continued)

Table 19.2 (continued)

Sl. No	Name of compound	Targeted protein/ enzyme/receptor	Nature of interaction	Mechanism of molecular interaction	Binding energy (kcal/ mol) or docking score	Software utilized	References
18.	Ellagic acid	Sigma 1 glutathione S-transferase (Sigma 1 GST)	Activation	Ellagic acid forms H-bond with amino acid residue Phe-144	−6.2 kcal/ mol	Schrodinger Maestro 9.3	Gawande et al. (2014)
19.	Genistein	Kir6.2 protein	Inhibitory	Genistein forms H-bonds with amino acid residues Arg-301, Gln-299, His-183 and Phe-183 of Kir protein	–	AutoDock Vina	Jagadeb et al. (2014)

of natural products as the alternative source of medicine by various pharmaceutical and cosmeceutical industries has been growing at an extensive rate. Hence, there is always a need to regulate and validate the efficacy, safety and quality of natural products by putting forward the standardization process. In this chapter, we have elaborated the application of in silico screening method incorporating virtual molecular docking technique, taking into account some potential active constituents from plants having positive interaction with a range of targeted proteins for pharmacological action. The output of the chapter will aid in understanding the interaction of active phytoconstituents with the target at the molecular level with a defined mechanism of binding property. Thus, in silico molecular docking simulation study will serve as an important standardization tool in future aspects of drug discovery and will minimize the problems associated with conventional standardization methods, thereby promoting ease of research and development with accurate and reproducible results.

Acknowledgement The author wished to acknowledge Shrimanta Shankar Academy (SSA) and Girijananda Chowdhury Institute of Pharmaceutical Science, Guwahati, Assam (India), for providing the necessary facilities of the central library for access to the Internet and online subscribed journals.

References

Abdul Manap AS, Wei Tan AC et al (2019) Synergistic effects of curcumin and piperine as potent acetylcholine and amyloidogenic inhibitors with significant neuroprotective activity in sh-sy5y cells via computational molecular modelling and in vitro assay. Front Aging Neurosci 11:206

Agarwal S, Mehrotra R (2016) An overview of molecular docking. JSM Chem 4:1042–1045

Al-Baghdadi OB, Prater NI, Van der Schyf CJ et al (2012) Inhibition of monoamine oxidase by derivatives of piperine, an alkaloid from the pepper plant *Piper nigrum*, for possible use in Parkinson's disease. Bioorg Med Chem Lett 22:7183–7188

Allen BK, Mehta S, Ember SWJ et al (2015) Large-scale computational screening identifies first in class multitarget inhibitor of EGFR kinase and BRD4. N Sci Rep 5:1–16

Armstrong MS, Morris GM, Finn PW et al (2010) ElectroShape: fast molecular similarity calculations incorporating shape, chirality and electrostatics. J Comput Aided Mol Des 24:789–801

Ashoor A, Nordman JC, Veltri D et al (2013) Menthol inhibits 5-HT_3 receptor–mediated currents. J Pharmacol Exp Ther 347:398–409

Atanasov AG, Waltenberger B, Pferschy-Wenzig E et al (2015) Discovery and resupply of pharmacologically active plant-derived natural products: a review. Biotechnol Adv 33:1582–1561

Atanasov AG, Waltenberger B, Pferschy-Wenzig EM et al (2016) Discovery and resupply of pharmacologically active plant-derived natural products: a review. Biotechnol Adv 33:1582–1614

Bai J, Zhao S, Fan X (2019) Inhibitory effects of flavonoids on P-glycoprotein in vitro and in vivo: food/herb-drug interactions and structure–activity relationships. Toxicol Appl Pharmacol 369:49–59

Berman HM, Westbrook J, Feng Z et al (2000) The protein data bank. Nucleic Acids Res 28:235–242

Bhaskar BV, Muni T, Babu C et al (2016) Homology modeling, molecular dynamics, and virtual screening of NorA efflux pump inhibitors of *Staphylococcus aureus*. Drug Des Devel Ther 10:3237–3252

Bhattacharjee B, Chatterjee J (2013) Identification of proapoptopic, anti-inflammatory, anti-proliferative, anti-invasive and anti-angiogenic targets of essential oils in cardamom by dual reverse virtual screening and binding pose analysis. Asian Pacific J Cancer Prev 14:3735–3742

Bucar F, Wube A, Schmid M (2013) Natural product isolation-how to get from biological material to pure compounds. Nat Prod Rep 30:525–545

Chen J, Lai L (2006) Further developments on receptor-based pharmacophore modeling. J Chem Inf Model 46:2684–2691

Cho HJ, Yoon IS (2015) Pharmacokinetic interactions of herbs with cytochrome p450 and p-glycoprotein. Evid Based Complement Alternat Med 2015:736431

Coan KE, Ottl J, Klumpp M (2011) Non-stoichiometric inhibition in biochemical high-throughput screening. Expert Opin Drug Discov 6:405–417

Colovic MB, Krstic DZ, Pasti TDL et al (2013) Acetylcholinesterase inhibitors: pharmacology and toxicology. Curr Neuropharmacol 11:315–335

Corson TW, Crews CM (2007) Molecular understanding and modern application of traditional medicines: triumphs and trials. Cell 130:769–774

Cragg GM, Newman DJ (2013) Natural products: a continuing source of novel drug leads. Biochim Biophys Acta 1830:3670–3695

da Silva Costa J, da Silva Lopes Costa K, Cruz JV et al (2018) Virtual screening and statistical analysis in the design of new caffeine analogues molecules with potential epithelial anticancer activity. Curr Pharm Des 24:576–594

da Silva JK, Silva JR, Nascimento SB et al (2014) Antifungal activity and computational study of constituents from Piper divaricatum essential oil against Fusarium infection in black pepper. Molecules 19:17926–17942

Dalmizrak O, Teralı K, Abdullah RK et al (2018) Mechanistic and structural insights into the in vitro inhibitory action of hypericin on glutathione reductase purified from baker's yeast. J Biochem Mol Toxicol 32:e22051

David B, Wolfender JL, Dias D (2015) The pharmaceutical industry and natural products: historical status and new trends. Phytochem Rev 14:299–315

Erukainure OL, Sanni O, Ijomone OM et al (2019) The antidiabetic properties of the hot water extract of kola nut (*Cola nitida* (Vent.) Schott & Endl.) in type 2 diabetic rats. J Ethnopharmacol 242:1–12

Farag MA, Ezzat SM, Salama MM et al (2016) Anti-acetylcholinesterase activity of essential oils and their major constituents from four Ocimum species. De Gruyter Online 71:393–402

Ferreira LG, dos Santos RN, Oliva G et al (2015) Molecular docking and structure-based drug design strategies. Molecules 20:13384–13421

Gawande ND, Subashini S, Murugan M et al (2014) Molecular screening of insecticides with sigma glutathione S-transferases (GST) in cotton aphid Aphis gossypii using docking. Bioinformation 10:679–683

Giersiefen H, Hilgenfeld R, Hillish A (2003) Modern methods of drug discovery: an introduction. In: Hillisch A, Hilgenfeld R (eds) Modern methods of drug discovery. Springer Basel AG, Switzerland, pp 1–18

Goldman BB, Wipke WT (2000) Quadratic shape descriptors. 1. Rapid superposition of dissimilar molecules using geometrically invariant surface descriptors. J Chem Inf Comput Sci 40:644–658

Gschwend DA, Good AC, Kuntz ID (1996) Molecular docking towards drug discovery. J Mol Recognit 9:175–186

Hameed A, Ashraf S, Khan MI et al (2018) Protein kinase A-dependent insulinotropic effect of selected flavonoids. Int J Biol Macromol 119:149–156

He Y, Zhu S, Wu C et al (2018) Bioactivity-guided separation of potential D_2 dopamine receptor antagonists from *Aurantii Fructus* based on molecular docking combined with high-speed counter-current chromatography. Molecules 23:3135

Hein M, Zilian D, Sotriffer CA (2010) Docking compared to 3D-pharmacophores: the scoring function challenge. Drug Discov Today Technol 7:e229–e236

Heinrich M (2010) Ethnopharmacology in the 21st century-grand challenges. Front Pharmacol 1:8

Henrich CJ, Beutler JA (2013) Matching the power of high throughput screening to the chemical diversity of natural products. Nat Prod Rep 30:1284–1298

Hu XM, Li RT, Zhang MM et al (2018) Phenanthroline-linked berberine dimer and fluorophore-tagged DNA conjugate for the selective detection of microRNA-185: experimental and molecular docking studies. Anal Chim Acta 1051:153–159

Hung TC, Jassey A, Liu CH et al (2018) Berberine inhibits hepatitis C virus entry by targeting the viral E2 glycoprotein. Phytomedicine 53:62–69

Institute of Medicine (US) Committee on Military Nutrition Research (2001) Caffeine for the sustainment of mental task performance: formulations for military operations, vol. 2, Pharmacology of caffeine. National Academies Press, Washington, DC. https://www.ncbi.nlm.nih.gov/books/NBK223808/

Islam MM, Sonu VK, Gashnga PM et al (2016) Caffeine and sulfadiazine interact differently with human serum albumin: a combined fluorescence and molecular docking study. Mol Biomol Spectrosc 152:23–33

Jagadeb M, Konkimalla VB, Rath SN et al (2014) Elucidation of the inhibitory effect of phytochemicals with Kir6.2 wild-type and mutant models associated in type-1 diabetes through molecular docking approach. Genomics Inform 12:283–288

Jarvis GE, Barbosa R, Thompson AJ (2016) Noncompetitive inhibition of 5-HT3 receptors by citral, linalool, and eucalyptol revealed by nonlinear mixed-effects modeling. J Pharmacol Exp Ther 356:549–562

Jeong CH, Bode AM, Pugliese A et al (2009) [6]-Gingerol suppresses colon cancer growth by targeting leukotriene A4 hydrolase. Cancer Res 69:5584–5591

Ji HF, Shen L (2009) Interactions of curcumin with the PfATP6 model and the implications for its antimalarial mechanism. Bioorg Med Chem Lett 19:2453–2455

Jones WP, Kinghorn AD (2012) Extraction of plant secondary metabolites. Methods Mol Biol 864:341–366

Kaboli PJ, Leong MPY, Ismail P et al (2018) Antitumor effects of berberine against EGFR, ERK1/2, P38 and AKT in MDA-MB231 and MCF-7 breast cancer cells using molecular modelling and *in vitro* study. Pharmacol Rep 71:13–23

Kaiser H (2008) Von der Pflanze zur Chemie-Die Frühgeschichte der"Rheumamittel". Z Rheumatol 67:252–262

Kinghorn AD, Pan L, Fletcher JN et al (2011) The relevance of higher plants in lead compound discovery programs. J Nat Prod 74:1539–1555

Kingston DG (2011) Modern natural products drug discovery and its relevance to biodiversity conservation. J Nat Prod 74:496–511

Kitchen DB, Decornez H, Furr JR et al (2004) Docking and scoring in virtual screening for drug discovery: methods and applications. Drug Discov 3:935–949

Krovat EM, Steindl T, Langer T (2005) Recent advances in docking and scoring. Curr Comput Aided Drug Design 1:93–102

Kruse PR (2007) Geschichte der Pharmazie. Vol II: Von der Frühen Neuzeit bis zur Gegenwart-by Rudolf Schmitz Centaurus 49, pp 182–183

Kumar S, Kumar S, Ram H (2019) Anti-aggregation property of Allicin by in vitro and molecular docking studies. J Exp Neurosci 13:1–5

Lee KE, Bharadwaj S, Yadava U et al (2019) Evaluation of caffeine as inhibitor against collagenase, elastase and tyrosinase using in silico and in vitro approach. J Enzyme Inhib Med Chem 34:927–936

Lengauer T, Rarey M (1996) Computational methods for biomolecular docking. Curr Opin Struct Biol 6:402–406

Mahajanakatti AB, Murthy G, Sharma N et al (2014) Exploring inhibitory potential of curcumin against various cancer targets by in silico virtual screening. Interdiscip Sci 6:13–24

Maldonado-Rojas W, Olivero-Verbel J (2011) Potential interaction of natural dietary bioactive compounds with COX-2. J Mol Graph Model 30:157–166

Markey SP (2007) Chapter 11—Pathways of drug metabolism. In: Principles of clinical pharmacology, 2nd edn, pp 143–162

Mateen S, Rehman Md T, Shahzad S et al (2019) Anti-oxidant and anti-inflammatory effects of cinnamaldehyde and eugenol on mononuclear cells of rheumatoid arthritis patients. Eur J Pharmacol 852:14–24

Mazumder MK, Choudhury S, Borah A (2019) An *in silico* investigation on the inhibitory potential of the constituents of pomegranate juice on antioxidant defense mechanism: relevance to neurodegenerative diseases. IBRO Reports 6:153–159

Meng EC, Shoichet BK, Kuntz ID (1992) Automated docking with grid-based energy evaluation. J Comput Chem 13:505–524

Nandeesh R, Vijayakumar S, Munnolli A et al (2018) Bioactive phenolic fraction of *Citrus maxima* abate lipopolysaccharide induced sickness behaviour and anorexia in mice: in silico molecular docking and dynamic studies of biomarkers against NF-κB. Biomed Pharmacother 108:1535–1545

Newman DJ, Cragg GM (2012) Natural products as sources of new drugs over the 30 years from 1981 to 2010. J Nat Prod 75:311–335

Ojeda-Montes MJ, Ardid-Ruiz A, Tomas-Hernandez S et al (2017) Ephedrine as a lead compound for the development of new DPP-IV inhibitors. Future Med Chem 9:2129–2146

Paarakh PM, Sreeram DC et al (2015) In vitro cytotoxic and *in silico* activity of piperine isolated from *Piper nigrum* fruits Linn. In Silico Pharmacol 3:9

Pedretti A, Marconi C, Bettinelli I et al (2009) Comparative modeling of the quaternary structure for the human TRPM8 channel and analysis of its binding features. Biochim Biophys Acta 1788:973–982

Petzer A, Pienaar A, Petzer JP (2013) The interactions of caffeine with monoamine oxidase. Life Sci 93:283–287

Pierce AC, Jacobs M, Stuver-Moody C (2008) Docking study yields four novel inhibitors of the protooncogene Pim-1 kinase. J Med Chem 51:1972–1975

Porter AG, Janicke RU (1999) Emerging roles of caspase-3 in apoptosis. Cell Death Differ 6 (2):99–104

Rollinger JM, Langer T, Stuppner H (2006) Integrated *in silico* tools for exploiting the natural products' bioactivity. Planta Med 72:671–678

Rollinger JM, Schuster D, Danzl B et al (2009) In silico target fishing for rationalized ligand discovery exemplified on constituents of *Ruta graveolens*. Planta Med 75:195–204

Sahoo M, Jena L, Daf S et al (2016) Virtual screening for potential inhibitors of NS3 protein of Zika virus. Genomics Inform 14:104–111

Sampath S, Subramani S, Janardhanam S et al (2018) Bioactive compound 1,8-cineole selectively induces G2/M arrest in A431 cells through the upregulation of the p53 signaling pathway and molecular docking studies. Phytomedicine 46:57–68

Scannell JW, Blanckley A, Boldon H et al (2012) Diagnosing the decline in pharmaceutical R&D efficiency. Nat Rev Drug Discov 11:191–200

Sertürner FW (1817) Über das Morphium, eine neue salzfähige Grundlage, und die Mekonsäure, als Hauptbestandteile des Opiums. Ann Phys 25:56–90

Shoichet BK, McGovern SL, Wei B et al (2002) Lead discovery using molecular docking. Curr Opin Chem Biol 6:439–446

Silva IT, Munkert J, Nolte E et al (2018) Cytotoxicity of AMANTADIG -a semisynthetic digitoxigenin derivative—alone and in combination with docetaxel in human hormone-refractory prostate cancer cells and its effect on Na^{+}/K^{+-}ATPase inhibition. Biomed Pharmacother 107:464–474

Singh S, Sharma DK (2011) *In silico* modeling in conjunction with natural products: paving the way for rational drug-design. Biotech Mol Biol Rev 6:88–93

Tan L, Li C, Chen H et al (2017) Epiberberine, a natural protoberberine alkaloid, inhibits urease of Helicobacter pylori and jack bean: susceptibility and mechanism. Eur J Pharm Sci 110:77–86

Teimouri M, Junaid M, Saleem S et al (2016) In-vitro analysis of selective nutraceuticals binding to human transcription factors through computer aided molecular docking predictions. Bioinformation 12:354–358

Totrov M, Abagyan R (2008) Flexible ligand docking to multiple receptor conformations: a practical alternative. Curr Opin Struct Biol 18:178–184

Truman AW, Kim KY, Levin DE (2009) Mechanism of Mpk1 mitogen-activated protein kinase binding to the Swi4 transcription factor and its regulation by a novel caffeine-induced phosphorylation. Mol Cell Biol 29:6449–6461

Vajragupta O, Boonchoong P, Morris GM (2005) Active site binding modes of curcumin in HIV-1 protease and integrase. Bioorg Med Chem Lett 15:3364–3368

Venkatraman V, Sael L, Kihara D (2009) Potential for protein surface shape analysis using spherical harmonics and 3D zernike descriptors. Cell Biochem Biophys 54:23–32

Verebova V, Belej D, Joniova J et al (2016) Deeper insights into the drug defense of glioma cells against hydrophobic molecules. Int J Pharm 503:56–67

Verma A, Kushwaha HN, Srivastava AK et al (2017) Piperine attenuates UV-R induced cell damage in human keratinocytes via NF-kB, Bax/Bcl-2 pathway: an application for photoprotection. J Photochem Photobiol B 172:139–148

Wang Y, Anderson M (2012) Chapter 22—intracellular signalling pathways in cardiac remodeling. Muscle. Fundamental biology and mechanisms of disease, vol. 1, pp 299–308

Wang J, Kollman PA, Kuntz ID (1999) Flexible ligand docking: a multistep strategy approach. Proteins 36:1–19

Wang W, Zhang W, Duan Y et al (2013) Investigation of the binding sites and orientation of caffeine on human serum albumin by surface-enhanced Raman scattering and molecular docking. Mol Biomol Spectrosc 115:57–63

Waszkowycz B, Clark DE, Gancia E (2011) Outstanding challenges in protein-ligand docking and structure-based virtual screening. Wiley Interdiscip Rev Comput Mol Sci 1:229–259

Wei BQ, Weaver LH, Ferrari AM et al (2004) Testing a flexible-receptor docking algorithm in a model binding site. J Mol Biol 337:1161–1182

Wojtukiewicz MZ, Hempel D, Sierko E et al (2016) Thrombin-unique coagulation system protein with multifaceted impacts on cancer and metastasis. Cancer Metast Rev 35:213–233

Wolber G, Langer T (2005) LigandScout: 3-D pharmacophores derived from protein-bound ligands and their use as virtual screening filters. J Chem Inf Model 45:160–169

Xing M, Akowuah GA, Gautam V et al (2017) Structure based design of selective phosphodiesterase 4B inhibitors based on ginger phenolic compounds. J Biomol Struct Dyn 35:2910–2924

Yeggoni DP, Rachamallu A, Kallubai M et al (2015) Cytotoxicity and comparative binding mechanism of piperine with human serum albumin and α-1-acid glycoprotein. J Biomol Struct Dyn 33:1336–1351

Zauhar RJ, Gianti E, Welsh WJ (2013) Fragment-based shape signatures: a new tool for virtual screening and drug discovery. J Comput Aided Mol Des 27:1009–1036

Zazeri G, Povinelli APR, Lima MF et al (2019) Experimental approaches and computational modeling of rat serum albumin and its interaction with piperine. Int J Mol Sci 20:E2856

Zenk MH, Juenger M (2007) Evolution and current status of the phytochemistry of nitrogenous compounds. Phytochemistry 68:2757–2772

Zhang H, Xu X, Chen L et al (2011) Molecular determinants of magnolol targeting both RXR-alpha and PPAR-gamma. PLoS One 6:e28253

Zhong W, Yang W, Qin Y et al (2019) 6-Gingerol stabilized the p-VEGFR2/VE-cadherin/β-catenin/actin complex promotes microvessel normalization and suppresses tumor progression. J Exp Clin Cancer Res 38:285

Zhou X, Li H, Shi Z et al (2017) Inhibition activity of a traditional Chinese herbal formula Huang-Lian-Jie-Du-Tang and its major components found in its plasma profile on neuraminidase-1. Sci Rep 7:15549

Standardization and Quality Evaluation of Botanicals with Special Reference to Marker Components

20

Khomendra Kumar Sarwa, Dipali Patel, Mithun Rudrapal, Soumya Bhattacharya, Swarnlata Saraf, Vishal Jain, Veenu Joshi, Ravindra Pandey, and Amber Vyas

Abstract

Botanicals or herbal drugs have gained wide popularity because of their easy availability and acceptance by patients. In earlier days, no standardized methods/techniques were available to test the quality efficacy and safety of marker components obtained from herbal sources. It was a big challenge in the formulation of herbal pharmaceutical products. In recent days, with the advancement in instrumentation and technology, it has become easier to identify and screen the marker compounds responsible for the bio-efficacy of herbal drugs. Therefore it

K. K. Sarwa (✉)
Department of Pharmacy, Government Girls Polytechnic, Raipur, Chhattisgarh, India
e-mail: khomendra.sarwa@gmail.com

D. Patel
Shri RawatpuraSarkar Institute of Pharmacy, Kumhari, Chhattisgarh, India
e-mail: dipkrish002@gmail.com

M. Rudrapal
Srikrupa Institute of Pharmaceutical Sciences, Siddipet, Telangana, India
e-mail: rsmrpal@gmail.com

S. Bhattacharya
Guru Nanak Institute of Pharmaceutical Science and Technology, Kolkata, India

S. Saraf · V. Jain · A. Vyas
University Institute of Pharmacy, Pt. Ravishankar Shukla University, Raipur, Chhattisgarh, India
e-mail: swarnlata_saraf@rediffmail.com; Vish_106@rediffmail.com; ambervyas@gmail.com

V. Joshi
Center for Basic Science, Pt. Ravishankar Shukla University, Raipur, Chhattisgarh, India
e-mail: vinu.jsh@gmail.com

R. Pandey
Columbia Institute of Pharmacy, Raipur, Chhattisgarh, India
e-mail: ravindraiop@gmail.com

S. C. Mandal et al. (eds.), *Evidence Based Validation of Traditional Medicines*,
https://doi.org/10.1007/978-981-15-8127-4_20

has given a new direction in the formulation of efficacious and safe and multi-component herbal medicine. Chemical marker-based and genetic level standardization can reduce the chances of adulteration, toxicity, and global market-based variation of botanical drugs. Our chapter will cover the latest developments in techniques and strategies that foster the evaluation of marker components in terms of their efficacy, safety, and quality as per WHO guidelines. The chapter would encompass various challenges that still exist and regulatory guidelines to be complied with in standardization for marker components to use herbal drugs with a high margin of safety and efficacy.

Keywords

Standardization · Quality · Botanicals · Herbal drugs · Adulteration · Marker · Bio-efficacy · Toxicity

20.1 Introduction

Botanical drugs are shared with large portions of a source of medicine. Its applicability has been part of accepted since ancient times as an important healthcare system. Botanical drugs have been an important natural source of conventional drugs available in the market. An application of the botanical drug is not limited to the traditional medicinal practice, but it is also accepted in modern medicinal practice, few examples of botanical drug which are used in allopathic system pilocarpine, such as digoxin, atropine, ephedrine, tubocurarine, vasaka, balsum of tolu, artemisinin, morphine, paclitaxel, etc. Moreover, botanicals are used as raw materials for a different range of products not only for herbal pharmaceuticals but also for dietary supplements, food industry, cosmetics, beverages, agrochemicals, etc. However, the phytotherapeutic agents or bioactive markers are believed to be the key components responsible for the medicinal importance as well as the industrial applicability of botanicals. As per the World Health Organization (WHO), botanical drugs are sharing about 80% part of the primary healthcare system. Popularity and demand for botanicals are highly increased in the last few decades, so pharmaceutical manufacturers show interest in botanicals. This increasing demands standardization which ensures quality, safety, efficacy, and availability of raw materials with a similar quality profile. The manufacturer involved in research and development tries to find out new lead compounds from botanical sources (Sarwa et al. 2014a, b).

20.2 Botanical Drugs/Botanical Products

Botanicals drugs mean a plant or plant parts used as an active ingredient for a pharmaceutical formulation that may contain the crude drug or extract or the isolated active ingredient. A botanical preparation means finished formulation in a suitable dosage form which is ready for marketing.

The therapeutic application of any pharmaceutical formulation is based on the added active principal. This active principle is called as an active pharmaceutical ingredient. This biomarker term is more precisely called as bioactive chemical markers (BCM). Any chemical compound from a botanical drug is considered as BCM if the overall therapeutic effect is based on these compounds which are differentiated as well as comparable to another chemical group. The therapeutic potential can be modulated by controlling the strength of these bioactive chemical markers, which suggests a set of promising quality assurance and tools for representing the pharmacological activity of botanical drugs (Panteghini 2004).

20.3 Standardization of Botanical Drugs/Botanical Products

Standardization is a method to check the presence of some predefined specifications in a product or a process. In general terminology, standardization is a process whether confirmation of identity and determination of purity has been done by morphological, microscopically, physical, chemical, and biological methods for any botanicals. This process has been carried out with respect to the prescribed set of standards, constant parameters, inherent characteristics, and definitive qualitative and quantitative values. Standardization process is involved in the determination of technical standards which is helpful for the reproduction of one set of results when needed. The process of standardization assured the quality, which will finally reflect the efficacy and safety of the botanicals as well as finished preparation.

The standardization of botanicals is required because of the following reasons:

1. Identification of the source
2. Quantification of biomarker
3. Identification of common adulterants
4. Determination of quality, safety, efficacy, and stability

Development of standards for specific botanicals drugs and preparation is a long-term process work out by experimentation. The experimental results are critically reviewed and finally prescribed as a set of parameters. A different agency is continuously working at national and international levels for setting standards for botanical drugs and preparation. The American Herbal Products Association (AHPA) is a national trade association working on botanical preparation that defines "standardization refers to the body of information and control necessary to produce material of reasonable consistency." The development of botanical finished product need to remove batch-to-batch variation. This is achieved by the standardization of botanical raw materials (Kunle et al. 2012; Kumari and Kotecha 2016). Hence, the process of standardization is a tool for determining identity, purity, and quality of botanical. The steps of the development of botanical preparation are summarized in Fig. 20.1.

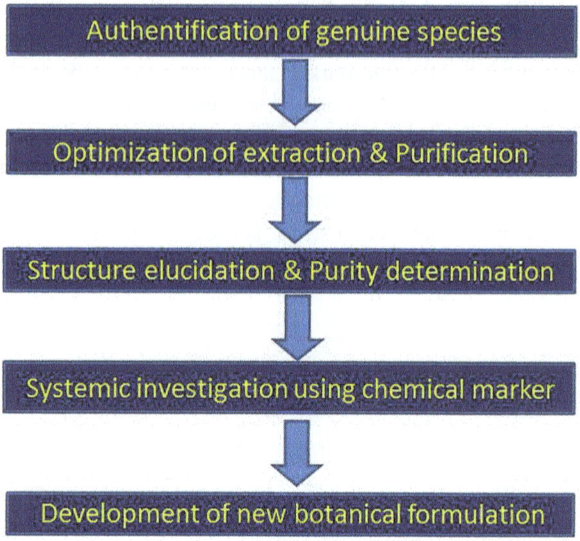

Fig. 20.1 Steps of development of botanical pharmaceuticals

20.3.1 Requirement of Standardization in Botanicals

The quality of botanical will affect the response of the finished product and ultimately applicability of the preparation. The applicability of standardization is presented in the following headings:

1. A dynamic revolution in the field of sophisticated instruments has to create a new horizon to set new quality parameters for traditional herbs.
2. The supply of genuine botanicals is a challenge without preset quality standards.
3. The geographical factor, storage condition, and agronomic variables will affect the quality of botanical.
4. A batch-to-batch variation is more common in botanicals as compared to synthetic drugs
5. Legal applicability of botanical drugs and preparation also demands a standard quality practice in a documented format.
6. The traditional practice of botanical drug is based on the assessment of response and need to be correlated with the quality of applied botanicals.

20.3.2 Standardization of Botanical/Polyherbal Formulation

Standardization of botanical preparation divided into a two-step process. In the first step, crude drug or raw botanical material is standardized. The main aim of this step is to confirm biochemical variation in the source, deterioration due to the storage, and remove substitution and adulteration.

This step is helpful in the selection of the best raw material for the finished botanical preparation. The information collected in this step is preserved throughout the self-life of the product and used as a reference at any time of the manufacturing procedure. The standardization of the botanical finished product is quite complex as compared to the synthetic drug preparations. The botanical finished pharmaceuticals may be classified into four categories, first preparations which contain single marker compound in pure isolated form like digoxin tablet (Lanoxin, Sun Pharma). The second category is containing more than one isolated compound. The example for this class is Migranil EC containing ergotamine and caffeine in tablet formulation. The third category is a product that doesn't contain an isolated compound but having an extract of botanicals. A The Himalaya Drug Company product Hadjod Tablet contains extract of Asthisamhara (*Cissus quadrangularis*). The fourth category is a product that contains raw botanicals like Sankat Mochan manufactured by L.P. Nagar & Company. The Sankat Mochan polyherbal formulation contains *Cinnamomum camphora*, *Trachyspermum ammi*, *Eugenia caryophyllus*, *Cinnamomum zylanicum*, *Foeniculum vulgare* oil, *Eucalyptus globulus* oil, and *Mentha arvensis* (Sarwa et al. 2017).

The botanical preparations containing isolated compounds are much simple by using chromatographic fingerprinting and genetic fingerprinting, but standardization of botanical preparation containing semipurified extract is more complex due to the interference of another available compound. A simultaneous fingerprint is required for this class of finished product to avoid batch-to-batch variation with an assuring the quality, efficacy, safety, and acceptability (Sarwa et al. 2017) To avoid variation in finished product, a good laboratory practice (GLP) and good manufacturing procedure (GMP) must be accepted. The acceptance of good agriculture practice (GAP) in the production of raw materials may use in the development of the standardized finished product.

A clinical trial is required to establish the relationship between active compound and therapeutic response. This correlation would promote clinical use as well as giving a direction to the standardization process. Sometimes a semipurified extract shows better clinical response than pure isolated compound (Sarwa et al. 2014a, b, 2016). The safety consideration will helpful to the determination of the application of botanical drugs as a prescription or as over the counter medicine. Standardization parameters reported in The Ayurvedic Pharmacopoeia of India are given in Tables 20.1 and 20.2.

20.4 Quality Variation in Botanicals

The botanicals are prone to quality variation, such variations are ecotypic, genotypic, and chemotypes variations. It is difficult to control all variables at the same level for all cultivators by which minimize the batch-to-batch variation, so the better opportunity is to develop standardization techniques by scaling up all products at the same level. Numbers of factors affect the plant chemical constituent profile. It is difficult

Table 20.1 Standardization parameters reported in The Ayurvedic Pharmacopoeia of India

Parameters
• **Test and Determination**
Microscopic identification
Determination of quantitative data
Limit tests
Microbial limit tests
Pesticide residues
Gas chromatography
Test for aflatoxin
Test for the absence of methanol
• **Physical tests and determinations**
• **Chemical tests and assays**

for manufacturers to control all these factors for making a similar product in different batches. Some of the factors are summarized below (Sampaio et al. 2016):

- Collection and harvesting time and procedure, storage conditions, and method
- Environmental conditions such as rainfall, humidity, sunlight, altitude, nature of soil, and temperature
- Manufacturing techniques and processes such as selection, size reduction, drying, extraction, evaporation, crystallization, purifying and extraction, etc.
- Selection of seed is important because genetic variability is also found in the seed which will affect the resultant product quality and quantity.
- Pollination, microbial attack, birds bite, and insect feeding may affect secondary metabolites of the botanical plant.
- Seasonal changes largely affect the chemical constituent profile. The chemical constituent composition is changed by diurnal variations
- The chemical of interest in botanicals is varying with stages of ripeness.

20.5 Types of Evaluation

Evaluation of botanicals drug is defined as a process of identification, quantification of a constituent of interest, and detection of an adulterant. The evaluation process necessary to be considered all quality attributes in the development of the standardization procedure (Kumari and Kotecha 2016). A standardization process is an extended form of older quality evaluation with adding some new parameters which are likely to be constant for particular botanicals or its preparation. Standardization covers all scientific class of investigation in which standard referral limit and the procedure have been set for a specific outcome. This procedure includes the development of physical standards, chemical standards, and clinical standards.

Table 20.2 Standardization parameters reported in The Ayurvedic Pharmacopoeia of India

Determination of quantitative data	Limit tests	Microbial limit tests	Pesticide residues	Physical tests and determinations	Chemical tests and assays
• Net content • Foreign matter • Determination of total ash • Determination of acid-insoluble ash • Determination of water-soluble ash 1.1.6. Determination of Sulphated ash • Determination of alcohol-soluble extractive • Determination of water-soluble extractive • Determination of ether-soluble extractive (fixed oil content) • Determination of moisture content (loss on drying) • Determination of volatile oil in drugs • Special processes used in alkaloidal assays • Thin layer chromatography (TLC) • Starch Estimation (Mont Gomery, 1957) • Sugar Estimation (Mont Gomery, 1957)	• Limit test for arsenic • Limit test for chlorides • Limit test for heavy metals • Limit test for Iron • Limit test for Lead • Limit test for Sulphates • Heavy metals by atomic absorption spectrophotometry	• Total aerobic microbial count • Tests for specified micro-organisms	• Qualitative and quantitative analysis of pesticide residues • Test for pesticides • Quantitative analysis	• Refractive index • Weight per milliliter and specific gravity determination of pH value • Determination of melting and congealing range – Determination of melting range – Determination of congealing range • Determination of boiling range • Determination of optical rotation • Determination of viscosity • Determination Total solids • Solubility in water • Determination of saponification value • Determination of iodine value • Determination of acid value • Determination of peroxide value • Determination of	• Estimation of Total Phenolics • Estimation of Total tannins • Estimation of sugars – Nelson–Somogyi photometric method – Reducing and non-reducing sugar • Estimation of curcumin by TLC densitometer • Determination of Aluminium • Determination of borax • Determination of calcium • Determination of copper • Determination of Iron (Fe) • Determination of magnesium • Determination of mercury • Determination of

(continued)

Table 20.2 (continued)

Determination of quantitative data	Limit tests	Microbial limit tests	Pesticide residues	Physical tests and determinations	Chemical tests and assays
• Fatty oil estimation • Protein Estimation • Method for alkaloid estimation				unsaponifiable matter • Detection of mineral oil (Holde's test) • Rancidity test (Kreis test) • Determination of alcohol content	silica (SiO_2) • Estimation of sodium and potassium by flame photometry • Determination of sodium chloride • Determination of Sulphur • Qualitative reactions of some radicals

20.5.1 Physical Evaluation

The physical evaluation develop standards which are based on a visual characteristic of materials. Identification, authentication, macroscopic and microscopic characteristic, sensor characteristic, pharmacognostic, etc. parameters may consider in their class. A microscopic analysis is the very oldest technique adopted for identification. Nowadays electron microscope is also applied for the evaluation process (Serrano et al. 2010). The estimation of quality based on size shape color order taste and texture is categorized as macroscopic evaluation. A further application of the microscope is involved in macroscopic evaluation. In this procedure, section cutting is very common to identify the cellular structure and few identified characters like a prism-shaped crystal of calcium oxalate and stone cells in some plants (AHPA 2019). The example of few physical standardization parameters like ash value, moisture content, melting point, optical rotation, and refractive index, etc. comes under this evaluation.

20.5.2 Chemical Evaluation

Chemical evaluation of the botanical includes phytochemical profile investigation and assessment of potency of an active chemical constituent of interest. A phytochemical investigation is an important aspect of standardization procedure because therapeutic application depends on chemical constituents. The determination of nature and quantity of chemical constituents of the botanicals come in chemical evaluation. These plant constituents are classified as a carbohydrate, glycoside, tannin, lipids, terpenoids and alkaloids, etc. Some other parameters like acid value, saponification value, sulfated ash, acetyl value, ester value, and methoxy determination also come under this evaluation. Chemical fingerprinting by both chromatographic techniques and spectrophotometric analysis is adopted in chemical evaluation which is extended to and genetic fingerprinting (AHPA 2019; Serrano et al. 2010).

20.5.3 Clinical Evaluation

Clinical evaluation is based on direct assessment of therapeutic responses on a living organism or isolated organs. The strength of the response is directly reflecting the qualitative as well as quantitative assessment of botanical with reference to their therapeutic applicability. In some cases, the microbial assay may also be adopted for effective assessment of the clinical response of botanical and their preparation. Pharmacokinetic and bioavailability parameter determination is also covered. A limit and standard procedure of determination of safety and toxicity are needed to be determined. Various safety parameters like maximum safe dose, short and long term toxicity, pesticide residue, microbial load, aflatoxin analysis, etc. are come under this heading. The various types of evaluation are summarized in Fig. 20.2.

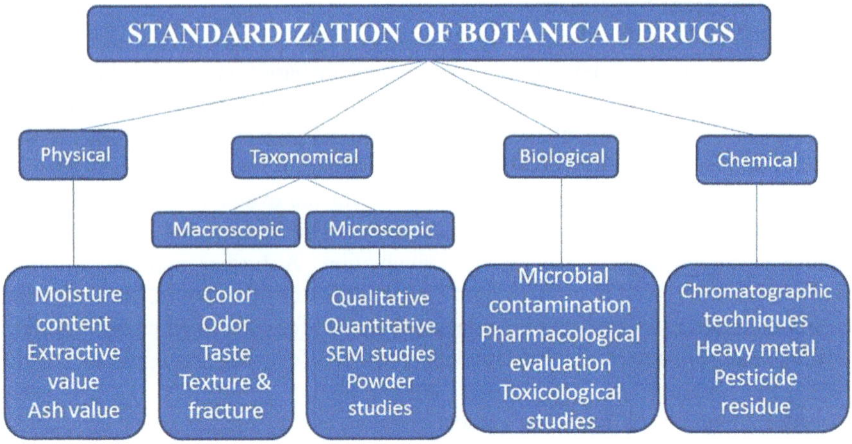

Fig. 20.2 Standardization of drug: physical, taxonomical, biological, and chemical

20.6 Markers Compounds: Key of Standardization of Botanical Drugs and Preparation

Markers are biochemical as well as phytoconstituents including both primary and secondary metabolites as well as other macromolecules like nucleic acid considered for quality control purpose may/may not possess therapeutic activity. The World Health Organization defines "markers as reference substances that are chemically defined constituents of herbal material. They may or may not contribute to the therapeutic activity. However, even when they contribute to the therapeutic activity, evidence that they are solely responsible for the clinical efficacy may not be available" (WHO 2017). The European Medicines Agency (EMA) defines "the chemical markers as chemically defined constituents or groups of constituents of an herbal medicinal product which are of interest for quality control purposes regardless of whether they possess any therapeutic activity (EMA)." The definition of these markers might be extended by added genetic materials. Now the markers are classified into two broad categories: first genetic markers that are used in the Genetic Fingerprinting Techniques. In these techniques, genetic materials like DNA are used for quality control purpose. Another is a Chromatographic Fingerprinting Technique in which chemical constituents are used for quality attributes. Any compound may serve as markers compound if isolated as single. Normally primary and secondary metabolites act as a marker compound for botanical drugs. They belong to group glycosides, terpenes, steroid, alkaloid, flavonoid, aromatic, and heteroaromatic frameworks. Similarly, glycosides having alcoholic, amide, ester, carbonyl, and olefinic acid functionalities are highly useful in the standardization. In the standardization process, chemical markers can be easily analyzed, quantified, and

Table 20.3 Characteristics of markers compound (WHO)

	Marker substances of constituents with known therapeutic activity	Marker substances of constituents with recognized pharmacological activities	Marker substances of characteristic constituents	Marker substances for toxic constituents
Purposes and function	Identification or quantification	Qualitative and quantitative measures in herbal medicine	Identification and quantification of herbal materials in herbal preparations and finished herbal products	Define maximum acceptable concentrations of toxic constituents in herbal materials, herbal preparations, or finished herbal products
Selection criteria	Readily available Easy to separate/ distinguish the marker analytically from other structurally similar herbal constituents Detectable and quantifiable with the available analytical instrumental method	Occur naturally in sufficient quantities in herbal materials A representative of the main therapeutic or pharmacological profiles of the herbal materials and finished products Detectable and quantifiable by available instrumental analytical methods	A marker should be specific for one plant. If not, the marker should be specific for a certain plant species, genus, and family Marker for quantification should be available in sufficient quantity for assay	A consequence of the composition of the herbal material or product, such a limit test is needed Defined upper tolerable limit for the mode of application and posology intended Toxicological evaluation including genotoxicity, mutagenicity, and carcinogenicity Analytical detection procedure for the established tolerable limits should be available

distinguished from similar constituents. So this parameter may easily adoptive for both crude botanical standardization and finished product quality assessment. Characteristics of markers compound as per WHO are summarized in Table 20.3.

20.7 Genetic Marking: DNA Fingerprinting

The term molecular marking, DNA marking, is a synonym used for genetic marking. All the living organisms having a genetic material and this deoxyribonucleic acid (DNA) is considered as the fundamental building constituents. The characteristics and physical features of botanicals depend on this DNA sequence. So this property is applied in the standardization of botanicals. The genetic marking is based on the identification of specific arrangements of DNA base-pair sequences in the cell. Molecular marking techniques are attractive to scientists because it is unique for each species. The genetic composition of botanical is not affected by age or physiological conditions or environmental factors. The important aspect of this fingerprinting is simplicity in procedure and applicability in both fresh as well as old samples. Another aspect of genetic fingerprinting does not restrict by physical form and quantity (Sucher et al. 2012).

Genetic marking can be further categorized into two classes as dominant and co-dominant. A basic difference in these two is the capability of analyzing loci with PCR reaction. Dominant markers are carried out an analysis of many loci at one time, and co-dominant can do analyze one locus at a time. The commonly used genetic marking techniques are AFLP (amplified fragment length polymorphism), microsatellite polymorphism, RAPD (random amplification of polymorphic DNA), RFLP (restriction fragment length polymorphism), SFP (single feature polymorphism), SNP (single-nucleotide polymorphism), STR (short tandem repeat), and VNTR (variable number tandem repeat) (Chial 2008). The scheme of genetic marking is presented as Fig. 20.3.

20.7.1 DNA Barcoding: Tool for Genetic Marker-Based Standardization

Authentication of raw botanicals materials is an initial step of standardization. Due to variation in geo-climatic level, species level, chemotypes level, cultivar level, and market-level adulteration, taxonomic or chemical marker-based authentication is a challenging job. To overcome this problem, DNA barcoding technique is used to find out the differences in very short regions (locus) of DNA (Balachandran et al. 2015). Generally, the process associated with a resolution ability of chloroplast (Cp) genomes including maturase K (*mat*K) and ribulose 1, 5-bisphosphate carboxylase/oxygenase large subunit (*rbc*L). For closely related species, combination of nuclear genome along with the chloroplast genome is also considered [nuclear internal transcribed spacer (*ITS*)]. Molecular standardization successfully applied in the determine the total substitution in herbal products like St. John's wort (*Hypericum perforatum*) substituted with *Senna alexandrina* (Fabaceae) and Ginkgo product with *Juglans nigra* (black walnut) (Newmaster et al. 2013). Although this technique is going to be an effective tool for standardization, still chemical constituent-based standardization cannot be avoided for botanicals because

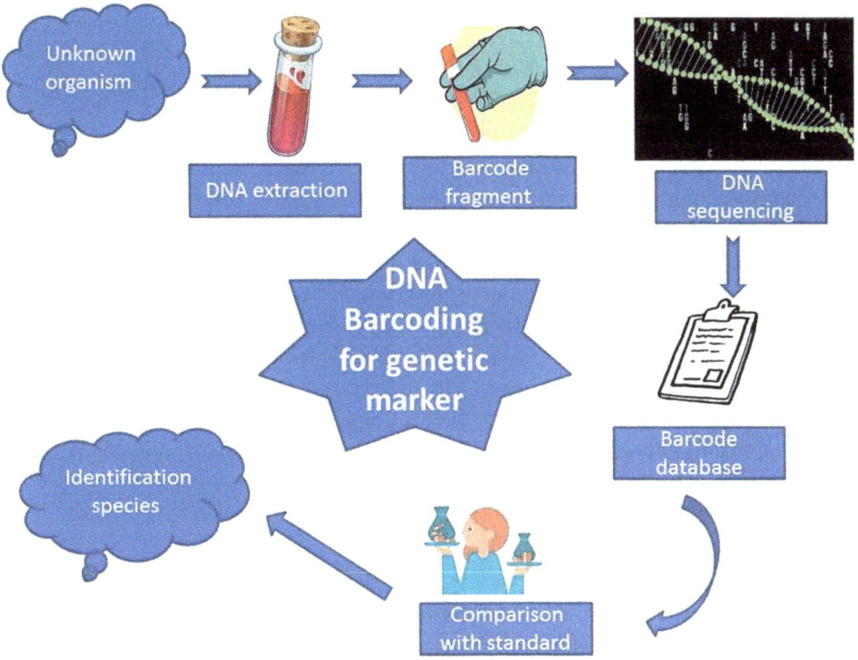

Fig. 20.3 Steps of genetic marking of botanicals drugs

Table 20.4 Example of botanicals with important DNA barcoding types

Botanical drugs	Potential chemicals constituent	Barcode for species authenticity
Datura stramonium (Solanaceae)	Hyoscine, hyoscyamine (tropane alkaloids)	*ITS2* or *psbA-trnH*
Datura metel (Solanaceae)	Hyoscine, hyoscyamine (tropane alkaloids)	*ITS2* or *psbA-trnH*
Datura innoxia (Solanaceae)	Hyoscine, hyoscyamine (tropane alkaloids)	*ITS2* or *psbA-trnH*
Boerhavia diffusa (Nyctaginaceae)	Punarnavine (alkaloid)	*ITS* and *ITS1*
Asparagus racemosus (Liliaceae)	Shatavarin (saponin glucoside)	*ITS2*
Catharanthus roseus (Apocynaceae)	Vincristine, vinblastine (indole alkaloid)	*matK*
Crocus sativus (Iridaceae)	Crocin, Picrocrocin (glycoside)	*psbA-trnH*

ITS internal transcribed spacer, *matK* matures kinase

pharmacological activities are based on the active principle of botanical. An example of botanicals with important DNA barcoding is mentioned in Table 20.4.

20.8 Role of Genetic Marketing in Botanical Drug

20.8.1 Genetic Variation/Genotyping

A chemical profile of botanical drugs is well affected by geographical condition. The commercial importance of botanical drugs is determined by presence of chemical constituents of interest. Germplasm technique is helpful in mapping and preventing botanical source for long time (Dhillon et al. 2007).

A product from different geographical sources is identified by using genetic marking. Mostly applied techniques are random amplified polymorphic DNA (RAPD) and random fragment length polymorphism (RFLP) for this purpose. A different accession of *Azadirachta indica, Allium schoenoprasum, Andrographis paniculata* and *Taxus wallichiana* is identified using RAPD. Random fragment length polymorphism (RFLP) and RAPD are applied in the identification of inter-species of different genera such as *Curcuma*, *Echinacea*, *Glycyrrhiza*, etc. (Li et al. 2019).

20.8.2 Authentication of Medicinal Plants

Genetic fingerprinting is widely used for authentication of the botanical source. Arbitrarily Primed PCR (AP–PCR), RAPD, RFLP, and sequence characterized amplified region (SCAR) are frequently applied techniques. These techniques are capable of distinguishing nearby species developed due to the mutation and/or alteration in genomic loci. Yu Ping Feng San is a Chinese herbal preparation. The RAPD techniques are successfully applied to authenticate of *Lycium barbarum* and other botanicals. Single RAPD primer technique is applied to identify the presence of three herbs *Astragalus membranaceus* Fisch Bge, *Atractylodes macrocephala* Koidz, and *Ledebouriellas eseloides* Wolff, in a single formulation (Zhang et al. 2001).

20.8.3 Detection of Substitution and Adulteration

DNA marking is well applied for identification of substitution and adulteration which is difficult by physical and chemical marking due to morphological resemblance as well as similarity in chemical profile. The DNA marking techniques like SCAR, RAPD, RFLP, and AP-PCR are applied for such purpose. *P. quinquefolius* (American ginseng) is used for substitution of *P. ginseng which is* identified and detected by DNA marking techniques (Baek et al. 2012).

20.8.4 Selection of Desirable Chemotypes

A genetic marking is applied in the selection of chemotypes. Few examples are *Cannabis sativa* for cannabinoid and *Acorus calamus* chemotypes differing in their essential oil composition by using RAPD and AFLP (Baum et al. 2001; Toth et al. 2020).

20.8.5 Standardization and Quality Assurance of Botanical Plant Materials

Genetic marking has shown applicability in quality assurance of commercially important botanical plants like *Atractylodes*, soybean, maize, ginseng, *Echinacea*, etc. (Datukishvili et al. 2015). This method is effective for the selection of the best botanical source. Ultimately this identification is responsible for the desired clinical effect. The application has been extended to differentiated germ and non-germ product of soybeans and capsicum products. Primers specific for inserted genes and crop endogenous genes are used for this purpose (Joshi et al. 2004).

20.9 Chemical Markers

Chemical makers are plant constituents either possess clinical activity or not but must be interesting for quality control purposes. Both primary and secondary metabolites, as well as other macromolecules such as proteins, polysaccharides, and nucleic acid, may serve as a chemical marker. Chemical marking is applied in wide area like the selection of botanical source, investigation of new resource, identification of substitute, determination of best collection and harvesting time, optimization processing and extraction procedure, development of purification method post-extraction, purity determination, shelf life assessment, and other quality control aspects (Li et al. 2008).

20.9.1 Types of Chemical Markers

Chemical makers are broadly classified into three groups, called active markers, analytical markers, and negative markers. The chemical constituents which possess pharmacological activity come under this heading. Few examples of these markers are Allin in *Allium sativum*, Ephedrine in *Ephedra sinensis*, and Silymarin in *Silybum marianum*. If the chemical constituents used only for analytical assessment without respect to their pharmacological activity, then it is called analytical markers. For example, the presence of alkyl amides bond is used for the authentication of *Echinacea* species. This bond is absent in the *Echinacea pallid* but present in the *Echinacea angustifolia* and *Echinacea purpurea*. The last is a negative marker, which is a chemical constituent that is allergic or toxic in properties that may interfere with the bioavailability of other therapeutic agents. An example of negative marker is ginkgolic acid found in *Ginkgo biloba* leaf extract.

20.9.2 Comparative Study Between Genetic Marking and Chemical Marking

For both genetic and chemical marking, uniqueness with a constant value of targeted chemical or DNA for a particular sample is a most crucial factor. The chemical compassion is varying with respect to physiological and environmental factors, while genetic marking system could not be affected by this factor. Another advantage of genetic marking is to provide an efficient and accurate authenticity of several samples simultaneously, whereas the conventional chemical-based methods usually longer duration. So in this respect, analysis of DNA markers is more reliable than chemical markers. Lots of work have been done in the field of applicability of molecular marking of botanical materials and preparation, but the biggest limitation of this technology is DNA fingerprint in some irrespective to plant part used. The applicability of genetic marking is limited based on the availability of nucleic acid. Normally the secondary metabolites like gum, latex, oils, etc. do not content cellular structure so difficult to standardized based on genetic fingerprinting.

20.10 Chemical Marking: Chromatographic Fingerprinting

Separation and identification is a step-by-step procedure of standardization. Both steps are simultaneously achieved by advanced chromatographic techniques for single botanical preparation as well as polyherbal formulations. The name of chromatographic techniques used in the fingerprinting of chemicals is paper chromatography, column chromatography, thin-layer chromatography, ion-exchange chromatography, size exclusion chromatography, gas chromatography, gas-liquid chromatography, high-pressure chromatography, high-performance thin-layer chromatography, supercritical fluid chromatography, and capillary electrophoresis. Similarly, the spectroscopy method is also used in the combination of chromatographic techniques like liquid chromatography-mass spectroscopy (LC-MS), gas chromatography- atomic absorption spectroscopy, and gas chromatography-mass spectroscopy (GC-MS). The concept behind combinational techniques is chromatography applied for the separation of components of the interest, and spectrophotometer is used to identify the compound.

20.11 Indian Official Book as a Reference for Marker-Based Standardization of Botanicals/Herbal Drugs

Indian Pharmacopoeia is an official book published by the Indian Pharmacopoeia Commission (IPC), Ghaziabad, Utter Pradesh India, under Ministry of Health and Family Welfare, Government of India, and is considered as one of the legally recognized books for the standardization of drugs and medicine in India (Prakash et al. 2017). *Indian Pharmacopoeia* 2014 and 2018 have given more and more emphasis on the standardization of herbal medicines on the basis of chemical marker along with their pharmacognostic parameters. Along with mentioning marker compounds and assay analytical techniques, they also provide internal phytochemicals reference sample (PRS) and botanical reference standard (BRS). A list of botanicals drugs with their marker for assay as per *Indian Pharmacopoeia* is mentioned in Table 20.5. Similarly, the *Ayurvedic Pharmacopoeia of India* is published by the Ayurvedic Pharmacopoeia Committee working under the Department of Ayurveda, Yoga, Naturopathy, Unani, Siddha, and homeopathy, under Ministry of Health and Family Welfare, Government of India, and prescribes the standards for Ayurveda formulation sale in Indian Market.

20.12 The Need of Fingerprinting

Nowadays high precious sophisticated instruments are available for fingerprinting, but still challenges are difficult to remove. Botanical preparation has been applied in the healthcare system since ancient times. An Indian Ayurveda system is considered as the oldest treatment methodology of the world, apply multiple botanical to single patients for a single disease, and sometimes prescribe single botanical for multiple diseases. In this situation difficult to identify, the marker compound over which therapeutic activity assessed. Complexity has been increased because single botanical also contains a variety of chemical constituents. However, a potential risk to the health of patients still needs to be considered. Thus, comprehensive characterizations of drug substances are important. A batch-to-batch variation is also needed to be addressed. With the help of the standardization process, a limit of fluctuation (range) can be decided. The fingerprinting concept also needs to link the Adjusted Efficacy Score based method developed by in vivo experiments (Yang et al. 2017). For getting approval for authorities, it is necessary to report the markers compound which is assessed in quality control test.

20.13 Conclusion

Botanical drugs have great market accessibility, traditional history, and potency to become lead molecules for modern medicine. Apart from medicine, botanical drugs can be used as new cosmeceuticals, nutraceuticals, perfumes, and pesticides. However, due to great variations of raw materials globally, it is difficult to reproduce for

Table 20.5 List of botanicals drug with their marker for assay (Indian Pharmacopoeia 2018)

Common name	Biological source	Category	Method of assay	Reference solution of marker for assay
Amra	The dried stem bark of *Mangifera indica* L. (Fam. Anacardiaceae)	Cardiotonic, astringent, nourishing tonic in hemoptysis	LC	Dissolve 10 mg of ***mangiferin*** *RS* in 10 mL of dimethylformamide and dilute to 100 mL with methanol
Arjuna	The dried stem bark of *Terminalia arjuna*. (Fam. Combretaceae)	Antihyperlipidaemic, antihypertensive, astringent, cardioprotective, hridroga	LC	0.1% w/v solution of ***arjungenin*** *RS* in *ethanol*
Artemisia	Dried leaves or the dried leaves and flowering tops of *Artemisia annua L.* (Fam. *Asteraceae*)	Antimalarial, antibacterial, cytotoxic, antihelmintic	HPTLC	0.1% w/v solution of ***artemisinin*** *RS* in *hexane*
Bakuci	Dried ripe fruits of *Psoralea corylifolia* Linn. (Fam. *Fabaceae*)	Skin problems, psoriasis	TLC	0.02% w/v solution of ***psoralen*** *RS* in *methanol*
Bala	Dried roots of *Sida acuta* Burm, f. ssp. Acuta	Astringent, diuretic, urinary infections	TLC	0.1% w/v solution of ***ecdysterone*** *RS* in *methanol*
Bassant	Whole or cut, dried flowering tops, or aerial parts of *Hypericum perforatum* (Fam. Hypericaceae)	Antidepressant, anxiolytic, sedative, antioxidant, anti-ulcerogenic	TLC	0.01% w/v solution of ***hypericin*** RS in methanol
Caraway oil	An essential oil obtained by steam distillation of ripe fruits(dried and crushed) of *Carum carvi* L.(Fam. Umbelliferae)	Carminative	GC	1. 2.0% w/v solution of ***limonene*** RS in ethanol (95%) 2. 2.0% w/v solution of ***carvone*** RS in ethanol (95%).
Cassia oil	An essential oil obtained by steam distillation of the leaves and young branches of *Cinnamomum cassia* Blume (Fam. Lauraceae)	Antidiarrheal, antiviral, antibacterial	GC	Dissolve 100 μL of *trans*-***cinnamic aldehyde*** RS, 10 μL *of* ***cinnamyl acetate*** *RS*, 10 μL of ***engenol*** *RS*, 20 mg of ***coumarin*** *RS* and 10 μL of *trans-2-* ***methoxycinnamaldehyde*** RS in 1 mL of *acetone*

Daruharidra stems	Cut dried stems of *Berberis aristata*, DC (Fam. Berberidaceae)	Hepatoprotective, anti-inflammatory, anticancer, amatisara	TLC	A 0.002% w/v solution of ***berberine hydrochloride** RS* in 0.1% *strong ammonia* solution in *methanol*
Dill seed oil	An essential oil obtained by stem distillation of the mature crushed seeds of Anethum sowa (Fam. Umbelliferae)		GC	1. A 2.0% w/v solution of ***cis-dihydrocarvone*** RS in *ethanol* (95%). 2. A 2.0% w/v solution of **L-*carvone*** *RS* in *ethanol* (95%)
Eucalyptus oil	An essential oil obtained by stem distillation of the leaves of *Eucalyptus globulus* (Fam. Myrtaceae)	Counterirritant, antiseptics, expectorant, antirheumatic	GC	1. A 2.0% w/v solution *of 1,8-**cineole** RS* in *ethanol* (95%) 2. A 2.0% w/v solution of ***alpha-pinene*** *RS* in *ethanol* (95%)
Garcinia	Dried deseeded fruit of *Garcinia cambogia* (Fam. Guttiferae)	Antiulcer, sthoulya, medohara	LC	1. A 0.8% w/v solution of ***hydroxycitric acid calcium salt*** *RS* in *dilute orthophosphoric acid* 2. A 0.1% w/v solution of ***citric acid*** *in dilute orthophosphoric acid*
Ginseng	Dried roots of *Panax ginseng* (Fam. Araliaceae)	Aphrodisiac, anticancer, adaptogenic, antidiabetic, and antiviral	LC	Dissolve 2.0 mg of **ginsenoside** (Rg1 and Rb1) in 10.0 mL of solvent mixture
Gokhru	Dried fruits of *Tribulus terrestris* L. (Fam. Zygophyllaceae)	Diuretic, anti-inflammatory, anthelmintic, potent aphrodisiac	LC	A 0.1% w/v solution of ***diosgenin*** RS in *methanol*
Haridra/ Haldi	Dried rhizomes of *Curcuma longa* Lnn. (Fam. Zingiberaceae)	Anti-inflammatory, galactagogue, stomachic, spasmolytic, visavikara	LC	A 0.01% w/v solution of ***curcumin*** *RS* in *methanol*
Hingu	Oleo-gum-resin obtained from rhizomes and roots *Ferula asafoetida* Regel. (Fam. Umbelliferae)	Anti-spasmodic, carminative, expectorant, laxative, sedative	LC	A 0.005% w/v solution of ***trans-ferulic acid*** *RS* in *ethanol*
Ipecac Tincture	Ipecac Tincture obtained from roots and rhizomes of *Cephaelis ipecacuanha* (Fam. Rubiaceae)	Emetic, antiamoebic	LC	A 0.01% w/v solution of ***emetine hydrochloride*** *heptahydrate RS* in 0.01 M *hydrochloric acid*

(continued)

Table 20.5 (continued)

Common name	Biological source	Category	Method of assay	Reference solution of marker for assay
Jatamansi	Dried rhizomes of *Valeriana jatamansi* (Fam. Valerianaceae)	Antispasmodic, tranquillizer	TLC/ HPTLC	0.001% w/v solution of ***valproic acid*** *RS* in *hexane*
Shatavari	Consists of the tuberous roots of *Asparagus racemosus* Willd. (Fam. Liliaceae)	Anti-inflammatory, immunomodulatory, neuroprotective, and antidysentery	TLC/ LC	1. A 0.008% w/v solution of ***shatavarin IV*** *RS* in *methanol* 2. A 0.1% w/v solution of **shatavarin IV RS** in methanol. Dilute suitably to prepare 0.0075–0.075% w/v solution
Punarnava	Consists of dried root of *Boerhaavia diffusa* Linn. (Fam. Nyctaginaceae)	Hepatoprotective, diuretic, antinephritic	LC	A 0.002% w/v solution of ***boeravinone B*** *RS* in methanol

formulations on commercial-scale uniformly. The safety and efficacy of botanicals are totally dependent upon the proper standardized quality of crude material. Proper agricultural, pharmacognostical, phytochemical, marker-based standardization, and DNA barcoding can generate a harmonized source of raw materials for commercial applications. Due to the diversity in chemical profile of botanical source it is necessary to perform toxicity marking at different level of finished good preparation. Different agencies are working on the standardization of botanicals focusing on a molecular level. In India, a new class of drug, "phytopharmaceuticals" is introduced which leads to proper validation of age-old Indian systems of medicine (ISM). In nutshell, standardization leads to more efficacious and safer botanical drugs for humans and animal health.

References

AHPA issues guidance on organoleptic analysis of herbal ingredients. https://www.nutraceuticalsworld.com/issues/2013-05/view_suppliers-corner/ahpa-issues-guidance-on-organoleptic-analysis-of-herbal-ingredients/. Accessed 1 Dec 2019

Baek SH, Bae ON, Park JH (2012) Recent methodology in ginseng analysis. J Ginseng Res 36(2):119–134. https://doi.org/10.5142/jgr.2012.36.2.119

Balachandran KRS, Mohanasundaram S, Ramalingam S (2015) DNA barcoding: a genomic-based tool for authentication of phytomedicinals and its products. Botanics 5:77–84

Baum BR, Mechanda S, Livesey JF, Binns SE, Arnason JT (2001) Predicting quantitative phytochemical markers in single Echinacea plants or clones from their DNA fingerprints. Phytochemistry 56(6):543–549

Chial H (2008) DNA fingerprinting using amplified fragment length polymorphisms (AFLP): no genome sequence required. Nat Educ 1(1):176

Datukishvili N, Kutateladze T, Gabriadze I, Bitskinashvili K, Vishnepolsky B (2015) New multiplex PCR methods for rapid screening of genetically modified organisms in foods. Front Microbiol 6:757. https://doi.org/10.3389/fmicb.2015.00757. eCollection 2015

Dhillon RS, Mohapatra T, Singh S, Boora KS, Singh K (2007) Assessment of genetic diversity in Azadirachtaindica based on DNA fingerprinting. Indian J Biotechnol 6(4):519–524

Indian Pharmacopoeia Commission (2018) Indian Pharmacopoeia. Indian Pharmacopoeia Commission, Ghaziabad. isbn: 9381238197

Joshi K, Chavan P, Warude D, Patwardhan B (2004) Molecular markers in herbal drug technology. Curr Sci 87(2):159–165

Kumari R, Kotecha M (2016) A review on the standardization of herbal medicines. Int J Pharma Sci Res 7(2):97–106

Kunle OF, Egharevba HO, Ahmadu PO (2012) Standardization of herbal medicines—a review. Int J Bifurc Chaos 4(3):101–112. https://doi.org/10.5897/IJBC11.163

Li S, Han Q, Qiao C, Song J, Cheng CL, Xu H (2008) Chemical markers for the quality control of herbal medicines: an overview. Chin Med 3:7. https://doi.org/10.1186/1749-8546-3-7

Li F, Fu C, Li Q (2019) A simple genome walking strategy to isolate unknown genomic regions using long primer and RAPD primer. Iran J Biotechnol 17(2):e2183. https://doi.org/10.21859/ijb.2183. eCollection 2019

Newmaster SG, Grguric M, Shanmughanandhan D, Ramalingam S, Ragupathy S (2013) DNA barcoding detects contamination and substitution in north American herbal products. BMC Med 11:222

Panteghini M (2004) Role and importance of biochemical markers in clinical cardiology. Eur Heart J 25(14):1187–1196

Prakash J, Srivastava S, Ray RS, Singh N, Rajpali R, Singh GN (2017) Current status of herbal drug standards in the Indian pharmacopoeia. Phytother Res 31(12):1817–1823. https://doi.org/10.1002/ptr.5933. Epub 2017 Oct 13

Sampaio BL, Edrada-Ebel R, Da Costa FB (2016) Effect of the environment on the secondary metabolic profile of Tithonia diversifolia: a model for environmental metabolomics of plants. Sci Rep 6:29265. https://doi.org/10.1038/srep29265

Sarwa KK, Das PJ, Mazumder B (2014a) A nanovesicle topical formulation of Bhut Jolokia (hottest capsicum): a potential anti-arthritic medicine. Expert Opin Drug Deliv 11(5):661–676. https://doi.org/10.1517/17425247.2014.891581. Epub 2014 Mar 24

Sarwa KK, Rudrapal M, Debnath M, Kumar A, Verma VK (2014b) Phytochemical and biological potential of *Cassia tora* Linn. Eur J Med Plants 4(8):946–963

Sarwa KK, Mazumder B, Suresh PK, Kaur CD (2016) Topical analgesic Nanolipid vesicles formulation of Capsaicinoids extract of Bhut Jolokia (Capsicum chinense Jacq): pharmacodynamic evaluation in rat models and acceptability studies in human volunteers. Curr Drug Deliv 13(8):1325–1338

Sarwa KK, Vishwakarma P, Suryawanshi V, Sahu T, Kumar L, Shree J (2017) An Ayurvedic formulation *Sankat Mochan*: a potent anthelmintic medicine. Drug Dev Therap 8(1):13–17

Serrano R, Silva G, Silva O (2010) Application of light and scanning electron microscopy in the identification of herbal medicines microscopy. In: Méndez-Vilas, Díaz J (eds) Science, technology, applications and education. Formatex, Guadalajara, pp 182–190

Sucher NJ, Hennell JR, Carles MC (2012) DNA fingerprinting, DNA barcoding, and next-generation sequencing technology in plants. Methods Mol Biol 862:13–22. https://doi.org/10.1007/978-1-61779-609-8_2

Toth JA, Stack GM, Cala AR et al (2020) Development and validation of genetic markers for sex and cannabinoid chemotype in Cannabis sativa L. GCB Bioenergy 12:213–222

WHO guidelines for selecting marker substances of herbal origin for quality control of herbal medicines. WHO technical report series, no. 1003, 2017, Annex 1. http://apps.who.int/medicinedocs/en/m/abstract/Js23240en/. Accessed 1 Dec 2019

Yang Z, Shao Q, Ge Z, Ai N, Zhao X, Fan X (2017) A bioactive chemical markers based strategy for quality assessment of botanical drugs: Xuesaitong injection as a case study. Sci Rep 7 (1):2410. https://doi.org/10.1038/s41598-017-02305-y

Zhang KY, Leung HW, Yeung HW, Wong RN (2001) Differentiation of Lyciumbarbarum from its related Lycium species using random amplified polymorphic DNA. Planta Med 67(4):379–381

Analytical Standardization of *Haridra* Formulation by UV-Vis Spectrophotometry and RP-HPLC

21

Panja Srikanth, Huzaifa Naaz, Mithun Rudrapal, and Manjunath S. Yalagatti

Abstract

In our study, analytical methods were developed for the determination of *Haridra* in the marketed formulation by UV-Vis spectrophotometry and RP-HPLC. The percentage of curcumin estimated in the marketed formulation complies with the standard (not less than 1.5% w/w of curcumin) specified in the official monograph of *Haridra*. The developed methods showed acceptable linearity, accuracy, precision, ruggedness, robustness, specificity, LOD, and LOQ. Results of validation studies were found satisfactory with % RSD values of less than 2%, which indicates good specificity, validity, and reliability of the developed methods. The methods proposed herein are claimed to be accurate and precise. Our study reports the standardization of *Haridra* by means of analytical determination of curcumin in the marketed formulation. The developed methods are proposed to be used for the routine analytical determination of *Haridra* in crude drugs, traditional preparations, and marketed formulations.

Keywords

Haridra · Herbal formulation · Curcumin · Analytical standardization · UV-Vis spectrophotometry · RP-HPLC

P. Srikanth · H. Naaz · M. Rudrapal (✉) · M. S. Yalagatti
Srikrupa Institute of Pharmaceutical Sciences, Siddipet, Telangana, India
e-mail: rsmrpal@gmail.com

S. C. Mandal et al. (eds.), *Evidence Based Validation of Traditional Medicines*,
https://doi.org/10.1007/978-981-15-8127-4_21

Abbreviations

ICH	International Council of Harmonization of Technical Requirements for Pharmaceuticals for Human use
IP	Indian Pharmacopoeia
LOD	Limit of Detection
LOQ	Limit of Quantification
RP-HPLC	Reverse Phase-High Performance Liquid Chromatography
RSD	Relative Standard Deviation
RT	Retention Time
UV-VIS	Ultraviolet-Visible
WHO	World Health Organization

21.1 Introduction

Haridra (Turmeric) is an herbal drug official in the Indian Pharmacopoeia (IP). It contains dried rhizomes of *Curcuma longa* Linn. belonging to the Zingiberaceae family (IP 2014). Since ancient times, the use of *Haridra* has been documented in various traditional systems of Indian medicine (*Ayurveda*, *Unani* and *Siddha*) for the treatment and management of a variety of human ailments (Noorafshan and Ashkani-Esfahani 2013). Various formulations of *Haridra* are now available in the market, which are used extensively as antioxidant, anti-inflammatory, and anthelmintic and other medications (Choudhary and Sekhon 2012). Curcumin (not less than 1.5% w/w) is the chief active principle of *Haridra* (IP 2014). It is chemically a diarylheptanoid [(1E,6E)-1,7-bis (4-hydroxy-3-methoxyphenyl)-1,6- heptadiene-3,5-dione, Fig. 21.1], belonging to the group of curcuminoids. It occurs as an intense yellow-colored crystalline solid sparingly soluble in water and freely soluble in methanol and ethanol (Manolova et al. 2014; Tonnesen and Karlsen 1985). Curcumin and its derivatives are attributed to be responsible for the medicinal potential and health benefits of *Haridra*. The curcumin is commercially used as an ingredient in nutraceuticals, dietary supplement, and cosmetics and as a flavor in foods and/or food products (Hewlings and Kalman 2017; Beevers and Huang 2011).

Since *Haridra* is an herbal drug, there is no sufficient standardized protocol (as per WHO guidelines) for its evaluation (quality, purity, and potency) in crude preparations or marketed formulations. Due to this fact, the analytical evaluation of *Haridra* is not well-established. Literature survey reveals availability of many

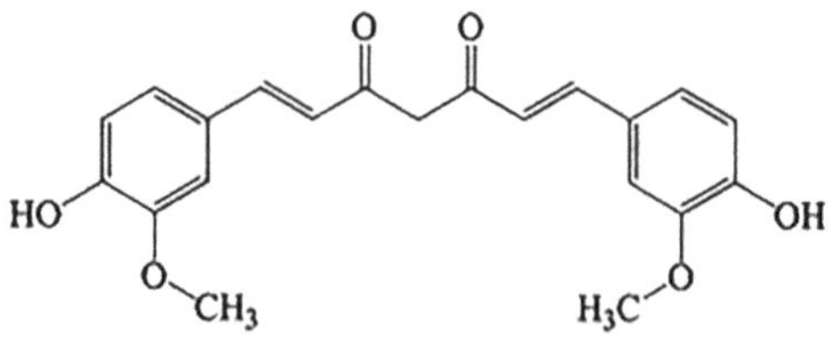

Fig. 21.1 Structure of curcumin

scientific methods, but they lack in consistency, adequacy, and reliability (Singh and Avupati 2017; Syed et al. 2015; Kachhadiya et al. 2014). In this context, it is necessary to develop some new analytical methods that can be applied widely for the standardization of *Haridra* in traditional and/or marketed preparations. In this study, our effort was to develop simple and specific analytical methods for the standardization of marketed formulation of *Haridra* by means of determination of its principle ingredient, i.e., curcumin by UV-Vis spectrophotometry and RP-HPLC.

21.2 Experimental

21.2.1 Chemicals

Curcumin was commercially procured from Yarrow Chem Products, Mumbai. *Haridra* formulation (capsules) was obtained from the local market. Methanol, acetonitrile, and water were of HPLC grade supplied by Merck Pvt. Ltd., Mumbai.

21.2.2 Instruments and Analytical Parameters

Spectrophotometric analyses were done on ELICO SL 244 Double Beam UV-Vis spectrophotometer with quartz cuvette of 1 cm width using methanol as solvent. Chromatographic analysis was performed on a Cyber Lab RP-HPLC equipped with a pump, manual sampler, and a UV detector. The chromatographic column was a Phenomenex Luna C18 column (250 mm × 4.6 mm i.d., 5 μm). The column temperature was kept constant at 55 °C. The mobile phase was a mixture of acetonitrile and 0.1% orthophosphoric acid in the ration of 85: 15% v/v. The chromatographic separation was done on an isocratic mode of operation at room temperature. The flow rate was 1.0 mL/min and the volume of injection was 20 μL. The time for chromatographic run was 6.0 min. The wavelength of UV detection was kept at 418 nm. Citizen Ultra Sonicator was used for sonicating the sample solutions. Digital balance (SHIMADZU AUX 220) was used for weighing.

21.2.3 Preparation of Standard and Sample Solutions

21.2.3.1 UV Method

To prepare the standard stock solution, accurately weighed 10 mg of curcumin was added into a volumetric flask of 10 mL, and the volume was adjusted up to 10 mL with methanol to get a stock concentration of 1000 μg/mL. Standard solutions were prepared with the concentration range between 1 and 5 μg/mL by further dilutions with methanol.

Powder of 20 capsules equivalent to the standard stock concentration was weighed and dissolved in a volumetric flask of 10 mL using methanol to get a sample stock concentration of 1000 μg/mL. The solution was mixed well by sonication for 20 min and filtered on a Whatman filter paper. Using the stock solution, a series of sample solutions (1–5 μg/mL) were made by serial dilution with methanol.

21.2.3.2 HPLC Method

To a 10 mL of volumetric flask, 10 mg of curcumin was transferred and dissolved with acetonitrile: 0.1% orthophosphoric acid (85: 15% v/v) to produce a standard stock concentration of 1000 μg/mL. Standard solutions of a series of concentrations in the range of 2–10 μg/ml was prepared by dilution with the same solvent system.

Powder of 20 capsules equivalent to the standard stock concentration was weighed and dissolved in a 10 mL of volumetric flask using acetonitrile/0.1% orthophosphoric acid (85:15% v/v) to produce a sample stock solution of 1000 μg/mL. The solution was mixed under sonication for 20 min and filtered with the help of Whatman filter paper. Using the stock solution, final sample solutions (2–10 μg/mL) were prepared by serial dilution with the same solvent system.

21.2.4 Assay of Curcumin in *Haridra* Formulation

21.2.4.1 UV Method

The percentage of curcumin in the test solution (5 μg/mL) of *Haridra* was estimated by absorbance ratio method. The absorbances of solutions were measured at the wavelength of 418 nm using methanol as a blank.

21.2.4.2 HPLC Method

The percentage of curcumin in the test solution (4 μg/mL) of *Haridra* was determined by calculating the peak area of curcumin. The wavelength of detection was at 418 nm.

21.3 Results and Discussion

21.3.1 Method Development

21.3.1.1 UV Method

Based upon solubility studies, methanol was selected as a suitable solvent for spectrophotometric measurements. To find out the wavelength of absorbance maxima (λ_{max}), a test solution of curcumin (5 μg/mL) was prepared and scanned between 300 and 500 nm of UV range against methanol as a blank. Relevant information regarding selection of suitable solvent system and determination of wavelength of absorption was recollected from available literature (Syed et al. 2015). A representative spectrum of curcumin is given in Fig. 21.2. The spectrum indicates a broad and well-defined peak at the λ_{max} of 418 nm.

21.3.1.2 HPLC Method

Several trial runs were performed using C_8 and C_{18} RP columns, various mobile phase compositions, and different flow rates for the separation of curcumin with

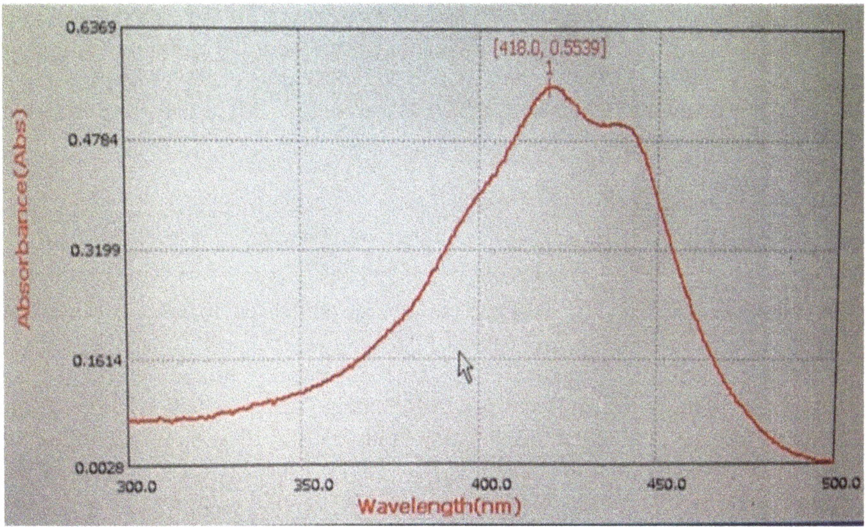

Fig. 21.2 Spectrum of curcumin (methanol) at λ_{max} 418 nm

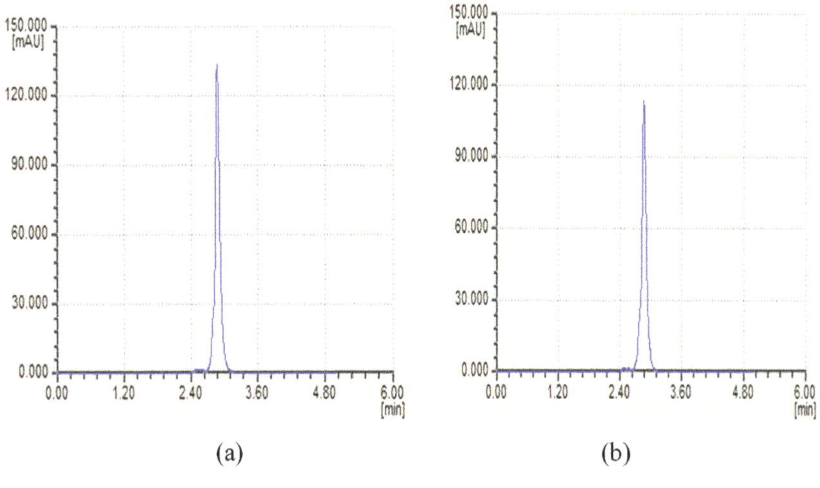

Fig. 21.3 Chromatogram of curcumin (**a**): test and (**b**): standard (RT = 2.88)

good chromatographic parameters (retention time, resolution, peak area, theoretical plates, tailing factor, etc.). A reverse-phase C_{18} column (250 mm × 4.6 mm, i.d. 5 μm) as a stationary phase with a mobile phase of acetonitrile/0.1% orthophosphoric acid (85:15% v/v) at a flow rate of 1.0 mL/min and a wavelength detection of 418 nm afforded the best separation with a sharp and well-resolved peak for curcumin. The optimized chromatograms obtained for the standard and test sample of curcumin are displayed in Fig. 21.3. Results indicate that test chromatogram matches with that of the standard chromatogram with the retention time

(RT) of 2.88 min. The analytical values of peak area (70568), theoretical plates (3508), and tailing factor (1.04) were found acceptable for the optimized chromatogram. The development and optimization of the method was achieved with the help of available literature (Joshi et al. 2018; Warule et al. 2017; Patil and Salunkhe 2012).

21.3.2 Method Validation

The validation of UV and RP-HPLC methods was performed by means of following analytical parameters according to the ICH (Q2 R1) guidelines: linearity and range, accuracy and percentage recovery, precision studies, ruggedness and robustness, limit of detection (LOD) and limit of quantification (LOQ), specificity, and system suitability (Gavali et al. 2016; Surya Kiran et al. 2016; Rudrapal et al. 2015; Green 1996; ICH 1996; Wegscheider 1996).

21.3.2.1 Linearity

The linearity was studied by analyzing the standard solutions of curcumin (1–5 μg/mL and 2–10 μg/mL in UV and HPLC, respectively) at 418 nm. The calibration curve (Fig. 21.4) was constructed between average absorbance (mean ± SD) and concentration for triplicate observations ($n = 3$). The calibration plot was obtained linear in the studied range of concentrations described above. The equations of regression analysis were obtained as follows: $y = 0.1444x - 0.0001$ ($r^2 = 0.999$) and $y = 11{,}280x + 1981.1$ ($r^2 = 0.999$) for UV and HPLC method, respectively, where y = absorbance or peak area, x = concentration of solution, and r^2 = the square of correlation coefficient. Results imply that the developed methods are linear for the specified analytical range.

21.3.2.2 Accuracy

To determine accuracy or percentage recovery studies, a known standard concentration of the drug was mixed to the sample solution to obtain three different concentrations. It was studied at three concentration levels, i.e., 50, 100, and

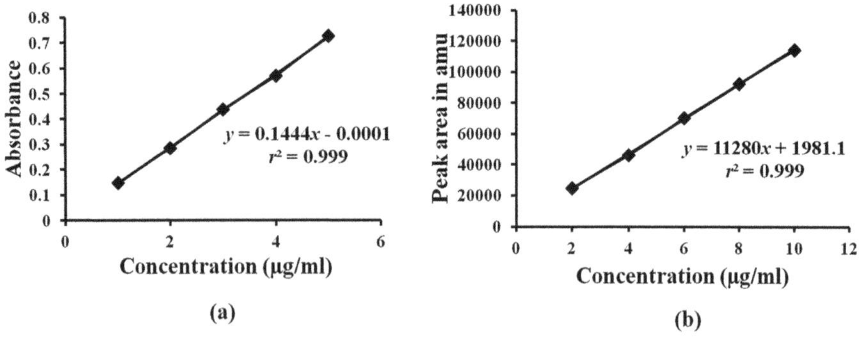

Fig. 21.4 Calibration curve of curcumin (**a**): in UV and (**b**): in HPLC

Table 21.1 Accuracy (% recovery) data

Recovery level	Concentration (μg/mL) Test (initial amount)	Standard (amount added)	Absorbance/ Peak area[a]	Mean % recovery	% RSD
UV method					
50%	1	1	0.2618 ± 0.008	98.49	0.810
100%	1	2	0.4296 ± 0.005	98.57	0.531
150%	1	3	0.5639 ± 0.004	99.01	0.419
HPLC method					
50%	2	2	42,669 ± 0.005	98.58	0.513
100%	2	4	65,264 ± 0.001	99.31	0.178
150%	2	6	86,105 ± 0.003	99.48	0.357

[a]Values are presented as mean ± RSD of three replicate observations ($n = 3$)

150%, in triplicate observations ($n = 3$). The percent recoveries were between 98.49 and 99.01% and 98.58 and 99.48% for UV and HPLC method, respectively. The percentage of curcumin recovery in the marketed formulation was within the acceptance range between 98 and 102%. From the results, it is apparent that both methods are accurate. Satisfactory RSD values with less than 2% were obtained. The % recovery data are summarized in Table 21.1.

21.3.2.3 Precision

Repeatability (intra-day precision) was studied by analyzing standard solutions for six replicate observations ($n = 6$) on the same day. Similarly, reproducibility (inter-day precision) was determined by conducting studies in the same laboratory but on different days under similar experimental conditions. The standard concentrations were 4 and 6 μg/mL for UV and HPLC method, respectively. Results of precision studies were also found to be acceptable. The % RSD values were below 2%, indicating good repeatability and reproducibility of both UV and HPLC methods. Results of precision studies are depicted in Table 21.2.

21.3.2.4 Ruggedness

Ruggedness was determined by the analysis of six samples ($n = 6$) of the standard solution by different analysts but in the same laboratory under similar experimental conditions. The standard concentrations were 4 and 6 μg/mL for UV and HPLC method, respectively. Results of ruggedness studies are displayed in Table 21.3. The % RSD values with 2% prove optimum ruggedness of the methods.

21.3.2.5 Robustness

The robustness of the methods was analyzed for six replicate studies ($n = 6$) of standard solution by changing the UV spectrometric measurements such as temperature and instrumental parameters in chromatographic analysis such as flow rate. The standard concentrations were 4 and 6 μg/mL for UV and HPLC method, respectively. Results of robustness studies presented in Table 21.4 indicate that both the

Table 21.2 Precision

Precision[a]	Repeatability (intra-day)		Reproducibility (inter-day)	
	Morning	Evening	Day 1	Day 2
UV method				
Absorbance[b]	0.5773 ± 0.004	0.5980 ± 0.018	0.5779 ± 0.136	0.0060 ± 0.010
%RSD	0.451	1.847	1.360	1.052
HPLC method				
Peak area[c,d]	69510.8 ± 0.004	69434.67 ± 0.003	70641.67 ± 0.003	70719.00 ± 0.003
%RSD	0.438	0.337	0.348	0.374

[a]Values are presented as mean ± RSD of six replicate observations ($n = 6$)
[b]Concentration was 4 μg/mL
[c]Concentration was 6 μg/mL
[d]RT was 2.88 min

Table 21.3 Ruggedness

Ruggedness[a]	Analyst 1	Analyst 2
UV method		
Absorbance[b]	0.5790 ± 0.009	0.5929 ± 0.011
%RSD	0.929	1.186
HPLC method		
Peak area[c,d]	72496.33 ± 0.003	72581.83 ± 0.002
%RSD	0.398	0.292

[a]Values are presented as mean ± RSD of six replicate observations ($n = 6$)
[b]Concentration was 4 μg/mL
[c]Concentration was 6 μg/mL
[d]RT was 2.88 min

Table 21.4 Robustness

Robustness[a]	Parameter	
UV method	Room temperature (29 °C)	Elevated temperature (35 °C)
Absorbance[b]	0.5727 ± 0.009	0.5901 ± 0.009
%RSD	0.9507	0.9023
HPLC method	Flow rate 0.8 mL/min	Flow rate 1 mL/min
Peak area[c,d]	114888.2 ± 0.006	88421.67 ± 0.003
%RSD	0.6970	0.36706

[a]Values are presented as mean ± RSD of six replicate observations ($n = 6$)
[b]Concentration was 4 μg/mL
[c]Concentration was 6 μg/mL
[d]RT was 2.88 min

methods are practically robust. The % RSD values of the methods determined under robustness conditions were below 2.0%.

21.3.2.6 LOD and LOQ

The limit of detection (LOD) and limit of quantification (LOQ) of the UV method were 0.054 and 0.180 μg/mL, respectively. For the HPLC method, the values were 0.068 and 0.207 μg/mL, respectively. The LOD and LOQ indicate that the developed methods are adequately sensitive for the precise determination of the component of interest, i.e., curcumin in the sample.

21.3.2.7 Specificity

The specificity of chromatographic method is meant to separate principal analytes from other components such as excipients, impurities, other active principles, degradants, etc.. About 20 μL of curcumin and related components were injected into the instrument, and their chromatograms were recorded. No peaks were observed in the chromatogram at the desired location other than the peak due to curcumin with the retention time of 2.88 min. Results of this study indicate that the developed HPLC method is free from interfering substances such as excipients and/or related components that are attributed to be present in the formulation. The proposed HPLC method is, therefore, claimed to be specific for the quantitative analysis of curcumin in the herbal formulation.

21.3.3 Assay

In both UV and HPLC methods, the %curcumin estimated in the marketed *Haridra* formulation were satisfactory as per the label claimed (Table 21.5). In the UV method, the %curcumin was estimated to be 99.46%, whereas it was 99.85% by HPLC method, at the tested concentration of 5 and 4 μg/mL, respectively. The assay chromatogram is represented in Fig. 21.5. The %curcumin estimated in the marketed formulation was found to be within limit (98–102%). The estimated amount of curcumin also complies with the standard specified (not less than 1% w/w of curcumin) in the official monograph of *Haridra* (IP 2014).

Table 21.5 Assay

Standard absorbance/ peak area[a]	Test absorbance/peak area[a]	Amount found (mg/mL)	% of curcumin estimated
UV method[b]			
0.7268 ± 0.0007	0.6756 ± 0.0002	4.973	99.46
HPLC[c]			
446,310 ± 0.057	43,219 ± 0.014	3.994	99.85

[a]Values are expressed as mean ± RSD of three replicate observations ($n = 3$)
[b]Concentration was 5 μg/mL
[c]Concentration was 4 μg/mL

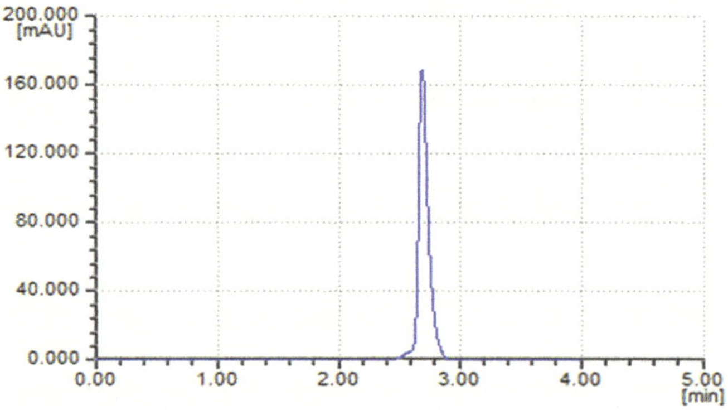

Fig. 21.5 Assay chromatogram of curcumin

21.3.4 System Suitability

System suitability was analyzed by introducing the standard solution (6 μg/mL) of curcumin in six replicate observations ($n = 6$). Results of system suitability studies (Table 21.6) were found within acceptable limit.

The summary of validation parameters are represented in Table 21.6. From validation studies, it is evident that the developed analytical methods are practically valid and thus useful. Results of validation studies described in the above sections are satisfactory for both the methods. In all validation experiments, the % RSD values are not more than 2%, indicating high reliability and validity of the methods. Furthermore, results of our present study are consistent with existing literature (Joshi et al. 2018; Baghel et al. 2017; Purohit et al. 2014; Patil and Salunkhe 2012). In HPLC, good separation and high resolution of the analyte peak proves satisfactory performance of the proposed analytical method. Moreover, higher percentage of recovery from accuracy studies and non-interference of excipients and/or related components assure that the HPLC method is quite specific for the estimation of curcumin in the *Haridra* formulation.

Table 21.6 Summary of validation parameters

Parameters		Result	
		UV method	HPLC method
Wavelength of detection (λ_{max}, nm)		418	418
Linearity	Beer's law limit (μg/mL)	1–5	2–10
	Slope	0.0001	1981.1
	Intercept	0.1444	11,280
	Coefficient of correlation	0.999	0.999
Accuracy or % recovery (%RSD)		<2.0	<2.0
Precision (% RSD)	Repeatability	<2.0	<2.0
	Reproducibility	<2.0	<2.0
Ruggedness (% RSD)		<2.0	<2.0
Robustness (% RSD)		<2.0	<2.0
LOD (μg/mL)		0.054	0.068
LOQ (μg/mL)		0.180	0.207
System suitability	Retention time (RT, min)	–	2.88
	Peak area	–	70,568
	Theoretical plates	–	3508
	Tailing factor	–	1.04
	Efficiency	–	36,490
	Asymmetry	–	1.201

21.4 Conclusion

In our study, analytical methods were developed for the standardization and/or evaluation of *Haridra* in the marketed formulation by UV-Vis spectrophotometry and RP-HPLC. The methods proposed herein are claimed to be simple, accurate, and precise. The developed methods are also reported to be highly valid, specific, and reliable. However, our study reports the standardization of *Haridra* by means of quantitative determination of curcumin in the marketed herbal formulation. Similarly, the crude preparations of *Haridra* can also be standardized by these proposed methods. Our developed methods, thus, can be used in routine practice for quality control analysis of *Haridra* in crude drugs, traditional preparations, and marketed formulations.

Acknowledgments Authors are thankful to the Management of Srikrupa Institute of Pharmaceutical Sciences for proving necessary laboratory facilities for carrying out this work.

References

Baghel US, Nagar A, Pannu MS, Singh D, Yadav R (2017) HPLC and HPTLC methods for simultaneous estimation of quercetin and curcumin in polyherbal formulation. Indian J Pharm Sci 79(2):197–203

Beevers CS, Huang S (2011) Pharmacological and clinical properties of curcumin. Bot Targets Ther 1:5–18

Choudhary N, Sekhon BS (2012) Potential therapeutic effect of curcumin—an update. J Pharm Educ Res 3(2):64–68

Gavali JB, Pradhan N, Waghmare N (2016) Current trends in analytical methods of medicinal plant drugs. Ayurpub 1(3):101–107

Green JM (1996) A practical guide to analytical method validation, analytical chemistry. Anal Chem 689:305A–309A

Hewlings SJ, Kalman DS (2017) Curcumin: a review of its' effects on human health. Foods 6:92. https://doi.org/10.3390/foods6100092

Indian Pharmacopoeia (IP) (2014) Indian pharmacopoeia (Vol. IV), Published by the Indian Pharmacopoeia Commission (IPC), Ministry of Health and Family Welfare, Government of India, New Delhi

International Conference on Harmonization (ICH) of Technical Requirements for the Registration of Pharmaceuticals for Human Use (1996) Validation of analytical procedures: methodology, ICH-Q2B, Geneva

Joshi S, Jadav VM, Kadam VJ (2018) Analytical method development and validation for simultaneous estimation of catechin and curcumin by HPLC in an Ayurvedic formulation. World J Pharm Sci 7(4):1693–1702

Kachhadiya KH, Shah JS, Maheshwari D (2014) Analytical method development and validation for simultaneous estimation of embelin and curcumin in bulk and polyherbal formulation. Int J Pharm Res Biosci 3(3):548–558

Manolova Y, Deneva V, Antonov L, Drakalska E, Momekova D, Lambov N (2014) The effect of the water on the curcumin tautomerism: a quantitative approach. Spectrochim Acta A Mol Biomol Spectrosc 132:815–820

Noorafshan A, Ashkani-Esfahani S (2013) A review of therapeutic effects of curcumin. Curr Pharm Des 19:2032–2046

Patil SJ, Salunkhe VR (2012) Simultaneous estimation of curcumin and quercetin in Ayurvedic proprietary medicine by UV spectrophotometry. Int J Res Ayurveda Pharm 3(2):267–271
Purohit PJ, Kapupara PP, Shah KV (2014) Development and validation for simultaneous estimation of curcumin and Gallic acid in different polyherbal formulations by HPLC. Res J Pharm Tech 7 (7):749–753
Rudrapal M, Oduri MU, Samidala NR, Surya Kiran BVVS, Junejo JA, Singh KD et al (2015) Development and validation of RP-HPLC method for simultaneous estimation of Olmesartan and hydrochlorothiazide in tablet dosage form. Orient J Chem 31(2):921–926
Singh A, Avupati VR (2017) Development and validation of UV spectrophotometric method for the estimation of curcumin in standardised polyherbal formulations. J Young Pharm 9(4):491–495
Surya Kiran BVVS, Sridhar N, Raghavendra M, Rudrapal M (2016) Development and validation of a RP-HPLC method for simultaneous estimation of minoxidil and aminexil in topical formulation. Asian J Chem 28(1):157–160
Syed HK, Liew KB, Loh GOK, Peh KK (2015) Stability indicating HPLC-UV method for detection of curcumin in *Curcuma longa* extract and emulsion formulation. Food Chem 170:321–326
Tonnesen HH, Karlsen J (1985) Studies on curcumin and curcuminoids. VI Kinetics of curcumin degradation in aqueous solution. Z Lebensm Unters Forsch 180(5):402–404
Warule PS, Patel VP, Gosavi SA (2017) Development and validation of UV spectrophotometric method for the estimation of curcumin in Ayurvedic formulation *Haridrakhand*. Int J Pharm Drug Anal 5(5):193–197
Wegscheider (1996) Validation of analytical methods. In: Guenzle H (ed) Accreditation and quality assurance in analytical chemistry. Springer, Berlin

Ethnobotanical Survey: The Foundation to Evidence-Based Validation of Medicinal Plants

22

Sunday O. Otimenyin

Abstract

Plants are vital to the survival of humans and animals. They supply the required nutrient and medicine to this group of living things. Plant seems to be the healthiest diet compared to diet of animal origins. Their benefits have been actualized over the year by the ability of humans and animals to recognize the difference between plants and their uses. Today much has been documented about plants and their food and health benefits. The ability to identify plants and their uses has been beneficial to the survival of mankind. Wrong identification can lead to poisoning and eventual death or to lack of expected activities and thus treatment failure and the consequence of treatment failure. Evidence that plant contents are dependent on the environment they are grown has been documented.

Research into plant usefulness and actions has been facilitated by taxonomical identification of the plant. This enhances documentation of plants and their actions. Plant location and distribution plays a vital role as it gives information about where to harvest the plants. This provides steady availability of the plants, a crucial requirement for their use in traditional medicine. Plant documentations brought to bear the plants that are about to go into extinction and prompt man to take positive actions, like cultivation of such plants for preservation. The process of identifying the plants and documenting their location and use is referred to as ethnobotanical survey. This is vital to evidence-based validation of traditional medicine. In this chapter, we have looked at the advantages of ethnobotanical survey and their impact on the validation of the uses of traditional medicine. The chapter revealed that such survey has enhanced plant's knowledge, their uses, and location documentation. It has also reduced duplication of studies and saved time

S. O. Otimenyin (✉)
Department of Pharmacology, Faculty of Pharmaceutical Sciences, University of Jos, Jos, Nigeria
e-mail: sundayotimenyin@gmail.com; sundayo@unijos.edu.ng

S. C. Mandal et al. (eds.), *Evidence Based Validation of Traditional Medicines*,
https://doi.org/10.1007/978-981-15-8127-4_22

and monies met for fruitful research. It has served as the foundation for evidence-based validation of traditional medicines.

Keywords

Ethnobotany · Ethnopharmacology · Evidence-based validation · Medicinal plant · Reverse pharmacology

Abbreviations

ATP	Adenosine Triphosphate
HIV	Human Immunodeficiency Virus
HPLC	High-Performance Liquid Chromatography
NMR	Nuclear Magnetic Resonance
Spp	Species
USA	United States of America

22.1 Introduction

Plants are vital to the survival of humans and animals. They supply the required nutrient and medicine to this group of living things. Plants seem to be the healthiest diet compared to diet of animal origins. Their benefits have been harnessed over the year by the ability of humans and animals to recognize the difference between plants and their importance or uses. Today much has been documented about plants, their food, and health benefits. The ability to identify plants and their uses has been beneficial to the survival of mankind. Wrong identification can lead to poisoning and eventual death or to lack of the expected effects or activities and thus treatment or use failure. Evidence that plant contents are dependent on the environment they are grown has been documented.

22.2 Ethnobotany

Ethnobotany as a discipline is currently oriented toward the exploration and documentation of new plant resources, collection of genetic materials, drug discovery, and product development. In recent years, the discipline of ethnobotany has become increasingly associated with the search for medicines. The search for herbal medicines and phytonutrients continues to gain more attention and expand rapidly across the world, with many people now resorting to these products for treatment of various health challenges in both developing and developed countries. Herbal remedies are available in drug stores, food stores, and supermarkets and on our streets (Höft et al. 1999; Peters 1996; Phillis and Gentry 1993).The surge in the utilization of herbal products has fueled research in this area and resulted in transition from classic herbal teas to standardized herbal products (Aumeeruddy-Thomas and Shengji 2003; Martin 1995; Davidson-Hunt 2000). The major reasons for seeking herbal therapy are the beliefs that it promotes healthier living and that it is natural and therefore safe. Others resort to herbal medicine when orthodox medicine has "failed," especially if the patient believes that there is spiritual involvement in the disease condition. Herbal medicines are, therefore, often viewed as a balanced and moderate approach to healing and recently represent a substantial proportion of the global drug market (Höft et al. 1999; Peters 1996). Another proof for the growing interest in herbal medicine industry is the novel basis of classical drug development. In 2007, China has collected 3563 extracts, 64,715 compositions, and 5000 single compounds from 3000 Chinese herbs together with about 130 kinds of chemical drugs (Elvine-lewis 2001). These findings follow the same trends from what is documented in most countries. As the global use of herbal medicinal products continues to grow and many more new products are introduced into the market, public health issues and concerns surrounding their safety are also increasingly recognized and overwhelming. Although some herbal medicines have promising potential and are widely used, many of their pharmacological and toxicological properties remain unverified. This makes knowledge of their potential toxicity and adverse effects very limited (Etkin 2001). It has become essential, therefore, to furnish the general public, including healthcare professionals, with adequate information based on scientific findings to facilitate better understanding of the risks associated with the use of these products and to ensure their safety, efficacy, suitability, and quality.

22.3 Basic Steps in Plant Research

Research into plant usefulness and actions has been facilitated by taxonomical identification of plants. This enhances documentation of plants, their distribution, their uses, and their actions. It also enhances the prescription or recommendation of specific plant to a patient or an individual. Plant location and distribution has also played a vital role as it gives information about where to harvest the plants for further scientific study. Plants that have not been identified scientifically cannot favor scientific research because of the difficulty in linking the discovery to the said plant. Knowledge of the local name of the plant is useful, but such identity cannot replace scientific name. Traditional names of plants are only useful in the area or region where such language is understood. This was what necessitated the introduction of botanical names for easy identification and global use. Knowledge of plants' scientific names or local names aids sourcing of the plants and provides steady availability of the plants, a crucial requirement for their use in traditional medicine and scientific research. Documentations have also brought to bear the plants that are about to go into extinct and prompted man to take positive conservative actions, like cultivation of such plants for preservation and/or the restriction of plant collection. The process of identifying the plants and documenting their location and use is referred to as ethnobotanical survey. This is vital to evidence-based validation of traditional medicine. In this chapter, we have looked at the advantages of ethnobotanical survey and their impact on the validation of the uses of traditional medicine. The chapter revealed that such survey has enhanced plant data collection, data base, their uses, and location documentation and pharmacological records on such plants. It has also reduced duplication of studies and saved time and monies met for fruitful research. It has served as the foundation for evidence-based validation of traditional medicines.

22.4 Plant Cycle

The role of plants in the life cycle (cycle of living things and non-living things) cannot be overemphasized. Plants have played a major role in the survival of man, animal, and plants. It is also known to be instrumental to the replenishment of key minerals in earth deposits. They are the builders of the basic building blocks of life. Plants assemble simple elements, like carbon, hydrogen, nitrogen, and other elements, into important building blocks required for normal body function and growth and for mineral deposits. They also provide enabling environment for nitrogen-fixing bacteria in plants and promotes the formation of nitrogen-containing compounds. The assembling process results in the generation or production of a wide range of chemical substances that are both beneficial and harmful to living things. Substances produced include food (carbohydrates and amino acids), medicines, perfumes and scents, and poisons (useful in warfare). The ability of plants to produce these products makes them as the sustainer of life. The identification of these roles of plants has made man to be pre-occupied with the selection and cultivation of important plants—a major pre-occupation of man. Records have shown that selected and cultivated plants are plants that are of benefits to man. Plants that are regarded as not beneficial are often weeded out—this has led to the extinction of some plants. Generally, as will be seen later in this chapter, all plants are of immerse benefit to mankind. Of notable importance are plants that hold soil together and prevent soil erosion and landscape destruction.

22.4.1 Importance of Plants

The importance of plants to the survival and sustenance of life cannot be overemphasized. Plants are utilized in all the endeavors of man. They are used in crude stage or processed stage. Animals, a minor source of food, survive on plants. They use the basic and complex chemicals produced by plants to produce other chemicals that are beneficial to man. Plants and animals depend on oxygen produced and released from plants for normal cell function. Plants have been found useful in many areas of life cycle and human endeavors. Important uses of plants are depicted in Table 22.1.

Table 22.1 Uses of plants

Flowering plants	Virtually all crop plants are flowering plants
	Beverage: teas, coffee, cocoa
	Cooking; oil from corn, canola, peanut, sesame, olive
	Spices and seasonings: cloves, vanilla, black pepper, cinnamon, nutmeg, saffron
	Alcohol, latex, adhesives
	Dyes, perfumes, saffron
	Drugs; morphine, aloe, strychnine, atropine, aspirin, digitalis, quinine
Ferns and fern	Horticultural; ornamental plants
	Food
	Art: basketry and weaving
	Coal; form coal deposit in forests
Conifers and conifer	Paper product; from wood pulp
	Turpentine, varnish, resin, adhesives, chewing gum, printing ink
	Cellulose fibers are processed into cellophane, rayon, and insulation
	Drugs: methamphetamines
	Horticultural; landscaping, Christmas trees, flowers
	Recreation: hiking, camping
Mosses and Moss	Fuel for cooking
	Herbal medicines; astringents, germicides, antiseptics, antibiotics
	Building material; as low-cost alternative
	Bioindicators of air pollution
	Horticulture: manures, mulch, and for propagation by "air layering"
	Water filtration and waste water treatment
	Arts and crafts

22.5 Plant Identity and Contents

Plants, like animals, have their genetic makeup. This determines their color, shape, and physical characteristics—a distinguishing element that aids in the identification of plants by taxonomists. The genetic composition of plants also determines the types of chemical agents they can produce or synthesize. The color and physical appearance helps in the identification of plant; it may also reveal the type of chemical agents that can be produced by such plant. Their identity also gives information of the type of chemical agents that can be found in them. This helps in distinguishing plants that are used as food, medicines, etc. Chemical agent found in the plants determines their uses. Plants are used as food because they contain carbohydrate, protein, and useful minerals. Carbohydrates are used as fuel by living things (Balick and Cox 1996). In man, it is fed into the Krebs cycle to produce ATP, an important element for normal cell function.

22.5.1 Genetics and Environment: Determinant of Plant Contents and Products

The gene in the cell of plants ensures plants are reproduced to specification. The genetic materials in plants are housed in the nucleus of each cell. It contains codes of the general and salient features of the plants. It can be used for proper plant identification and characterization but is seldom used because of the cost implications and facilities that may not be readily available or easily moved to field. The outputs of gene expression, like the leaf, flower, fruit, stem, and root shapes and types, are used for identification by taxonomists.

Several plants that have not been in existence before now are being produced via gene mutation and basic gene element rearrangement. Such mutation is fueled by change in the environment in which the plants are grown and arises as a survival tool in harsh environment. Genetic mutation, apart from changing the physical features of the plants, also impacts on the chemical agents formed by the plants.

The environment, to a greater extent, determines the types of biochemical agents that are produced by plants. As mentioned earlier, plant utilizes basic elements in the environment (particularly the soil) for the production/synthesis of the chemical principles found in them. This by implication means that the soil and air must contain these basic elements required for the synthesis of biochemical. This informed the variation of the type of chemicals found in the same species of plants across the globe. The full capacity of a plant to produce biochemical may not be realized if the basic elements needed for the formation of the said biochemical are not present in the environment. This implies that there might be a need to vary the growth environment if the yield of the needed agent from plant is low. Research into the basic elements needed for such agents to be formed may provide a clue to what is needed in the plant environment for optimum production of the required agents.

22.5.2 Local Identification of Plants

Proper identification of plants is crucial if the benefits of their contents are to be harnessed. The locals in a particular community identify plants by their shape (shape of leaves, fruits, stem, and roots), color, and smell and sometimes taste. This identification skill of the traditional people has been used by botanist for the identification of plant, since genetic identification is not easily accessible and costly. Local identification of plants by the traditional people is a skill inherited from their mentors and forefathers. This has led to the proper identification of plants that serves as food and/or medicines. Animals feed on plants and also eat plant for their medicinal purposes. They use the same features used by man to identify plants in the environment and determine the plants that are poisonous or edible or could be useful in the healing of their disease conditions. Plants that have resulted in the death of animals are often regarded as poisonous and avoided by man and animals.

22.5.3 Spiritual Identification

This is a subject of debate by scientists and non-scientists, but most ethnobotanical surveys have records of how the ancestors claim that dreams, spirits, and God reveal the curative herbs to them. This explains why their practice includes incantation, a belief that the locals hope it gives life and potency to their medicinal plant. What remains uncertain is the contribution of these claims to plant identification and uses by traditional people.

22.5.4 Taxonomical Identification of Plant

Taxonomists are the authority in the identification of plants in scientific environment. They use key features of the plants for plant identification. These features are not too different from the features used by the traditional people, but it is more refined and generally accepted. The taxonomist attains this height of plant identification after undergoing specific training in recognized university or institution. During ethnobotanical survey, plant herbarium is collected for plant identification. Collected herbarium is submitted to the taxonomist for onward identification of the plant. The voucher specimen is given a number after identification and stored in the herbarium for future reference. This practice has ease report submission and data storage in the data base of plants. The data bank avail the scientist the needed information on the plant or disease conditions they intend to research on. It also reveals the level of scientific probe that has been carried on the said plant. Twofold identification may be adopted as seen in most plant data base: the taxonomical identity (botanical name) and the local identity (local name) of the plants. Misidentification has been a major issue in data collection and storage, implying that effort must be made for proper identification if the information in the data base is to be reliable and credible. The identity of a plant is crucial if a plant that has been used by locals is to be validated for pharmacological activity. Plant identification either as local name or botanical name is also important for evidence of use of the medicinal plant. Correct plant identification is therefore crucial in evidence-based validation of medicinal plants.

22.6 Ethnopharmacology

Ethnopharmacology was first used in 1967 in a book titled *The Hallucinogens*. It is a relatively new field in pharmacology. It covers ethnobotany and the validation of the claims documented during ethnobotanical survey. As with ethnobotany, it covers the observation, collection, identification, description, and experimental investigation of the chemically active principles and the effects of the chemical agents found in traditional medicine preparation (Abdolbaset et al. 2006).

Ethnopharmacology is defined as "the interdisciplinary scientific evaluation and documentation of biologically active agents in medicinal plants" (Abdolbaset et al.

2006). Many drugs in use today were discovered through ethnopharmacological research. Many plants have been and are being evaluated with the aim of discovering new drugs. Knowledge of medicinal plants used in the management of disease conditions is an important element of traditional medical systems in many communities in the world, and these resources is a part of traditional knowledge of a culture (Heinrich et al. 2018). In some cultures, these practices or cultures are documented in their traditional pharmacopoeia. However, most of the traditional knowledge are not documented but are passed from generation to generation through apprenticeship. In this process, traditional knowledge are either modified or lost. The quest for modern education had negatively impacted the passage of traditional knowledge and culture in most African communities. This informs the need for ethnobotanist and ethnopharmacologists to act fast in the documentation of these rich cultures to avoid loss of knowledge that may lead to the discovery of life-saving drugs in the future. Some countries have well-developed pharmacopoeias, with research ongoing to validate the efficacy of the claims in such records. There exists the Asian, European, indigenous, and non-Western type of herbal medicines whose practices date back to thousands of years. Indigenous medical systems are the most diverse and are still practiced where such cultures are intact but are continuously evolving as contact with other cultures is continuous. The knowledge may reside exclusively with traditional healers, or may be known generally (Davidson-Hunt 2000). Ethnopharmacology is a field that yields collaboration among different researchers as well as the application of that knowledge to practical ends for both scientific and indigenous communities (Elvine-lewis 2001).

22.6.1 Imperatives of Ethnopharmacological Studies

Ethnopharmacological approach in the study of medicinal plants and cultural practices has yielded products of know pharmacological effects (Table 22.2). Such products have been integrated into modern orthodox medical practice or provided detailed understanding of its use in different cultures and/or the understanding of some physiological function of the human body. The study of the botanical origin of the arrow poison curare, its physiological effects, and the compound responsible for these effects has provided explanation for its use as poison arrow. The botanical source of curare was identified as the climbing vine *Chondrodendron tomentosum* Ruiz and Pavon; other species of the Menispermaceae (*Curarea* spp. and *Abuta* spp.) and Loganiaceae (*Strychnos* spp.) are also used in the production of curares of varying types (Abdolbaset et al. 2006). Ior and her team (Ior et al. 2017) reported medicinal plants used for the management of psychosis in Northern Nigeria. Their investigation into the claimed activity of these plants revealed that a number of plants have central nervous system effects (Otimenyin and Ior 2019). Figure 22.1 and Table 22.3 show the area of study and the plants claimed to have antipsychotic activity. Table 22.4 shows the results that validated some of the claims (Ior and Otimenyin 2019). Discovery of reserpine from a traditional medicinal plant (*Rauwolfia serpentina*) was a monumental breakthrough that led to the introduction of a

Table 22.2 Ethnobotanical approach outcomes

Drug/Agents	Plant	Action
Berberine	*Berberis vulgaris*	Management of bacillary dysentery
Atropine	*Atropa belladonna*	Anticholinergic
Asiaticoside	*Centella asiatica*	Vulnerary
Arecoline	*Areca catechu*	Anthelmintic
Anisodamine	*Anisodus tanguticus*	Anticholinergic
Anisodine	*Anisodus tanguticus*	Anticholinergic
Andrographolide	*Andrographis paniculata*	Hepatoprotective management of bacillary dysentery, hepatoprotective
Allyl isothiocyanate	*Brassica nigra*	Rubefacient
Agrimophol	*Agrimonia supatoria*	Anthelmintic
Ajmalicine, serpentine	*Rauwolfia serpentina*	Management of circulatory disorders
Agrimophol	*Agrimonia supatoria*	Anthelmintic
Aesculetin	*Fraxinus rhynchophylla*	Antidysentery
Aesculetin	*Aesculus hippocastanum*	Anti-inflammatory
Adoniside	*Aesculus hippocastanum*	Cardiotonic
Gossypol	*Gossypium species*	Male contraceptive
Glaucarubin	*Simarouba glauca*	Amoebicide
Gitalin	*Digitalis purpurea*	Cardiotonic
Etoposide	*Podophyllum peltatum*	Antitumor agent
Ephedrine	*Ephedra sinica*	Sympathomimetic, antihistamine
Emetine	*Cephaelis ipecacuanha*	Emetic, amoebicide
Digoxin	*Digitalis purpurea*	Cardiotonic
Digitalin	*Digitalis purpurea*	Cardiotonic
Digitoxin	*Digitalis purpurea*	Cardiotonic
Deslanoside	*Digitalis lanata*	Cardiotonic
Deserpidine	*Rauvolfia canescens*	Antihypertensive, tranquilizer
Danthron	*Cassia species*	Laxative
Curcumin	*Curcuma longa*	Choleretic
Cynarin	*Cynara scolymus*	Choleretic
Convallatoxin	*Convallaria majalis*	Cardiotonic
Codeine	*Papaver somniferum*	Analgesic, antitussive
Colchicine	*Colchicum autumnale*	Antitumor, antigout
Cocaine	*Erythroxylum coca*	Local anesthetic
(+)-Catechin	*Potentilla fragarioides*	Hemostatic
Chymopapain	*Carica papaya*	Proteolytic, mucolytic
Caffeine	*Camellia sinensis*	CNS stimulant
Betulinic acid	*Betula alba*	Anticancer
Bromelain	*Ananas comosus*	Anti-inflammatory, proteolytic
Monocrotaline	*Crotalaria sessiliflora*	Topical antitumor agent
Lapachol	*Tabebuia species*	Anticancer, antitumor

(continued)

Table 22.2 (continued)

Drug/Agents	Plant	Action
a-Lobeline	*Lobelia inflata*	Smoking deterrent, respiratory stimulant
Kheltin	*Ammi visage*	Bronchodilator
Lanatosides A, B, C	*Digitalis lanata*	Cardiotonic
Kainic acid	*Digenea simplex*	Ascaricide
Kawain	*Piper methysticum*	Tranquilizer
Irinotecan	*Camptotheca acuminate*	Anticancer, antitumor agent
Hyoscyamine	*Hyoscyamus niger*	Anticholinergic
Hemsleyadin	*Hemsleya amabilis*	Treatment for bacillary dysentery
Hydrastine	*Hydrastis canadensis*	Hemostatic, astringent
Glycyrrhizin	*Glycyrrhiza glabra*	Sweetener, treatment for Addison's disease
Gossypol	*Gossypium species*	Male contraceptive
Scopolamine	*Datura species*	Sedative
Santonin	*Artemisia maritima*	Ascaricide
Scillarin A	*Urginea maritima*	Cardiotonic
Salicin	*Salix alba*	Analgesic
Rotenone	*Lonchocarpus nicou*	Piscicide, Insecticide
Rhomitoxin	*Rhododendron molle (rhododendron)*	Antihypertensive, tranquilizer
Rorifone	*Rorippa indica*	Antitussive
Reserpine	*Rauwolfia serpentine*	Antihypertensive, tranquilizer
Rescinnamine	*Rauwolfia serpentine*	Antihypertensive, tranquilizer
Quisqualic acid	*Quisqualis indica*	Anthelmintic
Quinine	*Cinchona ledgeriana*	Antimalarial, antipyretic
nor-pseudoephedrine	*Ephedra sinica*	Sympathomimetic
Pseudoephedrine	*Ephedra sinica*	Sympathomimetic
Protoveratrines A, B	*Veratrum album*	Antihypertensives
Podophyllotoxin	*Podophyllum peltatum*	Antitumor, anticancer agent
Picrotoxin	*Anamirta cocculus*	Analeptic
Pilocarpine	*Pilocarpus jaborandi*	Parasympathomimetic
Phyllodulcin	*Hydrangea macrophylla*	Sweetener
Physostigmine	*Physostigma venenosum*	Cholinesterase inhibitor
Papain	*Carica papaya*	Proteolytic, mucolytic
Ouabain	*Strophanthus gratus*	Cardiotonic
Neoandrographolide	*Andrographis paniculata*	Treatment of dysentery
Noscapine	*Papaver somniferum*	Antitussive
Morphine	*Papaver somniferum*	Analgesic
Sennosides A, B	*Cassia species*	Laxative
Silymarin	*Silybum marianum*	Antihepatotoxic
Stevioside	*Stevia rebaudiana*	Sweetener

(continued)

Table 22.2 (continued)

Drug/Agents	Plant	Action
Strychnine	*Strychnos nux-vomica*	CNS stimulant
Teniposide	*Podophyllum peltatum*	Antitumor agent
Tetrahydropalmatine	*Corydalis ambigua*	Analgesic, sedative, tranquilizer
Yuanhuacine	*Daphne genkwa*	Abortifacient
Yohimbine	*Pausinystalia yohimbe*	Aphrodisiac
Tubocurarine	*Chondrodendron tomentosum*	Skeletal muscle relaxant
Valapotriates	*Valeriana officinalis*	Sedative
Trichosanthin	*Trichosanthes kirilowii*	Abortifacient
Theophylline	*Theobroma cacao and others*	Diuretic, bronchodilator
Topotecan	*Camptotheca acuminate*	Antitumor, anticancer agent

new drug for the management of some disease conditions in orthodox medicine. *Rauwolfia serpentina*, called Chotachand in Hindi, was reported to have been used by the local people of the Himalayan Mountains for snakebite. Ethnopharmacological survey revealed the claim by a local legend that in ancient times, mongooses feed on the plant before engaging in combat with cobra. This observation prompted the local people in that region to test the plant for activity in human bitten by snake and found that the shrub could serve as a potent antidote to snakebite. In Bihar Province of India, people use this plant to treat insanity, epilepsy, and insomnia (Balick and Cox 1996). Reserpine, a potent drug for hypertension, was later isolated from *Rauwolfia*.

Some plants have been reported to be of no value after scientific investigation. Traditional medical practitioners were advice not to use such drugs for the management of the unvaried disease conditions. Their insistence on the use of such plants has generated a lot of concerns in the scientific world. This has stimulated to further investigations that involved following up on the patients taking such plant to ascertain the claimed activity. Reports have shown that indeed some of such plants are effective. Lack of activity reported has been pinned to the type of solvents used during extraction processes and the fact that some plant constituents act in synergy to elicit their biological activity. These findings have reiterated the need to revisit plants that have been claimed to be devoid of pharmacological activity. Table 22.5 includes plants, ethnobotanical leads, verified biological activities, and herbal products produced from medicinal plants.

22.6.2 Research Driven by Ethnopharmacology

The volume of plants in the world is large. Conducting unfocused research is time-consuming and tasking and produces poor outcomes. Our biodiversity is evolving. This explains why it will be near impossible to screen all the plants that exist on

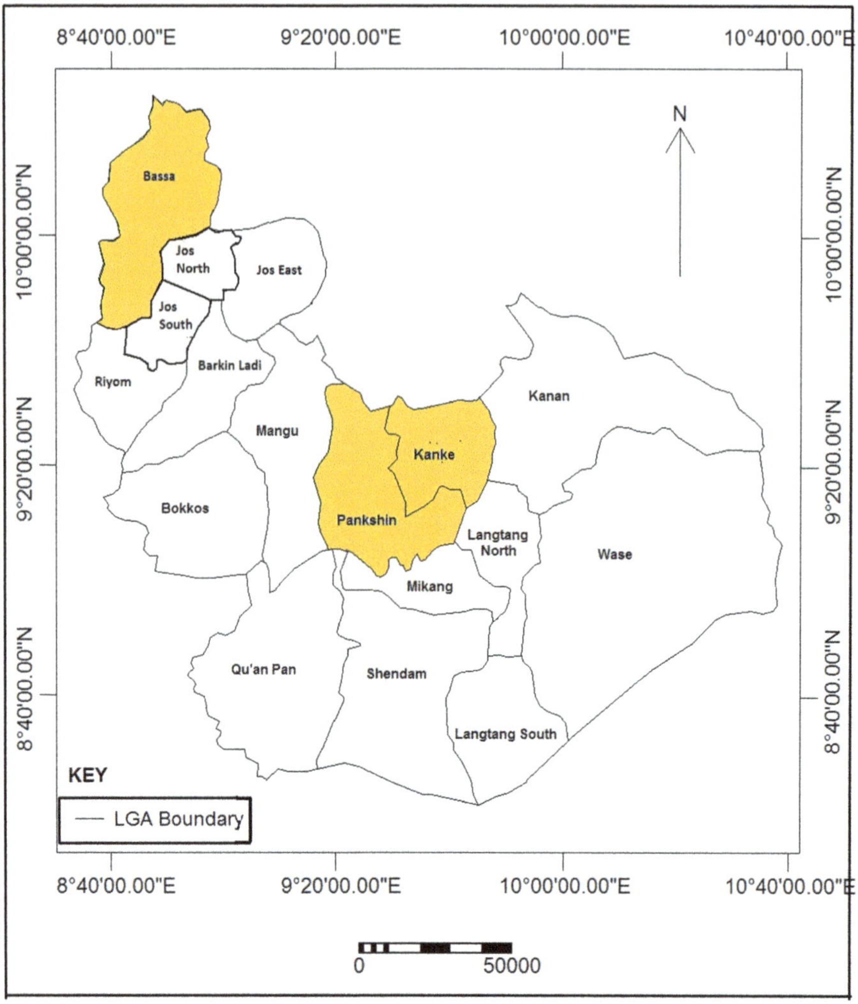

Fig. 22.1 Map of Plateau State showing study areas (Ior and Otimenyin 2019) Source: Geographic Information Systems (GIS) Laboratory, Department of Geography and Planning, University of Jos (2014)

earth. Research driven by folkloric use of plants saves cost and time and may be more productive. Ethnobotanical or ethnopharmacological survey documents folkloric plants and their uses. During such survey, it is possible to obtain first-hand evidence of individuals who have benefited from the plants. Beneficiaries can be interviewed to authenticate folkloric claims. The composition of ethnopharmacological survey team contributes to verification of folkloric claims. The team has qualified health practitioner who can confirm their diagnosis and the resolution of the disease condition. Positive evidence often gives hope and encourages further scientific study. Such evidence-based studies have been found

Table 22.3 Medicinal plants used in the treatment of CNS disorders in Bassa, Pankshin, and Kanke Area of Plateau State, Nigeria

S/N	Family	Botanical name	Local name	Voucher	Part used	Preparation	NC	Traditional uses	UV
1	Anacardiaceae	*Anacardium occidentale*	Kashu/fisaa	FHJ 255	Leaves, stem bark	Oral infusion	2	Aggression/ insomnia	0.03
2	Annonaceae	*Uvaria chamae*	Rikuki	FHJ 258	Leaves, root	Oral infusion/steaming	3	Aggression/ psychosis	0.04
3	Annonaceae	*Annona senegalensis*	Gwandan daji/ wut	ABU 90012	Leaves, stem bark	Oral decoction	3	Psychosis/ depression	0.04
4	Asparagaceae	*Asparagus africanus*	Turaakazomo	FHJ 251	Whole plant	Inhalation/incense	2	Hallucination/ depression	0.03
5	Asteraceae	*Vernonia amygdalina*	Shiwaka	ABU 595	Leaves	Oral decoction	2	Hallucinations	0.03
6	Bambuseae	*Oxytenanthera abyssinica*	Gora	FHJ 270	Leaves, stem bark	Oral infusion	2	Psychosis	0.03
7	Bignoniaceae	*Stereospermum kunthianum*	Sansami	FHJ 263	Stem bark	Oral infusion/bathing	2	Depression/ convulsion	0.03
8	Bignoniaceae	*Newbouldia laevis*	Aduruku	FHJ 277	Leaves	Oral infusion/steam bath	2	Psychosis/ insomnia	0.03
9	Bignoniaceae	*Spathodea campanulata*	Aduruku	FHJ 269	Leaves, root	Oral infusion/bathing	3	Hallucination	0.04
10	Burseraceae	*Boswellia dalzielii*	Hanno/ mwarmwar/ arrarabi	ABU 1314	Leaves, bark, root	Oral infusion/incense	3	Hallucinations/ aggression	0.04
11	Caesalpiniaceae	*Daniellia oliveri*	Maji	FHJ 264	Stem bark	Oral infusion/bathing	2	Hallucination	0.03
12	Caricaceae	*Carica papaya*	Gwanda	ABU 005	Leaves	Oral infusion	2	Psychosis	0.03
13	Chenopodiaceae	*Chenopodium ambrosioides*	Tafarnuwa turawa	ABU 1921	Leaves	Oral infusion/steam bath/ inhalation	4	Psychosis/ convulsion	0.06

(continued)

Table 22.3 (continued)

S/N	Family	Botanical name	Local name	Voucher	Part used	Preparation	NC	Traditional uses	UV
14	Combretaceae	*Guiera senegalensis*	Sabara	ABU 900165	Leaves	Maceration	4	Psychosis	0.06
15	Combretaceae	*Terminalia macroptera*	Baushe	FHJ 259	Leaves	Oral infusion/incense	2	Hallucinations	0.03
16	Crassulaceae	*Bryophyllum pinnatum*	Abomada	FHJ 254	Leaves	Oral decoction	2	Anxiety	0.03
17	Cucurbitaceae	*Momordica charantia*	Garafunii	FHJ 271	Fruits	Oral decoction	2	Psychosis	0.03
18	Cycadopsida	*Carissa edulis*	Lemun tsuntsu	ABU 900086	Leaves, root	Oral infusion	3	Psychosis	0.04
19	Euphorbiaceae	*Euphorbia hirta*	Yinfut	FHJ 261	Leaves	Inhalation/steam bath	2	Insomnia/depression	0.03
20	Euphorbiaceae	*Croton zambesicus*	Rim ase		Leaves	Oral infusion	4	Aggression/mania	0.06
21	Euphorbiaceae	*Jatropha curcas*	Bitadazuru, Dazugu, Mbilit	FHJ 267	Leaves	Oral infusion	2	Psychosis/aggression	0.03
22	Fabaceae	*Indigofera erecta*	Pinone		Leaves	Oral infusion	2	Psychosis/convulsion	0.03
23	Fabaceae	*Cassia singueana*	Senna	ABU 6855	Leaves	Oral infusion	2	Psychosis	0.03
24	Fabaceae	*Erythrina senegalensis*	Minjirya	ABU 7721	Leaves	Inhalation/steaming/bathing	2	Psychosis/convulsion	0.03
25	Lamiaceae	*Ocimum canum*	Dod- doyo	FHJ 275	Whole plant	Oral infusion/incense/bath infusion/steaming/incense	3	Psychosis/convulsion	0.04
26	Lamiaceae	*Thymus vulgaris*	Malaka		Whole plant	Inhalation/steaming	2	Hallucination	0.03

27	Lamiaceae	*Ocimum basil*	Wuzab	FHJ 276	Leaves, stem bark	Decoction/infusion	3	Aggression/ insomnia	0.04
28	Lamiaceae	*Clerodendrum capitatum*	Tabataab	FHJ 266	Leaves	Incense/inhalation	2	Psychosis	0.03
29	Loranthaceae	*Tapinanthus dodoneifolius*	Ndur/Kanci	ABU 6517	Leaves	Oral infusion	3	Psychosis	0.04
30	Malvaceae	*Sida cordifolia*	Banza	FHJ 253	Leaves	Oral infusion	2	Psychosis	0.03
31	Meliaceae	*Khaya senegalensis*	Tan/madaci	FHJ 252	Leaves, stem bark	Oral infusion	2	Psychosis	0.03
32	Myrtaceae	*Syzygium guineense*	Malmo	ABU 900295	Leaves	Oral infusion	3	Psychosis/ depression	0.04
33	Olacaceae	*Ximenia americana*	Chibolng/tsada	ABU 1612	Whole plant	Oral infusion	2	Aggression, depression	0.03
34	Poaceae	*Eleusine indica*	Juji	FHJ 260	Leaves	Oral infusion	2	Anxiety	0.03
35	Polygalaceae	*Securidaca longepedunculata*	Sannya	ABU 900141	Leaves, Root	Oral infusion/incense/bath	4	Psychosis	0.06
36	Rhamnaceae	*Ziziphus mucronata*	Magariyan kura	FHJ 257	Leaves	Oral decoction	2	Depression	0.03
37	Rubiaceae	*Nauclea latifolia*	Tafashiya/ Gilng	ABU 005	Leaves, bark, root	Oral decoction/inhalation	4	Psychosis/ aggression	0.06
38	Sapindaceae	*Paullinia pinnata*	Yatsa biyar	FHJ 256	Leaves	Oral infusion	3	Psychosis	0.04
39	Solanaceae	*Nicotiana tobacum*	Tiba/taba	ABU 1611	Leaves, root	Infusion/incense	2	Mania/ depression	0.03
40	Verbenaceae	*Lantana camara*	Kashin kuda/ yinfut	FHJ 273	Leaves, stem bark	Oral decoction	2	Mania/ depression	0.03
41	Verbenaceae	*Vitex doniana*	Dinnya	FHJ 272	Leaves	Oral decoction	3	Psychosis/ anxiety	0.04

(continued)

Table 22.3 (continued)

S/N	Family	Botanical name	Local name	Voucher	Part used	Preparation	NC	Traditional uses	UV
42	Vitaceae	*Cissus populnea*	Dafara	FHJ 268	Leaves Incense/inhalation	Incense/inhalation	2	Psychosis/depression	0.03

Key: *NC* number of citation, *UV* use value
Source: Ior and Otimenyin 2019

Table 22.4 Medicinal plants with validated antipsychotic activity

Treatment	Dose (mg/kg)	Number of climbing 10 min	Number of climbing 20 min	Number of climbing 30 min
Normal saline	10 mL/kg	5.52 ± 0.29	4.00 ± 0.10	3.11 ± 0.03
P. pinnata	800	1.53 ± 0.08*	1.72 ± 0.13*	1.11 ± 0.33*
O. canum	800	4.08 ± 0.22	3.21 ± 0.26	2.57 ± 0.12
D. oliveri	800	3.24 ± 0.12	2.44 ± 0.21	1.15 ± 0.17*
T. macroptera	800	1.95 ± 0.66*	1.80 ± 0.60*	1.25 ± 0.42*
U. chamae	800	2.34 ± 0.27*	2.52 ± 0.13	1.28 ± 0.17*
C. Zambesicus	800	2.83 ± 0.60*	2.67 ± 0.88	2.07 ± 0.17
J. curcas	800	1.85 ± 0.19*	2.14 ± 0.32*	1.81 ± 0.48*
S. campanulata	800	2.15 ± 0.29*	2.58 ± 0.29*	1.72 ± 0.44
C. edulis	800	4.53 ± 0.08	2.52 ± 0.13	2.28 ± 0.17
S. Kunthianum	800	4.21 ± 0.26	3.87 ± 0.43	2.67 ± 0.26
S. longepedunculata	200	2.21 ± 2.67*	2.67 ± 0.26*	1.52 ± 0.44
I. erecta	800	4.73 ± 0.55	3.34 ± 0.52	2.88 ± 0.55
N. latifolia	800	2.62 ± 0.32*	1.00 ± 0.63*	1.30 ± 0.83*
T. Ddodoneifolius	800	2.45 ± 0.43*	1.51 ± 1.43*	1.33 ± 1.31
B. dalzielii	800	1.66 ± 0.42*	1.98 ± 0.10*	0.60 ± 0.36*
L. camara	800	2.44 ± 0.30*	3.22 ± 0.57	0.45 ± 0.25*
C. ambrosioides	800	2.34 ± 0.48*	1.60 ± 0.60*	0.56 ± 0.56*
G. Senegalensis	800	2.64 ± 0.33*	3.12 ± 0.38	2.28 ± 0.62
N. laevis	800	2.50 ± 0.07*	1.00 ± 0.63*	1.30 ± 0.83*
V. doniana	800	2.75 ± 0.25	2.75 ± 0.31	3.07 ± 0.12
Chlorpromazine	4	0.05 ± 0.01*	0.02 ± 0.01*	0.05 ± 0.01*

Comparison of the effects of aqueous extracts of some collected plants on apomorphine-induced stereotypic climbing behavior in mice
Values are presented as mean ± SEM ($n = 5$); * $p < 0.05$ compared to control
Source: Ior and Otimenyin 2019

to be result-oriented and most likely to lead to the development of new or lead drugs (Abdolbaset et al. 2006). The interview sometimes reveals the side effects of the medicinal plant under investigation. Such findings can reveal unknown effects of the plant under investigation and result in the discovery of new drugs. It is also possible that the drug relieves only the symptoms as is the case with most analgesic and anti-inflammatory drugs. The relief of pain without managing the underlying cause of pain poses a great problem in the use of traditional medicine. The traditional medical systems have referral system that encourages referrals, especially if the patient is not responding to treatment or the practitioner lacks expertise in the area. Records also show that traditional medical system has specialists in different disease conditions; such specialists are located through referral system. This makes the traditional heritage of disease management rich and all encompassing. Most schools believe that the present orthodox medicine system copied the traditional medicine systems. There are known approaches for the selection of medicinal plants for study. Five

Table 22.5 Some medicinal plants with known pharmacological activities

Botanical name	Family	Product	Uses	Source
Capparis decidua	Capparidaceae	Rishta-capparis decidua in brine	Anti-inflammatory, antimicrobial, arthritis	Extract
Aloe barbadensis	Xanthorrhoeaceae	Aloe vera	No clear therapeutic use But used for its anti-inflammatory effects	Extracts in combination with other herbs extract
Balanites aegyptiaca	Balanitaceae	Balanites oil, Balanites fruit pulp	Antimyco-bacterial and anti-inflammatory, pain, arthritis, dysentery, and bilharzias	Extract
Dracaena cinnabari		Dracaena cinnabari		Dracaena cinnabari
Rosa canina	Rosaceae	Powder, supplements.	Anti-inflammatory, osteoarthritis	Dry fruit powder
Terminalia sericea	Combretaceae	Cosmelene®	Anti-inflammatory, pain, antimicrobial, diarrhea and swollen painful eyes, diabetes, and wounds	Root, bark
Prunus africana and Pygeum africanum	Rosaceae	Extract	Prostatic hypertrophy, allergies, inflammation, kidney diseases, malaria	Bark, extracts, Ttea
Camellia sinensis	Theaceae	Peso exacto, Sbelttix, hoodia slim	Anti-inflammatory, antiallergy, antihypercholesterolemic, antimicrobial, antioxidant	Extract
Acacia senegal, and Acacia seyal Del.	Mimosaceae	Capsules, cream Extracts and liquid extract	Antihypertensive, anti-inflammatory, and anti-platelet Aggregatory activity, antiarthritis	Acacia gum Acacia seyal
Albizia adianthifolia and Albizia julibrissin	Fabaceae	Aminogenesis	Anti-inflammatory, headaches, antioxidant, antitumor, ulcer, and wound healing	Roots, powdered bark, snuff

major approaches of plant selection for pharmacological screening are known, namely:

1. Random approach, which involves the collection of all plants found in the study area
2. Phytochemical targeting, which entails the collection of all members of a plant family known to be rich in bioactive compounds
3. The ethno-directed sampling approach, based on traditional medicinal uses of a plant
4. Chemotaxonomic approach
5. A method based on specific plant parts, such as seeds, flowers, leaves, etc. (Cotton 1996)

Khafagi and his research team (Khafagi and Dewedar 2000) investigated the antimicrobial activity of Sinai Peninsula plants collected randomly versus those collected based on ethno-medical uses. Their results revealed that about 83.3% of plants collected by ethno-directed approach showed antimicrobial activity, while only 41.7% of plants collected randomly showed antimicrobial action. Ethno-directed plant selection approach demonstrated higher percentage of test results of species with antimicrobial activity either way as compared to random selection approach. Both ethno-directed and random selection approach may serve as a useful strategy in the search for biologically active compounds with potential antimicrobial activity (Khafagi and Dewedar 2000). It is reasonable to conduct ethnobotanically directed research in order to optimize the search for novel pharmaceuticals. Random selection approach may be useful for the identification of the presence of bioactive compounds from plants with unknown folk medicinal use, but ethnobotanical approach is believed to be the best and is giving the highest chances to success (Sairafianpuor 2002). Ethno-directed approach reveals the possible uses of the plant and suggests screening methodology and extraction technique to be employed. Extraction methods determine the nature of chemicals that will be extracted from any medicinal plant. If the method used does not favor extraction of active principle in the plant, results obtained may show that the plant does not possess the claimed biological activities. This explains why, for the first study of evaluating plants for activity, it is advisable to choose the extraction methods that is same or close to the methods used by the traditional medical practitioner in the community where ethnobotanical survey was carried out. This saves time and may screen out the toxic principle in the plant.

22.6.3 Enriching Indigenous Medicinal Knowledge

Attention has been on traditionally used medicinal product with little efforts to add to the data base of the traditional knowledge of medicinal plants. In the past before the advent of medicine or pharmacology, medicinal knowledge is constantly being enriched by the discovery of new herbs that have healing prospects. Today attention has shifted to orthodox medicine and industrial or laboratory search for drugs. This has been helpful though, but attention should be given to the traditional methods of drug discovery and development of indigenous knowledge. Traditional data bases can be enriched by random study of plants for medicinal value. Plants that have not been documented or used by traditional medical practitioner as medicines can be evaluated for bioactivity. The study of the antimicrobial activity of Sinai Peninsula plants collected randomly shows that non-ethno-directed investigation can yield bioactive compounds (Bernard 1957).

22.6.4 Preservation of Ethnobotany Knowledge

Human living conditions have changed over decades, a change that has impacted on the environment. This change may be beneficial or adverse, but generally the change was intended for the betterment of humans and intended to favor survival and produce persistent effects that are beneficial to mankind (Laland et al. 2010; Levis et al. 2017; Sullivan et al. 2017). New generations always inherit modified environments that result from past decisions. This has affected the knowledge and current use of plants in our society (Lins Neto and Albuquerque 2018). Human activity has also impacted on the climate resulting in climate change (Ladio 2017). Climate change has affected the soil composition and the type of plants that can grow on such soil. Record has shown that plant used by a community of people depends on the types of diseases prevalent in their region. The tropics, especially Africa, have high incidence of malaria (Santoro et al. 2017). This always correlates with the type and number of plants used by the indigenous people living in this environment for the management of malaria (Albuquerque et al. 2019). Changes in the environment that favored the breeding of mosquitos may have contributed to the wealth of knowledge of plants used in the management of malaria. Ethnobotanical survey in these regions will unlock the wealth of knowledge of plants used in the management of prevalent disease condition in the region. The evidence gathered during such survey will direct scientific research in this direction and may result in the validation of the useful medicinal plants. The world today has artemisinin, a compound extracted from *Artemisia Annum*, as its commonly used antimalarial agent. This plant was used in china by the local people for the management of malaria. It has saved the world after the development of resistance to chloroquine.

22.6.5 Ethnobotanical Leads and Potential Drug Candidate

Plants produce economically important organic compounds such as oils, resins, tannins, rubber, gums, waxes, dyes, flavors and fragrances, pesticides, and pharmaceuticals (Eldeen et al. 2010). As indicated earlier in this chapter, many of the medicines used and/or currently in use such as aspirin, codeine, ipecac, pilocarpine, pseudoephedrine, quinine, reserpine, scopolamine, theophylline, vinblastine. etc. have been derived from medicinal plants based on ethnobotanical research programs (Balick and Cox 1996). It is estimated that among the total flowering plants occurring in the tropical regions of the world, only few were studied for their pharmaceutical potentials. Most plants contain chlorophyll-a, an agent that has been shown to possess anti-inflammatory activity (Subramoniam 2014). This may support the claim that all chlorophyll-a-bearing plants have pharmaceutical potential (Subramoniam 2014). Pharmaceutical companies are gradually turning their search for potent bioactive compounds to medicinal plants. They have produced a number of herbal preparations based on ethnobotanical leads for different purposes. Few examples of these products and their botanical sources are given in Table 22.2. Otimenyin and his research team (Otimenyin et al. 2013) validated the presence of pharmacologically active compound in an herbal preparation that contains *Mormordica balsamina*, an herbal mixture sold over the counter in Southern Nigeria. They also established the safety profiles of the medicinal preparations they validated (Otimenyin et al. 2013; Otimenyin and Uguru 2006), since safety is one of the major concerns in the use of medicinal plants.

22.6.6 Challenges of Ethnobotany and Herbal Medicine

It is believed that all medicinal plants are safe for use in the management of disease conditions. This belief is hinged on the fact that they are natural occurring and that most of these plants are found in our normal daily menus. This claim or assertion is baseless and false. Scientific examination of known herbal preparations has revealed that a sizable number of plants have unsafe pharmacological profiles. This may be dose-dependent or highly toxic. The increasing interest in medicinal plants has necessitated the screening of these plants, with the aim of ascertaining their safety profiles. Such study has led to the discovery of plants with deleterious effects on human health (Otimenyin and Uguru 2006). There exists a pool of research findings that reveals the hypersensitivity and damaging effects of plants on internal organs in the body. Herbs have been shown to be capable of producing a wide range of undesirable/adverse reactions, some of which are capable of causing serious injuries, life-threatening conditions, and even death. In many countries before now herbal medicines are registered without verifying their safety, the wealth of scientific evidence of the harmful effects of plants has led to the adoption of proof of safe toxicological profiles before registration of herbal medicines.

Plants if used as it is used by the traditional people may be safe for consumption. Research has shown that the method of preparation of herbal medicine impacts their safety. Some plants contain substances that can neutralize the toxic components in that plant, implying that separation of the constituents of the plant may result in obtaining a preparation that is harmful to human health. Taxol, also known as paclitaxel, an anticancer agent from *Taxus brevifolia* is an example. Toxicological study of the crude extract showed that the preparation is not toxic; the herbal medicine obtained with industrial method of preparation was shown to be toxic. This led to the development of more appropriate extraction/processing methods with less toxic products (Wilson et al. 2001).

The use of herbal medicines over the past decades represents approximately 40% of all healthcare services delivered in China, and the percentage of people who have used herbal medicines at least once in Australia, Canada, the USA, Belgium, and France is estimated to be between 48 and 75% (Ekor 2013). This increase in medicinal plant use raises concerns about the need for effective scientific evaluation of safety and toxicity of medicinal plants. There are a number of herbal products that have been scientifically evaluated for bioactivity and safety; some have undergone extensive clinical investigation and subjected to systematic review/meta-analysis. However, their development and large-scale production have been hampered due to the complex nature of the herbal products and the variation between manufacturer methods. Evaluation and documentation of efficacy and safety of medicinal plants should be extract-specific. This has become necessary because each manufacturer has different methods of extraction. As pointed out earlier, extraction procedure determines the contents of the extract and thus their efficacy and safety. Presently, finger printing of the extracts has been adopted for standardization purposes and to address changes that may occur from batch to batch. This involves subjecting the extract through mass spectrometry, HPLC, and NMR. The "picks" obtained from the standard extract that produced the desired effects are used as standard for other extracts, since it reveals the constituents present in the preparation. A number of cases of adverse effect and toxicity of herbal medicines have been reported. Yoyo "Cleanser" bitters, an herbal remedy that is found and used in Nigeria, has been reported to elevate plasma levels of liver enzymes and induce hypokalemia in rats that ingested it for 30 days consecutively (Ekor 2013). Efforts must be made to determine the safety profiles of all traditional medicines as this will ensure the safety of the large number of people that depends on herbal medicines for their health needs. This to my view should be the priority rather than the validation of the bioactive compound, though validation of bioactivity is not less important. It will reduce the number of toxicities related to medicinal plant use and sound a precautionary warning to the large populace that consume herbal medicines and encourage the use of the medicinal plants that are safe for consumption. The level of safety should be specified when reporting on the safety profiles of herbal remedies. Regulatory authorities in most countries have made this a priority for the registration of herbal remedies. Manufacturers are required by law to provide some evidence of safety of their preparation before drug regulatory agencies register their products.

22.7 Evidence-Based Validation of Medicinal Plants

We have established that plants have played key roles in the survival of man and animal. It took care of their food, health, and domestic needs. Ethnobotanical/ethnopharmacological survey provides documented evidence of the use of plants by a known community. Further survey does reveal the testimonies of the inhabitants in that community that have benefited from the practice of the use of a particular plant for the management of a known disease condition. Such documentation provides a guide to the scientific validation of medicinal plants. It saves time and is cost-effective compared to the practice of selection of plants at random for scientific evaluation. It also suggests the method of investigation of the plants by giving a clue to the disease conditions it is effective in remitting. This practice has added substantial knowledge to the medicinal plant library, has led to the discovery of new drugs and drug leads, and has contributed immensely to meeting the health needs of man. Okwuasaba et al. 1991 reported a survey of medicinal plants used as contraceptives by women in Jos, Plateau State of Nigeria. The survey led to the discovery of *Ricinus communis,* a plant seed used in Plateau State for the child spacing. Scientific evaluation of the plant revealed that it has contraceptive and hormonal effects. Many drugs have been discovered following this similar pattern of research; see Tables 22.2 and 22.4.

It is also possible to liaise with the traditional medical practitioners to clerk their patients: that is teach them the healthcare reporting systems and documentation process. Some of the clinical features to look out for in the disease being managed may also be revealed to them to enable them to manage and give proper records of their observation and outcome. This is imperative because it is not possible to stop the practice or discourage the patients from patronizing them. This practice is referred to as reverse pharmacology, an aspect of pharmacology that starts with human subjects that are presently using the herbal medicine for the evaluation of the effect of a known herbal medicine. Ethics is a major issue in this field and care must be taken if this method is to be used. In Nigeria, human immunodeficiency virus was a scourge, and many claim to have cure for the disease condition. Patients were forced to subject themselves to any form of treatment, since there was no orthodox medicine available at that time or the cost was not affordable. Efforts were taken to advice the traditional practitioners to keep records. For some who did, record of the success rates, the number of deaths, and the number that had serious side effects were extracted from their records. Such records were useful as it helped dissuade the herbalists from using their claimed remedy, when it showed that it was not effective.

22.7.1 Reverse Pharmacology

Human and clinical effects of medicinal plants and poisons started the entire enterprise of drug development and pharmaceutical industry. Bernard himself had said, "The most useful path for physiology and medicine to follow now is to seek to discover new facts instead of trying to reduce to equations the facts which science already possesses." (Saxena et al. 2015). At least in the early decades of the last century, scientists with medical backgrounds pursued the factual activity of plants in man and searched the mechanisms in animals and tissues (Raut et al. 2017). Striking observation can be made during ethnobotany/ethnopharmacological survey. During survey it is possible to see patients that are undergoing treatment for the claimed disease condition. This is possible because some of the herbal practitioner may have clinics and admission policies. An example of this setup is the traditional bonesetters in Nigeria. The health practitioner in the team may examine the patients and determine the progress of the patients. This might necessitate carrying out some

Table 22.6 Leads and indication from reverse pharmacology

Plants	Activity/lead	Indication
Enicostemma littorale	Antioxidant DNA protection Lipemic control	Type 2 diabetes mellitus
Zingiber officinale	Anti-inflammatory, Antiemetic, and anti-arthritic	Arthritis, nausea, vomiting
Curcuma longa	Anticancer	Cervical precancer
Semecarpus anacardium and Tinospora cordifolia	Chondroprotection	Arthritis
Nyctanthes arbor-tristis	Antiparasite Anticytokine	Malaria
Saraca asoca	Kidney problems, ovulatory dysfunction	Menorrhagia

medical investigations to ascertain the resolution of the disease being managed. Once a clear evidence of disease resolution has been documented, a para-clinical trial can be conducted to further support their claim. Once this has been established, the use of such herbal remedy is encouraged, while basic laboratory evaluation of the herbal remedy is initiated until the active principle in the herbal preparation is established. Research can progress to the synthesis of the active principle in the herbal preparation, in which case the active principle is derived from laboratory-based synthesis rather than obtaining it from the plant, thus saving the plant biodiversity. This type of drug discovery starts from bedside to benches and is known as reverse pharmacology. It is essentially a transdisciplinary quest for clinical significance of traditional medical practice (Saxena et al. 2015). Reverse pharmacology involves the investigation of the potential of widely used herbal remedies. Reverse pharmacology is a novel initiative that offers a paradigm shift in the new drug discovery process. The ingenuity of reverse pharmacology is meant for the integration of traditional remedies with the wisdom of robust documentation of safety and efficacy (Raut et al. 2017). Reverse pharmacology is staged as experiential knowledge/data, exploratory research, and relevant clinical/experimental studies. It can also be relevant for new uses of old drugs or for following up on a new unseen indication of a drug candidate. The clinically novel biodynamic actions of traditional remedies may open up new vistas in biomedicine and life sciences. The phyto-active molecules can also provide novel chemical scaffolds for structural modifications with defined drug targets. Hence, reverse pharmacology can play a dual role – it can inspire new drugs from traditional remedies and enrich the chemical repertoire of medicinal chemists for new chemical entities. Antimalarial have been developed through reverse pharmacology. Table 22.6 shows some drugs that were discovered through reverse pharmacology.

The scope of reverse pharmacology is immense: (1) to evaluate clinically the evidence of safety, efficacy, and quality of drugs/plants used in herbal medicine; (2) to discover new drugs from natural products already in use by humans; (3) to find new clinical facts and bedside biodynamic phenomena that may lead to new insights in human biology; (4) to overcome the current costly, drawn out, and attractive process of drug discovery/development; and (5) to complement the extant process by novel phyto-actives as chemical scaffolds for new chemical entities (Saxena et al. 2015). The origin of reverse pharmacology is often from the diverse and rich big data of traditional/modern literature, ethnobotanical, phytochemical, experimental, clinical, and anecdotal cases. Ethnobotany is an important area of drug discovery and is the foundation for drug discovery and validation of traditional medicinal products.

References

Abdolbaset G, Naghibi F, Mosaddegh M (2006) Ethnobotany, Ethnopharmacology and drug discovery. Iran J Pharm Sci Spring 2(2):109–118

Albuquerque UP, Nascimento ALB, Soldati GT, Feitosa IS, Campos JLA, Hurrell JA, Hanazaki N, de Medeiros PM, da Silva RRV, Ludwinsky RH, Júnior WSF, Reyes-García V (2019) Ten important questions/issues for ethnobotanical research. Acta Bot Bra 33(2):376–385

Aumeeruddy-Thomas Y, Shengji P (2003) Applied ethnobotany: case studies from the Himalayan region. People and plants working paper initiative, working paper series, UNESCO, Paris, p 24

Balick MJ, Cox PA (1996) Plants, people, and culture: the science of ethnobotany. Scientific American Library series 60, Scientific American Library, New York, pp. 2–3

Bernard C (1957) An introduction to the study of experimental medicine. Dover, New York, NY

Cotton CM (1996) Ethnobotany: principles and applications. John Wiley and Sons, Chichester

Davidson-Hunt I (2000) Ecological ethnobotany: stumbling toward new practices and paradigms. MASA J 16:1–13

Ekor M (2013) The growing use of herbal medicines: issues relating to adverse reactions and challenges in monitoring safety. Front Pharmacol 4:177. https://doi.org/10.3389/fphar.2013.00177

Eldeen I, van Staden J, Elgorashi E (2010) Ethnopharmacology of some African trees. VDM Verlag Dr. Muller, Germany, p 132. ISBN: 978-3-639-25373-3

Elvine-lewis M (2001) Should we be concerned about herbal remedies? J Ethnopharmacol 75:141–144

Etkin NC (2001) Perspective in ethnopharmacology: forging a closer link between bioscience and traditional empirical knowledge. J Ethnopharmacol 76:177–182

Heinrich M, Barnes J, Gibbons S, Williamson EM (2018) Fundamental pharmacognosy phytotherapy. Churchill, London

Höft M, Barik SK, Lykke AM (1999) Quantitative ethnobotany: applications of multivariate and statistical analyses in ethnobotany. People and plants working paper 6. UNESCO, Paris

Ior LD, Otimenyin SO (2019) Effect of selected medicinal plants on stereotypic behavior induced by apomorphine. In Press

Ior LD, Otimenyin SO, Okwori VA, Umar DM, Azila JJ (2017) Ethnobotanical survey of plants used in the management of mental illnesses in some selected local government areas of Plateau State, Nigeria. J Pharmacogn Phytother 9(10):146–156

Khafagi IK, Dewedar A (2000) The efficiency of random versus ethno-directed research in the evaluation of Sinai medicinal plants for bioactive compounds. J Ethnopharmacol 71:365–376

Ladio AH (2017) Ethnobiology and research on global environmental change: what distinctive contribution can we make? Ethnobiol Conserv 6:7. https://doi.org/10.15451/ec2017-07-6.7-1-8

Laland KN, Odling-Smee J, Myles S (2010) How culture shaped the human genome: bringing genetics and the human sciences together. Nat Rev Genet 11:137–148

Levis C, Costa FRC, Bongers F et al (2017) Persistent effects of preColumbian plant domestication on Amazonian forest composition. Science 355:925–931

Lins Neto EMF, Albuquerque UP (2018) Theories of niche construction and optimal foraging: weaknesses and virtues in understanding the early stages of domestication. Ethnobiol Conserv 7:7. https://doi.org/10.15451/ec2018-04-7.7-1-6

Martin GJ (1995) Ethnobotany: a methods manual. Chapman & Hall, London

Okwuasaba FK, Osunkwo UA, Ekwenchi MM, Ekpenyong KI, Onwukeme KE, Olayinka AO, Uguru MO, Das SC (1991) Anticonceptive and estrogenic effects of a seed extract of *Ricinus communis* var. minor. J Ethnopharmacol 34(2–3):141–145

Otimenyin SO, Ior LD (2019) Neuroprotective effect of ethyl acetate fraction of *Terminalia macroptera* leaf against ketamine-induced schizophrenia related behaviour in mice. In: Abstract of the 49th annual meeting of the Society for Neuroscience (SfN), Chicago, IL, 19–23 October 2019

Otimenyin SO, Uguru MO (2006) Acute toxicity studies, anti-inflammatory and analgesic activities of methanolic extract of the stem back of *Enantia chlorantha* and *Nauclea latifolia*. J Pharm Biores 3(2):111–115

Otimenyin SO, Olorunfemi PO, Sabo YS, Attah SH (2013) Sub-acute toxicity studies of the herbal cough mixture (Hamekof) in adult male Wistar rats. J Med Med Sci 4(11):27–30

Peters CM (1996) Beyond nomenclature and use: a review of ecological methods for ethnobotanists. In: Alexiades MN (ed) Selected guidelines for ethnobotanical research. The New York Botanical Garden, New York, pp 241–276

Phillis O, Gentry A (1993) The useful plants of Tambo Pata, Peru: I. statistical hypothesis tests with a new quantitative technique. Econ Bot 47:15–32

Raut AA, Chorghade MS, Vaidya ADB (2017) Reverse pharmacology. In: Innovative approaches in drug discovery I. https://doi.org/10.1016/B978-0-12-801814-9.00004-0

Sairafianpuor M (2002) Iranian medicinal plants and antiparasitic compounds: from ethnobotany to contemporary scientific evidence. Ph.D. Dissertation, Department of Medicinal Chemistry, Royal Danish School of Pharmacy, Copenhagen, Denmark, pp. 36–49

Santoro FR, Santos GC, Ferreira Júnior WS et al (2017) Testing an ethnobiological evolutionary hypothesis on plant-based remedies to treat malaria in Africa. Evol Biol 44:216–226

Saxena A, Saxena S, Singh S (2015) Reverse pharmacology: a new approach to drug development. Innor Int J Sci 2:7–12

Subramoniam A (2014) Present scenario, challenges and future perspectives in plant based medicine development. Ann Phytomed 3:31–36

Sullivan AP, Bird DW, Perry GH (2017) Human behaviour as a long-term ecological driver of non-human evolution. Nat Ecol Evol 1:65. https://doi.org/10.1038/s41559-016-0065

Wilson CR, Sauer J-M, Hooster SB (2001) Taxines: a review of the mechanism and toxicity of yew (Taxus spp.) alkaloids. Toxicon 39:175–185

Part III

Safety and Regulatory Issue of Traditional Medicine

Phytotherapeutics: The Rising Role of Drug Transporters in Herb-Drug Interactions with Botanical Supplements

23

Shruti Surendran, Pooja Dhurjad, and Satheeshkumar Nanjappan

Abstract

There is increase in the consumption of traditional medicines all over the globe despite of the lack of sufficient information on the safety. Researchers are showing interest in traditional medicines as precursor for pharmacological activity. Membrane transporter proteins play a crucial role in drug absorption, distribution, metabolism, and excretion, which determines the efficacy and safety. Herb-drug interaction (HDI) may involve multiple drug transporters in the body, which leads to alteration in the pharmacokinetics of substrate drugs, consequently affecting their efficacy and toxicity. In silico databases and in vitro approaches are there for screening HDIs; in vivo studies are the conclusive way to determine the clinical importance of HDIs. Evaluation of herbal product interaction is challenging due to variability in herbal product composition, uncertainty of the causative constituents, scant knowledge of causative constituent pharmacokinetics, pathological conditions, and genetic polymorphism. The current chapter is an extensive systematic review of the effect of transporters on ADME of traditional products, novel screening approaches of HDIs, challenges in the evaluation in the HDI, and future perspectives.

S. Surendran · P. Dhurjad
Drug Metabolism and Interactions Research Lab, Department of Pharmaceutical Analysis, National Institute of Pharmaceutical Education and Research (NIPER), Hyderabad, Telangana, India

S. Nanjappan (✉)
Drug Metabolism and Interactions Research Lab, Department of Pharmaceutical Analysis, National Institute of Pharmaceutical Education and Research (NIPER), Hyderabad, Telangana, India

Department of Natural Products, National Institute of Pharmaceutical Education and Research (NIPER), Kolkata, Kolkata, West Bengal, India
e-mail: satheesh.nanjappan@gmail.com

S. C. Mandal et al. (eds.), *Evidence Based Validation of Traditional Medicines*,
https://doi.org/10.1007/978-981-15-8127-4_23

Keywords

Drug transporters · Traditional medicines · Interaction

Abbreviations

ABC	ATP-binding cassette
ADME	Absorption, distribution, metabolism, and excretion
ATP	Adenosine triphosphate
BCRP	Breast cancer resistance protein
CHO	Chinese hamster ovary
DI	Drug interaction
DME	Drug-metabolizing enzyme
HDI	Herb-drug interactions
HEK293	Human embryonic kidney
LLC-PK1	Porcine kidney epithelial cells
MATE	Multidrug and toxin extrusion
MDCK II	Madin-Darby canine kidney II
MRPs	Multidrug resistance-associated proteins
NTCP	Sodium-dependent taurocholate co-transporting polypeptide
OAT	Organic anion transporters
OATP	Organic anion-transporting polypeptides
OCT	Organic cation transporters
P-gp	P-glycoprotein
SLC	Solute carrier family

23.1 Introduction

There is increase in the consumption of traditional medicines all over the globe despite the lack of sufficient information on the safety. Researchers are showing interest in traditional medicines as a precursor for pharmacological activity. This drift might be attributed to the myth among the people that traditional medicines are nontoxic as they are of natural origin and extensive usage history (Calitz et al. 2015). In addition, the common public have rising knowledge of health and easy access to herbal supplements (Stickel and Shouval 2015). Although herbals have been demonstrated to be efficacious, several herbal extracts may also have deadly fractions (Calitz et al. 2015; Stickel and Shouval 2015). Furthermore, herbal supplements may interact with concomitantly used drugs by modulating drug metabolism, interacting with transporters, or both and may consequently affect pharmacokinetics and pharmacodynamics, resulting in adverse drug reactions and altered efficacy (Li and Paxton 2013; Haefeli and Carls 2014; Domitrović and Potočnjak 2016).

Few decades earlier, drug-metabolizing enzymes (DME) were thought to be accountable for drug interactions (DI). However, recent studies have proved the importance of role of transporters in modulating the drug absorption, distribution, metabolism, and excretion (ADME) processes. Transporter-based interactions turned out to be of immense interest among the researchers and have been comprehensively recorded in the recent years. Drug transporters are present in several tissues and play an important role in transport of drugs into and out of the cells (Giacomini et al. 2010). Consequently, when drugs are concomitantly administered, there is a competition among them for the same transporter, which causes a significant change in the ADME of drugs. Therefore, the subsequent DI alters drug level in the blood and tissue, which changes the safety and efficacy profiles of the drug. In cases like these, the drug loses its effectiveness, or there is an increase in undesired adverse drug reactions, especially in the case of drugs with narrow therapeutic index. Hence, evaluation of the clinical significance of transporter-mediated DI has become one of the regulatory requirements during drug approval processes (Huang et al. 2008).

Transporters are expressed at either the apical membrane or the basolateral membrane of the polarized cells; they are classified based on their function, i.e., efflux or influx transporters (Giacomini and Sugiyama 2006). Among these transporters, solute carrier family (SLC) and the ATP-binding cassette (ABC) superfamily are actively involved in transporter-mediated DI. ABC superfamily transporters are chief active efflux transporters, which uses ATP hydrolysis-derived energy for the active cellular efflux of drugs (Poller et al. 2011; van Herwaarden et al. 2003). In contrast, the SLC transporters assist the cellular influx of drugs by facilitated diffusion or by co-transport of endogenous ions to provide the driving force (Hediger et al. 2004; Fredriksson et al. 2008). Few SLC transporters exhibited efflux properties or are bidirectional that is determined by the concentration gradient of the substrate and coupled ions across the membrane (Hediger et al. 2004). Many herbal products were found to show modulatory effect on the transporters, and some transporter-mediated herb-drug interactions (HDI) were clinically significant. Therefore, in the current chapter, we describe an extensive systematic review of the effect of transporters on ADME of traditional products, novel screening approaches of HDIs, challenges in the evaluation in the HDI, and future perspective.

23.2 Mechanism of Drug Transporter-Based Pharmacokinetic HDIs

In contrast with DME-based interactions, mechanism of transporter-based interaction is restricted, while the knowledge gap is beginning to narrow. The following are the biochemical mechanism underlying pharmacokinetic HDIs (Brantley et al. 2014a). Figure 23.1 represents the mechanism of HDIs.

1. **Inhibition:** Transporter proteins are susceptible to inhibition by two mechanisms, i.e., competitive reversible inhibition due to perpetrator obstructing the victim-binding site; Noncompetitive reversible inhibition due to change in the conformation of the transporter that decreases transporter activity due to the perpetrator. Inhibition of transporter activity in vivo can manifest as alteration in the systemic and tissue concentrations of the victim drug. The course of the alteration is influenced by the site of transporter expression and direction of flux.

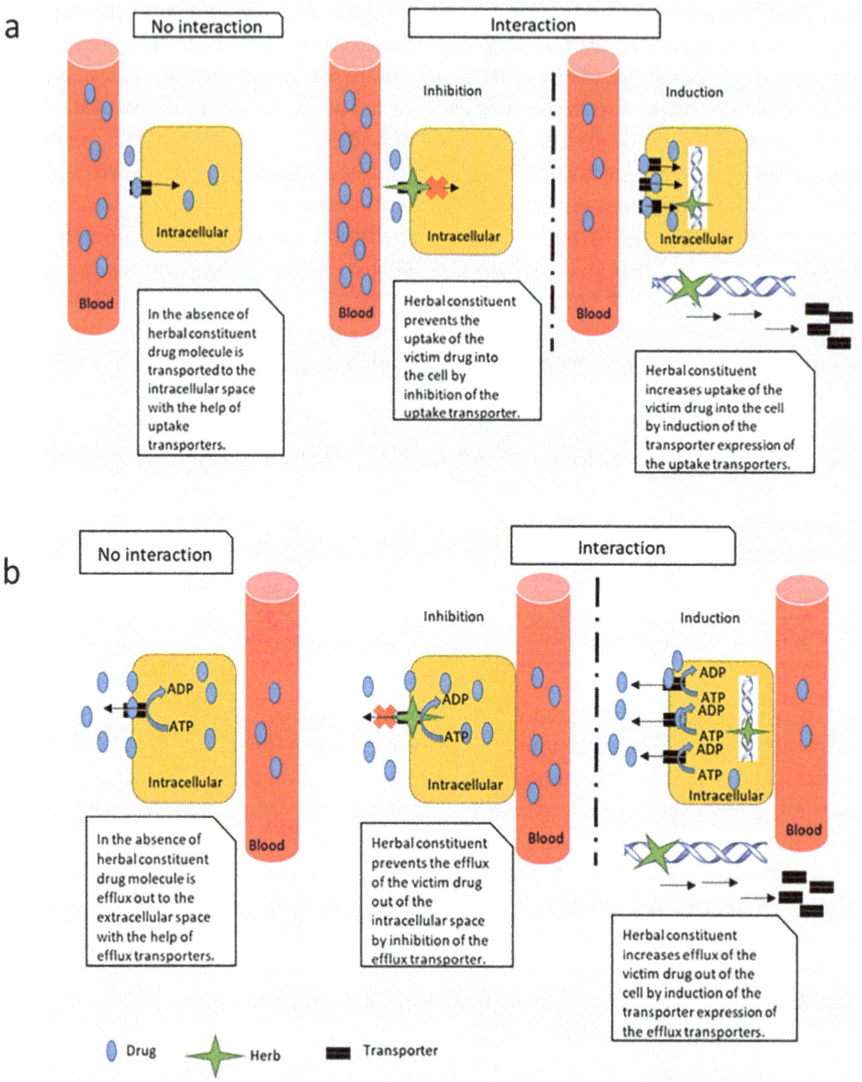

Fig. 23.1 Mechanism of HDIs: (**a**) mechanism of HDIs of uptake transporter, (**b**) mechanism of HDIs of efflux transporter

2. **Induction:** Apart from inhibition, HDIs can induce increase in transporter expression. Most common induction mechanism comprise increase in gene transcription or stabilization of mRNA or active protein (transporter), of which the receptor-mediated increase in the gene transcription is the chief mechanism for transporter induction. The perpetrator activates one or more nuclear receptors by binding to one of the ligand-binding site of the nuclear receptor, which causes a cascade of events leading to increase transcription and subsequent translation of

mRNA into protein. Similar to inhibition, induction of transporters exhibits as increased or decreased in circulating and tissue concentrations of the victim drug depending on the site of transporter expression and direction of flux.

23.3 Role of Drug Transporters in HDIs

Generally, ABC transporters include breast cancer resistance proteins (ABCG2 or BCRP), P-glycoprotein (MDR1 or P-gp), and multidrug resistance-associated proteins (MRPs), while the SLC transporters include organic cation transporters (OCTs), organic anion transporters (OATs), and organic anion-transporting polypeptides (OATPs). These efflux and uptake transporters are widely present in the apical and luminal membrane of epithelia of many organs (e.g., intestines, kidney, liver, brain, testis, and placenta) that relate to drug disposition. Figure 23.2 shows the distribution of transporters on different organs. To foresee potential HDIs or demarcate the basic mechanisms, it is vital to apprehend the tissue distribution of transporters and to recognize their substrates, inhibitors, and inducers. It has been worldwide documented that transporters leading the drug movement across the biological membranes are highly related with drug ADME (Scherrmann 2009).

23.3.1 Absorption

Transporters expressed on the apical membrane of enterocytes serve as a foremost line of barrier for absorption of orally administered drugs into the systemic

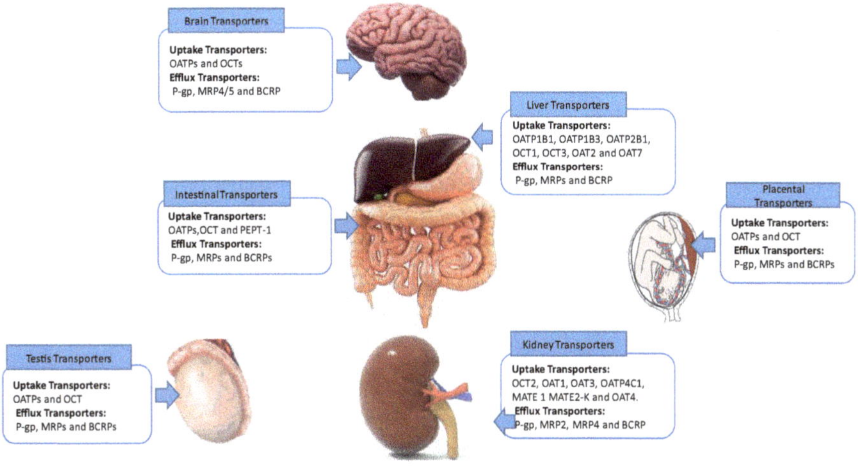

Fig. 23.2 Transporter distribution on different organs

circulation. Efflux transporters like MDR1, MRPs, and BCRPs and SLC transporters like OATPs and PEPT-1 transporters hinder the absorption of its substrates. Natural products that inhibit or induce transporters may significantly alter the intestinal absorption of the concomitantly administered drugs. Inhibition of the efflux transporter increases the bioavailability of the substrate drugs, whereas the induction of efflux transporter function may lead to restricted absorption of drug and vice versa for the SLC transporters. Many herbal extracts and phytotherapeutics modulate the efflux as well as the SLC transporters in the intestine. For example, milk thistle (*Silybum marianum* extract) is traditionally used for liver disorders like hepatitis and cirrhosis and gallbladder disorders. Milk thistle has been found to modulate the function of many transporters by inhibiting them like BCRPs, MRPs, and OATPs (Köck et al. 2013). A flavonoid from *Ginkgo biloba* (ginkgo), kaempferol inhibits OATP2B1 transporter (Mandery et al. 2010). Curcumin from *Curcuma longo*, which has been used for generations in the traditional medication, also inhibits P-gp, BCRP, and MRP2 (Wu et al. 2011). Piperine (*Piper nigrum* flavonoid) and garlic extract are inducers for BCRP and MRP2, respectively (Li et al. 2011; Berginc et al. 2010).

23.3.2 Distribution

Physiological barriers block the entry of possibly toxic xenobiotics. Transporters are broadly expressed on the physiological barriers including the blood-brain barrier, blood-testis barrier, and blood placenta. Transporters expressed in the brain mainly include BCRPs, P-gp, MRP4/5, OCTs, and OATPs. Some phytotherapeutics are substrates of these transporters. For therapeutic effect, few conventional drugs must move to the brain, but transporters present on the brain limit the exposure of the drug in the brain which leads to treatment failure. In some cases, co-administration of transporter modulator with drugs, which has poor brain bioavailability, is a favorable way to increase its brain concentration. For instance, natural flavonoids silymarin and quercetin improve the brain distribution of co-administered substrate by modulating the efflux transporters (Reddy et al. 2016). P-gp transporter present on the placenta plays a crucial role by protecting the fetus exposure to several harmful compounds (Iqbal et al. 2012). Modulation of the P-gp transporter on the placenta affects the distribution and subsequently the efficacy or the fetal toxicity of the P-gp substrates. Efflux transporters present on the testis protect them from unwanted exposure of the drugs and in some cases can be negative if the site of action is located behind the blood-testis barrier (Mruk et al. 2011). Nevertheless, the possible adverse effects should also be taken into consideration if an excess drug penetrates these physiological barriers.

23.3.3 Metabolism and Excretion

Liver is the major organ responsible for metabolism of drugs. The hepatic portal blood is separated from the basolateral membrane of the hepatocyte by fenestrated sinusoidal endothelial cell layer, which results in the direct contact of this basolateral membrane with the portal blood plasma. The SLC or the uptake transporters (OATP1B1, OATP1B3, OATP2B1, OCT1, OCT3, OAT2, and OAT7) are expressed on the basolateral membrane of the hepatocytes, while the ABC transporters (P-gp, ABCG2, and MRP2) are present on the canalicular membrane, which mediates the extrusion of the drugs and its metabolites into the bile (Giacomini et al. 2010; Roth et al. 2012). Other SLC transporters such as MATE1 (multidrug and toxin extruder 1) are expressed on the canalicular membrane, which shows the possible involvement to xenobiotic extrusion into the canaliculus. Kidneys are responsible for transporting drugs out of the blood into urine. Transporters like OCT2, OAT1, OAT3, and OATP4C1 are present on the basolateral side, MATE1 and MATE2-K are present on the luminal membrane of the proximal tubule cell, while OAT4, P-gp, MRP2, MRP4, and BCRP are present on the brush border membrane of the kidneys (Masereeuw and Russel 2010; van Aubel et al. 2002; Sakurai et al. 2004). Few examples of metabolism and excretion based HDIs are Epigallocatechin has inhibitory activity on OCT1, OCT2, MATE1, MATE2-K, OATP1B1, OATP1B3 and P-gp transporters (Knop et al. 2015). Ursolic acid affects the uptake of rosuvastatin in hepatocytes by inhibiting the transport of OATP1B1 (Hua et al. 2014). Berberine significantly reduces metformin elimination from the kidney by OCT2 inhibition (Wang et al. 2018). Rhein is a major constituent of Chinese herbal extract of Dahuang and inhibits OAT1/OAT3 and can lead to HDIs (Wu et al. 2016).

23.4 HDI: In Silico Tools, In Vitro and In Vivo Evaluation

23.4.1 In Silico

The approach of in silico tool is virtual screening to check if there is possibility of HDIs. Various in silico tools are used to assess the drug-drug interaction, but assessment of HDIs is difficult and still is the area to explore. There are different approaches used for evaluation of HDI (Ai et al. 2015). Different approaches for screening of transporter-based HDIs are shown in Fig. 23.3.

23.4.1.1 Hypothesis-Driven In Silico Approaches

The availability of 3D structure of transporters helps to study computational model-based HDIs in relation with transporters. This approach predicts HDIs at exact protein level and helps to understand the fundamental mechanism of interaction based on the previously collected data for HDIs. It includes four different approaches: pharmacophore modeling, machine learning methods, protein-based modeling, and hybrid approach. Generally, to study HDIs the pharmacophore

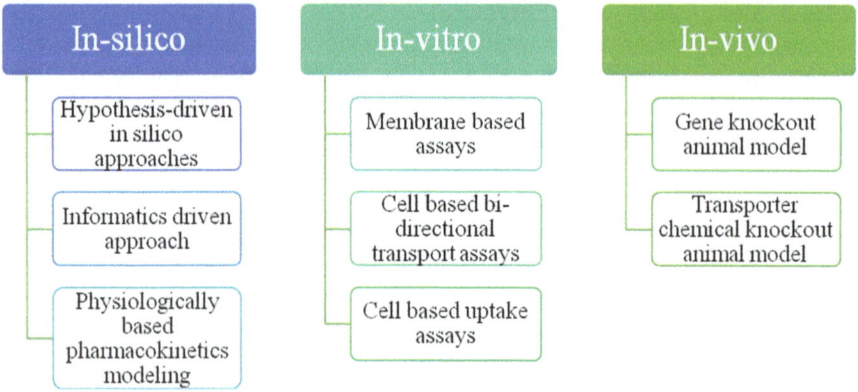

Fig. 23.3 Different approaches for screening of transporter-based HDIs

model is a more preferred approach. The pharmacophore approach deals with development of pharmacophores of enzymes, transporters, ion channels, and receptor followed by screening of plethora of compounds. Discovery Studio (Catalyst) provided by Accelrys, Inc.; Phase by Schrödinger, Inc.; and SYBYL (GALAHAD) by Tripos, Inc. are few of the examples of pharmacophore method (Ekins 2004; Toba et al. 2006; Zheng et al. 2009; Zhou et al. 2007).

23.4.1.2 Informatics-Driven Approach

There is enormous data available through various sources such as publications, pharmaceutical industry data, clinical study data, online data libraries, etc. This available data is extracted, compiled, and validated by using informatics tools. Informatics-driven approach is a more effective and relevant in silico approach as compared to others because it uses heterogeneous and huge experimental dataset for virtual screening. By using this approach, probable harmful interactions between anticoagulant or antiplatelet agents and herbal medicines have been identified and reported. The additive effect of anticoagulant or antiplatelet of herbal medicines, specifically danshen, dong quai, ginger, ginkgo, licorice, and turmeric, causes the major consequence of increased bleeding risks (Ai et al. 2015; Tsai et al. 2013).

23.4.1.3 Physiologically Based Pharmacokinetics Modeling (PBPK)

The herbal medicine contains various numbers of active compounds; physiologically based pharmacokinetics model has the capability to screen huge number of compounds at a same time, which is beneficial to study complex HDIs. However, the major drawback of PBPK approach in HDIs is the lack of data reported for standardization of herbal medicine, impact of herbal medicines on the modulation of enzymes or transporters, and also the variability in data reported for herbal medicines across the laboratories (Ai et al. 2015; Brantley et al. 2014a). Brantley et al. provide an example for prediction of HDIs between silibinin an herb and midazolam and warfarin the two drugs by using PBPK model approach. Though there was minimal

HDIs observed with no clinical relevance, still as a proof of concept, clinical study data was done which also showed minimal HDIs. Although in this case only one herbal constituent was used to study interaction, in future a complex mixture with more than one active constituents can be used to predict the interaction (Brantley et al. 2014b). The PBPK-generated in silico data is of regulatory relevance, and regulatory agencies emphasize on the use of PBPK model to study the HDI (Zheng et al. 2009; Ekins 2004; Zhao et al. 2012).

23.4.2 In Vitro

23.4.2.1 Membrane-Based Assays

These are high-throughput assays used to identify the substrates and inhibitors of the efflux pumps and uptake transporters. It involves ATPase assay and membrane vesicular transporter study. The ATPase assay involves calorimetric detection of released inorganic phosphate due to hydrolysis of ATP in presence of substrate. The ABC transporter directly interacts with the substrate or inhibitor and alters the ATPase activity, i.e., either inhibit or stimulate the transition state of ATPase reaction, and this altered activity shows the interaction of substrate with transporter. The altered activity of ATPase is monitored by release of inorganic phosphate by the enzyme. The drawback of ATPase assay is its limitations to differentiate between substrate and inhibitor as it is not a functional assay and there is huge variation in the experimental data; the chances of false positive are more in the case of ATPase assay (Bjornsson et al. 2003; Brouwer et al. 2013; Xia et al. 2007). The membrane transporters are ATP-dependent unidirectional pumps, which mediate the efflux transport of drugs from the cells. Examples of efflux transporters are P-gp, BCRP, and MRP, and uptake transporter is sodium-dependent taurocholate co-transporting polypeptide (NTCP). As these transporters are expressed on the apical membrane of cells, the inside-out vesicular membranes are used to study efflux mechanism, and they are considered as more reliable tool especially for ABC transporters. Isolated primary hepatocytes, isolated primary cells from other tissues, transfected cell lines, overexpressed cell lines, derived or immortalized cell lines, and recombinant cell lines are used as in vitro tools to study efflux transporters, and oocytes are more specifically used to study uptake transporter. Transfected cell lines are HeLa, human embryonic kidney (HEK293), Madin-Darby canine kidney II (MDCK II), V79 hamster, porcine kidney epithelial cells (LLC-PK1), and Chinese hamster ovary (CHO) cells. HEK293 and CHO cell lines are preferred to study uptake transporters, as they are easy to maintain and demonstrate low endogenous transporters. The advantages of membrane-based transporter study are high throughput, ease of experimentation, and easy maintenance of prepared membranes (Bjornsson et al. 2003; Giacomini et al. 2010; Glavinas et al. 2008; Keogh 2012; Kwatra et al. 2012; Wu et al. 2016).

23.4.2.2 Cell-Based Assays

(a) **Bi-directional transport assays with cell-based systems:** To study efflux or uptake transporters by cell-based system, polarized monolayer cells are used. Cells used to study interactions are intact cells (hepatocytes), cell lines (CaCo2 cell line), and recombinant cell lines (MDCK and LLC-PK1). To study bi-directional transport assay, the test analyte is added into the apical compartment (A) and buffer in basolateral compartment (B) which gives A→B transport. In addition, in corresponding wells, test analyte is added in basolateral compartment and buffer in apical compartment (A) which gives B→A transport. The permeability coefficient (Papp) is calculated for both the directions. If an efflux transporter is expressed on the apical cell membrane, then Papp A→B < Papp B→A. If an uptake transporter is expressed on the apical cell membrane, then Papp A→B > Papp B→A. The situation will be reversed if the transporters are expressed on the basolateral cell membrane. This assay is suitable to determine the transport of test analyte as only the analyte that is fluxed through the cell monolayer is measured (Bjornsson et al. 2003; Brouwer et al. 2013; Xia et al. 2007). CaCo2 cell lines are extensively used in cell-based assays; they are derived from heterogeneous human epithelial colorectal adenocarcinoma cells. It expresses P-gp, BCRP, and MRP efflux transporters and PEPT, OAT, OATP, and OCT uptake transporters. The major limitation of CaCo2 cell lines is its cultivation period which is 21 days. MDCK cell lines are used as alternative for CaCo2 cell line which is derived from dog kidney and has a shorter cultivation period of 3–5 days (Awortwe et al. 2014; Sun et al. 2008).

(b) **Uptake assays with cell-based systems:** The uptake assays are performed to measure the uptake of compound in the cells. The inhibitors of uptake transporter will reduce the uptake of compound; sometimes substrates act as competitive inhibitors because of more affinity than that compound. Suspended hepatocytes and sandwiched cultured hepatocytes are used to study the interaction because of uptake transporters. Organic anion-transporting polypeptides (OATP), organic cation transporters (OCT), and NTCP are the major transporters present on the basolateral membrane of heapotocytes (Liao et al. 2019; Bjornsson et al. 2003; Brouwer et al. 2013; Xia et al. 2007). Freshly isolated or cryopreserved suspended hepatocytes are utilized to study active hepatobiliary uptake transport as the presence of NTCP-mediated taurocholate and OATP-mediated estradiol-17β-glucuronide (E17βG) uptake has been reported by Shitara et al. (2003). The rate of hepatic uptake is determined by linear or dynamic regression analysis. The hepatocytes are pooled to avoid inter-individual variability due to differences in protein expression or genetic variation. Limitations of suspended hepatocytes are they cannot be used to study efflux transport due to internalization of canalicular membrane during isolation and identification of specific isoform responsible for uptake of compound is difficult as there is no specific substrate or inhibitor. As there is loss of polarity in suspended hepatocytes which disables their use as efflux transporter, sandwiched cultured hepatocytes are used which retain the polarity and are used to study efflux transport. The hepatocytes are sandwiched between the two layers of collagen gel. The sandwiched cultured hepatocytes retain more in vivo like properties, and there is existence of good in vitro and in vivo correlation (Bjornsson et al. 2003; Brouwer et al. 2013; Giacomini et al. 2010; Liao et al. 2019; Swift et al. 2010).

23.4.3 In Vivo

Gene knockout animal model and transporter chemical knockout animal model are used as in vivo tools to study transporter-based HDIs. Gene knockout mice are developed by eliminating the transporter gene of interest to study specific role of transporter. Chemical knockout animal model is developed by using the inhibitor of specific transporter; it gives detailed information of role of efflux transporter. The limitation of gene knockout model is there is gene deletion of a specific transporter, it may alter the expression of other transporters or enzymes, and also it will change the physiology of gene knockout animal. In the case of chemical knockout animal model, the selection of dose of an inhibitor is critical; also the specificity of inhibitor is of major concern (Agarwal et al. 2012; Klaassen and Lu 2007; Xia et al. 2007).

23.5 Challenges with Assessment of Transporter-Based HDIs

HDIs are mainly of pharmacokinetic origin, i.e., either DME based or drug transporter based. These interactions lead to variation in the victim drug's exposure or clearance due to induction of inhibition of DME and transporters by the herbal product (de Lima Toccafondo Vieira and Huang 2012; Tsai et al. 2013; Shi and Klotz 2012). The following point focuses on the challenges with assessment of HDIs.

23.5.1 Variation in Herbal Product Composition

Herbal products consist of multiple components, which vary in composition batch to batch and between manufacturers. The bioactive components of the herbal products are often plant-derived secondary metabolites produced from normal plant metabolism or as a defense reaction to environmental stress (Rousseaux and Schachter 2003). The percentage of the bioactive compound formation varies according to the growing conditions, i.e., temperature and rainfall (Rousseaux and Schachter 2003). Proper care should be taken to the composition of herbal products to ensure reproducibility within studies and to enable comparisons between studies.

23.5.2 Identification of the Constituents of Herbal Product

Modulation of transporters can be due to interaction between one or more constituents of herbal product. Hence, identification of the constituents in herbal product is necessary for precise prediction of HDI. Bioactivity-guided fractionation technique helps to elucidate the components of herbal product (Roth et al. 2011; Kim et al. 2011).

23.5.3 Pharmacokinetics of the Constituents of Herbal Product

Majority of the herbal product components have high first-pass metabolism and low bioavailability via modulation of various transporters, which results in low system concentration of the perpetrator component; as a result the concentration at site of interaction might be less (Peterson et al. 2019). Furthermore, on concomitant administration with the victim drug, it alters the clearance and system exposure of the victim drug. Henceforth, knowledge of the pharmacokinetics of the herbal product component is desirable for precise prediction of HDIs.

23.5.4 Interplay of Drug Transport and Drug Metabolism

The uptake of drug by the transporters present on the membrane is a precondition for their subsequent metabolism as DMEs are present intracellularly. Additionally, transfer of drugs or its metabolites also affects the DME kinetics, as kinetics is dependent on the concentration of the drugs or its metabolites present in the cell. Hence, variation in efflux or uptake of drug might also affect intracellular metabolism (Benet 2009; Fan et al. 2010). Vectorial transport is a process when uptake of drugs happens through the basolateral membrane and efflux of drugs or its metabolites happens through the apical membrane, as different families of transporters share a common substrate spectrum. For example, OATP family and MRP family or the OCT family and P-gp family have common substrate spectrum (Nies et al. 2008). Moreover, DMEs like phase I and phase II enzymes have wide substrate spectrum and accept the drugs transported by various transporter families. The DMEs and transporters work together, and any alteration to one of them might result in altered pharmacokinetics of drugs. Hence, when DMEs and transporters share a common substrate assessing the role and contribution of a single transporter in drug interaction might be challenging. In such cases, the selection and concomitant administration of the interacting drugs should be taken into account for planning a drug interaction study. For greater efficiency in drug development and avoidance of toxic interactions in clinics, quantitative prediction of DI consisting of both DME and transporters would give beneficial information. Hence, for designing the appropriate in vivo interaction studies and producing data for the informative labeling and effective use of medication, the development of standardized mechanism-based modeling and simulation is strongly needed (Huang and Lesko 2004; Zhang et al. 2010).

23.5.5 Transporter Polymorphism

Xenobiotic can produce a valuable effect or toxic effect in a particular patient. The nature and degree of the effect are dependent on the ADME of the drugs. The drug transporters have a vital role in the ADME of drugs. They control the passage of the drugs and their metabolites. Hence, polymorphism of their gene alters the ADME and eventually safety and efficacy of the drugs. Inter-individual variation in drug pharmacokinetics is due to genetic transporter polymorphism (Liu et al. 2015). After baicalin treatment, the AUC_{0-72} and $AUC_{0-\infty}$ of rosuvastatin were reduced, OATP1B1 haplotype ∗1b/∗15 (21.0 ± 20.6% and 23.9 ± 8.66%), ∗15/∗15 (9.20 ± 11.6% and 1.76 ± 4.89%), and ∗1b/∗1b (47.0 ± 11.0% and 41.9 ± 7.19%), respectively (Fan et al. 2008). Digoxin effluxes by MDR1 were significantly reduced in MDR1 gene variants of 2677G/893Ala and 2677T/893Ser with the treatment of *Dictamnus dasycarpus* and of 2677T/893Ser with the treatment of *Poria cocos* (Liu et al. 2015).

23.5.6 Pathological Conditions

Human disease can prompt a widespread effect on the expression and function of transporter proteins. These changes can modulate pharmacokinetics and ultimately pharmacological response. Numerous experimental data has shown that drug transporter expression is interrupted during pathophysiological conditions mainly due to reductions in gene expression of these transporters. Approximately 10% of the hospitalized cases are due to adverse effects caused by altered metabolism. The actual incidence may be much higher; many of the cases go unreported. Nevertheless, transporter expression and function in diseased states are not as well characterized (Atilano-Roque et al. 2016). The following point focuses on the transporter expression in liver, kidney, and brain diseases.

23.5.6.1 Liver Disease and Transporters

Literature database in human and rodent models shows that SLC and ABC transporter expression is altered in liver diseases. There is an increase in efflux transporters and decrease in uptake transporters in nonalcoholic steatohepatitis and cholestasis. However, alteration in MRP transporter expression is reported in hepatocellular carcinoma and viral hepatitis. The liver is the most commonly transplanted organ after the kidney. During transplantation procedure, liver ischemia contributes to post-operative hepatobiliary dysfunction. A study performed in a rat model of arterial liver ischemia demonstrated alterations in hepatobiliary transporter expression in response to hypoxia. At present, there is limited data on how transporter

Table 23.1 Altered expression of transporters in hepatic disease

Liver disease	Transporter	Alteration	References
Cholestasis	OATP1B1	Decrease in mRNA expression	Mennone et al. (2006), Chen et al. (2008), Chai et al. (2012), Canet et al. (2014), Trauner et al. (1997)
	OATP1B3	Decrease in mRNA expression	
	OCT1	Decrease in mRNA and protein expression	
	MRP3	Increase in mRNA and protein expression	
	MRP4	Increase in mRNA and protein expression	
	MRP2	Decrease in mRNA and protein expression	
HCV	OATP1B1	Decrease in mRNA expression	Nakai et al. (2008), Hanada et al. (2012), Kikuchi et al. (2010), More et al. (2013), Ros et al. (2003), Ogasawara et al. (2010), Kurzawski et al. (2012)
	OATP2B1	Decrease in mRNA expression	
	OAT2	Decrease in mRNA expression	
	MRP1/MRP1	Increase in mRNA and protein expression	
	MRP3	Increase in mRNA and protein expression	
	MRP4	Increase in mRNA and protein expression	
	BCRP	Decrease in mRNA expression	
	MRP2	Decrease in mRNA expression	
	P-gp	Increase in mRNA and protein expression	
HCC	OATP1B1	Decrease in mRNA and protein expression	Kullak-Ublick et al. (1996), Vavricka et al. (2004), Libra et al. (2006) Vander Borght et al. (2005), Tsuboyama et al. (2010), Schaeffeler et al. (2011), Heise et al. (2012), Ros et al. (2003), Zhao et al. (2010), Borel et al. (2012), Sukowati et al. (2012), Sun et al. (2010), Namisaki et al. (2014)
	OATP1B3	Decrease in mRNA and protein expression	
	OCT1	Decrease in mRNA and protein expression	
	MRP1	Increase in mRNA and protein expression	
	MRP3	Increase in mRNA expression	
	MRP4	Increase in mRNA expression	
	BCRP	Increase in mRNA and protein expression	

(continued)

Table 23.1 (continued)

Liver disease	Transporter	Alteration	References
	P-gp	Decrease in mRNA and increase in protein expression	
NASH	OATP1B1	Increase in protein expression	Aguilar-Olivos et al. (2015), Fisher et al. (2009), Okushin et al. (2016), Bedogni et al. (2005), Clarke et al. (2014b), Canet et al. (2014, 2015), Clarke et al. (2014a), Hardwick et al. (2011), Tanaka et al. (2012)
	OATP1B3	Decrease in protein expression	
	OAT2/ OAT2	Decrease in mRNA expression	
	MRP1	Increase in mRNA and protein expression	
	MRP3	Increase in mRNA and protein expression	
	MRP4	Increase in mRNA and protein expression	
	BCRP	Increase in mRNA and protein expression	
	MRP2	Increase in mRNA and protein expression	
	P-gp	Increase in mRNA and protein expression	
Cirrhosis	OATP1B1	Decrease in mRNA and protein expression	Zollner et al. (2003), Kojima et al. (2003), More et al. (2013), Takeyama and Sakisaka (2012), Ros et al. (2003)
	OATP1B3	Decrease in mRNA and protein expression	
	OATP2B1	Increase in mRNA and protein expression	
	MRP1	Increase in mRNA and protein expression	
	MRP3	Increase in mRNA and protein expression	
	MRP4	Increase in mRNA and protein expression	
	BCRP	Increase in mRNA and protein expression	
	P-gp	Increase in mRNA and protein expression	

protein expression and function can be altered with human liver transplantation. Table 23.1 shows the altered expression of transporters in liver disease.

23.5.6.2 Kidney Disease and Transporters

Variation in transporter expression and function has been reported in specific kidney diseases in comparison to normal kidneys. In acute kidney injury, OAT, OCTs, and

P-gp modulation has been proved. In glomerulonephritis and chronic kidney disease, mainly a variation in the efflux transporters has been reported. Variation in expression of transporters like PEPT, GLUT, and SGLT has been reported in diabetic nephropathy. Table 23.2 shows the altered expression of transporters in kidney disease.

23.5.6.3 Brain Pathologies and Transporters

Literature data suggests that there is a relationship between blood-brain barrier and ABC transporter expression and function. The current data shows that the expression of efflux transporters like P-gp and BCRP increases in diseases like brain tumor, epilepsy, amyotrophic lateral sclerosis, and ischemic stroke. Expression of uptake transporters like OATP1A2 and OATP2B1 decreases in brain tumor, and OATP expression increases in ischemic stroke. Table 23.3 shows the altered expression of transporters in brain disease.

Literature data on different pathophysiological conditions' influence on the expression of transporters are limited. To optimize the treatment course and to avoid drug interaction, a better understanding of the influence of disease state on transporter expression and function would be helpful. Development and application of modern methodologies such as physiologically based pharmacokinetics model and quantitative proteomics will help in advancement of our knowledge and understanding of influence of diseases in the functional changes of transporters and its impact on the pharmacokinetics and pharmacological response of drugs.

23.6 Conclusion

Treatment herbal product administration along with the prescribed drugs is increasing due to limitation in the conventional drugs, ultimately contributing to the increasing incidence of HDIs. Due to variability in the herbal product composition, uncertainty of the causative constituents, and limited information of the pharmacokinetics of the phytotherapeutics, mechanism of HDIs remains unpredictable, and evaluation of the interaction is still a big challenge (Brantley et al. 2014a). Majority of the phytotherapeutics are found to be multi-target modulators of ABC and SLC transporters. Most of the investigation with phytochemical has been done in vitro, and its outcome is dependent on the specific phytotherapeutics and its concentration and cell line and substrate used (Giacomini et al. 2010). Hence, to predict the clinical outcome from the in vitro data, various factors have to be considered like the concentration of the phytochemical in the gut, its solubility in the gut, and different transporter kinetics. However, it is often difficult to ascertain the in vivo effect of the phytotherapeutics is due to transporters solely as there are other factors involved like DMEs as many phytotherapeutics are its substrates and expression of the transporters that is up-/downregulated due to genetic polymorphism of transporters or the diseased state (Zhang et al. 2009). There is considerably less information about the metabolites of the phytotherapeutics, which can modulate the transporters in the liver or kidneys and alter the pharmacokinetics of the phytotherapeutics.

Table 23.2 Altered expression of transporters in kidney disease

Kidney disease	Transporter	Alteration	References
AKI	OAT1	Decrease in mRNA and protein expression	Ji et al. (2002), Matsuzaki et al. (2008), Schneider et al. (2007), Erman et al. (2014), Brandoni and Torres (2015), Huang et al. (2000)
	OAT3	Decrease in mRNA and protein expression	
	OCT1	Decrease in mRNA and protein expression	
	OCT2	Decrease in mRNA and protein expression	
	MRP2	Increase in mRNA and protein expression	
	MRP4	Increase in mRNA and protein expression	
CKD	P-gp	Increase in mRNA and protein expression	Lu and Klaassen (2008), Moe and Chen (2008), Lau et al. (2014), Dankers et al. (2013)
	BCRP	Decrease in mRNA and protein expression	
	MRP2	Increase in mRNA and protein expression	
	MRP3	Increase in mRNA and protein expression	
	MRP4	Decrease in mRNA and protein expression	
	MRP6	Decrease in mRNA and protein expression	
GN	P-gp	Decrease in mRNA and protein expression	Joy et al. (2014)
	BCRP	Decrease in mRNA and protein expression	
DN	P-gp	Increase in mRNA expression	Tramonti et al. (2006)
	BCRP	Increase in mRNA expression	

(continued)

Table 23.2 (continued)

Kidney disease	Transporter	Alteration	References
	PEPT1	Increase in mRNA and protein expression	
	PEPT2	Increase in mRNA expression	

Table 23.3 Altered expression of transporters in brain disease

Brain disease	Transporter	Alteration	Reference
Brain tumor	P-gp	Increase in mRNA and protein expression	de Vries et al. (2007), Adkins et al. (2013), Bronger et al. (2005)
	BCRP	Increase in mRNA and protein expression	
	OATP1A2, OATP2B1	Decrease in mRNA and protein expression	
Epilepsy	P-gp	Increase in mRNA and protein expression	Dombrowski et al. (2001), Tishler et al. (1995), Wen et al. (2008), Römermann et al. (2015)
	BCRP	Increase in mRNA and protein expression	
Amyotrophic lateral sclerosis	P-gp	Increase in mRNA and protein expression	Milane et al. (2010), Jablonski et al. (2014)
	BCRP	Increase in mRNA and protein expression	
Ischemic stroke	P-gp	Increase in mRNA and protein expression	Ji et al. (2013), Dazert et al. (2006), Thompson et al. (2014)
	BCRP	Increase in mRNA and protein expression	
	OATPs	Increase in mRNA and protein expression	

Different funding agencies like the National Institute of Health (NIH) and related organizations in other countries are encouraging the scientific evaluation related to the safety of the herbal products by giving financial support. This drift will endorse standardization of the herb medicines as that of conventional drugs, and the development might promote discovery of new and effective drugs from herbal extract.

Acknowledgments The authors are thankful to the Science and Engineering Research Board (No. EMR/2016/002098), New Delhi, India, for the financial support for this work.

References

Adkins CE, Mittapalli RK, Manda VK, Nounou MI, Mohammad AS, Terrrell TB, Bohn KA, Yasemin C, Grothe TR, Lockman JA (2013) P-glycoprotein mediated efflux limits substrate and drug uptake in a preclinical brain metastases of breast cancer model. Front Pharmacol 4:136

Agarwal S, Uchida Y, Mittapalli RK, Sane R, Terasaki T, Elmquist WF (2012) Quantitative proteomics of transporter expression in brain capillary endothelial cells isolated from P-glycoprotein (P-gp), breast cancer resistance protein (Bcrp), and P-gp/Bcrp knockout mice. Drug Metab Dispos 40(6):1164–1169

Aguilar-Olivos NE, Carrillo-Córdova D, Oria-Hernández J, Sánchez-Valle V, Ponciano-Rodríguez G, Ramírez-Jaramillo M, Chablé-Montero F, Chávez-Tapia NC, Uribe M, Méndez-Sánchez N (2015) The nuclear receptor FXR, but not LXR, up-regulates bile acid transporter expression in non-alcoholic fatty liver disease. Ann Hepatol 14(4):487–493

Ai N, Fan X, Ekins S (2015) In silico methods for predicting drug-drug interactions with cytochrome P-450s, transporters and beyond. Adv Drug Deliv Rev 86:46–60

Atilano-Roque A, Roda G, Fogueri U, Kiser JJ, Joy MS (2016) Effect of disease pathologies on transporter expression and function. J Clin Pharmacol 56:S205–S221

Awortwe C, Fasinu P, Rosenkranz B (2014) Application of Caco-2 cell line in herb-drug interaction studies: current approaches and challenges. J Pharm Pharm Sci 17(1):1–19

Bedogni G, Miglioli L, Masutti F, Tiribelli C, Marchesini G, Bellentani S (2005) Prevalence of and risk factors for nonalcoholic fatty liver disease: the Dionysos nutrition and liver study. Hepatology 42(1):44–52

Benet LZ (2009) The drug transporter–metabolism alliance: uncovering and defining the interplay. Mol Pharm 6(6):1631–1643

Berginc K, Milisav I, Kristl A (2010) Garlic flavonoids and organosulfur compounds: impact on the hepatic pharmacokinetics of saquinavir and darunavir. Drug Metab Pharmacokinet 25 (6):521–530

Bjornsson TD, Callaghan JT, Einolf HJ, Fischer V, Gan L, Grimm S, Kao J, King SP, Miwa G, Ni L (2003) The conduct of in vitro and in vivo drug-drug interaction studies: a PhRMA perspective. J Clin Pharmacol 43(5):443–469

Borel F, Han R, Visser A, Petry H, van Deventer SJ, Jansen PL, Konstantinova P, Réseau Centre de Ressources Biologiques Foie (French Liver Biobanks Network), France (2012) Adenosine triphosphate-binding cassette transporter genes up-regulation in untreated hepatocellular carcinoma is mediated by cellular microRNAs. Hepatology 55(3):821–832

Brandoni A, Torres AM (2015) Altered renal expression of relevant clinical drug transporters in different models of acute uremia in rats. Role of urea levels. Cell Physiol Biochem 36 (3):907–916

Brantley SJ, Argikar AA, Lin YS, Nagar S, Paine MF (2014a) Herb-drug interactions: challenges and opportunities for improved predictions. Drug Metab Dispos 42(3):301–317

Brantley SJ, Gufford BT, Dua R, Fediuk DJ, Graf TN, Scarlett YV, Frederick KS, Fisher MB, Oberlies NH, Paine MF (2014b) Physiologically based pharmacokinetic modeling framework for quantitative prediction of an herb-drug interaction. CPT Pharmacometrics Syst Pharmacol 3: e107

Bronger H, König J, Kopplow K, Steiner H-H, Ahmadi R, Herold-Mende C, Keppler D, Nies AT (2005) ABCC drug efflux pumps and organic anion uptake transporters in human gliomas and the blood-tumor barrier. Cancer Res 65(24):11419–11428

Brouwer KL, Keppler D, Hoffmaster KA, Bow DA, Cheng Y, Lai Y, Palm JE, Stieger B, Evers R, International Transporter Consortium (2013) In vitro methods to support transporter evaluation in drug discovery and development. Clin Pharmacol Ther 94(1):95–112

Calitz C, Du Plessis L, Gouws C, Steyn D, Steenekamp J, Muller C, Hamman S (2015) Herbal hepatotoxicity: current status, examples, and challenges. Expert Opin Drug Metab Toxicol 11 (10):1551–1565

Canet MJ, Hardwick RN, Lake AD, Dzierlenga AL, Clarke JD, Cherrington NJ (2014) Modeling human nonalcoholic steatohepatitis-associated changes in drug transporter expression using experimental rodent models. Drug Metab Dispos 42(4):586–595

Canet MJ, Merrell MD, Hardwick RN, Bataille AM, Campion SN, Ferreira DW, Xanthakos SA, Manautou JE, A-Kader HH, Erickson RP (2015) Altered regulation of hepatic efflux transporters disrupts acetaminophen disposition in pediatric nonalcoholic steatohepatitis. Drug Metab Dispos 43(6):829–835

Chai J, He Y, Cai S-Y, Jiang Z, Wang H, Li Q, Chen L, Peng Z, He X, Wu X (2012) Elevated hepatic MRP3/ABCC3 expression in human obstructive cholestasis is mediated through TNFα and JNK/SAPK signaling pathway. Hepatology 55(5):1485

Chen H-L, Liu Y-J, Chen H-L, Wu S-H, Ni Y-H, Ho M-C, Lai H-S, Hsu W-M, Hsu H-Y, Tseng H-C (2008) Expression of hepatocyte transporters and nuclear receptors in children with early and late-stage biliary atresia. Pediatr Res 63(6):667

Clarke JD, Hardwick RN, Lake AD, Canet MJ, Cherrington NJ (2014a) Experimental nonalcoholic steatohepatitis increases exposure to simvastatin hydroxy acid by decreasing hepatic organic anion transporting polypeptide expression. J Pharmacol Exp Ther 348(3):452–458

Clarke JD, Hardwick RN, Lake AD, Lickteig AJ, Goedken MJ, Klaassen CD, Cherrington NJ (2014b) Synergistic interaction between genetics and disease on pravastatin disposition. J Hepatol 61(1):139–147

Dankers AC, Mutsaers HA, Dijkman HB, van den Heuvel LP, Hoenderop JG, Sweep FC, Russel FG, Masereeuw R (2013) Hyperuricemia influences tryptophan metabolism via inhibition of multidrug resistance protein 4 (MRP4) and breast cancer resistance protein (BCRP). Biochim Biophys Acta 1832(10):1715–1722

Dazert P, Suofu Y, Grube M, Popa-Wagner A, Kroemer H, Jedlitschky G, Kessler C (2006) Differential regulation of transport proteins in the periinfarct region following reversible middle cerebral artery occlusion in rats. Neuroscience 142(4):1071–1079

de Lima Toccafondo Vieira M, Huang S-M (2012) Botanical-drug interactions: a scientific perspective. Planta Med 78(13):1400–1415

de Vries NA, Zhao J, Kroon E, Buckle T, Beijnen JH, van Tellingen O (2007) P-glycoprotein and breast cancer resistance protein: two dominant transporters working together in limiting the brain penetration of topotecan. Clin Cancer Res 13(21):6440–6449

Dombrowski SM, Desai SY, Marroni M, Cucullo L, Goodrich K, Bingaman W, Mayberg MR, Bengez L, Janigro D (2001) Overexpression of multiple drug resistance genes in endothelial cells from patients with refractory epilepsy. Epilepsia 42(12):1501–1506

Domitrović R, Potočnjak I (2016) A comprehensive overview of hepatoprotective natural compounds: mechanism of action and clinical perspectives. Arch Toxicol 90(1):39–79

Ekins S (2004) Predicting undesirable drug interactions with promiscuous proteins in silico. Drug Discov Today 9(6):276–285

Erman F, Tuzcu M, Orhan C, Sahin N, Sahin K (2014) Effect of lycopene against cisplatin-induced acute renal injury in rats: organic anion and cation transporters evaluation. Biol Trace Elem Res 158(1):90–95

Fan L, Zhang W, Guo D, Tan ZR, Xu P, Li Q, Liu YZ, Zhang L, He TY, Hu DL (2008) The effect of herbal medicine baicalin on pharmacokinetics of rosuvastatin, substrate of organic anion-transporting polypeptide 1B1. Clin Pharmacol Ther 83(3):471–476

Fan J, Chen S, Chow ECY, Sandy Pang K (2010) PBPK modeling of intestinal and liver enzymes and transporters in drug absorption and sequential metabolism. Curr Drug Metab 11(9):743–761

Fisher CD, Lickteig AJ, Augustine LM, Elferink RPO, Besselsen DG, Erickson RP, Cherrington NJ (2009) Experimental non-alcoholic fatty liver disease results in decreased hepatic uptake transporter expression and function in rats. Eur J Pharmacol 613(1–3):119–127

Fredriksson R, Nordström KJ, Stephansson O, Hägglund MG, Schiöth HB (2008) The solute carrier (SLC) complement of the human genome: phylogenetic classification reveals four major families. FEBS Lett 582(27):3811–3816

Giacomini KM, Sugiyama Y (2006) Membrane transporters and drug response. In: Goodman & Gilman's the pharmacological basis of therapeutics, vol 11, pp 41–70

Giacomini KM, Huang S-M, Tweedie DJ, Benet LZ, Brouwer KL, Chu X, Dahlin A, Evers R, Fischer V, Hillgren KM (2010) Membrane transporters in drug development. Nat Rev Drug Discov 9(3):215

Glavinas H, Méhn D, Jani M, Oosterhuis B, Herédi-Szabó K, Krajcsi P (2008) Utilization of membrane vesicle preparations to study drug–ABC transporter interactions. Expert Opin Drug Metab Toxicol 4(6):721–732

Haefeli WE, Carls A (2014) Drug interactions with phytotherapeutics in oncology. Expert Opin Drug Metab Toxicol 10(3):359–377

Hanada K, Nakai K, Tanaka H, Suzuki F, Kumada H, Ohno Y, Ozawa S, Ogata H (2012) Effect of nuclear receptor downregulation on hepatic expression of cytochrome P450 and transporters in chronic hepatitis C in association with fibrosis development. Drug Metab Pharmacokinet 27 (3):301–306

Hardwick RN, Fisher CD, Canet MJ, Scheffer GL, Cherrington NJ (2011) Variations in ATP-binding cassette transporter regulation during the progression of human nonalcoholic fatty liver disease. Drug Metab Dispos 39(12):2395–2402

Hediger MA, Romero MF, Peng J-B, Rolfs A, Takanaga H, Bruford EA (2004) The ABCs of solute carriers: physiological, pathological and therapeutic implications of human membrane transport proteins. Pflugers Arch 447(5):465–468

Heise M, Lautem A, Knapstein J, Schattenberg JM, Hoppe-Lotichius M, Foltys D, Weiler N, Zimmermann A, Schad A, Gründemann D (2012) Downregulation of organic cation transporters OCT1 (SLC22A1) and OCT3 (SLC22A3) in human hepatocellular carcinoma and their prognostic significance. BMC Cancer 12(1):109

Hua WJ, Hua WX, Nan FY, Jiang WA, Yan C (2014) The influence of herbal medicine ursolic acid on the uptake of rosuvastatin mediated by OATP1B1* 1a and* 5. Eur J Drug Metab Pharmacokinet 39(3):221–230

Huang SM, Lesko LJ (2004) Drug-drug, drug—dietary supplement, and drug—citrus fruit and other food interactions: what have we learned? J Clin Pharmacol 44(6):559–569

Huang Z-H, Murakami T, Okochi A, Yumoto R, Nagai J, Takano M (2000) Expression and function of P-glycoprotein in rats with glycerol-induced acute renal failure. Eur J Pharmacol 406(3):453–460

Huang SM, Strong JM, Zhang L, Reynolds KS, Nallani S, Temple R, Abraham S, Habet SA, Baweja RK, Burckart GJ (2008) New era in drug interaction evaluation: US Food and Drug Administration update on CYP enzymes, transporters, and the guidance process. J Clin Pharmacol 48(6):662–670

Iqbal M, Audette M, Petropoulos S, Gibb W, Matthews S (2012) Placental drug transporters and their role in fetal protection. Placenta 33(3):137–142

Jablonski MR, Markandaiah SS, Jacob D, Meng NJ, Li K, Gennaro V, Lepore AC, Trotti D, Pasinelli P (2014) Inhibiting drug efflux transporters improves efficacy of ALS therapeutics. Ann Clin Transl Neurol 1(12):996–1005

Ji L, Masuda S, Saito H, Inui K-I (2002) Down-regulation of rat organic cation transporter rOCT2 by 5/6 nephrectomy. Kidney Int 62(2):514–524

Ji BS, Cen J, He L, Liu M, Liu YQ, Liu L (2013) Modulation of P-glycoprotein in rat brain microvessel endothelial cells under oxygen glucose deprivation. J Pharm Pharmacol 65 (10):1508–1517

Joy MS, Roberts BV, Wang J, Hu Y, Hogan SL, Falk RJ (2014) A pilot study of leukocyte expression patterns for drug metabolizing enzyme and transporter transcripts in autoimmune glomerulonephritis. Int J Clin Pharmacol Ther 52(4):303
Keogh J (2012) Membrane transporters in drug development. In: Advances in pharmacology, vol 63. Elsevier, pp 1–42
Kikuchi R, McCown M, Olson P, Tateno C, Morikawa Y, Katoh Y, Bourdet DL, Monshouwer M, Fretland AJ (2010) Effect of hepatitis C virus infection on the mRNA expression of drug transporters and cytochrome p450 enzymes in chimeric mice with humanized liver. Drug Metab Dispos 38(11):1954–1961
Kim J-H, Chen C, Kong A-NT (2011) Resveratrol inhibits genistein-induced multi-drug resistance protein 2 (MRP2) expression in HepG2 cells. Arch Biochem Biophys 512(2):160–166
Klaassen CD, Lu H (2007) Xenobiotic transporters: ascribing function from gene knockout and mutation studies. Toxicol Sci 101(2):186–196
Knop J, Misaka S, Singer K, Hoier E, Müller F, Glaeser H, König J, Fromm MF (2015) Inhibitory effects of green tea and (−)-epigallocatechin gallate on transport by OATP1B1, OATP1B3, OCT1, OCT2, MATE1, MATE2-K and P-glycoprotein. PLoS One 10(10):e0139370
Köck K, Xie Y, Hawke RL, Oberlies NH, Brouwer KL (2013) Interaction of silymarin flavonolignans with organic anion-transporting polypeptides. Drug Metab Dispos 41 (5):958–965
Kojima H, Nies AT, König J, Hagmann W, Spring H, Uemura M, Fukui H, Keppler D (2003) Changes in the expression and localization of hepatocellular transporters and radixin in primary biliary cirrhosis. J Hepatol 39(5):693–702
Kullak-Ublick G-A, Beuers U, Paumgartner G (1996) Molecular and functional characterization of bile acid transport in human hepatoblastoma HepG2 cells. Hepatology 23(5):1053–1060
Kurzawski M, Dziedziejko V, Post M, Wójcicki M, Urasińska E, Miętkiewski J, Droździk M (2012) Expression of genes involved in xenobiotic metabolism and transport in end-stage liver disease: up-regulation of ABCC4 and CYP1B1. Pharmacol Rep 64(4):927–939
Kwatra D, Budda B, Vadlapudi AD, Vadlapatla RK, Pal D, Mitra AK (2012) Transfected MDCK cell line with enhanced expression of CYP3A4 and P-glycoprotein as a model to study their role in drug transport and metabolism. Mol Pharm 9(7):1877–1886
Lau WL, Liu S, Vaziri ND (2014) Chronic kidney disease results in deficiency of ABCC6, the novel inhibitor of vascular calcification. Am J Nephrol 40(1):51–55
Li Y, Paxton JW (2013) The effects of flavonoids on the ABC transporters: consequences for the pharmacokinetics of substrate drugs. Expert Opin Drug Metab Toxicol 9(3):267–285
Li S, Lei Y, Jia Y, Li N, Wink M, Ma Y (2011) Piperine, a piperidine alkaloid from Piper nigrum re-sensitizes P-gp, MRP1 and BCRP dependent multidrug resistant cancer cells. Phytomedicine 19(1):83–87
Liao M, Zhu Q, Zhu A, Gemski C, Ma B, Guan E, Li AP, Xiao G, Xia CQ (2019) Comparison of uptake transporter functions in hepatocytes in different species to determine the optimal model for evaluating drug transporter activities in humans. Xenobiotica 49(7):852–862
Libra A, Fernetti C, Lorusso V, Visigalli M, Anelli PL, Staud F, Tiribelli C, Pascolo L (2006) Molecular determinants in the transport of a bile acid-derived diagnostic agent in tumoral and nontumoral cell lines of human liver. J Pharmacol Exp Ther 319(2):809–817
Liu M-Z, Zhang Y-L, Zeng M-Z, He F-Z, Luo Z-Y, Luo J-Q, Wen J-G, Chen X-P, Zhou H-H, Zhang W (2015) Pharmacogenomics and herb-drug interactions: merge of future and tradition. Evid Based Complement Alternat Med 2015:321091
Lu H, Klaassen C (2008) Gender differences in mRNA expression of ATP-binding cassette efflux and bile acid transporters in kidney, liver, and intestine of 5/6 nephrectomized rats. Drug Metab Dispos 36(1):16–23
Mandery K, Bujok K, Schmidt I, Keiser M, Siegmund W, Balk B, König J, Fromm MF, Glaeser H (2010) Influence of the flavonoids apigenin, kaempferol, and quercetin on the function of organic anion transporting polypeptides 1A2 and 2B1. Biochem Pharmacol 80(11):1746–1753

Masereeuw R, Russel FG (2010) Therapeutic implications of renal anionic drug transporters. Pharmacol Ther 126(2):200–216

Matsuzaki T, Morisaki T, Sugimoto W, Yokoo K, Sato D, Nonoguchi H, Tomita K, Terada T, Inui KI, Hamada A (2008) Altered pharmacokinetics of cationic drugs caused by down-regulation of renal rat organic cation transporter 2 (Slc22a2) and rat multidrug and toxin extrusion 1 (Slc47a1) in ischemia/reperfusion-induced acute kidney injury. Drug Metab Dispos 36(4):649–654

Mennone A, Soroka CJ, Cai SY, Harry K, Adachi M, Hagey L, Schuetz JD, Boyer JL (2006) Mrp4−/− mice have an impaired cytoprotective response in obstructive cholestasis. Hepatology 43(5):1013–1021

Milane A, Fernandez C, Dupuis L, Buyse M, Loeffler J-P, Farinotti R, Meininger V, Bensimon G (2010) P-glycoprotein expression and function are increased in an animal model of amyotrophic lateral sclerosis. Neurosci Lett 472(3):166–170

Moe SM, Chen NX (2008) Mechanisms of vascular calcification in chronic kidney disease. J Am Soc Nephrol 19(2):213–216

More VR, Cheng Q, Donepudi AC, Buckley DB, Lu ZJ, Cherrington NJ, Slitt AL (2013) Alcohol cirrhosis alters nuclear receptor and drug transporter expression in human liver. Drug Metab Dispos 41(5):1148–1155

Mruk DD, Su L, Cheng CY (2011) Emerging role for drug transporters at the blood–testis barrier. Trends Pharmacol Sci 32(2):99–106

Nakai K, Tanaka H, Hanada K, Ogata H, Suzuki F, Kumada H, Miyajima A, Ishida S, Sunouchi M, Habano W (2008) Decreased expression of cytochromes P450 1A2, 2E1, and 3A4 and drug transporters Na+-taurocholate-cotransporting polypeptide, organic cation transporter 1, and organic anion-transporting peptide-C correlates with the progression of liver fibrosis in chronic hepatitis C patients. Drug Metab Dispos 36(9):1786–1793

Namisaki T, Schaeffeler E, Fukui H, Yoshiji H, Nakajima Y, Fritz P, Schwab M, Nies AT (2014) Differential expression of drug uptake and efflux transporters in Japanese patients with hepatocellular carcinoma. Drug Metab Dispos 42(12):2033–2040

Nies AT, Herrmann E, Brom M, Keppler D (2008) Vectorial transport of the plant alkaloid berberine by double-transfected cells expressing the human organic cation transporter 1 (OCT1, SLC22A1) and the efflux pump MDR1 P-glycoprotein (ABCB1). Naunyn Schmiedeberg's Arch Pharmacol 376(6):449–461

Ogasawara K, Terada T, Katsura T, Hatano E, Ikai I, Yamaoka Y, Inui KI (2010) Hepatitis C virus-related cirrhosis is a major determinant of the expression levels of hepatic drug transporters. Drug Metab Pharmacokinet 25(2):190–199

Okushin K, Tsutsumi T, Enooku K, Fujinaga H, Kado A, Shibahara J, Fukayama M, Moriya K, Yotsuyanagi H, Koike K (2016) The intrahepatic expression levels of bile acid transporters are inversely correlated with the histological progression of nonalcoholic fatty liver disease. J Gastroenterol 51(8):808–818

Peterson B, Weyers M, Steenekamp JH, Steyn JD, Gouws C, Hamman JH (2019) Drug bioavailability enhancing agents of natural origin (bioenhancers) that modulate drug membrane permeation and pre-systemic metabolism. Pharmaceutics 11(1):33

Poller B, Iusuf D, Sparidans RW, Wagenaar E, Beijnen JH, Schinkel AH (2011) Differential impact of P-glycoprotein (ABCB1) and breast cancer resistance protein (ABCG2) on axitinib brain accumulation and oral plasma pharmacokinetics. Drug Metab Dispos 39(5):729–735

Reddy DR, Khurana A, Bale S, Ravirala R, Reddy VSS, Mohankumar M, Godugu C (2016) Natural flavonoids silymarin and quercetin improve the brain distribution of co-administered P-gp substrate drugs. Springerplus 5(1):1618

Römermann K, Helmer R, Loescher W (2015) The antiepileptic drug lamotrigine is a substrate of mouse and human breast cancer resistance protein (ABCG2). Neuropharmacology 93:7–14

Ros JE, Libbrecht L, Geuken M, Jansen PL, Roskams TA (2003) High expression of MDR1, MRP1, and MRP3 in the hepatic progenitor cell compartment and hepatocytes in severe human liver disease. J Pathol 200(5):553–560

Roth M, Araya JJ, Timmermann BN, Hagenbuch B (2011) Isolation of modulators of the liver-specific organic anion-transporting polypeptides (OATPs) 1B1 and 1B3 from Rollinia emarginata Schlecht (Annonaceae). J Pharmacol Exp Ther 339(2):624–632

Roth M, Obaidat A, Hagenbuch B (2012) OATPs, OATs and OCTs: the organic anion and cation transporters of the SLCO and SLC22A gene superfamilies. Br J Pharmacol 165(5):1260–1287

Rousseaux CG, Schachter H (2003) Regulatory issues concerning the safety, efficacy and quality of herbal remedies. Birth Defects Res B Dev Reprod Toxicol 68(6):505–510

Sakurai Y, Motohashi H, Ueo H, Masuda S, Saito H, Okuda M, Mori N, Matsuura M, Doi T, Fukatsu A (2004) Expression levels of renal organic anion transporters (OATs) and their correlation with anionic drug excretion in patients with renal diseases. Pharm Res 21(1):61–67

Schaeffeler E, Hellerbrand C, Nies AT, Winter S, Kruck S, Hofmann U, van der Kuip H, Zanger UM, Koepsell H, Schwab M (2011) DNA methylation is associated with downregulation of the organic cation transporter OCT1 (SLC22A1) in human hepatocellular carcinoma. Genome Med 3(12):82

Scherrmann JM (2009) Transporters in absorption, distribution, and elimination. Chem Biodivers 6 (11):1933–1942

Schneider R, Sauvant C, Betz B, Otremba M, Fischer D, Holzinger H, Wanner C, Galle J, Gekle M (2007) Downregulation of organic anion transporters OAT1 and OAT3 correlates with impaired secretion of para-aminohippurate after ischemic acute renal failure in rats. Am J Physiol Renal Physiol 292(5):F1599–F1605

Shi S, Klotz U (2012) Drug interactions with herbal medicines. Clin Pharmacokinet 51(2):77–104

Shitara Y, Li AP, Kato Y, Lu C, Ito K, Itoh T, Sugiyama Y (2003) Function of uptake transporters for taurocholate and estradiol 17β-d-glucuronide in cryopreserved human hepatocytes. Drug Metab Pharmacokinet 18(1):33–41

Stickel F, Shouval D (2015) Hepatotoxicity of herbal and dietary supplements: an update. Arch Toxicol 89(6):851–865

Sukowati CH, Rosso N, Pascut D, Anfuso B, Torre G, Francalanci P, Crocè LS, Tiribelli C (2012) Gene and functional up-regulation of the BCRP/ABCG2 transporter in hepatocellular carcinoma. BMC Gastroenterol 12(1):160

Sun H, Chow EC, Liu S, Du Y, Pang KS (2008) The Caco-2 cell monolayer: usefulness and limitations. Expert Opin Drug Metab Toxicol 4(4):395–411

Sun Z, Zhao Z, Li G, Dong S, Huang Z, Ye L, Liang H, Qu J, Ai X, Zhang W (2010) Relevance of two genes in the multidrug resistance of hepatocellular carcinoma: in vivo and clinical studies. Tumori J 96(1):90–96

Swift B, Pfeifer ND, Brouwer KL (2010) Sandwich-cultured hepatocytes: an in vitro model to evaluate hepatobiliary transporter-based drug interactions and hepatotoxicity. Drug Metab Rev 42(3):446–471

Takeyama Y, Sakisaka S (2012) Hepatobiliary membrane transporters in primary biliary cirrhosis. Hepatol Res 42(2):120–130

Tanaka N, Matsubara T, Krausz KW, Patterson AD, Gonzalez FJ (2012) Disruption of phospholipid and bile acid homeostasis in mice with nonalcoholic steatohepatitis. Hepatology 56 (1):118–129

Thompson BJ, Sanchez-Covarrubias L, Slosky LM, Zhang Y, Laracuente ML, Ronaldson PT (2014) Hypoxia/reoxygenation stress signals an increase in organic anion transporting polypeptide 1a4 (Oatp1a4) at the blood–brain barrier: relevance to CNS drug delivery. J Cereb Blood Flow Metab 34(4):699–707

Tishler DM, Weinberg KI, Hinton DR, Barbaro N, Annett GM, Raffel C (1995) MDR1 gene expression in brain of patients with medically intractable epilepsy. Epilepsia 36(1):1–6

Toba S, Srinivasan J, Maynard AJ, Sutter J (2006) Using pharmacophore models to gain insight into structural binding and virtual screening: an application study with CDK2 and human DHFR. J Chem Inf Model 46(2):728–735

Tramonti G, Xie P, Wallner EI, Danesh FR, Kanwar YS (2006) Expression and functional characteristics of tubular transporters: P-glycoprotein, PEPT1, and PEPT2 in renal mass reduction and diabetes. Am J Physiol Renal Physiol 291(5):F972–F980

Trauner M, Arrese M, Soroka CJ, Ananthanarayanan M, Koeppel TA, Schlosser SF, Suchy FJ, Keppler D, Boyer JL (1997) The rat canalicular conjugate export pump (Mrp2) is down-regulated in intrahepatic and obstructive cholestasis. Gastroenterology 113(1):255–264

Tsai H-H, Lin H-W, Lu Y-H, Chen Y-L, Mahady GB (2013) A review of potential harmful interactions between anticoagulant/antiplatelet agents and Chinese herbal medicines. PLoS One 8(5):e64255

Tsuboyama T, Onishi H, Kim T, Akita H, Hori M, Tatsumi M, Nakamoto A, Nagano H, Matsuura N, Wakasa K (2010) Hepatocellular carcinoma: hepatocyte-selective enhancement at gadoxetic acid–enhanced MR imaging—correlation with expression of sinusoidal and canalicular transporters and bile accumulation. Radiology 255(3):824–833

van Aubel RA, Smeets PH, Peters JG, Bindels RJ, Russel FG (2002) The MRP4/ABCC4 gene encodes a novel apical organic anion transporter in human kidney proximal tubules: putative efflux pump for urinary cAMP and cGMP. J Am Soc Nephrol 13(3):595–603

Van Herwaarden AE, Jonker JW, Wagenaar E, Brinkhuis RF, Schellens JH, Beijnen JH, Schinkel AH (2003) The breast cancer resistance protein (Bcrp1/Abcg2) restricts exposure to the dietary carcinogen 2-amino-1-methyl-6-phenylimidazo [4, 5-b] pyridine. Cancer Res 63 (19):6447–6452

Vander Borght S, Libbrecht L, Blokzijl H, Nico Faber K, Moshage H, Aerts R, Van Steenbergen W, Jansen PL, Desmet VJ, Roskams TA (2005) Diagnostic and pathogenetic implications of the expression of hepatic transporters in focal lesions occurring in normal liver. J Pathol 207 (4):471–482

Vavricka SR, Jung D, Fried M, Grützner U, Meier PJ, Kullak-Ublick GA (2004) The human organic anion transporting polypeptide 8 (SLCO1B3) gene is transcriptionally repressed by hepatocyte nuclear factor 3β in hepatocellular carcinoma. J Hepatol 40(2):212–218

Wang H, Zhu C, Ying Y, Luo L, Huang D, Luo Z (2018) Metformin and berberine, two versatile drugs in treatment of common metabolic diseases. Oncotarget 9(11):10135

Wen T, Liu Y-C, Yang H-W, Liu H-Y, Liu X-D, Wang G-J, Xie L (2008) Effect of 21-day exposure of phenobarbital, carbamazepine and phenytoin on P-glycoprotein expression and activity in the rat brain. J Neurol Sci 270(1–2):99–106

Wu C-P, Ohnuma S, Ambudkar SV (2011) Discovering natural product modulators to overcome multidrug resistance in cancer chemotherapy. Curr Pharm Biotechnol 12(4):609–620

Wu X, Ma J, Ye Y, Lin G (2016) Transporter modulation by Chinese herbal medicines and its mediated pharmacokinetic herb–drug interactions. J Chromatogr B 1026:236–253

Xia CQ, Milton MN, Gan L-S (2007) Evaluation of drug-transporter interactions using in vitro and in vivo models. Curr Drug Metab 8(4):341–363

Zhang L, Zhang Y, Huang S-M (2009) Scientific and regulatory perspectives on metabolizing enzyme– transporter interplay and its role in drug interactions: challenges in predicting drug interactions. Mol Pharm 6(6):1766–1774

Zhang L, Reynolds KS, Zhao P, Huang S-M (2010) Drug interactions evaluation: an integrated part of risk assessment of therapeutics. Toxicol Appl Pharm 243(2):134–145

Zhao J, Yu B-Y, Wang D-Y, Yang J-E (2010) Promoter polymorphism of MRP1 associated with reduced survival in hepatocellular carcinoma. World J Gastroenterol 16(48):6104

Zhao P, Rowland M, Huang SM (2012) Best practice in the use of physiologically based pharmacokinetic modeling and simulation to address clinical pharmacology regulatory questions. Clin Pharmacol Ther 92(1):17–20

Zheng X, Ekins S, Raufman J-P, Polli JE (2009) Computational models for drug inhibition of the human apical sodium-dependent bile acid transporter. Mol Pharm 6(5):1591–1603

Zhou SF, Zhou ZW, Li CG, Chen X, Yu X, Xue CC, Herington A (2007) Identification of drugs that interact with herbs in drug development. Drug Discov Today 12(15–16):664–673

Zollner G, Fickert P, Silbert D, Fuchsbichler A, Marschall H-U, Zatloukal K, Denk H, Trauner M (2003) Adaptive changes in hepatobiliary transporter expression in primary biliary cirrhosis. J Hepatol 38(6):717–727

An Insight into Herb Interactions: Clinical Evidence-Based Overview

24

Namdeo Ajay Gajanan and Bandekar Sparsha Suresh

Abstract

Medicines are used for betterment of human health, and huge number of human population uses medicines for the same. But the desired effects of these drugs are obtained when taken in appropriate concentration called 'dose'. These same medicines can be ineffective at a subtherapeutic dose and toxic above it. Often herbs taken concomitantly with pharmaceutical drugs can pose a problem in gaining the desired effect by potentiating or inhibiting the drug action. This can prove to be life-threatening in severe cases. There is a need to identify the potential interactions and understand their mechanisms. Various clinical studies, case reports and review articles are available highlighting the prominent herb interactions. However summary of such clinical evidences for each main class of pharmaceutical drugs separately may provide for better understanding and further prevention of harmful interactions. An attempt is made to summarize potential herb-drug interactions for various therapeutic classes of prescription drugs.

Keywords

Drug · Herb · Interaction · Clinical · Evidence · Therapeutic

N. A. Gajanan (✉)
Department of Pharmacognosy, Poona College of Pharmacy, Bharati Vidyapeeth University, Erandwane, Pune, Maharashtra, India
e-mail: agnamdeo@gmail.com; ajay.namdo@bharatividyapeeth.edu

B. S. Suresh
Yashwantrao Bhonsale College of Pharmacy, Sawantwadi, Maharashtra, India
e-mail: sparshasb@gmail.com

S. C. Mandal et al. (eds.), *Evidence Based Validation of Traditional Medicines*,
https://doi.org/10.1007/978-981-15-8127-4_24

Abbreviations

CNS	Central nervous system
CVS	Cardiovascular system
CYP	Cytochrome P450
HDI	Herb-drug interaction

24.1 Introduction

Herbal remedy and traditional medicines have bloomed to a great extent in the late nineteenth century and is an interest worldwide till today. The increasing interest and an insight of equating the term 'herbal' or 'natural' to 'safety' have resulted in marked rise in consumption of herbal medication globally. Great percentage of population considers herbal therapy as a treatment which is safe and efficacious over a long run. The huge popularity acquired by the herbal therapy has led in use of herbs in the first or second line of choice of treatment. Moreover, there is a tremendous concomitant use of herbal remedies along with other therapeutic regimen. It has been a belief the natural products would afford an added advantage in treatment along with extended safety (Ekor, 2014).

However herbal products and drugs obtained from natural origin are not devoid of side effect and adverse reactions. Various manifestations such as rash, gastric discomfort, nausea, vomiting, allergies, headache, etc. have been associated with herbal drugs. Certain herbs can be even toxic or life-threatening to humans. The fact that herbs contain multiple components or active principles shouldn't be neglected, and this could be the possible reason for the adverse reaction especially in case of preparation containing multiple herbs due to the interaction of multiple components of the preparation (Yuan et al., 2016).

Multitude of patients considers herbs as an alternative of complementary medicine. Concomitant use of herbs with prescription medications could lead to interaction of the herbal component with conventional drug. This could alter the pharmacological effects of the drug and cause toxicological manifestations. In addition various evidences suggest that the patient does not disclose to physicians, about herbal remedies concomitantly used. So certain events of herb-drug interaction that occur are not even reported. Based on certain clinical reports, some of the herb drug interactions could be summarized.

24.2 General Consideration to Mechanisms of Herb-Drug Interactions

Interaction between herb and drug can lead to synergism (additive effect), thus potentiating the action of the pharmaceutical drug or antagonism (negative effect), thus diminishing the effect of drug administered, or may also lead to alteration of effects of one or the other drug.

Interactions occur at level of biotransformation pathways of the drug which results in changes in serum drug concentrations or drug levels at the site of action (Marchetti et al., 2007).

Herb-drug interactions can be broadly categorized into two major classes (Capasso et al., 2000; Izzo, 2012):

- Pharmacokinetic interactions: Those that alter the plasma drug concentration of drug.
- Pharmacodynamic interactions: The interactions are at receptors on target organs.

24.2.1 Pharmacokinetic Interactions

All those interactions that arise due to alteration in absorption, interference in distribution patterns of drug or competition at metabolic or excretory pathways can be categorized under this heading. These are more likely to occur due to induction or inhibition of drug-metabolizing enzymes such as cytochrome P450 or drug transporter proteins like P-glycoprotein. Both play a necessary role in drug bioavailability (Izzo, 2004).

24.2.1.1 Interaction with Cytochrome P450 (CYP) Enzymes

CYP enzyme families play an important role in metabolizing a wide array of drugs. It is expressed in the liver as well as in the intestine. There are various families of this enzyme, of which CYP1, CYP2 are CYP3 have a major role in drug metabolism. The subfamilies predominantly involved in drug metabolism are CYP1A2, CYP2A6, CYP2C9, CYP2C19, CYP2D6, CYP2E1, CYP3A4 and CYP3A5. The interactions with these enzymes result in alteration of absorption, distribution patterns or competitive changes in metabolic or excretory pathways. Different conventional medicines are metabolized by cytochrome P450 enzymes (phase I drug-metabolizing enzyme system); the interaction of herbs with this enzyme could lead to adverse events. There might be over-expression or under-expression of this enzyme leading to diminishing effect or potentiating effect of drug leading to inefficacy or toxicity problems (Izzo, 2004, 2012; Fasinu et al., 2012; Colalto, 2010; Zhou, 2008).

St. John's warts have shown to affect the blood concentrations of various drugs that are metabolized by cytochrome P450 enzymes (Izzo, 2004, 2012; Fasinu et al., 2012).

24.2.1.2 Interaction with P-Glycoprotein

Drug transporter proteins are functional in the pharmacokinetics of drugs at different levels according to its site of expression in the human body. Those at intestinal site will alter the absorption of drug into the bloodstream; those at excretory sites will affect the drug excretion. As a result it plays a role in modulation of the plasma drug concentration and bioavailability. Transporter protein, namely, P-glycoprotein, influences absorption, distribution and elimination of drug and might be responsible for interacting effects. This transporter protein influences pharmacokinetics of drug by limiting the cellular transport from the intestinal lumen into epithelial cells and by enhancing the excretion of drugs from hepatocytes and renal tubules into adjacent luminal space (Izzo, 2004, 2012; Fasinu et al., 2012; Zhang et al., 2009).

The pharmacokinetic profiles of various drugs transported by P-glycoprotein are altered by St. John's warts leading to potentiating or diminishing of the effect of drug (Izzo, 2004, 2012; Fasinu et al., 2012).

24.2.2 Pharmacodynamic Interactions

These interactions are due to different degrees of alteration of physiological environment or interaction with receptor sites by chemical entities. The interaction of herb at receptor level of the drug might be 'additive or synergetic' and potentiate the pharmacological/toxicological action of synthetic drugs or 'antagonistic', i.e. the herb may diminish the action of concomitantly administered drug. These can also be a result of interaction of drugs with hepatic enzymes (Izzo, 2004, 2012; Fasinu et al., 2012).

The anticoagulant effect of warfarin can be increased or diminished by coumarin containing herbs and vitamin K containing herbs, respectively (Izzo, 2004, 2012; Fasinu et al., 2012).

24.3 Clinical Significance and Evidence of Herb-Drug Interaction Study

Various study methods are available, which could help us to summarize that the herb-drug interaction is a comprehensive manner. But these have certain advantages as well as limitation. A comparative study of different types of methods can be used to come to some reliable conclusions. The reports of HDI studies are usually available for potential interactions, but it becomes a concern to identify the magnitude of clinical reliance of such interactions. The overall limitations of various methods could be summarized as variations caused in phytochemical composition of herbs due to seasonal variation, demographic changes of herbal drug, adulteration cause and its nature, poor authentication and characterization, methods employed in extracting and fractionating the herbals, etc. (Izzo, 2004, 2012; Fasinu et al., 2012).

In addition to these, the different study methods may attribute to certain specific limitations. Summary of different methods and certain limitations that these methods are prone to is stated in Table 24.1 (VanRoon et al., 2015).

Recently, evidences of HDI are derived through structured assessment procedures which gives due consideration to clinical evidence of potential adverse events, patient-specific risks and disease conditions which need to be considered for certain interaction. A system for hierarchical evidence-based structured assessment procedure of drug-drug interaction has been developed by Van Roon et al., which can be also applicable for HDI. This method can be used to identify well-established HDIs.

Table 24.1 Limitation to methods of studying HDIs

Method of study	Limitations
In vitro studies: These include the studies conducted using cell lines, tissues and organs or by employing certain metabolic enzymes	Reproducibility of results is difficult, clinical vs experimental variations; the in vivo changes and phenomena like plasma protein binding and correlating to clinical situation are poor
In vivo studies: This involves study conducted employing animals	Interpretation of results could be difficult due to species variation and changes in dosing patterns in comparison to physiological dose
Case reports: Patient history reports or diagnosis reports identifying HDI are considered ideal for better conclusions	Underreporting by patients, poor identification by physicians and difficulty in statistical evaluation of each herbal drug especially in the case of polyherbal preparations
Human studies: Human subjects are involved in these studies	Ethical constraints, costs involved and use of healthy subjects pose a problem in correlating it with pathological conditions and genetic variations

It categorizes evidences into four different levels as summarized below (Fasinu et al., 2012; VanRoon et al., 2015).

Level 1: Published theoretical proof or expert opinion on the possibility of HDI due to certain factors including the presence of known interacting phytochemicals in the herbs and structure-activity relationship

Level 2: Pharmacodynamic and/or pharmacokinetic animal studies, in vitro studies with a limited predictive value for human in vivo situation

Level 3: Well-documented, published case reports with the absence of other explaining factors

Level 4: Controlled, published interaction studies in patients or healthy volunteers with surrogate or clinically relevant endpoint

These levels of herb-drug evidence can be beneficial for better understanding and correlation with the clinical risk assessment of the various herb-drug interactions (Fasinu et al., 2012; VanRoon et al., 2015).

24.4 Classification of Herb-Drug Interactions According to Clinical Classes of Drugs

For better understanding of any study, it is necessary to classify the wide array of information available into feasible classes. Various herbs interact with the conventional drugs which may alter the therapeutic action of the drug and thus have a clinical manifestation. The main problem arising today is due to the underreporting of herbal products used by the patients, which finally affect the therapeutic regimen of the conventional drugs. An attempt is made to classify the herbal interactions with conventional drugs as per the therapeutic classes of pharmaceuticals.

The herbs may interact with various classes of synthetic drugs. Focus is laid only upon prime therapeutic classes that may have a clinical manifestation and on which relative clinical evidence is available.

Herb-drug interactions for following therapeutic classes have been illustrated.

- Cardiovascular drugs
- Central nervous system drugs
- Antiretroviral and anticancer drugs
- Hypoglycaemic drugs
- Immunosuppressants
- Miscellaneous drugs

This attempt of classification will ease the understanding of HDIs and further preventive measure can be laid upon during administration of those herbal preparations that might have a potential to interact with drug therapy of patient's need.

24.4.1 Interactions with Cardiovascular Drugs

Cardiovascular diseases are much prevalent in today's era, and variety of drugs is available in this class. Various cardiovascular drugs such as anticoagulants, cardiac inotropic drugs, antihyperlipidaemics and antihypertensive drugs interact differently with different herbs. The herbs that show potential interactions include St. John's wort, ginkgo, ginger, guar gum, garlic, boldo, fenugreek, devil's claw, dong quai, ginseng, danshen, cranberry, soya, green tea, goldenseal, etc. A summary of these interactions is provided in Table 24.2.

St. John's wort (*Hypericum perforatum*) is found to interact with a wide array of drugs belonging to different subclasses, and the mechanism is found to be induction of CYP enzyme or P-glycoprotein. The CYP enzyme induction ultimately leads to decreased blood concentrations of the drugs. *Ginkgo biloba* interacts with P-glycoproteins or has a pharmacodynamic interaction with resultant potentiation of drug action. Herbs like garlic, boldo, fenugreek, devil's claw, dong quai, ginseng, danshen, cranberry, etc. interact with blood thinners such as warfarin and potentiate their effects. Other herbs like soya and green tea have opposite effect on anticoagulant drugs. Exact mechanisms of these interactions are not known, but the anticoagulant properties of the herbs might be the underlying cause. Further there is scope for deducing the mechanism of interaction at the receptor level.

24.4.2 Interactions with Central Nervous System Drugs

Antidepressants, neuroleptics, antiepileptics, anxiolytics and antiparkinson drugs can be clubbed together as centrally acting drugs. Potentiating or inhibiting the effects of these drugs may prove to be life threatening. Herbs do interact with large

Table 24.2 Interaction with conventional cardiovascular drugs

Herb	Biological source	Drug	Result of interaction	Possible mechanism	References
St. John's wort	*Hypericum perforatum*	Digoxin	Decreased plasma digoxin concentration	Induction of P-glycoprotein involved in digoxin absorption or inhibition	Izzo, (2012), Wang et al. (2001), Zhou et al. (2004)
		Simvastatin	Decreased plasma digoxin concentration	Induction of P-glycoprotein and CYP enzymes	Izzo, (2012), Wang et al. (2001), Zhou et al. (2004)
		Atorvastatin	Reduced efficacy of rosuvastatin	Induction of P-glycoprotein and CYP3A4 enzyme	Izzo, (2012), Wang et al. (2001), Zhou et al. (2004)
		Rosuvastatin	Reduced efficacy of atorvastatin	Induction of P-glycoprotein and CYP3A4 enzyme	Izzo, (2012), Wang et al. (2001), Zhou et al. (2004)
		Verapamil	Decreased bioavailability of verapamil	Induction of intestinal CYP3A4	Izzo, (2012), Wang et al. (2001), Zhou et al. (2004)
		Nifedipine	Decreased blood concentrations	Induction of intestinal CYP3A4	Izzo, (2012), Wang et al. (2001), Zhou et al. (2004)
		Ivabradine	Decreased blood concentrations	Induction of intestinal CYP3A4	Izzo, (2012), Wang et al. (2001), Zhou et al. (2004)
		Talinolol	Decreased blood concentrations	Interaction with P-glycoprotein	Izzo, (2012), Wang et al. (2001), Zhou et al. (2004)
		Phenprocoumon	Decreased anticoagulant effect	Induction of CYP enzymes	Izzo, (2012), Wang et al. (2001), Zhou et al. (2004)
		Warfarin	Decreased anticoagulant effect	Induction of CYP enzymes	Izzo, (2012), Wang et al. (2001), Zhou et al. (2004)
Ginkgo	*Ginkgo biloba*	Aspirin	Spontaneous hyphema	Additive effect on platelet aggregation	Izzo and Ernst (2009)
		Ibuprofen	Over-anticoagulation	Antiplatelet properties	Izzo and Ernst (2009), Unger (2013)

(continued)

Table 24.2 (continued)

Herb	Biological source	Drug	Result of interaction	Possible mechanism	References
		Warfarin	Over-anticoagulation	Antiplatelet properties	Izzo and Ernst (2009), Unger (2013)
		Talinolol	Increased blood concentration	Interaction with P-glycoprotein	Izzo and Ernst (2009), Unger (2013)
Eleuthero or Siberian ginseng	*Eleutherococcus senticosus*	Digoxin	Elevated plasma digoxin concentration	Interaction with metabolic enzymes	Unger (2013)
Guar gum	*Cyamopsis tetragonoloba*	Digoxin	Decreased plasma digoxin concentration	Delay in gastric emptying	Izzo and Ernst (2009), Izzo (2012)
Wheat bran	–	Digoxin	Decreased plasma digoxin concentration	Trap digoxin in the gut	Izzo and Ernst (2009), Izzo (2012)
Pectin or oat bran	–	Lovastatin	Decreased absorption of lovastatin	Binding or trapping of lovastatin in gut	Izzo (2012)
Green tea	*Camellia sinensis*	Simvastatin	Intolerance, increased blood levels	Unknown mechanism	Izzo (2012)
Goldenseal	*Hydrastis canadensis*	Debrisoquine	Decreased urine recovery ratio	Inhibition of CYP2D6	Izzo and Ernst (2009), Izzo (2012)
Peppermint	*Mentha piperita*	Felodipine	Increased blood concentration	Inhibition of CYP3A4	Izzo (2012)
Magnolia berry	*Schisandra chinensis*	Talinolol	Increased blood concentration	Inhibition of P-glycoprotein	Izzo (2012)
Ginger	*Zingiber officinale*	Phenprocoumon	Over-anticoagulation	Additive effect on coagulation due to antiplatelet action of ginger	Izzo (2012)
Fenugreek	*Trigonella foenum-graecum*	Warfarin	Over-anticoagulation	Anticoagulant properties	Izzo (2012), Ge et al. (2014)

Boldo	*Peumus boldus*	Warfarin	Over-anticoagulation	Anticoagulant properties	Izzo (2012), Ge et al. (2014)
Dong quai	*Angelica sinensis*	Warfarin	Over-anticoagulation	Anticoagulant properties	Izzo (2012), Ge et al. (2014)
PC-SPES	–	Warfarin	Over-anticoagulation	Anticoagulant properties	Izzo (2012), Ge et al. (2014)
Danshen	*Salvia miltiorrhiza*	Warfarin	Over-anticoagulation	Antiplatelet properties	Izzo (2012); Ge et al. (2014)
Garlic	*Allium sativum*	Warfarin	Over-anticoagulation	Antiplatelet properties	(Izzo, 2012; Ge et al., 2014)
Ginseng	*Panax quinquefolius*	Warfarin	Over-anticoagulation	Antiplatelet properties	Izzo (2012), Ge et al. (2014)
Chamomile	*Matricaria recutita*	Warfarin	Over-anticoagulation	Synergistic/additive effect	Izzo (2012), Ge et al. (2014)
Curbicin	–	Warfarin	Over-anticoagulation	Contain high amount of vit E which antagonize vit K	Izzo (2012), Ge et al. (2014)
Cranberry	*Vaccinium macrocarpon*	Warfarin	Over-anticoagulation	Inhibit CYP enzymes	Izzo (2012), Ge et al. (2014)
Devil's claw	*Harpagophytum procumbens*	Warfarin	Over-anticoagulation	Unknown mechanism	Izzo (2012), Ge et al. (2014)
Chinese wolfberry	*Lycium barbarum*	Warfarin	Over-anticoagulation	Unknown mechanism	Izzo (2012), Ge et al. (2014)
Mango	*Mangifera indica*	Warfarin	Over-anticoagulation	Unknown mechanism	Izzo (2012), Ge et al. (2014)
Papaya	*Carica papaya*	Warfarin	Over-anticoagulation	Unknown mechanism	Izzo (2012), Ge et al. (2014)
Green tea	*Camellia sinensis*	Warfarin	Decreased anticoagulant effect	It contains antagonist vit K	Izzo (2012), Ge et al. (2014)
Chlorella	*Chlorella pyrenoidosa*	Warfarin	Decreased anticoagulant effect	It contains antagonist vit K	Izzo (2012), Ge et al. (2014)
Soya	*Glycine max*	Warfarin	Decreased anticoagulant effect	Unknown mechanism	Izzo (2012), Ge et al. (2014)

number of these drugs through both pharmacokinetic and pharmacodynamic manners. Induction of CYP enzymes and additive or synergistic effects have been found to be underlying cause of many such interactions; still some mechanisms of interactions need to be identified. Interactions with CNS drugs have been given in Table 24.3.

Same as that for the CVS drugs, the herb that prominently interacts with most drugs is St. John's worts. It either induces the CYP3A4 leading to decreased plasma drug concentration or affords an additive effect on neurotransmitter signaling, thus potentiating the drug action. Other herbs interacting with drugs include *Echinacea*, kava, evening primrose oil, psyllium, herbal diuretics, ginseng, Shankhapushpi, betel nut, ginkgo, danshen, garlic, goldenseal, passion flower, valerian, aloe vera, etc., showing variable mechanisms of action. Herbal interactions with CNS drugs can have deadly effects such as tremors, loss of seizure control, fatal seizures, etc. Therefore understanding and preventing such interactions are most important to reduce the prevalence of clinical manifestations.

24.4.3 Interactions with Antiretroviral and Anticancer Drugs

Herbs interacting with antiretroviral drugs may lead to either increased or decreased blood concentrations of such drugs via CYP3A4 enzyme induction, induction of transporter protein or an unknown mechanism which falls under the scope of further research. St. John's worts, garlic, ginseng, ginkgo and cat's claw have shown to interact with such drugs. St. John's worts also interact with anticancer drugs. These drugs with their herbal interactions have been condensed in Table 24.4.

24.4.4 Interactions with Hypoglycaemic Drugs

Herbs such as garlic and karela have been used to regulate the blood sugar levels. But there exist certain herbs that can interact with certain hypoglycaemic drugs and alter their potential. Herbs such as garlic, ginkgo, prickly pear cactus and St. John's worts have shown the interactions resulting in hyperglycaemia or hypoglycaemia through CYP enzyme induction or additive effect. These interactions are encapsulated in Table 24.5.

24.4.5 Interactions with Immunosuppressant Drugs

St. John's worts and red yeast rice have been found to interact with immunosuppressants like tacrolimus and cyclosporine. The possible mechanisms of interactions include induction of CYP enzyme or protein transporters, thus altering the drug concentration in blood and leading to organ rejection episodes in cases of organ transplants stabilized by immunosuppressant therapy. Herbs such as St. John's worts, red yeast rice, red avens and magnolia berry are herbs of interest

Table 24.3 Interaction of herbs with central nervous system drugs

Herb	Biological source	Drug	Result of interaction	Possible mechanism	References
St. John's wort	*Hypericum perforatum*	Amitriptyline	Decreased plasma levels of amitriptyline	Induction of both CYP2C19 and P-glycoprotein	Izzo, (2012), Wang et al. (2001), Zhou et al. (2004)
		Bupropion	Orofacial dystonia	Additive effect of dopaminergic transmission	Izzo, (2012), Wang et al. (2001), Zhou et al. (2004)
		Nefazodone	Serotonin syndrome	Additive effect on 5-HT signaling	Izzo, (2012), Wang et al. (2001)
		Paroxetine	Serotonin syndrome	Additive effect on 5-HT uptake	Izzo, (2012), Wang et al. (2001)
,		Serotonin reuptake inhibitors	Serotonergic syndrome	Synergistic effect on 5-HT uptake	Izzo, (2012), Wang et al. (2001)
		Eletriptan	Serotonergic syndrome	Synergistic effect on 5-HT uptake	Izzo, (2012), Wang et al. (2001)
		Fluoxetine	Serotonergic syndrome	Synergistic effect on 5-HT uptake	Izzo, (2012), Wang et al. (2001)
		Sertraline	Serotonin syndrome	Additive effect on 5-HT uptake	Izzo, (2012), Wang et al. (2001)
		Venlafaxine	Serotonin syndrome	Additive effect on 5-HT uptake	Izzo (2012), Wang et al. (2001)
		Alprazolam	Decreased plasma levels of alprazolam	Induction of CYP3A4	Izzo (2012), Wang et al. (2001), Zhou et al. (2004)
		Midazolam	Decreased plasma levels of Midazolam	Induction of CYP3A4	Izzo (2012), Wang et al. (2001), Zhou et al. (2004)
		Buspirone	Hypomania	Synergistic effect on 5-HT receptors	Izzo (2012)
		Quazepam	Decreased blood concentration	Induction of intestinal CYP3A4	Izzo (2012), Wang et al. (2001), Zhou et al. (2004)

(continued)

Table 24.3 (continued)

Herb	Biological source	Drug	Result of interaction	Possible mechanism	References
		Methadone	Decreased blood concentration	Induction of CYP3A4 and/or P-glycoprotein	Izzo (2012), Wang et al. (2001), Zhou et al. (2004)
		Mephenytoin	Increased urinary excretion of metabolites	Induction of CYP2C19	Izzo (2012), Wang et al. (2001), Zhou et al. (2004)
		Chlorzoxazone	Increased serum ratio	Induction of CYP2E1	Izzo, (2012), Wang et al. (2001), Zhou et al. (2004)
		Anaesthetics	Delayed emergence	Interaction with CYP enzymes	Izzo, (2012), Wang et al. (2001), Zhou et al. (2004)
		Ephedrine/ phenylephrine	Decreased responsiveness	Unknown mechanism	Izzo (2012), Wang et al. (2001), Zhou et al. (2004)
Echinacea	*Echinacea* spp.	Midazolam	Increased or decreased clearance	Inhibits intestinal CYP3A4, while it induces hepatic CYP3A	Izzo (2012)
		Caffeine	Reduction of caffeine oral clearance	Inhibition of CYP1A2	Izzo (2012)
Kava	*Piper methysticum*	Alprazolam	Semicomatose state	Additive effect on GABA receptors. And inhibition of CYP3A4	Izzo (2012)
		Paroxetine	Lethargic state	Unknown mechanism	Izzo (2012)
		Levodopa	Reduced efficacy of levodopa	Kava possesses dopaminergic antagonistic properties	Izzo (2012), Lundstrom et al. (2017)
		Chlorzoxazone	Decreased serum ratio	Interaction with CYP2E1	Izzo (2012)
Ginkgo	*Ginkgo biloba*	Trazodone	Coma	Possible mechanism	Izzo (2012), Unger (2013)
		Valproic acid	Fatal seizure	Unknown mechanism	Izzo (2012), Unger (2013)
		Phenytoin	Fatal seizure	Unknown mechanism	Izzo (2012), Unger (2013)
		Risperidone	Priapism	Unknown mechanism	Izzo (2012), Unger (2013)

Shankhapushpi	*Canscora decussata*	Phenytoin	Loss of seizure control	Unknown mechanism	Izzo (2012)
Evening primrose oil	*Oenothera biennis*	Fluphenazine	Seizures	Gamolenic acid from evening primrose oil lowers the seizure threshold	Izzo (2012), Lundstrom et al. (2017)
Ginseng	*Panax ginseng*	Phenelzine	Sleeplessness, tremor and headaches	Unknown mechanism	Izzo (2012), Lundstrom et al. (2017)
Betel nut	*Areca catechu*	Procyclidine	Rigidity, bradykinesia, jaw tremors	Antagonistic effect of arecoline	Izzo (2012), Lundstrom et al. (2017)
Passion flower	*Passiflora incarnata*	Lorazepam	Hand tremor, dizziness, throbbing and muscular fatigue	Additive CNS depressant effect	Izzo (2012), Lundstrom et al. (2017)
Valerian	*Valeriana officinalis*	Lorazepam	Hand tremor, dizziness, throbbing and muscular fatigue	Additive CNS depressant effect	Izzo (2012), Lundstrom et al. (2017)
Danshen	*Salvia miltiorrhiza*	Midazolam	Increased blood concentration	Induction of intestinal CYP3A4	Izzo (2012)
Goldenseal	*Hydrastis canadensis*	Midazolam	Increased blood concentration	Induction of intestinal CYP3A4	Izzo (2012)
Magnolia berry	*Schisandra sphenanthera*	Midazolam	Increased blood concentration	Inhibition of CYP3A4	Izzo (2012)
Aloe	*Aloe vera*	Sevoflurane	Blood loss	An additive effect on platelet function	Izzo (2012)
Garlic	*Allium sativum*	Chlorzoxazone	Decreased serum ratios	Inhibition of CYP2E1	Izzo (2012)
Psyllium, ispagula	*Plantago ovata*	Lithium	Decreased plasma lithium concentration	Hydrophilic psyllium may prevent lithium from ionizing	Izzo (2012)
Herbal diuretics	–	Lithium	Decreased plasma lithium concentration	Unknown mechanism	Izzo (2012)

Table 24.4 Interaction of herbs with anti-retroviral drugs and anticancer

Herb	Biological source	Antiretroviral drug	Result of interaction	Possible mechanism	References
St. John's wort	*Hypericum perforatum*	Imatinib	Decreased blood concentration drug	Induction of CYP3A4	Izzo, (2012), Wang et al. (2001), Zhou et al. (2004)
		Indinavir	Decreased blood concentration drug	Induction of CYP3A4	Izzo (2012), Wang et al. (2001), Zhou et al. (2004)
		Nevirapine	Decreased blood concentration drug	Induction of CYP3A4	Izzo (2012), Wang et al. (2001), Zhou et al. (2004)
Ginseng	*Panax ginseng*	Imatinib	Hepatotoxicity	Unknown mechanism	Izzo (2012)
Cat's claw	*Uncaria tomentosa*	Protease inhibitors (atazanavir, ritonavir and saquinavir)	Increased blood concentration	Unknown mechanism	Izzo (2012)
Garlic	*Allium sativum*	Ritonavir	Severe GI toxicity	Induction of CYP enzymes in gut wall	Izzo (2012), Fasinu et al. (2015)
		Saquinavir	Decreased blood concentration	Induction of P-glycoprotein in the gut	Izzo (2012), Fasinu et al. (2015)
Ginkgo	*Ginkgo biloba*	Efavirenz	Decreased blood concentration	Unknown mechanism	Izzo (2012), Fasinu et al. (2015)
Herb	**Biological source**	**Anticancer drug**	**Result of interaction**	**Possible mechanism**	**References**
St. John's wort	*Hypericum perforatum*	Irinotecan	Decreased blood concentration of active metabolite SN-38	Interaction with CYP3A4	Izzo (2012), Haefeli and Carls (2014)

which interact with this class of drugs. The list of these interactions has been enumerated in Table 24.6.

Table 24.5 Interaction of herbs with oral hypoglycaemic drugs

Herb	Biological source	Drug	Result of interaction	Possible mechanism	References
St. John's wort	*Hypericum perforatum*	Gliclazide	Decreased blood concentrations	Induction of CYP enzyme	Izzo (2012), Gupta et al. (2017)
Prickly pear cactus	*Opuntia polyacantha*	Glipizide and metformin	Hypoglycaemic effect	Additive effect	Izzo (2012), Gupta et al. (2017)
Garlic	*Allium sativum*	Chlorpropamide	Hypoglycaemia	Unknown mechanism	Izzo (2012), Gupta et al. (2017)
Ginkgo	*Ginkgo biloba*	Tolbutamide	Decreased blood concentrations	Unknown mechanism	Izzo (2012), Gupta et al. (2017)

24.4.6 Interactions with Miscellaneous Classes of Drugs

Few interactions have been found with other classes of drugs such as antipyretics, antiinflammatory, antiallergic, antiasthmatic, antihistamines, oral contraceptives, anthelmintics, antacids, antiemetics, antimalarial, antidiarrheal, antifungal, vitamin supplements, etc. St. John's worts have been found to interact with multiple classes of drugs. Other herbs include garlic, ginkgo, green tea, hibiscus, milk thistle, valerian, etc. The mechanisms of such interactions are unknown. These interactions are enlisted in Table 24.7.

24.5 Conclusion

Though it is believed that drugs obtained from natural origin are safe, it has been observed that many herbs interact with the conventional drugs causing adverse effects to different extent, even causing fatal manifestations. This has been evident from various in vitro and in vivo clinical studies as well as patient history reports. Evidences of such interactions are available to different level of confirmation, and only few interactions are known along with their mechanism of actions. The clinical significance of such interactions is yet to be identified. And many more herbal products need to be evaluated, to identify the incidences of herb-drug interactions.

It is of prime importance to create awareness among the patients of seriousness of herbal interactions in order to increase the patient compliance and reporting of adverse events by patients. Physicians should also be responsible to make patient history reporting, in order to prevent inadvertent impact on patient's health. This will also help in disclosure of other interactions which are not known so far.

Further, evidence-based study, to identify the underlying mechanisms of interaction, needs to be conducted. This will serve the purpose of evaluation of reliability of case studies of herb-drug interactions.

Table 24.6 Interaction of herbs with immunosuppressants

Herb	Biological source	Drug	Result of interaction	Possible mechanism	References
St. John's wort	*Hypericum perforatum*	Cyclosporine	Decreased blood concentrations	Induction of CYP3A4 and/or P-glycoprotein	Izzo (2012), Wang et al. (2001), Zhou et al. (2004)
Red yeast rice	Mould species of *Monascus purpureus*	Cyclosporine	Rhabdomyolysis	Red yeast rice contains statins, whose concentration is increased by inhibition of CYP3A4 by cyclosporine	Izzo, (2012)
Red avens	*Geum chiloense*	Cyclosporine	Increased blood concentration	Unknown mechanism	Izzo (2012)
Magnolia berry	*Schisandra sphenanthera*	Tacrolimus	Increased blood concentration	Inhibition of P-glycoproteins	Izzo (2012)

Table 24.7 Interaction of herbs with miscellaneous drugs

Herb	Biological source	Drug	Result of interaction	Possible mechanism	References
St. John's wort	*Hypericum perforatum*	Fexofenadine	Decreased blood concentration	Induction of P-gylcoprotein	Izzo, (2012), Wang et al. (2001), Zhou et al. (2004)
		Oral contraceptives	Changes in pharmacokinetics	Induction of intestinal CYP3A4	Izzo (2012), Wang et al. (2001), Zhou et al. (2004)
		Prednisolone	Manic episodes	Unknown mechanism	Izzo (2012), Wang et al. (2001), Zhou et al. (2004)
		Tibolone	Acute hepatitis	Unknown mechanism	Izzo (2012), Wang et al. (2001), Zhou et al. (2004)
		Theophylline	Changes in pharmacokinetics	Induction of hepatic enzymes	Izzo (2012), Wang et al. (2001), Zhou et al. (2004)
		Voriconazole	Decreased blood concentration	Unknown mechanism	Izzo (2012), Wang et al. (2001), Zhou et al. (2004)
Hibiscus	*Hibiscus sabdariffa*	Chloroquine	Reduced plasma concentration	Unknown mechanism	Izzo (2012)
		Paracetamol	Changes in pharmacokinetics	Unknown mechanism	Izzo (2012)
Garlic	*Allium sativum*	Paracetamol	Changes in pharmacokinetics	Unknown mechanism	Izzo (2012)
Ginkgo	*Ginkgo biloba*	Omeprazole	Reduction in blood levels	Unknown mechanism	Izzo (2012)
Milk thistle	*Silybum marianum*	Metronidazole	Decreased blood concentration	Unknown mechanism	Izzo (2012)
Valerian	*Valeriana officinalis*	Loperamide	Acute delirium	Unknown mechanism	Izzo (2012)
Liquorice	*Glycyrrhiza glabra*	Prednisolone	Increased blood concentration	Metabolic enzyme inhibition	Izzo (2012)
Green tea	*Camellia sinensis*	Folic acid	Decreased blood concentration	Unknown mechanism	Izzo (2012)

References

Capasso R, Izzo AA, Pinto L, Bifulco T, Vitobello C, Mascolo N (2000) Phytotherapy and quality of herbal medicines. Fitoterapia 71(suppl 1):S58–S65

Colalto C (2010) Herbal interactions on absorption of drugs: mechanisms of action and clinical risk assessment. Pharmacol Res 62:207–227

Ekor M (2014) The growing use of herbal medicines: issues relating to adverse reactions and challenges in monitoring safety. Front Pharmacol 10:4–177

Fasinu PS, .Bouic PJ, Rosenkranz B (2012) An overview of the evidence and mechanisms of herb drug interactions. Front Pharmacol 3(69):1–12

Fasinu PS, Gurley BJ, Walker LA (2015) Clinically relevant pharmacokinetic herb-drug interactions in antiretroviral therapy. Curr Drug Metab 17(1):52–64

Ge B, Zhang Z, Zuo Z (2014) Updates on the clinical evidenced herb-warfarin interactions. Evid Based Complement Alternat Med 2014:957362

Gupta RC, Chang D, Nammi S, Bensoussan A, Bilinski K, Roufogalis BD (2017) Interactions between antidiabetic drugs and herbs: an overview of mechanisms of action and clinical implications. Diabetol Metab Syndr 26:9–59

Haefeli WE, Carls A (2014) Drug interactions with phytotherapeutics in oncology. Expert Opin Drug Metab Toxicol 10(3):359–377

Izzo AA (2004) Herb-drug interactions: an overview of the clinical evidence. Fundam Clin Pharmacol 19(1):1–16

Izzo AA (2012) Interactions between herbs and conventional drugs: overview of the clinical data. Med Princ Pract 21:404–428

Izzo AA, Ernst E (2009) Interactions between herbal medicines and prescribed drugs: an updated systematic review. Drugs 69(13):1777–1798

Lundstrom K, Pham HT, Dinh LD (2017) Interaction of plant extracts with central nervous system receptors. Medicines (Basel) 4(1):12

Marchetti S, Mazzanti R, Beijnen JH (2007) Concise review: clinical relevance of drug–drug and herb–drug interactions mediated by the ABC transporter ABCB1 (MDR1, P-glycoprotein). Oncologist 12:927–941

Unger M (2013) Pharmacokinetic drug interactions involving *Ginkgo biloba*. Drug Metab Rev 45 (3):353–385

VanRoon EN, Flikweert S, le Comte M, Langendijk PN, Kwee-Zuiderwijk WJ, Smits P, Brouwers JR (2015) Clinical relevance of drug-drug interactions—a structured assessment procedure. Drug Saf 28:1131–1139

Wang Z, Gorski JC, Hamman MA, Huang SM, Lesko LJ, Hall SD (2001) The effects of St John's wort *(Hypericum perforatum)* on human cytochrome P450 activity. Clin Pharmacol Ther 70:317–326

Yuan H, Ma Q, Ye L, Piao G (2016) The traditional medicine and modern medicine from natural products. Molecules 21(5):559

Zhang W, Han Y, Lim SL, Lim LY (2009) Dietary regulation of P-gp function and expression. Expert Opin Drug Metab Toxicol 5:789–801

Zhou SF (2008) Drugs behave as substrates, inhibitors and inducers of human cytochrome P450 3A4. Curr Drug Metab 9:310–322

Zhou S, Chan E, Pan SQ, Huang M, Lee EJ (2004) Pharmacokinetic interactions of drugs with St John's wort. J Psychopharmacol 18(2):262–276

Pharmacovigilance of Herbal Medicines: An Overview

25

Saipriyanka Bhimaneni, Ruchika Sharma, Prasanta Dey, and Anoop Kumar

Abstract

Herbal medicine and its formulations play a main role in the treatment of different types of diseases such as diabetes, arthritis, memory disorders, liver disorders, etc. across the globe. The most common myth about these medicines is that these are totally safe and can be consumed without prescription. However, various reports in the literature have indicated the adverse drug reactions of these medicines. Pharmacovigilance is essential for developing reliable information on the safety of herbal medicines. The existing systems are successfully developed and implemented for synthetic medicines, but regarding herbal medicines, some modification is required to address the specific differences of medicinal herbs. This chapter starts with a review of safety concerns of various types of herbal medicines, pharmacovigilance methods of signal detection and causality assessment of herbal drugs as well as regulatory aspects. Finally, this chapter concludes with current challenges and future perspectives of herbal pharmacovigilance.

S. Bhimaneni
Department of Regulatory Toxicology, National Institute of Pharmaceutical Education and Research (NIPER)-Raebareli, Lucknow, Uttar Pradesh, India

R. Sharma
Department of Pharmacognosy, Indo-Soviet Friendship College of Pharmacy (ISF), Moga, Punjab, India

P. Dey
School of Pharmacy, Sungkyunkwan University, Suwon, Republic of Korea

A. Kumar (✉)
Department of Pharmacology and Toxicology, National Institute of Pharmaceutical Education and Research (NIPER)-Raebareli, Lucknow, Uttar Pradesh, India
e-mail: abitmesra@gmail.com

S. C. Mandal et al. (eds.), *Evidence Based Validation of Traditional Medicines*,
https://doi.org/10.1007/978-981-15-8127-4_25

Keywords

Herbal medicines · Adverse drug reactions · Pharmacovigilance

Abbreviations

ADRs	Adverse drug reactions
AIA	All India Institute of Ayurveda
ATC Classification System	Anatomical Therapeutic Chemical Classification System
CDSCO	Central Drugs Standard Control Organisation
COSTART	Coding Symbols for a Thesaurus of Adverse Reaction Terms
EU Directive	European Union Directive
EudraVigilance	European Union Drug Regulating Authorities Pharmacovigilance
HC	Health Canada
ICD9CM	International Classification of Diseases, ninth Revision, Clinical Modification
ICH	International Council for Harmonisation of Technical Requirements for Pharmaceuticals for Human Use
INN	International Nonproprietary Names
MedDRA	Medical Dictionary for Regulatory Activities
MHRA	Medicine and Healthcare products Regulatory Agency
PV	Pharmacovigilance
WHO ART	World Health Organization Adverse Reaction Terminology
WHO DDE	World Health Organization Drug Dictionary Enhanced
WHO	World Health Organization

25.1 Introduction

Pharmacovigilance is made up of two words, i.e. "Pharmakon" and "Vigilance". Pharmakon is a Greek word which means drug, whereas Vigilance is a Latin word which means to keep watch. In simple words, pharmacovigilance is the science of collecting, monitoring, assessing and evaluating information from health-care professionals (HCP) and patients on adverse effect of medications, biological products, vaccines, blood products, medical devices as well as herbals and traditional medicines. The World Health Organization (WHO) defined pharmacovigilance as "the science and activities related to the detection, assessment, understanding, and prevention of adverse drug effects or any other possible drug-related problems" (WHO 2004a). After completion of the phase 3 clinical trial, the sponsor submits the new drug application (NDA) to the respective regulatory authorities to get market approval. The NDA contains all the safety as well as efficacy data in animals (preclinical) and human (phase 1–phase 3 clinical trials). However, these all studies are not that much reliable because of limited number of patients studied in clinical trials and strict inclusion and exclusion criteria, for example, special group of population (children, pregnant women, etc.) are normally excluded. Thus, the long-term adverse drug reactions and adverse drug reactions in special population are difficult to detect. Therefore, to study rare ADRs which are in long latency and ADRs with special groups, the careful monitoring of drugs in post-marketing phase is necessary (Wester et al. 2008). The main objective of the pharmacovigilance is to monitor the safety of drugs continuously by evaluating adverse effects which are previously unrecognized during the clinical trial (Mann and Andrews 2002). Apart from pharmaceutical medicines, the safety of other pharmaceutical products such as vaccines, herbal products, blood products and medical devices is also monitored (Shaw et al. 2012).

Herbal medicinal products are used from the ancient time for the treatment of various disorders/diseases such as anxiety (Sarris et al. 2011); arthritis (Mills et al. 1996); depression (Sarris et al. 2011); high blood pressure (Mansoor 2001); viral infections (Liu et al. 2003); hormonal imbalance such as premenstrual tension, eczema, chest problems, insomnia and migraine; gastro-intestinal disorders (Chen et al. 2014); etc. These medicines are also consumed by special populations like pregnant women and children (Park et al. 2010). The term herbal medicine describes crude preparations which are obtained from different parts of a plant, herbal tinctures, usually supplied in the formulation as tablets, capsules and powders which can be available without a prescription (Barnes and Anderson 2003).

However, the safety of these herbal products is still unclear; many herbal practitioners assume that these medicines are completely safe, but in some experimental studies, some of these medicines are found to be toxic (Patel et al. 2012). Thus, there is a need to monitor the safety profile of herbals in a large group of population. This chapter starts with a review of safety concerns of various types of herbal medicines, pharmacovigilance methods of signal detection and causality assessment of herbal drugs as well as regulatory aspects. Finally, this chapter concludes with current challenges and future perspectives of herbal pharmacovigilance.

25.2 Need of Pharmacovigilance for Herbal Products

The safety of herbal products is still unknown. However, it is assumed that these products are safe as used from ancient times, but unfortunately, researchers have found the various safety concerns regarding the use of herbal products. The herbal products contain a mixture of plants that contain several hundreds of chemical constituents which are sometimes unknown also. Out of these hundreds of constituents, only some of the constituents have a therapeutic activity, whereas others have no therapeutic activity and sometimes also toxic (Barnes et al. 1998). In literature, various preclinical (animals) and clinical (human) studies have reported various adverse reactions related to herbal medicines. The dermatological problems such as photosensitivity and severe allergenicity have been observed after the use of St. John's wort herbal preparation (Ernst 2000). The hyper-reflexia due to the use of belladonna extract has been observed (Ulbricht et al. 2004). Liver toxicity in rats has been reported after the chronic use of kava kava (Sorrentino et al. 2006). Liver fibrosis is reported after the administration of *Hypericum* in rats (Gregoretti et al. 2004). The hepatotoxicity is reported due to kava kava (Denham et al. 2002), and urothelial cancer due to *Aristolochia* species (Cosyns 2003). The various ADRs related to herbal medicines are compiled in Table 25.1. The herbal products are commonly associated with hypersensitivity (Drain and Volcheck 2001), allergenicity, idiosyncratic, hypoglycaemia, abdominal pain, etc. (Kamsu-Foguem and Foguem 2014). The most common adverse reactions related to the herbal products as per the system organ class (SOC) have been compiled in Table 25.2. Thus, to avoid these all adverse effects which occurred due to herbal medicine, it is necessary to have proper pharmacovigilance system for the herbal medicines.

25.3 Classification of Adverse Drug Reactions

Adverse drug reactions are defined as "noxious and unintended response occurs at doses normally used in the population for the prophylaxis, diagnosis or therapy of disease, or modification of physiological function" (WHO 1972). An adverse effect can be termed as the unexpected or irrelevant toxic response of a drug caused due to the pharmacological response of drugs (Syed et al. 2015).

These adverse effects are mainly classified into six types (Drain and Volcheck 2001) as mentioned below:

- Augmented (Type A)
- Bizarre (Type B)
- Chronic (Type C)
- Delayed (Type D)
- End of treatment (Type E)
- Failure of treatment (Type F) (Edwards and Aronson 2000)

The brief classification of adverse drug reactions is compiled in Fig. 25.1.

25.4 Reasons for Safety Concerns of Herbal Products

The reasons for safety concerns of herbal products are due to presence of multiple constituents in the herbal preparations and also the percentage of chemical constituents which are not the same as throughout the plant for all time, and also

Table 25.1 List of herbal products and their traditional uses along with adverse effects

S. No	Medicinal plant	Traditional uses	Adverse effects	References
1.	*Pygeum africanum* (*Prunus africana*)	Prostate cancer, prostatitis, benign prostatic hyperplasia, other urinary tract infections, and aphrodisiac	Gastro-intestinal upsets	Kamsu-Foguem and Foguem (2014)
2.	Wild wisteria, violet tree (*Securidaca longipedunculata*)	Laxative, nervous system ailments (epilepsy), wounds, sores, coughs, venereal diseases, snakebites, bilharzias, headaches, fever related to malaria, erectile dysfunction or aphrodisiac, dysmenorrhoea, and abortion induction	Acute kidney (cortical necrosis, acute interstitial nephritis), injury, diarrhoea, dehydration, and collapse	Kamsu-Foguem and Foguem (2014)
3.	Madagascar rosy periwinkle (*Catharanthus roseus*)	Cancer chemotherapy	Medullary aplasia, leucopenia, incoordination of movements, convulsions, fatigue, mucositis, constipation, and neutropenia of short duration	Kamsu-Foguem and Foguem (2014)
4.	Red spinach (also known as Chinese spinach, Hon-toi-moi, Yin choy, Hsien tsai) or spleen amaranth (*Amaranthus dubius*)	Diuresis (high blood pressure, kidney infections, obesity, and the oedema associated with premenstrual syndrome (PMS) or traumatic injuries)	Hypotension, skin irritations, to extensive organ and tissue damage with death induction	Kamsu-Foguem and Foguem (2014)
5.	Licorice (*Glycyrrhiza glabra*)	Sore throat, cough, arthritis, and weight loss induction	Acute kidney injury (hypokalaemic nephropathy), amenorrhoea, pseudoaldosteronism, hypertension, heart failure, and rhabdomyolysis	Kamsu-Foguem and Foguem (2014)
6.	Betamethasone	Asthma, seasonal allergies, transfusion reactions	Corticosteroid-like side effects	Wal et al. (2011)
7.	Aristolochic acid	Traditional Chinese medicine used in manufacturing of	Kidney failure, mutagenicity, carcinogenicity	Wal et al. (2011)

(continued)

Table 25.1 (continued)

S. No	Medicinal plant	Traditional uses	Adverse effects	References
		cough, immune stimulation products		
8.	*Ginkgo biloba*	Alzheimer's disease, anxiety, depression	Prolonged prothrombin time, increased coagulation time, subcutaneous haematoma, intracranial haemorrhage	Wal et al. (2011)
9.	Myrrh (*Commiphora myrrha*)	Astringent, wound healing agent, antifungal, anti-bacterial	Liver, kidney toxicity, immune nephritis, tachypnoea	Wal et al. (2011)
10.	*Bidens pilosa* L.	Antihyperglycaemic, antihypertensive, antiulcerogenic, hepatoprotective, antipyretic, immunosuppressive and anti-inflammatory, anti-leukaemic, anti-malarial, anti-bacterial, antioxidant, antitumour and antifungal effects	Abnormal abortions to pregnant women	Kiguba et al. (2016)
11.	*Hoslundia opposita*	In post-partum to cleanse the uterine blood clots, heal vaginal lacerations	Abnormal abortions in pregnant women	Kiguba et al. (2016)
12.	Yoyo cleanser (*Entandrophragma utile* and *Anacardium occidentale*)	To cleanse body and blood infections	On chronic exposure lung tumours	(Ekor 2014)
13.	*Ephedra sinica*	Asthma, allergy, weight loss, dietary supplement	Severe cardiotoxicity, central nervous system disorders	(Ekor 2014)
14.	*Aconitum carmichaelii* and *Aconitum kusnezoffii*	Pain relief, stroke and heart failure, diarrhoea and diabetes	Cardiotoxicity, ventricular fibrillation, tachycardia	(Ekor 2014)
15.	*Tussilago farfara*	Acute and chronic cough	Hepatic veno-occlusive disease, cirrhosis	(Ekor 2014)
16.	Garlic (*Allium sativum*)	Hypertension, hypercholesterolaemia	Burning sensation in GIT nausea, diaphoresis and light-headedness, spinal epidural hepatoma	(Ekor 2014)

(continued)

Table 25.1 (continued)

S. No	Medicinal plant	Traditional uses	Adverse effects	References
17.	Kava (*Piper methysticum*)	CNS depressant, anxiolytic	Dizziness, disorientation, coma, numbness up on oral ingestion, liver injuries	(Ekor 2014)
18.	St. John's wort (*Hypericum perforatum*)	Against viral infections, antidepressant	Allergic reactions, headache, restlessness, fatigue, dry mouth, nausea, vomiting, constipation, photosensitivity. Hyperesthesia and a syndrome of dyspnoea hyperventilation with flushing headache, mydriasis, nausea, palpitations, tremor	(Ekor 2014)
19.	*Aloe vera*	Laxative, anti-inflammatory properties, inactivates HSV virus, antidiabetic properties, weight loss	Lowers blood sugar by stimulating insulin, diarrhoeal/abd. cramps, arrhythmia from hypokalaemia, contact dermatitis, stinging, soreness, acute hepatitis, renal failure, nephritis, abortifacient, may increase uterine bleeding, avoid in pregnancy and lactation	Sucich and Sanoski (2011)
20.	Ashwagandha (*Withania somnifera*)	Antidiabetic, anti-ageing, hyperlipidaemia, Parkinson disease, adaptogen	Lowers BP, blood sugar and testosterone, FSH levels increase WBC, platelet	Sucich and Sanoski (2011)
21.	Chamomile	Antispasmodic, anxiolytic, children diarrhoea, GI upset, infant colic, sedative effects	Decrease blood pressure, increase blood sugar, increase in heart rate, asthma	Sucich and Sanoski (2011)
22.	Fenugreek seed	Diabetes mellitus, hyperlipidaemia, loss of appetite, stimulates milk production. Induce child birth	Bloating, flatulence, diarrhoea, hypoglycaemia, hypokalaemia, miscarriage	Sucich and Sanoski (2011)

(continued)

Table 25.1 (continued)

S. No	Medicinal plant	Traditional uses	Adverse effects	References
23.	Ginger	Arthritis, muscle pain, pregnancy nausea, chemotherapy nausea, platelet aggregation	Nausea, arrhythmias, increase bleeding risk	Sucich and Sanoski (2011)
24.	Lavender	Anxiety, depression, insomnia, GI upset, alopecia, topical antiseptic, aromatherapy	Hypersensitivity, skin pigmentation, GI upset	Sucich and Sanoski (2011)
25.	Milk thistle (*holy thistle*) *Sylimarin*	Chronic hepatitis, liver cirrhosis, anti-cancer drug for breast, prostate cancer, cervical cancer	Decrease blood sugar, GI side effects H/A, insomnia, exacerbates hemochromatosis, arthralgia	Sucich and Sanoski (2011)

difference occurs between the same plants. The following factors are responsible for variations in chemical constituents of plants (Barnes et al. 1998):

- Inter- and intra-species variation in constituents.
- Harvesting time.
- Post-harvesting factors such as storage conditions, drying, etc.
- Variations exist between different manufacturer's products and preparations of the same herbal products.
- Quality issues such as contamination with soil, microorganism, pesticides, etc.

25.5 Regulatory Aspects

In most of the countries, the herbal medicines and other related nutraceutical products are widely introduced into the market without any mandatory toxicological studies. Many of these countries don't have an effective machinery for the regulation of herbs such as manufacturing practices, sound knowledge of the particular herbs and quality control standards, data related to reference drugs used for comparison, etc.

For the assessment of ADRs associated with herbal medicine, the following regulations are implemented worldwide:

- WHO International Drug Monitoring Programme for safety monitoring of herbal and traditional medicine in 2000, 2001 (World Health Organization 2004a, b, c, d).
- EU—The Directive on Traditional Herbal Medicinal Products (Directive 2004/24/EC) (Hitchen 2006).
- UK—Yellow Card Scheme for monitoring of ADRs (Ekor 2014).

Table 25.2 Common adverse effects regarding use of herbal medicines as per system organ class (SOC) (Patel et al. 2012)

No.	System organ class involved	Most frequently reported adverse reactions
1.	Endocrine disorders	Hypoglycaemia, Cushing's syndrome, adrenal insufficiency
2.	Central and peripheral nervous system disorders	Unconsciousness, drowsiness, giddiness, coma
3.	Body as a whole	General disorders, lethargy, fever, diaphoresis
4.	Liver and biliary system disorders	Jaundice, hepatitis, increased liver enzymes, hepatic failure
5.	Skin and appendage disorders	Rash, urticaria, maculopapular rash, Stevens-Johnson syndrome
6.	Gastro-intestinal system disorders	Vomiting, nausea, abdominal pain
7.	Psychiatric disorders	Confusion, abnormal mental state, insomnia
8.	Urinary system disorders, platelet	Acute renal failure, urine discolouration, urine incontinence
9.	Respiratory system disorders	Shortness of breath, wheeze, respiratory failure
10.	Metabolic and nutritional disorders	Hypokalaemia, hyponatraemia, diabetes mellitus
11.	Heart rate and rhythm disorders	Palpitations, tachycardia
12.	Cardiovascular disorders	Hypotension, hypertension, septic shock
13.	Platelet and bleeding, clotting disorders	Thrombocytopenia, increased prothrombin time
14.	White cell and reticuloendothelial system disorders	Eosinophilia, leukopenia, pancytopenia
15.	Musculo-skeletal system disorders	Rhabdomyolysis, myalgia, myopathy
16.	Red blood cell disorders	Anaemia, haemolysis, methaemoglobinaemia
17.	Reproductive disorders, female	Hypomenorrhoea, hypermenorrhoea, postmenopausal bleeding
18.	Vascular extracardiac disorders	Vasculitis, stroke, subdural haemorrhage
19.	Myo, endo, peripheral and valve disorders	Myocardial infraction, myocarditis, endocarditis
20.	Vision disorders	Blurred vision, blindness
21.	Collagen disorder	Systemic lupus erythematosus
22.	Hearing and vestibular disorders	Tinnitus
23.	Neoplasms	Hepatocellular carcinoma

- India—Ayurveda, Yoga & Naturopathy, Unani, Siddha, Sowa Rigpa and Homoeopathy (AYUSH), under the Ministry of Health and Family Welfare. The National Medicinal Plants Board (2000) by the Indian government to deal with the herbal medicine. All India Institute of Ayurveda (AIA) (Wal et al. 2011).
- USA FDA—The Dietary Supplement Health and Education Act (DSHEA), 1994 (Bent 2008).
- Russia—Federal Service for Surveillance in Healthcare under the Ministry of Health. It is governed by the Constitution of the Russian Federation, Federal

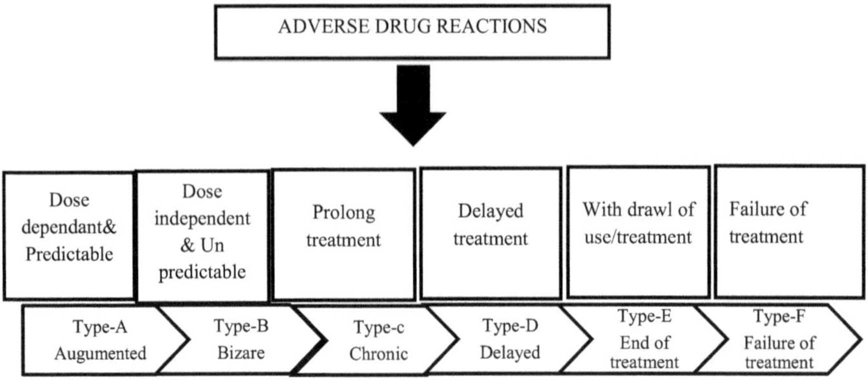

Fig. 25.1 Classification of adverse drug reactions (ADRs)

Constitutional Laws, federal laws, government of the Russian Federation and international treaties of the Russian Federation (Kumar et al. 2016).

25.6 Case Processing

For any herbal product, if any adverse drug reactions were found, then the case is processed in following mentioned steps:

- Data collection
- Case validity
- Triage of cases
- Entry of data into databases
- Quality assessment
- Medical coding
- Causality assessment
- Reporting of cases
- Aggregate reporting (PSURs, PBRERs, etc.)
- Submission of aggregate reports
- Regulatory action

The all above-mentioned steps have been described below.

25.6.1 Data Collection

The data related to ADRs of herbal products are collected from manufacturers, consumers, health-care reporters, patients, poison information centres, a consumer organization, clinical trials, spontaneous reporting, etc. These all sources are mentioned below in detail and compiled in Fig. 25.2.

Fig. 25.2 Sources of reporting of ADRs related to herbal medicines

25.6.1.1 Reports from the Consumers

The reporting of ADRs by the consumers is very much appreciated and essential for safety of medicines. This type of reporting could be useful for the identification of unknown ADRs or known effects of herbal medicine. The herbal medicines are mostly available for the public without prescription. Thus, reports of herbal medicines associated adverse drug reactions from physicians and health-care professionals will be less as compared to reports directly from the consumers. However, in allopathic medicines, most of the ADRs are obtained from physicians and health-care professionals as these medicines are mostly prescribed with prescription.

The reporting of ADRs by the consumers can be done in one form or any another form which includes an important information on patient, medication, ADRs, and reporter detail. Unfortunately, very less number of regulatory authorities is focusing on the collection of direct ADR reports from consumers. The Council for International Organizations of Medical Sciences (CIOMS) proposes a lot of policy approaches and practices to get the consumer reports. Further, the analysis of these reports is done in appropriate way.

25.6.1.2 Manufacturers

The manufacturers involved in the preparation of herbal medicine are also an important source of information to point out the ADRs associated with their products. However, very few regulatory authorities are focusing on this source of reporting.

25.6.1.3 National Poison Centres

National poison centres play vital role in pharmacovigilance of herbal products mainly in those countries where resources are limited.

25.6.1.4 Drug Information Centres

The data related to ADRs of herbal product can also be collected from drug information centres. These drug information centres must have a better communication with national pharmacovigilance centres to provide the information on behaviour of herbal products in large number of population.

25.6.1.5 Consumer Organizations

The consumer organization can also play an essential role in the collection of data related to ADRs of herbal medicines. These organizations receive complaints from the market regarding the use of any type of products. Thus, these organizations can also obtain the complaints regarding herbal medicine which can be useful for risk and benefit analysis (World Health Organization 2004a, b, c, d).

25.6.2 Validity of Case

The safety information of herbal products is reported in the standard form of Individual Case Study Report (ICSR). A valid case needs to have four elements which are mentioned below (Dhanalakshmi and Harikrishnan 2017):

- Patient/consumer identification
- List of medicines used by the patient
- Outcome of the suspected adverse reaction
- Reporter identification

The standard format of ADR reporting form is available at www.cdsco.gov.in under consumer section.

25.6.3 Triage of Cases

Next step is triage of cases which means prioritization of cases for reporting to regulatory authorities. The cases are assessed in the prioritization way such as serious adverse events are regarded as if any ADRs is fatal such as life-threatening adverse effect (may result to patient's death), disabilities are detected which leads to showing any effect on person's quality of life, or person leads to prolonged hospitalization or may have congenital abnormalities, i.e., adverse effect to the baby when a pregnant woman consumed. Depending on this, the adverse effects are categorized and listed on the bases of priority for further assessment (Kumar and Khan 2015).

25.6.4 Entry of Data in Databases

Next step is to enter the data in the safety databases, and these databases vary across the globe. For example, in the USA, Adverse Event Reporting System (AERs) is

Table 25.3 Common databases used in pharmacovigilance (Kumar and Khan 2015)

S. no	Country	Regulation	Software	ADR form
1.	USA	FDA	AERS	Med Watch 3500A
2.	Europe	EMEA	EudraVigilance	Yellow Card
3.	India	CDSCO	VigiFlow	Suspected Adverse Drug Reaction Form
4.	Canada	HC	Canada Vigilance	Canada vigilance reporting system

used, whereas in Europe, EudraVigilance is used. In India, VigiFlow is used and, in Canada, Canada vigilance. The sponsor or pharmaceutical companies are also using other software like Oracle AERS, PV-Works, Clintrace, etc. Some of these databases are mentioned below and compiled in Table 25.3.

25.6.4.1 VigiFlow

VigiFlow is the well-known database for the management of individual case safety reports (ICSRs). This database is widely used in national centres under WHO Programme for International Drug Monitoring (Chandel et al. 2014).

25.6.4.2 AERS (USA)

The Adverse Event Reporting System (AERS) is the US-based database which contains information about the adverse events. This database is designed for the safety surveillance programme of drug and therapeutics and other biological products by the FDA. The framework of AERS fully adheres to International Conference on Harmonisation (ICHE2B). The reports in AERS have been reviewed and evaluated by the departments such as Centre for Biologics Evaluation and Research (CBER) and Centre for Drug Evaluation and Research (CDER) to monitor the safety of products.

25.6.4.3 EudraVigilance (Europe)

EudraVigilance (European Union Drug Regulating Authorities Pharmacovigilance) is the data processing network and management system of Europe for reporting and evaluation of adverse drug reactions (Eudravigilance 2019).

25.6.4.4 Canada Vigilance (Canada)

In Canada, Health Canada is using Canada Vigilance Adverse Reaction Online Database for the reporting and analysis of ADRs related to medicinal products (Health Canada 2019).

25.6.4.5 ARISg Software

ARISg software is used by most of the pharmaceutical companies for the management of reporting of adverse drug events (Kumar and Khan 2015).

25.6.5 Quality Assessment

After the collection of reports, quality assessment is done. The selection of appropriate data is most important for the signal detection and assessment of ADRs. It's difficult or impossible to do causality assessment of reports which contain inappropriate information. The good quality reports should contain the following information:

- Patient information includes age, sex, date of birth, weight, BMI and diagnostic results due to the particular medicine.
- Drug which has been taken by patient. ADR information includes what they have faced, when the event happened and what was the event? How it affected patient life quality? How the event is managed? Such as by decreasing dose or by restricting its usage whether the event is reappeared again? Is there any supportive laboratory data?
- Information related to drug includes name of the drug, brand name, manufacturing address, dose, etc. Information is added to get a quality of report (Kumar and Khan 2015).

25.6.6 Medical Coding

After reviewing the data, the ADRs are converted into universal medical terminology with the help of different dictionaries of drug such as WHO ART, Med DRA, WHO DDE, ICD9CM, CONSTART, etc. These are internationally accepted medical dictionaries used by regulatory authorities and biopharmaceutical industries (Talbot and Aronson 2012). The most commonly used medical dictionary is Med DRA. This dictionary is organized in the form of system organ class (SOC) and divided into different terms for enlisting the ADRs such as high level group terms (HLGT), high level terms (HLT), preferred terms (PT) and lowest level terms (LLT) (Brown et al. 1999). Usually individual cases are coded in LLT level, and outputs of that counts are coded in PT level. The current version of Med DRA is 23.0 (September 2019). WHO ART is another drug dictionary maintained by UMC and used in the coding of medical events related to the drugs. The US FDA use COSTART for coding, filling and retrieving ADRs. The various types of dictionary used for the coding of adverse events are summarized in Table 25.3 (Qureshi 2012).

25.6.7 Causality Assessment

The causality assessment between the drug and the adverse event is done by using various methods. The most common methods are Naranjo scale and WHO UMC causality categories scale. The Naranjo scale assesses the adverse drug reactions in clear, transparent and consistent form by explaining the relationship between drug and adverse effect in the form of a probability scale. This scale contains set of

Table 25.4 Naranjo probability scale

Questions	Yes	No	Don't know
Are there any previous reports on this reaction?	+1	0	0
Did the adverse event appear after the suspect drug was administered?	+2	−1	0
Did the AR reduced when the drug was discontinued or specific antagonist administered?	+1	0	0
Did the adverse drug reaction reappear when drug was re-administered?	+2	−1	0
Are there alternate causes that could alone have caused the reaction?	−1	+2	0
Did the reaction reappear when other than drug-like placebo has given?	−1	+1	0
Is that drug detected in body fluids and found to be toxic?	+1	0	0
Was the reaction is dose dependent?	+1	0	0
Did the patient have similar reaction on previous exposure of similar molecules?	+1	0	0
Is that AR confirmed by any evidence?	+1	0	0

questions to find out causal relationship between the drug and adverse event. The scoring of Naranjo algorithm is done depending on the score which is obtained after addition of all points mentioned in Table 25.4: >9 gives definite ADR which means the drug is found to have adverse drug reactions, 5–8 score gives probable ADR and gives an idea that the drug may have adverse drug reaction on suitable population, 1–4 score gives possible ADR about the drug, and 0 is doubtful ADR. According to WHO UMC causality categories (Table 25.4), the different causality terms like certain, possible/likely, possible, unlikely, conditional or unclassified and inaccessible or unclassifiable are used depending upon the relationship between the drug and ADRs. The WHO UMC causality categories are summarized in Table 25.5 (Zaki 2011).

25.6.8 Reporting of Cases

After the causality assessment, case with ADR is reported to regulatory authorities (as per requirement). Nowadays, in most of the countries, cases with the ADR are reported through online system.

25.6.9 Aggregate Reporting

In aggregate reporting, the risk and benefit analysis of the particular product during the fixed interval time is done and submitted to respective regulatory authorities (Jarernsiripornkul et al. 2002). The periodic safety update reports (PSUR), periodic benefit risk evaluation report (PBRER), etc. are the examples of aggregate reporting (Sharma et al. 2020).

Table 25.5 WHO UMC causality categories

Causality term	Assessment criteria
Certain	Event or abnormalities in laboratory test data which depends on time relationship to drug administration Response to withdrawal (pharmacologically, pathologically) Event with specific medical disorder is recognized Rechallenge satisfactory if necessary
Probable or likely	Event or abnormalities in laboratory test data which depends on time relationship to drug administration Relation with the other disease or drugs Response to treatment such as withdrawal treatment Rechallenge not required
Possible	Event or abnormalities in laboratory test data which depends on time relationship to drug administration Can explain with the relationship with other drugs or disease Unclear information about the withdrawal
Unlikely	Event or abnormalities in laboratory test data which depends on time relationship to drug administration are improper but not impossible Relation with the other disease or drugs
Conditional or unclassified	Event or abnormalities in laboratory test data Need more data for proper assessment Additional data is under examination
Un accessible or unclassified	Report suggestion on AR Insufficient information to take action or judgment on particular drug Data is unable to supplement or verify

25.6.10 Submission of Aggregate Reports

As per the regulatory requirements, these reports are submitted.

25.6.11 Regulatory Action

On the basis of risk and benefit profile of herbal products, the following action can be taken by the regulatory authority:

- Possible actions are taken to minimize the risks related to the product.
- Advice is given to community (health-care professionals, consumer, etc.) regarding safety of the drugs.
- Modifications in the product such as dose, dosage, etc. to avoid ADRs.
- Restriction of the product from the market for the interest of public safety.
- Safety warnings and instructions sent to the manufacturers regarding the quality of the product in manufacturing which affects the safety.
- Label changing for the products.
- Marketing authorization withdrawal or cancellation of approval.
- Licence cancellation.

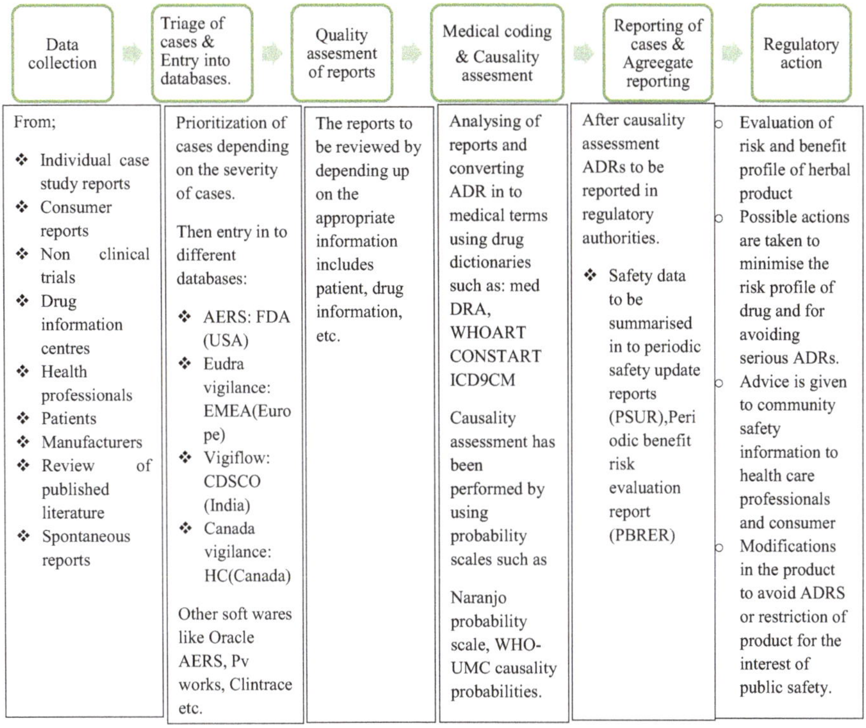

Fig. 25.3 Case reporting and management of ADRs

The overview of case reporting and management of ADRs related to herbal products is presented in Fig. 25.3.

25.7 Current Challenges

In case of pharmacovigilance of herbal products, the regulatory authorities and sponsor are facing various challenges. Some of the challenges are mentioned below and presented in Fig. 25.4.

25.7.1 Regulation

There are no uniform regulations for monitoring of ADRs associated with herbal products. The regulations may vary from country to country. These types of medicines are mostly available without prescription, and consumers directly get the medicine. These medicines are also not handled by qualified persons, and there are no proper regulations for proper quality control, proper distribution channels, etc. which can lead to increase in adverse effects of herbal medicines. Further, there is no

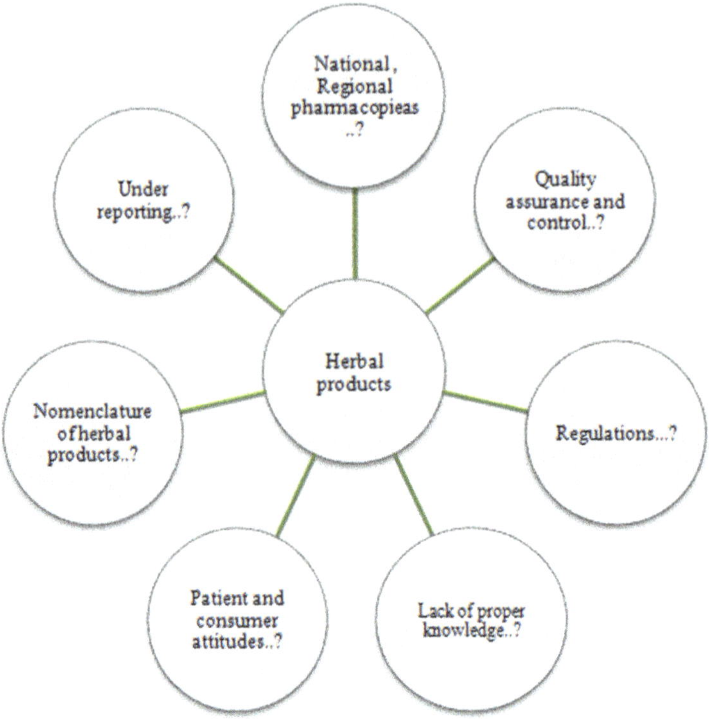

Fig. 25.4 Current challenges in pharmacovigilance of herbal products

proper regulation regarding import or export of herbal medicine. Thus, these types of products are directly coming into the market without proper regulations. National regulatory information on herbal products is different among all the countries, and there is a lack of sharing information such as safety monitoring, pharmacovigilance centres and also national regulatory authority regulation. Thus, there is need of proper regulatory guidelines regarding herbal products.

25.7.2 Quality Assurance and Control

The safety and efficacy of herbal products directly depend upon quality. Thus, quality control and assurance measures are required as per national and regional pharmacopoeias (Bandaranayake 2006). These pharmacopoeias provide the standard for herbal material and assurance of good manufacturing practice (GMP) for herbal products. However, these standards are not implemented in every country, and poor quality control measurement leads to poor quality of medicines, adverse drug reactions, adulterations with unknown potent substances, etc. (Sucich and Sanoski 2011).

Practically, the quality control of herbal medicine is difficult due to presence of multiple constituents. The quality of herbal medicine is depending on various factors such as cultivation of plant, seasonal variation, harvesting factors, etc. Thus, quality assurance and quality control of herbal medicine are one of the challenges in herbovigilance.

25.7.3 National and Regional Pharmacopoeias

The national and regional pharmacopoeias regulate the quality specifications and standards of the herbal materials and herbal preparations which include essential oils, powdered herbal materials, etc. The inclusion of herbal products in pharmacopoeia may depend on the availability of herbal products in a particular region or country. The availability of herbal products depends on obtaining original medicinal products. The ecological and environmental habitats play a major role in it. Therefore, if the same pharmacopoeia monograph name is given to a herbal material and the same name is enlisted in pharmacopoeia in the different form from the original medicinal product, then it's very difficult to analyse the ADRs associated with herbal product.

25.7.4 Lack of Proper Knowledge Regarding Herbal Medicines

The herbal medicine products are most common as traditional herbal medication which is used in different countries in different ways. These medicines are available in the market without prescription. The mode of administration of these medicines is also varying person to person. Most of the persons are also in dilemma whether they have to take it or not and assume these medicines are completely safe. However, all the herbal medicines are not safe completely. Thus, proper knowledge is required among physician as well as in public regarding use of herbal medicine.

25.7.5 Patient Consumer Attitudes Towards Herbal Medicine

Most of the people believe that herbal medicines are completely safe as these are prepared from the natural resources. The patient who uses herbal medicine along with the conventional medication could result in ADRs as in the often cases they will not report to physicians about their usage of herbal medicine along with the other medication. Thus, it is necessary to take interview of consumers by the health-care professionals and herbal medicine providers, respectfully, about what are all the medication they are taking, prescription medicine, herbal medicine, and any other health-care and self-care products used by patients to know the complete history usage of products. It is required to educate the entire stakeholder regarding proper use of herbal medicines (World Health Organization 2004a, b, c, d).

25.7.6 Nomenclature of Herbal Products

To provide consistency for pharmacovigilance of herbal products, the nomenclature of herbs also plays a major role due to having different naming of the same herbs in different countries. The same herb with different names among various countries may lead to confusion about the herbs. Thus, the pharmacovigilance of these types of herb is challenging task (Farah et al. 2006).

25.7.7 Underreporting

The underreporting of ADRs associated with herbal medicine is also one of the challenges. The public as well as physician and health-care professional don't know how to report ADRs related to herbal medicine. Thus, there is less number of reports related to ADRs of herbal medicine as actually exist (Kiguba et al. 2016).

If we can overcome these all above-mentioned challenges, then we can ensure the proper use of herbal medicines.

25.8 Future Perspectives

In the future, to achieve the effective pharmacovigilance system for herbal medicines, it is necessary to have proper regulatory authorities, sufficient quality assurance and quality control studies, proper national and regional pharmacopoeias, knowledge about herbal medicine to the public, changing attitudes of patient towards herbal medicine and use of proper scientific names such as nomenclature of herbs and awareness to the public in reporting of ADRs. The proper regulatory authorities will be helpful for prioritizing post-marketing surveillance and safety monitoring of herbal medicine. The strict actions are also required towards non-prescription medicines to strengthen the pharmacovigilance of herbal products. In this direction, the European Union (EU) of Directive on Traditional Herbal Medicinal Products has implemented some regulatory action for the manufacturers of herbal medicine. The UK also put a forward step to strengthen herbal pharmacovigilance to formulate a committee regarding safe use of herbal medicines. The Netherlands also ensure the pharmacovigilance of herbal products in future years by inventing common electronic health record which contains all the data about patient and herbal medicine, which can be accessed by community pharmacist. However, this system is probably applied to prescription medicine only. The Department of Health is increasing the use of technology advances by developing computerized record linkage to common electronic heath record to access the non-prescription medicine also. Italy also taken a forward step for ensuring drug safety by motivating the patients in reporting ADRs which improves patient care and clinical practice also (Russo and Sarro 2013).

The individual genetic pattern also affects the occurrence of adverse drug reactions related to herbal medicines in the population. Thus, pharmacogenomic studies will be helpful for ensuring the safety of herbal medicines. These studies will give the idea about importance of genetic factors in determining the individual susceptibility to ADRs. Further, these studies are also helpful in the optimization of treatment and potential reduction of ADRs on the basis of patient genotype (Munir and Park 2001).

To implement pharmacovigilance of herbal products in effective way, the proper scientific names for the herbs should be used which will be helpful for the physicians and health-care providers for reporting ADRs related to herbal medicines. Underreporting of the ADRs is the major challenge in pharmacovigilance. To increase the reporting of ADRs associated with herbal medicines, the awareness related to reporting of ADRs should be strengthen. Thus, it is necessary to educate the public regarding the ADRs associated with herbal medicines and importance of reporting of these ADRs. We hope that, in coming years, the pharmacovigilance of herbal medicines is also implemented in the same way as implemented for the synthetic medicines.

25.9 Conclusion

In conclusion, herbal medicines are not always safe. There are numbers of examples which raise the question on the safety of these medicines. Thus, the proper pharmacovigilance system is also required for the herbal medicines to ensure the public health.

Acknowledgments Authors are thankful to Dr. S.J.S. Flora, Director, NIPER, Raebareli, and Mr. Parveen Garg, Chairman, ISFCP, Moga, for their constant support and motivation. First author, Sai Priyanka Bhimaneni acknowledges the financial assistance received from the Department of Pharmaceuticals, Ministry of Chemicals & Fertilizers, Government of India, for carrying out this work.

References

Bandaranayake WM (2006) Quality control, screening, toxicity, and regulation of herbal drugs. In: Ahmed I, Aqil F, Owais M (eds) Modern phytomedicine: turning medicinal plants into drugs. Wiley-VCH Verlag GmbH & Co. KGaA, Weinheim, pp 25–57

Barnes J, Anderson AL (2003) In: Phillipson JD (ed) Herbal Medicines: A guide for healthcare professionals, 2nd edn. Pharmaceutical Press, London

Barnes J, Mills SY, Abbot NC, Willoughby M, Ernst E (1998) Different standards for reporting ADRs to herbal remedies and conventional OTC medicines: face-to-face interviews with 515 users of herbal remedies. Br J Clin Pharmacol 45:496–500

Bent S (2008) Herbal medicine in the United States: review of efficacy, safety, and regulation. J Gen Intern Med 23:854–859

Brown EG, Wood L, Wood S (1999) The medical dictionary for regulatory activities (MedDRA). Drug Saf 20:109–117

Chandel S, Jain N, Joshi A, Sonawane R, Sharma A (2014) Signal detection—an imperative activity of pharmacovigilance. Int J Pharm Sci Rev Res 28:95–100

Chen HY, Ma CH, Cao KJ, Ho JC, Ziea E, Wong VT, Zhang ZJ (2014) A systematic review and metaanalysis of herbal medicine on chronic obstructive pulmonary diseases. Evid Based Complement Alternat Med 2014:1. https://doi.org/10.1155/2014/925069
Cosyns JP (2003) Aristolochic acid and Chinese herbs nephropathy. Drug Saf 26:33–48
Denham A, McIntyre M, Whitehouse J (2002) Kava-kava—the unfolding story: report on a work-in-progress. J Altern Complement Med 8:237–263
Dhanalakshmi S, Harikrishnan N (2017) Pharmacovigilance of herbal drugs in poly arthritic patients. Int J Pharm 8:4396–4401
Edwards IR, Aronson JK (2000) Adverse drug reactions: definitions, diagnosis, and management. Lancet 356:1255–1259
Ekor M (2014) The growing use of herbal medicines: issues relating to adverse reactions and challenges in monitoring safety. Front Pharmacol 4:177. https://doi.org/10.3389/fphar.2013.00177
Ernst E (2000) Adverse effects of herbal drugs in dermatology. Br J Dermatol 143:923–929
Eudravigilance. pharmacovigilance in Europian Economic Area. Ema [online]. 2019. http://eudravigilance.ema.europa.eu/human/index.asp. Accessed 21 Jun 2019
Farah MH, Olsson S, Bate J, Lindquist M, Edwards R, Simmonds MSJ, Leon C, de Boer HJ, Thulin M (2006) Botanical nomenclature in pharmacovigilance and a recommendation for standardisation. Drug Saf 29:1023–1029
Gregoretti B, Stebel M, Candussio L, Crivellato E, Bartoli F, Decorti G (2004) Toxicity of Hypericum perforatum (St. John's wort) administered during pregnancy and lactation in rats. Toxicol Appl Pharmacol 200:201–205
Health Canada. 2019. www.hc-sc.gc.ca. Accessed 21 June 2019
Hitchen L (2006) Adverse drug reactions result in 250000 UK admissions a year. BMJ 332:1109
Jarernsiripornkul N, Krska J, Capps PAG, Richards RME, Lee A (2002) Patient reporting of potential adverse drug reactions: a methodological study. Br J Clin Pharmacol 53:318–325
Kamsu-Foguem B, Foguem C (2014) Adverse drug reactions in some African herbal medicine: literature review and stakeholders' interview. J Integr Med 3:126–312
Kiguba R, Ononge S, Karamagi C, Bird SM (2016) Herbal medicine use and linked suspected adverse drug reactions in a prospective cohort of Ugandan inpatients. BMC Complement Altern Med 16:145. https://doi.org/10.1186/s12906-016-1125-x
Kumar A, Khan H (2015) Signal detection and their assessment in pharmacovigilance. Open Pharm Sci J 2:66–73
Kumar A, Venkatesh MP, Kumar PTM (2016) Regulations and challenges of herbal medicines in Russia. Int J Ayurveda Herbal Med 6:2149–2169
Liu J, Manheimer E, Tsutani K, Gluud C (2003) Medicinal herbs for hepatitis C virus infection: a Cochrane hepatobiliary systematic review of randomized trials. Am J Gastroenterol 98:538–544
Mann RD, Andrews EB (2002) Pharmacovigilance. Wiley, Chichester
Mansoor GA (2001) Herbs and alternative therapies in the hypertension clinic. Am J Hypertens 14:971–975
Mills SY, Jacoby RK, Chacksfield M, Willoughby M (1996) Effect of a proprietary herbal medicine on the relief of chronic arthritic pain. A double-blind study. Rheumatology 35:874–878
Munir P, Park BK (2001) Genetic susceptibility to adverse drug reactions. Trends Pharmacol Sci 22:298–305
Park JJ, Kang M, Shin S, Choi E, Kwon S, Wee H, Nam B, Kaptchuk TJ (2010) Unexplained infertility treated with acupuncture and herbal medicine in Korea. J Altern Complement Med 16:193–198
Patel DN, Ling W, Liesbet WL, Mui-Mui T, Tan B, Zhang Q, Low MY, Chan CL, Koh HL (2012) Adverse events associated with the use of complementary medicine and health supplements: an analysis of reports in the Singapore pharmacovigilance database from 1998 to 2009. Clin Toxicol 50:481–485
Qureshi S (2012) Adverse event and drug coding in clinical research. J Clin Res Best Pract 8:1–4

Russo E, Sarro GD (2013) Pharmacovigilance in Calabria (Italy): local experiences resonate international relevance. J Pharmacol Pharmacother 4:S1

Sarris J, Panossian A, Schweitzer I, Stough C, Scholey A (2011) Herbal medicine for depression, anxiety and insomnia: a review of psychopharmacology and clinical evidence. Eur Neuropsychopharmacol 21:841–860

Sharma R, Kant K, Kumar A (2020) Overview of periodic safety update reports: where we reached? Appl Clin Res Clin Trials Regul Affairs 7(1):4–11. https://doi.org/10.2174/2213476X06666190823151514

Shaw D, Graeme L, Pierre D, Elizabeth W, Kelvin C (2012) Pharmacovigilance of herbal medicine. J Ethnopharmacol 140:513–518

Sorrentino L, Capasso A, Schmidt M (2006) Safety of ethanolic kava extract: results of a study of chronic toxicity in rats. Phytomedicine 13:542–549

Sucich M.I and Sanoski C. Herbal notes: complementary & alternative medicine pocket guide. 2011

Syed IA, Sulaiman SAS, Hassali MA, Lee CK (2015) Adverse drug reactions and quality of life in HIV/AIDS patients: advocacy on valuation and role of pharmacovigilance in developing countries. HIV AIDS Rev 14:28–30

Talbot J, Aronson JK (2012) Stephen's detection and evaluation of adverse drug reactions. Wiley-Blackwell, Chichester

Ulbricht C, Basch E, Hammerness P, Vora M, Wylie J, Woods J (2004) An evidence-based systematic review of belladonna by the natural standard research collaboration. J Herb Pharmacother 4:61–90

Wal P, Wal S, Gupta S, Sharma G, Rai AK (2011) Pharmacovigilance of herbal products in India. J Young Pharm 3:256–258

Wester K, Jonson AK, Spigset O, Druid H, Hagg S (2008) Incidence of fatal adverse drug reactions: a population based study. Br J Clin Pharmacol 65:573–579

World Health Organization. International drug monitoring: the role of national centres, report of a WHO meeting [held in Geneva from 20 to 25 September 1971]. World Health Organization; 1972

World Health Organization. WHO pharmacovigilance indicators: a practical manual for the assessment of pharmacovigilance systems. 2004a

World Health Organization. WHO guidelines on safety monitoring of herbal medicines in pharmacovigilance systems. 2004b

World Health Organization. WHO guidelines on safety monitoring of herbal medicines in pharmacovigilance systems. 2004c, p 17–19

World Health Organization. WHO guidelines on safety monitoring of herbal medicines in pharmacovigilance systems. 2004d, p 22

Zaki S (2011) Adverse drug reaction and causality assessment scales. Lung India 28:152

Herbal Drugs: Efficacy, Toxicity, and Safety Issues 26

Mithun Rudrapal and Dipak Chetia

Abstract

Herbal drugs are being in use for the management of human health and for prevention as well as to cure human diseases since ancient civilization. In recent days, the use of herbal drugs has been increased significantly in various forms such as herbal formulations, dietary supplements, and nutraceuticals in the global market. This growing demand undoubtedly proves the therapeutic claims of herbal drugs as biomedicines and/or functional foods. However, the safe use of herbal products/herbal medicines is still challenging due to the toxicity and regulatory issues. This review discusses toxicity-related and safety issues of herbal medicinal products, factors responsible for, and suitable remedial measures. Some challenges associated with monitoring the safety of herbal drugs are also discussed to ensure their effectiveness for adequate protection of public health and the relevant regulatory issues.

Keywords

Herbal drugs · Traditional medicine · Quality · Efficacy · Toxicity · Safety · Standardization · Validation

M. Rudrapal
Srikrupa Institute of Pharmaceutical Sciences, Siddipet, Telangana, India
e-mail: rsmrpal@gmail.com

D. Chetia (✉)
Department of Pharmaceutical Sciences, Dibrugarh University, Dibrugarh, Assam, India
e-mail: dchetia@dibru.ac.in

S. C. Mandal et al. (eds.), *Evidence Based Validation of Traditional Medicines*,
https://doi.org/10.1007/978-981-15-8127-4_26

Abbreviations

CAM	Complementary and alternative medicines
CVS	Cardiovascular system
DSHEA	Dietary Supplement Health and Education Act of 1994
FDA	Food and Drug Administration
GC-MS	Gas chromatography-mass spectrometry
GMP	Good manufacturing practices
HMPs	Herbal medicinal products
LC-MS	Liquid chromatography-mass spectrometry
TM	Traditional medicine
WHO	World Health Organization

26.1 Introduction

Herbal drugs have been used to maintain human health and to prevent and/or cure human diseases since ancient civilizations. Herbal medicinal preparations predate a recorded history with their usage in traditional Greek, Egyptian, Indian, and Chinese medicine for a variety of therapeutic purposes (Ang-Lee et al. 2001; Fabricant and Farnsworth 2001). In developing world, herbal drugs are increasingly used as a part of traditional medical practices in addition to complementary and alternative medicines (CAM). In recent years, the consumption of herbal medicines has been increased in developed countries including Europe, the United States, and Australia (Fabricant and Farnsworth 2001; Jamshidi-Kia et al. 2018). According to the estimates of World Health Organization (WHO), about 80% of the world's population relies mainly on traditional medicines (TM) including botanicals and herbal drugs for their primary healthcare needs (Ganesan 2008; Farnsworth et al. 1985). Over the past few decades, the use-consumption of botanicals such as phytomedicines, herbal drugs, dietary supplements, nutraceuticals, cosmetics, and other products has been dramatically increased across the globe (Jamshidi-Kia et al. 2018). This is because of the fact that botanical drugs are naturally derived and, therefore, believed to be safe. Also, these are locally available and cheap as compared to conventional Western medicines (Cragg et al. 1997; Kong et al. 2003). Herbal medicines are, therefore, considered as a rational approach of alternative therapeutic measure for diseases of human and animals. Owing to the growing demand, the sales and trades of herbs and herbal drugs/drug products have been dramatically increased with significant proportion in the global market. In a recent report, a huge expenditure (about 100 billion in US dollars) has been accounted for the use of herbal drugs as primary remedies at home and also the over-the-counter medicines (Cordell and Colvard 2012; Pal and Shukla 2003).

As the global consumption of herbal medicines is continuously increasing with the introduction of many new herbal products into the market, public health faces challenges. Herbal drugs sector, therefore, requires a more detailed guideline from the regulatory bodies to ensure the quality, efficacy, and safety of herbal medicines.

Although herbal medicines possess promising level of therapeutic efficacy, they lack valid scientific and clinical evidences to comply their efficacy and safety profile. It is a challenging task to develop scientifically validated and standardized herbal drug products. Some guiding protocols have been developed for the standardization and/or quality evaluation of herbal drugs. However, practical issues related to their safety and toxicity still exist. In this chapter, the therapeutic efficacy, toxicity issues, and safety concerns that are commonly encountered for the effective and safe use of herbal drugs/medicines are discussed. This review also discusses some significant challenges and relevant regulatory issues associated with monitoring of safety of herbal drugs to ensure their overall effectiveness for adequate protection of public health.

26.2 Herbal Drugs and Related Terms

26.2.1 Terms and Definitions

The following terms and definitions related to herbal drugs/herbal medicines/herbal products are discussed as per WHO guidelines (WHO 2004) on traditional medicines.

Herbs: A herb is a plant or plant part generally used for its flavor, scent, or health benefits. Herbs include crude plant materials such as leaves, flowers, fruit, seeds, stems, bark, wood, roots, rhizomes, or other plant parts, which may be whole plant, fragmented part, or powdered form of the plant parts.

Herbal materials: Herbal materials include herbs, fresh juices, gums, fixed oils, essential oils, resins, and dry powders of herbs. In some countries, the herbal materials may be subjected for processing by various local procedures, such as roasting, steaming, and fermentation.

Herbal preparations: These are the finished herbal products and/or herbal formulations and may include grounded or powdered herbal materials, herbal extracts, tinctures, and volatile or fatty oils of herbal materials. Such preparations are produced by means of extraction, fractionation, distillation, purification, concentration, and other processes/techniques like physical or biological procedures.

Herbal products: Herbal products are herbal medicines derived from plants or plant parts or herbs. They are used to improve health and physical/mental well-being and may also be used for other therapeutic claims. Such products are available as powders, tablets, capsules, teas, extracts, decoction and fresh or dried plants, and so on. The three most common types of herbal medicines available globally are *Ayurvedic* medicine (Indian), Chinese herbal medicine (Chinese), and Western medicine (European).

Herbal formulations: Herbal formulations are the finished labeled herbal products that contain active ingredients such as aerial or underground plant parts or other plant material or combinations of plant parts (e.g., leaves, flowers, barks, or roots), either in the crude form or as processed plant preparations (fresh or dried).

Herbal drug/herbal medicine: A plant or plant part or an extract or mixture of these which is used in the form of medicinal preparation to treat illness and promote health is known as **herbal drug** or **herbal medicine**. It is also called **botanical medicine**, **phytomedicine**, or **phytotherapy**. Herbal medicine includes herbs, herbal materials, herbal preparations, and finished herbal products or herbal formulations.

According to the definition of WHO, herbal drugs contain plant parts as active ingredients or plant materials in the crude form or processed state along with certain excipients, such as solvents, diluents, binders, preservatives, etc. The active chemical principles responsible for the biological action of herbal drugs are unknown. Other important characteristics of herbal medicines are their diverse therapeutic uses and wide acceptance by the population. Formulation with active chemical ingredients or any combination with isolated constituents is not considered to be herbal medicines.

Botanicals: Botanicals include herbal materials (plant or their component parts as discussed above), algae, fungi, and/or their combinations. They are intended for the prevention of illness and promoting healing. **Botanical drugs** are the botanicals intended for use as drugs with therapeutic claim. Herbs or herbal medicines fall into the general category of botanicals. **Botanical products** are diverse and classified as herbal medicines, dietary supplements, nutraceuticals, phytopharmaceuticals, food, cosmetics, and beverages.

Phytomedicines: They are also known as **phytotherapeutic agents**. They are standardized preparations of herbals drugs consisting of complex mixtures of one or more herbs or plant materials. Phytomedicines are used in the treatment of various diseases.

Polyherbal formulation (PHF): It is the use of more than one herb in an herbal medicinal preparation. The concept is found in *Ayurvedic* and other traditional medicinal systems where multiple herbs in a particular ratio may be used in the treatment of a diverse range of human illness. The concept behind polyherbalism is to exert maximal therapeutic efficacy with reduced toxicity. It is insufficient to achieve the desired therapeutic effects by the active phytochemical constituents of plants.

26.2.2 Herbal Drugs vs. Conventional Drugs

In developing countries, many of the prescription drugs available today are botanical drugs (drugs derived from plants). Herbs or botanicals are the natural chemicals within a plant. Either the extract is administered in its original form, sometimes combined with other herbal extracts, or it is refined or purified. On the other hand, a conventional drug is derived from a chemical that is synthesized or isolated in a laboratory, even though it may have originally derived from plant. The conventional drugs are based on active pharmaceutical ingredients. Unlike herbal drugs, the use of conventional drugs must be approved by the Food and Drug Administration (FDA)

Table 26.1 Differences between herbal drugs and conventional drugs (Cordell and Colvard 2012; Yuan et al. 2016)

Herbal drugs	Conventional drugs
Herbal drugs are finished, labeled medicines containing pharmacologically active parts of a plant or plants, either in the crude state or processed form (physically modified during processing)	Conventional drugs are either pharmacological single entities which have been derived by chemical synthetic procedures or single chemical derivatives of naturally occurring pharmacologically active substances and isolated from plant, fungus, bacteria, or animal
They are prescribed often as chemically complex remedies and administered in the form of multicomponent formulations	They are rather prescribed as a single agent or more often combined with other conventional drugs. This is done either to enhance the desired clinical effect and/or to alleviate the unwanted side effects of the principal drug
They are usually administered by oral route or in the topical dosage form to cure diseases (acute/chronic) and are available as crude preparations (fresh or dried plants, teas, or extracts) or as powders, tablets and capsules, and many others	They are given to patients by oral route topically or by parenteral administration to treat disorders in chronic or acute state and are usually available as powders, tablets, capsules, liquids, and many other pharmaceutical formulations
Herbal therapy is usually combined with advices related to lifestyle such as changes in diet, exercise, etc. This is supposed to enhance the potency of the herbal drugs by boosting the patient's innate self-healing processes or immune system and natural defense system	Conventional therapy is used to alleviate the patient's disease causing symptoms, viz., the relief of pain, depression, digestive disorders, and movement problems, and restore the presenting signs to normal. There is, however, no real or urgent imperative to rectify underlying causes of the disorder, whether lifestyle, environmental, or behavioral in origin
Herbal products contain many active chemical principles which can act in combination or synergistically. The basic pharmacological response may be different from when one active ingredient acts alone, as the primary mechanism may be intensified/potentiated by individual mechanism	Conventional drugs are used generally as single chemical or therapeutic entity. It is mainly responsible for the principal pharmacological response or therapeutic indication with well-defined mechanism of action at biological matrix
Herbal products usually have several different pharmacological/therapeutic actions, and parameters that predominate the therapeutic efficacy are dosage employed, the part of plant used, and also the presence of other active components	A single conventional drug may only have one major direct action on a particular receptor site. Indeed, it is more specific for one type of receptor involved which is in turn believed to be the preferred outcome of research for new therapeutic agents

(Yuan et al. 2016). Table 26.1 describes some important differences between herbal drugs and conventional medicines.

26.3 Efficacy, Toxicity, and Safety Issues

Herbal medicinal products (HMPs) are being released into the market without performing any standardized efficacy, safety, or toxicity evaluation studies. This is because of less availability of appropriate guiding protocols and/or methodologies for the standardization and evaluation (analytical/biological) of herbal drugs. It is, therefore, difficult to execute standardization, evaluation, or quality control tests on herbal drugs in order to ascertain their quality, efficacy, and safety profile. Due to this reason, herbal products available at consumer level produce, in many cases, some serious health hazards like adverse drug reactions and/or toxic effects.

26.3.1 Efficacy

Herbal medicines usually contain a diverse range of biologically active components. Herbal medicinal preparations contain up to 150 active constituents (Cordell and Colvard 2012; De Smet 1995). It is believed that herbal remedies generally work due to biochemical action of the whole plant material and/or mixtures of plant parts, and not due to only one particular active ingredient (Roberts and Tyler 1997). The therapeutic activity of herbal medicines does not depend on a particular component or active ingredient that is present in the herbal drug product and/or herbal formulation. It is firmly believed that isolated active principle has weaker efficacy than that of entire plant material and/or crude plant extracts (Fetrow and Avila 2004). There are many scientific studies on the efficacy of certain herbal medicines in human disease indications, but they do not possess valid clinical evidences (Ernst 2005).

26.3.2 Toxicities and Adverse Effects

Clinical evidences suggest that adverse effects or toxicities of herbal drugs are relatively less intense as compared to conventional medicines or synthetic drugs (Parle and Bansal 2006; Zhang et al. 2015). In clinical practice, the occurrence of herbal drug toxicities is also not very routine. The incidence of side effects is normally less severe which affects a small population (Drew and Myers 1997). Herbal medicines are, therefore, generally considered safer and less harmful over synthetic drugs. However, some herbs, for instance, comfrey, ephedra, and kava, can cause heart attacks, strokes, and hepatotoxicity (Ekor 2014; D'Arcy 1993). Most toxicities of herbal drugs involve the skin, liver, GI tract, and heart. Adverse effects related to herbal drugs are mainly attributed to be due to predictable toxic effects including allergic reactions, overdosage, and herb-drug interactions. One more attribute is that problems due to preparation and formulation such as improper identification of plant species, lack of protocol for standardization, inadequate good manufacturing practices, adulteration and contamination of plants/plant parts, improper preparations and dosage, etc. (Ekor 2014; George 2011).

26.3.3 Herb-Drug Interactions

With the increasing consumption of herbal drugs, the incidence of herb-drug interactions has attained a serious health concern worldwide. According to global regulatory authorities such as Therapeutic Goods Administration and US Food and Drug Administration (US FDA), the increasing issue of herb-drug interactions is believed to be an important aspect as far as the safety of herbal drugs is concerned (Bland 1998; Coxeter et al. 2004). Herbal drugs can interact with conventional over-the-counter drugs. Herbal medicine may also interact with other herbal preparations causing alteration of therapeutic response of both the drugs (Bushra et al. 2011; Herr et al. 2002). In a research, it has been reported that about 50% of patients consuming herbal medicines are cancer patients due to the increasing incidence of toxicities of cancer medicine (Izzo and Ernst 2011). Herbal remedies can either potentiate or antagonize some drug from other conventional therapies (Abubakar et al. 2015). However, common remedial measure is to avoid the use of a drug or an herb that is associated with potential toxic effects (Fugh-Berman and Ernst 2001; Otles and Senturk 2014).

26.3.4 Evidence of Clinical Efficacy and Toxicity

Scientific and clinical evidences available in literature support the efficacy, toxicities, and herb-drug interactions for herbal drugs. Evidences may be indirectly based on some preclinical efficacy studies of the herbal drugs and/or herbal drug's active constituents in in vitro models and in vivo experimental animals. Evidences are directly based upon some clinical observations such as spontaneous clinical reports or case studies, drug metabolism studies, and drug interactions in patients and healthy individuals (Mukherjee et al. 2006). Preclinical data (in vitro or in vivo) may provide required evidence about the clinical efficacy, toxicities, and herb-drug interactions of herbal drugs. Controlled clinical studies may put light on clinical efficacy, possible adverse effects, and drug interactions of herbal medicines. Case reports may also be a valid evidence of certain clinically important herb-drug interactions (Zhi-Xu et al. 2014). Herbal drugs are generally studied by well-controlled double-blind trials to ensure their efficacy and safety. Scientific evidences from randomized and controlled double-blind clinical trials are thus available for certain herbal medicines (Boullata and Nace 2000; Ekor 2014). The efficacy of some common herbs and herbal drugs (used in traditional medical practices), their adverse responses, and interactions are depicted in Tables 26.2 and 26.3.

26.4 Efficacy, Toxicity, and Safety Assessment

The multicomponent character of herbal drugs renders their evaluation efficacy comparatively more tedious and complex than that of isolated active chemicals. In this case, the entire plant extract is considered as the bioactive principle. The

Table 26.2 Efficacy of some common herbs and herbal drugs (Ernst 2005; Staines 2011)

Herbal drug	Effect/clinical efficacy[a]
Ginger (*Zingiber officinale* L.)	Inhibits platelets aggregation (antiplatelet)
Garlic (*Allium sativum* L.)	Decreases low-density protein cholesterol (in hypercholesterolemia) and beneficial in hypertension and some disturbances of CVS (anti-lipidemic and anti-hypertension) and inhibition of progression of certain cancer (anticancer)
Ginkgo (*Ginkgo biloba* L.)	Useful in dementia and tinnitus
Ginseng (*Panax ginseng* C.A. Meyer)	Improves physical performance, beneficial in diabetes and cognitive dysfunction, helps in immunomodulation, and acts to inhibit herpes simplex infection
Aloe vera (*Aloe barbadensis* L.)	Useful as antidiabetic and anticoagulant medication
Licorice (*Glycyrrhiza glabra*)	Causes salt and water retention
Feverfew (*Tanacetun parthenium*)	Used to cure migraine headache
Chamomile (*Matricaria chamomilla*)	Used as a carminative, anti-inflammatory agent, and antispasmodic
Valerian (*Valeriana officinalis*)	Used as a sedative-hypnotic to induce sleeping
Kava (*Piper methysticum*)	Used as an anxiolytic
Silybium marianun (milk thistle)	Used for repairing liver cirrhosis
Echinacea (*Echinacea purpurea*)	Used as anti-inflammatory and immunostimulant

[a]Strength of evidence: No trial data available/uncertain efficacy in almost all cases

evaluation of efficacy is generally achieved by selecting the key active ingredient of the herbal drugs and/herbal drug products, or, if such an ingredient is not known, a suitable marker component is preferably selected (Mukherjee et al. 2013). There are some prescribed guidelines used to investigate the safety and/or efficacy of herbal drugs (Akerele 1993; WHO 2005a). However, following the procedure prescribed in the WHO guidelines, limited standardization of herbal drugs could be achieved, where one or two ingredients can be standardized and rest of ingredients may remain unattended. It could largely affect both the efficacy and safety of herbal drugs/drug products. Therefore, complete product characterization and/or quality control tests are essential for the overall assessment of standardization of herbal drugs (Fig. 26.1). It could also be followed to regulate as well as establish further the quality standards of crude herbal drugs and/or finished herbal preparations.

Table 26.3 Some common adverse effects of herbal drugs and herb-drug interactions (Staines 2011; Tasneem 2012; Thomas et al. 2012)

Herbal drug(s)	Conventional drug(s)/ drug class	Interactions/effects
Garlic, ginger, gingko, feverfew, saw palmetto, guarana, *Passiflora*, cranberry, evening primrose oil	Anticoagulants (warfarin), anti-platelets (aspirin)	Increase anticoagulant effects of warfarin
Ginseng, green tea, St. John's wort, chamomile		Decrease anticoagulant activity of warfarin
Garlic, ginseng, karela, eucalyptus, fenugreek, St. John's wort	Antidiabetic/ hypoglycemic agents (insulin, sulfonylureas, biguanides, etc.)	Hypoglycemia
Ephedra, licorice		Hyperglycemia
Licorice	Diuretics (spironolactone)	Antagonism of diuretic effect
St. John's wort	Antiretrovirals, digoxin, theophylline, cyclosporin, oral contraceptives	Decreases clinical effect
Ginseng	Estrogens, corticosteroids	Additive effects
Licorice	Oral contraceptives	Increases the risk of fluid accumulation, blood pressure
Ginseng		Additive estrogenic effects
Feverfew, garlic gingko	Nonsteroidal anti-inflammatory drugs	Inhibition of herbal effect, increase risk of bleeding
Evening primrose oil, gingko, ephedra	Anticonvulsants	Lower seizure threshold, increase risk of seizures, decrease drug effect
Licorice, hawthorn, cascara, cassia, aloe, senna	Digoxin, laxatives	Increase risk of hypokalemia
Ginseng		Elevates digoxin levels
St. John's wort		Decreases digoxin levels
St. John's wort, gingko	Antidepressants	Increase serotonin levels
Valerian, ginseng, St. John's wort	Opioid analgesics	Additive CNS depression, reduce analgesic effectiveness
Kava, hawthorn, hops, St. John's wort, valerian	Anesthetics or sedative-hypnotics (benzodiazepines)	Additive-sedative effects, reduce effectiveness
Ephedra, guarana, licorice, ginseng	Anti-hypertensives	Hypertension
St. John's wort		Reduces effectiveness of some anti-hypertensives
Echinacea, licorice, St. John's wort	Immunosuppressants	Reduce immunosuppressant levels
Garlic, St. John's wort, milk thistle, gingko ginseng, *Echinacea* – used for only short duration	Anti-HIV drugs	Decrease concentration of protease inhibitor, increase the risk of antiretroviral resistance

(continued)

Table 26.3 (continued)

Herbal drug(s)	Conventional drug(s)/ drug class	Interactions/effects
Cranberry, gingko, St. John's wort	Proton-pump inhibitors (PPIs)	Reduce effectiveness of PPIs
Dandelion, fennel	Antibiotics	Decrease effectiveness of fluoroquinolones
Cinnamon		Decreases effectiveness of tetracyclines
St. John's wort		Increases risk of phototoxicity with tetracyclines

Fig. 26.1 Quality, efficacy, and safety cycle

26.4.1 Physicochemical Assessment

As a general consideration, herbal medicine must of acceptable quality and be safe as per the prescribed standards and guidelines. The quality control may be done by assessing identity, purity or assay, and other physical chemical or biological parameters (Shulammithi et al. 2016). Standardization of herbal drugs are, therefore, essential for the final assurance of their efficacy and safety. Standardization of botanical drugs is defined as the process of documenting a set of guidelines/ standards and qualitative as well as quantitative parameters which assure the quality aspects including stability, efficacy, safety, and toxicity of herbal drugs (Bijauliya et al. 2017). The standardization of botanicals is done by the identification of botanicals, identification of active component(s), identification of common adulterants or contaminants, and systematic quality evaluation of crude botanicals (Shukla et al. 2009).

However, despite the use of modern analytical tools and techniques, scientific validation of herbal drugs still remains to be a challenging task. The efficacy is defined by the active ingredients (i.e., phyto-constituents) of herbal drugs and/or herbal formulations. Compared with synthetic molecules, the physicochemical determination of herbal drugs, particularly in terms of thorough phytochemical investigations, involves a more tedious and rigorous procedure due to the multi-ingredient nature of herbal products. In most of the cases, phyto-constituents are present at negligible amount and also vary in content that may interfere with the quality control process (Bauer and Tittel 1996). Moreover, phytomedicines are mixtures of many components which make them difficult to characterize. In the absence of active chemical principles, suitable marker component(s) can be used for analytical studies to establish and/or assess the efficacy profile of herbal drugs (Nikam et al. 2012). The quality of herbal products has to be assured at all stages of manufacturing starting from cultivation, harvesting and procurement of raw materials, herbal materials processing, and finally production of finished herbal formulations (Calixto 2000). The fixation of criteria for quality of herbal drugs is, therefore, based on evidenced-based scientific investigations of the crude or raw plant materials, plant preparations, and finished products.

The source and processing of raw materials play a significant role to assure the quality and stability of herbal drugs. Some atmospheric and other factors such as the temperature, exposure to light, availability of water, nutrients, harvesting period and time, method of collection, plant part(s) collected, processing for drying, packing condition, storage and transportation of raw material, etc. can considerably influence the quality and consequently the therapeutic potential of herbal materials (Patra et al. 2012). Apart from these, others factors related to processing such as the extraction methods, production processes, and quality control process can also influence the quality assessment of herbal drugs. In quality control tests, the botanical description including genus and species, plant profile, plant part(s) used, and active ingredients should be systematically depicted. Type(s) of herbal preparation, sensory parameters, physical constants, moisture content, extractive values, ash values, and presence of contaminants or adulterants should be thoroughly determined to identity and also to assess purity of herbal drugs. Microbiological contamination, impurities and foreign substances, such as toxic heavy metals, pesticide(s) or pesticide residues, aflatoxins, and radioactive contaminants are also required to be determined (Nikam et al. 2012; Shulammithi et al. 2016). The application of appropriate modern analytical tools and techniques such as UV-Vis spectroscopy, TLC, HPTLC, HPLC, GC, mass spectrometry (MS), GC/MS, LC/MS, etc. is necessary in order to establish certain standards for the qualitative as well as quantitative evaluation of herbal dugs/preparations.

26.4.2 Biological Assessment

Biological standardization includes screening for potency, toxicity, or safety in various experimental models such as in vitro techniques, screening using cell lines, and in vivo studies in animals. The efficacy screening is preliminarily done for various pharmacological evaluations such as anti-inflammatory, antiulcer, hepatoprotective, antidiabetic, antitumor, insecticidal, antiparasitic, neuropharmacological, cardioactive property, and so on in statistically well-defined animal models. For antimicrobial efficacy of plant materials, the screening is carried out using methodologies described under WHO guidelines for the assay of antimicrobial herbal drugs (Patra et al. 2012; Schulz et al. 2001). In clinical stage, clinical trials are generally carried out to assess the therapeutic efficacy as well as toxicity using outcomes of several clinical, laboratory, or diagnostic studies (Ekor 2014).

Toxicity investigations of herbal drugs involve several techniques such as in vivo cell culture, in vivo technique using animal models, and micro-array techniques. Toxicity studies are usually carried out in laboratory animals to assess various levels of toxicities such as acute, subacute, and chronic toxicities. In assessing so, the dose chosen is very important. Carcinogenicity, hepatotoxicity, cardiotoxicity, immunogenicity, and also reproductive toxicity are generally investigated (Nasri and Shirzad 2012). However, there are some clinical evidences of the occurence of serious physiological adverse effects after the consumption of certain herbal products. In many cases, the toxicities have been attributed mainly to be due to the inherent toxicity of plant constituents and ingredients, manufacturing malpractices, and contaminations or adulterations (George 2011). In addition to adverse effects, some serious drug-drug and drug-food interactions have also been investigated from animal studies and clinical trials for certain commonly used herbal medicines (Bushra et al. 2011). Due to the inadequate evidences and reports, the biological evaluation of toxic effects and herb-drug still, however, requires thorough phytochemical, toxicological, and pharmacological investigations.

26.5 Safety Monitoring and Challenges

Safety is a fundamental requirement as far as quality control of herbal drugs is concerned. With the enormous global usage of herbal drugs, toxicity and safety issues are the main risk factors associated with herbal medicines (WHO 2004). The safety has, therefore, become a matter of great concern in the domain of public health. Increasing cases of adverse events and toxicities urge thorough toxicological investigation on herbal drugs along with active pharmacovigilance program. According to WHO Programme for International Drug Monitoring, the safety monitoring is an inevitable requirement for herbal drugs together with thorough toxicity and safety assessment (Rodrigues and Barnes 2013; Shetti et al. 2011).

Adverse effects of herbal drugs are commonly attributed to be due to several factors such as the use of the wrong plant species, adulteration of herbal materials, admixture with toxic elements or other contaminants, overdosage of drugs, improper use of herbal medicines, and simultaneous use of herbal drugs with other medications (Saxena et al. 2008; Ekor 2014). Some challenges that are commonly encountered for monitoring safety of herbal drugs (George 2011; Zhang et al. 2015; Wu et al. 2008; Warude and Patwardhan 2005) are enumerated below:

- Lack of proper knowledge on herbal medicines: Medicine providers such as doctors, nurses, and pharmacists may have little knowledge regarding the use of herbal remedies and detailed knowledge on their health implications. It is suggested to provide them with adequate training since most patients are not acquainted proper usage of medicines. In fact, educating the healthcare professionals, herbal practitioners, herbal medicine providers, and patients or consumers is necessary for combating serious risks resulting from the herbal medicine misuses.
- Patient or consumer attitudes toward herbal medicines: Herbal medicine consumers or patients generally believe that herbal drugs are absolutely safe and they do not possess any potential risk of toxicities. Patients who take herbal medicines and conventional medicines together often do not inform so to their doctors. It creates some unavoidable herb-drug interactions.
- Difficulty in standardization of herbal drugs: For safe and effective use of herbal medicines, proper standardization is necessary for achieving consistency in chemical and biological properties of herbal materials. However, herbal drugs seldom meet the quality standards because of several hurdles encountered during standardization and/validation studies described in above sections.
- Inadequate quality of herbal preparations: If an herbal product or formulation is effective, quality assurance is needed to be ensured that the product has the expected effects. Even in the absence of data on efficacy, quality assurance is important. The quality is an important determinant of safety as well.
- Adulteration: Some of the common adulterants that are found in plant materials or botanicals are other plant materials, botanicals, heavy metals, microbial residues, and pesticides.

26.6 Regulatory Status

Currently, several regulatory guidelines for the use of herbal medicines exist, which include prescription medicines, traditional remedies, and dietary supplements. In fact, the regulatory situation for herbal drugs varies from country to country. As per regulations of food and medicines, a botanical preparation may be used as a food, a dietary supplement, a functional food, and an herbal medicine. Herbal preparations used for diagnosis, cure, mitigation, treatment, or prevention of diseases are considered and regulated as drugs (Bandaranayake 2006; Warude and Patwardhan 2005). Botanical products and/or natural products are made marketed as dietary supplements as per the regulation of Dietary Supplement Health and Education Act of 1994 (DSHEA 1994). A dietary supplement is defined as a product that is intended to supplement the normal diet and contains some ingredients of diet. The dietary ingredients may include vitamins, minerals, herbs, or other botanical preparations (Ekor 2014; Ernst 2011). To perform toxicities, the FDA assures certain guidelines for an herbal product or dietary ingredient whether it is toxic or not safe for use. WHO has set regulatory guidelines (Fig. 26.2) for evaluating the quality, efficacy, and safety of herbal medicines (WHO 2005b). Pharmacopoeial

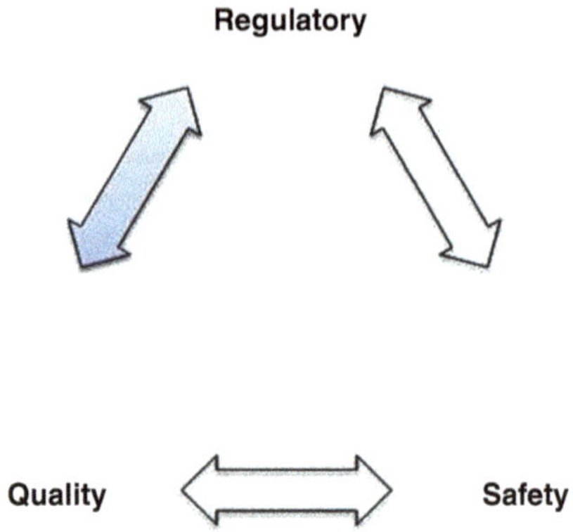

Fig. 26.2 Regulatory aspects

monographs have set guidelines for the quality assessment of herbal medicines. WHO has also urged the governments to establish regulatory mechanisms to control the quality and safety of herbal products (WHO 2004).

However, there are several increasing regulatory concerns in relation to research applications and commercialization of herbal medicines, particularly in developing countries. Though several regulatory guidelines are available, among various traditional medicinal practices, herbal medicines represent some unique challenges in research and regulations. Challenges often encountered are those related to regulatory requirements, safety monitoring, and inadequate or poor knowledge about traditional, complementary/alternative, and herbal medicines (Calixto 2000; Wu et al. 2008). Therefore, the development of regulatory standards for the analysis of herbal drugs is necessary on a global and/or regional scale. To address the safety issues, the development of new regulations for herbal therapies is, therefore, urgently required.

26.7 Conclusion

In recent times, the global consumption of herbal drugs is increasing. Issues relating to toxicity and safety of herbal drugs are also becoming a growing concern. It is believed to be due to the misbelief that herbal medicines are safe because of their natural origin. Such attributes are due to cultural aspects, inadequate scientific studies (standardization and validation protocol), and the weak regulatory mechanism of herbal drugs in most of the countries.

In this chapter, toxicity issues and major safety concerns related to herbal medicines are briefed. Since safety continues to be a major challenging issue, it is, therefore, imperative to create increased awareness program for understanding and monitoring the safety as well as toxicity of herbal medicines. To do so, it is urgently required to gather sufficient knowledge of possible harmful effects and/or adverse reactions and their mechanism of action and possible interactions with conventional drugs and functional foods associated with the use of herbal remedies. Regulatory policies are required to be standardized and strengthened on a global scale. Regulatory authorities may implement appropriate measures that ascertain the quality and safety of herbal products before getting released into the market. It is also important to create adequate awareness among the general public, healthcare providers, policymakers, and manufacturers with sufficient information for understanding of the risks or toxicities linked with the consumption of herbal drugs. Moreover, monitoring the safety of herbal medicine would require effective collaboration among botanists, phytochemists, pharmacologists, toxicologists, and other major stakeholders. However, appropriate scientific and regulatory measures must be implemented for promoting the rational and safe use of herbal medicines and/or to protect health of public in general by ensuring that all herbal drugs are safe and efficacious and of acceptable quality.

References

Abubakar AR, Chedi BAZ, Mohammed KG, Haque M (2015) Drug interaction and its implication in clinical practice and personalized medicine. Natl J Physiol Pharm Pharmacol 5:343–349

Akerele O (1993) Summary of WHO guidelines for the assessment of herbal medicines. HerbalGram 28:13–19

Ang-Lee MK, Moss J, Yuan CS (2001) Herbal medicines and perioperative care. J Am Med Assoc 286:208–216

Bandaranayake WM (2006) Quality control, screening, toxicity, and regulation of herbal drugs modern Phytomedicine. In: Ahmad I, Aqil F, Owais M (eds) Turning medicinal plants into drugs. Wiley-VCH Verlag GmbH & Co. KGaA, Weinheim, pp 25–57

Bauer R, Tittel G (1996) Quality assessment of herbal preparations as a precondition of pharmacological and clinical studies. Phytomedicine 2:193–198

Bijauliya R, Alok S, Chanchal DK, Kumar M (2017) A comprehensive review on standardization of herbal drugs. Int J Pharm Sci Res 8:3663–3677

Bland SE (1998) Drug-food interactions. J Pharm Soc Wisconsin 14:S28–S35

Boullata JI, Nace AM (2000) Safety issues with herbal medicine. Pharmacotherapy 20:257–269

Bushra R, Aslam N, Khan AY (2011) Food-drug interactions. Oman Med J 26:77–83

Calixto JB (2000) Efficacy, safety, quality control, marketing and regulatory guidelines for herbal medicines (phytotherapeutic agents). Braz J Med Biol Res 33:179–189

Cordell GA, Colvard MD (2012) Natural products and traditional medicine: turning on a paradigm. J Nat Prod 75:514–525

Coxeter PD, McLachlan AJ, Duke CC, Roufogalis BD (2004) Herb-drug interactions: an evidence based approach. Curr Med Chem 11:1513–1525

Cragg GM, Newman DJ, Snader KM (1997) Natural products in drug discovery and development. J Nat Prod 60:52–60

D'Arcy PF (1993) Adverse reactions and interactions with herbal medicines. Part 2—drug interactions. Adverse Drug React Toxicol Rev 12:147–162

De Smet PAGM (1995) Health risks of herbal remedies. Drug Saf 13:81–93

De Smet PAGM (1997) The role of plant derived drugs and herbal medicines in healthcare. Drugs 54:801–840

Drew A, Myers SP (1997) Safety issues in herbal medicine: implications for the health professions. Med J Aust 166:538–541

Ekor M (2014) The growing use of herbal medicines: issues relating to adverse reactions and challenges in monitoring safety. Front Pharmacol 4:1–10

Ernst E (2005) The efficacy of herbal medicine—an overview. Fundam Clin Pharmacol 19:405–409
Ernst E (2011) Herbal medicinal products: an overview of systematic reviews and meta-analyses. Perfusion 14:398–404
Fabricant DS, Farnsworth NR (2001) The value of plants used in traditional medicine for drug discovery. Environ Health Perspect 109:69–75
Farnsworth NR, Akerele O, Bingel AS, Soejarto DD, Guo Z (1985) Medicinal plants in therapy. Bull World Health Organ 63:965–981
Fetrow CW, Avila JR (2004) Professional's handbook of complementary and alternative medicines. Lippincott Williams & Wilkins, Philadelphia
Fugh-Berman A, Ernst E (2001) Herb-drug interactions: review and assessment of report reliability. Br J Clin Pharmacol 52:587–595
Ganesan A (2008) The impact of natural products upon modern drug discovery. Curr Opin Chem Biol 12:306–317
George P (2011) Concerns regarding the safety and toxicity of medicinal plants—an overview. J Appl Pharm Sci 1:40–44
Herr SM, Ernst E, Young VSL (2002) Herb-drug interaction handbook. Church Street Books, New York
Izzo AA, Ernst E (2011) Interactions between herbal medicines and prescribed drugs: a systematic review. 15:2163–2175
Jamshidi-Kia F, Lorigooini Z, Amini-Khoei H (2018) Medicinal plants: past history and future perspective. J Herbmed Pharmacol 7:1–7
Kong JM, Goh NK, Chia LS, Chia TF (2003) Recent advances in traditional plant drugs and orchids. Acta Pharm Sin 24:7–21
Mukherjee PK, Harwansh RK, Bahadur S, Banerjee S, Kar A (2006) Evidence based validation of Indian traditional medicine—way forward. World J Tradit Chin Med 2:48–61
Mukherjee PK, Harwansh RK, Bahadur S (2013) Standardization and chemoprofiling of herbal medicine through marker analysis. Pharm Rev 58:161–163
Nasri H, Shirzad H (2012) Toxicity and safety of medicinal plants. J Herbmed Pharmacol 2:21–22
Nikam P, Kareparamban J, Jadhav A, Kadam V (2012) Future trends in standardization of herbal drugs. J Appl Pharm Sci 2:38–44
Otles S, Senturk A (2014) Food and drug interactions: a general review. Acta Sci Pol Technol Aliment 13:89–102
Pal SK, Shukla Y (2003) Herbal medicine: current status and the future. Asian J Cancer Prev 4:281–288
Parle M, Bansal N (2006) Herbal medicines: are they safe? Nat Prod Rad 5:6–14
Patra KC, Pareta SK, Harwansh RK, Kumar KJ (2012) Traditional approaches towards standardization of herbal medicines—a review. J Pharm Sci Technol 2:372–379
Roberts JE, Tyler VE (1997) Tyler's herbs of choice. The therapeutic use of phytomedicinals. The Haworth Press, New York
Rodrigues E, Barnes J (2013) Pharmacovigilance of herbal medicines: the potential contributions of ethnobotanical and ethnopharmacological studies. Drug Saf 36:1–12
Saxena A, Tripathi KP, Roy S, Khan F, Sharma A (2008) Pharmacovigilance: effects of herbal components on human drugs interactions involving cytochrome P450. Bioinformation 3:198–204
Schulz V, Hänsel R, Tyler VE (2001) Rational phytotherapy. A physician's guide to herbal medicine. Springer, Berlin
Shetti S, Kumar CD, Srivastava NK (2011) Pharmacovigilance of herbal medicines: current state and future directions. Pharmacogn Mag 7:69–73
Shukla SS, Saraf S, Saraf S (2009) Approaches towards standardization and quality assessment of herbals. J Res Edu Indian Med 15:25–32
Shulammithi R, Sharanya M, Tejaswini R, Kiranmai M (2016) Standardization and quality evaluation of herbal drugs. IOSR J Pharm Biol Sci 11:89–100

Staines SS (2011) Herbal medicines: adverse effects and drug-herb interactions. J Malta Coll Pharm Pract 11:38–42

Tasneem TSS (2012) Drug-herb interactions. Int J Pharmaceut Chem Biol Sci 3:75–78

Thomas AS, Varughese P, Shirwaikar A, Shirwaikar A (2012) Herb-drug interactions: a review. Hygeia J D Med 4:33–40

Warude D, Patwardhan B (2005) Botanicals: quality and regulatory issues. J Sci Ind Res 64:83–92

WHO (2004) WHO guidelines on safety monitoring of herbal medicines in pharmacovigilance systems. World Health Organization, Geneva

WHO (2005a) WHO global atlas of traditional, complementary and alternative medicine. In: Ong CK, Bodek G, Grundy CB, Kin S (eds) . World Health Organization, Geneva

WHO (2005b) National Policy on traditional medicine and regulation of herbal medicines—report of a WHO global survey. World Health Organization, Geneva

Wu SK, Ghantous H, Birnkrant DB (2008) Current regulatory toxicology perspectives on the development of herbal medicines to prescription drug products in the United States. Food Chem Toxicol 46:2606–2610

Yuan H, Ma Q, Ye L, Piao G (2016) The traditional medicine and modern medicine from natural products. Molecules 21:1–18

Zhang J, Onakpoya IJ, Posadzki P, Eddouks M (2015) The safety of herbal medicine: from prejudice to evidence. Evid Based Complement Alternat Med 2015:316706

Zhi-Xu H, Thach C, Zhou SF (2014) Clinical herb-drug interactions as a safety concern in pharmacotherapy. J Pharmacol Drug Metab 1:1–3

Herbal Drug Patenting

27

Kalpana Patil and Raju Wadekar

Abstract

Herbal drugs are increasing overall noticeable quality because of their particular points of interest. Creating nations have begun investigating the ethnopharmacological methodology of medication disclosure and have started to document patents on herbal medicines. Worldwide, herbal drug manufacturing units are congregating to these bazaars in an endeavour to shelter a clamp for their products. Several research institutes were set up and indulged to bring herbal products, which include herbal medicine, to the international arcade. As herbal drug market magnifies and progresses, the role of intellectual property protection right becomes significant in its existence. Analysts and experts must build up notoriety and push one advance step towards selectiveness utilizing rights for insurance. IP incorporates licences, trademarks, exchange privileged insights and copyrights. A great many people may accept there is no IP assurance for home herbal drugs/medicines thereof. Truth be told, this isn't accurate. The most well-known types of IP insurance for home-grown drugs are exchange mysteries and trademarks.

Keywords

Intellectual property · Patent · Geographical indication · Biopiracy · Trademark · Copyright

K. Patil
Department of Pharmacognosy, KLE's College of Pharmacy, Belagavi, Karnataka, India
e-mail: kalpanapatil@klepharm.edu

R. Wadekar (✉)
Department of Pharmacognosy and Phytochemistry, Shri Vile Parle Kelavani Mandal's Institute of Pharmacy, Dhule, Maharashtra, India
e-mail: raviraj_333@rediffmail.com

S. C. Mandal et al. (eds.), *Evidence Based Validation of Traditional Medicines*,
https://doi.org/10.1007/978-981-15-8127-4_27

Abbreviations

ABS	Access and benefit-sharing
BDA	Biological Diversity Act
CBD	Convention on Biological Diversity
CGRFA	Commission on Genetic Resources for Food and Agriculture
COICA	Coordinating Body of Indigenous Organisations of the Amazon Basin
COP	Conference of Parties
CSIR	Council for Scientific and Industrial Research
EPO	European Patent Office
FPO	Farmer Producers Organization
GI	Geographical indication
GO	Government of India (I)
IPR	Intellectual property right
IUPGRFA	International Undertaking on Plant Genetic Resources for Food and Agriculture
NBA	National Biodiversity Authority
PBRs	Plant breeders' rights
PGRs	Plant genetic resources
PPVFR	Protection of Plant Varieties and Farmers' Rights Act
SIPO	State Intellectual Property Office
TCM	Traditional Chinese medicine
TK	Traditional knowledge
TKDL	Traditional Knowledge Digital Library
TRIPs	Trade-Related Aspects of Intellectual Property Rights
UPOV	International Convention For The Protection Of New Varieties Of Plants
WIPO	World Intellectual Property Indicators
WTO	World Trade Organization

27.1 Introduction

According to the World Trade Organization survey, the global trade of herbal drugs has been increased at 60 billion US dollars. In today's era, pharmaceutical industries were developing enormously on the basis of the discovered medicinal plants and products. The pharmaceutical companies neither recognize nor give any compensation for those who created/reported their intellectual knowledge of traditional medicines, and on the contrary, pharmaceutical companies retain the economic benefits derived from these traditional medicines (Aguilar 2001). Thus it is necessary to create awareness and constitute policies and law for the protection of the intellectual property of the inventors and proprietor of this traditional knowledge by granting patent to the newly invented product or process through the traditional knowledge of medicines such as herbal drugs/extracts, formulations, phytoconstituents, etc. (Bayram 2005). Intellectual property (IP) defined as the form of invention, skill, knowledge, information, design or process is neither known nor obvious to others. It should be a novel and self-owned defined commodity. An intellectual property is owned by an individual or by a group of inventors of either an institute or an organization. The owner of an intellectual property owes all rights and laws whether to sell, practices, give away, trade or commerce, etc. (Bhat 1996). Nevertheless, several private and government research organizations have developed globally to explore the new horizons and avenues in the field of natural product research and herbal medicines so that the herbal products get a place into the international market. As characteristic medication extends and develops, licensed innovation (IP) assurance is assuming a significant work in its presence. Innovators and scientists have to develop a reputation and recognition for their intellectually developed or discovered products/molecules and must step forward for the exclusivity by the use of intellectual property (IP) protection. Intellectual Property Rights or protections can be of different types based on the inventions to be protected, for example, patents granted for product and process, whereas trademarks, trade secrets and copyrights for the protection of process design, etc. Surprisingly, most of the inventors believe that there is no intellectual property rights or protections in or for the herbal drugs/medicines. However, in fact, the trade secrets and trademarks are the main IP protection for herbal medicines and products thereof (Boyd 1996). In this chapter, the efforts have been made to comprehend the jurisdiction/laws/rights regarding IPR for herbal drugs and also for the traditional knowledge of herbal medicines.

27.2 Definition and Introduction of Patent

Patent is defined as a legal intellectual protection right granted to the original and first discoverer or "inventor" of a new intellectual property granted by a government, for a defined period of time to prohibit others from selling or using or making (Albert and Jason 2004).

The grantor allows a patent "skilled in the art" to the inventor if the claimed is novel unobvious and useful to others (Albert and Jason 2004).

A patent application is in a form of a written document which describes "claimed invention and discloses sufficient details and comprehensiveness of an art" so as to use or practise the invention. Thereafter, the inventor may submit the application form where patent protection is desired and available in its respective country. The patent issued may be failed in case, where the requirement of the invention utility, novelty and unobviousness deficient. Hence, to avoid such circumstances, it is always suggested to provide detailed information about the novelty and applicability of the invention (Boyd 1996). Unfortunately, the mechanism for the protection of intellectual wisdom of traditional knowledge and indigenous people is still lacking, and, therefore, the ethnic tribes strongly believe that their ethnobotanical/ethnopharmacological knowledge and bio-resources is the subject of biopiracy. Due to lack of guidelines or laws in protection of their traditional intellectual property including traditional medicine, it was necessary to take immediate action and calls for the establishment of a sui generis system (*A "sui generis" framework just signifies "one that is of its own sort". For this situation it alludes to the production of another national law or the foundation of global standards that would manage the cost of security to licensed innovation managing hereditary assets—or biodiversity—and the biotechnology that may result. It additionally alludes to a law that may secure manifestations, developments, models, drawings, plans and advancements contained in pictures, figures, images, petroglyphs, workmanship, music, history and other conventional masterful articulations*) for the protection of ethnic knowledge. It was thought worthwhile and suggested by the experts that sui generis system be separate from the existing IPR system which should be designed to protect inventions, practices and innovations associated with biological resources (Ganguli 2000). However, sui generis system protocols, paradigms and structured methodologies are yet to be finalized. Recently, about 95% of patents in the world are held in developed countries. Perhaps, there is a strong possibility to modify and to implement new guidelines and laws for the intellectual property ownership for those who seek to protect knowledge of traditional herbal medicine (Cheeptham and Chantawannakul 2001).

27.3 Convention on Biological Diversity (CBD) and TRIPS Agreement on Traditional Knowledge

In 1992, the Convention on Biological Biodiversity was held in Rio de Janeiro wherein individuals accredited the rule about the traditional knowledge of the herbal medicines that shall be the sole property of the respective nation and the government shall have rights to utilize them as commercial products for trade and sale. Be that as it may, most nations in the creating scene have not so far authorized enactment to actualize the goals intended in the convention. Therefore, the guidelines laid down on the basis of accepted norms for the transfer of indigenous resources used for the research and development or for commercial production in the form of bilateral and multilateral agreements (Cordell 2000).

The CBD played a major role that assigns rights of biodiversity not only to indigenous communities but to individuals as well and asserts their rights to protect this knowledge. Two articles of this convention are particularly relevant (Okan and Mine 2007).

1. Article 8 (j): The article states that, the respective nation shall preserve, conserve and actively involved in the sustainable biodiversity and utilisation of the innovations, knowledge and practices of ethnic/local and indigenous communities including their traditional lifestyle. "The government has equal responsibility to preserve and endorse the application with the authorization and association of the holders of such acquaintance, practices and creations and support the unbiased sharing of the remuneration arising from the exploitation of such knowledge, innovations and practices" (Bodeker 2000).

2. Article 18.4: "To encourage and cooperate to develop models for the sustainable development and use of technologies, including traditional and indigenous technologies (Bernaid 2002)".

There are two significant global conventions, viz. the World Trade Organization's Trade-Related Aspects of Intellectual Property Rights (TRIPs) and the Convention on Biological Diversity (CBD), that have to bear on protected innovation rights in indigenous information frameworks. The outing is a link between national understandings in advancing the harmonization of national IPR systems. These two conventions express that licence will be accessible for any creations, items and procedures, given a creative advance, fit for modern era of technology, whereas protected innovation rights is a specific part of property covering "everything that radiates the activity from the human mind". Thus, the patents, 'plant breeders' rights, geographical indications, trade secrets, trademarks and copyrights are covered under the major intellectual property rights (Correa 1998). Usually, in all legal systems, companies and individuals invest and acquire rights on property such as sale and lease and gain economic returns from their investments. Therefore, to protect intellectual property of traditional knowledge, it is necessary to separate the existing system of intellectual property and design sui generis system associated with access to genetic resources (King et al. 1996). Moreover, sui generis laws adopted by the countries like Brazil, Costa Rica, India, Peru, Panama, the Philippines, Portugal, Thailand and the United States for the protection of at least some aspects of traditional knowledge and also at present play a crucial role for accomplishing security and benefits for traditional knowledge holders (Okan and Mine 2007).

Although under existing intellectual property rights some aspects of traditional medicine may be protected sin form of patents still some part of ethnobotanical information need to be protected and thus sui generis system has been developed and accepted. The regimens in updation of sui generis system are regularly, explicitly in light of contemplations of value. If trail blazers in the "formal" arrangement of advancement get pay through IPRs, equity necessitates that holders of conventional information be comparably treated, hence the need for the development of appropriate legislation on intellectual property rights that recognize and protect indigenous knowledge. Whereas, inventive insurance of licensed innovation of conventional drugs as indicated by their disposition ought to be created which are the fundamental necessities and states of protected innovation (Walden 1998).

27.4 Intellectual Property Rights for the Protection of Traditional Herbal Medicine and Herbal Medicinal Products and in the Natural Product Drug Discovery

The cost of drug discovery and development is a costly affair of more than thousand millions to several billion dollars. IPR ensures thoughts and demeanour inside inventions and procedures. Patents provide protection to the inventor for its invention may be a product/process for the period up to 20 years (Walden 1998). The pharmaceutical products can be protected by patent based on the following parameters:

- Drug discovery/patent for discovering new chemical components
- Novel process patent in producing the products
- Trademark/copyright (Zhang 2000)

27.4.1 Patents for New Chemical Components from Natural Products

In nature, species of plants, animals, marine organisms and microorganisms synthesize enormous variety of primary and secondary metabolites of therapeutic significance (Rocha et al. 2001). Roughly 33% of the drugs on the planet are characteristic items of their subordinates. Notwithstanding major logical and mechanical advancement in combinatorial science, drugs got from regular items despite everything that makes a gigantic commitment to tranquilize disclosure (Rocha et al. 2001). A plant serves as a solar power biochemical laboratory which synthesizes thousands of primary and secondary metabolites of therapeutic importance (Udgaonkar 2004). Innovative methods such as extraction, isolation and purification and structural identification for the drug discovery and development from natural products were employed to find a new natural product and its useful biological properties of the new lead; synthesis of new analogues or derivatives (Boyd 1996). The plant-derived or natural product-derived compounds discovered unreservedly insignificant revelation which may not be patentable. However, if the natural product appropriately portrayed with respect to its method of production, synthesis and its novelty as a potent therapeutically active compound in the supreme feeling recently perceived presence, at that point the substance might be patentable. For example: Microorganism derived natural products such as proteins, amino acids, neurotoxins, etc. (Walden 1998; Ganguli 2000). It is also stated further that the value of patent is placed on a compound with novelty in its structural appearance, derivatives, sight specificity structure-activity relationships and uniqueness in its mechanistic pharmacology. Modification in various stages of its development and discovery with the intention to reduce its toxicity profile and overall of pharmacokinetic and pharmacodynamics (Cordell 2000).

27.4.2 Patents for Known Phytoconstituents from Natural Products

To patent the known phytoconstituents from the natural product, the conditions must be appropriate such as the following: the lead molecules should possess defined pharmacological activity. However, company would purchase only those plant-derived entities having potential and promising pharmacological activity. The traditional medicine likely to patent, if the old medicine found to have novelty in its activity which is not mentioned in the traditional text but found effective as drug lead. For instance, if betulinic acid exhibits target-specific anticancer activity against melanoma cells and also possesses in vivo antiproliferative and cancer chemopreventive activities could be patentable. Similarly, a semisynthetic approach in drug discovery applied on a known compound and can be modified with a view to enhancing potency and reducing toxicity or modifying solubility can be patentable. Also in case of new techniques derived for the isolation and purification of known compounds from nature and synthesis of useful new analogues or derivatives could be patentable (Kartel et al. 2006).

27.4.3 Patent Rules for Traditional Herbal Medicines and Herbal Medicinal Products

Value-added products as well as active phytoconstituents and active extracts derived obtain from the various cultivated and wild grown medicinal plants and traditionally utilised herbs. Thus, ethnobotanical or ethnomedicinal plants may serve as a precursor for the screening of natural products of therapeutic importance. However, this traditional medicine-derived plant components need to be analysed scientifically and systematically to complement its use as a medicine. The endeavours of present-day industry are to exploit traditional medicines and to use it for the conventional information, especially phytopharmaceuticals (Correa 2002).

The herbal formulations derived from the traditional medicines are prepared in various forms such as tinctures, concentrates, extracts, oils and powdered natural products. They incorporate various techniques and procedures such as soaking or warming home-grown products in mixed drinks as well as nectar, etc. Practical this process cannot be eligible for patent. The extracts are prepared with different solvents (nonpolar to polar) and from different parts of the same plant (stem, leaves, roots, fruits) each with an exclusive right for a different set of applications. Procedure's or patent process have generally been seen as being of a less financial incentive than "item" patent, incompletely in light of the fact that they are hard to police, while new specialized systems can likewise regularly be found to deliver a specific item. An alternative is a sweeping 'item by procedure's patent case, which offers ascend to the most productive imposing business model rights. Scientists or organizations may likewise guarantee patent innovation rights over natural assets as well as customary information, after marginally adjusting them. Instances of these incorporate patents gave identified with the neem tree, kava, barbasco. Protection can be given to the use of technology, including biotechnology, to develop a useful product using traditional knowledge (Zhang 2000).

27.5 Classification of Patent Protection for the Herbal Medicine Inventions

27.5.1 Patent on Formulation of Herbal Medicine

Preparing a novel dosage form containing herbal medicine using new technology to achieve synergistic or significant therapeutic effect.

27.5.2 Combination Patents on Herbal Medicine

It may include new or alternative herbal drug/herbal preparation targeted for same indication, and patent is granted in the following circumstances:

(a) The combination of herbal medicines prepared from several crude drugs and the composition is newly prepared and completely different from the original one or by changing ration of its constituents.
(b) The combination of novel herbal medicines should report new indications other than.

27.5.3 Patents on Process of Herbal Medicine

Novelty in the process or method of preparation herbal medicines can be patented. Following are the inventive key steps may get patentable such as:

(a) To process such as cultivation, collection, drying and sampling of crude drugs
(b) To process and prepare pharmaceutical formulations using herbal drugs or herbal drug preparations
(c) To separate, extract and isolate phytoconstituents from known herbal drugs which are used in the procedure of pharmaceutical production
(d) To formulate and standardize solid/liquid herbal extracts or powders or tinctures from herbal drugs,
(e) To process crude drug to produce granules and drying process of standardized extracts
(f) To develop novel methods and process to prevent moisture and microbial contamination of standardized extracts
(g) To develop process for removal of trace amounts of contaminant substances such as pesticides, toxins and surfactants from plant products
(h) To develop novel process and techniques to increase the production of standardized extracts from herbal drugs
(i) To invent new procedures to decrease the cost of preparation of standardized extracts from herbal drugs
(j) To enhance the purity of the extract separated from herbal drugs

(k) To develop new techniques and model to eliminate the side effects of herbal medicine (Bayram 2005)

27.6 New Indication of Herbal Drug and Herbal Medicine

If the herbal drug or medicine is found to possess new indication such as novel therapeutic activity in its extract form or in formulation preparation under such circumstances, patent can be granted. For instance, herbal medicine could have an obscure impact of relieving bosom malignant growth found by a researcher. In such a case, the new sign can be secured by patent law. The use of new technology for the formulation of new dosage form of the herbal drug showing new indication to achieve significant therapeutic effect can be patentable (Kartal 2007).

27.7 Prerequisites for Patent of Herbal Medicines/Drugs

27.7.1 Validation

Validation of the unknown phytochemicals in herbal medicines is the mandate for each batch of drug formulation. To achieve validation of the unknown compounds, the quality of product may be identified. Comparative study claims and use of standard drug may also be patentable.

27.7.2 Clinical and/or Animal Studies

The herbal medicines should be tested for all clinical and animal studies to ensure its acute toxicity profile as well as to find out the therapeutic dose. Thus the effective dose must be calculated and may be patented.

27.7.3 New Formulations

In case of new formulations and their new route of administration and drug delivery is discovered, then the patent protection is necessary. For that all the relevant data must be provided, which can prove the efficacy of such modifications.

27.7.4 Active Ingredients

All types of active phytoconstituents present in herbal medicine, which may be simple or are complex in nature, are patentable.

27.7.5 Second-Generation Products

In this case, if the natural product should be improved product from the native product. Such type of second generation product may be patentable. These second-generation products are chemically decoded or modified by some reactions and change in any functional group. Also the bioactivity or target specific activity of such molecules must be known.

27.7.6 Authentication Kits

Authentication of herbal drugs and medicines thereof is of prime importance as the malpractices of adulteration are prevalent. The authentication of plant sample can be done by sophisticated techniques like DNA fingerprinting and protein analysis. However, if the plant material is found to be authenticated by means of these methods, the inventor may subsequently decide whether to keep patent or to keep it for trade secret (Albert and Jason 2004).

However, in India the certain legislative measures are active and considered to protect IPR of the biological source and traditional knowledge. Patent protection to the biological materials and their control of biopiracy are sought under the Biodiversity Bill, thereby resulting in the financial and nonfinancial benefits contributing to the final product. A biological resources can be obtained by an Indian for commercial purpose after giving proper and prior information to the State Biodiversity Board concerned (Patwardhan 2005). An extensive procedure for ensuring herbal drug knowledge, think about the network, national, local and worldwide measurements. The more grounded the incorporation and coordination between each level, the more probable the general viability. Numerous people groups, nations, and, what's more, provincial association are attempting to address these levels individually. National laws are the prime factor for accomplishing security for TK holders. Countries like Brazil, Costa Rica, India, Peru, Panama, the Philippines, Portugal, Thailand and the United States have all embraced sui generis laws that secure probably some part of TK (WIPO 2005). The Convention on Biological Diversity is working on implementation of Article 8(j) and related provisions: development of elements of sui generis systems for the protection of the knowledge, innovations and practices of indigenous and local communities. As per the choice VIII/5 E (27 June 2006); Parties and Governments, indigenous and nearby networks, and non-legislative associations are mentioned to impart to the Executive Secretary data on existing sui generis frameworks and their perspectives on the definitions contained in archive UNEP/CBD/WG8J/4/7 addition II, no later than 30 June 2007 to guarantee they are considered in the readiness of documentation for the following gathering of the Working Group on Article 8(j). Parties and governments are also requested to report on initiatives to adopt local and national sui generis models and to share experiences through the Clearing-House Mechanism (Notification from Secretariat of the Convention on Biological Diversity 2006).

The implementation of the Convention on Biological Diversity, in particular Article 8(j), ensures that traditional knowledge will receive adequate protection. The laws, priorities and definition of concerns, aims and objectives of the protection to be sought at the international level to delineate interests more clearly in different international organizations such as the CBD, WTO and WIPO (The Inter-national Dimension of TK Protection 2006). The WIPO work is based on the core principles that should underpin the protection of TK. This serves as an opportunity to a potential establishment for worldwide lawful advancement as exact approach and administrative choice for improved assurance of TK through adjusted or extended ordinary licensed innovation (IP) frameworks, or through independent sui generis frameworks (Timmermans 2003).

A traditional Chinese medicine (TCM) patent information is based on the principle built up by the State Intellectual Property Office (SIPO) of PR China. The database has just been put to use in the patent assessment office in SIPO since April 2002. There are 29 pursuit fields in the database that fall into four classes: bibliographic data, subject file terms, utilizes/impacts and TCM recipes. Nevertheless, the database provided by TCM accessible to Chinese examiners contains approximately 30,000 records in the bibliographic database and more than 60,000 TCM formulae. Till November 2005, 1761 items in the English demo version are made publicly available and also contain 1995 items in the Chinese demo version. This type of database is provided as an efficient way for patent examiners or other users to search TCM-related patent documents. Thus, every state's patent office should establish a traditional medicine patent database of each community in each country. Furthermore, in order to obtain a more powerful traditional medicine patent database, they should be merged into one central database system (Yanhuai and Yanling 2004).

At the University of Illinois, Chicago, a policy has been developed on contracts for collecting natural product samples for drug discovery which includes royalty sharing and other important provisions. For example, the providing organization for their endeavours and their ability, and perceive a commitment to offering a sovereignty to the establishment in the occasion that medicates revelation, with ensuing commercialization, owes its source, to a material that it gave and repay by the Glaxo. The new commercialized agricultural products such as bio-fertilizers and pesticides will also come under the power of IPR and become monopolized. In such cases patent granted for the neem extracts which is used as biopesticide to some US and Japanese companies. Thus, before filing the patent, it is mandatory to submit a publication related to the invention issued at least a year ago from the date of filing of patent application (Timmermans 2003). Patents on Ayurvedic herbal extracts are listed on Table 27.1 (https://ipindiaservices.gov.in/publicsearch).

Table 27.1 Patent on Ayurvedic/herbal extracts

Sl. No.	Patent No.	Title	Main herbs
1.	155/ MUM/ 2008	Herbal extract and Ayurvedic composition for the treatment of diabetes	*Momordica charantia*
2.	1734/ KOL/ 2007	A process for preparing an Ayurvedic medicament effective against leukaemia and carcinoma of the lung and intestine	Lime, asafoetida and black jeera
3.	1938/ DEL/ 2006	A process for preparation of Ayurvedic composition for treatment of hepatic disorder	Kagzi nimbu (lemon) [*Citrus medica (aurantifolia)*], Sarjika [*Salsola kali* Linn., *Fagonia cretica* Linn., Barilla] and Varat (Kapard) *Cypraea moneta*
4.	1623/ MUM/ 2006	An Ayurvedic composition for oral consumption in treatment of heart diseases and hypertension	Arjuna, *Terminalia arjuna*; *Ajamoda, Apium graveolens*; Punarnava, *Boerhavia diffusa*; Rasona, *Allium sativum*; Shigru, *Moringa oleifera*; Draksha, *Vitis vinifera*; Pippali, *Piper longum*; Guduchi, *Tinospora cordifolia*; and Triphala
5.	228/ CHE/ 2006	A process for the preparation of Ayurvedic tooth powder/paste	*Azadirachta indica*, menthol, thymol, camphor and gall nut
6.	3207/ DEL/ 2005	An Ayurvedic composition for joining fractured bone and as anti-inflammatory and process for preparation thereof	*Cissus quadrangularis, Pterocarpus marsupium* heartwood, buffalo/cow milk and *Chenopodium murale* (kurund)
7.	726// CHE/ 2005	An Ayurvedic medicine for curing viral hepatitis	*Luffa aegyptiaca, Luffa cylindrica* or *Luffa acutangula* or *Luffa operculata* and seeds of *Cuminumcyminum*
8.	313/ CHE/ 2005	An unique combination of Ayurvedic compounds for correcting a rare form of mulleri and ysgenesis	*Asoka, Asana* and Bilwa, *Shorea robusta* Gaertn, *Pinus roxburghii* Sargent, *Cyperus rotundus* Linn., *Sida rhombifolia, Cyperus rotundus* Linn., *Gmelina asiatica* Linn., *Nardostachys jatamansi* DC, *Randia dumetorum* Linn., *Kaempferia galangal* Linn., *Teramnus labialis* Spreng., *Phaseolus trilobus, Inula racemosa* Hook. f., *Cinnamomum zeylanicum, Syzygium aromaticum* Merr., *Parmelia* (Sangejaranath), *Crocus sativus* Linn., *Cinnamomum camphora* T. Nees & Ebem
9.	146/ MUM/ 2005	An Ayurvedic herbal hair oil composition and preparation thereof	Jatamansi (*Nardostachys jatamansi* (D.Don) DC.), Amla, Brahmi (*Gratiola*), Bhringraj (*Eclipta alba*) Thistles, Nagamothra (*Cyperus rotundus*), Kapur kachari (*Hedychium*

(continued)

Table 27.1 (continued)

Sl. No.	Patent No.	Title	Main herbs
			spicatum) and Kavath (*Feronia elephantum*)
10.	31/ MUM/ 2005	An Ayurvedic herbal composition for treatment of cancer/skin diseases and process of making thereof	*Murraya paniculata* (Bhutkati), *Latanacamara* (ghanera or Tantani), *Terminalia* (Kinjal), *Toalia asiatica* (Jangli-Mirch) *chawat* or *dhundari* and teakwood

27.8 Protection of Geographical Indications

Geographical indication implies a sign which recognizes farming or common or fabricated merchandise as starting or produced in a specific region "where a given quality, notoriety or another trait of such products is basically owing to its geological origin". Medicinal plants beginning from specific locales numerous a period have unmistakable restorative properties and, subsequently, are qualified for enlistment and assurance under the Geographical Indications of Goods (Registration and Protection) Act, 1999. The approved clients of an enlisted geographical indication (GI) just are qualified for the rights under the Act which are the rights to utilize the sign and to get alleviation thereof in the event of encroachment. An application for enrolment is to be made by "a relationship of people or makers or any association or authority set up by or under any law speaking to the premiums of the makers of the concerned goods". The enlisted assortments have the benefit of telling premium value in the market. A model is Navara rice from Kerala which is an enrolled assortment. It is used in Ayurveda for specific treatments like *Panchakarma* and in various treatments for arthritis (Dutfield 2003). The List of medicinal/aromatic plants registered as geographical indications is presented in Table 27.2 and patent on thirty medicinal plants in India in Table 27.3.

27.9 Farmers' Right and Breeders' Right

Plant genetic resources (PGRs) are forming the building blocks for continuous development of a food and nutritionally secure society. Furthermore, plants have numerous utilizations, including feed, fibre, medication and modern applications. Ranchers may have almost no comprehension of the logical premise of hereditary decent variety, yet they surely comprehend its vital significance to agribusiness and the requirement for advancing fluctuation in rural practices. The self-governance that each rancher practises in choosing, sparing and keeping up seed for re-planting has been principal for the agronomic change of plant species into crops and their further choice. Any sensible valuation of PGR created by ranchers could all around run into trillions of dollars, which is commonly higher than the worth that cutting-edge plant

Table 27.2 Registered geographical indications of medicinal/aromatic plants and allied products

Plant/allied product	State	Reg. No.	Reference
Mysore *Agarbathi*	Karnataka	11	http://ipindiaservices.gov.in/publicsearch/. Accessed on 20 April 2015
Mysore Sandalwood oil	Karnataka	23	http://ipindiaservices.gov.in/publicsearch/. Accessed on 20 April 2015
Mysore Sandal soap	Karnataka	24	http://ipindiaservices.gov.in/publicsearch/. Accessed on 20 April 2015
Mysore Betel leaf	Karnataka	28	http://ipindiaservices.gov.in/publicsearch/. Accessed on 20 April 2015
Nanjanagud banana	Karnataka	29	http://ipindiaservices.gov.in/publicsearch/. Accessed on 20 April 2015
Mysore Jasmine	Karnataka	37	http://ipindiaservices.gov.in/publicsearch/. Accessed on 20 April 2015
Udupi Jasmine	Karnataka	38	http://ipindiaservices.gov.in/publicsearch/. Accessed on 20 April 2015
Hadagali jasmine	Karnataka	39	http://ipindiaservices.gov.in/publicsearch/. Accessed on 20 April 2015
Navara rice	Kerala	40	http://ipindiaservices.gov.in/publicsearch/. Accessed on 20 April 2015
Coorg Green Cardamom	Karnataka	57	http://ipindiaservices.gov.in/publicsearch/. Accessed on 20 April 2015
Naga tree tomato	Nagaland	220	http://ipindiaservices.gov.in/publicsearch/. Accessed on 20 April 2015
Sikkim large cardamom (*Amomum subulatum*)	Sikkim	222	http://ipindiaservices.gov.in/publicsearch/. Accessed on 20 April 2015
Assam Karbi Anglong Ginger	Assam	226	http://ipindiaservices.gov.in/publicsearch/. Accessed on 20 April 2015
Khasi Mandarin	Meghalaya	231	http://ipindiaservices.gov.in/publicsearch/. Accessed on 20 April 2015

reproducing has contributed. In any case, PGRs were treated as the "legacy of humankind" and were shared openly among countries, till the worries for the preservation of organic decent variety were raised by the Convention on Biological Diversity (CBD), which came into power in 1993. The preservation and manageable usage and access to natural assorted variety were considered as national sway by CBD. Therefore, numerous issues with respect to the privileges of the conservers, clients, reproducers, ranchers and protected innovation have risen. During 2001, huge advancements have occurred concerning the acknowledgement of the privileges of raisers, ranchers and nearby networks assortment is thorough, permitting no adaptability for ranchers and negligible adaptability for reproducers, contingent upon the locale. The essential basic rule of IPR on plant assortments is the acknowledgement of human advancement in building up another plant assortment through choice, with or without recombination, which is novel and unmistakable from the previous assortments. Dissimilar to the developments that are made in numerous non-organic spaces, life structures, for example, crop assortments, are not

Table 27.3 Patent on 30 medicinal plants in India

Plant/herb	Number of patent applications published	Number of patents granted	References
Lemon	65	16	http://ipindiaservices.gov.in/publicsearch/. Accessed on 11 Dec 2019
Neem	173	47	http://ipindiaservices.gov.in/publicsearch/. Accessed on 11 Dec 2019
Aloe vera	185	43	http://ipindiaservices.gov.in/publicsearch/. Accessed on 11 Dec 2019
Terminalia bellirica	123	25	http://ipindiaservices.gov.in/publicsearch/. Accessed on 11 Dec 2019
Turmeric	103	16	http://ipindiaservices.gov.in/publicsearch/. Accessed on 11 Dec 2019
Ginger	86	19	http://ipindiaservices.gov.in/publicsearch/. Accessed on 11 Dec 2019
Pepper	80	17	http://ipindiaservices.gov.in/publicsearch/. Accessed on 11 Dec 2019
Garlic	59	13	http://ipindiaservices.gov.in/publicsearch/. Accessed on 11 Dec 2019
Sandal/ Chandan	50	29	http://ipindiaservices.gov.in/publicsearch/. Accessed on 11 Dec 2019
Arjuna	47	16	http://ipindiaservices.gov.in/publicsearch/. Accessed on 11 Dec 2019
Amla	47	12	http://ipindiaservices.gov.in/publicsearch/. Accessed on 11 Dec 2019
Almond	23	12	http://ipindiaservices.gov.in/publicsearch/. Accessed on 11 Dec 2019
Poppy	18	05	http://ipindiaservices.gov.in/publicsearch/. Accessed on 11 Dec 2019
Camphora	14	04	http://ipindiaservices.gov.in/publicsearch/. Accessed on 11 Dec 2019
Senna	06	03	http://ipindiaservices.gov.in/publicsearch/. Accessed on 11 Dec 2019

(continued)

Table 27.3 (continued)

Plant/herb	Number of patent applications published	Number of patents granted	References
Jasmine/ *Jasminum*	25	07	http://ipindiaservices.gov.in/publicsearch/. Accessed on 11 Dec 2019
Nutmeg/Jati	26	03	http://ipindiaservices.gov.in/publicsearch/. Accessed on 11 Dec 2019
Vetiver/ Khus	09	04	http://ipindiaservices.gov.in/publicsearch/. Accessed on 11 Dec 2019
Rose geranium	94	19	http://ipindiaservices.gov.in/publicsearch/. Accessed on 11 Dec 2019
Patchouli	07	02	http://ipindiaservices.gov.in/publicsearch/. Accessed on 11 Dec 2019
Chamomilla	06	02	http://ipindiaservices.gov.in/publicsearch/. Accessed on 11 Dec 2019
Basil/Tulsi	60	09	http://ipindiaservices.gov.in/publicsearch/. Accessed on 11 Dec 2019
Lavender	10	01	http://ipindiaservices.gov.in/publicsearch/. Accessed on 11 Dec 2019
Rosemary	15	01	http://ipindiaservices.gov.in/publicsearch/. Accessed on 11 Dec 2019
Mint	57	16	http://ipindiaservices.gov.in/publicsearch/. Accessed on 11 Dec 2019
Mucuna/ Kawanch	28	06	http://ipindiaservices.gov.in/publicsearch/. Accessed on 11 Dec 2019
Jamun	10	01	http://ipindiaservices.gov.in/publicsearch/. Accessed on 11 Dec 2019
Celastrus	11	02	http://ipindiaservices.gov.in/publicsearch/. Accessed on 11 Dec 2019
Babchi/ *Psoralea*	20	01	http://ipindiaservices.gov.in/publicsearch/. Accessed on 11 Dec 2019
Musli	19	04	http://ipindiaservices.gov.in/publicsearch/. Accessed on 11 Dec 2019

totally imagined but however are constantly made from previous living things and engendered by regular procedures. Along these lines, the making of another assortment has two segments: the utilization of prior assortments and the information required to choose another assortment by recombining the previous ones or by different procedures. Value requests that the acknowledgement of advancements made on the recently reproduced assortments ought to likewise incorporate the comparatively inventive segment put resources into the source assortments (e.g. plant hereditary assets). The last basically speaks to the far more prominent total scholarly information sources contributed by ages of cultivating networks over a significant stretch. The way that those networks need character and institutional support, in contrast to the current business plant raisers, does not imply that they are given less significance or acknowledgement for their scholarly information sources. While IPRs on plant assortments are maintained, the interest with the expectation of complimentary access to assortments created by ranchers, without the instalment of sovereignties material to assortments secured by licensed innovation (IP), can be viewed as a twofold standard concerning rights. Additionally, the allowing of restrictive rights over the seed or proliferating material of an IP-shielded assortment denotes a defining moment from the conventional unlimited right ranchers had appreciated over seeding. This limitation on the seed of a patent-shielded assortment denotes a defining moment from the conventional unhindered right ranchers had delighted in over seeding. This limitation on the seed of a patent-secured assortment is thorough, permitting no adaptability for ranchers and negligible adaptability for raisers, contingent upon the ward (Brahmi et al. 2004).

IPR that is allowed to reproducers of plant assortments is alluded to as plant raisers' privileges (PBRs). The International Convention for the Protection of New Varieties of Plants (UPOV Convention) is the most punctual framework for plant assortment assurance and is at present clung to by 69 nations. PBRs permit a plant raiser to bar others from the creation, preparing, loading, conveyance, promoting, deal, fare and import of engendering material of a secured assortment for a predetermined number of years. It additionally permits the reproducer to authorize such rights to other people and to get eminences created from the approved utilization of the proliferating material. These rights may in certain nations likewise incorporate reaped material, for example, cut blossoms, organic products or foliage of the ensured assortment, in situations where the raisers don't have sensible chances to practise their privileges over the planting materials. The lawful space accessible to ranchers concerning the seed of a secured assortment under such a framework for plant varietal insurance appears as ranchers' privileges, together with PBRs, or that of the ranchers' benefit inside PBRs. The UPOV Convention of 1978 had created the special case to PBRs as a private and nonbusiness activity on engendering material of the secured assortment, which permitted—as indicated by specific translations—ranchers to utilize the spreading material got to as the result of the gather acquired by planting. The UPOV rendition of 1991, be that as it may, makes reference to the ranchers' benefit which restrains the exclusion to specific harvests (by the decision of the part state) and just to seed delivered by ranchers for planting back on their own possessions, in this way significantly diminishing the customary ranchers' privileges for use. While perceiving PBRs on plant assortments, the administrative Commission on Genetic Resources for Food and Agriculture (CGRFA) likewise affirmed ranchers' privileges in a goal on the understanding of the International Undertaking on Plant Genetic Resources for Food and Agriculture (IUPGRFA) in 1989. Its essential target for perceiving ranchers' privileges was to guarantee that ranchers would keep on adding to the preservation and reasonable utilization of PGR for reinforcing the worldwide nourishment and wholesome security (Brahmi et al. 2004).

The Protection of Plant Varieties and Farmers' Rights Act (PPVFR Act) tries to address the privileges of plant reproducers and ranchers on an equivalent balance. It asserts the need for perceiving and ensuring the privileges of ranchers as for the commitment they make in saving, improving and making PGR accessible for the advancement of new plant assortments. The PPVFR Act additionally regards it similarly important to secure PBRs to animate venture for innovative work, both in general society and private part, for the advancement of new plant assortments. Under the Act, PBRs permit reproducers to hold selective rights to create, sell, advertise, appropriate, import or fare the engendering material of an enlisted assortment (Brahmi et al. 2004). The PPVFR Act perceives the different jobs played by ranchers in developing, moderating, creating and choosing assortments. As to creating or choosing assortments, the Act alludes to the worth added by ranchers to wild species or conventional assortments through choice and distinguishing proof of their helpful qualities. As needs are, ranchers' privileges incorporate the jobs of ranchers as clients, conservers and reproducers. Farmers are granted nine specific rights, which are briefly described below.

27.9.1 Right 1: Access to Seed

Farmers were qualified to spare, use, sow, resow, trade, offer or sell their homestead produce, including seed of ensured assortments, in a similar way as they were qualified before the PPVFR Act came into power. Be that as it may, farmers will not be qualified to sell marked seed of an assortment ensured under this Act. The Act doesn't determine the amount of seed that farmers can spare from a yield developed in their own homesteads from a secured variety.

27.9.2 Right 2: Benefit-Sharing

All Indian legitimate substances who give PGR to raisers to growing new varieties, including farmers, will get a decent amount of the advantages from the business gains of the enrolled assortments. Out of all the national plant assortment security laws enacted since 2001, the PPVFR Act primarily coordinates an arrangement for access and benefit-sharing (ABS) alongside PBRs. Legitimate promotion of the hereditary asset utilized inbreeding isn't tended to in the Act; this falls rather under the Biological Diversity, 2002. In any case, the PPVFR Act requires a reproducer to make a sworn declaration on the geological inception of the hereditary assets utilized in the family of the new variety, and how they were gotten too.

27.9.3 Right 3: Compensation

Registered seed must be sold with full disclosure including the details of their agronomic performance under recommended management conditions. If the farmer found that the seed sold to him fails to provide the expected performance under recommended management conditions, then the farmer shall claim compensation from the breeder through the office of the PPVFR Authority.

27.9.4 Right 4: Rational Seed Price

Farmers reserve the privilege to get to seed of enlisted assortments at a sensible cost. At the point when this condition isn't met, the reproducer's elite directly over the assortment is suspended under the arrangement concerning mandatory authorizing, and the raiser is committed to permit the seed creation, dispersion and deals of the assortment to an able legitimate substance. A large portion of the laws for plant assortment assurance have arrangements on mandatory authorizing of secured assortments to guarantee sufficient seed supply to farmers, and a few of them additionally utilize out of line evaluating as reason for necessary permitting.

27.9.5 Right 5: Farmers' Recognition and Reward for Contributing to Conservation

Farmers who have been occupied with PGR protection and yield improvement, and who have caused considerable commitments in giving hereditary assets to edit improvement, to get acknowledgement and awards from the national quality store. The quality reserve gets assets from the execution of the Act, which thusly is supplemented by commitments from national and universal associations. The utilisation of the reserve are reserved to help the preservation and reasonable utilization of PGR, and along these lines, it very well may be viewed as a national identity to the worldwide benefit-sharing asset working inside the ITPGRFA. Since 2007, the plant genome guardian angel grant, related to the national quality reserve, has been remunerating cultivating networks and individual ranchers for their commitment to in situ protection on ranch and to the choice of PGR.

27.9.6 Right 6: Registration of Farmers' Varieties

The Indian PPVFR Act takes into account the enlistment of existing ranchers' assortments that satisfy prerequisites for uniqueness, consistency, security and section, but however excludes that of curiosity. This privilege furnishes ranchers with a coincidental open door for a restricted timeframe, from the second when a yield animal variety is remembered for the harvest portfolio under the PPVFR Act for enlistment. When enrolled, these assortments are qualified for all PBRs.

27.9.7 Right 7: Prior Authorization for the Commercialization of Essentially Derived Varieties

At the point when farmers' assortments, regardless of whether surviving or new, are utilized by an outsider as source material for the advancement of a basically determined assortment, the ranchers need to give earlier approval to its commercialization. Such a procedure can permit ranchers to arrange the terms of approval with the raiser, which may incorporate sovereignties, one-off instalments, advantage sharing and so on.

27.9.8 Right 8: Exemption from Registration Fees for Farmers

Under the PPVFR Act, ranchers have the benefit of being totally absolved from paying any sort of charges or different instalments that are regularly payable for assortment enlistment; tests for uniqueness, consistency and strength (DUS), and different administrations rendered by the PPVFR Authority; just as for lawful procedures identified with encroach coach different causes.

27.9.9 Right 9: Farmer Protection from Accidental Infringement

On the off chance that a farmer can some way or another demonstrate under the watchful eye of the court that the person in question didn't know about the presence of any rights at the hour of encroachment on any such rights, as pointed in the PPVFR Act, the individual in question won't be charged. This arrangement is made with regard to the hundreds of years old unreasonable rights that the ranchers had over the seed everything being equal, the novel idea of the PPVFR Act, and the low lawful education of farmers.

Along these lines, the arrangement for reasonable and fair advantage sharing from plant reproducers utilizing the PGR monitored by ranchers is a significant motivating force for connecting preservation with occupation advancement, during the time spent network biodiversity the executives (CBM). So as to guarantee the proceeded with commitments of ranchers through CBM, ranchers over the world must be conceded liberal rights over the seed of each assortment that they create. These rights ought to be joined by the arrangement of money related, good and specialized help, through the use of CBM rehearses inside seed frameworks including those utilizing current innovations. Such rights and bolster must be given as a need and should be equivalent to the select rights that have been conceded to plant raisers, who scrounge on the PGR that once started and keep on being preserved and created by those ranchers (Benjamin et al. 2007).

27.10 Biopiracy

The indigenous individuals of the world have monstrous information on their surroundings, in light of hundreds of years of living near nature. Conventional information (TK)- framework held by indigenous networks, regularly identifying with their encompassing common habitat like agriculture information, logical information, specialized information, natural information, restorative information, "articulations of fables" as music, movie, tune, handwork plans, stories, craftsmanship, biodiversity protection, nourishment methods, tradition-based artistic works, information, and all other customary-based advancements, Healing information (Example: Aspirin, got from salicylic corrosive found in meadowsweet and the salivation of the vampire bat of South America opens stopped up courses quicker) and so on. The key qualities of conventional information to protected and transmitted in a customary setting from age to age relates to a specific conventional or indigenous individuals or community are not static, yet rather advance as networks react to new difficulties and requirements are aggregate/individual in nature. Around 370 million indigenous and innate individuals all around the globe are the genuine caretaker and holders of conventional information, and up to 80% of the total populace relies upon customary medication for its essential social insurance (Battiste and Henderson 2000). This information is fundamental for the most unfortunate sections of society and by the TK additionally forestalls land and soil corruption, fisheries exhaustion, biodiversity disintegration and deforestation. Roughly 90% of the world's biodiversity is packed in tropical and sub-tropical districts inside creating nations, explicitly in Mexico, Brazil, India, Indonesia, Australia and the Democratic Republic of Congo. An expected 90 percent of the world's biodiversity exists in the regions of indigenous people groups. India brags a wide assortment of verdure which is enhanced in nature. The faunal and flower lavishness has been perhaps the greatest resource of India. Accessible information place India in the tenth situation on the planet and fourth in Asia in plant decent variety. With this India has for some time been a casualty of biopiracy (Gollin 1999).

Bio-robbery alludes to the procedure through which the privileges of indigenous societies to these assets and information are deleted and supplanted by restraining infrastructure rights. Biopirates, privateers of life, are looting another sort of riches, that of biodiversity and the conventional information and strategies of provincial and indigenous people groups. Biopiracy can be characterized as, "the misappropriation and commercialization of hereditary assets and conventional information on the country and indigenous individuals". Pharmaceutical biopiracy is a term utilized for the most part to describe the legitimate practice by pharmaceutical organizations abusing the indigenous individuals' conventional information on medication. India and other creating nations are rich in bio-assets, and TK are the most loved targets and casualties of biopiracy. Famous market analyst and Nobel Prize winner Joseph

E. Stiglitz remarks on the World Trade Organization's Trade-Related Aspects of Intellectual Property Rights (TRIPs) Agreement—what we were not completely mindful of was another threat, what has come to be named bio-robbery, worldwide organizations protecting traditional medicines and nourishments. It isn't just that they try to bring in cash from "assets" and information that legitimately has a place with the creating nations, however in this manner, they suppress residential firms that have since quite a while ago gave the items. Biopirates are for the most part pharmaceutical, corrective and agri-nourishment firms. Biopiracy of hereditary assets and hereditary materials is likewise noted. It was drawn on biodiversity problem areas so as to make apparently "inventive" items and assurance their restraining infrastructure on them through the patent framework. Such misappropriation of TK brings about the award of a patent for the innovation to the "first-to-document" (the pharmaceutical organization) instead of the "first-to-design" (the indigenous network). It includes making benefits from unreservedly accessible normal items (plants, seeds, leaves and so forth.), by duplicating methods utilized day by day for ages by nearby people groups so as to take care of or deal with themselves. Biopirates don't give any benefit or appropriate advantage to nearby networks and TK holders. The issue of intellectual property rights (IPRs) is subsequently getting firmly identified with the issue of bio-robbery and scholarly theft of western-style IPR systems. The absence of lawful insurance of our organic and social legacy has made the indigenous networks of the third world defenceless against bio-robbery and scholarly theft as in the instances of neem and so on. The current patent systems dependent on western ideal models are one sided for enormous transnational corporations with interests cutting across pharmaceuticals and agri-synthetic concoctions (Turner 1996). Citations on biopiracy across the globe are discussed in Table 27.4.

With this backdrop of practicing biopiracy and to bring out some concrete solution to stop biopiracy, and to give justice to the guardians of the traditional knowledge of medicines, it was thought worthwhile to form international treaties which can deal with biopiracy. There are two international treaties that help protect peoples and biodiversity from biopirates.

1. Convention on Biological Diversity (CBD) Evolved from the Rio Earth Summit (1992)
2. Nagoya Protocol on Biodiversity, Negotiated in Japan (2010)

Prior to these two bargains, in 1994, one endeavour was made through the United Nations Draft Declaration on the Rights of Indigenous People for the security of indigenous information or TK. In this assertion Article 29 commands for the control, acknowledgement to full proprietorship and assurance of their social and licensed innovation. Moreover, the assertion likewise incorporates acknowledgement and insurance of their hereditary assets, seeds, meds, writing, plans, visuals and performing expressions. Accordingly, in 1989, The Indigenous and Tribal Peoples (Convention 169) perceived the yearnings of these people groups to practise authority over their own organizations, lifestyles and monetary turn of events and to keep

Table 27.4 Citations on biopiracy across the globe

Sr. No.	Name of country	Name of drugs	References
1	India	**Neem:** Several US and European patents on neem seed were received by W.R. Grace and Co. received in the year 1992 to 1995. Howerer, the neem patent was evoked by the European Patent Office on the basis of biopiracy	Marcia and De (2003)
		Turmeric: The Council for Scientific and Industrial Research (CSIR) and other government authorities from India challenged the patent for turmeric obtained by Dr. Suman K. Das and Dr. HariHar P. Cohly, faculty at the University of Mississippi, for its efficacy in the treatment of various untreatable conditions in humans. This challenge, often described in the press as a complex legal battle fought by CSIR, provides an interesting insight into patent reexamination and revocation practices at the end of the twentieth century	Emily (1999)
		Phyllanthus niruri: Crude drugs like *Phyllanthus niruri* which are well-known for their potential hepatoprotective activity in the traditional system of medicine. Patent has been granted to the Fox Chase Cancer Centre of Philadelphia, USA, for the manufacture of a medicine for treating hepatitis B	Miriam (2001)
		Recently, various well-known herbs and its preparations such as Ashwagandha, Neem, Tulsi, Brahmi, etc. were discovered in the West, known to have significant medicinal value without adverse side effects. However, pharmaceutical industries along with academic institutions have conflicts with western agencies for exploiting the traditional Ayurvedic practitioners over the intellectual property rights of their owned herbal products. Thereby, indulged in the significant economic loss of the traditional knowledge by such biopiracy practices. It was also observed that a patent for the antidiabetic properties of Karela, Jamun and Brinjal was granted by the US government to the nonresidential Indians, namely, Onkar S. Tomer and Kripanath Borah and their colleague Peter Gloniski. However, traditionally these substances are well-known for their antidiabetic activity and were used enormously in day-to-day diet in Indian cuisine. The practice of biopiracy is a sensible problem getting deep and systemic. Thus attention is required for a change	
2	Africa	***Banisteriopsis caapi* Mort.:** "Ayahuasca" a well-known drink obtained from the bark of the plant *B. caapi* Mort. from the Amazon vicinity used by the tribes of the Amazon basin as a religious ritual for healing and to diagnose the illness. A patent on the alleged variety of this plant was granted to an American Loren Miller obtained US patent (No. 5, 751 issued in 1986), The patent claimed that Da Vine represented a new and distinct variety of *B. caapi* Mort. primarily because of the flower colour. Thereafter, members of the Coordinator of the Indigenous Organizations of the Amazon Basin (COICA) protested	Homsy et al. (1999)

(continued)

Table 27.4 (continued)

Sr. No.	Name of country	Name of drugs	References
		against the patent as Miller could not have discovered it and should not have been granted such rights, which, in effect, appropriated indigenous and the drug is already used in TRM and cultivated for that purpose for generations. Nevertheless, on 3 November 1999, USPTO revoked and reexamined this patent, and the inventor convinced to USPTO the claims he had made on the said species. Because reconfirmation of the original claims made by the inventor was found appropriate, therefore patent rights were restored in favour of the inventor on 17 April 2001	
		Rosy Periwinkle: African local knowledge of herbs and their preparations have been inappropriately used by the pharmaceutical drug developers and Co. For instance, vinca is used as potential drug candidate for the treatment of cancer as it gives two molecules, i.e. vincristine and vinblastine. By knowing this fact, Michael Brown, an anthropologist, fought for several years for the rights of the native peoples and for giving back their ownership, dignity, knowledge and respect for their local knowledge of this herb	Michael (2003)
		Rooibos Tea: "red bush" is the rich source of various phyto-elements and herbal products including tea derived from this plant which is native to South Africa. It was noticed that the trademark purchased by the US company was abandoned for its further use	Amin et al. (2011)
3	South America	**Quinine:** Quinine isolated in the year 1820 is derived from the bark of *Cinchona* which was indigenous to the peoples of Andes for the cure of malaria. The indigenous knowledge of the drug was passed on to the Jesuit priests by the Peruvian in the year 1630. After, the disclosure of this traditional knowledge, a couple of pharma industries indulged in the developing and manufacturing of quinine and its synthetic derivatives for the treatment of arrhythmia	Walter and Veena (2011)
		Tamate: Another example of biopiracy is the drug called tamate obtained from tomato containing lycopene. This fruit was used as an anticancer drug by the aboriginal Amazonian Indians in Ecuador. However, the active ingredient of the tamate, lycopene, was discovered and developed by multinational pharma industry and made the formulation out of it for the treatment of cancer	Walter and Veena (2011)

up and build up their characters, dialects and religions inside the structure of the states in which they live. The show explicitly shielded the privileges of these people groups identified with the regular assets relating to their properties. In Article 15(1), these rights remember that of interest for the utilization, the board and protection of these assets.

27.10.1 Convention on Biological Diversity (CBD)

A global network structured the Convention on Biological Diversity (CBD) (1992) as another instrument to forestall the loss of biodiversity around the world. Understanding the significance of indigenous knowledge (IK) and the target of the CBD with respect to profit sharing, the Conference of Parties to the CBD, in its sixth gathering (COP 6) in 2002, embraced the Bonn Guidelines on Access to Genetic Resources and Fair and Equitable Sharing of the Benefits Arising out of their Utilization. It gives the rules to creating authoritative, managerial or strategy quantifies on access and advantage sharing. Articles 15 and 8 (j) of this show are significant and characterize the lawful standards and measures for bio-exchange, including the utilization of characteristic local fixings and frequent customary information concerned. CBD sets up three correlative goals, for example, the protection of biodiversity, its economical use and the reasonable and impartial sharing of the advantage emerging from their utilization which is otherwise called the "access and benefit-sharing (ABS)" component. The CBD perceives the estimation of the "information, developments and practices of indigenous and neighbourhood networks" for the protection and economical utilization of organic decent variety. The fundamental issue is that of access and benefit-sharing (ABS). ABS is actualized under Article 15 of CBD under which national governments are required to set up local laws and arrangements to permit access to hereditary assets. It means to destroy the wrongs of biopiracy just as to ensure the interests of TK holders.

27.10.2 The Nagoya Protocol

The Nagoya Protocol on Access to Genetic Resources and the Fair and Equitable Sharing of Benefits is a milestone bargain that was concocted remembering the expanding loss of biodiversity on Earth. The Nagoya Protocol indicates the methods by which the CBD can be applied. The Nagoya Protocol, particularly the ABS condition, calls for frameworks to be set up. These frameworks are required to drive the expenses brought about by pharmaceutical organizations during the medication revelation stage. The Nagoya Protocol could adversely affect the pharmaceutical business. Numerous organizations feel that the access and benefit-sharing statement will expand item improvement costs and confuse the medication revelation stage. As indicated by the convention, associations should pay a lot of their income and sovereignties to indigenous networks and host nations for the medication they create from hereditary assets. The amended patent framework will likewise add to the expense of medication advancement. Because of the guidelines and guidelines set somewhere around the Nagoya Protocol, associations would need to execute joint licences with the networks from whom they source assets.

27.10.3 India Fights Against Biopiracy

India is one of the 17 uber biodiversity nations with 2.4% of the worldwide land territory and records for 7–8% of the recorded types of the world, making it increasingly inclined to biopiracy. The turmeric case, in which India prevails with regard to upsetting a patent allowed by the United States Patent and Trademark Office on turmeric powder, was a milestone in the fight against "bio-theft". It was the principal case in which a third world nation prevailing in its issue with a remote patent in light of the fact that it depended on conventional information known to the nation for ages. In 1995, two exile Indians at the University of Mississippi Medical Center (Suman K. Das and Harihar P. Cohly) conceded a US patent (no. 5,401,504) on the utilization of turmeric in wound mending. The Council of Scientific and Industrial Research (CSIR), New Delhi, India, documented a reconsideration case with the USPTO testing the patent on the grounds of existing of earlier craftsmanship. CSIR contended that turmeric has been utilized for a huge number of years for mending wounds and rashes and, in this manner, its restorative use was not a novel innovation. Their case was upheld by narrative proof of conventional information, including antiquated Sanskrit content and a paper distributed in 1953 in the *Journal of the Indian Medical Association*. Regardless of intrigue by the patent holders, the USPTO maintained the CSIR protests and dropped the patent. The award of the patent for the fungicidal impact of neem oil in 1994 is another model. The European Patent Office (EPO) allowed a patent (No. 436257) to the US Corporation W.R. Elegance Company and the US Department of Agriculture for a strategy for controlling parasites on plants by the guide of hydrophobic extricated neem oil. In 1995, a gathering of universal NGOs and delegates of Indian ranchers recorded lawful resistance against the patent. They submitted proof that the fungicidal impact of concentrates of neem seeds had been known and utilized for a considerable length of time in Indian agribusiness to secure yields and, in this manner, can't be licensed. In 1999, the EPO discovered that as per the proof, all highlights of the current case were unveiled to general society preceding the patent application, and the patent was not considered to include an innovative advance. The patent allow on neem was denied by the EPO in May 2000.

27.11 Learning from such Bitter Experiences: Indian Government Framed Following Acts and Projects to Fight Against Biopiracy

27.11.1 Biological Diversity Act (2002), India

In India, empowering arrangements have been made for securing the conventional information in the Biodiversity Bill 2000. Area 36(iv) accommodates security of information on nearby individuals identifying with biodiversity through measures, for example, enrolment of such information and advancement of a sui generis framework. For guaranteeing fair sharing of benefits arising from the utilization of natural assets and related information, Areas 19 and 21 specify earlier endorsement of the National Biodiversity Authority (NBA) before their entrance. Segment 6 states that anyone looking for any sort of protected innovation rights on an exploration dependent on natural asset or information acquired from India needs to get an earlier endorsement from the NBA. The NBA will force benefit-sharing conditions. Area 18 (iv) specifies that one of the elements of NBA is to take measures to restrict the award of IPRs in any nation outside India on any organic asset acquired from India or information related to such natural assets. India's Biological Diversity Act, 2002 (BDA) looks to carry out a few responsibilities, including to manage access to natural assets to make sure about the fair offer in benefits emerging out of the utilization of organic assets and related TK to moderate and reasonably utilize natural assorted variety; to regard and ensure TK of nearby networks to protect sharing of advantages with neighbourhood individuals as conservers of organic assets and TK holders. The Biological Diversity Rules of 2004 further recognizes benefit-sharing strategies, for example, joint endeavours, innovation move, item improvement, training, mindfulness raising, institutional limit building and investment assets and states that applications will be resolved dependent upon the situation (James 2016).

27.12 Protection of Plant Varieties (PPV) and Farmers' Rights Act (2001)

The Indian enactment for the Protection of Plant Varieties and Farmers' Rights Act, 2001, additionally recognizes that the preservation, investigation, assortment, portrayal, assessment of plant hereditary assets for nourishment and agribusiness are basic to meet the objectives of national nourishment and dietary security as likewise for economic improvement of horticulture for the people in the present and the future.

27.13 Traditional Knowledge Digital Library (TKDL)

India has taken different activities with respect to the security of conventional information under licensed innovation rights, including the Traditional Knowledge Digital Library (TKDL), to ensure its customary information and to forestall award of wrong licences. A community-oriented undertaking among CSIR and the Department of AYUSH, Ministry of Health and Family Welfare, TKDL is a ready Indian exertion to help forestall misappropriation of customary information having a place with India at International Patent Offices. By recording the conventional information, lawfully, it becomes an open space information. Under the patent law, this implies it is viewed as earlier craftsmanship and consequently isn't patentable. Such a put-down account, in a structure effectively open to patent workplaces around the globe, would give every single such office a record of India's earlier craftsmanship. Patent analysts could easily check this database and reject any patent application that might be a minor duplicate of customary information. Being in document structure, it is satisfactory to patent workplaces that insist on a setup account of earlier craftsmanship, as in the United States. To this degree, it would forestall instances of "bio-robbery". Around the time the TKDL was built up in 2001, the TKDL master bunch evaluated that, yearly, approximately 2000 licences identifying with Indian restorative frameworks were as a rule incorrectly allowed by patent workplaces around the globe. TKDL hence empowers crossing out/withdrawal of wrong patent applications concerning India's customary information at zero expense and in scarcely any week's time. In any case, biopiracy is an emerging logical aggravation in the pharmaceutical business. It can be a market locally just as all-around verifiable truths, acquired information, conventional information, network astuteness and so on, so as to investigate new chance and cost sparing in pharmaceuticals innovative work. CBD and Nagoya Convention have attempted to streamline the contention between bio-privateers and unique asset bearers by proposing administrative understanding which despite everything should be overhauled and reimagined. In India, NBDA and TKDL are two beginning up activities to balance the biopiracy (James 2016).

27.14 Trademark and Copyright

A trademark is a word, expression, image or potential structure that recognizes and recognizes the wellspring of the products of one gathering from those of others. A service mark is a word, expression, image as well as a structure that recognizes and separates the wellspring of assistance as opposed to merchandise. A few models include brand names, mottos and logos. The expression "trademark" is frequently utilized from a general perspective to allude to the two trademarks and administration marks. Trademark rights originate from real "use". Accordingly, a trademark can keep going perpetually inasmuch as you keep on utilizing the imprint in business to show the wellspring of merchandise and enterprises. A trademark enrolment can likewise keep going everlastingly—insofar as you. A trademark is an interesting imprint (e.g. the name of the item introduced on signs or bundles) which can be utilized to secure a homegrown medication's notoriety. Makers distinguish a governmentally enlisted trademark with a circumnavigated R image, for example, "®". They offer the end client an assurance of value, administration and item picture. This is the explanation that numerous coffee shoppers pick McDonald's over other drive-through joints. The trademark close to the McDonald's name guarantees shoppers that the cheeseburgers will taste the equivalent in Beijing as they do in New York. On account of regular or homegrown meds, a trademark can convince the purchaser that the item is sold by a similar organization that makes other top-notch wellbeing nourishment items. This is especially significant on the grounds that loads of common items bear comparable names or bundling. For instance, there are in excess of 60 *Ginkgo biloba* items now available. A trademark must be particular and must be utilized persistently so as to be considered substantial. Non-utilization of the trademark in the market would set up by all appearance instances of deserting.

Trademarks likewise should be used at the time of its application submission. A candidate may record a plan to go through the trademark to 3 years preceding genuine use, however, the candidate must give evidence that the imprint is utilized inside the timespan showed. It is certainly worth investing the energy to pick the correct trademark for an item. Any trademark candidate needs to put forth an attempt to comprehend the US culture and market and pick an imprint that will be adequate to the end client. On account of regular or natural medication, this end client could be a common individual keen on ensuring or improving his/her wellbeing, a clinical specialist, a medication store, an oversaw human services model and so on. Along with these lines, the notoriety of conventional information can be defended to a limited degree by trademark framework; however it won't ensure the substance of such information. It will guarantee cautious security against demonstrations of going off non-certifiable items or administrations. Such utilization of imprint can unquestionably set up item devotion and shield against loss of notoriety coming about because of the utilization of the assignment of conventional information for subsidiaries items. Especially similar to the utilization of trademarks significantly after the termination of licences, especially if there should arise an occurrence of pharmaceutical licences, to draw out item devotion. On the occasion, if a patent denies the indigenous network from selling the item, they could enlist the trademark and consequently permit out the utilization of the trademark so as to permit organizations to guarantee validity. Existing techniques could be performed on items and affirmed by a network as a strategy for increasing the value of an item with the possibility to gather sovereignties on the items sold. While, a land sign distinguishes merchandise as starting in a region or district, or region in a region, where a given quality, notoriety, or another trait of the products are inferable from its geological beginning. Like trademarks, when related with an item, it decidedly ascribes a known quality to the item that is related with a particular topographical area (James 2016).

On account of advertising restorative plant items and even in selling the plants, all things considered, trademarks can be utilized. These imprints can be utilized by singular merchants and furthermore by aggregate gatherings. So as to guarantee the quality, confirmation stamps additionally can be enrolled and applied to explicit therapeutic plant items. Ayurvedic firms, for example, Dabur and Zandu, use trademarks. Restorative plant cultivars and dealers can utilize the course of trademarks. "India Organic" is an accreditation trademark conceded by the Spices Board based on consistency with the National Standards for Organic Production. The logo has an explicit shading plan. Therapeutic plants by their very nature so as to keep up the quality must be created naturally and, along these lines, could utilize this, on the off chance that they fulfil the conditions determined. Another affirmation mark is the FPO mark granted by the Ministry of Food Processing Industries which is utilized for prepared natural product items. It guarantees that the item is fabricated in a sterile ‹food-safe› condition.

The AGMARK is utilized on farming items to guarantee that they comply with a lot of guidelines affirmed by the Directorate of Marketing and Inspection, an office of the Government of India. These imprints ought not to be utilized on merchandise without the endorsement of the equipped affirming specialists. Else it might draw in punishments and fines as the instance of encroachment of geographical indication rights. Therefore, numerous IPRs can be utilized for getting insurance and for receiving rewards on account of therapeutic plants. In any case, these conceivable outcomes have not been investigated by the Indian cultivars ideally. In numerous examples, more than one IPR can be abused. The absence of information and mindfulness could be viewed as the primary explanation behind the equivalent. While common society bunches have been dynamic in battling against bio-theft, the positive course additionally should be investigated. This is especially the situation with numerous fragrant plants, for example, sandalwood, lavender (from Kashmir Valley, Himachal Pradesh and Uttarakhand) and so on. On account of items sent out to different nations, there is an additional peril, since, if certain parts or characteristics of those items have been protected or even trademarked in those nations, it might cause the exporters much avoidable difficulties.

27.15 Copyright

It is a selective right given by law to the makers of proficiency, melodic and creative universes, movies and records. These rights are managed lawful security to forestall unlawful re-assurance of such works. The object of the Copyright Act is to shield the essayist and the craftsman from unlawful re-creation, counterfeiting, robbery and impersonation. In any case, the creation of an indistinguishable thing through autonomous inventive research isn't disallowed, and there can be no risk for encroachment in such cases. The law fundamentally is worried about the negative right of forestalling the copyright of physical material, existing in the field of writing and craftsmanship. The copyright empowers the creator to guarantee initiation of the work just as to control or guarantee harms in regard to bending or different adjustments in the work or whatever other activity which is biased to his respect or notoriety comparable to that work. While if there should be an occurrence of licences, plans and trademarks, the rights can be procured uniquely by enlistment, in the event of copyright enrolment isn't essential and it stays alive naturally. Like all the licensed innovation rights, duplicate rights insurance as acquired by the residential law of a nation and the sufficiency of the current copyright Act, 1957.

Copyright insurance can in all likelihood be stretched out to legends. There are numerous nations which secure old stories as protected innovation under the copyright law. For instance, Angola ensures as old stories, all abstract, masterful and logical works made by creators dared to begin in specific areas or ethnic networks, went from age to age, secretly or on the whole or by different methods, and establishing one of the essential components of the customary social legacy. Benin Law on the Protection of Copyright, 1984, further broadens the extent of legends and secures all abstract, masterful, strict, logical, mechanical and different customs and creations made by the national networks gave from age to age and subsequently comprising the essential components of the national social legacy. Under Article 10 of the Benin Copyright Law, legends have a place with native to the national legacy. Copyright Law, 1990, of Cameroon covers and secures society stories, people verse, mainstream melodies, and instrumental music, society moves and shows, just as imaginative articulations, customs, and creations of well-known craftsmanship inside the articulation old stories. A few different nations likewise have sanctioned extraordinary arrangements to ensure their one-of-a-kind legend. For example, Congo Law on Copyright and Neighbouring Rights, 1982; 213 Ghana Copyright, 1985; 214 Law Adopting Provisions on Copyright and Neighbouring Rights in the Revolutionary People's Republic of Guinea, 1980; 215 the Copyright Act, 1966 of Kenya; 216 Liberian Act Adopting another Patent, Copyright and Trademark Law, 1972; 217 Dahir (Act) Relating to the Protection of Literary and Artistic Works, 1970, of Morocco; 218 Law Governing Copyright, 1983, of Rwanda; 219 the Copyright Act 1973 (as changed in 1986) of Senegal; 220 Law on Protection of Copyright and Neighbouring Rights, 1986 (Zaire); 221 Code of Intellectual Property Act (Sri Lanka), 1990; 222 Barbados Copyright Act, 1982; 223 Law on the Protection of Copyright, Folklore and Neighbouring Rights (Togo), 1991. It is clear from the above conversation that the abstract, aesthetic and strict, logical, mechanical and different customs and creations made by the national or ethnic networks or obscure or unidentified creators, yet gave from age to age can fantastically be given lawful acknowledgement and security under the copyright law as protected innovation. A wide scope of customs which structure some portion of social legacy can be ensured under this choice. It is very well may be contended that this sort of assurance would furnish conventional and indigenous networks with lawful intends to forestall any demonstrations that mutilate the paternity privileges of the network or influence their honesty as TK holders. Be that as it may, in this viewpoint, distinguishing proof of the creator and term of security will be worried in the Copyright Law. In spite of the fact that copyright obliges the idea of joint initiation, copyright can't be vested over the whole clan or network since copyright doesn't perceive network proprietorship. Copyright doesn't perceive unending insurance too. In such a circumstance until we build up a sui generis framework or the idea of network possession in the Indian Copyright Law, the standard law must win (Anantha 2013).

27.15.1 Copyrights for Herbal-Related Matter

- Copyrights are also registrable.
- It includes specific text matter, advertisement materials, jingles, music elements, documentaries, product promotion literatures, clinical trial reports, books, PIL's, training manuals, etc.
- State that the material is copyrighted and who owns the same.
- Provisions for reproductions to be cited.
- Photocopying for personal use is okay.
- Is a matter of severe discussion and legal battles in pharmaceuticals area for new drugs where the published data of the new drug belongs to the innovators and cannot be cited by the generic applicant.

- Large firms have policy of the same under data management.
- Exact copying to be avoided, and if needed and done provide correct and full references to the original copyrighted material.
- Many of us also quote from published material, many of which are copyrighted. Doing so without giving reference and also doing so where "exact parts/paras/ sentences" are copied may turn out as infringement of copyrights. Copying may also turn out to become cases of "plagiarisms". One needs to be careful in this area, an area where many of us do not have good awareness, let alone competency.
- A few years ago, to check if any copyrighted material is being used, it was very difficult and time-consuming as it had to be done by manually searching "potential areas"; with the advent of IT tools, it is much easier now and can be done as easily as "a click of the mouse". Examples of sites called "Turnitin" and similar cites are available to trace if any copyrighted material is used in any submission. Most such sites may need to be subscribed to and accessed with login IDs and passwords. In one paper that I was reviewing, I was given access to this site, and it was wonderful to get so much information, which almost lists which sentences have been lifted from which publication (Anantha 2013).

27.16 Protecting Traditional Knowledge Under the Indian Copyright Law

The Indian Copyright Law as such doesn't accommodate insurance of articulation of legends or assurance of customary information on indigenous individuals; anyway a surmising can be drawn from Section 31A of the Indian Copyright law, which secures the unpublished Indian work. The question that emerges is whenever in all actuality, regardless of whether copyright law would be adequate for the insurance of customary information? Some of the major drawbacks in protecting traditional knowledge with copyright are as follows:

27.16.1 Authorship

Under the Indian Copyright Law, insurance is given to the creator or proprietor of the work. Customary knowledge is network claimed information and is commonly evolved and developed from generation to generation. In such a case following the creator of the conventional information isn't just troublesome but practically unthinkable.

27.16.2 Protection for Limited Time

Indian copyright insurance is time restricted as it is only allowed for a specific term of 60 years. Customary knowledge is objective, and it ought to have interminable security as opposed to constrained insurance.

27.16.3 Fixed Form

To secure any work with copyright under the Indian Copyright Law, it is necessitated that the work must be available in an unmistakable structure. A fixed type of customary information is elusive. In the greater part of the cases, customary information is ignored ages in a network in the type of stories. These accounts are once in a while accessible in fixed structure (Section 17 of the Copyright Act 1957).

References

Aguilar G (2001) Access to genetic resources and protection of traditional knowledge in the territories of indigenous peoples. Environ Sci Policy 4:241–256

Albert WC, Jason CL (2004) Intellectual property protection of natural products. Asia Pacific Biotech News 8(10):540–545

Amin R, Blumenthal Mand, Wayne S (2011) Rooibos tea trademark dispute settled 68 (2005). Rooibo tea: noting that '(r)ooibos, otherwise known as red bush. http://www.herbalgram.org/herbalgram/articleview.asp?a2891

Anantha DB (2013) Intellectual property rights for herbal products. Pharmabiz.com

Battiste M, Henderson JY (2000) Protecting indigenous knowledge and heritage: a global challenge. Purich Publishing Ltd., Saskatoon

Bayram S (2005) Patent research on 15 medicinal plants which are significant for phytotheraphy and have preparations in the European market. J Ethnol 51:59–73

Benjamin C, Bhavani S, Richard B (2007) Bias from Farmers Self Selection in genetically modified crop productivity estimates: evidence from Indian data. J Agri Eco 58:24–36

Brahmi P, Sanjeev S, Dhillon BS (2004) The protection of plant varieties and farmers rights act of India. Curr Sci 86(3):10

Bhat MG (1996) Trade-related intellectual property rights to biological resources: socioeconomic implications for developing countries. Ecol Econ 19:205–217

Bodeker G (2000) Indigenous medical knowledge: the law and politics of protection—this study was presented at the Oxford Intellectual Property Research Centre Seminar in St. Peter's College, Oxford on 25th January 2000

Boyd MR (1996) The position of intellectual property rights in drug discovery and development from natural products. J Ethnol 51:17–27

Cheeptham N, Chantawannakul P (2001) Intellectual property management and awareness at the university level in the biotechnology era: a Thai perspective. World Patent Inf 23:373–378

Cordell GA (2000) Biodiversity and drug discovery—a symbiotic relationship. Phytochemistry 55:463–480

Correa CM (1998) Implementing the TRIPS agreement in the patents field: options for developing countries. J World Intellect Prop 1(1):75–99

Correa CM (2002) Protection and promotion of traditional medicine implications for public health. In: Developing countries. University of Buenos Aires, by support of Rockefeller Foundation, pp 13–52

Dutfield G (2003) Protecting traditional knowledge and folk-lore, issue paper 1. ICTSD and UNCTAD Project on IPRs and Sustainable Development, Geneva

Emily M (1999) The neem tree patent: international conflict over the commodification of life. J Int Comp Law Review, 279

Ganguli P (2000) Intellectual property rights: mothering innovations to markets. World Patent Inform 22:43–52

Gollin MA (1999) Legal and practical consequences of biopiracy. Diversity 15:7–9

Homsy J, Katabira E, Kabatesi D et al (1999) Evaluating herbal medicine for the management of *Herpes Zoster* in human immunodeficiency virus-infected patients in Kampala, Uganda. J Altern Complement Med 5(6):553–565

James TC (2016) IPR issues related to medicinal and aromatic plants (herbs & their allied products). J Tradit Folk Pract 2, 3, 4(1):7–17

Kartel NT, Sych NY, Tsyba NN, Strelta VV (2006) Effect of combined activation on the preparation of high porous active carbon from granulated post-consumer polyethylenelecephtalate. Appl Surf Sci 252(23):8062–8066

King SR, Carlson TJ, Moran K (1996) Biological diversity, indigenous knowledge, drug discovery and intellectual property rights: creating reciprocity and maintaining relationships. J Ethnol 51 (1–3):45–57

Marcia E, De G (2003) Bio-piracy: the appropriation of indigenous peoples' cultural knowledge. J Intellect Prop Law, 179

Michael F (2003) Who owns native culture? Economic law review 136 suggesting that assertions about rosy periwinkle conceal a much more complex story by pointing out the potential difficulty of disentangling proprietary claims originating in folk traditions

Miriam L (2001) Quinn, "protection for indigenous knowledge: an international law analysis" 14 St. Thomas Law Review, 287

Notification from Secretariat of the Convention on Biological Diversity 2006 Decision VIII/5 E on Article 8(j) and related provisions: Development of elements of *sui generis* systems for the protection of the knowledge, innovations and practices of indigenous and local communities. www.biodiv.org/doc/notifications/2006/ntf-2006-073-8j-en.pdf

Okan A, Mine GO (2007) Traditional medicine and intellectual property rights. J Fac Pharm Ankara 36(2):135–151

Patwardhan B (2005) Ethnopharmacology and drug discovery. J Ethnol 100:50–52

Rocha AB, Lopes RM, Schwartsmann G (2001) Natural products in anticancer therapy. Curr Opin Pharmacol 1:364–369

Timmermans K (2003) Intellectual property rights and traditional medicine: policy dilemmas at the interface. Soc Sci Med 57:745–756

Turner DM (1996) Natural product source material use in the pharmaceutical industry: the Glaxo experience. J Ethnol 51:39–44

Udgaonkar S (2004) The protection of medicinal plants in India. http://envis.frlht.org.in/sangeeta.htm

Walden I (1998) Preserving diversity: the role of property rights. In: Swanson TM (ed) Intellectual property rights and biodiversity conservation. Cambridge University Press, Cambridge, pp 176–197

Walter HL, Veena R (2011) Ethics and practice in ethnobiology: analysis of the international cooperative biodiversity group project in Peru (Unpublished manuscript at 2, on file with the wash univ school of law for interdisciplinary studies). http://law.wustl.edu/centeris/confpapers/PDFWrdDoc/lewisramani.pdf

WIPO (2005) Intellectual property and traditional knowledge booklet no. 2, WIPO Publication no. 920

Yanhuai L, Yanling S (2004) China traditional Chinese medicine (TCM) patent database. World Patent Inform 26:91–96

Zhang X (2000) Traditional medicine and its knowledge. UNCTAD expert meeting on systems and national experiences for protecting traditional knowledge, innovations and practices. Geneva, 30 October–1 November. World Health Organization (WHO), Geneva

Safety and Regulatory Issues on Traditional Medicine Entrusted Drug Discovery 28

Divya Sankaramourthy, Kavimani Subramanian, and Sudha Rani Sadras

Abstract

Traditional medicine refers to the therapeutic approach of diagnosing, treating or preventing ailments with the aid of knowledge acquired through ancient practices and beliefs using phytocompounds and animal- or mineral-based compounds, applied singularly or as a combination. Around 70–95% people living in Asian, African, Latin American and Middle East countries have been utilizing indigenous knowledge as traditional medical practice for health care. Different practices are adopted in different parts of the world for traditional medicines and are given in the form of dietary supplements, health foods, functional foods, phytoprotectants, over-the-counter medicines, etc. Remarkably, the interest in traditional medical systems has increased globally owing to increased cost of modern medicines associated with development of drug resistance as well as lack of treatment to several chronic and emerging diseases. Nevertheless, all medicines have their own adverse effects and indigenous medicines are no exception. Most of the reported side effects of traditional medicines arise due to failures in identifying precise plants, adopting good manufacturing practices, standardizing products and preventing contamination of products and also because of incorrect preparations. Along with this, certain other factors including soil quality, temperature, light, plant age and time of harvest also influence the quality of traditional medicines. Keeping these flaws in mind, if meticulous scientific approach with focussed attention to ensure quality, safety and efficacy

D. Sankaramourthy · K. Subramanian (✉)
Department of Pharmacology, Mother Theresa Post Graduate and Research Institute of Health Sciences, Gorimedu, Puducherry, India
e-mail: kavimanis@yahoo.co.in

S. R. Sadras
Department of Biochemistry and Molecular Biology, School of Life Sciences, Pondicherry University, Kalapet, Puducherry, India

S. C. Mandal et al. (eds.), *Evidence Based Validation of Traditional Medicines*,
https://doi.org/10.1007/978-981-15-8127-4_28

of traditional medicines is adopted, it will certainly benefit mankind to overcome the consistently increasing threat of drug resistance for modern medicine. The current chapter presents discussion on the safety and regulatory issues associated with practice of traditional medicines and appraise the operational approaches towards entrusted drug discovery.

Keywords

Indigenous medicine · Traditional medicine · Safety · Regulatory issues · Ayurveda · Siddha · World Health Organization

Abbreviations

GACP	Good agricultural and collection practice
GMP	Good manufacturing practice
TM	Traditional medicine
WHO	World Health Organization

28.1 Introduction

As defined by the WHO, "Traditional medicines include herbal medicines composed of herbs, herbal materials, herbal preparations, and finished herbal products, that contain as active ingredients parts of plants, or other plant materials, or combinations thereof. Traditional medicines may also use animal parts and or minerals" (WHO 2002). Traditional medicine applies the therapeutic approach of diagnosing, treating or preventing ailments with the aid of knowledge acquired through ancient practices and beliefs using phytocompounds and animal- or mineral-based compounds singularly or as a combination. They are practised in different ways around the world in the form of home remedies, dietary supplements, health foods, functional food, phytoprotectants (World Health Organization 2003a, b, c), over-the-counter medicines, etc. (World Health Organization 2005). However, increasing costs of modern medicines, lack of treatment to several chronic and emerging diseases and increasing drug resistance had also expanded the global interest to traditional medicine.

Though the importance of traditional medicine is valued everywhere, nevertheless, the indigenous medicines also have their own adverse effects. One such ideal example is liquorice (*Glycyrrhiza glabra*), which is used worldwide conventionally for treating sore throat, cough, arthritis and weight loss induction. But it is said that prolonged use of liquorice induces acute kidney injury (hypokalaemic nephropathy), amenorrhoea, pseudo-aldosteronism, hypertension, heart failure and rhabdomyolysis (Foguema and Foguemb 2014). Another plant named *Catharanthus roseus* or rosy periwinkle, used in Eastern Africa for its anticancer properties, also causes adverse effects like medullary aplasia, leucopenia, incoordination of movements, convulsions, fatigue, mucositis, constipation and neutropenia (Morel and Talbot 2010).

Furthermore, majority of side effects of traditional medicines also arise from factors such as improper plant identification, poor manufacturing practices, lack of product standardization, contamination of products, substitution or incorrect preparations or dosage. The quality of preparation of traditional medicines was also influenced by environmental factors such as light, temperature, soil quality, period and time of harvest, age of the plant, etc. This demands future research to ascertain the quality, safety and efficacy of traditional medicine.

28.2 Significance of Traditional Medicines: An Overview

The existence and practice of traditional medicines were evident from 1552 BC (Joachim et al. 1890), whereas the rise of modern medicines took place in the middle of the twentieth century starting from commercial production of penicillins (Sevillano and Ramos 2007). Thereby, since ancient times, there is an indispensable place for traditional medicine in several countries. They are practised as diverse therapies throughout the world under indigenous names such as traditional European medicine, traditional Chinese medicine, traditional Korean medicine, traditional African medicine, Ayurveda, Siddha, Unani, ancient Iranian medicine, Persian medicine, Islamic medicine, Muti and Ifa. However, all these traditional medicines centre around the balance of mind, body and environment with an emphasis on health and natural way of living. Around 70–95% of the Asian, African, Latin American and Middle East population depend on traditional medicines for their primary health care (WHO 2002; World Health Organization 2003a, b, c; World Health Organization 2004). It is a well-known fact that nearly 25% of drugs currently in market like aspirin, artemisinin, ephedrine and paclitaxel are formulated from bioactive compounds isolated from traditionally used medicinal herbs. Undeniably, modern anticancer drugs such as vincristine and vinblastine carry their origin from ancient medicinal plant *Catharanthus roseus* or rosy periwinkle which is indigenous to Madagascar (Newman and Cragg 2007). In that way, the knowledge acquired from traditional medicines and their contribution to the development of modern medicines is irrefutable. Ayurveda, one of the world's oldest medical systems, uses natural products in combination for remedy. One such well-known medicine in Ayurveda is guggul. Guggul, an oleo-gum-resin acquired from *Commiphora mukul* tree, is known for its antibacterial, anti-inflammatory (Gujral et al. 1960), hypolipidaemic and anticoagulant (Mester et al. 1979) activities and is recommended in Ayurveda for treating various ailments such as inflammatory diseases, muscle spasms, urinary infections, skin diseases, obesity, rheumatoid and osteoarthritis (Gujral et al. 1960).

Many such traditional medicines are being used in various countries for recuperating from several health conditions including sleeplessness, headaches, hyperlipidaemia, increased blood pressure, diabetes mellitus, skin diseases, infections of respiratory and gastrointestinal tracts, menopause, etc. (Calixto 2000;

World Health Organization 2004). Moreover, it has been reported that countries like Brazil have used traditional medicines for treating 89% of reported cancer cases (Holtz 2007). This reflects the significance of traditional medicines which are used worldwide under the labels of alternative medicines, complementary medicines, etc.

28.3 Challenges Faced by Traditional Medicines and Its Regulatory Issues

28.3.1 Challenges Related to Safety and Efficacy Assessment

Traditional medicines are generally administered as combinations of medicinal herbs with animal products, minerals, etc. Each medicinal plant may possess several bioactive compounds, and hence, a medical formulation might comprise of even more numbers of unknown constituents. As the combinational treatment fails to identify or isolate the active constituents responsible for the pharmacological effect, there is a lack of specificity in the treatment resulting in scepticism about the efficacy of traditional medicines. Inadvertently, treating an ailment with such combinations may increase the incidence of toxic components along with therapeutic ingredients, thereby questioning the safety of traditional medicines. Potential adverse effects such as acute nephropathy, vomiting, diarrhoea, hepatitis, hypertension, cardiac failure, duodenal ulcer, convulsions, liver failure, sexual dysfunction, hypoglycaemia, hypotension, skin irritations and confusions have been documented worldwide upon the use of indigenous medicinal plants for therapeutic purposes (Luyckx 2012; Morel and Talbot 2010; Moolla and Viljoen 2008; Brendler and Wyk 2008; Jain and Singh 2013; Halberstein 2012). In that way, research on safety and efficacy demands primary importance for the continued development of traditional medicines.

28.3.2 Challenges Associated with Ensuring Quality of Herbal Medicines

The therapeutic potential of traditional medicines is closely associated with its quality, safety and efficacy. Extrinsic factors such as failure to identify correct plant species, adopting substandard manufacturing protocols, inadequacies in standardization of products, adulteration of products, inappropriate preparation of dosage, etc. will influence the quality of preparation of medicine. Additionally, other factors such as agroclimatic conditions like light, temperature, soil quality, plant age, harvest time etc. may also affect the quality and consistency of the medicinal product (Marcus and Grollman 2002; Peters et al. 2003; Li et al. 2008). It becomes extremely challenging to perform quality control checks for each of the raw material used in the formulation. However, good manufacturing practice (GMP) stipulates several inevitable methods for quality control of raw materials that include precise recognition of appropriate medicinal plants, implementing suitable methods for their proper storage, cleaning, etc.

28.3.3 Challenges Associated with Safety Monitoring of Indigenous Medicines

World Health Organization reports have indicated that several of the indigenous medicines sold in the market for remedial purpose were found to be frequently contaminated with toxic substances such as heavy metals, mycotoxins, pesticides and alternate plant species (Teschke et al. 2013). Preparation of traditional medicines with misidentified plant species, adulterated plant materials, undeclared medicines and toxic contaminants may also imperil the consumers to adverse effects. Overdosage, misuse of herbal medicines by consumers or health-care providers and simultaneous use of traditional medicines with other medicines may also lead to hazardous effects. Since majority of traditional medicines are most commonly used for self-care purpose, there is an impending necessity to create awareness among the consumers on the proper use of traditional medical practices.

28.3.4 Dearth of Knowledge Update Among National Drug Authorities on Traditional Medicines Affects Development of National Policies on Traditional Medicines

It has been speculated that a deficit in the depth of knowledge on traditional medicines prevailing among the national drug authorities and inadequacies in suitable assessment methods have greatly impacted formatting or updating of national policies on traditional medicines.

The shortcomings of modern medicine are its symptomatic treatment approach that reveal therapeutic gap between complete relief and cure of the disease. Though the treatment approach in modern medicines has helped mankind in handling several life-threatening diseases such as cancer, tuberculosis, AIDS (Manosuthi et al. 2006), etc., it ails with unavoidable adverse drug reactions and development of multi-drug resistance (Tanwar et al. 2014). Moreover, modern medicines have failed to provide complete cure for iatrogenic diseases such as Parkinsonism, Alzheimer's disease, stroke, etc. On the other hand, knowledge from traditional medical practices on the naturalistic cure for the above-said diseases remains to be unveiled. Nevertheless, several traditional medicines were used for centuries as nature's cure for several diseases such as *Mucuna pruriens* for Parkinsonism (Sathiyanarayanan and Arulmozhi 2007), *Gymnema sylvestre* for diabetes (Tiwari et al. 2017), *Vinca rosea* for cancer (Vipasha et al. 2016), *Artemisia afra* for tuberculosis (Buwa and Afolayan 2009) and ashwagandha for stroke (Mishra and Singh 2000). Hence, gaining sufficient knowledge from historic roots of older traditional medical systems and implementing rigorous research efforts to ascertain the hidden therapeutic benefits of traditional practices will benefit mankind enormously to handle life-threatening diseases.

28.3.5 Challenges Faced by Countries in Formation of Regulations for Herbal Medicines and WHO Support

Invariably, various countries have faced persistent challenges in developing and standardizing their traditional medicines that differ by culture, history and varied applications. In case of conventional pharmaceuticals, though it may take 10–12 years to develop a conventional drug (a single pure compound) and cost several hundreds of million dollars (WHO 2002; Sucher and Carles 2008), the money invested could be augmented after patenting and marketing the product. On the other hand, developing a traditional medicine has more difficulties as it involves longer periods of time for steady research to optimize the product and even more money. In addition, natural sources like plants and minerals used for developing the indigenous drug could not be patented. Even though major proportion of traditional medicines are from plants, it is difficult to hold trademarked rights over a tree or flower where anyone can grow the plant for similar medicinal purpose (McIntyre 1999). Hence, the money spent for standardizing and developing a traditional medicine cannot be redeemed easily.

Based on the regulations applied to food and medicine in each country, the categorization of medicinal plants may also differ under the classification of functional foods, dietary supplements, herbal medicines, etc. Thereby, lack of uniformity in development of drug regulations for traditional medicines worldwide made the WHO to develop WHO traditional medicine strategy (WHO 2002). The four major objectives of WHO traditional medicine strategy were framing a policy; enhancing safety, efficacy and quality; ensuring access; and promoting rational use.

A resolution on traditional medicines was adopted at the 56th World Health Assembly in May 2003 which requested the WHO to support member states to provide internationally acceptable guidelines, technical standards and evidence-based information in framing policy and regulations to govern the safety and efficacy and quality of traditional medicines. In turn, the WHO also came out with numerous supportive documents such as policies on research and evaluation of traditional medicines, good agricultural and collection practice (GACP), good manufacturing practice (GMP) and the duty of pharmacists (World Health Organization 1998a, b; World Health Organization 2000; World Health Organization 2003a, b, c; World Health Organization 2007).

28.4 WHO Global Survey on Regulation of Traditional Medicines

In 1994, the WHO requested various countries to provide information regarding the regulatory affairs with respect to herbal medicines, and only 52 out of 191 countries responded positively, and based on the details obtained, the WHO published a review in 1998 on the regulatory policies adopted worldwide for herbal medicines (World Health Organization 1998a, b). Subsequently in 2001, the WHO prepared a global survey questionnaire with items like whether those countries practicing traditional and alternative medicines would require technical guidance and support from the WHO, whether the policies and regulations on traditional medicine were reviewed, etc. From the acquired positive responses, the WHO created a global database representing 74% of its member states. This served as a primary reference line to all the member states to collect, share and update the information on national policies and regulations on traditional and alternative medicines. In addition, the database provides comprehensive information on laws and regulations, regulatory requirements, monographs and pharmacopeia but limits its access to national health authorities. However, efforts have been taken to expand and update the database in the future by conducting regular surveys and data-gathering activities. The WHO also plans to update the outcomes of the exercises for developing WHO's Traditional Medicine Strategy. In 2009, a resolution on traditional medicine (WHA 62.13) was put forward by the World Health Assembly emphasizing the national governments and member states to share their knowledge on traditional medicines while formulating national policies and regulations with the help of conventional and traditional practitioners (World Health Organization 2009). The resolution also fosters to expand the safe and effective use of traditional medicines and their inclusion in national health-care systems through pertinent research.

28.5 National Policy on Traditional Medicines

The following key elements are considered to be essential while constituting national policy on traditional medicine as per the WHO survey on traditional medicine strategy (WHO 2002). The requirements include a classic definition on tradition medicine and stipulation for devising laws and regulations as well as for inclusion of intellectual property right issues. The proposed laws and regulations on traditional medicine can also be included in the national policy.

The global survey conducted by the WHO exposed that 32% (45 out of 141) of member states possessed a national policy on traditional medicine and 64% (90 out of 141) of member states did not frame any policies while the remaining failed to respond. In addition, 44 out of 45 countries with national policy have also mentioned their year of issue, and 51 (54%) out of 90 member states without national policy have specified their progressive status towards development of a national policy.

Laws and regulations on traditional medicine form the preliminary stage of legislative procedures. The law comprehends legal conditions required for systematizing traditional medicines in accordance with national traditional medicine policy. Legal procedures for traditional medicine field concerning licensing of manufacturers, medical practitioners, sales practice, production of primary stuffs required for the manufacture of end products and counselling of professionals are covered by law. The administrative and technical goals of law are established by a secondary tool termed as regulation. The regulation encompasses the obligatory rules related to the responsibilities of licensed practitioners, traditional medicine manufacturers and penal sanctions for those who violate the law. However, the survey on national policy on traditional medicine enquired about the existence or promulgation of national law and regulation in each member state following the indicators for monitoring national drug policies (Brudon Jacobowicz et al. 1999). Consequently, it has been evident that nearly 54 (38%) countries reported (year: 1987–2003) to have laws and regulations provided the composition and inclusiveness of laws and regulations vary from one country to the other. On the contrary, 60% of countries (80) had failed to develop laws and regulations, and about 2% of countries (3) had failed to report.

Organizing national programmes, setting up of national office and establishment of expert committee and national research institutes are also considered to be important for maintaining national policy and traditional medicine drug development. The national programmes are usually conducted by the Ministry of Health in order to accomplish the goals in time with national policy or legislation. The WHO survey on national policy on traditional medicine have revealed that nearly 40 (28%) member states have created national programmes on TM, whereas 93 (66%) countries haven't organized national programmes. Furthermore, 31 (33%) countries have indicated that the development of national programme were in progress.

A national office is a governing body associated with national authority and is established for addressing the issues related to traditional medicines. The results of the WHO survey on national policy on traditional medicine revealed that 75 (53%) countries have national office, 61 (43%) countries do not have national office and 5 (4%) countries haven't responded to the question. Among 61 countries without national office, 19 (31%) have indicated that the process of setting up national office was in progress.

Similarly, expert committees structured by national governments are considered to be indispensable for revising, policy-making and propounding practical commendations concerning traditional medicines. The WHO survey on national policy on traditional medicine exposed that 61 (43%) countries positively responded to have an expert committee, whereas 72 (51%) countries responded "No" for an expert committee and 8 (6%) countries where nonresponders.

In the same way, national research institutes also play an essential role towards drug development of traditional medicines. The WHO survey on national policy on traditional medicine indicated that only 28% of its member states have developed a national research institute for traditional medicine for the period of 1970–2003 which is expected to increase over a period of time.

28.6 Regulatory Status of Traditional Medicines

The regulatory statuses of traditional medicines were enquired among the member states in order to classify them under separate regulatory categories. The member states were requested to broadly classify their traditional medicines under the following regulatory categories such as prescription medicines (drugs that are legally dispensed under physician's prescription), over-the-counter medicines (drugs that do not require medical prescription to be dispensed), self-medication only (drugs that do not require medical supervision during administration), dietary supplements (nutritional add-ons such as vitamins, minerals, herbs or amino acids), herbal medicines under separate regulatory category (medicines derived from plant origin), health foods (foods marketed under definite health claims), functional foods (super foods of extra nutrient value which claims physiological benefits, also known as nutraceuticals), etc.

The survey results indicated an overwhelming response with more than 130 member states providing information. As each drug was categorized in more than one class, about 97 responses were recorded under the over-the-counter medicines category, 50 as prescription medicines, 47 as dietary supplements, 40 as self-medication only, 25 as herbal medicines, 15 under health foods, 9 under functional foods and 12 under excluded (other) categories. Of 130 member states, 23 did not show proper regulatory status for herbal medicines, while 13 countries showed the categories included in their legislation for herbal medicines as herbal remedies; supportive medicines; homeopathic, bioactive, probiotic substances; etc.

28.7 Regulation on Claims, Pharmacopeial Database and Monographs for Herbal Medicines

28.7.1 Claims

Claims are the assertions or statements labelled on a product assuring promising results upon usage. The probabilities of claims on traditional medicines were queried in the WHO global survey. If so, the member states were asked to mention the possible category of claim on their traditional medicines, specified in accordance with their law or regulation. Definitions on various forms of claims were also furnished in the survey. The claims include medical claims (specifying treatment, cure or prevention of a disease or modification of a physiological condition), health claims (specifying definite health benefits, nutritional function or recommended dietary practice), nutritional content claims (specifying nutritional composition of a product) and functional claims (specifying beneficial effect on the body).

The results of the survey revealed that 103 (73%) countries sell their traditional medicines with appropriate claims, while 34 (24%) countries do not permit selling with claims. Furthermore, the responding (103) countries came up with chosen categories of claims to their traditional medicines. Of that, 87% were medical claims, 60% health claims, 48% nutritional content claims and 38% structure/functional claims. Six other member states also mentioned other claims such as cultural, traditional and cosmetic use and against witchcraft and accidents. Exceptionally the survey also exposed five member states picking "No claim can be made according to the law" category.

28.7.1.1 Pharmacopeial Database

A pharmacopoeia is an official book published by the authority of a government or a medical or pharmaceutical society (Holmes 1911). It serves as a drug reference standard containing details about the drugs used in the medical practice, composition, physical and chemical properties and mode of preparation (Churchill Livingstone 1989). In addition, assay methods for ensuring purity, preservation of quality and biological potency as well as for identifying active elements may also be included in pharmacopoeia.

When the WHO global survey on traditional medicine queried about the existence of national pharmacopoeia, 34 (24%) countries responded "yes" on having a national pharmacopoeia, while 104 countries (74%) reported "no", and subsequently, of the 104 countries, 26 have specified that the development of national pharmacopoeia were in process.

28.7.2 Monographs for Herbal Medicines

Monographs are the reference standards comprising detailed research-based information on safety, efficacy and quality control of widely used medicinal plants. They are generally prepared for supporting the manufacture of herbal formulations, for developing new monographs on unknown medicinal plants and for easy sharing of knowledge among the member states (World Health Organization 1999). In order to set the standards high, the WHO initially provided a comprehensive reference by publishing monographs on various medicinal plants for the benefit of drug regulatory authorities, physicians, traditional health practitioners, pharmacists, manufacturers, research scientists and general public.

In a survey conducted by the WHO among the member countries regarding the strategy adopted for traditional medicines, the countries were requested to provide information on the prevalence of national monographs. In response to this, 46 countries reported existence of national monographs, 84 reported absences of national monographs, while another 25 countries indicated positive move towards preparation of monographs.

28.7.3 Manufacture of Indigenous Medicines

In order to meet safety and quality standards while manufacturing indigenous medicines, the WHO global survey on national policy of traditional medicine asked the member states to choose their opinion on regulatory requirements by providing set of options in the survey form. The choices given were the need for adherence to guidelines provided in pharmacopoeias or monographs, adopting GMP rules alike conventional pharmaceuticals or special GMP rules, no necessity for regulatory guidelines and other requirements.

The response from the member countries regarding the need for regulatory requirements was varied wherein the choice of adopting similar GMP rules for herbal medicines as used for conventional medicines was chosen by 73 countries followed by adherence to information in pharmacopoeias or monographs by 59 countries, the choice of special GMP rules was picked by 30 countries, while the choice of no regulatory requirements needed was selected by 28 countries. Interestingly, six other member states also delivered additional regulatory requirements essential for manufacturing indigenous medicines such as implementation of hygienic practices and few other improved elements of GMP like need for documentation, licensing, packing, etc.

28.7.4 Regulatory Requirements for Safety Assessments

Traditional medicines even though they are of natural origin may not always be safer for human consumption and held to be non-toxic (Calixto 2000). Unavoidable adverse effects and drug interactions such as food-drug interaction and drug-drug interaction may even bound to happen upon inappropriate use of traditional medicines. Thereby, safety assessments for traditional medicines are highly recommended. For that reason, the WHO global survey on traditional medicines requested the member states to choose among the options provided for assessment of safety of traditional medicines. The choices given were applying the same safety requirements as used for conventional pharmaceuticals, unique requirements or no requirements. The survey results revealed that 82 member states have special requirements for herbal medicines, 57 member states use the same regulatory requirements as for conventional pharmaceuticals and 28 member states do not follow any safety assessment strategy for traditional medicines. Furthermore, the member countries which chose unique requirements were asked to select the relevant category of special requirements or to mention if any other special requirements exist. As a result, out of 82 member states with special requirements, 66 indicated that their laws and regulations employ the regulatory requirement of traditional use without proven harmful effects. Fifty-three countries stated that their regulatory requirements were designed based on the references to documented scientific research on similar products. Twenty-one countries picked the option as other requirements and came up with documentary evidences on clinical studies, bibliography, screening of herbs not suitable for consumption, screening of toxic elements, radioactive and heavy metals, well-established use, traditional literature review and toxicological studies.

28.7.5 Registration System for Herbal Medicines and Essential Drug List

As of conventional pharmaceuticals are concerned, each drug product must be registered under relevant state authority before its release into the market. The chief aim of registration system is to ensure the consumers that the registered drugs supplied are of good quality and safer to use. Thereby, the WHO global survey on traditional medicine was curious to know about the existence of registration system for herbal medicines among member countries. Results of the survey have shown that nearly 85 countries have proper registration system, while 54 countries do not have a registration system for herbal medicines. The countries which have registered their herbal medicines were also asked to provide the number of registered herbal medicines. Accordingly, 64 countries provided the number of registered herbal medicines which approximately ranged from 1 to 10,000. It was also evident that several countries weren't able to mention the exact number of registered herbal medicines due to implementation of new registration system.

An essential drug list comprises of all the drugs accepted for public use (Jacobowicz et al. 1999). In general, the essential drug list represents drugs that are commonly denoted by short scientific names indicating active ingredient of the drug known as international nonproprietary name (INN). Moreover, the essential drug list must also be approved by the Ministry of Health before its release into the public sector.

The WHO global survey on traditional medicine questioned the member states regarding the addition of herbal medicines in essential drug list. The results of the survey have indicated that only 22 (16%) countries have included their herbal medicines in national essential countries, whereas the remaining 111 (78%) countries haven't included their herbal medicines in the essential drug list. The maximum number of essential herbal drugs (1242) was reported by China. An average of 165 essential herbal drugs was reported by other member states.

28.7.6 Sales and Postmarketing Surveillance

The WHO global survey on national policy of traditional medicine aimed to fulfil the regulatory requirements essential for traditional drugs among the member states and to maintain its uniformity worldwide. In that way, postmarketing surveillance is the final stage of regulatory requirement for a drug product after the drug is released into the market for sale. Postmarketing surveillance is the practice of monitoring and documenting the adverse drug reactions of a drug product after its clinical approval and release into the market.

The member states were inquired about the prevalence of postmarket survey system for herbal drugs. The survey results revealed that nearly 59 countries have postmarketing surveillance system for herbal drugs and nearly 90% countries have indicated that they have a nationwide system for monitoring damaging side effects of herbal drugs. Another 77 countries indicated that they do not have a postmarketing survey mechanism but of that 44 countries (58%) mentioned that such a system was in development stage.

The survey also enquired the member states regarding the modes of procurement of herbal medicines by consumers whether as prescription drugs from registered physicians, over-the-counter drugs, no restrictions prevailed for sales of herbal medicines, etc. A total of 137 countries responded the survey where 101 countries reported as sales in pharmacies as over-the-counter drugs, about 70 countries indicated that no restrictions were followed in selling herbal drugs, 48 countries reported as sales in pharmacies as prescription drugs and 30 countries informed as sales by licensed practitioners. About 22 countries declared the option as sales by other ways which included selling methods such as peddling in markets; door-to-door selling; sales by unregistered physicians, ethnic groups, traditional therapists, supermarkets and food markets; mail order; multilevel marketing systems; etc.

The survey results also indicated a steady progress in annual sales and demand for herbal medicines from 1999 to 2001 in nine representative countries picked from various parts of the world, namely, Bhutan, Canada, the Czech Republic, the Islamic Republic of Iran, Madagascar, Malaysia, Pakistan, Sudan and Sweden. Stimulatingly, there has been 40% increase in the sales value of herbal drugs observed in these countries (World Health Organization 2005).

28.8 Conclusion

For the entrusted drug discovery of traditional medicines, continued research on safety and efficacy of the formulated product is of prime importance. Production of quality-assured indigenous medicines by good manufacturing practice and safety monitoring of traditional medicines entrusting proper education to consumers about the product have become a worldwide necessity for the development of traditional medical system (Sen et al. 2011). The WHO has played an indispensable role, addressing various regulatory issues faced by member states concerning the development of traditional medicines. This had set a way to progress to uniformity and standardization of regulatory requirements among the member states assuring quality, safety and efficacy of the manufactured product. Standardization of preparation methods, growing, harvesting and processing of raw materials is also essential for production of uniform grade products (Noller et al. 2001). Commencement of postmarketing surveying practice among the member states ensures the resulting step of quality in evaluating the adverse effects, herb-herb or herb-drug interactions, etc.

A gradual improvement is being speculated among the member states towards formulating national standards, policies and regulations governing the production and use of traditional medicines. This in turn encourages good practice among the manufacturers and traditional medicine practitioners assuring safety to consumers. Concurrent sharing of available information, knowledge and model guidelines among the member states will ease the process of framing national policies for traditional medicine and also help to attain greater international standards. Therefore, the member states would be enabled to maintain the regulatory standards while formulating traditional medicines.

References

Brendler T, Wyk VBE (2008) A historical, scientific and commercial perspective on the medicinal use of Pelargonium sidoides (Geraniaceae). J Ethnopharmacol 119:420–433
Buwa LV, Afolayan AJ (2009) Antimicrobial activity of some medicinal plants used for the treatment of tuberculosis in the Eastern Cape Province, South Africa. Afr J Biotechnol 8:6683–6687
Calixto JB (2000) Efficacy, safety, quality control, marketing and regulatory guidelines for herbal medicines (phytotherapeutic agents). Braz J Med Biol Res 33:179–189
Foguema BK, Foguemb C (2014) Adverse drug reactions in some African herbal medicine: literature review and stakeholders interview. Integr Med Res 3:126–132
Gujral ML, Sareen K, Tangri KK, Amma MKP, Roy AK (1960) Anti-arthritic and anti inflammatory activity of gum (Balsamodendron mukul Hook). Indian J Physiol Pharmacol 4:267–273
Halberstein RA (2012) Botanical medicines for diuresis: cross-cultural comparisons. Chapter 1. Stud Nat Prod Chem 37:1–41
Holmes, Edward Morell (1911) Pharmacopoeia In Chisholm, Hugh (ed.). Encyclopædia Britannica. Cambridge University Press Cambridge 21 (11): 353–355
Holtz C (2007) Global health care: issues and policies. Jones and Bartlett Publishers, Sudbury
Jacobowicz BP, Rainhorn JD, Reich MR (1999) Indicators for monitoring national drug policies, a practical manual, 2nd edn. WHO, Geneva, pp 1–250. WHO/EDM/PAR/99.3
Jain S, Singh SN (2013) Metabolic effect of short term administration of *Hoodia gordonii*, a herbal appetite suppressant. S Afr J Bot 86:51–55
Joachim H, Berlin, G, Reimer (1890) Papyros Ebers. The first complete translation from the Egyptian
Li S et al (2008) Chemical markers for the quality control of herbal medicines: an overview. Chin Med 28(3):7
Livingstone C (1989) Churchill medical dictionary. Churchill Livingstone, New York
Luyckx VA (2012) Nephrotoxicity of alternative medicine practice. Adv Chronic Kidney Dis 19:129–141
Manosuthi W, Chottanapand S, Thongyen S, Chaovavanich A, Sungkanuparph S (2006) Survival rate and risk factors of mortality among HIV/tuberculosis coinfected patients with and without antiretroviral therapy. J Acquir Immune Defic Syndr 43:42–46
Marcus D, Grollman A (2002) Botanical medicines: the need for new regulations. N Engl J Med 347(25):2073–2075
McIntyre M (1999) Protecting the availability of herbal medicines. J Altern Complement Med 5 (2):109–113
Mester L, Mester M, Nityanand S (1979) Inhibition of platelet aggregation by "guggulu" steroids. Planta Med 37:367–369
Mishra LC, Singh BB (2000) Scientific basis for the therapeutic use of Withania somnifera (aswagandha): a review. Altern Med Rev 5:334–346
Moolla A, Viljoen AM (2008) 'Buchu'—Agathosmabetulina and Agathosmacrenulata (Rutaceae): a review. J Ethnopharmacol 119:413–419
Morel A, Talbot D (2010) Critical evaluation of vinflunine in the treatment of refractory metastatic urothelial carcinoma. Open Access J Urol 2:99–108
Newman DJ, Cragg GM (2007) Natural products as sources of new drugs over the last 25 years. J Nat Prod 70:461–477

Noller BN et al (2001) Global forum on safety of herbal and traditional medicine, Gold Coast, Australia. J Altern Complement Med 7(5):583–601
Peters D et al (2003) Time for a new approach for reporting herbal medicine adverse events. J Altern Complement Med 9(5):607–609
Sathiyanarayanan L, Arulmozhi S (2007) Mucuna pruriens A comprehensive review. Pharmacognosy Rev 1:157–162
Sen S, Chakraborty R, Biplab D (2011) Challenges and opportunities in the advancement of herbal medicine: India's position and role in a global context. J Herb Med 1:67–75
Sevillano D, Ramos C (2007) The birth of the greatest invention of the 20th century. Rev Esp Quimioter 20(3):354–358
Sucher NJ, Carles MC (2008) Genome-based approaches to the authentication of medicinal plants. Planta Med 74(6):603–623
Tanwar J, Das S, Fatima Z, Hameed S (2014) Multidrug resistance: an emerging crisis. Interdiscip Perspect Infect Dis 2014:541340
Teschke R, Eickhoff A, Wolff A, Frenzel C, Schulze J (2013) Herbal hepatotoxicity and WHO global introspection method. Ann Hepatol 12(1):11–21
Tiwari P, Ahmad K, Baig MH (2017) Gymnema sylvestre for diabetes: from traditional herb to future's therapeutic. Curr Pharm Des 23(11):1667–1676
Vipasha S, Hardeep K, Tarun K, Tullika M (2016) Traditional Indian herb *Catharanthus roseus* used as cancer treatment: a review. Int J Pharmacogn Phytochem Res 8(12):1926–1928
WHO Traditional medicine strategy 2002–2005 (2002) (document WHO/EDM/TRM/2002.1). Geneva, World Health Organization, pp 1–59
World Health Organization (1998a) Regulatory situation of herbal medicines: a worldwide review. Geneva (document WHO/TRM/98.1)
World Health Organization (1998b) The role of the pharmacist in self-care and self-medication: report of the Fourth WHO Consultative Group on the Role of the Pharmacist. The Hague, The Netherlands, Geneva, pp 26–28
World Health Organization (1999) WHO monographs on selected medicinal plants, vol 1. WHO, Geneva, pp 1–2
World Health Organization (2000) General guidelines for methodologies on research and evaluation of traditional medicine. Geneva (WHO/EDM/TRM/2000.1)
World Health Organization (2003a) Joint United Nations Programme on HIV/AIDS, AIDS Epidemic Update. Geneva
World Health Organization (2003b) Report of the regional meeting on traditional medicine and herbal medicines. WHO, Geneva, pp 18–20
World Health Organization (2003c) WHO guidelines on good agricultural and collection practices (GACP) for medicinal plants. WHO, Geneva
World Health Organization (2004) Guidelines on registration of traditional medicines in the WHO African region. Brazzaville, Regional Office for Africa. (AFR/TRM/04.1)
World Health Organization (2005) National policy on traditional medicine and regulation of herbal medicines: report of a WHO global survey. WHO, Geneva
World Health Organization (2007) WHO guidelines on good manufacturing practices (GMP) for herbal medicines. WHO, Geneva
World Health Organization (2009) Resolution WHA 62.13. Traditional medicine. WHO, Geneva

Pharmacovigilance of Herbal Medicine: An Evolving Discipline 29

Gaurav Ranjan, Rakesh K. Gupta, and Sugato Banerjee

Abstract

With increase in popularity and use of herbal products both as medicine and nutraceuticals throughout the world, there is an urgent need to monitor their adverse drug reactions. Pharmacovigilance is the discipline of monitoring, reporting and evaluating adverse drug reaction for medicinal products including herbal medicines. However, the current model of pharmacovigilance, its activities and associated tools have been developed with respect to conventional medicines. Applying these methods to monitor the safety of herbal medicinal products provides new challenges due to the diverse nature, regulatory requirements and usage of these products. In this chapter we discuss the importance and regulations associated with monitoring adverse drug reactions of herbal medicine. We also discuss the current challenges and future opportunities in this evolving field of importance.

Keywords

Spontaneous reporting · Cohort event monitoring · Natural products · Vigilance · Safety · Herbals

G. Ranjan · R. K. Gupta
Department of Pharmaceutical Sciences and Technology, Birla Institute of Technology, Mesra, Ranchi, Jharkhand, India

S. Banerjee (✉)
Department of Pharmacology and Toxicology, National Institute of Pharmaceutical Education and Research, Kolkata, West Bengal, India
e-mail: sugato.banerjee@niperkolkata.edu.in

S. C. Mandal et al. (eds.), *Evidence Based Validation of Traditional Medicines*,
https://doi.org/10.1007/978-981-15-8127-4_29

Abbreviations

ADR	Adverse drug reaction
AE	Adverse effect
CEM	Cohort event monitoring
GMP	Good manufacturing practice
HIV	Human immunodeficiency virus
HPLC	High-performance liquid chromatography
ICSR	Individual case study report
IR	Infrared
MS	Mass spectroscopy
NMR	Nuclear magnetic resonance
OTC	Over-the-counter
WHO	World Health Organization

29.1 Introduction to Pharmacovigilance

We all have heard about the thalidomide tragedy. There may be argument that the history of pharmacovigilance was much earlier, but the era of modern pharmacovigilance begins from the thalidomide tragedy (Waller 2010). The concept of pharmacovigilance was not much used earlier, and the testing and evaluation of medicines were entirely dependent on pharmaceutical companies in the late 1990s. As far as the thalidomide tragedy was concerned, iniquitous allegation of safety in pregnancy was questioned because of its use as a sedative in pregnant women (Baker 1960). It was seen that drug was having teratogenic effect, bringing forth a variety of abnormalities (Gregg 1941), especially limb defects called phocomelia (Warkany and Pediatr 1944). Overall, about 12,000 foetuses were afflicted, notably in Europe where the medicine was initially sold (Sullivan 1900). A sequence of three cases linked with thalidomide was divulged in *The Lancet* in the 1960s, and the trouble was finally identified, and the medicine was withdrawn from the market (McBride 1961). In the initial 1960s, the only effective mechanism of drawing attention was the publication of adverse drug reaction. Thalidomide seemed to produce nonlethal but teratogenic and adverse effect, which lead individuals to ask about the reason of impaired infants after birth (Botting and Botting 2015). This question was central to subsequent developments. It is not possible to predict all the adverse and harmful effects which may be caused by medicines, but limiting them is now easily achievable. An important lesson learnt from phocomelia was that drug safety problem cannot be taken for granted and these entire incidents should be taken earnestly, quite literally in this case. The recent system called pharmacovigilance which is prevalent now started to develop after the thalidomide tragedy (Evans 2000).

29.1.1 Definition

According to the World Health Organization, pharmacovigilance is defined as 'The science and activities relating to the detection, assessment, understanding and prevention of adverse effects or any other drug-related problems' (WHO 2004). Though the term pharmacovigilance has been defined in a lot of many ways, still the definition given by the WHO seems to be the appropriate one as there is a clear implication of 'risk management'. This basic concept is recently developed in pharmaceuticals, but it was already applied to many parts of modern life. The ultimate aim of pharmacovigilance is to enhance the proper use of medicine which will conclusively lead to better patient care and public healthcare (Mann and Andrews 2002).

29.1.2 Important Lessons Learnt from the Thalidomide Tragedy

- Testing of drugs should be mandatory prior to marketing.
- Government rules and regulations should be implicated for medicines.
- Adverse effect of drugs needs to be identified.
- The relationship between safety and marketing claims should be properly established.
- Preventing irrelevant use of drugs in pregnancy.

The major lesson learnt as an upshot of the thalidomide tragedy with respect to pharmacovigilance was that effective systems for hazard detection are required.

29.1.3 Importance of Pharmacovigilance

A medicine or drug product after being released for marketing needs to be studied for its safety. Once the medicines are being released in the market, patients with different ailments and different dietary pattern and tradition and patients already using several other drugs take them, so they may get affected and react in a different way. There may be a slight difference between different brands of medicines because of the alteration in ingredients used and their manufacturing. The toxicity, hazardous effect and adverse drug reactions allied with herbal and natural products need to be regulated in each country. One country's data or information received for adverse effect may not be applicable or significant to other country's inhabitants (Kshirgsagar 2005). There are many cases found in which chances of adverse effect befalls in a particular habitat. In order to avoid patient suffering unnecessarily or to use medicine in a proper, safe and efficient way, a regulating system is vital for each country which should be supported by physicians, pharmacist, nurses and other healthcare professionals. The responsibility to improve the safety profile of drugs through proper monitoring is led by the Department of Health's Essential Drug Programme and Drug Regulatory. Adverse drug reaction should be testified and reported on a routinely basis by the authority through the Drug Regulatory Authority's national pharmacovigilance programme.

29.1.4 Scope and Purposes of Pharmacovigilance

The process of pharmacovigilance has time and again been reflected to begin when a medicine is certified for the administration in conventional practice in earlier days. However, recently all safety-related activity is usually considered to embrace before individuals are first exposed to a new medicine (Maldonado et al. 2014). The endmost tenacity of pharmacovigilance is to diminish and minimize the potential for damage or harm that is linked with all active medications (De Ponti 2014). Even though data and information about all types of adverse drug reactions are gathered, the foremost emphasis is on detecting and preventing all the serious harmful and hazardous side effects. Thus, adverse drug reactions must fulfil any of the following conditions:

1. Lethal or fatal
2. Extends hospitalization
3. Deadly or life threatening
4. Results in chronic debility

Furthermore, all congenital deformities are considered harmful and serious, and the meaning of 'serious' permits the solicitation of medical verdict such that a response may be considered severe, even if there is no clear indication of criteria met. Non-serious or mild reactions are significant to every patients and healthcare professionals involved in their treatment, but they can generally be accomplished clinically, and their impression is negligible when possible potential benefits and risk are calculated. Thus, pharmacovigilance can be regarded as a part of healthcare system which aids in subsidizing the incidence of harmful effects and also helps in achieving all possible measures which promote rational and safety use of drugs and provide specific precautions against known health risks.

29.2 Pharmacovigilance Methods

Pharmacovigilance is the process of monitoring, collecting, reporting and analysing the adverse reaction caused by any drug molecule or medicine that is being administered by an individual for therapeutic purposes.

Various methods are used for pharmacovigilance process such as:

1. Spontaneous reporting
2. Intensified adverse drug reaction reporting
3. Targeted reporting
4. Cohort event monitoring
5. Electronic health record mining

29.2.1 Spontaneous Reporting

This method includes voluntary submission of ICSR by healthcare professionals, patients, pharmaceutical companies, etc. to the National Pharmacovigilance Centre (Bate and Evans 2009). This method includes patient and healthcare professional (Van Puijenbroek et al. 2002). When the patient observes any undesirable medical event which may occur after the administration or exposure of any medicine, then he reports to the healthcare professional, and the healthcare professional fills up the ICSR and submit it to the National Pharmacovigilance Centre. The patient may also report independently (Bégaud et al. 1993).

Advantages

1. Easiest method to establish
2. Covers whole population
3. Can detect new, rare, serious ADR which may not be detected during preclinical and clinical trail
4. Least expensive

Disadvantages

1. Reporting bias
2. Only suspected ADRs are reported
3. Delayed ADRs are difficult to detect
4. Denominator unknown
5. Under reporting is common

29.2.2 Intensified Reporting

It is the extension of spontaneous reporting system. It is used in early post-marketing phase to improve reporting of adverse drug events.

ADR of medicines reported by this method includes medicines which contain the following:

(a) New active substance
(b) Biological medicine applied after 1 January 2011
(c) Those medicines for which marketing authorization holder ID is required to carry out a post-authorization safety pass

29.2.3 Targeted Reporting

This method is used in a defined population to check the occurrence of known adverse drug reaction which is associated with a specific medicine. It may be done for specific medicines and specific ADRs, for example, TSR UGANDA. Targeted reporting method was used in UGANDA to check the adverse drug reaction of tenofovir which causes renal toxicity on HIV-infected patients.

Advantages
1. Can utilize existing ADR reporting infrastructure
2. Denominator known
3. Target specific medicines of interest
4. Possible to implement monitoring programme that targets specific issues of concern

Disadvantages
1. Captures only suspected ADRs or known toxicities
2. It relies on diagnostic capability of reporter
3. May limit reporting only to specific ADRs
4. Under-reporting

29.2.4 Cohort Event Monitoring (CEM)

This method is used in definite group of population to monitor particular adverse effect caused by the drugs (Layton et al. 2014). This method includes monitoring of adverse effect from the ignition of the therapy. It is the real-time monitoring of adverse events associated with one or more monitored medicines in a defined group of patients over a period of time from the starting of treatment (Pal et al. 2013). Its objective is to collect the required information about the safety profile of the chemical entity which is newly discovered when the drug enters the market. It document's all clinical events and not just the probable adverse effects. It is a programme limited to a time frame to compliment pharmacovigilance activities and not intended to alter spontaneous reporting (Dodoo et al. 2014) (Fig. 29.1).

Advantages
1. It helps to record complete profile of ADR/AE for medicine of interest.
2. It helps in the assessment of risk and detection of risk factors.
3. It can detect inefficacy of medicines.
4. It helps in characterizing known reactions.

Disadvantages
1. It is costlier and labour intensive than spontaneous or targeted reporting.
2. It is a new technique to healthcare professionals and pharmacovigilance centres and requires training.

Fig. 29.1 Cohort event monitoring

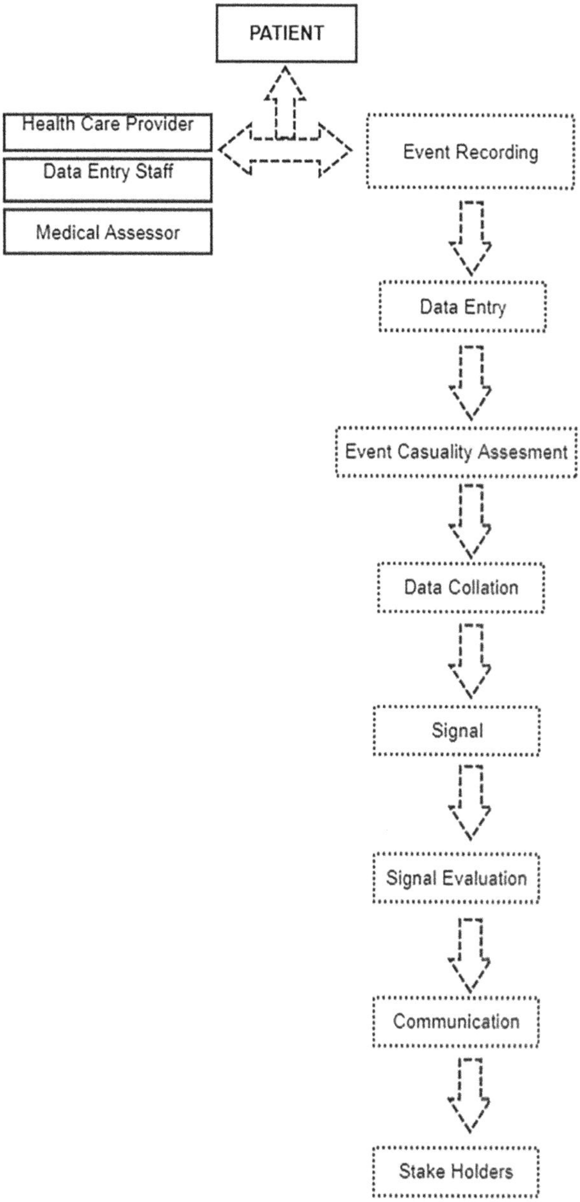

29.2.5 Electronic Health Record Mining

This method uses existing health records for pharmacovigilance activities. Health records cover large populations and give the detailed information of medical histories of all types of patients (Jensen et al. 2012).

The clinical information includes:

1. Prescriptions
2. Laboratory test results
3. Hospital referrals
4. Admission
5. Signs and symptoms

29.3 Herbal Medicine

According to the World Health Organization (WHO), herbal medicines are defined as finished, labelled medicinal products that contain active ingredients, aerial or underground parts of the plant or other plant material or combinations. The WHO has fixed precise guiding principle for the assessment of the quality, safety and efficacy of herbal drugs. As estimated by the World Health Organization, approximately 78% of the total population currently takes herbal medicine for primary healthcare and disease diagnosis (Mosihuzzaman and Choudhary 2008). Formal definitions and other terms used in relation to herbal medicines are included in Table 29.1.

Herbal drugs are progressively being used up by the individuals without prescription. Herbal medicines are recognized for almost 4000 years ago. Various plant parts such as leaves, stems, roots, seeds, bark, flowers, fruits and their extracts have been used over the periods. Herbal preparations have extended widespread adequacy and acceptability as therapeutic agents which include anti-diabetic, anti-fertility,

Table 29.1 Formal definitions and other terms used in relation to herbal medicines

Term	Definition
Herbal substance (herbal drug)	*Herbal substance may be defined as all whole, uncut, disjointed or cut flora, plant parts, fungi, algae, lichen in an unrefined or crude, generally dried form. Sometimes fresh materials are also included under it. Various exudates which are not exposed to a specific treatment may be considered as herbal substances. Herbal substances are accurately defined by the part of the plant used and its scientific name in accordance to the binomial system*
Herbal drug preparation	*Preparations attained by exposing herbal materials to treatments like fractionation, concentration, purification, fermentation, distillation, maceration expression, or extraction. These comprise comminuted or powdered herbal substances, expressed juices, tinctures, extracts, essential oils and treated exudates*
Herbal medicinal product	*Herbal medicinal products are demarcated as any pharmaceutical product which solely comprises a single active constituent or consist of more than one active ingredient in combination. Vitamins or minerals may be present and well-documented confirmation for safety, only if their action is auxiliary with that of the herbal active constituents concerning the specific claimed suggestions*

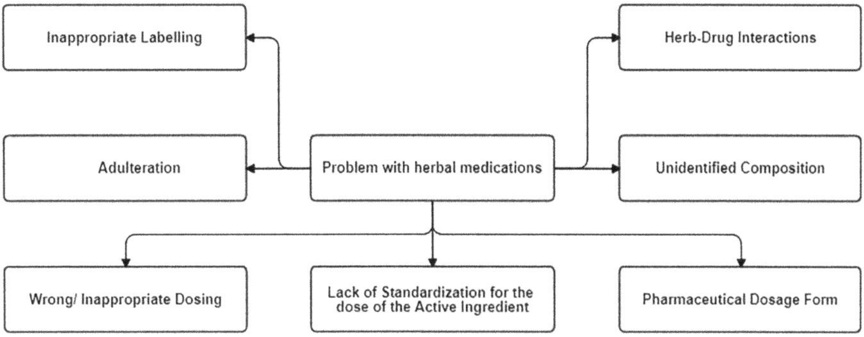

Fig. 29.2 Problems associated with herbal medicines

anti-ageing, antimicrobial, antiarthritic, sedative, anti-HIV, vasodilatory, hepatoprotective, antiasthmatic, antidepressant, anti-anxiety, antispasmodic, analgesic and anti-inflammatory and for the treatment of cirrhosis, acne, gall stones, hypertension, chronic fatigue, impotence, menopause, migraine, obesity, Alzheimer's disease and learning and memory-enhancing activities. These drugs have undergone thousands of years of human testing. Few drugs have been obsolete due to their toxicity, harmful and noxious effects, whereas some medicines have been altered or combined with supplementary herbs to counterbalance ADR (Delgoda et al. 2004). Many herbal product experienced modifications in their pharmacological uses.

Herbal supplements are used as a substitute and/or adjunct to their prescribed medicine by the patients. Herbal drugs are favoured by individuals, for the reason that they are natural and are supposed to be safer and have minor side effects than synthetic drugs. In the United States, herbal drugs are utilized by 20% of the inhabitants. Plants comprehend a number of active constituents that produce physiological effect in the system.

If an herbal drug is claimed to have advantageous effect on the human health, then it should be capable of altering the physiological system, i.e. exert a pharmacological response. Consequently, it may be expected to retain side effects as well. For instance, herbal tea made from *Digitalis purpurea* leaves will show toxic and harmful effect although it is natural and herbal product. Herbal supplements must be commercially obtainable after being standardized according to a specific active constituent. Therefore, there are several issues which arise during the use of these herbal drugs including side effects which may be due to drug-drug interaction or contaminants. Because of all the directly above cited explanations, there are numerous case reports of adverse effects of herbal supplements, specifically in combination. Nevertheless, many cannot be recognized due to inadequate information about the content of the mixture or combination (Toklu 2016) (Fig. 29.2).

29.4 Safety of Herbal Medicines

Generally, there is a misconception that natural products are safe, but the true fact is that many herbal products are toxic and may produce serious side effects. The safety of the drug is the major concern in herbal products and herbal medicines. With herbal medicines flooding the market, the necessity for the development of the system for quality control is essential.

Herbal medicines may also possess side effect like other medicines, which may be of an adverse nature. Adverse event may also arise from the following.

The mistaken use of the wrong species of medicinal plants (Chavez et al. 2006):

1. Incorrect dosing
2. Error in the use of herbal medicines both by the healthcare providers and consumers
3. Interactions with other medicines and use of products contaminated with potentially (Ernst 2000)
4. Hazardous substances such as toxic metals
5. Pathogenic microorganisms and agrochemical residues (Naithani and Kakkar 2006).

For example, herbal products were found to contain 0.1–0.3 mg of betamethasone per capsule after some patients developed corticosteroid-like side effects (Kumari et al. 2006).

29.5 Adverse Drug Reactions of Herbal Drugs

There may be a slight possibility that herbal remedies can cause adverse drug reactions (Baldwin 1987). *Ginkgo biloba* (Sparreboom et al. 2004) which is a common herb causes spontaneous bleeding. Another example can be *Hypericum perforatum* (Bennett Jr et al. 1998) which causes allergic reactions, fatigue, gastrointestinal disturbances, dizziness, confusion and photosensitivity (Rey and Walter 1998). *Capsicum annuum* is responsible for adverse drug reactions like myocardial infarction, hypertension and cardiac arrhythmias. Anxiety is caused by the commonly used plant *Ephedra*. *Vitex agnus* (chaste tree fruit) causes known adverse drug reactions like diarrhoea and headache. Liver toxicity is caused by *Piper methysticum* (Boullata and Nace 2000).

Drug Interactions*:* Drugs with a narrow therapeutic index should not be encouraged for herbal use like cyclosporine, phenytoin, procainamide, digoxin, theophylline, warfarin, etc. Drugs having narrow therapeutic index when used along with herbal products may either result in increased adverse effect or be less effective. *Ginkgo*, which is a remedy for Alzheimer's disease, when used along with aspirin causes increased bleeding. Synergism is shown by ginseng when used with monoamine oxidase inhibitors. Anxiolytic kava shows synergism with benzodiazepines. The reduction of plasma levels of drugs like cyclosporine, theophylline and oral contraceptives is caused by St. John's Wort which is an antidepressant. Ancient physicians permitted the use of heavy metals in medicines but in limited concentrations (Stephens 1998). Nowadays toxicity is reported when there is use of heavy metals in traditional drugs. Heavy metals responsible for causing toxicity are silver, arsenic, mercury, lead, copper and gold. The patient henceforth must cautiously use herbal drugs with modern medicines as there are high chances of adverse drug reactions (De Smet et al. 1997).

29.6 Standardization of Herbal Medicines

Standardization refers to the code of conduct which represents the necessity for a comprehensive quality management system in order to ensure proper designing and manufacturing of medicine (Waldesch et al. 2003). Complex mixture of various ingredients or mixture of herbals is used according to Ayurveda system, so it is necessary to standardize herbals. In these cases, the component which is actually responsible for the effect is unknown.

The physicochemical properties like boiling point, melting point, optical rotation and pre-formulation data followed by MS, NMR, IR and other analytical data should be considered as validated markers along with structural elucidation as these are the most important aspect for standardization. Safety, strength, purity, identity and quality of herbal drugs should be placed under GMP procedures (Kamboj et al. 2012). The assessment of plant preparation, crude plant material and finished products is necessary for assuring the quality of herbal medicines. For products which are imported, confirmation of regulatory status in the country where it originated is required. The pharmaceutical products which are exported should be certified under the WHO scheme. Monographs have been provided by several pharmacopoeias which state the standard and parameter of various herbs and also products made out of these herbs. Several pharmacopoeias contain monograph for herbal products to maintain a standard quality irrespective of nations like the British Herbal Pharmacopoeia, Pharmacopoeia Committee, British Herbal Compendium, Japanese Standards for Herbal Medicines and the Ayurveda Pharmacopoeia of India. Quality control and parameters for several Ayurvedic medicines have been recommended by the Government of India. The chemical and physical stability of marketed product should be evaluated under specific storage conditions, and the shelf life of the product should be determined. Toxicological studies are an important parameter for the safety of herbal drugs. Pharmacological and clinical effects of active ingredients are responsible for the efficacy of herbal medicines. Chromatography or instrumental analysis is done for the qualitative and quantitative standardization of polyherbal products (Patra et al. 2010).

29.7 Stability Testing of Herbal Medicines

The stability of herbal medicines is difficult because the whole herb is considered as an active ingredient irrespective of constituents with known therapeutic activity. Factors influencing the quality of herbal products are environmental factors such as oxygen, moisture, temperature, light, other ingredient or excipient in dosage form, microbial contamination, trace metal contamination and particle size of drug leaching from container. The main aim of stability evaluation is to provide proof how the quality of herbal products varies with time under such factors and also to establish proper storage conditions and shelf life. The product must be of desired quality throughout its storage period which is ensured by stability testing. Stability studies are conducted under natural atmospheric conditions and on at least three production batches. Stability studies which are conducted under accelerated atmospheric conditions of humidity, light and temperature are known as short-term stability, and the shelf life of the product is obtained by the data produced. Dosage forms in packed container should also be subjected to stability testing. Proper data and shelf life is established with modern analytical techniques like HPLC and spectrophotometry and by introducing proper guidelines. This will help in the acceptance of herbal medicines globally (Sachan et al. 2010).

29.8 Pharmacovigilance of Herbal Medicines

Developed and developing countries widely use herbal medicines, but recently there are various safety concerns influencing public health. Surveillance and monitoring are still needed for herbal products even though they are considered harmless in order to determine the potential risk factors (Awodele et al. 2013). Data which are already published indicates that risk factors are majorly due to contaminant or any added drug. Extensive and continuous monitoring of safety of these products must be carried out as there is limited knowledge about the components of herbal medicines and lack of quality control and effects in humans and their heterogeneous nature (Gulmez 2013). The WHO International Drug Monitoring Programme promotes herbal safety monitoring issues like classification, assessment of efficacy, pharmacovigilance and control of advertisements of herbal products which are covered under the regulation of herbal medicines which the WHO guidelines takes into account. The pharmacovigilance of herbal products is challenging because they are usually available from outlets where healthcare professionals are absent, so mostly they are purchased in conventional OTC environment. Different methods for pharmacovigilance are passive surveillance under which we have spontaneous and stimulated reporting and active surveillance by sentinel sites, registries, drug event monitoring, comparative observational studies by case-control study, survey study, targeted clinical investigations by investigating drug-drug interaction and food drug interactions (Toklu and Mensah 2016).

To monitor the safety of herbal medicines, four complementary actions are needed:

1. Identification of adverse events.
2. Risk management.
3. Establish guidelines to measure adverse effect and prevent its occurrence.
4. Discussion about the safety and the risk benefit ratio of using herbal medicine.
5. Safety monitoring of herbal medicines can be done by analysing the following types of reports (Rodrigues and Barnes 2013) (Zhang et al. 2012).

The common source of information on adverse events and reactions to medicines are clinical trials and spontaneous reports.

Reports from Healthcare Professionals

Generally, herbal medicines are bought without prescription from the retail pharmacy without any prior post-marketing surveillance; hence, the necessity of the nurses and community pharmacist is important.

Reports from Consumers

The adverse reactions reported by the consumer should be noted seriously, as they contribute as markers of the unknown effects of herbal medicine.

Reports from Manufacturers

The manufacturers are a reliable source of information of the adverse events related with their products; in many countries the reporting of adverse events are mandatory by the manufacturers.

Problems Associated with Herbal Medicines May Be Reported as Toxicity to the Following:

1. National poison centres: Poison centres are the major source for the pharmacovigilance and safety associated with herbal medicines in areas where resources are limited and no pharmacovigilance centres are established.
2. Drug information centres: They are the primary place of contact to provide the requisite clinical information. They could be the first point of contact and may provide a wealth of clinical information.
3. Consumer organizations: They obtain the required information from the list of complaints regarding the products.
4. Clinical trials and clinical studies: They can also be a point of information for safety monitoring of herbal products.

29.9 Regulatory Position of Herbal Medicines

Every country has its own legal documentation of herbal medicine, i.e. it varies from country to country (Saito 2000). In folklore medicines, the knowledge of herbs and their use in traditional medicine is quite prevalent in developing countries, but the problem with these countries is that they don't have any legislative measures to incorporate these traditionally used herbal drugs in drug regulation (WHO 1998). Authorization of herbal drugs in most of the countries is centred on traditional herbal references, only if they are known to be safe when used to treat minor ailments (Mukherjee 2003). In the current scenario, claims are being made to diagnose more severe ailments with herbal drugs for which no traditional knowledge exists. Therefore, regulatory necessities for herbal drugs are essential to certify the quality, safety and efficacy to aid specific signs; scientific and clinical evidence must be learned and acquired. Submission of clinical trial data and toxicity data depends upon the nature, quality of herbs and market accessibility. The governing requirements of herbal drugs vary from country to country (Morgan 2002). There are some countries which accept traditional, knowledge-based evidence, while some countries consider herbal medications as hazardous or of doubtful value (Calixto 2000).

References

Awodele O, Daniel A, Popoola TD et al (2013) A study on pharmacovigilance of herbal medicines in Lagos West Senatorial District, Nigeria. Int J Risk Saf Med 25(4):205–217

Baker JM (1960) The effects of drugs on the foetus. Pharmacol Rev 12:37–90

Baldwin CA (1987) What pharmacists should know about feverfew. Pharm J 239:237–238

Bate A, Evans SJ (2009) Quantitative signal detection using spontaneous ADR reporting. Pharmacoepidemiol Drug Saf 18(6):427–436

Bégaud B, Chaslerie A, Fourrier A et al (1993) Methodological approaches in pharmacoepidemiology. Applications to spontaneous reporting, 1st edn. Elsevier Science Publishers, Amsterdam

Bennett DA Jr, Phun L, Polk JF et al (1998) Neuropharmacology of St. John's wort (Hypericum). Ann Pharmacother 32:1201–1208

Botting JH, Botting RM (2015) The history of thalidomide. Animals and medicine: the contribution of animal experiments to the control of disease. Open Book Publishers, London

Boullata JI, Nace AM (2000) Safety issues with herbal medicine. Pharmacotherapy 20:257–269

Calixto JB (2000) Efficacy, safety, quality control, marketing, and regulatory guidelines for herbal medicines. Braz J Med Biol 33:179–189

Chavez ML, Jordan MA, Chavez PI (2006) Evidence-based drug–herbal interactions. Life Sci 78:2146–2157

De Ponti F (2014) New approaches in pharmacovigilance in the pharmacogenomic era: a call for papers. J Pharmacovigilance 2:e122

De Smet PAGM, Keller K, Hansel R et al (1997). Adverse effects of herbal drugs, 3rd edn Springer, Heidelberg 129(3), 137–145

Delgoda R, Ellington C, Barrett S et al (2004) The practice of polypharmacy involving herbal and prescription medicines in the treatment of diabetes mellitus, hypertension and gastrointestinal disorders in Jamaica. West Indian Med J 53:400–405

Dodoo AN, Fogg C, Nartey ET et al (2014) Profile of adverse events in patients receiving treatment for malaria in urban Ghana: a cohort-event monitoring study. Drug Saf 37(6):433–448

Ernst E (2000) Possible interactions between synthetic and herbal medicinal products. Part 1: a systematic review of the indirect evidence. Perfusion 13:2200–2211
Evans SJW (2000) Pharmacovigilance: a science or fielding emergencies. Stat Med 19:3199–3209
Gregg NM (1941) Congenital cataract following German measles in the mother. Epidemiol Infect 107(1):iii–xiv
Gulmez SE (2013) The new pharmacovigilance legislation and impact on observational studies. Marmara Pharm J 17:61–64
Jensen PB, Jensen LJ, Brunak S (2012) Mining electronic health records: towards better research applications and clinical care. Nat Rev Genet 13:395–405
Kamboj A, Vallisuta O, Olimat SM (2012) Drug discovery research in pharmacognosy. In: Vallisuta O (ed) Analytical evaluation of herbal drugs, pp 23–60
Kshirgsagar N (2005) The pharmacovigilance system in India. Drug Saf 28:647–650
Kumari B, Madan VK, Kathpal TS (2006) Monitoring of pesticide residue in fruits. Environ Monit Assess 123:407–412
Layton D, Shakir SA, Andrews EB et al (2014) Prescription event monitoring (PEM)—the evolution to the new modified PEM and its support of risk management. In: Mann's pharmacovigilance, 3rd edn. Wiley-Blackwell, New Jersey, pp 359–384
Maldonado C, Vazquez M, Guevara N et al (2014) Use of STOPP/START criteria to perform active pharmacovigilance in the elderly. J Pharmacovigilance 2:146
Mann RD, Andrews EF (eds) (2002) Pharmacovigilance, 2nd edn. Wiley, Chichester
McBride WG (1961) Thalidomide and congenital abnormalities. Lancet ii:1358
Morgan C (2002) Medicine of the gods: basic principles of ayurvedic medicine, 2nd edn. Mandrake of Oxford, Oxford
Mosihuzzaman M, Choudhary MI (2008) Protocols on safety, efficacy, standardization, and documentation of herbal medicine (IUPAC technical report). Pure Appl Chem 80 (10):2195–2230
Mukherjee PK (2003) Exploring botanicals in Indian systems of medicine regulatory perspectives. Clin Res Regul Aff 20:249–264
Naithani V, Kakkar P (2006) Estimation of organochlorine pesticide residue in two popular spices extensively used as herbal tea ingredients in India. Bull Environ Contam Toxicol 76:429–435
Pal SN, Duncombe C, Falzon D et al (2013) WHO strategy for collecting safety data in public health programmes: complementing spontaneous reporting systems. Drug Saf 36(2):75–81
Patra KC, Pareta SK, Harwansh RK et al (2010) Traditional approaches towards standardization of herbal medicines. JPST 2(11):372–379
Rey JM, Walter G (1998) Hypericum perforatum (St John's wort) in depression: pest or blessing? Med J Aust 169:583–586
Rodrigues E, Barnes J (2013) Pharmacovigilance of herbal medicines: the potential contributions of ethnobotanical and ethnopharmacological studies. Drug Saf 36:1–12
Sachan AK, Sachan NK, Kumar S et al (2010) Evaluation and standardization of essential oils for development of alternative dosage forms. Eur J Sci Res 46(2):194–203
Saito H (2000) Regulation of herbal medicines in Japan. Pharmacol Regul 41:515–519
Sparreboom A, Cox MC, Acharya MR et al (2004) Herbal remedies in the United States: potential adverse interactions with anticancer agents. J Clin Oncol 22:2489–2503
Stephens MD (1998) Detection of new adverse drug reactions. MacMillan Publishers, Scotland
Sullivan WC (1900) The children of the female drunkard. Med Temp Rev 1:72–79
Toklu HZ (2016) Pharmacovigilance of herbal medicine: herbavigilance. Adv Pharmacoepidemiol Drug Saf 5:208
Toklu HZ, Mensah E (2016) Why do we need pharmacists in pharmacovigilance systems? Online J Public Health Inform 8(2):e193
Van Puijenbroek EP, Bate A, Leufkens HG et al (2002) A comparison of measures of disproportionality for signal detection in spontaneous reporting systems for adverse drug reactions. Pharmacoepidemiol Drug Saf 11(1):3–10

Waldesch FG, Konigswinter BS, Remagen HB (2003) Herbal medicinal products scientific and regulatory basis for development quality assurance and marketing authorization. Medpharm Stuttagart and CRC Press, Washington, DC
Waller P (2010) An Introduction to Pharmacovigilance, 2nd edn. Blackwell Publishing, New Jersey
Warkany J, Pediatr J (1944) Congenital malformations induced by maternal nutritional deficiency. J Pediatr 25:476–480
World Health Organization. Programme on Traditional Medicine (1998) Regulatory situation of herbal medicines: a worldwide review. World Health Organization. https://apps.who.int/iris/handle/10665/63801
World Health Organization (2004) WHO guidelines on safety monitoring of herbal medicines in pharmacovigilance systems. World Health Organization, Geneva
Zhang L, Yan J, Liu X et al (2012) Pharmacovigilance practice and risk control of traditional Chinese medicine drugs in China: current status and future perspective. J Ethnopharmacol 140:519–525

Toxicity Studies Related to Medicinal Plants 30

Sankhadip Bose, Rana Datta, and Ward G. Kirlin

Abstract

Medicinal herbs are extensively used for therapy in many parts of the world. This traditional knowledge has been passed on from one generation to the other, based on their long-term clinical use. Little clinical and experimental data is available for most herbal medicaments. Herbs are normally considered to be safe. They are obtained from living species. Yet many adverse cases have been reported, especially on long-term use. These include toxic effects or allergic reactions. Frequently herbal drugs tend to interact with one another. Some plants produce toxic constituents for defence purposes such as *Aconitum columbianum*, *Digitalis purpurea* and *Hyoscyamus niger*. Reasons for herbal drug toxicity are improper identification/authentication, improper labelling or standardization or contamination with fungal toxins, such as aflatoxin. Accordingly, the World Health Organization has prescribed study guidelines for toxicity investigation of herbal medicines, including acute toxicity testing, long-term toxicity testing, local toxicity tests and special toxicity tests. Guidelines have also been put forward by Organization for Economic Co-operation and Development in this regard. As traditional and herbal medicines are increasingly used, scientific validation of its folkloric usage has become necessary.

S. Bose (✉)
Bengal School of Technology, Hooghly, West Bengal, India
e-mail: sankha.bose@gmail.com

R. Datta
Gupta College of Technological Sciences, Asansol, West Bengal, India

W. G. Kirlin
Department of Pharmacology and Toxicology, Morehouse School of Medicine, Atlanta, GA, USA

S. C. Mandal et al. (eds.), *Evidence Based Validation of Traditional Medicines*,
https://doi.org/10.1007/978-981-15-8127-4_30

Keywords

Herbal · Standardization · Toxicity · Guidelines

Abbreviations

%	Percentage
°C	Degree centigrade
BW	Body weight
cm	Centimetre
FTIR	Fourier transform-infrared radiation
g	Gram
g/kg	Gram/kilogram
HPTLC	High-performance thin-layer chromatography
mg	Milligram
mL	Millilitre
mm	Millimetre
mM	Millimolar
MTD	Maximum tolerated dose
nm	Nanometre
NOAEL	No-observed-adverse-effect level
OECD	Organization for Economic Co-operation and Development
PO	Per oral
UV	Ultraviolet
μg	Milligram

30.1 Introduction

Medicinal plants have been used for therapeutics since time immemorial (Cragg and Newman 2001). Across ages and cultures, they have influenced healthcare practice. Traditional medicines mostly consist of herbal drugs. Not only for the treatment of acute ailments, herbs have also been beneficial in chronic disorders. Hence it is very important in the modern era to incorporate the herbal remedies as part of the national drug policy, to enable them to be a part of mainstream modern therapeutics. More than 80% of the world population still resort to herbal formulations for initial treatment. It is time to strengthen drug regulatory measures involving them.

Even WHO recognizes that usefulness of most herbal drugs has been enhanced, when scientists started isolating the active phytoconstituents. It established the traditional usage scientifically to the world (Farnsworth 1984). Herbal remedies are increasingly becoming popular, as most people believe them to be free of side effects. For chronic ailments they offer a useful alternative to allopathic drugs. Figure 30.1 shows the data that represents the usage trends of herbal drugs in a few less developed countries.

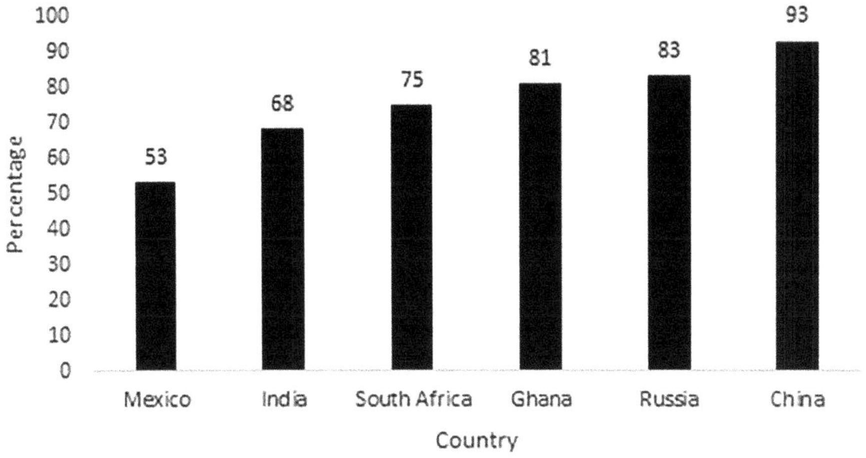

Fig. 30.1 Use of traditional medicine in middle-income countries

Long-term clinical usage is the main rationale for herbal drug usage. Little scientific and experimental data is available in most cases, pertaining both to safety and efficacy (Zhu et al. 2002). This traditional information of usage of herbal medicines has been passed on across generations. At times incorrect documentation and perpetuation of knowledge leads to alteration of the original usage formulation or application. With the upsurge in usage of herbs as an alternative to allopathic medicine, it has become very important to standardize them. Validation of traditional knowledge will greatly enhance their acceptance among the developed world nations. Though considered safe by most people, herbs do possess adverse effects at times, which occasionally may be life-threatening (Whitto et al. 2003). Adulteration or use of substituent possesses another challenge to use of herbs. These in addition to direct toxic actions and allergic reactions make it necessary to evaluate the therapeutic profile of herbs, before they are used for clinical practice. Toxicity of many phytoconstituents aggravate on isolation. This offers another grave challenge to the scientists (Bent and Ko 2004).

Undesirable changes upon usage of herbal drugs are its adverse effects. Though death-mortality is considered the ultimate adverse effect, altered/unwanted physiological status pertaining to drug usage may also necessitate withdrawal of herbal drugs from therapeutic regimen. Altered body-organ weight and altered levels of enzyme are also indications of drug toxicity (Duffus and Worth 2006). Hence all the herbal drugs should be stringently tested for efficacy and safety. The guidelines of testing should be the same as used for synthetic drug moieties (Talalay and Talalay 2001). Many herbal drugs, otherwise toxic, may be used for therapeutics if they are used within tolerable limits. Preliminary acute toxicological studies help elucidate the appropriate doses to be used for therapeutics. Acute toxicity studies are essential to prevent overdose of herbal drugs, which otherwise may lead its withdrawal from therapeutics. LD_{50} or lethal dose 50 is determined after acute, single exposure.

Herbal drugs on most occasions are taken in orally. Thus the adverse effect of the drug on the GIT should be ascertained wherever possible. Other body parts include the lungs and the skin that may serve as an entry point of herbal toxicant. Airborne herbal toxicants, specially associated with the manufacturing process of herbal drugs, need to be ascertained (Ozbek et al. 2004). The skin rarely serves as an entry point. For herbs to be used topically, however, skin irritation tests must also be performed.

Single-dose toxicity studies are not very useful. Always the studies should be done using multiple doses (Alvarez et al. 2004). The results and data obtained from the acute toxicity study help to determine the doses to be used for repeated dose studies. Pharmacological studies of biological activity in animals and in humans may also help in part in the dose selection procedure (Da Silva et al. 2002). Most toxicologists believe that the doses used for preclinical chronic toxicity testing should be higher than the clinical doses to be used for human (Datta and Bose 2017; Feres et al. 2006).

The results of the toxicity study guides as to whether a herbal drug can be used for clinical use. Toxicological studies are broadly of three types—acute, subacute and chronic (Baki et al. 2007). This classification is based on the duration of exposure of animals to drugs. Toxicity of a herbal drug is a function of the dose and the toxic properties of the phytoconstituents. Both these factors are equally important, and their relation greatly determines the doses to be used for therapeutics in clinical practice (Hayes 2001).

Extensive preclinical toxicological data is required to support the safety profile of a herbal drug, especially if it is to qualify the stringent clinical trials guidelines, like any other synthetic molecule. Two-week toxicological study (usually rodents) elucidates significant toxicological profile of a herbal drug. Along with this toxicokinetics studies are also necessary to determine the no-observed-adverse-effect level (NOAEL) of a herbal drug. There is a threshold dose beyond which the clinical signs of toxicity of a herbal drug may not be measurable (Talalay and Talalay 2001). NOAEL estimation greatly depends on this threshold dose. It helps to ascertain the maximum permissible daily intake of herbal drugs. Even the save limits of food additives and pesticides are related to NOAEL (Duffus and Worth 2006).

30.2 Toxicity of Herbs

Most herbal drugs are safe or possess minimal toxicity. At times they possess toxic phytoconstituents which are not identified or written on the label (Barrett et al. 1999). These may be intentionally or unintentionally added. Intentional adulteration is done to maximize economic gains, though it severely compromises the toxicity and safety profile of a herbal drug. Misidentification of a plant or contamination with pesticides/heavy metals may account for unintentional compromise with the quality of a herbal drug (Asif 2012).

Most modern system of allopathic medications has side effects. These unwanted effects are elucidated based on clinical use, either during clinical trials or post-

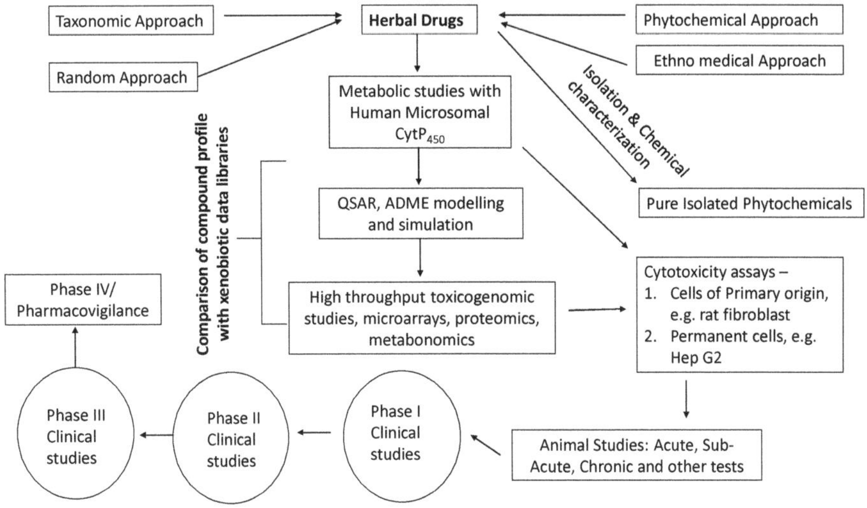

Fig. 30.2 Schematic representation of screening of herbal drugs for potential toxicities

marketing surveillance. Precautions need to be taken for pregnant mothers. Lactating and breastfeeding mothers offer an additional risk of harming their infants. This information should be made available to doctors as standard references. Herbal drug formulations should also meet strict quality control measures. Uniform standards should be laid down in official formularies to ensure product and quality uniformity. It is a big challenge to standardize herbal drugs and formulation using sophisticated modern analytical techniques, in a reproducible manner. Figure 30.2 outlines the pathway normally followed for the screening of herbal drugs for potential toxicities.

Herbal drugs frequently lack standardized references. Quality control and validation are very crucial. It is a challenge to modern era scientists. One batch may be different from the other, based on time of collection. Ample references exist in literature that show that expression of phytoconstituents can vary in herbs based on the time of collection. Also if a herb is known to possess toxicity, the manufacturer may not reveal it to the consumers (Barrett et al. 1999). WHO and the regulatory authorities of various countries are constantly working in this direction. Most plant exposures are unintentional. Many people try out herbal treatment for self-medication before actually visiting the physician. This is mainly to reduce the healthcare costs. Thus the herbal industry has rapidly evolved in the recent time. The definition of herbal drugs is confusing at times. Though literally speaking herbs refer to drugs obtained from plants, herbal formulations also occasionally do possess animal and/or mineral products to synergize the therapeutic efficacy of plant materials (Asif 2012).

30.2.1 Intrinsic Adverse Effects

These are also known as type A adverse effects. These are normally predictable in nature. The intensity of type A adverse effects is related directly to dose. As the dose of the causative agent increases, the chances and intensity of adverse effects also increase. The adverse effect normally occurs due to an increased pharmacological response of the drug. Since these adverse effects are predictable, mortality does not normally occur due to these adverse effects. It is easier to manage and monitor these adverse effects. These types of adverse effects are usually identified during the preclinical trials and the clinical trials. Thus herbal drugs that may show these types of adverse effects at therapeutic doses are usually not allowed to come to the market. Acetaminophen is often cited as a prototype candidate possessing dose-dependent adverse effect. It produces fulminant hepatic failure at higher-dose levels. However significant hepatic dysfunction is usually not observed at doses below 4 g/ day in healthy patients (Feres et al. 2006). Repeated administration of herbal drugs, especially those used for chronic disorders, can increase the risk of developing type A adverse effects in several folds. Work done by many researchers have established that chronic exposure of hepatotoxic herbal just like medications, just like acetaminophen, overtime increases the chances of development of intrinsic hepatotoxic reactions (Baki et al. 2007).

30.2.2 Extrinsic Adverse Effects

These types of adverse reactions are also referred to as type B or idiosyncratic drug reactions. These are in general unpredictable in nature, making them even more concerning. These are expressed in certain individuals as a combination of unique factors. Being rare in nature, the actual incidence of idiosyncratic reactions is difficult to estimate. But most clinicians believe that the expression rate ranges from 1 in 1000 to 1 in 200,000, depending on the agent. Liver injury caused by antibiotics and certain classes of NSAIDs is considered as type B adverse reactions. High degree of variable presentation of idiosyncratic reactions makes them extremely difficult to predict. Thus they are not usually identified during the preclinical or the clinical trials. Only post-marketing surveillance may be considered an effective tool to track and monitor type B adverse drug effects. These are further subdivided into immunologic and non-immunologic type B adverse drug reactions. Usually these types of reactions due to herbal drugs are not related to the dose; but also examples do exist where dose levels can predict the response up to certain extent (Bunchorntavakul and Reddy 2013).

30.3 Possible Reasons for Toxicity of Herbal Medicines

30.3.1 Self-medication

Herbal drugs can be easily purchased from the market without prescription. Often traditional treatments based on plants at hand such as neem and tulsi are taken as initial therapy by many patients before consulting the doctors. Herbal drugs are also frequently advertised as miracle treatments of many disorders people are shy of consulting a physician. Many people adopt herbal medicines when they have lost hope on allopathic system of medicine, or are not being able to tolerate their side effects. Many patients take herbal remedies along with their regular allopathic medications without informing their physicians (Barrett et al. 1999). People also regularly take herbal remedies prophylactically to stay healthy (Andrade et al. 2007). All these reasons may contribute to self-medication of herbal drugs and remedies.

30.3.2 Untrained/Unqualified Practitioners

In many countries of the world, untrained or unqualified practitioners are prescribing herbal remedies. Though laws and regulations exist in many countries, it is very difficult to implement those to herbal practitioners. Medical practitioners having vast knowledge of human body and the drug pharmacology and years of experience are usually allowed to practice herbal therapy in developed countries of the world (Barrett et al. 1999). Degrees in alternative systems of medicines are being offered by many reputed universities of the world. Still almost half of the actual herbal practitioners do not have a formal degree or training. They acquire these knowledge and information from their elderly and forefathers. These practitioners are themselves not aware of the untoward adverse effects a herbal drug may produce (Pai and Nahata 2000).

30.3.3 Substandard Product

Many herbal drugs are marketed without assuring uniform standards. They do not meet the prerequisite standards. The reason for this being inadequate testing before the marketing of the finished products. Herbal formulations at times do not contain the active ingredients as claimed in the labels. Incorrect identification and authentication of the plant sample may result in adulteration. The abovementioned errors, whether deliberate or not, may even lead to total absence of the bioactive phytoconstituents. Deliberate adulteration may also lead to substandard quality of herbal products. Improper storage of the plant sample of the finished product may also lead to loss of efficacy (Cragg and Newman 2001). Extraneous materials that are not claimed in the labels may also be present, for example, earthy materials, minerals and heavy metals. Presence of toxins or pesticide residues, employed in plant cultivation, may also account for the presence of toxic principles in the finished herbal product (Maffè et al. 2013).

30.3.4 Improper Intake

Doses of conventional allopathic medications are fixed after years of testing and trials. Age or body weight is the major parameters that govern the dose fixation of the allopathic medications (Cragg and Newman 2001). But no such stringent testing or regulatory procedures are normally followed for the herbal drugs. At times herbal remedies are consumed as dietary supplements; no definite dose regimen is being outlined. Also incomplete/inappropriate packaging and failure to provide a cup or spoon for accurate dosing of herbal drugs may contribute to inaccurate intake. The pharmaceutical companies involved in the manufacturing and marketing of herbal drugs frequently misguide the patients by claiming their product to be free of side/adverse effects. So people at times are inclined to overdose themselves to get quick relief from their ailments. Some people continue to take such drugs over months and years, without realizing the cumulative toxicity the active herbal drugs can have on the human body.

30.3.5 Herbal-Drug Interactions

Most people believe herbal drugs to be safe and free of side effects. Research has shown that herbal drugs may also contain harmful ingredients. They can also interact with other allopathic medications, dietary supplements and even food. Thus it becomes very necessary to identify all possible interactions of herbal remedies (Nielsen and Brant 2002). Rigorous research is usually not carried out for herb-allopathic drug interactions. Drugs with narrow therapeutic index may aggravate the problem. Many studies with the drugs warfarin and digoxin have highlighted the issue (Farnsworth 1984). These interactions can even lead to therapeutic failure and in extreme cases loss of life. For example, it has been found that the drug St. John's wort reduces the drug levels of cyclosporine in the body. This eventually will lead to transplant rejection in many cases. We now know many of the predisposing and protective factors that determine whether or not an interaction occurs, but in practice, it is still very difficult to predict what will happen when an individual patient is given two potentially interacting medicines (Bent and Ko 2004). Table 30.1 outlines a list of some of herbal drugs that have been found to cause toxicity.

30.4 Preclinical Toxicity Testing

Limited scientific data is available regarding herbal drugs. Their safety and efficacy are not well-documented in the literature. Thus many doubt their therapeutic usage as an alternative system of medicine. Stringent standards have been set by the

Table 30.1 Potential toxic effects associated with some common herbal medicines marketed for different indications

Plant source	Parts used	Pharmacological use	Toxicity	References
Panax ginseng (ginseng)	Roots	Relieves stress, promotes mental and physical activity	Central nervous system stimulation, hypertension, skin eruptions	Chan and Fu (2007)
Hypericum perforatum (St. John's wort)	Aerial parts	Antidepressant, mood stabilizer	Highly potent cytochrome P450 enzyme inducer which affects drug metabolism. Also causes hepatotoxicity and nephrotoxicity in pregnancy and lactation	Gregoretti et al. (2004)
Piper methysticum (kava kava)	Roots	Sedative, anxiolytic	Hepatotoxic, cytochrome P450 enzyme inhibitor	Gow et al. (2003)
Ginkgo biloba (ginkgo)	Leaves	Impotence, vertigo, circulatory disorders, improves mental alertness	Gastric irritability, spontaneous bleeding	Sierpina et al. (2003)
Salvia miltiorrhiza (danshen)	Exterior taproot	Angina pectoris, antihyperlipidemic, ischemic stroke	Bleeding, anticoagulant effects	Wang (2010)
Crataegus oxycantha (hawthorn)	Flowers, roots, berries	Mild to moderate congestive heart failure	Cardiac arrhythmias, lowered blood pressure	Rothfuss et al. (2001)
Symphytum officinale (comfrey)	Leaves	Anti-inflammatory, anti-diarrhoel and treatment of thrombophlebitis	Hepatotoxicity, carcinogenicity	Stickel and Seitz (2010)
Glycyrrhiza glabra (liquorice)	Roots	Antiulcer, anti-inflammatory, antihypertensive	Hypokalaemic myopathy, pseudoaldosteronism, thrombocytopenia	Celik et al. (2012)
Larrea tridentata (chaparral, creosote bush)	Leaves and twigs	Blood thinner, weight loss, antioxidant, anticancer, antiarthritis	Carcinogenic, nephrotoxic, hepatotoxic	Allard et al. (2013)
Urginea maritima (squill)	Bulbs	Antiarthritic, bronchial expectorant	Symptoms resembling digitalis toxicity	Tuncok et al. (1995)
Ephedra (mahuang)	Stem	Promotes weight loss and mental and physical alertness	Cardiotoxicity, thyrotoxicosis, seizures	Woolf et al. (2005)
Senna occidentalis (senna)	Seeds	Laxative	Skeletal and cardiac muscle degeneration, hepatotoxicity, neurotoxicity	Barbosa-Ferreira et al. (2005)
Aloe vera (aloe)	Leaves	Wound healing, laxative	Cytogenetic toxicity	Verma et al. (2012)

prominent regulatory agencies like the US Food and Drug Administration (US FDA) (USFDA 2006) and the Committee on Safety of Medicines, United Kingdom (CSMUK) (Barnes 2003). Thus scientists are trying to establish a toxicity and safety profile of the herbal drugs. The Organization for Economic Co-operation and Development (OECD) test guidelines have been globally accepted for toxicity testing in animals. They provide a uniform and harmonized approach.

Many scientists have been trying to isolate the active principles from the plant sources, so that large-scale production is possible after determination of its structure. But it has been found that many chemical compounds exert toxicity if isolated from the plants, because in the plants, possible untoward effect may be balanced or neutralized by the presence of other constituents (Rao 1985). Contamination of herbal drugs by pesticides, herbicides or naturally occurring toxins and microbes or adulteration by means of synthetic substitutes are also causes for concern (Sharma 2000).

Several factors have to be kept in mind while adopting OECD guidelines for toxicity testing of herbal drugs:

1. ***Lack of standardization***—Herbal drugs often lack uniform standards and pharmacopoeial monographs. Identity of herbs, composition of polyherbal formulations and dose may differ depending on the practitioner (Chopra and Doiphode 2002). Thus OECD guidelines ask for material characterization. It is a prerequisite to preclinical and clinical drug testing. Only then can the polyherbal formulations be validated. Use of marker compound detection by instrumental techniques such as HPLC and HPTLC may offer a solution to this problem.
2. ***Use of vehicles for drug administration***—On most occasion herbal drugs need a vehicle. The vehicles may be honey, clarified butter or curd. These carriers may modify the therapeutic efficacy or the toxicity of the herbal drugs. The adjuvants and vehicles may modify the toxicokinetic properties (Raghunathan 1976). Thus vehicle control groups in preclinical testing designs. Along with it a suitable and robust statistical tool can help enhance the predictive significance of the data generated.
3. ***Polyherbal formulations***—Most herbal drugs consist of many active ingredients. A 'single holistic entity' approach should better be adopted than a 'reductionist approach' during the toxicity testing design. Efforts should be made to use the whole extract of the herbal drugs as far as possible, instead of trying to isolate the individual constituents.

30.4.1 General Tests

Standard guidelines have been put forward by OECD (1998a, b) to harmonize and provide uniformity to testing of herbal drugs. Some factors that may influence such tests include:

Preparation of herbal drug: Different dose forms may be formulated for drug administration. Tablets, capsules or creams are the dosage forms commonly prepared for toxicity testing. It is important that the dosage form provides uniform and predictable dosage administration in the test species. Before actual drug administration, it is important to standardize the herbal drug. Quantitative content uniformity is an important criterion the scientists should be careful of. The dose and the route of administration are related to the proposed clinical usage of the herbal drug.

Animal welfare: The animals used for preclinical studies need to be humanely treated. Appropriate housing conditions should be maintained. The preclinical design and testing protocol must be approved by an appropriate Animal Ethics Committee. To account for intraspecies variation in drug response, multiple animal species should be tested on. The protocol should include both rodents and non-rodents for better extrapolation of preclinical data into clinical studies. If the animals seem to be suffering during repeated dosing, an informed decision should be made by the investigator as to whether to humanely euthanize the animal before the end of the study period.

Regulatory requirements: An independent animal ethics committee should review and scrutinize all animal experimentation protocols. The justification of the test procedures and animal usage should be appropriately explained. The regulations may differ from one country to another.

30.4.2 Acute Toxicity Testing

It is designed to measure the toxicological response of the test animals to a single (or rarely multiple) dose of the test herbal drug, over a period of not more than 24 h (OECD 2001a, b, c, d). This test is also used to calculate median lethal dose (LD_{50}) of a substance (OECD 2019; OECD 2006a, b). For preclinical testing two rodents species should be used wherever possible. Mice and rats are usually selected for the preliminary studies. To study the effect of gender on the toxicological response of a herbal drug, both male and female test organisms are included in the study design. Animals are administered by the same route as that of the intended route in the humans (Nandy and Datta 2012). The maximum cutoff limit is at least ten times the human therapeutic dose. If single-dose administration is not possible, the dose may be divided and administered within a course of 24 h. The toxic signs should be carefully noted. Time to onset of these adverse reactions and their severity should be monitored. Additional groups of animas should be included to study the reversibility of toxic signs, once the herbal drug has been withdrawn from the regimen. These observations are usually made after 14 days of administration of herbal drugs (Lorke 1983). Changes in colour of the skin, eyes and mucous membranes should also be noted. Tremors, convulsions, diarrhoea or lethargy in movement are noted. Moribund animals are humanely killed. Autopsy of all the dead animals should be done. Histopathological studies are also done of the major organs after the death of the animals.

Table 30.2 The duration of period of exposure in repeated toxicity study in animals

Serial number	Duration of proposed human administration	Period of exposure in toxicity study	References
1	Single dose	2 weeks	WHO (1993)
2	More than 2 weeks to less than 4 weeks	4 weeks	WHO (1993)
3	More than 4 weeks to less than 12 weeks	12 weeks	WHO (1993)
4	More than 12 weeks to less than 24 weeks	24 weeks	WHO (1993)
5	More than 24 weeks	Same as that of expected period of use of trial drug	WHO (1993); WHO (2002)

30.4.3 Repeated Dose Toxicity Study

The herbal drugs are repeatedly administered over a period, ranging from weeks to a few months. If the period of exposure is 28–30 days (OECD 1995), the term 'subacute' is used. If the period of exposure ranges from 2 to 3 months (60–90 days) (OECD 1998a, b), the term 'subchronic' toxicity can be used. Deleterious effects on organs and haematological and biochemical indices are explored. Herbal drugs are tested on two mammalian species, one of which should be a non-rodent. To study the effect of gender on the toxicological response of a herbal drug, both male and female test organisms are included in the study design. Animals are administered by the same route as that of the intended route in the humans (Table 30.2). Three dose levels are used for the study. The highest dose selected produces some observable toxicity (selected on basis of MTD calculated in acute toxicity study). The lowest should not produce any observable toxicity. A middle dose should be selected between the two. Also a vehicle control should be included in the testing. The parameters evaluated at the end of study period include body weight, food and water intake. Additional groups of animas should be included to study the delayed occurrence and reversibility of toxic signs (if any), once the herbal drug has been withdrawn from the regimen. These observations are usually made after 14 days of administration of herbal drugs. Biochemical parameters to study the effect of herbal drug on liver, kidney and haematological parameters are studied. At the end of study period, the animals are sacrificed; autopsies are performed; and histopathological studies of the major organs are done.

30.4.4 Reproduction and Developmental Toxicity Studies

These tests are usually carried out on rodents, which are mice and rats. If additional animal species are required for the study, rabbits may be considered. Few researchers also include zebra fish and roundworm for preliminary reproductive studies to minimize the number of animals involved. In this test the animals are dosed repeatedly with the herbal drug (Datta and Bose 2018). The test animals are dosed throughout the period of mating. Drug administration should proceed even after delivery to study the adverse effects of the herbal drug (if any) on the reproductive organs. The offspring born are carefully observed for signs of anomaly. Often toxicological reactions result in spontaneous abortions. Additionally birth defects may also occur in the newborn, indicating toxic end points. The route of administration is the same as the intended route of administration in humans. The OECD guidelines 414 (OECD 2001a, b, c, d) and 416 (OECD 2001a, b, c, d) outline the reproduction and developmental toxicity studies. Three segments of these tests are:

Segment I—Female fertility study
Segment II—Teratogenicity study
Segment III—Perinatal study

30.4.5 Special Toxicity Tests

30.4.5.1 Genotoxicity Test

Herbal drugs at times cause damage to the genetic makeup of an organism. Such alterations are evaluated by genotoxicity tests. These tests may be done both in vitro and in vivo. Mutations induced by herbal drugs or any alteration of the genetic materials are evaluated (WHO 1993). The genotoxicity study must be completed before the Phase III clinical trials. They are not required before the Phase I and Phase II trials.

1. Ames test with *S. typhimurium*—It is used to assess the potential of a herbal drug to induce genetic mutations in bacteria.
2. Chromosomal damage due to a herbal drug may be evaluated using in vitro mammalian cell cultures.
3. Chromosomal damage due to a herbal drug may be evaluated using in vitro mouse lymphoma tk assay.
4. Chromosomal damage due to a herbal drug may be evaluated using in vivo models involving rodent. Here effects of the drugs are tested on hematopoietic cells.

30.4.5.2 Carcinogenicity Test

These tests are usually performed on rodents (rats). Alternatively mice may be used for these tests, provided the use is justified after the scrutiny of the Animal Ethics Committee. The selection of proper animal strain is important (WHO 1993). It should not have a very high or a very low incidence of spontaneous tumours. These tests are done at two levels:

(a) Preliminary test
(b) Full-scale study

Initially the preliminary tests are done to study the initial response of the test organism to the herbal drug. These studies help to determine the dose levels to be used for subsequent studies. A preliminary study for carcinogenicity requires the involvement of at least two animal species. Both male and female animals should be incorporated in the study group. For example, a group of rodents containing at least ten male and females would suffice the purpose. While for non-rodents, a lower number of animals may be considered in each study group. For example, three animals per sex per group would be a rational number. The duration of drug administration should be at least 90 days. Usually the same route to be used for human trials is used in the animal species.

The period may be longer if the drug tends to cause delayed or cumulative effects. During the observation period, the animals are monitored for food intake, water intake and body weight changes. Necropsy and histopathological examinations are done at the end of the study period.

In the full-scale carcinogenicity studies, the test animals are administered the highest dose that was tested in the preliminary investigation. A full-scale study for carcinogenicity requires the involvement of at least two animal species. Both male and female animals should be incorporated in the study group. Herbal drugs are administered for 24 months in rats and for 18 months in mice. During the observation period, the animals are monitored for food intake, water intake and body weight changes. Also blood biochemistry and haematological examination should be done for the animals. Necropsy and histopathological examinations are done at the end of the study period.

A test substance is considered to be positive for carcinogenicity when any of the following types of response have been observed in the carcinogenicity study:

1. If the test animals develop any tumour, which was not seen in the control animals.
2. If the rate of incidence of tumours in test animals is more.
3. More of the organs of the test animals are affected by the tumour growth.
4. The tumours appear early in the test animals.

30.4.6 Local Toxicity Tests for Topical Preparations

30.4.6.1 Dermal Toxicity Study

Dermal toxicity study of herbal drugs is done on rat or rabbit. Daily topical application of the dosage form of the herbal drugs allows detection of dermal toxicity (if any). To apply the drug formulation, at least 10% of total body surface area is shaved (WHO 1993). Precaution is taken so as not to irritate the area during the shaving process. To explore the toxicity profile, the herbal drug should be applied in at least three different concentrations. Erythema or oedema is considered as dermal toxicity. At the end of the study, histological examination is done of the tissues. Tissues are collected from the site of administration and observed under a microscope to see the changes in the cellular level.

30.4.6.2 Photoallergy or Dermal Phototoxicity

Guinea pigs are the most suitable test animal for exploring photoallergy or dermal phototoxicity of herbal drug formulations. It is also referred to as Armstrong or Harber test. Usually a pretest or preliminary examination is done using a lesser number of animals. Most acceptable protocols need drug application to at least eight animals for preliminary evaluation. Herbal formulations should be applied at four different concentrations. If the herbal drug is formulated as a patch, application becomes easier. The patches should be applied for 2 h $\pm$ 15 min. After application of herbal patches, the test animals are exposed to UV radiation. The strength of the radiation should be 10 J/cm^2. Additional animals are dosed, but without exposing them to the UV rays. The highest dose of the herbal drug that does not produce any irritation in the test animal is thereafter determined. The site of application is observed for signs of toxicity, including oedema and erythema formation. The toxic signs are also graded based on the severity of expression. Observations are usually taken after 24 and 48 h of herbal drug exposure (WHO 1993). In the study design, positive control groups should also be included. Such animals are dosed with musk or psoralen.

30.4.6.3 Vaginal Toxicity Test

These tests are done on rabbit or dog as test animals. The herbal drugs are formulated as a pessary, cream or ointment (WHO 1993). They are applied topically in the vaginal mucosa of the animals. The minimum drug treatment period is 7 days. The maximum drug treatment period is 30 days. Vaginal swelling and histological studies are done to evaluate vaginal toxicity.

30.4.6.4 Rectal Tolerance Test

These tests are done on rabbit or dog as test animals. Six to ten animals per dose group should be taken. Formulation in volume comparable to human dose (or the maximum possible volume) should be applied once or several times daily, per rectally, to achieve administration of multiples of daily human dose. The minimum duration of application is 7 days (more according to clinical use), subject to a maximum of 30 days. For the ease of application, a suitable suppository size may be formulated with the herbal drug. The concentration of drug in the suppository is much higher than the therapeutic dose to be used in humans (WHO 1993). At the end of the study period, the animals are observed for signs of pain. Presence of the blood in the faeces is also indicative of rectal discomfort and toxicity. Microscopical examination of histological tissues provides conclusive evidence of toxicity to rectal tissue.

30.4.6.5 Neurotoxicity Test

Certain herbal drugs may cause neurological effects and convulsions. Acute or subacute administration may cause cerebrovascular accident or encephalopathy (WHO 1993). the presence of high amounts of metals in herbal drugs may also lead to neurotoxicity.

30.4.6.6 Allergenicity/Hypersensitivity Test

To explore antigenic potential of herbal drugs, GPMT or LLNA tests may be used (WHO 1993).

GPMT stands for guinea pig maximization test. A preliminary test is performed before performing the main GPMT test. The objective of the preliminary test is to determine the minimum irritant dose and the maximum safe dose. Four doses of the herbal formulation are administered by intradermal route to eight guinea pigs. The group should consist of four animals of each gender. Of these two of each sex are administered the Freund's adjuvant. Thus the minimum dose that produces signs of irritation is determined (WHO 1993). In the main study, a total of 12 guinea pigs are tested on. Six animals of each gender are used. One test and one control group should be used. It is preferable to have one more positive control group. Intradermal induction (day 1) coupled with topical challenge (day 21) should be done. If the initial dosing does not produce any response, the herbal formulation is re-administered after 7–30 days. This re-challenge often revokes the hypersensitization reactions typical of many allergens. Erythema and oedema are considered as toxicological responses. These are scored based on the severity of the response produced.

LLNA tests stands for local lymph node assay. Mice are the most suitable animal for performing this test. Contrary to the other test protocols, here animals of a single sex/gender are used. Herbal drugs are administered on the skin of the ear. Drugs are evaluated at three dose levels. The maximum dose administered should not produce any irritation in the test animals. Additionally a vehicle control animal group should be incorporated in the test design. In each testing group, there should be a minimum of six mice (WHO 1993). The Herbal drug formulations are applied on the ear skin of mice for 3 days consecutively. The mice are sacrificed on day 5. The draining auricular lymph nodes should be dissected out 5 h after I.V. administration of 3H-thymidine or bromodeoxyuridine (BrdU). Increase in 3H-thymidine or BrdU incorporation in the lymph nodes is used as the criterion for evaluation of results.

30.5 Replacement of Toxicity Testing Using New Technologies

30.5.1 Computerized Modelling

Bioinformatics and systems biology are increasingly being used to predict toxicity of herbal drug molecules. Computer simulation is used speeding the rate of discovery. Wet lab work is also significantly curtailed in the process. The quality of the data generated in such experiments is a major issue. The methods should be of predictive value in assessing in vivo toxicity and in exploring the mechanisms involved. Data from tumour cell lines cannot always be applied to in silico models. Morphology and signalling pathways of cancer cells may vary from normal cells. Quantitative in vitro to in vivo extrapolations are crucial. This helps to predict the systemic toxicity of any herbal drug. They can reduce animal experimentations. Also human toxicity potential can be made before clinical trials (Daneshian 2012).

30.5.2 Microarray Technology

The method allows the simultaneous semiquantitative measurement of the transcriptional activity of thousands of genes in a biological sample. A microarray is like a 'multiplex lab-on-a-chip.' Microarrays are generated by immobilizing cDNAs, PCR products or cloned DNA. This immobilization is done on solid supports. Most frequently the commercially used supports consist of glass slides or silicon films. Microscopic DNA spots contain the specified DNA (Wishart 2007). These are also referred to as DNA microarrays or probes. Differences in gene expressions are determined by detection techniques. Fluorescent or radioactive-labelled RNA probes are frequently used that may be hybridized. Toxicological screening of herbal drugs is a major area where this method can be applied (Schena et al. 1995).

30.5.3 Genomics and Transcriptomics

The discipline deals with the study of the genomes of organisms. Entire DNA sequences of various organisms are determined, to study complex disease mechanisms. Many diseases are caused by interplay of genetic and environmental factors. Genomics helps to unearth the fine-scale genetic mapping. The information present in a cell is based in its DNA or RNA. Thereafter the genes encoded in DNA are copied into mRNA. Translation helps in production of functional proteins. Quantitative measure of mRNA can be done by transcriptomics (Wilkins et al. 1997).

Herbal drugs with potential toxicity can be characterized by these methods. Predictive safety evaluation becomes more meaningful by the application of these technologies. Complex toxicological mechanisms can also be determined by genomics as the pattern of gene expression changes during toxicity. Overtime it can help in exploration of biomarkers for the diagnosis of toxicant exposure (Buchwald 1984).

30.5.4 Proteomics

Automated and high-throughput techniques facilitate the analysis of body fluids or tissue samples (Srivastava 2008). Structure and function of proteins are studied in proteomics. Isoelectric focusing and SDS-PAGE allow separation of proteins. Matrix-assisted laser desorption/ionization and protein mass fingerprinting allow identification of proteins. It is possible to detect and identify even very low levels of proteins (Wilkins et al. 1997). The genome of an organism is fixed, but its proteome may vary. Disease states or environmental conditions may lead to altered protein expression. Proteomics allows the peptide data to be connected directly to nucleotide sequences and gene information. 2D-gel electrophoresis helps in exploring toxic mechanisms. When altered protein expression takes place after exposure to a herbal toxicant, the biochemical pathways are altered.

30.5.5 Metabonomics/Metabolomics

The study of chemical processes involving metabolites is referred to as metabonomics. The study of metabolite profiles greatly helps in identifying toxicity due to herbal drugs (Nicholson and Lindon 2008). These products of cellular responses are greatly altered in production once an organism encounters toxicants. The focus is to identify the metabolic changes that take place in an organism (Wishart 2009).

30.5.6 Lipidomics

It is the study of pathways and networks of cellular lipids. Changes in lipid profile are the target of the study. The profile at the levels of cell, tissue or organism may be studied, once an organism encounters herbal toxicants. The lipid profile within an organism is referred to as 'lipidome.' Many diseases such as obesity and atherosclerosis are associated with lipid profile abnormalities. Instrumental techniques such as mass spectrometry, NMR and fluorescence spectrometry may be used for these lipidomics studies (Han and Gross 2005).

30.5.7 In Vitro Models

These models attempt to mimic the in vivo conditions. Controlled testing conditions provide an alternative to whole organism studies (Kandárová and Letašiová 2011). To some toxicologists the predictive value of these studies is arguable. Especially chronic toxicological effects of herbal drugs cannot be evaluated by these study methods. However cell culture or tissue assays may be designed to assess toxicity in studies discussed above (Guillouzo 1998).

30.5.7.1 Cell Cultures

Bioinformatics and systems biology are increasingly being used to predict toxicity of herbal drug molecules. Computer simulation is used speeding the rate of discovery. Wet lab work is also significantly curtailed in the process. The quality of the data generated in such experiments is a major issue. The methods should be of predictive value in assessing in vivo toxicity and in exploring the mechanisms involved. Data from tumour cell lines cannot always be applied to in silico models. Morphology and signalling pathways of cancer cells may vary from normal cells. Quantitative in vitro to in vivo extrapolations are crucial. This helps to predict the systemic toxicity of any herbal drug. They can reduce animal experimentations. Also human toxicity potential can be made before clinical trials (Daneshian 2012).

Cell cultures provide an effective alternative to in vivo methods for toxicity testing. Interaction of herbal toxicants with cultured cells can be studied under microscope. Both qualitative and quantitative changes can be studied to ascertain toxicological changes. Higher growth of cells in culture medium can speed up the studies, minimizing time consumption. It is possible to study the effect of drugs on both animal and human cell lines. Specific organ cells can be cultured in appropriate medium to study organ-specific toxicity of herbal drugs (Guillouzo 1998). A major drawback of these in vitro studies is low proliferation of a few cell lines. Also these methods do not allow designing of chronic studies. As repeated herbal drug administration is not possible. Variable results may be obtained from animal cell cultures, compared to human ones. Interspecies variation is very common in biological and preclinical studies. This reduces the predictability of the data generated significantly. Though these studies can provide preliminary toxicological data about herbal toxicants, such as toxic concentrations or effects on specific types of cells, detailed in vivo studies are required from drug regulatory point of view.

30.5.7.2 Target Organ/Tissue Assays

These may be used to predict toxicity of herbal drugs mostly on the eye or skin. The ocular toxicity assays are bovine corneal opacity and permeability (BCOP) test (OECD 2013a, b, c, d), isolated rabbit eye test (IRE) (OECD 2013a, b, c, d) and chicken enucleated eye test (CEET). Skin tests include skin irritation, skin corrosion, skin sensitization, skin penetration and photo-toxicity tests.

Living corneas of cows are used in BCOP test. Here the herbal formulations are applied directly to the epithelial surface of corneas. The corneas are obtained from slaughtered animals (OECD 2013a, b, c, d). If the drug affects the cornea adversely, the opacity is altered. Even the drug can hamper the permeability. These physiological alterations are monitored by various modern analytical optical devices.

In the isolated rabbit eye test (IRE), herbal drugs are applied to the eyes of test rabbits. Drugs are considered toxic if it produces irritation in the eyes of the test animals. End point of the assay is analysed by slit-lamp biomicroscopes (OECD 2013a, b, c, d). In the chicken enucleated eye test (CEET), the herbal drugs to be evaluated are applied to the isolated eyes of sacrificed chickens. Noncorrosive chemicals are identified by these methods. A score of 0 to 4 is given for varying degree of erythema and oedema formation. Herbal drugs available in liquid formulations are applied undiluted for the study. Solid samples may be dissolved and diluted with a minimum amount of solvent.

Skin irritation tests widely used are EPISKIN (reconstructed human epidermis test, OECD 439) (OECD 2013a, b, c, d), EpiDerm, pig ear and PREDISKIN.

EPISKIN is a 3D model of human skin. The model contains a reconstructed epidermis. Even the stratum corneum is functional, which offers an additional advantage to study the local effects of a herbal formulation. After application to the skin, MTT assay is used to evaluate toxicity. Cell viability after toxicant exposure is the end point of the assay method (Dreher et al. 2002). EpiDerm contains cultured human keratinocytes. It offers an in vitro skin irritation testing platform (Kubilus et al. 2004).

Irritant and non-irritant herbal drugs may be distinguished using the pig ear test. An increase in trans-epidermal loss of the non-perfused pig ear is considered the end point. The skin corrosion tests approved by OECD for the evaluation of herbal drugs include Corrositex (membrane barrier test, OECD 435) (OECD 2006a, b), transcutaneous electrical resistance (TER) (OECD 430) (OECD 2013a, b, c, d) and SkinEthic (reconstituted human epidermal model, OECD 431) (OECD 2014).

The rate of penetration of a herbal drug into a skin membrane maybe predicted by Corrositex. It is a protein membrane that simulates the human skin. The membrane encounters a colour change on potential toxicant exposure. Transcutaneous electrical resistance helps to identify potential corrosive herbal drugs or formulations. Loss of integrity of skin layers is the end point. This is estimated by the transcutaneous electrical resistance of the skin layers. In SkinEthic assay human epidermal keratinocyte cultures are used. Herbal drugs are applied topically, and the loss of cell viability is evaluated. This may be due to the release of cytokines from the cells (de Brugerolle et al. 1999).

In murine local lymph node (LLNA) assay (OECD 429) (OECD 2002), live mice are used. Increase in thymidine (3H-TdR) incorporation is the end point. The test estimates sensitization of a herbal drug in mice as a function of proliferative activity induced in lymph nodes.

3T3 NRU photo-toxicity test (OECD, 432) and ECVAM pre-validated EpiDerm Phototoxicity Test are used to evaluate phototoxicity (OECD 2004a, b). The test allows high-throughput testing. Poorly water-soluble compounds cannot be evaluated by this test method. The test utilizes the mouse fibroblast 3T3 cell line. Cytotoxicity is evaluated using the neutral red uptake (NRU) assay. On the other hand, ECVAM uses reconstructed human skin model, and the cytotoxicity is determined by MTT assay.

OECD Test Guideline 428 describes skin penetration tests of herbal drugs and other potential toxicants (OECD 2004a, b). Diffusion of herbal drugs into and across excised skin gives an indication of the end point. Ex vivo human, pig skins or reconstructed skin models may be used for the test. Radiolabelled or nonradiolabelled herbal test materials are evaluated by quantitating the amount of chemical that crosses the skin to test topically applied pharmaceutical formulations.

30.6 Clinical Trials of Herbal Medicines: World and Indian Scenario

According to the World Health Organization (WHO), herbals drugs are used more than allopathic medicines. Among the alternative systems of medicines prevalent in the society, Herbal formulations constitute a major portion (Bauer and Tittel 1996). With regard to safety and efficacy assessment of these drugs, nonsystemic clinical trials approach poses a great challenge. Standardization and variability in composition aggravates the problem. Standardization and variability in composition aggravates the problem, though the problem has been addressed to quite some extent with the discovery of the modern sophisticated analytical instruments. However lack of suitable marker compounds often jeopardizes the standardization protocols. The Western Pacific Regional Office of WHO took a significant step in this direction. In the year 1992, they tried to lay down the general principles to evaluate herbal medicines (WHO 1993). The urgent need to devise protocols for clinical testing of herbal drugs was laid down. Guidelines for standardization of herbal drugs were also outlined. Studies related to both pharmacology and toxicology of herbal drugs were elaborated in the protocol.

Some of the issues that influence the designing of clinical trial protocols of herbal formulations are:

- Adulteration, which may lead to non-uniform dosing
- Herbal drug-food interactions and herbal drug-allopathic medication interactions
- Toxicological data
- Dose fixation in preclinical models and extrapolation to clinical trials

Ethical practice of clinical research is of utmost importance in clinical trials. The same stringent principles should be adopted both for synthetic and natural molecules. The principal investigator should be sufficiently trained and skilled to honour the interests of the participating subjects. They should identify and recognize the adverse effects (if any) at the earliest possible in the best interests of the patients. If the herbal drugs tend to worsen any preexisting illness of the participating subject, the same should be promptly attended to. Ethics committees must apply the same vigilant attitude towards herbal studies as they do towards conventional treatment protocols.

The Drugs and Cosmetics Act of 1940 and the Drugs and Cosmetics Rules of 1945 regulate the commerce and usage of traditional medicines in India. The Government of India recognized traditional Indian medicinal system in 1959. State drug control authorities give the necessary licence for the manufacture of traditional medicines. The Drugs and Cosmetics Act has prescribed the books, which should be followed while compounding these traditional medicines. Pharmacopoeia committees of Ayurveda, Siddha and Unani drugs have been constituted (Chakravarty 1993). The safety and efficacy guidelines for the herbal drugs were put forward by the Indian government in 1993. The authorizing application of a new herbal product should be prepared according to these guidelines. Herbal drugs may be classified into three categories:

- Category 1: Drugs in use for more than 5 years
- Category 2: Drugs in use for less than 5 years
- Category 3: New herbal medicines

Central Drugs Standard Control Organisation (CDSCO), India, issued Good Clinical Practice guidelines in 2001 for herbal drugs (CDSCO 2001). It outlines the approaches that should be adopted for the clinical trial of herbal drugs.

According to the guidelines:

- The Drug Controller General of India should lay down the rules for clinical evaluation of herbal drugs. Such drugs may thereafter use in allopathic hospitals.

- All New Chemical Entities should be evaluated for repeated dosing of toxicological data. This should include exploration of toxicological effects after acute, subacute and chronic administration.
- An extract or an isolated compound should be regarded as an NCE if its proposed use is not mentioned in traditional literature.
- Clinical trials should be done after identification of markers and standardization of herbal drugs.
- Good manufacturing practices should be practiced for formulation of herbal drugs.
- During the Phase I of clinical trials, MTD should be estimated.
- If proposed duration of use is more than 3 months, 4- to 6-week toxicity studies in at least two animals should be done before Phase 2 clinical trials.
- Ethical principles should be followed, and informed consent should be taken from the participants of the clinical trials.
- During clinical evaluation of herbal drugs, a competent person should be a coinvestigator. He should be an expert in the field of Ayurveda, Siddha or Unani medicine.

In March 2013, the Department of AYUSH, Ministry of Health & Family Welfare, Government of India, introduced Good Clinical Practices guidelines for the conduction of clinical trials for Ayurveda, Siddha and Unani medicines. Drugs and Cosmetics Rule 158 B was introduced in August 2010, to further outline GCP guidelines in ASU (Ayurvedic, Siddha and Unani) drug research (AYUSH 2013).

Evidence of safety and efficacy of herbal medicines in clinical trials is limited. Most data are available in animal models only. Efficacy end points are at times variable for herbal medicines, complicating the designing of clinical trials. Different extracts of the same plant may vary in composition, and factors such as time of collection and extraction methods can give diverse results in clinical trials.

30.7 Summary and Path Forward

In the modern era, toxicity testing of herbal drugs has considerably evolved. Focus on alternative screening strategies bypassing animal usage has increased considerably. Identification of unique approaches to toxicity testing to satisfy regulatory needs remains challenging.

Animal welfare and usage have drawn considerable attention of the scientists recently. Though alternative strategies have been developed, it is impossible to predict long-term toxicity by bypassing animal models. Adoption of '3R' can help usage of animals in toxicity testing of herbal drugs. Replacement, reduction or refinement of studies using animals might help animals, but the goal is to use the best model with as few animals as necessary for a meaningful study. This might improve the studies, but it is more of an ethical consideration. In in vitro toxicity testing using cell lines, both human and animal can help achieve the objective to a great extent.

Next-generation sequencing technology, toxicogenomics and proteomics have been found to be of great value in predicting toxicity of herbal drugs. Suitable data libraries (genomic signatures of isolated and identified phytochemicals) should be developed to exploit these tools further. Modern analytical techniques such as HPLC, HPLC-MS, HPTLC and HPLC-NMR have helped to isolate the marker compounds from herbal drugs and helped in standardization and toxicity testing.

More so, the integration of recent biotechnological innovations undoubtedly brings about significant advances in predicting and determining herbal medicine safety.

Acknowledgement The authors (S. Bose and R. Datta) are indebted to Drug Development Diagnostics & Biotechnology Department, Indian Institute of Chemical Biology, Kolkata, West Bengal, India, to help explore the in vitro and cell-line toxicity studies of plant extracts in 2018 as presented in the book chapter.

References

Allard T, Wenner T, Greten HJ, Efferth T (2013) Mechanisms of herb induced nephrotoxicity. Curr Med Chem 20(22):2812–2819

Alvarez L, Gil AG, Ezpeleta O, Jalon-Garcıa JA, Cerain L (2004) Immunotoxic effects of Ochratoxin A in Wistar rats after oral administration. Food Chem Toxicol 42(5):825–834

Andrade RJ, Robles M, Fernández-Castañer A, López-Ortega S, López-Vega MC, Lucena MI (2007) Assessment of drug-induced hepatotoxicity in clinical practice: a challenge for gastroenterologists. World J Gastroenterol 13(3):329–340

Asif M (2012) A brief study of toxic effects of some medicinal herbs on kidney. Adv Biomed Res 1:44

AYUSH (2013) Good Clinical Practices Guidelines for clinical trials in Ayurveda, Siddha & Unani. Department of AYUSH, Ministry of Health & Family Welfare, Government of India, New Delhi. http://ayush.gov.in/sites/default/files/5110899178-Final%20Book%2028-03-13_0.pdf. Accessed 14 Aug 2019

Baki MA, Khan A, Al-Bari MAA, Mosaddik A, Sadik G, Mondal K (2007) Sub-acute toxicological studies of Pongamol isolated from *Pongamia pinnata*. Res J Med Med Sci 2(2):53–57

Barbosa-Ferreira M, Dagli ML, Maiorka PC, Gorniak SL (2005) Sub-acute intoxication by *Senna occidentalis* seeds in rats. Food Chem Toxicol 43(4):497–503

Barnes J (2003) Pharmacovigilance of herbal medicines: a UK perspective. Drug Saf 26 (12):829–851

Barrett B, Keifer D, Rabago D (1999) Assessing the risks and benefits of herbal medicine: an overview of scientific evidence. Altern Ther 5(4):40–49

Bauer R, Tittel G (1996) Quality assessment of herbal preparations as a precondition of pharmacological and clinical studies. Phytomedicine 2(3):193–198

Bent S, Ko R (2004) Commonly used herbal medicines in the United States: a review. Am J Med 116(7):478–485

Buchwald M (1984) Use of cultured human cells for biochemical analysis. Clin Biochem 17 (3):143–150

Bunchorntavakul C, Reddy KR (2013) Review article: herbal and dietary supplement hepatotoxicity. Aliment Pharmacol Ther 37(1):3–17

CDSCO (2001) Guidelines for clinical trials on pharmaceutical products in India. Good Clinical Practices Central Drugs Standard Control Organization Directorate General of Health Services Ministry of Health & Family Welfare, Government of India. http://www.sgpgi.ac.in/sop/GCP-%20Indian.pdf. Accessed 14 Aug 2019

Celik M, Karakus A, Zeren C, Demir M, Bayarogullari H, Duru M, Al M (2012) Licorice induced hypokalemia, edema and thrombocytopenia. Hum Exp Toxicol 31(12):1295–1298

Chakravarty BK (1993) Herbal medicines. Safety and efficacy guidelines. Reg Aff J 4:699–701

Chan P, Fu PP (2007) Toxicity of *Panax ginseng*—an herbal medicine and dietary supplement. J Food Drug Anal 15(4):416–427

Chopra A, Doiphode VV (2002) Ayurvedic medicine: core concept, therapeutic principles, and relevance. Med Clin North Am 86(1):75–89

Cragg MG, Newman DJ (2001) Natural product drug discovery in the next millennium. Pharm Biol 39(1):8–17

Da Silva J, Herrmann SM, Heuser V, Peres W, Marroni NP, Gonzalez-Gallego J, Erdtmann B (2002) Evaluation of the genotoxicity affect of rutin and quercetin by comet assay and micronucleus test. Food Chem Toxicol 40(7):941–947

Daneshian M (2012) A roadmap for the development of alternative (non-animal) methods for systemic toxicity testing. ALTEX 29(1):3–91

Datta R, Bose S (2017) A reproductive toxicological evaluation of 90 days repeated exposure of *Pterospermum acerifolium* (L) Willd in rodents. J Pharm Sci Tech 7(2):1–55

Datta R, Bose S (2018) A reproductive and developmental study following 28 days repeated oral exposure of leaves of *Pterospermum acerifolium* (L.) Willd in Wistar albino rats. Int J Pharm Sci Res 9(12):5246–5252

de Brugerolle d F, Picarles V, Chibout S, Kolopp M, Medina J, Burtin P, Ebelin ME, Osborne S, Mayer FK, Spake A, Rosdy M, De Wever B, Ettlin RA, Cordier A (1999) Predictivity of an in vitro model for acute and chronic skin irritation (SkinEthic) applied to the testing of topical vehicles. Cell Biol Toxicol 15(2):121–135

Dreher F, Patouillet C, Fouchard F, Zanini M, Messager A, Rouget R, Cottin M, Leclaire J, Beech-Kieffer F (2002) Improvement of the experimental setup to assess cutaneous bioavailability on human skin models: dynamic protocol. Skin Pharmacol Appl Skin Physiol 15S1:31–39

Duffus JH, Worth HGJ (2006) Introduction to toxicology. In: Fundamental toxicology, 2nd edn. RSC publishing, Cambridge

Farnsworth NR (1984) The role of medicinal plants in drug development. In: Krogsgaard-Larson P, Brogger CS, Munksgaard H (ed). Natural products and drug development. 2nd edn. Alfred Benzon Symposium, Copenhagen, p 17–30

Feres CAO, Madalosso RC, Rocha OA, Leite JPV, Guimaraes TMDP, Toledo VPP, Tagliati CA (2006) Acute and chronic toxicological studies of *Dimorphandra mollis* in experimental animals. J Ethnopharmacol 108(3):450–456

Gow PJ, Connelly NJ, Hill RL, Crowley P, Angus PW (2003) Fatal fulminant hepatic failure induced by a natural therapy containing kava. Med J Aust 178(9):442–443

Gregoretti B, Stebel M, Candussio L, Crivellato E, Bartoli F, Decorti G (2004) Toxicity of *Hypericum perforatum* (St. John's wort) administered during pregnancy and lactation in rats. Toxicol Appl Pharm 200(3):201–205

Guillouzo A (1998) Liver cell models in *in vitro* toxicology. Environ Health Perspect 106S2:511–532

Han X, Gross RW (2005) Shotgun lipidomics: electrospray ionization mass spectrometric analysis and quantitation of cellular lipidomes directly from crude extracts of biological samples. Mass Spectrom Rev 24(3):367–412

Hayes AW (2001) Principles and methods of toxicology. Raven Press, New York

Kandárová H, Letašiová S (2011) Alternative methods in toxicology: pre-validated and validated methods. Interdiscip Toxicol 4(3):107–113

Kubilus J, Hayden PJ, Ayehunie S, Lamore SDK, Servattalab C, Bellavance KL, Sheasgreen JE, Klausner M (2004) Full thickness EpiDerm (TM): a dermal-epidermal skin model to study epithelial mesenchymal interactions. Altern Lab Anim 32S1A:75–82

Lorke D (1983) A new approach to practical acute toxicity testing. Arch Toxicol 54(4):275–287

Maffè S, Paffoni P, Laura Colombo M, Davanzo F, Dellavesa P, Cucchi L, Zenone F, Paino AM, Franchetti Pardo N, Bergamasco L, Signorotti F, Parravicini U (2013) Herbs and cardiotoxic effects. G Ital Cardiol 4(6):445–455

Nandy S, Datta R (2012) Acute and sub-acute toxicity studies of methanolic leaves extract of *Pterospermum acerifolium* (L.) Willd in rodents. Int J Pharm Life Sci 3(3):1519–1529

Nicholson JK, Lindon JC (2008) Systems biology: metabonomics. Nature 455(7216):1054–1056

Nielsen E, Brant J (2002) Chemotherapy induced neurotoxicity: assessment and interventions for patients at risk. Am J Nurs 102:16–19

OECD (1995) Test guideline 407. Repeated dose 28-day oral toxicity study in rodents. In: OECD guidelines for the testing of chemicals. Organisation for Economic Cooperation and Development, Paris. https://ntp.niehs.nih.gov/iccvam/suppdocs/feddocs/oecd/oecdtg407-2008.pdf. Accessed 5 Aug 2019

OECD (1998a) Principles of good laboratory practice. OECD Environmental Health and Safety Publications. https://www.oecd.org/chemicalsafety/testing/good-laboratory-practicleglp.htm. Accessed 4 Aug 2019

OECD (1998b) Test guideline 408. Repeated dose 90-day oral toxicity study in non-rodents. In: OECD guidelines for the testing of chemicals. Organisation for Economic Cooperation & Development, Paris. https://www.oecd.org/chemicalsafety/testing /Revision-OECD-TG408-repeated-dose-90-day-oral-toxicity-study-in-rodents.pdf. Accessed 12 Aug 2019

OECD (2001a) Guidelines for the testing of chemicals: test no. 414: prenatal developmental toxicity study. https://ntp.niehs.nih.gov/iccvam/suppdocs/feddocs/oecd/oecd_gl414.pdf. Accessed 12 Aug 2019

OECD (2001b) Guidelines for the testing of chemicals: test no. 416: two-generation reproduction toxicity study. http://www.oecd.org/chemicalsafety/risk-assessment/1948466.pdf. Accessed 12 Aug 2019

OECD (2001c) Test guideline 420. Acute oral toxicity—fixed dose procedure (FDP). In: OECD guidelines for the testing of chemicals. Organisation for Economic Cooperation and Development, Paris. https://ntp.niehs.nih.gov/iccvam/suppdocs/feddocs/oecd/oecd_gl420.pdf. Accessed 5 Aug 2019

OECD (2001d) Test guideline 423. Acute Oral toxicity—acute toxic class method. In: OECD guidelines for the testing of chemicals. Organisation for Economic Cooperation and Development, Paris. https://ntp.niehs.nih.gov/iccvam/suppdocs/feddocs/oecd/oecd_gl423.pdf. Accessed 5 Aug 2019

OECD (2002) Test guideline 429. Skin sensitisation: local lymph node assay. https://www.oecd-ilibrary.org/environment/test-no-429-skin-sensitisation_9789264071100-en. Accessed 13 Aug 2019

OECD (2004a) Test guideline 432. *In vitro* 3T3 NRU phototoxicity test. https://ntp.niehs.nih.gov/iccvam/suppdocs/feddocs/oecd/oecdtg432-508.pdf. Accessed 13 Aug 2019

OECD (2004b) Test guideline 428. Skin absorption: *in Vitro* method. https://ntp.niehs.nih.gov/iccvam/suppdocs/feddocs/oecd/oecdtg428-508.pdf. Accessed 13 Aug 2019

OECD (2006a) Test guideline 425. Acute oral toxicity—up and down procedure. https://ntp.niehs.nih.gov/iccvam/suppdocs/feddocs/oecd/oecdtg425.pdf. Accessed 13 Aug 2019

OECD (2006b) Test guideline 435. *In vitro* membrane barrier test method for skin corrosion. https://ntp.niehs.nih.gov/iccvam/suppdocs/feddocs/oecd/oecdtg435.pdf. Accessed 13 Aug 2019

OECD (2013a) Test guideline 430. *In vitro* skin corrosion: transcutaneous electrical resistance test (TER). https://ntp.niehs.nih.gov/iccvam/suppdocs/feddocs/oecd/oecd-tg430-2013-508.pdf. Accessed 13 Aug 2019

OECD (2013b) Test guideline 437. Bovine corneal opacity and permeability test method for identifying ocular corrosives and severe irritants. https://ntp.niehs.nih.gov/iccvam/suppdocs/feddocs/oecd/oecd-tg437-2013-508.pdf. Accessed 13 Aug 2019

OECD (2013c) Test guideline 438. Isolated chicken eye test method for identifying ocular corrosives and severe irritants. https://ntp.niehs.nih.gov/iccvam/suppdocs/feddocs/oecd/oecd-tg438-2013-508.pdf. Accessed 13 Aug 2019

OECD (2013d) Test guideline 439. *In vitro* skin irritation: reconstructed human epidermis test method. https://ntp.niehs.nih.gov/iccvam/suppdocs/feddocs/oecd/oecd-tg439-2013-508.pdf. Accessed 13 Aug 2019

OECD (2014) Test guideline 431. *In vitro* skin corrosion: human. Skin model test. http://www.oecd.org/env/ehs/testing/Draft%20Revised%20Test%20Guideline%20431%20In%20Vitro%20Skin%20Corrosion%20Reconstructed%20Human%20Epidermis%20Test%20Method.pdf. Accessed 13 Aug 2019

OECD (2019) OECD guidelines for the testing of chemicals. Organisation for Economic Cooperation and Development, Paris. https://ntp.niehs.nih.gov/iccvam/suppdocs/feddocs/oecd/oecdtg425.pdf. Accessed 5 Aug 2019

Ozbek H, Ozturk M, Ozturk A, Ceylan E, Yener Z (2004) Determination of lethal doses of volatile and fixed oils of several plants. Eastern J Med 9(1):4–6

Pai VB, Nahata MC (2000) Cardiotoxicity of chemotherapeutic agents: incidence, treatment and prevention. Drug Saf 22(4):263–302

Raghunathan K (1976) Pharmacopoeial standards for Ayurvedic formulations. Central Council for Research in Indian Medicine Homeopathy (CCRIMH), Publ. 15, New Delhi

Rao SKR (1985) Encyclopedia of Indian medicine: historical perspective. Popular Prakashan, Bombay

Rothfuss MA, Pascht U, Kissling G (2001) Effect of long-term application of *Crataegus oxycantha* on ischemia and reperfusion induced arrhythmias in rats. Arzneimittelforsch 1(1):24–28

Schena M, Shalon D, Davis RW, Brown PO (1995) Quantitative monitoring of gene expression patterns with a complementary DNA microarray. Science 270(5235):467–470

Sharma DC (2000) India raises standards for traditional drugs. Lancet 356(9225):231

Sierpina VS, Wollschlaeger B, Blumenthal M (2003) Ginkgo biloba. Am Fam Physician 68 (5):923–926

Srivastava S (2008) Move over proteomics, here comes glycomics. J Proteome Res 7(5):1799

Stickel F, Seitz HK (2010) The efficacy and safety of comfrey. Public Health Nutr 3(4A):501–508

Talalay P, Talalay P (2001) The importance of using scientific principles in the development of medicinal agents from plants. Acad Med 76(3):238–247

Tuncok Y, Kozan O, Cavdar C, Guven H, Fowler J (1995) *Urginea maritima* (squill) toxicity. Clin Toxicol 33(1):83–86

USFDA (2006) Guidance for industry on complementary and alternative medicine products and their regulation by the Food and Drug Administration. Draft guidance U.S. Department of Health and Human Services Food and Drug Administration

Verma A, Gupta AK, Kumar A, Khan PK (2012) Cytogenetic toxicity of *Aloe vera* (a medicinal plant). Drug Chem Toxicol 35(1):32–25

Wang BQ (2010) *Salvia miltiorrhiza*: chemical and pharmacological review of a medicinal plant. J Med Plant Res 4(25):2813–2820

Whitto PA, Lau A, Salisbury A, Whitehouse J, Evans CS (2003) Kava lactones and the kava-kava controversy. Phytochemistry 64(3):673–679

WHO (1993) Research guidelines for evaluating the safety and efficacy of herbal medicines. http://apps.who.int/medicinedocs/en/d/Jh2946e/. Accessed 18 Aug 2019

WHO (2002) WHO expert committee on specifications for pharmaceutical preparations. Geneva. https://apps.who.int/iris/bitstream/handle/10665/42424/WHO_TRS_902.pdf;jsessionid=E83777A3295F5030C33BBEE49267C977?sequence=1. Accessed 13 Aug 2019

Wilkins MR, Williams KL, Appel RD, Hochstrasser DF (1997) Proteome research: new frontiers in functional genomics. Springer, Berlin

Wishart DS (2007) Improving early drug discovery through ADME modelling: an overview. Drugs R D 8(6):349–362

Wishart DS (2009) Computational strategies for metabolite identification in metabolomics. Bioanalysis 1(9):1579–1596

Woolf AD, Watson WA, Smolinske S, Litovitz T (2005) The severity of toxic reactions to ephedra: comparisons to other botanical products and national trends from 1993-2002. Clin Toxicol 43 (5):345–355

Zhu M, Lew KT, Leung P (2002) Protective effect of plant formula on ethanol induced gastric lesions in rats. Phytother Res 16(3):276–280

Herb-Drug Interactions

31

R. Vimalavathini, R. Shri Hari Subhashri, and S. Kavimani

Abstract

Modern drugs are being more replaced by herbal medicines especially for chronic and severe diseases for their fewer side effects, better patient adherence, cost-effectiveness and vast availability. However, an unexpected herb-drug interaction (HDI) can amplify or block the action of drugs which is a detrimental issue to be considered. Some of HDIs may even be fatal. For example, concomitant administration of warfarin with *Ginkgo biloba and Allium sativum* leads to haemorrhage, precipitation of mania in depression patients consuming antidepressants and *Panax ginseng* together, induction of hypertension on concurrent administration of tricyclic antidepressants with *Pausinystalia yohimbe* and therapeutic failure of phenytoin on simultaneous administration with shankhapushpi. Our current knowledge on HDI is limited because of the poor reporting by patients and the failure on part of practitioner to recognise HDIs. Also reports on HDIs are vague since these herbs contain many phytoconstituents and hence difficult to interpret. HDI may be of pharmacokinetic or pharmacodynamics in nature. Pharmacokinetic interactions occur mainly by inhibiting or inducing drug-metabolising enzymes and transporters, which adversely affect absorption, distribution, metabolism and excretion of concurrently administered victim drugs. Pharmacodynamic HDIs such as synergism, additive, agonist and antagonist effect are also known to occur. This chapter provides an insight into the significance of pharmacokinetic and pharmacodynamic interactions that occur on concomitant administration of herbs and conventional drugs.

R. Vimalavathini · R. Shri Hari Subhashri · S. Kavimani (✉)
Department of Pharmacology, College of Pharmacy, Mother Theresa Post Graduate and Research Institute of Health sciences, Puducherry, India
e-mail: kavimanis@yahoo.co.in

S. C. Mandal et al. (eds.), *Evidence Based Validation of Traditional Medicines*,
https://doi.org/10.1007/978-981-15-8127-4_31

Keywords

Herb-drug interactions · Absorption · Cytochrome P450 · Drug transporters · Synergism · Antagonism

Abbreviations

ATP	Adenosine triphosphate
BCRP	Breast cancer resistance protein
CYP	Cytochrome
HDIs	Herb-drug interactions
MRP	Multidrug resistance-associated protein
OATP	Organic anion-transporting polypeptides
OCT	Organic cation transporters
P-gp	P-glycoprotein
SLC	Solute carrier transporters
UGT	Uridine diphospho-glucuronosyltransferase

31.1 Introduction

Modern drugs are being more replaced by herbal medicines especially for chronic and severe diseases for their easy accessibility, fewer side effects, better patient adherence, and cost-effectiveness. China and India rank top on the use of medicinal plants in the world. Europe and the United States are major importers of herbal product in the world. Medicinal plants are used as food, drugs and raw materials for the production of drugs (Santosh et al. 2016). In fact most potent drugs such as digitalis, morphine, atropine and many anticancer agents were derived from herbs (Yaheya and Ismail 2009). Currently there is a global transition from allopathic to herbal remedies for reasons such as inconsistent increase of human population, scarcity of allopathic drugs in economically poor countries, high price and severe adverse effects associated with Western drugs (Ramawat and Goyal 2008).

Herbal drugs contain many active ingredients and when taken along with pharmaceutical drugs may interact with each other in body leading to pharmacokinetic and pharmacodynamic alterations. This raises the issue of potential herb drug interactions (HDIs). Many reports of HDIs are vague since herbs contain many phytoconstituents and are difficult to interpret. HDI may be of pharmacokinetic or pharmacodynamics in nature and unexpected HDI_S may amplify or block the drug action, and it is mandatory to identify and understand the mechanisms underlying HDIs in order to prevent adverse interactions and improve beneficial HDI_S (Brantley et al. 2014).

Major hindrance in identification of the phytoconstituents responsible for HDIs is periodic variation in the phytochemical composition of herbal materials; poor characterisation of the herbal ingredients; lack of content uniformity; poor documentation of the presence and nature of adulterants present; variations in methodologies such as extraction; deficit in data regarding the bioavailability of herbal components; interactions with many other herbal constituents; patient-related factors such as the dose, schedule and route of administration; pharmacogenetics involved in drug absorption, metabolism, distribution, excretion and drug action; underreporting by patients to healthcare providers; and failure to recognise HDI by practitioners (Venkataramanan et al. 2006). For few herbal medicines like St. John's wort and milk thistle, their constituents have been characterised (Brantley et al. 2014).

31.2 Mechanisms Involved in HDIs

Mechanisms of HDIs are divided into pharmacokinetic and pharmacodynamic interactions. Inhibition or induction of drug-metabolising enzymes and/or transporters leads to pharmacokinetic interactions, which affect absorption, distribution, metabolism and excretion of concurrently administered drugs. This alteration of plasma drug concentrations would be of clinical significance for drugs with narrow therapeutic index (warfarin, digoxin), steep dose effect profile and disease condition. Pharmacodynamic HDIs such as synergistic, additive, agonist and antagonistic effect also occur. Some of HDIs may even be fatal. For example, concomitant administration of warfarin with *Ginkgo biloba*, *Pausinystalia yohimbe and Allium sativum* leads to haemorrhage, precipitation of mania in depression patients consuming antidepressants and *Panax ginseng* together, induction of hypertension on concurrent administration of tricyclic antidepressants with and therapeutic failure of phenytoin on simultaneous administration with shankhapushpi.

31.2.1 Pharmacokinetic HDIs

Pharmacokinetic drug interaction can cause drug, toxicity, or treatment failure. Drug toxicity is due to inhibition of drug-metabolising enzymes and transporters which affects metabolism and clearance of these drugs, while therapeutic failure is due to induction of drug-metabolising enzymes and transporters leading to faster drug metabolism (Santosh et al. 2016). Pharmacokinetic interactions are observed more for antiepileptic and antibiotics which have narrow therapeutic index and hence require continuous therapeutic drug monitoring for better efficacy and safety (Shord et al. 2009).

31.2.1.1 HDIs at Absorption Level

Pharmacokinetic-based HDIs lead to variations in absorption, distribution, metabolism and excretion of the drug. To add on physical and chemical properties of drugs such as partition coefficient, molecular size and degree of ionisation, modulation of drug transporters adversely affects the amount of drug absorbed. Some herbal medicine alters the pH of the GIT which in turn can alter dissolution and absorption of pH-dependent drugs, namely, ketoconazole and itraconazole. Anthranoid-containing plants *Cassia senna*, *Rhamnus purshiana*, *Rheum officinale*, psyllium and guar gum disturb gut epithelium by inhibiting Na+/K+ ATPase and augment the activity of nitric oxide synthase leading to decreased drug absorption by altering the intestinal water and salt absorption (Mamindla et al. 2016).

31.2.1.2 HDIs at Distribution Level

HDIs at distribution level may be due to altered protein binding or at induction or inhibition of drug transporters. *Salvia miltiorrhiza* interfere with warfarin protein binding and hence increase anticoagulant effect of warfarin (Borse et al. 2019). Toxicity of drugs such as warfarin and carbamazepine is increased due to displacement from plasma protein by herbs such as meadow sweet and black willow (Mamindla et al. 2016).

31.2.1.3 HDIs at Metabolism Level

HDIs at metabolism are due to cytochrome P450 (CYP) enzymes and drug transporters. Though mostly found in the liver, they are widely distributed in most tissues. Of this CYP2B6, -2C9, -2C19, 2D6, and -3A4 isoenzymes metabolise most drugs, and CYP3A4 alone accounts for metabolism of 60% of drugs. Phytoconstituents can cause liver injury by oxidative stress, mitochondrial injury and apoptosis leading to hepatitis, liver failure, vaso-occlusive disorders, liver cirrhosis, fibrosis, cholestasis, necrosis and steatosis which can in turn affect drug metabolism. HDIs due to metabolism occur due to induction or inhibition of drug-metabolising enzymes by concurrent herbal medicines and are reversible after discontinuing the administration of herbal drugs.

Induction is a slow process involving gene activation and increased expressions of the drug-metabolising enzyme leading to therapeutic failure. It can also affect stabilisation of mRNA or protein. Enzyme induction culminates in subtherapeutic effect of the drug (Gorman 2012; Brantley et al. 2014). For instance, hyperforin, an active ingredient of St. John's wort, acts as an agonist of pregnane X receptor-modulating CYP3A4 expression, leading to decrease in bioavailability of irinotecan and imatinib. Subtherapeutic effect of midazolam was observed when co-administrated with *Echinacea purpurea*, due to CYP3A4 induction in the liver (Murtaza et al. 2017).

Inhibition of metabolic enzymes occurs when herbal medicines decrease the expression or activities of metabolic enzymes causing an increase in bioavailability of the concomitant drug and toxicity. Enzyme inhibition can result in reversible or irreversible loss of activity. Reversible inhibition is again classified as competitive, non-competitive and uncompetitive inhibition. In competitive inhibition the herbal constituent and drug compete with each other for binding to the enzyme, and most HDIs are of competitive inhibition type. Non-competitive inhibition occurs when the herb brings about an allosteric modulation of the enzyme, thus preventing the binding of the drug to the enzyme. In uncompetitive inhibition herbal medicine binds to the enzyme-drug molecule complex, thus perturbing its affinity and activity. Irreversible inhibition occurs when the herbal constituent mediates enzymatic degradation and can be revived only after new enzymes are synthesised. However, the subsequent pharmacodynamic effect depends on other drug and patient-related factors (Fasinu et al. 2012; Shord et al. 2009).

Furanocoumarins in *Citrus paradisi* juice and *Citrus sinensis* inhibit CYP3A4 enzyme, thus affecting the bioavailability of cyclosporine, terfenadine, midazolam and felodipine. Also studies show that *Ginkgo biloba* can inhibit human CYP-1A2, 209, 2C19, 2D6, and 3A4 in vitro. Also in rats it increased the plasma levels of diltiazem by inhibition of CYP3A enzyme (Nowack 2008). Green tea inhibits CYP3A4, contributing to increased plasma levels of simvastatin. Resveratrol on concomitant administration with diclofenac and carbamazepine significantly affected their plasma levels by inhibition of CYP2C9 and CYP3A4, respectively (Murtaza et al. 2017). *Curcuma xanthorrhiza* have high content of xanthorrhizol and can inhibit glucuronidation of lovastatin and lorazepam that are primarily metabolised by uridine diphospho-glucuronosyltransferase (UGT) (Salleh et al. 2016).

31.2.1.4 HDIs at Elimination Level

Renal route is the major routes of drug excretion. HDIs at kidneys are due to variations of urinary pH, glomerular filtration, tubular reabsorption and active secretion. *Arctostaphylos uva ursi*, *Taraxacum officinale*, *Urtica dioica*, *Solidago virgaurea*, *Equisetum arvense*, *Medicago sativa*, *Juniperus communis*, *Levisticum officinale*, *Petroselinum crispum*, and *Asparagus officinalis* exhibited diuretic property and thus can increase the renal elimination of other drugs (Borse et al. 2019; Liu et al. 2011).

31.2.1.5 HDIs Due to Drug Transporters

Drug transporters are widely distributed in body and play significant role in pharmacokinetic and pharmacodynamic perspectives of drug. The effect of drug transporters on drug depends on their expression and activity which in turn can be modulated by many endogenous and exogenous factors (Ishikawa et al. 2004). Variation of drug action by herbal constituents may be due to interaction with one or more drug transporters through inhibition or induction of drug transport. Inhibition of drug transporter leads to altered bioavailability of drugs pertaining to its toxicity. Induction of drug transporter can lead to therapeutic failure (Ondieki et al. 2017). The most commonly affected drug transporters are discussed herewith:

P-Glycoprotein

P-gp is an active transmembrane efflux pump and affects many drugs (digoxin, vinblastine) and endogenous substrates (steroids, lipids, bilirubin, bile acids). Phytochemicals can also modulate P-gp by inhibition of adenosine triphosphate (ATP) binding, hydrolysis or coupling of ATP-hydrolysed molecules, thus preventing transport of P-gp substrates. Induction of P-gp will decrease the bioavailability of drugs causing therapeutic failure, and inhibition of P-gp leads to drug toxicity by increasing the bioavailability of drugs. Therefore, the inhibition or induction of P-gp by concurrently administered herbs may result in interactions that can adversely affect therapy (Cho and Yoon 2015).

Studies show that *Piper nigrum*, *Curcuma longa*, *Morinda lucida* and *Carica papaya* inhibit P-gp efflux pump, leading to digoxin toxicity. Herbs such as *Momordica charantia*, *Citrus aurantium L*, *Rosmarinus officinalis*, *Zingiber officinale* and *Camellia sinensis* inhibit P-gp efflux pump, culminating in vinblastine toxicity. Also *Citrus paradise* and *Mangifera indica* inhibit P-gp efflux when combined with calcium channel blockers and increase the plasma concentration of these drugs. *Hypericum perforatum*, *Morus nigra* and *Glycyrrhiza glabra* induce P-gp efflux when treated with cyclosporine decreasing the bioavailability of cyclosporine. *Hypericum perforatum* when combined with digoxin and sulphonylureas induce P-gp efflux leading to their subtherapeutic effect. *Allium sativum* induce P-gp efflux pump when co-administered with saquinavir, thus decreasing saquinavir activity (Subhashri et al. 2018).

Chemotherapeutic failure and poor prognosis in cancer patients have been attributed to high P-gp expression levels. Herbal modulation of P-gp expression or activity has been exploited to overcome chemotherapeutic resistance (Cho and Yoon 2015). Studies reveal that administration of P-gp modulators helps to overcome this chemotherapeutic drug resistance. Studies using first- and second-generation P-gp modulators proved to be too toxic to patients. This was followed by the study of third-generation modulators of low toxicity. Currently fourth-generation P-gp modulators of herbal origins such as flavonoids and curcuminoids are being extensively studied (Amin 2013).

Breast Cancer Resistance Protein

BCRP is widely distributed in the body and responsible for export of bile acids, estrone, doxorubicin, daunorubicin, topotecan, rosuvastatin and protease inhibitors (Griffin et al. 2011). Studies reveal rhinacanthin-C from *Rhinacanthus nasutus* inhibits intestinal BCRP efflux transporters and thus increases bioavailability of topotecan, paclitaxel, digoxin, indinavir and rosuvastatin (Dunkoksung et al. 2019). Also flavonoids such as apigenin, genistein, biochanin A and kaempferol from *Silybum marianum* increase intracellular concentration of mitoxantrone by inhibition of BCRP (Mamindla et al. 2016; Fasinu et al. 2012).

Multidrug Resistance-Associated Protein

MRP are organic anion pumps, being expressed widely in the liver, kidneys and intestine. MRP substrates include glutathione conjugates and anionic drugs (Griffin et al. 2011). Studies show that curcumin by modulating MRP pump increased the absorption of baicalein (Li et al. 2016).

Solute Carrier Transporters (SLC)

Organic cation transporters (OCT) are present mainly in the kidneys and liver and transport small cations, carnitine and acetylcholine (Griffin et al. 2011; Lu et al. 2014). Flavonoids in herbal medicine have the potential to modulate this OCT transport. Organic anion-transporting polypeptides (OATP) are distributed in the liver, kidney, intestine and placenta. They transport thyroid hormones, bile acids, steroids, imatinib, fexofenadine, methotrexate, statins and HIV protease inhibitors. The phytoconstituents of pomegranate such as oleanolic acid, ursolic acid and gallic acid inhibit OATP, thus affecting the bioavailability of rosuvastatin (Li et al. 2014). Flavonoids present in grape juice can inhibit OATP affecting the activity of celiprolol and fexofenadine. The plasma levels of atenolol and fexofenadine are decreased due to inhibition of OATP by apple and orange juice in human subjects (Yu et al. 2017).

31.2.2 Pharmacodynamic HDIs

Pharmacodynamic interactions cause changes in pharmacological responses that may result in stimulation or opposition of the pharmacological activity of a concurrently administered drug manifesting as additive, synergistic, agonist or antagonistic actions. HDIs alter the affinities for receptor sites which precipitate pharmacodynamic toxicity or antagonistic effects (Fasinu et al. 2012). The problem in predicting such HDIs is that herbal preparation contains several components with unknown activity and also herbal medicine can modulate the effects of concomitantly administered drugs through concurrent effects on the same or different targets (Mamindla et al. 2016).

31.2.2.1 HDIs: Additive

The total effect of two or more drugs is equal to the sum of their individual effect, it is called additive effect. Additive activity on vitamin K epoxide reductase with *Matricaria recutita*, *Medicago sativa*, ginger, ginkgo, garlic, ginseng and Danshen may augment haemorrhage in patients on chronic warfarin therapy. Ephedrine-like alkaloids increase the actions of monoamine oxidase inhibitor (Mamindla et al. 2016).

31.2.2.2 HDIs: Synergism

Synergism occurs when one drug is enhanced or facilitated by another drug. Thus they provide favourable effects either by amplifying drug efficacy or diminishing adverse effects. For example, combined therapy of garlic with ACE inhibitor exhibited better anti-hypertensive effect. Similarly patients co-administered with silymarin and desferrioxamine showed favourable effects on thalassemia patients (Fasinu et al. 2012).

31.2.2.3 HDIs: Antagonism

One drug opposing or inhibiting the action of another is antagonism. *Hypericum perforatum* decreases anticoagulant activity of warfarin. Arecoline reduces the effect of anti-parkinsonism drugs such as flupenthixol, phenothiazine and anticholinergic such as procyclidine. *Piper methysticum* counteract L-dopa by dopamine antagonism. Phytoconstituents of Trikatu interacts with diclofenac and decreases its bioavailability and anti-inflammatory actions. *Aspilia africana* when co-administered with artemisinin and chloroquine in malaria antagonised their effects (Mamindla et al. 2016).

31.3 Pharmacogenomics in HDIs

Genetic polymorphism is intersubject variability in drug response due to pharmacokinetic and pharmacodynamic reasons and may affect drug efficacy and toxicity. Genetic polymorphisms can occur at the level of drug receptors, enzymes and transporters. St. John's wort when co-administered with nifedipine, gliclazide, mephenytoin and caffeine exhibited genetic polymorphism by inducing expression of CYP3A4 and CYP2C9 gene. St. John's wort induce OATP transporter gene, thereby decreasing the plasma concentration of repaglinide. Baicalin induces OATP transporter gene, thereby decreasing the bioavailability of rosuvastatin (Liu et al. 2015).

31.4 Evaluation of HDIs

HDIs are evaluated by in vivo, in vitro and in silico models. In vitro studies deal with disposition of herbs by drug-metabolising enzymes and transporters. Drug transport activity is determined by using cell lines such as Caco-2, Madin-Darby Canine Kidney, primary or immortalised human hepatocytes or microsomes and sandwich-cultured hepatocytes. Of this inhibition, experiments depend on cell fractions, while induction experiments are conducted on intact cells and by evaluation of RNA or protein expression (Brantley et al. 2014; Shord et al. 2009). The limitations of in vitro studies are poor sink condition and poor association to clinical situation. In vivo studies are mostly conducted in mice and rats, but interspecies variability is of major concern. The advantages are dose, and bioavailability of components is considered. Human studies are conducted in healthy volunteers or patients, and hence the results are directly extrapolative on interactions (Shord et al. 2009). The limitations are ethical issues, genetic polymorphism and population size and cost (Fasinu et al. 2012).

31.5 Regulation of Herbal Medicine

Unlicensed preparations account for over 80% of herbal medicine sales, and regulatory agencies have limited ability to restrict herbal pharmacotherapy because of wide range of herbal products. Various regulatory agencies have developed stringent regulatory guidelines for addressing market availability and safety of herbal medicine. In the United Kingdom, many herbal medicinal products remain unregistered since the licencing fee is high and limited data on efficacy and safety profile of the product is available. In Germany herbal medicines have special licencing procedures. In France more than 200 herbs have been officially accepted as phytomedicines. Australia has developed regulatory policies for herbal market for various non-Western herbs. The European Herbal Practitioner and National Institute of Medical Herbalist have been set up to integrate herbalists (Pal and Shukla 2003).

31.6 Conclusion

In conclusion, many reports have documented pharmacokinetic and pharmacodynamic interactions with therapeutic drugs which may precipitate toxicity or therapeutic failure. Identifying the mechanism of HDI is important, and when mixing of herbs with pharmaceutical drugs is inevitable, dose adjustment or discontinuation of therapy may be essential to avert obnoxious HDIs. Both patients and healthcare professionals should be educated on these medical implications of HDIs in order to curtail it.

References

Amin ML (2013) P-glycoprotein inhibition for optimal drug delivery. Drug Target Insights 7:27–34. https://doi.org/10.4137/DTI.S12519

Borse SP, Singh DP, Nivsarkar M (2019) Understanding the relevance of herb–drug interaction studies with special focus on interplays: a prerequisite for integrative medicine. Porto Biomed J 4(2):pe15. https://doi.org/10.1016/j.pbj.0000000000000015

Brantley SJ, Argikar AA, Lin YS, Nagar S, Paine MF (2014) Herb-drug interactions: challenges and opportunities for improved predictions. Drug Metab Dispos 42(3):301–317. https://doi.org/10.1124/dmd.113.055236

Cho HJ, Yoon IS (2015) Pharmacokinetic interactions of herbs with cytochrome P450 and p-glycoprotein. Evid Based Complement Alternat Med 2015:1–10. https://doi.org/10.1155/2015/736431

Dunkoksung W, Vardhanabhuti N, Siripong P, Jianmongkol S (2019) Rhinacanthin-C mediated herb-drug interactions with drug transporters and phase I drug-metabolizing enzymes. Drug Metab Dispos 47(10):1040–1049. https://doi.org/10.1124/dmd.118.085647

Fasinu PS, Bouic PJ, Rosenkranz B (2012) An overview of the evidence and mechanisms of herb–drug interactions. Front Pharmacol 3:1–19. https://doi.org/10.3389/fphar.2012.00069

Gorman GS (2012) Mechanisms and implication of drug-herbal interactions. J Bioequiv Availab 4 (4):xiii–xiv. https://doi.org/10.4172/jbb.10000e16

Griffin L, Annaert P, Brouwer K (2011) Influence of drug transport proteins on pharmacokinetics and drug interactions of HIV protease inhibitors. J Pharm Sci 100(9):3636–3654. https://doi.org/10.1002/jps.22655

Ishikawa T, Hirano H, Onishi Y, Sakurai A, Tarui S (2004) Functional evaluation of ABCB1 (P-glycoprotein) polymorphisms: high-speed screening and structure-activity relationship analyses. Drug Metab Pharmacokinet 19(1):1–14. https://doi.org/10.2133/dmpk.19.1

Li Z, Wang K, Zheng J, Cheung FS, Chan T, Zhu L et al (2014) Interactions of the active components of *Punica granatum* (pomegranate) with the essential renal and hepatic human solute carrier transporters. Pharm Biol l52(12):1510–1517. https://doi.org/10.3109/13880209.2014.900809

Li B, Zhao B, Liu Y, Tang M, Lue B, Luo Z et al (2016) Herb-drug enzyme-mediated interactions and the associated experimental methods: a review. J Tradit Chin Med 36(3):392–408

Liu CX, Yi XL, Si DY, Xiao XF, He X, Li YZ (2011) Herb-drug interactions involving drug metabolizing enzymes and transporters. Curr Drug Metab 12(9):835–849. https://doi.org/10.2174/138920011797470083

Liu MZ, Zhang YL, Zeng MZ, He FZ, Luo ZY, Luo JQ et al (2015) Pharmacogenomics and herb-drug interactions: merge of future and tradition. Evid Based Complement Alternat Med 2015:1–8. https://doi.org/10.1155/2015/321091

Lu X, Chan T, Xu C, Ng WV, Zhu L, Zhou F (2014) The interactions of herbal compounds with human organic anion/cation transporters. J Pharmacogenom Pharmacoproteom 5(5):1–8. https://doi.org/10.4172/2153-0645.1000142

Mamindla S, Prasad KVSRG, Koganti B (2016) Herb-drug interactions: an overview of mechanisms and clinical aspects. IJPSR 7(9):3576–3586

Murtaza G, Ullah N, Mukhtar F, Nawazish S, Muneer S, Mariam (2017) Phytotherapeutics: the emerging role of intestinal and hepatocellular transporters in drug interactions with botanical supplements. Molecules 22(10):1–20. https://doi.org/10.3390/molecules22101699

Nowack R (2008) Review article: cytochrome P450enzyme, and transport protein mediated herb-drug interactions in renal transplant patients: grapefruit juice, St John's Wort—and beyond! Nephrology (Carlton) 13(4):337–347. https://doi.org/10.1111/j.1440-1797.2008.00940.x

Ondieki G, Nyagblordzro M, Kikete S, Liang R, Wang L, He X (2017) Cytochrome P450 and P-glycoprotein-mediated interactions involving African herbs indicated for common noncommunicable diseases. Evid Based Complement Alternat Med 2017:1–18. https://doi.org/10.1155/2017/2582463

Pal SK, Shukla Y (2003) Herbal medicine: current status and the future. Asian Pac J Cancer Prev 4(4):281–288

Ramawat KG, Goyal S (2008) The Indian herbal drugs scenario in global perspectives. In: Ramawat K, Merillon J (eds) Bioactive molecules and medicinal plants. Springer, Berlin, pp 325–347. https://doi.org/10.1007/978-3-540-74603-4_18

Salleh NA, Ismail S, AbHalim MR (2016) Effects of *Curcuma xanthorrhiza* extracts and their constituents on phase II drug-metabolizing enzymes activity. Pharmacognosy Res 8(4):309–315. https://doi.org/10.4103/0974-8490.188873

Santosh J, Farhat PM, Manojkumar P (2016) Comprehensive review on: interaction between herbal drugs & allopathic drugs. Int J Res Pharm Pharm Sci 1(1):63–66

Shord SS, Shah K, Lukose A (2009) Drug-botanical interactions: a review of the laboratory, animal, and human data for 8 common botanicals. Integr Cancer Ther 8(3):208–227. https://doi.org/10.1177/1534735409340900

Subhashri SHR, Sridevi R, Monisha P, Shabna E, Vimalavathini R, Kavimani S (2018) Herbal drug interactions mediated by P-glycoprotein—an overview. EJPMR 5(4):270–272

Venkataramanan R, Komoroski B, Strom S (2006) In vitro and in vivo assessment of herb drug interactions. Life Sci 78(18):2105–2115. https://doi.org/10.1016/j.lfs.2005.12.021

Yaheya M, Ismail M (2009) Herb-drug interactions and patient Counseling. Int J Pharm Pharm Sci 1(1):151–161

Yu J, Zhou Z, Tay-Sontheimer J, Levy RH, Ragueneau-Majlessi I (2017) Intestinal drug interactions mediated by OATPs: a systematic review of preclinical and clinical findings. J Pharm Sci 106(9):2312–2325. https://doi.org/10.1016/j.xphs.2017.04.004

Plant-Based Traditional Herbal Contraceptive Use in India: Safety and Regulatory Issues

32

Susmita Dey, Dola Das, Arijit Chakraborty, Shubhadeep Roychoudhury, Birupakshya Paul Choudhury, Arun Paul Choudhury, and Subhash C. Mandal

Abstract

Contraception is essential in preventing unwanted pregnancy in addition to controlling population outgrowth. Control of population explosion has remained an important issue across the developing world, and socioeconomic conditions have been associated with the choice of contraceptive methods. A large proportion of rural population still practice traditional herbal medicines; hence the evaluation of contraceptive and/or anti-fertility agents assumes huge significance not only for their efficacy but also for their safety. On an average, women in the age group of 21 and 28 years use contraception in India with traditional methods being the primary choice particularly in rural settings. Estrogen and progesterone content of modern-day contraceptives has certain risk factors, which has triggered the debate over the need of traditional contraceptives. A lot of traditional contraceptive methods employ the use of plants and/or plant-based products. Although the plant-based products offer a cheap, accessible, and women-friendly approach toward contraception, yet safety and regulatory measures needed to contain the chemical and functional integrity of the compounds have remained questionable at large. Also, proper dosages of such products, if not well

S. Dey · A. Chakraborty
Department of Zoology, Royal Global University, Guwahati, India

D. Das
Lerner Research Institute, Cleveland Clinic, Cleveland, OH, USA

S. Roychoudhury (✉) · B. P. Choudhury
Department of Life Science and Bioinformatics, Assam University, Silchar, India
e-mail: shubhadeep1@gmail.com

A. P. Choudhury
Department of Obstetrics and Gynaecology, Silchar Medical College and Hospital, Silchar, India

S. C. Mandal
Department of Pharmaceutical Technology, Jadavpur University, Kolkata, India

S. C. Mandal et al. (eds.), *Evidence Based Validation of Traditional Medicines*,
https://doi.org/10.1007/978-981-15-8127-4_32

monitored, may result in potential toxicity and unwanted medical hazards. Furthermore, the risk of sexually transmitted diseases (STDs) may never be overruled while employing traditional methods that might put the women's health at jeopardy. Moreover, validation of such herbal and plant-based traditional products for possible development of contraceptive drug, proper evaluation of their safety, and regulatory issues coupled with associated toxicity and side effects, if any, is mandatory. In this regard, safety pharmacology applies the principles of proper drug testing for safe human administration. It deals with quality control and testing for efficacy which includes toxicity studies supported by well-documented preclinical data, dose escalation studies, regulatory review, and mandatory clinical trials (Phases 1, 2, & 3). A lot of the traditional herbal products do not follow the required guidelines of drug development and medication usage. Central Drug Standard Control Organization (CDSCO) is the Indian national regulatory organization for pharmaceuticals and medical devices which is equivalent to the Food and Drug Administration (USA), European Medicines Agency (European Union), Pharmaceuticals and Medical Devices Agency (Japan), and Medicines and Healthcare products Regulatory Agency (UK). Many traditional contraceptives in usage in rural India as folk medicines have not yet been tested for safety and efficacy and assessed by CDSCO safety guidelines/licensing procedures. Although drug testing and approval by CDSCO will enhance cost of medication, yet it is mandatory for the long-term medical cost and health of individuals exposed to untested/unlicensed drugs.

Keywords

Traditional medicine · Herbal contraceptive · Safety pharmacology · Regulation · CDSCO

Abbreviations

CDER	Center for Drug Evaluation and Research
CDSCO	Central Drugs Standard Control Organization
EMA	European Medicines Agency
FAERS	FDA Adverse Event Reporting System
FDA	Food and Drug Administration
HA	Hypothalamic amenorrhea
IND	Investigational New Drug
IUD	Intrauterine device
LAM	Lactational amenorrhea method
LPD	Luteal phase dysfunction
MHRA	Medicines and Healthcare products Regulatory Agency
PAHO	Pan American Health Organization
PCOS	Polycystic ovarian syndrome
PMDA	Pharmaceuticals and Medical Devices Agency

PSTC	Predictive Safety Testing Consortium
STD	Sexually transmitted disease
TGA	Therapeutic Goods Administration
WHO	World Health Organization
WTO	World Trade Organization

32.1 Introduction

The world population continues to rise, and there is an urgent need to increase the choice of family planning methods available to couples. As such, continued research to develop safe, effective, affordable, and reversible contraceptives is crucial. Indeed, the past few decades have brought about great advances in hormonal and non-hormonal contraceptive studies on both sexes alike (Sitruk-Ware et al. 2013). However, long-term use of systemic hormone-based contraceptives might result in adverse side effects or undesired pathological circumstances (i.e., increase the incidence of cancer, hypertension, obesity, and/or blood clots, etc.) (Marnach et al. 2013).

Contraceptive method is the best protection against unwanted pregnancy. It can be defined as the global prevention of contraception via various devices, chemicals, drugs, or surgical procedures. To control the population, contraception has been a basic need for the mankind. The prime aim of contraception lies in allowing a couple maximum comforts, freedom to choose expansion of family at an appropriate time, and have no fear of unwanted pregnancy. Certain contraceptive methods also prevent sexually transmitted diseases (STDs) apart from reducing the vulnerability of adolescents against the risk of unwanted pregnancy outcome(s).

Energy metabolism is closely linked to human reproduction. Balanced nutrition is not only required during the pregnancy, but they also play a vital role during the adolescent age as well as the preconception period (Templeton 2000). Poor lifestyle habits corresponding to altered body mass index, physical activity, and nutritional intake can affect fertility (Chavarro et al. 2007). Malnutrition in poor countries is very common due to inadequate availability of food and lack of awareness; while in developed countries, inappropriate food habits as well as lifestyle choices are the major reasons for undernutrition in females which can culminate into reproductive disorders (Hassan and Killick 2004).

Studies have shown that decrease in food intake, prolonged undernutrition, and negative energy balance can suppress the reproductive function (Knuth et al. 1977). When there is prolonged food scarcity or energy imbalance, growth and reproduction are compromised over essential body functions in order to ensure the individual's survival. Altered nutritional status has also been encountered in polycystic ovarian syndrome (PCOS) and hypothalamic amenorrhea (HA) (Chavarro et al. 2007), which are common causes of infertility. Body fat ratio is higher in PCOS patients as compared to normal ones irrespective of their normal weight. Approximately 4–9% of healthy reproductive age group women also suffer the potential risk of luteal phase dysfunction (LPD) (Lenton et al. 1984). The LPD makes the endometrium less receptive for implantation which may lead to early pregnancy loss and sometimes infertility (Smith 1988). Under such conditions of malnutrition and associated disorders, preventing pregnancy through proper use of contraception might be the only answer to prevent further health complications.

32.2 Types of Contraceptive Methods

Over the past decades, there has been an increasing need for custom-made, sexual health services for adolescents. Societal change led by urbanization and modernization has directed to flexibilities in family ties leading toward premarital sexual activity (Muniruzzaman 2017). Younger ages of menarche combined with delayed average age of marriage culminate in a longer time period between the onset of sexual maturity and marriage. An important source of information to the adolescents on responsible sexual behavior gets eliminated with decline in intergenerational relationships. As such, adolescents tend to learn on this subject from peers or from the mass media and social networking sites. This puts them at a risk to most likely depend on cheap traditional contraceptives. Sex education and information on proper use of contraceptives are of utmost importance to achieve unwanted adolescent pregnancies. There are many different types of contraceptives, and some are very effective. They may be classified as given below.

1. **Traditional methods**: These may include lactational amenorrhea method (LAM), withdrawal, rhythm method, etc.
2. **Modern methods**: These include male condom, female condom, oral contraceptive pills, injectables, emergency contraceptive pills, etc.
3. **Surgical methods**: These include intrauterine devices (IUD), female sterilization (tubectomy), and male sterilization (vasectomy).
4. **Other methods**: These include hormonal implants, barrier/chemical method, and the use of diaphragm and spermicides in some of the developed countries.

In general, any method other than sterilization that is appropriate and physiologically safe to healthy adults is also considered as safe and appropriate for post-pubertal adolescents. However, informed contraceptive decision-making includes more than just medical safety.

Female condoms, although not a new idea, offers a great tool for contraception and potential protection against sexually transmitted diseases. Recent advancements in the design and use of polyurethane to make female condoms have made them widely available to willing users. Polyurethane, being a stronger material compared to latex is less likely to rupture or to have the pinholes common in natural latex condoms. Female condoms offer certain advantages over male condoms. One primary advantage is that it can be inserted in advance allowing more sexual impulsiveness. The female condom can provide greater protection against STDs due to its ability to cover both the internal and external genitalia (Prudhomme et al.

2005). Additionally, the female condom does not need to be immediately removed after ejaculation, thereby providing adequate space for greater intimacy after intercourse. Finally, it is absolutely safe for individuals with latex allergy.

Pregnancy is rare among breastfeeding women with LAM. This method comprises of the informed use of breastfeeding as a contraceptive method by a woman who is still amenorrheic and who is not feeding her baby with supplements, for up to 6 months after delivery (Tietze and Lincoln 1987). Breastfeeding delays return of fertility and menstrual cycle, thereby providing a natural and physiological method of contraception. Users of this method are thought to have 98% protection from pregnancy (Kennedy et al. 1989). It is unquestionably cost-effective with greater health benefit, as breastfeeding alone provides the infant with adequate nutrition and fluid intake through the first 6 months. Breast milk is considered a healthier option than its substitutes for infants in low-resource settings, as suitably practiced in countries like Egypt (Khella et al. 2004).

The withdrawal method or coitus interruptus is another effective method of contraception. This involves withdrawal of penis from the vagina and woman's external genitals before ejaculation with the objective to prevent sperm from entering the vagina. However, this is not an effective form of birth control (Handelsman and Waites 2006). Sperm may enter the vagina if withdrawal is not properly timed or if pre-ejaculation fluid contains sperm. The withdrawal method does not offer protection from sexually transmitted infections. Rhythm method, also known as calendar method, is another natural method of family planning. This requires tracking the menstrual history to predict when ovulation will possibly occur. This helps to determine when the person is most likely to conceive. If the female is hoping to get pregnant, the rhythm method is one of the possible ways to determine indulging in intercourse. Similarly, if one is hoping to avoid pregnancy, rhythm method may be used to determine the exact days to avoid unprotected sex. Rhythm method is based on detecting ovulation and the fertile period in a woman. Ovulation occurs approximately 14 days before the start of next menstrual cycle (Wilcox et al. 2000). Throughout a menstrual cycle, there are approximately 6 days of fertile interval during which conception will likely occur if intercourse happens (Bilian et al. 2010). The fertile window can comprise of 5 days before and on the day of ovulation. Rhythm method requires persistence and careful record keeping. Couples interested in getting pregnant have only a limited period of time for fertilization to happen due to limited viability of the ovum and sperm (Colombo and Masarotto 1996). The ovum usually loses its viability 10–24 h post-ovulation (Stanford et al. 2002). Life span of sperm is usually 24–48 h and depends on the internal environment of the female reproductive tract. Under conditions where estrogenic cervical mucus supports the life span of sperm, their fertilizing capacity can last from 3–7 days in the periovulatory period (Pallone and Bergus 2009).

32.3 Limitations to Contraception in India and the Availability of Herbal Contraceptives

There are certain limitations to contraception in India due to a number of factors as noted below.

32.3.1 Inadequate Cognitive Capacity

Adolescents in their preadolescence phase do not understand properly the risks involved in sexual indulgence. Until they move from concrete to abstract, they are unlikely to engage themselves in correct sexual decision-making. Often, such individuals are quite ignorant of the fact that their non-deductive decision-making might put their life in turmoil (Molina and González 2012).

32.3.2 Egocentric Thinking

A kind of egocentric belief makes some adolescents feel that they entirely depend upon natural methods to avoid pregnancy. This makes them more inclined toward taking chances rather than using contraceptives (Whaley 1999).

32.3.3 Inadequate/Lack of Knowledge of Contraceptives

Low uptake of contraceptives is often attributed to lack of proper knowledge of contraceptive methods. A study on abortion-seeking unmarried girls reported that 88% women did not know the link between sexual relations and pregnancy (Chabbra and Nuna 1994).

32.3.4 Anxiety

Less knowledge about sexual information leads to anxiety and becomes an obstruction in use of contraception which results in contraceptive failure. Moreover, cognitive problems related to socially disapproved activity often leads to anxiety and prevents adolescents from seeking reliable advice related to contraception (Furstenberg et al. 1981).

32.3.5 Misinformation

The misinformation about safe periods most of the times becomes the cause for failure to use contraception. Though the sexually active adolescents in urban areas have knowledge about birth control and contraception, such knowledge may not always ensure proper use of birth control methods during coital activity (Gandotra and Das 1996).

32.3.6 Perception of Dangers of Contraceptives

Some adolescents believed that condoms and IUDs could move upward in the women's body and cause complications or pills cause permanent infertility (Olson and Rollins 1982). It has been noted that sometimes beliefs are distinct and distant from actual information on birth control and availability of methods/devices.

32.3.7 Faulty Use and Storage

Condoms are often stored in cupboards or under mattresses in order to hide from kids and family members. Due to pressure, heat, and humidity, the quality of condoms may deteriorate.

32.3.8 Poor Quality of Contraceptives

Due to poor quality and defective packaging or storage, contraceptives may fail leading to the disappointment of the user. Estrogen and progesterone content of contraceptive pills has certain risk factors (Ness et al. 2000), which has triggered the need of traditional contraceptives. Therefore, there is a quest for contraceptive drug development from herbal products used by indigenous and ethnic communities for ages. Such plant-based agents could effectively replace the pills having untoward side effects. Furthermore, the reversible herbal contraceptives are non-invasive and are believed to be relatively safe. Therefore, development of newer contraceptives with the ethno-pharmacological background using the reverse pharmacology approach is perhaps the way forward (Gurib-Fakim 2011). In India, there are many references of such plants and/or plant products with contraceptive and/or anti-fertility properties (Table 32.1).

32.4 Efficacy

The main aim of contraception is to ensure a couple with a full comfort without fear of an unwanted pregnancy and to bestow the couple with sufficient freedom to have offspring when desired. The primary emphasis lies on ensuring maximum comfort and privacy at minimum cost and without any side effects. The need of young couples is not termination of fertility but proper management and spacing out pregnancies. This requires the use of reversible and non-invasive methods. Reversibility rate of many of the modern-day contraceptives are very low and so less accepted. Reversibility is a very good factor for a novel contraceptive. Steroid-

Table 32.1 Plants used traditionally in India having potential for development of herbal contraceptives

Sl. No	Scientific name	Parts used	Solubility	Effect	Author/year
1	*Abrus precatorius*	Seeds	Ethanolic	Abortifacient	Priya et al. (2012)
2	*Acalypha indica*	Seeds	Alcoholic	Anti-estrogenic	Kong et al. (1985)
3	*Achillea millefolium*	Flowers	Water	Contraception	Raj et al. (2011)
4	*Achyranthes\ aspera*	Whole stem and root	Ethanolic	Anti-implantation and abortifacient	Pokharkar et al. (2010)
5	*Adhatoda vasica*	Leaves	Methanolic	Anti-implantation and abortifacient	Pokharkar et al. (2010)
6	*Aegle marmelos*	Leaf	Alcoholic	Contraception	Shah et al. (2009)
7	*Afrormosia laxiflora*	Stem bark	Ethanolic	Anti-gonadotropic, estrous cycle blocker	Pokharkar et al. (2010); Gediya et al. (2011)
8	*Ailanthus excelsa*	Leaf, stem, bark	Ethanolic	Anti-implantation	Shah et al. (2009); Raj et al. (2011)
9	*Alangium salvifolium*	Stem bark	Ethanolic	Abortifacient, anti-implantation	Priya et al. (2012)
10	*Albizia lebbeck*	Seeds, roots, pods	Ethanolic	Anti-fertility	Raj et al. (2011)
11	*Allium cepa*	Bulb	Ethanolic	Anti-implantation	Priya et al. (2012)
12	*Aerva lanata*	Aerial parts	Ethanolic	Anti-implantation	Pokharkar et al. (2010)
13	*Aloe vera*	Latex	Ethanolic	Anti-ovulatory	Gediya et al. (2011)
14	*Arctium lappa*	Leaves and roots	Alcoholic	Abortifacient	Shah et al. (2009)
15	*Ardisia solanacea*	Plants excluding roots	Water extract	Spermicidal	Pokharkar et al. (2010)
16	*Aristolochia tagala*	Whole plant	Ethanolic	Anti-implantation	Narwaria et al. (1994)
17	*Artemisia africana*	Leaf	Water	Abortion	Pathak et al. (2005)

(continued)

Table 32.1 (continued)

Sl. No	Scientific name	Parts used	Solubility	Effect	Author/year
18	*Artemisia vulgaris*	Whole plant	Water	Disrupts spermatogenesis	Priya et al. 2012
19	*Aspilia africana*	Leaves	Alcoholic	Anti-ovulatory	Priya et al. (2012)
20	*Austropenckia populnea*	Pods	Alcoholic	Anti-implantation & abortification	Shah et al. (2009)
21	*Azadirachta indica*	Seed oil	Ethanolic	Anti-androgenic	Pokharkar et al. (2010)
22	*Ballota undulate*	Leaves, flower	Alcoholic	Anti-implantation	Priya et al. (2012)
23	*Mentha arvensis*	Leaves	Water	Contraception	Ahmad et al. (2011)
24	*Mentha longifolia*	Leaves	Water	Contraceptive	Kaur et al. (2011)
25	*Melia azedarach*	Seed	Water	Anti-implantation	Azmeera et al. (2012)
26	*Momordica cymbalaria*	Root	Methanolic	Anti-implantation	Pokharkar et al. (2010)
27	*Mondia whitei*	Root	Methanolic	Anti-spermatogenic	Kaur et al. (2011)
28	*Nelumbo nucifera*	Seeds	Methanolic	Anti-estrogenic	Kaur et al. (2011)
29	*Oxalis corniculata*	Whole plant	Methanolic	Estrogenic	Priya et al. (2012)
30	*Piper longum*	Roots, leaves, fruits	Methanolic	Anti-fertility	Keshri et al. (1996)
31	*Piper nigrum*	Fruit powder	Water	Contraception	Panda et al. (2011)
32	*Piper betel*	Petiol	Methanolic	Anti-estrogenic	Singh and Singh (2010)
33	*Punica granatum*	Fruits	Alcoholic	Anti-implantation	Pokharkar et al. (2010)
34	*Plantago ovata*	Seed	Methanolic	Abortion	Panda et al. (2011)
35	*Polygonum hydropiper*	Root, powder	Water	Anti-estrogenic	Singh and Singh (2010)
36	*Physalis alkekengi*	Plants	Water	Anti-implantation	Keshri et al. (1996)
37	*Phyllanthus amarus*	Whole plant	Methanolic	Contraception	Panda et al. (2011)
38	*Pterocarpus erinaceus*	Stem bark	Ethanolic	Anti-gonadotropic	Pokharkar et al. (2010)
39	*Plumbago zeylanica*	Root	Ethanolic	Abortifacient	Ravichandran et al. (2009)

(continued)

Table 32.1 (continued)

Sl. No	Scientific name	Parts used	Solubility	Effect	Author/year
40	*Pergularia daemia*	Twig	Ethanolic	Anti-implantation, late abortifacient	Choudhary et al. (2013)
41	*Quassia amara*	Bark, leaves	Benzene	Contraception	Ravichandran et al. (2009)
42	*Striga orobanchioides*	Plant	Ethanolic	Anti-implantation	Choudhary et al. (2013)
43	*Sesbania sesban*	Seeds	Chloroform	Anti-implantation	Pokharkar et al. (2010)
44	*Ruta graveolens*	Root, plant powder	Aqueous	Contraception	Ravichandran et al. (2009)
45	*Rumex steudeli*	Root	Acetone	Contraception	Singh and Singh (2010)
46	*Cassia fistula*	Fruits, bark aqueous	Aqueous	Anti-implantation, estrogenic	Panda et al. (2011)
47	*Areca catechu*	Fruit/seed	Ethanolic	Anti-ovulatory	Shrestha et al. (2010)
48	*Ocimum sanctum*	Leaves	Benzene	Anti-fertility	Pandey and Madhuri (2010)
49	*Plumbago rosea*	Leaves	Ethanolic	Anti-ovulatory	Sheeja et al. (2009)
50	*Trichosanthes cucumerina*	Whole plant	Ethanolic	Anti-ovulatory	Kage et al. (2009)
51	*Hibiscus rosa sinensis*	Root, flower	Benzene	Anti-implantation, anti-ovulatory	Vasudeva and Sharma (2008)
52	*Achyranthes aspera*	Roots	Ethanolic	Anti-ovulatory, anti-implantation	Vasudeva and Sharma (2007)
53	*Dysoxylum binectariferum*	Stem bark	Ethanolic	Anti-implantation	Keshri et al. (1996)
54	*Balanites roxburghii*	Bark, fruit, seed, leaves	Ethanolic	Abortifacient, anti-implantation	Singh and Singh (2010)
55	*Lantana camara*	Leaves	Hydroalcoholic	Female infertility	Mello et al. (2005)
56	*Derris brevipes*	Leaf, root	Ethanolic	Anti-implantation, anti-fertility	Badami et al. (2003)
57	*Tinospora cordifolia*	Stem	Aqueous	Infertility	Choudhary et al. (2013)

(continued)

Table 32.1 (continued)

Sl. No	Scientific name	Parts used	Solubility	Effect	Author/year
58	*Lepidium sativum*	Mature plant	Methanolic	Anti-ovulatory	Pande et al. (2002)
59	*Calotropis procera*	Root	Aqueous	Anti-ovulatory	Circosta et al. (2001)

based contraceptive pills may develop unwanted side effects, such as obesity, cardiovascular disease, and cancer of the breast and uterus and hence not recommended for long-term use.

Estradiol and progesterone used in contraceptive pills are metabolized in the gastrointestinal tract and liver resulting in side effects like abdominal cramping, bloating, changes in menstrual periods, breast pain, and liver toxicity.

Due to better compatibility and lesser side effects, herbal contraceptives are now more acceptable. The development of new fertility regulating drugs from plants is an attractive proposition. Herbal contraceptive also offers alternatives to those women who lack access to modern contraceptives, basically, the women in the rural areas in developing nations with very high fertility rates. Ethnic plant products of proven birth control efficacy are expected to be of higher acceptability and are likely to generate greater confidence and safety with reference to public health.

32.5 Case Study

Spermicides have also been reported to be of similar effect in some studies. Reddy et al. (2004) reported that nisin, a known antimicrobial peptide, can have a safe and effective contraception in rabbits both in vitro and in vivo. Nisin, at a concentration of 400 mg/mL, was found to be spermicidal in vitro. It was also observed that the effect was dose and time dependent. Intravaginal application of 1 mg nisin inhibited conception in rabbits (in vivo studies). It was further observed that repeated application in rabbits (50 mg per animal per day) for 14 consecutive days did not cause any damage to vaginal epithelium or local inflammation.

32.6 Safety and Regulatory Issues

Safety pharmacology is an important aspect of drug development not only from inception in the laboratory to the launch of a finished product in the market but also for the life cycle optimizations of the product. Regulatory agencies across the globe in many countries take the prime responsibility to ensure that a drug product is safe for human use and is used as recommended and that there are no toxic side effects. Regulatory affairs cover not only the aspects associated with the product development and manufacturing but also registration of the product, market viability, and life cycle monitoring or optimization. Safety monitoring and putting safety as the first rule of thumb during the process of drug development are critical steps to determine the efficacy and long-term success of a drug product (Gewin 2009; Yao et al. 2013; Sanghi and Tiwle 2014). Different countries worldwide have different agencies that monitor the drug development process, and approval of these agencies is mandatory for labeling and marketing of drugs. Some of established agencies are Food and Drug Administration (FDA, USA), Medicines and Healthcare products Regulatory Agency (MHRA, UK), Therapeutic Goods Administration (TGA, Australia), Central Drug Standard Control Organization (CDSCO, India), European Medicines Agency (EMA, European Union), Health Canada (Canada), Swiss Medics (Switzerland), and Pharmaceuticals and Medical Devices Agency (PMDA, Japan). In addition, World Health Organization (WHO), Pan American Health Organization (PAHO), and World Trade Organization (WTO) also play an important role in pharmaceutical and medical device research and development, licensing, manufacturing, marketing, and intellectual property protection. In essence the prime objective of such organizations is to regulate proper research and safety measures of drugs and devices used in healthcare industry (Cronenwett et al. 2007). Here, we highlight the functional aspects of FDA and compare facts with CDSCO with the objective to focus on improving safety measures and regulatory affairs during the process of marketing and use of traditional herbal-based therapies/ medications.

All drugs and medical devices available in the US market have to get an approval or licensed through FDA. FDA functions under tight regulation and surveillance in the development and marketing of healthcare products and continues to work on safety evaluation of drugs and devices post-marketing. The objective is to support drug safety and performance throughout the life cycle of drugs (Ciociola et al. 2014). Surprisingly, in certain cases, new emerging safety profiles have casted original risk assessment data in doubt. A company interested in launching a new product and has at that point collected sufficient preclinical data is required to put in an Investigational New Drug application (IND) with FDA (Young et al. 1988). Following the application, the Center for Drug Evaluation and Research (CDER) has 30 days to review the IND and approve for clinical trials. A clinical trial primarily comprises of 3–4 phases (Mahan 2014). Phase 1 is a safety evaluation and dose escalation study mostly on healthy subjects. Positive data from Phase I is needed to proceed to Phase II which comprises 100–200 subjects. Phase 3 allows the testing of drugs in larger number of subjects. Success of Phase III will determine if the drug is ready to seek FDA approval for marketing and labeling. In certain cases, investigators may decide or FDA may mandate a Phase IV clinical trial of a new drug (Hermans 2014). Drug safety science is another functional aspect of FDA which continues to bring new insights into a disease process and information related to the cellular and molecular incorporation of the drug in question. FDA employs high end software and tools to help CDER's staff during product evaluation. Jumpstart is a software platform used by CDER/FDA during early assessment of drug product, data quality, and safety measures in consideration. FDA mandates that all clinical trial data be deposited at the Janus clinical trial repository and made available to public. Post-marketing drug surveillance is an important aspect of FDA. FDA utilizes a passive and an active system to monitor safety of drugs already available in the market. FDA Adverse Event Reporting System (FAERS) is the passive system that employs an electronic database allowing spontaneous submission of adverse events. FDA also utilizes the sentinel system (active system) to monitor post-marketing drug safety and efficacy. In addition, MedWatch is routinely used to report and collect clinically relevant safety information on marketed drugs or devices. Predictive Safety Testing Consortium (PSTC) was also established with a mission to scientifically validate new biomarkers for the detection, evaluation, and monitoring of adverse events induced by drugs. FDA also works to regulate or monitor misuse of drugs. Working toward drug product efficacy and safety FDA regularly updates public on new drug safety information through drug safety communications (Lynch 2019). Figure 32.1 summarizes the drug approval and safety regulatory process followed by FDA.

Central Drug Standard Organization (CDSCO) is the Indian national regulatory authority that functions under the Directorate General of Health Services, Government of India. CDSCO oversees compliant conduction of clinical trials, approval of new drugs, manufacturing standards, and quality of imported drugs. CDSCO works in coordination with the State Drug control organization with the objective to establish uniformity in the enforcement of Drug and Cosmetic Act (CDSCO website). New investigational drug or foreign imported drugs have to get approval from CDSCO before their launch in market. Agencies interested in either can apply online, make due payments, and upload all required data, following which the review committee will review facts and data and decide whether or not to grant permission. During the process all steps can be monitored online. All permissions and licenses are organized through a comprehensive database (SUGAM). According to the CDSCO Clinical trial guidelines, sponsors, investigators, regulator,and ethics committee are solely responsible for the proper design and ethical conduction of trials. Investigators are limited to working on a maximum of three trials at the same time, whilemanufacturers, sponsors, and clinical research organizations are advised to provide compensation for any post-trial adverse event. Clinical trials of foreign approved drugs in Indian population can only be waived under extreme conditions of emergency or epidemic. Generic or bio-similar products marketed in other countries with satisfactory report for over 4 years can be approved in India and require only short trials. Under conditions where two or more countries remove a certain drug from their market on grounds of safety and efficacy, the continued marketing of that particular drug will require stringent monitoring by CDSCO. In addition, CDSCO has established many drug testing laboratories throughout India that engages in testing and analytical quality control of new, imported drugs and cosmetics that require CDSCO approval. Certificate of accreditation from these laboratories is required for drug approval and quality control. Although CDSCO functions to regulate drugs and devices permitted or licensed to be in the Indian market and maintains strict regulatory protocols, proper safety evaluation post-marketing is not as well monitored and regulated as in the United States or other European nations. An initiative by the central/regional drug testing laboratories to test the traditional herbal contraceptives used particularly in rural India, for safety and efficacy, will be

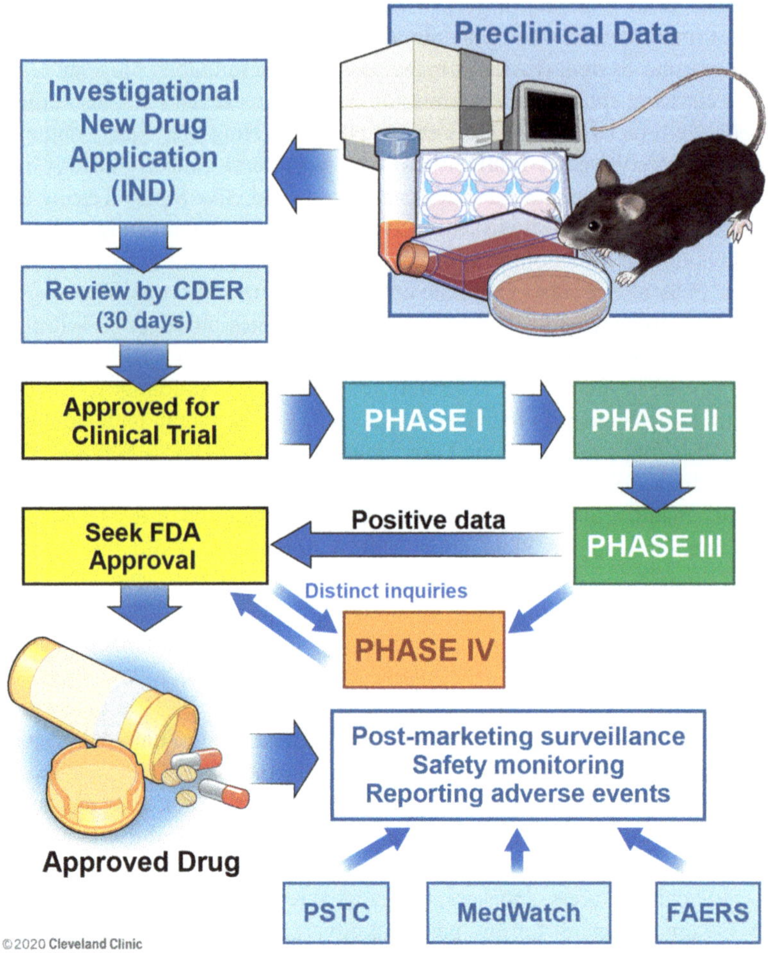

Fig. 32.1 Schematic representation of drug approval and safety regulatory processes

an important step toward providing cheap and safe contraceptive means to the population.

32.7 Conclusions

The use of contraceptives is a useful method for safe sex without unwanted pregnancies in the short term while also contributing to a long-term effect on population control. Lack of information, fear to discuss with elders, and social stigma often result in the use of cheap and traditional plant-based contraceptive methods, particularly in rural India. Although these herbal traditional means of contraception have delivered some success toward achieving the goal of avoiding unwanted pregnancies, yet their role in protection against sexually transmittable diseases is not known. Besides, some of these drugs may not be as effective or may result in adverse side effects. Since a lot of these traditional herbal products have not gone through stringent manufacturing, safety and regulatory measures, or approval by CDSCO, their efficacy, safe dosage, and adverse effects are not properly documented or imparted to end users, thereby imposing a greater health hazard. Although cheap and readily available to low-income sectors, their safety and efficacy are questionable, and the use of these drugs should be approached with great concern. CDSCO and its regional drug testing laboratories can play a fundamental role in evaluating the safe dosage, adverse effects, or any other regulatory issues pertaining to the use of such traditional plant-based contraceptives. If properly manufactured with all safety measures evaluated and considered, these plant-based contraceptive methods can produce significant result and be readily available for use particularly in rural areas.

References

Ahmad S, Jamal Y, Mannan A (2011) Review of some medicinal plants with anti-fertility activities. Unani Res 1:24–28

Azmeera M, Elumalai A, Eswaraiah MC, Mathangi N (2012) An updated review on anti-fertility plants. Int J Pharm 2:4–6

Badami S, Aneesh R, Sankar S, Sathishkumar MN, Suresh B, Rajan S (2003) Antifertility activity of *Derris brevipes* variety *coriacea*. J Ethnopharmacol 84:99–104

Bilian X, Heng Z, Shang-chun W, Xiao-ping J, Chang-hai H, Wen-qi S, Zhen-wu L (2010) Conception probabilities at different days of menstrual cycle in Chinese women. Fertil Steril 94:1208–1211

Chabbra R, Nuna C (1994) Abortion in India: an overview. Veerendra Printers, New Delhi

Chavarro JE, Rich-Edwards JW, Rosner BA, Willett WC (2007) Diet and lifestyle in the prevention of ovulatory disorder infertility. Obstet Gynecol 110:1050–1058

Choudhary N, Siddiqui MB, Azmat S, Khatoon S (2013) *Tinospora cordifolia*: ethnobotany, phytopharmacology and phytochemistry aspects. Int J Pharm Sci Res 4:891–899

Ciociola AA, Cohen LB, Kulkarni P, Kefalas C, Buchman A, Burke C, Cain T, Connor J, Ehrenpreis ED, Fang J, Fass R (2014) How drugs are developed and approved by the FDA: current process and future directions. Am J Gastroenterol 109:620–623

Circosta C, Sanogo R, Occhiuto F (2001) Effects of *Calotropis procera* on oestrous cycle and on oestrogenic functionality in rats. IL Farmaco 56:373–378

Colombo B, Masarotto G (1996) Menstrual cycle fecundability study group, daily fecundability: first results from a new data base, vol 94. B.R. Publishing Corporation, New Delhi, p 114

Cronenwett LR, Bootman JL, Wolcott J, Aspden P (eds) (2007) Preventing medication errors. National Academies Press, Washington, DC, pp 409–446

Furstenberg FF, Lincoln R, Menken JA (1981) Teenage sexuality, pregnancy, and childbearing. University of Pennsylvania Press, Philadelphia, pp 280–300

Gandotra MM, Das NP (1996) Factors influencing choice of a contraceptive and the reason for its discontinuation. In: Khan ME, Cernada G (eds) Spacing as an alternative strategy: India's Family Welfare Programme. B. R Publishing Corporation, New Delhi, pp 95–114

Gediya S, Ribadiya C, Soni J, Shah N, Jain H (2011) Herbal plants used as contraceptives. Int J Curr Pharm Rev Res 2:47–53

Gewin V (2009) Safety first! Careers in drug safety and toxicology. https://www.sciencemag.org/features/2009/03/safetyfirstcareersdrugsafetyandtoxicology. Accessed 5 Nov 2019

Gurib-Fakim A (2011) Traditional roles and future prospects for medicinal plants in health care. Asian Biotechnol Dev Review 13:77–83

Handelsman D, Waites G (2006) Traditional methods. In: Schill WB, Comhaire F, Hargreave TB (eds) Andrology for the clinician. Springer, Berlin

Hassan MA, Killick SR (2004) Negative lifestyle is associated with a significant reduction in fecundity. Fertil Steril 81:384–392

Hermans L (2014) Conducting a clinical safety review in drug development regulatory environment. https://www.cambridgecognition.com/blog/entry/conducting-a-clinical-safety-review-in-drug-development. Accessed 5 Nov 2019

Kage DN, Malashetty VB, Seetharam YN, Suresh P, Patil SB (2009) Effect of ethanol extract of whole plant of *Trichosanthes cucumerina* var. *cucumerina* L. on gonadotropins, ovarian follicular kinetics and estrous cycle for screening of antifertility activity in albino rats. Int J Morphol l27:173–182

Kaur R, Sharma A, Kumar R, Kharb R (2011) Rising trends towards herbal contraceptives. J Nat Prod Plant Resour 1:5–12

Kennedy KI, Rivera R, McNeilly AS (1989) Consensus statement on the use of breastfeeding as a family planning method. Contraception 39:477–496

Keshri G, Singh MM, Lakshmi V, Kambhojposycoital VP (1996) Contraceptive efficacy of the seeds of *Nigella sativa* in rats. Indian J Physiol Pharmacol 39:59–62

Khella AK, Fahim HI, Issa AH, Sokal DC, Gadalla MA (2004) Lactational amenorrhea as a method of family planning in Egypt. Contraception 69:317–322

Knuth UA, Hull MG, Jacobs HS (1977) Amenorrhea and loss of weight. Br J Obstet Gynaecol 84:801–807

Kong YC, Nag KH, Wat KH, Wong A, Saxana IF, But PPH, Chang HT (1985) Yuechukene, a novel anti-implantation indole alkaloids from *Murraya paniculata*. Planta Med 51:304–307

Lenton EA, Landgren BM, Sexton L (1984) Normal variation in the length of the luteal phase of the menstrual cycle: identification of the short luteal phase. BJOG Int J Obstet Gynaecol 91:685–689

Lynch SS (2019) Drug efficacy and safety. https://www.merckmanuals.com/professional/clinicalpharmacology/conceptsinpharmacotherapy/drug-efficacy-and-safety. Accessed 4 Nov 2019

Mahan VL (2014) Clinical trial phases. Int J Clin Med 5:1374–1383

Marnach ML, Long ME, Casey PM (2013) Current issues in contraception. Mayo Clinic Proc 88:295–299

Mello FB, Jacobus D, Carvalho K, Mello JR (2005) Effects of *Lantana camara* (Verbenaceae) on general reproductive performance and teratology in rats. Toxicon 45:459–466

Molina RC, González EA (2012) Teenage pregnancy. Endocrine Dev 22:302–331

Muniruzzaman MD (2017) Transformation of intimacy and its impact in developing countries. Life Sci Res Policy 13:10

Narwaria A, Khosa RL, Dhar SK (1994) Experimental studies on *Artemisia vulgaris*—a possible antifertility drug. Ancient Sci Life 14:10–15

Ness RB, Ann Grisso J, Klapper J, Schlesselman JJ, Silberzweig S, Vergona R, Morgan M, Wheeler JE (2000) Risk of ovarian cancer in relation to estrogen and progestin dose and use characteristics of oral contraceptives. Am J Epidemiol 152:233–241

Olson L, Rollins J (1982) Psychological barriers to contraceptive use among adolescent women. Pregnancy in adolescence: needs, problems, and management. Van Nostrand, New York

Pallone SR, Bergus GR (2009) Fertility awareness-based methods: another option for family planning. J Am Board Family Med 22:147–157

Panda SK, Padhi LP, Mohanty G (2011) Antibacterial activities and phytochemical analysis of *Cassia fistula* (Linn.) leaf. J Adv Pharm Technol Res 2:62–67

Pande D, Malik S, Bora M, Srivastava AP (2002) A rapid protocol for *in vitro* micropropagation of *Lepidium sativum* Linn. An enhancement in the yield of Lepidine. In Vitro Cell Dev Biol 38:451–455
Pandey G, Madhuri S (2010) Pharmacological activities of *Ocimum sanctum* (tulsi): a review. Int J Pharm Sci Rev Res 5:61–66
Pathak AK, Mallurwar VR, Kondalkar AK, Soni S (2005) A review of plants with anti-fertility activity. Nig J Nat Prod and Med 9:4–10
Pokharkar RD, Saraswat RK, Kotkar S (2010) Survey of plants having antifertility activity from Western Ghats area of Maharashtra state. J Herbal Med Toxicol 4:71–75
Priya G, Saravanan K, Renuka C (2012) Medicinal plants with potential antifertility activity—a review of sixteen years of herbal medicine research (1994-2010). Int J Pharmtech Res 4:481–494
Prudhomme M, Boucher J, Delberghe P, Christman H, Leroux MC (2005) Female condom or male condom: offer a choice! Gynécol Obstét Fertil 33:891–897
Raj A, Singh A, Sharma A, Singh N, Kumar P, Bhatia V (2011) Antifertility activity of medicinal plants on reproductive system of female rat. Int J Bioeng Sci Technol 2:44–50
Ravichandran V, Arunachalam G, Subramanian N, Suresh B (2009) Contraception and its significance in traditional system of medicines. Int J Pharm Sci 1:1–21
Reddy KV, Aranha C, Gupta SM, Yedery RD (2004) Evaluation of antimicrobial peptide nisin as a safe vaginal contraceptive agent in rabbits: *in vitro* and *in vivo* studies. Reproduction 128:117–126
Sanghi DK, Tiwle R (2014) Role of regulatory affairs in a pharmaceutical industry. IJPRR 4:127–131
Shah GM, Khan MA, Ahmad M, Zafar M, Khan AA (2009) Observations on antifertility and abortifacient herbal drugs. Afr J Biotechnol 8:1959–1964
Sheeja E, Joshi SB, Jain DC (2009) Antiovulatory and estrogenic activity of Plumbago rosea leaves in female albino rats. Indian J Pharmacol 41:273–277
Shrestha J, Shanbhag T, Shenoy S, Amuthan A, Prabhu K, Sharma S, Banerjee S, Kafle S (2010) Antiovulatory and abortifacient effects of *Areca catechu* (betel nut) in female rats. Indian J Pharmacol 42:306–311
Singh OM, Singh TP (2010) Phytochemistry of *Solanum xanthocarpum*: an amazing traditional healer. J Pharm Res 8:98–109
Sitruk-Ware R, Nath A, Mishell DR Jr (2013) Contraception technology: past, present and future. Contraception 87:319–330
Smith JF (1988) Influence of nutrition on ovulation rate in the ewe. Aust J Biol Sci 41(1):27–36
Stanford JB, White GL Jr, Hatasaka H (2002) Timing intercourse to achieve pregnancy: current evidence. Obstet Gynecol 100:1333–1341
Templeton A (2000) Infertility and the establishment of pregnancy—overview. Br Med Bull 56:577–587
Tietze SL, Lincoln R (1987) The effect of breastfeeding on the rate of conception. In: Fertility regulation and the public health. Springer, New York, pp 333–338
Vasudeva N, Sharma SK (2007) Estrogenic and pregnancy interceptory effects of *Achyranthes aspera* Linn. root. Afr J Tradit Complement Altern Med 4:7–11
Vasudeva N, Sharma SK (2008) Post-coital antifertility activity of *Hibiscus rosa-sinensis* Linn. roots. Evid Based Complement Alternat Med 5:91–94
Whaley AL (1999) Preventing the high-risk sexual behavior of adolescents: focus on HIV/AIDS transmission, unintended pregnancy, or both? J Adolesc Health 24:376–382
Wilcox AJ, Dunson D, Baird DD (2000) The timing of the "fertile window" in the menstrual cycle: day specific estimates from a prospective study. BMJ 321:1259–1262
Yao B, Zhu L, Jiang Q, Xia H (2013) Safety monitoring in clinical trials. Pharmaceutics 5:94–106
Young FE, Norris JA, Levitt JA, Nightingale SL (1988) The FDA's new procedures for the use of investigational drugs in treatment. JAMA 259:2267–2270

Traditional Medicine Stability and Pharmacokinetic Issue

33

Parmandeep Kaur, Sunidhi Sharma, and Diptiman Choudhury

Abstract

Phytomedicine including Ayurveda, Siddha, Naturopathy, Homeopathy, Chinese traditional medicine, Unani, etc. has a long history of therapeutic application in humans and animals. However, the efficacy of phytomedicine depends not only on the pharmacokinetics but also on the shelf life of the formulations as well. Many of the compounds/herbal formulations due to short shelf life show different pharmacological responses at different time points. But unfortunately for many formulations, stability of the compounds and relative response to its pharmacokinetics have not been well established. Further many of the herbal formulations are a raw extract of one or multiple plants where the exact composition of the formulations is not known; therefore their stability and pharmacokinetic parameters largely remain unexplored. Here in this chapter different methods of studying the stability of the formulations and effect of stability on the pharmacokinetics have been discussed to benefit the people working in the field of traditional and translational medicine.

Keywords

Traditional medicines · Pharmacokinetics · Drug absorption and distribution · Drug stability · Therapeutic index · Drug storage and packaging

P. Kaur · S. Sharma · D. Choudhury (✉)
School of Chemistry and Biochemistry And Thapar Institute of Engineering and Technology-Virginia Tech (USA) Center for Excellence in Material Sciences, Thapar Institute of Engineering and Technology, Patiala, India
e-mail: diptiman@thapar.edu

S. C. Mandal et al. (eds.), *Evidence Based Validation of Traditional Medicines*,
https://doi.org/10.1007/978-981-15-8127-4_33

Abbreviations

°C	Degree Celsius
ADR	Adverse drug reactions
API	*Ayurvedic Pharmacopoeia of India*
APIs	Active pharmaceutical ingredients
FDCs	Fixed-dose combination
FPPs	Finished pharmaceutical products
GACP	Good agricultural and collection practices
GC-MS	Gas chromatography-mass spectrometry
GLC	Gas liquid chromatography
GMP	Good manufacturing practice
HDPE	High-density polyethylene
HPLC	High-performance liquid chromatography
HPTLC	High-performance thin-layer chromatography
h	Hours
IR	Infrared spectroscopy
JAMA	Journal of American Medical Association
LD50	Lethal dosage 50
NLN	Natural lipid-based nanoparticles
PVP	Polyvinylpyrrolidone
RH	Relative humidity
TLC	Thin-layer chromatography
UV-VIS	Ultraviolet visible spectroscopy
WHO	World Health Organization

33.1 Introduction

From ancient times, Indians, Egyptians, Chinese, and people from other countries have used various plant parts and products as herbal products. In the indigenous medical texts, about 25,000 plant-based formulations are written, and the modern pharmacopeias are known to have at least 25% drugs that are derived from plants or are the synthetic analogs of the prototype compounds that are isolated from the plants. Since the Vedic era, plants and its parts have been used in medicines, and about 500 plants have been reported for medicinal uses. WHO had estimated that for primary health care, about 81% of the world population uses herbal medicines.

Phytomedicines are known to have a short shelf life. Those get deteriorated easily when not stored properly due to factors like environmental conditions including humidity, temperature, sunlight, and pollution, leading to the loss of active components having pharmaceutical value. Plants contain a variety of components such as lecithins, saponins, polyphenols, alkaloids, flavonoids, phenols, etc. which act as therapeutic agents in herbal medicine. When the herbal extracts are formed, these various components come to a single phase which can lead to chemical reactions between them causing the production of some toxic metabolites, which in turn can cause serious effects on the body if consumed (Johnson et al. 2011). According to the new Indian government norms, the quality of the herbal product should be within the permissible limits. Before marketing, the stability of the herbal medication should be tested for its active product and its related toxicity. Based on the toxicity study, the safety of herbal medicine is assured (Hussin 2001). Stability testing is a very critical step in the case of herbal formulations as the whole formulation is regarded as the potent medicine, regardless of the other components present and their effects on the organism. The rules for herbal drug preparation and its marketization differ from country to country according to the native herbal medicine preparations and practices. Of the total herbal medicines practiced so far, only a few of them have been tested for their safety and efficacy. There are certain guidelines published by the WHO defining the criteria for evaluating herbal medicines. There are other regulatory bodies or national agencies established by the different countries for assurance of the safety of herbal medicines (Calixto et al. 2000).

For the safety and effectiveness of herbal medicines, pharmacovigilance and pharmacognosy of the herbal drug are important. Pharmacovigilance is the discipline that is related to the collection of information, its monitoring and researching, and then evaluation of the information collected from the patients or health service providers. The word pharmacovigilance comes from Greek and Latin words *pharmakon* (Greek for drug) and *vigilare* (Latin for to keep watch). Pharmacovigilance strongly emphasizes on the adverse drug reactions (ADR) caused by any pharmaceutical compound. It precisely evaluates the risk factors and beneficial factors associated with any potent drug (Nainwal and Nainwal 2010). Pharmacognosy means a complete systematic study of herbal drug, which includes scientific name, the source of the drug, its chemical composition, and its harvesting and storage. Pharmacovigilance identifies the hazards related to a drug and protects the people from unnecessary harm (Krishnaiah 2010; Nainwal and Nainwal 2010). It identifies and does an assessment of the drug and its interactions with other components or drugs including pet medications (Routledge 1998; Nainwal and Nainwal 2010).

In previous times, all the standards of herbal medicine were optimized by the herbal medicine practitioner. Nowadays, it is well determined by the herbal pharmacopoeia. Pharmacopoeia is a book that contains all the guidelines laid down by the government to monitor the quality of herbal medicines. It is published by the governmental authority or the pharmaceutical society.

Internationally many pharmacopeias provide the standards for herbal medicinal products like:

- *The Ayurvedic Pharmacopoeia of India* (API)
- *Japanese Standards for Herbal Medicine*
- *Pharmacopoeia Committee*
- *The United States Herbal Pharmacopoeia*
- *British Herbal Compendium*
- *British Herbal Pharmacopoeia*

- *Chinese Herbal Pharmacopoeia*
- *European Pharmacopoeia*

There are different pharmacopoeia monographs to maintain the quality of the herbal preparations. For example, *The Ayurvedic Pharmacopoeia of India* (API) recommends the standards for the 80 Ayurveda herbal formulations that are commonly used (Patel 2011).

It is a general perception of the people that herbal preparations are safe to use and do not possess any side effects, but it is completely untrue. In 1991 and 1992, in Brussels, Belgium, it was reported that 30 women were treated with a Chinese herbal preparation to reduce weight. In that herbal formulation, one of the herbs was mistakenly considered as non-toxic. They died of renal failure due to the presence of aristolochic acid in that herbal preparation. Therefore, pharmacovigilance for natural formulations is very important and should be practiced regularly to avoid such a situation in the future (Ernst 1998). Different analytical techniques are employed to evaluate herbal drug formulations such as chromatography and spectroscopic techniques. As herbal formulations contain a lot of contents which makes it very difficult to analyze each component, therefore sometimes markers are used. Markers are the chemically defined components that are used for the control purpose, even if they do not have any medicinal or therapeutic value. For the quality, purity, and identifying the kind of herbal preparation, its physical parameters, its moisture content, ash content, adulterants' presence, and other solvent residues need to be checked. The correct identification and quality of herbal formulation are of supreme importance in quality control of phytoformulations (Shulammithi et al. 2016).

The demand for herbal plants is increasing day by day around the world in the form of drugs, food supplements, beverages, and cosmetics. When compared to well-defined synthetic drugs, these herbal medicines show marked differences like:

- The unknown active principles.
- Stability, quality control, and standardization are feasible but not easy.
- The quality and availability of raw materials are of concern.
- It's difficult to prove the efficacy and safety of well-controlled double-blind clinical and toxicological studies.
- Have a wide range of therapeutic use.
- Suitable for chronic treatments.
- With herbal medicines the undesirable side effects were known to be less frequent, but it was revealed with the help of well-controlled randomized clinical trials that they also exist.
- Generally the cost is less than synthetic drugs.
- Known to be effective, economical, and relatively less toxic.
- These are easily available in the neighborhood.

Because herbal medicines are being used directly from natural sources, it may contain impurities, heavy metals, biological contaminants, toxic substances, and hazardous foreign materials and therefore must be checked before consumption, or it could cause serious health problems.

Phytotherapeutic agents do not possess immediate pharmacological action and therefore are not used for emergency treatment. Combinations of homeopathic preparations along with the chemically defined active substances are not considered to be herbal medicines even though they contain plants.

Herbal products that are used in prevention, diagnosis, or treatment of diseases are classified as drugs depending on a country and its current legislation. The marketing of botanical products in the form of dietary supplements is done in few nations like the USA. While in some nations, the drugs from the herbal preparations need to be registered and have to be examined to show that they are safe and have clinical efficacy.

Some programs have been established to investigate the safety and efficacy of herbal medicines as the WHO guidelines have initially recommended for the evaluation of herbal medicines. Even though clinical trials of herbal drugs are possible, placebo-controlled double-blind trials have done for the herbal medicines as it has been shown by the specialized literature's survey, and this has been confirmed by the meta-analysis done recently by the reviews that have been published in prestigious medical journals like the Journal of the American Medical Association (JAMA), the Annals of Internal Medicine, the Lancet, and the British Journal of Clinical Pharmacology, and the British Medical Journal. For the explanation of discrepancies like this, many factors play an important role, like a deficiency in quality control and standardization of the herbal drugs that are used in clinical trials; using various dosages of herbal medicines; patients not selected properly and improper randomization in many studies; because of the different aromas, tastes, etc., there is a challenge in establishment of suitable placebos; for the achievement of statistical significance, there is an insufficient number of patients in most trials; and herbal medicines have a large number of variations in the duration of treatments. To verify the efficacy and safety that may have been lacking in herbal drugs as a function of the above difficulties, some have been studied appropriately, and well-controlled double-blind clinical trials are done.

33.2 WHO Guidelines for Herbal Drug Stability

Herbal medicines have been defined by the World Health Organization as the finished labeled medicinal products that contain an aerial or underground part of the plant, an active ingredient or other plant material, or combinations. According to WHO, around 80% of the world population depends on traditional medicines for their primary health-care needs as these herbal drugs are comparably cheaper and easily available because of the presence of rich agroculture conditions, but its careless utilization may result in threatening the sustainability of many of plant species.

The Drugs and Cosmetics Act of 1940 and The Drugs and Cosmetics Rules of 1945 govern traditional medicines. The Drugs and Cosmetics Act was amended in

1959 by the Government of India, and it incorporated drugs that were obtained from traditional medicines. In 1993, the guidelines for the safety and efficacy of these herbal drugs were formed by an expert committee, and the procedures of which were devised by the Drug Controller General of India for the allopathic drugs, and it was stated that all the traditional and herbal products should follow these guidelines before undergoing clinical trials for therapeutic conditions.

WHO guidelines of the year 1993 asserted that both the regulatory bodies and the manufacturers should take the responsibility for the quality of herbal products and drugs. The responsibility to make guidelines on several aspects of data evaluation, dossiers, and quality assurance and the evaluation of the agreement of the post-marketing product along with the specifications that were given by the producers and in compliance with GMP was given to the regulatory authorities.

In the WHO Summit, 2000, numerous suggestions were given for the cultivation of raw materials of herbal formulations and quality control. Some of these suggestions comprised the development of a sub-committee on GACP that will promote good-quality herbal medicine's availability to the market by giving advice and training to small farmers and producers. Incentives were given to the producers of botanical raw materials which will help to promote the GACP implementation, and that could involve giving help in the form of logistic and technical support for the selection of a suitable site for agricultural production, selecting pesticides and fertilizers, providing seedlings and seeds, and giving advice on primary processing and machinery used for harvesting.

The license issuing for wholesalers, manufacturers, assemblers, and importers of herbal medicinal must be ensured by the National Drug Regulatory Authority. Depending on what type of business is involved, the dealers of imported herbal medicinal products can ask for one or more of the licenses.

Detailed knowledge is required for the registration of the herbal medicinal products like the list of ingredients, details of the assembler(s) and manufacturer (s), complete product formula for the imported herbal-based medicinal products, manufacturer's certificate or license that is provided by the drug regulatory authority, dosage form, mode of administration, dosage, it's adverse effects, precautions, warnings, indications, duration of use, product brand name, contraindications, and major drug interactions and, if feasible, the expiry date of product, date of manufacturing, storage condition, and lot/batch number.

Herbal product's pharmacovigilance centers were expected to collect and evaluate the information regarding the efficacy and safety of herbal products by observing the adverse drug reactions. Only 4 out of the 11 member countries in the SE Asia region have national systems for observing the safety of traditional medicinal products (Parveen et al. 2015).

33.2.1 WHO Guideline Goals

1. Strengthening of the research in testing of efficiency and safety of herbal medicines

2. Promoting and strengthening the fair use of herbal medicines

33.2.2 WHO Guideline Objectives

1. Ensuring the efficiency and safety of herbal medicines that are used in the health-care systems of countries within the region and elsewhere in the world
2. Providing research criteria for evaluation of the efficiency and safety of herbal medicines to propose basics for the states to make research guidelines of their own for the study of herbal medicines
3. Facilitating the exchange of research information and experience so that a source of reliable data can be stored for the affirmation of herbal medicines

33.2.3 WHO Guidelines for Stability Testing of Pharmaceutical Products Having Well-Established Drug Substances in Conventional Dosage Forms

The Committee reviewed and adopted the advised modifications of the storage conditions for climatic zone IV to read with a relative humidity of 65% (±5%) and 30 °C (±2 °C) for the real-time stability studies. In cases the storage conditions and special transportation were not following these criteria, extra study data explaining these conditions were required to be made available.

33.2.4 Stability Testing of Active Pharmaceutical Ingredients and Finished Pharmaceutical Products

WHO in the Technical Report Series No. 953 (2009) published the guidelines on stability testing of active pharmaceutical ingredients (APIs) and finished pharmaceutical products (FPPs) that aimed to make a clear picture about the core stability of the data packages that were needed for the registration of APIs and FPPs. Another thorough discussion, as well as feedback on the report and on the numerous sections of the existing guidelines, was done in July 2016 in Copenhagen during the joint meeting with the Prequalification of Medicines Team Assessment Group and the Medicines Quality Assurance Group on regulatory guidance for multisource products, and timely revisions of the text was concluded.

The main objective of stability testing was in rendering evidence regarding the quality and how it varies in accordance with the time when under the influence of a different kind of environmental factors like light, humidity, and temperature. The stability testing also involved the study of product-related factors that influence its quality, for example, packaging materials, container-closure systems, and the interaction of the APIs with excipients. In the case of fixed-dose combination FPPs (FDCs), the interactions between two or more APIs were considered. Due to this stability testing, the shelf life for the FPP or the retest period for the APIs can be established, and suitable storage conditions can be recommended. In exceptional cases like that of unstable APIs, shelf life is given.

33.2.5 Guidelines for Active Pharmaceutical Ingredient

Knowledge about the stability of active pharmaceutical ingredients (APIs) is an essential part of the methodical approach towards stability evaluation. The shelf life or the retest period specified by the API(s) producer should be acquired using the stability testing data.

Stress Testing: API(s) stress testing could help in identifying the suitable degradation products that can help in validating the stability-indicating power of the analytical procedures that have been used and establishing the degradation pathways along with the intrinsic stability of the molecule.

The following strategies can be used for an active pharmaceutical ingredient: when feasible, it would be acceptable to give the relevant data that is being published in the scientific literature for supporting the identified degradation pathways and products, and in case no published data is available, stress testing should be performed.

Stress testing can be executed with a single slot of the API(s), and it should include effects of humidity (75% relative humidity (RH) or higher), temperature (in 10 °C increments above the temperature used for accelerated testing), and, wherever suitable, photolysis and oxidation of the API(s). The susceptibility of the API(s) towards hydrolysis over a justified pH range when present in solution or suspension should also be evaluated during the testing.

The aim of stress testing is not to degrade the API(s) completely but to recognize the primary degradation products. The conditions investigated should cause small extent degradation, typically with 10–30% loss of API(s) to occur when concluded by an assay as compared to non-degraded API(s). Such a target should be chosen so that secondary products do not get generated, but only some degradation occurs.

Batch Selection: Stability study data for at least three primary batches of the APIs should be given, and these API(s) batches should be fabricated at a pilot-scale by using the identical synthesis route as that of production batches. A production method and a scheme that mimic the final process should be used in case of production batches. The whole quality of the API(s) batches that are placed on stability studies should represent the material quality that has to be formed on the production scale.

Container-Closure System: A container-closure system should be used for the stability studies carried out on the API(s) packaged in a way either that is the same or that simulates the packaging that has been proposed for storage and distribution.

Specification Stability Studies: Testing of different stability-indicating attributes like those that are sensitive to change during storage and are expected to affect the safety, efficacy, and/or quality should be included, and the biological, chemical, microbiological, and physical attributes should also be covered. The employment of certified stability-indicating analytical procedures should be done.

Frequency of Testing: In the case of long-term studies, the testing frequency should be sufficient enough to establish the API(s) stability profile. For APIs that have recommended shelf life or retest period of at least 12 months, the testing frequency at long-term storage conditions should normally be every 3 months for the first year, every 6 months for the second year, and yearly thereafter throughout the recommended shelf life or retest period. A minimum of three time points from a 6-month study, including the initial and final time points like 0, 3, and 6 months, is recommended for the accelerated storage conditions.

Storage Conditions: An API(s) in general should be assessed under storage conditions with suitable tolerances, which tests its thermal stability and, if suitable, its sensitivity to moisture. The lengths of studies chosen and the storage conditions should be adequate to cover the shipment and storage. Monitoring and recording of the storage conditions should be done; however, short-term environmental variations because of the storage facility's opening and closing of the doors are accepted as unavoidable. Any effect of excursions that affects the stability data due to equipment malfunction should be evaluated, addressed, and notified. If the registration application is being evaluated, then the additional data gathered during this period should be submitted to the authorities along with the data in response to outstanding questions. To assess the effect of short-term excursions outside the label storage conditions that might happen during shipping, data from the intermediate storage condition and the accelerated storage condition can be used. In case long-term studies are performed at 60% RH ± 5% RH/25 °C ± 2 °C and at the accelerated storage condition if some "significant change" happens at any time during 6 months' testing, additional tests should be conducted at the intermediate storage conditions and then assessed against the notable change criteria. The testing during the intermediate storage condition should incorporate all long-term tests unless it is otherwise justified. At least 6 months' data from a 12-month study at the intermediate storage condition should be incorporated in the initial application.

Stability Commitments: A commitment should be made in case the proposed shelf life or retest period is not covered on the primary batches of available long-term stability data that had been granted at the time of approval and should also continue with the stability studies even after post-approval to firmly establish the shelf life or retest period. A post-approval commitment is unnecessary if the submission covers the proposed shelf life or retest period of long-term stability data on the three production batches. If not then one of the following commitments should be made: in case stability study data on less than three production batches is included in the submission form, then a commitment should be made so that these studies continue through the suggested retest period, and additional production batches should be made in long-term stability studies up to a total of at least three through the suggested shelf life or retest period; in case stability study data on three production batches is included in the submission form, a commitment should be made such that these studies continue through the suggested shelf life or retest period; and in case stability data on production batches are not included in the submission form, a commitment should be made so that the first three production batches should be placed on long-term stability studies through the suggested shelf life or retest period. The stability protocol should be the same for both, in case of long-term studies for the stability commitment and that for the primary batches unless otherwise it is scientifically justified.

Evaluation: The stability study aims to establish a shelf life or a retest period that is appropriate to all forthcoming API(s) batches that are fabricated under similar circumstances. The amount of variability in the case of individual batches affects the confidence if a future production batch will have similar specifications throughout the shelf life or retest period that was assigned. It is apparent from the data that with so little variability and so little degradation, the requested shelf life or retest period will be imparted. It is unnecessary to do statistical analysis under those circumstances. One proposal for the data analysis for the quantitative attribute which is assumed to vary with time is to determine the time during which 95% of the one-sided confidence limit for the mean curve will intersect at the acceptance criterion. If small batch-to-batch variability is shown in the analysis, it would be beneficial to combine all the data to make it one overall estimate which can be performed firstly by applying suitable statistical tests like P values in case the level of significance of rejection is greater than 0.25, to the zero time intercepts for individual batches and slopes of the regression lines. If it is unsuitable for data to combine from several batches, then the overall shelf life or retest period should depend on how much minimum time a batch can be supposed to reside in the acceptance criterion. The character of the degradation relationship would help in determining if the data should be modified for the linear regression analysis. Normally, this relationship can be expressed by a cubic, linear, or quadratic function on a logarithmic or an arithmetic scale. The choice of model should be supported by a chemical and/or physical rationale and should also acknowledge the amount of available data (using the parsimony principle). To examine the quality of fit of data on all the batches, statistical methods should be employed. Long-term data from the long-term storage condition with limited extrapolation can be taken beyond the perceived range to prolong the shelf life or retest period that can be engaged if verified, and this justification should be based on the consequences of examining under accelerated conditions, what is known about the degradation mechanism, the goodness of fitting of any of the mathematical models, batch size, and occurrence of supporting stability results. Nevertheless, this extrapolation implies that beyond the observed data, the same degradation relationship will proceed to apply. The evaluations should cover the levels of degradation products and other stability-indicating attributes along with the assay.

Statements and Labeling: Based on API(s) stability evaluation, a storage statement should be set up for the label display. Specific instructions should be provided where it is suitable, particularly in case of APIs which cannot tolerate the freezing or temperature excursions. Words like "room temperature" or "ambient conditions" should be avoided. Using the stability information, a retest period should be procured, and the retest date should be shown on the label of the container if suitable. An API(s) batch after the retest period that is intended for usage in the production of an FPP could be retested, and if it is in agreement with the specifications, it could be immediately used (within 30 days). In case the retested batch is found compliant, then the batch does not get extra time in comparison to the time that was established for the retest period. Nevertheless, retesting of a batch of API(s) can be done multiple times, and different portions of the batch can be used after every retest, as long as it remains compliant with the specification. It is more relevant to establish a shelf life rather than a retest period for the APIs that are known to be labile (like certain antibiotics).

Ongoing Stability Studies: The API(s) stability should be observed according to an appropriate and continuous program that will allow the detection of stability issues like the variation in the number of degradation products. The ongoing stability program aims to observe the API(s) and determine if the API(s) within the terms supporting the storage conditions that are designated on the label within the shelf life or the retest period will persist or can be expected to persist in all the future batches. The ongoing stability program should be reported in a written protocol, and the results should be displayed and made available on-site in the form of a formal report. The protocol should extend up to the end of the shelf life or retest period and also include the following parameters: number and sizes of different batch(es); appropriate chemical, biological, microbiological, and physical test parameters along with the reference to the attached specifications or the acceptance criteria; testing frequency; reference to test methods; description of the storage conditions; other applicable parameters that are specific to the API(s); and the container-closure system(s) description. Unless none of the production batches of API(s) is produced during a year, at least one production batch per year should be added to the stability monitoring program and at least every 6 months in the first year and then annually to confirm the stability.

33.2.6 Guidelines for Finished Pharmaceutical Product

The designing of finished pharmaceutical products (FPPs) for the stability studies should be done based on the understanding of the behavior and properties of the active pharmaceutical ingredients (APIs) and also on the experience gained during similarly marketed formulations, pre-formulation studies, and investigational FPPs.

Another essential part of stress testing, the photostability testing, should be done on a minimum of one primary batch of the FPP. Some additional stress testing like the freeze-thaw studies in case of liquid products or cyclic studies in case of semisolid products with a particular type of dosage form would be appropriate.

Data from stability studies from at least three primary batches of each of the proposed strength of FPP having new APIs should be provided for the FPPs. One of the three batches can be smaller, but two out of the three batches should at least be of pilot-scale batches. FPP data having existing APIs should be given for not less than two batches of a minimum of pilot scale, or otherwise of an uncomplicated three FPP like non-sterile solutions or immediate-release solid FPPs, at least one batch of a minimum of pilot scale, and a second batch which could be smaller like in case of solid oral dosage forms, 25,000 or 50,000 tablets or capsules of recommended FPP strength. The primary batches should have the same preparedness and should be packaged in the same container-closure system as it was intended for marketing. The primary batch production process should assume such as to be applied in case of production batches, and the product should be of the same quality and should match the specification as it was proposed for marketing. In case the primary batch size used is smaller than a pilot scale, then discussion or data is required which will verify that the smaller batch represents the proposed production size, in addition to its method of manufacture and formulation. Wherever possible, the FPP batches should be produced using different API(s) batches, and the stability studies should be done on every individual dosage form and container type, strength, and size of the FPP except in case matrixing or bracketing is applied.

Container-Closure System: The stability testing should be performed on the dosage form that is packaged inside the primary container-closure systems initially intended for marketing. In case the secondary container-closure system has some protective properties, the packaging of the product is in a semi-permeable container, and the ingredients from the secondary packaging can transfer into the product or in case the labeling asks the product to be stored in the primary and secondary packaging, for example, "store tablets in blisters in the provided cartons," then the secondary packaging can become a part of the stability samples' packaging system. Any available evaluation performed on the FPP outside its immediate container or in some other packaging materials can form a beneficial part in the stress testing in case of the dosage form, or it can be regarded as the supporting information, respectively.

Specification: Stability-indicating attributes of the FPP should also be included in the case of stability studies like those that are sensitive to alteration during storage and are expected to affect the safety, efficacy, and/or quality. The evaluation should include the chemical, microbiological, biological, and physical attributes, functionality tests like in case of a dose delivery system, and preservative content like the antimicrobial or the antioxidant preservatives. Stability-indicating and full validation should be done by analytical procedures. The outcomes of validation studies will tell up to what extent and whether replication should be performed.

Frequency of Testing: For products having an intended shelf life of a minimum of 12 months, the testing frequency should usually be every 3 months for the first year, every 6 months for the second year, and thereafter annually throughout the intended shelf life in case of long-term storage conditions. In case of the accelerated storage conditions, at least three time points from a 6-month study, including the initial and final time points like 0, 3, and 6 months, are usually recommended. If it is expected that the outcomes from accelerated testing are expected to propose a significant change criterion, then either the testing should be increased by the addition of samples during the last time point or a fourth time point is included in the study scheme. If the evaluations at the intermediate storage condition during the accelerated storage condition are known as outcomes of significant change, then at least four time points from a 12-month study are recommended including the initial and final time points like 0, 6, 9, and 12 months. t0 is considered as the initial date of storage, and the stability time points should be described as dates in accordance with t0. For example, if 1 January 2020 is t0, then the time point of 1 month will correspond either to 31 January 2020 or to 1 February 2020, and for every time point, withdrawing and evaluation of samples should be according to the protocol. It should be noted that the testing/evaluations should be finished as soon as possible and any deviation from the protocol should be recorded/documented and justified.

Storage Conditions: The stability of the medicinal product should be demonstrated throughout its expected shelf life in the climatic conditions that are common in the target countries. If the same requirements are applied to the other markets, like in case the stability studies are performed at the storage conditions for the countries of I/II climatic zone and the products are supplied in countries of III/IV climatic zones, it could probably lead towards the substandard products. Generally, an FPP should be tested under storage conditions along with the designated tolerances that will evaluate the thermal stability and its moisture sensitivity or the potential in case there is a solvent loss. The lengths of studies chosen along with the storage conditions should be enough to incorporate the storing, shipping, and subsequent application with due consideration to the climatic conditions during which the product is proposed to be marketed. During storage, the orientation of the product, i.e., whether it should be upright, inverted, or on the side, along with the justification for the orientation may be needed to include in the protocol in place where the contact of the product along with the closure system would be required to alter the product's stability contained like semisolids and liquids or in case there has been a variation in the container-closure system. The additional testing at the intermediate storage condition, accelerated storage condition should be performed and tested for a significant change criterion in case the long-term studies are performed at 60% RH $\pm$ 5% RH/25 °C $\pm$ 2 °C and "significant change" happens at any time in 6 months. In such a case, at least 6-months' data from a 12-month study should be included in the initial application at the intermediate storage condition. A "significant change" in case of an FPP is usually described as follows: a variation in the API(s) initial content of 5% or more as identified by an assay or a failure in meeting the acceptance criterion for potency when immunological or biological procedures are used; the acceptance criterion of any of its degradation product exceeds; if it fails to satisfy the acceptance criterion for physical attributes, appearance, and functionality test like phase separation, color, caking, resuspendability, hardness, and dose delivery per actuation. Nevertheless, some variations in the physical attributes could be considered under accelerated conditions like creams melting, suppositories softening, and partial loss of adhesion in the case of transdermal products; the acceptance criterion for pH fails to satisfy; or the acceptance criterion fails to satisfy for the dissolution of 12 dosage units.

The stability commitments are as follows:

- The primary batch stability commitment is a commitment that should be performed to continue the post-approval stability studies throughout the recommended shelf life when the long-term stability data that is available for the primary batches does not comprise the recommended shelf life that was granted during the time of approval.
- In the production batch stability commitment, a commitment should be performed to place up to a total of at least three, the next production batches,

and long-term stability studies should be done throughout the recommended shelf life and on the accelerated studies for a period of 6 months if data from stability studies on less than three production batches are included in the submission.
- For the ongoing stability commitment, an ongoing stability program is needed for each product to observe its shelf life and in determining that the product remains and will remain in the specifications according to the storage conditions that are there on the label. The stability protocol used should be the same for both studies on commitment batches and that for the primary batches unless it is justified scientifically.

Evaluation: For the evaluation and presentation of the stability information, a systematic approach should be used that must have the results from the chemical, microbiological biological, and physical tests, and some particular attributes of the dosage form like the solid oral dosage form's dissolution rate should also be included. A summary of extra knowledge and an understanding gained from modeling, supporting studies, and the predictive tools regarding stability should be included wherever suitable to support knowledge gained from the primary stability program. The stability study aims to establish the label storage and shelf life instructions that are suitable for all future batches of the FPP that are produced and packaged in similar circumstances based on testing of a minimum number of FPP batches. The amount of variability in individual batches affects the assurance that throughout its shelf life, the future production batch will persist within its specifications. Usually, it is not necessary to do statistical analysis in case the results show very little variability and very little degradation. Another approach to analyzing the quantitative attribute data which varies from time to time is to determine the time period during which, for the mean curve, 95% of the one-sided confidence limit intersects with the acceptance criterion, and if a small batch-to-batch variability is shown in the analysis, it would be beneficial for the data to combine in one overall estimate. This can be performed firstly by employing suitable statistical tests like P values having value more than 0.25 for the level of significance of rejection to the zero time intercepts and the regression line slopes for the individual batches. If combining data from different batches is unsuitable, then the overall shelf life should be based on how much minimum time can a batch be expected to remain within the acceptance criterion.

Labeling and Statements: Based on the FPP stability evaluation, a storage statement for the label should be established. Specific instructions should be given wherever suitable particularly in the case where freezing cannot be tolerated by the FPPs. Words like "room temperature" or "ambient conditions" should be avoided, and there should be a direct connection between the FPPs' demonstrated stability and the storage statements on the label. On the label of the container, an expiry date should be written. Packaging of FPPs should be done in containers that will assure protection from the deterioration and FPP stability. In the case of inferior or inadequate packaging, a storage statement should not be used as compensation. In cases the results of the stability testing show some limiting factors, additional labeling statements can be used.

In-Use and Hold-Time Stability: The testing for in-use stability aims in providing information regarding the labeling on the development; utilization period for multi-dose products after they are opened, reconstituted or solution dilution; and for the storage conditions, for example, in case an antibiotic injection is supplied in the form of powder for the reconstitution or hygroscopic (moisture-sensitive) solid oral FPP present in a large format multi-dose containers like 500 tablets are present in high-density polyethylene (HDPE) bottle. A 30-day in-use time period is usually accepted without any additional supporting data. The tests should be designed considering any reconstitution or dilution before usage and the filling volume of the container to mimic the use of the FPPs in actual practice. The suitable amount should be removed at intervals similar to those during practice using the withdrawal methods that are commonly described and used in the product literature. The FPP's chemical, microbial, and physical properties that are sensitive to alteration during storage can be determined during the proposed in-use shelf life. Testing on the FPP's final amount that remained in the container should be conducted at the end of the proposed in-use shelf life and during intermediate time points. The effectiveness and content of preservatives are required to be studied for specific parameters like in the case of semisolids and liquids. The testing of a minimum of two batches of at least a pilot scale should be conducted on the primary batches of the diluted or reconstituted FPP or for the solid oral FPP, as a part of the stability studies during the recommended in-use period for the initial and final time points. In the case of bulk products like coated tablets, before final packaging, consideration should be given to hold-time studies, and similar considerations should be applied to the intermediates which are used and stored for periods surpassing 30 days.

Variations: Additional stability studies are needed once the registration of FPPs has been done and whenever alterations are made that can change the API(s) or FPP stability. The expected change will have an effect on the stability, and the quality characteristics of APIs and/or FPPs should be investigated by the applicant. Based on the experience and knowledge gained on APIs and FPPs, the design and scope of the stability studies for the alterations can be studied. For guidance, the available variation guidelines should be advised for the expectations concerning the stability requirements to approve variations to the APIs and FPPs. For the specific regulatory authority or region, the requirements of the guidelines predominate for an assigned region, but in case such guidelines are absent, then the "WHO Prequalification Team: Medicines guidelines" can be implemented, and depending upon the change, there is a requirement of either a commitment to perform such as test or the stability study data. Variations that need supporting data must incorporate certain modifications to the retest period of APIs or in the storage conditions and, in case of the formulation of FPPs, fabrication process, shelf life, container-closure system, storage conditions, and in-use period. There is a need for a commitment to support the variations for the stability studies in case of alterations in the API(s) certificate of the manufacturing process, manufacturing site, or prequalification and certificate of suitability or specific variation in the FPP batch size, container-closure system, or manufacturing site. The outcomes of these stability studies should be reported to the concerned regulatory authorities along with the appropriate specifications specified in the region variation guidelines.

Ongoing Stability Studies: After a marketing authorization has been allowed, the FPP stability should be observed properly in respect to the continuous program that will allow the detection of any stability issue like the variation in levels of dissolution profile or degradation products that are correlated to the formulation present in the container-closure system through which it is marketed. The ongoing stability program aims in monitoring the product over its shelf life and in determining that the product will remain, or can be assumed to remain, in accordance with the specifications that are given on the label of the storage conditions. In the ongoing stability program, results should be formalized and described in a written protocol in the form of a report. For an ongoing stability program, the protocol for the shelf life period should prolong till the end and should include the following parameters: different batch sizes and the number of batches per strength, and in case of different batch size, it should be recorded; appropriate chemical, biological, microbiological, and physical test parameters along with the reference to the attached specifications or the acceptance criterion; information of the container-closure system(s); reference to test methods; testing frequency usually in 6-months' time points and annual time points for ongoing studies is satisfactory; storage condition description like the standardized conditions for long-term testing as told in the guidelines and consistency with the labeling of the product should be used; and other applicable parameters that are specific to the FPP. The protocol of the initial long-term stability study could vary from the ongoing stability program as it was submitted in the marketing authorization dossier only if it is documented and justified in the protocol. The testing frequency and the number of batches should provide an adequate amount of data to allow for the trend analysis. A minimum of one batch per year of the product produced in every primary packaging type and every strength should be incorporated in the stability program unless it is otherwise justified or none of the product is manufactured in that year. The principle of matrixing designs and bracketing could be utilized if in the protocol it is justified scientifically. Any verified significant change should be reported to the appropriate accountable authorities immediately, and consultation should be considered with the appropriate accountable authorities for the possible impact on batches on the market.

33.3 Stability-Related Issues with Herbal Drugs

The various factors responsible for the physical and chemical stability of the drug depend on the following (Fig. 33.1):

33.3.1 Physical Instability

Herbal medicines show a small half-life period due to the degradation of its components when they come to direct contact with the environment. Microbial

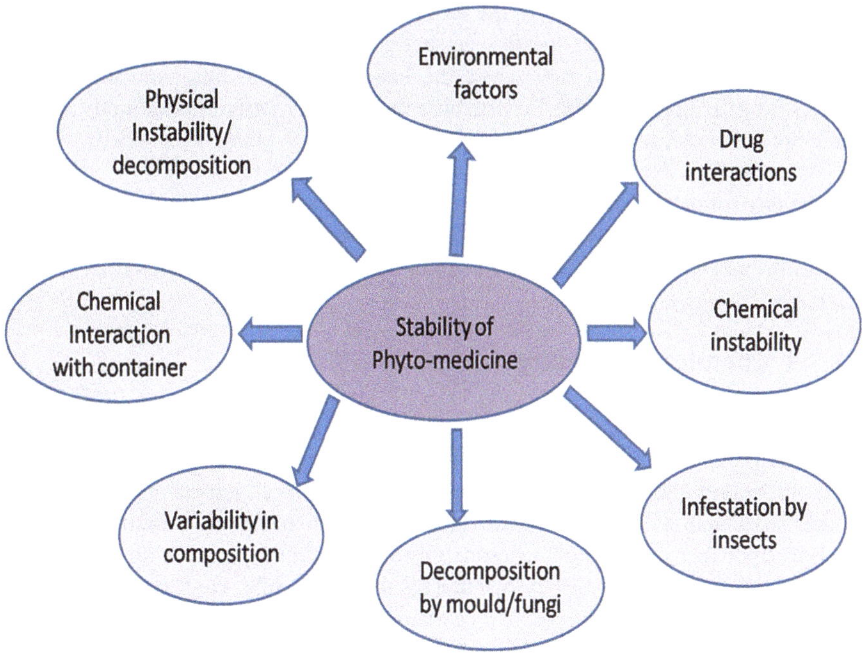

Fig. 33.1 The various factors on which the chemical and physical stability of the drug depends

contamination is very common in herbal medications which causes instability of the formulation. One should choose wisely the container generally earthen pots, glass vessel, or bone china vessels to preserve the herbal formulations and to prevent the deterioration of the medicine by the component of the container itself. Another factor that is responsible for the physical instability of herbal formulation is its volatile components that are generally aromatic in nature such as essential oils. Essential oils get volatile with time resulting in the loss of its therapeutic activity. This is the reason that the species that contain essential oils lose their effect if stored for long times, and the loss is rapid if not stored properly. Examples include cardamom, cinnamon, etc. that lose their active ingredients if not properly stored. Microorganism and insect infestation can strongly affect the plant secondary metabolites and lead to change in the chemical composition of the drug. To avoid instability of the phytoformulation, factors such as volatility of components, microbial contamination, insect infestation, and deliquescent or hygroscopic nature of the components should be taken into consideration and should be properly answered (Thakur et al. 2011).

33.3.2 Environmental Conditions

The environmental conditions play a magnificent role in the storage of herbal drugs, for example, moisture content, temperature, harvesting, processing methods, and time of collection of the herb. In the case of essential oils, during harvesting from flowers, one should collect the flowers before sunrise, and the sample should be dried in the shade as sun rays cause the loss of essential oils from the flowers. Excessive heat results in the loss of pharmaceutically active compounds. Higher moisture content causes an increase in the chance of being attacked by fungus, molds, or insects (Nainwal and Nainwal 2010; Thakur et al. 2011).

The environment conditions also count for the genetic variability of species, soil contents, day length, seasonal variations of the area from which particular herb that affects the activity, concentration, and stability of active phytoformulation.

33.3.3 Chemical Instability

The half-life of different compounds in phytoformulations is different during storage which affects the concentration of active ingredients of the herbal formulations. Microbial contamination might be present at the time of harvesting of the herb, which can cause a change in chemical composition over time. Additional factors such as hydrolysis, oxidation, photon sensitivity, humidity, and high temperature enhance the deterioration of phytochemicals during storage. High moisture content makes the formulation more prone to be attacked by fungi and other microorganisms that can cause the hydrolysis of many active compounds and cause chemical instability. Sometimes the product itself contains native enzymes that result in the instability of the herbal formulation with time (Nainwal and Nainwal 2010; Thakur et al. 2011).

Chemical evaluation can be done to know the chemical composition and its stability. Chemical analysis for the resins can be done using the acid value or sulfated ash. For the balsams, the acid value, saponification value, and better values can be evaluated, and for the evaluation of the volatile oils, acetyl and ester value can be evaluated. Various chromatographic techniques, HPTLC techniques, and UV-VIS spectrophotometry can also be employed to test for the chemical constituents and their stability (Patel 2011).

33.3.4 Variability in Composition and Inconsistency

The composition of the herbal formulation depends upon different factors; the composition and amount can be different due to harvesting methods, processing, regional changes, soil composition, and pollution levels. Even the same species grown in a different geographical area may have variability in composition due to environmental differences and soil composition. All the mineral and organic nutrition of the plant is provided by the soil that is different in different geographical areas and has a different composition of micro- and macronutrients that causes a change in the chemical composition of herbal formulations. These factors count for physical parameters that are responsible for the variability in composition and inconsistency. On the other hand, genetic factors also lead to variability in composition. In species or sub-species differences, a small genetic change can lead to huge differences in the components and consistency of herbal drugs. Herbal formulations are a mixture of many components, and each constituent has a different half-life and different percentage in composition. Therefore it becomes very difficult to assess the exact composition and consistency of a single component in herbal formulations. When a phytoformulation for pharmaceutical benefit is formed, it contains many components. In case of phytoformulations, the half-life, chemical activity, percentage composition, and consistency are different for all the different components, which leads to loss of precision in the final product formed. Hence it is difficult to determine the stability of the final herbal formulations (Thakur et al. 2011).

A study was done on the genus *Ocimum* of Lamiaceae family having geographical distribution in tropical and subtropical regions of Asia, Africa, and South America. The species of *Ocimum* genus are the source of essential oils containing a mixture of monoterpenoids, sesquiterpenoids, triterpenoids, and phenolic compounds as their chemical components (Ogendo et al. 2008; Ghasemi Pirbalouti et al. 2013; Hussain et al. 2017).

The essential oils isolated from various spices are well recognised for their anti-allergic, antifungal, antimicrobial, and immunity-booster properties. Henceforth, these essential oils are widely used in food, pharmaceutical, cosmetics, and perfume industries (Makri et al. 2008; Dambolena et al. 2010).

In Iran, the analysis of essential oil was done on *Ocimum* which was obtained from 28 accessions (plant parts obtained from single species collected from the same location at the same time), and this was investigated using GC-MS, and three different chemotypes were found showing genetic variation in the same species in the same geographical area (Moghaddam et al. 2017). The inconsistency and variability in composition of herbal formulations making them difficult to standardize.

33.3.5 Drug Interactions, Deterioration, Decomposition, and Storage

Patients should be restrained from using herbal products with a narrow therapeutic index, for example, digoxin, cyclosporine, procainamide, phenytoin, warfarin, theophylline, etc. All the drugs having a narrow therapeutic index when used in combination with herbal products may either be less effective or have enhanced adverse effects. A general depiction of herb-drug interaction is shown in Fig. 33.2. For example, *Ginkgo biloba* was found to be good for patients having Alzheimer's dementia as it was found to increase blood flow in the brain, but if it is taken with aspirin, it causes increased bleeding, showing a synergic effect with aspirin. Ginseng has versatile uses but shows synergism with monoamine oxidase inhibitor. Kava has been identified with the properties to cure anxiety disorders, but it shows a synergic effect when taken along with benzodiazepines, a well-known drug used to relieve anxiety. Also, the famous St. John's wort, a flowering plant wide grown in various

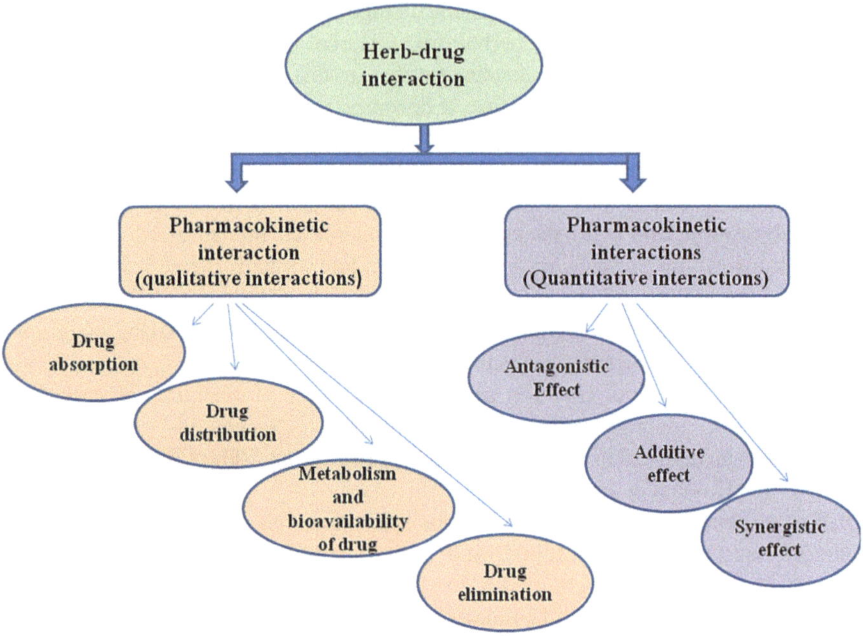

Fig. 33.2 Herb-drug interaction

parts of the globe is used to treat depression causes the reduced plasma levels of warfarin drug (an anticoagulant drug) and oral contraceptives (Hussin 2001).

Heavy metals are quite common in traditional medicinal formulations. Heavy metals such as copper, lead, arsenic, mercury, gold, and silver added in the herbal formulation traditionally are toxic and causes enormous heath issues. Patients are advised not to use herbal drugs in combination with modern medicines as there are high chances of drug synergies and enhanced risk of adverse drug reactions (Nainwal and Nainwal 2010).

As composition of herbal formulations are unknown for many traditional medicines may lead to undesirable sideeffects due to drug-drug interactions. It is very often that people believe that herbal formulations are safe and do not have adverse effects, and sometimes they take the herbal preparation with another drug, especially allopathic drugs. Some herbal formulations have agnostic and some have antagonistic effects on each other which can be even fatal for the person. Due to drug-drug interaction many herbal formulations should not be used simaltaneously such as Mentat, an antiatheritic drug and Celery root, a drug used for monipose treatment. Other examples are Gingko biloba with fish oil supplement, Melatonin with antidepressants, (such as benzodiazepines, narcotics, etc), St. John's wort. with antidepressants.

The other environmental factors such as moisture content or microbial contamination can result in the interaction of herbal formulation with the container or the intraspecific interactions of different components of the same herbal formulations.

Therefore the stability of the herbal formulation is different as compared to the stability of its individual components.

Natural medicines consisting of plant parts or other naturally available compounds contain a lot of microorganisms on their surface. Faulty methods of processing of herbal medicine can cause contamination of the herbal formulation with microorganisms especially sporulating bacteria and fungi. The plant parts are tested for the total number of microbes and for the presence of various gram-positive and gram-negative bacteria and fungus. It is well included in the *European Pharmacopoeia* that *Escherichia coli* and *Salmonella* species should not be present in herbal preparations. Microorganisms' growth, high moisture levels at the time of harvesting of the drug, or the faulty packaging practices can cause the loss or deterioration of the active component in herbal formulations. Normally there are hundreds of components in a single herbal formulation, and it is quite obvious that there are intra-chemical interactions between these components, which can change the composition and activeness of the herbal formulation. Further, the different components have different stability parameters which can lead to loss of effectiveness of the drug or cause the decomposition of that herbal drug (Thakur et al. 2011).

33.4 Effect of Drug Stability on Pharmacokinetics

Pharmacokinetics is the study of the movement of the drug within the body; it tells about what the drug does to the body. Pharmacokinetics of herbal drug includes the five basic steps similar to that of all the conventional medicinal therapy as shown in the Fig. 33.3.

Whenever a person takes the drug orally, the drug moves through the GI tract to the intestinal region where the maximum of the drug gets absorbed. During the movement through the gastrointestinal tract, the drug encounters different enzymes and pH conditions that vary from pH 1.5 to pH 7.4, which largely affects the stability of the herbal formulations. Factors responsible for the plasma concentration of drug are its dosage concentration, absorption, distribution, metabolism, half-life of the drug in plasma, and its excretion.

As the herbal formulations generally contain crude extracts from plants, each component has a different dosage concentration, absorption, and distribution in the body according to its chemical nature. Pharmacokinetics of the herbal preparation or herbal drug helps to evaluate the efficacy and its related toxicity.

In South Africa, a study of pharmacokinetics was done on *Hypoxis* extract having 50–55% of hypoconid, a di-glucoside, primarily found in corms of many *Hypoxis* species. In vitro studies showed that it is non-toxic to cancer cells for up to the concentration of 100 mg/mL. It gets hydrolyzed into aglucone and rooperol having a cytotoxic concentration of 2–10 mg/mL. When *Hypoxis* extract dosage of 1600–3200 mg was administered orally to lung cancer patients, it was observed that neither rooperol nor hypoxoside was seen in the blood. Only rooperol conjugates were measured along with glucuronide sulfate as major metabolites, and diglucuronide and disulfate as minor metabolites were obtained. Large inter-

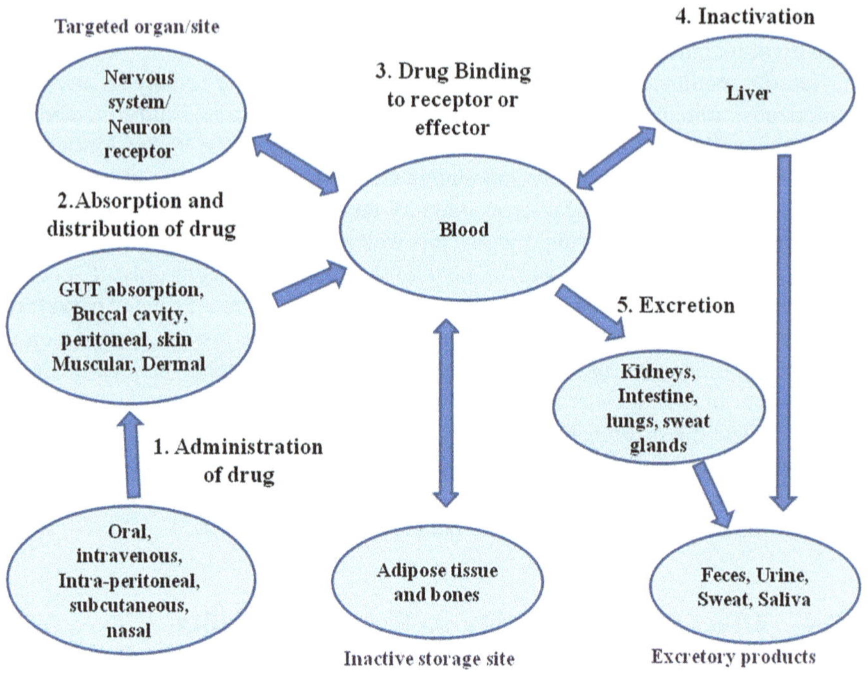

Fig. 33.3 Five steps depicting the pharmacokinetics of the herbal drug within the body, viz., administration, absorption, distribution and binding, inactivation, and finally excretion from the body

patient variation in the concentration was seen in the concentration-time relationship of the major metabolite which may be due to metabolizing capacity. The lag phase for the metabolites was also different for different patients. Some patients showed enterohepatic recirculation immediately after taking the *Hypoxis* extract orally. Hence the individual dosage of the herb varies according to the metabolism of the individual.

33.5 Tools for Dealing with Natural Product Instability

33.5.1 Physical Parameters

The consistency of the herbal formulations is very important for defining its therapeutic effect and toxicity. Instability of phytomedicines may take place due to interaction of components, as the composition may vary based on environmental and geographical factors. Further it may also alter based biological factors such as pest or microbial infestation. Other internal factors may include pesticides, insecticide content, presence of radioactivity or heavy metal poisoning, and alpha toxins which might be due to soil pollution. To make a standard herbal medicine, a suitable analytical method should be applied (Thakur et al. 2011).

The herbal preparations should be stored carefully by considering all the physical parameters like temperature and exposure to sunlight and other parameters as discussed earlier to ensure the stability. There are several factors which influence the natural product instability and make herbal medicines different from allopathic medicines like:

1. Herbal drugs contain multiple components.
2. The active component in most of the cases is unknown.
3. Reference compounds or reference analytical methods are generally not available.
4. Plant contents are chemically variable due to genetic diversity and geographical differences.
5. Different sources cause a difference in the quality and standard of herbal medicine.
6. Method of harvesting, storage, transportation, and processing methods also count for the instability and should be addressed to maintain the quality and activeness of phytoformulations (Folashade et al. 2012).

Stability Testing of Herbal Medicines: Generally the whole herb is regarded as the active substance, even if the therapeutic value of each component is not known. Due to multi-component, it becomes a tedious job to define the stability of herbal medication. Stability testing of a herbal formulations ensures its quality, effectiveness, and shelf life. Stability testing is done in batches for the proposed period of storage denoting its shelf life. Long-term stability testing is generally done under natural atmospheric conditions, while short-term stability data is collected under accelerated atmospheric conditions like humidity, temperature, and light. The data obtained from both studies are used to predict the shelf life of the herbal product. The stability testing is done on the dosage form, packed in the container prior to marketization, and it is generally done with the help of instruments such as HPLC (high-performance liquid chromatography) or spectrophotometric study and also by following proper guidelines of pharmacopoeia for safety and effectiveness of herbal drugs which lead to standard marketable herbal formulation, hence improving the global market and worldwide acceptance of the herbal product (Kumari 2016; Shulammithi et al. 2016).

33.5.2 Impurity Profile

The impurity profile of herbal drugs is important for the assessment of the stability of phytomedicine, related toxicity, and its effectiveness. The impurity profile gives a tool especially for the assessment of a new drug. A degradted product of unstable subject can be utilized to make reference librabry and can be utilized as referenecs for future studies. This reference library acts as a reference for the routine testing of impurity profile of the respective substance. From here we can discover the nature of

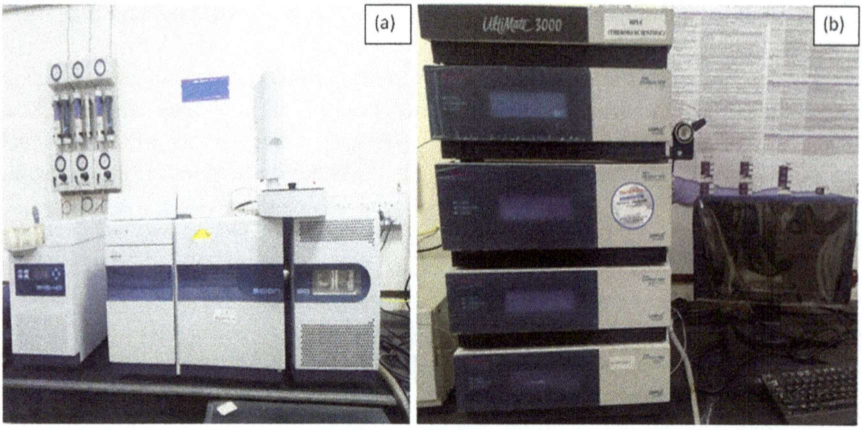

Fig. 33.4 Analytical methods that can be used to impurity profiling of herbal drugs. (**a**) HPLC. (**b**) GC-MS

the impurity using analytical methods like HPLC, TLC, GC-MS, and electrophoresis (Fig. 33.4).

The strategy is chosen for the assessment of the total impurity profile of the drug which includes all the physical and chemical evaluation starting from the selection and handling of crude plant material; generally the following quality parameters are chosen for the assessment of the marketable finished product:

1. Identification of the right variety and checking for adulterants.
2. Removal of unnecessary plant material, other than the drug source.
3. Analysis of ash values to analyze the identity and purity of the drug.
4. Analysis of moisture content for proper storage.
5. Calculation of extractive values to judge the weights of chemical constituents present in the crude drug.
6. Determination of crude fibers to calculate the woody material and to judge the purity of the drug.
7. Qualitative analysis of the crude drug with the help of phytochemical analysis to detect and extract the active components of the drug. It involves the botanical identification of the herb, the use of an appropriate solvent to extract the drug, purification of the drug, and characterization of the active ingredient of pharmacological importance.
8. Chromatographic analysis of the crude drug by using the major constituents as markers through gas liquid chromatography (GLC) and high-performance thin-layer chromatography (HPTLC). HPTLC provides a valuable tool for the quality assessment of the extracts of botanical origin in a very efficient and convenient way.

9. Evaluation of presence of potential toxic contents (for examples microbial toxins, pesticides, heavy metals, etc) to ensure the safety of the drugs.
10. Biological assays and LD50 value (lethal dosage 50) evaluation on the living systems for ensuring the safety of the herbal medicine (Folashade et al. 2012).

All the guidelines and approaches for the herbal drug are for the efficacy, stability, and safety of herbal medications and to address the problems related to herbal medicines. The various techniques are employed to render it safe and effective for the people. The herbal drugs are screened through various stages and become standardized marketable products only when they meet the criteria laid by the Pharmacopoeia of that nation or as guided by WHO.

33.5.3 Identification and Quantification of Components

WHO gives a strong emphasis on the quantitative and qualitative tools for the characterization of herbal samples. Quantification and identification of all the metabolites in the herbal preparation are important to define the stability and efficacy of the herbal formulation. Fingerprint profile through IR spectroscopy with chemometric data processing enables to get the details of the total metabolites present in the herbal formulations. This technique enables the inspection at a time and gives acceptance or rejection for the classification of material. If the main active component having the pharmacological importance in the herbal preparation is known, it becomes easy and logical to quantify that compound, and the herbal preparation could be standardized to its components. In the photo preparations where the active component is not known, a marker that is specific for that plant is used for the analysis of the active component of the herbal medicine (Thakur et al. 2011; Shulammithi et al. 2016).

The most reliable method of quantitative analysis should always be done along with the qualitative analysis generally done with chromatographic tools. The correct identification and purity are the two utmost things needed for the standardization of herbal preparation. The primary aim of these methods is to quantify the active compound having the pharmacological activity (Folashade et al. 2012).

Correct identification and purity of the herb are the preliminary things to be done before assigning it to be used as medicine. All the quality aspects of the herb should be critically examined. Sometimes outbreaks of some phyto-diseases could lead to wrong identification of the herb. The exemplary case was seen in the 1990s when a South American product was marketed which was labeled as "Paraguay Tea." It was the reason behind the outbreak of anticholinergic poisoning in New York. After a series of chemical investigations, it was found that a different metabolite was present in the product which was normally not part of the metabolites present generally in the plants from which the Paraguay Tea was used to be made of.

33.5.4 Storage Conditions

To enhance the stability of the herbal preparation protecting it from the environmental factors are essential. Therefore storing the preparation in clean, dry and shady area away from the sunlight is generally preferred. The container should be properly airtight and moisture-free and should not chemically interact with the herbal formulation stored in it. Proper storage influence the shelf life of the drugs directly (Thakur et al. 2011). Therefore the container for storage should be made up of inert materials (such as glass) which do not change in environmental influence.

33.6 Approaches to Address the Problem of Instability of Natural Products

Some of the major challenges like the astringent taste, hygroscopicity, uncontrolled release, physiological instability, poor absorption and bioavailability, gastric and/or intestinal irritation, and high-dose-induced toxicity because of unspecific delivery restrict the widespread applications of bioactive compounds. Due to this, encapsulation is an efficient method to overcome these problems and to intensify the bioactivity of these natural products.

33.6.1 Nanoparticle Coating

Nanoparticle coatings have been used widely for antimicrobial properties shown by the bioactive compounds. Ag, Au, Se, and Cu metal-based nanoparticles have also exhibited potent antimicrobial properties. Nowadays, green chemistry is being used for the synthesis of metal-based nanoparticles of biological compounds of plants, algae extracts, yeast, fungi, and bacteria (Amini 2019). In phytoformulation research, development of polymeric nanoparticles (for example nanospheres and nanocapsules, liposomes, proliposomes, nanoemulsion, solid lipid nanoparticles, etc.) has been found to exhibit many benefits for herbal drugs, such as improvement in bioavailability, solubility, stability and phamodynamics (Musthaba et al. 2009). Improved stability of liquid dosage compositions of stable nanoparticulate drugs was found to be more than the conventional dosage forms (Bosch et al. 2004).

Geranyl cinnamate loaded in polycaprolactone nanoparticles that were prepared using solvent evaporation technique and are known for their antibacterial activity showed good colloidal stability for over 60 days and was not released even after 72 h. To release the active compound, triggers like the oil phase was required (Zanetti et al. 2019) (Fig. 33.5). Using green chemistry, silver nanoparticles (AgNPs) were synthesized, and natural extracts of dark (D) or white (W) seeds of *Salvia hispanica* L. were used as a reducing-stabilizing agent. These were studied for their antimicrobial activity against gram-positive (*E. coli*) and gram-negative (*S. aureus*) bacteria; it was observed that the nanoparticles that were synthesized

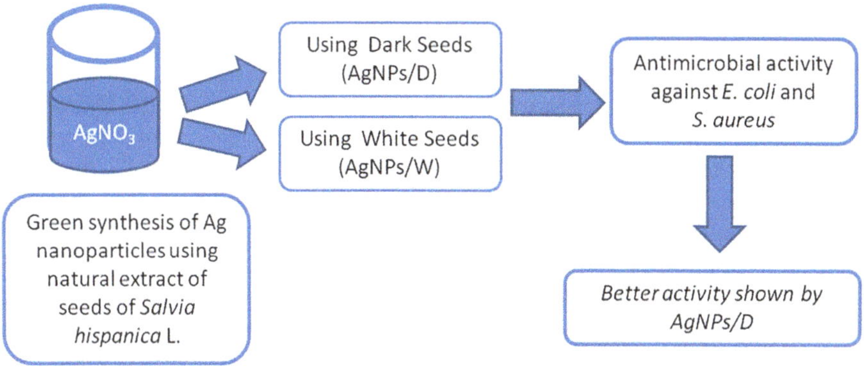

Fig. 33.5 Silver nanoparticle green synthesis using a natural extract of dark or white *Salvia hispanica* L. seeds and the study of their antibacterial property

using dark seeds (AgNPs/D) showed more antibacterial activity than the ones prepared using white seeds (AgNPs/W) (Hernández-Morales et al. 2019).

33.6.2 Cellulose Coating

Polysaccharides of different physicochemical properties can easily be obtained from plant parts at a cheaper cost and can be changed chemically as per requirement. Because of intramolecular hydrogen bonding and high molecular weight, polysaccharides can be used for the formation of biosystems that are non-toxic, complex, stable, easily biodegradable, and highly hydrophilic. Chemical and/or biochemical modifications can be done to enhance the bioadhesiveness and biostability of the tissues (Lu et al. 2019).

Plant materials that are rich in polysaccharides have been utilized successfully in the altered release dosage forms in the form of matrix formers and have been examined for its application in the formation of solid oral dosage forms. Some polysaccharides having the property of gelation can be utilized in the controlled delivery of herbal formulations that have enhanced storage stability. For effective drug delivery, several polymers of plant origin are being used, and some others are being examined for the designing of dosage forms as the excipients.

Polysaccharides like alginate, carrageenan, glucomannan, guar gum, gum Arabic, and locust bean gum that are obtained from plants have displayed remarkable potential as carrier materials in controlled release dosage forms of the matrix type in the form of beads, cross-linked hydrogel microparticles, and tablets. Pectin, rosin, and inulin have been examined for its properties in film formation, while some other plant polysaccharides have been modified physically or chemically for the development of starch and cellulose derivatives. The semi-synthetic polymers are used for the formation of conventional dosage forms and are being investigated for its application in drug delivery systems (Beneke et al. 2009).

33.6.3 Lipid Coating

Polar, amphiphilic, and non-polar regions are present in the liposomal system of the phospholipid bilayer compartment, and this can encapsulate both the hydrophilic and the lipophilic compounds. Improved biological activities were observed on inclusion of the natural compound into a nanoliposome when compared to the native compound.

A low-linolenic acid-base compound is used by the long-chain fatty acid which is derived from plant for incorporation in many formulations to improve the product stability as it enhances the shelf life of the product and decreases the requirement of hydrogenation by the oil-based formulations. Some of the plant extracts are also encapsulated in lipid-based formulations for better stability and encapsulation.

Alkaloid extract of *Solanum lycocarpum* was loaded in natural lipid-based nanoparticles (NLN), and these particles have shown high stability, better encapsulation efficiency, and enhanced anti-proliferative and anti-tumor activity. These NLN-AE particles have also been reported to induce apoptosis in bladder cancer cells when compared to alkaloid extract alone (Fig. 33.6). When alkaloid extract was encapsulated in NLN particles, cell viability was decreased up to 5.4 times showing better results as anti-tumor particles (Carvalho et al. 2019).

33.6.4 Protein Coating

Chaperonin-like Sp1 protein can be used for the formation of therapeutic protein by forming a fusion product and complex that has improved thermal stability, with the core comprising of at least one protein that is biologically active and the coating comprising of a micronized product obtained from leguminous plants (Thakur et al. 2011). The *Azadirachta indica* (neem) extract loaded in albumin nanoparticles was prepared using the coacervation process and bovine serum albumin cross-linked with glutaraldehyde. These particles showed better antimicrobial properties when treated with cotton fabrics than neem extract alone. The antimicrobial activity was retained up to 25 washes, whereas the neem extract-treated fabrics retained the activity only up to 10 washes (Ahmed et al. 2012).

33.6.5 Suspension and Emulsion

A study on herbal extracts prepared by dissolving emulsifiers (such as olein oil 7 (heptaethylene) glycol monooleate, methylcellulose, and Tween 20) in ethoxypolysiloxane and water shows that the viscosity decreases in case of the emulsions that contained methylcellulose. when the storage stability in case of oil-in-water emulsions and the size and number of globules during the storage was studied. The emulsions having Tween 20 had a similar globule size and number and increased viscosity after storage. The emulsion containing olein oil 7 (heptaethylene) glycol monooleate was the least stable. The ability to subdue the formation of

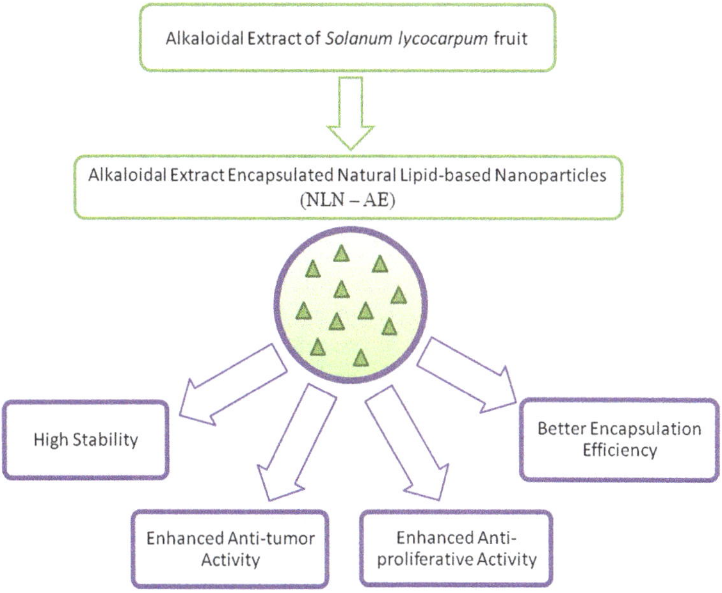

Fig. 33.6 Effect of alkaloid extract of *Solanum lycocarpum* when loaded in natural lipid-based nanoparticles on RT4 bladder cancer cell line

aqueous solutions having plant extract by emulsions was preserved during the storage period (Thakur et al. 2011).

A nanoemulsion formulation consists of a therapeutically active aqueous phase that is captured in oil phase chosen from one or more essential oils and utilized in the topical treatment of skin disorders like acne and some others like psoriasis, aging, scarring, and eczema and showed enhanced percutaneous penetration, increased efficacy, low skin irritation, and great thermodynamic stability that ensures longer shelf life and reservoir effect that favors localization of drug in the skin and enables controlled drug delivery of the stated therapeutic agent (Thakur et al. 2011).

The suspension stability depends on the viscosity and some other factors. Xanthan gum can improve the stability of the formulation by enhancing the viscosity. Cellulose derivatives, in comparison to xanthan gum, have lead to decreased stability due to lesser viscosity. Nonionic surfactant and the suspending agent in the presence or absence of propylene glycol help in stabilizing the sparingly soluble, sparingly insoluble, or the insoluble plant (Thakur et al. 2011).

Spray microcapsule formulations are used in nutraceutical manufacturing and in the production of cosmetic products like anti-wrinkle creams as these are sufficient in stabilizing compounds that are of natural origin. Microcapsules are formed by a consecutive water-in-oil-in-water type of microencapsulation process, and its formulation comprises of dispersion of microcapsules in the continuous water phase that contains oil drops, and in the middle of every oil phase drop, there would be a dispersion of water or water-dispersible material or aqueous extract or water-soluble material. Encapsulation of the oil drops is done with the help of a polymerizable material that is of a natural origin. Microcapsules like these are suitable for spray-drying that can be used as a gel, dry powder, cream, lyophilized, any liquid form, and self-emulsifiable powder. The active compounds that are incorporated in the microcapsules are useful in health and other biological purposes.

33.6.6 Use of Antioxidants

Antioxidants like polyphenols, *Ginkgo biloba*, *Glycyrrhiza*, fatty acids, and esters of flavanol are important as they help in lessening oxidation. The formulations are damaged more easily due to the formation of free radicals. Here, the antioxidants behave as scavengers of free radicals and thereby enhance the stability of liquid herbal formulations and other products.

During the preparation of linctuses, the physical properties such as viscosity, pH, and shelf life are determined with the help of formulations that are glycerin-based. Using these glycerin-based formulations, the value of viscosity and specific gravity becomes higher, and formulation properties are more stable on storage (Iwu et al. 2009).

It has been shown in studies that the disrupted plant cells get transformed genetically to express the immunogens or some other polypeptides which further form the immunoprotective compounds, biologically active proteins which show robustness and stability and are helpful in pharmacological and vaccine developments. The functional proteins or antigens get accumulated in the membrane part of the plant cell and the cell wall of cytoplasm, and these get released in the form of particles in the form of physical or mechanical disruption or by some other methods that form immunoprotective compounds (Miller et al. 2009).

33.6.7 Tablet and Pellets

The chewable tablets have benefits of better dispersion, fast dissolution, little disintegrating time, larger stability and bioavailability, faster absorption in the human body, convenient administration, and good taste when compared to the already present dosage forms having the same component. For example, when a chewable tablet is formed having a mixture of the fruit of *Gardenia jasminoides*, *Flos lonicerae*, and *Herba artemisiae* Scopariae, it has better bioavailability and stability when compared to the similar injection and liquid formulations (Xu et al. 2015).

In case of a stable liquid preparation, a solution comprises an active ingredient which is coated with a water-soluble cellulose derivative and has a water content of 10–80%. This liquid preparation preserves the water-unstable active ingredient in it, and a hard capsule can be formed from it, and this also helps in masking the unpleasant smell or taste.

For the preparation of a visually stable tablet that contains medicine which is a volatile liquid, damp-proof coating is performed after the volatile liquid active medicine is included and moisturized. It can be done by the following procedure: the volatile liquid active medicine is made which is then incorporated in β-cyclodextrin and moisturized with colloidal solution of silicon dioxide; this is mixed with the other additives and then pulverized and pressed in fixed size for the preparation of dried granules; this is then mixed with lubricants and disintegrating agents and later re-pressed to form a tablet which is then damp-proof coated with a coating material (e.g., isopropane of polyvinylacetaldimethylaminoacetate), and later sugar coating is done on it.

33.6.8 Powder in Oil

The powders contain oil compounds and base powders that are prepared by drying of water-in-oil emulsions in which emulsifiers contain oil phase and a water-soluble dispersible active substance is contained in the aqueous phase. The active substances like plant extracts and flavors are stabilized in the powders in oil. A powder preparetion using pinedex 3 and arabic gum (starch hydrolyzate) are useful for masking the taste of the extract.

33.6.9 Miscellaneous

Derivatives of carvacrol, thymol, caffeic acid, N-(3-oxo-dodecanoyl)-l-homoserine lactone, and pterostilbene, which are naturally occurring compounds and known for their strong anti-inflammatory properties, are either chemically unstable or have poor solubility problems. For these compounds, liposomal formulations have been employed, and it was found that lipophilic 3-oxo-C_{12}-homoserine lactone and stilbene derivatives when loaded into the liposomal lipid bilayer showed an efficiency of 50–70% and inhibited tumor growth by approximately 70% in a murine tumor model. This showed that simple solubilization techniques can have important therapeutic benefits (Coimbra et al. 2011). The compounds that were solubilized by the liposomes allowed intravenous administration without using solvents. There was an improvement in the chemical instability of resveratrol due to liposome encapsulation which prevented the inactivation of *cis-trans* isomerization. By the derivatization into a phenyl ester, caffeic acid was encapsulated stably without any chemical degradation. Liposomal delivery of natural products had shown an increased protective effect against cancer, hepato-/neurotoxicity, oxidative stress, inflammation, hyperlipidemia, and microbial disease (Islam Shishir et al. 2019).

Plant extracts collected through aqueous extraction could be stabilized by utilizing a water-soluble and stabilizing agent like polyvinylpyrrolidone (PVP) that was capable of chelating the extract. Nutritional supplement's dosages have a huge amount of stable ascorbic acid, herbal extracts, and vitamins for mammals which were prepared using improved compositions and methods and have shown prolonged shelf life without substantial degradation. The processes involved heating the mixture of humectants and ascorbic acid to higher temperatures and agitation to stabilize the ascorbic acid at chosen water activity. Therefore, vitamin C solution was prepared that comprised of citric acid, vitamin C, glycerine, and water. It was inferred that a large amount of glycerin, i.e., about 46%, and temperature higher than 135 °F, which is a detrimental temperature, are required in order to obtain a low level of water activity. Heat exposure of about 6 min was required to achieve complete solubilization of many herbal preparetions. But heat treatment sometime damage many components of the herbal formulation such as vitamin C.

33.7 Summary

Herbal medicines generally consist of several components, and the whole preparation is regarded to be as medicine. The components of the drug may interact with each other and may have differences in shelf lives. Further composition of phytochemicals in a drug formulation distinctively varies based on the geographical distribution and harvesting time period of the plant species. These ambiguities may cause alteration of pharmacokinetics in different batches as well as affect the shelf life of the herbal medicine. The study of pharmacokinetics and the stability of the drug can help to make the use of herbal preparations safe and can save people from herbal drug's toxicity. Moreover, the guidelines for the safety of herbal drugs laid by the local government and the WHO should be followed by herbal drug practitioners who will make the herbal preparations more reliable and acceptable worldwide. Identification and quantification of metabolites are necessary for the characterization of herbal samples and to define its stability and efficacy. Different techniques like spectroscopic and chromatographic techniques can be used for the same. Knowing the active component of the herbal products will result in easy and logical quantification and standardization of the components for herbal medicines. These herbal medicines prepared must be stored in proper conditions that are moisture-free and in an inert environment. Different stability parameters of different components can result in loss of effectiveness of the drug or can cause the decomposition of the herbal drug. Therefore, to increase the efficacy of herbal drugs, the active substance can be delivered by advanced methodologies like using various colloidal systems such as microgels, emulsions, nanoemulsions, microemulsions, liposomes, solid lipid nanoparticles, and biopolymer nanoparticles to encapsulate the natural bioactive compounds, which will make the drug absorption more efficient.

References

Ahmed H, Rajendran R, Balakumar C (2012) Nanoherbal coating of cotton fabric to enhance antimicrobial durability. Appl Chem 45:7840–7843

Amini SM (2019) Preparation of antimicrobial metallic nanoparticles with bioactive compounds. Mater Sci Eng C 103:109809

Beneke CE, Viljoen AM, Hamman JH (2009) Molecules polymeric plant-derived excipients in drug delivery. Molecules 14:2602–2620. https://doi.org/10.3390/molecules14072602

Bosch H, Hilborn M, Hovey D, et al (2004) Liquid dosage compositions of stable nanoparticulate active agents. US Patent 0258757A1

Calixto JB, Federal U, Catarina DS (2000) Efficacy, safety, quality control, marketing and regulatory guidelines for herbal medicines (phytotherapeutic agents). Braz J Med Biol Res 33:179–189

Carvalho IPS, Miranda MA, Silva LB et al (2019) In vitro anticancer activity and physicochemical properties of Solanum lycocarpum alkaloidic extract loaded in natural lipid-based nanoparticles. Colloids Interface Sci Commun 28:5–14. https://doi.org/10.1016/j.colcom.2018.11.001

Coimbra M, Isacchi B, Van Bloois L et al (2011) Improving solubility and chemical stability of natural compounds for medicinal use by incorporation into liposomes. Int J Pharm 416:433–442. https://doi.org/10.1016/j.ijpharm.2011.01.056

Dambolena JS, Zunino MP, López AG et al (2010) Essential oils composition of Ocimum basilicum L. and Ocimum gratissimum L. from Kenya and their inhibitory effects on growth and fumonisin production by Fusarium verticillioides. Innov Food Sci Emerg Technol 11:410–414. https://doi.org/10.1016/j.ifset.2009.08.005

Ernst E (1998) Harmless herbs? A review of the recent literature. Am J Med 104:170–178

Folashade O, Omoregie H, Ochogu P (2012) Standardization of herbal medicines—A review. 4:101–112. https://doi.org/10.5897/IJBC11.163

Ghasemi Pirbalouti A, Mahdad E, Craker L (2013) Effects of drying methods on qualitative and quantitative properties of essential oil of two basil landraces. Food Chem 141(3):2440–2449

Hernández-Morales L, Espinoza-Gómez H, Flores-López LZ et al (2019) Study of the green synthesis of silver nanoparticles using a natural extract of dark or white Salvia hispanica L. seeds and their antibacterial application. Appl Surf Sci 489:952–961. https://doi.org/10.1016/j.apsusc.2019.06.031

Hussain AI, Ali S, Chatha S et al (2017) Chemical composition and biological activities of essential oil and extracts from Ocimum sanctum chemical composition and biological activities of essential oil and. Int J Food Prop 20:1569–1581. https://doi.org/10.1080/10942912.2016.1214145

Hussin AH (2001) Adverse effects of herbs and drug-herbal interactions. Malaysian J Pharm 1:39–44

Islam Shishir MR, Karim N, Gowd V et al (2019) Liposomal delivery of natural product: a promising approach in health research. Trends Food Sci Technol 85:177–200

Iwu M, Okunji C, Tchimene M et al (2009) Stability of cough linctus (Streptol) formulated from named medicinal plant extracts. Chem Pharm Bull (Tokyo) 57:229–232. https://doi.org/10.1248/cpb.57.229

Johnson JJ, Nihal M, Siddiqui IA et al (2011) Enhancing the bioavailability of resveratrol by combining it with piperine. Mol Nutr Food Res 55:1169–1176. https://doi.org/10.1002/mnfr.201100117

Krishnaiah YSR (2010) Pharmaceutical technologies for enhancing oral bioavailability of poorly soluble drugs. J Bioequivalence Bioavailab 2:28–36. https://doi.org/10.4172/jbb.1000027

Kumari R (2016) A review on the standardization of herbal medicines. Int J Pharma Sci Res 7:97–106

Lu X, Chen J, Guo Z et al (2019) Using polysaccharides for the enhancement of functionality of foods: a review. Trends Food Sci Technol 86:311–327

Makri O, Kintzios S, Kintzios S (2008) Ocimum sp. (Basil): botany, cultivation, pharmaceutical properties, and biotechnology Ocimum sp. (Basil): properties, and biotechnology. 6475. https://doi.org/10.1300/J044v13n03

Miller T, Lincoln NE, Matthew J, et al (2009) Stable immunoprophylactic and therapeutic compositions derived from transgenic plant cells and methods for production. US patent 0087448A1

Moghaddam M, Farhadi N, Ranjbar M (2017) Variability in essential oil content and composition of Ocimum ciliatum accessions from Iran: evidence for three chemotypes. Int J Food Prop 20:1489–1500. https://doi.org/10.1080/10942912.2017.1352599

Musthaba S, Ahmad S, Ahuja A et al (2009) Nano approaches to enhance pharmacokinetic and pharmacodynamic activity of plant origin drugs. Curr Nanosci 5:344–352. https://doi.org/10.2174/157341309788921453

Nainwal P, Nainwal P (2010) Pharmacovigilance of herbal medicines: an intangible approach. IJPSR 1:60–65

Ogendo JO, Kostyukovsky M, Ravid U, Matasyoh JC, Deng AL, Omolo EO, Kariuki ST, Shaaya E (2008) Bioactivity of Ocimum gratissimum L. oil and two of its constituents against five insect pests attacking stored food products. J Stored Prod Res 44(4):328–334

Parveen A, Parveen B, Parveen R, Ahmad S (2015) Challenges and guidelines for clinical trial of herbal drugs. J Pharm Bioallied Sci 7(4):329–333

Patel NM (2011) Compendial testing method on herbal crude drug—a review. Int J Pharm Technol 1:49–52

Routledge P (1998) 150 years of pharmacovigilance. Lancet 351:1200–1201

Shulammithi R, Sharanya M, Tejaswini R, Kiranmai M (2016) Standardization and quality evaluation of herbal drugs standardization and quality evaluation of herbal drugs. IOSR J Pharm Biol Sci 11(5)

Thakur L, Ghodasra U, Patel N, Dabhi M (2011) Novel approaches for stability improvement in natural medicines. Pharmacogn Rev 5:48–54

Xu XD, Shao SX, Jiang HP et al (2015) Warburg effect or reverse Warburg effect? A review of cancer metabolism. Oncol Res Treat 38:117–122. https://doi.org/10.1159/000375435

Zanetti M, Mazon LR, de Meneses AC et al (2019) Encapsulation of geranyl cinnamate in polycaprolactone nanoparticles. Mater Sci Eng C 97:198–207. https://doi.org/10.1016/j.msec.2018.12.005

Pharmacovigilance: Methods in Developing the Safety and Acceptability of Traditional Medicines 34

Pougang Golmei, Pratap Kalita, S. S. Agrawal, Abdul Baquee Ahmed, Saikat Sen, and Raja Chakraborty

Abstract

Traditional medicines are widely recognized and used as therapeutic agents such as anti-inflammatory, cough remedies, antipyretic, etc. The myth with public concerning traditional medicine is that it is safe to consume as it has been bestowed by our forefathers and it can be consumed without prescription which have led to self-medication which in turn often led to disappointing side effects and end results. There is alarming need for increasing awareness to develop pharmacovigilance practices with regard to herbal and traditional medicines. The present status of pharmacovigilance and its tools have been developed in relation to drugs which are synthetic in nature. With recent several prominent drugs withdrawn from the market, the pharmaceutical agencies have raised the issue of pharmacovigilance. Early uncovering of signals from clinical and pharmacovigilance studies has now been adapted by main companies in order to detect and identify the association of risk with medicinal products and to effectively manage the risk by applying health hazard management plans during the life cycle of the product. Applying the methods to monitor the safety of traditional medicine is an exclusive challenge. A strong, well-defined system for monitoring adverse events is in a position for evaluating the safety of the drugs. In

P. Golmei (✉) · S. S. Agrawal
Department of Pharmacology, Delhi Pharmaceutical Science & Research University, New Delhi, India
e-mail: pougolmei@gmail.com

P. Kalita · S. Sen
Faculty of Pharmaceutical Science, Assam Down Town University, Guwahati, Assam, India

A. B. Ahmed
Girijananda Chowdhury Institute of Pharmaceutical Science, Guwahati, Assam, India

R. Chakraborty
NEF College of Pharmacy, Lokra, Guwahati, Assam, India

S. C. Mandal et al. (eds.), *Evidence Based Validation of Traditional Medicines*,
https://doi.org/10.1007/978-981-15-8127-4_34

today's world, there is an immense need to understand and implement pharmacovigilance with more activity of clinical research conducted all over the globe.

Keywords

Adverse effects · Traditional medicine · Herbal medicine · Pharmacovigilance · Medicinal products

34.1 Introduction

Patient safety is a fundamental principle in the provision of traditional medicines and herbal products for health care and a critical component of quality control. The consumption of traditional medicines worldwide is enormous, so that, in terms of population exposure alone, it is essential to identify the risks associated with their use. Safety of traditional medicines is therefore an important public health issue. Medicines from traditional origin are frequently used in conjunction with other medicines, and it is essential to understand the consequences of such combined use and monitor whether any adverse effects are arising or not. This can be achieved most readily within existing pharmacovigilance systems (Geneva; WHO 2004a, b).

Pharmacovigilance is the science and activities relating to the detection, assessment, understanding and prevention of adverse effects of drugs or any other possible drug-related problems. Recently its concern has widened to include herbals, traditional and complementary medicines, blood products and biological (Geneva; WHO 2004a, b). The purpose of pharmacovigilance is to detect, assess, understand and prevent the adverse effects or any other possible drug-related problems which are not only confined to chemical drugs but also extended to herbal, traditional and complementary medicines, biologicals, vaccines, blood products and medical devices.

In traditional systems of medicine, the medicinal plants play a foremost role and constitute their backbone. Adverse events may also ascend from the misuse of the wrong species of medicinal plants, incorrect dosing, errors in the use of herbal medicines by consumers and health-care providers, interactions with other medicines and use of products contaminated with potentially hazardous substances, such as pathogenic microorganisms, and agrochemical residues, toxic metals (Wal et al. 2011).

34.2 Traditional Medicines

Traditional medicine refers to the knowledge, skills and practices based on the beliefs, theories and experiences indigenous to different cultures, used in the maintenance of health and in the diagnosis, prevention, improvement or treatment of physical and mental illness. Traditional medicine is often termed alternative or complementary medicine in many countries. Herbal treatments are the most popular form of traditional medicine. One third of the population lacks access to essential medicines, and the provision of safe and effective traditional as well as alternative remedies could become a significant way of increasing admittance to health-care services. The traditional use of herbal medicines has long historical basis, on the basis of which they are widely acknowledged to be safe and effective (WHO 2004a, b).

Traditionally made herbal formulations have reached widespread acceptability as therapeutic agents such as cough remedies, hepatoprotective and antidiabetic, etc. Herbal medicines are traditionally considered harmless since these belong to natural sources. However, this is not accurate as there are numerous case reports of adverse reactions of herbal drugs mentioned in published literatures. Even though most traditional therapies are presumed to be safe, there is still the problem of how to assess and quantify the possibility of very rare adverse events. A serious event is sufficient to tip the scales in contradiction of the use of the alternative medicine therapy (Gadre 1910). Increasing uses of these drugs are growing concerns about the safety of Ayurvedic medicines (Trikramji 2001).

Regulation and standardization are not satisfactory in spite of the growing popularity of herbal products. In 1994, Dietary Supplement Health and Education Act (DSHEA) were passed to protect access by consumers to safe dietary supplements. Thus, all herbs, vitamins, botanicals and minerals are identified as dietary supplements (Bass and Raubicheck 2000). In India, the Mashelkar committee on pharma industry was recommending formulations of national policy to bring over-the-counter drugs and herbal drugs under one umbrella to maintain standards of safety and efficacy (Geneva; WHO 2000).

Another traditional form of medicine, the Unani medicine, is an ancient form of medicine originated from Greece. It is more commonly practised in Indian subcontinent and has an age-old concept and principles of drug management. It has drugs from natural identity and source. It may pose threat if not prepared, isolated, identified and administered properly (Khan et al. 2008). According to Unani medicine, correct reasoning in the method of preparation of drugs, including a rationality underlying combination of various medicinal plants, animal products, minerals, etc., is given in various Unani formularies. Therefore, correctives to drugs are used since a long time to lessen some undesirable effects, which the basic and the adjuvant ingredients may produce in a normally prescribed combination of both single and compound crude form of drugs which are sometimes toxic and are processed and purified in various ways before use. In spite of the fact that every drug used in traditional systems of medicine may have some side effects, the aim of the precautions taken by well-informed and experienced physician was noticeably to avoid any adverse drug reactions (Khan et al. 2008).

34.3 Pharmacovigilance

Pharmacovigilance (PV) is well-defined as 'the study of the safety of marketed drugs under the practical conditions of clinical usage in large communities' (Mann and Andrews 2002). The objective is to encompass safety monitoring and distinguish adverse events that have previously been unrecognized despite evaluation in clinical trials. While these procedures were established for monitoring pharmaceutical medicines, they are also used for other evaluation of the safety and efficacy of other medicinal products including traditional, herbals, vaccines, blood products and even medical devices.

34.3.1 Aims of Pharmacovigilance

The thalidomide tragedy incident highlights the life-threatening status of effective drug monitoring systems for all medicines. The pharmacovigilance programmes are aiming (Geneva; WHO 2004a, b):

- To progress public health and safety with respect to use of medicines.
- To progress patient care and well-being in relation to the use of medicines and medical and paramedical interventions.
- To encourage education, understanding and clinical training in pharmacovigilance.
- To perform the valuation of harm, benefit, effectiveness and risk of medicines, inspiring their rational and safe use.

The main purposes of pharmacovigilance are comprise of exhibiting the efficacy of drugs by intensively monitoring their adverse effect profile for many years from the lab to the pharmacy; tracing for any extreme effects of drugs improving public health and safety in relation to the use of medicines, encouraging the safe, rational and cost-effective use of drugs and endorsing understanding, education and clinical training in pharmacovigilance; and effective communication to the generic public. In addition, providing information to consumers, practitioners and regulators on the effective use of drugs along with designing programmes and procedures for collecting and analysing reports from patients and clinicians conclude to the objectives of pharmacovigilance studies (Tuffs 2002).

The fact of plant having an important role in the development of modern medicines is undeniable. Directly or indirectly approximately 60–70% of modern medicines is available in the present market derived from plant source (Geneva; WHO 2003). Due to the lack of clinical trials for most herbal medicinal products, post-marketing pharmacovigilance becomes a critical source of safety information. However, the assessment of adverse reactions associated with herbal medicinal products offers unique challenges in the quantity and quality of available information (Trikramji 2001; Shastri 1994).

34.4 Adverse Drug Reaction of Herbal Origin and Drugs

Herbal products unlike conventional drugs are not measured nor regulated for purity and potency. Thus impurities, contamination of mineral, undeclared adulteration and batch-to-batch discrepancy are few examples of the constant threats in their usage and can also justify many of the reports on their toxic effects (Kinsel and Straus 2003; Ernst 2002).

Extensively, ADRs-reported issues associated with *Aristolochia* and ephedra have shown that herbal medicinal products can give results to toxicity in human beings. The most common adverse effects reported are hepatic and renal problems. However, it is difficult to identify the causative agent associated with the ADRs encountered because traditional herbal preparations often contain multiple ingredients (Bowdler 1997) (Tables 34.1 and 34.2).

According to the database of WHO, which elaborates that there are large number of suspected herbal case reports and some of the most commonly reactions reported among them are rash, pruritis, urticaria, nausea, vomiting, headache, abdominal pain, fever, etc. (Geneva; WHO 2004a, b).

In one of the case reports, two botanical varieties of *Piper betle Linn* (betel leaf), the Mysore variety and Ambadi variety, had different actions for lowering the activity of enzymes of the intestine and stimulation of enzyme activity, respectively (Prabhu et al. 1995). Another report shows that two patients prescribe with phenytoin 300 mg/day had well-controlled seizures later presented with sudden loss of seizure control and due to which a strict history was conducted and revealed that they had started taking "SHANKHAPUSHPI." Administration of single drug dose does not lead to any change in phenytoin levels but decreased its antiepileptic effect (Jawadekar et al. 1992).

Table 34.1 List of specific herbal drugs and their adverse interactions

Sl no.	Herb	Drugs	Adverse effect	Reference
1	*Ginkgo biloba*	Drugs such as aspirin, warfarin, ticlopidine clopidogrel, dipyridamole, garlic, vitamin E	With aspirin-retards aspirin absorption	Kubde (2016)
2	Psyllium seed	Coumarin derivatives	Retards absorption of drugs	Kubde (2016)
3	Devil's claw (*Harpagophytum*)	Warfarin	May be addictive in nature	Elvin-Lewis (2001)
4	Echinacea	Aspirin, rosuvastatin, escitalopram, atorvastatin, esomeprazole, montelukast, levothyroxine, alprazolam, cetirizine	Toxic side effects	Nyambega et al. (2018)

Table 34.2 List of herbs with suspected or known adverse effects

Sl no.	Herbal drugs	Adverse effects	Reference
1	St. John's wort	Gastrointestinal disturbances, allergic, reactions, fatigue, dizziness, confusion, dry mouth, photosensitivity	Kubde (2016)
2	*Ginkgo biloba*	Bleeding	Boullata and Nace (2000)
3	Ephedra (ma huang)	Hypertension, insomnia, arrhythmia, nervousness, tremor, headache, seizure, cerebrovascular event, myocardial infarction, kidney stones	Kubde (2016)
4	Kava (*P. methysticum*)	Sedation, oral and lingual dyskinesia, torticollis, oculogyric crisis, exacerbation of Parkinson's disease, painful twisting movements of the trunk, rash	Nyambega et al. (2018)
5	*Aristolochia* sp. (found use in Chinese medicine)	Kidney toxicity, carcinogenicity	Kubde (2016)

34.5 Approaches of Monitoring Traditional Medicine

Modern drug depends only on the perception of disease and suffering to be treated by intervening a drug or potion, to the patient. Still, evidence gathering is that a wide variety of apparently unrelated ailments can be due to abnormal reactions to consumed food or other things and that a cure can be affected by simply suppressing those ingredients. A struggle also arises because symptoms are often vague rather than specific and chronic rather than acute. Grievances of irritability, fatigue, depression, headache, muscle and joint pain or gastrointestinal disturbances may well be dismissed as being of emotional origin or psychological (Khan et al. 2008). Basically, diagnosis is very time-consuming, tough and involved skilled personnel, which can be presented only by one who has engaged in special awareness. Even though, traditional remedies (called indigenous remedies in Europe and United States) are widely used by the public continuously as there is a perception among the consumers that these remedies are naturally made and therefore have no ill-effects and are very safe to be taken by human being.

Explanations for herbal medicine may be varied, like risks associated with parenteral use, and are greater because all drugs are prepared for internal purpose or for external application. Adulteration of herbals with pharmaceutical drugs is a problem in many countries. In the present scenario, when adverse drug reactions (ADR) monitoring is being done on wide scale and in a well-maintained way, there is still a very low reporting of adverse drug reactions of herbs. The traditional medicine also varies from country-to-country permissible status and approval mechanism and when it comes to risks linked with its irrational use are also larger (Khan

Table 34.3 Selective examples of clinical effectiveness based on meta-analyses of clinical traditional medicine therapy research

Therapy	Medical condition	Conclusion	Reference
St. John's wort	Mild depression	As effective as synthetic anti-depressants, more effective than placebo	Berner et al. (2008)
Kava kava	Anxiety	More effective than placebo	Nyambega et al. (2018)
Glucosamine sulphate	Osteoarthritis	Decreased pain and increased function	Jorg Jerosch (2011)
Horse chestnut	Chronic venous insufficiency	Effective in reducing leg pain and swelling for short term	Pittler and Ernst (2012)

et al. 2008). Table 34.3 includes few examples of clinical effectiveness based on meta-analyses of clinical traditional medicine therapy research.

34.6 Need of Pharmacovigilance

In view of all medicines, it is anticipated that traditional form of medicines also has side effects, whose consequence may be of an adverse nature.

- It has been broadly acknowledged that development of medicines is a process of multifaceted nature which also comprises traditional and herbal origin and their products. Once medications are released for the consumer, it proceeds down the protected and secure scientific setting of clinical study and is free for consumption by all. Thereafter, utmost medicines will only have been tested for its efficacy and short-term safety on a restricted number of cautiously selected people, be it of herbal or traditional origin of medicines (Geneva; WHO 2000).
- Henceforward, pharmacovigilance is certainly a necessity in drugs of various origins including the traditional source of medicine which comprise of fortifying the initial discovery of any new adverse reactions or patient's subdivisions of extraordinary sensitivity and taking up convincing actions so as to manage such risks.
- Additionally, it is of important matter that the new medicine that is available in markets is strictly observed after being sold for their safety and effectiveness in real-life environments.
- Supplementary evidence furthermore is generally required in terms of its usage in precise groups like elderly people, pregnant women and children when it comes to safety and its efficacy of chronic routine usage in adjunction to other drugs (Geneva, WHO 2002).
- Large number of people do not inform their physician about their use of herbal medicines which can seriously lead to unexpected results of laboratory test (Dasgupta 2003).

- It was observed that there is decreased plasma concentration of simvastatin after co-administration with St. John's wort (Sugimoto et al. 2001).

34.7 Safety Monitoring of Medicine with Special Reference to Herbal Origin

Clinical trials and spontaneous reports are among the common sources of information on adverse events and reactions to medicines. In some countries, adverse reaction reporting by physicians is mandatory; such reports are regarded as spontaneous (Shetti et al. 2011).

34.7.1 Sources of Reports

It has been recommended by the Council for International Organizations of Medical Sciences (CIOMS) working group V that a common controlling principle which is emphasise on the quality of a report and not on its source and therefore the value of a report does not lie in who made the report but in the thoroughness and care with which it is prepared, documented, received, recorded, followed-up, clarified and analysed (The Council for International Organizations of Medical Sciences 2001).

34.7.2 Reports from Health-Care Professionals

Internationally, adverse drug reaction reporting systems in the post-marketing safety surveillance setting depend primarily on voluntary reporting by health-care professionals, preferably those directly associated with the care of the patient/consumer, i.e. the patient's primary health-care provider or specialist (Geneva; WHO 2004a, b).

34.7.3 Reports from Consumers

Consumers should be appreciated positively for their concern regarding probable adverse effects in the use of herbal medicines and products in health care. Their reports on adverse reactions should be acknowledged as a serious source of data, which can in return help in identification of indications for unknown effects of herbal medicines and products (Geneva; WHO 2004a, b).

34.7.4 Reports from Other Sources

Any problems allied with herbal medicines may be reported from various different sources, namely, clinical trials/studies, National Poisons Centre, drug information centres and consumer organizations (Geneva; WHO 2004a, b).

34.8 Herbal Products Targeted for Safety Monitoring

It is useful to think of herbal products so as to have maximum exposure, in the following categories:

1. Regulatory status.
   - Herbal medicines in the prescription medicines category and non-prescription medicine's category.
   - Other herbal products intended for use in health care.
2. Registration/marketing status.
   - Herbal medicines undergoing the new drug development process: In clinical trials prior to national drug regulatory approval and under post-marketing safety surveillance.
   - Herbal medicines undergoing re-evaluation under the current protocol: In clinical trials and under post-marketing safety surveillance.
   - Other herbal products marketed for health care, such as dietary supplements (Geneva; WHO 2004a, b).

34.8.1 Reporting of Suspected Adverse Reactions

A. The main players who should be reporting an adverse reaction are:
   - Health professionals (physicians, pharmacists and nurses) should report to the national pharmacovigilance centre.
   - Patients/consumers usually report to their physicians or providers of herbal medicines, but they may also report directly to the national pharmacovigilance centre, consumer organizations or manufacturers.
   - Manufacturers should be directly report to the national pharmacovigilance centre or national regulatory authority (Geneva; WHO 2004a, b).

B. A complete report should contain information as follows:
   - Where it is permitted by the country health information privacy code, and with appropriate confidentiality, some form of identification of the patient/consumer in order to avoid duplications and facilitate follow-up.
   - Age, sex and a brief medical history of the patient (when relevant).
   - Details of suspected product(s) if known: species name (Latin binomial name and common vernacular name of medicinal plant) and/or brand or ingredient

name(s), including the part of medicinal plant used, preparation methods, manufacturer, country of origin, batch number, expiry date and provider.

- Details of drug administration: dose and quantity supplied, dosage form, route, start-stop dates including the indication or reason for use.
- Data of adverse reaction: date of onset (or duration from first administration to onset of event), description with symptoms and signs, severity and seriousness, results of clinical investigations and tests, course and outcome.
- Risk factors, e.g. age, impaired renal function, previous exposure to the herbal medicine(s) concerned, previous allergies, drug misuse or abuse and the social use of drugs.
- All other medicines used (including self-medication), with administration details.

  Full name and address of the reporter (which is to be considered confidential) are to be used only for data verification, completion and case follow-up (Gneneva; WHO 2004a, b).

C. How to Report an Adverse Reaction:
   For all medicines which include traditional medicines (herbal source), a single reporting form should be used. A familiarized form for health-care providers is previously included in a national pharmacovigilance system, which will help in facilitating a report. A standard printed or electronic reporting form will be desirable for usage and to ensure that forms are available extensively. A sample of the herbal product and its packaging should be submitted with the report whenever feasible. Reports should be acceptable to receive by telephone, letter or e-mail (Geneva; WHO 2004a, b).

D. Recording and Coding the Identity of Herbal Medicines:
   For reports to be submitted to Uppsala Monitoring Centre (UCM), it is desirable to use a standardized classification and identification for transmitting. Coding of adverse reactions/events should be compatible to herbal medicines with that for other medicines. Therefore, Uppsala Monitoring Centre suggests the use of the WHO Drug Dictionary (WHO-DD), as it has been structured to store, classified information on the names of herbal products and their ingredients in the same way as similar information on other medicines. Herbal medicines typically contain variety of ingredients, and identifying them all is not possible. Then such cases should be recorded and referred to Uppsala Monitoring Centre, which will contribute in identifying the product. The product information will be added (if not recorded earlier) in the global WHO database, along with the existing information (Geneva; WHO 2004a, b).

34.9 Challenges in Monitoring the Safety of Herbal Medicines

34.9.1 Regulation, Quality Assurance and Control

National registration and regulation of traditional medicines vary from one country to another. Where medicines are regulated, they may be categorized as either prescription or non-prescription medicines. Traditional-based herbal products may also be categorized other than as medicines. Likewise, the regulatory status of a particular traditional or herbal product may differ from one country to another. The framework generally includes the involvement of qualified medicine providers and distributors of respective constituents. Regulatory status consequently determines distribution or access route of medicinal products (The Council for International Organizations of Medical Sciences 2001).

34.9.2 Quality Assurance and Control

National quality specification and standards for herbal materials, good manufacturing practices (GMP), labelling and licencing outlines should be regulated in places of the country where herbal medicines are manufactured, imported and marketed. These measures are vital for ensuring the safety and efficacy of traditional medicines. Negligence in regulations as well as quality control may lead to high prevalence of adverse reactions particularly resulting from adulteration with assumed potent substances and/or mixing with possibly hazardous substances and residues (Geneva; WHO 2004a, b).

34.9.3 Appropriate Use

According to respective country's national health-care delivery system and judicial agenda, various health-care professionals are oblige as qualified providers of traditional medicines and its products. Those countries where herbal medicines are classified as prescription medicines, dispensers and prescriber other than physicians, pharmacists and dentists are left occasionally from existing reporting systems (Geneva; WHO 2004a, b).

34.9.4 Action Required

It is very essential to create an atmosphere of trust to enable the sharing of knowledge about the use and safety of traditional and herbal medicines so that the prescribers are effectively involved. Due to lack of knowledge, providers of medicines, such as physicians, nurses and pharmacists, may have less training and understanding of how this traditional origin medicines affect the health of their particular patients, who may be taking other medicines also be it prescription or non-prescription. An appropriate knowledge base is also relevant to diagnostic and treatment decision-making (Geneva; WHO 2004a, b).

34.10 Challenges in Pharmacovigilance of Traditional Medicine

34.10.1 The Complexity of Herbal Products (with Regards to the Following)

- Lack of clinical trial data: Systematic clinical trial data for several traditional and herbal medicinal products is not available continuously unlike conventional medicines.
- Chemical complexity: Remedies and preparations of herbal and traditional medicine are chemically rich mixtures containing several hundreds of complex constituents, and the resultant effects are likely to be contributing to a group of correlated constituents rather than a single constituent.
- Non-uniformity or non-standardization: The constituents of a plant are usually not uniform all over a plant, and it may be noted that some parts of the plant can be poisonous. The detailed profile of constituents is expected to vary among different batches of herbal products, and factors such as drying, storage, environment, time of harvesting and processing can affect their variability. This makes it difficult to determine toxicology pharmacokinetics and pharmacodynamics and to suggest which ingredient of the plant may cause a safety concern (WHO 2017).
- Quality assurance and control: In comparison with conventional pharmaceutical products, traditional and herbal medicines are made from ingredients of herbal origin which are often obtained from various commercial and geographical sources, resulting in ambiguous condition. Also, techniques and procedures used in its manufacturing and quality control measure are very unlike from those used for conventional medicines.
- Due to lack of technical skill personnel, manpower and amenities to assess and dealt with the problem, especially in identifying adulterated, substandard and contaminated incorrect medicinal plants, which is frequently observed with traditional medicine products.

34.10.2 Botanical Nomenclature

The nomenclature of crude plants is not constant. In various manuscripts the names are in Latin, consisting of two parts, one indicating the plant part and another related to the scientific name, e.g. *Digitalis folium*.

34.10.3 Difference in Product Regulation

The differences on classification of traditional and herbal medicines, inadequate data and lack of admittance to reliable information support such as product name, part use, etc., for analysing the products which is of very concerned.

34.10.4 Safety Monitoring

Most of the health-care providers are undertrained on safety monitoring of medicines including traditional medicine (pharmacovigilance methods), which results into zero or very less reports (WHO 2017).

34.11 Measures for Improvising the Pharmacovigilance of Traditional Medicines

With special reference to usage of herbal medicine, the contemporary practice of contemporary and alternative medicine (CAM) is massively different from what it was envisioned and practise in ancient times. In urban scenario, the patient, health-care provider, symptoms, plants, formulations, etc. have drastically changed, thus making safety and pharmacovigilance a sweltering issue (Gogtay et al. 2002). Roughly vital steps that need to be taken in order to establish safety and efficacy of these interventions can be enumerated as below:

(a) Physician's training: Well-organized training to enhance the physician's capability to recognize and report adverse events.
(b) Public awareness and education: Increasing public awareness as well as education can certainly put weightage to improve pharmacovigilance of traditional medicine in the future.
(c) Conducting randomized, controlled clinical studies: The gold standard for establishing the safety and efficacy of any interventions remains to conduct randomized, controlled clinical trial/studies (RCT) which will definitely help in recognizing type A adverse events easily, though some uncertain blocks that remain are blinding, placebo, appropriate standard, evaluation paradigms, etc.
(d) Incorporation of safety monitoring of herbal medicines as part of routine safety monitoring.
(e) Appropriate stringent regulations regarding quality and purity of traditional products should be of favourable steps be it herbal or animal sources.
(f) Another approach can be introduction of product information leaflet for herbal medicines for better understanding of the source, consumption method as well as possible side effects.

34.12 Conclusion

As a potential source of therapeutics aids, traditional medicines (herbal medicines) have accomplished a significant role in health care system all over the world for combating diseased condition in human beings and for maintaining proper health. Several types of drugs are continuously increasing with advancement, but the financial resources for health-care services still persist to be of limited, and as a result, rational drug management has become an increasingly vital topics as to make optimal use of the drug budget and to provide highest possible standard of health services which is safe and free/less of adverse drug reactions. There is no systematic data on the incidence of traditional medicine-associated adverse effects. One of the difficulties has been that there are many complex issues relating to adverse event detection with traditional products. These include the problems of products with multiple ingredients, drugs of multiple systems of medicine, misclassification of names, poor standardization of products, clinical trial conducted very less, variation in manufacturing processes, contamination, adulteration and misidentification of herbs. In particular, rare adverse events and delayed effects may not be readily identified despite traditional use, countering the argument that many herbal remedies are safe because of previous traditional use. Adopting guidelines for rational use of these traditional medicines are essential to be outlined such as general guidelines for methodologies on research and evaluation of traditional medicines. National pharmacovigilance systems should be closely connected to national drug regulatory systems. A national safety monitoring programme for herbal medicines should be running conjunction with an active national drug regulatory system for its proper functioning, with the intentions and potential to respond to signals originating from reports of adverse effects of herbal medicines and to take appropriate regulatory measures.

34.13 Future Aspects

The education among providers of traditional/herbal medicines, health-care professionals and consumers is vital for the prevention from misapplication of herbal medicines of probable serious risks. Health-care professionals and providers of traditional medicines should ask patients directly, respectfully and persistently what other medicines they are taking, including prescription medicines, herbal medicines and other health products for self-care.

The following are ways to improve pharmacovigilance application in traditional and herbal medicines:

1. In case of herbal formulations, making it a mandatory step for reporting of adverse reactions to regulatory.
2. Including undergraduate and postgraduate level, the concept of pharmacovigilance into the syllabus of herbals.

3. It will be essential to train herbal experts personally in the field of pharmacovigilance and include them in reporting and in valuation of the adverse reactions.
4. Specifically, when it comes to treatment with narrow therapeutic indices, being vigilant by all health-care professionals for potential interactions between prescription medications and herbals.

References

Bass IS, Raubicheck CJ (2000) Marketing dietary supplements. Food and Drug Law Institute, Washington DC

Berner M, Kriston L, Linde K et al (2008) St John's wort for major depression. Cochrane Database Syst Rev 4:CD000448

Boullata JI, Nace AM (2000) Safety issues with herbal medicine. Pharmacotherapy 20:257–269. https://doi.org/10.1592/phco.20.4.257.34886

Bowdler J (1997) Effective communications in pharmacovigilance. The Erice report. Birmingham

Dasgupta A (2003) Review of abnormal laboratory test results and toxic effects due to use of herbal medicines. Am J Clin Pathol 120(1):127–137

Elvin-Lewis M (2001) Should we be concerned about herbal remedies. J Ethnopharmacol 75:141–164

Ernst E (2002) Toxic heavy metals and undeclared drugs in Asian herbal medicines. Trends Pharmacol Sci 23:136–139

Gadre GK (1910) AshtangHridaya. 13thAdhyaya, 2nd edn. Aryabhushan Printing Press, Pune, p 65

Gogtay NJ, Bhatt HA, Dalvi SS, Kshirsagar NA et al (2002) The use and safety of non-allopathic Indian medicines. Drug Saf 25:1005–1019

Jawadekar YM, Shah KN, Kshirsagar NA et al (1992) A study of phenobarbital and dilantin in neonatal seizures. Indian J Pediatr 59:729. https://doi.org/10.1007/BF02859409

Jerosch J (2011) Effects of glucosamine and chondroitin Sulfate on cartilage metabolism in OA: outlook on other nutrient partners especially Omega-3 fatty acids. Int J Rheumatol 2011:Article ID 969012. https://doi.org/10.1155/2011/969012

Khan RA, Latif A, Rahman SZ et al (2008) Importance of pharmacovigilance in Unani system of medicine. Indian J Pharmacol 40(Suppl SI):17–20

Kinsel JF, Straus SE (2003) Complementary and alternative therapeutics: rigorous research is needed to support claims. Annu Rev Pharmacol Toxicol 43:463–484

Kubde S (2016) Adverse drug reactions and pharmacovigilance of herbal medicines in India. IJGP 10(1)

Mann RD, Andrews EF (eds) (2002) Pharmacovigilance. Wiley, Chichester

Nyambega RG, Sharma PK, Dutta S et al (2018) Need of PV in Ayurvedic medicines. Int J Green Pharm 12(1):S21

Pittler MH, Ernst E (2012) Horse chestnut seed extract for chronic venous insufficiency. Cochrane Database Syst Rev 11:CD003230. https://doi.org/10.1002/14651858.CD003230

Prabhu MS, Platel K, Saraswati G, Srinivasan K et al (1995) Effect of orally administered betel leaf (Piper betle Linn.) on digestive enzymes of pancreas and intestinal mucosa and on bile production in rats. Indian J Exp Biol 33(10):752–756

Shastri PK (1994) Rasantarangini. 2nd Adhyaya, 11th edn. Sri Jainendra Press, New Delhi, pp 22–24

Shetti S, Kumar CD, Sriwastava NK, Sharma IP (2011) Pharmacovigilance of herbal medicines: current state and future directions. Pharmacogn Mag 7(25):69–73

Sugimoto K, Ohmori M, Tsuruoka S, Nishiki K et al (2001) Different effects of St John's wort on the pharmacokinetics of simvastatin and pravastatin. Clin Pharmacol Ther 70(6):518–524

The Council for International Organizations of Medical Sciences (2001) Current challenges in pharmacovigilance: pragmatic approaches. The Council for International Organizations of Medical Sciences, Geneva. https://cioms.ch. Accessed 3 Nov 2019

Trikramji AJ (2001) CharakSamhita. 8th Adhyaya, 5th edn. Chaukhambha Sanskrit Sansthan, Varanasi, p 276

Tuffs A (2002) Three out of four Germans have used complementary or natural remedies. Br Med J 325:990

Wal P, Wal A, Gupta S et al (2011) Pharmacovigilance of herbal products in India. J Young Pharm 3:256–258

WHO (2000) General guidelines for methodologies on research and evaluation of traditional medicine. World Health Organization, Geneva. https://apps.who.int. Accessed 27 Dec 2019

WHO (2002) The importance of pharmacovigilance—safety monitoring of medicinal products. World Health Organization, Geneva. https://apps.who.int. Accessed 28 Dec 2019

WHO (2003) WHO guidelines on good agricultural and collection practices (GACP) for medicinal plants. World Health Organization, Geneva. https://apps.who.int. Accessed 28 Dec 2019

WHO (2004a) Guidelines on developing consumer information on proper use of traditional, complementary and alternative medicine. World Health Organization, Geneva. http://apps.who.int/medicinedocs. Accessed 28 Dec 2019

WHO (2004b) WHO guidelines on safety monitoring of herbal medicines in pharmacovigilance systems. World Health Organization, Geneva. https://apps.who.int. Accessed 27 Dec 2019

WHO (2017) Pharmacovigilance for traditional medicine products: why and how? World Health Organization, Southeast Asia. https://apps.who.int. Accessed 16 Dec 2019

Zoopharmacognosy (Plant-Animal Interaction) 35

Shiv Shankar Shukla, Sachin Kumar Jain, Gunjan Kalyani, Beena Gidwani, Ravindra Kumar Pandey, Rupesh Pandey, and Amber Vyas

Abstract

As humans we have taken several things as slender. In case of any ailment, there is a visit to the doctor, in order to have some control and cure. However, such competence does not exist for floras and faunas. This is due to the fact that they have adapted toward the progression and have attained an innate association to the soils, minerals, algae, and supplementary remedies that environment proffers them in order to restore their health from the ailments and diseases through which they are suffering. Zoopharmacognosy also known as animal self-medication had been worn out by several animals, which is comprehensive of our genuine, connubial escorts and their untamed relatives, for the treatment of a diverse ailments and diseases. Alongside the path of progression, and with the employment of trial-and-error methodology, animals have acquired the knowledge of nature's reward by which they can heal themselves. With the vigorous participation of their insight and perception, animals inquire about the beneficial and

S. S. Shukla (✉)
Department of Pharmaceutical Analysis and Quality Assurance, Columbia Institute of Pharmacy, Raipur, Chhattisgarh, India

Columbia Institute of Pharmacy, Raipur, CG, India
e-mail: shivpharma007@gmail.com

S. K. Jain
IPS Academy College of Pharmacy, Indore, MP, India
e-mail: sachin225819@rediffmail.com

G. Kalyani · B. Gidwani · R. K. Pandey
Columbia Institute of Pharmacy, Raipur, CG, India

R. Pandey
KSCP, Swami Vivekanand Subharti University, Meerut, UP, India

A. Vyas
University Institute of Pharmacy, Raipur, India

S. C. Mandal et al. (eds.), *Evidence Based Validation of Traditional Medicines*,
https://doi.org/10.1007/978-981-15-8127-4_35

helpful plant secondary metabolites and by employing the diverse methodology such as by smelling the delicate scent of pure essential oils and by recognizing assured colors. Plants fabricate these nonnutritious materials either for defending themselves or for the attraction of pollinators and carrying out of certain supplementary functions that have been undiscovered. At present, the scientists have discovered that the secondary plant metabolites are beneficial to the animals as well as humans. The modern medicine system and traditional folk medicine system substantiate on the principle which embraces trial-and-error methodology. Owing to the environmental alterations and diverse anthropogenic activities, many of the essential imperative and fundamental species have been gone astray. The present chapter thus furnishes an outline of the repercussions pertaining to the biodiversity and furthermore relating to the conservation strategies. Expert opinion taking care of the critical points and look in has also been discussed.

Keywords

Zoopharmacognosy · Humans · Animals · Environment · Essential oil

35.1 Introduction

Self-medicating accomplishment has drawn together a hasty curiosity of the experts pertaining to the diverse fields such as ethnobotanists, chemical ecologists, conservationists, and physicians as well. Scientists/researchers from various restraints have investigated the prospect and opportunity that quite a lot of plant species, soil, insects, and fungi have been employed and utilized as "medications" as a matter of course against controlling and curing of the illness which is referred to as preventive medicine usage and/or to mitigate the disagreeable symptoms which are referred to as curative or therapeutic medicine approach.

It is an imperative measure to note that the systematic and methodical study of animal self medication is not allied to affirm that animals stand with an innate 'astuteness' by which they unblemished get acquire for the good to them. Self-medication approaches are endurance dexterity which works on by natural hodge-podge. In numerous cases self-medication could be instigated by a craving to instantaneously sponsor the disagreeable stipulations. Some species, predominantly great apes, confirmed the rationale of their medication, and as a result, the term "zoopharmacognosy" was coined to obscure the course by which feral animals selectively employ and utilize specific plants demeanor with medicinal properties for the healing and preclusion of diverse ailments and diseases (Balza et al. 1989).

Therefore, "**zoopharmacognosy**" refers to the progress by which animal do self-medication, by deciding and exploiting the plants along with the soil as well as insects for the control and cure of ailments and diseases. This term was introduced by Dr. Eloy Rodriguez, who was a biochemist and professor at Cornell University. The word is blend of zoo ("animal"), pharma ("drug"), and gnosy ("knowing"). From times immemorial, people have evidenced the observations that an animal in fact discovers the curative potential of plant species and thus medicates with the natural tinge. This is in accordance with the name of many herbs which surmise to their exploitation, such as dog grass (*Agropyron repens*), catnip (*Nepeta cataria*), horny goat weed (*Epimedium* sp.), etc. which advocates for their credible use by animals as indicated in their common name. However, these sorts of inspections are for the most part being unexplored by science. Tales and stories pertaining to animal self-medication are merely premeditated to enlighten and empathize the herbal lore rather than validating it on the scientific basis.

As per the Chinese folklore, centuries ago a farmer in the Yunnan district discovered a snake near his shelter. Afraid for his life, he ridiculously killed the snake. After few days of this incident, the same snake returned. Again he killed the snake, but every time the same happened. The farmer in turn followed the wounded snake, which made a slow progress into a clump of weeds. The snake started feeding on the weeds, which cured the injuries caused. The plant which is responsible for the curing the wounds in the story was identified as *Panax ginseng*, and it is the major constituent in the Chinese herbal formulation "Yunnan bai yao," which is actually a white powder that is used for the cuts and stems exterior flow of the blood straight away. It was then utilized during the Vietnam War, when soldiers were wounded (Barrera and Rodriguez 1990).

35.2 The Significance of Studying Zoopharmacognosy

As nature's medicine breakfront moderate every day, for the animals, with whom we crack this earth, and for ourselves, we have no time to abuse and must work to defend environment. Thoughtful consequence of zoopharmacognosy and its compensation to animals and humans will help congeal the inevitability in which mankind must uphold species-rich habitat and preserve biodiversity. Not only is it vital to conserve our earth's assorted species and ecosystems for their beauty and for their involvement to our own expedition for knowledge but most prominently for the indecision we hold, of not knowing what the effects of trailing it might entail (Fig. 35.1).

There must be a multidiscuiplinary approach for creature of drug, which will enhance creature's and our lives from side to side give the associate expected to recognize why certain species go cleared out and the methods we can take to deflect this, advance unique therapeutic colleague and furthermore embrace the potential disclosure of new fixes and prescriptions for our own species. Just with circumstance shared modification activities will zoopharmacognosy be fit to detach the numerous repayment to be aggregated by natural life biologists, open strength experts, pharmaceutical organizations, preservationists, and the wide-extending open (Clark and Mason 1985).

35.3 Domestic Livestock Benefits

Further than zoopharmacognosy's sizeable benefits, the feasible discovery of recent clinical cures, it also offers a whole realm of other guarantee. May be even greater giant than the medicinal potential, the thoughtful animal self-medicinal drug can

Fig. 35.1 Red-and-green macaws (*Ara chloroptera*) consumption clay to aid digestion

assist humans uphold their meals supply. We stay in a global world where meals, for the most part, isn't any long farmed but mass-produced in a similar fashion to that of a congregation line. However, nature does now not have characteristic as such. Rather, nature functions as a web of assorted interlinking, overlapping, and complicated relationships, which do now not dash on a linear aircraft as discovered on a manufacturing facility floor. Nonetheless, maximum of our meals is produced in this way, in the shape of monoculture or manufacturing facility farms, producing lots of livestock, chickens, and swine. So far, in nature, plants do not develop fruitfully in a monoculture, or in exhausted soil, nor do animals preserve on wholesome within confined, enclosed spaces, at the same time as standing in their individual feces. Even though we experience we've mastered the technology of mass-produced food, all we've consummate is growing a gadget of extra work, greater inputs, and a less humane technique than that of the past. More time must be spent fertilizing fields, spraying insecticides, and deep-tilling the earth (Clark and Mason 1987).

Considering the aspect of the agricutural enterprise more money must be spent on antibiotics and veterinarians to make wonderful the fitness of animals in what most crucial might regard as no longer to fit stay in farms. Even though the agriculture industry can also take years to increase into more humane and greater sustainable, there is want for the future for the duration of zoopharmacognosy. Pragmatic evidence suggests that sick, debilitated animals, when given the selections of certain flowers, soils, minerals, among other substances, will self-medicate and heal themselves. Provided more studies happen and collaboration among scientists and farmers develops, the solutions to be had via applying zoopharmacognosy ideas could shop the farming enterprise billions of dollars. When animals are given the choice to self-medicate, farmers will no longer deliver antibiotics to all their animals, together with the healthful ones, as they do at the present. Not simplest will they shop cash and skip these reserves onto the customer; in addition they will keep away from the extra buildup of bacterial confrontation to antibiotics, which will be dangerous for animal populations as properly as human beings (Clark and Mason 1988).

35.3.1 Chewing Plants

Huffman is one of the pioneers of zoopharmacognosy, on account of his clarification in 1987 of a creature, the chimp, endeavoring to restore to well-being herself. Charmed by her moment recovery and inquisitive about the reason for her disease, Huffman investigated the chimp's manure and discovered the intestinal parasite *Oesophagostomum stephanostomumto* is a decent number likely edification for her side effects. Likewise, he build up lower levels of the worm in one increasingly female chimp's discharges 20 h after she ate the severe substance from a Vernonia tree, when trouble from looseness of the bowels. Huffman and his associates separated a totally new class of mixes from the essence, one of which, vernonioside B1, was found to procure antiparasitic, antitumor, and antibacterial properties (Clark and Smeraski 1990) (Fig. 35.2).

Fig. 35.2 Leaf-eating habit of chimpanzee

35.3.2 Fur-Rubbing Behavior

Mary Baker, an anthropologist at the University of California, opined that white-colored Capuchin monkeys (Cebus capucinus) open the products with specific types of Citrus plants by tearing, mashing, scouring and squeezing into their place of hide. They also tore stems, leaves, and seed units from *Clematis dioica*, *Piper marginatum*, and *Sloanea terniflorastems*, blended with salivation, and energetically focused on them too. These botanicals have optional mixes with helpful and creepy-crawly repulsing uniqueness. Pastry specialist likewise examined that hide scouring conduct turns into extra constant when temperatures and mugginess increment during the blustery period. This might be because of the ensuing increase in the danger of bacterial or contagious diseases. North American earthy-colored bears (*Ursus arctos*) bite the foundation of *Ligusticum porteri*, making a glue of the plant with spit, stroke on their countenances. *Ligusticum porteri* contains coumarins-fragrant natural aggravates that may ward off creepy crawlies when topically applied (File et al. 1976) (Fig. 35.3).

35.3.3 Eating Bacteria for Digestion

The folivorous are those herbivore animals that feed on leaf, and utilize the specific microorganisms in the yield to crush down difficult to-process verdant plant

Fig. 35.3 Fur-rubbing habit of monkey

substance. Literature survey shows that the fledgling's gut microscopic organisms also offset harmful auxiliary mixes beginning in the plants it eats (Huffman and Seifu 1989).

35.3.4 Consumption of Soil

"Geophagy" is depicted as purposefully expending soil, stones, and rock by herbivorous and omnivorous warm-blooded animals, flying creatures, reptiles, and creepy crawlies. This kind of behavior is experiential and determined with regard to self-drug in Japanese macaques (*Macaca mulatta*), mountain gorillas (gorilla), chimpanzees (container troglodytes), and African elephants. Geophagy is elective as a way to keep up gut pH, to meet wholesome necessities for minerals, to persuade long for sodium, to detoxify before devoured plant auxiliary metabolites, and to battle intestinal difficulties like looseness of the bowels (Itokawa et al. 1983).

35.4 Zoonotic Disease Prevention

As the human populace ends up off the impact points of the swine flu pandemic, we are familiar with too well how negative creatures can very quickly convert into unfortunate people. Zoonotic maladies, those which are passed across species limitations, have flooded the features and presented extreme terrorizing to human well-being lately (Janzen 1978). Despite the fact that AIDS might not have been restricted from side to side the comprehension and importance of zoopharmacognosy standards and disclosures, there are such huge numbers of zoonotic ailments that may have been forestalled. Obliging creature self-medicine conduct and guaranteeing that creatures, familial and wild, have permission to the regular cures they require may have the option to put off comparative wellbeing pandemics, for example, Mad Cow infection, Bird Flu and Swine Flu later on. If zoopharmacognosy is contemplated extra with the relationship of scientists, creature conduct specialists, drug specialists, veterinarians, and human well-being concern experts, people may be fit to turn away cataclysmic zoonotic illness transmissions or if nothing else discover prescriptions to mitigate them.

35.4.1 Wild Remedies for Reproduction

Creatures may have endless supply of approaches to control copy, and researchers accept ebb and flow disclosures are just a glimpse of something larger. As indicated by World Wildlife Fund researcher Holly Dublin, African elephants (*Loxodonta africana*) search for specific types of tree, presumably to prompt work. Dublin followed a pregnant elephant for over a year in East Africa and explored that the elephant followed a severely uniform eating regimen, an example of every day conduct holding up close to the end of development. The elephant strolled 17 miles in a single day a few more than her standard three and a tree of the Boraginaceae family from leaves to trunk! After 4 days she brought forth a sound calf. The University of Wisconsin anthropologist Karen Strier starts that, at various occasions, muriqui monkeys (*Brachyteles arachnoides*) of Brazil make a special effort to eat leaves of *Apuelia leiocarpa* and *Platypodium elegans* and the product of *Enterolobium contortisiliquim* (monkey's ear). The initial two plants have isoflavonoids which are compounds practically equivalent to estrogen. Ingesting the leaves may bring estrogen step up in the body, accordingly lessening ripeness. Then again, eating monkey's ear may raise the monkey's odds of getting pregnant on the grounds that the plant contains an ancestor to progesterone (the "pregnancy hormone") called stigmasterol (Kapundu and Penders 1988).

35.4.2 Antimicrobial Property of Plant

As indicated by scholar John Berry at Cornell University, sweet red products of Aframomum angustifolium, forces' antimicrobial impacts basically represent a stomach related danger to the ordinary, well populace of microorganisms found in the gorilla's gut. Utilization of products such as wild ginger and antibacterial mixes in the plant can for the present harm these microorganisms, thus around upsetting the gorilla's stomach-related structure on the off-chance that they aren't before an ordinary piece of the eating routine. Affirmation uncovered that the gorilla's microbiota has urbanized protection from the organically dynamic segments of the plant in territories where it is generally eaten in an adjustment (Kawabata and Nishida 1991).

35.4.3 Antimicrobial Lining in the Nests

The leaves of wild carrot (*Daucus carota*, Umbelliferae) impressively lessen the quantity of fowl bugs (*Ornithonysus sylviarum*) in starling homes. The dusty-footed wood rodents (*Neotoma fuscipes*) put inlet foliage in the locale of their resting homes, and it has been tentatively demonstrated that the inclusion of narrows foliage altogether lessens the bug larval endurance. The wood ants, *Formica paralugubris*, frequently incorporate enormous amounts of set conifer gum into their homes. By making pitch-free and tar-rich test homes, it was built up that the coordinated gum restrains the improvement of pathogenic smaller-scale life form inside subterranean insect homes (Kokwaro 1976).

35.4.4 Plants as Stimulants

Chacma monkeys (*Papio ursinus*) in South Africa are known to eat a little amount of leaves of explicit plants, which are notable for their energizer property. The plants were accessible everywhere throughout the day, and their availability isn't confined; however, monkeys benefited from them discontinuously and just in little amounts. These plants were not under mandrill's standard eating regimen and are gathered as "euphoric." The use of such plants isn't straightly identified with any ailment or an ailing state however shows energizer action. These incorporate *Croton megalobotrys* (Euphorbiaceae), *Euphorbia avasmontana* (Euphorbiaceae), *Datura innoxia*, and *D. stramonium* (Solanaceae). In any case, there are no trial contemplates announcing the plant's particular pharmacological advantages (Kupchan et al. 1969).

35.4.5 Self-Medication as Adaptive Plasticity: Augmented Eating of Plant Toxins by Parasitized Caterpillars

Self-medication is a down-to-earth restorative and self-versatile conduct in response to infection or parasitism. Tainted creatures could change their scavenging to remember therapeutic substances for their eating regimens. They see self-drug as a sort of versatile pliancy, which are typically described by ecologically initiated changes in conduct or phenotype through a person's life expectancy that recuperates its conjecture for perseverance and propagation. Versatile pliancy is totally expected when there is an obvious exchange off in the versatile thought of uncommon phenotypes under perceptibly extraordinary environmental conditions. Thusly, we envision creatures to interface in self-medicine when it is versatile in the event of ailment or parasitism, yet not to associate in such conduct without ailment or parasitism because of its wellness cost (Lwande et al. 1985) (Fig. 35.4).

35.5 Impacts on the Organic Food Revolution

Bacterial and other microorganism cause infections that cause genuine wellbeing dangers, and they cannot be ignored proportionally due to the risk of malignant growth. Since the world depends on a ton of seriously fake medication, manures, and pesticides, one got the opportunity to address whether driving this unnatural way may bring about hazardous physical state angle impacts like malignant growth. The

Fig. 35.4 Self-medication by caterpillar and butterfly

natural food upheaval has aforesaid started an answer for requests of people's inadequate to a more beneficial, more secure substitute to plant created, concoction soaked nourishments. Zoopharmacognosy will energize the natural affiliation even extra through particular and commitment options in contrast to pesticides and genetically changed monoculture. Enquiry shows that honey bees are utilizing explicit blossoms to self-sedate and help battle unwellness. Since honey bees are critical to regularly yield well-being as pollinators, its fundamental that ranchers support a solid supplier of blossoms, on that honey bees trust. This kind of connection guarantees self-medicine anticipating honey bees yet as different species, which can advantage from the protection of a great deal of encompass or arrangement of species (Mcgrew et al. 1989).

Thereof, if ranchers would repulse creepy crawlies, reasonable recognition of plants should be possible, having explicit plant optional metabolites that normally forestall bugs. Insightful and finding about creature self-prescription will most likely encourage the deciding of such plants, since creatures conjointly need them to protect themselves from creepy crawlies and thusly the maladies they bear. In the event that plants, that grip comparative properties, it tends to be learned through Zoopharmacognosy investigation, and therefore planted among crops, which will diminish the use of pesticides discernibly and initiate the security of food. Any nonheritable animal self-medicine associate can be shared over the cultivating network to convey enormous scope-cultivating activities, likewise as profiting little scope and local kinds of cultivating (Gompper and Hoylman 1993).

35.6 Biodiversity and Conservation Repercussions

Regrettably the species-rich tropics are not impervious to outsized extent farming and ranching. It has only been recently when the world has seen an emergent trend of great degree of operational activities in the tropics; it is therefore intimidating to the authenticity of the world's tropical rainforests. Even though tropical rainforests do not swaddle much of the surface of earth, they are the abode to most of species on the earth.

However, there are some threats to the tropical rainforests which is due to the following reasons, as cut and burn agriculture, legal and banned logging, cattle ranching, removal, and climate; therefore humans must make an attempt in order to defend for the better management of the leftover plant species in the ecosystems. The research have shown that plants manage to give the medicinal benefits; it is awfully indispensable that they must be preserved and studied as many as feasible before they are lost gradually. Changes in weather in particularly elevate the danger for the perpetuation of tropical plant species, since the tropics transform somewhat diminutive in warmth; the species that flourish there have advanced to stay alive in very explicit temperature circumstances.

However, with the change in weather, the temperature of earth rises up, and heating effect increases with it; the species unable to adapt with these changes will become extinct. Those medicinal plants that are able to withstand the changes in the

Fig. 35.5 Mushrooms and lichens start the disintegration of a fallen tree

temperature emit stronger aromas which are all due to the heating system climate change, which can perplex animals that rely on the medicinal plants for their self-medication. This is due to the positive contribution between the raised concentrations of secondary plant metabolites and the elevated temperatures. Though the change in weather is not under the human control, zoopharmacognosy researchers can toil simultaneously with conservationists in order to safeguard the biodiversity in the tropics (Nishida 1990). In the study of zoopharmacognosy and secondary plant metabolites, both researchers along with the conservationists can contribute to the healthy and holistic ecosystems, as their association will assist and help the local forest dwelling and other farming communities to have an advantage from the zoopharmacognosy and for their protection as well.

Regrettably for the maintenance efforts, there is an equal requirement of international collaboration, self-indulgence. For these abovementioned reasons, humans must overpower their own egotistical ambitions and act hastily to bring to a standstill for the destruction of very significant ecosystems, such as the tropical rainforests, before it is too late for the species and communities as well, relying on them (Hemmes et al. 2002). (Fig. 35.5).

Zoopharmacognosy supports the fortification efforts for the sake of ensuring continued animal self-medication as well as for the discovery of human cures for ailments and diseases. There should be the efforts to preserve the species for the inherent value; furthermore, there is also a gigantic financial gain, if the pharmaceutical companies and farming industry could better appreciate nature's remedies (Rodriguez et al. 1985). There will be profits for major pharmaceutical companies, along with a source of income to the rural communities as well, and it will also consent the people to their advanced understanding for the animal needs which will assure better and more humane treatment of animals.

Earth's weather is not constant over the time, and there is continued observance of the variations. Plant species will either die out or will adapt against their disappearance from the ecosystems. The existing ecological concern and trepidation, however, lies not in the fear that we are causing transformation in our natural world but in the fact that the rate of these changes has been considerably hastened by human movement and far exceeds the environment's capability to get well. Our never-ending craving for natural resources, physical space, and energy and our short nervousness for anything other than our own species have resulted in a disturbing rate of habitat destruction and loss of species, putting both biodiversity and humans at risk. Human aptitude and the competence to problem-solve have delayed our defense efforts of natural resources and species, simply because we think we will be able to come up with a speedy fix in the future. Yet, indisputably, the human community has defeat wonderful hurdles when faced with worldwide dilemmas (Takasaki and Hunt 1987).

35.7 Expert Opinion: Critical Points and Look in

Currently, the scientists revealed and did their expertise in medical miracles leading to the discovery of penicillin and immunization processes for several deadly diseases such as tuberculosis, measles, mumps, etc.

Zoopharmacognosy shall learn to light on novel solutions for our ever-increasing medicinal needs. From miscellaneous estimation and accepting of animals' self-medication performance, new cures can be discovered for our medical necessitates. The approach for this is to safeguard and preserve the habitats since we do not completely appreciate the consequence of destroying incalculable ecosystems and species. Continued observance and knowledge about the animal behavior can escort to the appreciation of holistic surroundings for them as well as ourselves (Wrangham and Goodall 1989).

Zoopharmacognosy can enlighten the path that is required for the liability of our delinquency, overdo, and ignorance about the species in this world. Once unchained, the probable reimbursement of zoopharmacognosy and its affirmative connotation for conservation can then revolutionize the humanity. John Muir, a Scottish-American preservationist, on one occasion cleverly said, "One learns, that the world, although made, is yet being made. That this is still the morning of making." One would simply anticipate creating a world, where we and all other species can thrive.

35.8 Future Prospects

Babu Kalunde reported for the lessons of animal self-medication and ethnomedicinal observation which are pragmatic and outstandingly lead to approaching source of medicine. An insight into the behavior of animals pertaining to the use of plants and products will lead to some novel pathways for new strategies.

The rationale behind this is to incorporate the pragmatic results into local health apprehension and farm animal's supervision systems so that close by accessible plants can be fittingly used for the benefit of all.

References

Balza F, Lopez I, Rodriguez E, Towers GHN (1989) Dithiacyclohexadiene and thiophenes from *Ambrosia chamissonis*. Phytochemistry 28(12):3523–3524

Barrera E, Rodriguez E (1990) Studies on the African plant *Ficus exasperata* eaten by wild chimpanzees *Pan troglodytes*, (Schweinfurthii). Society for the Advancement of Chicanos and Native Americans in Science (SACNAS) Phoenix, Arizona

Clark L, Mason JR (1985) Use of nest material as insecticidal and antipathogenic agents by the European starling. Oecologia 67:169–176

Clark L, Mason JR (1987) Olfactory discrimination of plant volatiles by the European starling. Anim Behav 35:227–235

Clark L, Mason JR (1988) Effect of biologically active plants used as nest material and the derived benefit to starling nestlings. Oecologica (Berl) 77:174–180

Clark L, Smeraski CA (1990) Seasonal shifts in odor acuity by starlings. J Exp Zool 255:22–29

File SK, McGrew WC, Tutin EG (1976) The intestinal parasites of a community of feral chimpanzees, *Pan troglodytes* (Schweinfurthii). J Parasitol 62(2):259–226

Gompper ME, Hoylman AM (1993) Grooming with Trattinnickiaresin: possible pharmaceutical plant use by coatis in Panama. J Trop Ecol 9:533–540

Hemmes RB, Alvarado A, Hart BL (2002) Use of California bay foliage by wood rats for possible fumigation of nest-born ectoparasites. Behav Ecol 13:381–385

Huffman MA, Seifu M (1989) Observations on the illness and consumption of a possibly medicinal plant *Vernonia amygdalina* (Del.) by a wild chimpanzee in the Mahale Mountains National Park, Tanzania. Primates 30(1):51–63

Itokawa H, Takeya K, Mihara K, Mori N, Hamanaka T, Sonobe T, Iitoka Y (1983) Studies on the antitumor cyclic hexapeptides, obtained from *Rubia* radix. Chem Pharm Bull 37(6):1670–1672

Janzen DH (1978) Complications in interpreting the chemical defenses of trees against tropical arboreal plant-eating vertebrates. In: The ecology of arboreal Folivores. Smithsonian Institution Press, Washington D.C., pp 73–84

Kapundu M, Penders A (1988) Structure du Kotschyioside, Saponside Novueau Isole de *Aspilia kotschyi* (Sch. Bip.). Oliv Bull Soc Chim Belg 97(5):329–342

Kawabata M, Nishida T (1991) A preliminary note on the intestinal parasites of wild chimpanzees in the Mahale Mountains, Tanzania. Primates 32(2):275–278

Kokwaro JO (1976) Medicinal plants of East Africa. East Africa Literature Bureau, Nairobi

Kupchan SM, Hemingway RJ, Karim A, Werner D (1969) Tumor inhibitors. XLVII Vernodalin and vernomygdin, two new cytotoxic sesquiterpene lactones from *Vernonia amygdalina*. J Org Chem 34(2):3908–3911

Lwande W, Macfoy C, Okecj M, Delle Monache F, Marini Bettolo G (1985) Kaurenoic acids from *Aspilia pluriseta*. Fitoterapia 65(2):126–128

Mcgrew WC, Tutin GEG, Collins DA, File SK (1989) Intestinal parasites of sympatric *Pan troglodytes* and *Papio* spp. at two sites; Gombe (Tanzania) and Mt. Assirik (Senegal). Am J Primatol 17:147–158

Nishida T (1990) A quarter century of research in the Mahale Mountains: an overview. In: Nishida T (ed) The chimpanzees of the Mahale Mountains: sexual and life history strategies. University of Tokyo Press, Tokyo, pp 3–35

Rodriguez E, Aregullin M, Nishida T, Uehara S, Wrangham RW, Abramowski Z, Finlayson A, Towers GHN (1985) Thiarubrine-A, a bioactive constituent of *Aspilia* (Asteraceae) consumed by wild chimpanzees. Experientia 41:419–420

Takasaki H, Hunt K (1987) Further medicinal plant consumption in wild chimpanzees. Afr Stud Monogr 8(2):125–128

Wrangham RW, Goodall J (1989) Chimpanzee use of medicinal leaves. In: Heltne PG, Marguart LA (eds) Understanding Chimpanzees. Harvard University Press, Cambridge, MA, pp 22–37

Significance of Stability and Pharmacokinetic Issues in Traditional Medicine

36

Lavanya Bolla, Navitha Reddy Gangireddy, Devi Naga Jyothi Bale, and Satheeshkumar Nanjappan

Abstract

Herbal medicine has become the backbone of the traditional medicine in recent years. There is an increase in demand in the global market and international attention for herbal medicines because of the proved scientific significance of traditional medicines commercially. In an herbal drug, separate active constituents or entire herb is used as an herbal medicine. Because of the presence of many active ingredients, stability of the herbal products becomes difficult throughout the storage period which may change the chemical composition of active ingredients. So, stability testing has become an important component in the development of herbal products to determine the altered therapeutic activity of the formulation ingredients. Furthermore, this complexity of the natural products leads to various pharmacokinetic interactions results in altered absorption, distribution, metabolism and excretion of drugs when co-administered with western drugs. These issues need to be analyzed adequately to avoid the adverse effects of traditional drugs. In this chapter, importance of stability and pharmacokinetics-related issues, and the measures taken to avoid any adverse effects will be discussed in detail.

L. Bolla · N. R. Gangireddy · D. N. J. Bale
Drug Metabolism and Interactions Research Lab, Department of Pharmaceutical Analysis, National Institute of Pharmaceutical Education and Research (NIPER), Hyderabad, Telangana, India

S. Nanjappan (✉)
Drug Metabolism and Interactions Research Lab, Department of Pharmaceutical Analysis, National Institute of Pharmaceutical Education and Research (NIPER), Hyderabad, Telangana, India

Department of Natural Products, National Institute of Pharmaceutical Education and Research (NIPER), Kolkata, Kolkata, West Bengal, India
e-mail: satheesh.nanjappan@gmail.com

S. C. Mandal et al. (eds.), *Evidence Based Validation of Traditional Medicines*,
https://doi.org/10.1007/978-981-15-8127-4_36

Keywords

Herbal medicine · Stability of herbal drugs · Pharmacokinetics and pharmacodynamics interactions

Abbreviations

ABC	ATP binding cassette
ASU	Ayurveda, Siddha, and Unani
AUC	Area under curve
AYUSH	Department of Ayurveda, Yoga & Naturopathy, Unani, Siddha and Homoeopathy
CE-MS	Capillary electrophoresis-mass spectrometry
C_{max}	Maximum concentration of a drug in the systemic circulation
CNS	Central nervous system
CRISM	Centre for Research on Indian Systems of Medicine
CYP	Cytochrome P-450 enzyme family
D&C Act	Drugs and Cosmetics Act
DPPH	2,2-Diphenyl-1-picrylhydrazyl
EU	European Union
FDA	Food and Drug Administration
GACP	Good Agricultural and Collection Practices
GC-MS	Gas chromatography-mass spectrometry
HDI	Herb-Drug Interactions
HM	Herbal medicine
HPLC	High-performance liquid chromatography
HSA-DB-L	Human Serum Albumin
K3G	Kaempferol 3-O-gentiobioside
L-AA	Levo-ascorbic acid
LC-MS	Liquid chromatography-mass spectrometry
MAO	Monoamine oxidase
MRP	Multidrug resistance protein
NMR	Nuclear magnetic resonance spectroscopy
NSAIDS	Nonsteroidal anti-inflammatory drugs
OAT	Organic anion transporters
OATP	Organic anion transporting polypeptides
OCT	Organic cation transporters
PBPK modelling	Physiologically based pharmacokinetic modelling
PEG	Polyethylene glycol
P-gp	P-glycoprotein (transporter protein)
PK	Pharmacokinetics
Poly-PK	Poly-pharmacokinetics
PPO	Phenylperoxidase

QC	Quality Control
RH	Relative humidity
RSA	Radical scavenging activity
SLC	Solute carrier transporters
SLN	Solid-lipid nanoparticles
$T_{½}$	Half-life of a drug
$T_{1/2}$	Half-life
TAA	Total ascorbic acid
UGT	UDP glucuronosyltransferase
US	United States
USFDA	United States Food and Drug Administration
WHO	World Health Organisation

36.1 Introduction

Traditional medicines are generally herbal extracts, and these constitute the main component in curing the ailments. As having less side effects and more acceptance in the patient population, the use of herbal medicine (HM) is variably increased in the past decades because they offer treatment for a wide variety of diseases. The use of these medicines increased mainly due to the valuable experiences and observations given by the people who use them. Important aspect in this boost and usage of HM includes cost-effectiveness and wide intake in the people as a status of being a natural product/extract (Builders 2018; Kamboj 2000).

36.1.1 Herbal Medicine

Herbal medicine which is also regarded as a phytomedicine includes active natural products that are made up of herbs. These natural products and the chemical characteristics of them differ from several factors. So, the dose may vary from person to person and with respect to the community sometimes. These HMs have a very long history and are under practice based on the beliefs, theories and experiences of different times. Some medications may contain bee and fungal extracts, insect parts, few minerals like bentonite and kaolin for the betterment of the health and also for the management of several diseases.

HM is being propagated as the popular kind of traditional medicine in the health-care sector; but there is a need to test the efficacy of the HMs. HM offers number of health benefits yet can show some adverse effects. These pharmacological and adverse effects are shown by HMs majorly connected to their secondary metabolites. Unknown of these facts, public think that natural products/HMs are inherently safe. But HMs do have a risk of carrying some side effects; this suggests intensified research should be carried out on herbal extracts (Firenzuoli and Gori 2007). Containing many active ingredients in the extracts, HM stability is one of the most important elements that need to be maintained in the development of HMs. Next aspect that is to be considered is the pharmacokinetic interactions that are mainly caused by the complexity of the natural herbal product and also with the co-administration of western allopathic drugs.

36.1.2 Common Herbal Medicines and Chemical Constituents

Differing from conventional medicine, HMs use part or whole of the plant – for treating acute and chronic diseases. The main enigma of the herbal product is about the composition. Their chemical composition alters based on factors like anatomical part of the herb (leaf, root, flower, or seed), storage conditions and type of ground the herb is collected from, humidity and so on. Same products from manufacturers tend to produce extracts with different concentration of the constituents. This irregularity in the production of HMs from batch to batch gives chance of difference in pharmacological activity: both pharmacokinetic and pharmacodynamic changes (Firenzuoli and Gori 2007; Kamble et al. 2018). Raw plant materials (bulk herbs) are used to prepare a wide variety of herbal dosage forms. Most popular are the extracts and powders in oral dosage forms. Freeze-dried herbs and tinctures are consumed well too. Some herbal drugs and their uses are mentioned in Table 36.1.

The chemical composition of the HMs may show adverse and side effects: people are still aware about introducing a natural ingredient into the body. This vegetable origin unknown to the people may cause some effects like sequestrenes, salicylic

Table 36.1 Herbal drugs and their uses (Falzon and Balabanova 2017; Vickers et al. 2001)

Herb	Scientific name	Dosage form available	Used for
Garlic	*Allium sativum*	Oil-filled capsules, enteric-coated tablets	Hypertension, hypercholesterolemia, aphrodisiac
Ginkgo	*Ginkgo biloba*	Solid and liquid extracts	Intermittent claudication and dementia
Echinacea	*Echinacea purpurea*	Liquid extracts and tinctures	Short-term immune system stimulation
Ginseng	*Panax* spp.	Elixirs, tablets and capsules	Anti-inflammation
St. John's wort	*Hypericum perforatum*	Tinctures, tablets and capsules	Mild depression, gastritis
Turmeric	*Curcuma longa*	Capsules	Anti-inflammatory effect
Kava	*Piper methysticum*	Powders, tablets and capsules	Stress and anxiety
Devil's claw	*Harpogophytumprocumbens*	Capsules and gel	Anti-inflammatory effect
Marigold	*Calendula officinalis*	Creams and herbal tea	Hemollient and healer (only topic use)
Saw palmetto	*Serenoa repens*	Capsules	Benign prostatic hyperplasia

Table 36.2 Herb-conventional drug interactions (Boullata and Nace 2000; Vickers et al. 2001)

Herb	Conventional/western drugs	Interactions observed
St. John's wort	Tetracyclins, CNS stimulants	Increased phototoxicity
Ginseng	MAO inhibitors	Tremors, headache and mania
Echinacea used above 8 weeks	Anabolic steroids, ketoconazole	Hepatotoxicity
Kava kava	Benzodiazepines	More sedation
Saw palmetto	Estrogen	Herbal effect is increased
Fever few	NSAIDS	Herbal effect inhibition
Liquorice	Spironolactone	Diuretic antagonism
Shankapushpi	Phenytoin	Drug inhibition
Kelp	Thyroxin	Herbal iron content interferes with thyroid replacement
Ginkgo	Anticoagulants, NSAIDS	Risk of excess bleeding

glucosides produce allergic reactions, bioavailability of some drugs are improved or reduced by the use of grapefruit juice and St. John's Wort. Some herbal extracts may also be carcinogenic. Special attention or care should be taken before taking HMs in pregnancy and should be done under medical supervision. Some herbs and conventional drugs when taken together may lead to side effects: herb-drug interactions depicted in Table 36.2.

In this chapter we brief about the stability issues and pharmacokinetic interactions with their regulatory perspectives related to HMs.

36.2 Stability-Related Issues

Herbal products/medicines involve many specific markers or constituents for the cure of a disease. Any changes in the components of HM/changes in their content will affect the therapeutic activity. In herbal formulations, the herb in it is entirely treated as the active ingredient. The contents of the HM that belong to different chemical classes undergo molecular changes by the exposure of light, humidity, pH, heat and various reactions due to storage, manufacture and transport. The stability report alone of the active ingredient will not serve the purpose. Stability data in the presence of other substances in herbal formulations is to be provided with actual fingerprint chromatograms (EMEA GOS 2006; Bansal et al. 2016; Nanjappan et al. 2018).

Some interactions are very complex among different constituents and make the final product very active/less active or toxic. For example, tannins, lignins and polysaccharides tend to develop mono-dentate and bi-dentate complexes with heavy metals (Martin-Dupont et al. 2006). Non-covalent interaction seen in paeoniflorin and glycyrrhizin increased the water-solubility of tannins with glycosides. These kinds of reactions probably change the chemical detail of the HMs during their storage which alters the therapeutic activity. Proper analytical data

Table 36.3 Stability testing conditions (WHO 2009)

Type of stability study	Storage condition	Duration to be covered before submission	Testing frequency
Long-term stability	25 ± 2 °C/60 ± 5% RH or 30 ± 2 °C/65 ± 5% RH or 30 ± 2 °C/75 ± 5% RH	12 months	0, 3, 6, 9, 12, 18, 24 months and annually
Intermediate stability	30 ± 2 °C/65 ± 5% RH	6 months	0, 6, 9, 12 months
Accelerated stability	40 ± 2 °C/75 ± 5% RH	6 months	0, 3, 6 months

is to be provided focusing the drug is unaltered during its shelf life. Generally, HMs are formulated as either tablets, capsules, or oral solutions. Tests/parameters distinct to these formulations are also to be additionally performed like dissolution, hardness, disintegration, viscosity, suspendability and microbial contamination. These tests should be done according to the pharmacopoeia protocols (Table 36.3) in HM stability period (Atmakuri and Dathi 2010; Bansal et al. 2016). Stability data studies are good to be performed on at least three production batches during shelf life period of the herbal medicines. Stability should be done on the herbal dosage packing in the container which is to be released in the market (Atmakuri and Dathi 2010).

36.2.1 Shelf Life

Shelf life determination is same as for the chemical APIs, but special care is taken for the nature of herbal product involved. In known therapeutic active herbal products, the variation in components should not exceed ±5% to that of initial value. If a proper reason is given to justify the changes caused, the acceptance limit is up to ±10%. Tests like water content and fingerprint chromatograms should be added. Due to the climatic changes, harvesting behaviour and biological variance, the marker added to the herbal product is to be highly given importance during the stability study. Different types of approaches available for assessing the shelf life are biological assays, assay of markers added and chromatographic profiling of both control and stability samples. Expiry date provided for a herbal medicine should have some stability data which supports its shelf life (World Health Organization 2006).

36.2.2 Recent Literary Work in Herbal Medicine Stability Testing

Many stability reports in the recent decades showed a wide difference in the conditions applied for the assessment of shelf life. In this current book chapter, we have identified reports that showed (1) physical changes, (2) chemical changes, (3) biological and (4) physico-chemical and biological changes in different stability conditions. These changes help to know the challenges faced and the kind of approaches to be developed during stability testing.

36.2.2.1 Physical Changes during Herbal Medicines/Products Assessment

Physical changes in herbal medicines include change in formulation colour, viscosity and few specific variations respective to that of dosage form. Neem (*Azadirachta indica*) cream at a stability condition of 25–30 °C, 12 months is found to be stable (Aremu et al. 2009). Topical gel (2.5% and 5%) of *Clerodendron infortunatum* was prepared which has its use in skin diseases, inflammation and asthma. The topical gel filled in collapsed tubes was subjected to stability testing according to ICH guidelines. The storage conditions were 25 °C ± 2 °C/60% ± 5% RH, 30 °C ± 2 °C/65% ± 5% RH and 40 ° C ± 2 °C/75% ± 5% RH for a period of 3 months and studied for appearance, viscosity, pH and spreadability. The gels with lower concentration (2.5%) showed better stability than compared to 5% topical gel formulation. The stability conditions didn't show much variation on viscosity and spreadability either (Das et al. 2011).

Akhtar and co-workers developed a topical skin cream of sea buckthorn (*Hippophae rhamnoides*) for anti-sebum secretion effects. This cream being a w/o emulsion is analysed for the stability parameters in a storage period of 4 weeks. The storage was in 8 ± 0.1 °C, 25 ± 0.1 °C and 40 ± 0.1 °C with 75% RH. Physical stability was estimated by subjecting the cream to centrifugation and by making a note of pH changes. Centrifugation studies showed slight phase separation for the formulation preserved at 40 °C after 21 days period. pH was reasonably reduced till the final stage of the storage period (Akhtar et al. 2010). Topical gels of fenugreek (*Trigonella foenum-greacum*) developed for the anti-inflammatory action was tested for stability for a time period of 3 months. The stability conditions maintained were 40 ± 2 °C and 75 ± 5% RH for 90 days. The colour was green to that of the initial preparation. The formulation was stable showing no changes in viscosity and drug content for a period of 90 days (Jyothi and Koland 2016).

36.2.2.2 Chemical Changes during Herbal Medicines/Products Assessment

Chemical changes cause major effect on the activity of the drug. So, these changes must be properly studied and estimated with proper storage specifications. *Andrographis paniculata* is used traditionally for liver complaints and as immunostimulant. The dried sample extract when kept under 25 ± 2 °C, 60 ± 5% RH and 30 ± 2 °C, 60 ± 5% RH storage conditions showed an insignificant reduction the andrographolide content. The same is the case in storage of 5 ± 2 °C for 3 months. The ambient condition for storage of *Andrographis paniculata* dried leaves is decided as 30 ± 2 °C, 60 ± 5%RH. This attempt declares no special container is required to preserve the quality of dried *Andrographis paniculata* (Ibrahim and Chong 2008). Birch leaves and passion flowers which are rich in flavonoid content were tested for stability. The samples stored at 40 °C and 75% RH for 6 months showed a significant decrease in the total flavonoid content.

Here the birch leaves showed degradation of flavonoid glycosides into sugar and aglyca-moieties (Heigl and Franz 2003). Sea buckthorn which consists of vitamin C as the primary vitamin within it is evaluated for the storage capacity of the berries and juice. The study estimates the effect of temperature on total ascorbic acid content (TAA). The content was estimated with different temperatures, viz. -20 °C, 6° to 25 °C and also 40 °C. The TAA content was decreased to half of its percentage at elevated temperatures like 40 °C after 7 days. L-ascorbic acid (L-AA) is majorly prone to chemical and enzymatic oxidation. L-AA showed good aqueous stability in the pH working range of 3.0–4.5 (Gutzeit et al. 2008).

Piper sarmentosum, medicinal plant used in phytopharmaceuticals, has many benefits to cure ailments. It has pharmacological activities like anti-amoebic, anti-neoplastic and anti-malarial (Hussain et al. 2012). The ethanolic extracts of this *Piper sarmentosum* were studied for the stability by HPLC. The accelerated stability was done by the addition of the selective markers like pellitorine, sarmentine and sarmentosine. The samples were studied for 0, 1, 2, 4 and 6 months. Three different temperatures selected are 30 °C/60% RH, 40 °C/75% RH and 60 °C/85% RH for 6 months. The integrity of the samples was good in 30 °C/60% RH showing less than 10% of content lost in contest to the markers. The reduction in the concentration was likely found to be 25% and 40% at 40 °C/75% RH and 60 °C/85% RH. This indicating the higher temperatures and humidity caused the loss of activity of *Piper sarmentosum*. These samples should be stored in room temperatures (Khalid et al. 2011). *Stevia rebaudiana* leaves are known for its commercial sweetener purpose due to the presence of steviol glycosides. These glycosides stability was determined in leaves and also the commercial sweetener Truvia. The methanolic extracts of the steviol and its glycosides were stored at 4° and −20 °C for 30 days. The samples at 4 °C showed a good stability result, whereas samples stored at −20 °C reported a decrease of 3.6% in all the analytes (Gardana et al. 2010).

36.2.2.3 Biological Activity Assessment during Stability Testing

Antioxidant property of *Aegle marmelos* with leaves extracted with ethanol, methanol and water was subjected to pH studies. The pH selected was 4, 7 and 9 and the incubation was 24 h. The ethanol and water extracts at pH 4 showed good radical scavenging activity (RSA) initially for 30 min. This eventually declined at the 24 h. time point. The antioxidant extracts at pH 7 showed high RSA till 24 h making the samples stable in neutral environment. The ability to scavenge the 2,2-diphenyl-1-picrylhydrazyl (DPPH) radicals was completely lost at pH 9 and may be due to the structural denaturation of antioxidant composite (Reddy and Urooj 2013). *Echinacea purpurea* formulations like creams and gels are checked for the antioxidant activity under the storage condition of 4, 30 and 40 °C for 6 months. Linear regression analysis was used for the antioxidant activity purpose. The antioxidant activity didn't last for more than 4 months in 30 °C. The gel components for antioxidants decreased far than that of cream. The gels are incorporated with water which causes oxidation of the bioactive compounds. 0.1% w/w of α-tocopherol or disodium edetate were added to improve the stability of *Echinacea purpurea* creams and gels, but the results seemed not satisfactory (Yotsawimonwat et al. 2010).

Aqueous garlic extract studied for the antibacterial activity at different ranges of pH 6, 7, 8 and 9. The initial pH 5.8 of *Allium sativum* aqueous extracts showed good pathogenic activity against *Escherichia coli*, *Salmonella typhi* and *Bacillus*. But as the pH increases from 5.8 to 9, the antibacterial activity was eventually lost (Durairaj et al. 2009).

36.2.2.4 Herbal Products Evaluated for Biological along with Physico-Chemical Changes

Allium sativum is well known for its antimicrobial and also anti-thrombotic activities. The instabilities in allyl 2-propenylthiosulphinate (allicin), a major constituent of *Allium* were evaluated. This was done in various solutions like water and different percentages of ethanol. The biological half-life $T_{1/2}$ (17 days) was high and good compared to chemical $T_{1/2} = 12$ days. The allicin was seen to be more stable in 20% ethanolic extract but unstable in vegetable oil (Fujisawa et al. 2008). The abundant flavonoid glycoside found in *Cassia alata* leaves, kaempferol 3-O-gentiobioside (K3G) when sundried and heat dried and then tested for anti-inflammatory effects on COX-1, showed least inhibitory activity (Moriyama et al. 2003). *Withania somnifera* has its therapeutic uses in cancer, stress and arthritis. Dried extracts (powder) of *Withania* were tested for physicochemical stability in correlation with biological effects in accelerated and real-time storage conditions for 6 months. Test conditions included 30 ± 2 °C and 65 ± 5% relative humidity (RH) for the real-time testing and 40 ± 2 °C with 75 ± 5% RH for the accelerated testing. Moisture content and size distribution were assessed under physical parameters while chemical stability by HPLC for the determination of withanolides. The physical form of the Withania powder changed to clump in five-month sample of real time study and at third month of accelerated storage. Moisture content was found to increase in accelerated and real time storage case. The particle size range was from 20 to 140 μM. HPLC analysis of withaferin A and withanolide A showed a percentage of 90.40 and 92.82 in real-time storage. In accelerated storage, these percentages went to 63.60 and 88.86. Immunomodulatory effect of levamisole (reference drug) and Withania spray-dried extract was measured in biological stability which showed a high decline in accelerated condition (Patil et al. 2010).

36.2.3 Challenges Faced in Current Scenario of Herbal Medicine Stability

The quality, efficacy and safety of the herbal products are to be maintained throughout their shelf life. The main challenges faced nowadays by herbal products stability testing are the biochemical variation in raw materials, selection of specific markers, chemical complexity of the composition and the enzyme activity. The herbal drugs are mixture of various heterogeneous chemical components. The development of chromatographic finger printing is quite difficult with the divergent chemical composition. The exposure to light, humidity, heat and air may trigger new chemical reactions and changes the amount/levels of components present. These chemicals exist in tiny amounts in an herb where the extraction becomes difficult, and also many constituents together produce their individual pharmacological action leading to synergistic or antagonistic action. To avoid this, identification of chemical markers is important with respect to their biological activities in parallel to the control sample during the stability testing of herbal drugs. The other quality attribute of stability is the microbial contamination of the herbal drug product. The microbial enzymes not only cause hazardous effects by the product but also degrade the components present. Biochemical variation occurs by the herbs when collected from different climates (Bansal et al. 2016).

Commercial formulations of *Hypericum perforatum* were meant to have 0.3% hypericin of the product. But the study showed it was somewhere around lowest of 7.72% and a highest of 38.57% of the label claim (Shah et al. 2005). Such studies prove that chemical composition varies from batch to batch of same herbal formulation. The most common phenomenon observed in herbals is the post-harvest browning. The reasons behind these are the phenyl peroxidase (PPO), phenylalanine ammonia lyase and/or peroxidase present which imparts brown colour to the product by oxidising the phenolic compound to a quinine (giving the brown colour). Enzymatic activities should be a must check in the processing of the HMs (Adams 2010; Bansal et al. 2016).

36.2.4 Advances in Improvement of Stability

Many components in herbal products and natural medicines are causing a common problem that they react with each other, further leading to stability issues in the formulation. Different studies to deal with herbal products stability are (a) physical parameters testing, (b) complete study of impurity profile, (c) quantification of all metabolites and (d) properly maintained storage conditions. Instability issues can be dealt with techniques mentioned here like- preparation of nanoparticle coating to enhance the shelf life. Coating of herbal formulations with nanoparticles prevents them for oxidation and environment degradation and improves the shelf life (Thakur et al. 2011). Solid-lipid nanoparticles (SLN) have gained good importance in the improvement of the physicochemical stability and provide good protection against degradation of labile drugs (Martins et al. 2012; Kolenyak dos Santos et al. 2013). SLN were incorporated into curcumnoids. In the absence of light, the stability samples showed 91%, 96% and 88% of curcumin, demethoxycurcumin and bisdemethoxycurcumin, respectively. SLN are a good choice in improving the stability of curcumnoids (Tiyaboonchai et al. 2007). Camptothecin, anti-tumour agent from *Camptotheca acuminata*, is an active lactone. This was incorporated into PEGylated liposomes by the addition of 3,5-bis (dodecyloxy) benzoic (PO)-polyethylene glycol-containing liposomes. The surface was overlaid with human serum albumin (HSA-DB-L). These studies performed for this anti-tumour action showed that HSA-DB-L improves the stability of camptothecin (Watanabe et al. 2008). The instabilities caused by liposomes can be replaced by the usage of proliposomes. Proliposomes convert into liposomes upon the addition of water.

This was applied for silymarin which improved its stability along with bioavailability and also encapsulation efficiency (Yan-yu et al. 2006).

Phytosomes gained good popularity in the twentieth century for formulating herbals. These are used in the nanosized range as controlled and sustained release systems. The phosphatidylcholine chains form chemical bonds showing good stability (Sharma et al. 2016). Soft gel complex of curcumin powder was prepared with soy phosphatidylcholine by solvent evaporation method. These curcumin-loaded phytosomes were filled in soft gelatin capsules to assess the stability. The samples were kept in 25 °C and 65% RH for 3 months. The zeta potential between the aqueous and curcumin phytosome complex varied from -38.8 ± 3.48 mV compared -35.4 ± 2.45 mV indicating a good stability (Allam et al. 2015). Topical herbal formulations help in the therapy of skin disorders like psoriasis, scarring and eczema. These are nano-emulsions drug-based aqueous phase tapped in essential oils. These promote good thermodynamic stability with long shelf life (Chaudhary and Naithani 2009). Storage stability can be increased by the addition of β-cyclodextrin to the herbal liquid compound (Thakur et al. 2011). Silver is always known for its prominent role. Silver nanoparticles prepared for the antibacterial traditional Chinese medicines showed good role in the upliftment of stability. The conjugation of these silver nanoparticles improved the potency against *Pseudomonas aeruginosa*, *Staphylococcus epidermidis* and *Staphylococcus aureus* (Sun et al. 2014). Tannins when microencapsulated with β-cyclodextrin, sodium alginate and ethylcellulose have shown better stability in light and heat conditions compared to normal tannins (Wu et al. 2006). Many a times long-chain fatty acid derivatives from plants are incorporated into herbal formulations by the use of linoleic acid-base compound. This reduces the hydrogenation of oil-based products and improves the stability character of the drugs and increases its shelf life (Wilkes 2009).

In addition to all the above examples, traditional herbal and natural medicinal active ingredients are formulated into promising delivery systems like transferosomes, ethosomes, polymer drug-conjugates, dendrimers, microemulsions, nanostructured lipid carriers and hybrid nanocarriers (Fig. 36.1). These are able to improve both the stability and also biological activity of the natural herbal product (Liu and Feng 2015).

36.3 Pharmacokinetic Issues

The usage of herbs for an ailment dates back to centuries, but in recent times alternate treatments to allopathic medicine have been gaining increased popularity for a wide range of ailments like common cold to cancer and even metabolic disorders. Due to the fact that herbal drugs having more than one component and some are being taken as food supplements along with the prescribed drugs, there is a chance for potential herb-drug interactions. With the multiple components in a single formulation, the herbal drug acts on multiple targets in the body that produces multiple responses which are interdependent and makes the interaction study quite difficult to perform. Due to increased popularity, there is an increased scope of

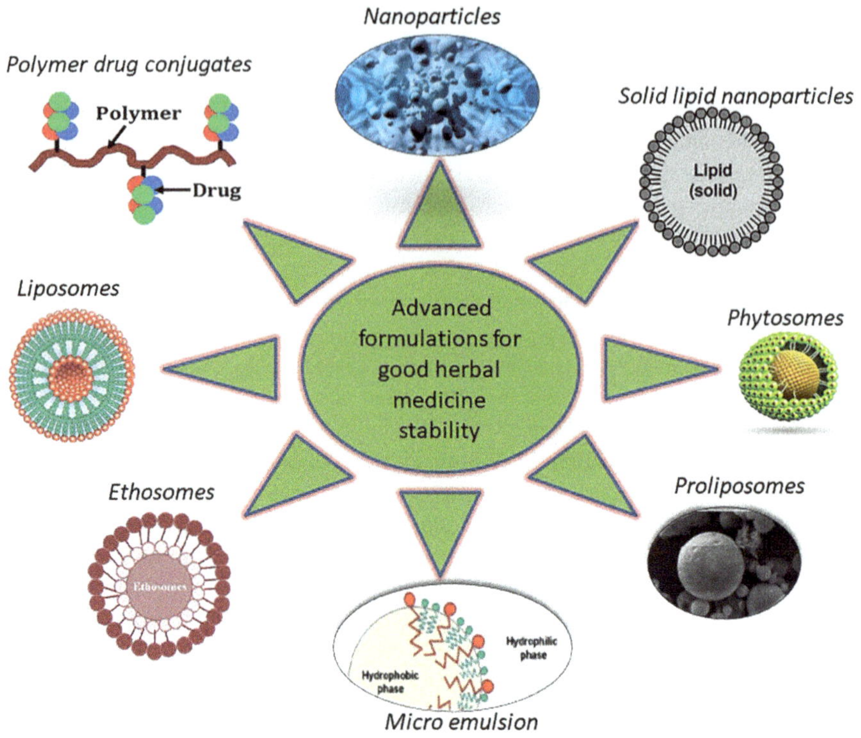

Fig. 36.1 Advanced formulations for good stability of herbal medicine

performing clinical studies for elucidating herb-drug interactions. The interactions become potential or life threatening in case of drugs with low therapeutic index. In general, the possible types of interactions are pharmacokinetic and pharmacodynamic, but the mostly reported ones are pharmacokinetic interactions. It is said to be pharmacodynamic herb-drug interaction when the herbal drug antagonises or synergises the pharmacological effect of the drug. In case of pharmacokinetic interactions, there is alteration in metabolism or transport of a drug resulting in changes in absorption, distribution, metabolism and excretion leading to the altered pharmacokinetics. These interactions are clinically more significant because of the changes it brings to area under curve (AUC), $T_{1/2}$ and $C_{\max}$ of the simultaneously administered drug. The two main mechanisms behind these pharmacokinetic interactions are metabolism-mediated and transporter-mediated interactions.

36.3.1 Metabolism-Mediated HDI

In this type of HDI, phase I and phase II enzymes are involved. Phase I enzymes are nothing but CYP family, and phase II enzymes are non-CYP family like UGTs. The herbal medicines induce or inhibit the metabolic enzymes of the drug. The inhibition occurs when the herbal medicine competitively or non-competitively decreases the expression or the activity of the enzyme. Induction of enzyme involves the gene activation and increase in protein or gene level of a particular enzyme which makes it a much slower process. However, the induction or inhibition processes are reversible and the enzyme levels reach normal when the herbal medicine administration is stopped. The inhibition of CYP enzymes increases the plasma concentrations of the drug and induction leads to the therapeutic failure with low plasma concentrations of the drug.

36.3.2 Transporter-Mediated HDI

These are less reported than metabolism-mediated HDI but are equally important. To know about the transporters mediated HDI, it is essential to know about the types and functions of these transporters. There are two super families of drug transporters: (i) ATP-binding cassette [ABC] and (ii) solute carrier transporters [SLC]. The ABC transporters include P-glycoprotein, multidrug resistance protein (MRP) and breast cancer resistance protein (BCR), whereas SLC include organic anion transporters (OATs), organic cation transporters (OCTs) and organic anion transporting polypeptides (OATPs) that mediate the uptake of their substrates. Unlike metabolic enzymes which are mainly expressed in the liver and intestine, these transporters are expressed in most of the tissues and involve in absorption, distribution and excretion and also involved in oral bioavailability, hepatobiliary and intestinal and renal excretion of drugs. These transporters also cause changes in pharmacokinetics of drugs administered along with herbal medicine that have the potential to induce or inhibit the transporter for which the drug is a substrate. Herb-drug interactions because of the alteration in metabolism and transport enzymes are mentioned in Table 36.4.

The pharmacokinetic interactions can be identified by different kinds of models like in vitro models, animal in vivo models and human in vivo models. Of them in vitro models are widely used because of the ease of operation and practical applicability. Generally cryopreserved human liver cells and microsomes expressing recombinant enzymes are used as in vitro models. And mice and rats are used as in vivo models. Sometimes probes are also used to determine the pharmacokinetic interactions. Studies performed in human healthy volunteers and patients provide clinically relevant data from which systemic concentrations and elimination of the drug can be known (Meng and Liu 2014).

However, the wide number of metabolites present in herbal drugs makes it difficult for effective pharmacological evaluation and for drug development. The complexity of herbal drugs and that of biological samples provides a challenge to measure the concentration-time profiles of herbal drugs quantitatively through analytical approach that is just out of the scope in present traditional study. The study of herbal pharmacokinetics along with multiple metabolites is called poly-pharmacokinetics is only seen in a few studies and needs to be concentrated in order to establish a safety profile for an herbal drug/formulation. Due to drug-drug

Table 36.4 Herb-drug interactions because of the alteration in metabolism and transport enzymes (Meng and Liu 2014)

Herbal drug	Enzymes altered	Transporters altered	Drugs affected
Ginkgo biloba	CYP2B6, CYP2C19, CYP3A4	P-gp, OATP	Efavirenz, midazolam, talinolol, fexofenadine, omeprazole
Ginseng	CYP1A2, CYP2D6, CYP2E1, CYP3A	P-gp	Midazolam, debrisoquine
Garlic	CYP1A2, CYP2D6, CYP2E1, CYP3A4	P-gp	Chlorzoxazone, saquinavir
Black cohosh	CYP1A2, CYP2E1, CYP3A4	P-gp	Debrisoquine
Echinacea	CYP1A2, CYP3A	P-gp	Midazolam
Milk thistle	CYP1A2, CYP2C9, CYP2D6, CYP2E1, CYP3A4	UGT1A, P-gp	Losartan, metronidazole, talinolol
Kava	CYP2E1, CYP1A2, CYP2D6, CYP3A4	P-gp	Chlorzoxazone
St. John's wort	CYP2B6, CYP2C19, CYP2D6, CYP2E1, CYP3A4	P-gp	Alprazolam, amitriptyline, bupropion, chlorzoxazone, cimetidine, ciclosporin, digoxin, fexofenadine, warfarin, imatinib, mephenytoin
Goldenseal	CYP3A4, CYP2D6, CYP3A5		Caffeine, chlorzoxazone, digoxin, indinavir, midazolam, ciclosporin

interactions, the PK of a given compound differs from single component assay to multicomponent assay. With large number of variables and complicated network of metabolites, it is technically quite challenging to identify a metabolite and changes occurring in it in the metabolite pool. The metabolic fate of a chemical in human body can be first determined by the structure and varied by diet, dose, gut microbiota, lifestyle, genetic differences, environment and routes of exposure as well as other xenobiotics intentionally or unintentionally. To simplify all this, in silico prediction coupled with in vitro screening and identification and in vivo validation were introduced for xenobiotic metabolism studies. However, extrapolations among different platforms (in vitro/in vivo/in silico) are not robust or reliable; the xenobiotic metabolite studies require extra attention.

Recent advances in hyphenated techniques like LC-MS, GC-MS, CE-MS and NMR made it easier to identify small metabolites simultaneously. Metabolomics is another approach of analysis of metabolites. This particular omics branch can be used in many ways; some of them are for determining how a biological system reacts to a stimulus, identifying metabolic pathways and differentiating the pathways of endogenous and xenobiotic metabolites, etc. It untangles the interlinked pathways of both endo- and exogenous metabolites and can analyse hundreds and even thousands of variables. It can be used to form a new platform in PK studies that can measure multiple herbal drugs in vivo and to identify alterations in the human body upon exposure to an herbal drug. Presently utilising metabolomics to study the safety and efficacy of herbal drugs is the main focus in present research. There are advances in the aspect of metabolic fate in case of single xenobiotics, whereas multicomponent drug systems with complicated and highly variable composition provide a challenge to conduct studies to know how the herbal drug works inside the body.

There are two strategies generally followed to determine the herb-drug interactions; they are (a) classic strategy and (b) metabolomics-based strategy.

1. Classic strategy: It is a recently discussed knowledge-based technique. A number of metabolites were determined by this technique with the advancement of bioanalytical techniques. This strategy can reveal intra-herb interactions and robust and reliable but time consuming.
2. Metabolomics-based strategy: the first study under this category is NMR-based study on human nutritional intervention. This gave rise to new study called 'Nutrikinetics' that is a combination of latest technologies and methodologies that gives a complete note on the fate of a metabolite in the body. This has been used in many studies to know the fate of metabolites exposed to diet.

Metabolomics with multivariate statistical analysis can address all the challenges faced in poly-PK. The results we get are reliable and many factors and variables that are faced in case of multicomponent herbal drugs are taken into account. Focus in this area further helps to increase the efficacy of the analysis. Elucidation of complete pharmacokinetic parameters is essential to reduce complications, toxicity and costs and to increase the quality of patient's life.

36.3.3 Challenges Faced with the Complexity of Metabolism of Herbal Drugs

To develop protocols for in vivo metabolomics, there are many challenges that need to be addressed. Some of them are (1) multicomponent system and the difficulty to standardise it, (2) overlap of herbs with daily diet, (3) co-metabolism in the gut by microbes that varies and gives unreliable data, (4) differentiating endo- and exogenous metabolites and (5) similar chemical structure of different constituents with interlinked metabolic pathways. From these one can say that research in this area is still in its infancy and needs to be explored (Lan et al. 2013).

The composition of the herbal drug is not constant and subjected to variation based on many factors like climatic changes, rainfall, temperature, growing conditions, etc. To achieve reproducibility, the composition of the formulation has to be more focussed on. The detailed screening of the constituents is done only for some herbs like Milk thistle, St. John's wort, etc. and needs to be expanded to every possible herb. Some techniques like bioactivity-guided fractionation are used to determine the constituents of herbs. The variable experimental protocols across laboratories and lack of standardisation of herbal drugs give low-quality and variable results. Due to multiple constituents in a single drug, the static equations used for extrapolations are not amenable which renders the requirement of more sophisticated techniques like physiologically based pharmacokinetic modelling (PBPK) and simulation studies (Brantley et al. 2014).

Although the studies on these interactions are increasing, the area is still limited by lack of long-term in vivo studies and lacks clear mechanisms of the interactions. The pharmacodynamic end points must be thoroughly investigated to draw a clinical relevance of these interactions. This enables us to take clear decisions regarding patient's safety. These types of interactions can find out by systematic study during drug discovery and development stage and further continued by clinical trials. There is a need to develop a protocol to conduct clinical trials to determine risks and any lethal interactions. The composition variability is the biggest challenge in herb-drug interaction study which is altered by species of the plant, geographical area, plant part used, collection methods, post-harvesting treatment, etc. This demands standardisation and harmonisation of herbal drug standards.

Another emerging challenge is polymorphism in genotypes related to the metabolic pathways of drugs and herbs. Extreme criteria are needed to assess the adverse effects that may occur when herb and drug are taken simultaneously during the development phase. Very little scientific data is available on interactions of herbs with the drugs which need to be the present focuses in order get clinically significant and quality data. The interactions reported are less because the patients do not report them to their health care professionals. There should be awareness brought among the consumers regarding the usage of herbs and the potential interactions that they cause when taken along with allopathic drugs (Tarirai et al. 2010).

There is a need for pharmacovigilance to assess risks, quality and safety in case of herb-drug interactions which is not tested or not subjected to approval by regulatory agencies. Because of the underreporting of the incidents by the consumers, it is difficult to keep a track on number of interactions occurring. It is the national pharmacovigilance centre's responsibility to bring awareness about this herb-drug interactions and drug safety among the consumers, health professionals, regulatory authorities and suppliers of the herbal medicine (Skalli and Bencheikh 2012).

Herbal practitioners and manufacturing companies do not require clinical trials for endorsement and marketing internationally. And also, there are no strict regulation requirements during regulatory approvals. This may be the cause of lack of proper protocols for clinical trials for HDI (Parveen et al. 2015). Quality assurance, health and safety are global concerns regarding standards of herbal drugs and regulatory requirements. The documentation and traceability of every herbal drug entering into local or international market is lacking and has to be maintained. Research regarding the cultivation, drying, various processing methods, storage practices and their effect on the herbs not only benefit the consumer but also achieve stable herbal products (Street et al. 2008).

36.3.4 Regulatory Perspective

In India, under the Ministry of Health & Family Welfare, Department of Ayurveda, Yoga & Naturopathy, Unani, Siddha and Homoeopathy (AYUSH) was established in 1995 which regulates the herbal medicines. Drugs and Cosmetics Act 1940 and Drugs and Cosmetics Rules of 1945 governs the manufacture, import, sale and distribution of drugs and cosmetics. In 1959, the govt. of India included the drugs derived from traditional Indian medicine. Without state drug control, authorities license no products derived from traditional medicine should be manufactured. Proprietary medicines and drugs with patents should have ingredients mentioned in the books recognised by the above regulatory systems. Special committee for ayurvedic, siddha and unani medicines advises the govt. regarding any issues in traditional medicine. Pharmacopoeias for all these systems are being prepared by the pharmacopoeia committees. Indian government appointed an expert committee to develop guidelines for safety and efficacy of herbal drugs in 1993 which were intended to be included in the D & C Act and Rules. The herbal medicines other than those authorized by the regulatory agencies are not allowed to be manufactured or marketed except for those drugs that are manufactured in compliance with the formulae taken from the authoritative books for herbal medicine. For marketing authorisation for a new drug, the manufacturer must submit the safety and efficacy data. The requirement for submission of clinical trial and toxicity data differs according to the nature of the herb and its availability. The herbal drugs are classified into three categories based on their nature and availability. Category 1: Already in use for more than 5 years, Category 2: In use for less than 5 years and Category 3: New medicines. This classification depends on whether the drug has processed or unprocessed plant parts or presence of any poisonous substance (World Health Organization 1998).

The WHO produced a large amount of data on how to handle traditional plant material and their products from collecting the material to clinical trials. These guidelines are facilitating the work of academicians, industries and regulatory authorities concerned with the trade and regulation of the herbal drugs. The WHO published a collection of test procedure to assess the purity, identity, content of the herbal drugs in the name "Quality Control Methods for Medicinal Plant Materials" to engage national laboratories in drug QC. In 2003, guidelines on good agricultural and collection practices (GACP) for medicinal plants and in 2007, 'WHO guidelines for assessing quality of herbal medicines with reference to contaminants and residues', were published. Regional and national guidelines for GACP were developed in European Union, China and Japan so as to ensure that the irrigation water and soil used for cultivation of herbal drugs are having within the limits of heavy metals and free from herbicides, pesticides and toxic substances. In Germany a concept was introduced to ensure the consistency of herbal drugs known as 'phytoequivalence' in which a fingerprint is constructed for an herbal constituent and compared with reference (Ameh et al. 2010; Sahoo et al. 2010).

The main reason for fall back of traditional medicine internationally is that no efficacy and safety studies are required for marketing approval according to D & C Act, and in United States all the traditional medicines are being marketed under Dietary Supplement Health and Education Act of 1994, as dietary supplements which doesn't require any safety and toxicity data. Under this act the manufacturers are not allowed to claim any health benefit towards any disease. By this, the value of the traditional Indian medicine is not being recognised fully and properly. For producing or selling dietary supplements, the industries do not need to register with USFDA or have to take approval. This has been a loophole for the wide marketing in United States for herbal drugs. The FDA can take any action against safety issues only after the product being released into the market. Most of the manufacturers favourably export the traditional medicine to United States because of the relaxed requirements, and no requirement of submitting any scientific evidence for the health claims for dietary supplements which require a lot of money.

While this is the case in United States, European Union requires a lot of data to be submitted for registration and application for marketing authorization. The data need to be submitted are description of manufacturing methods, contraindications, qualitative and quantitative data of the herbal drug constituents, adverse reactions, posology, dose, route of administration, etc. The product is allowed into the market only if it passes the full regimen including safety and toxicity data which is an expensive issue making EU less favourable for the industries although the demand for herbal medicine is equally increasing in both areas.

Different countries have different pharmacopoeial standards, the comparison to which produces differences in permissible limits of heavy metals, plant-specific parameters, microbial contamination, pesticides, quality standards, etc. There are country-specific standards and regional guidelines evolving which have become bottlenecks to comply with for manufacturers for trading of herbal drugs. To resolve this issue, it is necessary to support the profession and strengthen it in other countries in order to increase the value for Indian herbal medicine worldwide. For this the government of India has taken necessary steps by organising major events, exhibitions, providing technical support to universities for the research, exchange of scholars and funding research. An Indo-US Centre for Research on Indian Systems of Medicine (CRISM) which provides scientific validation and spread the information on ASU (Ayurveda, Siddha, and Unani) medicines through collaborative research was set up in National Center for Natural Products Research, University of Mississippi. Through this mission, it was intended that the scientific acceptance of the Indian traditional medicine will be increased.

To overcome the problem of difficulty in identification of components in multicomponent herbal drug systems, 'marker-based assessment' was introduced. It was already well established in developed countries along with chemical fingerprinting to assure the identity and quality of raw materials and active ingredients which has recently been introduced in India. In this approach, for identification and quality assessment, one or more compounds are being selected by natural product analysts as markers. But no guidelines are provided regarding 'marker-based assessment' in Indian GMP. One of the surveys regarding the quality issues in herbal medicines revealed that the poor quality of herbal medicines is the result of lack of sufficient regulatory guidelines in various aspects off production. Manufacturers feel that there should be further development and elaboration of guidelines on quality control of herbal drugs. Good storage practices are also very important in herbal drug manufacturing other than QC and GACP. National Medicinal Plants Board, India, developed guidelines on good agricultural practices and good field collection practices specific to India in line with GACP developed by the WHO. However, many people are not aware of these guidelines, and there is a need to spread awareness about these guidelines. Even though some companies are aware of them, they find it impractical because the growers are unaware of those practices and also the operational cost is more. There is a need to establish government certified raw material supply centres in every state in order to supply standardised and certified raw materials to produce quality herbal medicines.

As a first attempt in elaboration of guidelines, in D & C Act, fourth amendment a rule was introduced for the guidelines for evaluation of ASU drugs. Another drawback in the Indian herbal drug industry is that the non-uniformity in drug registration timeline which differs in different states. There is an urgent need to establish a unified system in the country with defined timelines and unified protocols (Sahoo and Manchikanti 2013).

36.4 Conclusion

Vast numbers of herbal medicines are used to improve health by using different forms such as phytoformulations, nutraceuticals and functional food. Plants are the main source of treatment against most of the diseases and in future can dominate the synthetic drug market. Because of the increased use of these herbal drugs, it is essential to have clear and complete picture on pharmacokinetics of the herbal drugs. The pharmacokinetic profile of the herbal drugs acts as a blueprint and is used to determine the clinical and toxicological value of the herbs. Since there are increasing number of patients using these herbal drugs, clinicians must have a clear picture about the interactions that a particular herb can cause. Either dose adjustments or discontinuation of the therapy is to be suggested if any herb is found to be interacting with a particular herb (Mehta et al. 2015; Mamindla et al. 2016).

Not only about the interactions but also the factors affecting the interactions must be known. This can lead to better understanding of the interactions thereby leading to better rationale behind the administration. However these studies are limited to in vivo animal models and since the extrapolations are not robust and reliable; there is a need to expand these studies to clinical research to better understand the mechanisms of the interactions and to elucidate the metabolic fate of the herbal constituents (Sun et al. 2019).

Because of the less strict regulations and lack of standardised raw materials, the labels identified on the herbal formulations are incorrect or incomplete. The composition may change from batch to batch. This can be overcome by forming strict guidelines and bringing awareness on different regulatory practices like GACP, WHO guidelines, etc. in order to get consistent herbal drugs. There is also need for well-designed clinical studies and making it mandatory to submit clinical safety and toxicity data to acquire manufacturing and marketing approvals (Shi and Klotz 2012).

Acknowledgements Authors are thankful to the Science and Engineering Research Board (No. EMR/2016/002098), New Delhi, India, for the financial support for this work.

References

Adams JB (2010) Effect of enzymatic reactions on color of fruits and vegetables. In: Enzymes in fruit and vegetable processing. CRC Press, pp 33–58

Akhtar N, Khan BA, Mahmood T, Parveen R, Qayum M, Anwar M (2010) Formulation and evaluation of antisebum secretion effects of sea buckthorn w/o emulsion. J Pharm Bioallied Sci 2(1):13

Allam AN, Komeil IA, Abdallah OY (2015) Curcumin phytosomal softgel formulation: development, optimization and physicochemical characterization. Acta Pharma 65(3):285–297

Ameh SJ, Obodozie OO, Inyang US, Abubakar MS, Garba M (2010) Current phytotherapy-a perspective on the science and regulation of herbal medicine. J Med Plants Res 4(2):072–081

Aremu O, Femi-Oyewo M, Popoola K (2009) Repellent action of Neem (Azadirachta indica) seed oil cream against Anopheles gambiae mosquitoes. Afr Res Rev 3(3)

Atmakuri LR, Dathi S (2010) Current trends in herbal medicines. J Pharm Res 3(1):109–113

Bansal G, Suthar N, Kaur J, Jain A (2016) Stability testing of herbal drugs: challenges, regulatory compliance and perspectives. Phytother Res 30(7):1046–1058

Boullata JI, Nace AM (2000) Safety issues with herbal medicine. Pharmacotherapy 20(3):257–269

Brantley SJ, Argikar AA, Lin YS, Nagar S, Paine MF (2014) Herb–drug interactions: challenges and opportunities for improved predictions. Drug Metab Dispos 42(3):301–317

Builders PF (2018) Introductory chapter: introduction to herbal medicine. In: Herbal medicine. IntechOpen

Chaudhary M, Naithani V (2009) Novel herbal topical formulations for treatment of acne and skin disorders. Indian Pat Appl Publ 28

Das S, Haldar PK, Pramanik G (2011) Formulation and evaluation of herbal gel containing Clerodendron infortunatum leaves extract. Int J Pharm Tech Res 3(1):140–143

Durairaj S, Srinivasan S, Lakshmanaperumalsamy P (2009) In vitro antibacterial activity and stability of garlic extract at different pH and temperature. J Biol 5(1):5–10

EMEA GOS (2006) Test procedures and acceptance criteria for herbal substances 1. Herbal preparations and 2. Herbal medicinal products 3. Traditional Herbal Medicinal Products, European Medicines Agency

Falzon CC, Balabanova A (2017) Phytotherapy: an introduction to herbal medicine. Prim Care 44 (2):217–227

Firenzuoli F, Gori L (2007) Herbal medicine today: clinical and research issues. Evid Based Complement Alternat Med 4(S1):37–40

Fujisawa H, Suma K, Origuchi K, Kumagai H, Seki T, Ariga T (2008) Biological and chemical stability of garlic-derived allicin. J Agric Food Chem 56(11):4229–4235

Gardana C, Scaglianti M, Simonetti P (2010) Evaluation of steviol and its glycosides in Stevia rebaudiana leaves and commercial sweetener by ultra-high-performance liquid chromatography-mass spectrometry. J Chromatogr A 1217(9):1463–1470

Gutzeit D, Baleanu G, Winterhalter P, Jerz G (2008) Vitamin C content in sea buckthorn berries (Hippophae rhamnoides L. ssp. rhamnoides) and related products: a kinetic study on storage stability and the determination of processing effects. J Food Sci 73(9):C615–C620

Heigl D, Franz G (2003) Stability testing on typical flavonoid containing herbal drugs. Die Pharmazie- Int J Pharm Sci 58(12):881–885

Hussain K, Hashmi FK, Latif A, Ismail Z, Sadikun A (2012) A review of the literature and latest advances in research of Piper sarmentosum. Pharm Biol 50(8):1045–1052

Ibrahim M, Chong G (2008) Stability of andrographolide in Andrographis paniculata under selected storage conditions. Int J Eng Technol 5(1):60–65

Jyothi D, Koland M (2016) Formulation and evaluation of an herbal anti-inflammatory gel containing Trigonella Foenum Greacum seed extract. Int J Pharm Sci 8(1):41–44
Kamble B, Athawale N, Gugale A, Advankar A, Ghagare A, Kanukuntla P (2018) Herbal Drug Development: Challenges and Opportunities. Plant-and Marine-Based Phytochemicals for Human Health: Attributes, Potential, and Use: 3
Kamboj VP (2000) Herbal medicine. Curr Sci 78(1):35–39
Khalid H, Zhari I, Amirin S, Pazilah I (2011) Accelerated stability and chemical kinetics of ethanol extracts of fruit of piper sarmentosum using high performance liquid chromatography. Iran J Pharm Res 10(3):403
Kolenyak dos Santos F, Helena Oyafuso M, Priscila Kiill C, Palmira Daflon-Gremiao M, Chorilli M (2013) Nanotechnology-based drug delivery systems for treatment of hyperproliferative skin diseases-a review. Curr Nanosci 9(1):159–167
Lan K, Xie G, Jia W (2013) Towards polypharmacokinetics: pharmacokinetics of multicomponent drugs and herbal medicines using a metabolomics approach. Evid Based Complement Alternat Med 2013:819147
Liu Y, Feng N (2015) Nanocarriers for the delivery of active ingredients and fractions extracted from natural products used in traditional Chinese medicine (TCM). Adv Colloid Interf Sci 221:60–76
Mamindla S, Prasad K, Koganti B (2016) Herb-drug interactions: an overview of mechanisms and clinical aspects. Int J Pharm Sci Res 7:3576–3586
Martin-Dupont F, Gloaguen V, Guilloton M, Granet R, Krausz P (2006) Study of the chemical interaction between barks and heavy metal cations in the sorption process. J Environ Sci Health A 41(2):149–160
Martins S, Costa-Lima S, Carneiro T, Cordeiro-da-Silva A, Souto E, Ferreira D (2012) Solid lipid nanoparticles as intracellular drug transporters: an investigation of the uptake mechanism and pathway. Int J Pharm 430(1–2):216–227
Mehta P, Shah R, Lohidasan S, Mahadik K (2015) Pharmacokinetic profile of phytoconstituent (s) isolated from medicinal plants—a comprehensive review. J Tradit Complement Med 5 (4):207–227
Meng Q, Liu K (2014) Pharmacokinetic interactions between herbal medicines and prescribed drugs: focus on drug metabolic enzymes and transporters. Curr Drug Metab 15(8):791–807
Moriyama H, Iizuka T, Nagai M, Miyataka H, Satoh T (2003) Antiinflammatory activity of heat-treated Cassia alata leaf extract and its flavonoid glycoside. Yakugaku Zasshi 123(7):607–611
Nanjappan S, Paul D, Bolla L (2018) Assessing herb–drug interactions of herbal products with therapeutic agents for metabolic diseases: analytical and regulatory perspectives. In: Studies in natural products chemistry, vol 59. Elsevier, pp 283–322
Parveen A, Parveen B, Parveen R, Ahmad S (2015) Challenges and guidelines for clinical trial of herbal drugs. J Pharm Bioallied Sci 7(4):329
Patil D, Gautam M, Jadhav U, Mishra S, Karupothula S, Gairola S, Jadhav S, Patwardhan B (2010) Physicochemical stability and biological activity of Withania somnifera extract under real-time and accelerated storage conditions. Planta Med 76(05):481–488
Reddy VP, Urooj A (2013) Antioxidant properties and stability of Aegle marmelos leaves extracts. J Food Sci Technol 50(1):135–140
Sahoo N, Manchikanti P (2013) Herbal drug regulation and commercialization: an Indian industry perspective. J Altern Complement Med 19(12):957–963
Sahoo N, Manchikanti P, Dey S (2010) Herbal drugs: standards and regulation. Fitoterapia 81 (6):462–471
Shah AK, Avery BA, Wyandt CM (2005) Content analysis and stability evaluation of selected commercial preparations of St. John's wort. Drug Dev Ind Pharm 31(9):907–916
Sharma P, Verma S, Misri P (2016) Global need for novel herbal drug formulations. Int J Phcog Phytochem Res 8(9):1535–1544
Shi S, Klotz U (2012) Drug interactions with herbal medicines. Clin Pharmacokinet 51(2):77–104

Skalli S, Bencheikh RS (2012) Safety monitoring of herb-drug interactions. Drug Saf 35 (10):785–791

Street R, Stirk W, Van Staden J (2008) South African traditional medicinal plant trade—challenges in regulating quality, safety and efficacy. J Ethnopharmacol 119(3):705–710

Sun S, Wang Y, Wu A, Ding Z, Liu X (2019) Influence factors of the pharmacokinetics of herbal resourced compounds in clinical practice. Evid Based Complement Alternat Med 2019:1983780

Sun W, Ding Qu YM, Chen Y, Liu C, Zhou J (2014) Enhanced stability and antibacterial efficacy of a traditional Chinese medicine-mediated silver nanoparticle delivery system. Int J Nanomedicine 9:5491

Tarirai C, Viljoen AM, Hamman JH (2010) Herb–drug pharmacokinetic interactions reviewed. Expert Opin Drug Metab Toxicol 6(12):1515–1538

Thakur L, Ghodasra U, Patel N, Dabhi M (2011) Novel approaches for stability improvement in natural medicines. Pharmacogn Rev 5(9):48

Tiyaboonchai W, Tungpradit W, Plianbangchang P (2007) Formulation and characterization of curcuminoids loaded solid lipid nanoparticles. Int J Pharm 337(1–2):299–306

Vickers A, Zollman C, Lee R (2001) Herbal medicine. West J Med 175(2):125

Watanabe M, Kawano K, Toma K, Hattori Y, Maitani Y (2008) In vivo antitumor activity of camptothecin incorporated in liposomes formulated with an artificial lipid and human serum albumin. J Control Release 127(3):231–238

WHO (2009) Stability testing of active pharmaceutical ingredients and finished pharmaceutical products. WHO technical report series, no. 953. World Health Organization, Geneva

Wilkes RS (2009) Food compositions incorporating stearidonic acid. Google Patents

World Health Organization (1998) Regulatory situation of herbal medicines: a worldwide review. WHO, Geneva

World Health Organization (2006) Supplementary guidelines on good manufacturing practices for the manufacture of herbal medicines. WHO, Geneva

Wu C, Guo L-J, Zhang Y (2006) Study on encapsulation of tannin for improving its stability. Chem World 12

Yan-yu X, Yun-mei S, Zhi-peng C, Qi-neng P (2006) Preparation of silymarin proliposome: a new way to increase oral bioavailability of silymarin in beagle dogs. Int J Pharm 319(1–2):162–168

Yotsawimonwat S, Rattanadechsakul J, Rattanadechsakul P, Okonogi S (2010) Skin improvement and stability of Echinacea purpurea dermatological formulations. Int J Cosmet Sci 32 (5):340–346

Part IV

Pharmaceutical Excipients from Nature

37 Pharmaceutical Formulation Development Based on the Polymers Obtained from Edible Plants: An Excellent Approach for the Betterment in Health-Care Services

Basudev Sahana and Kuhu Bhaduri

Abstract

Pharmacy is a professional health science that deals with the techniques and chemical sciences of preparing and dispensing drugs to ensure the safe and effective use of medicines. Pharmaceutical scientists have been engaged and much more interested in the field of pharmaceutical product development based on new molecule invention through synthesis or isolation from herbal sources, chemical modification of drugs or alternate use of drugs and use of different excipients obtained from plant sources especially polymers for proper administrable drug formulation development and to minimize the toxicity or health hazards of synthetic drugs and additives. The design and development depends upon the polymers for both the conventional and controlled release drug delivery systems, but the quality of drugs and polymers can affect the safety for applications. Formulation development based on the herbal drugs and polymers obtained from edible plants is the current interest in pharmaceutical formulation design and development. Edible drugs and polymers are playing a major role in several ways of herbal formulation design for potential improvement of efficacy like modern medicines. The present study is highlighting the edible and biocompatible herbal polymers and drugs utilized in mechanized controlled release and long-acting formulations for better health-care services.

Keywords

Edible polymers · Mucoadhesive · Biodegradable · Biocompatible · Nanoparticles

B. Sahana (✉) · K. Bhaduri
Department of Pharmaceutics, Institute of Pharmacy, Jalpaiguri, Govt. of West Bengal, Jalpaiguri, West Bengal, India
e-mail: sahanabasu33@rediffmail.com; bsahana_29@yahoo.com

S. C. Mandal et al. (eds.), *Evidence Based Validation of Traditional Medicines*,
https://doi.org/10.1007/978-981-15-8127-4_37

37.1 Introduction

Pharmaceutical product development based on **herbal constituents** and **excipients** is an important and emerging field for the betterment in health-care systems for modern human society (Hoover 1975) **to minimize toxicity and avoid unwanted health hazards.**

37.1.1 Pharmacy or Pharmaceutical Science

It is the branch of sciences which deals with drugs, chemical modification of drugs, better efficacious new molecule invention through synthesis or isolation and design and development of formulations for delivering the drug in proper administrable forms with the help of some nondrug components or additives according to the need of the patients. Pharmacy is the science and technique of preparing and dispensing drugs. It is a health professional science that deals with the techniques and chemical sciences to ensure the safe and effective use of drugs or medicines. The pharmacy practice includes compounding and dispensing medications and modern health-care services including clinical services, reviewing medications for safety and efficacy and providing drug information. **Pharmacists** are specialized personnel engaged in procurement, handling, testing, manufacturing, quality testing, storing, dispensing and distributing of drug products or medicines (Hoover 1975). They are the experts on drug therapy and primary health professionals who optimize use of medication and minimize toxicity for the benefit of the patients.

37.1.2 Pharmaceutics

This is the formulation science main branch of pharmacy which deals with the drugs, excipients and drug formulation design. Investigation of herbal and chemical drug molecules which are having sufficient pharmacologic or therapeutic response and studies regarding those molecules prior to the formulation for safe, economic, legal, efficacious and stable dosage form development are the major working fields of pharmacy. Pharmaceutical formulation design and development has been performed through the analysis of their physicochemical properties, drug-excipient compatibility studies and other pre-formulation testings to achieve a safe, efficacious, quality and stable dosage form. The safe, effective, legal and rational use of drug through dispensing a quality and stable dosage form to the patients is the major part of pharmaceutics. Biopharmaceutics and pharmacokinetics deal with formulation design according to the need of biosystems with proper dosage regimen and dosage interval adjustment. Clinical pharmacokinetics deals with individual dosage requirements and dosage intervals with selection of proper drug or drugs for optimization of drug efficacy and minimization of drug hazards or toxicity (Hoover 1975; Liberman and Lachman 1980).

37.1.3 Drugs

Drugs are the agents or substances used in prevention, control, diagnosis, mitigation, treatment and cure of disease or ailments through the applications or administration in the body of human beings or animals. Drugs are very rarely used in their pure forms and applied in proper administrable forms (medicines) with the help of some nondrug components used for giving proper shape, size, mass or volume of drug ingredient. Proper administrable dosage forms or medicines contain both active pharmaceutical ingredients (API) or drug components and nondrug components or additives or excipients or adjuncts or pharmaceutical aids. Drugs are used for disease prevention (BCG vaccines act to prevent tuberculosis), control (streptomycin and rifampicin act to control tuberculosis), diagnosis (barium meal X-ray helps to diagnose gastric or peptic ulcer), mitigation or relieving (paracetamol is used to relieve fever and pain) and treatment and cure (antibiotics and other drugs are used to treat and cure diseases) (Hoover 1975; Liberman and Lachman 1980).

37.1.4 Additives

Additives are nondrug components used to make a proper administrable dosage form for delivering the drug as **form stabilizers** or **form givers** such as suspending or emulsifying agents that help in the formation of suspension or emulsion; as **form disintegrators** like disintegrating agents which are used in tablets to break down the tablet in biological systems; as **solvents** (water, alcohol) which are used in solution preparation or monophasic liquid preparations and as **vehicles** (water) used in dispersion systems for carrying or dissolving the drugs in suspension or colloidal dispersion systems; as **bases** (white or yellow soft paraffin) used as carriers of drugs in semisolid preparation in which drug or drugs are incorporated or distributed evenly; as **diluents** used to dilute the drug as well as make up the volume or mass for proper shape or size of medicines; as **formulation stabilizers** which include **preservatives** (sodium benzoate), **antioxidants** (tocopherol) and **buffers** (citrate or phosphate) used to improve the stability of the product; as **organoleptic** additives for the improvement of the essence using **flavouring** agents (ethyl vanillin) or appearance using **colouring** agents (amaranth) or acceptability using **sweetening** agents (sucrose) of the product; and as **manufacturing additives** which are used to improve the process of manufacturing (lubricant or glidant in tablet manufacturing) (Raymond et al. 2012).

37.1.5 Polymers

Polymers are excellent dosage form devising materials (Jain 2006; Sahana and Bhaduri 2019a) used widely in conventional and controlled release pharmaceutical formulations of all categories such as solid, semisolid and liquid dosage forms. Conventional dosage forms used commonly and conveniently such as tablets, capsules, gels, solutions, colloidal dispersions, suspensions and emulsions are required to make use of polymers in their formulation. Polymers have been used in solubility enhancement of poorly soluble drugs; in suspension formulations as suspending, thickening and dispersants; and in emulsion preparations as emulsifying agents. Tablets have been prepared using polymers as binders and disintegrants as well as coating materials. Capsule shells are prepared with gelatin (nonvegcapsule) and semi-synthetic cellulosic polymers (vegcapsules). Semi-synthetic polymers like HPMC, HPMC k4, are used as gelling agents to prepare clear gels. Controlled release drug delivery systems have been devised through specially mechanized way totally based on polymers obtained from natural, semi-synthetic and synthetic sources and especially biodegradable and biocompatible polymers and polymers obtained from edible plants or vegetables. Controlled release formulations have been designed specially in such a way that makes the drug available for the target, providing the predetermined sufficient release rate and prolonging duration to produce the desired effect (Sahana and Bhaduri 2019a; Ali et al. 2009).

37.2 Formulation Development

Pharmaceutical formulation development is the most important and emerging area for pharmaceutical scientists and technologists engaged in the pharmaceutical research laboratory and industry materials (Jain 2006). They have constantly been engaged in devising dosage forms with the objective of maintaining the effective or therapeutic blood levels of drugs in the biological systems for prolonged period without causing dose dumping. Specially designed dosage form with controlled drug release at a predetermined rate and spatial placement of the drug formulations in the site of action or absorption fulfils the goal for providing an effective amount of drug to the proper site in the body to promptly achieve action and then maintain the desired drug concentration for a long time. Polymeric materials based drug delivery systems are now very much important and glorious fields among the several controlled drug delivery strategies such as specially devised micro- and nanoparticles, lyposomes and niosomes, matrices, pellets, floating systems, microemulsions, colloidal dispersions, liquid crystals, solid dispersions, nanosuspensions, transdermal systems, cyclodextrin inclusion complexes, osmotic pumps, etc. (Jain 2006; Sahana and Bhaduri 2019b; Ali et al. 2009). The spatial placement for enhanced residence period of dosage forms and long-term action can be achieved from the special formulation based on mucoadhesive (Nagai and Konishi 1987) polymers interacting with the mucus membrane that lines organs and body cavities such as the mouth, gut, rectum, genital area, nose and eye lids (Nagai and Konishi 1987).

37.2.1 Polymers in Devising Controlled Release Drug Delivery Systems

The study lays emphasis mainly on the polymers used in devising controlled release drug delivery systems (CRDDS) design and development with natural polymers (Sahana and Bhaduri 2019b; Mahajan et al. 2013; Sahana 2002). Specially designed dosage form using different grades of polymers with controlled release of drug at a predetermined rate and spatial placement of the drug formulations in the site of action or absorption fulfils the goal of providing an effective amount of drug to the proper site in the body to promptly achieve action and then maintain the desired drug concentration for a long time. Polymers are basic ingredients for designing and formulating the CRDDS which can provide extended duration of action and thus assure greater patient compliance (Jain 2006; Sahana and Bhaduri 2019b; Ali et al. 2009).

37.2.2 Advantages of Controlled Release Drug Delivery Systems

Mainly improvement in patient compliance, reducing dosing frequency, more consistent and prolonged therapeutic effect, reduction in health-care cost, decreased intensity of adverse effect and toxicity, better drug utilization, controlled rate and site of release and a greater selectivity of pharmacological activity (Jain 2006; Ali et al. 2009).

37.2.3 Disadvantages of Controlled Release Drug Delivery Systems

Increased variability among dosage units, stability problems, toxicity due to dose dumping, increased cost, more rapid development of tolerance and need for additional patient education and counselling (Jain 2006; Ali et al. 2009).

37.2.4 Rationale of Controlled Release Drug Delivery Systems

Ideally the optimization of therapeutic efficacy and safety of controlled release formulation is attained as a result of providing a nearly constant pharmacological response, thereby avoiding the normal peak and valley pattern associated with multiple dosing of conventional drug product. Thus the basic rationale of controlled drug delivery systems is to optimize the physicochemical, pharmacokinetic and pharmacodynamic properties of drug for maximizing the efficacy and minimizing the side effects by using the smallest quantity of drug administered by the most suitable route. Physicochemical characteristics such as solubility, diffusion, molecular weight, pKa, apparent partition coefficient and drug stability and pharmacokinetic characteristics such as elimination half-life, area under curve (AUC), total clearance, apparent volume of distribution (V_d), mean steady-state concentration, mean residence time, first pass effect, intrinsic absorption rate constant and dosage form index can be altered or changed and optimized for the development of controlled release formulations. Controlled release formulations have been designed in several ways of control such as dissolution controlled release, diffusion controlled release, both dissolution and diffusion controlled release, ion exchange resins, pH-independent formulations, pH-dependent controlled formulations, osmotically controlled release, altered density formulations, pro-drugs and delayed-release systems (Jain 2006; Ali et al. 2009).

37.3 Herbal Formulations

Herbal drugs (Table 37.1) and excipients (Table 37.2) have been utilized in new formulation development of both allopathic and Ayurvedic systems of medicines (Andrew et al. 2001). Presently, Indian traditional system of medicine Ayurvedic is becoming much more popular and acceptable as less hazardous for health. Herbal polymers and other excipients (Table 37.2) specially obtained from edible plants or vegetables have been becoming the excellent dosage form devising materials to minimize the hazards of synthetic materials.

Natural products (Table 37.1) specially obtained from edible plants have been utilized in the treatment of human diseases. Presently about 80% of people in developing countries especially in India has been using traditional medicines based on herbal drugs obtained from edible plants for their primary health care as India is a large repository of medicinal plants (Chopra et al. 1956). Formulations based on herbal drugs are currently in demand, and their popularity is increasing day by day due to toxicity and side effects of allopathic medicines. This led to the sudden increase in the number of herbal drug manufactures (Agarwal 2005). A number of scientific investigations have highlighted the importance and the contribution of many plant families, i.e. Asteraceae, Liliaceae, Apocynaceae, Solanaceae, Caesalpiniaceae, Rutaceae, Piperaceae and Sapotaceae, used as medicinal plants (Verma and Singh 2008). Medicinal plants play a vital role for the development of new drugs by extracting, isolating, standardizing and analysing the safety studies of the actual bioactive compounds.

37.3.1 Natural Polymers Obtained from Edible Plants for Formulation Development

Polymer molecule is made by linking many small units (monomers) together to form a large molecule. Polymers have been successfully employed in the solid, liquid and semisolid pharmaceutical formulations and specially used in the CRDDS. The polymers are mainly of three types: natural, semi-synthetic and synthetic. The natural polymers are obtained from the nature (plant or animal tissue, marine and mine sources), semi-synthetic polymers are chemically modified form of natural

Table 37.1 Medicinal plants used in formulations in better health-care systems

Sl. no	Name and identifying Picture	Parts used	Constituents	Uses	Formulations
01	*Acorus calamus* 	Rhizome	Volatile oil containing beta-asarone and alpha-asarone, saponins, lectins, sesquiterpenoids, lignans, steroids	To treat digestive disorders and pain	 Hifenac gel
02	*Aloe vera* *Aloe barbadensis miller* 	Dried leaf	Vitamins A (beta-carotene), C and E, which are antioxidants. *Anthraquinones* aloin and emodin, *hormones* auxins and gibberellins, cholesterol, campesterol, β-sitosterol and lupeol (Surjushe et al. 2008)	• Aloin and emodin act as analgesics, antibacterials and antivirals. Antioxidant accelerates the healing of burns, reduces dental plaque and constipation. May also improve skin and prevent wrinkles. Lowers blood sugar level..	 Patanjali Aloe Vera Gel
03	*Argemone mexicana* 	Fruit	*Argemone oil* or *katkar oil* contains the toxic alkaloids sanguinarine and dihydrosanguinarine. Four quaternary isoquinoline alkaloids, dehydrocorydalamine, jatrorrhizine, columbamine and oxyberberine	Treatment of several ailments including tumour, warts, skin diseases, inflammation rheumatism, jaundice, leprosy, microbial and fungal infections and malaria (Singh et al. 2010)	Extract cream and ointment (Alzomor et al. 2016) 

(continued)

Table 37.1 (continued)

Sl. no	Name and identifying Picture	Parts used	Constituents	Uses	Formulations
04	*Adhatoda vasica* 	Whole plant	Alkaloids vasicine and vasicinone	Bronchodilatory activity. Exhibits strong respiratory and uterine-stimulant activity and moderate hypotensive activity (Kumar et al. 2010)	*Vasaka* juice (*Vasa swarasa*) (Soni et al. 2008)  Vasaka leaf powder
05	*Curcuma amada* 	Rhizome	Phenolics, terpenoids and other bioactive constituents were hydroocimene, ocimene and myrcene (Aradhya and Singh 2011)	Antioxidant, antibacterial, antifungal, anti-inflammatory, anticancer, antidepressant, anti-tubercular and platelet aggregation inhibitory activities	Cooling effect, aggravates Vata.  Amba haldi powder
06	*Cinnamomum* *Cinnamomum zeylanicum* Blume (Family: Lauraceae)	Inner bark and leaf	Rich in monoterpenoids and phenylpropanoids. Cinnamaldehyde, eugenol, benzyl benzoate and linalool are the main components of the leaf oil (Niranjan and Prakash 2008)	Soothe an upset stomach, clear up urinary tract infections. Cinnamon helps people with diabetes metabolize sugar better (Mallavarapu and Rao 2007)	Cinnamon extract capsules 

07	*Curcuma longa* 	Rhizome	Curcumin, demethoxycurcumin and bisdemethoxycurcumin collectively known as curcuminoids (3–6%) are major polyphenolic compounds in turmeric rhizomes (Niranjan and Prakash 2008)	Exhibits antiparasitic, antispasmodic, anti-inflammatory, anticarcinogenic and gastrointestinal effects. Anti-HIV, antibacterial, antioxidant, nematocidal, antiparasitic, antispasmodic and anticarcinogenic activities	Good antifungal activity Khattak et al. (2005)  Turmeric curcumin powder
08	*Curcuma aromatic* 	Rhizome	Alpha-curcumene (ar-curcumene), beta-curcumene, D-camphor, alpha- and beta-turmerone	Used for its anti-inflammatory, wound healing, anti-tumour, anticancer and repellent activities (Sikha et al. 2015)	 *Curcuma aromatica* powder

(continued)

Table 37.1 (continued)

Sl. no	Name and identifying Picture	Parts used	Constituents	Uses	Formulations
10	*Turmeric Garcinia indica* 	Fruit	Rich source of compounds including garcinol, xanthochymol, isoxanthochymol and hydroxycitric acid. These are flavonoids, benzophenones, xanthones, lactones and phenolic acids	Garcinol has been studied for its anticancer, anti-ulcer, anti-oxidative and antiglycation activity	 Kokum butter
11	*Cassia lanceolata* 	Leaves	Senna contains four active crystalline anthraquinone glycosides, sennosides A, B,C and D	Laxative	Tab., capsules, chewable tab., Senna syrup 
12	*Gloriosa superba* 	Tuber and seed	Colchicine is the major compound isolated from the seed of this plant, and another important compound is gloriosine (Sarin et al. 1974)	Antithrombotic/anticoagulant	 Gloriosa superba-kalappai kilangu-seed

13	*Glycyrrhiza glabra* 	Root	Glycyrrhizin is comprised of a triterpenoid aglycone, glycyrrhetic acid conjugated to a disaccharide of glucuronic acid, glabridin, glabrene and glycyrrhizinic acid (Kaur et al. 2018)	Anti-ulcer activity Corticosteroid activity	 Yashtimadhu herbal powder
14	*Juniperus communis* 	Fruit	Monoterpenes alpha- and beta-pinene, sabinene, limonene, terpinen-4-ol, alpha-terpineol, borneol, geraniol, myrcene, camphene, camphor, alpha-eudesmol and many others	Juniper has been used for indigestion and as a steam inhalant in the management of bronchitis	 Juniper young shoot extract
15	*Withania somnifera* 	Vegetable rennet	Alkaloids (tropine, pseudotropine, withasomnine)	Used as tonic, aphrodisiac and sedative. Used in insomnia and blood pressure	 Ashwagandha extract
16	*Myrica nagi* 	Bark	Myricetin, myricitrin and glycosides	Used to prepare pickle and refreshing drinks. Cough-relieving product	 Koflyn

(continued)

Table 37.1 (continued)

Sl. no	Name and identifying Picture	Parts used	Constituents	Uses	Formulations
18	*Strychnos nux-vomica*  	Bark and seed	Strychnine, brucine. Minor alkaloids present in the seeds are protostrychnine, vomicine, n-oxystrychnine, pseudostrychnine, isostrychnine, chlorogenic acid and a glycoside	Spinal cord stimulant, bitter stomachic	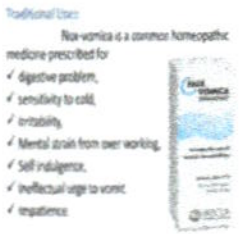 Nux Vomica Homaccord
19	*Piper longum* 	Fruit	Piperine is the major and active constituent	Used to improve appetite and digestion and treat stomachache, heartburn, indigestion, intestinal gas, diarrhoea, cholera and lung problems including asthma, bronchitis and cough	 *Piper longum* powder
20	*Phyllanthus amarus* 	Fruit	The plant contains high levels of saponins and tannins with low content of cyanogenic glycosides (Danladi et al. 2018)	Hepatoprotective effect, antidiabetic and hyperlipidemic activity, hyperuricemic effect, antiplasmodial activity	 *Phyllanthus niruri* Bhumyamlaki

21	*Rubia cordifolia* 	Madder root	Anthraquinone, terpenes, glycosides	Used for blood purifier activity and anticancer, astringent, antiacne, anti-inflammatory, anti-microbial, antidysentric, antiseptic, nephroprotective, antirheumatic and hepatoprotective activities	 Manjistha capsules
22	*Ricinus communis* 	Seed	80% of glyceride of ricinoleic acid	Purgative	 Pure castor oil
23	*Symplocos racemosa* 	Bark	Flavanol, glycosides like symplocoside, symposide, leucopelargonidin 3-O-glucoside, etc.	Treatment of diarrhoea, dysentery, liver diseases, uterine disease, leprosy, liver complaints, ophthalmia and conjunctivitis	 Organic lodhra capsules

(continued)

Table 37.1 (continued)

Sl. no	Name and identifying Picture	Parts used	Constituents	Uses	Formulations
24	*Rauvolfia serpentina* 	Root	Indole alkaloids (reserpine, rescinnamine)	Treatment of hypertension, sedative	Deserpidine tablets  Pure mother tincture
25	*Swertia chirata* 	Whole plant	Xanthones, secondary metabolites such as flavonoids, iridoid glycosides and triterpenoids	Antimalarial, hypoglycaemic, febrifuge, etc. *Swertia chirata* is a bitter tonic, carminative, laxative, anti-pyretic, antiperiodic, anti-inflammatory, stomachic and anthelmintic	Ayurvedic Kadu kirayata powder  Sage Chirata capsule
26	*Ocimum sanctum* 	Leaf and essential oil	Volatile oil (eugenol)	Antibacterial, stimulant expectorant, aromatic carminative, flavouring agent and used in cold	 Tulsi capsules

27	*Terminalia chebula* 	Bark and seed	Tannins (chebulanin, chebulagic acid), terpenoids (arjunic acid, arjunolic acid, terminolic acid), flavonoids (rutin, quercetin, etc.)	Kidney and liver dysfunctions. *Terminalia chebula* showed promising anti-microbial and anti-viral potentials	 Powder
28	*Tylophora purpurea* 	Root	Alkaloid tylophorine	Bronchial asthma and allergic rhinitis	Leaf extract  Tylophora Plus Capsules
29	*Zingiber officinale* 	Rhizome	Volatile oil, pungent principle (gingerol)	Spice, stimulant aromatic carminative	Ginger extract 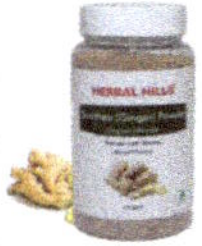 Ginger oil
30	*Vinca rosea* 	Leaf, seed and stem	Indole alkaloids (vinblastine, vincristine)	Used in the treatment of cancer	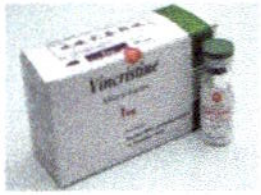 Oncovin vial

(continued)

Table 37.1 (continued)

Sl. no	Name and identifying Picture	Parts used	Constituents	Uses	Formulations
31	*Wedelia calendulacea* 	Leaf and root	Petals and pollen contain triterpenoid esters, carotenoids, flavoxanthin and auroxanthin. Leaves and stems contain other carotenoids and beta-carotene	The leaves are used in dyeing grey hair and promoting the growth of hair. Useful in coughs, cephalalgia, skin diseases and alopecia	Monalisa hair oil 
32	*Cumin* (*Cuminum cyminum*) 	fruit	Phenols (salicylic acid, gallic acid, cinnamic acid, hydroquinone, resorcinol, p-hydroxybenzoic acid, rutin, coumarine, quercetin) were isolated from seeds (Roberts and Martens 2016)	Stimulant, carminative and astringent and useful in dyspepsia and diarrhoea	 Cumin seed essential oil
33	*Plantago ovata* 	Seed coat	Mucilage	Used as laxative	 Psyllium husk capsules

All the figures are taken from the Internet site

polymers, and the synthetic polymers have been made synthetically in laboratory or industry (Raymond et al. 2012; Jain 2006).

1. Natural polymers—starch, cellulose, mucilage, pectin, gelatin, gum etc.
2. Semi-synthetic polymers—cellulose nitrate/acetate, ethyl cellulose, CMC, HPC and HPMC.
3. Synthetic polymers—polyethylene, polyvinyl chloride (PVC), nylon and PVA

Presently pharmaceutical formulations based on the use of natural polymers (Sahana 2002; Sahana and Bhaduri 2019b) have been becoming more popular and attractive because of their non-toxicity and potential biodegradability and biocompatibility. They have versatile advantages like cheap, safe and devoid of side effects, easy availability, flexibility in polymer chain, degrading in biological fluids, quickly eliminated by the kidney, forming a strong non-covalent bond with mucin-epithelial cell surfaces, soothing action and non-irritant nature. Although the natural polymers obtained from edible plant sources are very much advantageous for formulation development, some disadvantages are also present which should be taken under consideration: easy to microbial degradation, batch-to-batch variation, unreserved rate of hydration, heavy metal adulteration, sometimes immunogenic in nature, viscosity of the formulation reduced during the storage and most of the natural polymers are seasonal.

Excipient selection is based on the drug-excipient compatibility within an acceptable range for a given drug product. Drug-excipient compatibility studies have been usually performed for detecting the potential incompatibilities that can alter physical, chemical, microbiological or therapeutic properties of the drug in the formulations. Natural herbal polymers selection is too tough through the analysis of their physicochemical properties to achieve a quality, safe, efficacious and stable dosage form.

Natural excipients (Table 37.2) have certain advantages over other synthetic and semi-synthetic excipients. So these excipients are getting preference day by day. Potential characteristics of them help the scientists to formulate better formulation. But, while working with these excipients or using them in dosage form, it should pass all the regulatory requirements. Experiments and researches should be done with natural excipients in order to get the most safe and suitable one for pharmaceutical applications (Sahana 2002; Sahana and Bhaduri 2019b). Natural polymers can act numerously as excipients in pharmaceutical preparations.

37.3.2 Polymers Obtained from Edible Plants Used in Better Health-Care Systems

Edible plants like vegetables that are used in medicinal purposes containing mucilage, gum, pectin or other polymers and their utility for formulation development for better health-care services are highlighted in this work.

Table 37.2 Natural polymers and their pharmaceutical applications

Sl. No	Name and sources	Chemical nature	Identifying picture	Uses
1	Acacia gum: *Acacia senegal* and *Vachellia (Acacia) seyal*	Gum		Used as suspending agent, emulsifier, binder, demulcent and emollient and in osmotic drug delivery system
2	*Aegle marmelos* gum *(AMG)*	Fruit		Solubility enhancer
3	*Artocarpus heterophyllus* Lam	Fruit		Used as mucoadhesive polymer in the preparation of nasal gel containing antiemetic drug domperidone
4	Banana powder *(dehydrated)*	Fruit		Used as polymer in the treatment of gastric ulcer and diarrhoea
	Cashew gum *Anacardium occidentale* L.	Exudate gum		Binding agent in paracetamol tablet, gelling agent in aceclofenac topical gel formulation

5	Cellulose Green plants and algae	Polygalactoside 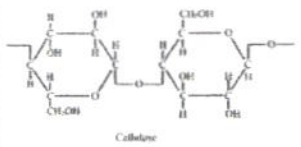 $(C_6H_{10}O_5)_n$		Binder, filler, diluents, thickening and viscosity imparting agent, compressibility enhancer, mucoadhesive agent
6	*Dillenia indica*	Fruit extract		*Oxytocin mucoadhesive nasal gel*
7	*Delonix regia*	Seed extracted polysaccharide		*Sustained release tablet containing antipsychotic drug*
7	Dextran	Polygalactoside 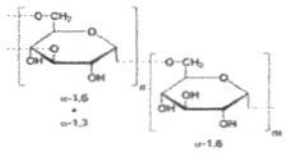		Colon-specific drug delivery

(continued)

Table 37.2 (continued)

Sl. No	Name and sources	Chemical nature	Identifying picture	Uses
8	Fenugreek seed Trigonella foenum-graecum L.	Mucilage		Nasal gel containing benzodiazepine derivative drug diazepam
9	*Ficus carica*	Fruit mucilage		Nasal in situ gel of midazolam
10	Guar um *Cyamopsis tetragonoloba*	Galactomannan polysaccharide 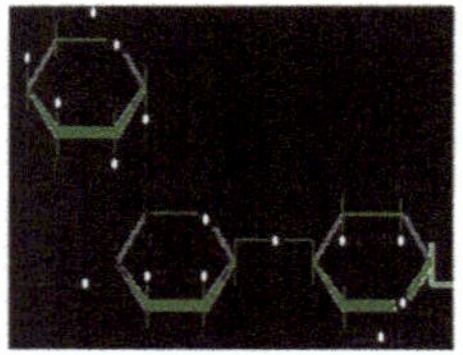		Used as binding, thickening, emulsifying, disintegrating agent and as laxative. Used in colon targeted drug delivery, cross-linked microspheres formulating agent. Thickening and stabilizing properties useful in the diltiazem SR tab
11	Gum agar *Gelidium cartilagineum*, *Gracilaria confervoides* and related red algae	Polygalactoside 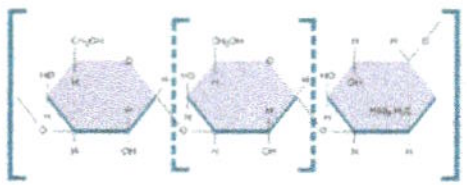		Used as suspending, emulsifying agent, lubricant, gelling agent, laxative, surgical lubricant; disintegrates, bacterial culture media

12	*Gum karaya (Sterculia urens)*	*Gum*		Gastric retentive formulations
13	Gum *odina* *Odina wodier*	Gum		Used as a suspending agent, emulsifier, mucoadhesive agent and coating material and used as prebiotic and probiotic
14	*Hibiscus rosa-sinensis* mucilage	*Mucilage*		Utilized as thickeners, suspending agent, water retention agent and disintegrants
15	*Ispaghula husk Plantago ovata*	*Mucilage*	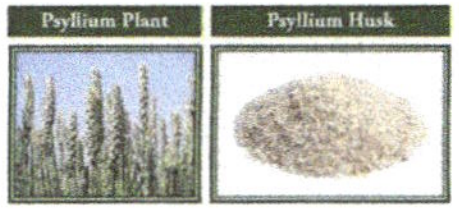	Used in the treatment of ulcerative colitis

(continued)

Table 37.2 (continued)

Sl. No	Name and sources	Chemical nature	Identifying picture	Uses
16	Jute leaf extract Corchorus olitorius L.	*Mucilage*		Gastrointestinal mucoadhesive tablets
17	*Lepidium sativum mucilage*	Mucilage	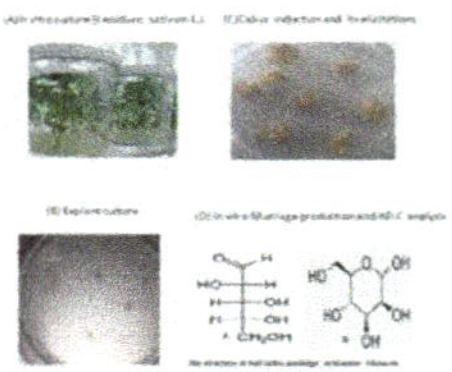	Utilized as thickeners and suspending agent
18	*Leucaena* seed gum	*Seed gum*		Used as emulsifier, suspending agent, binder, disintegrating agent
19	Locust bean gum *(Carob gum)* *Ceratonia siliqua* L. plant seed	Polygalactoside 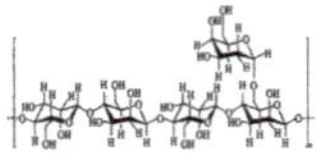		Used as gelling and thickening agent in food industry. Utilized as bioadhesive, to control drug release; and it enhances solubility of poorly water-soluble drugs

20	*Mangifera indica* gum *(MIG)(Mango)*	Gum powder		Tablet retardant polymer, sustained release of diclofenac sodium, utilized as disintegrant, binder, suspending agent and emulsifying agent
21	*Mango peel pectin* Pectin extracted from citrus fruits	D-galacturonic acid		Mucoadhesive agent specially for colon-specific drug delivery, controlled release formulation, patch and transdermal drug delivery nanoparticulate drug formulations
22	*Soy polysaccharide*	Cotyledon		Super disintegrants, used in nutritional products
23	*Soya seed extract* (*Glycine max*)	*Glycin* Seed extract		Soybeans are a globally important crop, providing oil and protein

(continued)

Table 37.2 (continued)

Sl. No	Name and sources	Chemical nature	Identifying picture	Uses
24	Starch *Solanum tuberosum* (potato), *Zea mays* (corn), *Oryza sativa* (rice)	Polysaccharide 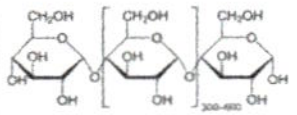		Binder, disintegrating agent, aiding drug delivery, film coating material, mucoadhesive agent
25	Tamarind seed gum	Seed extracted polysaccharide		Matrix tablet containing diclofenac sodium (Sikha et al. 2015). Used as binding, coating and gelling agent
25	Tragacanth powder Astragalus adscendens, Astragalus gummifer Astragalus brachycalyx and Astragalus tragacantha	Polysaccharides		Suspending agent, emulsifier, demulcent, emollient, sustained release formulating agent as drug carrier
26	*Trigonella foenum-graecum* L.	Seed		Diazepam *mucoadhesive* nasal gel

27	Vine spinach leaf extract Basella alba L. Basella rubra L.	Mucilage		Gastrointestinal mucoadhesive tablets
28	*Xanthan gum/Gellan gum* *Pseudomonas elodea/* *Xanthomonas campestris*	Exocellular polysaccharide 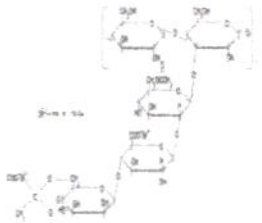		Metoclopramide hydrochloride in situ nasal gel for antimigraine activity. Metoprolol tartrate SR tab

All the figures are taken from the Internet site

37.3.2.1 Polymers Obtained from Natural Sources

Polysaccharide-based polymers: agarose, alginates, alginic acid, carrageenan, cellulose, cyclodextrins, dextran, hyaluronic acid, lectin, microcrystalline cellulose, pectin, polysialic acid, soluble starch.

37.3.2.2 Agarose

Agarose is a complex group of polysaccharides derived from seaweed, extracted from the agarocytes of Rhodophyceae, marine algae found predominately in the Pacific and Indian Oceans. It is used as a matrix to encapsulate cells for cartilage tissue engineering (Malviya et al. 2010; Buschmann et al. 1992).

37.3.2.3 Alginate

Alginate is a group of naturally occurring anionic polysaccharides derived from brown algae cell walls (Szekalska et al. 2016); seaweed (Lee and Mooney 2012) including *Macrocystis pyrifera*, *Laminaria hyperborea* and *Ascophyllum nodosum* (Nagai and Konishi 1987); and several bacteria strains such as *Azotobacter* and *Pseudomonas* (Remminghorst and Rehm 2006). Alginates are useful polymer as viscosity-increasing agents, thickeners and suspension and emulsion stabilizers in food and pharmaceutical industry (Repka and Singh 2009). ALG is useful for its mucoadhesive properties to prepare buccal, nasal, ocular, gastrointestinal and vaginal mucosa tissue (Martin-Villena et al. 2013).

37.3.2.4 Carrageenans

Carrageenans are high-molecular-weight polysaccharides made up of repeating galactose units obtained from several species of red seaweeds. It is useful for tissue engineering, wound healing, growth factor, drug delivery systems, immobilization of enzymes and encapsulation of cells for in vivo delivery (Aramwit 2016). The gelling, thickening, stabilizing and strong interaction with protein properties of carrageenans are widely useful in the food industry, dairy and meat products.

37.3.2.5 Cellulose

Cellulose is the most abundant polysaccharide found in nature, a polymer of long, unbranched chains of β-D-glucose, oriented with CH_2OH groups alternating above and below the plane of the cellulose molecule. The absence of side chains in cellulose molecules brings them close to each other to form rigid structures. Cellulose is mainly used to produce paper and paperboard (Strydom et al. 2011).

37.3.2.6 Chitosan

Chitosan is a natural polycationic copolymer consisting of glucosamine and N-acetyl glucosamine units, mostly obtained by deacetylation of chitin derived from the exoskeleton of crustaceans. Chitosan is having valuable properties of biocompatibility and biodegradability for devising nanoparticulate pharmaceutical formulations (Rani et al. 2010).

37.3.2.7 Cyclodextrins

Cyclodextrins are naturally available in three forms as alpha, beta and gamma consisting of six, seven and eight D-glucopyranose residues, respectively, useful for carrying drug materials. It can alter physical, chemical and biological properties of active molecules through the formation of complexes. Cyclodextrin-tamoxifen citrate complex is utilized for enhancing the solubility of tamoxifen citrate (Mondal et al. 2018). The different hydrophilic, hydrophobic and ionic derivatives of cyclodextrin are used as excellent devising agents for novel drug carriers to administer drugs in various routes such as oral route of drug delivery (prednisolone, to reduce gastric ulceration) and nasal drug delivery (midazolam, to enhance bioavailability of drug) (Tiwari et al. 2010).

37.3.2.8 Dextran

Dextran is an extracellular bacterial homopolysaccharide (Seymour and Knapp 1980) used for devising sustained release dosage forms of proteins, vaccines, and drugs for its water solubility, biocompatibility and biodegradability properties.

37.3.2.9 Hyaluronic Acid

Hyaluronic acid is a major component of the skin involved in tissue repairing and is found throughout the body in various tissues and fluids and binds to specific cell surface receptors. Hyaluronic acid is the only non-sulphated glycosaminoglycan that consists of repeating units of *N*-acetyl-d-glucosamine and d-glucuronic acid and is degraded in the presence of hyaluronidases. Hyaluronic acid hydrogels are readily fabricated as microspheres, sponges and fibres depending on the intended application (Burdick and Stevens 2005).

37.3.2.10 Lectins

Lectins are cell-agglutinating proteins of nonimmune origin that bind mono- and oligosaccharides reversibly and with high specificity. They also precipitate polysaccharides and glycoproteins. Bean lectins agglutinate erythrocytes owing to their ability to bind with cell-surface glycoproteins and glycolipids. Lectins are useful in controlled release drug delivery systems and in targeted drug delivery systems.

37.3.2.11 Microcrystalline Cellulose

Microcrystalline cellulose is playing an important role in formulation design of solid dosage forms as multifunctional excipients such as compressibility enhancer, binder in granulations, thickeners and flow enhancer (Wan and Prasad 1988).

37.3.2.12 Pectin

Pectin is a high-molecular-weight carbohydrate polymer which is present in virtually all plants where it contributes to the cell structure. The gelling properties of pectin have been known for centuries, but the isolation of commercial pectin only started at the beginning of the twentieth century. Pectin is used as viscosity-improving agent and as suspending agent used in mucoadhesive drug delivery systems.

37.3.2.13 Polysialic Acid

Polysialic acid is a linear polymer of sialic acid, covalently bound to proteins as a post-translational modification. (1) It is widely expressed in nature in bacterial capsules, fish, sea urchin eggs, embryonic tissues, amphibians, animal and human brains and in a variety of cancers. (2) Polysialic acid, a homopolymer of α-2,8-linked sialic acid, is one of the carbohydrates expressed on neural precursors in the embryonic and adult brain. Polysialic acid, synthesized by two polysialyltransferases (ST8SiaII and ST8SiaIV), mainly modulates functions of the neural cell adhesion molecule. Polysialic acid-based micelles for encapsulation of hydrophobic drugs will be helpful for better therapeutic efficacy (Bader et al. 2011).

37.3.2.14 Soluble Starch

Soluble starch is a branched polysaccharide composed of two substances: amylose and amylopectin. Natural starch contains 10–20% amylase and 80–90% amylopectin. Amylose forms a colloidal dispersion in hot water, whereas amylopectin is completely insoluble. Starches are hydrolysed to simple sugars using acids or enzymes as catalysts. It plays an important role in pharmaceutical formulation development as binder, disintegrant and strong film-forming agent (Jane 1995).

37.4 Herbal Polymers Utilized in Modern Pharmaceutical Formulations

Herbal polymeric materials are very much advantageous for controlled release drug delivery systems and retardant mucoadhesive formulations in several ways such as edible and biodegradable, degrading in biological fluids with progressive release of dissolved or dispersed drugs, surgical removal of mucoadhesive device after releasing the drug is not required, diffusion controlled delivery systems which are excellent means of achieving predetermined rates of drug delivery. New oxytocin-containing nasal gel formulation has been developed using natural mucoadhesive agent obtained from the fruit of *Dillenia indica* (Ketousetuo and Bandyopadhyay 2007). Nasal gel containing benzodiazepine derivative drug diazepam and natural mucoadhesive agent obtained from fenugreek seed (*Trigonella foenum-graecum* L.) was developed (Datta and Bandyopadhyay 2005). In situ nasal gel of midazolam with natural mucoadhesive material obtained from mucilage of *Ficus carica* fruit (Basu and Bandyopadhyay 2010) and antimigraine drug metoclopramide HCl with mucoadhesive material xanthan gum have been developed and characterized (Parmar and Lumbhani 2012). Nasal gel containing antiemetic drug domperidone (Sabale et al. 2012) with natural polymer obtained from mucilage of jackfruit tree has been developed and mucoadhesive tablets of metformin using jackfruit latex has been developed.

37.4.1 Herbal Mucoadhesive Polymers Used in Formulation Development

The uses of two natural mucoadhesive agents, jute leaf extract and vine spinach leaf extract, and two synthetic mucoadhesive agents, carbopol 934 and carbopol 940, for the development of gastrointestinal mucoadhesive tablets for lingering the residence period of the formulations were highlighted for the comparison of herbal and synthetic polymers. The 1.0%w/v solution of those mucoadhesive agents was used as binder for granulations of the tablets, and the 3.0%w/v solutions were used as coating agents of the adhesive tablets of barium sulphate (Sahana 2002). The barium sulphate oral mucoadhesive tablets were administered to the rabbit. The adhesive tablets of barium sulphate adhered for a long time to the mucosal absorptive membrane of the gastrointestinal tract of the rabbit, which was observed by X-ray plate analysis of the rabbit's gastrointestinal tract after specific time of interval of tablet administration. Barium sulphate adhered up to 8 h in the small intestine; that is why it may be concluded that any kind of drug will be absorbed from the small intestine for a long time up to 8 h or more. It was also observed that the adhesive action of the two leaf extracts is much longer than the two synthetic polymers. This will be the potential alternative of extending the drug action for a long time through long-term absorption.

37.4.2 Herbal Novel Drug Delivery Systems: Phytosomes or Phytolipids

Drug delivery systems for the betterment in drug actions through improved bioavailability due to high lipophilicity are formulated with lipophilic complexes of components of plant origin like *Silybum Marianum, Ginkgo biloba, ginseng, etc.* with natural herbal phospholipid (Semalty et al. 2009).

37.4.3 Novel Ayurvedic Medicines

Herbal drugs or purified phytochemicals have been incorporated in novel drug delivery systems in Indian Ayurvedic medicines to combat serious diseases. Herbal novel drug delivery systems (Semalty et al. 2009) such as buccal tablet, mouth dissolving tablets, sustained release formulations, mucoadhesive drug delivery systems, transdermal dosage forms, micro- and nanoparticulate drug delivery systems, etc. based on herbal drugs and herbal excipients have a lot of potential for the betterment in health-care systems.

37.5 Future Scope

Herbal excipients and herbal drugs are most useful in the betterment of health-care services by avoiding the health hazards of synthetic materials. In the near future, it will be the potential alternative of modern medicine through the applications of special techniques in designing and fabrication of herbal medicines to achieve safe, efficacious, stable and legal quality formulations.

37.6 Conclusion

Presently the novel drug delivery systems based on herbal drugs and excipients have been utilized widely all over the world for their better therapeutic value as they have less adverse effects as compared with synthetic drugs and excipients. Research on natural herbal drugs and excipients is the current interesting field, and the use of herbal formulation is the potential alternative in place of synthetic and allopathic formulations. Medicinal herbs are potential sources of therapeutic aids, and the excipients obtained from herbal sources specially from edible plants have a great and significant role in designing and development of special formulations such as mucoadhesive drug delivery systems, controlled release drug delivery systems, micro- and nanoparticulate formulations, etc. which have been uprising the pharmaceutical sciences for the betterment in health-care services. Herbal formulations need to overcome the problems regarding dosing and dosing interval, clinical pharmacokinetic studies, bioassays for biological standardization, pharmacological and toxicological evaluation, investigation of their sites of absorption, toxicity and safety evaluation and legal and regulatory aspects of herbal drugs.

References

Agarwal A (2005) Critical issues in quality control of herbal products. Pharma Times 37(6):9–11

Ali J, Khar RK, Ahuja A (2009) Dosage form design. Birla Publications Pvt. Ltd. ISBN 10: 8186270752, ISBN 13: 9788186270752

Alzomor AK, Moharem AS, Noman NM, Alafifi J, Mokred H, Alaregi N (2016) Extraction, formulation and evaluation of Argemone Mexicana Linn leaves as antimicrobial cream and ointment. Int J Pharm Res BioSci 5(4):37–51

Andrew V, Catherine Z, Roberta L (2001) Herbal medicine. West J Med 175(2):125–128. https://doi.org/10.1136/ewjm.175.2.125

Aradhya SM, Singh L (2011) Mango ginger (Curcuma amada Roxb.) a promising spice for phytochemicals and biological activities. RS Policegoudra. J Biosci 36(4):739–748

Aramwit P (2016) Wound healing biomaterials, vol 2. Woodhead publishing, pp 3–38

Bader RA, Silvers AL, Zhang N (2011) Polysialic acid-based micelles for encapsulation of hydrophobic drugs. Biomacromolecules 12(2):314–320

Basu S, Bandyopadhyay AK (2010) Development and characterization of Mucoadhesive in situ nasal gel of midazolam prepared with Ficus carica mucilage. AAPS PharmSciTech 11 (3):1223–1231
Burdick JA, Stevens MM (2005) Biomaterials artificial organs and tissue engineering. Biomedical hydrogels Science Direct:107–115
Buschmann MD, Gluzband YA, Grodzinsky AJ, Kimura JH, Hunziker EB (1992) Chondrocytes in agarose culture synthesize a mechanically functional extracellular matrix. J Orthop Res 10 (6):745–758
Chopra RN, Nayar SL, Chopra IC (1956) In glossary of Indian medicinal plants, vol 1. Council of Scientific and Industrial Research, New Delhi, p 197
Danladi S, Idris MA, Umar II (2018) Review on pharmacological activities and phytochemical constituents of Phyllanthus niruri (Amarus). J Phytopharmac 7(3):341–348
Datta R, Bandyopadhyay AK (2005) Development of a new nasal drug delivery system of diazepam with natural mucoadhesive agent from Trigonella foenum-graecum L. J Sci Ind Res 64:973–977
Hoover JE (ed) (1975) Remington's pharmaceutical sciences. Mack Publishing Co
Jain NK (ed) (2006) Pharmaceutical product development. CBS Publishers and Distributers, New Delhi
Jane J (1995) Starch properties, modifications, and applications. J Macromol Sci 32(4):751–757
Kaur R, Kaur H, Dhindsa AS (2018) Glycyrrhiza glabra: a phytopharmacological review. International journal of pharmaceutical sciences and research, projected impact factor 0.83, citescore: 0.27
Ketousetuo K, Bandyopadhyay AK (2007) Development of oxytocin nasal gel using Mucoadhesive agent obtained from the fruits of Dillenia indica L. ScienceAsia 33:57–60
Khattak S, Saeed-ur-Rehman, Ullah Shah H, Ahmad W, Ahmad M (2005) Biological effects of indigenous medicinal plants Curcuma longa and Alpinia galanga. Fitoterapia 76:254–257. [PubMed]
Kumar KPS, Bhowmik D, Chiranjib TP, Kharel R (2010) Indian traditional herbs Adhatoda vasica and its medicinal application. J Chem Pharm Res 2(1):240–245
Lee KY, Mooney DJ (2012) Alginates: properties and biomedical applications. Prog Polym Sci 37:106–126
Liberman HA, Lachman L (1980) Pharmaceutical Dosage Forms: Marcel Dekker
Mahajan P, Kaur A, Aggarwal G, Harikumar SL (2013) Mucoadhesive drug delivery system: a review. J Drug Dev Res 5(1):11–20
Mallavarapu GR, Rao BRR (2007) Chemical constituents and uses of Cinnamomum zeylanicum Blume. In: Jitrovetz L, Dung NX, Varshney VK (eds) Aromatic plants from asia their chemistry and application in food and therapy, 1st edn. Har Krishan Bhalla & Sons, Dehradun
Malviya R, Srivastava P, Bansal V, Sharma PK (2010) Formulation, evaluation and comparison of sustained release matrix tablets of diclofenac sodium using natural polymers as release modifier. Int J Pharm Bio Sci 1(2):1–8
Martin-Villena MJ, Fernandez-Campos F, Calpena-Campmany AC, Bozal-de Febrer N, Ruiz-Martınez MA, Clares-Naveros B (2013) Novel microparticulate systems for the vaginal delivery of nystatin: development and characterization. Carbohydr Polym 94(1):1–11
Mondal L, Mukherjee B, Chakraborty S, Bhattacharya S, Ehsan I, Sengupta S, Pal MM (2018) Comparison of enhanced solubility profile analysis of thermodynamic parameters and pharmacokinetic profile related to tamoxifen citrate solubilisation. Nov Appro Drug Des Dev 3(5)
Nagai T, Konishi R (1987) Buccal/gingival drug delivery systems. J Control Release 6:353–360
Niranjan A, Prakash D (2008) Chemical constituents and biological activities of turmeric (Curcuma longa L.)—a review. J Food Sci Technol 45(2):109–116
Parmar V, Lumbhani AN (2012) Development and evaluation of ion- dependent in- situ nasal gelling systems of metoclopramide hydrochloride as an antimigraine model drug. Int J Latest Res Sci Technol 1(2):80–89
Rani M, Agarwal A, Negi YS (2010) Review- chitosan based hydrogel polymeric beads – as drug delivery system. Bioresources 5(4):2765–2807

Raymond CR, Sheskey PJ, Cook WG, Fenton ME (2012) Handbook of pharmaceutical excipients. Pharmaceutical Press, London

Remminghorst U, Rehm BHA (2006) Bacterial alginates: from biosynthesis to applications. Biotechnol Lett 28(21):1701–1712

Repka MA, Singh A (2009) Alginic acid. In: Handbook of pharmaceutical excipients, 6th edn. Pharmaceutical Press, London, pp 20–22

Roberts JJ, Martens PJ (2016) Biosynthetic polymers for medical applications. Engineering biosynthetic cell encapsulation systems, Science Direct, Woodhead Publishing Series in Biomaterials, pp 205–239

Sabale V, Patel V, Paranjape A (2012) Isolation and characterization of jackfruit mucilage and its comparative evaluation as a mucoadhesive and controlled release component in buccal tablets deetu. J Pharm Sci 2(2):61–69

Sahana B (2002) Design and development of oral mucoadhesive drug delivery systems with some natural mucoadhesive agents. Jadavpur University

Sahana B, Bhaduri K (2019a) Polymers: excellent formulations devising agent. Int J Pharm Res Sch 8(2):1–15

Sahana B, Bhaduri K (2019b) Special approach to linger residence period if drug formulations for enhancement of bioavailability and therapeutic activity based on mucoadhesive polymers. Int J Pharm Res Sch 8(1–4):1–17

Sarin YK, Jamwal PS, Gupta BK, Atal CK (1974) Colchicine from the seeds of Gloriosa superba. Curr Sci 43:87–90

Semalty A, Semalty M, Rawat BS, Singh D, Rawat MS (2009) Pharmacosomes: the lipid-based new drug delivery system. Expert Opin Drug Deliv 6(6):599–612

Seymour FR, Knapp RD (1980) Structural analysis of dextran from strains of Leuconostoc and related genera that contain 3-O-glucosylatedd-glucopyranosyl residues at the branched points, or in consecutive, linear position. Carbohydr Res 81:105–129

Sikha A, Harini A, Hegde Prakash L (2015) Pharmacological activities of wild turmeric (Curcuma aromatica Salisb): a review. J Pharmacogn Phytochem 3(5):01–04

Singh S, Singh TD, Singh VP, Pandey VB (2010) Quaternary alkaloids of Argemone mexicana. Pharm Biol 48(2):158–160

Soni S, Anandjiwala S, Patel G, Rajani M (2008) Validation of different methods of preparation of *Adhatoda vasica* leaf juice by quantification of Total alkaloids and Vasicine. Indian J Pharm Sci 70(1):36–42

Strydom SJ, Otto DP, Liebenberg W, Lvov YM, Villiers MM (2011) Preparation and characterization of directly compactible layer-by-layer nanocoated cellulose. Int J Pharm 404(1–2):57–65

Surjushe A, Vasani R, Saple DG (2008) Aloe vera: a short review. Indian J Dermatol 53 (4):163–166. https://doi.org/10.4103/0019-5154.44785. PMCID: PMC2763764.PMID: 19882025

Szekalska M, Puciłowska A, Szymanska E, Ciosek P, Winnicka K (2016) Alginate: current use and future perspectives in pharmaceutical and biomedical applications -review article. Int J Polymer Sci 2016:1–17

Tiwari G, Tiwari R, Rai AK (2010) Cyclodextrins in delivery systems: applications. J Pharm Bioallied Sci 2(2):72–79

Verma S, Singh SP (2008) Current and future status of herbal medicines. Vet World 1(11):347–350

Wan LSC, Prasad KPP (1988) Effect of microcrystalline cellulose and cross-linked sodium carboxymethylcellulose on the properties of tablets with methylcellulose as a binder. Int J Pharm 41:159–167

Evaluation of Bioactivity of Green Nanoparticles Synthesized from Traditionally Used Medicinal Plants: A Review

38

Moumita Chatterjee and Ambarish Mukherjee

Abstract

Contemporarily the worldwide growth of interest in chemically synthesized nanoparticles (NPs) for use in therapeutic applications is noteworthy in spite of the broad range of toxicity imposed by them in vertebrates and invertebrates so far tested. Further trend of advancement aimed toward getting rid of toxicity for which green synthesis of metal nanoparticles was indispensable as a cost-effective and better environment-friendly alternative. Interestingly the 'green' chemical techniques involving traditionally used medicinal plants for synthesizing nanoparticles are getting more and more accepted far and wide for being simple, one-step, cost-effective, environment-friendly, and relatively reproducible to handover more stable materials. So far there have been important publications evaluating rationally the efficacy of green nanoparticles synthesized from plants used traditionally in ethnomedicine over the antibiotics and anticancer chemotherapeutic drugs in use that are not only very costly and toxic but also with the possibilities of facing drug resistance problems. The present review provides information on green nanoparticles synthesized from plants, their bioactivity, and ethnomedicinal uses.

Keywords

Ethnomedicine · Green nanoparticles · Bioactivity · Antibiotics · Chemotherapeutic drugs

M. Chatterjee (✉)
Institute of Wood Science and Technology, Bangalore, Karnataka, India
e-mail: moumitataxonomy@gmail.com

A. Mukherjee
Department of Botany, UGC Centre for Advanced Study (Phase II), The University of Burdwan, Burdwan, West Bengal, India
e-mail: amukherjee.bu@gmail.com

S. C. Mandal et al. (eds.), *Evidence Based Validation of Traditional Medicines*,
https://doi.org/10.1007/978-981-15-8127-4_38

Abbreviations

AgNP	Silver nanoparticle
AuNP	Gold nanoparticle
CuNP	Copper nanoparticle
NPs	Nanoparticles
ZnNP	Zinc nanoparticle

38.1 Introduction

Realizing the importance of traditional knowledge at the disposal of ethnic communities regarding medicinal use of plants in one hand and modern technology regarding patient-friendly therapeutic use of green nanoparticles on the other, the present authors decided to review the publications of concern for innovative contemplation of medical scientists working hard for conveying benevolence to mankind.

Understanding medicinal potential of a plant in the surroundings of man has been the subject addressed by him almost since the dawn of his consciousness as a part of his struggle for existence. The direct "man-plant" relationship formed the basis for discovery of different types of species serving as source of food, medicine, fodder, fuel, fiber, floss, timber, and other phytoresources essential for sustenance of life. Concomitantly with sociocultural advancement, science and technology started exploring the chemical weapons produced by plants to survive and reveal their pharmacological effects and bioactivities. While doing so, series of secondary metabolite were identified belonging to such categories as alkaloids, phenols, flavonoids, tannins, terpenoids and glycosides, etc. that can thrill, kill, and heal man. Scientific endeavors started establishing their pharmacological effects that paved the pathway for isolation, attenuation, characterization, and therapeutic proving and hence resulted in discovery of active principles constituting medicines suitable for use against certain ailments/diseases.

The advent of nanotechnology in medicine in the recent past has made us very hopeful regarding the exciting therapeutic possibilities offered by different types of nanoparticles (NPs) that subsequent to synthesis have dimensions less than 100 nm. This review aims to highlight the bioactivity of nanoparticles synthesized from plants to the ethnomedicinal uses which have been literally documented.

38.1.1 Ethnomedicine

Ethnomedicine is a discipline to include an age-long indigenous therapeutic system in vogue among folk and ethnic communities, thus referring to the study of traditional medical practices with the cultural interpretation of health, illness and diseases, as well as their healthcare and healing practices (Krippner 2003). It is a complex multidisciplinary system that includes the plants usage, spirituality, and the natural environment which lead to the healing source for people for millennia.

In India, ethnomedicine has been nurturing age long the codified traditional systems like Unani, Ayurveda, and Siddha and also enabling development of modern medicine. Since the ethnomedicinal knowledge has been getting orally transmitted through generations with progressive extinction keeping parity with modernization, its documentation for validation has been presently prioritized. Since medicinal uses of plants documented in Ayurvedic literature have been responding positively to pharmacological evaluation and clinical trials for validation, WHO and UNESCO have resolved to nurture this Indian Traditional System of Medicine on global basis. According to Uniyal et al. (2002), study discovery of multiple vital drug of modern age is linked with the Great Sage Charak knowledge. This resolution has evoked immense rejuvenation of interest in ethnomedicine all over the world for scientific validation of ethnopharmacy-based medicines and their standardization with optimum proportions of active ingredients (secondary metabolites) for inclusion in modern herbal-healthcare system. The health-conscious people in modern societies have been showing involvement especially with Ayurvedic medicines for their prophylactic, palliative, and immunomodulatory actions, amelioration of refractory diseases, and antioxidant activities (Mukherjee 2012).

Presently the uses of traditional medicines of plant origin for primary health care are about 80%. Users of traditional medicine for health care have a good numerical representation being 40% of the populations in China and Cambodia as well and 71% in Chile. India as one of the mega-biodiversity countries of the world is having rich vegetation with a wide diversity of species having medicinal value. Rural people depend extensively on wild plants as sources of food, medicine, fodder, and fuel for which they have been trying to work out methods of resource management, which may be basic to the conservation of some of the world's major habitats. In India, about 65% of the populations in rural areas are known to use medicinal plants to fulfill their needs for primary health care. There have been important publications concerning ethnomedicinal plants of India (Ahirwar et al. 2010; Anonymous 1976, Bhattacharya 2004; Biswas and Mukherjee 2017; Borgohain et al. 2016; Borthakur 1976a, b, 1981, 1992, 1996a, b; Borthakur and Goswami 1995; Borthakur and Sarma 1996; Borthakur and Nath 2007; Borthakur et al. 2004; Chopra and Simon 2002; Das and Pandey 2007; Das et al. 2010; Gautam et al. 1998; Jain 1991, 1994, 2004; Jain and Jain 2015; 2016a; b; Jain and Srivastava 1999; Kapoor 1990; Kaushik and Dhiman 1999; Kumar 2014; Lalramnghinghlova 2001; Lalramnghinghlova and Jha 1997; Nadkarni 1992; Nath 2015; Naik 2004; Nair and Mohanan 1998; Rana 2003; Sahu et al. 2015; Satyavati and Gupta 1987; Shah 1992; Swami Brahmananda 2002; Zafar 1999). Ethnobotanical investigations have also been revealing edible species capable of preventing and curing various diseases and disorders with restoration of health. The nutritional and medicinal importance of wild edible plants started subsequently getting revealed (Jain et al. 1977; Arora and Pandey 1996; Arjariya and Rawat 2005; Devi and Salam 2016). Efficacy of ethnomedicines, especially of India, has been driving most of the corporate world crazy for profitable acquisition of the crude drugs along with the indigenous knowledge about their use, thus eventually indirectly sensitizing India to arouse consciousness about her own knowledge on plant resources (Mukherjee and Banerjee 2019).

38.1.2 Nanotechnology in Medicine

The use of nanotechnology in medicine offers some challenging therapeutic prospects. In recent years, NPs have emerged as promising agents in modern medicine, with applications in drug delivery, diagnosis, cell repair, and wound treatment, and as antibacterials. In biosciences, organic dyes are getting replaced by nanoparticles since organic dyes require high photostability as well as high multiplexing properties. Nanotechnology promises important role in agriculture and food industry by innovation of such new techniques as "precision farming techniques" augmenting the ability of plants to absorb nutrients and make more efficient and targeted use of inputs. The technology has been facilitating detection and control of diseases, imparting the ability to withstand environmental pressures and providing effective systems for processing, storage, and packaging of products.

Conventionally nanoparticles can be synthesized by various physical, chemical, and biological methods. The production of nanoparticles using plant extract (green nanoparticle) is a better alternative to the conventional methods. These particles have no harmful effects on environment. Moreover, green nanoparticles have a wide range of applications, viz., as detectors, catalysts, surface-coating agents, anticancer substances, antibacterial, antifungal, etc.

The success of nanoparticles in drug release is rooted in its ultra-dimensions. Drug release is known to be affected by particle size. Smaller particles have a larger surface area-to-volume ratio; therefore, most of the drug associated with small particles would be at or near the particle surface, leading to faster drug release. In contrast, larger particles with large cores allow more drugs encapsulation per particle and give slower release rate (Redhead et al. 2001). Thus, particle size is a vital issue for regulating drug release rates.

Some of the most studied metallic nanoparticles include silver (Ag) (Ajitha et al. 2015; Feng et al. 2013), gold (Au) (Moreira dos Santos et al. 2012), zinc (Zn) (Fakhari et al. 2019; Ezealisiji et al. 2019), and copper (Cu) (Chung et al. 2017).

Plant extracts may act both as reducing agents and stabilizing agents in the synthesis of nanoparticles (Jirovetz et al. 2003). Compared with other NPs, CuNP synthesis has drawn attention since their useful properties are achievable at costs lower than those of silver and gold (Han et al. 2006). Synthesis of gold nanoparticles (AuNP) using plant extract is useful for two reasons, firstly it has almost no environmental effects, and secondly it can be used to produce large quantities of nanoparticles, in spite of high cost, gold nanoparticles have revolutionized the field of medicine due to their efficacy in widespread applications in targeted drug delivery, imaging, diagnosis, and therapeutics. Moreover, their extremely small size, high surface area, stability, non-cytotoxicity, and tunable optical, physical, and chemical properties (Thakor et al. 2011; Huang and El-Sayed 2010) add immensely to their credentials. Among different types of nanoparticles, zinc nanoparticles (ZnNPs) are versatile semiconductors in the sense that they can display significant optical transparency and luminescent properties in UV-visible (UV-vis) regions (Cao et al. 2011) and excellent potential as chemical and thermal stable compound (Nunes et al. 1999). Easy accessibility is the main advantage of synthesize silver nanoparticles (AgNP) using plant extracts. Moreover, AgNPs are rendered safety (non-toxicity) in most cases by the concerned plants which have different types of metabolites that can help in the reduction of silver ions and operate the synthesis at rates faster than those operated by microbes (Anandan et al. 2019). Green silver nanoparticles synthesized by Anandan et al. (2019) from the aqueous leaf extract of *Dodonaea viscosa* exhibited antibacterial and anticancer activities. Incidentally green silver nanoparticles synthesized from aqueous leaf extracts of *Elephantopus scaber* L. were seen to be with antibacterial and anticancer properties (Das 2019).

38.2 Publications Reviewed

This review is based on the usage of 26 species of plants in ethnomedicine (Table 38.1) and in synthesis green nanoparticles for use in modern medicine (Table 38.2). So far the ethnomedicinal uses of these plants are concerned on how wide the range of their remedial applications is (Fig. 38.1). It can well be presumed that they must be equipped with variety of secondary metabolites keeping parity with the evolutionary status and the environments in which they grow. Selection of these plants for resolving their ethnomedicinal use was rather indirect. Firstly, these plants were selected on the basis of their efficacy in synthesis of green NPs and in exhibiting bioactivity. The trend of researches on green NPs is to work out their efficacy as antibacterial/viral/fungal agents and antioxidant, cytotoxic, and anticancerous so that mankind can be served with more friendly and effective remedies than those contemporarily used. Leaves are mostly used in synthesis of green nanoparticles. Literature reveals that different plant parts are useful for synthesis of green NPs (Fig. 38.2) involving leaf (17 plant species), root (3 species), seed (2 species), peel (2 species), fruit (1 species), and gall (1 species). For a few plants, viz., *Abutilon indicum*, *Acorus calamus*, *Carica papaya*, *Acalypha indica*, *Nelumbo nucifera*, *Memecylon edule*, *Melia dubia*, *Psoralea corylifolia*, and *Vitex negundo*, same plant parts were used in both ethnomedicine and green NPs, whereas in case of *Artocarpus lakoocha* and *Carissa edulis*, different plant parts were used. Analysis of therapeutic potential of green NPs in the context of the concerned

Table 38.1 Ethnomedicinal potential of selected species of plants

Name of the plant	Ethnomedicinal potential	Reference
Abutilon indicum	Leaf paste is taken orally to relieve body pain and to cure piles	Prakshanth et al. (2006)
	The fruit decoction mixed with ammonium chloride is used against hemorrhagic septicemia	Ali (1999)
Acorus calamus	The various extract of this is traditionally used for the antidiabetic, antiproliferative, immunosuppressive, hypolipidemic, mitogenic, and anticarcinogenic activity toward human lymphocytes	Umamaheshwari and Rekha (2018)
Argyreia nervosa	Traditionally the plant is used in the treatment of gonorrhea, strangury, and chronic ulcers. The leaves are used to prevent conception and antiphlogistic, used externally in the treatment of ringworm, eczema, itch, and other skin diseases. It is also used as a local stimulant and rubefacient	Padhi et al. (2013)
Acalypha indica	Extract of fresh leaves may be employed in scabies and other skin diseases, with lime and onion. It is a good stimulating application in rheumatism. Powder of dry leaves is used in bed sores	Saha and Ahmed (2011)
Carica papaya	Papaya leaf when dried and cured like a cigar is smoked by asthmatic persons. Fresh papaya leaves infusion is used by person to expel or destroy intestinal worms. Fresh green papaya is also used to remedy colic, a certain stomach disorder, or cramp. Outer part of the papaya roots decoction is able to cure dyspepsia	Yogiraj et al. (2014)
Cymbopogon citratus	It is used in herbal medicine for a wide range of applications based on its antibacterial, antifungal, antiprotozoal, anticarcinogenic, anti-inflammatory, antioxidant, cardioprotective, antitussive, antiseptic, and antirheumatic activities. It has also been used to inhibit platelet aggregation, treat diabetes, dyslipidemia, gastrointestinal disturbances, anxiety, malaria, flu, fever, and pneumonia, as well as in aromatherapy	Christopher et al. (2014)
Coccinia indica	It has remarkable effect in reducing the amount of sugar in the urine of patients suffering from diabetes mellitus	Wealth of India (1992)
Citrus sinensis	It has been used traditionally to treat ailments like constipation, cramps, colic, diarrhea, bronchitis, tuberculosis, cough, cold, obesity, menstrual disorder, angina, hypertension, anxiety, depression, and stress	Milind and Chaturvede (2012)
Musa paradisiaca	Treatment of depression, anemia, and constipation	Sampath Kumar et al. (2012)
Melia dubia	The leaf is a good source of essential oil camphene and exhibited bacteriostatic and fungistatic activities against wide range of human pathogens. The leaf and	Chauhan et al. (2018)

(continued)

Table 38.1 (continued)

Name of the plant	Ethnomedicinal potential	Reference
	bark extract has antifeedant, insecticidal, and anti-larval property and anti-inflammatory, anticancer, and hepato-protective activities	
Memecylon edule	The lotion prepared from leaf was used for conjunctivitis and ophthalmia. Decoction of leaf was given internally for treating gonorrhea and urticarial antipyretic in Thai medicine	Srinivasan et al. (2014)
Nelumbo nucifera	Lotus leaves are used against diarrhea, high fever, hemorrhoids, leprosy weakness, skin inflammation, body heat imbalance, hematemesis, epistaxis, hemoptysis, hematuria, and metrorrhagia	Subzar (2014)
Plumbago zeylanica	Act as stimulant, digestant, expectorant, laxative, abortifacient, and also treating muscular pain and rheumatic disease	Vishnukanta and Rana (2011)
Psoralea corylifolia	In India, seed powder is mixed with haratala bhasma (yellow arsenic) and cow urine to make paste. This paste is used to treat leukoderma lesions. In another formulation, the mixture of powdered seeds with buttermilk has been used externally for treating ringworm and scabies. In Ayurveda, the seeds are used in the form of paste and as an ointment for external as well as internal use for treatment of different conditions such as alopecia, inflammation, leukoderma, leprosy, psoriasis, and eczema	Fiaz et al. (2017)
Vitex negundo	Decoction of leaves of *Vitex negundo* or boiled leaves can be applied locally to reduce inflammation	Kabeeruddin et al. (2007)
Carum copticum	Traditionally, it has been used in the past for various therapeutic effects including bloating, fatigue, diarrhea, abdominal tumors, abdominal pain, respiratory distress, and loss of appetite	
Syzygium cumini	Charaka used seeds, leaves, and fruits in decoctions for diarrhea and the bark as an astringent. According to Sushruta the fruit can be used in obesity problem, vaginal discharges, and menstrual disorders	Nair and Subramanian (1962)
	The bark decoction is potent mouthwash and gargle method for curing spongy gums, stomatitis, relaxed throat, and other mouth diseases. In Ayurveda, its bark is acrid, sweet, digestive, astringent to the bowels, anthelmintic, and also used for treating sore throat, asthma, bronchitis, thirst, biliousness, blood impurities dysentery, and ulcers	Kirtikar and Basu (1975)
	The leaf juice of amra, jambu, and amalaka mixed with honey and goat milk is used for curing diarrhea with the blood. Leaf juice also taken orally for treating diabetes. Fresh juice of *S. cumini* leaves is taken orally for curing stomach pain	Bhandary et al. (1995)

(continued)

Table 38.1 (continued)

Name of the plant	Ethnomedicinal potential	Reference
Hovenia dulcis	In East Asia, it is a traditional herbal medicine used for curing liver problem and also acts as alcoholic poisoning detox agent	Gadelha et al. (2005)
Dendropanax morbifera	The leaves, stems, and roots of the plant have been traditionally used for treating various disease, viz., dysmenorrhea, infectious diseases, and migraine problem	Bae (2000)
Rhus chinensis	Historically, gallnuts have been used by both Western and eastern cultures as a traditional medicine for various body disorders, as an astringent in painful hemorrhoids, an antiphlogistic for inflammatory conditions, and also used for treating diarrhea, dysentery, toothache, and dental caries	Larew (1987), Sariozlu and Kivanc (2011)
Carissa edulis	Aqueous extract of root used as anti-plasmodial agent in traditional malaria therapy in Meru and Kilifi districts of Kenya	Kiriraa et al. (2006)
Linum usitatissimum	Traditionally been used for curing diarrhea as well as infections of Gastrointestinal track	Palla et al. (2015)
Suaeda maritima	Traditionally, different parts of the plant have been used as an ethnomedicine for curing various ailments	Nayak et al. (2018)
Cassytha filiformis	In Taiwan, it has beneficial medicinal value against the gonorrhea, kidney ailments, and also the diuretic. In Africa it was used to cure cancer and African trypanosomiasis	Mythili et al. (2011)
Artocarpus lakoocha	Bark powder is applied for curing any wound and also used to draw out purulent matter from any abscess. The stomach and liver disease can be cured by the seed and bark of the plant. The juice and seeds from this plant are used as purgative, and the bark is used as astringent. The root is used as refresher, whereas the leaves are used for curing dropsy	Hossain et al. (2016)
Cassia fistula	According to the Unani and Ayurvedic medicines, this plant is potent agent for curing liver troubles, tuberculous glands, and skin diseases and also used for treating hematemesis, pruritus, leucoderma, and diabetes as has been suggested	Ali (2014) Dutta and De (1980)

publications (Fig. 38.3) reveals them as antioxidants (5 species), antimicrobials (18 species), anticancerous (3 species), antileukemic (1 species), and larvicidal (1 species).

Table 38.2 Bioactivity of green nanoparticles synthesized from ethnomedicinal plants

Name of the plant	Plant parts used	Bioactivity type	Reference
Abutilon indicum	Leaf	Antioxidant activity	Sadeghi and Gholamhoseinpoor (2015)
Acorus calamus	Rhizome	Antioxidant and antibacterial, cytotoxic activities	Nakkala et al. (2014)
Argyreia nervosa	Seed	Antibacterial and antifungal activities	Thombre et al. (2014)
Acalypha indica	Leaf	Antibacterial activity	Krishnaraj et al. (2010)
Carica papaya	Fruit	Antibacterial activity	Jain et al. (2009)
Cymbopogon citratus	Leaf	Antibacterial and antifungal activities	Masurkar et al. (2011)
Coccinia indica	Leaf	Antibacterial activity	Kumar et al. (2013)
Citrus sinensis	Peel	Antibacterial activity	Kaviya et al. (2011)
Musa paradisiaca	Peel	Antibacterial and antifungal activities	Bankar et al. (2010)
Melia dubia	Leaf	Anticancer activity	Kathiravan et al. (2014)
Memecylon edule	Leaf	Antioxidant activity	Elavazhagan and Arunachalam (2011)
Nelumbo nucifera	Leaf	Larvicidal activity	Santhoshkumar et al. (2011)
Plumbago zeylanica	Root	The effects of quantitative biofilm inhibition and disruption assays of *Acinetobacter baumannii*, *Staphylococcus aureus*, and *Escherichia coli* biofilms	Salunke et al. (2014)
Psoralea corylifolia	Seed	Antimicrobial activity	Sunita et al. (2014)
Vitex negundo	Leaf	Antibacterial activity	Zargar et al. (2011)
Carum copticum	Root	Antibacterial activity	Esmaeili and Ghobadianpour (2016)
Syzygium cumini	Leaf	Antitubercular and antimycobacterial activities	Singh et al. (2016)
Hovenia dulcis	Leaf	Antibacterial activity	Salunke et al. (2016)
Dendropanax morbifera	Leaf	Anticancer activity	Wang et al. (2016)
Rhus chinensis	Gall	Antibacterial activity	Patil et al. (2016)

(continued)

Table 38.2 (continued)

Name of the plant	Plant parts used	Bioactivity type	Reference
Carissa edulis	Leaf	Antioxidant and antibacterial activities	Fowsiya et al. (2016)
Linum usitatissimum	Leaf	Antibacterial and antifungal activities	Ghaedi et al. (2016)
Suaeda maritima	Leaf	Antileukemic activity	Rajendran et al. (2016)
Cassytha filiformis	Leaf	Anticancer, antifungal, and antibacterial activities	Jena et al. (2016)
Artocarpus lakoocha	Leaf	Antibacterial activity	Kaewkod et al. (2016)
Cassia fistula	Leaf	Antioxidant, antimicrobial, and cytotoxic activities	Mohanta et al. (2016)

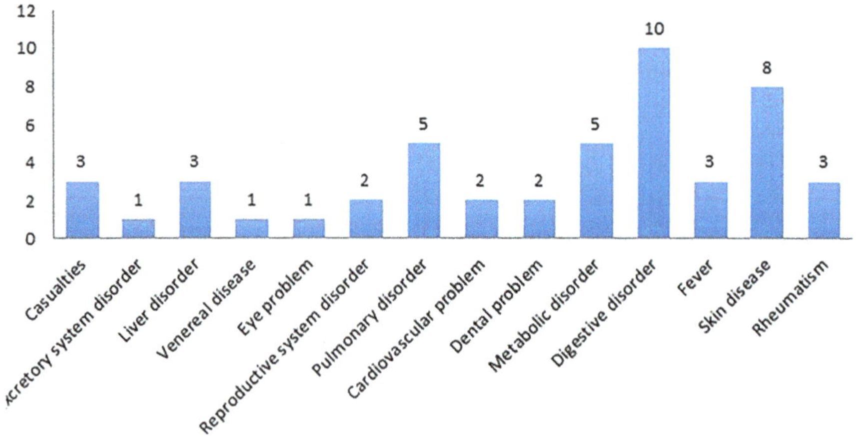

Fig. 38.1 Ethnomedicinal potential of selected plants showing number of species vis-à-vis ailments cured

38.3 Discussion

Since time immemorial the plant resources in India have been evoking profound interest in her people to discover various medicinal plants and their uses and develop indigenous system of medical science that still survives today in Indian traditions. Such a scientific concern of the people covered even the exotic plants that immigrated and naturalized in India in the recent past and discovered many of their hitherto unknown uses. Synthesis of nanoparticles (NPs) by various chemical reactions got global attention due to their wide range of bioactivity despite the fact

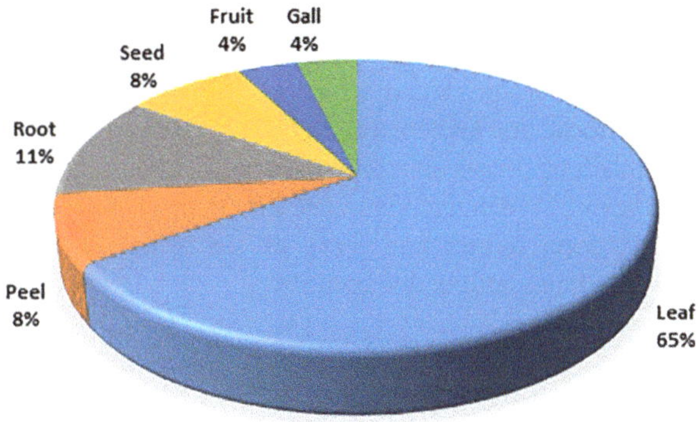

Fig. 38.2 Different plant parts used in synthesis of green NPs

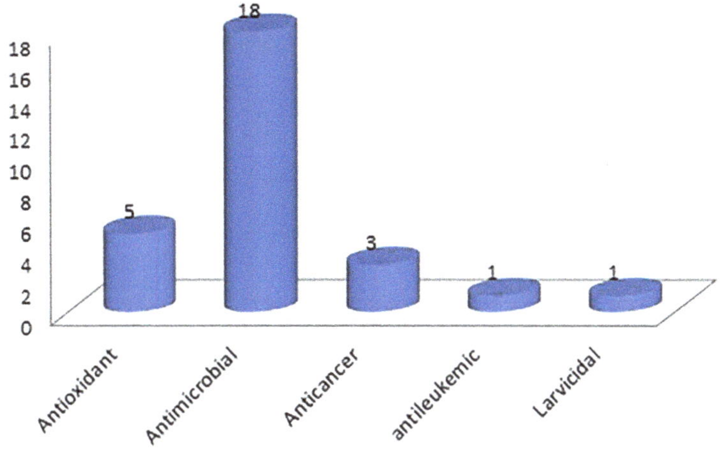

Fig. 38.3 Curative properties of green NPs showing number of species vis-à-vis bioactivity

that they possess a broad range of toxicity in invertebrates and vertebrates. The "green" (environment-friendly) techniques in chemistry and chemical technologies become widely accepted to overcome problems associated with environmental concerns (Thuesombat et al. 2014). Researchers have explored the use of silver nanoparticles as carriers for delivering small drug molecules or large biomolecules to specific targets. Once the NP has had sufficient time to reach its target, which could be potentially triggered by an internal or external stimulus, and then provide high concentrations at specific target sites and could minimize side effects. NPs production per year is estimated to increase in next few years for its profound role in field of high sensitivity biomolecular detection, catalysis, biosensors, agriculture, and medicine; it has been acknowledged to have bactericidal effects along with the antifungal, anti-inflammatory, and anticancer activities (Abalkhil et al. 2017; Afrah et al. 2018).

Antibiotics and anticancer chemotherapeutic drugs are very costly and toxic, besides there being a drug resistance problem. Consequently, looking for alternative medicines is essential. NPs produced with an eco-friendly approach is efficient, boasts limited side effects, and less costly. From the present review, one can develop the idea of giving much emphasis on ethnomedicinally important plants for selection of species for synthesis of green nanoparticles. The therapeutic potential of a species used among ethnic communities can serve as the criteria for strengthening its candidature for use in nanoparticle therapeutics.

38.4 Summary and Path Forward

The green AgNPs prepared from plants should be seriously considered as an antibacterial and anticancer drug. Our investigations in the future will focus mainly on answering specific questions, and the knowledge gained in ethnomedicine is likely to enhance our understanding of the efficacy of nanoparticles in biological and biotechnological applications for restoration and sustenance of life worldwide.

Acknowledgments Both the authors are grateful to UGC Centre for Advanced Study (Phase II), Department of Botany of the University of Burdwan for providing scientific facilities; they extend gratitude to Institute of Wood science and Technology.

References

Abalkhil TA, Alharbi SA, Salmen SH, Wainwright M et al (2017) Bactericidal activity of biosynthesized silver nanoparticles against human pathogenic bacteria. *Biotechnology &amp.* Biotechnological Equipment 31(2):411–417

Afrah E, Mohammed ID, Al-Qahtani A, al-Mutairi A, Al-Shamri B, Aabed K et al (2018) Antibacterial and cytotoxic potential of biosynthesized silver nanoparticles by some plant extracts. Nano 8:382

Ahirwar R, Khan A, Mishra A, Parna SK, Shrivastava IN, Singh G et al (2010) Ethnomedicinal plants of district Anuppur used in treatment of renal calculi (kidney stones). Indian J Appl Pure Biol 25(2):251–255

Ajitha B, Ashok Kumar Reddy Y, Sreedhara Reddy P et al (2015) Green synthesis and characterization of silver nanoparticles using *Lantana camara* leaf extract. Mater Sci Eng C 49:373–381

Ali M (2014) *Cassia fistula* Linn: a review of phytochemical and pharmacological studies. Int J Pharm Sci Res 5(6):2125–2130

Ali ZA (1999) Folk veterinary medicine in Moradabad District (Uttar Pradesh), India. Fitoterapia 70:340–347

Anandan M, Poorani G, Boomi P, Varunkumar K, Anand K, Anil Chuturgoon A, Saravanan M, Prabua HG et al (2019) Green synthesis of anisotropic silver nanoparticles from the aqueous leaf extract of *Dodonaea viscosa* with their antibacterial and anticancer activities. Process Biochem 80:80–88

Anonymous (1976) Medicinal plants of India, vol 1. ICMR, New Delhi

Arjariya A, Rawat A (2005) An ethnobotanical survey of certain wild edible plants of Chhatarpur district (M.P.). Flora Fauna 11(1):97–100

Arora RK, Pandey A (1996) Wild edible plants of India: diversity, conservation and use. National Bureau of Plant Genetic Resources, New Delhi

Bae K (2000) The medicinal plants of Korea. Kyohak Publishing, Seoul
Bankar A, Joshi B, Kumar AR et al (2010) Banana peel extract mediated novel route for the synthesis of silver nanoparticles. Colloids Surf A Physicochem Eng Asp 368:58–63
Bhandary MJ, Chandrashekar KR, Kaveriappa KM et al (1995) Medical ethnobotany of the siddis of Uttara Kannada district, Karnataka, India. J Ethnopharmacol 47(3):149–158
Bhattacharya SK (2004) Handbook of aromatic plants. Pointer Publications, Jaipur
Biswas S, Mukherjee A (2017) A census of ethnomedicinal plants used in Bankura District, West Bengal. Res J Pharm, Biol Chem Sci 8(1):424–431
Borgohain MP, Teron R, Tamuli AK et al (2016) Ethnomedicinal plants of the Tai Phake tribe of upper Brahmaputra Valley in Assam, India. Pleione 10(1):123–133
Borthakur SK (1976a) Less known medicinal uses of plants among the tribes of KarbiAnglong (Mikir Hills), Assam. Bull Bot Surv India 18(1–4):166–171
Borthakur SK (1976b) Traditional weaving implements among the Mikirs (the Karbis) of Assam. Bull Bot Surv India 11:46–50
Borthakur SK (1981) Plants in folklore and folktales of the Karbis (Mikirs) of Assam. In: Jain SK (ed) Glimpses of Indian ethnobotany. Oxford and IBH Publishing Co, New Delhi, pp 170–181
Borthakur SK (1992) Native phytotherapy of child and women diseases from Assam. Fitoterapia 63:483–488
Borthakur SK (1996a) Wild edible plants in markets of Assam, India-an ethnobotanical investigation. In: Jain SK (ed) Ethnobiology in human welfare. Deep Publications, New Delhi, pp 31–34
Borthakur SK (1996b) Post natal care of women in traditional system in Assam. Ethnobotany 8:51–55
Borthakur SK, Choudhury BT, Gogoi R et al (2004) Folklore hepato-protective herbal recipes from Assam in Northeast India. Ethnobotany 16:76–82
Borthakur SK, Goswami N (1995) Herbal remedies from Dimoria of Kamrup district of Assam in Northeastern India. Fitoterapia 66(4):333–340
Borthakur SK, Nath KK (2007) Ethnobotanical aspects of weeds from Darrang district of Assam. Ethnobotany 19:82–87
Borthakur SK, Sarma UK (1996) Ethnoveterinary medicine with special reference to cattle prevalent among the Nepalis of Assam, India. In: Jain SK (ed) Ethnobiology in human welfare. Deep Publications, New Delhi, pp 197–199
Cao B, Liu R, Huang ZX, Ge S et al (2011) Flash synthesis of flower-like ZnO nanostructures by microwave-induced combustion process. Mater Lett 65:160
Chauhan RS, Jadeja DB, Thakur NS, Jha SK, Sankanur MS et al (2018) Selection of candidate plus trees (CPTs) of Malabar Neem (*Meliadubia* Cav.) for enhancement of farm productivity in South Gujarat, India. Int J Curr Microbiol App Sci 7(5):3582–3592
Chopra D, Simon D (2002) The Chopra Centre herbal hand book. Rider, London
Christopher EE, Ernest EA, Daniel NE et al (2014) Phytochemical constituents, therapeutic applications and toxicological profile of *Cymbopogon citratus* Stapf (DC) leaf extract. J Pharmacogn Phytochem 3(1):133–141
Chung IM, Abdul Rahuman A, Marimuthu S, Kirthi AV, Anbarasan K, Padmini P, Rajakumar G et al (2017) Green synthesis of copper nanoparticles using Eclipta prostrata leaves extract and their antioxidant and cytotoxic activities. Exp Ther Med 14(1):18–24
Das AP, Ghosh C, Sarker A, Biswas R, Biswas K, Choudhury D et al (2010) Preliminary report on the medicinal plants from three MPCAs in Terai and Duars of West Bengal, India. Pleione 4 (1):90–101
Das AP, Pandey AK (eds) (2007) Advances in ethnobotany. Bishen Singh Mahendra Pal Singh, Dehra Dun
Das M (2019) Ph.D. Thesis: Understanding the medicinal potential of *Elephantopus scaber* L. The University of Burdwan, West Bengal, India
Devi MR, Salam S (2016) Wild edible plants used by the Monsang Naga tribe of Manipur, India. Pleione 10(1):90–96

Dutta A, De B (1980) Seasonal variation in the content of sennosides and rhein in leaves and pods of *Cassia fistula*, Ind. J Pharmacol Sci 60:388–390
Elavazhagan T, Arunachalam KD (2011) *Meme cylonedule* leaf extract mediated green synthesis of silver and gold nanoparticles. Int J Nanomedicine 6:1265–1278
Esmaeili A, Ghobadianpour S (2016) Antibacterial activity of *Carum copticum* extract loaded $MnFe_2O_4$ nanoparticles coated with PEGylated chitosan. Ind Crop Prod 91:44–48
Ezealisiji KM, Siwe-Noundou X, Maduelosi B et al (2019) Green synthesis of zinc oxide nanoparticles using *Solanum torvum* (L) leaf extract and evaluation of the toxicological profile of the ZnO nanoparticles–hydrogel composite in Wistar albino rats. Int Nano Lett 9:99
Fakhari S, Jamzad M, Fard HK et al (2019) Green synthesis of zinc oxide nanoparticles: a comparison. Green Chem Lett Rev 12(1):19–24
Feng A, Wu S, Chen S, Zhang H, Shao W, Xiao Z et al (2013) Synthesis of silver nanoparticles with tunable morphologies via a reverse nano-emulsion route. Mater Trans 54(7):1145–1148
Fiaz A, Khan GN, Hassan BA, Muhammad H et al (2017) *Psoralea corylifolia* L: ethnobotanical, biological, and chemical aspects: a review. Phytother Res:1–19
Fowsiya J, Madhumitha G, Al-Dhabi NA, Arasu MV et al (2016) Photocatalytic degradation of Congo red using *Carissa edulis* extract capped zinc oxide nanoparticles. J Photochem Photobiol 162:395–401
Gadelha APR, Vidal F, Castro TM, Lopes CS, Albarello N, Coelho MGP, Figueirede SFL, Monteiro-Leal LH et al (2005) Susceptibility of *Giardia lamblia* to *Hoveniadulcis* extracts. Parasitol Res 97:399–407
Gautam AL, Rana R, Srivastava V, Ray Choudhuri SP, Singh BR et al (1998) Prospects of medicinal plants. Indian Soc Plant Genet Res, New Delhi
Ghaedi M, Yousefi-Nejad M, Safarpoor M, Hashemi S, Goudarzi A, Tyagi I et al (2016) Investigation of phytochemical and antimicrobial properties of *Linum usitatissimum* in presence of ZnO/Zn(OH)2 nanoparticles and extraction of euphol from *Euphorbia microsciadia*. Desalin Water Treat 57(43):20597–20607
Han WK, Choi JW, Hwang GH, Hong SJ, Lee JS, Kang SG et al (2006) Fabrication of cu nanoparticles by direct electrochemical reduction from CuO nanoparticles. App Surf Sci 252:2832–2838
Hossain MF, Islam MA, Akhtar S, Numan SM et al (2016) Nutritional value and medicinal uses of monkey Jack fruit (*Artocarpus lakoocha*). Int Res J Biol Sci 5(1):60–63
Huang X, El-Sayed MA (2010) Gold nanoparticles: optical properties and implementations in cancer diagnosis and photothermal therapy. J Adv Res 1:13–28
Jain A, Jain SK (2016a) Indian ethnobotany: bibliography of 21st century (2001–2015). Scientific Publisher, Jodhpur
Jain D, Daima HK, Kachhwaha S et al (2009) Synthesis of plant-mediated silver nanoparticles using papaya fruit extract and e valuation of their antimicrobial activities. Dig J Nanomater Biostruct 4:557–563
Jain SK (1991) Dictionary of Indian folk medicine and ethnobotany. Deep Publications, New Delhi
Jain SK (1994) Ethnobotany and research on medicinal plants in India. In: Derek J, Chadwick JM (eds) Ethnobotany and the search of new drugs, CIBA foundation symposium, 185. John Wiley and Sons, Chichester, UK, pp 153–168
Jain SK (2004) Manual of ethnobotany. Scientific Publishers, Jodhpur
Jain SK, Hajra P, Shanpru R et al (1977) A survey of wild edible plants in bazaars of Meghalaya. Bull Meghalaya Sci Soc 2:29–34
Jain SK, Jain V (2015) Evolution of ethnobotany in India during last sixty years: a brief overview. Ethnobotany 27:1–6
Jain SK, Srivastava S (1999) Dictionary of ethnoveterinary plants of India. Deep Publications, New Delhi
Jain V, Jain SK (2016b) Compendium of Indian folk medicine and ethnobotany (1991–2015). Deep Publications, New Delhi

Jena S, Singh RK, Panigrahi B, Suar M, Mandal D et al (2016) Photo-bioreduction of Ag+ ions towards the generation of multifunctional silver nanoparticles: mechanistic perspective and therapeutic potential. J Photochem Photobiol 164:306–313

Jirovetz L, Buchbauer G, Shafi MP, Leela NK et al (2003) Analysis of the essential oils of the leaves, stems, rhizomes and roots of the medicinal plant *Alpinia galanga* from southern India. Acta Pharma 53(2):73–82

Kabeeruddin AM et al (2007) Makhzanul Mufredat. Idara Kitab-ul-shifa, New Delhi, pp 266–267

Kaewkod T, Tragoolpua K, Tragoolpua Y et al (2016) Encapsulation of *Artocarpus lacucha* Roxb. Extract in alginate chitosan nanoparticles for inhibition of methicillin resistant *Staphylococcus aureus* and bacteria causing skin diseases. Chiang Mai J Sci 43(5):946–958

Kapoor LD (1990) Handbook of Ayurvedic medicinal plants. CRC Press, Boca Raton

Kathiravan V, Ravi S, Kumar SA (2014) Synthesis of silver nanoparticles from *Melia dubia* leaf extract and their in vitro anticancer activity. Spectrochim Acta A Mol Biomol Spectrosc 130:116–121

Kaushik P, Dhiman AK (1999) Medicinal plants and raw drugs of India. Bishen Singh Mahenderpal Singh, Dehradun

Kaviya S, Santhanalakshmi J, Viswanathan B et al (2011) Biosynthesis of silver nanoparticles using *Citrus sinensis* peel extract and its antibacterial activity. Spectrochim Acta A Mol Biomol Spectrosc 79:594–598

Kiriraa PG, Rukungab GM, Wanyonyia AW, Muregib FM, Gathirwab JW, Muthaurab CN, Omarc SA, Tolob F, Mungaid GM, Ndiege IO et al (2006) Anti-plasmodial activity and toxicity of extracts of plants used in traditional malaria therapy in Meru and Kilifi Districts of Kenya. J Ethnopharmacol 106:403–407

Kirtikar KR, Basu BD (1975) Indian medicinal plants, vol II. Periodical Experts, New Delhi, pp 1052–1053

Krippner S (2003) Models of ethnomedicinal healing. In: Ethnomedicine Conferences, Munich, 26–27 April and 11–12 October

Krishnaraj C, Jagan E, Rajasekar S et al (2010) Synthesis of silver nanoparticles using *Acalypha indica* leaf extracts and its antibacterial activity against water borne pathogens. Colloids Surf B-Biointerfaces 76:50–56

Kumar S (2014) Ethnobotanical studies in India. Deep Publications, New Delhi

Kumar S, Daimary RM, Swargiary M et al (2013) Biosynthesis of silver nanoparticles using *Premna herbacea* leaf extract and evaluation of its antimicrobial activity against bacteria causing dysentery. Int J Pharm Biol Sci 4(4):378–384

Lalramnghinghlova H (2001) Ethnobotanical interpretations and future prospects of ethnobotany in the north-East India. SciVision 1:24–31

Lalramnghinghlova H, Jha LK (1997) Forest resources: an overview. In: Jha LK, Nangia SB (eds) Natural resource management, vol, vol 1. A.P.H. Publishing Corporation, New Delhi, pp 203–255

Larew HG (1987) Oak galls preserved by the eruption of mount vesuvius in A.D. 79, and their probable use. Econ Bot 41(1):33–40

Masurkar SA, Chaudhari PR, Shidore VB et al (2011) Rapid biosynthesis of silver nanoparticles using *Cymbopogon citratus* (Lemon grass) and its antimicrobial activity. Nano-Micro Lett 3 (3):189–194

Milind P, Chaturvede D (2012) Orange: Range of benefits. Int Res J Pharm 3:59–63

Mohanta YK, Panda SK, Biswas K, Tamang A, Bandyopadhyay J, De D et al (2016) Biogenic synthesis of silver nanoparticles from *Cassia fistula* (Linn.): in vitro assessment of their antioxidant, antimicrobial and cytotoxic activities. IET Nanobiotechnol 10(6):438–444

Moreira dos Santos M, João Queiroz M, Baptista PV et al (2012) Enhancement of antibiotic effect via gold: silver-alloy nanoparticles. J Nanopart Res 14(5):859–866

Mukherjee A (2012) Herbal remedies: aspects and prospects with special reference to India. In: Maiti G, Mukherjee SK (eds) Multidisciplinary approaches in angiosperm systematics, vol 2. University of Kalyani, Kalyani, West Bengal, pp 576–588

Mukherjee A, Banerjee M (2019) Rejuvenation of interests in herbal remedies as elixir of life. In: Sen S, Chakraborty R (eds) Herbal medicine in India. Springer, Singapore

Mythili S, Gajalakshmi S, Sathiavelu A, Sridharan TB et al (2011) Pharmacological activities of *Cassytha filiformis*: a review. Asian J Plant Sci Res 1(1):77–83

Nadkarni P (1992) Indian medicinal plants. Orient Longman, Hyderabad

Naik VN (2004) Identification of common Indian medicinal plants. Scientific Publishers, Jodhpur

Nair AG, Subramanian S (1962) Chemical examination of flowers of *Eugenia jambolana*. J Sci Ind Res 21(9):457

Nair CKN, Mohanan N (1998) Medicinal plants in India. Nag Publications, Delhi

Nakkala JR, Mata R, Kumar Gupta A et al (2014) Biological activities of green silver nanoparticles synthesized with *Acorus calamus* rhizome extract. Eur J Med Chem 85:784–794

Nath M (2015) Some less known medicinal uses of plants among Dimsa tribe of Barak valley, Assam (North-East India). Ethnobotany 27:49–52

Nayak B, Roy S, Roy M, Mitra A, Karak K et al (2018) Phytochemical, antioxidant and antimicrobial screening of *Suaedamaritima* L (Dumort) against human pathogens and multiple drug resistant bacteria. Indian J Pharm Sci 80(1):26–35

Nunes JP, Fernandes B, Fertunato E, Vilarinho P, Martins R et al (1999) Performances presented by zinc oxide thin films deposited by spray pyrolysis. Thin Solid Films 337:176–179

Padhi M, Mahapatra S, Panda J, Mishra NK et al (2013) Traditional uses and Phytopharmacological aspects of Argyreia nervosa. J Adv Pharm Res 4(1):23–32

Palla AH, Khan NA, Bashir S, Ur-Rehman N, Iqbal J, Gilani AH et al (2015) Pharmacological basis for the medicinal use of *Linum usitatissimum* (flaxseed) in infectious and non-infectious diarrhea. J Ethnopharmacol 160:61–68

Patil MP, Rokade AA, Ngabire D, Kim GD et al (2016) Green synthesis of silver nanoparticles using water extract from galls of *Rhus chinensis* and its antibacterial activity. J Clust Sci 27 (5):1737–1750

Prakshanth V, Neelam S, Padh H, Rajani M et al (2006) Search for antibacterial and antifungal agents from selected Indian medicinal plants. J Ethnopharmacol 107:182–188

Rajendran N, Subramaniam S, Raja MR, Brindha P, KarMahapatra S, Sivasubramanian A et al (2016) Plant phenyl-propanoids-conjugated silver nanoparticles from edible plant *Suaedamaritima* (L.) dumort. Inhibit proliferation of K562-human myeloid leukemia cells. Artif Cells Nanomed Biotechnol:1–7

Rana AK (2003) Indian medicinal and aromatic plants 1 & 2. Indian Forester 129:1–297

Redhead HM, Davis SS, Illum L et al (2001) Drug delivery in poly (lactide-co-glycolide) nanoparticles surface modified with poloxamer 407 and poloxamine 908: in vitro characterisation and in vivo evaluation. J Control Release 70:353–363

Sadeghi B, Gholamhoseinpoor F (2015) A study on the stability and green synthesis of silver nanoparticles using *Ziziphora tenuior* (Zt) extract at room temperature. Spectrochim Acta A Mol Biomol Spectrosc 134:310–315

Saha R, Ahmed A (2011) Phytochemical constituents and pharmacological activities of *Acalypha indica* Linn: a review. Int J Pharm Sci Res 2(8):1900–1904

Sahu V, Bajpai R, Upreti DK et al (2015) Traditional use of lichens of Rudrabhishek (spiritually ceremony): a Hindu practice to worship Lord Shiva in India. Ethnobotany 27:92–96

Salunke BK, Sawant SS, Kim BS et al (2016) Enhancement of antibacterial effect by biosynthesized silver nanoparticles with antibiotics. J Nanosci Nanotechnol 16(7):7191–7194

Salunke GR, Ghosh S, Santosh Kumar R et al (2014) Rapid efficient synthesis and characterization of silver, gold, and bimetallic nanoparticles from the medicinal plant *Plumbagozeylanica* and their application in biofilm control. Int J Nanomedicine 9:2635–2653

Sampath Kumar KP, Bhowmik D, Duraivel S, Umadevi M et al (2012) Traditional and medicinal uses of Banana. J Pharmacogn Phytochem 3(12):51–63

Santhoshkumar T, Rahuman AA, Rajakumar G et al (2011) Synthesis of silver nanoparticles using *Nelumbo nucifera* leaf extract and its larvicidal activity against malaria and filariasis vectors. Parasitol Res 108:693–602

Sariozlu NY and Kivanc M (2011) "Gallnuts (*Quercus infectoria* Oliv. And *Rhus chinensis* Mill.) and their Usage in Health," Nuts and Seeds in Health and Disease Prevention 505–511
Satyavati GV, Gupta KA (1987) Medicinal plants of India, vol 1&2. ICMR, New Delhi
Shah S (1992) The selected Indian medicinal plants. Chemixil, Madras
Singh R, Nawale L, Arkile M, Wadhwani S, Shedbalkar U, Chopade S et al (2016) Phytogenic silver, gold, and bimetallic nanoparticles as novel antitubercular agents. Int J Nanomedicine 11:1889–1897
Srinivasan R, Natarajan D, Shivakumar MS et al (2014) Antimicrobial and GC-MS analysis of *Memecylon edule* leaf extracts. Int J Curr Pharm Rev Res 5(1):1–13
Subzar AS (2014) Ethno-medicinal uses and pharmacological activities of lotus (*Nelumbo nucifera*). J Med Plants Stud 2(6):42–46
Sunita D, Tambhale D, Parag V et al (2014) Facile green synthesis of silver nanoparticles using *Psoralea corylifolia*. Seed extract and their in-vitro antimicrobial activities. Int J Pharm Biol Sci 5(1):457–467
Swami Brahmananda (2002) Common medicinal plants of India. Dominant Publishers, New Delhi
Thakor AS, Jokerst J, Zavaleta C, Massoud TF, Gambhir SS et al (2011) Gold nanoparticles: a revival in precious metal administration to patients. Nano Lett 11:4029–4036
Thombre R, Parekh F, Patil N et al (2014) Green synthesis of silver nanoparticles using seed extract of *Argyreia nervosa*. Int J Pharm Biol Sci 5(1):114–119
Thuesombat P, Hannongbua S, Akasit S, Chadchawan S et al (2014) Effect of silver nanoparticles on rice (*Oryza sativa* L. cv. KDML 105) seed germination and seedling growth. Ecotoxicol Environ Saf 104(1):302–309
Umamaheshwari N, Rekha A (2018) Sweet flag: (*Acorus calamus*)-an incredible medicinal herb. J Pharmacogn Phytochem 7(6):15–22
Uniyal SK, Awasthi A, Rawat GS et al (2002) Current status and distribution of commercially exploited medicinal and aromatic plants in upper Gori valley, Kumaun Himalaya, Uttaranchal. Curr Sci 82(10):1246–1252
Vishnukanta, Rana AC (2011) *Plumbago zeylanica*: a phytopharmacological review. Int J Pharm Sci Res 2(2):247–255
Wang C, Mathiyalagan R, Kim YJ, Castro-Aceituno V, Singh P, Ahn S et al (2016) Rapid green synthesis of silver and gold nanoparticles using *Dendropanax morbifera* leaf extract and their anticancer activities. Int J Nanomedicine 11:3691–3701
Wealth of India (1992) A dictionary of Indian raw materials and India products, vol 4. Raw Material, New Delhi, p 312
Yogiraj V, Goyal PK, Singh Chauhan C, Goyal A, Vyas B et al (2014) *Carica papaya* Linn: an overview. Int J Herb Med 2(5):01–08
Zafar R (1999) Medicinal plants in India. CBS Publications, New Delhi
Zargar M, Hamid AA, Bakar FA et al (2011) Green synthesis and antibacterial effect of silver nanoparticles using *Vitex negundo* L. Molecules 16:67–76

Physicochemical, Micromeritics, Biomedical, and Pharmaceutical Applications of Assam Bora Rice Starch

39

Pratap Kalita, Abdul Baquee Ahmed, Abhinab Goswami, Saikat Sen, and Raja Chakraborty

Abstract

In current times the utilization of the natural starches in various aspects like as pharmaceutical, biomedical, food materials, etc. due to their uniqueness properties. Assam bora rice starch is a promising biopolymer as well as bio-carrier in different aspects. Natural starch as well as modified starch was being used abundantly as natural polymers in different drug delivery systems. The prime advantages of using natural starches are low cost, easily available, renewability, etc. Starch obtained from Assam bora rice is having potential applications in pharmaceutical industries such as excipients and binder and disintegrates in tablet formulations, as an active drug carrier in nano, microformulations, and controlled drug delivery formulations. Assam bora rice starch as a native starch is having maleate particles which are often used for drug delivery carrier in the field biomedical application. The reasons behind this were due to their non-toxicity, biocompatibility, and inexpensive nature. In this chapter, we specially focus on Assam bora rice starch to profile its different

P. Kalita (✉)
Faculty of Pharmaceutical Sciences, Assam Down Town University, Guwahati, Assam, India

Assam Science and Technology University, Guwahati, Assam, India
e-mail: Pratapkalita0@gmail.com

A. B. Ahmed
Girijananda Chowdhury Institute of Pharmaceutical Science, Guwahati, Assam, India

A. Goswami
Regional Institute of Paramedical And Nursing Sciences, Aizawl, Mizoram, India

S. Sen
Faculty of Pharmaceutical Sciences, Assam Down Town University, Guwahati, Assam, India

R. Chakraborty
NEF College of Pharmacy, Guwahati, Assam, India

S. C. Mandal et al. (eds.), *Evidence Based Validation of Traditional Medicines*,
https://doi.org/10.1007/978-981-15-8127-4_39

applications along with physicochemical and powder flow properties which drastically affect the compactness of the starch granules.

Keywords

Assam bora rice · Bora rice starch · Pharmaceutical aspects · Biomedical applications

Abbreviations

ABR	Assam Bora Rice
ABRS	Assam Bora Rice Starch
API	Active Pharmaceutical Ingredients

39.1 Introduction

Starch, a polysaccharide formed in the cytoplasm of the plant's cells, forms the principal carbohydrate reservoir in plants (Moorthy 2002). In its natural form, starch is semicrystalline in nature also known as grains or granules. The size, shape, and structure of starch granules are characteristic of the botanical source. Starch is a biocompatible and biodegradable natural polymer, white, soft, and smooth, and possesses viscosity modifying properties; these attributes greatly attract investigators to evaluate its potential as a pharmaceutical excipient in various drug delivery systems (Tester et al. 2004). When native starch is modified, new attributes such as enhanced flow, disintegration, direct compression, and formation of stable gels in hot and cold water are introduced. These enhanced attributes make starch not only applicable to conventional but also to novel drug delivery systems.

Assam bora rice (ABR) also known as Bora Saul (*Oryza sativa*) is a traditional waxy or glutinous rice variety of Assam state of India (Sharma 2014). Traditionally ABR is grown by the farmers to meet up their domestic consumptions. ABR is also required to prepare a number of food items during and after the festival and religious ceremonies; the food items include pithas (biscuit-like confectionaries), chira (flaked rice), hurum (expanded wax rice), and sandoguri (fried rice powder). Apart from its usage in food items, some communities of Assam state also prepare high quality of rice beer out of bora rice. Because of its waxy or glutinous and instant cooking property, many companies outside of the state are attracted for the preparation of instant and packet food, bakery components, etc. (Deka et al. 2014).

Assam bora rice mainly consist of amylopectin (>99%), while the maximum amylose content of ABR is 0.257% (Kumar 2011). Contrary to the linear structure of amylose, amylopectin has a highly branched structure, consisting of three types of branch chains. The chains present in amylopectin can be classified as A-, B-, and C-. A-chains are those which are linked to other chains (B- or C-) by their reducing ends through α-D-(1 → 6) linkages, but are not branched themselves. B-chains are those linked to another B-chain or a C-chain, but B-chains are branched by A-chain or other B-chains at O-6 of a glucosyl unit. Each amylopectin has only one C-chain, which carries the sole-reducing end of the molecule (Finar 1975). The high content of amylopectin in bora rice attributes to its adhesive properties, which can be utilized as a pharmaceutical excipient in conventional and novel drug delivery vehicles (Sharma et al. 2013). Several investigations on bora rice as pharmaceutical excipient or as a drug delivery vehicle involve the use of flour or isolated starch. These includes the use of Assam bora rice starch (ABRS) as directly compressible excipient for tablet formulation (Sharma et al. 2011; Rajak and Bhuyan 2011), use as mucoadhesive and bioadhesive drug delivery vehicles for colon targeting (Ahmad et al. 2012a, b, c). Apart from the use as pharmaceutical excipients, ABRS is investigated as plasma volume expander (Rajak et al. 2014).

Based on the previous studies, it can be concluded that the starch content in ABR is very high (Sharma et al. 2011). The reports also conclude the huge potential of the ARBS in the application as pharmaceutical excipients, biomedical field, and in industry. Therefore, we have design to profile the various utilization, physiochemical characterization, and micrometrics of ABRS in a form of review.

39.2 Extraction of Starch from ABR

Alkali extraction of protein is the most common method of preparation of ABRS (Deka et al. 2014; Rajak and Bhuyan 2011; Ahmad et al. 2012a, b, c). The method involved stepping of ABR with 0.2–0.4% sodium hydroxide solutions with regular stirring for every 6 h, followed by replacement of the sodium hydroxide solution for every 18–24 h. The steeped rice then blended with two parts of sodium hydroxide solution to each part of steeped rice, which results in a milky solution. The starch suspension thus formed was diluted and allowed to settle, which is followed by filtration and drying at temperature of 40°–50 °C. The dried ABRS is then pulverized and pass through sieves. Applications and properties of Assam bora rice starch and flour are given in Fig. 39.1.

39.3 Physiochemical Properties

39.3.1 Chemical Structure

Chemically starch consists of two main structural units or components: amylose and amylopectin. Amylose is a linear polymer with glucose units arranged in α (1–4) linkage; it constitutes 15–20% of starch, with a degree of polymerization upto 6000, and has a molecular mass of 105–106 g/mol. Amylopectin is a larger branched molecule with α (1–6) linkage which results in a compact, branched structure. Amylopectin has a molecular weight of 107–109 g/mol with a degree of polymerization of two million (Rodrigues and Emejea 2012; Builders and Arhewoh 2016). ABRS is waxy starch which contains virtually all amylopectin (more than 99%). The

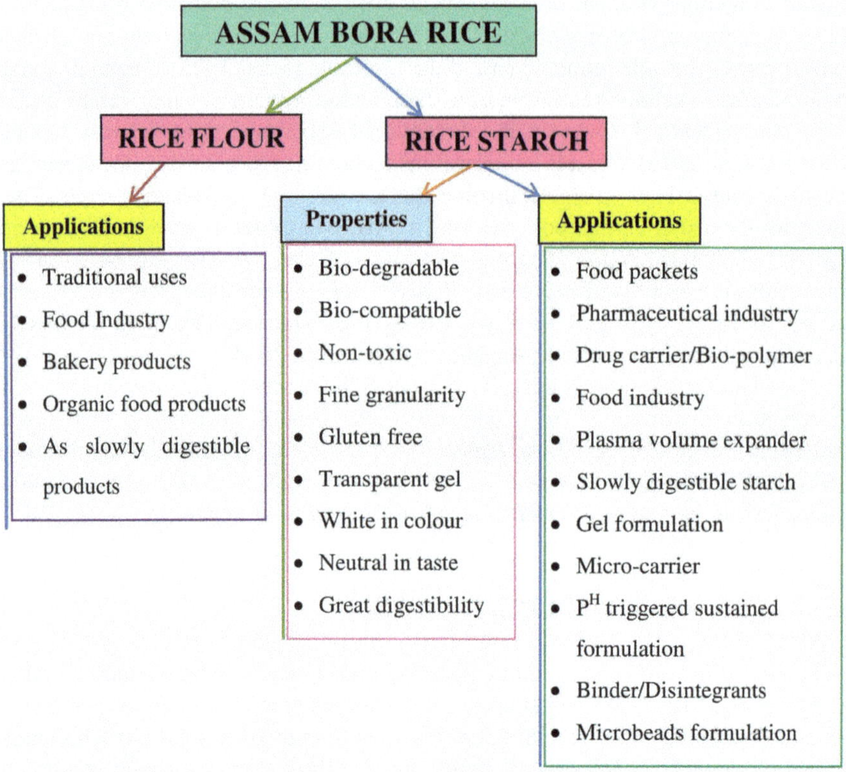

Fig. 39.1 Applications and properties of Assam bora rice starch and flour

amylose content of ABRS is 0.257% along with trace amount of starch lipids, proteins, and fibers (Kumar 2011). Amylose possesses a linear structure, while amylopectin has a highly branched structure with branch chains. The chains present in amylopectin can be classified as A-, B-, and C-. A-chains are those which are linked to other chains (B- or C-) by their reducing ends through α-D-(1 → 6) linkages, but are not branched themselves. B-chains are those linked to another B-chain or a C-chain, but B-chains are branched by A-chain or other B-chains at O-6 of a glucosyl unit. Each amylopectin has only one C-chain, which carries the sole-reducing end of the molecule (Finar 1975). This indicates a differences in certain function properties of starches, such as crystallinity, viscosity, shear resistance, gelatinization characteristics, solubility, swelling, and retrogradation as compared with native starch which consists of 10–15% of amylose content.

39.3.2 Morphology

Granule morphology is an important parameter as it is helpful in identifying and differentiating the starch obtained from different botanical sources. Starch granules can be classified into following categories: large (>25 mm), medium (10–25 mm), small (5–10 mm), and very small (5 mm) (Manek et al. 2005). Starch granules of rice are the smallest granule produced by the plants and have an average size of 3–8 μm (Rajak and Bhuyan 2011). The starch granules of rice are irregular and polygonal in shapes. ABRS starch granules have an average size of 21.75 μm; however, most of the particles falls in the size range of 8–12 μm (Kumar 2011). Owing to the small size rice starch granules, several reports have described the possibility of use as directly compressible excipient for the formulation of tablets (Rajak and Bhuyan 2011).

39.3.3 Solubility

ABRS has a cold water solubility of 1.01 $\pm$ 0.08 g/dm^3; this indicates the low cold water solubility of ABRS (Rajak and Bhuyan 2011). The low cold water solubility of ABRS can be attributed to its high content of amylopectin (>99%). However the solubility in water can be increased, when the dispersion is heated to a certain value also known as gelatinization temperature (Builders and Arhewoh 2016). Gelatinization is a property, by virtue of which starch undergoes an enormous swelling, increased viscosity, translucency, and solubility. These changes often involve the breakage of the hydrogen bonds within the starch granules and allow water to enter the granules to make it swell (Shimelis et al. 2006). DSC study suggests that the gelatinization of ABRS is 64 °C, which complies with that of the waxy rice (64–71 °C); however, complete gelatinization does not occur below 76–78 °C up to 45 min of heating; this may be due to water acting as a plasticizer and mobility enhancer in the starch/water system by depressing the thermal transition (Sachan et al. 2012).

39.3.4 Viscosity

The viscosity of waxy starch such as ABRS is higher than that of non-waxy starches. This is due to the presence of high amylopectin content which imparts high degree of branching and very high weight and average molecular weight. Bhattacharya et al. investigated the viscosity of ABRS using cone plate type Brookfield programmable viscometer (Sachan et al. 2012). The results indicated that the viscosity increases with an increase in concentration of starch.

39.3.5 Organoleptic Properties

ABRS being waxy rice have higher amylopectin content than amylose; this increase in amylopectin gives ABRs its characteristic dull milky white color (Sachan and Bhattacharya 2006).

39.4 Micrometricsproperties of ABRS

Micromeritics is the science and technology of small particles. It includes the study of fundamental and derived properties of individual and collection of particles. The knowledge and control of the particles are an important aspect in pharmaceuticals (Sinko and Singh 2006).

39.4.1 Particle Size and Shape

The size and shape of particles are important parameters as they govern every and every manufacturing step which includes mixing, granulation, drying, milling, coating, and compression. Particle size is an important parameter for solid dosage form and biphasic liquid dosage form as it influence the efficacy and stability of the formulation. Similarly particle shape plays an important role in particle size distribution, the shape of a particle can be symmetrical or asymmetrical and therefore difficult to measure by a meaningful diameter. Particle size is usually expressed as the equivalent spherical diameter to correlate with the size of particles to that of a sphere with the same diameter, surface area, and volume. The measurement of size of irregular particles depends upon which type of method is applied (Lowell and Shields 2013; Maheshwari et al. 2018).

39.4.2 Flow Properties

Flow property is an important characteristic of powder for the pharmaceutical manufacturing process. Manufacturing of solid dosage forms such as tablets involves the filling of tablets die with powder or granules based on volume; therefore the flow property determines the various crucial parameters of tablets such as weight, hardness, and content uniformity. Similarly the manufacturing of capsules involves the filling of powder or granules in the empty capsule shells. Therefore assessments of flow properties are important prior to the manufacturing of pharmaceuticals. There are various methods available to measure the powder flow (Shah et al. 2008). The compendial methods include measurement of the angle of repose (USP 2007a, b), bulk density, tapped density (USP 2007a, b), Carr's compressibility index (Carr 1965), or Hausner ratio (Hausner 1967). These methods suffer from various limitations including reproducibility, performance conditions, and predictability. For instance, powder flow measurement by the angle of repose has limitations for very cohesive powder which doesn't flow through the funnel, and vibrating the funnel introduces inherent variability in measurement technique. The other compendial methods of powder flow characterization which includes measurements of bulk and tap density which is semi-quantitative (Shah et al. 2008).

Prakash et al. extracted starch from ABR by alkali extraction method using 0.4% caustic soda and evaluated its flow properties (Ahmad and Bhattacharya 2010). Evaluation parameters were true density, bulk density, tapped density, Hausner's ratio, Carr's compressibility index, and angle of repose. Their findings indicated that starch from Assam bora rice is not porous and is poor-flowing powder; this is reflected by the low bulk and tapped densities (0.31 ± 0.06 and 0.46 ± 0.01, respectively); and the poor-flowing property of ABRS was further confirmed by the values of Hausner ratio (1.47 ± 0.06), Carr's compressibility index (31.93 ± 2.88%), and angle of repose (40.31 ± 1.52°). Similar results were reported describing the poor flow property of ABRS (Sharma et al. 2011; Rajak et al. 2014).

However, Ahmad et al. reported good flowing property of ABRS. The good flowing property is indicated by the higher value of tapped and bulk densities as compared to the other reports (Ahmad et al. 2012a, b, c). The value of Hausner's ratio (1.18 ± 0.001), Carr's compressibility index (15.92 ± 1.23), and angle of repose (23.02° ± 1.01°) also indicated the good flow property. The moisture content of ABRS was found to be low. The difference in the results in flow property of ABRS obtained by different investigators may be attributed to the extraction procedure. M.Z. Ahmad et al. used 0.01 M sodium hydroxide solution for the extraction of starch rather than the use of 0.2–0.4% solution. Therefore, it may be concluded that optimal modification in the extraction procedure can lead to improvement in the flow properties of starch which is necessary requirement for pharmaceutical excipients.

39.5 Application of ABRS

39.5.1 Pharmaceutical Excipients

Excipients are the substances other than the active pharmaceutical ingredients (API) which are used in pharmaceutical formulations with the aim to provide stability, safety, and integrity to the pharmaceutical formulations. Native and modified starch has been widely used as pharmaceutical excipients. Excipients based on starch are widely used owing to their low cost, biocompatibility, safety, and product quality. Based on the properties, starch has been also evaluated as novel drug delivery carriers.

39.5.1.1 ABRS as Binder

Binders are pharmaceutical excipients which are added to tablet formulation to increase plasticity as well as to increase interparticulate bonding strength in tablets. Binders also increase the degree of consolidation or compactions while decreasing the brittle fracture tendency during tableting. Rajak et al. prepared paracetamol tablets (400 mg) using ABRS as a tablet binder and evaluated the tablets against tablets prepared by using gelatin as tablet binder (Shah et al. 2008). The binder concentrations in both cases used were 2%, 4%, 6%, and 8% w/w. The evaluations of the prepared tablets were conducted by measuring weight variation, tablet hardness, friability, disintegration, and dissolution study. From the results obtained, it can be concluded that the tablets prepared by using ABRS at a binder concentration of 8% w/w show more optimum results; also the gelatin and ABRS showed comparative effectiveness as binders in paracetamol tablets. Therefore, ABRS could be used as a better binder in tablet formulations comparable to the gelatin powder.

39.5.1.2 ABRS As Directly Compressible Agents

Directly compressible excipients are pharmacologically inactive, non-medicinal substances which may be compacted under force with no difficulty and which may do so even when mixed with active drug substances without significant change in tablet quality, in addition to their good compression property, some of the directly compressible agents also possess good flow property and high bulk density (Sharma et al. 2011). The advantages of the direct compression method of tablets preparation lead to the extensive use of the directly compressible excipients; this is supported by the fact that about 50% of worldwide tablet production is made by this method (Sharma et al. 2011).

Sharma et al. explored the potential application of ABRS as a directly compressible agent in tablet formulations (Sharma et al. 2011). The atorvastatin chloride tablets were prepared by using unmodified ABRS, acid, and thermally modified ABRS, and spray-modified ABRS and MCC were used for the preparation of different tablet formulations.

The results indicate that spray modified and acid-modified showed good hardness, and the results obtained after evaluation of all the prepared tablets on the based on weight variation, friability, drug uniformity, disintegration time, and dissolution were comparable with the tablets prepared by using MCC as directly compressible agent. Similar investigation was performed by Rajak et al. by preparing ketotifen tablets using ABRS and comparing it with tablets prepared by using pregelatinized starch. The results obtained after evaluating the ABRS as directly compressible excipients were found to be comparable with that of the tablets prepared by using pregelatinized tablets. The above studies serve as basis for further evaluation of ABRS as directly compressible excipients.

39.5.2 ABRS as Plasma Volume Expander

Plasma volume is a crucial and important factor in maintaining hemodynamic and tissue oxygenation. Decrease in the amount of circulating blood may be the result of acute bleeding, sepsis, or any kind of polytrauma, in such circumstances rapid restoration of normal blood volume and sufficient filling of vascular components have to be guaranteed, else the situation may turn fatal. Plasma volume expanders serve in restoring the deficient volumes; these mainly consist of polymeric substances which are isotonic to the blood. Plasma volume expanders work by the virtue of increased osmotic pressure. Most of the plasma volume expanders are not physiologically homogeneous, besides they are insufficiently excreted in urine, which leads their accumulation in body organs (Kulicke et al. 1993; Kulicke and Heinze 2006; Dewachter et al. 1992).

ABRS is mainly consisting of amylopectin, which is similar in compatible with physiological tissues, besides being unaffected by the presence of the amylase. Bhattacharya et al. evaluated the potential of ABRS as plasma volume expander by polymer analysis (Ahmad et al. 2012a, b, c). ABRS was characterized by determining the FTIR spectra, degree of branching by H1 NMR, osmotic pressure by internal measurement technique, establishment of Mark-Houwink relationship, and determination of molecular weight-viscosity relationship. From their findings it can be concluded that ABRS posses the required characteristics to be potentially used as plasma volumes expanders, owing to its biocompatibility to physiological tissues.

39.5.3 Advanced Drug Delivery

ABRS was investigated as novel colon-targeted mucoadhesive microspheres for the delivery of 5-fluorouracil to the colon (Ahmad et al. 2012a, b, c). The microspheres were formulated by a double emulsion evaporation method using Tween-80 as an emulsifying agent. The prepared microspheres were characterized by evaluating their size, shape, surface morphology, size distribution, drug incorporation efficiency, and in vitro and in vivo drug release studies. From the release studies, it can be concluded that the release of 5-fluorouracil is sufficiently less in the stomach and small intestine as compared to the colon; this results in reduced systemic side effect of 5-fluorouracil; and this is supported by the decreased level of liver enzymes, serum albumin, creatinine, leucocytopenia, and thrombocytopenia in animals receiving 5-fluorouracil-loaded mucoadhesive microsphere.

ABRS was also investigated as novel colon-targeted bioadhesive microsphere loaded with metronidazole (Ahmad et al. 2012a, b, c). The microspheres were prepared by double emulsion evaporation techniques. The prepared microspheres were obtained with a uniform spherical shape, with sufficiently high retention time. In vitro drug release study in different physiological environment indicates sufficiently high drug release in target site. The drug release pattern was also found to be over a longer period of time. In vitro bacterial inhibition studies show an equivalent or higher zone of inhibition against bactericides fragile than that of marketed formulations. In vivo studies showed that the ABRS microspheres remained attached to the GIT and release the drug once they reach the colon, the release of drug from the microsphere is due to microbial digestion of ABRS by microbial flora residing in the colon, and the results obtained from in vivo studies comply with that of the results obtained from the in vitro drug release studies. These studies show the potential application of ABRS as drug delivery system for colon targeting.

Sharma et al. investigated the use of natural polysaccharides for controlled release of metformin hydrochloride from the microspheres by the emulsion solvent diffusion method (Sharma et al. 2013). Polysaccharides used in the microsphere were obtained from *Dillenia indica* L., *Abelmoschus esculentus* L., and Assam bora rice starch. The prepared microspheres were then evaluated for their particle size, drug entrapment, swelling index, mucoadhesive strength, and in vitro drug release studies. Mean particle sizes of different batches of prepared microspheres were found to be within the range from 77 $\pm$ 1 to 190 $\pm$ 10 μm, as shown in the microspheres which were found to be uniformly sized and spherical with maximum drug entrapment efficiency of 84.7 $\pm$ 2.3%. Drug release from the prepared microspheres was found to diffusion (non-Fickian) controlled as evident from the correlation coefficient obtained from drug release kinetic study. The controlled release of drug from the microsphere may be attributed to higher content of amylopectin present in ABRS rice and pectic polysaccharides in *Dillenia indica* L. and *Abelmoschus esculents* L. extracts.

39.6 Conclusion

The applications of native starch are increasing exponentially in different fields including pharmaceutical, food industry, film preparation, etc. The unique property of starch granules of ABRS is remarkable. ABRS is also a good candidate to use as slowly digestible starch (SDS) which can be a valuable co-product to incorporate in food products as well as supplements. The long chain of amylopectin structure abundant in starch granules plays the ideal role to provide the SDS property. Natural as well as modified form of ABRS is used as biopolymer, bio-carrier, micro carrier, nano carrier, etc. in the bunch of drug delivery systems.

39.7 Future Challenges

Native starch is considered as promising biopolymer in drug delivery systems. Assam bora rice starch has different effective properties. In this review, we specially focus on Assam bora rice starch to profile its different applications along with physicochemical and powder flow properties. But some research work on Assam bora rice starch still need to be establish in different aspects. Future investigations required seeing the different chemical modification along with the length of the amylose and amylopectin chains distribution in native starch system. The advance physical and chemical approaches are required to explore the better functional properties and molecular strategies of Assam bora rice starch in future research.

References

Ahmad MZ, Bhattacharya A (2010) Isolation and physicochemical characterization of Assam bora rice starch for use as a plasma volume expander. Curr Drug Deliv 7:162–167

Ahmad MZ, Akhter S, Ahmad I et al (2012c) *In vitro* and *in vivo* evaluation of Assam bora rice starch-based bioadhesive microsphere as drug carrier for colon targeting. Expert Opin Drug Deliv 9:141–149

Ahmad MZ, Akhter S, Anwar M et al (2012b) Compatibility and compressibility studies of Assam Bora rice starch. Powder Technol 224:281–286

Ahmad MZ, Akhter S, Anwar M et al (2012a) Assam Bora Rice starch based biocompatible mucoadhesive microsphere for targeted delivery of 5-fluorouracil in colorectal cancer. Mol Pharm 9:2986–2994

Builders PF, Arhewoh MI (2016) Pharmaceutical applications of native starch in conventional drug delivery. Starch/Stärke 68:1–10

Carr RL (1965) Evaluating flow properties of solids. Chem Eng 72:69–72

Deka SD, Sarmah B, Sharma S (2014) Conservation and utilization of indigenous rice varieties for a sustainable livelihood security. Int J Environ Res Dev 4:291–296

Dewachter P, Laxenaire MC, Donner M et al (1992) *In vivo* rheological study of plasma substitute. Ann Fr Anesth 11:516–525

Finar IL (1975) Organic Chemistry 1: The fundamental principles, 6th Ed

Hausner HH (1967) Friction conditions in a mass of metal powder. Int J Powder Metal 3:7–13

Kulicke WM, Heinze T (2006) Improvement in polysaccharides for use as blood plasma expanders. Macromol Symp 231:47–59

Kulicke WM, Rossner D, Wiebke K (1993) Characterization of hydroxyethyl starch by polymer analysis for use as plasma volume expander. Starch/Starck 4:445–450

Kumar N (2011) Fabrication and Biopharmaceutical Characterization of microparticulate Controlled drug Delivery System. Composed of Assam *Bora Rice* as Possible Mucoadhesive Backbone. Dissertation, Dibrugarh University

Lowell S, Shields JE (2013) Powder surface area and porosity. Springer Science & Business Media

Maheshwari R, Todke P, Kuche K, Raval N, Tekade RK (2018) Micromeritics in pharmaceutical product development. In: Dosage form design considerations. Academic Press, pp 599–635

Manek RV, Kunle OO, Emeje MO et al (2005) Physical, thermal and sorption profile of starch obtained from Tacca leontopetaloides. Starch/Stärke 57:55–61

Moorthy SN (2002) Physiochemical and functional properties of tropical tuber starches: a review. Starch/Stärke 54:559–592

Rajak P, Bhuyan B (2011) Role of *Assam Bora* Rice starch as direct compressible tablet excipient. World J Pharm Pharm Sci 2:4150–4158

Rajak P, Nath LK, Bhuyan B (2014) Application of Assam bora rice starch as a binder in formulation of paracetamol tablets. Int J Pharm Pharm Sci 6:118–120

Reis AV, Guilherme MR, Moia TA et al (2008) Synthesis and characterization of a starch-modified hydrogel as potential carrier for drug delivery system. J Polym Sci Polym Chem 46:2567–2574

Rodrigues A, Emejea M (2012) Recent applications of starch derivatives in nanodrug delivery. Carbohydrate 87:987–994

Sachan NK, Bhattacharya A (2006) Evaluation of Assam Bora Rice starch as a possible natural mucoadhesive polymer in the formulation of microparticulate control drug delivery systems. J Assam Sc Soc 47:34–41

Sachan NK, Ghosh SK, Bhattacharya A (2012) Differential scanning calorimetric analysis of drug-polymer interaction in glutinous Rice based microbeads. Asian J Chem 24:2207–2209

Shah RB, Tawakkul MA, Khan MA (2008) Comparative evaluation of flow for pharmaceutical powders and granules. AAPS Pharm Sci Tech 9:250–258

Sharma HK (2014) Bora Rice: a promising pharmaceutical. Rice Res: Open Access 3:1–2

Sharma HK, Lahkar S, Nath LK (2013) Formulation and in vitro evaluation of metformin hydrochloride loaded microspheres prepared with polysaccharide extracted from natural sources. Acta Pharma 63:207–220

Sharma HK, Mukherjee A, Nath LK (2011) Evaluation and comparison of treated-untreated Assam bora Rice flour for use as directly compressible agent. Int J Curr Biomed Phar Res 1:173–177

Shimelis E, Meaza M, Rakshit S (2006) Physico-chemical Properties, Pasting Behavior and Functional Characteristics of Flours and Starches from Improved Bean (*Phaseolus vulgaris* L.) Varieties Grown in East Africa. Agricultural Engineering International: the CIGR E.J., Manuscript FP 05 015. VIII (2006). https://cigrjournal.org/index.php/Ejounral/article/view/629/623. Accessed 30 Oct 2019

Sinko PJ, Singh Y (2006) Micromeritics. In: Martin's physical sciences and pharmaceutical sciences: physical chemical and biopharmaceutical principles in pharmaceutical science, 6th edn. Lippincott Williams & Wilkins, a Wolters Kluwer business, China, pp 442–491

Tester RF, Karkalas J, Qi X (2004) Starch-composition, fine structure and architecture. J Cereal Sci 39:151–165

USP. <1174> Powder flow. USP30 NF 25 (2007a). https://www.drugfuture.com/pharmacopoeia/usp35/PDF/0801-0804%20%5b1174%5d%20POWDER%20FLOW.pdf accessed on 21 Nov 2019

USP. <616> Bulk density and tapped density. USP30 NF 25 (2007b). https://www.usp.org/sites/default/files/usp/document/harmonization/gen-chapter/bulk_density.pdf accessed on 21 Nov 2019

Natural Excipients in Pharmaceutical Formulations

40

Pradeep Singh, Garima Mishra, and Subas Chandra Dinda

Abstract

Pharmaceutical excipients are the backbone of pharmaceutical formulations. These are inert substances added to contribute physicochemical stability in vitro as well as in vivo to the active pharmaceutical ingredients during formulation development. Excipients significantly affect the quality of a finished pharmaceutical dosage forms. They are intended to be used to facilitate the pharmacokinetic as well as pharmacodynamic profiles of the medicament. Furthermore, excipients also influence the physical properties of medicinal products including solubility, consistency, weight and volume, release rate, etc. which are required for administration of adequate dose of active drug in a particular dosage formulation. Currently, numerous synthetic excipients have been explored in pharmaceutical formulations such as tablets, capsules, novel drug delivery systems, etc. However, they are found to have some unwanted side effects in vivo after administration. Therefore, there is a quest to have some excipients from natural origin with minimum side effect or toxic manifestations.

Nature is enriched with a wide array of precious components which directly or indirectly are beneficial for maintaining the good health of living organisms.

The original version of this chapter was revised by updating affiliation for Subas Chandra Dinda. The correction to this chapter can be found at https://doi.org/10.1007/978-981-15-8127-4_51

P. Singh (✉) · G. Mishra
Department of Pharmacy, College of Health Sciences, Debre Tabor University, Debre Tabor, Amhara, Ethiopia
e-mail: pradeep2682@gmail.com; gp_nmr2002@yahoo.co.in

S. C. Dinda
Department of Pharmaceutics, College of Pharmacy, Teerthanker Mahaveer University, Moradabad, UP, India
e-mail: subasdinda@gmail.com

S. C. Mandal et al. (eds.), *Evidence Based Validation of Traditional Medicines*,
https://doi.org/10.1007/978-981-15-8127-4_40

Natural excipients such as binders, diluents, sweeteners, colorants, preservatives, film formers, etc. and their derivatives are found abundantly in plants, animals, and minerals. Since natural ingredients are inert, biocompatible and biodegradable with minimum toxic effects, as well as cost-effective, hence pharmaceutical industries are showing their interest for these excipients to be used in formulating novel drug formulations, cosmetics, as well as in food products.

This book chapter entitled "Natural Excipients in Pharmaceutical Formulations" has been compiled to explore the extensive information about the excipients of natural origin to be utilized as bulking, binding, film forming, lubricating, sweetening, preservative, coloring agent, etc. in design and development of various pharmaceutical dosage forms.

Keywords

Excipients · Gums and mucilage · Polysaccharides · Colorants · Dosage form

Abbreviations

3,6-AG	3,6-Anhydro-D-galactose
AGP	Andrographolide
ALG	Alginic acid
CG	Carrageenan
CPM	Chlorpheniramine maleate
CR	Copal resin
EC	Ethyl cellulose
GalA	D-galacturonic acid
LA	Lignosulfonic acid
LBG	Locust bean gum
LCCs	Lignin–carbohydrate complexes
LER	Lignin epoxy-modified resin
NDDS	Novel drug delivery systems
NPs	Lignin nanoparticles
SA	Sanguinarine
Sclg	Scleroglucan

40.1 Introduction

Pharmaceutical formulations are composed of many components including active pharmaceutical ingredient(s) along with other excipients such as binders, diluents, disintegrants, etc. These ingredients contribute a key role in maintaining the quality, safety, and efficacy of the formulation to be manufactured. Several studies report that the extensive use of synthetic excipients in developing formulations poses a vexed problem due to their toxicity, incompatibility, nonavailability, and huge cost. The emergence of natural excipients in pharmaceutical field led to the need for novel ways to prepare safer and effective medicines for the patients. However, several challenges have been found to be associated with chemicals derived from natural sources including their structural complexity in mixture and productivity which may differ according to the geographical location and season in addition with slow and expensive isolation and purification processes (Lam 2007; McChesney et al. 2007).

There are tremendous sources including plants, animals, marines, etc. to obtain natural excipients. In addition, a wide range of derivatives of natural excipients can also be prepared through modification techniques such as grafting and cross-linking. In recent years, the research has been focused on plant-derived ingredients like polysaccharides, gums, and mucilages and volatile oils. These plant-based polymers have potential applications in pharmaceutical preparations such as solid and liquid dosage forms, controlled and sustained drug delivery, osmotic drug delivery, micro-encapsulation, injectables, buccal films, and implants. These ingredients are employed in the form of binders, disintegrators, diluents, emulsifiers, suspending agents, coloring agent, flavoring agents, gelling agents, bioadhesives, and viscosity enhancers in formulations (Sakarkar 2012).

In this context, the book chapter highlighted brief information collected from various research and review articles, regarding the naturally derived biopolymers employed in food, cosmetics, and pharmaceutical domains along with their biomedical applications.

40.2 Classification of Natural Excipients

Natural excipients are commonly classified according to their origin, chemical nature, and applications.

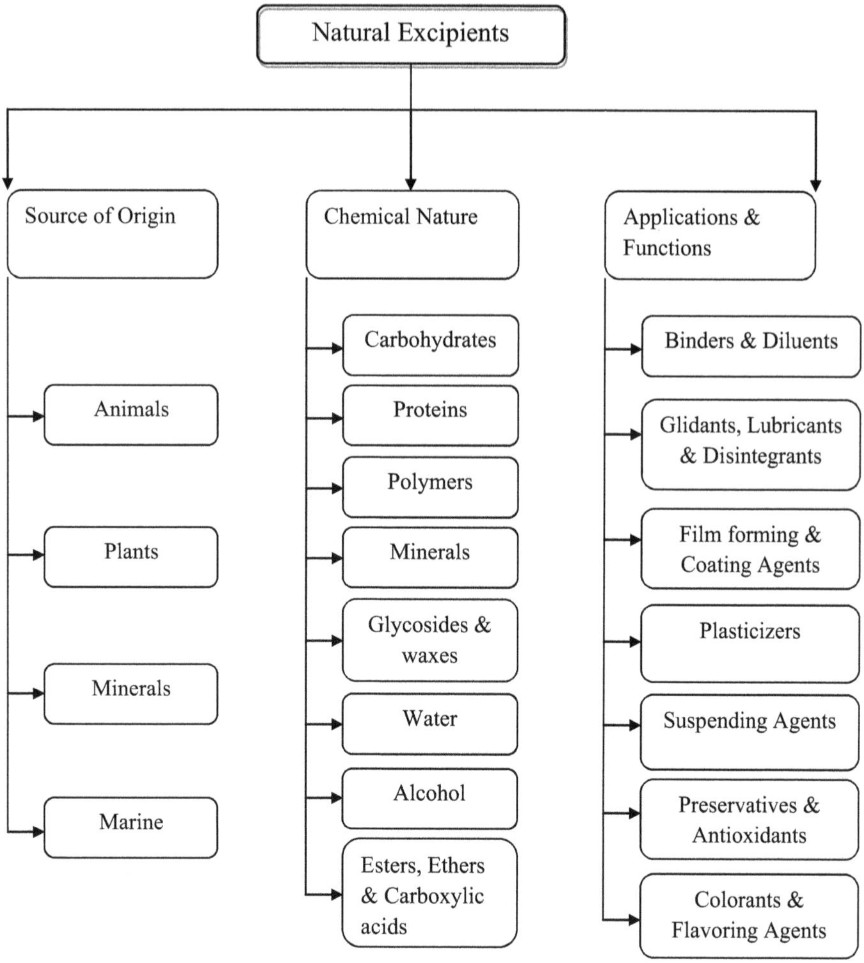

40.2.1 Starch

Starch is the major dietary source of carbohydrate in plants. It is present in root, tubers, leaves, stem, fruits, seeds, legumes, and cereals of higher plants (Hu et al. 2015). Biochemically, starch is synthesized from infinite numbers of glucose molecules by photosynthesis (Teramoto et al. 2003). Cereals (40–90%), roots (30–70%), tubers (65–85%), legumes (25–50%), and fruits are the rich sources of starch [Santana and Meireles 2014].

Chemically, starch is a polysaccharide which constitutes exclusively a large number of D-glucopyranose units joined together by α-1,4 and α-1,6 glycosidic bonds. It is primarily made up of two inherent incompatible components: amylose (15–30%), a linear chain polymer, and amylopectin (70–85%), a branched chain

Table 40.1 Modified starches and their applications

Materials	Applications	Reference
Dextrin, hydroxyethyl starch, cyclodextrin	Solid dispersions of phenylbutazone and nalidixic acid	Chowdary and Rao (1994), Schmidt and Broegmann (1988)
Maltodextrin, hydrogenated maltodextrins	Vitamin C effervescent tablets	Schmidt and Broegmann (1988)
Dialdehyde starch	Sustain release dosages forms of isoniazid and isonicotinic acid	Hiraku and Nagai (1986)
Pregelatinized starch, short-chain amylase like amylodextrin	Control drug delivery systems	Herman and Remon (1989); Te Wierik et al. (1996); Te Wierik et al. (1993)

polymer (Robyt 2008). Despite tremendous use of starch in a variety of industries like cosmetics, textile, food, and paper, it has also achieved a great position in pharmaceutical field. Native starch is employed as binder, diluents, and disintegrants in solid dosage forms. Moreover, starch finds its novel applications in the area of nasal, periodontal, film-based drug delivery systems (Hartesi et al. 2016). Nowadays, chemical modification of native starch is carried out to alter the physicochemical properties like stability, flowability, and compressibility (Table 40.1). The most commonly used modified starch in pharmaceutical industry is pre-gelatinized starch that is marketed under the name of starch 1500 (Alcazar-Alay and Meireles 2015).

40.2.2 Celluloses

Cellulose, a polysaccharide, is the most important natural polymer of plants. It is the structural or skeleton component of plant cell walls. Cotton seed fibers are the major source of pure cellulose. Other cellulose-producing plants are hemp, jute, corn, flasks, rice, wheat straw, and sisal. Currently, cellulose has also been isolated from bacteria *Acetobacter xylinum* by fermentation of glucose from corn syrup. Cellulose is a water insoluble fiber. Humans cannot digest cellulose in smaller components (Lavanya et al. 2011).

Cellulose has fascinating structure consisting of linear chain of β-1,4-linked D-glucose units. It has parallel arrangements of crystallites and crystallite strands, which are the basic elements of the fibers (Krässig 1996). Supramolecular structure of cellulose reveals the two main regions: amorphous (low ordered) and crystalline (high ordered) (Fainberg and Mikhailov 1966). Presently, cellulose and their derivatives (Table 40.2) of pharmaceutical importance can be prepared by chemical processes like esterification, etherification, and cross-linked or graft copolymerization (Marques-Marinho and Vianna-Soares 2013).

Table 40.2 Cellulose and their derivatives of pharmaceutical importance

Chemical class	Examples	Reaction	Applications
Native cellulose	Microcrystalline cellulose	Controlled hydrolysis of cellulose with mineral acid	An adsorbent, suspending agent, capsule diluent, and bulking agent
Cellulose ether derivatives	Methyl cellulose (MC)	Etherification by methyl group	Emulsifying agent (1–5%), suspending agent (1–2%), capsule disintegrants, viscosity enhancer, topical creams and gel formulations, syrup preparations
	Ethyl cellulose (EC)	Etherification by ethyl group	Viscosity enhancer, flavoring agent. Oral and topical preparations like gels, lotions, creams, thickening agents stabilizer in emulsions
	Hydroxyethyl cellulose	Etherification by hydroxyethyl group	Suspending agent, thickening agent viscosity enhancer, cosmetics preparation
	Hydroxypropyl cellulose	Incorporation of hydroxypropyl group	Suspending agent, thickening agent viscosity enhancer, topical formulations
	Hydroxypropyl methyl cellulose	O-methylated and O-(2-hydroxypropylated derivative)	Foaming agent, emulsifying agent, dispersing agent, thickening agent, Controlled and sustained release agent
	Carboxymethyl cellulose (CMC)	Polycarboxymethyl ether of cellulose	Capsule disintegrant, emulsifier, suspending agent, gel-forming agent
Cellulose ester derivatives	Cellulose acetate	Acetylation of hydroxy group	Capsule diluent, filler and taste-masking agent
	Cellulose acetate phthalate (CAP)	Obtained by reaction of phthalic acid and partial ester of cellulose	Enteric coating agents (0.5–9%), delayed and controlled release systems

40.2.3 Hemicelluloses

Hemicelluloses are the heterogeneous group of polysaccharide in woody plants and cereals. These are present in plants in association with lignin and cellulose. Hemicellulose offers rigidity to the cell wall and acts as adhesive between single cells in plants (Muchlisyam et al. 2015). The composition of hemicellulose varies according to the type of plant. Structurally, they are composed of both linear and branched monomeric sugar moieties such as xylans (β-1,4-linked D-xylose units), mannans (β-1,4-linked D-mannose units), arabinans (α-1,5-linked L-arabinose units), and galactans (β-1,3-linked D-galactose units) (Belgacem and Gandini 2008). Various types of hemicelluloses and their sources with composition are as depicted in the Table 40.3. Hemicelluloses are potentially applied in pharmaceutical field in

Table 40.3 Types of hemicelluloses and their sources with composition

Type of hemicellulose	Source	Composition (%)
Arabinoglucuronoxylans	Softwood (non-woody materials)	7–10
Galactoglucomannans	Softwood	20
Glucomannans	Harwood	2–5
Arabinogalactans	Larch wood	5–35
Glucuronoxylans	Hardwood	15–30
Arabinoxylans	Brans and annual plant	Variable

developing drug formulation excipients such as binders, thickeners, stabilizers, and disintegrators (Ma et al. 2017). In recent years, hemicellulose-based hydrogels have been designed and evaluated for novel drug delivery system. Besides, the hydrogels can be applied in contact lenses. The current investigations on hemicellulose-based hydrogels demonstrated the blending of hemicellulose with chitosan and polyvinyl alcohol to create hybrid hydrogels responsible for antibacterial function (Hu et al. 2018).

40.2.4 Chitins and Chitosan

Chitin is a natural biopolymer and comprised of β-(1 → 4)-linked 2-acetamido-2-deoxy-β-D-glucose (N-acetylglucosamine). It serves as structural component in the exoskeleton of arthropods as well as cell walls of fungi and yeast. Commercially, two marine crustaceans, namely, crab and shrimp shells are the major sources of chitin (Rinaudo 2006; Younes and Rinaudo 2015).

Chitosan is a polycationic polysaccharide, derived from chitin through deacetylation process. Recent study involved the isolation of chitosan from fungal strains, *Aspergillus niger*, *Rhizopus oryzae*, *Lentinus edodes*, *and Pleurotus sajor-caju,* and two yeasts, *Zygosaccharomyces rouxii* and *Candida albicans* by fermentation technique (Pochanavanich and Suntornsuk 2002).

It is a linear polymer consisting of S (1–4)-linked 2-acetamido-2-deoxy-S-D-glucopyranose and 2-amino-2-deoxy-S-D-glycopyranose units. Molecular weight and degree of deacetylation are the most important characteristics of chitosan which have determining effects on solubility, film-forming capacity, and optical and structural properties (Hon 1996; Dutta et al. 2004). Notably, the presence of primary amino group in chitosan imparts promising characteristics such as cationic nature, mucoadhesion, controlled release, in situ gelation, permeation enhancement, and antimicrobial properties (Ali and Ahmed 2018). The multiple applications of chitin and chitosan (Fig. 40.1) have been identified in pharmaceutical as well as biomedical field (Younes and Rinaudo 2015).

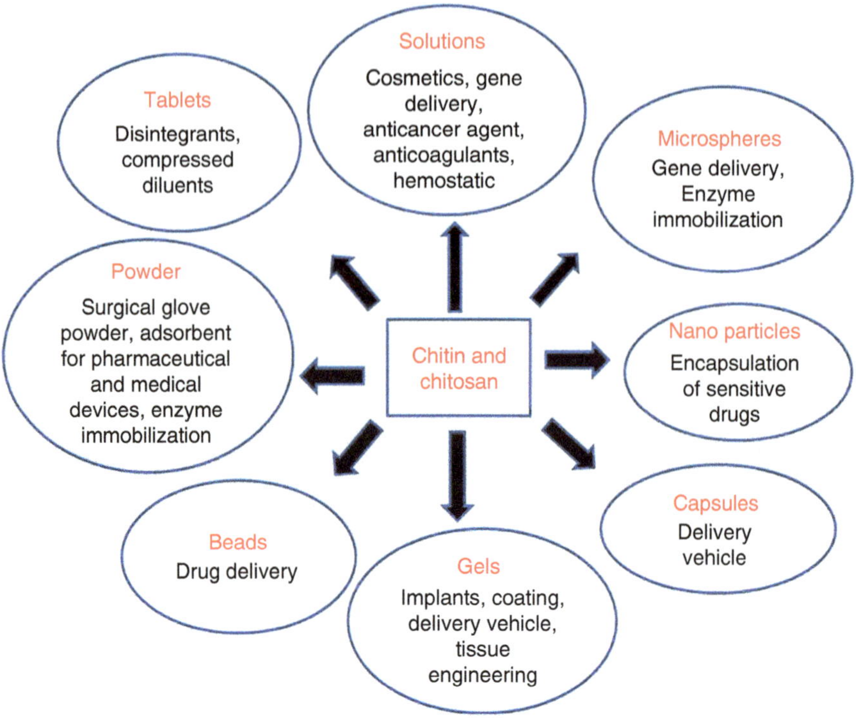

Fig. 40.1 Pharmaceutical and biomedical applications of chitin and chitosan

40.2.5 Lignins

Lignins are the naturally occurring biopolymers and extensively found in secondary cell walls of the plants. Lignin in combination with cellulose provides mechanical strength to the plant cell walls. Lignin, a highly complex in structure, consists of randomly cross-linked phenylpropanoid units (monolignols) such as cumaryl, coniferyl, and sinapyl alcohol. The monolignol units when connected with the lignin polymer are called guaiacyl (G), syringyl (S), and p-hydroxyphenyl (H) units.

Phenylpropane units are randomly cross-linked to each other with a variety of chemical bonds such as βO-4-aryl ether linkages, α-O-4-aryl ether, 4-O-5-diaryl ether, β-5-phenylcoumaran, 5–5-biphenyl, β-1-(1,2-diarylpropane), and β-β-(resinol) (Agrawal et al. 2014; Vanholme et al. 2010; Witzler et al. 2018). Interestingly, lignin is a nonsugar macromolecule found in the plants. At present, lignin and its derivatives are broadly investigated in numerous pharmaceutical and medical domains as described in the Table 40.4, and pharmaceutical perspectives of lignins are presented in Table 40.5.

Table 40.4 Chemically derived lignins and their biomedical applications

Ingredients	Source	Uses	References
Native lignins	*Acacia nilotica wood* (Mimosaseae)	Oxidative stress and diabetes, mouth, bone, and skin cancer	Barapatre et al. (2016); Kalaivani and Mathew (2010)
Modified alkali lignin	Microbial biotransformation by the ligninolytic fungi *Aspergillus flavus* and *Emericella nidulans*		Quesille-Villalobos et al. (2013)
Lignosulfonic acid (LA)	Sulfite pulping of softwood	α-Glucosidase inhibitory activity, raw material in vanilla flavor, vanillin, anti-HIV activity	Andrei et al. (2011); Fargues et al. (1996); Hasegawa et al. (2015)
Lignin–carbohydrate complexes (LCCs)	Pinecones and pine nut shells, *Lentinus edodes mycelia* and *Sasa senanensis* Rehder leaves extracts	Anti-UV activity (cosmetic sun care products)	Sakagami et al. (2016)
	Pinecone LCC from *Prunella vulgaris*	Anti-herpes activity	Zhang et al. (2007)
	LCCs from *P. anisum*	Antiviral activities	Lee et al. (2011)
Sulfonated lignins	–	Anticoagulants	Monien et al. (2006)
Sulfated low-molecular-weight lignins	–	Anticoagulants	Henry and Desai (2014)
Sulfated β-*O*-lignins	–	Anticoagulation and antiplatelet actions	Mehta et al. (2016)

40.2.6 Pectin

Pectin is a heteropolysaccharide and widely extracted from the cell walls of higher plants. Citrus fruits, apple, plume, gooseberry, oranges, cherries, and grapes are the rich sources of pectin. It is mainly composed of D-galacturonic acid (GalA) units linked together by α-(1–4) glycosidic linkage (Mukhiddinov et al. 2000). Pectin has promising applications in pharmaceutical industry (Sriamornsak 2006). These are used as binding, thickening, emulsifying, suspending, and stabilizing agents in many dosage forms. Pectin hydrogels have been exploited as binder in tablet dosage forms and controlled release matrix tablet formulations. Sungthongjeen et al. (1999) studied HM-pectins in controlled release matrix formulations. Various pharmaceutical applications of pectin have been briefly mentioned in Fig. 40.2.

Table 40.5 Pharmaceutical perspectives of lignins

Formulation	Use	Reference
Lignins nanoparticles (NPs)	Controlled release of different herbicides and pesticides	Fernández-Pérez et al. (2007); Pereira et al. (2003)
	Stabilizers of cosmetic	Frangville et al. (2012)
Water-dispersed lignin NPs	Stabilize Pickering emulsions	Lievonen et al. (2016)
Lignin-based NPs from *A. tequilana* Weber bagasse	Sun protection	Gutiérrez-Hernández et al. (2016)
Lignin coated with silver NPs	Antibacterial effect	Klapiszewski et al. (2015)
Superabsorbent hydrogels prepared by xanthan and lignin epoxy-modified resin (LER) mixture, cross-linked with epichlorohydrin	Drug delivery of bisoprolol fumarate	Răschip et al. (2015)

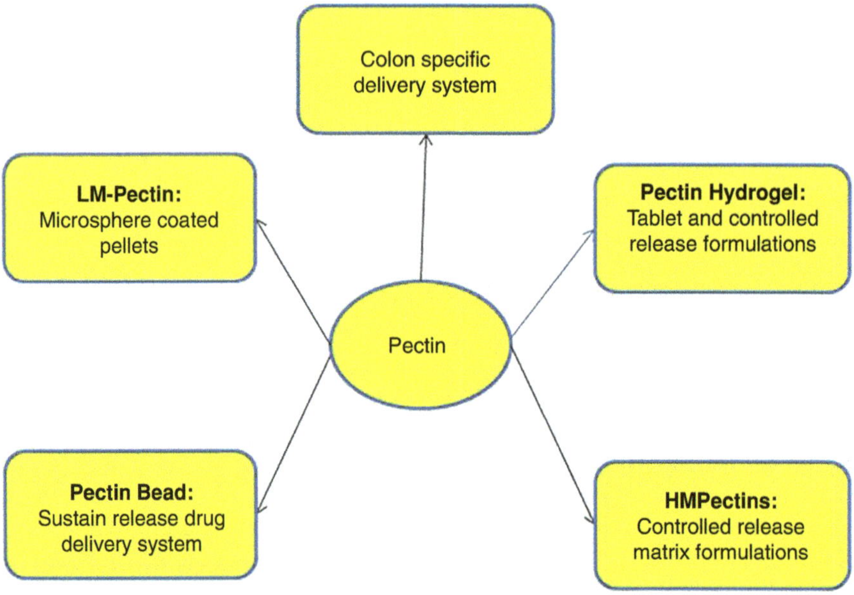

Fig. 40.2 Various applications of pectin

40.2.7 Alginates (Alginic Acid)

Alginates are the marine-derived polysaccharides comprising linear chain of β-d-mannuronic acid and α-l-guluronic acid residues connected with α and β- 1 → 4 glycosidic bonds. Alginates are widely found in brown seaweeds such as *Laminaria*, *Ascophyllum*, *Macrocystis*, *Sargassum*, *Durvillaea*, *Lessonia*, and *Turbinaria*

Table 40.6 Pharmaceutical applications of alginates and their derivatives

Types	Pharmaceutical applications
Alginic acid	Tablet binder and disintegrant, sustained release and release-modifying agent, taste-masking agent, thickener, suspending and viscosity-increasing agent, stabilizer
Sodium alginate	Suspending and viscosity-increasing agent, tablet and capsule disintegrant, tablet binder, stabilizer, sustained release agent, diluent in capsule formulation, thickener
Calcium alginate	Tablet disintegrant
Ammonium alginate	Color diluent, emulsifier, film former, humectants
Propylene glycol alginate	Stabilizer, emulsifier, suspending and viscosity-increasing agent

Table 40.7 Current investigations on chemically modified alginates and their applications

Alginates material	Active substance	Dosage form	Application
Oxidized-NaALG	Flurbiprofen	Beads	Sustained oral delivery
Sulfated ALG		Microspheres	Reduction of secretion inflammatory cytokines, improvement of the biocompatibility
Propylene glycol ALG	Lysozyme	Microparticles	Protein encapsulation with a sustained release
β-Cyclodextrin-ALG conjugate	Ondansetron	Gel	Controlled drug delivery systems
α-Cyclodextrin-ALG conjugate	*Sphingomonas cloacae*	Bead	Immobilization of bacteria
Galactosylated ALG	Hepatocytes	Microcapsules	Cell carrier with mechanical stability and selective permeability
NaALG-co-polyacrylamide	Famotidine	Hydrogel	Sustained release gastroretentive carrier
Starch–poly (acrylic acid)-NaALG	Diclofenac sodium	Hydrogel beads	Matrices for the oral drug delivery

species. Of these, *Laminaria hyperborea*, *Laminaria digitata*, *Laminaria japonica*, *Ascophyllum nodosum*, and *Macrocystis pyrifera* are the most common marine algae. In addition, bacterial alginates extracted from *Azotobacter* and *Pseudomonas* have also been reported (Lee and Mooney 2012). Alginic acid and its salts are reported to serve as tablet binders and disintegrants, stabilizers in emulsions, and suspending agents (Table 40.6). Recently, chemically modified alginates prepared by different reactions (Table 40.7) have been extensively investigated as excipients in drug delivery systems (Szekalska et al. 2016).

40.2.8 Inulin

Inulin is a heterogeneous polysaccharide and widely distributed in plants like chicory, dalia, Jerusalem artichoke, onion, and garlic. Inulin is composed of β-D-(2 → 1) fructosyl fructose (fractan) units with a starting α–D-glucose unit. The chain length in inulin molecules varies from 2–60 fractan units (Roberfroid 2007). Presence of β (2 → 1) linkage is the unique characteristic of inulin and is responsible for reduced calorific values and dietary fiber effect. Inulin is a popular ingredient of food industry. Recently, inulin and its derivatives have also been extensively used in pharmaceutical industry as excipients. It is being potentially used in drug delivery systems due to its rapid water solubility, stability, and low friability against gastric and intestinal enzymes (Akhgari et al. 2006; Shahwar et al. 2012).

40.2.9 Carrageenans

Carrageenans are the naturally occurring linear and anionic polysaccharides containing both sulfated and nonsulfated residues (Table 40.8). High-molecular-weight carrageenans are composed of alternate units of D-galactose and 3,6-anhydro-D-galactose (3,6-AG) linked by α-1,3 and β-1,4-glycosidic linkage (Michel et al. 1997). Carrageenans are extracted from a number of red seaweeds especially from *Chondrus crispus*, *Eucheuma*, *Gigartina stellata*, *Iridaea*, *Hypnea*, *Solieria*, *Agardhiella*, and *Sarconema* belonging to the family Rhodophyceae. There are mainly three classes of carrageenans designated as kappa (κ), iota (ι), and lambda (λ). They differ in their chemical composition, solubility, viscosity, and gel-forming property with potassium and calcium ions (Necas and Bartosikova 2013). Carrangeenans are widely explored in pharmaceutical arena as a gelling agent/viscosity-enhancing agent for controlled drug release. The current investigations declared the use of carrangeenan as polymer matrix in oral extended-release tablets, as a novel extrusion aid for the production of pellets and as a carrier/stabilizer in micro–/nanoparticles systems (Li et al. 2014). Rodriguez et al. (2014) examined two carrageenan (CG)-based formulations, namely, PC-515 gel and the lubricant Divine 9 for in vitro and in vivo anti-HPV activity against HPV16, 18, and 45 pseudoviruses (PsVs). Another study was designed to formulate the bioadhesive films containing

Table 40.8 Types of carrageenans and their characteristics

Type	Source	Chemical composition		Solubility	Gelling property
		Ester sulfate content	3,6-AG content		
Kappa (κ)	*Kappaphycus alvarezii*	25–30%	28–35%.	Soluble in hot water	Form hard or brittle gels
Iota (ι)	*Eucheuma denticulatum*	28–30%	25–30%.	–	Forms soft and elastic gels
Lambda (λ)	*Gigartina*	32–39%	Nil	Soluble in hot and cold water	No gel-forming ability

κ-carrageenan along with other excipients and ibuprofen as a model drug for buccal drug delivery system (Kianfar et al. 2011). In addition, the blending of κ carrageenan with another polysaccharide chitosan in the presence of cross-linker tripolyphosphate (TPP) led to the development of stable nanoparticles (Khan et al. 2017).

40.2.10 Scleroglucan

Scleroglucan (Sclg), a well-known branched exopolysaccharide, is produced from filamentous fungi of the genus *Sclerotium* predominantly from *Sclerotium rolfsii* (Farina et al. 1998). It is a nonionic, high-molecular-weight homoglucan which is composed of a backbone of a linear chain of β-d-(1–3)-glucopyranosyl and β-d-(1–6)-glucopyranosyl units as a side chain for every three main residues. This water soluble homopolysaccharide on complete hydrolysis liberates D-glucose units (Farina et al. 1998; Rinaudo and Vincendon 1982). Scleroglucan possesses numerous unique physicochemical properties like water solubility, and viscosity, and, interestingly, it exhibits great stability at high temperature and pH. Therefore, it is globally considered as multipurpose compound employed in food, oil, and pharmaceutical industries. Recent investigation on scleroglucan demonstrated the development of a novel spray dried formulation for dry powder inhaler loaded with model drug andrographolide (AGP) which enhanced the in vitro and in vivo lung deposition and pulmonary antihypertensive activity (Mali et al. 2017). In another study, physical hydrogel prepared from high-carboxymethylated derivative of scleroglucan (Scl-CM_{300}A) was evaluated for topical drug delivery system using three different bioactive agents, namely, betamethasone, diclofenac, and fluconazole (Paolicelli et al. 2017). Moreover, physical hydrogels containing various conc. of carboxymethyl scleroglucan in the presence of calcium ion, loaded with diclofenac drug, were investigated as modified drug delivery system for topical use (Corrente et al. 2009). Brief summary of scleroglucan and its chemical derivatives (Coviello et al. 2005) with other pharmaceutical applications has been given in Table 40.9.

Table 40.9 Scleroglucan and its chemical derivatives with other pharmaceutical applications

Ingredient	Applications
Native scleroglucan	Slow release matrix, sustain release tablets, oral dosage forms like benzamide, theophylline
Aldehyde and carboxylated derivatives	Sustain release tablets
Cross-linked scleroglucan	Hydrogels preparations for controlled release delivery
Co-cross-linked scleroglucan	Hydrogels preparations for sustained release tablets

40.2.11 Rosin

Rosin or colophony is a naturally occurring oleoresin extracted from pine trees such as *Pinus soxburghui*, *Pinus longifolium*, and other species such as *Pinus palustris* Miller, *Pinus linnae*, etc. Chemically, it is predominantly comprised of tricyclic diterpene carboxylic acids, namely, abietic and pimaric acids. Presence of hydrophobic hydrophenanthrene ring is the salient feature of rosin that imparts film-forming property. Pharmaceutically, rosin and rosin-based polymers have been potentially used as film formers, coating materials, microencapsulation, and matrix materials in tablet dosage forms as well as drug delivery systems (Fulzele 2003; Yadav et al. 2016). Rosin and its derivatives have been investigated for film-forming property using diclofenac as model drug for sustained drug release (Kim and Fassihi 1997; Pathak and Dorle 1990). Further, rosin in combination with other polymers like polyvinylpyrrolidone and dibutyl phthalate was used to design matrix-type transdermal drug delivery system to prolong the drug release (Satturwar et al. 2005). Some investigations demonstrated the use of rosin polymers in the preparation of controlled release formulations with bioactive agent carbofuran as insecti-nematicide (Choudhary et al. 2006). Moreover, nanoparticles were designed using rosin as excipient and hydrocortisone for controlled release system (Lee et al. 2005). Of late, rosin nanoparticles were synthesized and studied for colloidal stability against numerous factors such as solvent nature, temperature, and pH (Kozlova et al. 2017).

40.2.12 Gums and Mucilages

Gums and mucilages are the translucent and amorphous substances obtained from plants, animals, and microbial sources. Both are composed of polysaccharide units in association with uronic acids. Gums are the breakdown products of cell walls and soluble in water. However, mucilages are the intracellular products of metabolism and exist in the form of slimy mass when combined with water.

Gums and mucilages and their derivatives play a pivotal role in food and pharmaceutical industries. Majority of gums and mucilages have been investigated as key ingredients in many formulations as binders, disintegrants, emulsifying agent, sustaining agents, film-forming agents, hydrogels, and thickening agents. Currently, drug delivery research involving novel drug delivery systems opened a new window to develop gums and mucilage-based formulations for targeting specific sites of delivery (Hamman et al. 2015; Jania et al. 2009).

40.2.12.1 Okra Gum

Okra gum is extracted from the fresh fruits of *Abelmoschus esculentus* (Family: Malvaceae). The gum is composed of a mixture of sugars galactose, galacturonic acid, and rhamnose and small portion of glucose, mannose, arabinose, and xylose. Okra polysaccharide has been reported as a popular excipient in many formulations. Okra polysaccharide was used to prepare sustained release matrix tablets and gastric floating matrix tablets of metformin hydrochloride (Chodavarapu 2011; Tawari 2018). Mucoadhesive gel for nasal drug delivery was prepared using okra mucilage as gel-forming agent and rizatriptan benzoate as model drug (Sharma 2013). In another study, okra mucilage has been investigated as sustained release matrix excipient on in vitro release of lamivudine drug (Palei et al. 2016).

40.2.12.2 Albizia Gum

Albizia gum is extracted from different species of *Albizia* plant such as *Albizia zygia*, *Albizia procera*, *Albizia lebbeck*, and *Albizia stipulata* belonging to the family Leguminosae. This naturally occurring polymer is comprised of β-1-3-linked D-galactose units with some ß1–6-linked D-galactose residues. *Albizia* gum has been reported to be used as emulsifier, binding agent in food and pharmaceutical industry (Ashton et al. 1975; Choudhary and Pawar 2014). In a study, it was found that *Albizia zygia* gum is used as binder and produced tablets with higher mechanical strength in comparison to gelatin (Odeku 2005). Furthermore, microbeads were prepared from *Albizia* gum for the controlled release of ibuprofen (Odeku et al. 2014). Controlled release matrix tablets of paracetamol were developed using *Albizia stipulata* gum as novel excipient. Recent investigation on *Albizia* gum has shown the formulation of floating tablets of cefuroxime axetil for controlling the release of the drug (Swapna et al. 2019; Veenus Seelan et al. 2016).

40.2.12.3 Locust Bean Gum (LBG)

Locust bean is a neutral polysaccharide obtained from the seeds of carob tree *Ceratonia siliqua* (family: Leguminosae). The gum is primarily composed of D-galacto-Dmannoglycan, pentane, proteins, and cellulose. It is slightly soluble in cold water and devoid of gel-forming ability. Hot solutions of LBG when mixed with other hydrocolloids like agar, xanthan, and carrageenan form gel on cooling below the gelling temperature. This natural polymer has diverse range of pharmaceutical applications comprising of production of solid monolithic matrix systems, microparticles, nanoparticles, implants, films, beads, injectable systems, as well as viscous liquid and gel formulations. The gum is widely used as thickener, stabilizer, emulsifier, and gelling agent and also serves as good candidate for controlled drug delivery systems (Beneke et al. 2009; Dionísio and Grenha 2012). A study on locust bean gum has shown the preparation of novel colon targeted drug delivery of mesalamine which is used in the treatment of ulcerative colitis (Sirisha et al. 2018). This gum has also been investigated for its superdisintegrant property in nimesulide oral dispersible tablets (Malik et al. 2011).

40.2.12.4 Tara Gum

Tara gum is derived from the endosperm of seed of *Caesalpinia spinosa* (family: Leguminosae or Fabaceae). The gum is a galactomannan polymer constituting a linear chain of (1 → 4)-β-D-mannopyranose units linked with α-D-galactopyranose units through (1 → 6) linkage. The ratio of mannose to galactose is 3:1 (Final Assessment Report: Application A546 Tara Gum as a Food Additive 2006). *Tara* gum is soluble in cold water and produces viscosity less than gaur gum. *Tara* gum has been investigated as an excellent carrier in production of gastroretentive controlled release tablets (Shin et al. 2006) and as emulsifying agent (Zeng et al. 2007) for various drugs including ciprofloxacin, metformin, and clozapine.

40.2.12.5 Khaya Gum

Khaya gum is a branched polysaccharide and obtained from the incised trunk of the tree *Khaya grandifoliola*, family Meliaceae. The gum consists of D-galactose, L-rhamnose, D-galacturonic acid, and 4-O-methyl-D-glucoronic acid (Aspinall and Bhattacharjee 1970). *Khaya* gum exhibits good binding capacity in tablet formulations and emulsifying property. Several investigations have been carried out on *Khaya* gum as a controlled release agent for paracetamol and indomethacin drugs (Odeku and Fell 2004). Besides, colon targeted drug delivery system using budesonide as active drug and *Khaya* gum as coating material was investigated (Prabhu et al. 2010).

40.2.12.6 Honey Locust Gum

Honey locust gum, commonly known as thorny locust, is obtained from the seeds of a leguminous tree *Gleditsia triacanthos* belonging to the family Leguminosae. The seeds are rich in carbohydrates, proteins, fats, and fibers. Honey locust gum has been investigated as sustaining material in matrix tablets containing theophylline as active drug (Üner and Altinkurt 2004).

40.2.12.7 Guar Gum

Guar gum is a nonionic, water soluble, and gel-forming polysaccharide obtained from the seeds of *Cyamopsis tetragonolobus* (family Leguminosae). It is commonly known as cluster bean, guaran, *Cyamopsis* gum, and guarkernmehl in German. The gum is predominantly composed of high-molecular-weight polysaccharides of galactomannans. Chemically, guar galactomannans constitute a linear chain of (1 → 4)-linked β-D-mannopyranosyl units with (1 → 6)-linked α-D-galactopyranosyl residues as side chains. Guar gum on complete hydration produces a viscous colloidal dispersion. Guar gum is used as binder and disintegrating agent in solid dosage forms, and it is also used as emulsifier, suspending agent, and stabilizer in liquid and topical formulations (Mudgil et al. 2014). In recent years, guar gum has gained its importance as carrier in development of three-layer matrix tablets for controlled drug delivery systems (Chavda et al. 2012; Krishnaiah et al. 2002). Further studies have shown that guar gum is successfully evaluated for the preparation of microspheres for oral and colon targeted drug delivery systems (Shukla et al. 2012).

40.2.12.8 Hakea Gum

Hakea gum is an arabinogalactan type of polysaccharide derived from the exudates of the plant *Hakea gibbosa* (family: Proteaceae). The gum has partial solubility in aqueous medium. The chemical composition of gum is comprised of glucuronic acid, galactose, arabinose, mannose, and xylose in the ratio of 12:43:32:5:8 (Peter et al. 1993). This arabinogalactans gum has been highlighted as a sustain release as well as mucoadhesive component in buccal tablets containing chlorpheniramine maleate (CPM) as a model drug (Alur et al. 1999).

40.2.12.9 Hupu Gum

Hupu gum, also called kondagogu gum, kolha gum, and silk cotton, is the dried gummy exudates obtained from the tree *Cochlospermum religiosum* (family: Bixaceae). Structurally, the gum consists of several carbohydrate units like rhamnose, galacturonic acid, glucuronic acid, β-D galactopyranose, α-D-glucose, β-D-glucose, galactose, arabinose, mannose, and fructose. Pharmaceutically, hupu gum has been considered as a novel excipient in designing controlled release tablets of freely soluble drugs like diltiazem hydrochloride, partially soluble drugs like rifampicin and sparingly soluble drugs like diclofenac sodium. In a study, hupu gum as a carrier in combination with other polymers revealed excellent dissolution characteristics for poorly soluble drugs like rofecoxib (Vadlamudia et al. 2014). Apart from that, hupu gum was successfully employed in colon-specific drug delivery system, development of sustain release matrix tablets, nanoparticles, and mucoadhesive microcapsules (Balambhaigari et al. 2012; Bhowmik et al. 2013; Krishna and Murthy 2010; Lankalapalli et al. 2014).

40.2.12.10 Tamarind Gum

Tamarind gum is a galactoxyloglucan polysaccharide obtained from the endosperm of the seed of *Tamarindus indica* (Family: Fabaceae). It is a high-molecular-weight polymer consisting of a (1Ñ4) β-D-glucan backbone substituted with side chains of α-D-xylopyranose and β-D-galactopyranosyl linked (1Ñ2)-α-D-xylo-pyranose linked (1Ñ6) to glucose units. The monomer units glucosyl/xylosyl/galactosyl are found in the ratio of 3:2:1 (Chawananorasest et al. 2016). Its high viscosity, broad pH tolerance, noncarcinogenicity, biocompatibility, mucoadhesive nature, and gel-forming ability offer a prime role in pharmaceuticals (Chandramouli et al. 2012; Khanna 1997; Sattle and Agrawal 2012). It is regarded as a novel excipient for oral-controlled drug release, ocular drug delivery systems, and in the design of sustained release drug delivery systems and dosage forms (Singh et al. 2011). Recent investigation on TSP demonstrated its sustained release property from matrix tablets developed using diclofenac sodium as standard drug (Mahavarkar et al. 2016). Tamarind seed polysaccharide (TSP) serves as vehicle in ophthalmic preparations (Sahoo et al. 2010). It has mucomimetic and mucoadhesive ability to form hydrogels. Tamarind seed polymer is an attractive candidate in the form of binder for tablet dosage forms (Kulkarni et al. 1998).

40.2.12.11 Cashew Gum

Cashew gum (CG) is an exudate heteropolysaccharide obtained from Brazilian plant *Anacardium occidentale* belonging to the family Anacardiaceae. The gum is arabinogalactan consisting of a variety of sugar molecules galactose (72–73%), glucose (11–14%), arabinose (4.6–5%), rhamnose (3.2–4%), and glucuronic acid (4.7–6.3%) (De Paula et al. 1998). Enormous numbers of studies have been performed on cashew gum as new excipient to formulate tablets, hydrogels, microspheres, and nanoparticles for controlled drug delivery system (Dias et al. 2016; Gowthamarajan et al. 2012; Guilherme et al. 2009; Okoye et al. 2012; Soares et al. 2014). In addition, cashew gum polysaccharide has been used as binder, diluents, coating agent, viscosity enhancer, and drying agent. A study on cashew gum demonstrated its diluent property in tablet formulation (Ana Paula de Sá et al. 2018; Kumar et al. 2012). In another study, a formulation named an Orabase gel containing cashew gum polysaccharide was evaluated for the treatment of periodontitis (Filho et al. 2018).

40.2.12.12 Damar Gum

Damar gum is a naturally occurring triterpenoid resin obtained from *Shorea* spp. including *Shorea wiesneri*, *S. javanica*, *S. lamellata*, and *S. retinodes* (family Dipterocarpaceae). This whitish to yellowish color gum is composed of α-resin (40%), β-resin (22%), dammarol acid (23%), and water (2.5%). Besides these constituents, a polymeric fraction, composed of polycadinene, has also been reported in the gum. It has been extensively used as emulsifier and stabilizer in food and cosmetics industries. Pharmaceutically, damar gum alone or in combination with copal gum has been employed as sustained release matrix forming materials in tablet formulations (Fulbandhe et al. 2012; Morkhade et al. 2006a, b; Prasanna et al. 2012). Presently, bilayer gastric retentive floating tablets have been formulated using neem gum, damar gum, and copal gum and ranitidine HCl and clarithromycin as standard drugs to prolong the gastric residence time and bioavailability (Rajeswari et al. 2017).

40.2.12.13 Sandarac Gum

Sandarac gum is a resinous substance collected by incision from the stem of coniferous tree *Callitris quadrivalvis* (family: Pinaceae). Till date, sandarac gum has not been investigated for medicinal and pharmaceutical purposes with few exceptions. A comparative study on sandarac gum with ethyl cellulose (EC) illustrated that the gum can be used as promising matrixing material for formulating coated pellets for sustain release of the drug (Khobragade et al. 2016).

40.2.12.14 Gum Copal

Gum copal is a naturally occurring polymer derived from the plant *Bursera bipinnata* (family Burseraceae). Additionally, the gum can also be obtained from the plants belonging to the families Caesalpiniaceae and Araucariaceae. Gum copal obtained from both families differs in its chemical composition. Copal resin (CR) consists of agathic acid (diterpenoid) and related lobdane compounds along with cis-communic acid, trans-communic acid, polycommunic acid, sandaracopimaric acid, agathalic acid, monomethyl ester of agathalic acid, agatholic acid, and acetoxy agatholic acid. CR obtained from Leguminosae family contains copalic acid, pimaric acid, isopimaric acid, dehydro-dehydroabietic acid, dehydroabietic acid, and abietic acid (Osete-Cortina and Domenech-Carbo 2005). The gum has both medicinal and pharmaceutical significance. It is recommended to use in the treatment of headache, fever, burns, and stomach ache and as binder in preparation of dental products (Dutton et al. 1993; Whitmore 1980). Moreover, gum copal has been investigated as film-forming agent for sustained release and colon targeted drug delivery systems (Morkhade et al. 2006a, b).

40.2.12.15 Bahera Gum

Bahera gum is a natural gum extracted from the bark of plant *Terminalia bellirica* Roxb. belonging to family Combretaceae. The gum primarily consists of tannins such as ß-sitosterol, gallic acid, ellagic acid, ethyl gallate, galloyl glucose, and chebulaginic acid. It has potential use in pharmaceutical and cosmetic industries as emulgent. Medicinally, it is used as demulcent and purgative (Kokate et al. 2009). Also, the gum has been used as additive in water-based natural rubber adhesive formulations (Saha et al. 2005). Recently, bahera gum has been evaluated as excipient in developing microcapsules of famotidine for sustained release microencapsulated drug delivery system by ionic gelation technique (Nayak et al. 2008).

40.2.12.16 Fenugreek Mucilage

Fenugreek mucilage is a gummy substance naturally obtained from the seeds of the plant *Trigonella foenum-graceum* (family: Leguminosae). However, mucilage in aqueous media has poor solubility, but when in contact with fluid, it generates viscous and glutinous mass when comes in contact with fluid (Petropoulos 2002). Fenugreek seeds contain major amount of mucilage (26%), protein (22%), fatty acids, minerals, and fibers (Ahmad 2017). This mucilaginous substance is composed of different monomers such as mannose, galactose, and xylose. Fenugreek mucilage has been used and investigated as gelling agent, disintegrant, and granulating agent and as control release material for tablet formulation of propranolol hydrochloride (Avachat et al. 2007; Kumar et al. 2009; Kuppusamy et al. 2002; Nokhodchi et al. 2008).

40.2.13 Natural Dyes in Pharmaceuticals

Colorants are the group of additives which impart distinctive elegance/appearance to the pharmaceutical dosage forms particularly in tablets, capsules, oral liquids, ointments, and cosmetic preparations. Pharmaceutical preparations are colored to enhance the acceptability to the patient and identification of similar looking products as well as stability purposes (Allam and Kumar 2011). Despite the tremendous use of synthetic colorants, there is a great demand of natural colors in pharmaceuticals and nonpharmaceutical industries. Natural colorants can be obtained from a variety of sources such as plants, animals, minerals, insects, and microbial flora (Chengaiah et al. 2010; Joshi et al. 2003). The past studies demonstrated that only 0.5% from 70% plants have been explored and investigated as coloring agents (Mohd-Nasir et al. 2018). Here, the most commonly used natural colorants derived from different sources including plants, microbes, and animals have been discussed.

40.2.13.1 Natural Dyes Derived from Plants

Curcumin

Curcumin, chemically known as diferuloyl methane, is isolated from the rhizomes of plant *Curcuma longa* Linn. (turmeric). It consists of two hydrophobic aromatic rings with o-methoxy phenolic groups, joined by a seven carbon linker consisting of an α,β-unsaturated β-diketone moiety. Curcumin, a turmeric constituent, is mainly responsible for its yellow color (Gilda et al. 2010; Pyrzanowska et al. 2010; Soni et al. 2011). Curcumin is insoluble in water, readily soluble in polar solvents (DMSO, methanol, ethanol, acetonitrile, chloroform, ethyl acetate), and sparingly soluble in hydrocarbon solvents like cyclohexane and hexane. Curcumin exhibits wide range of therapeutic applications including anti-inflammatory, antioxidant, antimicrobial, antiparasite, antispasmodic, and anticancer activities. In addition, it is also used as coloring agent in food, cosmetics, as well as pharmaceutical industries. Curcumin has been investigated as an excellent candidate for skin care preparations due to its antioxidant, antiseptic, anti-inflammatory, and antiaging activities (Priyadarsini 2014; Arct et al. 2014). Arct et al. (2014) investigated skin-coloring properties of curcumin-based cosmetic emulsion at various concentrations.

Crocin

Crocin is a carotenoid extracted from the stigmas of *Crocus sativus* Linn. (saffron) and the fruits of *Gardenia jasminoides* Ellis (Pham et al. 2000). Besides crocin, the other constituent of saffron are crocetin, picrocrocin, and safranal. Crocin is a water-soluble pigment and imparts golden yellow coloration to saffron. Structurally, it is primarily composed of mono- and di-glycosylic esters of polyene dicarboxylic acid. The natural dye obtained from saffron finds its wide applications in food, cosmetic, textile, and pharmaceutical industry. It has been investigated for several medical properties such as antiproliferative, antioxidant, anticancer, anticoagulant, and anti-Alzheimer (Sedaghati and Abbaszadeh 2018).

β-Citraurin and Violaxanthin

Both β-citraurin and violaxanthin are carotenoid pigments isolated from the peel of Valencia orange fruit, scientifically named as *Citrus sinensis* (Szabolcs 1980). β-citraurin, a C30 apocarotenoid, is responsible for the reddish orange color of citrus fruits (Ma et al. 2013). It is degradation product of cryptoxanthin and zeaxanthin. Violaxanthin is a natural xanthophylls dye and biosynthesized from zeaxanthin by epoxidation (Gonzalez-Jorge et al. (2016).

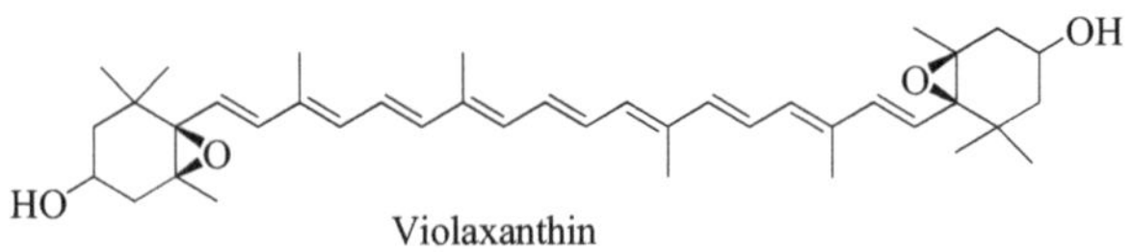

Violaxanthin

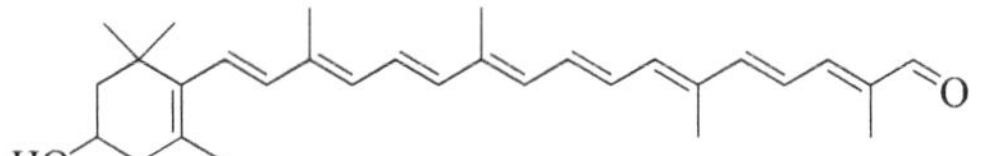

Citraurin

Lutein

Lutein is a member of xanthophyll family and widely found to occur in fruits, flowers, and vegetables. This yellow-colored pigment has also been isolated from the flowers of plant *Tagetes erecta* L. (marigold flowers) (family: Asteraceae). Structurally, it is composed of C40 isoprenoid units containing ten conjugated double bonds. The flowers make about 80% of lutein diesters of dry matter. It is considered as a powerful antioxidant. It has been reported to be used as food additive and in dermatology and cosmetology areas (Philip and Berry 1976; Šivel et al. 2014; Vasudevan et al. 1997).

Carthamin

Carthamin, a red natural dye, is obtained from the plant *Carthamus tinctorius* L. (safflower). The plant also contains carthamidin, arctigenin, tacheloside, N-feruloyl tryptamine, N-feruloylserotonin, steroids, flavonoids, and polyacetylenes. Carthamin is a water-insoluble pigment, while carthamidin is responsible for producing water-soluble yellow dye. Apart from food coloring and flavoring agent, carthamin has been investigated for uterine stimulating, coronary dilating, and hypotensive properties. It also has the cytotoxic, antigenic, and antiplatelet activities (Azami et al. 2019; Kizil et al. 2008; Takahashi et al. 1982).

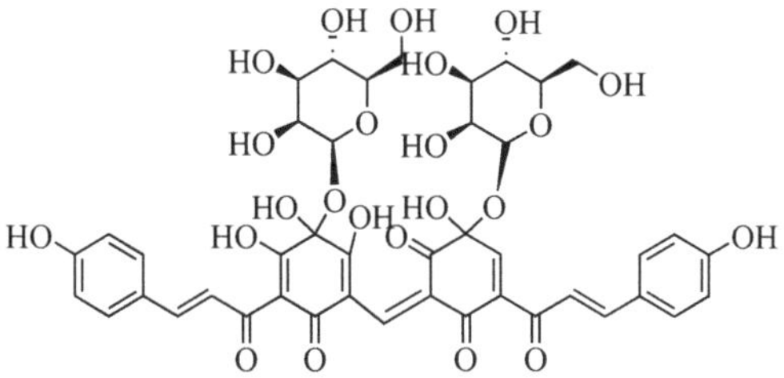

Carthamin

Bixin and Norbixin

Bixin and norbixin are naturally occurring pigments derived from the outer resinous coatings of the seeds of *Bixa orellana* Linn. (Annato) (family: Bixaceae). These carotenoid pigments are responsible for red-orange color. The plant pigments are largely used as colorants in food and pharmaceuticals. Bixin is oil-soluble component and suitable for preparation of fat- and oil-based formulations such as emulsions and other oral dosages forms. Norbixin, a water-soluble compound, is found suitable in food preparations (Dinda and Mukharjee 2008; Satyanarayana et al. 2003).

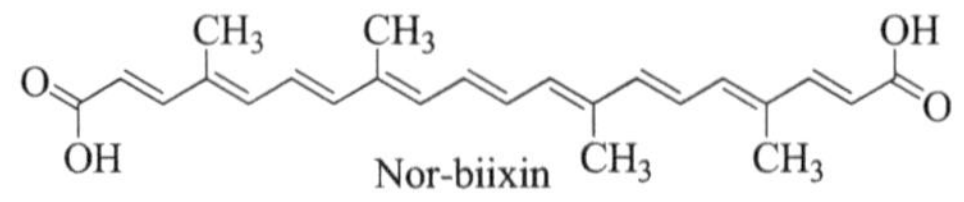

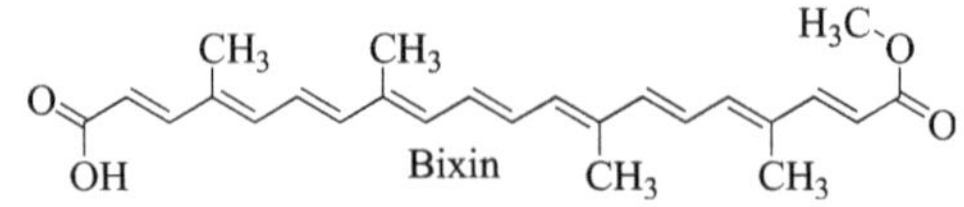

Lycopene

Lycopene is the chief constituent of fresh ripens fruits of plant *Solanum lycopersicum* (tomato). The pigment principally contributes red color to the tomato fruits. Chemically, it is polyunsaturated open linear chain of aliphatic hydrocarbons containing 11 conjugated and two unconjugated double bonds. Lycopene exists in both cis and trans forms. The red color of lycopene is due to trans-isomeric form. It has attracted much attention as colorant in food and pharmaceutical industries. The presence of double bonds in the pigment offers numerous therapeutic properties such as antioxidant, antiulcer, anticancer, and anti-neuroinflammatory activities (Kumar et al. 2017; Story et al. 2010).

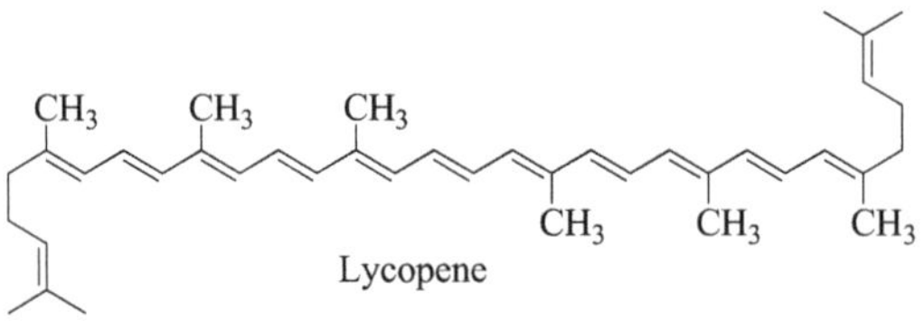

Capsanthin and Capsorubin

Capsanthin and capsorubin are the natural pigments principally responsible for red color of the fruits of *Capsicum annuum* L. (paprika, sweet peppers). Paprika constituents make about 60% of total carotenoids of paprika fruits. These are water insoluble and stable to heat and pH variation but deteriorate in light (Cantrill 2008).

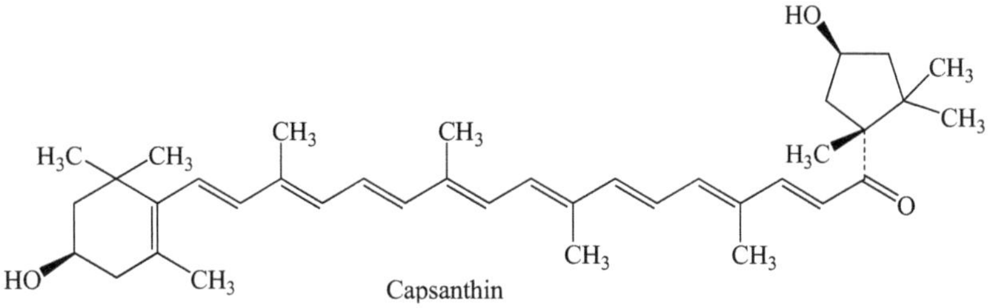

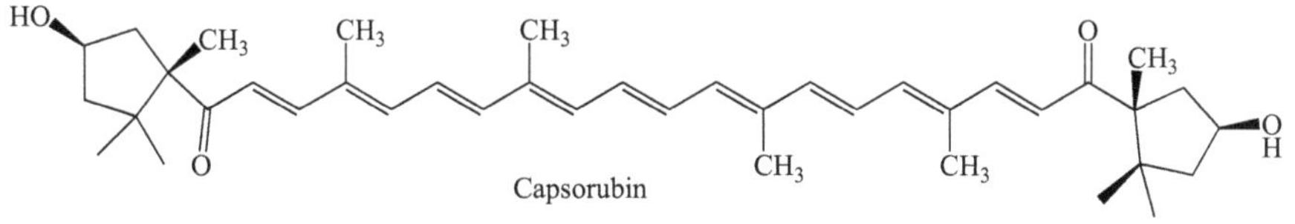

Lawsone

Lawsone, a red-orange-colored natural hair dye, is obtained from the fresh and dried leaves of the plant *Lawsonia inermis* L. (henna) (family: Lythraceae). It is extensively used in textile and cosmetic industry as coloring agent. Structurally, lawsone is 2-hydroxy-1,4-naphthoquinone present at a concentration of 1.0–1.4% w/w (Hasan et al. 2015).

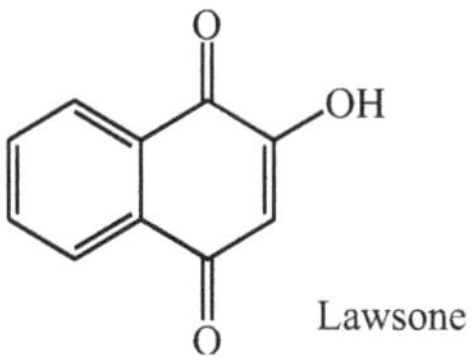

Indigo Dyes

Indigo dye is extracted from various plants such as *Indigofera tinctoria*, *Persicaria tinctoria*, *Strobilanthes cusia*, and *Isatis tinctoria.* Pure indigo is slightly soluble in water. The reduced form of white indigo on oxidation produces a substance of deep blue color. The isomer of blue indigo named indirubin (red color pigment) is a pharmacologically active molecule (Stasiak et al. 2014). In addition, indigo dye has also been isolated from the fruits and flowers of *Couroupita guianensis* (cannonball tree). The plant contains indigotin and indirubin as chief constituents which are used in dyeing of cotton fabrics (Tayade and Adivarekar 2014).

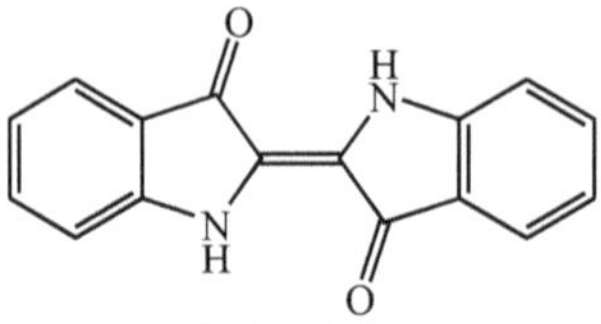

Indigotin

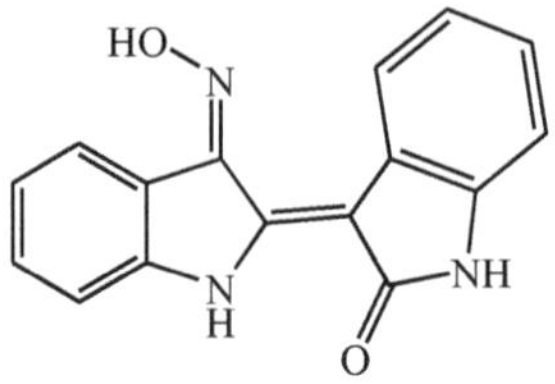

Indirubin

Alizarin and Purpurin

Alizarin and purpurin are the two important natural pigments containing anthraquinone ring in their structures. These pigments are largely found in the roots and tubers of *Rubia tinctorum* (Common Madder). Recently, some other anthraquinone-type compounds named lucidin primeveroside, ruberythric acid, and lucidin-ω-ethyl ether have also been isolated from the plant (Henderson et al. 2013). Both the pigments are present in the form of glycoside. Alizarin and purpurin have been investigated for mutagenic and antimutagenic along with antimicrobial properties. These coloring matters are widely exploited in food, textile, and cosmetic industries (Uysal et al. 2019). Alizarin is converted into an insoluble red dye in the presence of mordants like alum and alkali.

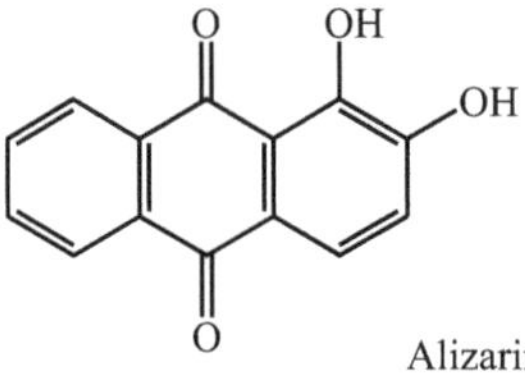

Alizarin

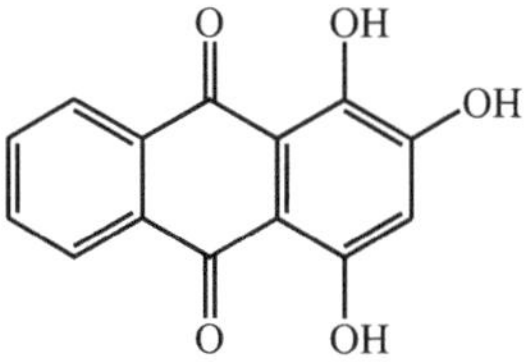

Purpurin

Catechin

Catechin is a naturally occurring polyphenolic compound with excellent antioxidant properties. It is present in several medicinal plants such as *Camellia sinensis* (green tea), *Agrimonia eupatoria* (agrimony), *Areca catechu* (betel nut), *Fragaria* sp. (strawberry), *Hordeum vulgare* (barley), *Olea europaea* (olive), *Piper betle* (betel leaf), and *Ribes nigrum* (black currant) (Duke 1992; Ghayur 2007). It was originally isolated from *Acacia catechu* belonging to the family Fabaceae. Chemically, it is a monomeric flavonoid belonging to the flavan-3-ol family. In nature, it exists as a mixture of (−)-epicatechin and (+)-catechin. This phytochemical is mainly responsible for brown coloration. Interestingly, a novel hair dye named catechinone has been synthesized from catechin and tyrosinase enzyme which gives reddish orange color to the hairs (Yasunaga 2012). Catechin is potentially used in pharmaceutical formulations and cosmetics (Gadkari and Balaraman 2015).

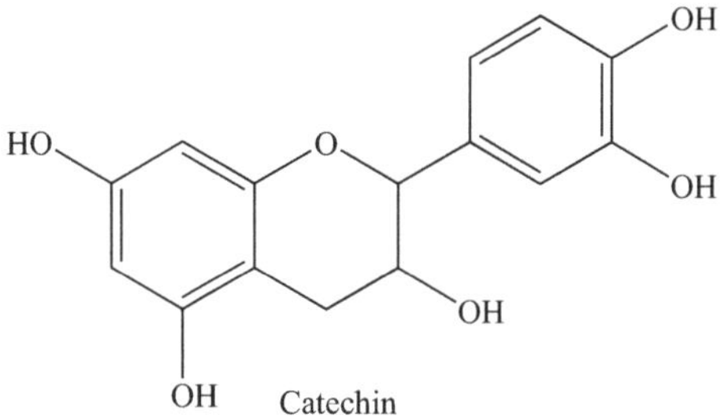

Catechin

Sanguinarine

Sanguinarine (SA), a benzophenanthridine alkaloid, is naturally drawn from many plants including *Sanguinaria canadensis* (bloodroot), *Poppy fumaria*, *Bocconia frutescens*, *Chelidonium majus*, *Macleaya cordata*, and *Bocconia frutescens* (Govender and Gathiram 2008). Chemically, it is designated as 13-methyl (1,3) benzodioxolo(5,6-c)-1,3-dioxolo(4,5)phenanthridinium. Medicinally, SA is known to display cytotoxic and cytostatic effects on various human cancer cells including prostate cancer, pancreatic carcinoma, colon cancer, breast cancer, lung cancer, promyelocytic leukemia, and bone cancer (Singh and Sharma 2018). Sanguinarine (1%, colorless alkaloid yielding red salts) is also used as coloring matter in various industries. Berberine is yellow-colored constituent isolated from the blood root plant.

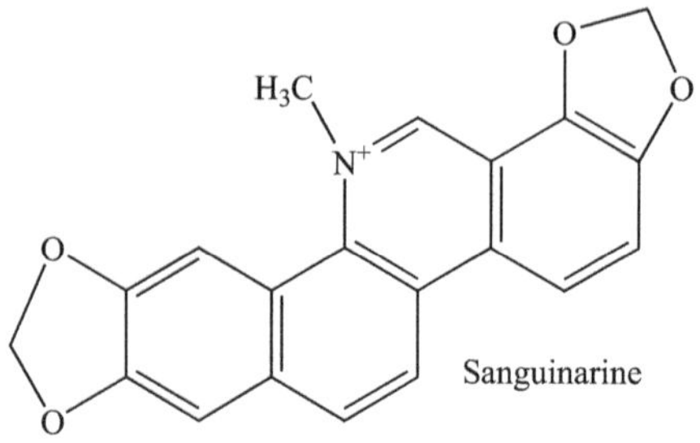

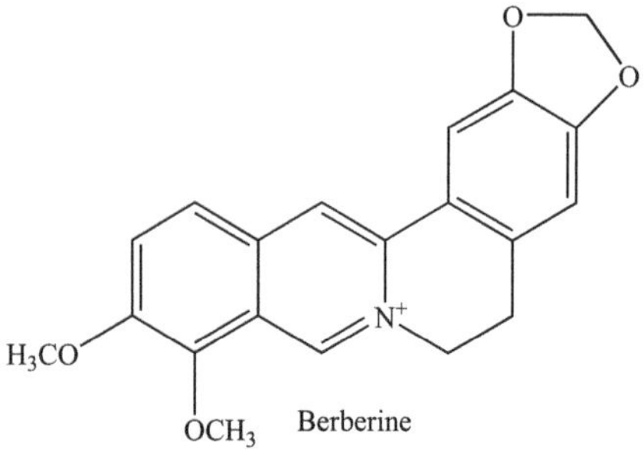

40.2.13.2 Natural Colorants Derived from Animals/Insects

Carminic Acid (Cochineal Dye)

Carminic acid consists of an anthraquinone aglycone part linked to a *C*-D-glucopyranose unit. It is extracted from the dried bodies of female cochineal insect, *Dactylopius coccus*. The scaly insect is largely found in Mexico and Central America. It is a red color pigment which turns black-red in aqueous solution and yellow-violet in acids. The dye has multiple applications in food, cosmetics, pharmaceuticals, and textile industries (Borges et al. 2012). In recent years, researchers have isolated a novel biopigment designated as spiroketalcarminic acid from the cochineal extract (Ito et al. 2017).

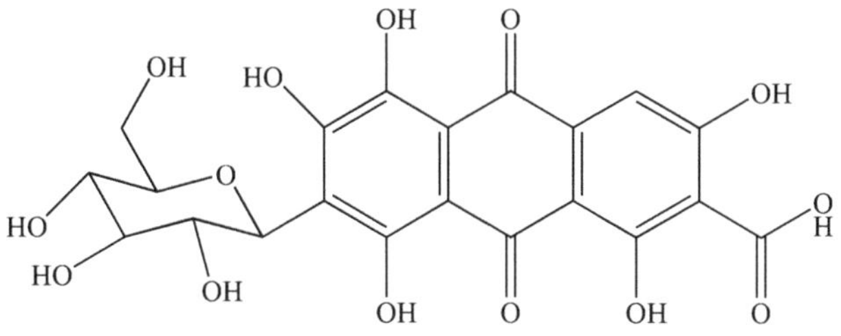

Carminic acid

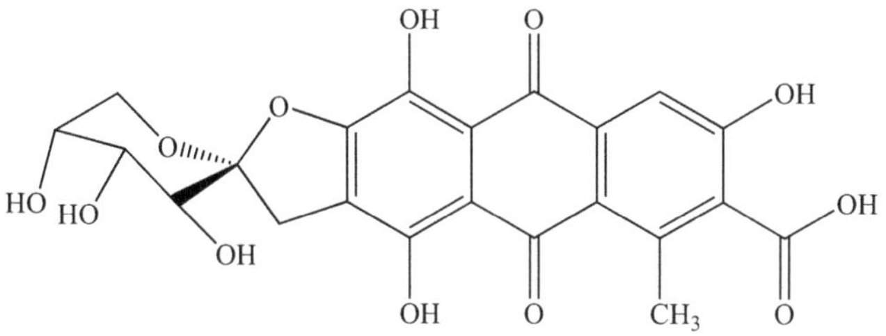

Spiroketalcarminic acid

Lac Dye

Lac dye, also known as lake and lacca, is a natural colorant that can be produced by washing the tiny insects *Kerria lacca* (family: Tachardiidae). Lac dye is a mixture of anthraquinone derivatives designated as laccaic acid A, B, C, D, and E. The red color of dye is due to laccaic acid. The methylated derivative of lac dye has been investigated for antifungal and antibacterial activity. Its main use is with foods, cosmetics, and pharmaceutical applications (Srivastava et al. 2017).

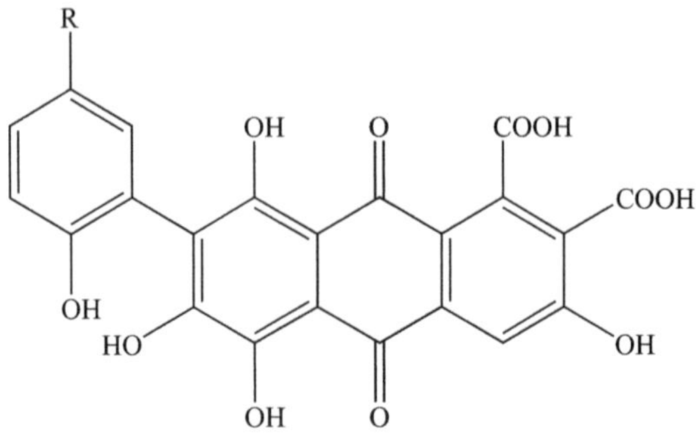

R= $CH_2CH_2NHCOCH_3$ = Laccaic acid A

R= CH_2CH_2OH = Laccaic acid B

R= $CH_2CH(NH_2)COOH$ = Laccaic acid C

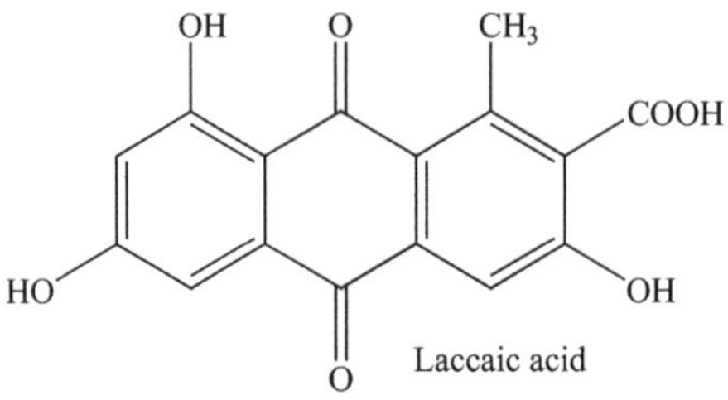

Laccaic acid

Kermesic Acid

Kermesic acid is a natural colorant obtained from the female kermes insect *Kermes vermilio* found on the tree kermes oak (*Quercus coccifera* L). It imparts crimson red color and used as colorant in textile and pharmaceutical industries.

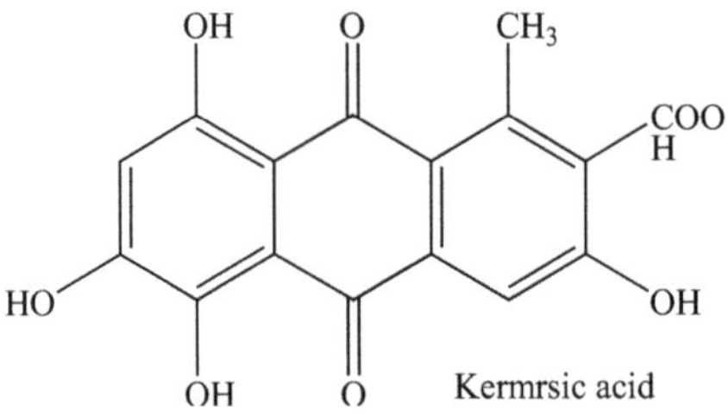

Kermrsic acid

Tyrian Purple

Tyrian purple, also known as royal purple, shellfish purple, derived exclusively from marine shellfish of the Muricidae and Thaisidae families. Chemically, it consists of major component named 6,6′-Dibromoindigo imparting purple color to the dye (Cooksey 2001).

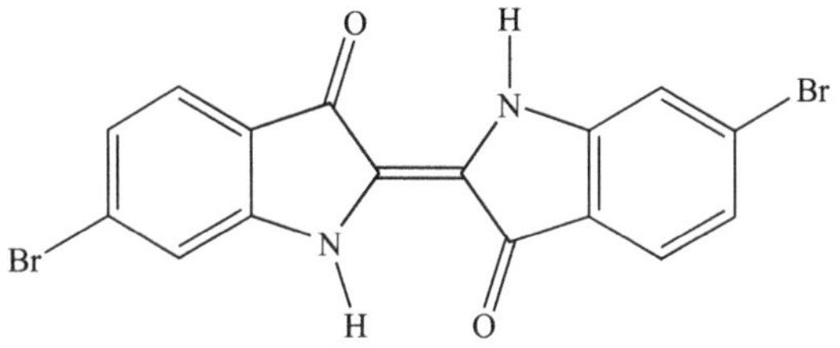

Tyran purple

40.2.13.3 Natural Dyes Derived from Microorganisms

In addition to the plant sources, a variety of microorganisms have been recognized as potential source of pigments including carotenoids, melanins, flavins, quinines, monascins, violacein, etc. These microbial dyes not only are used as colorants in various foods and cosmetic industry but also display anti-inflammatory, antioxidant, antimicrobial, and anticancer activities. Some of the coloring agents obtained from microorganisms are briefly described in Table 40.10 (Tuli et al. 2015).

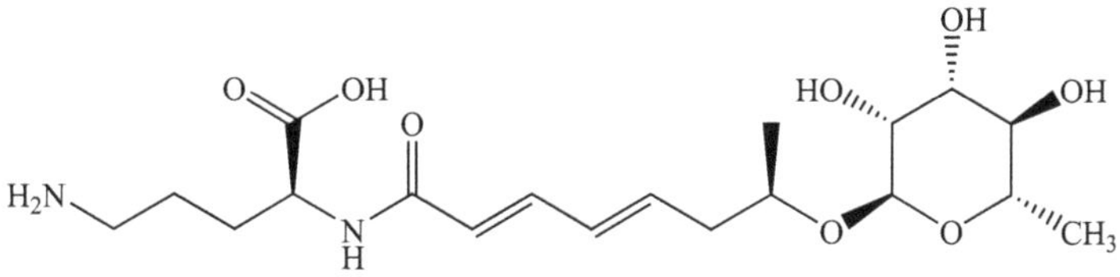

Granadaene

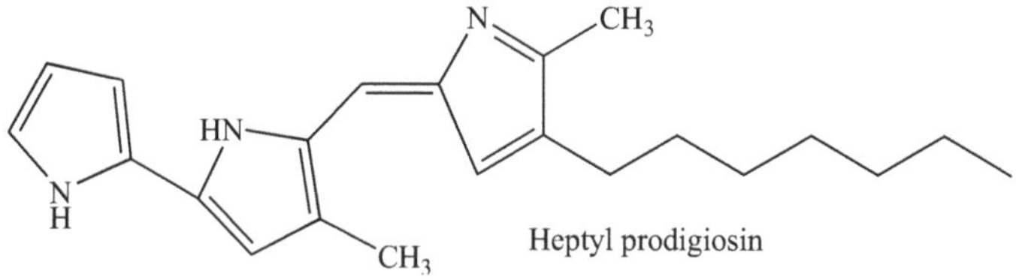

Heptyl prodigiosin

Table 40.10 Coloring agents obtained from microorganisms

Pigment	Source	Color
Ankaflavin	*Monascus* spp.	Yellow
Anthraquinone	*Penicillium oxalicum*	Red
Astaxanthin	*Haematococcus pluvialis*, *Euphausia pacifica* (Pacific krill), *Euphausia superba* (Antarctic krill), *Pandalus borealis* (shrimp), *Xanthophyllomyces dendrorhous*	Pink red
Canthaxanthin	*Bradyrhizobium* spp.	Orange-red
Granadaene	*Streptococcus agalactiae*	Orange-red
Heptyl prodigiosin	*Proteobacteria*	Red
Indigoidine	*Corynebacterium insidiosum*	Blue
Lycopene	*Fusarium*, *Blakeslea trispora*	Red
Naphthoquinone	*Cordyceps unilateralis*	Deep red
Procynin	*Pseudomonas* spp.	Blue, green
Riboflavin	*Ashbya gossypii*	Yellow
Rubrolone	*Streptomyces echinoruber*	
Staphyloxanthin	*Staphylococcus aureus*	Golden
Undecylprodigiosin	*Streptomyces*	Red
Violacein	*Janthinobacterium lividum*, *Chromobacterium violaceum*	Purple
Xanthomonadin	*Xanthomonas oryzae*	Yellow
Zeaxanthin	*Staphylococcus aureus*, *Paracoccus zeaxanthinifaciens*	Yellow
β-Carotene	*Dunaliella salina*	Orange

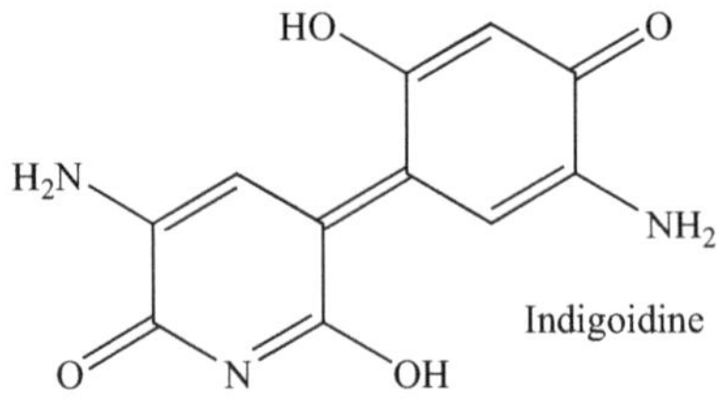

Indigoidine

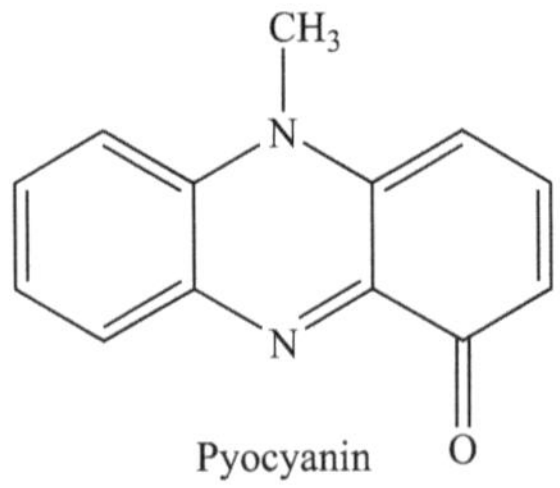

Pyocyanin

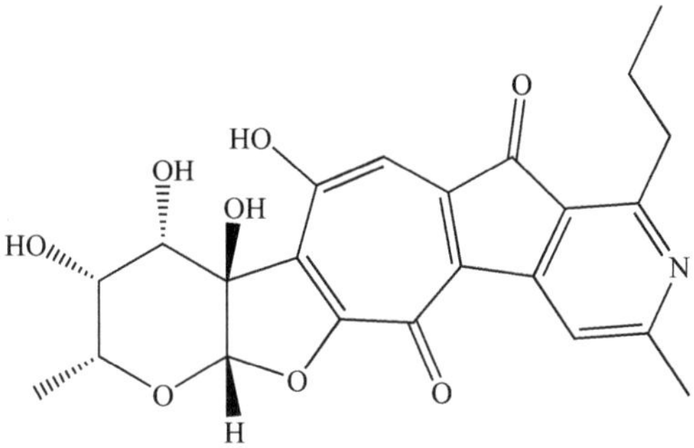

Rubrolone

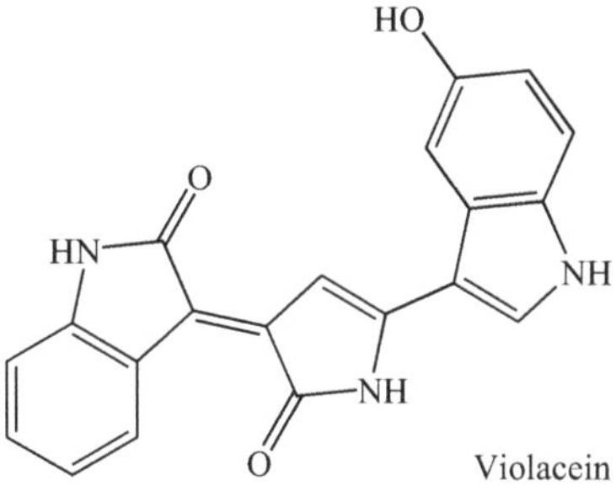

Violacein

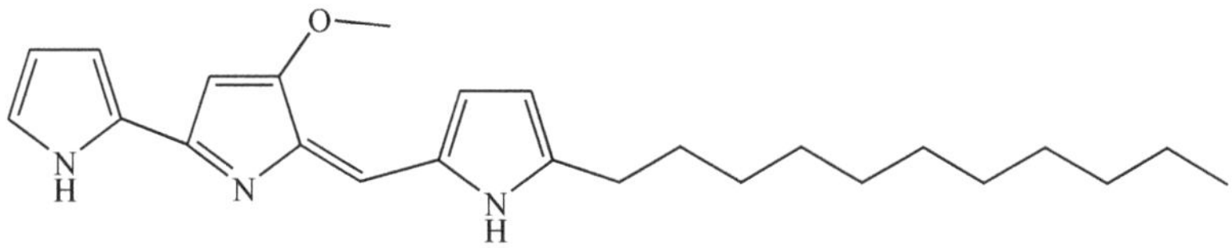

Undecylprodigiosin

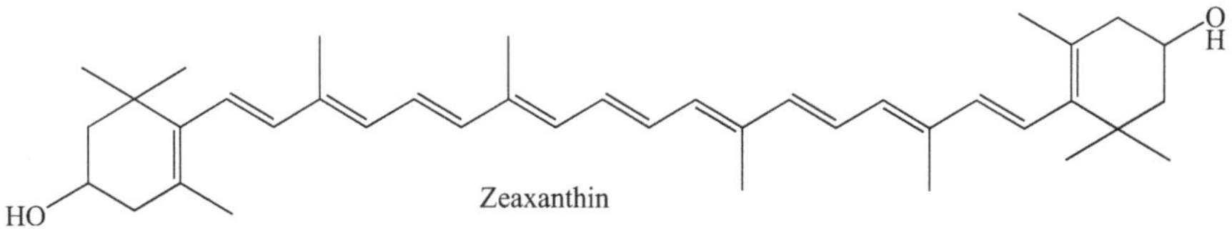

Zeaxanthin

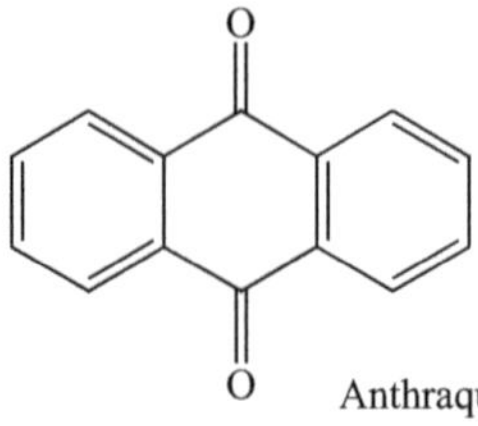

Anthraquinone

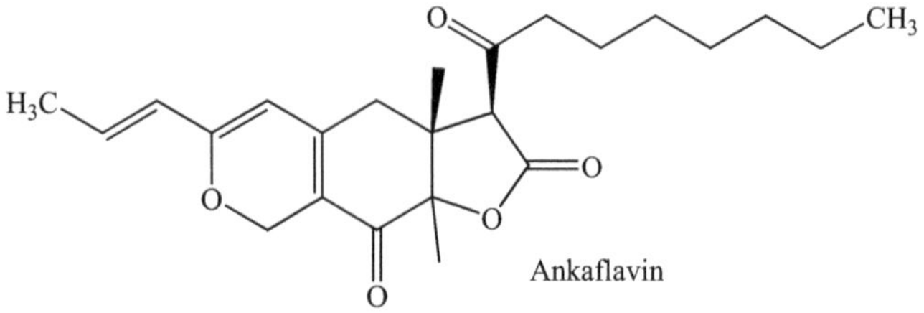

Ankaflavin

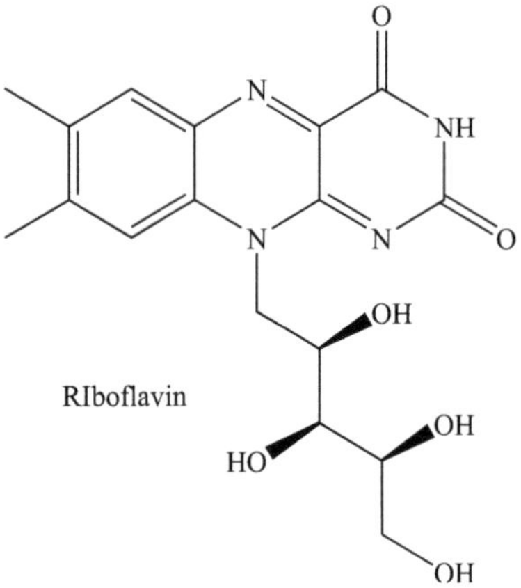

RIboflavin

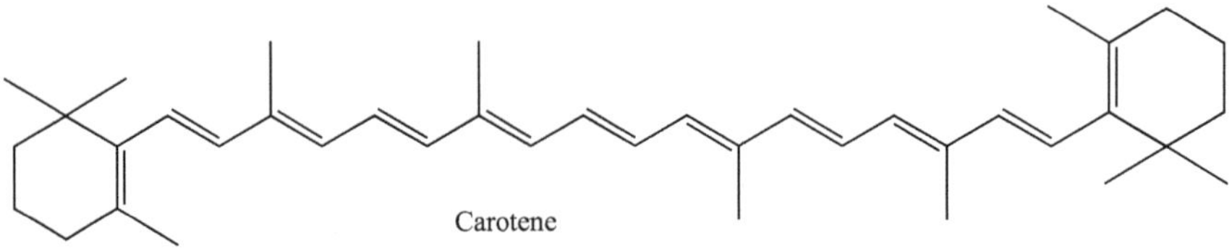

Carotene

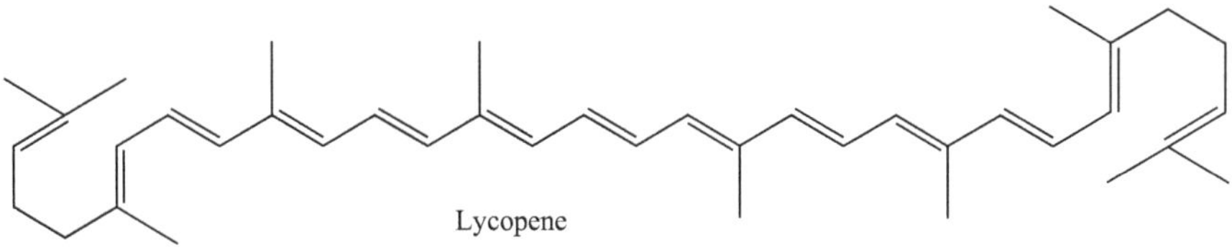

Lycopene

40.2.13.4 Natural Dyes Derived from Minerals

The most commonly used dyes derived from mineral sources are chrome yellow, chrome green, chrome orange, manganese brown, iron buff, Prussian blue, etc. Mineral colors are basically inorganic compounds which are insoluble in water and precipitated onto the fiber by double decomposition. Red lead (Sindur), laminated red earth (Gem), cinebor (Sangraj), ultramarine (Lajerd), zinc white (Sajeda), etc. are the well-known minerals used as natural dyes.

40.3 Conclusion

Conclusively, pharmaceutical excipients derived from natural sources have attracted a great attention for researchers in developing a wide array of convectional dosage forms as well as novel drug delivery systems (NDDS). In recent times, chemically modified natural excipients have been extensively used to overcome the disadvantages of their native forms. They offer a putative role not only in pharmaceutical domain but also in other areas like food, textile, and paper industries. Pharmaceutically, natural excipients are being preferred over synthetic to ensure the administration of accurate and desired amount of drug at the specific target at the right time. In addition, they also enhance the bioavailability, stability, patient acceptance, and confirm the safety as well as efficacy during administration. In future, a lot of research on these excipients is required to explore their numerous features and potential role in drug delivery in order to obtain a better pharmaceutical formulation.

References

Agrawal A et al (2014) Derivatives and applications of lignin–an insight. Scitech J 1:30–36

Ahmad D (2017) Physicochemical properties of Chemurgic-fenugreek (Trigonella foenum-graecum) herb of different origin. Der Pharma Chem 9:102–106

Akhgari AF et al (2006) Permeability and swelling studies on free films containing inulin in combination with different polymethacrylates aimed for colonic drug delivery. Eur J Pharm Sci 28:307–314

Alcazar-Alay SC, Meireles MAA (2015) Physicochemical properties, modifications and applications of starches from different botanical sources. Food Sci Technol 35(2):215–236

Ali A, Ahmed S (2018) A review on chitosan and its nanocomposites in drug delivery. Int J Biol Macromol 109:273–286

Allam KV, Kumar GP (2011) Colorants the cosmetics for the pharmaceutical dosage form. Int J Pharma Pharma Sci 3:3–21
Alur HH et al (1999) Evaluation of the gum from Hakea gibbosa as a sustained-release and mucoadhesive component in buccal tablets. Pharm Dev Technol 4:347–358
Ana Paula de Sá P et al (2018) Evaluation of the application of cashew gum as an excipient to produce tablets. Polímeros 28:302–308
Andrei G, Lisco A, Vanpouille C, Introini A, Balestra E, Van den Oord J et al (2011) Topical tenofovir, a microbicide effective against HIV, inhibits herpes simplex virus-2 replication. Cell Host Microbe 10:379–389
Arct J, Ratz-lyko A, Mieloch M et al (2014) Evaluation of skin colouring properties of *curcuma longa* extract. Indian J Pharm Sci 76:374–378
Ashton WA et al (1975) Physical properties and applications of aqueous solutions of Albizia zygia gum. J Sci Food Agric 26:697–704
Aspinall GO, Bhattacharjee AK (1970) Plant gums of the genus Khaya. Part IV. J Chem Soc:365–369
Avachat A, Gujar KN, Kotwal VB et al (2007) Isolation and evaluation of fenugreek seed husk as a granulating agent. Indian J Pharm Sci 69:676–679
Azami K, Hayashi T, Kusumi T et al (2019) Total synthesis of Carthamin, a traditional natural red pigment. Angew Chem Int Ed Engl 58:5321–5326
Balambhaigari RY et al (2012) Design and assessment of colon specific drug delivery of mefenamic acid using modified pulsincap technique and hupu gum. Indo Am J Pharm Res 2:1026–1038
Barapatre A et al (2016) In vitro evaluation of antioxidant and cytotoxic activities of lignin fractions extracted from *Acacia nilotica*. Int J Biol Macromol 86:443–453
Belgacem MN, Gandini A (eds) (2008) Monomers, polymers and composites from renewable resources. Elsevier, Amsterdam
Beneke C et al (2009) Polymeric plant-derived excipients in drug delivery. Molecules 14:2602–2620
Bhowmik A et al (2013) Design and development of atenolol matrix tablet employing natural and synthetic polymers. J Appl Pharm 3:103–108
Borges ME, Tejera RL, Díaz L et al (2012) Natural dyes extraction from cochineal (*Dactylopius coccus*)-new extraction methods. Food Chem 132:1855–1860
Cantrill R (2008) Paprika extract: Chemical and Technical Assessment. In: Joint FAO/WHO Expert Committee on Food Additives. http://www.fao.org/fileadmin/templates/agns/pdf/jecfa/cta/69/Paprika_extract.pdf
Chandramouli Y et al (2012) Tamarind seed polysaccharides (TSP)-an adaptable excipient for novel drug delivery system. Int J Pharm Pract Drug Res 2:57–63
Chavda HV et al (2012) Preparation and in vitro evaluation of guar gum based triple-layer matrix tablet of diclofenac sodium. Res Pharm Sci 7:57–64
Chawananorasest K et al (2016) Extraction and characterization of tamarind (tamarind indica L.) seed polysaccharides (TSP) from three difference sources. Molecules 21:775
Chengaiah B, Rao KM, Kumar KM et al (2010) Medicinal importance of natural dyes- a review. Int J Pharm Tech Res 2:144–154
Chodavarapu NP (2011) Formulation and evaluation of Abelmoschus esculentus mucilage based metformin hydrochloride floating matrix tablets. IJPT. ONLINE 3(2):2725–2745
Choudhary G et al (2006) Development of controlled release formulations of Carbofuran and evaluation of their efficacy against Meloidogyne incognita. J Agric Food Chem 54:4727–4733
Choudhary PD, Pawar HA (2014, 2014) Recently investigated natural gums and Mucilages as pharmaceutical excipients: an overview. J Pharm 9. https://doi.org/10.1155/2014/204849
Chowdary KPR, Rao PVV (1994) Dissolution, Bioavailability and Ulcerogenic Studies on Solid Dispersions of Indomethacin in Water Soluble Cellulose Polymers. Drug Dev Ind Pharm 20:799-813.
Cooksey CJ (2001) Tyrian purple: 6,6′-Dibromoindigo and related compounds. Molecules 6:736–769

Corrente F et al (2009) Physical carboxymethylscleroglucan/calcium ion hydrogels as modified drug delivery systems in topical formulations. Molecules 14(8):2684–2698
Coviello T et al (2005) Scleroglucan: a versatile polysaccharide for modified drug delivery. Molecules 10:6–33
De Paula R et al (1998) Characterization of Anacardium occidentale exudate polysaccharide. Polym Int 45:27–35
Dias SFL, Nogueira SS, França Dourado F, Guimarães MA, Oliveira Pitombeira NA, Gobbo GG et al (2016) Acetylated cashew gum-based nanoparticles for transdermal delivery of diclofenac diethyl amine. Carbohydr Polym 143:254–261
Dinda SC, Mukharjee B (2008) Annatto seed color as natural coloring agent in oral dosage forms. Int J Pharma Sci Tech 1:10–14
Dionísio M, Grenha A (2012) Locust bean gum: exploring its potential for biopharmaceutical applications. J Pharm Bioallied Sci 4:175–185
Duke JA (1992) Handbook of phytochemical constituents of GRAS herbs and other economic plants. CRC Press, London
Dutta PK et al (2004) Chitin and chitosan: chemistry, properties and applications. J Sci Ind Res 63:20–31
Dutton FB, Summitt JB, Chan DCN et al (1993) Effect of a resin lining and rebonding in the marginal leakage of amalgam restoration. J Dent 21:52–56
Fainberg EZ, Mikhailov NV (1966) Molecular and supramolecular structure of cellulose fibres. Polym Sci USSR 8:1805–1812
Fargues C et al (1996) Kinetics of vanillin production from Kraft lignin oxidation. Ind Eng Chem Res 35:28–36
Farina JI et al (1998) High scleroglucan production by Sclerotium rolfsii: influence of medium composition. Biotechnol Lett 20:825–831
Fernández-Pérez M et al (2007) Controlled-release formulations of cyromazine-lignin matrix coated with ethylcellulose. J Environ Sci Health B 42:863–868
Filho MD, Medeiros JVR, Vasconcelos DFP, Silva DA, Leódido ACM, Fernandes HF et al (2018) Orabase formulation with cashew gum polysaccharide decreases inflammatory and bone loss hallmarks in experimental periodontitis. Int J Biol Macromol 107:1093–1101
Final Assessment Report: Application A546 Tara Gum as a Food Additive (2006) Food Standards Australia New Zealand. http://www.foodstandards.gov.au/code/applications/documents/FAR_A546_Tara_gum.pdf. Accessed 09 August 2019
Frangville C et al (2012) Fabrication of environmentally biodegradable lignin nanoparticles. Chem Phys Chem 13:4235–4243
Fulbandhe VM et al (2012) Evaluation of release retarding property of gum Damar and gum copal in combination with hydroxypropyl methylcellulose. Indian J Pharm Sci 74:189–194
Fulzele SV (2003) Studies of the biodegradation and in vivo biocompatibility of novel biomaterials. Eur J Pharm Sci 20:53–61
Gadkari PV, Balaraman M (2015) Catechins: sources, extraction and encapsulation: a review. Food Bioprod Process 93:122–138
Ghayur MN (2007) Antispasmodic, bronchodilator and vasodilator activities of (+)-catechin, a naturally occurring flavonoid. Arch Pham Res 30:970
Gilda S, Kanitkar M, Bhonde R et al (2010) Activity of water-soluble turmeric extract using hydrophilic excipients. LWT-Food Sci Technol 43:59–66
Gonzalez-Jorge S, Mehrshahi P, Magallanes-Lundback M, Lipka AE, Angelovici R, Gore MA et al (2016) Zeaxanthin Epoxidase activity potentiates carotenoid degradation in maturing seed. Plant Physiol 171:1837–1851
Govender MT, Gathiram P (2008) Review Sanguinarine. Cardiovasc Ther 26:75–83
Gowthamarajan K et al (2012) Development of buccal tablets for curcumin using Anacardium occidentale gum. Carbohydr Polym 88:1177–1183

Guilherme AM, Clara MWS, Durcilene AS et al (2009) Microspheres of chitosan/carboxymethyl cashew gum (CH/CMCG): effect of chitosan molar mass and CMCG degree of substitution on the swelling and BSA release. Carbohydr Polym 77:217–222

Gutiérrez-Hernández JM et al (2016) Use of Agave tequilana-lignin and zinc oxide nanoparticles for skin photoprotection. J Photochem Photobiol B Biol 163:156–161

Hamman H et al (2015) Use of natural gums and Mucilages as pharmaceutical excipients. Curr Pharm Des 21:4775–4797

Hartesi B, Sriwidodo AM et al (2016) Starch as pharmaceutical excipient. Int J Pharm Sci Rev Res 41(2):59–64

Hasan MM, Nayem KA, Azim AYMA et al (2015) Application of purified Lawsone as natural dye on cotton and silk fabric. J Text 2015:1–7

Hasegawa Y et al (2015) Lignosulfonic acid-induced inhibition of intestinal glucose absorption. J Nutr Sci Vitaminol 61:449–454

Henderson RL et al (2013) Isolation and extraction of lucidin primeveroside from Rubia tinctorum L. and crystal structure elucidation. Phytochemistry 95:105–108

Henry BL, Desai UR (2014) Sulfated low molecular weight Lignins, allosteric inhibitors of coagulation proteinases via the heparin binding site, significantly alter the active site of thrombin and factor Xa compared to heparin. Thromb Res 134:1123–1129

Herman J, Remon JP (1989) Modified starches as hydrophilic matrices for controlled oral delivery. II. In vitro drug release evaluation of thermally modified starches. Int J Pharm 56:65–70.

Hiraku O, Nagai T (1986) Characterization and evaluation of dialdehyde starch as an erodible medical polymer and a drug carrier. Int J Pharm 30:133–141.

Hon DNS (1996) Chitin and chitosan: medical applications. In: Dumitri S (ed) Polysaccharides in medicinal application. Marcel Dekker, New York, pp 631–649

Hu A, Jiao S, Zheng J, Li L, Fan Y, Chen L et al (2015) Ultrasonic frequency effect on corn starch and its cavitation. LWT-Food Sci Technol 60:941–947

Hu L et al (2018) Hemicellulose-based hydrogels present status and application prospects: a brief review. Open J Forest 8:15–28

Ito Y, Harikai N, Ishizuki K et al (2017) Spiroketalcarminic acid, a novel minor Anthraquinone pigment in cochineal extract used in food additives. Chem Pharm Bull 65:883–887

Jania GK et al (2009) Gums and mucilages: versatile excipients for pharmaceutical formulations gums and mucilages. Asian J Pharm Sci 4:308–322

Joshi V, Attri D, Bala A et al (2003) Microbial pigments. Indian J Biotechnol 2:362–369

Kalaivani T, Mathew L (2010) Free radical scavenging activity from leaves of *Acacia nilotica* (L.) wild. *ex* Delile, an Indian medicinal tree. Food Chem Toxicol 48:298–305

Khan AK, Saba AU, Nawazish S, Akhtar F, Rashid R, Mir S et al (2017) Carrageenan based Bionanocomposites as drug delivery tool with special emphasis on the influence of ferromagnetic nanoparticles. Oxidative Med Cell Longev 2017:1–13. https://doi.org/10.1155/2017/8158315

Khanna M (1997) Polyose from seeds of *Tamarindus indica* of unique property and immense pharmaceutical use. Trends Carbohydr Chem 4:79–81

Khobragade DS et al (2016) Evaluation of gum sandarac as a novel release controlling polymer for sustained release matrix pellets. Int J Adv Pharm 5:40–45

Kianfar F et al (2011) Formulation development of a carrageenan based delivery system for Buccal drug delivery using ibuprofen as a model drug. J Biomater Nanobiotechnol 2:582–595

Kim H, Fassihi R (1997) New ternary polymeric matrix system for control drug delivery of highly soluble drugs. Pharm Res 14:1415–1421

Kizil S, Çakmak O, Kirici S et al (2008) A comprehensive study on safflower (carthamus tinctorius L.) in semi-arid conditions. Biotechnol Biotechnol 22:947–953

Klapiszewski L, Rzemieniecki T, Krawczyk M, Malina D, Norman M, Zdarta J et al (2015) Kraft lignin/silica-AgNPs as a functional material with antibacterial activity. Colloids Surf B Biointerfaces 134:220–228

Kokate CK et al (2009) Drugs containing tannins in: Pharmacognosy. Nirali Prakashan, Pune, pp 9.5–9.6
Kozlova EA et al (2017) Synthesis and investigation of rosin nanoparticles. In: Saratov Fall Meeting 2016: Fourth International Symposium on Optics and Biophotonics. Saratov, Russian Federation. doi.org/https://doi.org/10.1117/12.2269313
Krässig H (1996) Cellulose, polymer monographs, vol 11. Gordon and Breach Science Publishers, Amsterdam, pp 6–42
Krishna RR, Murthy TEGK (2010) Preparation and evaluation of mucoadhesive microcapsules of glipizide formulated with gum kondagogu: *In-vitro* and *in-vivo*. Acta Pharm Sci 52:335–344
Krishnaiah YS et al (2002) Three-layer guar gum matrix tablet formulations for oral controlled delivery of highly soluble trimetazidine dihydrochloride. J Control Release 81:45–56
Kulkarni D et al (1998) Performance evaluation of tamarind seed polyose as a binder and in sustained release formulations of low drug loading. Indian J Pharma Sci 1:50–53
Kumar A et al (2012) Cashew gum a versatile Hydrophyllic polymer: a review. Curr Drug Ther 7:2–12
Kumar R, Patil S, Patil MB et al (2009) Isolation and evaluation of Disintegrant properties of fenugreek seed mucilage. Int J Pharm Tech Res 1:982–996
Kumar VNP, Elango P, Asmathulla S et al (2017) A systematic review on lycopene and its beneficial effects. Biomed Pharmacol J 10:2113–2120
Kuppusamy G, Kulkarni GT, Muthukumar AB et al (2002) Evaluation of fenugreek mucilage as gelling agent. Int J Pharma Excip:16–19
Lam KS (2007) New aspects of natural products in drug discovery. Trends Microbiol 15:279–289
Lankalapalli S et al (2014) Preparation and evaluation of vancomycin polyelectrolyte complex nanoparticles. Indian J Nanosci 2:10–18
Lavanya D, Kulkarni PK, Dixit M et al (2011) Sources of cellulose and their applications–a review. Int J Drug Formulation Res 2:19–38
Lee CM et al (2005) Rosin nanoparticles as a drug delivery carrier for the controlled release of hydrocortisone. Biotechnol Lett 27(19):1487–1490
Lee JB et al (2011) Antiviral and immunostimulating effects of lignin-carbohydrate-protein complexes from Pimpinella anisum. Biosci Biotechnol Biochem 75:459–465
Lee KY, Mooney DJ (2012) Alginate: properties and biomedical applications. Prog Polym Sci 37:106–126
Li L et al (2014) Carrageenan and its applications in drug delivery. Carbohydr Polym 103:1–11
Lievonen M, Valle-Delgado JJ, Mattinen ML, Hult EL, Lintinen K, Kostiainen MA et al (2016) Simple process for lignin nanoparticle preparation. Green Chem 18:1416–1422
Ma G, Zhang L, Matsuta A, Matsutani K, Yamawaki K, Yahata M et al (2013) Enzymatic formation of β-citraurin from β-cryptoxanthin and Zeaxanthin by carotenoid cleavage dioxygenase4 in the flavedo of citrus fruit. Plant Physiol 163:682–695
Ma R et al (2017) Preparation and characterization of hemicellulose-based printable films. Cellulose Chem Technol 51:939–948
Mahavarkar RV et al (2016) Formulation and evaluation of tamarind seed polysaccharide matrix tablet. Pharm Biol Eva 3:241–255
Mali AJ et al (2017) In vitro and in vivo performance of novel spray dried Andrographolide loaded Scleroglucan based formulation for dry powder inhaler. Curr Drug Deliv 4:968–980
Malik K et al (2011) Locust bean gum as superdisintegrant-formulation and evaluation of nimesulide orodispersible tablets. Polim Med 41:17–28
Marques-Marinho FD, Vianna-Soares CD (2013) Cellulose and its derivatives use in the pharmaceutical compounding practice. InTech, pp 141–162. https://doi.org/10.5772/56637
McChesney JD, Venkataraman SK, Henri JT (2007) Plant natural products: Back to the future or into extinction? Phytochemistry 68:2015–2022
Mehta AY et al (2016) Allosterism-based simultaneous, dual anticoagulant and antiplatelet action. Allosteric inhibitor targeting the glycoprotein Ibα and heparin-binding site of thrombin. J Thromb Haemost 14:828–838

Michel AS et al (1997) Physico-chemical properties of carrageenan gels in presence of various cations. Int J Biol Macromol 21:195–200
Mohd-Nasir H, Abd-Talib N, Mohd-Setapar SH, Wong LP, Idham Z, Casillas AC et al (2018) Natural colorants from plants for wellness industry. Int J Pharm Sci Res 9:836–843
Monien BH et al (2006) Novel chemo-enzymatic oligomers of cinnamic acids as direct and indirect inhibitors of coagulation proteinases. Bioorgan Med Chem 14:7988–7998
Morkhade DM, Fulzele SV, Satturwar PM et al (2006b) Gum copal and gum Damar: novel matrix forming materials for sustained drug delivery. Indian J Pharm Sci 68:53–58
Morkhade DM et al (2006a) Gum copal and gum Damar: novel matrix forming materials for sustained drug delivery. Indian J Pharm Sci 68:53–56
Muchlisyam et al (2015) Hemicellulose: isolation and its application in pharmacy. In: Thakur VK, Thakur MK (eds) Handbook of sustainable polymers: processing and applications, 1st edn. Pan Stanford Publishing, Singapore, pp 305–339
Mudgil D et al (2014) Guar gum: processing, properties and food applications-a review. J Food Sci Technol 51:409–418
Mukhiddinov ZK et al (2000) Isolation and structural characterization of a pectin homo and ramnogalacturonan. Talanta 53:171–176
Nayak BS, Nayak UK, Patro KB et al (2008) Design and evaluation of controlled release Bhara gum microcapsules. Res J Pharm Tech 1:433–437
Necas J, Bartosikova L (2013) Carrageenan: a review. Vet Med-Czech 58:187–205
Nokhodchi A, Nazemiyeh H, Khodaparast A et al (2008) An *in-vitro* evaluation of fenugreek mucilage as a potential excipient for oral controlled-release matrix tablet. Drug Dev Ind Pharm 34:323–329
Odeku OA (2005) Assessment of Albizia zygia gum as a binding agent in tablet formulations. Acta Pharma 55:263–276
Odeku OA, Fell JT (2004) Evaluation of khaya gum as a directly compressible matrix system for controlled release. J Pharm Pharmacol 56:1365–1370
Odeku OA et al (2014) Formulation and in vitro evaluation of natural gum-based microbeads for delivery of ibuprofen. Trop J Pharm Res 13:1577–1583
Okoye EI et al (2012) Evaluation of LD50 of cashew gum and the comparative study of its functionality in cotrimoxazole granule and tablet formulations. Br J Pharmacol Toxicol 3:156–164
Osete-Cortina L, Domenech-Carbo MT (2005) Analytical characterization of diterpenoid resins present in pictorial varnishes using pyrolysis-gas chromatography-mass spectrometry with on line trimethylsilylation. J Chromatogr A 1065:265–278
Palei NN et al (2016) Formulation and evaluation of lamivudine sustained release tablet using okra mucilage. J Appl Pharm 6:69–75
Paolicelli P, Varani G, Pacelli S, Ogliani E, Nardoni M, Petralito S et al (2017) Design and characterization of a biocompatible physical hydrogel based on scleroglucan for topical drug delivery. Carbohydr Polym 174:960–969
Pathak YV, Dorle AK (1990) Rosin and rosin derivatives as hydrophobic matrix materials for controlled release of drugs. Drug Des Deliv 6:223–227
Pereira FM et al (2003) Estimation of solubility effect on the herbicide controlled-release kinetics from lignin-based formulations. Appl Biochem Biotechnol 108:913–919
Peter FK et al (1993) Molecular structures of gum exudates from Hakea species. Phytochemistry 34:709–713
Petropoulos GA (2002) Fenugreek: the genus Trigonella. In: Petropoulus GA (ed) Botany. Taylor and Francis, London, pp 9–17
Pham TQ, Cormier F, Farnworth E et al (2000) Antioxidant properties of Crocin from *Gardenia jasminoides* Ellis and study of the reactions of Crocin with linoleic acid and Crocin with oxygen. J Agric Food Chem 48:1455–1461
Philip T, Berry JW (1976) Process for the purification of lutein-fatty acid esters from marigold petals. J Food Sci 41:163–164

Pochanavanich P, Suntornsuk W (2002) Fungal chitosan production and its characterization. Lett Appl Microbiol 35:17–21

Prabhu P, Ahmed N, Harish NM, Ahmed MG, Charyulu RN, Satyanarayana D et al (2010) Investigation of colon specificity of novel polymer Khaya gum. Indian J Pharm Educ Res 44:56–65

Prasanna JL et al (2012) Influence of formulation variables on the *in vitro* dissolution of gum Damar based matrix tablets. Int J Pharm Sci Res 3:4520–4527

Priyadarsini KI (2014) The chemistry of Curcumin: from extraction to therapeutic agent. Molecules 19:20091–20112

Pyrzanowska J, Piechal A, Blecharz-Klin A, Lechner M, Skórzewska A, Turzynska D et al (2010) The influence of the long-term administration of *Curcuma longa* extract on learning and spatial memory as well as the concentration of brain neurotransmitters and level of plasma corticosterone in aged rats. Pharmacol Biochem Behav 95:351–358

Quesille-Villalobos AM et al (2013) Phenolic compounds, antioxidant capacity, and in vitro α-amylase inhibitory potential of tea infusions (*Camellia sinensis*) commercialized in Chile. CyTA J Food 11:60–67

Rajeswari S et al (2017) Development, formulation and evaluation of a bilayer gastric retentive floating tablets of ranitidine HCl and clarithromycin using natural polymers. Int J Pharm Pharm Sci 9:164–177

Răschip IE et al (2015) In vitro testing of xanthan/lignin hydrogels as carriers for controlled delivery of bisoprolol fumarate. Rev Med Chir Soc Med Nat Iasi 19:1189–1194

Rinaudo M (2006) Chitin and chitosan: properties and applications. Prog Polym Sci 31:603–632

Rinaudo M, Vincendon M (1982) 13C NMR structural investigation of scleroglucan. Carbohydr Polym 2:135–144

Roberfroid MB (2007) Inulin-type fructans: functional food ingredients. J Nutr 137:2493S–2502S

Robyt J (2008) Starch: structure, properties, chemistry, and enzymology. In: Fraser-Reid BO, Tatsuta K, Thiem J (eds) Glycoscience. Springer, Berlin, Heidelberg, pp 1437–1472

Rodriguez A et al (2014) *In vitro* and *in vivo* evaluation of two carrageenan-based formulations to prevent HPV acquisition. Antivir Res 108:88–93

Saha S, Singha N, Chattopdhaya RN et al (2005) Studies of *Terminalia bellirica* (Bahera), a natural gum, as an additive in a water-based adhesive composition. J Adhes Sci Technol 19:1349–1361

Sahoo S et al (2010) Tamarind seed polysaccharide: a versatile biopolymer for Mucoadhesive applications. J Pharm Biomed Sci 8:1–12

Sakagami H, Sheng H, Okudaira N, Yasui T, Wakabayashi H, Jia J et al (2016) Prominent anti-UV activity and possible cosmetic potential of lignin-carbohydrate complex. In Vivo 30:331–339

Sakarkar DM (2012) Natural gums and Mucilage's in NDDS: applications and recent approaches. Int J Pharm Tech Res 2:800

Santana AL, Meireles MAA (2014) New starches are the trend for industry applications: a review. Food Public Health 4(5):229–241

Sattle A, Agrawal S (2012) Solubility enhancement potential of tamarind seed polysaccharide as pharmaceutical excipient. Int J Pharm Biol Arch 3:456–459

Satturwar PM et al (2005) Evaluation of polymerized rosin for the formulation and development of transdermal drug delivery system: a technical note. AAPS PharmSciTech 6:E649–E654

Satyanarayana A et al (2003) Chemistry, processing and toxicology of annatto (*Bixa orellana* L.). J Food Sci Technol 40:131–141

Schmidt PC, Broegmann B (1988) Effervescent tablets: choice of a new binder for ascorbic acid. Acta Pharm Technol 34: 22–26.

Sedaghati S, Abbaszadeh F (2018) Extraction of saffron Crocin as a natural pharmaceutical source with crystallization method. Int J Pharma Med Biol Sci 7:84–87

Shahwar I et al (2012) Application and use of inulin as a tool for therapeutic drug delivery. Biotechnol Genet Eng Rev 28:33–46

Sharma N (2013) Development of Abelmoschus esculentus (Okra)-based Mucoadhesive gel for nasal delivery of Rizatriptan benzoate. Trop J Pharm Res 12:149–153

Shin HJ et al (2006) Gastric-retentive controlled release mono-matrix tablet WO/2006/088305. 24 Aug 2006

Shukla RK et al (2012) Formulation development and characterization of guar gum microspheres for colon delivery. Nanosci Nanotech—Asia 2:28. https://doi.org/10.2174/2210681211202010028

Singh N, Sharma B (2018) Toxicological effects of Berberine and Sanguinarine. Front Mol Biosci 5:21

Singh R et al (2011) Extraction and characterization of tamarind seed polysaccharide as a pharmaceutical excipient. Pharm J 3:17–19

Sirisha VNL et al (2018) A novel approach of locust bean gum microspheres for colonic delivery of mesalamine. Int J App Pharm 10:86–93

Šivel M, Kráčmar S, Fišera M et al (2014) Lutein content in marigold flower (*Tagetes erecta* L.) concentrates used for production of food supplements. Czech J Food Sci 32:521–525

Soares PAG, Bourbon AI, Vicente AA, Andrade CAS, Barros W Jr, Correia MTS et al (2014) Development and characterization of hydrogels based on natural polysaccharides: Policaju and chitosan. Mater Sci Eng C 42:219–226

Soni H, Patel SS, Mishra K et al (2011) Qualitative and quantitative profile of curcumin from ethanolic extract of *Curcuma longa*. Int Res J Pharm 2:180–184

Sriamornsak P (2006) Chemistry of pectin and its pharmaceutical uses: a review. Silpakorn Univ Int J 3:206–228

Srivastava S et al (2017) Antimicrobial efficacy of methylated lac dye, an anthraquinone derivative. Indian Microbiol 57:470–476

Stasiak N et al (2014) Modern industrial and pharmacological applications of indigo dye and its derivatives-a review. Acta Pol Pharm 71:215–221

Story EN, Kopec RE, Schwartz SJ et al (2010) An update on the health effects of tomato lycopene. Annu Rev Food Sci Technol 1:189–210

Sungthongjeen S et al (1999) Studies of pectins as potential hydrogel matrices for controlled release drug delivery. Drug Dev Ind Pharm 25:1271–1276

Swapna R et al (2019) Formulation *In-Vitro* and *In-Vivo* evaluation of cefuroxime Axetil floating tablets using natural gums. Int J Life Sci Pharma Res 9:37–53

Szabolcs PMJ (1980) β-Citraurin epoxide, a new carotenoid from Valencia orange peel. Phytochemistry 19:633–637

Szekalska M et al (2016) Alginate: current use and future perspectives in pharmaceutical and biomedical applications. Int J Polym Sci 2016:1–17. https://doi.org/10.1155/2016/7697031

Takahashi Y, Miyasaka N, Tasaka S, Miura I, Urano S, Ikura M et al (1982) Constitution of two coloring matters in the flower petals of Carthamus Tinctorius L. Tetrahedron Lett 23:5163–5166.

Tawari PD (2018) Formulation and evaluation of metformin sustained release matrix tablet using okra (Abelmoschus Esculentus L.) polysaccharide. Res Pharm 2:13–19

Tayade PB, Adivarekar RV (2014) Extraction of indigo dye from *Couroupita guianensis* and its application on cotton fabric. Fashion Text 1:16

Te Wierik GHP, Bergsma J, Arends-Scholte W, Boersma T, Eissens AC, Lerk CF (1996) A new generation starch products as excipient in pharmaceutical tablets: I. Preparation and binding properties of high surface area potato starch products. Int J Pharm 134:27–36.

Te Wierik GHP, Eissens AC, Bergsma J et al (1993) Preparation, characterization and pharmaceutical application of linear dextrins: IV. Drug release from capsules and tablets containing amylodextrin. Int J Pharm 98:219–224.

Teramoto N, Motoyama T, Yosomiya R et al (2003) Synthesis, thermal properties & biodegradability of propyl-etherified starch. Eur Polym J 39:255–261

Tuli HS et al (2015) Microbial pigments as natural color sources: current trends and future perspectives. J Food Sci Technol 52:4669–4678

Üner M, Altinkurt T (2004) Evaluation of honey locust (*Gleditsia triacanthos* Linn.) gum as sustaining material in tablet dosage forms. Farmaco 59:567–573

Uysal A et al (2019) Elucidation of biological properties of some commercial anthraquinones: mutagenic/antimutagenic and antimicrobial activity approaches. J Res Pharm 23:208–216
Vadlamudia HC et al (2014) Assessment of Hupu gum for its carrier property in the design and evaluation of solid mixtures of poorly water soluble drug—Rofecoxib. Curr Drug Deliv 11:62–71
Vanholme R et al (2010) Lignin biosynthesis and structure. Am Soc of Plant Biologists. https://doi.org/10.1104/pp.110.155119
Vasudevan P et al (1997) Tagetes: a multipurpose plant. Bioresour Technol 62:29–35
Veenus Seelan T et al (2016) Design and development of *Albizia stipulata* gum based controlled release matrix tablets in cancer therapeutics. Int J Biol Macromol 92:972–980
Whitmore TC (1980) Utilization, potential, and conservation of *Agathis*, a genus of tropical Asian conifers. Econ Bot 34:1–12
Witzler M, Alzagameem A, Bergs M, Khaldi-Hansen BE, Klein SE, Hielscher D et al (2018) Lignin-derived biomaterials for drug release and tissue engineering. Molecules 23:1885
Yadav BK et al (2016) Rosin: recent advances and potential applications in novel drug delivery system. J Bioact Compat Pol 31:111–126
Yasunaga H (2012) Hair dyeing by using Catechinone obtained from (+)-Catechin. JCDSA 2:158–163
Younes I, Rinaudo M (2015) Chitin and chitosan preparation from marine sources: structure, properties and applications. Mar Drugs 13:1133–1174
Zeng H et al (2007) Controlled-release emulsion compositions. Tech Rep, Penwest Pharmaceuticals, New York
Zhang Y, But PP, Ooi VE, Xu HX, Delaney GD, Lee SH et al (2007) Chemical properties, mode of action, and in vivo anti-herpes activities of a lignin-carbohydrate complex from *Prunella vulgaris*. Antivir Res 75:242–249

Part V

Plants for Future

A Wonder Plant *Withania*: Pharmacological and Chemical Perspectives

41

Surjeet Verma, Namrita Lall, and Debra Meyer

Abstract

Plants belonging to *Withania* genus are well-known medicinal plants and are commonly used as herbal medicine in the Ayurvedic health system. These plants belong to the family Solanaceae and comprise of 23 known species distributed in North Africa, the Middle East, Asia, the Mediterranean, and the Canary Islands. However, only six species, namely, *Withania somnifera* (L.) Dunal, *Withania coagulans* (Stocks) Dunal, *Withania adpressa* Coss, *Withania aristata* (Aiton) Pauquy, *Withania frutescens* (L.) Pauquy, and *Withania obtusifolia* Täckh, are extensively investigated with regard to their pharmacological properties and bioactive constituents. Therefore, in this chapter, the traditional uses of the six *Withania* species and the pharmacological validations of the plants and plant-derived constituents are discussed. Applications of the bioactive constituents in drug discovery are explained by discussing the semisynthetic modifications and structure-activity relationships.

S. Verma
Department of Plant and Soil Sciences, Faculty of Natural and Agricultural Sciences, University of Pretoria, Pretoria, South Africa

N. Lall (✉)
Department of Plant and Soil Sciences, Faculty of Natural and Agricultural Sciences, University of Pretoria, Pretoria, South Africa

School of Natural Resources, College of Agriculture, Food and Natural Resources, University of Missouri, Columbia, MO, USA

College of Pharmacy, JSS Academy of Higher Education and Research, Mysuru, Karnataka, India
e-mail: namrita.lall@up.ac.za

D. Meyer
Department of Biochemistry, Office of the Executive Dean, Faculty of Science, University of Johannesburg, Johannesburg, South Africa

S. C. Mandal et al. (eds.), *Evidence Based Validation of Traditional Medicines*,
https://doi.org/10.1007/978-981-15-8127-4_41

Keywords

Withania species · Traditional uses · Biological activities · Bioactive withanolides · Structure activity relationship

Photo credit: Piouswatson 2008

41.1 Introduction

W. somnifera and *W. coagulans* are economically popular and medicinally significant species of the genus *Withania*. These species are widely used and cultivated in several regions of the world. *W. somnifera* and *W. coagulans* are the most extensively investigated species with regard to the biological activities, bioactive constituents, and the scientific validation of the traditional usages. Other species of *Withania* are less popular, and very few scientific studies have been done on them.

The medicinal values of *Withania* species are well-documented in the Indian Ayurvedic system of traditional medicines and are also important in the traditional medicine of Africa and Asia (Mirjalili et al. 2009). However, around 80% of the total literatures available on *Withania* are based on the studies done on *W. somnifera* only. This species has diverse pharmacological properties and is the species responsible for the economic and medicinal importance of the genus. It is also known as "Ashwagandha" or "Indian ginseng" and is native to India and North Africa. *W. coagulans* is the second most explored plant of the genus and is known as "Indian cheese maker" as its fruit has a milk-coagulating characteristic. It is also well-known for its antidiabetic effect. The major constituents responsible for its antidiabetic effect are withanolide glucosides (Pramanick and Srivastava 2015). A number of phytochemicals have been isolated from different *Withania* species. The major phytochemical constituents of these plant species are "withanolides," a class of naturally occurring steroids based on an ergostane skeleton.

The purpose of this book chapter is to provide a brief description of the traditional usage of six selected species of *Withania*, their biological activities, their bioactive constituents, and their structure-activity relationship (SAR). In addition, this book chapter also summarizes the scientific work already done on these species for the validation of their traditional usages and to identify the gaps which still need to be addressed by further research.

41.2 *Withania* Species

In this chapter, only six species will be discussed, which have been investigated in detail for therapeutic potential and the presence of bioactive compounds. Those species are, namely:

- *Withania adpressa* Coss.
- *Withania aristata* (Aiton) Pauquy.
- *Withania frutescens* (L.) Pauquy.
- *Withania obtusifolia* Täckh.
- *Withania coagulans* (Stocks) Dunal.
- *Withania somnifera* (L.) Dunal.

41.3 Traditional Usage

In traditional medicine, *W. adpressa* is used to treat food poisoning by traditional healers of Morocco (Bakrim et al. 2018), while *W. aristata* has long been used as a scarring agent, antispasmodic, to treat rheumatism, eye diseases and otitis, insomnia, constipation, and urinary pathologies in folk medicine of the Canary Islands. Ingestion of the fruit strongly stimulates urine production, making it useful against hydropesia (Martin-Herrera et al. 2007). *W. frutescens* is used by the traditional healers of Morocco for the treatment of dysentery (Bouzidi et al. 2013), and *W. coagulans* fruit is reported to be sedative, emetic, and diuretic. They have been reported to be effective in chronic liver disorder, dyspepsia, flatulent coli, other intestinal infections, asthma, biliousness, and strangury. In some parts of the Indo-Pak subcontinent, the berries are used as a blood purifier, and the smoke of the plant is inhaled for the relief of toothache. In Northwestern parts of India, traditional practitioners use dry fruits of this species for the treatment of diabetic patients. It is also used for the treatment of ulcers, rheumatism, nephronia, bronchitis, and degenerative disease (Huang et al. 2009; Pramanick and Srivastava 2015). Unfortunately, no traditional usage of *W. obtusifolia* has been documented.

W. somnifera (commonly known as Ashwagandha in Hindi, a native language of India) is used traditionally to boost energy, strength, and stamina and for the treatment of numerous ailments including hepatic, cardiovascular, immunological, neurological, and metabolic disorders such as diabetes (Jonathan et al. 2015). It is also used for treating cold and coughs, ulcers, emaciation, diabetes, conjunctivitis, epilepsy, insomnia, senile dementia, leprosy, Parkinson's disease, nervous disorders, rheumatism, arthritis, intestinal infections, bronchitis, asthma, impotence, and as a suppressant of HIV (human immunodeficiency virus)/AIDS (acquired immunodeficiency syndrome). It might also be beneficial against diseases like tuberculosis (TB), chronic upper respiratory diseases, and HIV due to its strong immunostimulatory effect. Its usage as a blood tonic, especially in gynecological disorders including anemia and irregular menstruation, has been documented (Umadevi et al. 2012).

41.4 Biological Activities

41.4.1 *W. adpressa*

There are two reports on the biological effect of *W. adpressa*: one on antifungal activity (Askarne et al. 2012) and the other on cytotoxicity (Abdeljebbar et al. 2009). The dichloromethane, ethyl acetate, and methanol extracts from leaves of this plant exhibited cytotoxicity against the human laryngeal cancer cell line (Hep-2), human colorectal adenocarcinoma cell line (HT-29), human rhabdomyosarcoma cell lines (RD), Madin-Darby canine kidney (MDCK) cells, and Vero cancer cell lines with half maximal inhibitory concentration (IC_{50}) values in the range of 0.61–25 μg/mL. The leaf powder exhibited poor inhibitory potential against *Penicillium digitatum (Pers.:Fr.) Sacc.*

41.4.2 *W. aristata*

The dichloromethane extract of *Withania aristata* and its chemical components (withanolides) has shown interesting antiproliferative/cytotoxic activity against human cervical cancer cell line (HeLa), human lung cancer cell line (A549), human breast cancer cell line (MCF-7), and Vero cells (Llanos et al. 2010, 2012). The *W. aristata's* water and aqueous methanolic extracts exhibited potent diuretic activity in rats at a dose of 100 mg/kg body weight, with significant excretion of sodium and potassium ions (Martin-Herrera et al. 2007; Benjumea et al. 2009).

41.4.3 *W. frutescens*

The ethanolic extract of *W. frutescens* leaves has shown protective and curative action against carbon tetrachloride (CCl_4)-induced hepatotoxicity in rats (Montilla et al. 1990). It has been reported that the dichloromethane fraction of the methanolic extract of the leaf and isolated compounds "withanolides" possessed moderate cytotoxicity against human liver cancer cell line (HepG2) and HT-29 cancer cell lines (Bouzidi et al. 2013).

41.4.4 *W. obtusifolia*

Although *W. obtusifolia* (WO) extracts have not been evaluated for any biological activity, few cytotoxic withanolides have been isolated from the leaf ethanolic extract (Fig. 41.6). The activity of those withanolides is discussed in Sect. 41.5.4.

41.4.5 *W. coagulans*

Various organic, hydro-organic, and aqueous extracts of the different parts of *W. coagulans* (WC) along with the isolated compounds have shown interesting biological properties. The alcoholic extract and total alkaloid fraction of WC fruit (WCF) showed significant anti-inflammatory effect in subacute inflammation produced by formalin and acute inflammation produced with egg albumin. It also reduced the granulation tissue formation (Budhiraja et al. 1977). Methanolic extract of WC dried fruit (WCFE) showed significant anticancer activity against human breast cancer cell line MDA-MB-231. The cell viability was reduced to about 50% at 40 μg/mL concentration of methanolic extract in 50% DMSO. However, WCFE did not exhibit significant cytotoxicity against Vero cells at 170 and 200 μg/mL (Ahmad et al. 2017). The methanolic extracts of WC leaf stalk and fruit have shown remarkable activity (IC_{50}, 0.96–4.73 and 0.69–6.69 μg/mL, respectively) against human immortal cancer (HeLa), breast cancer (MCF-7), rhabdomyosarcoma cancer (RD), and rat glioma 2 (RG2) and INS-1 cancer cell lines (Maqsood et al. 2018).

The combination of aqueous and chloroform extracts of WCF showed highly significant ($p < 0.01$) reduction in blood glucose (55%), triglyceride, total cholesterol, LDL, and VLDL (very low density lipoprotein) levels and a highly significant ($p < 0.01$) increase in HDL (high-density lipoprotein) level in diabetic rats. The rats were administered with a dose of 1 g/kg body weight (b.wt) for 14 days (Hoda et al. 2010). Administration of an aqueous extract of WCF (1 g/kg per oral) to high-fat diet-induced hyperlipidemic rats for 7 weeks significantly reduced elevated serum cholesterol, triglycerides, and lipoprotein levels. It also showed hypolipidemic activity in triton-induced hypercholesterolemia (Hemalatha et al. 2006). Antidiabetic activity WCF extracts in diabetic rats have also been reported few other groups (Datta et al. 2013; Hemalatha et al. 2004; Jaiswal et al. 2009). A withanolide isolated from the alcoholic extract of WCF produced a moderate fall of blood pressure in dogs (34 ± 2.1, mm Hg) which was blocked by atropine but not by mepyramine or propranolol. In rabbit, it produced myocardial-depressant effects, but in perfused frog heart, it produced mild positive inotropic and chronotropic effects (Budhiraja et al. 1983).

The hydroalcoholic fraction of methanolic extract of WC fruit showed a significant increase in the rate of wound healing on topical application of 10% w/w ointment and oral administration (500 mg/kg, per oral) in streptozotocin-induced diabetic rats (Prasad et al. 2010). The ethyl acetate extract of WCF showed antimicrobial activity against *Enterobacter aerogenes* (28 mm) at a concentration of 150 mg/20disc, while methanolic extract showed the best activity against *Klebsiella pneumonia* (21 mm) at a concentration of 250 mg/20 discs (Sudhanshu et al. 2012). The antimutagenic effect of WCF extract has also been documented. Administration of WCF methanolic extract at the dose of 500, 1000, and 1500 mg/kg body weight significantly prevented the micronucleus formation in dose-dependent manner in bone marrow cells of mice as compared to cyclophosphamide treated group (Mathur and Agrawal 2011). WCF ethanolic extract exerted nephroprotective effect in Charles Foster albino (150–200 g) male rats at a dose of 400 mg/kg, per oral (Sharma et al. 2017). As mentioned in Sect. 41.3, *W. coagulans* is used as a diuretic agent in traditional folklore medicine. The diuretic potential of WCF aqueous extract was evaluated by Dabheliya et al. 2010. A significant ($p < 0.001$) increase (79.12% and 71.02%) in the urine volume was observed in Albino Wistar rats at a dose 500 and 750 mg/kg body weight. Thus, the traditional use of WC as a diuretic agent has been justified.

41.4.6 *W. somnifera*

The *W. somnifera* (WS) is known for various biological effects, including anti-stress, antibiotic, antioxidant, antiaging, anticonvulsant, nootropic, antiparkinson, cardiovascular, immunomodulatory, antihyperglycemic, hypolipidemic, stimulating sexual behavior, anticarcinogenic, and supportive toward osteoarthritis, hypothyroid, and anti-inflammatory activities (Kaur et al. 2013). *W. somnifera* leaf and root methanolic extracts have been reported to increase glucose uptake in myotubes and adipocytes in a dose-dependent manner. The leaf extract was found to be more active than the root extract. Leaf but not the root extract increased insulin secretion in basal pancreatic beta cells but not in stimulated cells (Jonathan et al. 2015). The root powder of WS has shown hypoglycemic, diuretic, and hypocholesterolemic effects on human subjects. Decrease in blood glucose was comparable to that of an oral hypoglycemic drug daonil (glibenclamide) after 30 days of treatment (Andallu and Radhika 2000). The aqueous alcoholic extract of WS root reduced the rate of cell division in the mammary carcinomas developed in female transgenic (MMTV/Neu) mice (Khazal et al. 2014). The alcoholic extract of WS (400 mg/kg b.wt) along with paclitaxel (33 mg/kg b.wt) provided stabilization of membrane-bound ATPase enzymes levels and decreased lipid peroxidation against benzo(a)pyrene-induced lung cancer in male Swiss albino mice (Senthilnathan et al. 2006). Other in vivo studies have shown that alcoholic extract of WS root could also be useful in the treatment of colon and brain or glioblastoma multiforme (GBM) cancers (Chang et al. 2016; Muralikrishnan et al. 2010). However, no evidence suggests that WS exerts similar effect in humans.

Stress has been reported to be a causative factor for male infertility. Administration of the root powder of WS (5 g/day) for 3 months to the human subjects resulted in a reduction in stress and thus the infertility (Mahdi et al. 2011). A stress hormone (cortisol) level was substantially ($p = 0.0006$) reduced in chronically stressed adults by administration of 300 mg of root extract twice a day for 60 days. The anxiety and insomnia were thus reduced by 69%, on average, compared with 11% in the placebo group (Chandrasekhar et al. 2012). The aqueous root extract blocked the stress pathway in the brain of the rats by regulating chemical signalling in the nervous system (Candelario et al. 2015).

It has a powerful effect on reproductive organs by increasing sperm count and its motility and by increasing testosterone levels. In a study a total of 180 infertile male patients were administered WS root powder at the rate of 5 g/day for a 3-month period. This therapy resulted in the repair of disturbed concentrations of lactate, alanine, citrate, glycerophosphocholine (GPC), histidine, and phenylalanine in seminal plasma and recovers the quality of semen of post-treated compared to pre-treated infertile men (Gupta et al. 2013). In another study, WS root powder (5 g/day) is administered orally for 3 months with milk to the infertile human subjects. The treatment resulted in improvement in sperm count and motility and recovery of the seminal plasma levels of antioxidant enzymes, vitamins A, C, and E, and corrected fructose. Moreover, treatment also significantly increased serum T (serum testosterone) and LH (luteinizing hormone) and reduced the levels of follicle-stimulating hormone (FSH) and prolactin (PRL) (Ahmad et al. 2010). *W. somnifera* is also used as a health tonic; the people who administered 6–10 g of pulverized root per day significantly gained muscle strength and size and loosed total- and low-density lipoprotein (LDL) cholesterol (Raut et al. 2012). The subjects who consumed the 300 mg of aqueous extract of Ashwagandha root twice a day lose the body fat by more than two times as compared to the placebo group after 30 days (Wankhede et al. 2015). Hypocholesterolemic and antioxidant effects of WS root powder have also been observed in male albino rats by treatment with 0.75 and 1.5 g/rat/day (Visavadiya and Narasimhacharya 2007).

W. somnifera root powder exerted protective effect on inflammatory markers and insulin resistance in fructose-fed rats. The rats treated with WS root powder (62.5 mg/g per diet) were found with significantly ($p < 0.05$) low glucose, insulin, homeostasis model assessment for insulin resistance (HOMA-R), interleukin (IL)-6, and tumor necrosis factor (TNF)-α level (Noshahr et al. 2015). The WS methanolic extract has also been reported to exert anti-inflammatory effect through alleviation of formalin-induced nociception in mice (Orrù et al. 2016). Treatment with WS aqueous root extract (WSAq) resulted in a dose-dependent reduction in arthritic index, autoantibodies, and C-reactive protein ($p < 0.05$) with maximum effect at dose of 300 mg/kg body weight in collagen-induced arthritic (CIA) rats. The oxidative stress in CIA rats was also ameliorated by treatment with different doses of WSAq, which was evidenced by a decrease in lipid peroxidation and glutathione-*S*-transferase activity and an increase in the glutathione content and ferric-reducing ability of plasma ($p < 0.05$) (Khan et al. 2015). A regular consumption of tea fortified with WS enhanced the natural killer (NK) cell activity in human volunteers. NK cells are important aspect of the (early) innate immune response to infections (Bhat et al. 2010). A major change in immune cell activation was observed after 96 h in human volunteers who consumed 6 mL of an Ashwagandha aqueous alcoholic root extract twice a day. A significant increase was observed in the expression of CD4 (cluster of differentiation 4) on $CD3^+$(cluster of differentiation 3) T cells after 96 h. $CD56^+$ NK (natural killer) cells were also activated after 96 h which was evidenced by expression of the CD69 (cluster of differentiation 69) receptor (Mikolai et al. 2009).

A methanol:chloroform (3:1) extract of WS root reversed the β-amyloid$_{1-42}$-induced toxicity in HIV-associated neurocognitive disorders in human neuronal cells (Kurapati et al. 2013). The aqueous leaf extract of WS suppressed the acute effects of sleep loss on learning and memory impairments in sleep-deprived Wistar rats administered with a dose of 140 mg/kg body weight for 15 consecutive days (Manchanda et al. 2017). Administration of aqueous-methanolic extract of WS root (WSRE) prevented hypobaric hypoxia-induced memory impairment and neurodegeneration in Sprague Dawley rats after 21 days. In addition it also reduced the nitric oxide, corticosterone, oxidative stress, and AchE activity in hippocampal region. Administration of sodium nitroprusside (SNP) along with WSRE supplementation during hypoxic exposure enhanced corticosterone levels and increased the number of pyknotic cells (Baitharu et al. 2013).

An almost complete amelioration of spatial memory deficits in temporal lobe epilepsy by enhancing antioxidant system and restoring altered N-methyl-daspartate (NMDA) receptor density epileptic rats was observed after treatment with an aqueous WS root extract (100 mg/kg body weight/day) and withanolide A (10 μmol/kg body weight/day) for 15 days (Soman et al. 2012). The healthy men, who took 500 mg of aqueous extract of root and leaf daily, experienced significant improvements in their task performance, in reaction times, choice discrimination, digit symbol substitution, digit vigilance, and card sorting tests. However, no effect was observed in the finger tapping test (Pingali et al. 2014). Thus, all these proved health benefits of WS, documented previously (Spritzler 2018), and validate its traditional usages for treating various ailments.

41.5 Bioactive Constituents

The major bioactive constituents isolated from the *Withania* species are the "withanolides." Hence, under this chapter, only withanolides and its associated biological activities will be discussed. The withanolides are a group of naturally occurring steroids (containing 28 carbons) based on an ergostane skeleton in which C-26 and C-22 or C-26 and C-23 are oxidized in order to form δ- or γ-lactones or lactols. The withanolides containing δ-lactone or δ-lactol are known as type-A withanolides, while those containing γ-lactone or γ-lactol are called type-B withanolides (Fig. 41.1). Withaferin A (WA) was the first withanolide isolated from *Withania somnifera* in 1965, and it possesses significant and diverse biological activities. Over the past 53 years, a number of withanolides have been isolated and characterized from *Withania* as well as from other species (Chen et al. 2011). Similar to withaferin A, all of the withanolides contain oxidized ring A and B systems

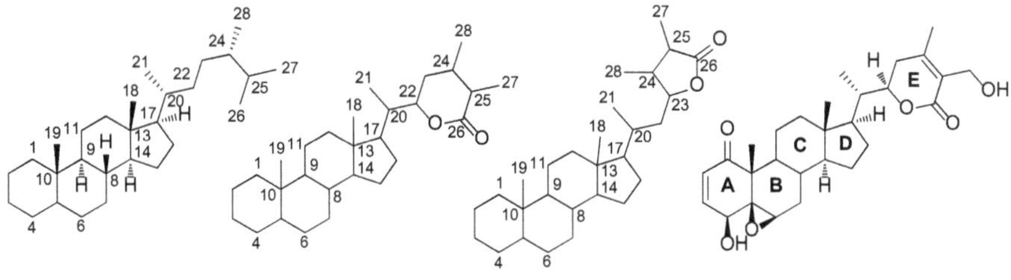

Fig. 41.1 Ergostane skeleton, types of withanolides, and structure of withaferin A

(Glotter 1991). The oxidation patterns of ring A and B can be categorized into eight major groups, i-viii (Fig. 41.2).

41.5.1 Cytotoxic Constituents of *W. adpressa*

Three withanolides, (17*S*,20*S*,22*R*)-14α,15α,17β,20β-tetrahydroxy-1-oxowitha-2,5,24-trienolide (**1**), withanolide F (**2**), and withanolide J (**3**), and two withanolide glycosides, coagulin L (**4**) and wadpressine (**5**) along with withaferin A (**WA**), were isolated from the dichloromethane and ethyl acetate fractions of the methanolic extract of the *W. adpressa* leaves (Fig. 41.3). Withanolide F and withaferin A exhibited good antiproliferative activity against multiple myeloma cancer stem cells (MM-CSCs), derived from the bone marrow of a multiple myeloma (MM) patient and the tumor plasma cells (RPMI 8226). The IC_{50} values for **2** and **WA** against MM-CSCs were found to be 5.3 and 0.33 μM, respectively, while against RPMI 8226 cells were found to be 0.1 and 0.17 μM, respectively. These compounds also inhibited necrosis factor NF-κB significantly, with IC_{50} values of 1.2 and 0.05 μM, respectively, as compared to the IC_{50} value for control drug (parthenolide), 0.53 μM. The other compounds possessed a moderate inhibition potential of cell proliferation in RPMI 8226 cells (IC_{50}, 3.2–7.1 μM).

Nicotiflorin, a flavonoid glycoside isolated from the methanolic extract of *W. adpressa*, exhibited a moderate-free radical-scavenging activity with IC_{50}, 35.3 μM in the 2,2-diphenyl-1-picryl-hydrazyl-hydrate (DPPH) assay, and with IC_{50}, 41.3 μM in nitric oxide (NO) assays. The methanol, dichloromethane, and ethyl acetate extracts, semi-pure fractions, and compounds **1–3** exhibited potent cytotoxicity against the other cancer cell lines, viz., Hep2, HT29, RD, Vero, and MDCK by inducing apoptosis (Abdeljebbar et al. 2009; Bakrim et al. 2018).

41.5.2 Diuretic and Cytotoxic Constituents of *W. aristata*

Withaferin A, witharistatin (**6**), and its mixture, isolated from the water extract of *W. aristata*, possessed significant diuretic effect in rats with a diuretic index of

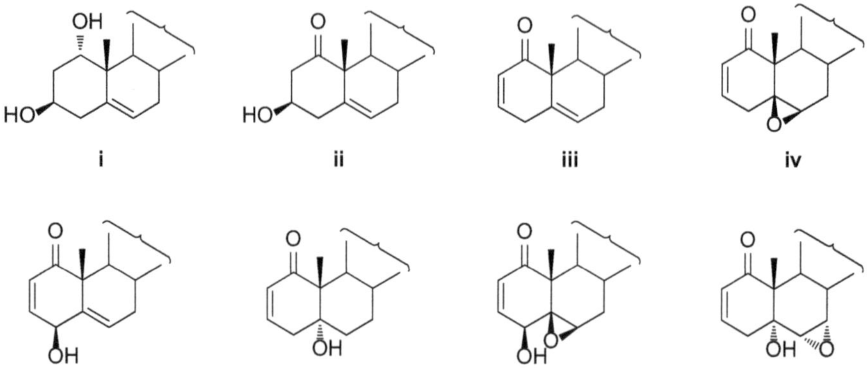

Fig. 41.2 Eight major patterns of the ring A and B systems in withanolides

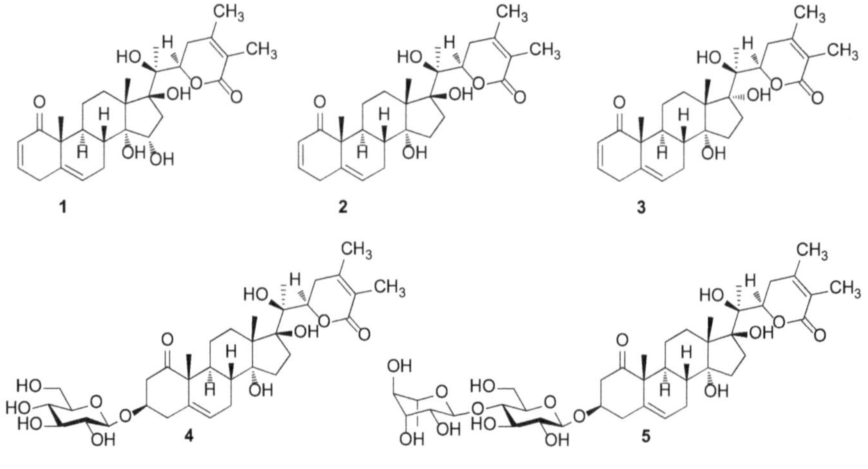

Fig. 41.3 Structures of compounds 1–5

2.43–2.59 (3.56 for the reference drug hydrochlorothiazide) at the dose of 10 mg/kg body weight (Benjumea et al. 2009). It has been reported that several other withanolides isolated from the dichloromethane extract of *W. aristata* leaves and semisynthetic derivatives (Fig. 41.4) possessed interesting antiproliferative/cytotoxic activities against HeLa, A-549, MCF-7, and Vero cells (Llanos et al. 2010, 2012). The withanolides 27-*O*-acetyl-withaferin A (**7**), 27-*O*-acetyl-viscosalactone B (**8**), 4*β*-hydroxy-1-oxo-5*β*,6*β*-epoxywitha-2,24-dienolide (**9**), 4*β*-hydroxy-1-oxo-5*β*,6*β*-epoxy-22*R*-witha-2,14,24-trienolide (**10**), 27-*O*-(*tert*-butyldimethylsilyl) withaferin A (**11**), 27-*O*-(*tert*-butyldimethylsilyl)-4-dehydroxy-4-oxo-withaferin A (**12**) and (4*S*,20*S*,22*R*)-27-acetoxy-4-*p*-bromobenzoyloxy-1-oxowitha-2,5,16,24-tetraenolide (**13**) along with withaferin A and witharistatin (**6**) exhibited potent antiproliferative/cytotoxic activity (IC_{50}, < 10 μM).

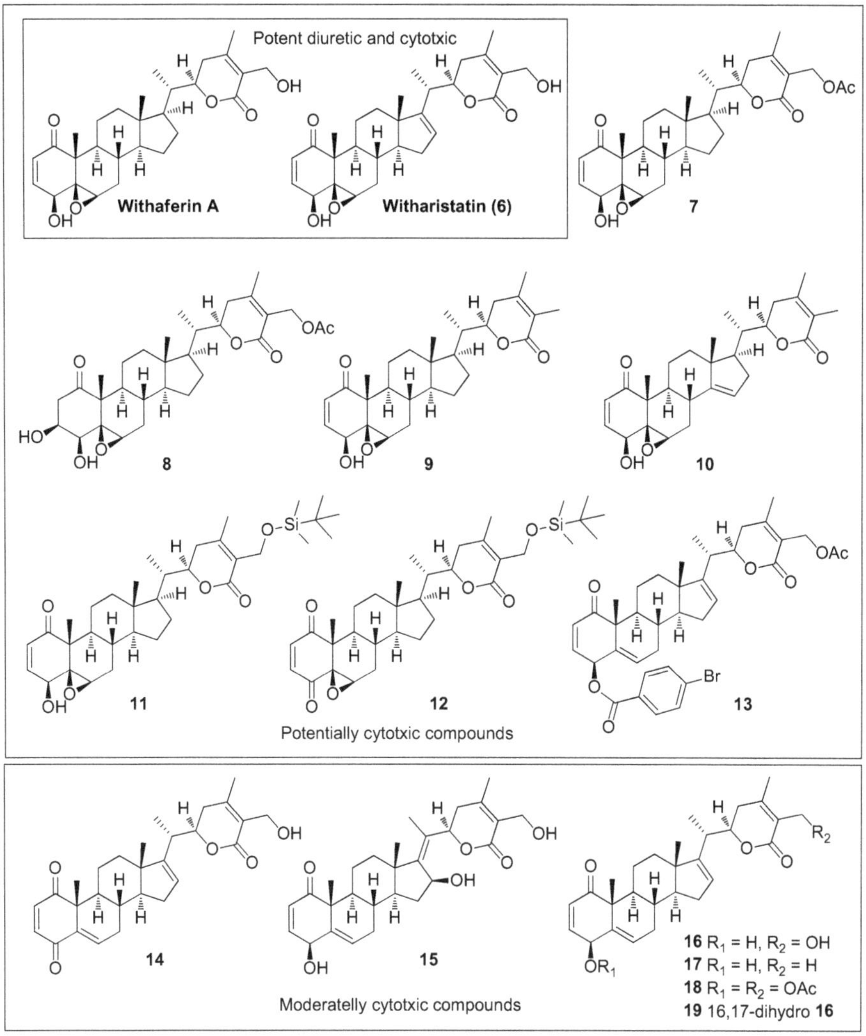

Fig. 41.4 The bioactive compounds from *W. aristata* (6–19)

Compounds **11** (IC_{50}, 0.7–1.5 μM) and **12** (IC_{50}, 0.5–5.3 μM) were found to be even more potentially active than the reference drug mercaptopurine (IC_{50}, 1.4–67.6 μM). These compounds are semisynthetic derivatives of withaferin A. The withanolides (20*S*,22*R*)-27-hydroxy-1,4-dioxo-witha-2,5,16,24-tetraenolide (**14**), (4*S*,22*R*)-4,16,27-trihydroxy-1-oxo-witha-2,5,17(20),24-tetraenolide (**15**), (4*S*,20*S*,22*R*)-4,27-dihydroxy-1-oxo-witha-2,5,16,24-tetraenolide (**16**), (4*S*,20*S*,22*R*)-4-hydroxy-1-oxo-witha-2,5,16,24-tetraenolide (**17**), (4*S*,20*S*,22*R*)-4,27-diacetoxy-1-oxo-witha-2,5,16,24-tetraenolide (**18**), and 4*β*,27-dihydroxy-1-oxo-witha-2,5,24-trienolide (**19**) possessed moderate activities (IC_{50}, 10–20 μM).

41.5.3 Cytotoxic Constituents of *W. frutescens*

Three cytotoxic withanolides, 4*β*,17*α*,27-trihydroxy-1-oxo-22*R*-witha-2,5,24-trienolide (**20**), 5*β*,6*β*-epoxy-4*β*,17*α*,27-trihydroxy-1-oxowitha-2,24-dienolide (**21**), and 2,3-dihydrowithaferin A-3*β*-*O*-sulfate (**22**) were isolated from the dichloromethane fraction of the methanolic extract of *W. frutescens* leaves (Fig. 41.5). Among these three compounds, **22** exhibited the strongest cytotoxicity against HT29 cancer cell lines (IC_{50}, 1.780 μM), comparable to that of 5-fluorouracil, a reference drug (IC_{50} of 0.81 μM). The other compounds (**20**,**21**) were moderately active, IC_{50}, 13.18, and 25.13 μM, respectively (Bouzidi et al. 2013).

41.5.4 Cytotoxic Constituents of *W. obtusifolia*

Seven withanolides including "withaferin A" have been identified from the ethanolic extract of the leaves of the *W. obtusifolia* (Fig. 41.6). The compounds obtusifonolide (**23**), sitoindoside IX(**24**), and 6*α*-chloro-5*β*-hydroxywithaferin A (**25**) and withaferin A exhibited potent cytotoxicity against SW-620 (human colon cancer) cell lines (IC_{50}, 0.3–7.3 μM). However, isowithanone (**26**), 2,3-dihydro-3-ethoxywithaferin A (**27**), and daturataturin A (**28**) were found to not be active against the tested cancer cell lines (Alali et al. 2014).

41.5.5 Cytotoxic, Immunosuppressive, and Antidiabetic Constituents of *W. coagulans*

The *W. coagulans* is the second most explored species after *W. somnifera*. An array of withanolides (including some of those isolated from the previous species) was isolated from the different plant parts and different extracts of the plant (Khodaei et al. 2012). The withacoagulin G (**29**), withacoagulin H (**30**), withacoagulin I (**31**), 27-hydroxywithanolide F or ajugin E (**32**), withanolide F (**2**), (20*R*, 22*R*-14*α*, 20*α*)-dihydroxy1-oxowitha-2,5,16,24 tetraenolide (**33**), withacoagulin (**34**), withanolide H (**35**), and withanolide G (**36**) were isolated from $CHCl_3$-MeOH extracts of

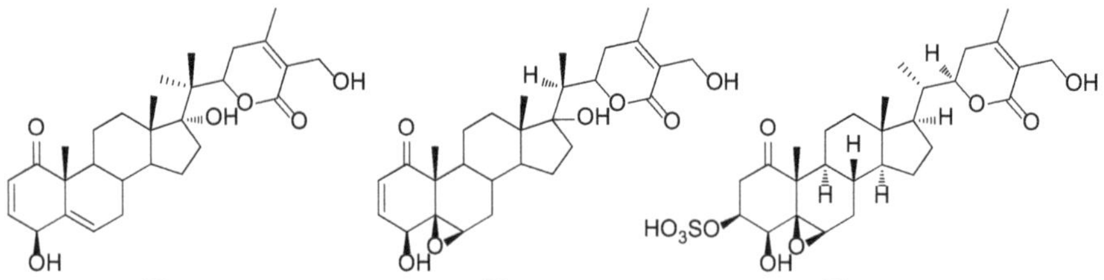

20 21 22

Fig. 41.5 Structures of compounds 20–22

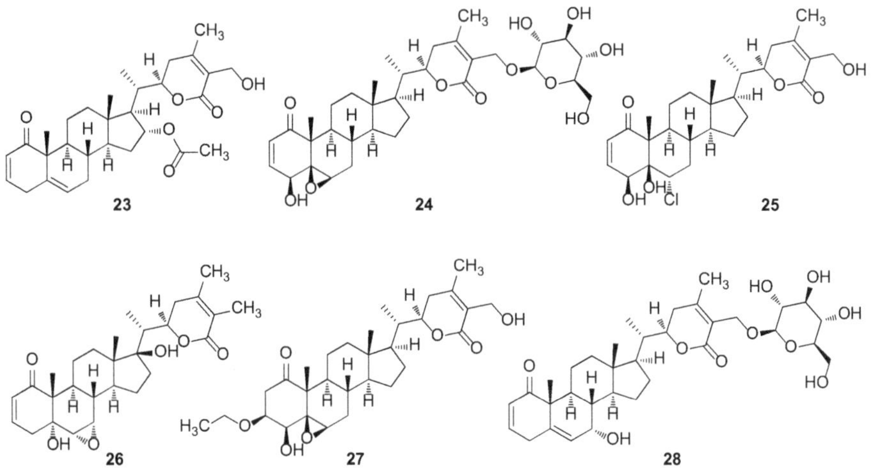

Fig. 41.6 Structures of compounds 23–28

W. coagulans (Fig. 41.7). These compounds inhibited nitric oxide production in lipopolysaccharide-activated murine macrophage RAW 264.7 cells with IC_{50} values of 1.9–38.2 μM. Compounds **29** and **30** were found to be the most active ($IC_{50,}$ 3.1 and 1.9 μM, respectively). These compounds also exhibited inhibition of tumor necrosis factor-α (TNF-α) and induced nuclear factor-kappa B (NF-κB) activation with IC_{50} values of 1.60–12.4 μM (Ihsan-ul-Haq et al. 2013).

Five other withanolides, **1**, **2**, **4,** coagulin C (**37**), and 17*β*-hydroxywithanolide K (**38**), isolated from *W. coagulans* fruits along with two semisynthetic derivatives of **4** (**39** and **40**), have shown significant inhibition on postprandial rise in hyperglycemia, post-sucrose load in normoglycemic rats, as well as streptozotocin-induced diabetic rats. The median effective dose of the compound **4** was found to be around 25 mg/kg body weight. The compound **4** also showed significant fall on fasting blood glucose profile and improvement in the glucose tolerance and antidyslipidemic activity in diabetic (db/db) mice (Maurya et al. 2008; Akanksha et al. 2010). The compounds **1**, **2**, and **4** were isolated from the *W. coagulans*, before their isolation from the *W. adpressa*. Some other withanolides, isolated from ethanolic extract of the aerial part of *W. coagulans*, have shown immunosuppressive activity by inhibiting T- and B-cell proliferation. The compounds **2**, **38**, **34**, withanolide L (**41**), and **1** potentially inhibited B- and/or T-cell proliferation (IC_{50}, 1.66–9.73 μM). The withacoagulin A (**42**), withacoagulin C (**43**), withacoagulin D (**44**), and withacoagulin E (**45**) and (**3**) exhibited moderate inhibitory activity against T- and B-cell (IC_{50}, 10.0–19.0 μM) proliferation (Huang et al. 2009).

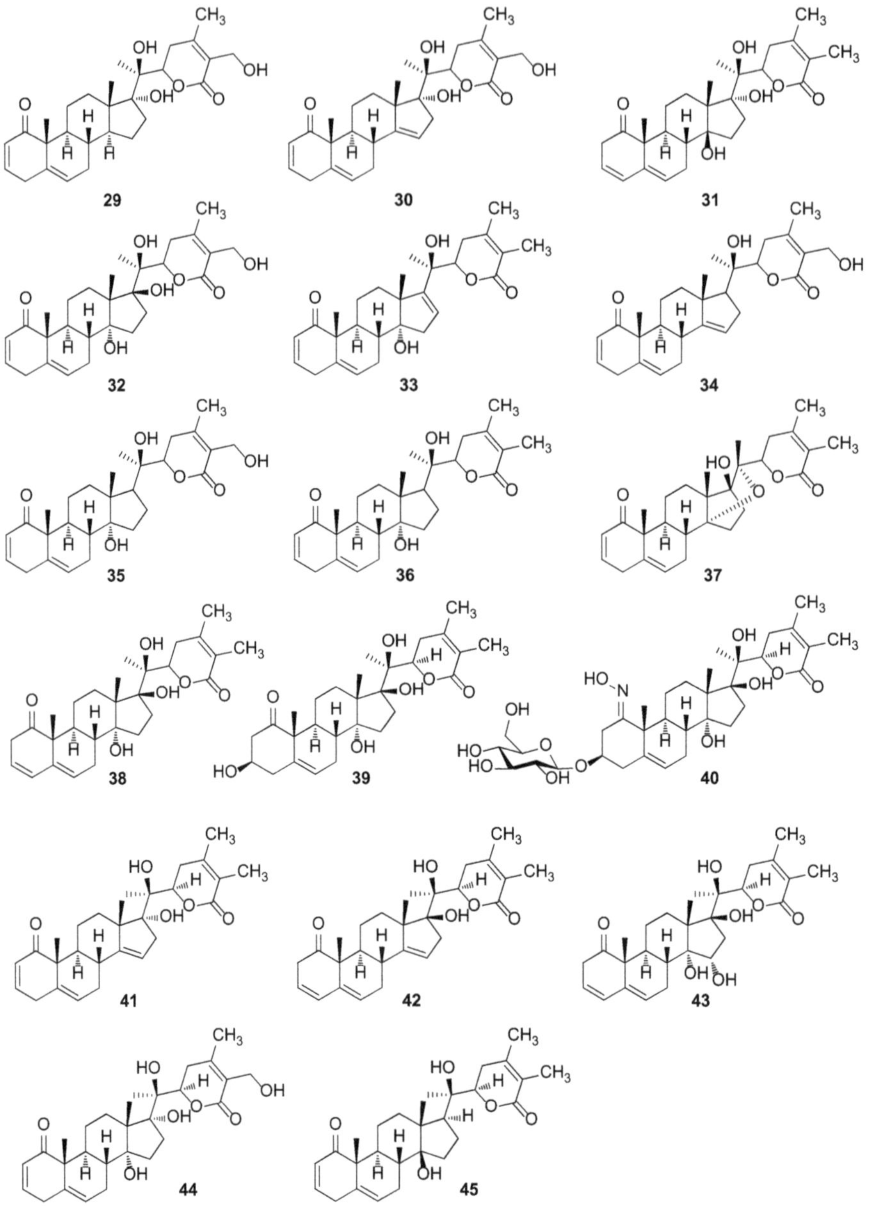

Fig. 41.7 Structures of compounds 29–45

41.5.6 Bioactive Constituents of *W. somnifera*

More than 100 withanolides have been isolated and characterized from the *W. coagulans* and *W. somnifera*, and the biological activities of these withanolides have been evaluated by several research groups independently and reviewed by Chen et al. (2011). Under this section, only some selected bioactive withanolides identified by bioassay guided isolation will be discussed.

A chlorinated steroidal lactone, 27-acetoxy-4*β*,6*α*-dihydroxy-5*β*-chloro-1-oxowitha-2,24-dienolide (**46**), a diepoxy withanolide, 5*β*,6*β*,14*α*,15*α*-diepoxy-4*β*,27-dihydroxy-1-oxowitha-2,24-dienolide (**47**), and withaferin A, isolated from the methanolic extract of the aerial part *of W. somnifera*, exhibited growth inhibition and cytotoxic activity against human lung cancer cell line NCI-H460 (Choudhary et al. 2010). The GI_{50} (the concentration for 50% of maximal inhibition of cell proliferation) values for these compounds were found to be 6.2, 1.5, and 0.18 μg/mL, while the LC_{50} (lethal concentration 50) values were found to be 95.6, 8.3, and 0.45 μg/mL, respectively. Seven withanolide glycosides, withanosides I, II, III, IV, V, VI, and VII (**48–54**), along with withaferin A were isolated form the methanolic extract of the root of *W. somnifera* (Fig. 41.8). The withanolide VI and withaferin A exhibited inhibitory activity for tachyphylaxis to clonidine in isolated guinea-pig ileum at 10 μM concentration (Matsuda et al. 2001).

Four other compounds, (20*S*,22*R*)-3*α*,6*α*-Epoxy-4*β*,5*β*,27-trihydroxy-1-oxowitha-24-enolide (**55**), withanolide A (**56**), (20*S*,22*R*)-4*β*,5*β*,6*α*,27-tetrahydroxy-1-oxowitha-2,24-dienolide (**57**), and coagulin Q (**58**) along with withanoside IV (**51**) and withanoside VI (**53**) of a methanolic extract of *W. somnifera* root, showed significant neurite outgrowth activity at a concentration of 1 μM on a human neuroblastoma SH-SY5Y cell line (Zhao et al. 2002). Two dimeric thiowithanolide, ashwagandhanolide **59** and **60,** have also been isolated from the methanolic extract of the root (Fig. 41.9). Compound **59** displayed growth inhibition against human gastric (AGS), breast (MCF-7), central nervous system (SF-268), colon (HCT-116), and lung cancer (NCI H460) cell lines, with IC_{50} values in the range 0.43–1.48 μg/mL. Compound **60** showed anticancer activity against human gastric (AGS), breast (MCF7), central nervous system (SF-268), and colon (HCT-116) cancer cell lines, with IC_{50} values in the range 0.74–3.63 μM. In addition, these compounds also inhibited lipid peroxidation and the activity of the enzyme cyclooxygenase-2 in vitro. It inhibited lipid peroxidation by 65% and the COX-2 enzyme by 60% at a concentration of 100 μg/mL. Further, compounds **59** and **60** completely suppressed TNF-induced NF-κB activation at a 100 μM concentration (Mulabagal et al. 2009; Subbaraju et al. 2006). Withaferin A is also believed to promote the formation of reactive oxygen species (ROS) inside cancer cells and disrupt their function. According to reports, it has also been documented to induce apoptosis in cancer cells by making them less resistant to drugs (Nishikawa et al. 2015).

The enzyme inhibitory activity against cyclooxygenase-2 has also been displayed by a few other withanolide glycosides and withanolides, isolated from the leaves of *W. somnifera*. The active compounds noted were 27-*O*-*β*-D-glucopyranosylphysagulin D (**61**), viscosalactone B (**62**), physagulin D (1 → 6)-*β*-D-glucopyranosyl-(1 → 4)-*β*-D-glucopyranoside (**63**), 27-*O*-*β*-D-glucopyranosyl viscosalactone B (**64**), physagulin D (**65**), withaferin A, 2,3-dihydrowithaferin A (**66**), 4-(1-hydroxy-2,2-dimethylcyclo-propanone)-2,3-dihydrowithaferin A (**67**),

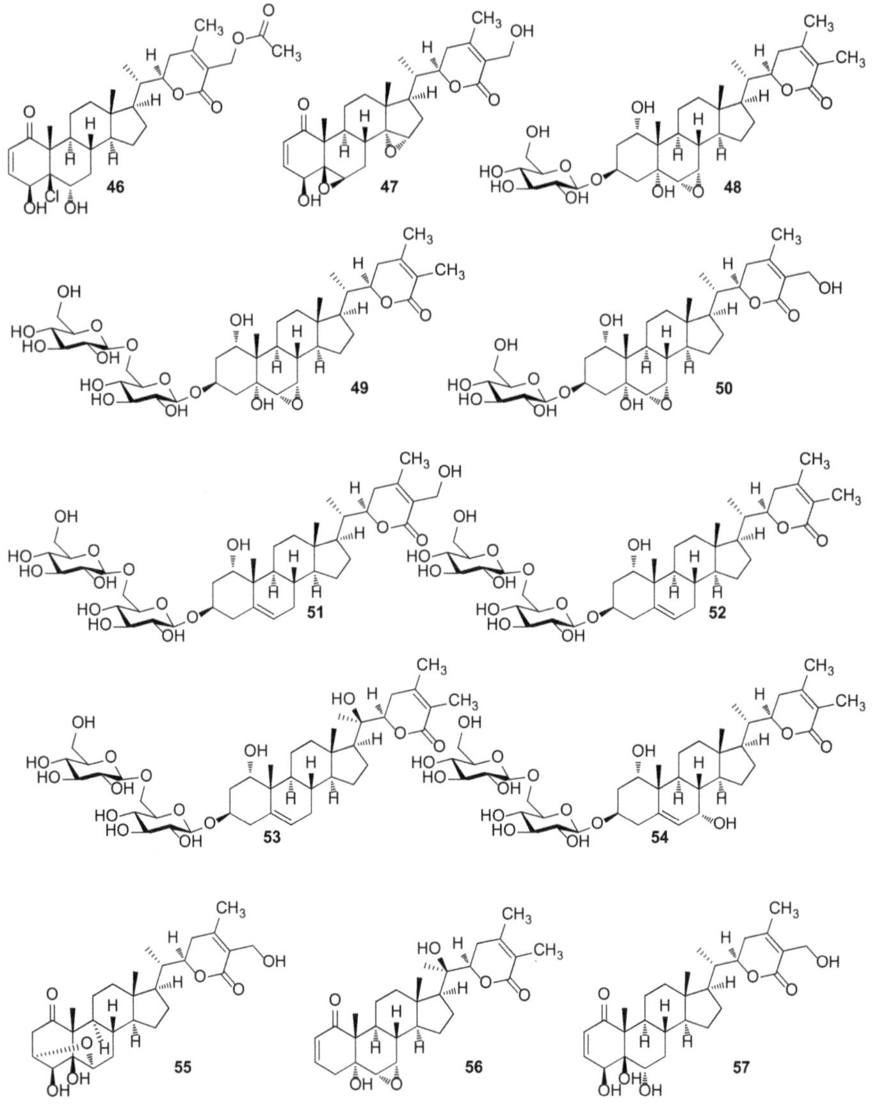

Fig. 41.8 Structures of compounds 46–57

24, and withanoside IV (**51**). These compounds selectively inhibited COX-2 enzyme by up to 40% and lipid peroxidation up to 55% at 100 μg/mL concentrations (Jayaprakasam and Nair 2003). A few withanolides isolated from the methanolic extract of the whole plant material inhibited cholinesterase enzyme. Compounds 17α-hydroxywithaferin A (**68**) and 6α,7α-epoxy-5α,20β-dihydroxy-1-oxowitha-2,24-dienolide (**56**) along with compound **10** and withaferin A displayed inhibitory

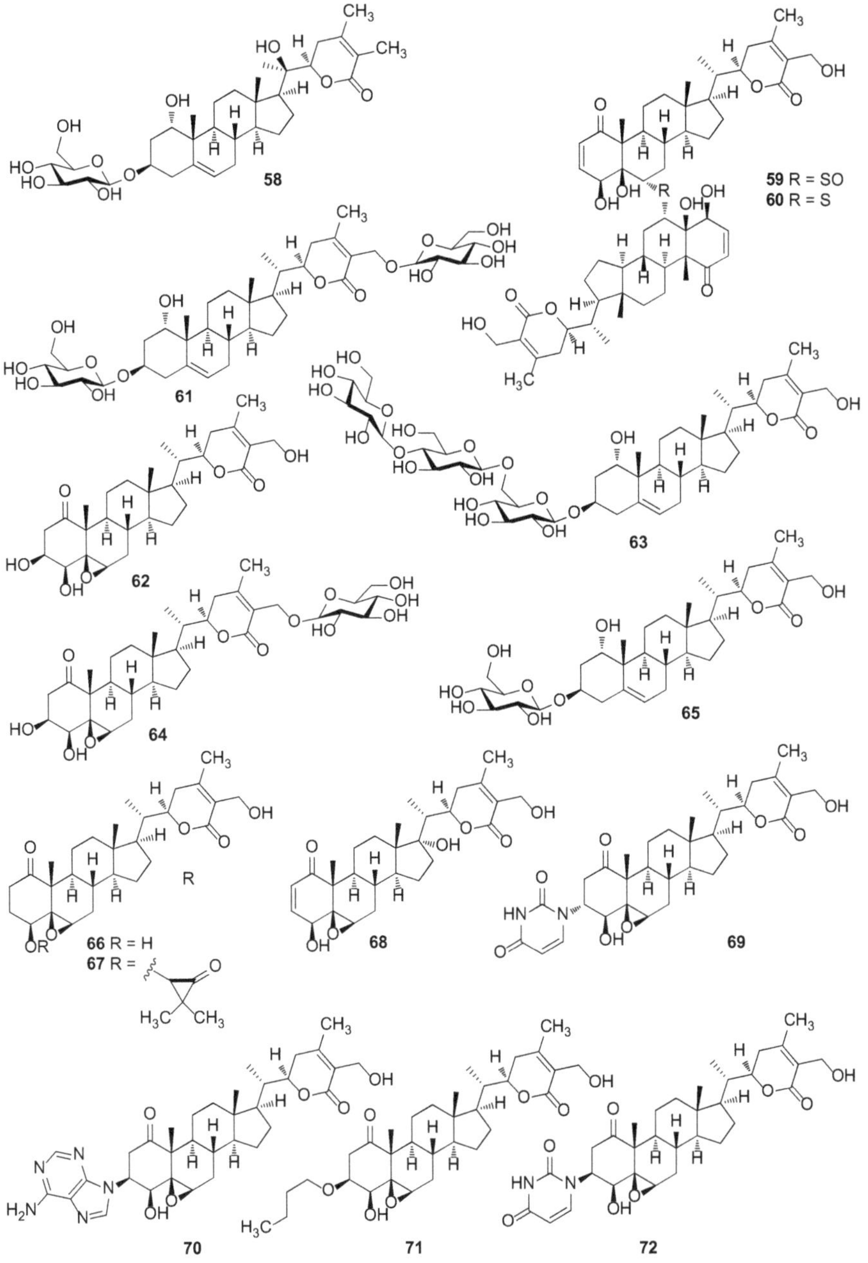

Fig. 41.9 Structures of compounds 58–72

potential against acetylcholinesterase (AChE) with IC_{50} values in the range of 50.0–161.5 μM.

However, only compounds **10** and **65** and withaferin A inhibited butyrylcholinesterase (BChE) with (IC_{50}) values in the range of 62.5–500 μM (Choudhary et al. 2004). The neurite outgrowth and cholinesterase inhibitory activities are related to the neurodegenerative diseases like Alzheimer's and Parkinson's diseases. The activities of these compounds against these diseases validate the traditional usages of *W. somnifera* for a memory boosting effect.

Some unusual withanolides (**69–71**) along with **22** are isolated from the methanolic extract of aeroponically grown *W. somnifera*. Compounds, 3*α*-(uracil-1-yl)-2,3-dihydrowithaferin A (**69**), 3*β*-(adenin-9-yl)-2,3-dihydrowithaferin A (**70**), and 3*β*-*O*-butyl-2,3-dihydrowithaferin A (**71**) and **22,** exhibited inhibition of cancer cell proliferation/survival, disruption of cytoskeletal organization, and induction of the cellular heat-shock response parallel to those displayed by withaferin A (Xu et al. 2009, 2011; Wijeratne et al. 2014). The detailed bioactivity will be discussed in the section structure activity. The beta-epimer of **69**, 3*β*-(uracil-1-yl)-2,3-dihydrowithaferin A (**72**) is a synthetic analogue obtained by the reaction of adenine with withaferin A.

41.6 Structure-Activity Relationships (SAR) in Withanolides

41.6.1 Cytotoxicity

From the biological activities of the withanolides isolated from different *Withania* species, it is evident that most of the compounds exhibited potent anticancer/cytotoxicity. A careful analysis of the structure of the withanolides with type-7 ring system and cytotoxic activity revealed that the enone group in ring A and epoxide group fuse with ring A and B at 5,6-position are essential for the cytotoxicity of the withanolides. The 2,3-dihydro or other analogues/derivatives produced from the addition of the chemical entities to the double bond at 2,3-position exhibited reduced cytotoxic activity (Fig. 41.10). For example, the IC_{50} value for the cytotoxicity of withaferin A (WA) against HeLa (human cervix carcinoma), A-549 (human lung carcinoma), MCF-7 (human breast adenocarcinoma), and Vero (African green monkey kidney) cell lines was found to be in the range of 0.6–2.3 μM, while that of 2,3-dihydrowithaferin A were found to be between 5.4 and 19.3 μM. Thus the latter has been found to be about ninefold less active than the WA (Llanos et al. 2012). Similarly, the IC_{50} value for the cytotoxicity of WA against myeloma (H929) cells was found to be 0.25 μM, whereas that of **69** and **70** was observed to be 0.54 and 0.61 μM, respectively (Wijeratne et al. 2014). The compounds **69** and **70** are the addition product of the withaferin A with uracil and adenine, respectively. The uracil and adenine derivatives are well-known for their cytotoxic effects. In fact the 5-fluorouracil is a medication used to treat cancer (Thedford et al. 1989; Takechi et al. 1997; Parker et al. 1999). Even then, their addition to withaferin A reduces its cytotoxicity by more than twofold.

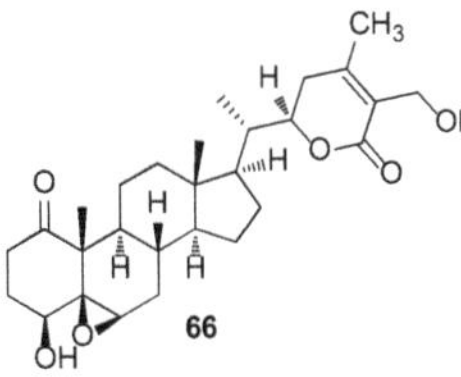

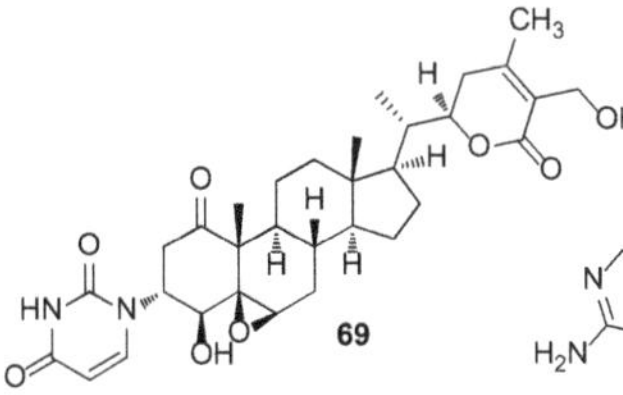

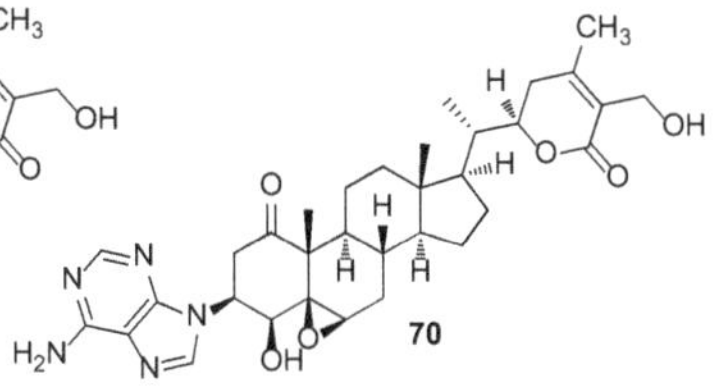

IC_{50} (HeLa, A-549, MCF-7, Vero), 5.4-19.3 μM

IC_{50} (H929), 5.25 μM

IC_{50} (H929), 8.74 μM

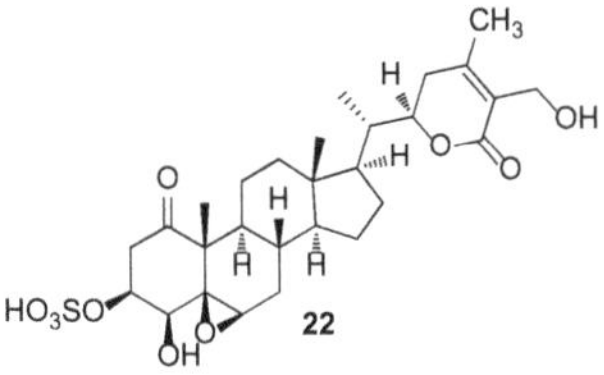

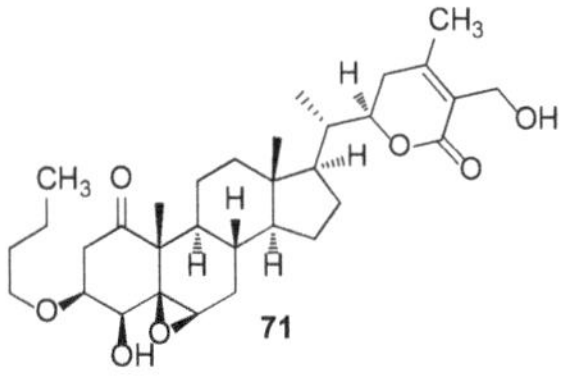

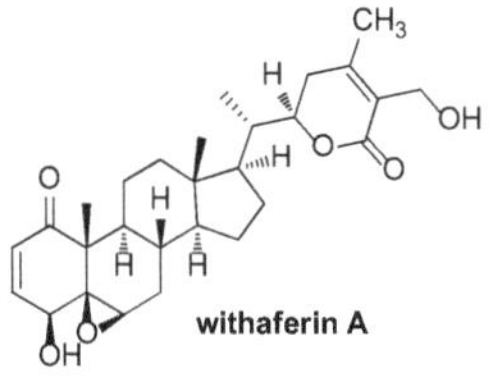

IC_{50} (H929), 5.25 μM

IC_{50} (H929), 8.74 μM

IC_{50} (H929), 0.25 μM; IC_{50} (HeLa, A-549, MCF-7, Vero), 0.6-2.3 μM

Fig. 41.10 Withanolides active against myeloma (H929) cells, reduction of the double bond at C-2 position in withaferin A decreased the cytotoxicity

The cytotoxicity of the addition products of sulfuric acid (**22**) and *n*-butanol (**71**) was found to be much lower (21- and 35-fold; IC_{50}, 5.25 and 8.74 μM, respectively) than the WA (Fig. 41.10). This result indicated the role of the enone group in cytotoxicity of the withanolides. The conversion of the epoxide ring into thioepoxide or cleavage of the epoxide ring to produce hydroxyl groups resulted into the 4- to 25-fold reduction in the cytotoxicity of withaferin A and other withanolides (Fig. 41.11) against human cancer cell lines WRL68 (liver cancer cells), CaCO-2 (colon cancer cells), PC-3 (prostate cancer cells), and MCF7 (hormone-dependent breast cancer cells) (Joshi et al. 2014). Thus, according to this result, one can conclude that the 5*β*,6*β*-epoxide ring is also essential for the cytotoxicity of the withanolides.

The acetylation of the hydroxyl group in some withanolides resulted in significant improvement in the cytotoxicity against myeloma (H929) cells. The benzoylation and silylation of the hydroxyl groups also resulted in significant enhancement in the activity against HeLa, A-549, MCF-7, and Vero cancer cells (Fig. 41.12). The benzoylation of 4-hydroxyl group with 4-bromobenzoyl group in **16**, isolated from *W. aristata*, enhanced the cytotoxicity up to three- to sevenfold (Llanos et al. 2010). Hence, the above concept can be applied in the semi-synthesis of the other withanolide derivatives with improved biological activities. Protection of the 27-OH group with tertiarybutyldimethylsilyl (TBDMS) group also enhanced cytotoxicity by three- to tenfold against the tested cancer lines.

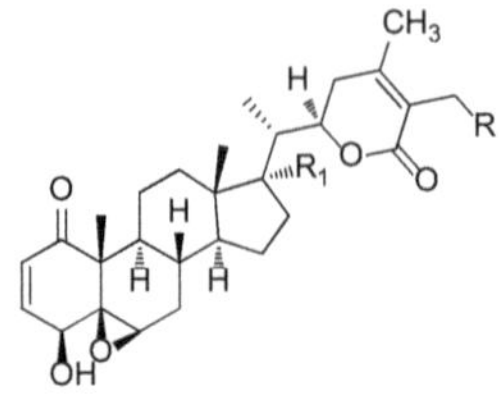

Compounds	IC_{50} (μg/mL)
WA, R = OH, R_1 = H	3.8-5.4
9, R = R_1 = H	4.6-6.2
68, R = R_1 = OH	4.8-6.8
73, R = H, R_1 = OH	4.8-6.7

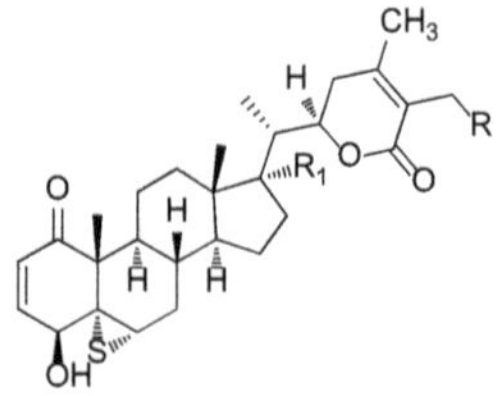

Compounds	IC_{50} (μg/mL)
WAa, R = OH, R_1 = H	77 to >100
9a, R = R_1 = H	84 to >100
68a, R = R_1 = OH	88 to >100
73a, R = H, R_1 = OH	87 to >100

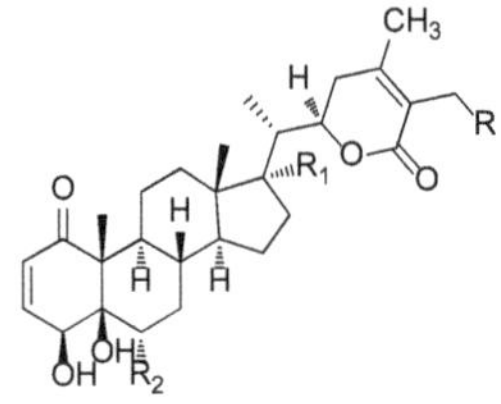

Compounds	IC_{50} (μg/mL)
WAb, R = OH, R_1 = H, R_2 = $NH(CH_2)_3CH_3$	100 to >100
9b, R = R_1 = H, R_2 = $NH(CH_2)_3CH_3$	98 to >100
68b, R = R_1 = OH, R_2 = $NH(CH_2)_3CH_3$	100 to >100
73b, R = H, R_1 = OH, R_2 = $NH(CH_2)_3CH_3$	98 to >100

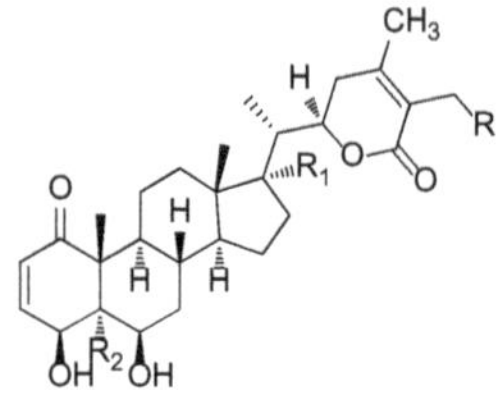

Compounds	IC_{50} (μg/mL)
WAc, R = OH, R_1 = H, R_2 = $NH(CH_2)_3CH_3$	100 to >100
9c, R = R_1 = H, R_2 = $NH(CH_2)_3CH_3$	100 to >100
68c, R = R_1 = OH, R_2 = $NH(CH_2)_3CH_3$	100 to >100
73c, R = H, R_1 = OH, R_2 = $NH(CH_2)_3CH_3$	100 to >100

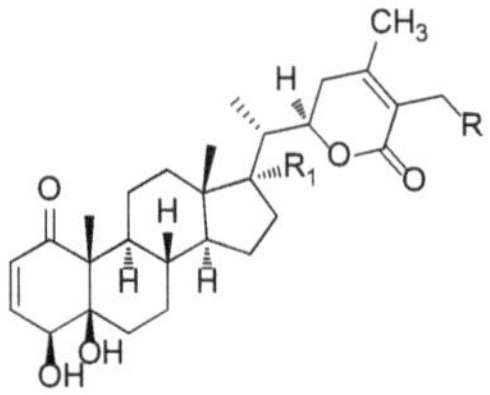

Compounds	IC_{50} (μg/mL)
WAd, R = OH, R_1 = H	55.5-59.0
9d, R = R_1 = H	17.0-64.0
68d, R = R_1 = OH	20.0-70.0
73d, R = H, R_1 = OH	22.0-72.0

Fig. 41.11 Reduction in the cytotoxicity of withanolides against WRL-68, Caco-2 PC-3, and MCF-7 cancer cells on modification in the epoxide ring

41.6.2 Cyclooxygenase-2 Inhibition

The cylooxigenase-2(COX-2) inhibition potential of the withanolides isolated from *W. somnifera* has been reported by Jayaprakasam and Nair (2003) and Chen et al. (2011). The only inactive compound contained the saturated lactone group in ring E (**79**).

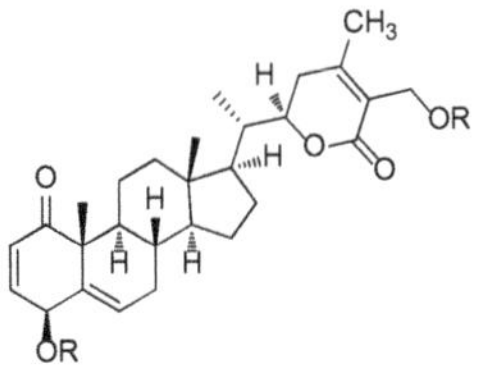
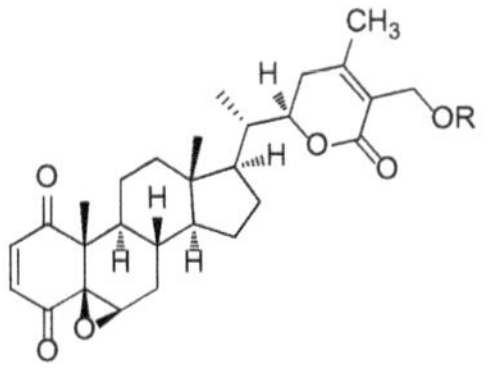
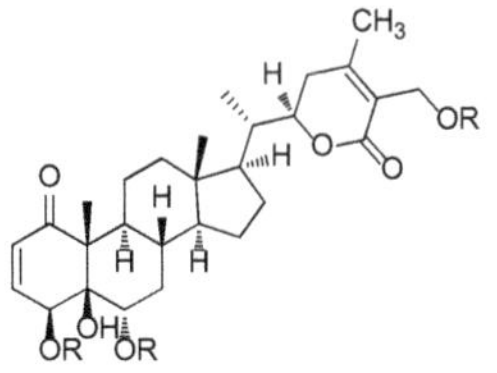

Compounds	IC_{50} (μM)
74, R = H	5.09
74a, R = Ac	1.63

Compounds	IC_{50} (μM)
75, R = H	1.16
75a, R = Ac	0.46

Compounds	IC_{50} (μM)
57, R = H	>10
57a, R = Ac	2.29

*cytotoxicity against myeloma (H929) cells

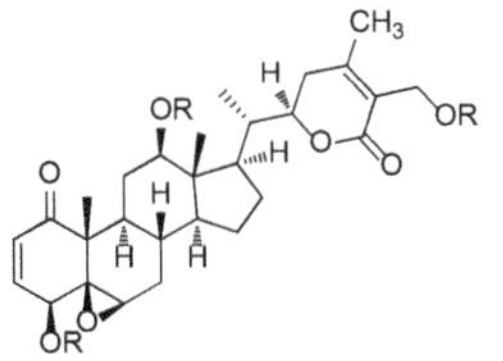
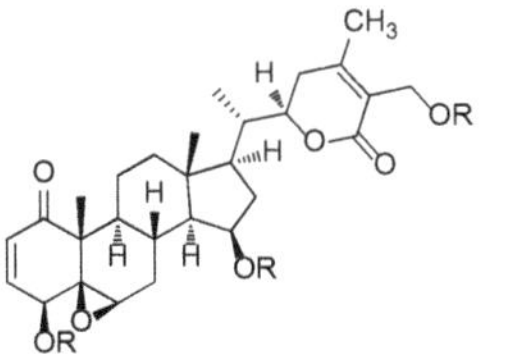

Compounds	IC_{50} (μM)
76, R = H	4.30
76a, R = Ac	0.18

Compounds	IC_{50} (μM)
77, R = H	3.83 μM
77a, R = Ac	0.24 μM

*cytotoxicity against myeloma (H929) cells

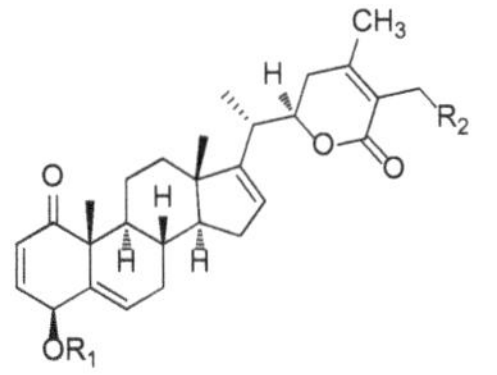
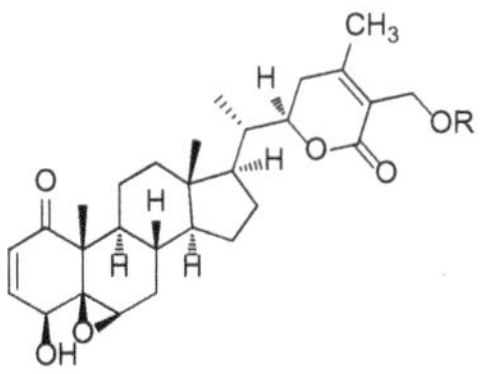
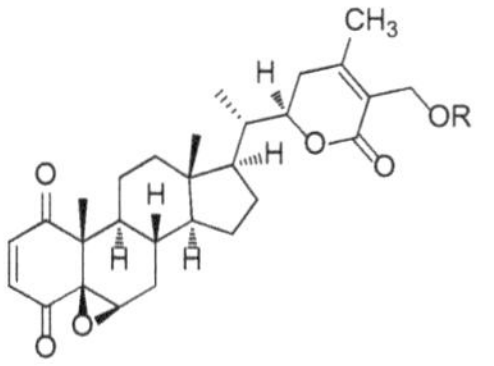

Compounds	IC_{50} (μM)
16, R_1 = H, R_2 = OH	10.8-25.2
13, R_1 = *p*-Br-Bz, R_2 = OAc	2.8-3.6

Compounds	IC_{50} (μM)
WA, R = H	1.3-6.6
11, R = TBDMS	0.7-1.7

Compounds	IC_{50} (μM)
75, R = H	3.2-25.4
12, R = TBDMS	0.5-5.3

*cytotoxicity against HeLa, A-549, MCF-7 and Vero cancer cells

Fig. 41.12 Improvement in cytotoxicity of withanolides on acetylation, benzoylation, and silylation of hydroxyl groups

Thus by comparing the structures and activity of the compounds, it can be concluded that the *α,β*-unsaturated lactone in ring A is essential for the cyclooxigenase-2 inhibitory activity, whereas the 5,6-epoxy and enone group are not essential (Fig. 41.13). The dimeric withanolides **59** and **60** also inhibited COX-2 enzyme by 60% and 100 μg/mL concentration.

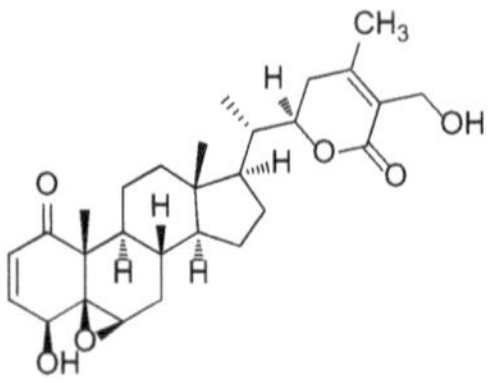

Compound **WA** % Inhibition 39

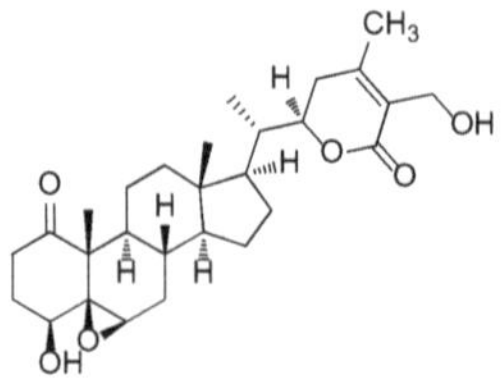

Compound **66** % Inhibition 27

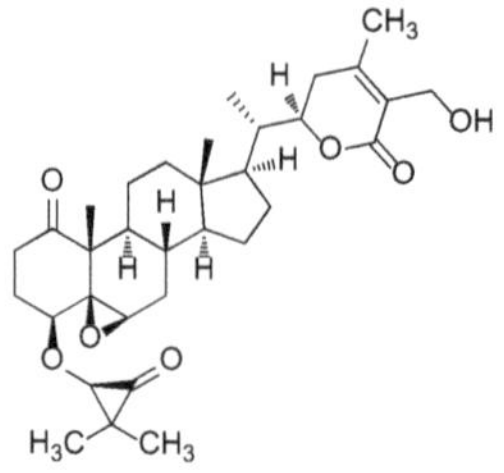

Compound **67** % Inhibition 28

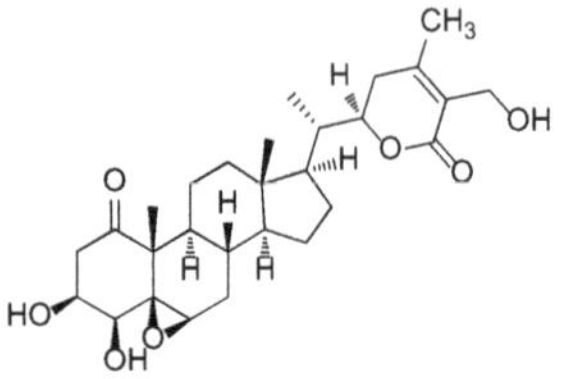

Compound **62** % Inhibition 35

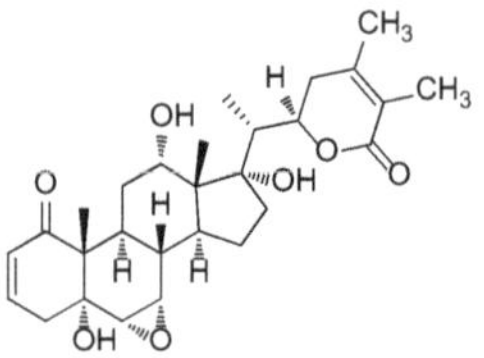

Compound **78** % Inhibition 46

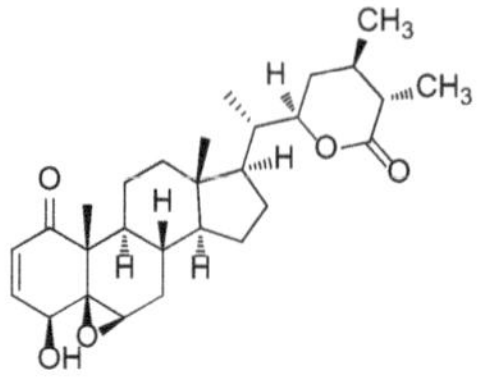

Compound **79** % Inhibition Inactive

Fig. 41.13 Percentage inhibition of COX-2 enzyme by withanolides at 100 μg/mL concentration

41.6.3 Acetylcholinesterase Inhibition

Choudhary et al. (2005) reported the cholinesterase inhibition by withanolides, isolated from *W. somnifera*. The withanolides having 5,6-epoxy ring were found to be more active against acetylcholinesterase (AChE) than the one without the epoxy ring at 5,6-position (Fig. 41.14). However, no such structure activity relationship could be observed with butyrylcholinesterase (BChE).

41.6.4 Neurite Outgrowth Activities

As mentioned earlier, several withanolides and withanolide glycosides, isolated from *W. somnifera*, exhibited neurite outgrowth activities in human SH-SY5Y cells. Withanolide A (**56**) is the most interesting compound and has been synthetically modified by Liffert et al. (2013) for the neurite outgrowth SAR study. On the basis of the results, it can be concluded that the enone group in ring A of withanolide A is not essential for the neurite outgrowth activities. The substitution of 1-oxo group by N-hydroxylimine group resulted in improvement in the activity. However, substitution with N-alkoxyimine resulted in a decrease in the activity. Similarly, replacement of the enone group by enol groups also resulted in the enhancement of the activity (Fig. 41.15).

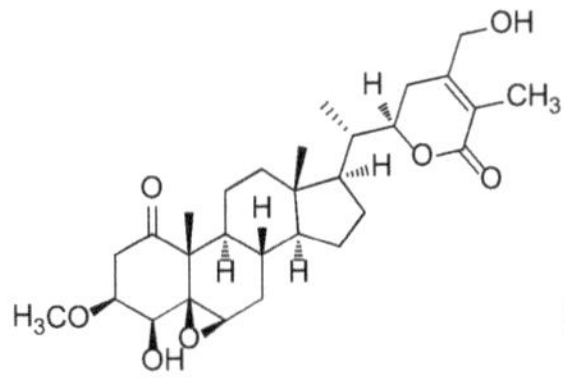

Compound	IC_{50} (μM)
80	25.2

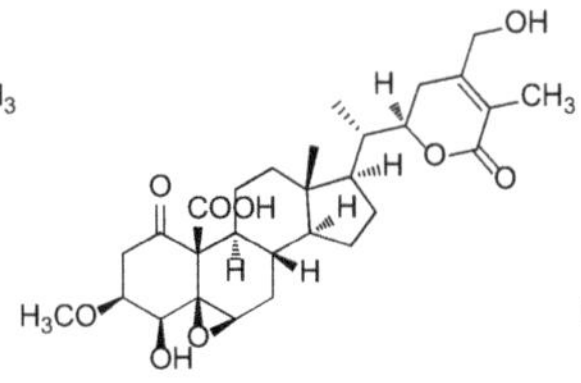

Compound	IC_{50} (μM)
81	22.1

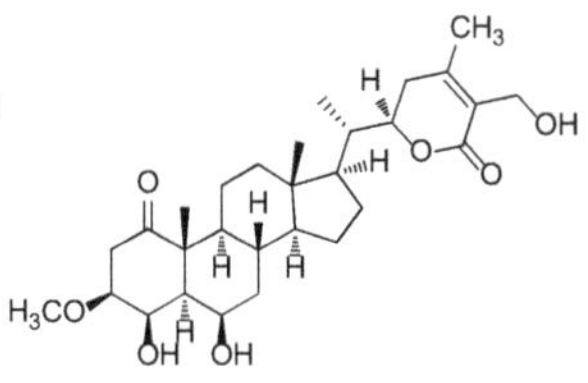

Compound	IC_{50} (μM)
82	49.2

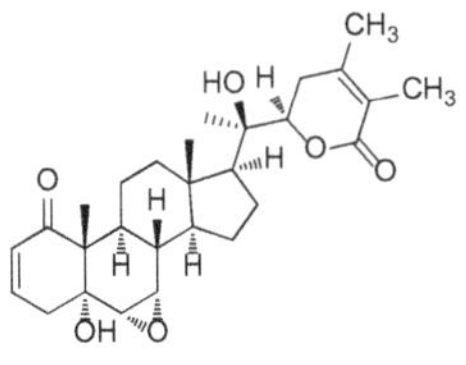

Compound	IC_{50} (μM)
56	45.2

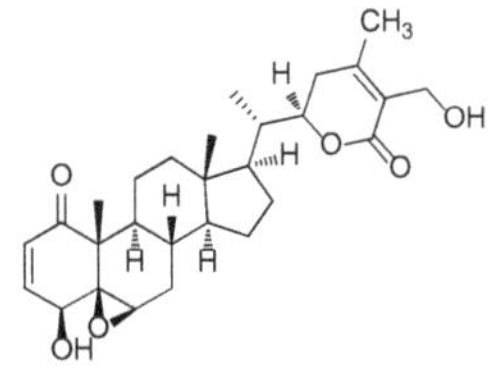

Compound	IC_{50} (μM)
WA	20.5

Fig. 41.14 Acetylcholinesterase inhibitory activity of the withanolides isolated from *W. somnifera*

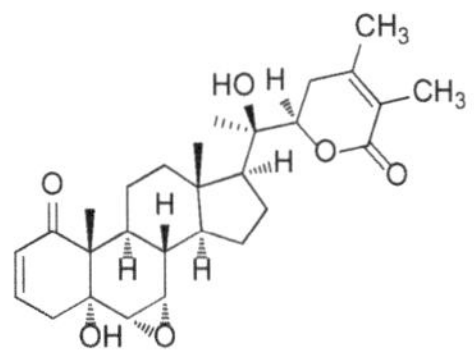

Compound	% growth
56	48

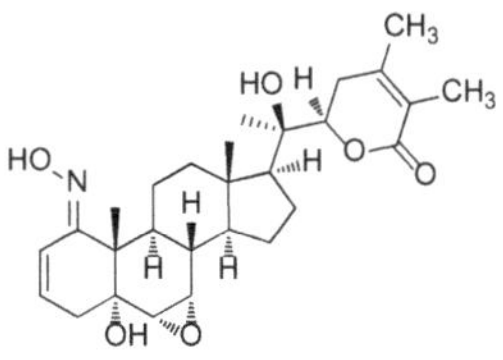

Compound	% growth
83	49

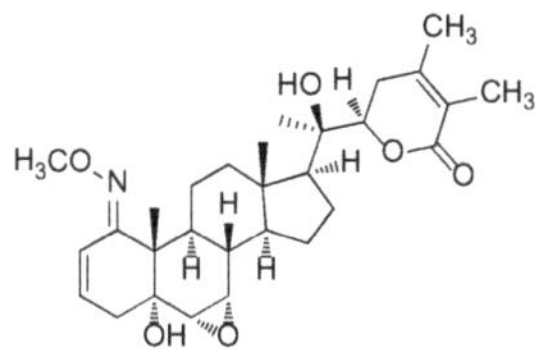

Compound	% growth
84	38

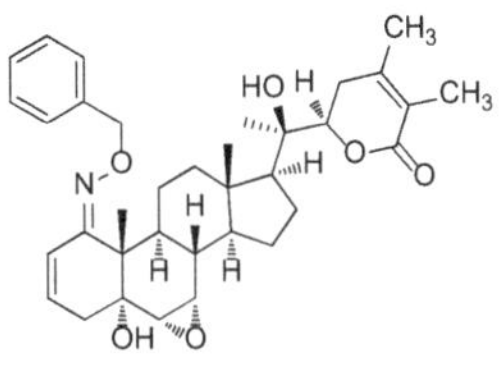

Compound	% growth
85	41

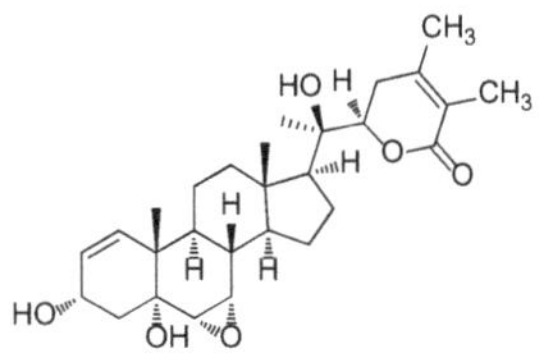

Compound	% growth
86	49.5

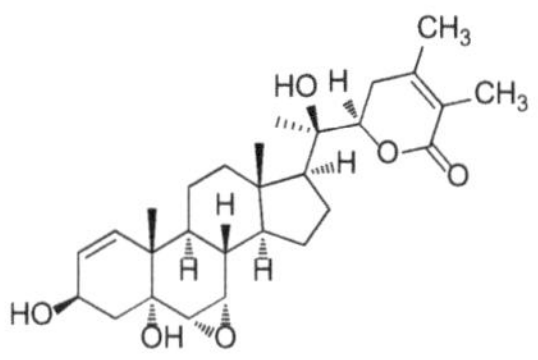

Compound	% growth
87	50

Fig. 41.15 Neurite outgrowth activity of withanolide A (**56**) and its synthetic derivatives

41.7 Summary and Future Prospects

The antifungal activity of *W. adpressa* has been reported for the leaf of this plant, but other parts have to be investigated. No chemical constituents have been identified which are responsible for this antifungal activity. Similarly, only five cytotoxic withanolides (**1–5**) have been isolated from the dichloromethane fraction of the methanolic extract of the *W. adpressa* leaf (Fig. 41.1). However, no work has been carried out on the phytochemical investigation of the other parts of this plant as well as on other fractions of the methanolic extract of the leaves. Hence, the other parts, viz., stem, root, fruits, and flower, and other fractions, viz., *n*-hexane and *n*-butanol of the methanolic extract of the leaf, require chemical investigation of the compounds responsible for the cytotoxicity and other biological activities.

Only diuretic and cytotoxic constituents have been isolated from the dichloromethane and aqueous extract of the *W. aristata leaves*; hence, the chemical constituents of the stem, root, bark, fruits and flowers, etc. still need to be identified. Further, the other extracts, viz. *n*-hexane, chloroform, ethyl acetate, *n*-butanol, acetone and methanol, etc. should also be prepared and can be investigated for chemical constituents and biological activities. Despite the report on hepatoprotective activity of the ethanolic extract of the leaf of *W. frutescens*, the hepatoprotective compounds still need to be identified. The phytochemical constituents are only isolated from the dichloromethane fraction of the methanolic extracts of the leaves. However, other parts of the plants and other extracts need to be investigated in detail.

Although *W. coagulans* and *W. somnifera* are explored and being used widely for many purposes across the world, most of the biological activity of the plants and compounds is tested in vitro and/or on animals. Several newly discovered biological effects (other than that in traditional medicine) still need to be tested in the humans. Similarly, several withanolides have been isolated from both the plants, but all were not evaluated for their biological potentials. Further, a number of chemical components have already been isolated from these two plants, and there is very little probability of finding new compounds. One can still isolate novel compounds by changing experimental methods. For example, Xu et al. 2009 has recently isolated three novel withanolides from the aeroponically frown *W. somnifera*. Hence, it is possible that one can find other novel compounds from the same as well as other species grown under different kinds of stress conditions.

A number of traditional usages have been known for the six plant species. However, none of the traditional usages have been clinically validated, and there is no solid evidence to justify their medicinal benefits and adverse effects if it is taken together with prescription drugs. The plants are not fully investigated for the biological activities and bioactive compounds to validate all the traditional uses. The isolated compounds can be further derivatized (synthetically modified) for structure activity relationships associated with different biological activities, which may lead to the discovery of more potent lead molecules than the isolated ones.

References

Abdeljebbar LH, Benjouad A, Morjani H, Merghoub N, El Haddar S, Humam M et al (2009) Antiproliferative effects of Withanolides from *Withania adpressa*. Therapie 64(2):121–127

Ahmad MK, Mahdi AA, Shukla KK, Islam N, Rajender S, Madhukar D et al (2010) *Withania somnifera* improves semen quality by regulating reproductive hormone levels and oxidative stress in seminal plasma of infertile males. Fertil Steril 94(3):989–996

Ahmad R, Fatima A, Srivastava AN, Khan MA (2017) Evaluation of apoptotic activity of *Withania coagulans* methanolic extract against human breast cancer and Vero cell lines. J Ayurveda Integr Med 8:177e183

Akanksha, Srivastava AK, Maurya R (2010) Antihyperglycemic activity of compounds isolated from Indian medicinal plants. Ind J Exp Biol 48:294–298

Alali FQ, Amrine CSM, El-Elimat T, Alkofahi A, Tawaha K, Gharaibah M et al (2014) Bioactive withanolides from *Withania obtusifolia*. Phytochem Lett 9:96–101

Andallu B, Radhika B (2000) Hypoglycemic, diuretic and hypocholesterolemic effect of winter cherry (*Withania somnifera*, Dunal) root. Indian J Exp Biol 38(6):607–609

Askarne L, Talibi I, Boubaker H, Boudyach EH, Msanda F, Saadi B et al (2012) In vitro and in vivo antifungal activity of several Moroccan plants against Penicillium italicum, the causal agent of citrus blue mold. Crop Prot 40:53–58

Baitharu I, Jain V, Deep SN, Hota KB, Hota SK, Prasad D et al (2013) *Withania somnifera* root extract ameliorates hypobaric hypoxia induced memory impairment in rats. J Ethnopharmacol 145(2):431–441

Bakrim WB, El Bouzidi L, Nuzillard J-M, Cretton S, Saraux N, Monteillier A et al (2018) Bioactive metabolites from the leaves of *Withania adpressa*. Pharm Biol 56(1):505–510

Benjumea D, Martín-Herrera D, Abdala S, Gutiérrez-Luis J, Quinones W, Cardona D et al (2009) Withanolides from *Whitania aristata* and their diuretic activity. J Ethnopharmacol 123:351–355

Bhat J, Damle A, Vaishnav PP, Albers R, Joshi M, Banerjee G (2010) In vivo enhancement of natural killer cell activity through tea fortified with Ayurvedic herbs. Phytother Res 24 (1):129–135

Bouzidi LE, Mahiou-Leddet V, Sok-S B, Larhsini M, Abbad A, Markouk M et al (2013) Cytotoxic withanolides from the leaves of Moroccan *Withania frutescens*. Pharm Biol 51(8):1040–1046

Budhiraja RD, Bala S, Garg KN (1977) Pharmacological investigations on fruits of *Withania coagulans* Dunal. Planta Med 32:154–157

Budhiraja RD, Sudhir S, Garg KN (1983) Cardiovascular effects of a withanolide from *Withania coagulans*, dunal fruits. Ind J Physiol Pharmacol 27(2):129–134

Candelario M, Cuellar E, Reyes-Ruiz JM, Darabedian N, Feimeng Z, Miledi R et al (2015) Direct evidence for GABAergic activity of *Withania somnifera* on mammalian ionotropic GABAA and GABAρ receptors. J Ethnopharmacol 171:264–272

Chandrasekhar K, Kapoor J, Anishetty S (2012) A prospective, randomized double-blind, placebo-controlled study of safety and efficacy of a high-concentration full-spectrum extract of Ashwagandha root in reducing stress and anxiety in adults. Indian J Psychol Med 34 (3):255–262

Chang E, Pohling C, Natarajan A, Witney TH, Kaur J, Xu L et al (2016) AshwaMAX and Withaferin a inhibits gliomas in cellular and murine orthotopic models. J Neuro-Oncol 126:253–264

Chen L-X, He H, Qiu F (2011) Natural withanolides: an overview. Nat Prod Rep 28:705–740

Choudhary MI, Hussain S, Yousuf S, Dar A, Mudassar A-u-R (2010) Chlorinated and diepoxy withanolides from *Withania somnifera* and their cytotoxic effects against human lung cancer cell line. Phytochemistry 71:2205–2209

Choudhary MI, Nawaz SA, Zaheer-ul-Haq LMA, Ghayur MN, Jalil S et al (2005) Withanolides, a new class of natural cholinesterase inhibitors with calcium antagonistic properties. Biochem Biophys Res Commun 334:276–287

Choudhary MI, Yousuf S, Nawaz SA, Ahmed S, Atta-ur-Rahman (2004) Cholinesterase inhibiting withanolides from *Withania somnifera*. Chem Pharm Bull 52(11):1358–1361

Dabheliya J, Khan SA, Joshipura M, Vasoya M, Patel S, Vijaya S (2010) Diuretic potential of aqueous extract of fruits of *W. coagulans* Dunal in experimental rats. Int J Pharm Pharm Sci 2 (4):51–53

Datta A, Bagchi C, Das S, Mitra A, Pati AD, Tripathi SK (2013) Antidiabetic and antihyperlipidemic activity of hydroalcoholic extract of *Withania coagulans* Dunal dried fruit in experimental rat models. J Ayurveda Integr Med 4:99–106

Glotter E (1991) Withanolides and related ergostane-type steroids. Nat Prod Rep 8(4):415–440

Gupta A, Mahdi AA, Shukla KK, Ahmad MK, Bansal N, Sankhwar P et al (2013) Efficacy of *Withania somnifera* on seminal plasma metabolites of infertile males: a proton NMR study at 800MHz. J Ethnopharmacol 149:208–214

Hemalatha S, Wahi AK, Singh PN, Chansouria JPN (2004) Hypoglycemic activity of *Withania coagulans* Dunal in streptozotocin induced diabetic rats. J Ethnopharmacol 93:261–264

Hemalatha S, Wahi AK, Singh PN, Chansuria JPN (2006) Hypolipidemic activity of aqueous extract of *Withania coagulans* Dunal in albino rats. Phytother Res 20:614–617

Hoda Q, Ahmad S, Akhtar M, Najmi AK, Pillai K, Ahmad SJ (2010) Antihyperglycaemic and antihyperlipidaemic effect of poly-constituents, in aqueous and chloroform extracts, of *Withania coagulans* Dunal. In experimental type 2 diabetes mellitus in rats. Hum Exp Toxicol 29 (8):653–658

Huang C-F, Ma L, Sun L-J, Ali M, Arfan M, Liu J-W et al (2009) Immunosuppressive Withanolides from *Withania coagulans*. Chem Biodivers 6:1415–1426

Ihsan-ul-Haq YUJ, Chai X, Park E-J, Kondratyuk TP, Simmons CJ et al (2013) Biologically active withanolides from *Withania coagulans*. J Nat Prod 76:22–28

Jaiswal D, Rai PK, Watal G (2009) Antidiabetic effect of *Withania coagulans* in experimental rats. Ind J Clin Biochem 24(1):88–93

Jayaprakasam B, Nair MG (2003) Cyclooxygenase-2 enzyme inhibitory withanolides from *Withania somnifera* leaves. Tetrahedron 59:841–849

Jonathan G, Rivka R, Avinoam S, Lumir H, Nirit B (2015) Hypoglycemic activity of withanolides and elicitated *Withania somnifera*. Phytochemistry 116:283–289

Joshi P, Misra L, Siddique AA, Srivastava M, Kumar S, Darokar MP (2014) Epoxide group relationship with cytotoxicity in withanolide derivatives from *Withania somnifera*. Steroids 79:19–27

Kaur N, Niazi J, Bains R (2013) A review on pharmacological profile of *Withania somnifera* (Ashwagandha). Res Rev J Bot Sci 2(4):6–14

Khan MA, Subramaneyaan M, Arora VK, Banerjee BD, Ahmed RS (2015) Effect of *Withania somnifera* (Ashwagandha) root extract on amelioration of oxidative stress and autoantibodies production in collagen-induced arthritic rats. J Complement Integr Med 12(2):117–125

Khazal KF, Hill DL, Grubbs CJ (2014) Effect of *Withania somnifera* root extract on spontaneous estrogen receptor-negative mammary cancer in MMTV/Neu mice. Anticancer Res 34:6327–6332

Khodaei M, Jafari M, Noori M (2012) Remedial use of withanolides from *Withania coagulans* (stocks) Dunal. Adv Life Sci 2(1):6–19

Kurapati KRV, Atluri VSR, Samikkannu T, Nair MPN (2013) Ashwagandha (*Withania somnifera*) reverses β-amyloid$_{1\text{-}42}$ induced toxicity in human neuronal cells: implications in HIV-associated neurocognitive disorders (HAND). PLoS One 8(10):e77624

Liffert R, Hoecker J, Jana CK, Woods TM, Burch P, Jessen HJ et al (2013) Withanolide a: synthesis and structural requirements for neurite outgrowth. Chem Sci 4:2851–2857

LLanos GG, Araujo LM, Jiménez IA, Moujir LM, Bazzocchi IL (2012) Withaferin A-related steroids from *Withania aristata* exhibit potent antiproliferative activity by inducing apoptosis in human tumor cells. Eur J Med Chem 54:499–511

LLanos GG, Araujo LM, Jiménez IA, Moujir LM, Vázquez JT, Bazzocchi IL (2010) Withanolides from *Withania aristata* and their cytotoxic activity. Steroids 75:974–981

Mahdi AA, Shukla KK, Ahmad MK, Rajender S, Shankhwar SN, Singh V et al (2011) *Withania somnifera* improves semen quality in stress-related male fertility. Evid Based Complement Alternat Med 2011:576962

Manchanda S, Mishra R, Singh R, Kaur T, Kaur G (2017) Aqueous leaf extract of *Withania somnifera* as a potential neuroprotective agent in sleep-deprived rats: a mechanistic study. Mol Neurobiol 54(4):3050–3061

Maqsood M, Qureshi R, Ikram M, Ahmad M, Jabeen B, Asi MR et al (2018) In vitro anticancer activities of *Withania coagulans* against HeLa, MCF-7, RD, RG2, and INS-1 cancer cells and phytochemical analysis. Integr Med Res 7:184–191

Martin-Herrera D, Abdala S, Benjumea D, Perez-Paz P (2007) Diuretic activity of *Withania aristata*: an endemic Canary Island species. J Ethnopharmacol 113:487–491

Mathur D, Agrawal RC (2011) Evaluation of in vivo antimutagenic potential of fruits extracts of *Withania coagulans*. Der Pharma Chemica 3(4):373–376

Matsuda H, Murakami T, Kishi A, Yoshikawa M (2001) Structures of withanosides I, II, III, IV, V, VI, and VII, new withanolide glycosides, from the roots of Indian *Withania somnifera* Dunal. and inhibitory activity for tachyphylaxis to clonidine in isolated Guinea-pig ileum. Bioorg Med Chem 9:1499–1507

Maurya R, Akanksha J, Singh AB, Srivastava AK (2008) Coagulanolide, a withanolide from *Withania coagulans* fruits and antihyperglycemic activity. Bioorg Med Chem Lett 18:6534–6537

Mikolai J, Erlandsen A, Murison A, Brown KA, Gregory WL, Raman-Caplan P et al (2009) *In vivo* effects of Ashwagandha (*Withania somnifera*) extract on the activation of lymphocytes. J Altern Complement Med 15(4):423–430

Mirjalili HM, Fakhr-Tabatabaei SM, Bonfill M, Alizadeh H, Cusido RM, Ghassempour A et al (2009) Morphology and withanolide production of *Withania coagulans* hairy root cultures. Eng Life Sci 9(3):197–204

Montilla MP, Cabo J, Navarro MC, Risco S, Jimknez J (1990) The protective and curative action of *Withania frutescens* leaf extract against CCl_4-induced hepatotoxicity. Phytother Res 4 (6):212–215

Mulabagal V, Subbaraju GV, Rao CV, Sivaramakrishna C, DeWitt DL, Holmes D et al (2009) Withanolide sulfoxide from Ashwagandha roots inhibits nuclear transcription factor-kappa-B, cyclooxygenase and tumor cell proliferation. Phytother Res 22:987–992

Muralikrishnan G, Dinda AK, Shakeel F (2010) Immunological investigations immunomodulatory effects of *Withania somnifera* on azoxymethane induced experimental colon cancer in mice. Immunol Investig 39:688–698

Nishikawa Y, Okuzaki D, Fukushima K, Mukai S, Ohno S, Ozaki Y et al (2015) Withaferin a induces cell death selectively in androgen-independent prostate cancer cells but not in normal fibroblast cells. PLoS One 10(7):e0134137

Noshahr ZS, Shahraki MR, Ahmadvand H, Nourabadi D, Nakhaei A (2015) Protective effects of *Withania somnifera* root on inflammatory markers and insulin resistance in fructose-fed rats. Rep Biochem Mol Biol 3(2):62–67

Orrù A, Casu MA, Tambaro S, Marchese G, Casu G, Ruiu S (2016) *Withania somnifera* (L.) Dunal root extract alleviates formalin-induced nociception in mice: involvement of the opioidergic system. Behav Pharmacol 27(1):57–68

Parker WB, Shaddix SC, Rose LM, Shewach DS, Hertel LW, Secrist JA III et al (1999) Comparison of the mechanism of cytotoxicity of 2-chloro-9-(2-deoxy-2-fluoro-β-D-arabinofuranosyl)adenine, 2-chloro-9-(2-deoxy-2-fluoro-β-D-ribofuranosyl)adenine, and 2-chloro-9-(2-deoxy-2,2-difluoro-β-D-ribofuranosyl)adenine in CEM cells. Mol Pharmacol 55:515–520

Pingali U, Pilli R, Fatima N (2014) Effect of standardized aqueous extract of *Withania somnifera* on tests of cognitive and psychomotor performance in healthy human participants. Pharm Res 6 (1):12–18

Piouswatson, 2008. Ashwagandha roots. https://commons.wikimedia.org/wiki/File: Aswagandhaherb11.jpg. Public Domain. Accessed Aug 2019

Pramanick DD, Srivastava SK (2015) Pharmacognostic evaluation of *Withania coagulans* Dunal (Solanaceae) - an important ethnomedicinal plant. Biosci Discov 6(1):06–13

Prasad SK, Kumar R, Patel DK, Hemalatha S (2010) Wound healing activity of *Withania coagulans* in streptozotocin-induced diabetic rats. Pharm Biol 48(12):1397–1404

Raut AA, Rege NN, Tadvi FM, Solanki PV, Kene KR, Shirolkar SG et al (2012) Exploratory study to evaluate tolerability, safety, and activity of Ashwagandha (*Withania somnifera*) in healthy volunteers. J Ayurveda Integr Med 3(3):111–114

Senthilnathan P, Padmavathi R, Magesh V, Sakthisekaran D (2006) Stabilization of membrane bound enzyme profiles and lipid peroxidation by *Withania somnifera* along with paclitaxel on benzo(a)pyrene induced experimental lung cancer. Mol Cell Biochem 292:13–17

Sharma S, Joshi A, Hemalatha S (2017) Protective effect of *Withania coagulans* fruit extract on cisplatin-induced nephrotoxicity in rats. Pharm Res 9(4):354–361

Soman S, Korah PK, Jayanarayanan S, Mathew J, Paulose CS (2012) Oxidative stress induced NMDA receptor alteration leads to spatial memory deficits in temporal lobe epilepsy: ameliorative effects of *Withania somnifera* and Withanolide a. Neurochem Res 37:1915–1927

Spritzler F (2018) 12 Proven health benefits of Ashwagandha. Healthline.com (https://www.healthline.com/nutrition/12-proven-ashwagandha-benefits)

Subbaraju GV, Vanisree M, Rao CV, Sivaramakrishna C, Sridhar P, Jayaprakasam B et al (2006) Ashwagandhanolide, a bioactive dimeric thiowithanolide isolated from the roots of *Withania somnifera*. J Nat Prod 69:1790–1792

Sudhanshu MS, Rao N, Menghani E (2012) Phytochemical and antimicrobial activity of *Withania coagulans* (stocks) Dunal (fruit). Int J Pharm Pharm Sci 4(4):387–389

Takechi T, Uchida J, Fujioka A, Fukushima M (1997) Enhancing 5-fluorouracil cytotoxicity by inhibiting dihydropyrimidine dehydrogenase activity with uracil in human tumor cells. Int J Oncol 11:1041–1044

Thedford R, Leyimu EO, Thornton DL, Mehta R (1989) Cytotoxicity of N6-cycloalkylated adenine and adenosine analogs to mouse hepatoma cells. Expl Cell Biol 57:53–59

Umadevi M, Rajeswari R, Rahale CS, Selvavenkadesh S, Pushpa R, Sampath KP (2012) Traditional and medicinal uses of *Withania somnifera*. Pharma Innov 1(9):102–110

Visavadiya NP, Narasimhacharya AV (2007) Hypocholesteremic and antioxidant effects of *Withania somnifera* (Dunal) in hypercholesteremic rats. Phytomedicine 14(2–3):136–142

Wankhede S, Langade D, Joshi K, Sinha SR, Bhattacharyya S (2015) Examining the effect of *Withania somnifera* supplementation on muscle strength and recovery: a randomized controlled trial. J Int Soc Sports Nutr 12:43

Wijeratne EMK, Xu Y, Scherz-Shouval R, Marron MT, Rocha DD, Liu MX et al (2014) Structure–activity relationships for withanolides as inducers of the cellular heat-shock response. J Med Chem 57:2851–2863

Xu Y, Gao S, Bunting DP, Gunatilaka AAL (2011) Unusual withanolides from aeroponically grown *Withania somnifera*. Phytochemistry 72:518–522

Xu Y, Marron MT, Seddon E, McLaughlin SP, Ray DT, Whitesell L et al (2009) 2,3-Dihydrowithaferin A-3β-*O*-sulfate, a new potential prodrug of withaferin a from aeroponically grown *Withania somnifera*. Bioorg Med Chem 17:2210–2214

Zhao J, Nakamura N, Hattori M, Kuboyama T, Tohda C, Komatsu K (2002) Withanolide derivatives from the roots of *Withania somnifera* and their neurite outgrowth activities. Chem Pharm Bull 50(6):760–765

Immunomodulators and Phytodrugs

42

Deepa S. Mandlik and Ajay G. Namdeo

Abstract

A strong, healthy immune system is the basis of good health. Immunity is the balanced state of having adequate biological defense to fight against infection, disease, or unwanted biological invasion while having tolerance to avoid allergy and autoimmune diseases. Immune responses occur due to effective interaction between innate (natural and nonspecific) and acquired (adaptive and specific) components of the immune system. The interaction of phagocyte and microorganism in the immunological system is a protective pathway but when unreasonably or improperly organized can damage the body and contribute to the development of several long-term inflammatory conditions like allergies, autoimmune disease, rheumatic arthritis, carcinoma, and many more. Over the past years, remarkable changes have occurred in the immune system due to exposure to chemicals, drugs, or environmental pollutants. Numerous factors perform a major part in alterations of the immunological system such as age, sex, stress, genetic variability, alcohol/drug abuse, lifestyle, malnutrition, and environmental pollution. In clinical practice, immunomodulators can be classified into three main categories such as immunostimulants, immunosuppressants, and immunoadjuvants. Majority of the immunomodulators and immunosuppressive agents used in medicinal practice are toxic and have detrimental adverse effects on the host body. Hence, there is a rising need for the usage of herbal drugs that

D. S. Mandlik
Department of Pharmacology, Bharati Vidyapeeth University, Poona College of Pharmacy, Pune, India
e-mail: deepa_ingawale@yahoo.com

A. G. Namdeo (✉)
Department of Pharmacognosy, Bharati Vidyapeeth University, Poona College of Pharmacy, Pune, India
e-mail: agnamdeo@gmail.com; ajay.namdeo@bharatividyapeeth.edu

S. C. Mandal et al. (eds.), *Evidence Based Validation of Traditional Medicines*,
https://doi.org/10.1007/978-981-15-8127-4_42

leads to modulation of immunological system which will prevent the infections rather than to treat the immunological disorders. Phytoconstituents such as alkaloids, glycosides, flavonoids, lactones, tannins, saponins, phenolic compounds, polysaccharides, and diterpenoids of herbal origin have been confirmed for their immunostimulating characteristics. Therefore, the exploration of herbs derived potent and safe immunomodulator drugs is gaining more attention. Hence, the current book chapter will focus attention on summary of extensively scrutinized phyto-immunomodulator plants such as *Boerhaavia diffusa*, *Withania somnifera*, *Allium sativum*, *Azadirachta indica*, *Acacia catechu*, *Curcuma longa*, *Ficus benghalensis*, *Tinospora cordifolia*, *Murraya koenigii* and *Ocimum sanctum.* These plant-derived immunomodulators drugs have displayed their action by acting on cellular and humoral immune system in variety of animal models, and we need to explore their role in clinical trials also.

Keywords

Immunity · Immune system · Cellular immunity · Humoral immunity · Immunomodulators · Plant-derived immunomodulators

Abbreviations

COX	Cyclo-oxygenase enzyme
ERK-1/2	Extracellular signal-regulated kinase-1/2
IFN-γ	Interferon-γ
IL-10	Interleukin-10
IL-12	Interleukin-12
IL-1α	Interleukin-1α
IL-1β	Interleukin-1β
IL-6	Interleukin-6
IL-8	Interleukin-8
MMP-2	Matrix metalloproteinase-2
MMP-9	Matrix metalloproteinase-9
NF-κB	Nuclear factor κB
PAMPs	Pathogen-associated molecular patterns
PBMCs	Peripheral blood mononuclear cells
RBCs	Red blood cells
STAT4	Signal transducer and activator of transcription 4
TLR6	Toll-like receptor 6
TNF-α	Tumor necrosis factor-α
VEGF	Vascular endothelial growth factor
WBC	White blood cell

42.1 Introduction

The scientist Edward Jenner was known for the idea of "immunomodulation." Balanced homeostatic state of body is maintained by body's immune system, but pathogenic organisms such as allergens, bacteria, viruses, and parasitic agents repeatedly attacks on it. Additionally, the immune system was weakened by various factors such as chronic stress, diseases, environmental conditions, and lifestyle-related changes and interrupt the balanced homeostasis state of host. The disturbed immune system can be repaired by using antibiotics and conventional chemotherapeutic agents, but excessive use of these drugs and chemical agents has more harmful effects on the immune system. So, for minimizing the side effects of these chemotherapeutic agents, the usage of herbal medicines has widely enhanced in the last few years. The management of immune system with plant-derived immunomodulators offers a safer alternative to conventional chemotherapeutic agents (Tzianabos 2000).

42.1.1 Immunity

It is stated that immunity is body's capability to fight against huge amounts of injurious microbes, facilitating the body to inhibit diseases, organ damage, or tissue damage. The causative factors of immunity trigger consist of various external stimuli, immunization, and previous infection. The moment, at which the outside matter enters the body, the combined and synchronized reaction of immune cells in response to foreign matters establishes the immune reaction (Baxter 2007). The immunological structure is not restricted to any particular body part. The stem cells are produced in the bone marrow, and on maturation they travel to various body parts. On maturation, an immune cell exhibits specific effect on human body. Two mechanisms of immune system like cellular immunity and humoral immunity plays important role in fighting against invading microorganisms (Ford and Roach 2009).

42.1.2 Immune Systems

Immune system has been divided in two extensive categories depending on their functions such as innate and adaptive immune system (Vesely et al. 2011). The chief mediators of immune system are cytokines, interleukins, complement, acute phase proteins, macrophages, monocytes, and neutrophils. Pathogen-associated molecular patterns (PAMPs) are the different moieties expressed by pathogens to identify the existence of microorganism in host body. The sensors in the host body are recognized as pattern recognition receptors that identify the PAMPs. After recognition of PAMPs by pattern recognition receptors, a sequence of immune responses get quickly activated through the generation of different types of cytokines, chemokines, and interferons (Parkin and Cohen 2001). The components of innate immunity system include antigen-presenting cells and macrophages that plays crucial role in

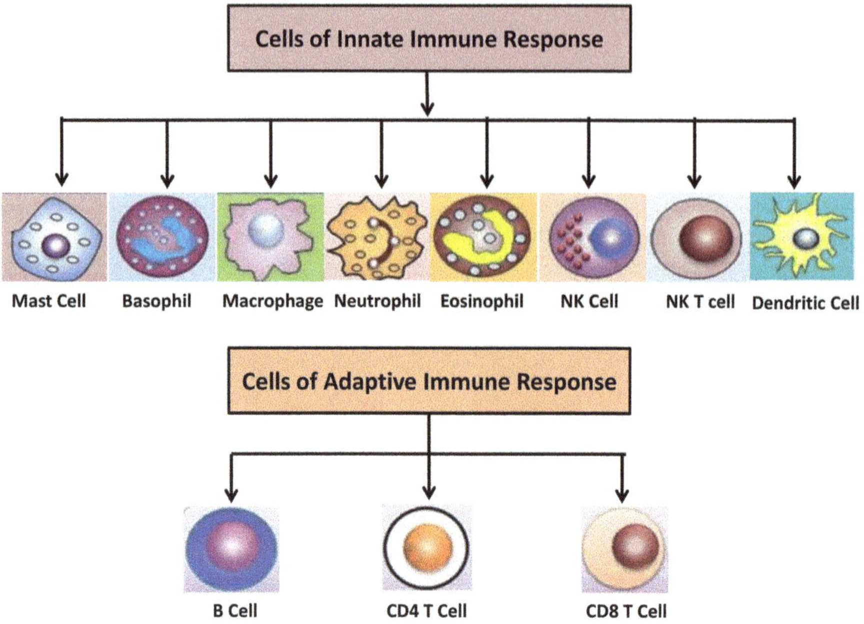

Fig. 42.1 Different cells involved in innate and adaptive immune responses

secretion of cytokines, production of nitric oxide, antibody-dependent cell-mediated cytotoxicity, antigen recognition, and phagocytic process. The stimulation of T and B cells is carried out by dendritic cells (Moradali et al. 2007). The third main part of innate immunity is component system (Oh et al. 2000). Adaptive immunity is developed by the generation of pathogen-specific lymphocytes by gene reorganization method. The acquaintance of host body to antigen for production of adaptive immune response that develops in 30 days but lasts for entire life span is known as active immunity. The adaptive immunity consists of T and B lymphocyte-mediated antigen-specific reactions. The infectious microorganisms are deactivated by combining with the antibodies. Antibodies lead to stimulation of variety of complement proteins that result in opsonization and destruction of microorganisms by phagocytic cells (Spellberg and Edwards Jr (2001). The several cells associated with innate and adaptive immune responses are represented in Fig. 42.1.

42.1.3 Immunomodulators

Several exogenous and endogenous factors affect on the efficiency of immune system and resulted in either immunosuppression or immunostimulation. The molecules of biological or synthetic source proficient of modulation, suppression, and stimulation of any component of immune system are recognized as immunomodulators, immuno-restoratives, or immuno-augmentors. In autoimmune

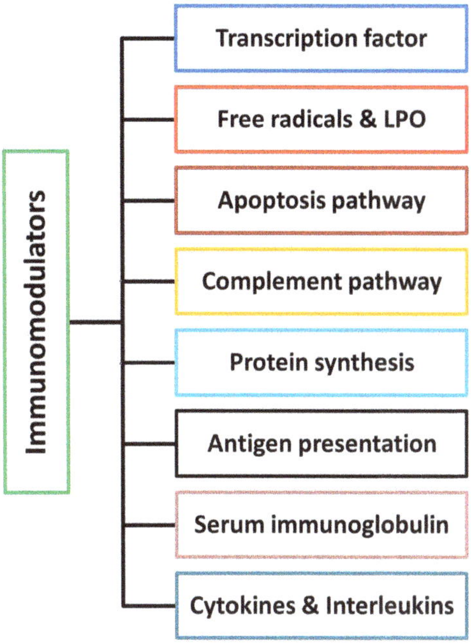

Fig. 42.2 Various targets of functioning of immunomodulator

disorders, a number of chemically synthesized compounds and monoclonal antibodies are being used as immunomodulators (Puri et al. 1994). Immunomodulators are the drugs that can modify cellular actions such as protein synthesis, apoptosis, antigen recognition, release of immune mediators, and transcription factors (Fig. 42.2). Immunomodulating agents are commonly classified into immunostimulants, immunosuppressants, and immunoadjuvants in medicinal field. Immunostimulants are the agents that enhance the level of immune response by enhancing the body's resistance power against infectious diseases wherein they act by both the responses like innate and adaptive. Immunostimulants modulate the insufficiency of immune system as observed in acquired immunodeficiency syndrome disease. The resistance against autoimmunity, allergy, infection, and cancer is enhanced by immunostimulants (Song et al. 2016). Immunosuppressant drugs exhibit a central role in suppressing the immune system to restore normality. They are utilized to control the pathological immune reaction subsequent to organ transplantation, treatment of allergic reactions, autoimmune infections, and infection associated immunopathology. Immunoadjuvants are the agents that can boost the efficiency of vaccines such as Freund's adjuvant (El-Sheikh 2008). The overall classification of immunomodulators and their uses are indicated in Fig. 42.3.

Hence, immunomodulators with surplus safety and efficacy are still needed to evaluate. So, association of several adverse effects with the use of synthetic medicines and immunomodulating agents of natural origin would be the proposed targets to substitute them in the treatments.

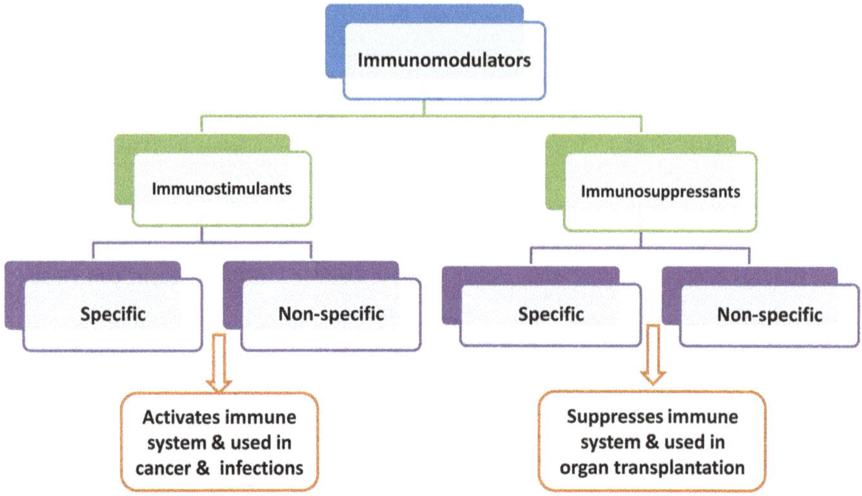

Fig. 42.3 Classification of immunomodulators and uses

42.2 Plant-Derived Immunomodulator Drugs

Abundant research work have been explored on Rasayana and Non-Rasayana plants for elucidation of phytoconstituents from them and understanding their role on the immunological system of the body. Ayurveda-Indian traditional medicine system defines the usage of numerous ayurvedic plants to fortify the body's defense system and helps to improve the general human well-being (Mukherjee et al. 2012). Rasayana herbal plants have revitalizing properties and boost the immune system against various diseases. As well as Non-Rasayana medicinal plants also display immunostimulatory, anti-inflammatory, and anti-pathogenic properties (Kumar et al. 2012). Rasayana and Non-Rasayana plants with immunomodulatory activities are presented in Table 42.1.

Since from the ancient times, medicinal herbs have been utilized for the management of several microbiological and general day-to-day disorders (Kalia 2011). As per World Health Organization, around 3/4 of the world's populace depends on plant-derived remedy for the treatment of numerous diseases (Kumar et al. 2012). Phyto-immunomodulatory agents can enhance the body's immune response against infectious microorganism by the activation of immunological system (Agarwal and Singh 1999). An immunostimulatory herb performs a crucial part for the management of inflammation, infectious diseases and immunodeficiencies by exerting their effect on different cells through the release of cytokines and interleukins (Jandu et al. 2017). The possible mechanism of action of phyto-immunomodulators could be as immunostimulators, immunopotentiators, or immunoadjuvants to enhance the antigen-specific immune response. Phyto-immunomodulators exhibit their

Table 42.1 List of Rasayana and Non-Rasayana plants in Ayurveda

Sr. no.	Rasayana plant	Sr. no.	Non-Rasayana plants
1	*Acorus calamus*	1	*Andrographis paniculata*
2	*Allium sativum*	2	*Aristolochia indica*
3	*Aloe vera*	3	*Bauhinia variegata*
4	*Anacyclus pyrethrum*	4	*Bombaxmalabarium*
5	*Asparagus racemosus*	5	*Butea monosperma*
6	*Azadirachta indica*	6	*Butea superba*
7	*Bacopa monnieri*	7	*Calotropis procera*
8	*Boerhavia diffusa*	8	*Catharanthus roseus*
9	*Butea monosperma*	9	*Gymnema sylvestre*
10	*Centella asiatica*	10	*Hibiscus esculentus*
11	*Chlorophytum borivilianum*	11	*Jasminum sambac*
12	*Clitoria ternatea*	12	*Lawsonia inermis*
13	*Commiphora mukul*	13	*Luffa cylindrical*
14	*Convolvulus pluricaulis*	14	*Mangifera indica*
15	*Curculigo orchioides*	15	*Melia azadirachta*
16	*Curcuma longa*	16	*Mentha spicata*
17	*Cynodon dactylon*	17	*Nardostachys jatamansi*
18	*Glycyrrhiza glabra*	18	*Ocimum canum*
19	*Hemidesmus indicus*	19	*Ocimum sanctum*
20	*Phyllanthus emblica*	20	*Picrorhiza kurroa*
21	*Piper longum*	21	*Piper betle*
22	*Sida spinosa*	22	*Saraca indica*
23	*Terminalia bellirica*	23	*Viscum album*
24	*Tinospora cordifolia*	24	*Vitex negundo*
25	*Withania somnifera*	25	*Zingiber officinale*

immunomodulatory action through the modulation of various body immune cells such as T cells, B cells, macrophages, and dendritic cells (Dwivedi et al. 2017).

42.2.1 *Boerhavia diffusa* (Common Names: Tarvine, Punarnava)

The common names of *B. diffusa* are tarvine or punarnava belonging to the family *Nyctaginaceae*. Punarnavine, boerhavia acid, boeravinone, palmitic acid, and ecdysteroid are the foremost phytoconstituents isolated from it. Punarnavine and syringaresinol are the two important immunostimulants obtained from the *B. diffusa* roots. Aqueous extracts of *B. diffusa* have showed considerable increase in the number of leukocytes and decreased the death rate in mouse model of abdominal sepsis induced by *E. coli*. It also exhibited reduction in stress prompted elevation in blood glucose and cholesterol levels. The alkaloid fraction of *B. diffusa* was found to lessen the delayed type of hypersensitivity reactions and stabilizes the plasma cortisol level in rodents (Mungantiwar et al. 1999). It is also been famous for its adaptogenic property. Punarnavine, syringaresinol, and quercetin are the molecules may be responsible for the immunostimulant potential of *B. diffusa*. Mehrotra et al. (2003) have reported immunosuppressive properties of *B. diffusa* alcoholic extracts that showed inhibition of human natural killer cells, generation of nitric oxide in macrophages and reduced proinflammatory cytokines level in human PBMCs (Mehrotra et al. 2003). The treatment of RAW 264.7 cell line with chloroform and ethanolic extract of *B. diffusa* on lipopolysaccharides stimulation convinced production of nitric oxide, inhibited production of proinflammatory cytokines in PBMCs and phytohemagglutinin-induced proliferation (Pandey et al. 2005).

B. diffusa methanol extract has inhibited melanoma metastasis in mice along with decreasing the levels of blood parameters (Leyon et al. 2005). In a different research study, levamisole was considered as a standard drug while performing phagocytic activity of macrophages that was enhanced with *B. diffusa* extract treatment (Sumanth and Mustafa 2007). The activity of natural killer cells, interferon-γ (IFN-γ) and interleukin-2 (IL-2) and antibody-dependent cellular cytotoxicity was enhanced with *B. diffusa* extract treatment. The proinflammatory cytokines level like IL-1β, IL-6, and TNF-α was abridged on treatment with punarnavine (Manu and Kuttan 2007). In the anticancer activity, punarnavine decreased matrix metalloproteinase and vascular endothelial growth factor expressions in lung melanoma (Manu and Kuttan 2009). The methanolic extract of whole plant decreased the viability of MCF-7 cell lines in G0-G1 phase (Sreeja and Sreeja 2009). The ethanolic extract prepared by using *B. diffusa* leaf and root part have reported anticancerous activity against Hela and U-87 tumor cell lines (Srivastava et al. 2005).

42.2.2 *Withania somnifera* (Common Names: Indian Ginseng, Ashwagandha)

It is commonly known as Indian ginseng or ashwagandha belonging to the family Solanaceae. In preclinical studies, *W. somnifera* has shown numerous pharmacological activities such as anti-inflammatory, anti-tumor, cardio-protective, anti-microbial, anti-stress, neuro-protective and anti-diabetic properties. The active phytoconstituents of *W. somnifera* include alkaloids (withanine, withananine, withasomnine, somniferine, etc.), steroidal lactones (withanolides, withaferins), and saponins (Mishra et al. 2000).

The administration of *W. somnifera* extract in mice has prevented the myelosuppression in mice induced by various immunosuppressant agents along with elevation in the blood cells number and body weight index (Ziauddin et al. 1996). Agarwal et al. have studied the cyclophosphamide-induced immunosuppression in mice in which the study result showed considerable elevation in the blood cells count (Agarwal et al. 1999). The immunological potential of mice was enhanced with increase in white blood cells, alpha-esterase-positive cells, bone marrow cellularity, circulating antibody titer, and spleen cells. The type IV hypersensitivity reaction and phagocytic activity of macrophages was improved on treatment with ashwagandha extract (Davis and Kuttan 2000). Immun-21, polyherbal formulation of *W. somnifera* stated considerable immunomodulatory potential in AIDS patients (Singh et al. 2001). Ashwagandha has shown immunostimulatory activity on lymphocyte that resulted in reduction in tumor growth in cancer cell cultures (Jayaprakasam et al. 2003). The immunosuppressive activity was exhibited by withaferin A and withanolide E on mice T and B lymphocytes as well as thymocytes (Aggarwal et al. 2012). In another research work, immunomodulatory potential of *W. somnifera* was evaluated against doxorubicin induced immunosuppression in rats. The results of the study stated significant elevation in blood cells count, alpha-esterase-positive cells, spleen, cells and bone marrow cellularity (Rizvi et al. 2016). The immunomodulatory activity of ashwagandha and withaferin A was also tested on zinc oxide nanoparticles that facilitated toxicity in mice. The nanoparticle-induced toxicity was found to abridge on *W. somnifera* extract and withaferin A administration in mice with decrease in gene expression and restoration of phagocytic potential (Kumar et al. 2019).

42.2.3 *Allium sativum* (Common Names: Garlic)

A. sativum is commonly known as garlic or lahsun of family Amaryllidaceae. It is used as a spice and condiment in a variety of food preparations due to its immunomodulation activity. Allicin, diallyldisulfide, S-allylcysteine, and diallyltrisulfide are the important phytoconstituents present in the *A. sativum* (Wang and Ng 1999). The ethanolic extract of garlic cloves was found to be effective against cancerous cell lines and natural killer cells (Kyo et al. 1998). The extract of *A. sativum* exerts an immunomodulatory effect on lymphocyte and monocytes multiplication by upregulation of IL-10 and downregulation of TNF-α expressions followed by lipopolysaccharide activation (Hodge et al. 2002). The bioactive compounds, allitridin or diallyltrisulfide, from garlic exhibit antiviral activity against murine cytomegalovirus by inhibiting regulatory T cells (Yi et al. 2005). Lectin isolated from *A. sativum* activates basophils and mast cells thereby acting as powerful mitogens and immunomodulator agent. Lectins and agglutinins were the immunoproteins isolated from *A. sativum* that are popular due to their mitogenic activity same as that of concanavalin A and phytohemagglutinin (Clement et al. 2010). Fructooligosaccharides obtained from the ethanolic extract of garlic exhibited mitogenic potential and phagocytic activity that was found to be comparable mannan and zymosan mitogenic agents (Chandrashekar et al. 2011). The treatment of ethanolic extract of garlic has inhibited the formation of histamine in rat basophilic cell culture. The ethanolic extract of garlic on oral administration, in Leishmania-infected murine macrophages, has considerably declined the immunoglobulin-E-mediated hypersensitivity reactions and proinflammatory cytokines such as IL-12, INF-γ, and inducible nitric oxide synthase (Gharavi et al. 2011). Allicin stimulates the growth of dendritic cells by elevating the expression of costimulatory molecule (CD40) in animal malarial model (Feng et al. 2012). The administration of allicin in *Plasmodium yoelii*-infected mice enhances the macrophage activation and activates the CD41 T-cells (Feng et al. 2012).

Diallyldisulfide obtained from garlic has lessened the inflammatory cytokines levels and nitric oxide formation in rodent macrophages and leukemia monocyte cell lines (Shin et al. 2013). Fresh *A. sativum* extracts enhance the activation and proliferation of CD81 T cells and exhibited type-IV hypersensitivity reaction (Ebrahimi et al. 2013). Allicin, an active constituent obtained from garlic has elevated the levels of monocyte chemoattractant protein-1 and growth response protein-1 in lipopolysaccharide-stimulated adipocytes. *A. sativum* extract in low doses augmented IL-10 expressions with decrement in the expressions of proinflammatory cytokines (Quintero-Fabian et al. 2013). Additional bioactive compounds of garlic such as allitridin, S-allyl-L-cysteine, and diallyl sulfide inhibited the expressions of proinflammatory cytokines (IL-6, IL-12, TNF-α, IL-1β, MCP-1) by inhibition of NF-κB factor (Kim et al. 2013). Treatment of *Labeo rohita* with combination of *A. sativum* and minerals oil augments serum antibody titer, serum agglutination, and hemolytic activity against *Aeromonas hydrophila* infection (Dash et al. 2014).

42.2.4 *Azadirachta indica* (Common Names: Neem)

The common name of *A. indica* is neem belonging to the family Meliaceae. Different phytoconstituents present in *A. indica* are quercetin, azadirachtin, nimbin, nimbinin, nimbidin, nimbanene, 6-desacetylnimbinene, nimbandiol, nimbolide, and nimbiol. The extract obtained from neem trees has been widely used in health management since from olden times and possesses a variety of health-promoting properties (Sudhir et al. 2010). It is also known for anti-microbial, hepatoprotective, hypoglycemic, hypolipidemic, hypotensive, antipyretic, cancer, skin diseases, digestive disorders and acquired immunodeficiency syndromes (Santhosh and Navartnam 2013).

The treatment of dried leaves powder of *A. indica* in broiler chickens resulted in significant enhancement of antibody titers against new castle disease virus antigen thereby showing its effect on humoral and cell-mediated immune responses (Sadekar et al. 1998). Neem leaf preparation plays important role as an adjuvant for the generation of antigen-specific antiserum in mice (Baral and Chattopadhyay 2004). In another research studies, administration of *A. indica* extract in Balb/c mice enhanced the activation of peripheral blood lymphocytes, T cells, and macrophages. It also exhibited significant effect in sarcoma model of Balb/c mice after the treatment with *A. indica* extracts (Belska et al. 2006). The nimbolide extracted from the *A. indica* leaves has exhibited anti-proliferative and anti-apoptotic activities through the upregulation and downregulation of specific genes. Hence, *A. indica* strengthens the body's immune system by acting as an anticancer agent (Kumar et al. 2006). The neem leaf glycoprotein significantly improved the IFN-γ and downregulated CXCR3B expression that causes lymphocyte death with decrement in the chemotactic movement of PBMCs in the direction of tumor (Chakraborty et al. 2008). Durrani et al. have investigated the growth-promoting and immunomodulatory effects of *A. indica* leaves infusion in broiler chicks. The study results showed that neem infusion effectively enhanced the growth performance and antibody titer in broiler chicks (Durrani et al. 2008). Neem leaf glycoprotein also causes T-cell stimulation and release of IFN-γ and can successfully stimulate oral and erythroleukemia cancerous cells (Bose et al. 2009). Treatment with watery extracts of neem considerably improved tumor antigen presentation by macrophages and lymphocytes (Tsang et al. 2011).

42.2.5 *Acacia catechu* (Common Names: Black Catechu, Cachou)

A. catechu is commonly known as black catechu or kattha belonging to the family of Fabaceae. *A. catechu* has been commonly used for the treatment of many diseases in Ayurveda. Its heartwood extract is used for the treatment of cough, asthma, stomatitis, bronchitis, skin reactions, colic pain, diarrhea, dysentery, boils, and sores. *A. catechu* exerts its immunomodulatory effect by acting on cell-mediated and humoral immune responses (Sunil et al. 2018).

The compounds isolated from *A. catechu* exert anti-viral effect against human immunodeficiency virus-1 strains (Marquez et al. 2005). Its bark extracts also exert anti-HIV activity by inhibition of viral protease enzyme and blocking Viral Tat protein interaction to its HIV-1 promoter sequence of long terminal repeats (Modi et al. 2013). The administration of water extract of black catechu (5 and 50 mg/kg) exhibited an increase in neutrophil adhesion with optimum rise in the hemagglutination titer values, phagocytic index, and serum immunoglobulin levels and decrease in the death ratio thus showing protective effect against cyclophosphamide-induced neutropenia (Ismail and Asad 2009). Methanolic and hexane extract of *A. catechu* bark exerts anti-proliferative, anticancer, and cytotoxic activity against several cancerous cell cultures. Hence, it would be helpful in finding out effective anticancer drug moiety. But, it is not effective on human peripheral lymphocytes (Nadumane and Nair 2011). Different extracts of black catechu have protective action against chemical-induced hepatic damage (carbon tetrachloride, 7,12-Dimethylbenz[a]anthracene), squamous cells, and breast cancers (Monga et al. 2011). Treatment of alcoholic extract of *A. catechu* in breast carcinoma cell culture exhibited improvement in Bax/Bcl2 ratio leading to the activation of caspases with 50% cytotoxic activity (Ghate et al. 2014). The immunomodulatory effects of *A. catechu* extracts was studied in mice. The study results showed that enhancement in the spleen antibody producing cells with reduction in type-IV hypersensitivity reaction (Sunil et al. 2018).

42.2.6 *Curcuma longa* (Common Names: Turmeric)

C. longa is commonly known as turmeric or haldi and belongs to the family of Zingiberaceae. It has wide spectrum of pharmacological activities such as anti-inflammatory, antiseptic, wound-healing, blood purifying, anti-tumor, anti-diabetic, anti-bacterial, neuro-protective, and immuno-modulatory and anticarcinogenic properties. The chief phytoconstituents extracted from turmeric are curcuminoids that contains curcumin, demethoxycurcumin, and bisdemethoxycurcumin and turmerones (He et al. 1998). The bioactive constituents of *C. longa*, curuminoid, inhibit cyclo-oxygenase-2 enzyme and nuclear factor κB (NF-κB) thereby acting as a potent anti-inflammatory agent (Plummer et al. 1999). The human T-cell maturation prompted by antigens is also affected by curcumin (Cipriani et al. 2001). In chronic mild stress model of rats, treatment of *C. longa* extracts improved the activity of natural killer cells and proinflammatory cytokine levels (Xia et al. 2006). Curcumin was found to downregulate the expression of various proinflammatory cytokines (TNF-α, IL-1, IL-2, IL-6, IL-8, IL-12) and chemokines through the inactivation of transcription factor such as NF-κB. Moreover, curcumin at small doses augment the antibody responses, suggesting its important role in treating variety of chronic diseases may be due to modulation of immune system (Jagetia and Aggarwal 2007). In experimental autoimmune encephalomyelitis model, curcumin has declined the production of IL-12 and signal transducer and activator of transcription 4(STAT4) (Fahey et al. 2007). The aqueous extracts of *C. longa* containing polysaccharides devoid of curcuminoids that possess mitogen properties and improved the spleen cells count as compared to lipopolysaccharide and concanavalin-A. Also it enhanced the levels of inflammatory cytokines in inactivated macrophages and spleen cells (Chandrasekaran et al. 2013). The molecular mechanism of curcumin for the management of carcinomas and inflammatory diseases would be combinations of multiple signaling pathways such as NF-κB and STAT3 signaling (Deguchi 2015). Several preclinical and clinical trials have discovered immunomodulatory actions of curcumin by showing its effect on immune cells and mediators involved in the immune response (Abdollahi et al. 2018).

42.2.7 *Ficus benghalensis* (Common Names: Banyan Tree)

The common name of *F. benghalensis* is banyan tree belonging to the family Moraceae. The root extract of *F. benghalensis* has been used in traditional medicine system for boosting the immune system. The numerous *Ficus* species have been widely used in conventional medicines as an astringent, hypotensive, vermicide, and anti-dysentery agent (Trivedi et al. 1969). The main phytoconstituents extracted from banyan tree are glucosides and flavonoids (Bhattacharjee 2008). The aqueous and methanolic extracts of *F. benghalensis* possess immunostimulatory potential as well as amplify the engulfing ability of PBMCs. The lymphocytes proliferation along with liberation of cytokines from the immune cells was enhanced by *F. benghalensis* extracts (Gabhe et al. 2006). The aqueous extract of *F. benghalensis* roots was administered in rats for evaluating immunostimulatory activity, using in vitro polymorphonuclear function test, hemagglutination, and hypersensitivity reactions (Khan et al. 2008). Immunostimulatory effect of the roots powder of *F. benghalensis* was studied in fish models (*Channa punctatus*). The study results showed that aspartate aminotransferase and alanine aminotransferase levels remained nearly same, whereas superoxide dismutase, total serum protein, nitric oxide, immunoglobulin, serum lysozyme, phagocytotic index have enhanced significantly (Verma et al. 2012). The hydro-alcoholic leaves extract of *F. benghalensis* has considerably enhanced the engulfing ability of human granulocytes that resulted in phagocytosis of microorganisms by leukocytes, with scavenging of free radicals and lessening of oxidative burden. Therefore, extracts of *F. benghalensis* showed immunomodulatory activity by phagocytosis of killed *C. albicans* and antioxidant property (Bhanwase and Alagawadi 2016).

42.2.8 *Tinospora cordifolia* (Common Names: Guduchi)

The family of *T. cordifolia* is Menispermaceae and reported for extensive range of immunological properties such as anti-inflammation, anti-allergic, anti-pyretic, anti-hepatotoxic, anti-diabetic, and anti-bacterial properties with minimal toxicity. Compound-D-glucan, cordifolioside A, cordifolioside B, and syringin are the chief phytoconstituents present in the guduchi plant (Nair et al. 2006). The engulfing capacity of human granulocytes was found to be enhanced with the treatment of *T. cordifolia* aqueous and ethyl acetate extracts. *T. cordifolia* extract has also stimulated the proliferation of stem cells and bone marrow cells along with increase in the number of antibody producing cells. Hence, indicating the role of *T. cordifolia* extract in the enhancement of humoral immune response. The guduchi extract also reported with anti-tumor activity with reduction in tumor progression as compared with the cyclophosphamide treatment (Mathew and Kuttan 1999). Lysozymes act as a strong microbicidal agent against gram-positive bacteria. The treatment with *T. cordifolia* extract has increased the quantity of lysosomes that lead to microbial activation of macrophages (Shimada et al. 2008). The immunomodulatory activity of *T. Cordifolia* was confirmed by evaluating its activity on macrophages and by release of lysosomes and nitric oxide (More and Pai 2011). The immunomodulatory activity of *T. cordifolia* extracts (ethyl acetate, water fractions, and hot water extract) was investigated in relation to phagocytosis and reactive oxygen species production in human neutrophil cells. The results of study indicated that all the three extract exhibited significant immunomodulatory activity with increase in the percentage of phagocytosis, nitric oxide, and reactive oxygen species generation. The immunostimulatory activity of *T. cordifolia* may be associated with the synergistic action of active constituents like syringin and cordifolioside A (Sharma et al. 2012). The water and methanolic extracts of *T. cordifolia* were tested against *Salmonella typhimurium* by the broth dilution and agar well diffusion assay methods. Treatment with methanolic extracts of *T. cordifolia* resulted in increased survival and reduced bacterial load in *S. typhimurium*-infected mice. Furthermore, aqueous and methanolic extracts of *T. cordifolia* treatment reduced the liver inflammation and rescued the levels of antioxidant enzymes in *S. typhimurium*-infected mice (Alsuhaibani and Khan 2017).

42.2.9 *Murraya koenigii* (Common Names: Curry Leaves)

M. koenigii is commonly known as curry leaves belonging to the family Rutaceae. Due to the aromatic quality of curry leaves, it is regularly used in cooking. Carbazole alkaloid present in the curry leaves are responsible for its antioxidant, anti-inflammatory, and anticancer activities. Other active phytoconstituents in curry leaves are koenine, koenidine, koenimbine, cyclomahanimbine, murrayastine, tetrahydromahanimbine, murrayazoline, girinimbin, iso-mahanimbin, carbazole alkaloids, mahanimbine, etc. The production of nitric oxide from macrophages was significantly improved by the methanolic extracts of *M. koenigii.* In addition to that, the phagocytic index was also increased in carbon-clearance test. The treatment of methanolic extract has significantly exhibited the humoral antibody response to ovalbumin with boosting of humoral response and immuno-stimulatory effect on B cells. The extract of *M. koenigii* does not show stimulatory effect on cell-mediated immunity (T-cells) as there was no significant difference in delayed types of hypersensitivity (Shah et al. 2008). The reduction in lipid peroxidation in ethanol-induced liver toxicity was found to be reduced by *M. koenigii* leaf extracts (Sathaye et al. 2011). In cadmium-induced oxidative stress, the cardiac tissue was protected by the treatment of aqueous extracts of *M. koenigii* in rats due to its antioxidant activity (Mitra et al. 2012). The administration of *M. koenigii* alcoholic extracts in rodents has significantly abridged anti-inflammatory effect on paw edema in rats induced by carrageenan (Bhandari 2012). The anti-inflammatory and chemopreventive effect of *M. koenigii* aqueous extract was studied on 4 T1 breast cancer cell-challenged mice. The aqueous extract of *M. koenigii* was found to reduce the tumor size and lung metastasis. Moreover, it decreased the level of inflammation-related cytokines, genes, and nitric oxide (Yeap et al. 2015).

42.2.10 *Ocimum sanctum* (Common Names: Holy Basil or Tulsi)

O. sanctum is commonly known as holy basil or tulsi belonging to the family Lamiaceae. It is known for analgesic, anti-inflammatory, immuno-stimulatory activity and has numerous therapeutic applications. Tulsi is a scavenger of free radicals and found to be effective against a variety of cancer. Ursolic acid, the main constituent of *O. sanctum*, has anti-inflammatory, anticancerous activity and inhibits cyclo-oxygenase enzyme (Prashar et al. 1998). The treatment of mice with water and alcoholic extracts of *O. sanctum* exhibited anti-tumor properties with reduction in size of tumors (Adhvaryu et al. 2008). The immunomodulatory potential of *O. sanctum* leave extract may be due to the existence of active constituents like ursolic acid, salrigenin, caryophyllen, oleanolic acid, eugenol, and methyleugenol (Mukherjee et al. 2005). In clinical trials in bovines, the aqueous extract of *O. sanctum* leaf has revealed immunomodulatory activity with improved number of granular and agranular white blood cells with reduction in bacterial count (Mukherjee et al. 2005). In *S. typhimurium* infection, the treatment of *O. sanctum* extract had lessen various cytokines levels such as TNF-α, IL-2, and IFN-γ along with the activation of macrophages that inhibits the bacterial growth (Goel et al. 2010). The GABAergic-mediated immune responses were regulated by the oil extracted from *O. sanctum* seeds (Vaghasiya et al. 2010). The treatment of fish with *O. sanctum* leaves extract has revealed immuno-stimulatory activity by inducing innate immunity and increasing phagocytes, red blood cells (RBCs), white blood cells (WBCs), and lymphocytes numbers in fish. Therefore, it could be utilized to boost the immune power for the treatment of several parasitic infections (Nahak and Sahu 2014). In another research investigation, treatment of *L. Rohita* (species of fish) with *O. sanctum* leaf extract has demonstrated a general immunological reaction against *A. hydrophila* infection that showed increase in hematological, biochemical, lysosomal, and total immunoglobulins activity (Das et al. 2015).

42.3 Conclusion

Immunomodulation by using herbal medicinal plants for the treatment of variety of diseases can provide a great substitute to conventional chemotherapy, especially when host defense immune system is impaired in situations like autoimmune disorders. The light has been thrown on numerous plant-based phytoconstituents since from many years for their immunomodulation property. Variety of disease conditions can be cured by herbal immunomodulators instead of chemotherapeutic agents. The invention and identification of immunomodulatory agents of herbal source have the ability to neutralize the adverse effects and high price of conventional drugs. This book chapter focuses on the importance of herbal plants as creators of immunomodulatory agents with possible applicability in preclinical and clinical health studies. There is a great potential for the discovery of more specific immunomodulators which mimic or antagonize the biological effects of cytokines and interleukins, and the refinement of assays for these mediators will create specific and sensitive screens.

Acknowledgments We are immensely grateful to Dr. K. R. Mahadik, Principal, Poona college of Pharmacy, Bharati Vidyapeeth (Deemed to be) University, Pune for his constant support and motivation throughout the manuscript.

References

Abdollahi E, Momtazi AA, Johnston TP et al (2018) Therapeutic effects of curcumin in inflammatory and immune-mediated diseases: a nature-made jack-of-all-trades? J Cell Physiol 233:830–848

Adhvaryu MR, Reddy NM, Vakharia BC (2008) Prevention of hepatotoxicity due to anti tuberculosis treatment: a novel integrative approach. World J Gastroenterol 14:4753–4762

Agarwal R, Diwanay S, Patki P et al (1999) Studies on immunomodulatory activity of *Withania somnifera* (Ashwagandha) extracts in experimental immune inflammation. J Ethnopharmacol 67:27–35

Agarwal SS, Singh VK (1999) Immunomodulators: a review of studies on Indian medicinal plants and synthetic peptides. Part I: medicinal plants. Proc Indian Natl Sci Acad 65:179–204

Aggarwal R, Diwanay S, Patki P et al (2012) Studies on immunomodulatory activity of *Withania somnifera* (Ashwagandha). J Appl Pharm Sci 2:170–176
Alsuhaibani S, Khan MA (2017) Immune-stimulatory and therapeutic activity of *Tinospora cordifolia*: double-edged sword against salmonellosis. J Immunol Res 1787803:1–9
Baral R, Chattopadhyay U (2004) Neem (*Azadirachta indica*) leaf mediated immune activation causes prophylactic growth inhibition of murine Ehrlich carcinoma and B16 melanoma. Int Immunopharmacol 4:355–366
Baxter D (2007) Active and passive immunity, vaccine types, excipients and licensing. Occup Med 57:552–556
Belska NV, Guriev AM, Danilets MG et al (2006) Enhancement of immune responses to neem leaf extracts (*Azadirachta indica*) correlates with antineoplastic activity in BALB/c-mice. In Vivo 20:247–251
Bhandari PR (2012) Curry leaf (*Murraya koenigii*) or cure leaf: review of its curative properties. J Med Nutr Nutraceuticals 1:92
Bhanwase AS, Alagawadi KR (2016) Antioxidant and immunomodulatory activity of hydroalcoholic extract and its fractions of leave of *Ficus benghalensis* Linn. Pharm Res 8:50–55
Bhattacharjee S (2008) Handbook of medicinal plants, vol 118. Pointer Publishers, Jaipur
Bose A, Chakraborty K, Sarkar K et al (2009) Neem leaf glycoprotein induces perforin-mediated tumor cell killing by T and NK cells through differential regulation of IFNγ signaling. J Immunother 32:42–53
Chakraborty K, Bose A, Pal S et al (2008) Neem leaf glycoprotein restores the impaired chemotactic activity of peripheral blood mononuclear cells from head and neck squamous cell carcinoma patients by maintaining CXCR3/CXCL10 balance. Int Immunopharmacol 8:330–340
Chandrasekaran CV, Kannan S, Gururaja JRE et al (2013) Immune stimulatory and anti-inflammatory activities of *Curcuma longa* extract and its polysaccharide fraction. Pharm Res 5:71–79
Chandrashekar PM, Prashanth KVH, Venkatesh YP (2011) Isolation, structural elucidation and immunomodulatory activity of fructans from aged garlic extract. Phytochemistry 72:255–264
Cipriani B, Borsellino G, Knowles H et al (2001) Curcumin inhibits activation of Vγ9Vδ2 T cells by phosphoantigens and induces apoptosis involving apoptosis-inducing factor and large scale DNA fragmentation. J Immunol 167:3454–3462
Clement F, Pramod SN, Venkatesh YP (2010) Identity of the immunomodulatory proteins from garlic (*Allium sativum*) with the major garlic lectins or agglutinins. Int Immunopharmacol 10:316–324
Das R, Raman RP, Saha H et al (2015) Effect of *Ocimum sanctum* Linn (Tulsi) extract on the immunity and survival of *Labeo rohita* (Hamilton) infected with *Aeromonas hydrophila*. Aquac Res 46:1111–1121
Dash P, Yadav S, Sahoo PK (2014) Immunoadjuvant effect of garlic (*Allium sativum*) mineral oil suspension on immunity and resistance to *Aeromonas hydrophila* infection in rohu, *Labeo rohita*. Int Aquat Res 6:167–173
Davis L, Kuttan G (2000) Immunomodulatory activity of *Withania somnifera*. J Ethnopharmacol 71:193–200
Deguchi A (2015) Curcumin targets in inflammation and cancer. Endocr Metab Immune Disord Drug Targets 15:88–96
Durrani FR, Chand N, Jan M et al (2008) Immunomodulatory and growth promoting effects of neem leaves infusion in broiler chicks. Sarhad J Agric 24:655–659
Dwivedi VP, Bhattacharya D, Yadav V et al (2017) The phytochemical Bergenin enhances T helper-1 responses and anti- mycobacterial immunity by activating the MAP kinase pathway in macrophages. Front Cell Infect Microbiol 7:149–158
Ebrahimi M, Hassan ZM, Mostafaie A et al (2013) Purified protein fraction of garlic extract modulates cellular immune response against breast transplanted tumors in BALB/c mice model. Cell J 15:65–75

El-Sheikh ALK (2008) Renal transport and drug interactions of immunosuppressants [thesis]. Radbound University, Nijmegen, p 62

Fahey AJ, Adrian RR, Constantinescu CS (2007) Curcumin modulation of IFN-beta and IL-12 signalling and cytokine induction in human T cells. J Cell Mol Med 11:1129–1137

Feng Y, Zhu X, Wang Q et al (2012) Allicin enhances host pro-inflammatory immune responses and protects against acute murine malaria infection. Malar J 11:268

Ford MS, Roach SS (2009) Introductory clinical pharmacology, 27th edn. Lippincott Williams and Wilkins, Philadelphia, PA, pp 567–568

Gabhe SY, Tatke PA, Khan TA (2006) Evaluation of the immunomodulatory activity of the methanol extracts of *Ficus benghalensis* roots in rats. Indian J Pharmacol 38:271–275

Gharavi MJ, Nobakht M, Khademvatan SH et al (2011) The effect of garlic extract on expression of INFγ and inos genes in macrophages infected with Leishmania major. Iranian J Parasitol 6:74

Ghate NB, Hazra B, Sarkar R et al (2014) Heartwood extract of *Acacia catechu* induces apoptosis in human breast carcinoma by altering bax/bcl-2 ratio. Pharmacogn Mag 10:27–33

Goel A, Kumar S, Bhatia AK (2010) Effect of *Ocimum sanctum* on the development of protective immunity against *Salmonella typhimurium* infection through cytokines. Asia Pac J Trop Med 3:682–686

He XG, Lin LZ, Lian LZ (1998) Liquid chromatography_electrospray mass spectrometric analysis of curcuminoids and sesquiterpenoids in turmeric (*Curcuma longa*). J Chromatogr A 818:127–132

Hodge G, Hodge S, Han P (2002) *Allium sativum* (garlic) suppresses leukocyte inflammatory cytokine production in vitro: potential therapeutic use in the treatment of inflammatory bowel disease. Cytometry A 48:209–215

Ismail S, Asad M (2009) Immunomodulatory activity of Acacia catechu. Indian J Physiol Pharmacol 53:25–33

Jagetia GC, Aggarwal BB (2007) Spicing up of the immune system by curcumin. J Clin Immunol 27:19–35

Jandu JJ, Neto M, Zagmignan RN et al (2017) Targeting the immune system with plant lectins to combat microbial infections. Front Pharmacol 8:671

Jayaprakasam B, Zhang Y, Seeram NP et al (2003) Growth inhibition of human tumor cell lines by withanolides from *Withania somnifera* leaves. Life Sci 74:125–132

Kalia A (2011) Textbook of industrial Pharmacognosy. CBS Publishers & Distributors Pvt, New Delhi

Khan T, Tatke P, Gabhe S (2008) Immunological studies on the aerial roots of the Indian banyan. Indian J Pharm Sci 70:287–291

Kim SR, Jung YR, An HJ et al (2013) Antiwrinkle and anti-inflammatory effects of active garlic components and the inhibition of MMPs via NF-κB signaling. PLoS One 8:e73877

Kumar D, Arya V, Kaur R et al (2012) A review of immunomodulators in the Indian traditional health care system. J Microbiol Immunol Infect 45:165–184

Kumar J, Mitra MD, Hussain A et al (2019) Exploration of immunomodulatory and protective effect of *Withania somnifera* on trace metal oxide (zinc oxide nanoparticles) induced toxicity in Balb/c mice. Mol Biol Rep 46:2447–2459

Kumar S, Suresh PK, Vijayababu MR et al (2006) Anti-cancer effects of ethanolic neem leaf extract on prostate cancer cell line (PC-3). J Ethnopharmacol 105:246–250

Kyo E, Uda N, Suzuki A et al (1998) Immunomodulation and antitumor activities of aged garlic extract. Phytomedicine 5:259–267

Leyon PV, Lini CC, Kuttan G (2005) Inhibitory effect of *Boerhaavia diffusa* on experimental metastasis by B16F10 melanoma in C57BL/6 mice. Life Sci 76:1339–1349

Manu KA, Kuttan G (2007) Effect of punarnavine, an alkaloid from *Boerhaavia diffusa*, on cell-mediated immune responses and TIMP-1 in B16F-10 metastatic melanoma bearing mice. Immunopharmacol Immunotoxicol 29:569–586

Manu KA, Kuttan G (2009) Anti-metastatic potential of Punarnavine, an alkaloid from *Boerhaavia diffusa* Linn. Immunobiology 214:245–255

Marquez N, Sancho R, Bedoya LM et al (2005) Mesuol, a natural occurring 4-phenylcoumarin, inhibits HIV-1 replication by targeting the NF-κB pathway. Antivir Res 66:137–145
Mathew S, Kuttan G (1999) Immunomodulatory and anti-tumour activities of *Tinospora cordifolia*. Fitoterapia 70:35–43
Mehrotra S, Mishra KP, Maurya R et al (2003) Anticellular and immunosuppressive properties of ethanolic extract of *Acorus calamus* rhizome. Int Immunopharmacol 3:53–61
Mishra LC, Singh BB, Dagenais S (2000) Scientific basis for the therapeutic use of *Withania somnifera* (ashwagandha): a review. Altern Med Rev 5:334–346
Mitra E, Ghosh AK, Ghosh D et al (2012) Protective effect of aqueous curry leaf (*Murraya koenigii*) extract against cadmium-induced oxidative stress in rat heart. Food Chem Toxicol 50:1340–1353
Modi M, Dezzutti CS, Kulshreshtha S et al (2013) Extracts from *Acacia catechu* suppress HIV-1 replication by inhibiting the activities of the viral protease and tat. Virol J 10:309
Monga J, Chauhan CS, Sharma M (2011) Human epithelial carcinoma cytotoxicity and inhibition of DMBA/TPA induced squamous cell carcinoma in Balb/c mice by *Acacia catechu* heartwood. J Pharm Pharm 63:1470–1482
Moradali MF, Mostafavi H, Ghods S et al (2007) Immunomodulating and anticancer agents in the realm of macromycetes fungi (macrofungi). Int Immunopharmacol 7:701–724
More P, Pai K (2011) Immunomodulatory effects of *Tinospora cordifolia* (Guduchi) on macrophage activation. Biol Med 3:134–140
Mukherjee PK, Nema NK, Venkatesh P et al (2012) Changing scenario for promotion and development of Ayurveda-way forward. J Ethnopharmacol 143:424–434
Mukherjee R, Dash PK, Ram GC (2005) Immunotherapeutic potential of *Ocimum sanctum* (L) in bovine subclinical mastitis. Res Vet Sci 79:37–43
Mungantiwar AA, Nair AM, Shinde UA et al (1999) Studies on the immuno-modulatory effects of *Boerhaavia diffusa* alkaloidal fraction. J Ethnopharmacol 65:125–131
Nadumane V, Nair S (2011) Evaluation of the anticancer and cytotoxic potentials of *Acacia catechu* extracts in vitro. J Nat Pharm 2:190
Nahak G, Sahu R (2014) Immunostimulatory effects of *Ocimum sanctum* Linn. Leaf extracts in *Clarias batrachus* Linn. Asian J Pharm Clin Res 7:15
Nair PR, Melnick SJ, Ramachandran R et al (2006) Mechanism of macrophage activation by (1, 4)-α-D-glucan isolated from *Tinospora cordifolia*. Int Immunopharmacol 6:1815–1824
Oh SR, Kinjo J, Shii Y et al (2000) Effects of triterpenoids from *Pueraria lobata* on immunohemolysis: beta-D-glucuronic acid plays an active role in anti-complementary activity in vitro. Planta Med 66:506–510
Pandey R, Maurya R, Singh G et al (2005) Immunosuppressive properties of flavonoids isolated from *Boerhaavia diffusa* Linn. Int Immunopharmacol 5:541–553
Parkin J, Cohen B (2001) An overview of the immune system. Lancet 357:1777–1789
Plummer SM, Holloway KA, Manson MM et al (1999) Inhibition of cyclo-oxygenase 2 expression in colon cells by the chemopreventive agent curcumin involves inhibition of NF-κB activation via the NIK/IKK signalling complex. Oncogene 18:6013–6020
Prashar R, Kumar A, Hewer A et al (1998) Inhibition by an extract of *Ocimum sanctum* of DNA-binding activity of 7,12-dimethylbenz [a] anthracene in rat hepatocytes in vitro. Cancer Lett 128:155–160
Puri A, Saxena R, Saxena RP et al (1994) Immunostimulant activity of Nyctanthes arbor-tristis L. J Ethnopharmacol 42:31–37
Quintero-Fabian S, Ortuno-Sahagun D, Vazquez-Carrera M et al (2013) Alliin, a garlic (*Allium sativum*) compound, prevents LPS-induced inflammation in 3T3-L1 adipocytes. Mediat Inflamm 381815:1–11
Rizvi TF, Razauddin M, Rahman SR et al (2016) Immunomodulatory effect of Ashwagandha against doxorubicin toxicity. Eur J Pharma Med Res 3:463–467

Sadekar RD, Kolte AY, Barmase BS et al (1998) Immunopotentiating effects of *Azadirachta indica* (Neem) dry leaves powder in broilers, naturally infected with IBD virus. Indian J Exp Biol 36:1151–1153

Santhosh VK, Navartnam V (2013) Neem prehistory to contemporary medicinal uses to humankind. Asian Pac J Trop Biomed 3:505–519

Sathaye S, Bagul Y, Gupta S (2011) Hepatoprotective effects of aqueous leaf extract and crude isolates of *Murraya koenigii* against in vitro ethanol induced hepatotoxicity model. Exp Toxicol Pathol 63:587–591

Shah AS, Wakade AS, Juvekar AR (2008) Immunomodulatory activity of methanolic extract of *Murraya koenigii* (L) Spreng leaves. Indian J Exp Biol 46:505–509

Sharma U, Bala M, Kumar N et al (2012) Immunomodulatory active compounds from *Tinospora cordifolia*. J Ethnopharmacol 141:918–926

Shimada J, Moon SK, Lee HY et al (2008) Lysozyme M deficiency leads to an increased susceptibility to *Streptococcus pneumoniae* induced otitis media. BMC Infect Dis 8:134

Shin IS, Hong J, Jeon CM et al (2013) Diallyl-disulfide, an organosulfur compound of garlic, attenuates airway inflammation via activation of the Nrf-2/HO-1 pathway and NF-kappaB suppression. Food Chem Toxicol 62:506–513

Singh B, Saxena AK, Chandan BK et al (2001) Adaptogenic activity of a novel, withanolide-free aqueous fraction from the roots of *Withania somnifera* dun. Phytother Res 15:311–318

Song L, Xiong D, Hu M et al (2016) Immunopotentiation of different adjuvants on humoral and cellular immune responses induced by HA1-2 subunit vaccines of H7N9 influenza in mice. PLoS One 11:e0150678

Spellberg B, Edwards JE Jr (2001) Type 1/type 2 immunity in infectious diseases. Clin Infect Dis 32:76–102

Sreeja S, Sreeja S (2009) An in vitro study on anti-proliferative and anti-estrogenic effects of *Boerhaavia diffusa* L. extracts. J Ethnopharmacol 126:221–225

Srivastava R, Saluja D, Chopra M (2005) Isolation and screening of anticancer metabolites from *Boerhaavia diffusa*. Indian J Med Res 151:S19

Sudhir PK, Mishra D, Ghosh P (2010) Biological action and medicinal properties of various constituent of *Azadirachta indica*. Ann Biol Res 1:24–34

Sumanth M, Mustafa SS (2007) Anti-stress, adoptogenic and immuno-potentiating activity roots of *Boerhaavia diffusa* in mice. Int J Pharmacol 3:416–420

Sunil MA, Sunitha VS, Radhakrishnan EK et al (2018) Immunomodulatory activities of Acacia catechu, a traditional thirst quencher of South India. J Ayurveda Integr Med:1–8

Trivedi CP, Shinde S, Sharma RC (1969) Preliminary phytochemical and pharmacological studies on *Ficus racemosa* (Gular). Indian J Med Res 57:1070–1074

Tsang JYS, Li D, Ho D et al (2011) Novel immunomodulatory effects of adiponectin on dendritic cell functions. Int Immunopharmacol 11:604–609

Tzianabos AO (2000) Polysaccharide immunomodulators as therapeutic agents: structural aspects and biologic function. Clin Microbiol Rev 13: 523-533.

Vaghasiya J, Datani M, Nandkumar K et al (2010) Comparative evaluation of alcoholic and aqueous extracts of *Ocimum sanctum* for immunomodulatory activity. Int J Pharm Biol Res 1:25–29

Verma VK, Rani KV, Sehgal N et al (2012) Immunostimulatory response induced by supplementation of Ficus benghalensis root powder, in the artificial feed the Indian freshwater murrel, *Channa punctatus*. Fish Shellfish Immunol 33:590–596

Vesely MD, Kershaw MH, Schreiber RD et al (2011) Natural innate and adaptive immunity to cancer. Annu Rev Immunol 29:235–271

Wang HX, Ng TB (1999) Natural products with hypoglycemic, hypotensive, hypocholesterolemic antiatherosclerotic and antithrombotic activities. Life Sci 65:2663–2677

Xia X, Pan Y, Zhang WY et al (2006) Ethanolic extracts from *Curcuma longa* attenuates behavioral, immune, and neuroendocrine alterations in a rat chronic mild stress model. Biol Pharm Bull 29:938–944

Yeap SK, Abu N, Mohamad NE et al (2015) Chemopreventive and immunomodulatory effects of Murraya koenigii aqueous extract on 4T1 breast cancer cell-challenged mice. BMC Complement Altern Med 4:306
Yi X, Feng F, Xiang Z et al (2005) The effects of allitridin on the expression of transcription factors T-bet and GATA-3 in mice infected by murine cytomegalovirus. J Med Food 8:332–336
Ziauddin M, Phansalkar N, Patki P et al (1996) Studies on the immunomodulatory effects of Ashwagandha. J Ethnopharmacol 50:69–76

Combined Effects of Plant Extracts on Ovarian Cell Functioning

43

Sushmita Nath, Attila Kadasi, Roland Grossmann,
Adriana Kolesarova, Anupam Das Talukdar, Alexander V. Sirotkin,
and Manabendra Dutta Choudhury

Abstract

Ovarian granulosa cell (GC) functions decide the fate of an ovarian follicle to establish fecundity and regular ovarian cycle. Literature depicts the combined effects of two or more herbal components are likely to be more prominent than the effect of a single ingredient. Based on this, the current study was designed to evaluate the combined effects of aqueous extract (RA) of *Ricinus communis* L. and curcumin (Cur) of *Curcuma longa* L., ethnopharmacologically reputed plants and porcine ovarian granulosa cell functions, in vitro. Cytotoxicity was determined by using real-time cell analyser. Combination ofAQ+Cur enhanced cyclin-B1 accumulation at low concentrations and significantly upregulated caspase-3 accumulation, evaluated by immunocytochemistry, and significantly

S. Nath (✉) · A. D. Talukdar · M. D. Choudhury
Department of Life Science and Bioinformatics, Assam University, Silchar, India
e-mail: sushmitanath84@gmail.com

A. Kadasi
Department of Animal Genetics and Reproduction, Research Institute of Animal Production Nitra, Luzianky, Slovak Republic

Department of Animal Physiology, Slovak University of Agriculture, Nitra, Slovak Republic

R. Grossmann
Department of Functional Genomics and Bioregulation, Institute of Animal Science, Federal Research Centre, Neustadt, Germany

A. Kolesarova
Department of Animal Physiology, Slovak University of Agriculture, Nitra, Slovak Republic

A. V. Sirotkin
Department of Animal Genetics and Reproduction, Research Institute of Animal Production Nitra, Luzianky, Slovak Republic

Department of Zoology and Anthropology, Constantine the Philosopher University, Nitra, Slovak Republic

S. C. Mandal et al. (eds.), *Evidence Based Validation of Traditional Medicines*,
https://doi.org/10.1007/978-981-15-8127-4_43

decreased progesterone (P4) and inhibited testosterone (T) release, as determined by the enzyme immunoassay. The resulted observations suggested the possible combined effect of AQ and Cur on GC cell functions via steroid output.

Keywords

Granulosa cell functioning · Steroid output · *Ricinus communis* · *Curcuma longa*

43.1 Introduction

For the last five decades, research on contraception is at the forefront, still there is no groundbreaking discovery of perfect contraception (Brache et al. 2012; Sitruk-Ware and Nath 2013). Till date, no plant-based contraceptive is available commercially (Singh et al. 2012), even though herbal contraceptives are described more safe and effective, yet less researched (Khillare and Shrivastav 2003). In spite of diverse contraceptive efficacy of medicinal plants, still no plant-based contraceptive is available in the market to practically employ herbs for fertility regulation (Singh et al. 2012). Emphasis is given to develop safe non-hormonal alternatives from biologically active substances of traditional medicinal plants. Most of the synthetic-hormonal contraceptive, structurally related either to progestogen or androgen acts potently on ovulation but with side-effects (Sitruk-Ware and Nath 2011). To curve these adverse side-effects, individual herbal alternatives acting on the aforesaid targets are already reported. Supported literature reveals that out of various medicinal plants *Ricinus communis* L. (R) and curcumin (Cur), folklore potent female contraceptive (Ammon and Martin 1991; Eigner and Scholz 1999; Ross 2001) has individual effect on both progestogen and androgen (Okwuasaba et al. 1991; Salhab et al. 1999; Kadasi et al. 2012).

Folklores use leave, seeds, roots and stem bark of R for fertility regulation (Kalita et al. 2011; Saharia and Sarma 2011). Seeds alter ovarian cyclicity and have anticonceptive, anti-implantation and abortifacient activities in rodents (Okwuasaba et al. 1991; Salhab et al. 1999). It decreases chances of pregnancy (Salhab et al. 1997; Salhab et al. 1998), induces long-term contraception (Meena and Rao 2010) and inhibits follicular development and ovulation by acting directly on ovary (Goncim et al. 2010). But scientific studies did not pay attention to R stem bark for contraceptive efficacy. Whereas *Nyishi* (Daffla) tribes of Arunachal Pradesh (India) use young twigs of the plant as locally applied abortifacient (Srivastava and Nyishi 2010), but its mechanism remains unknown. In our recent study for the first time, we have documented the potential contraceptive efficacy of aqueous extract of R stem bark on human spermatozoa (Nath et al. 2013).

On the other hand, Cur, the main active constituent of the rhizome *Curcuma longa* L., is a traditional contraceptive agent (Thakur et al. 2009, Niazi et al. 2010). Cur has contraceptive, anti-implantational, antiovulatory, antigonadotropic, antioestrogenic and antifertility effect (Purohit and Bhagat 2004; Ghosh et al. 2011; Yadav and Jain 2010) in rodents. Cur is also established as a potent human spermatozoa inhibitor (Rithaporn et al. 2003), lowers the androgen level (Purohit

1999) and inhibits testosterone-5-reductase (Liao et al. 2001). Its intravaginal use acts as contraceptive (Naz 2011). Kadasi and co-workers (Kadasi et al. 2012) demonstrated Cur's fertility regulatory activity on porcine ovarian granulosa cell functions – follicular development and steroidogenesis, as reproductive regulators affect fecundity by targeting ovarian cell steroidogenesis, proliferation and apoptosis. As ovarian progestogens, androgens and oestrogens are known regulators of ovarian folliculogenesis and oogenesis (Conneely et al. 2007; Sirotkin 2011; Qin et al. 2011).

It was cited that the combined effect of two or more herbal compounds likely to be more rigorous than their cumulative effect of single ingredient (Keith et al. 2005; Parasramka and Gupta 2012). Therefore, combined effect of R and Cur may have more rigorous efficacy towards contraception, as in recent times the concept of new combinations is getting priority in pharmacology of medicinal plants research. Moreover, their combined effects on various ovarian parameters are yet to be experimented. With this hypothesis, the current experiments were designed to investigate the direct influence of aqueous extract of R stem bark (RA) and RA in combination with Cur on ovarian granulosa cell functions (proliferation, apoptosis) and steroidogenesis. Furthermore, we discussed the effect of RA on Cur's activity by comparing present and reported data.

43.2 Materials and Methods

43.2.1 Preparation and Storage of Plant Extract

Fresh plant (voucher specimen number 2502, Assam University Herbarium, Department of Life Science and Bioinformatics) was collected from Cachar, India. After shade drying, the sectioned stem barks were powdered, and then aqueous extract (RA) was prepared in double distilled water (Nath et al. 2013).

43.2.2 Isolation, Preparation and Culture of Granulosa Cells from Ovaries

Granulosa cells were collected from the ovaries of non-cycling pubertal Slovakian White gilts (*Sus scrofa domestica* L.) without any visible reproductive abnormalities, from local abattoir of Animal Production Research Centre Nitra in accordance with Slovak Government guidelines. Ovaries were transported to the laboratory in containers of 4 °C. After washing them in sterile 0.9% physiological saline, follicular fluid containing granulosa cells (GC) were aspirated from ovarian follicles, centrifuged at 200xg for 10 min and washed in sterile DMEM/F12 1:1 medium and then resuspended in DMEM/F12 1:1 medium supplemented with 10% FCS and 1% antibiotic-antimycotic solution. Cells at a concentration of 10^6 cells/mL were counted with the help of haemocytometer (Sirotkin et al. 2008). GC (1×10^6 cells/mL) seeded in 24-welled culture plates (Becton Dickinson Labware, Bedford, MA,

2 mL/well) and Lab-Tek 16-welled chamber slides (Nunc Inc., International, Naperville, USA, 200 μL/well). These plates and slides were incubated at 37.5 °C under 5% CO_2 in humidified air until a 60–75% confluent adhering monolayer was formed. Bottom-attached cells were supplemented with renewed medium of previous composition.

43.2.3 Cytotoxicity Evaluation Using Real-Time Cell Analyser (RTCA)

The xCELLigence real-time system (Roche Applied Science, Germany) (Bird and Kirstein 2009; Ozturk et al. 2014) was used to evaluate the real-time cytotoxicity of RA and Cur on GC. The isolated GC were seeded as 50,000 cells/well into RTCA E-Plates 16, cultured in sterile medium of above-mentioned composition and were monitored by using xCELLigence DP real-time system of RTCA (by impedance measuring and a voltage of 20 mV) at 50 mins intervals in incubation for 24 ± 2 h at 37.5 °C under 5% $CO_{2.}$. Post 24 h, medium was replaced, experimental cells were treated with RA (1, 10 and 100 μg/mL) and Cur (1, 10 and 100 μg/mL) separately and again monitored for 200 ± 10 h. At the end of the experiment, the impedance was expressed as cell index (CI); the cell viability parameter and the resulted status were determined by using the RTCA software 1.2.

43.2.4 Processing Cultured Granulosa Cells

Experimental GC in 24-well plates were cultured with RA (0.01, 0.1, 1, 10 and 100 μg/mL) and RA + Cur (0.01, 0.1, 1, 10 and 100 μg/mL) separately for 24 h. Whereas, the in 16-welled plates, chamber slides were cultured with RA (0.01, 0.1, 1, 10, 100 μg/mL) and RA + Cur (0.01, 0.1, 1, 10, 100 μg/mL) in 4:1. After 48 h in culture, the media from the 24-well plates were aspirated and frozen at −70 °C for further immunoassay for steroidogenesis studies. Meanwhile, medium from chamber slides removed, and cells were washed in ice-cold PBS (pH 7.5), fixed in paraformaldehyde (4% in PBS, pH 7.2-7.4; 60mins) and kept at 4 °C for further immunocytochemical analysis.

43.2.5 Immunocytochemistry

Cyclin B1 and caspase-3 (intracellular signalling molecules) were detected in GC plated on chamber slides, using immunocytochemistry (Osborn and Isenberg 1994). As per manufacturer's instruction the ImmunoCruz Staining System and primary mouse monoclonal antibodies against cyclinB1 and caspase3 (dilution 1:500). Visualisation of the primary antibody binding sites was achieved with a corresponding secondary polyclonal antibody against mouse IgGs, labelled with HRP (dilution 1:1000) and DAB (10%). Chamber slides were stained with

HRP/DAB, coated with Glycergel and mounted with a coverslip, and then the cells containing specific immunoreactivity were counted under the light microscope.

43.2.6 Enzyme Immunoassay (EIA)

Concentrations of progesterone (P_4) and testosterone (T) were determined by using EIA, using antisera against steroids produced by the Institute of Animal Science, Neustadt, Germany. P_4 concentrations were measured as per standard protocol (Prakash and Meyer 1987). Antiserum cross-reacted ($\leq$0.1%) with dihydrotestosterone, testosterone and 17β-hydroxyprogesterone, and the EIA sensitivity was found to be 12.5 pg/mL. T concentrations were assayed according to Münster (Munster 1989); the sensitivity of the EIA was found to be 10 pg/mL. The cross-reactivity of T antiserum was with dihydrotestosterone ($\leq$96%), androstenedione ($\leq$3%), progesterone ($\leq$0.01%), cortisol ($\leq$0.02%) and corticosterone ($\leq$0.001%).

43.2.7 Statistical Analysis

Four culture wells represented each experimental group. Presented data are the mean values collected from three separate experiments executed on separate days using separate pools of ovaries. At least 100 $\times$ 10 (totally $\geq$1000 cells) cells per chamber of chamber slides were calculated to obtain the proportion of cells containing specific immunoreactivity. Rates of hormonal secretion (determined by EIA) were calculated from per 10^6 cells per day. Wilcoxon-Mann Whitney multiple-range test (Sigma Plot 11.0 software, Systat Software, GmbH, Erkhart, Germany) has been used, and differences from the control at $p < 0.05$ were considered as significant.

43.3 Results

43.3.1 Cytotoxicity Evaluation Using Real-Time Cell Analyser (RTCA)

Cell index (CI) data (Fig. 43.1) of 200 $\pm$ 10 h of treatment with RA and Cur at different dose levels obtained from RTCA analyser showed the viability of GC in culture medium. Cells at dose level of 1 μg/mL (for both RA and Cur) are having more CI than control. On the other hand, CI at dose level 10 μg/mL of RA is almost same as that of control. Though, at dose level 10 μg/mL (Cur) and 100 μg/mL (RA) CI value is lower than control, but the cells were still viable. On the contrary CI value at dose concentration 100 μg/mL (Cur) appeared significantly lower in comparison to control.

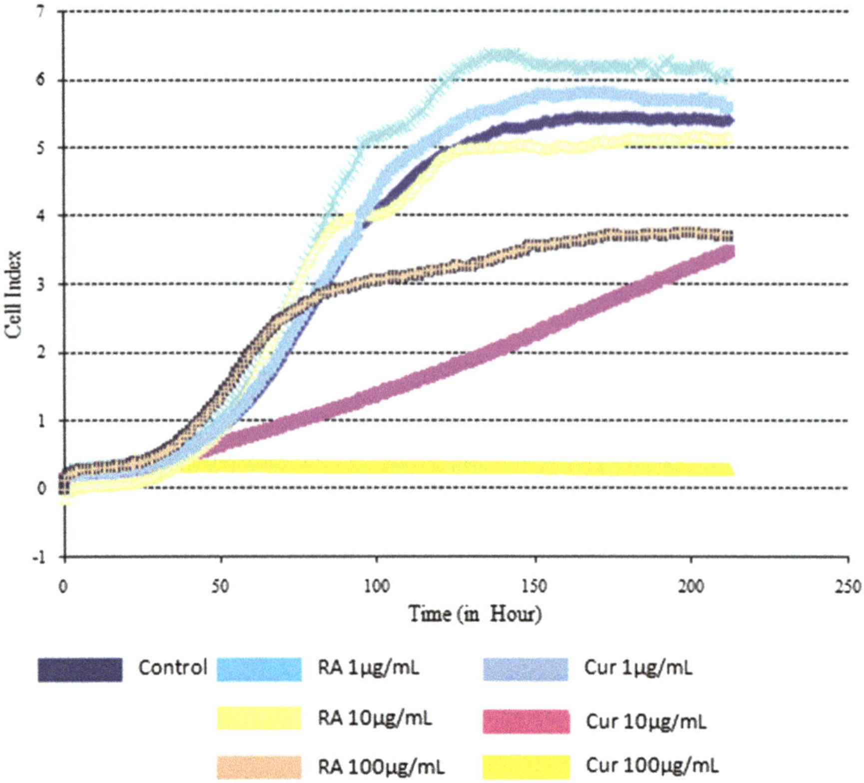

Fig. 43.1 RTCA profiling and plot of cell index (CI) data of RA and Cur on porcine ovarian granulosa cells for entire 200 ± 10 h of the experiments from E-Plates

43.3.2 Immunocytochemistry and Enzyme Immunoassay (EIA)

Isolated porcine GC was viable in culture medium, secreted steroid hormones and accumulated intracellular signalling substances. Immunocytochemistry revealed the accumulation of cyclin B1 and caspase-3 within the GC, while EIA demonstrated the release of P_4 and T. These parameters were influenced by the treatments given at various concentrations.

43.3.2.1 Effect of RA on Proliferation (Cyclin B1) and Apoptosis (Caspase-3) Markers

RA at 0.1 µg/mL enhanced the percentage of cyclin B1 (Fig. 43.2) accumulation significantly ($p < 0.05$) in comparison to the control. At dose level 1 to 100 µg/mL, RA inhibited the cyclin B1 expression by causing apoptosis instead of proliferation, whereas, at dose 0.01, 0.1 and 1 µg/mL, RA enhanced the caspase-3 accumulation, but at dose level 10 and 100 µg/mL, there was 100% cell death (Fig. 43.3).

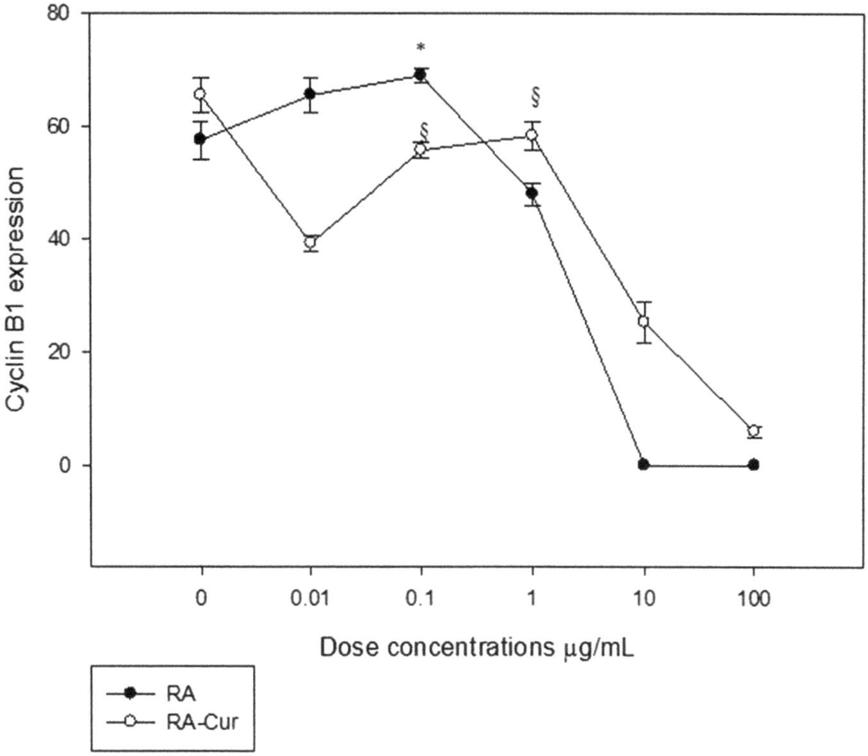

Fig. 43.2 Effect of RA and RA + Cur on cyclinB1 accumulation. Values are means ± SEM. * (RA), § (RA + Cur) denote values significantly different ($p < 0.05$) from control group. Immunocytochemistry

43.3.2.2 Combined Effect of RA and Cur on Proliferation (Cyclin B1) and Apoptosis (Caspase-3) Markers

Application of RA alone significantly enhanced and inhibited both cyclin B1 and caspase-3 accumulations in a dose-dependent manner. RA + Cur showed significant influence on cyclin B1 expression and rigorous effect on caspase-3 accumulation. Combination of RA + Cur (Fig. 43.2) upregulated cyclin B1 accumulation at doses 0.1 and 1 μg/mL significantly ($p < 0.05$). Whereas dose concentrations 0.01, 10 and 100 μg/mL downregulated the cyclin B1 accumulation. But when this combination is tested on caspase-3 accumulation (Fig. 43.3), it emerged significant ($p < 0.05$) upregulation of apoptosis by causing 100% cell death in all the dose levels (0.01–100 μg/mL).

43.3.2.3 Effect of RA on Steroidogenesis

RA decreased (0.1 to100 μg/mL) the P_4 release level in comparison to the control (Fig. 43.4). RA inhibited (at dose concentration 10 μg/mL) as well as decreased (at dose concentrations 0.1, 1 and 100 μg/mL) the T secretion in comparison to

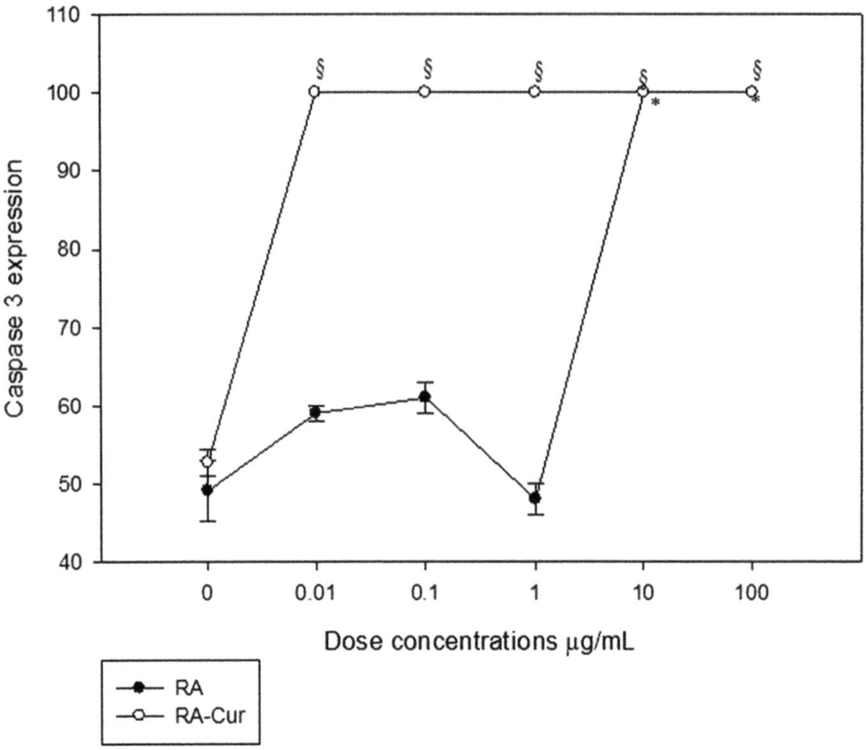

Fig. 43.3 Effect of RA and RA + Cur on caspase-3 accumulation. Values are means ± SEM. * (RA), §(RA + Cur) denote values significantly different (*p* < *0.05*) from control group. Immunocytochemistry

control (Fig. 43.5) significantly ($p < 0.05$), while lower dose 0.01 μg/mL did not affect it.

43.3.2.4 Combined Effect of RA and Cur on Steroidogenesis

In this series of experiment, RA + Cur significantly ($p < 0.05$) decreased (0.01, 0.1, 1 and 10 μg/mL) and inhibited (100 μg/mL) P_4 release level (Fig. 43.4). On the other hand, this RA + Cur combination has totally inhibited the T release (Fig. 43.5) in all the five dose levels (0.01–100 μg/mL) in comparison to control suggesting cumulative effect.

43.4 Discussion

In RTCA analysis the cells were viable for 200 ± 10 h under treatment condition at different dose concentrations of RA and Cur. The real-time CI curve suggests that these plant materials (at assigned doses) can be used on porcine ovarian granulosa cell in vitro. Formation of adherent GC monolayer, release of P_4 and T and

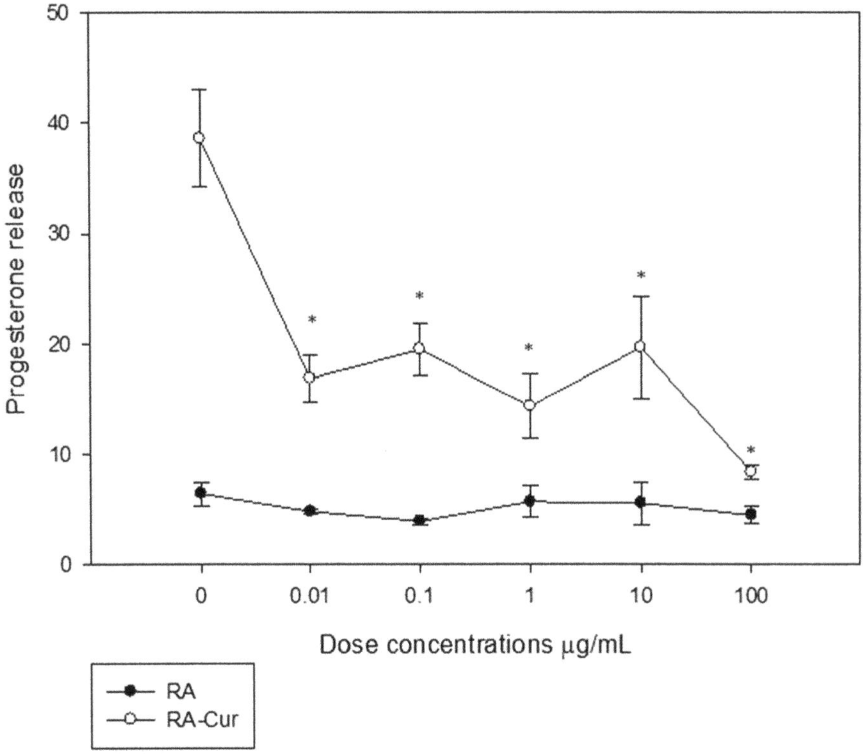

Fig. 43.4 Effect of RA and RA + Cur on progesterone (P_4) release. * (RA), § (RA + Cur) denotes significant ($p < 0.05$) differences compared to control group. Differences between groups were assessed by t-test. Values represent the mean ± SEM. EIA

expression of cyclin B1 (marker of cell proliferation) indicated the viability of cells used in the current experiments, whereas expression of caspase-3 (apoptosis marker) suggests the occurrence of apoptosis in these cells. Hence, the cells used were viable, functionally active and acceptable to examine the treatments affecting proliferation, apoptosis and secretory activity. These results resemble with earlier reports (Kolesarova et al. 2011; Kadasi et al. 2012).

Proliferation and apoptosis of ovarian cells are processes that determine ovarian folliculogenesis and ovulation (Sirotkin 2011). This is the first demonstration of the combined effect of R stem bark extract and Cur on basic ovarian cell functions – proliferation and apoptosis—and its direct involvement in the regulation of ovarian function. Current sets of results demonstrated that RA alone and RA + Cur by enhancing proliferation marker (cyclin B1) accumulation could act in the G_2 phase of the cell cycle (Gavet and Pines 2010) and by inhibiting cyclin B1 expression in a dose-dependent manner negatively affects the S-phase of the cell cycle. It was reported (Kadasi et al. 2012) that Cur alone suppresses proliferation in porcine GC, but the present study showed that RA + Cur regulated the proliferation in a

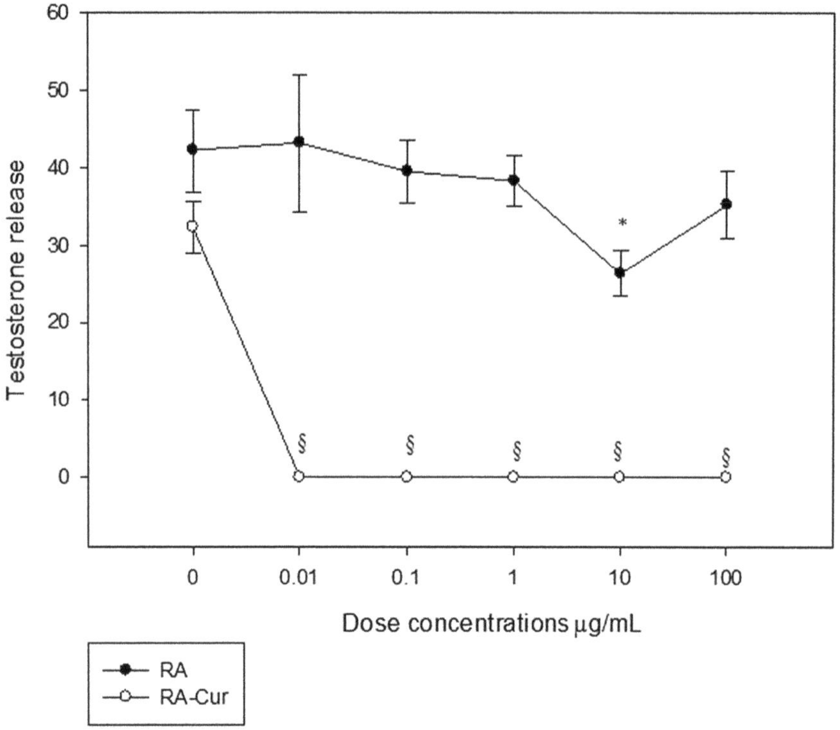

Fig. 43.5 Effect of RA and RA + Cur on testosterone (T) release. * (RA), § (RA + Cur) denotes significant ($p < 0.05$) differences compared to control group. Differences between groups were assessed by *t*-test. Values represent the mean ± SEM. EIA

dose-dependent manner. This suggests the antagonistic effect of RA on Cur's proliferation suppressive efficacy at lower doses and protagonistic effect of RA on Cur at highest dose level. RA alone and RA + Cur have upregulated the apoptotic peptide caspase-3 accumulation which indicates promotion in granulosa cell apoptosis. Earlier study (Kadasi et al. 2012) suggested that Cur acts as apoptotic agent (Aktas et al. 2012). The results obtained from this study also showed that due to the influence of RA, Cur's apoptotic activity become more active, and the combination is acting as apoptosis enhancer showing cumulative effect. Thus, RA can modify the Cur's effect on ovarian cell functions—proliferation and apoptosis. Therefore, the dose-dependent combined effect of RA + Cur on both ovarian cell proliferation and apoptosis demonstrates the combination's direct influence on ovarian follicle (growth and atresia) and an interrelationship between these two antagonistic mechanisms defining ovarian cell turnover (Sirotkin 2011).

Furthermore, P_4 is an important regulator of ovarian functions, folliculogenesis, ovarian cell proliferation, luteinisation and pregnancy maintenance (Sirotkin 2011). In the present study, RA reduced the P_4 release level which corresponds with previous reports on rodents (Okwuasaba et al. 1991; Salhab et al. 1999). RA + Cur combination significantly mitigated the P_4 release level, whereas Cur alone acts as P_4 enhancer (Kadasi et al. 2012). Thus RA has negatively affected the Cur's P_4 release enhancing activity. Hence, current observations demonstrate the influence of R on ovarian functions via alteration in P_4 output. RA alone at higher doses reduces as well as inhibited T release dose-dependently, thus affecting ovarian folliculogenesis, oocyte maturation, ovulation, follicular atresia and fecundity. On the other hand, Cur is reported as T release enhancer in porcine (Kadasi et al. 2012), but RA + Cur combination in the current study prohibited the T release. Thus, here also RA has negatively affected Cur's activity on androgen. Therefore this combination in turn can also affect ovarian folliculogenesis, follicular atresia, oocyte maturation, ovulation and fecundity via T, a known regulator of these processes (Sirotkin 2011; Walters et al. 2008). Hence, along with cell proliferation and apoptosis, RA can also alter the Cur's effect on ovarian steroidogenesis.

43.5 Conclusion

The comparison of real-time CI data and accumulation of cyclin B1, caspase-3 and progesterone-testosterone releases demonstrate that both RA and Cur can affect ovarian functions. Combination of RA + Cur has directly affected the (a) cell proliferation, (b) cell apoptosis and (c) progestogen and androgen release when Cur is used at a higher dose. Reversibly same combination with lower doses of Cur neutralised the cell proliferation suppressing activity of RA. Present report also suggests that RA can change the Cur's reported efficacy on granulosa cell function and RA + Cur combination can act as a fertility regulator. Since all the mentioned parameters are under female reproductive cycle, therefore, results of the findings suggest a possible female herbal contraceptive mechanism of action of RA + Cur combination.

Acknowledgement The authors are thankful to Slovak Academic Information Agency (SAIA, project no. SAIA 185-2011) and National Scholarship Programme of Government of Slovakia summer session 2011, Slovak Agency for Promotion of Research and Development (APVV, projects no. 0137-10 and 0854-11) and Operational Programme Research and Development funded from the European Regional Development Fund (project no. 26220220176) for their financial support to carry out the experiments. We are thankful to Mrs. Katarina Tothova and Ing. Zofia Kuklova (Animal Production Research Centre, Nitra-Luzianky) for their technical help to carry out the experiments and to Mrs. Iris Stelter (Institute of Animal Science, Neustadt, Germany) for performing the hormone EIA. We would also like to extend our deep sense of gratitude to University Grants Commission, Government of India. Facility for literature search provided by DeLCON available in Bioinformatics Centre, Assam University and Lab facility provided by Assam University Biotech Hub are sincerely acknowledged.

References

Aktas C, Kanter M, Kocak Z (2012) Antiapoptotic and proliferative activity of curcumin on ovarian follicles in mice exposed to whole body ionizing radiation. Toxicol Ind Health 28:852–863

Ammon HPT, Martin AW (1991) Pharmacology of *Curcuma longa*. Planta Med 57:1–7
Bird C, Kirstein S (2009) Real-time, label-free monitoring of cellular invasion and migration with the xCELLigence system. Nat Methods 6:v–vi
Brache V, Sitruk-Ware R, Williams A et al (2012) Effects of novel estrogen-free, progesterone receptor modulator contraceptive vaginal ring on inhibition of ovulation, bleeding patterns and endometrium in normal women. Contraception 85:480–488
Conneely OM, Mulac-Jericevic B, DeMayo F et al (2007) Reproductive functions of progesterone receptors. Recent Prog Horm Res 57:339–355
Eigner D, Scholz D (1999) *Ferula asa-foetida* and *Curcuma longa* in traditional medical treatment and diet in Nepal. J Ethnopharmacol 67:1–6
Gavet O, Pines J (2010) Progressive activation of CyclinB1-Cdk1 coordinates entry to mitosis. Dev Cell 18:533–543
Ghosh AK, Das AK, Patra KK (2011) Studies on antifertility effect of rhizome of *Curcuma longa* Linn. AJPLS 4:349–353
Goncim HY, Mador ES, Ogunranti JO (2010) *Ricinus communis Var Minor* inhibits follicular development and possibly ovulation in human subjects as shown by ultrasound follicle tracking. Clin Med Insights Reprod Health 4:35–38
Kadasi A, Sirotkin AV, Maruniakova N et al (2012) The effect of curcumin on secretory activity, proliferation and apoptosis of the porcine ovarian granulosa cells. JMBFS 2:349–357
Kalita JC, Chakrabarty A, Tanti B (2011) Assessment of antifertility activity of some traditionally used plants by different ethnic communities in three districts of Assam, India. J Herb Med Toxicol 5:65–72
Keith CT, Borisy AA, Stockwell BR (2005) Multicomponent therapeutics for networked systems. Nat Rev Drug Discov 4:71–78
Khillare B, Shrivastav TG (2003) Spermicidal activity of *Azadirachta indica* (neem) leaf extract. Contraception 68:225–229
Kolesarova A, Capcarova M, Marunikova N et al (2011) Deoxynivalenol-induced animal ovarian signaling: proliferation and apoptosis. JMBFS 1:323–332
Liao S, Lin J, Dang MT et al (2001) Growth suppression of hamster flank organs by topical application of catechins, alizarin, curcumin and myristoleic acid. Arch Dermatol Res 293:200–205
Meena AK, Rao MM (2010) Folk herbal medicines used by the Meena community in Rajasthan. Asian J Tradit Med 5:19–31
Munster E (1989) Entwicklung von enzymimmunologischenMessverfahren auf MikrotitrationsplattenzurBestimmung von Testosteron und ProgesteronimBlutplasma. Dissertation, University of Hohenheim
Nath S, Dutta Choudhury M, Roychoudhury S et al (2013) Male contraceptive efficacy of *Ricinus communis* L. extract. J Ethnopharmacol 149:328–334
Naz RK (2011) Can curcumin provide an ideal contraceptive? Mol Reprod Dev 78:116–123
Niazi J, Poonia P, Gupta V et al (2010) Pharmacotherapeutics of *Curcuma longa*- a potent patent. IJPPR 1:24–33
Okwuasaba FK, Osunkwo UA, Ekwenchi MM et al (1991) Anticonceptive and estrogenic effects of seed extract of *Ricinus communis* var. *minor*. J Ethnopharmacol 34:141–145
Osborn M, Isenberg S (1994) Cell biology: a laboratory handbook. Academic Press, New York
Ozturk F, Toy E, Hatunoglu E et al (2014) Real-time cell analysis of cytotoxicity of orthodontic cements on gingival fibroblasts. J Orthod Res 2:32–37
Parasramka MA, Gupta SV (2012) Synergistic effect of garcinol and curcumin on antiproliferative and apoptotic activity in pancreatic cancer cells. J Oncol. https://doi.org/10.1155/2012/709739
Prakash BS, Meyer HH (1987) Development of a sensitive enzyme immunoassay (EIA) for progesterone determination in unextracted bovine plasma using the second antibody technique. J Steroid Biochem 28:623–627
Purohit A (1999) Antifertility efficacy of *Curcuma longa* (50% E to H extract) with special reference to serum biochemistry and fertility test. Anc Sci Life 18:192–194

Purohit A, Bhagat M (2004) Contraceptive effect of *Curcuma long* (L.) in male albino rat. Asian J Androl 6:71–74
Qin Y, Zhao Z, Sun M et al (2011) Association of basal serum testosterone levels with ovarian response and in vitro fertilization outcome. Reprod Biol Endocrinol 9:9
Rithaporn T, Monga M, Rajasekaran M (2003) Curcumin: a potential vaginal contraceptive. Contraception 68:219–223
Ross IA (2001) Medicinal plants of the world: chemical constituents, traditional and modern medicinal uses. Humana Press Totowa, New Jersey
Saharia S, Sarma CM (2011) Ethno-medicinal studies on indigenous wetland plants in the tea garden tribes of Darrang and Udalguri district, Assam, India. NeBIO 2:27–33
Salhab AS, Al-Tamimi SO, Maha MNG et al (1998) The abortifacient effects of castor bean extract and ricin-a chain in rabbits. Contraception 58:193–197
Salhab AS, Issa AA, Alhougog I (1997) On the contraceptive effect of castor beans. Pharm Biol 35:63–65
Salhab AS, Shomaf MS, Maha MNG et al (1999) Effects of castor bean extract and ricin A-chain on ovulation and implantation in rabbits. Contraception 59:395–399
Singh KK, Parmar S, Tatke PA (2012) Contraceptive efficacy and safety of HerbOshield™ vaginal gel in rats. Contraception 85:122–127
Sirotkin AV (2011) Growth factors controlling ovarian functions. J Cell Physiol 226:2222–2225
Sirotkin AV, Chrenek P, Darlak K et al (2008) Some endocrine traits of transgenic rabbits. II. Changes in hormone secretion and response of isolated ovarian tissue to FSH and ghrelin. Physiol Res 57:745–751
Sitruk-Ware R, Nath A (2011) Metabolic effects of contraceptive steroids. Rev Endocr Metab Disord 12:63–75
Sitruk-Ware R, Nath A (2013) Characteristics and metabolic effects of estrogen and progestins contained in oral contraceptive pills. Best Pract Res Clin Endocrinol Metab 27:13–24
Srivastava RC, Nyishi C (2010) Traditional knowledge of *Nyishi (Daffla)* tribe of Arunachal Pradesh. IJTK 9:26–37
Thakur S, Bawara B, Dubey A et al (2009) Effect of *Carum carvi* and *Curcuma longa* on hormonal and reproductive parameter of female rats. Int J Phytomed 1:31–38
Walters KA, Allam CM, Handelsman DJ (2008) Androgen actions and the ovary. Biol Reprod 78:380–389
Yadav R, Jain GC (2010) Post-coital contraceptive efficacy of aqueous extract of *Curcuma longa* rhizome in female albino rats. Pharmacologyonline 1:507–517

Pterocarpus santalinus: A Wonder Gift of Nature

44

Tuhin Kanti Biswas, Shrabana Chakrabarti, Biswajit Auddy, Tanmoy Mondal, Srikanta Pandit, and Tapan Seal

Abstract

Pterocarpus santalinus Linn. (family Fabaceae), commonly known as Raktachandan or red sandalwood in India, is a rare plant, principally found in the southern part of India. The plant is described *Ayurveda* for multipurpose biological activities like anti-diabetic, wound healing, anti-inflammatory, antipyretic, aphrodisiac, anti-haemorrhagic, etc. Phytochemical investigations of this plant show that it contains substances, for example, alkaloids, phenols, saponins, glycosides, flavonoids, triterpenoids, sterols and tannins. What's more, heartwood contains isoflavone, glucosides and two enemies of tumour lignans, viz. savinin and calocedrin. Biological studies of the plant reveal that it has a potent wound healing property in animal model. Besides, hepatoprotective activity in rodents was also found to be perceptible, and protection against oxidative stress-induced DNA damage, investigated on the basis of comet assay, has also been noticed with this medicinal plant.

T. K. Biswas (✉) · S. Pandit
Research Unit, J. B. Roy State Ayurvedic Medical College and Hospital, Kolkata, India
e-mail: biswastuhin@rediffmail.com

S. Chakrabarti · B. Auddy
Chigurupati Technologies Pvt. Ltd., Hyderabad, India

T. Mondal
School of Genetic Studies, MaulanaAbulKalam Azad University of Technology (MAKAUT), Kolkata, India

T. Seal
Plant Chemistry Department, Botanical Survey of India, AJC Bose Indian Botanic Garden, Howrah, India

S. C. Mandal et al. (eds.), *Evidence Based Validation of Traditional Medicines*,
https://doi.org/10.1007/978-981-15-8127-4_44

Keywords

Pterocarpus santalinus · Toxicity study · Chemical analysis · Free radical scavenging activity · Wound healing · Hepatoprotective · Comet assay

Abbreviations

ABTS	2,2′-Azino-bis(3-ethylbenzothiazoline-6-sulfonic acid)
DMSO	Dimethyl sulfoxide
DNA	Deoxyribonucleic acid
DPPH	1,1-Diphenyl-2-picrylhydrazyl
EDTA	Ethylenediaminetetraacetic acid
EtOH	Ethanol
FBS	Foetal bovine serum
HBSS	Hanks Buffered Salt Solution
OTM	Olive tail moment
RNA	Ribonucleic acid
RPMI	Roswell Park Memorial Institute
SDS	Sodium dodecyl sulphate

44.1 Introduction

In the process of evolution of the universe for several thousands of years, numerous biosyntheses occur in the nature, some of which were represented as gifts for the mankind. These ranges of gifts included medicinal plants, lower and higher animals and non-animated minerals and metals. The most important use of natural products is highlighted as healthcare ingredients, next to food. Canvassers of natural products in the world have explored many treasures for using them to control and cure many diseases. These campaigners have also observed that a single natural product may have diverse biological activities. This information is noted in various languages in various countries under various traditional systems of medicine, and *Ayurveda* is such a wing, established in India about 5000 years back, which described this diversity of a single natural agent.

Many natural products are described in different classical texts of Ayurveda for therapeutic purpose, and medicinal plant is a major resource among these. A review reveals that a total number of 1587 medicinal plants are narrated in different texts of Ayurveda for various therapeutic purposes, either in single or in combined form. Among these 341 medicinal plants are described in *Charaka Samhita*, 395 in *Sushruta Samhita* and a maximum of 902 in Ayurvedic text *Ashtanga Hridayam* (Vassou et al. 2016). In reference to a large number of plants described in Ayurveda, only 395 medicinal plants are included in Ayurvedic Pharmacopoeia of India as most of these plants are not available now. Amusing descriptions of 500 medicinal plants

Fig. 44.1 The plant *Pterocarpus santalinus*

are found in *Charaka Samhita* categorized into 50 groups on the basis of their therapeutic mode of action (Biswas 2018).

Pterocarpus santalinus Linn. (family Fabaceae), commonly known as *Raktachandan*, is one such available medicinal plant, now enlisted in the rare category, which is descried in Ayurveda for its various therapeutic activities. *P. santalinus* is a small to medium measured deciduous tree having a thick, round crown arriving at a stature of 10–15 m with a circumference of around 90–160 cm (Fig. 44.1). The plant is known as red sandalwood or rubywood, found specifically in the southern parts of the Eastern Ghats of India. The heartwood of the plant is deep red in colour (Fig. 44.2a) and described in different texts of Ayurveda for its use in various therapeutic purposes like controlling vomiting, thirst, diabetes, haemorrhage, different diseases of eyes, wounds, fever, helminthiasis (intestinal worms), diverse psychological diseases, etc. (Biswas and Mukherjee 2003). The bark is regularly dull earthy coloured in shading with rectangular plates and profoundly fissured when developed and radiates a red shading gum with various pink

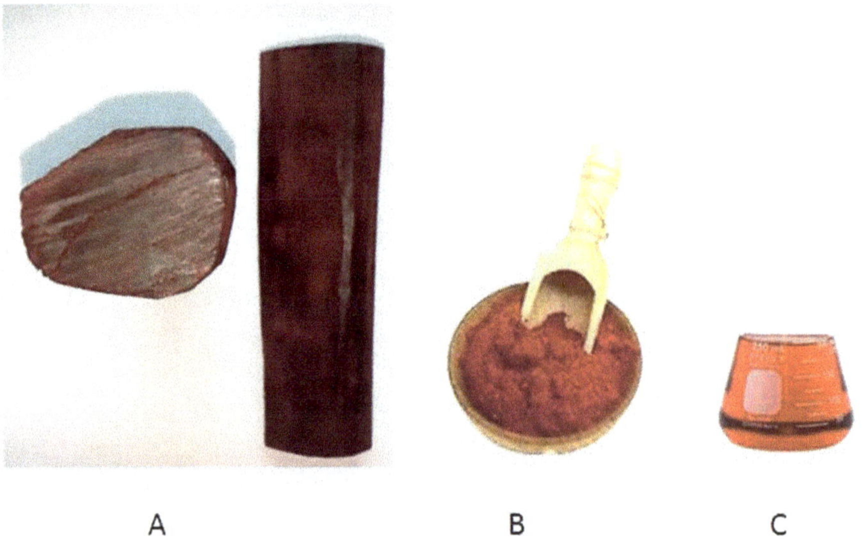

Fig. 44.2 The heartwood of *Pterocarpus santalinus* (**a**), powder (**b**) and ethanol solution (**c**)

streaks on bursting. The plant has tremendous demand in the international market, particularly in Japan, for its wavy red grained internal structure to prepare a three-stringed special musical instrument called as 'shamisen'. This typical structure of wood of the plant enables to prepare the best quality of such acoustic musical instrument. In China, the plant is used since the tenth century for the manufacture of high-value furniture (Arunkumar and Joshi 2014). Apart from its classical uses as mentioned in Ayurveda, there are a number of evidences for its different therapeutic uses by traditional healers in different parts of India. The traditional Santal group of folklores uses this plant for the treatment of burn wound (*Pachiari ghao*) (Bodding 1986). The whole plant of *P. santalinus* is used by Yerukula and Irula tribes of Chittoor district in Andhra Pradesh for the healing of ulcer (Vedavathy et al. 1997); the stem bark extract is used by a tribal group of the Western Ghats in Shimoga region of Karnataka state for treating diabetes, fever and snake bite and specially for jaundice (Manjunatha 2006); and various tribes in coastal Karnataka use red sanders as an anti-inflammatory agent for the treatment of herpes zoster (Bhandari and Chandrashekar 2011). The principal red pigment of red sanders contains the chemical constituents santalins 'A' and 'B' (Fig. 44.3), which are traditionally important for its cosmetic value; particularly it is applied topically as a popular home remedy used in southern India as lightening agent against post-acne and other facial scars. The alcoholic extract of the plant is used as a natural colouring agent for providing nutrition is cosmetic products and with the added benefit of antioxidant potential (Prakash and Majeed 2008). Interestingly, it is also mythologically reported that the stem of the plant is religiously used for the worship of lord Shiva, Durga and other goddess according to Hindu rituals in India (Sarma 2015).

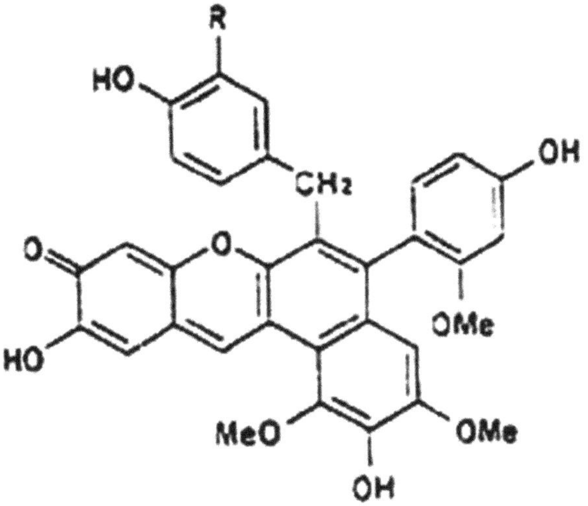

Fig. 44.3 Chemical structure of colouring pigments of *P. santalinus*(Source Prakash and Majeed 2008)

RH = OH ... santalin A
R = OMe ... santalin B

Scientific appraisal on this endangered plant is extremely limited. There is a report of the presence of carbohydrates, anthocyanins, phenols, triterpenoids, saponins, tannins, steroids, flavonoids, glycosides and glycerides in the plant (Bhange et al. 2016), but no such detailed information is available till date. On the other hand, the scientific information on the pharmacological and/or clinical activities of the plant other than its wound healing property is least. An updated scientific screening of the plant was performed, and fascinating findings were revealed with this plant on different biological fields as well as chemical analysis, which are being depicted in the current essay.

44.2 Scientific Screening

44.2.1 Collection and Identification of the Plant

The dried heartwood of *P. santalinus* was procured from an authorized vendor and was identified by a competent authority of the Botanical Survey of India. The heartwood was powdered finely with a mesh size of 80 (μm) and was separated into two parts, of which one part was kept as crude powder and other for further chemical extraction and analysis.

44.2.2 Toxicity Study of *P. santalinus*

Acute toxicity study including CNS effect were observed by conventional pharmacological method with aqueous extract of the powder of the plant on Swiss albino mice in three doses as 500 mg kg^{-1} b.w., 1 g kg^{-1} b.w. and 2 g kg^{-1} b.w. on oral and intra-peritoneal routes. No toxicities or mortalities were observed with these doses, indicating its safety aspect (Biswas et al. 2004a, 3(3)).

44.2.3 Preparation of Ointment

Ointment of 15% concentration (w/w) was prepared with white soft petroleum jelly (mp 60 °C) with crude powder and 5% concentration (w/w) with aqueous ethanol extract powder. These formulations were meant for the wound healing study in animal as well as clinical patients of non-healing wounds.

44.2.4 Chemical Screening of Ethanol Extract

44.2.4.1 Preparation of Plant Extract

Five gramme (5 g) of heartwood of *P. santalinus* was powdered and extricated (Fig. 44.2b, c) twice with 20 ml 80% fluid ethanol (EtOH) with constant stirring for 18–24 h at surrounding temperature. The concentrates gained from the first and the resulting extractions were blended and diluted to 50 ml, and aliquot was investigated for their all-out phenolic, flavonoid and flavanol content, reducing power and free radical scavenging limit.

44.2.4.2 Quantification of Total Phenolic, Flavonoid and Flavanol Contents

The measure of all-out phenolic substance of unrefined concentrate of *P. santalinus* was resolved by the Folin-Ciocalteu technique (Singleton and Rossi 1965). Total flavonoids in *P. santalinus* were assessed utilizing the strategy of Ordonez et al. (2006), and all-out flavanols in the plant were evaluated utilizing the technique of Kumaran and Karunakaran (2006).

44.2.4.3 Estimation of Reducing Power

The capacity of the concentrates to reduce iron (III) was evaluated by using the strategy of Oyaizu (1986). For this reason extricates (100 μl) of plant were added in phosphate buffer (2.5 ml, 0.2 M, pH 6.6) and 1% potassium ferricyanide (2.5 ml). The blend was incubated at 50 °C for 20 min. Aliquots of 10% trichloroacetic acid (2.5 ml) were added to the mixture, which was then centrifuged at 3000 rpm for 10 min. The upper layer of the solution (2.5 ml) was mixed in with distilled water (2.5 ml) followed by addition of ferric chloride (0.5 ml, 0.1%). The absorbance was estimated at 700 nm. The reducing power is given in ascorbic acid comparable (AAE) in milligramme per gramme (mg/g) of dry material utilizing the accompanying condition dependent on the calibration curve: $y = 0.0023x - 0.0063$, $R^2 = 0.9955$ where y was the absorbance and x was the ascorbic acid proportional (mg/g).

44.2.4.4 Evaluation of DPPH Free Radical Scavenging Activity

The free radical rummaging movement of the plant extract and butylated hydroxytoluene (BHT) as positive control was resolved utilizing the steady radical DPPH (1,1-diphenyl-2-picrylhydrazyl) (Blois 1958). Aliquots (20–100 ml) of the tried example were set in test tubes, and 3.9 ml of newly arranged DPPH arrangement ($25\ \text{mg L}^{-1}$) in methanol was included in each test tube and mixed. Thirty minutes later, the absorbance was estimated at 517 nm (UV-noticeable spectrophotometer Shimadzu UV-1800). The ability to rummage the DPPH radical was determined, utilizing the accompanying condition:

$$\text{DPPH scavenged } (\%) = \{(\text{Ac} - \text{At})/\text{Ac}\} \times 100$$

Here Ac is the absorbance of the control response, and At is the absorbance in nearness of the example of the concentrates. The antioxidant capacity of the concentrate was communicated as IC50. The IC50 value was characterized as the fixation in mg of dry material per ml (mg/ml) that represses the arrangement of DPPH radicals by half. Each value was resolved from regression equation. Values are introduced as mean ± standard error mean of three recreates. The all-out phenolic content, flavonoid content, flavanol content, reducing power and IC50 estimation of each plant material were determined by utilizing linear regression examination.

44.2.4.5 Determination of ABTS Radical Cation Scavenging Activity

The 2,2′-azino-bis(3-ethylbenzothiazoline-6-sulfonic corrosive) (ABTS) radical cation (ABTS.+)-searching movement was estimated by the strategy depicted by Miller (Miller and Rice-Evans 1997). ABTS was dissolved in water to a 7 mM focus. The ABTS radicals were created by including 2.45 mM potassium per sulphate (last concentration). The completion of radical formation was acquired in obscurity at room temperature for 12–16 h. This arrangement was then diluted with ethanol to modify its absorbance at 734 nm to 0.70 ± 0.02. To decide the scavenging activity, 1 ml of diluted ABTS. + solution was added to 10 μl of plant concentrate (or water for the control), and the absorbance at 734 nm was estimated 6 min after the underlying mixing, utilizing ethanol as blank. The level of inhibition was determined by the equation:

$$\text{ABTS scavenged } (\%) = (A_{\text{cont}} - A_{\text{test}})/A_{\text{cont}} \times 100$$

where A_c and A_s are the absorbencies of the control and of the test sample, respectively. From a plot of concentration against % inhibition, a linear regression analysis was performed to determine the IC_{50} value of the sample.

44.2.4.6 Estimation of Phenolic Acids and Flavonoids in the 80% Aqueous Ethanol Extract of *P. santalinus* by HPLC

HPLC examinations for the measurement of phenolic acids and flavonoids in the concentrate were performed following the strategy of Seal et al. (2017) utilizing Dionex Ultimate 3000 fluid chromatograph including a diode array detector (DAD) with 5 cm stream cell and with Chromeleon system manager as data processor. Separation was accomplished by a reversed phase Acclaim C18 column (5 micron molecule size, 250 × 4.6 mm). 20 μL of test was brought into the HPLC section. The strategy was approved by the USP and ICH rules (ICH-Q2A 1995; ICH-Q2B 1996). HPLC chromatograms were distinguished utilizing a photo diode array UV detector at three unique frequencies (272, 280 and 310 nm) as indicated by absorption maxima of analysed compounds. Each compound was recognized by its retention time and by spiking with measures under similar conditions. The evaluation of phenolic acids and flavonoids in the example extricates was done by the measurement of the integrated peak area, and the substance was determined using the calibration curve by plotting peak area against concentration of the respective standard sample. The data were described with convergence limit in triplicate. A representative HPLC chromatogram of the mixture of all standard phenolic acids and flavonoids recorded at 280 nm is delineated in Fig. 44.4.

The quantification of phenolic acids and polyphenolic compounds (e.g. gallic acid, protocatechuic acid, gentisic acid, chlorogenic acid, *p*-hydroxybenzoic acid, vanillic acid, caffeic acid, syringic acid, *p*-coumaric acid, ferulic acid, sinapic acid, salicylic acid and ellagic acid, catechin, rutin, myricetin, quercetin, naringin, apigenin and kaempferol) in the 80% aq. ethanol extract of *P. santalinus* were carried out by HPLC studies. The quantity of all phenolic acids and flavonoids in the investigated plant has been expressed as μg/mg dry plant material and data described in Table 44.1.

The HPLC chromatogram of the heartwood of *P. santalinus* (Fig. 44.5) showed the presence of protocatechuic acid (0.72 ± 0.007), catechin (8.16 ± 0.033), chlorogenic acid (1.89 ± 0.03), vanillic acid (6.47 ± 0.04gm), caffeic acid (2.36 ± 0.02), syringic acid (0.81 ± 0.02), p-coumaric acid (0.78 ± 0.02), ferulic acid (0.52 ± 0.01), sinapic acid (0.22 ± 0.02), naringin (3.53 ± 0.03), rutin (5.67 ± 0.04), ellagic acid (0.68 ± 0.02 mg/100 gm), myricetin (12.58 ± 0.07), quercetin (2.14 ± 0.04), naringenin (1.09 ± 0.007), apigenin (1.50 ± 0.02) and kaempferol (94.78 ± 0.07) (Table 44.2).

44.2.4.7 Antioxidant Activities of the Heartwood of *P. santalinus*

In this assay, estimation of all-out phenolic, all-out flavonoid and absolute flavanol content, decreasing force, ABTS and DPPH techniques were utilized to assess the in vitro cell reinforcement exercises of the 80% aq. ethanol concentrate of *P. santalinus*, and the result has been introduced in Table 44.3. The outcome indicated that the all-out phenolic content in the plant was 49.48 ± 4.62 mg GAE/gm dry concentrate. The flavonoid and flavanol sums that were identified in the heartwoods were 39.61 ± 0.41 mg/g and 91.80 ± 2.38 mg/g dry concentrate, respectively. The reducing power (AAE) of the 80% aq. ethanol extract was

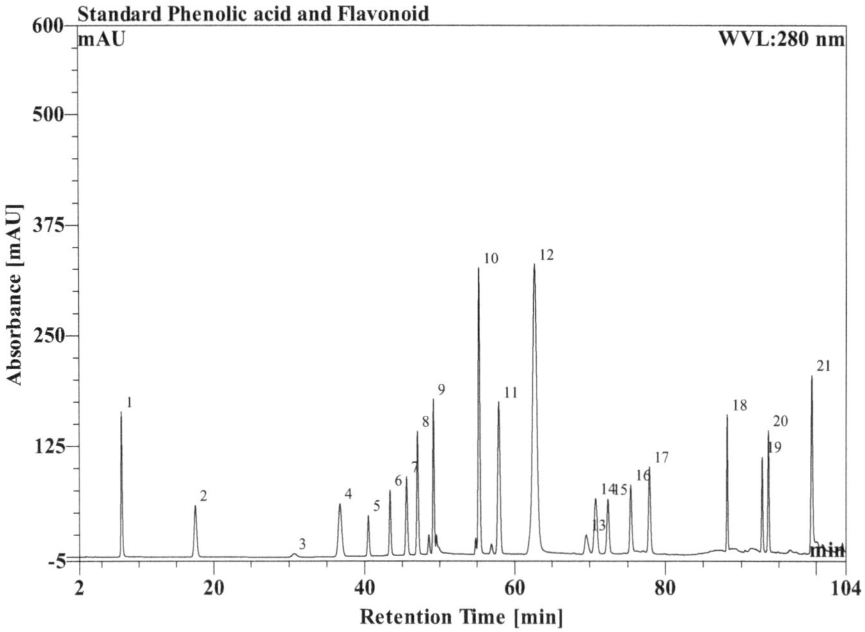

Fig. 44.4 HPLC chromatogram of mixture of standard phenolic acids and flavonoids. 1 gallic acid, 2 protocatechuic acid, 3 gentisic acid, 4 p-hydroxybenzoic acid, 5 catechin, 6 chlorogenic acid, 7 vanillic acid, 8 caffeic acid, 9 syringic acid, 10 p-coumaric acid, 11 ferulic acid, 12 sinapic acid, 13 salicylic acid, 14 naringin, 15 rutin, 16 ellagic acid, 17 myricetin, 18 quercetin, 19 naringenin, 20 apigenin, 21 kaempferol

70.21 ± 2.51 mg/g. The investigation likewise uncovered that the ABTS radical searching action indicated higher antioxidant capacities with IC50 esteem 0.02 ± 0.001 mg dry concentrate than DPPH measure (IC500.06 ± 0.001 mg dry concentrate).

44.2.5 Evaluation of Wound Healing Property

44.2.5.1 Study of Wound Healing Efficacy with Crude Ointment

Wound healing property was performed in Charles Foster rats ~~of~~ with ointment prepared from *P. santalinus* crude drug (15% w/w) to compare the efficacy. The efficacy of the ointment from *P. santalinus* crude drug (15% w/w) was evaluated in 8 mm full-thickness punch biopsy wound model by means of a specially designed instrument Acuderm (Fig. 44.6), and effect was observed on the basis of physical parameters like graphical measurement of wound contraction size (mm^2) (Fig. 44.7), healing periods (days) and tensile strength (g) as well as molecular markers like estimation of tissue DNA, RNA, protein and hydroxyproline. The study was conducted in comparison with untreated control, vehicle control (ointment base

Table 44.1 Phenolic acid and flavonoid content by HPLC μg/mg dry extract

Phenolic acids/flavonoids	Amount μg/mg dry extract	Phenolic acids/flavonoids	Amount μg/mg dry extract	Phenolic acids/flavonoids	Amount μg/mg dry extract
Gallic acid	ND	Caffeic acid	2.36 ± 0.02	Rutin	5.67 ± 0.04
Protocatechuic acid	0.72 ± 0.007	Syringic acid	0.81 ± 0.02	Ellagic acid	0.68 ± 0.02
Gentisic acid	ND	p-Coumaric acid	0.78 ± 0.02	Myricetin	12.58 ± 0.07
p-Hydroxybenzoic acid	ND	Ferulic acid	0.52 ± 0.01	Quercetin	2.14 ± 0.04
Catechin	8.16 ± 0.033	Sinapic acid	0.22 ± 0.02	Naringenin	1.09 ± 0.007
Chlorogenic acid	1.89 ± 0.03	Salicylic acid	ND	Apigenin	1.50 ± 0.02
Vanillic acid	6.47 ± 0.04	Naringin	3.53 ± 0.03	Kaempferol	94.78 ± 0.07

Each value in the table was obtained by calculating the average of three experiments, and data are presented as mean ± SEM

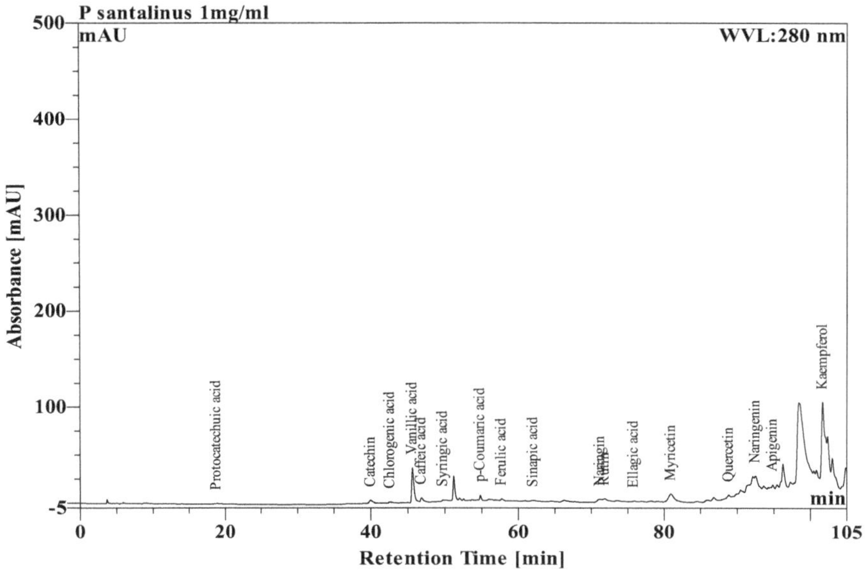

Fig. 44.5 HPLC chromatogram of heartwoods of *P. santalinus* showing phenolic acids and flavonoids

petroleum jelly with mp 60 °C) and standard comparator human placental extract. Tissues were also processed for histopathological findings with haematoxylin and eosin (H&E) stain. Interestingly, it was observed that the wound healing efficacy of the ointment prepared from the crude powder envisaged significant reduction of wound contraction size (mm^2) within a time period of 9 days which was significantly ($p < 0.05$) better than untreated control (16 days), vehicle control (15 days) or human placental extract (12 days). There was also significant ($p < 0.05$) generation of tensile strength (421 g) which was better than all the other treated groups. This evidence was supported with highly significant ($p < 0.001$) synthesis of molecular markers like DNA (2.45 mg/g), RNA (2.30 mg/g), protein (24.48 mg/g) and hydroxyproline 6.04 mg/g) (Table 44.4). The result of molecular marker was found to be quite comparable with human placental extract. The biological character of wound healing process was reflected with histological study with *P. santalinus* study showing potent epithelialization, neoangiogenesis and collagenases (Fig. 44.8) (Biswas et al. 2004a, 3(3)).

Clinical study of crude ointment was performed owing to the promising result of the experimental model, and the ointment was thereafter applied on selected patients of non-healing wounds. Efficacy of the wound healing property of the crude ointment of *P. santalinus* was observed on the basis of the physical parameters like measurement of wound contraction size (mm^2) and arbitrary scoring for assessment of granulation and epithelialization (Table 44.5) (Biswas et al. 2004b, 3(4)). Promising wound healing property of the ointment was observed in six patients as observed on the basis of physical and haematological parameters. In the treatment

Table 44.2 Phenolic acid and flavonoid content by HPLC μg/mg dry extract

Phenolic acids/flavonoids	Amount μg/mg dry extract	Phenolic acids/flavonoids	Amount μg/mg dry extract	Phenolic acids/flavonoids	Amount μg/mg dry extract
Gallic acid	ND	Caffeic acid	2.36 ± 0.02	Rutin	5.67 ± 0.04
Protocatechuic acid	0.72 ± 0.007	Syringic acid	0.81 ± 0.02	Ellagic acid	0.68 ± 0.02
Gentisic acid	ND	p-Coumaric acid	0.78 ± 0.02	Myricetin	12.58 ± 0.07
p-Hydroxybenzoic acid	ND	Ferulic acid	0.52 ± 0.01	Quercetin	2.14 ± 0.04
Catechin	8.16 ± 0.033	Sinapic acid	0.22 ± 0.02	Naringenin	1.09 ± 0.007
Chlorogenic acid	1.89 ± 0.03	Salicylic acid	ND	Apigenin	1.50 ± 0.02
Vanillic acid	6.47 ± 0.04	Naringin	3.53 ± 0.03	Kaempferol	94.78 ± 0.07

Each value in the table was obtained by calculating the average of three experiments, and data are presented as mean ± SEM

Table 44.3 Antioxidant properties of the heartwood of *P. santalinus*

Parameters	Amount (mg/g) dry extract
Total phenolic content (gallic acid equivalent)	49.48 ± 4.62
Total flavonoid content (rutin equivalent)	39.61 ± 0.41
Total flavanol content (quercetin equivalent)	91.80 ± 2.38
Reducing power (ascorbic acid equivalent	70.21 ± 2.51
DPPH radical scavenging activity (IC_{50}mg dry extract)	0.06 ± 0.001
ABTS radical scavenging activity (IC_{50}mg dry extract)	0.02 ± 0.001

Each value in the table was obtained by calculating the average of three experiments, and data are presented as mean ± SEM

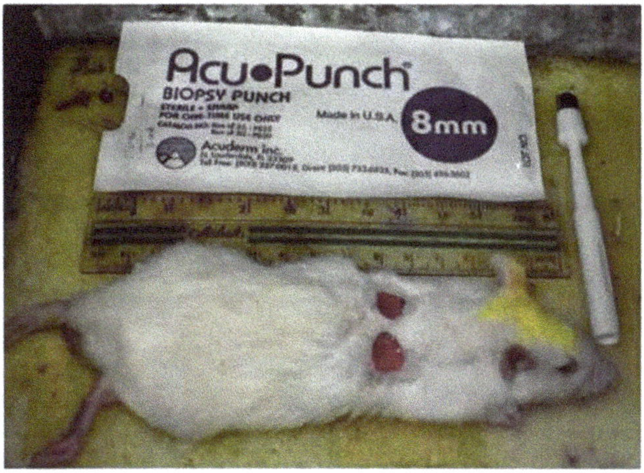

Fig. 44.6 Full-thickness punch wound of 8 mm^2 diameter

protocol, all the patients' wounds were primarily dressed with 70% ethanol followed by turpentine oil. No antibiotic was applied to any of the patients in tropical or systemic routes. A specific case study on a patient codified as R.S., 60 years, male had a deep seated wound of 15 mm^2 area over the lateral malleolus of left foot, was treated with the ointment on every alternate day, showed complete healing without any mal-union through healthy epithelialization and granulation along with decrease of elevated ESR (37.5% reduction) indicating its role in controlling the inflammatory procedure. The patient was treated with the ointment for 30 consecutive days and showed complete healing (Fig. 44.9). The haematological investigation showed 13.63% increase of lymphocyte, indicating promotion of collagen fibres as lymphocyte is considered as the pre-cursor of collagen (Biswas et al. 2004b, 3(4)).

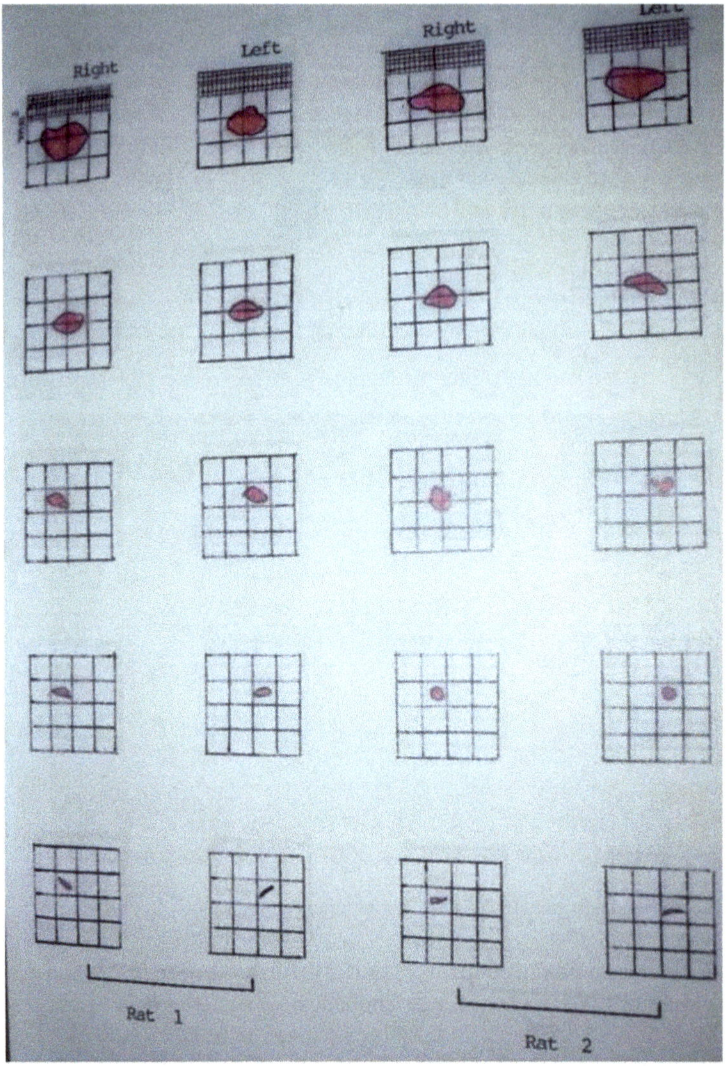

Fig. 44.7 Graphical measurement of wound contraction in different days

44.2.5.2 Study of Wound Healing Efficacy with 5% EtOH Extract Ointment

Wound healing efficacy of 5% ointment of *P. santalinus* ethanol extract was evaluated both in 8 mm full-thickness punch wound model in Wistar rats (inflicted with Acuderm instrument) and in patients of chronic non-healing wounds. Methods for infliction of wounds in rats were followed in the same way as done earlier. It was observed that there was significant ($p < 0.05$) reduction of wound size (1.30 mm^2) within a period of 8 days with *P. santalinus* 5% ointment, which was better than the

Table 44.4 Wound healing effect of crude ointment (15% w/w) *Pterocarpus santalinus* Linn. on 8 mm full-thickness punch wound in CF rats

Treatment groups	Values of physical parameters (mean ± SE)			Values of molecular markers (mean ± SE)			
	Wound size (mm^2)	Healing period (days)	Tensile strength (g)	DNA (mg/g)	RNA (mg/g)	Protein (mg/g)	Hydroxyproline (mg/g)
*P. santalinus*crude ($n = 8$)	2.45 ± 0.08*	09 ± 0.37*	421 ± 34.57*	2.45 ± 0.02**	2.30 ± 0.02**	24.48 ± 0.45**	6.04 ± 0.02 **
Untreated control ($n = 8$)	3.37 ± 0.25	16 ± 1.24	301 ± 24.08	1.01 ± 0.04	0.47 ± 0.01	09.09 ± 0.04	1.56 ± 0.01
Vehicle control ($n = 8$)	3.58 ± 0.43	15 ± 0.93	313 ± 25.14	1.25 ± 0.03	1.01 ± 0.02	12.30 ± 0.05	1.99 ± 0.03
Human placental extract ($n = 8$)	2.84 ± 0.05*	12 ± 0.63*	386 ± 34.54*	2.30 ± 0.03**	2.13 ± 0.02**	23.73 ± 0.19	4.69 ± 0.01 **

'*n*' indicated number of animal in each group. *$p < 0.05$. **$p < 0.001$. Vehicle control group consisted white soft petroleum jelly

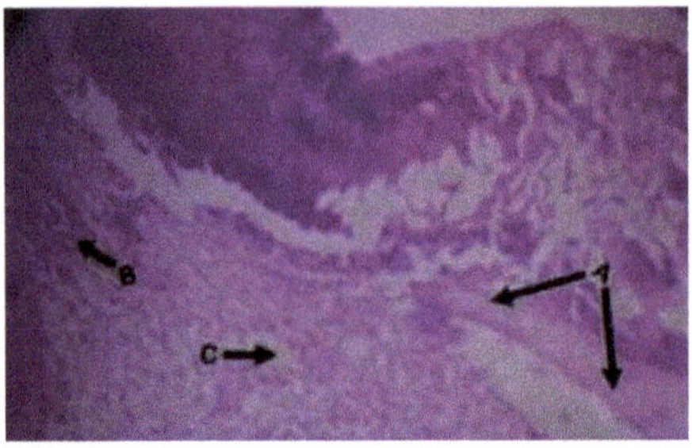

(40X)

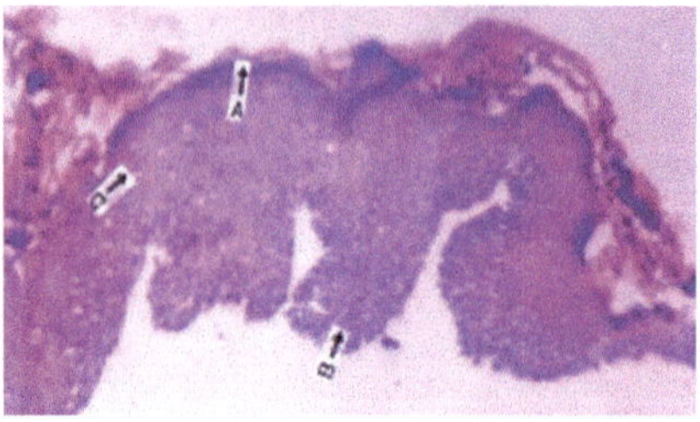

(10X)

Fig. 44.8 Histological characteristics of *P. santalinus* crude ointment (15%w/w)-treated healed wound tissues (A = epithelialization, B = collagenases, C = neoangiogenesis)

vehicle control (petroleum jelly) or framycetin topical ointment-treated wounds in rats (Fig. 44.10a, b, c). The sequential change of hydroxyproline (mg/g) of wound tissues collected on days 6, 9 and 11 with this treatment was performed by a special method of sealing in neutral glass ampules with 6 N HCl followed by boiling in water bath at 100 °C temperature for a period of 2 h and thereafter neutralizing the aliquot with 5 N NaOH. Estimation of hydroxyproline was done against blank and standard hydroxyproline in spectrophotometer at 557 nm after a series of chemical reactions (Woessner 1961). Results of sequential change of hydroxyproline (mg/g) of wound tissues collected on days 6, 9 and 11 were 5.28, 10.72 and 14.44, which were promisingly significant ($p < 0.001$) than other treatment groups and quite comparable with framycetin-treated group (Table 44.5). Hydroxyproline is the fundamental building block of collagen, composed of three major proteins proline, hydroxyproline and glycine, remaining in a triple helix manner and is the direct molecular marker for understanding the potentiality of wound healing efficacy. Results of animal study definitely support the potent wound healing efficacy of 5% ointment of *P. santalinus* ethanol extract. Results of physical and biochemical studies were well evidenced by the study of protein SDS-polyacrylamide gel electrophoresis (Bailey et al. 1979). This was done with wound tissues of *P. santalinus*-treated animals collected on days 6, 9 and 11 and compared with standard marker collagen IV (14–44 kDa). Interestingly it was observed that proteins of different molecular weights were generated with the treatment of 5% ointment of *P. santalinus* ethanol extract on days 6, 9 and 11, indicating its potentiality in the formation of wound base as well as wound tissues (Fig. 44.11).

The 5% ointment of *P. santalinus* ethanol extract was thereafter applied on patients of chronic non-healing wounds of lower extremity like diabetic foot ulcers and varicose vein ulcer. Conventional anti-diabetic and antibiotic therapies were administered therefore to the patients for controlling diabetic state and to prevent secondary infections. A total 12 patients were selected and were categorized into two groups. Six patients received 5% ethanol extracted *P. santalinus* ointment, and rest six received vehicle control (petroleum jelly) as comparator to evaluate effect the base of the ointment. There was significant ($p < 0.05$) reduction of wound size

Table 44.5 Wound healing effect of ethanol extracted ointment (5% w/w) of *Pterocarpus santalinus* Linn. on 8 mm full-thickness punch wound in Wistar rats

	Values of physical parameters (mean ± SE)		Hydroxyproline (mg/g)		
Treatment groups	Wound size (mm^2)	Healing period (days)	Day '6'	Day '9'	Day '11'
P. santalinus ethanol extracted5% ointment ($n = 6$)	1.30 ± 0.04*	08 ± 0.34**	5.28 ± 0.67*	10.72 ± 0.94*	14.44 ± 1.09**
Vehicle control ($n = 6$)	2.15 ± 0.22	14 ± 0.78	3.38 ± 0.42	8.03 ± 0.65	12.31 ± 0.80
Framycetin ointment ($n = 6$)	1.95 ± 0.19	10.50 ± 0.57*	6.53 ± 0.99	12.13 ± 1.35*	15.56 ± 1.56**

'*n*' indicated number of animal in each group. *$p < 0.05$. **$p < 0.001$. Vehicle control group consisted white soft petroleum jelly

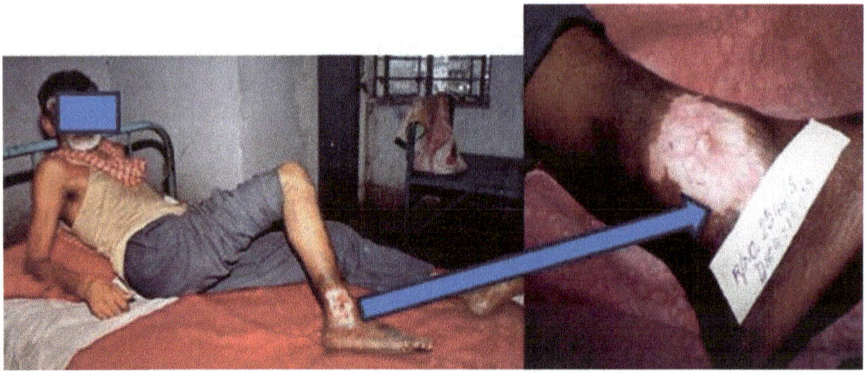

Fig. 44.9 Clinically treated non-healing wounds with crude *P. santalinus* crude ointment (15%w/w) on a patient of deep right lateral malleolus wounds

(50.12%) with good-quality epithelialization and granulation as detected by arbitrary scoring system in *P. santalinus*-treated group with respect to vehicle control (Table 44.6). This evidence was supported by haematological parameters like significant reduction of ESR and increase of lymphocyte count (Table 44.7). Manifestation of wound is an outcome of inflammatory process of human system where there is involvement of vascular response, cellular response and repair. Amelioration of vascular and cellular response is indicated by reduction of ESR, while repair of wounds is mediated by lymphocyte markers. It is reposted that lymphocytes initiates the immune response during inflammatory procedure and thereby converted into fibroblast which is the sourced of collagen (Rubin and Farber 1996). Healing of wound is chiefly based on qualitative and quantitative synthesis of collagen which is potentially performed by 5% ethanol extracted *P. santalinus* ointment. The efficacy of the ointment prepared from 5% EtOH extract of *P. santalinus* was exemplified with chronological improvement of a patient codified as C.B., 45 years, female, having diabetic foot ulcers showing the potent closure of wounds as measured by induration measurement scale (Fig. 44.12), which heal within a period of 22 days (Figs. 44.13 and 44.14) without any complication.

44.2.6 Hepatoprotective Effect of EtOH Extract of *P. santalinus*

Advancement of civilization results in amplified exposure of human body to numerous foreign chemicals known as xenobiotics. These potentially hazardous compounds are subjected to chemical alterations within the body in a safer, water-soluble and excretable molecule. The liver plays the central pivotal role for such crucial activities. In spite of enormous regenerative capacity, continuous injuries to hepatocytes and bile duct cells often lead to drug-induced liver injury. Furthermore, the enhanced regenerating and proliferative response of hepatocytes makes the liver susceptible to carcinogenesis. Carbon tetrachloride (CCl_4) has long been used as a

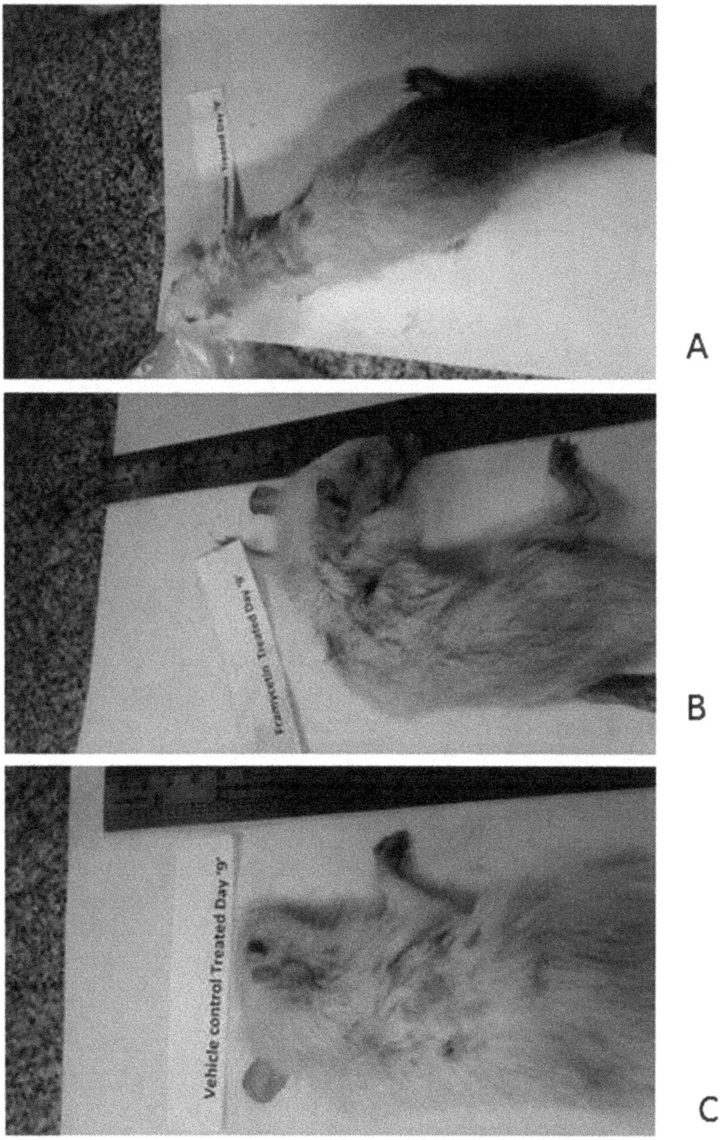

Fig. 44.10 Effect of 5% ethanol extract ointment of *P. santalinus*, framycetin ointment and vehicle control on 8 mm punch wound on day 9 in Wistar rats

standard hepatotoxin. It is evident that metabolic activation of CCl_4 into.CCl_3 and.Cl kicks off lipid peroxidation. Scientific reports suggest that some herbal extracts could ameliorate the damage caused by CCl_4-induced oxidative stress by altering the levels of increased lipid peroxidation and improving the decreased activities of

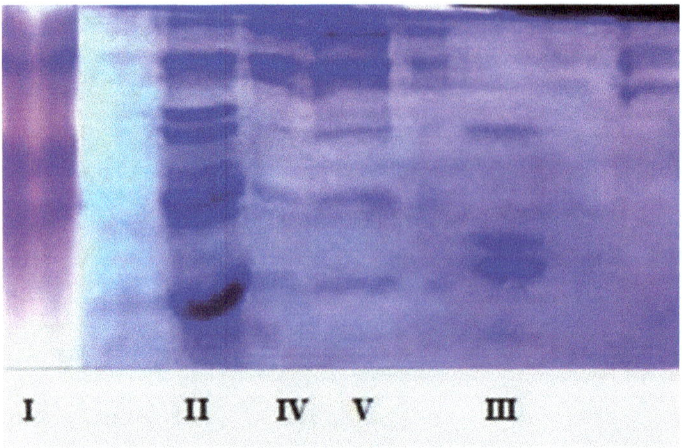

Fig. 44.11 Study of SDS-PAGE with wound tissues of different days treated with 5% ethanol extract *P. santalinus ointment* in Wistar rats (Lanes: I = collagen IV, II = normal skin, III = day 6, IV = day 9, V = day 11)

Table 44.6 Arbitrary scoring system for assessment of wound index, granulation and epithelialization in clinical patients of wounds

Sl	Wound index	Incidences of granulation and epithelialization	Score
1	Complete healing of wound with scar formation	Complete formation of epithelial and granular tissue (100%)	5
2	Incomplete but healthy progress of healing	Satisfactory progress of formation of epithelial and granular tissue (75%)	4
3	Healing process is delayed but initiation for granulation and epithelial tissue	Moderate formation of granulation and epithelial tissue (50%)	3
4	Wound is healthy but healing process is yet to start	Initiation of process of formation of epithelial and granular tissue (25%)	2
5	Unhealthy wound and non-initiation of healing process	No granular or epithelial tissue formation (0%)	1
Total			15

Here, the higher the score, the better the healing

endogenous antioxidant enzymes (viz. SOD, CAT), as well as increasing hepatic GSH.

In the present study, the hepatoprotective effect of EtOH extract of *P. santalinus* was evaluated on Wistar rats (body weight 160 ± 20 g) which were acclimatized under laboratory condition for a fortnight before starting the experiments. They were provided with standard diet and water ad libitum. The animals were divided into three groups, each group having six rats. Group I was treated as normal control, group II was treated as CCl_4 control in which rats were injected with CCl_4 in paraffin oil (1:1) at the dose of 2 ml^{-kg} b.w. through i/p route, and group III received hydro-alcoholic (30:70) extract (EtOH) of *P. santalinus* through oral route at a dose of

Table 44.7 Effect of ethanol extract ointment of *Pterocarpus santalinus* Linn. (5% w/w) in patients of chronic non-healing wounds

Treatment group	Physical criteria (mean ± SE)								Haematological parameters (mean ± SE)			
	Wound area (mm^2)		Wound index		Granulation		Epithelialization		ESR (mm)		Lymphocytes (%)	
	Day '0'	Day '15'	Day '0'	Day '15'	Day '0'	Day '15'	Day '0'	Day '15'	Day '0'	Day '15'	Day '0'	Day '15'
P. santalinus ($n = 06$)	16.54 ± 3.68	8.25* ± 1.76	0.14 ± 0.02	3.10* ± 0.15	0	3.96* ± 0.19	0	3.96* ± 0.19	14.75 ± 3.18	05.00 *±	18.00 ± 3.54	23.00* ± 4.07
Vehicle ($n = 06$)	25.00 ± 4.34	21.26 ± 4.03	2.00 ± 0.14	2.00 ± 0.14	0	0.66 ± 0.02	0	0.54 ± 0.02	16.00 ± 3.50	14.00 ± 3.12	19.00 ± 3.88	18.00 ± 3.74

'*n*' indicates number of patient in each group. *$p < 0.05$

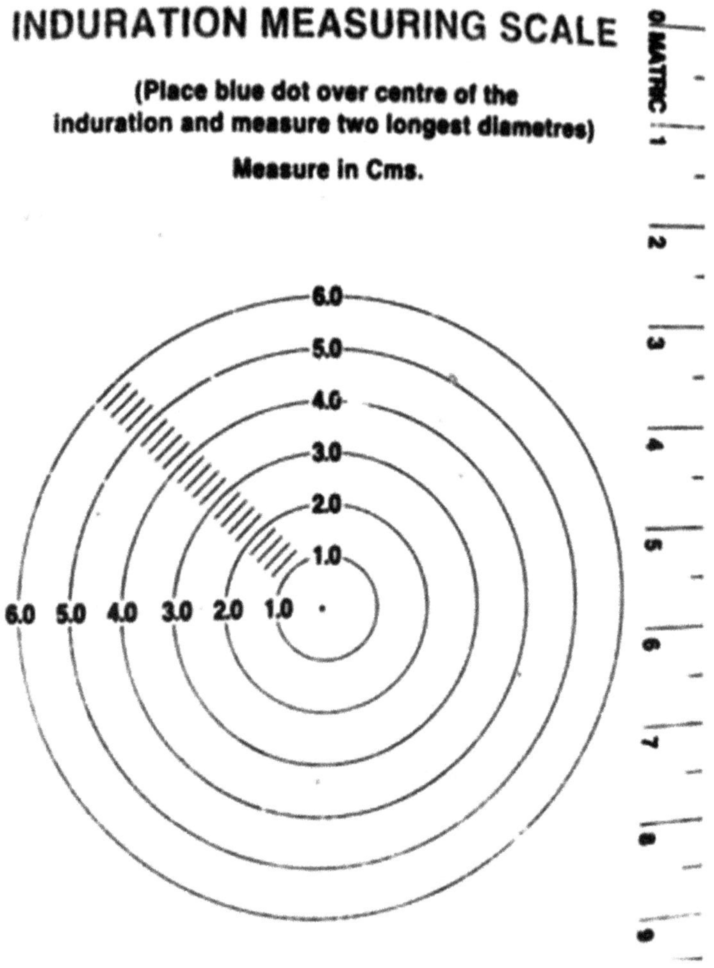

Fig. 44.12 Wound contraction size measurement scale in clinical patients

500 mg^{-kg} b.w. for 2 days prior to CCl_4 administration (day 3), and the extract was continued for another 3 days. On day 6, twenty-four (24) hours after the last i/p injection of CCl_4, blood samples from all the groups of rats were drawn from the heart after being anaesthetized under a light diethyl ether. Animals of all the groups were sacrificed thereafter, and liver tissues were collected, homogenized (10% w/v) in normal PBS (pH 7.2) and stored at −20 °C for biochemical analysis. Serum biochemical parameters like ALT, AST and ALP were estimated according to the method of commercially available kit (BioLab Diagnostics, India). From the liver homogenate, malondialdehyde (Uchiyama and Mihars 1978), GSH (Teitze 1969), SOD (Misra and Fridovich 1972) and catalase (Beers Jr. and Sizer 1952) were estimated.

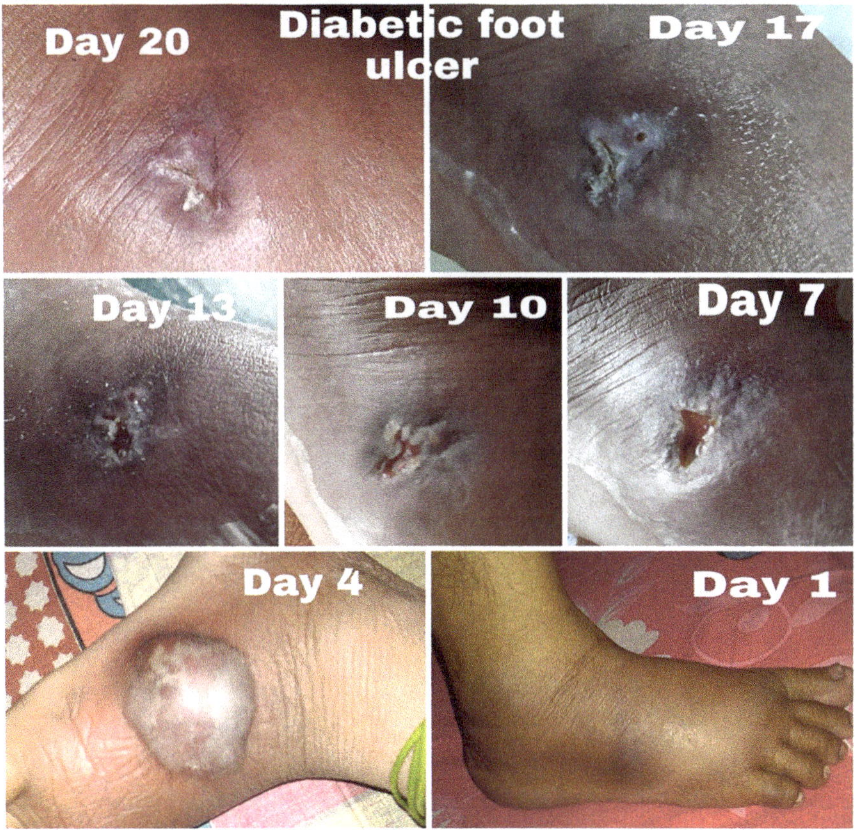

Fig. 44.13 Effects of EtOH extract ointment (5% w/w) of *P. santalinus* on different days in diabetic foot ulcer

The hepatoprotective effects of *P. santalinus* extract on serum ALT, AST and ALP activity are shown in Table 44.8. In the CCl_4 group, serum ALT, AST and ALP activity was significantly increased in comparison with that of control group, whereas these values were 22–35% decreased in drug-treated groups. GSH, a non-enzymatic endogenous antioxidant, is an important biomolecule for combating chemically induced toxicity. The level of GSH was substantially decreased in the CCl_4 group compared to the control group. Pre-treatment with *P. santalinus* reduced the extent of CCl_4-induced hepatic GSH depletion by 35% (Table 44.9). Lipid peroxidation is considered to be one of the principal causes of CCl_4-induced liver injury. Pre-treatment of rat with *P. santalinus* effectively inhibited CCl_4-induced hepatotoxicity (73.8%, $p < 0.02$), as shown by the reduced level of hepatic MDA formation, an index of the chain reaction of lipid peroxidation (Table 44.9). SOD is an extremely effective antioxidant enzyme. The increased production of free radicals caused by the administration of CCl_4 was a major cause of the reduced SOD activity.

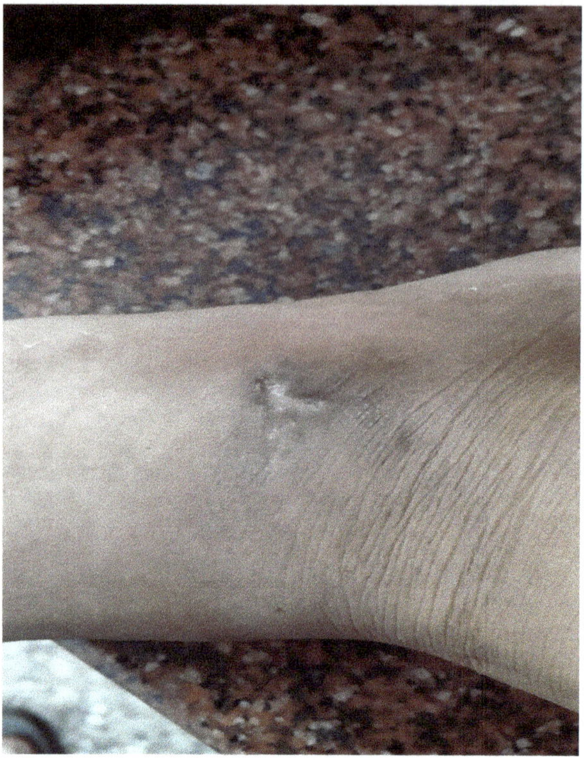

Fig. 44.14 Complete healing of diabetic foot ulcer on day 22 with EtOH extract ointment (5% w/w) of *P. santalinus*

Table 44.8 Changes in serum hepatic markers in CCl_4 and *P. santalinus* Linn. in rats

Groups	ALT (KU/ml)	AST (KU/ml)	ALP (KU/ml)
Normal control	60.42 ± 7.47	66.83 ± 13.86	80.17 ± 13.26
Only CCl_4	268.50 ± 11.52	480.83 ± 35.82	219.83 ± 22.10
CCl_4 + drug (500 mg/kg)	221.42 ± 11.95*	385.17 ± 23.56*	170.67 ± 10.82

*$p < 0.05$ when compared with CCl_4 group

Table 44.9 Changes in hepatic lipid peroxidation, GSH contents, CAT and SOD activities in CCl_4 and *P. santalinus* Linn. in rats

Groups	GSH (mU/mg tissue)	CAT (mU/mg tissue)	SOD (mU/mg tissue)	MDA (nM/mg tissue)
Normal control	1.789 ± 10.16	131.65 ± 5.81	101.91 ± 6.06	56.41 ± 6.52
Only CCl_4	1.364 ± 0.09	95.53 ± 4.76	41.64 ± 5.57	113.55 ± 10.33
CCl_4 + drug (500 mg/kg)	1.502 ± 0.10	116.63 ± 6.64*	83.47 ± 9.65*	71.37 ± 5.67*

*$p < 0.05$ when compared with CCl_4 group

The data showed that the SOD activity significantly decreased in rats treated with CCl_4 compared with the control group. Pre-treatment with *P. santalinus* significantly (69.4%, $p < 0.02$) prevented the decrease in SOD activity (Table 44.9). A significant decrease in hepatic CAT activity was observed in the CCl_4 control when compared with the control group, but the decrease induced by CCl_4 was significantly (58.4%, $p < 0.05$) ameliorated by the pre-treatment of *P. santalinus* extract (Table 44.9).

44.2.7 Evaluation of DNA Damage Protection Role of *P. santalinus* EtOH Extract (Comet Assay)

The experimentation was done on fresh goat livers which were perfused in PBS (pH 7.4) with collagenase, and the liver was then minced in minute pieces, and cells were separated using cell strainer having 40 μl pore size (Genetix cell strainer, S. Korea). The cells were then washed with HBSS and centrifuged at 800 g to eliminate fine debris. The purity of hepatocytes was examined by phase-contrast microscopy. The isolated cells were taken in an Eppendorf tube containing 0.5 ml RPMI and 10% FBS. Cells were then pre-incubated with with/without *P. santalinus* extract at a concentration of 100 μg/ml at 37 °C for 30 min, and then 10 μl 1.5% CCL_4 was added to treated and only CCL_4 groups and incubated for another 30 min at 37 °C in a CO_2 incubator. 20 μl cell suspensions in 90 μl of 0.5% low melting point agarose (LMPA) were used for preparation of slides and solidified at 4 °C on a slide pre-coated with a film of 1% normal melting point agarose (NMPA). Two slides were prepared for each sample. After solidification, slides were put in cold lysis buffer (2.5 M NaCl, 100 mM EDTA, 10 mM Tris buffer, 10% DMSO, Triton X-100 0.8%, pH 10) for 1 h, followed by alkylation with electrophoresis buffer (1 mM EDTA, 0.3 N NaOH, pH 13.0) for 20 min for unwinding of DNA. Then, electrophoresis was performed for 30 min at 25 V/300 mA, and electrophoresis slides were neutralized (three times) and stained with ethidium bromide solution (20 mg/ml). The stained nuclei were visualized under fluorescent microscopy and photographed. Olive tail moment (OTM) of individual stained nuclei was calculated using CaspLab® comet assay software. The lower the value of olive tail moment, the higher the level of DNA protection of the test extract.

The study revealed that there was about 50.39% prevention of DNA damage in CCl_4 along with *P. santalinus*-treated group with respect to CCl_4 alone (Table 44.10). Results are illustrated in Fig. 44.15 to envisage the mean olive tail moment (whose magnitude reflects the extent of DNA strand breaks per nucleus) for all groups including the untreated control and vehicle control group in which most cells presented with no comet. DNA was tightly compressed and maintained the circular nature of the normal nucleus (Fig. 44.15a, b). Fig. 44.15c of CCl_4 group; the profile of the nuclear DNA in this group was altered with the appearance of a

Table 44.10 Changes in DNA damage in CCl_4- and *P. santalinus* Linn.-treated isolated hepatocytes

Groups	OTM
Normal control	4.53 ± 1.41
Vehicle control	5.22 ± 1.91
Only CCl_4 (1.5%)	14.41 ± 3.81
CCl_4 + drug (100 μg/ml)	7.30 ± 2.04

Data are in mean ± SEM

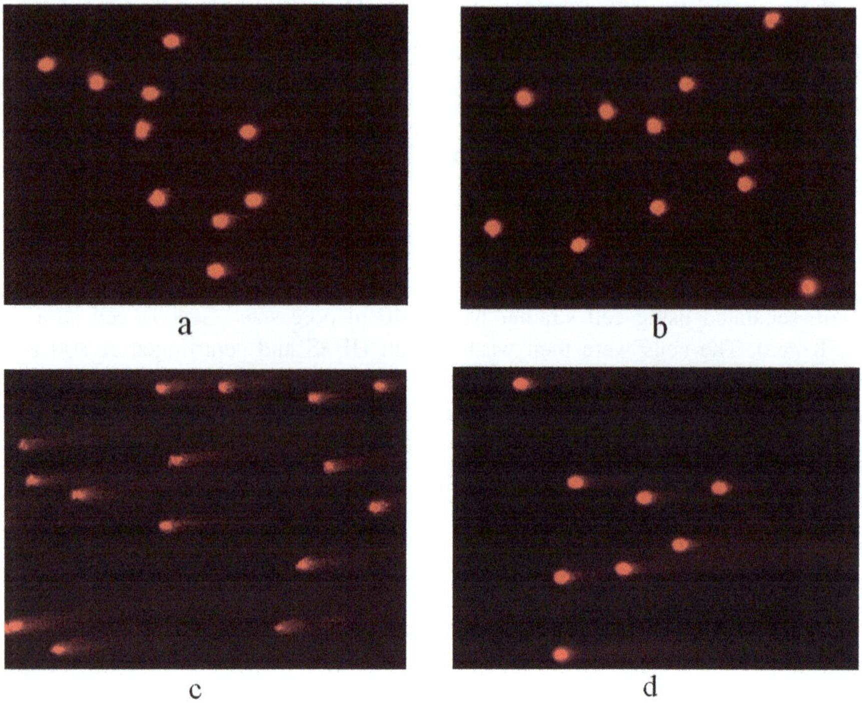

Fig. 44.15 Mean olive tail moment in different groups subjected to comet assay where **a** = normal control, **b** = vehicle control, **c** = only CCL_4 and **d** = CCL_4 along with *P. santalinus*-treated group

fluorescent streak extending from the nucleus. Image 15d of the treated group showed less damage of liver cells. Thus, treatment with PS effectively alleviated the DNA damage, induced by CCl_4.

44.3 Summary and Path Forward

Pterocarpus santalinus, commonly known in India as Raktachandan, is considered as a holy plant not only for its use in ritualistic activity but also for its multipurpose therapeutic applications. Interestingly many natural products are described in *Ayurveda* for their diverse therapeutic applications, and *P. santalinus* is such an important example. The description regarding the diverse therapeutic use of *P. santalinus* is available in different classical texts of *Ayurveda*, and Bhavaprakasha Samhita of the eleventh century (Sengupta et al. 2000) described it vividly as follows:

रक्तचन्दनम आख्यातं रक्ताङ्गं क्षुद्रचन्दनम।
तिलपर्णम रक्तसारं तत प्रवाल फलं स्मृतम।।
रक्तं शीतं गुरु स्वादु छर्दितृष्णाश्रपित्तहृत।
तिक्तं नेत्रहितं वृष्यं ज्वरव्रणबिषापहम।।

The verse narrates the potent role of the plant in the amelioration of diseases like vomiting manifested due to different pathological conditions, thirst disorders, diseases of Pitta source like hepatic disorders, diseases of eye, wounds and various systemic toxicities. The narration can be translated in modern scientific language as its role in the eradication of noxious substances from the body like oxidative free radicals, maintenance of physiological status of the hepatic biochemistry, healing of wounds, etc. The current research work was planned to establish the theories of the *Ayurveda* in modern parlance.

A detailed phytochemical investigation of the plant was performed with 80% aq. ethanol concentrate of *P. santalinus* which demonstrated strong antioxidant agent exercises utilizing DPPH and ABTS measure. The IC50 estimation of DPPH examined was seen as higher than that of ABTS measure. The complete phenolic segment displayed cell reinforcement action through adsorption and balance of the free radicals, while flavonoid and flavanol indicated antioxidant action through rummaging or chelating process. The antioxidant agent exercises of the extractive arrangement speak to a significant parameter to assess the organic property of the plant. In this way, it is important to describe and evaluate the significant mixes like phenolic acids and flavonoids present in the plant and furthermore to approve the technique for division and distinguishing proof of dynamic constituents. The HPLC investigation indicated the nearness of protocatechuic acid which is a broadly disseminated normally happening phenolic acid. It has auxiliary closeness with gallic acid, caffeic acid, vanillic acid and syringic acid which are notable antioxidant agent mixes. A decent measure of these phenolics identified in this plant may be liable for the solid cell reinforcement properties of the plant and in this manner help in the avoidance and treatment of different oxidative pressure-related illnesses, for example, neurodegenerative and hepatic sicknesses (Kakkar and Bais 2014). It is likewise detailed that because of the nearness of p-coumaric acid, the plant is accepted to have cancer prevention agent conduct in this way lessening the development of cancer-causing nitrosamines in the stomach (Ramadoss Karthikeyan et al. 2015). All these logical foundations are very much bolstered with the experimentation portrayed in the ebb and flow investigate work. The HPLC examination of the ethanol concentrate of *P. santalinus* showed the nearness of good measure of naringin and rutin. Rutin is a phenolic compound with glycosidic linkage. It is accounted for to display huge pharmacological exercises, including antioxidation, hostile to irritation, against diabetic and so on (Moghaddasian et al. 2013).

Biological investigation of the plant for wound healing, hepatoprotective and DNA damage protecting effect was performed on the basis of the chemical screening, and it is envisaged that *P. santalinus* has potent wound healing property which was observed with ointment prepared by both crude plant (15% w/w) and EtOH extract (5% w/w). The wound healing potentiality of ointment prepared from EtOH extract was found to be better than the crude part, but both of them showed potentiality in this perspective. The healing of wound is a natural phenomenon, but the natural process of healing may lack quality, aesthetics and duration. This entire goal was fulfilled by this plant due to its prompt synthesis of hydroxyproline, its anti-inflammatory mechanism and the proliferative property of lymphocytic tissues, which are the fundamental sources of collagen tissue formation. The antioxidative and free radical scavenging activities of the phenolic and polyphenolic compounds like gallic acid, protocatechuic acid, gentisic acid, chlorogenic acid, *p*-hydroxybenzoic acid, vanillic acid, caffeic acid, syringic acid, *p*-coumaric acid, ferulic acid, sinapic acid, salicylic acid and ellagic acid and catechin, rutin, myricetin, quercetin, naringin, apigenin and kaempferol are solely responsible for the healing property as well as hepatoprotective and DNA damage protection activity. The hepatoprotective role of the plant supports the maintenance pathway of different metabolic activities followed by the augmentation of biological synthesis of beneficial microelements like growth factors to accelerate the healing process and other biological achievements. The study of DNA damage protection is an added accomplishment which indicates the safety aspects of the plant on the one hand and role of *P. santalinus* in succeeding other biological outcomes like wound healing or hepatoprotective values on the other. It can be concluded that the idea of *Ayurveda* in considering single natural agent for multiple biological effects is contemplated well with the present view of research work to explore the wonder gift.

44.4 Validation of the Plant

Pterocarpus santalinus is a rare plant of the world which was first identified in *Ayurveda* a long way back for its medicinal use in varied fields. Identification of the plant and its specific parts for specific therapeutic uses definitely favours credential of Ayurveda regarding citation of this valuable wealth in various classical texts. In spite, many advanced research works were carried out with different plants of Ayurveda but it is interesting that a very few scientific personalities approaches with this rare plant to evaluate its biological properties. The plant *P. santalinus* is now becoming endangered from this land because of the wide cutting, uprooting and environmental threats. There is great global demand of about 3000 tonnes per annum of the plant for its deep red heartwood (red gold), and the plant is majorly distributed in Deccan eco-region of India (UNEP-WCMC 2017, Technical Report). Conservation of the plant is urgently needed not only for its cosmetic value but also for its tremendous biological property to save human life from different ailments. Germination of the plant in wide spectrum at the suitable geographical area like Deccan eco-region of India is advocated to prevent its destruction. It is reported with one such in vitro experimentation that 100% germination of the plant can be done in 3% sucrose solution (Padmalatha and Prasad 2007). The time has now come to apply the outcome of the scientific results of laboratory tests to the land for preserving this wonder gift.

References

Arunkumar AN, Joshi G (2014) Pterocarpus santalinus (red sanders) en endemic, endangered tree of India: current status, improvement and the future. J Tropic Forest Environ 4(2):01–10

Bailey AJ, Sims TJ, Duance VC et al (1979) Partial characterization of a second basement membrane collagen human placenta—evidences for the existence of two type IV collagen molecules. FEBS Lett 99(2):361–366

Beers RF Jr, Sizer IW (1952) J Biol Chem 195:133–140

Bhandari MJ, Chandrashekar KR (2011) Herbal therapy of herpes in the ethnomedicine of coastal Karnataka. Indian J Tradit Knowl 10(3):528–532

Bhange P, Bhangade SL, Mehere AP (2016) A review on red sandal wood. Int J Phytother 6 (1):9–13

Biswas TK (2018) Drug discovery from Ayurveda: mode of approach and applications. In: Mandal SC, Mandal V, Konishi T (eds) Natural products and drug discovery—an integrated approach. Elsevier, Amsterdam., Section I, pp 3–26

Biswas TK, Maity LN, Mukherjee B (2004a) Wound healing potential of *Pterocarpus santalinus* Linn: a pharmacological evaluation. Int J Lower Extrem Wounds 3(3):143–150

Biswas TK, Maity LN, Mukherjee B (2004b) The clinical evaluation of *Pterocarpus santalinus* Linn ointment on lower extremity wounds- a preliminary report. Int J Lower Extrem Wounds 3 (4):227–232

Biswas TK, Mukherjee B (2003) Plant medicine of Indian origin for wound healing activity: a review. Int J Lower Extrem Wounds 2(1):25–39

Blois MS (1958) Antioxidant determination by the use of a stable free radical. Nature 181:1199–1200

Bodding PO (1986) List of Santal prescriptions. In: Studies in Santal medicine and connected folklores. The Asiatic Society, Calcutta, pp 161–393

ICH-Q2A (1995) Text on validation of analytical procedures, March

ICH-Q2B (1996) Validation of analytical procedures: methodology, November

Kakkar S, Bais S (2014) A review on Protocatechuic acid and its pharmacological potential. ISRN Pharmacol 26:1–9

Kumaran A, Karunakaran RJ (2006) Antioxidant and free radical scavenging activity of an aqueous extract of *Coleus aromaticus*. Food Chem 97:109–114

Manjunatha BK (2006) Hepatoprotective activity of *Pterocarpus santalinus* L. f., an endangered plant. Indian J Pharmacol 38:25–28

Miller N, Rice-Evans C (1997) Factors influencing the antioxidant activity determined by the ABTS radical cation assay. Free Radic Res 26:195–199

Misra HP, Fridovich I (1972) The role of superoxide anion in the autoxidation of epinephrine and a simple assay for superoxide dismutase. J Biol Chem 247:31–70

Moghaddasian B, Eradatmand AD, Alaghemand A (2013) Simultaneous determination of rutin and quercetin in different parts of *Capparis spinosa*. Bull Env Pharmacol Life Sci 2:35–38

Ordonez AAL, Gomez JG, Vattuone MA et al (2006) Antioxidant activities of Sechium edule (Jacq.) swart extracts. Food Chem 97:452–458

Oyaizu M (1986) Studies on product on browning reaction prepared from glucose amine. Jpn J Nutr 44:307–315

Padmalatha K, Prasad MNV (2007) Seed germination studies in *Pterocarpus santalinus* L.f.—an endangered and endemic medicinal plant, and relevance of conservation. Seed Sci Biotechnol 1 (2):32–34

Prakash L, Majeed M (2008) Red sandal wood extract. In: Natural Actives Lend Colour to Cosmetics and More, Sabinsa Corporation, 20 Lake Drive, East Windsor, NJ 08520, USA, pp. 9–10

Ramadoss Karthikeyan SI, Devadasu C, SrinivasaBabu P (2015) Isolation, characterization and RP-HPLC estimation of p-coumaric acid from methanolic extract of Durva grass (*Cynodon dactylon* Linn.) (Pers.). Int J Anal Chem 20(15):1–7

Rubin E, Farber JL (1996) Repair, regeneration and fibrosis. In: Pathology, 2nd edn. J B Lippincott Co, Philadelphia, PA, pp 215–325

Sarma J (2015) Study on traditional worship plants in Hindu religion from Nalbari and Sonitpur districts of Assam. Int J Scientific Res Public 5(5):1–5

Seal T, Chaudhuri K, Pillai B (2017) Nutraceutical and antioxidant properties of Cucumis hardwickii Royle: a potent wild edible fruit collected from Uttarakhand, India. J Pharmacogn Phytochem 6(6):1837–1847

Sengupta K, Sensharma K, Bhattacharyya S (2000) Raktachandana: Karpuradivarga. In: Bhavaprakash, Deepayana, vol I. Purvakhanda, Kolkata, pp 195–207

Singleton VL, Rossi JA (1965) Colorimetry of total phenolics with Phosphomolybdic-phosphotungstic acid reagents. Am J Enol Vitic 16:144–158

Teitze F (1969) Enzymatic method for the quantitative determination of nanogram amounts of total and oxidized glutathione: applications to mammalian blood and other tissues. Anal Biochem 27:502–522

Uchiyama M, Mihars M (1978) Determination of malondialdehyde precursor in tissues by thiobarbituric acid. Ann Biochem 86:271–278

UNEP-WCMC (2017), *Pterocarpus santalinus:* India, In: Technical Report, Report on species/ country combination selected for review by the Plants Committee following CoP16, CITES Project no. A-498, UNEP World Conservation Monitoring Centre, Cambridge, UK, 22–32

Vassou SL, Nithaniyal S, Parani RM (2016) Creation of reference DNA barcode library and authentication of medicinal plant raw drugs used in Ayurvedic medicine. BMC Complement Alternet Med 16(1):186–191

Vedavathy S, Sudhakar A, Mrudula V (1997) Tribal medicinal plants of Chittoor. Anc Sci Life 26:307–331

Woessner JF (1961) The determination of hydroxyproline in tissue and protein samples containing small proportion of this amino acids. Arch Biochem Biophys 93:440–447

Phytochemicals and Investigations on Traditionally Used Medicinal Mushrooms

45

Wangkheirakpam Sujata, Sanjram Nomita Devi, and Subhash C. Mandal

Abstract

Mushrooms degrade matter to produce metabolite for sustaining life. For degradation they produce enzymes and certain metabolites. Both the enzymes and the metabolites are of use for human. The metabolites have medicinal applications. The traditional use and the scientific work on some traditionally used medicinal mushrooms have been discussed here. Some of the mushrooms are *Auricularia delicata*, *A. polytricha*, *A. auricular*, *Agaricus blazei*, *Coprinus comatus*, *Cordyceps* spp., *Fomes fomentarius*, *Fomitopsis pinicola*, *Ganoderma lucidum*, etc. *Auricularia delicata* has been used in the traditional medicine of Manipur, India, for dysentery and liver healing therapy. A scientific investigation done by one of the authors on traditionally used *Auricularia* species showed its hepatoprotective activity. The compound isolated from the ethyl acetate extract was chlorogenic acid. It is known to have hepatoprotective activity. This has been a good example illustrating that traditional medicine is a good lead to drug discovery.

Keywords

Traditional use of mushrooms · Scientific reason · Bioactive compounds · *Maibaron* · Puyas

W. Sujata (✉) · S. N. Devi
Chemistry Department, NIT Manipur, Langol, India
e-mail: wangkheimayums@yahoo.com

S. C. Mandal
Division of Pharmacognosy, Department of Pharmaceutical Technology, Jadavpur University, Kolkata, India

S. C. Mandal et al. (eds.), *Evidence Based Validation of Traditional Medicines*,
https://doi.org/10.1007/978-981-15-8127-4_45

45.1 Introduction

Over the past decades, mushrooms which have therapeutic property have been used as nutraceuticals and health supplements. These mushrooms are important areas of research for medical and pharmacological researchers (Guillamon et al. 2010; Kwok et al. 2005). In different places of the world, there are mushrooms which have been studied for their traditional medicinal use. The therapeutic knowledge of the traditionally used medicinal mushrooms has been transmitted orally from one generation to the next medicine man and woman across many places of the world. Such knowledge of use of medicinal mushrooms has been documented in different parts of the world as their medicine books in ancient texts like "Vedas." Roman and Chinese traditional medicinal use mentions of mushrooms having property of providing strength and life force. In olden times people considered mushrooms as the gifts of god Osiris. There are records of people of China, India, and Iran using medicinal mushrooms (Sharma 2003; Lowy 1971). Mushrooms play an important role in the degradation of organic matters on earth. Otherwise the earth would have been a huge heap of waste. Fungi degrade organic waste and manage the waste. For the purpose, it produces many enzymes to perform the degradation. Enzymes and certain metabolites are produced to protect themselves and sustain their livelihood. Such compounds have been useful for treating many ailments. Two *Auricularia* species have been studied for their chemical constituents and bioactivity. Both species are edible mushrooms. Chlorogenic acid which has hepatoprotective effect was present in both species. It augmented the traditional medicinal use of the species in liver healing in Manipur. The data can be a good lead in the discovery of drugs from the hepatoprotective effect of *Auricularia* sp. (Sujata and Joshi 2017; Sujata et al. 2018). As from the past, the traditional medicinal use has been good leads to new drugs. Metabolites, proteins, and enzymes from mushrooms can be leads to new drugs. A good example of protein being a good lead is exendin-4, a natural peptide of 39 amino acids. It is isolated from a lizard and used for the treatment of diabetes (Minghan 2011). Enzymes are also an important bioactive which have therapeutic use. In some diseases, especially inheritable genetic disorders, there may be a deficiency or even a total absence of one or more enzymes. There are disease conditions for which excessive activity of an enzyme may be the cause. Measurements of the activities of enzymes in blood plasma, erythrocytes, or tissue samples are important in the diagnosis of many diseases. Many drugs exert their biological effects through interactions with enzymes. Extracellular enzymes or exoenzymes are enzymes from mushrooms. They are synthesized inside the cell and then secreted outside the cell. They break down complex macromolecules into smaller units, which are to be taken up by the cell for growth and assimilation (David and Michael 2004; Sinsabaugh 1994). These enzymes degrade complex organic matter such as cellulose and hemicellulose into simple sugars. Host enzyme-producing organisms use the products as a source of carbon, energy, and nutrients (Burns et al. 2013). Some of the enzymes produced by fungus are cellulase, lignolytic, xylanase, hemicellulase, exoglucanase, laccases, manganese peroxidases, fibrinolytic enzymes, etc. Here the medicinal use of the mushrooms producing fibrinolytic enzymes is of interest. Most of the mushrooms have medicinal properties; herein in this chapter, the authors are focusing on the mushrooms which have been used traditionally for medicinal purpose because of the presence of different metabolites, proteins, and enzymes discussed. The scientific intervention and the results will be correlated with the traditional uses. It is a well-known fact that traditional medicinal system all around the world has been good guides for drug discovery. And in the far-flung rural places, such medicines are the only way to combat illness. There are many diseases for which there are no appropriate therapies, so a study on traditional mushroom is of good importance. *Auricularia delicata*, *A. polytricha*, *A. auricular*, *Agaricus blazei*, *Copernicus comatus*, *Cordyceps* spp., *Fomes fomentarius*, *Fomitopsis pinicola*, *Ganoderma lucidum*, etc. are some mushrooms that are being studied in this chapter. The mushrooms had traditional medicinal use, and their bioactivity was found out in many in vitro and in vivo experiments. To illustrate its traditional use as a guide for drug discovery, a case study on *A. delicata* conducted by the author will be discussed. Table 45.1 includes the list of mushrooms which have been used as traditional medicine and validated with scientific studies.

45.2 Traditional Mushrooms and their Scientific Studies

45.2.1 *Auricularia delicata, A. polytricha,* and *A. auricular*

Auricularia is a genus comprising edible macro-fungi. The *Auricularia* species grows on fresh wood or decaying tree trunks. *Auricularia* spp. of family *Auriculariaceae* are locally known as Uchina and are consumed as a dish widely by patients in the course of therapy by local traditional healers in Manipur, India, from olden times for the treatment of diarrhea, dysentery, hypertension, constipation, and liver ailments. Chlorogenic acid was identified from the *Auricularia* species (Sujata and Joshi 2017). Ethyl acetate extract of *Auricularia delicata* had hepatoprotective property (Sujata et al. 2018).

A. auricula-judae was considered as a medicinal mushroom in the Chinese ancient medicine book. It was effective in the treatment of chronic bronchitis. Over that it was consumed as an anti-aging food in ancient China. The DPPH radical scavenging assay suggests that the extract of *A. auricular* has very high antioxidant activity. The extract of *A. auricular* had a strong superoxide anion scavenging activity. The results showed that the extract of *A. auricular* has excellent antioxidant activity. The extract had antioxidant activity and promoted the biosynthesis of procollagen, a precursor of collagen in HaCaT cells. In addition, the expression of *HAS-3* (hyaluronic acid synthase), which is a moisturizing factor, was increased in HaCaT cells. These results support the use of *A. auricular* taken as beauty food in traditional use (Young-Ji et al. 2018).

Table 45.1 Mushrooms which have been used as traditional medicine and validated with scientific studies

Sl no.	Mushrooms	Traditional use	Scientific study or validation	Bioactives
1	*Auricularia delicata* *A. polytricha* *A. Auricular*	In Manipur, India, a traditional medicine used for diarrhea, dysentery, hypertension, constipation, and liver ailments A Chinese traditional medicine used for chronic bronchitis Anti-aging agent	Hepatoprotective activity of *A. delicata* Promoted the biosynthesis of procollagen, a precursor of collagen in HaCaT cells	Chlorogenic acid (Sujata and Joshi 2017; Sujata et al. 2018; Young-Ji et al. 2018)
2	*Agaricus blazei*	Used for treating cancer, type 2 diabetes, high cholesterol, hardening of the arteries (atherosclerosis), and hepatitis B and preventing heart disease, weakened bones (osteoporosis), and stomach ulcers in Japan and China	Anticancer and immunostimulant activities	Polysaccharide (Hang et al. 2013; Gonzaga et al. 2005; Kawagishi et al. 1990; Shu and Wen 2003)
3	*Coprinus comatus*	Used as digestives and in constipation and hemorrhoids in China	Antidiabetic and anticancer effect	Vanadium and ergothioneine, respectively (Ivo 2009; Jianzhe and Mao 1989; Badalyan et al. 2003; Bailey et al. 1984; Dijkstra and Wiken 1976; Ershova et al. 2001; Fan et al. 2006; Gu and Leonard 2006; Han et al. 2006; Han et al. 2003; Li et al. 2001a, b; Li and Xiang 2005; List 1957; Luo et al. 2004; Wu et al. 2003; Yang et al. 2003; Yilmaz et al. 2006)
4	*Cordyceps* spp.	Used in India, China, Nepal, and Bhutan and Tibetan medicine system	Enhanced endurance and resistance to fatigue	Ashok and Kailash (2011); Li et al. (2001a, b); Wang et al. (1995)

(continued)

Table 45.1 (continued)

Sl no.	Mushrooms	Traditional use	Scientific study or validation	Bioactives
5	*Fomes fomentarius*	The Chinese apply *Fomes fomentarius* in the treatment of cancer of the throat, stomach, and uterus. Further to increase the immunity of the body	Enhances blood circulation, regulates blood sugar, and lowers blood pressure. It extends life by dealing with malignant cells that cause debilitating diseases in our body	Wasser et al. (2005); McMeekin (2005)
6	*Ganoderma lucidum*	Used in popular traditional medicine in Asia	Anticancer activity	Polysaccharides and triterpenoids (Wasser et al. 2005; McMeekin 2005; Wachtel-Galor et al. 2004; Wachtel-Galor et al. 2011; Tomasi et al. 2004)
7	*Fomitopsis pinicola*	In folk medicine it has been an anti-inflammatory, anti-hemorrhagic, and antimicrobial agent	Apoptosis effect on ROS and migration arrest on matrix metalloproteinase	Ergosterol (Keller et al. 1996; Keller et al. 1996; Zjawiony 2004; Rogers 2011; Artem et al. 2018)
8	*Grifola frondosa*	A traditional Chinese medicine used in cancer treatment; curing of palsy, nerve pain, and arthritis; immune stimulation; and regulation of homeostasis	Inhibit carcinogenesis, metastasis, and tumor growth	D-fraction (Nik and Tom 2006)
9	*Hericium erinaceus*	Used in Chinese and Japanese medical systems to fortify the spleen and nourish the gut and also as an anticancer drug	Effective in Alzheimer's disease and Parkinson's disease	Polysaccharides, polyketides, and terpenoids (Kevin et al. 2017; Cui et al. 2016; Liu et al. 2015; Thongbai et al. 2015; Wang et al. 2015; Nagano et al. 2010; Wong et al. 2011)
10	*Lentinula edodes*	Used in China and Japan for their high immunomodulatory property in the treatment of cancer	Anticancer effect and good nutraceuticals	Lentinan (Kuppusamy et al. 2009; Medical Mushroom 2009a, b, c, d)
11	*Monascus purpureus*	Used in China and in Asia as a nutraceutical food to improve blood	Lower cholesterol by inhibiting HMG-CoA (5- hydroxy-3-methylglutaryl-	Monacolins (Thorne Research Inc. 2004; Li et al. 1998; Patrick and Uzick 2001;

(continued)

Table 45.1 (continued)

Sl no.	Mushrooms	Traditional use	Scientific study or validation	Bioactives
		circulation by decreasing cholesterol and triglyceride levels	coenzyme A) reductase, the rate-limiting step for cholesterol synthesis in the liver	Smith and Olive 2003; Liu et al. 2006; Klimek et al. 2009)
12	*Piptoporus betulinus*	Used in north European countries for clearing parasitic worms, soothing the nerves, and eliminating fatigue	It possess antibacterial, anti-fatiguing, immuno-enhancing, and soothing properties for wounds. The fractions significantly decrease viability, proliferation, and migration of tumor cells, effects the stimulatory effect of IGF. And it was nontoxic to the normal cells	Medical Mushroom (2009a, b, c, d), Marta et al. (2009)
13	*Pleurotus ostreatus*	Used in Chinese, Korean, and Japanese cuisines Lower cholesterol in the body	Antimicrobial, antiviral, antitumor, anti-human immunodeficiency virus (HIV), anti-mutagenic, antineoplastic, anti-lipidemic, antioxidant, hyperglycemic, hypotensive, anti-inflammatory, hypocholesterolemic, immunomodulatory, hepatoprotective, and anti-aging activities	Statins, lovastatins, and pleuromutilin (Medical Mushroom 2009a, b, c, d; (Kunjadia et al. 2014; Yashvant et al. 2012)
14	*Polyporus umbellatus*	A traditional Chinese medicine used for treating edema and promotion of diuretic processes (promoting urination)	Antitumor, anticancer, antioxidant, free radical scavenging, immune system enhancement, and antimicrobial activities	Yang and Zhonghua (1991), Ar et al. (2015), Zhao (2013), Ar et al. (2015)
15	*Poria cocos*	Used in traditional Chinese and Japanese medicine for its promising effects on diuretic, sedative, and tonic areas	The essential impacts were on certain diseases such as rheumatoid arthritis, psoriasis, autoimmune uveitis,	Polysaccharides and triterpenoids (Ríos 2011)

(continued)

Table 45.1 (continued)

Sl no.	Mushrooms	Traditional use	Scientific study or validation	Bioactives
			septic shock, and bronchial asthma	
16	*Trametes versicolor*	Used in Chinese and Japanese medicines	It has antioxidants activity, immune-boosting, immune function in cancers cells, enhance the efficacy of some cancer treatments, enhancement of gut health, HPV clearance, reducing inflammation, antibacterial activities, improving athletic performance, improving insulin resistance of human	Polysaccharopeptides (Medical Mushroom 2009a, b, c, d; Christopher 2004; Rajarathnam and Shashirekha 2003; Huaiqian and Lijuan 2019; Jillian 2018)

45.2.2 *Agaricus blazei* (Royal Sun Agaricus)

Agaricus mushroom is a fungus originating in Brazil. Recently grown in China, Japan, and Brazil for commercial purposes, the mushroom extract is used as medicine. The extracts of mushroom are traditionally used as a food additive and tea in Japan. *Agaricus* mushroom is used for treating cancer, type 2 diabetes, high cholesterol, hardening of the arteries (atherosclerosis), and hepatitis B. It is also helpful for digestive problems such as ulcerative colitis and Crohn's disease. It has remarkable property of reducing side effects caused by cancer therapy. Moreover, it helps in preventing heart disease, weakened bones (osteoporosis), stomach ulcers, etc. (Hang et al. 2013).

Agaricus blazei is used traditionally in Japan. Many researchers of Japan studied about the anticancer and immunostimulant activities with the extract of *Agaricus* mushroom. The consumption of this mushroom extract showed improvement in the diseased condition. The necessity to investigate the mechanism is of utmost importance. But there are lesser investigations on clinical studies. Data showed that polysaccharide is the main component of *A. brasiliensis* which was responsible for its antitumor activity (Gonzaga et al. 2005; Kawagishi et al. 1990; Shu and Wen 2003).

45.2.3 *Coprinus comatus* (Shaggy Mane)

Coprinus comatus has been recorded to be used in eastern folk medicines in China as a digestive. It is commonly known as shaggy mane since it is hairy. It is also use in the treatment of constipation and hemorrhoids. Chinese scientists used this mushroom and demonstrated the inhibition of the growth of malicious tumors in connective and supportive tissue (Bailey et al. 1984; Han et al. 2003; Ivo 2009; Jianzhe and Mao 1989). Shaggy mane has high blood sugar-lowering effects effective in the treatment of diabetes. The vanadium content of the mushroom was responsible for the property. It is reported that the insulin target cells are sensitized as the insulin-producing ß-cells are protected and regenerated in the pancreas (Badalyan et al. 2003). Researchers have identified some potent bioactive compounds useful for breast cancer from the water extract of *Coprinus comatus*. This has a great significance because there is no effective therapy for estrogen-independent (ER-) breast cancer (Badalyan et al. 2003; Bailey et al. 1984; Dijkstra and Wiken 1976; Ershova et al. 2001; Fan et al. 2006; Gu and Leonard 2006; Han et al. 2006; Han et al. 2003; Li et al. 2001a, b; Li and Xiang 2005; List 1957; Luo et al. 2004; Wu et al. 2003; Yang et al. 2003; Yilmaz et al. 2006). Further it has been reported to have hypoglycemic effects, anti-nematode activity, and antioxidant activity. Ergothioneine was responsible for the antioxidant activity. The hypoglycemic activity was proved with in vivo experiments with investigations on mice. It also showed metabolic effects in the weight gain in mice (Badalyan et al. 2003; Bailey et al. 1984; Dijkstra and Wiken 1976; Ershova et al. 2001; Fan et al. 2006; Gu and Leonard 2006; Han et al. 2006, 2003; Li et al. 2001a, b; Li and Xiang 2005; List 1957; Luo et al. 2004; Wu et al. 2003; Yang et al. 2003; Yilmaz et al. 2006).

45.2.4 *Cordyceps* Spp. (Caterpillar Fungus)

The fungus *Cordyceps sinensis* has been used in traditional Chinese medicine. It has been found to be used in Tibetan medicine also (Ashok and Kailash 2011). Its uses are also there in folk practices of Sikkim and other parts of India, China, Nepal, and Bhutan. It is a powerful antioxidant. The *Cordyceps* mushroom also has bioactives showing antioxidant properties (Li et al. 2001a, b).

In vivo investigation was conducted on mice, with a double-blind, placebo-controlled trial, for *Cordyceps* for property of enhanced endurance and resistance to fatigue. After 3 weeks of administration, the groups given CS-4 were able to swim significantly longer than the control groups. The results of the study were dose-dependent with one group on a higher dose showing a 30% increase in endurance and a second group showing a 73% increase in endurance. CS formulation would enhance lactate clearance and allow athletes greater anaerobic physical performance (Wang et al. 1995).

45.2.5 *Fomes fomentarius* (Tinder Fungus)

Fomes fomentarius has been used in the ancient Indian medicine as a diuretic agent. It is also known as tinder fungus. It is also used as a laxative to stimulate bowel movement. The fungus is also used to stabilize the nerves. The Chinese apply *Fomes fomentarius* in the treatment of cancer of the throat. The Chinese use it too in the treatment of cancer of the stomach and the cancer of the uterus (Wasser et al. 2005).

Fomes fomentarius or tinder is used to increase the immunity of the body. Investigations on the fungus have successful results such as enhancing blood circulation, regulating blood sugar, and lowering blood pressure. It enhances the medicinal use of the mushroom. It extends life by dealing with malignant cells that cause debilitating diseases in our body (McMeekin 2005).

45.2.6 *Ganoderma lucidum* (Reishi, Lingzhi)

Ganoderma lucidum is also known as reishi or lingzhi. Lingzhi has been used as a powerful medicinal mushroom over 2000 years. It has been documented in ancient scripts of China (Wasser et al. 2005). The proliferation of *G. lucidum* images in art began in 1400 AD, and they are associated with Taoism in China (McMeekin 2005).

Ganoderma species is a popular traditional medicine in Asia till date. Their use is spreading throughout the world (Wachtel-Galor et al. 2011). Fifty-eight basidiomycete mushrooms of *Ganoderma* species were studied for their in vitro anticancer activities. *G. lucidum* was shown to be the most effective in treating cancer cells. It was found out that *G. lucidum* induced cell cycle arrest and apoptosis in various human and rodent tumors (Wachtel-Galor et al. 2004; Tomasi et al. 2004). There were reports on antioxidant activities helping in the prevention of cancer and other chronic diseases. The mechanism was that antioxidants are protecting the cellular components from oxidative damage and thus decreasing the risk of mutations and carcinogenesis. It protects immune cells by maintaining immune surveillance and responses. The main source of their antioxidant activity is particularly from the polysaccharides and triterpenoids present in *G. lucidum* (Benzie and Wachtel-Galor 2009).

45.2.7 *Fomitopsis pinicola* (Red Belted Polypore)

Fomitopsis pinicola is known as red belted polypore. It has fruiting bodies which are considered nontoxic mushrooms in Europe (Keller et al. 1996). It has been also used in Korean folk medicine as a major hemostatic and anti-inflammation agent (Keller et al. 1996; Zjawiony 2004). It is used as tonic to decrease the inflammation of the digestive tract and increase the resistance to cancers. It is used by the Cree Indians as an anti-hemorrhagic agent to stop bleeding. From the ancient records in folk medicine, it has been used as an anti-inflammatory, anti-hemorrhagic, and antimicrobial agent (Rogers 2011).

F. pinicola has potent anticancer bioactives. The effects of different mushroom derivatives and their mechanisms of action are studied. In vivo experiments were carried out in mouse models for anticancer activity. It contains high reactive oxygen species (ROS) and downregulates matrix metalloproteinase. Extracts and ergosterol isolated from the mushroom had an apoptosis effect on ROS and migration arrest on matrix metalloproteinase (Artem et al. 2018).

45.2.8 *Grifola frondosa* (Maitake, Hen of the Woods)

Grifola frondosa has been called as "maitake" in Japan meaning as dancing mushroom as the obtainers usually dance when they find the mushroom. It is also known as "hen of the woods" because of the similarities of the texture and the meat of chicken. It is also referred to as "sheep head" in Western Pennsylvania and eastern Ohio (Nik and Tom 2006). As a well-documented traditional Chinese medicine, *Grifola frondosa* is documented for its great medicinal uses. Some of the uses of *Grifola frondosa* are in cancer treatment; curing of palsy, nerve pain, and arthritis; immune stimulation; and regulation of homeostasis.

Many bioactives were isolated from the extracts of *Grifola frondosa*. The most countable and useful of these activities is the antitumor activity of *Grifola*. Some of the extracts of this mushroom have been known to inhibit carcinogenesis, metastasis, and tumor growth. In Japan *Grifola* was prepared as a treatment for many cancer patients. Most researches on this mushroom have shown antitumor activities in mice and using in vitro studies with cancer cell lines. Studies of the crude extracts of this mushroom are carried out. Potent mixture of bioactives of complex branched polysaccharides called D-fraction was isolated. The D-fraction extract of *Grifola frondosa* is the hot-water-extractable, acid-insoluble, alkali-soluble fraction. This D-fraction extract has been used as an anticancer agent (Nik and Tom 2006).

45.2.9 *Hericium erinaceus* (Lion's Mane)

Hericium erinaceus has a long well-documented record in the history of traditional Chinese medicines. It is mostly known as lion's mane. It is an edible mushroom. It is also used in many regions of Asia as food and medicine. It grows on old or dead broadleaved trees (Kevin et al. 2017). In China the fruiting body of *Hericium erinaceus* is called hóu tóu gū ("monkey head mushroom") (Cui et al. 2016). In Japan it is known as yamabushitake ("mountain monk mushroom"). Traditionally *Hericium erinaceus* has been served as a good medicine in Chinese and Japanese medical systems. It has been reported to be used to fortify the spleen and nourish the gut and also as an anticancer drug (Liu et al. 2015).

Bioactive compounds consisting of high molecular weight compounds particularly polysaccharides and low molecular weight compounds, as polyketides and terpenoids, were isolated and characterized from *H. erinaceus* (Thongbai et al. 2015; Wang et al. 2015). It has extensive effectiveness in Alzheimer's disease and Parkinson's disease. *H. erinaceus* works against Alzheimer's disease in different ways (Nagano et al. 2010).

Lowers the cerebral αβ plaque burden.

Increases the NGF mRNA expression.
Decreases the plaque-activated microglia and astrocytes.
Increases the acetylcholine and choline acetyltransferase concentrations.
Increases the lipoxin A4 (LXA4) in the brain.
Decreases Tau tangles.

Oral administration of low-dose HEM (10.76 or 21.52 mg/day) is used in an animal model of Parkinson's disease. It has shown a significant improvement in oxidative stress and dopaminergic lesions in the striatum and substantia nigra after 25 days (Wong et al. 2011).

45.2.10 *Lentinula edodes* (Shiitake)

Lentinula edodes has been used in practicing traditional medicine in many Asian countries mainly in China and Japan for its high immunomodulatory property (Kuppusamy et al. 2009). It is commonly known as shiitake in Japan. A complex carbohydrate known as lentinan was isolated and identified from the mushroom. Shiitake has been used in the natural and traditional treatment of cancer in Japan because of its lentinan content. It is reported to be a source of selenium, an antioxidant that is said to prevent cancer (Medical Mushroom 2009a, b, c, d).

The National Cancer Institute has been using selenium in the hope of coming up with a cure to cancer. Meanwhile, the Japanese pharmaceutical company, Ajinomoto, is already using lentinan extracted from shiitake mushrooms to treat stomach cancers. Recently, other countries are also using it as injectable medication to fight cancer. It was reported that the mushroom is low in sodium, is low in glucose, and is a rich source of fiber. Hence, shiitake is ideal for diabetics and other invalids. Shiitake (*Lentinula edodes*) is recommend for lowering serum cholesterol (Medical Mushroom 2009a, b, c, d).

45.2.11 *Monascus purpureus* (Red Yeast Rice)

Monascus purpureus usually grows on the fermented product of rice. It has been used for medicinal purpose in Chinese cuisines to improve blood circulation for centuries (Thorne Research Inc. 2004) In Asian countries, red yeast rice is a dietary staple and is used to make rice wine, as a flavoring agent, and to preserve the flavor and color of fish and meat (Li et al. 1998).

This fermented rice product is used as a medicinal food to improve blood circulation by decreasing cholesterol and triglyceride levels (Patrick and Uzick 2001; Smith and Olive 2003). The supplement contains varying amounts of natural monacolins as a result of the different strains of *Monascus purpureus* used in fermentation (Liu et al. 2006). Monacolins lower cholesterol by inhibiting HMG-CoA (5-hydroxy-3-methylglutaryl-coenzyme A) reductase, the rate-limiting step for cholesterol synthesis in the liver. It has been used in the therapy for hyperlipidemia (Klimek et al. 2009).

45.2.12 *Piptoporus betulinus* (Birch Polypore)

The *Piptoporus betulinus* grows in cold climate on the host, the birch tree, for which it is known as birch polypore. Birch is very common in the Arctic and North European countries. In the past, *P. betulinus* was used to clear parasitic worms from the stomach and the digestive system. Traditionally it was added to tea where it acted as a laxative. Tea made by brewing with this mushroom is used for soothing the nerves or to eliminate fatigue (Medical Mushroom 2009a, b, c, d).

P. betulinus is used as an anti-parasitic and antimicrobial agent in the treatment of wounds, rectal cancer, and stomach diseases. Tea from this mushroom had antibacterial, anti-fatiguing, immuno-enhancing, and soothing properties for wounds. In vitro anticancer activity of fraction isolated from *P. betulinus* has been investigated. This mushroom has an effect on cell proliferation, motility, and viability in a range of cancer and normal cells. Fractions of extracts of dried fruiting bodies of *P. betulinus* were tested for anticancer evaluation in human lung carcinoma (A549), colon adenocarcinoma (HT-29), and rat glioma (C6) cell cultures. Further cytotoxic effect was tested on human skin fibroblasts (HSF), bovine aorta endothelial cells (BAEC), models of rat oligodendrocytes (OLN-93), hepatocytes (Fao), rat astroglia, and mouse neurons (P19). The fractions significantly decrease viability, proliferation, and migration of tumor cells, affects the stimulatory effect of IGF. And it was nontoxic to the normal cells (Marta et al. 2009).

45.2.13 *Pleurotus ostreatus* (Oyster Mushroom)

Pleurotus ostreatus is also known as oyster mushroom. It helps in lowering cholesterol in the body. The effect is due to the statin and lovastatin content of the mushroom. It is made as dishes on Chinese, Korean, and Japanese menus. The young mushrooms are only used for dishes. When it gets older, it gets tougher and smells exacerbated. Arabitol present in the mushroom can cause gastrointestinal upset to some people. *P. ostreatus* contains an antibiotic, pleuromutilin. Pleuromutilin was bactericidal on various bacteria including the bacterium *Salmonella* in 1950 (Medical Mushroom 2009a, b, c, d).

Ethanolic extracts of the mushroom showed antimicrobial activities against gram-positive and gram-negative bacteria. In vitro antifungal activities have been found. A nutraceutical formulation with nutritive, medicinal, and antimicrobial properties of *P. ostreatus* has been made (Kunjadia et al. 2014). The therapeutic values found in the extracts of oyster mushrooms were antimicrobial, antiviral, antitumor, anti-human immunodeficiency virus (HIV), anti-mutagenic, antineoplastic, anti-lipidemic, antioxidant, hyperglycemic, hypotensive, anti-inflammatory, hypocholesterolemic, immunomodulatory, hepatoprotective, and anti-aging activities (Yashvant et al. 2012).

45.2.14 *Polyporus umbellatus* (Umbrella Polypore)

Polyporus umbellatus is also known as Zhuling (sclerotium) in China. It is known as Chorei-maitake (fruiting body) in Japan (Yang and Zhonghua 1991). It is used in folk traditional medicine. IT belongs to the family *Polyporaceae*. The important part of this mushroom is underground sclerotia. Sclerotium is the hard dark resting body of certain fungi, consisting of a mass of hyphal threads. The body can remain dormant for long periods. Traditional Chinese medicine describes these sclerotia as Zhuling, a crude bioactive commonly used for treating edema and the promotion of diuretic processes (promoting urination) (Yang and Zhonghua 1991); Ar et al. 2015).

Many investigations on *P. umbellatus* mushroom led to the finding of a taxon containing many bioactives showing its antitumor, anticancer, antioxidant, free radical scavenging, immune system enhancement, and antimicrobial activities. With the promising medicinal value, it has been used as a main ingredient in many medicinal products. It is used as food supplement for providing many medicinal effects. This mushroom has been cultivated in large scale under natural and industrial conditions to meet the demand (Zhao 2013; Ar et al. 2015).

45.2.15 *Poria cocos* (Hoelen, Poria Mushroom)

Poria cocos (*Polyporaceae*) is a type of saprophytic fungus that widely grows in diverse species of pines. It is also known as hoelen and poria mushroom. It produces the bioactive sclerotium also known as fu-ling or hoelen. It has been used in traditional Chinese and Japanese medicine for its promising effects on diuretic, sedative, and tonic areas. Anti-inflammatory activities of acute and chronic inflammation were reported. It has been widely used in the preparation of many Asian medicine as a major constituent (Ríos 2011). Major phytochemical compounds, polysaccharides, and triterpenoids were present in *P. cocos*. Investigations reported the inhibitory effects of triterpenes on phospholipase A (2) (PLA (2)). Inhibitory effects of *P. cocos* on phospholipase A led to the secretion of different cytokinins from human peripheral blood monocytes. The essential impacts were on certain diseases such as rheumatoid arthritis, psoriasis, autoimmune uveitis, septic shock, and bronchial asthma. Polysaccharides potentiated the immune response. The polysaccharides from the mushroom potentiated the immune system by enhancing the secretion of immune stimulators and suppressed the secretion of immune suppressors. Furthermore, bioactives showed antitumor activity against different cancer cell lines. These mushrooms have the capacity to constrain angiogenesis by controlling both NF-κB and the induction of NF-κB/Rel translocation in the cancer cells (Ríos 2011).

45.2.16 *Trametes versicolor* (Turkey Tail)

Trametes versicolor is also known as turkey tail. It is a polypore mushroom found in many places of the world for its famous taste and medicinal property. From old times it has been used traditionally in Chinese and Japanese medicines. In Chinese medicine it is known as yun zhi, and in the Japanese medicine, it is kawaratake. This mushroom has a thick leather-like cap; it grows and develops different colors. It has been found to be grown as brownish or darkish brown with blackish areas having algae over the cap. It has a triangular shape and a round-shaped cap (Medical Mushroom 2009a, b, c, d).

T. versicolor contains the polysaccharide Krestin (PSK). It is known for having high molecular weight fractions. It is adjuvant approved in Japan in 1977. In vitro and in vivo studies have shown effects on enhancement of immune functions, antiviral effects, cholesterol-regulating effects, and major effect on colorectal and stomach cancers (Christopher 2004; Rajarathnam and Shashirekha 2003; Huaiqian and Lijuan 2019). It has many biological activities such as antioxidants, Immune-Boosting Polysaccharopeptides, Immune Function in Cancers cells, Enhance the Efficacy of some Cancer Treatments, Enhancement of Gut Health, HPV clearance, reducing inflammation, antibacterial activities, improving athletic performance, improving insulin resistance of our human body (Jillian 2018).

45.3 Case Study on the Hepatoprotective Activity of *Auricularia delicata*

There are species from *Auricularia* which have been studied by the author: *A. delicata* and *A. polytricha*. These species have been used in traditional medicinal practice by the medicine man of Manipur from olden times for the treatment of diarrhea, dysentery, hypertension, constipation, and liver ailments. The traditional medicinal knowledge of Manipur is well documented in the medicine book of the Manipuri *Maibaron*. *Maibaron* is one of the Puyas of the Manipuris where their cultural heritage is being documented (Sujata and Joshi 2017).

Both species have been studied for their elemental content and identification and estimation of bioactives. Elemental analysis was performed with dried samples of the two species. The technique used was EDX spectrometry. The elements were identified and quantitated, viz., C, N, O, Na, Mg, Al, Si, P, S, Cl, K, Ca, and Fe. Same types of elements have been present in both species but in different quantities. Essential elements such as K and Ca were present in favorable amount.

Chemical investigations were performed with HPLC. Extractions with different solvents were performed with the dried mushroom. Different types of standards such as chlorogenic acid, ferulic acid, coumaric acid, etc. were taken and screened with the respective method for their presence in different extracts. Chlorogenic acid peak was detected in the ethyl acetate extracts of both species. To further confirm the presence of CGA, standard CGA was procured, and the ethyl acetate extract was

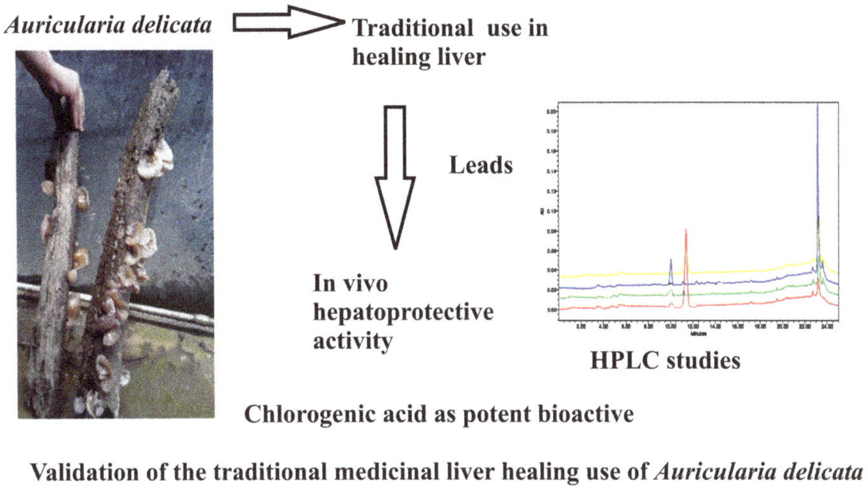

Fig. 45.1 Validation of traditional medicinal use of *Auricularia delicata*

co-injected with standard CGA. The CGA peak of the extract was spiked which confirmed the presence of CGA (Sujata et al. 2018).

Further there was investigation on the liver healing property which is used in the Manipuri traditional therapy system. The antioxidant and antimicrobial properties of the extracts of *Auricularia delicata* were conducted. Based on the results of the antimicrobial and antioxidant activity of the extracts, potent extracts were selected. As it is an experiment with animal sacrifice, assay-guided fractionation wasn't followed. Assay-guided fractionation in such experiment would involve sacrifice of a large number of animals. The experiments were performed under the permission and guidance of proper ethics committee. Mice were used for acute toxicity study. The experiment for liver healing was performed on rat model.

It was found that the ethyl acetate extract had optimum hepatoprotective effect. In our earlier experiments, it was investigated that chlorogenic acid is present in the ethyl acetate extract. Chlorogenic acid has been reported to have hepatoprotective activity.

This is a very good illustration where the scientific investigations were guided by the traditional medicinal practice of the past in Manipur, India. It is scientifically proved to a level with the presence of active components such as chlorogenic acid as expected. The liver healing traditional belief was proved in rat model (Sujata and Joshi 2017; Sujata et al. 2018; Sujata 2018). Figure 45.1 shows the validation of the traditional medicinal use of *Auricularia delicata*.

As many other experiments, this is another experiment with mushrooms which showed that traditional therapies are good guides for drug discovery.

45.3.1 Validation, Identification, and Determination of Pharmacognostic Character of Mushroom

Validation, identification, and determination of pharmacognostic properties of medicinal mushrooms are carried out for drug discovery and formulations of nutraceuticals. As for any natural products, mushrooms are collected from specific locations recording longitude, latitude, and elevation. The species are identified with appropriate taxonomic rules under an institute. Herbaria are submitted for the purpose. Compounds or extracts from the mushrooms are extracted, separated, and identified. Spectroscopic data of compounds and extracts are profiled. HPLC, LC-MS, and GC-MS fingerprints are recorded. Experiments are repeated in a controlled manner with the compounds or extracts. The desired bioactivity is studied with respect to its traditional use. In vitro experiments such as on antioxidant activity are performed with standard methods. Experiments are performed under controlled conditions with standards. In vivo experiments includes antioxidant, hepatoprotective, anti-aging activities, antitumor, anticancer, immune system enhancement, anti-human immunodeficiency virus (HIV), anti-mutagenic, antineoplastic, anti-lipidemic, hyperglycemic, hypotensive, anti-inflammatory, hypocholesterolemic and antimicrobial activities. Experiments with rat or mice model are performed with authentic permission from ethics committee, and clinical trials are then performed for further validation. NBA permission is sought if any products are prepared from such mushrooms. Clinical trials are performed for products. In the final stage, FDA approval is sought to get authentic product of desired bioactivity (Sujata and Joshi 2017; Sujata et al. 2018; Sujata and Warjeet 2019).

45.4 Results and Conclusion

In all the cases of the medicinal mushrooms which have been used in different places of the world, the scientific result has proved its medicinal use. Most of the mushrooms had high antioxidant activity. Many extracts had bioactivity. D-fraction was one active fraction which had medicinal effects. Polysaccharopeptides, triterpenoids, terpenoids, and polysaccharides were reported. Vanadium was also reported as one active element along with other elements like potassium and calcium. The mushrooms had bioactives such as chlorogenic acid, ergothioneine, ergosterol, polyketides, lentinan, monacolins, statins, lovastatins, and pleuromutilin.

Our study is focused on medicinal mushrooms in folk medicine. The medicinal properties which are recorded are results compiled after many hit and trial use. Such natural products are good leads for bioactives and fractions for therapy of many diseases. Here *Auricularia delicata* which is a traditional medicine of Manipur, India, has been validated with scientific investigations for its hepatoprotective activity.

Acknowledgments The authors acknowledge the Department of Chemistry, National Institute of Technology, Manipur, India, for providing the infrastructure for our studies to complete this chapter.

References

Ar B, Sylvie R, Darbhe JB et al (2015) *Polyporus umbellatus*, an edible-medicinal cultivated mushroom with multiple developed health-care products as food, medicine and cosmetics: a review. Cryptogam Mycol 36(1):3–42. https://doi.org/10.7872/crym.v36.iss1.2015.3

Artem B, Margarita Y, Valeriia M et al (2018) Medicinal mushrooms as an attractive new source of natural compounds for future cancer therapy. Oncotarget 9(49):29259–29274. https://doi.org/10.18632/oncotarget

Ashok KP, Kailash CS (2011) Traditional uses and medicinal potential of *Cordyceps sinensis* of Sikkim. J Ayurveda Integr Med 2(1):9–13. https://doi.org/10.4103/0975-9476.78183

Badalyan CM, Gasparyan AV, Garibyan NG (2003) Investigation of the antioxidant activity of some basidial macromycetes. Mikol Fitopatol 37(5):63–68

Bailey CJ, Turner SL, Jakeman KJ et al (1984) Effect of *Coprinus comatus* on plasma glucose concentrations in mice. Planta Med 50(6):525–526

Benzie IFF, Wachtel-Galor S (2009) Biomarkers of long-term vegetarian diets. Adv Clin Chem 47:169–220

Burns RG, DeForest JL, Marxsen J, Sinsabaugh RL, Stromberger ME, Wallenstein MD, Weintraub MN, Zoppini A (2013) Soil enzymes in a changing environment: current knowledge and future directions. Soil Biol Biochem 58:216–234. https://doi.org/10.1016/j.soilbio

Christopher RH (2004) Medicinal value of Turkey tail fungus *Trametes versicolor* (L.:Fr.) Pilat (Aphyllophoromycetideae). A literature review. Int J Med Mushrooms 6(3):195–218. https://doi.org/10.1615/IntJMedMushr.v6.i3.10

Cui F, Gao X, Zhang J et al (2016) Protective effects of extracellular and intracellular polysaccharides on hepatotoxicity by *Hericium erinaceus* SG-02. Curr Microbiol 73 (3):379–385

David LN, Michael MC (2004) Lehninger principles of biochemistry, 4th edn. W H Freeman and Co, New York, NY, pp 191–225

Dijkstra FY, Wiken TO (1976) Studies on mushroom flavours, Flavour compounds in Coprinus comatus. Z Lebensm Unters Forsch 160(3):263–269

Ershova EY, Efremenkova OV, Zenkova VA et al (2001) The revealing of antimicrobial activity of strains of the genus *Coprinus*. Mikol Fitopatol 35(6):32–37

Fan J, Zhang J, Tang Q et al (2006) Structural elucidation of a neutral fucogalactan from the mycelium of *Coprinus comatus*. Carbohydr Res 341(9):1130–1134

Gonzaga MLC, Ricardo NMPS, Heatley F et al (2005) Isolation and characterization of polysaccharides from *Agaricus blazei* Murill. Carbohydr Polym 60(1):43–49

Gu YH, Leonard J (2006) In vitro effects on proliferation, apoptosis and colony inhibition in ER-dependent and ER-independent human breast cancer cells by selected mushroom species. Oncol Rep 15(2):417–423

Guillamon E, Garcia-Lafuente A, Lozano M et al (2010) Edible mushrooms: role in the prevention of cardio-vascular diseases. Fitoterapia 81(7):715–723

Han C, Xing F, Jiang F et al (2003) A study on co-effects of *Coprinus comatus* fermentation liquid and sodium vanadate on the process of inhibiting Ascension of blood glucose in mice. Edible Fungi of China 22(1):39–40

Han C et al (2006) Hypoglycemic activity of fermented mushroom of *Coprinus comatus* rich in vanadium. J Trace Elem Med Biol 20(3):191–196

Hang W, Zhiming F, Chunchao H (2013) The medicinal values of culinary- medicinal Royal Sun Mushroom (*Agaricus blazei* Murrill). Evid Based Complement Alternat Med 2013:842619. https://doi.org/10.1155/2013/842619

Huaiqian D, Lijuan Z (2019) *Coriolus versicolor* polysaccharopeptide as an immunotherapeutic in China. Prog Mol Biol Transl Sci 163:361–381

Ivo B (2009) Moderne Mykotherapie. Alpha One AG. ISBN, 3000258809. Berlin, German

Jianzhe Y, Mao X (1989) Icons of medicinal fungi from China. CRC Press, Boca Raton, FL

Jillian K (2018) 5 immune boosting benefits of turkey tail mushroom. In: Nutrition. Healthline. DIALOG. https://www.healthline.com/nutrition/turkey-tail-mushroom. Accessed on 26 Aug 2019

Kawagishi H, Kanao T, Inagaki R et al (1990) Formolysis of a potent antitumor (1→ 6)-β-d-glucan-protein complex from *Agaricus blazei* fruiting bodies and antitumor activity of the resulting products. Carbohydr Polym 12(4):393–403

Keller AC, Maillard MP, Hostettmann K (1996) Antimicrobial steroids from the fungus *Fomitopsis pinicola*. Phytochemistry 41:1041–1046. https://doi.org/10.1016/0031-9422(95)00762-8

Kevin S, Elizabeth S, Aravind B (2017) Neurological activity of Lion's mane (*Hericium erinaceus*). J Restor Med 6:19–26. https://doi.org/10.14200/jrm.2017.6.0108

Klimek M, Pharm D, Shan W et al (2009) Safety and efficacy of red yeast Rice (*Monascus purpureus*) as an alternative therapy for hyperlipidemia. P T 34(6):313–327

Kunjadia PD, Nagee A, Pandya PY et al (2014) Medicinal and antimicrobial role of the oyster culinary-medicinal mushroom *Pleurotus ostreatus* (higher Basidiomycetes) cultivated on banana agro wastes in India. Int J Med Mushrooms 6(3):227–238

Kuppusamy UR, Chong YL, Mahmood AA et al (2009) *Lentinula edodes* (shiitake) mushroom extract protects against hydrogen peroxide induced cytotoxicity in peripheral blood mononuclear cells. Indian J Biochem Biophys 46(2):161–165

Kwok Y, Ng KF, Li CC et al (2005) A prospective, randomized, double blind, placebo-controlled study of the platelet and global hemostatic effects of Ganoderma lucidum (Ling-Zhi) in healthy volunteers. Anesth Analg 101(2):423–426

Li C, Zhu Y, Wang Y et al (1998) *Monascus purpureus* fermented rice (red yeast rice): a natural food product that lowers blood cholesterol in animal models of hypercholesterolemia. Nutr Res 18:71–81

Li S, An L, Zhang H (2001a) Effects of polysaccharide from *Coprinus comatus* on activity of serum lysozyme in Kunming mouse, China. Edible Fungi of China 20(4):36–38

Li SP, Li P, Dong TT et al (2001b) Anti –oxidant activity of different types of natural cordyceps sinensis and cultured cordyceps mycelia. Phytomedicine 8(3):207–212

Li Y, Xiang H (2005) Nematicidal activity of *Coprinus comatus*. Acta Phytopathol Sin 35 (5):456–458

List PH (1957) Occurrence of ergothioneine in shaggy-mane, *Coprinus comatus*. Arch Pharm Ber Dtsch Pharm Ges 290/62(11):517–520

Liu J, Du C, Wang Y et al (2015) Anti-fatigue activities of polysaccharides extracted from *Hericium erinaceus*. Exp Ther Med 9(2):483–487

Liu J, Zhang J, Shi Y et al (2006) Chinese red yeast rice *(Monascus purpureus*) for primary hyperlipidemia: a meta-analysis of randomized controlled trials. Chin Med 1(4):1–13

Lowy B (1971) New records of mushroom stones from Guatemala. Mycologia 63(5):983–993

Luo H, Mo MH, Huang XW et al (2004) *Coprinus comatus*: a basidiomycete fungus forms novel spiny structures and infects nematodes. Mycologia 96(6):1218–1224

Marta L, Ewa L, Jozef K et al (2009) Anticancer effect of fraction isolated from medicinal *Birch Polypore* mushroom, P. karst (2009) (Aphyllophoromycetideae): In vitro studies: *Piptoporus betulinus* (bull.: Fr.). Int J Med Mushrooms 11(4):361–364. https://doi.org/10.1615/IntJMedMushr.v11.i4.20

McMeekin D (2005) The perception of Ganoderma lucidum in Chinese and Western culture. Mycologist 18:165–169

Medical Mushroom (2009a) Medical Mushroom. Net. Johanneshov. http://www.medicalmushrooms.net/piptoporus-betulinus/. Accessed on 12 Aug 2019

Medical Mushroom (2009b) Medical Mushroom. net. Johanneshov. http://www.medicalmushrooms.net/lentinula-edodes-shiitake/. Accessed on 10 Aug 2019

Medical Mushroom (2009c) Medical Mushroom. net. Johanneshov. http://www.medicalmushrooms.net/pleurotus-ostreatus-oyster-mushroom/. Accessed on 10 Aug 2019

Medical Mushroom (2009d). Medical Mushroom.net. Johanneshov. http://www.medicalmushrooms.net/trametes-versicolor/. Accessed on 22 Aug 2019

Minghan WA (2011) Metabolic syndrome: underlying mechanisms and drug therapies. Wiley & Sons Editor, Inc, Hoboken, NJ, p 264

Nagano M, Shimizu K, Kondo R et al (2010) Reduction of depression and anxiety by 4 weeks *Hericium erinaceus* intake. Biomed Res 31:231–237

Nik Z, Tom V (2006) *Grifola frondosa*, the hen of the Woods, a.k.a. Sheepshead or Maitake. In: Tom Volk's Fungi. University of Wisconsin. Available via DIALOG. https://botit.botany.wisc.edu/toms_fungi/nov2006.html. Accessed on 07 Aug 2019

Patrick L, Uzick M (2001) Cardiovascular disease: C-reactive protein and the inflammatory disease paradigm: HMG–CoA reductase inhibitors, alpha- tocopherol, red yeast rice, and olive oil polyphenols, a review of literature. Altern Med Rev 6(3):248–270

Rajarathnam S, Shashirekha MN (2003) Mushrooms and truffles/use of wild mushrooms. In: Benjamin CP (ed) Encyclopedia of food sciences and nutrition, 2nd edn. Elsevier, Amsterdam Switzerland, pp 4048–4054

Ríos JL (2011) Chemical constituents and pharmacological properties of *Poria cocos*. Planta Med 77(7):681–691. https://doi.org/10.1055/s-0030-1270823

Rogers RD (2011) The fungal pharmacy: the complete guide to medicinal mushrooms and lichens of North America. North Atlantic, Berkeley, CA

Sharma N (2003) Medicinal uses of macrofungi. Ethnobotany 15:97–99

Shu CH, Wen BJ (2003) Enhanced shear protection and increased production of an anti-tumor polysaccharide by *Agaricus blazei* in xanthan-supplemented cultures. Biotechnol Lett 25 (11):873–876

Sinsabaugh RS (1994) Enzymic analysis of microbial pattern and process. Biol Fertil Soils 17 (1):69–74. https://doi.org/10.1007/BF00418675

Smith DJ, Olive KE (2003) Chinese red rice–induced myopathy. South Med J 96(12):1265–1267

Sujata W (2018) Traditional and folk medicine as target for drug discovery in natural products & drug discovery. In: Mandal SC, Konishi (eds) An integrated approach. Elsevier, Amsterdam Switzerland, pp 29–56

Sujata W, Joshi DD (2017) Screening and comparison of two edible macrofungi of Auricularia sp. Curr Sci 112(3):460–463

Sujata W, Joshi DD, Geeta DL et al (2018) Hepatoprotective effect of Auricularia delicata (Agaricomycetes) from India in rats: biochemical and Histopathological studies and antimicrobial activity. Int J Med Mushrooms 20(3):213–225

Sujata W, Warjeet L (2019) Novel method of separation and isolation of indirubin from strobilanthes flaccidifolius leaves. 327026 IPR patent, 033109274, 9 December 2019

Thongbai B, Rapior S, Hyde KD et al (2015) *Hericium erinaceus*, an amazing medicinal mushroom. Mycol Prog 14(91):1–23

Thorne Research Inc. (2004) *Monascus purpureus* (red yeast Rice). Altern Med Rev 9(2):208–210

Tomasi S, Lohezic-Le DF, Sauleau P et al (2004) Cytotoxic activity of methanol extracts from Basidiomycete mushrooms on murine cancer cell lines. Pharmazie 59:290–293

Wachtel-Galor S, Buswell JA, Tomlinson B et al (2004) *Lingzhi* polyphorous fungus. In: Herbal and traditional medicine, molecular aspects of health. Marcel Dekker Inc, New York, pp 179–228

Wachtel-Galor S, John Y, John A et al (2011) *Ganoderma lucidum* (Lingzhi or Reishi) a medicinal mushroom. In: Benzie IFF, Wachtel-Galor S (eds) Herbal medicine: biomolecular and clinical aspects, 2nd edn. CRC Press/Taylor & Francis, Boca Raton, FL

Wang K, Bao L, Qi Q et al (2015) Erinacerins C-L, isoindolin-1-ones with alpha- glucosidase inhibitory activity from cultures of the medicinal mushroom *Hericium erinaceus*. J Nat Prod 78 (1):146–154

Wang ZX, Wang XM, Wang TZ et al (1995) Current status of pharmacological studies on Cordyceps sinensis and Cordyceps hyphae. Chung-Kuo Chung His I Chieh ho Tsa Chih 15:255–256

Wasser SP, Coates P, Blackman M et al (2005) Encyclopedia of Dietary Supplements, New York: Marcel Dekker Reishi or Lingzhi (Ganoderma lucidum) p 680–690

Wong KH, Naidu M, David P et al (2011) Peripheral nerve regeneration following crush injury to rat peroneal nerve by aqueous extract of medicinal mushroom *Hericium erinaceus* (bull.: Fr) Pers. (Aphyllophoromycetideae). Evid Based Complement Alternat Med 580752:1–10. https://doi.org/10.1093/ecam/neq062

Wu L, Wu Z, Lin Q et al (2003) Purification and activities of an alkaline protein from mushroom *Coprinus comatus*. Wei Sheng Wu Xue Bao 43(6):793–798

Yang DA, Zhonghua WKZZ (1991) Inhibitory effect of Chinese herb medicine Zhuling on urinary bladder cancer. Chin J Surg 29(6):393–395. 399

Yang X, Wan M, Mi K et al (2003) The quantification of (1, 3)-β-glucan in edible and medicinal mushroom polysaccharides by using limulus G test. Mycosystema 22(2):296–302

Yashvant P, Ram N, Singh VK et al (2012) Medicinal properties of *Pleurotus* species (*Oyster* mushroom): a review. World J Fungal Plant Biol 3(1):01–12. https://doi.org/10.5829/idosi.wjfpb.2012.3.1.303

Yilmaz N, Solmaz M, Turkekul I et al (2006) Fatty acid composition in some wild edible mushrooms growing in the middle Black Sea region of Turkey. Food Chem 99(1):168–174

Young-Ji C, In-Sun P, Mi-Hee K et al (2018) The medicinal mushroom Auricularia auricula-judae (bull) extract has antioxidant activity and promotes procollagen biosynthesis in HaCaT cells. Nat Prod Res 33(22):3283–3286

Zhao YY (2013) Traditional uses, phytochemistry ,pharmacology, pharmacokinetics and quality control of *Polyporus umbellatus*. J Ethnopharmacol 149(1):35–48

Zjawiony JK (2004) Biologically active compounds from Aphyllophorales (polypore) fungi. J Nat Prod 67:300–310. https://doi.org/10.1021/np030372w

A Review on Investigational Studies of Marine Macroalgae *Spongomorpha indica* L

46

Mounika V. V. Donela, Ganga Rao Battu, Ramadevi Devarakonda, and Subhash C. Mandal

Abstract

Plants are being used for disease treatment, prevention, cure, and many other medicinal purposes long before the prehistoric era as traditional medicines. The use of natural products for the discovery of many novel drugs has been increasing over the globe. Along with terrestrial plants, marine plant species too have a role in drug discovery and serves as lead compounds. Marine plants showed potent action in tumor and cancer cells and further produced many novel compounds in natural products of chemistry. Thus, marine samples became attractive to pharmaceutical companies in order "re-discovery" of compounds from relatively well-characterized terrestrial sources and because of the realization that classes of molecular structures not found in terrestrial but found in the marine environment. The present review aims to provide a detailed note on new marine macroalgae of Chlorophyceae class, *Spongomorpha indica* L. that is also accepted as *Acrosiphonia orientalis* and its important aspects related to ethnobotany and ethnopharmacology. Identification of important elements such as K (802.5), I (215.5), etc., Compounds isolation (such as 2,5-Di-O-methyl-L-arabinose, arabinose, 2,4-Di-O-methyl-L-arabinose, etc.) for the assessment of usefulness of the marine macroalgae and important pharmacological activities like anti-inflammatory, antioxidant, antibacterial, antifungal, and antimicrobial activities. In overall, this review concludes the reports on phytochemical and

M. V. V. Donela (✉) · G. R. Battu · R. Devarakonda
Department of Pharmacognosy & Phytochemistry, Andhra University College of Pharmaceutical Sciences, Andhra University, Visakhapatnam, Andhra Pradesh, India
e-mail: mounikadonela2010@gmail.com

S. C. Mandal
Division of Pharmacognosy, Department of Pharmaceutical Technology, Faculty of Engineering & Technology, Jadavpur University, Kolkata, India
e-mail: scmandal1963@gmail.com

S. C. Mandal et al. (eds.), *Evidence Based Validation of Traditional Medicines*,
https://doi.org/10.1007/978-981-15-8127-4_46

pharmacological aspects of *Spongomorpha indica* L. and brief information on its genus *Spongomorpha* and species related to it. Marine algae are rich in vitamins, minerals, and secondary metabolites. Thus, provides the information on importance of studies on marine algae those have the medicinal, molecular, and other values that are not found in terrestrial but found in the marine environment only.

Keywords

Spongomorpha · Ulotrichaceae · Isolated compounds · Uses · Elemental analysis, and Biological activities

Abbreviations

%	Percentage
°C	Degree centigrade
cm	Centimeters
DPPH	2,2-diphenyl-1-picrylhydrazyl
g	Grams
HIV	Human Immunodeficiency Virus
m	Meters
mg/ml	Milligrams per milliliter
ml/L	Milliliter per liter
nm	Nanometers
ppm	Parts per million
ppt	Parts per thousand
RNA	Ribonucleic acid
USD	United States Dollar
WHO	World Health Organization
α	Alpha

46.1 Introduction

Plants have been used for medicinal purposes long before the prehistoric era as traditional medicines for the prevention and treatment of diseases. In ancient, Unani manuscripts, Egyptian papyrus, and Chinese writings described the uses of plants. They became important lead compounds for many semi-synthetic, total synthetic modifications and creating novel compounds for drug discovery having pharmacologically active compounds with many blockbuster drugs being derived directly or indirectly from plants. In current demand with synthetic chemistry in drug discovery and manufacturing of drugs, the contribution of plants to disease treatment and prevention became still enormous. Even at the dawn of the twenty-first century, 11% of the 252 drugs considered as basic and essential by the WHO were exclusive of flowering plant origins (Veeresham 2012).

For many years, pharmacognosy only focused on the investigation and identification of medically important terrestrial plants and animals, although many marine organisms were concerned with the naturally occurring substances of medicinal value from the marine. Marine samples became attractive to pharmaceutical companies in order "re-discovery" of compounds from relatively well-characterized terrestrial sources and because of the realization that classes of molecular structures not found in terrestrial but found in the marine environment. Interest in the development of marine natural products has been increasing, and internationally novel compounds have been isolating from marine sources; hence the research is continuing for drug discovery of novel compounds from marine sources (Russell 2009).

Marine plants belong to different types of families in which Ulotrichaceae family in green algae consists of nearly 13 types of genus and 14 species of Chlorophyta classification. This classification is well known for its antioxidant and tumor cell activities as well as uses in traditional Chinese systems (Rangaiah et al. 2010).

Marine natural products have been found to be an important source of drugs and drug leads. The resurgence of natural products those in the marine world initiated drug discovery of novel compounds. There is a growing interest in marine natural products for their secondary metabolites. Marine natural products have secondary metabolites as well as enzymes, lipids, and heteropolysaccharides derived from marine sources (Gudbjarnason 1999).

This field of research receives the attention of investigators in various fields, marine biology, marine ecology, biochemistry, chemistry, and pharmacology. Most of the natural products of interest to the pharmaceutical industry are secondary metabolites, and several such compounds, derived from marine invertebrates, have been in clinical trials as experimental anticancer drugs (Gudbjarnason 1999).

Many of the researches activities involved in the discovery of novel compounds from the marine environment fall into the area of marine biotechnology. This is one component of research in development for drug discovery, a process in which pharmaceutical companies invest over USD 20 billion per annum (Russell 2009).

Marine environment has proven to be very rich source of extremely potent compounds that have demonstrated significant activities in anti-tumor, anti-inflammatory, immunomodulatory, analgesia, allergy, and anti-viral assays. There are now significant numbers of very interesting molecules that come from marine sources or have been synthesized as a result of knowledge gained from a prototypical compound that are either in or approaching clinical trials in cancer, analgesia, allergy, and cognitive diseases. A substantial number of other potential agents are following in their wake in pre-clinical trials in these and in other diseases (Newman and Cragg 2004).

Approximately 16,000 marine natural products have been isolated from marine organisms and reported in 6800 publications. In addition to these publications, there are approximately another 9000 publication that covers synthesis, reviews, biological activity studies, ecological studies, etc. on the subjects of marine natural products. Several of the compounds isolated from marine source exhibit biological activity (Bhakuni and Rawat 2005).

During the 35 years, more than 130 diverse secondary metabolites, including alkaloids, terpenoids, peptides, lipids, and steroids, were discovered to have biological activities such as antimicrobial, RNA cleavage activating, cytotoxic, and anti-HIV effects. For instance, steroids and tri-terpenoids were mainly discovered in South China Sea, whereas alkaloids and lipids were mostly isolated from Japanese Sea (Wu et al. 2018).

Marine algae are distributed all over the world like India, China, Japan, Atlantic, Iceland, etc. It is having antioxidant, anti-inflammatory, antimicrobial, antifungal, and antibacterial activities. Isolated compounds are sulfated polysaccharides. They are high in vitamin C content than in any other green algae (Newman and Cragg 2004).

The common interest for scientists and industries around the globe is in biotechnological development and the exploitation of marine metabolites of high economic values in China and the ninth century A.D. in China (Keshini et al. 2016).

In the nineteenth century or so, several marine macroalgae have been used as natural fertilizers in several countries. In the late twentieth century, around 1980, the natural blue dye phycocyanin from the microalgae *Spirulina* sp. was mainly used as colorants for ice creams and cosmetics. At the very beginning of the twenty-first century, the array of marine algae and their derivatives are gaining increasing recognition worldwide (Keshini et al. 2016).

Marine algae are being widely used both at the molecular and organismal levels as food and nutraceuticals, animal and fish feed, biofertilizers, bioplastics, pharmaceuticals, cosmeceuticals, fluorophores, food colorants and textile dyes, biofuels, as well as phycoremediation (Keshini et al. 2016).

Marine algae have primary and secondary metabolites that are used as food, food additives, and raw materials for industries such as vegetables oils, fatty acids (used for making soap and detergents) and carbohydrates (sugar, starch, pectin and cellulose). Edible seaweeds are rich in bioactive compounds such as soluble dietary fibers, proteins, peptides, minerals, vitamins, and polyunsaturated fatty acids (Khalid et al. 2018).

Active compounds include sulfated polysaccharides; phlorotannins, carotenoids (e.g., fucoxanthin), minerals, peptides, sulfolipids, etc. are proven with benefits against degenerative metabolic disease (Khalid et al. 2018).

Isolated compounds from marine algae includes the following:

- Brown Algae: Dieckol, 6,6′-bieckol, dioxinodehydroeckol, fucoidan, fucoxanthin, fucoxanthinol, sargachromanol E, laminarin, fucodiphloroethol G, diphlorethohydroxycarmalol, ascophyllan, bis(2,3 dibromo-4,5 dihydroxybenzyl) ether, sarg A, tuberatolide B etc. (Alves et al. 2018).
- Red Algae: Lophocladines B, polyether triterpenoid dehydrothyrsiferol, eucheumaserra agglutinin, sulfated carrageenan, laurenditerpenol, bromophycolide A, elatol, ϒ carrageenan, mertesene, pheophorbide A, thyrsiferol, porphyran, etc. (Alves et al. 2018).
- Green Algae: Nigricanosides A, caulerpin, caulerpenyne, siphonaxanthin, clerosterol, etc. (Alves et al. 2018).

Seaweeds are regarded as good reservoirs of compounds with numerous biological and biomedical activities such as anticancer, anti-obesity, antidiabetic, antibacterial, antihypertensive, antihyperlipidemic, antioxidant, antifungal, antimicrobial, anti-inflammatory, anticoagulant, antiestrogenic, immunomodulatory, thyroid stimulating, neuroprotective and tissues healing properties (Khalid et al. 2018), neurodegenerative disorders, cardiovascular disorders, anti-infertility, antiviral/HIV, HIV transfusion, antiallergy, skin disorders, metalloproteinase, inhibition activity, and bone-related diseases (Thomas and Kim 2013).

Compared to the terrestrial plants and animal-based food, seaweeds are rich in some health-promoting molecules and materials such as dietary fiber, ω-3 fatty acids, essential amino acids, and vitamins A, B, C and E which are essential for cosmeceutical product development (Thomas and Kim 2013).

The biochemical composition of seaweed provides an excellent choice of biologically active components with a broad range of physiological and biochemical characteristics, many of which are rare or absent in other taxonomic groups. Hence many novel compounds are obtained from marine sources in which most compounds are used for anticancer activity (Thomas and Kim 2013).

Marine environment has demonstrated to be interesting source of compounds with uncommon and unique chemical features on which the molecular modeling and chemical synthesis of new drugs can be based with greater efficacy and specificity for the therapeutics (Alves et al. 2018). In addition marine algae are considered as sea vegetables not only for consumption but also as an alternative medicine since ancient times for skin related diseases. In other words, the marine environment is off many folds richer in its biodiversity, thereby making marine organisms and their metabolites unique. The majority of the investigation on the metabolites derived from brown algae (Thomas and Kim 2013).

On the chemo-ecological point of view, future work would be interesting to focus on discovery of targeted bioactive molecules such as highly cytotoxic tri-terpenoids from certain sea areas. Another outlook worth to mention is the possible biogenetic relationship between some metabolites in this genus. It could inspire biomimetic synthesis and compound interconversion, to scale up the available amounts for further pharmacological and toxicological evaluation. Further, generation of some metabolites could also be used for another direction future study (Wu et al. 2018).

46.1.1 Identification of Algae

Marine plants are known as marine algae. The study of algae is called phycology. They are identified based on their macroscopic, microscopic, and habitat characteristic. Similar types of algae are identified based on sexual systems, reproduction organs, type of reproduction, germinating zoospores, zygospore, pyrenoids, nuclei, rhizoids, etc. There are several types of living organisms present in oceans and seas that include phytoplankton, kelps, seagrasses, mangroves, corals, waterwheel plants, sea anemones, open brains, sponges, algae, bioluminescence plants and organisms, etc. (PRI 2019).

There are basically two categories of plants living in our ocean's waters (PRI 2019):

Floating plants: They can be spotted near the surface of the water.
Rooted plants: They can be found in shallow waters near the shore.
Other plants: They can be found in deep waters.

Algae can be generally categorized into two large sub-groups (Keshini et al. 2016):

1. Microalgae: Cyanophyta—Blue green algae.
   Pyrrophyta—Dinoflagellates.
   Chrysophyta—Diatoms and golden brown algae.
   Chlorophyta—Microscopic green algae.
2. Macroalgae: Chlorophyceae—Green algae.
   Rhodophyceae—Red algae.
   Phaeophyceae—Brown algae.

Spongomorpha indica L. (Family, Ulotrichaceae) is one of the macroalgae that was accepted as *Acrosiphonia orientalis* that distributed in the intertidal rocky habitats of tropical and subtropical seawaters of the globe. It is obtained from a parent plant *Spongomorpha kutzing* (Rangaiah et al. 2010). It is used as fodder, biogas production, fertilizers, dietary supplement, water waste treatments, etc. It is rich in vitamin C content, which acts as an antioxidant, than other chlorophyta plants (Rajubabu et al. 2017).

Historically, Thivy and V. Visalakshmi identified and scientifically named the plant as *Spongomorpha indica* in 1963. The specimen accepted in the World Register of Marine Species as *Acrosiphonia orientalis* given by (J. Agardh) P. C. Silva in 1996 (Guiry and Guiry 2019).

Seaweed is also known as brown, red, green seaweed, algae, kelp, egg wrack, kombu/konbu, sea spaghetti, wakame, nori, dulse/dillisk, sea lettuce, sea grass, carrageenan, and Irish moss (McHugh 2003).

46.1.2 Description of Family Ulotrichaceae (Taylor 1960)

- Family Ulotrichaceae belongs to green algae in the order Ulotrichales.
- Algae consisting of filaments with or without an attaching base.
- Filaments are cylindrical, typically unbranched, uniseriate cells with thin to gelatinous walls.
- Chloroplast consists of a parenchymatous layer with cells. It consists of a lateral plate-like or a band like chromatophores with or without pyrenoids.
- Asexual reproduction by filament fragmentation or zoospores and sexual reproduction by diplohaplontic takes places.

46.2 Botanical Description of *Spongomorpha indica* L.

Taxonomic classification of seaweed *Spongomorpha indica* L. (Guiry and Guiry 2019) given in Table 46.1. Vernacular names for word seaweed mentioned in Table 46.2.

46.2.1 Distribution and Habitat

Spongomorpha indica L. is a green macroalgal seaweed, which is distributed in the intertidal rocky habitats especially mid to low littoral regions of tropical and subtropical seawaters of the globe. They grow on hard rocks and rarely as epiphytes. It is a perennial weed which grows throughout the year within the intertidal zones with depth 1.1 m (10 cm = 1 m) and distance from the shore 10 m (2 cm = 1 m) consisting latitudes 17°14′13″ and 17°41' N and longitudes 83°16′25″ and 21°30″. Seawater conditions should be maintained as follows (Manjula 2015):

Salinity: 30–33 ppt.
p^H: 7.2–7.6.
Dissolved oxygen: 7–7.6 ml/L.
Temperature: 26–37 °C.
Water temperature: 21–26 °C.

Table 46.1 Taxonomic classification of seaweed *Spongomorpha indica* L.

Empire	Eukaryota
Kingdom	Plantae
Sub-kingdom	Viridiplantae
Phylum	Chlorophyta
Sub-phylum	Chlorophytina
Class	Ulvophyceae
Order	Ulotrichales
Family	Ulotrichaceae
Genus	*Spongomorpha*
Species	*Spongomorpha indica*

Table 46.2 Name of seaweed in different languages

Languages	Names
Telugu	Samudrapu paci
Hindi	Samudreesivaar
English	Seaweed
Marathi	Samudraparyatana
Guajarati	Sivida
Tamil	Katarpaci
Malayalam	Katalijalam
Punjab	Samudaritalaha
Kannada	Kadalakale

46.3 Morphology and Uses of *Spongomorpha indica* L.

46.3.1 Morphology and Distribution

Spongomorpha indica L. is seaweed that is classified as macroalgae with sub-classification as chlorophyta (green algae). It is a medium-sized sea plant with height 3–5 cm and width 5–7 cm having unbranched filaments or uniseriate branched filaments attached by descending, often branched, multicellular rhizoids as irregular branched clusters with thallus. Branches spreading in young plants, intertwined in older plants, forming secondary tufts. Branching intercalary, irregular, usually single per node, arising subterminally from upper cell pole with oblique cross wall at axial cell having rounded tips, rarely hooked. It may or may not be having other weeds and epiphytes attached to it. It is dark seaweed green in color (Setchell and Gardner 1920).

Spongomorpha distributed in temperate, cold temperate to arctic waters of North Atlantic and adjacent waters, growing in littoral zone, often in rock pools or epiphytically on larger algae. Cell wall thin in fast-growing plants, pecto-cellulosic in nature. Chemical compositions are unknown. X-ray diagrams indicating randomly arranged substances (Setchell and Gardner 1920). Images of seaweed regarding the habitat distribution and specimen images are given in Figs. 46.1–46.3. Figure 46.4 includes the uses of seaweed by their extraction methods and other uses.

Fig. 46.1 Picture of seaweed at the site of intertidal rocky habitat

Fig. 46.2 Picture of seaweed attached to rock

Fig. 46.3 Picture of seaweed

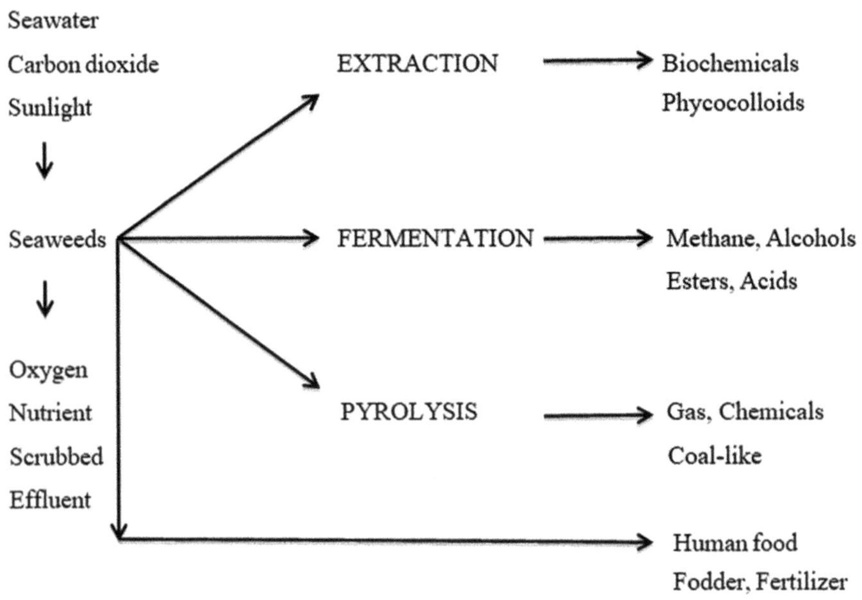

Fig. 46.4 Uses of seaweed by their extraction methods and other uses (Indergaad 1983)

Thallus: It composed of short intercalary, sometimes isodiametric cells in the lower parts of the thallus and longer cells in the upper part. Cells are uninucleate (Taylor 1957).

Filaments: They may be branched or unbranched uniseriate filaments, which either have hooks or rhizoids that cause the branches to become tangled. This gives the algae a matted rope-like appearance. It is also said to resemble Rastafarian hair (Taylor 1957).

46.3.2 Uses, Interaction, Toxicology of Seaweed

46.3.2.1 Recommended Dose

Clinical trials have used an oral dosage range of 4–12 g seaweed daily up to 2 months (Teas et al. 2009).

46.3.2.2 Interactions

Patients taking warfarin and consuming a large quantity of food containing seaweed may experience a change in international normalized ratio because of seaweed's high vitamin C, E, K, and B contents (Bartle et al. 2001).

46.3.2.3 Toxicology

Excessive intake of dried seaweed may result in increased serum thyroid-stimulating hormones. There have been case reports of yellowing of the skin with excessive seaweed consumption (Nishimura et al. 1998).

46.3.2.4 Uses of Seaweed *Spongomorpha*

Industrial Uses (Sekar 2015; Taylor 1957)

- Alginates, carrageenan, etc. extraction.
- Rich in elements like K, Na, Ca, Mg, Zn, Cu, Cl, S, P, Vi, Co, Mn, Se, Br, I, Ar, Ir, F.
- Metabolically required minerals.
- Therapeutically get tracer elements.
- Dry weight minerals.
- Preparation of agar, alginates, carrageenan.
- Biogas production.
- Gelling, stabilizing, and thickening agents in food.
- Confectionary.
- Dairy.
- Textiles and papers industries.
- Paint and varnish industries.

Ethnobotanical Uses (Wu et al. 2018)

- Wound healing.
- Wound dressing.
- Anti-inflammatory.
- Allergy.
- Anti-viral.
- Analgesia.
- Cognitive disorders.
- Anticancer.
- Skin disorders, etc.
- Tuberculosis.
- Arthritis.
- Colds and influenza.
- Worm infestation.

Traditional Uses (McHugh 2003)

- Cosmetic preparations.
- Traditional medicine.
- Seaweed oils.
- Water waste treatments.
- Soil conditioners.
- Livestock as fodder.
- Organic fertilizers.
- Dietary supplements.
- Foliar feeding for plants.

Medicinal Uses (Manjula 2015; Stein and Borden 1984)

- Dietary source of essential medicines.

- Cholesterol reduction.
- Appetite suppression.
- Heart problems.
- Mineral and vitamins supplements.
- Thyroxin production.
- Vermifugal agents.
- Inhibit herpes simplex virus.
- Carrageenan's used as anti-viral agents.
- Bone replacement therapy.
- Sexual desirability.

46.4 List of Accepted Species in Other Names

Some *Spongomorpha* species were accepted as *Acrosiphonia* based on common vegetative characters; however nuclear number provides stable features for generic distinction. A genus is redefined because of the presence and absence of heteromorphic life history (*Spongomorpha* with and *Acrosiphonia* without heteromorphic life history). This is the diversity among the endophytic sporophytic vesicles. The reason seems reasonable since some species have distinguishable sporophytes (Hoek 1963). The list of species was mentioned in Table 46.3.

Table 46.3 *Spongomorpha* species with accepted names (Guiry and Guiry 2020)

Plant names	Accepted names
Spongomorpha breviarticulata	*Acrosiphonia duriuscula*
Spongomorpha cartilaginea	*Acrosiphonia duriuscula*
Spongomorpha arcta	*Acrosiphonia centralis*
Spongomorpha congregata	*Spongomorpha aeruginosa*
Spongomorpha effusa	*Acrosiphonia effusa*
Spongomorpha flagellata	*Acrosiphonia flagellata*
Spongomorpha hemisphaerica	*Acrosiphonia hemisphaerica*
Spongomorpha indica	*Acrosiphonia orientalis*
Spongomorpha lanosa	*Spongomorpha aeruginosa*
Spongomorpha ochotensis	*Spongomorpha mertensii*
Spongomorpha oxyclada	*Camontagnea oxyclada*
Spongomorpha pallida	*Spongomorpha aeruginosa*
Spongomorpha saxatilis	*Acrosiphonia saxatilis*
Spongomorpha sonderi	*Acrosiphonia sonderi*
Spongomorpha spinescens	*Acrosiphonia spinescens*
Spongomorpha uncialis	*Spongomorpha aeruginosa*

46.5 List of Accepted Species of *Spongomorpha* Genus (Guiry & Guiry 2020)

1. *Spongomorpha aeruginosa.*
2. *Spongomorpha cincinnata.*
3. *Spongomorpha coalita.*
4. *Spongomorpha conjuncta.*
5. *Spongomorpha indica.*
6. *Spongomorpha glacialis.*
7. *Spongomorpha hystrix.*
8. *Spongomorpha pacifica.*
9. *Spongomorpha spiralis.*
10. *Spongomorpha vernalis.*
11. *Spongomorpha mertensii.*

Accepted names and description of seaweed regarding the chemical constituents and biological activities along with references were mentioned in Table 46.4.

46.6 Isolated Molecules

A sulfated heteropolysaccharide $[\alpha]_D$+ 59° was isolated from a green seaweed *Spongomorpha indica* L. by extraction with ammonium oxalate. Studies showed that the polysaccharide is a complex and multilinked polymer containing arabinose in both furanose and pyranose forms. The polymer is composed of arabinose, xylose, galactose, and glucose in the ratio 8.9:1:12:1. The compounds isolated from seaweed *Spongomorpha indica* L. by ammonium oxalate method were 2,5-Di-O-methyl-L-arabinose, 2,3,4-Tri-O-methyl-L-arabinose, arabinose, 2,3,5-Tri-O-methyl-L-arabinose, 2,3,4,6-Tetra-O-methyl-D-glucose, 3,5-Di-O-methyl-L-arabinose, 2,4-Di-O-methyl-L-arabinose, 6-O-methyl-D-galactose, 2,3,6-Tri-O-methyl-D-glucose, 2,3,6-Tri-O-methyl-D-galactose, 2,3,4-Tri-O-methyl-D-xylose, 2-O-methyl-L-arabinose, and 2,3,4,6-Tetra-O-methyl-D-galactose (Rao et al. 1991) (Fig. 46.5).

46.7 Elemental Analysis

Elemental analysis is a process where a sample of some material (e.g., soil, water, bodily fluids, minerals, chemical compounds, etc.) is analyzed for its elements and sometime isotonic composition. Elemental analysis can be quantitative (determining how much of each are present) and qualitative (determining what elements are present). Elemental analysis falls within the analytical chemistry. It is necessary to detect elemental analysis in materials especially food and chemical compounds to check the major and minor elements present in the samples. Trace elements, toxic elements such as heavy metals, and their implications have been dealt in a serious

Table 46.4 *Spongomorpha* species that have bioactive compounds and biological activities with references

S. no.	Seaweed name	Accepted name	Chemical constituents	Biological activities	References
1	*Spongomorpha sonderi*	*Acrosiphonia centralis*	Polysaccharides (includes mono and oligo) Sulfides and Sulfonium compounds	Antibacterial Antifungal Anti-mycobacterial	Flewelling et al. (2013), McKinnell and Perecival (1962)
2	*Spongomorpha indica*	*Acrosiphonia orientalis*	Sulfated heteropolysaccharides Vitamin C	Antifungal Antibacterial Antimicrobial Anti-inflammatory Antioxidant	Rao et al. (1991), Sarojini and Sharma (1999), Satyalakshmi et al. (2018), Rangaiah et al. (2010)
3	*Spongomorpha cartilaginea*	*Acrosiphonia duriuscula*	cDNA of cdc2	Regulating kinase activity	Kato et al. (2001)
4	*Spongomorpha arcta*	*Acrosiphonia sonderi*	Polysaccharides (includes mono and oligo) Sulfides and Sulfonium compounds	Glycosidase activity Antibacterial Antifungal Anti-mycobacterial	Flewelling et al. (2013), McKinnell and Perecival (1962)
5	*Spongomorpha sonderi*	*Acrosiphonia sonderi*	Polysaccharides (includes mono and oligo)	Antibacterial Antifungal Anti-mycobacterial	Flewelling et al. (2013), McKinnell and Perecival (1962)
6	*Spongomorpha coalita*	–	Fatty acids	Antiviral Antibacterial Antifungal	Hudson et al. (1999), Purshotam and Abhishek (2009)
7	*Spongomorpha hystrix*	–	–	Antibacterial	Salem et al. (2011)

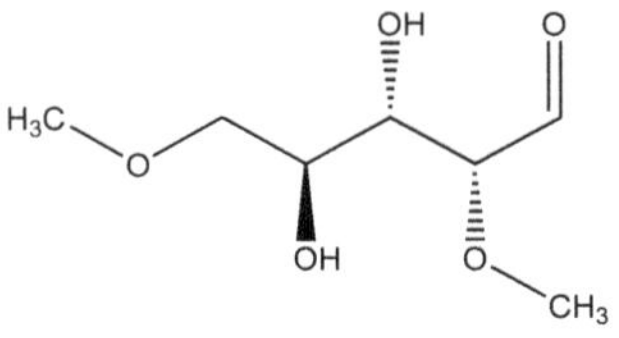

2,5-Di-O-methyl-L-arabinose

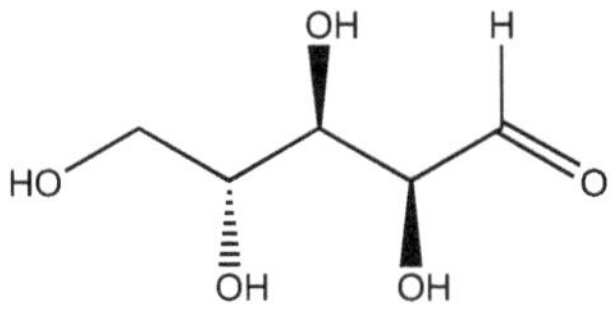

Arabinose

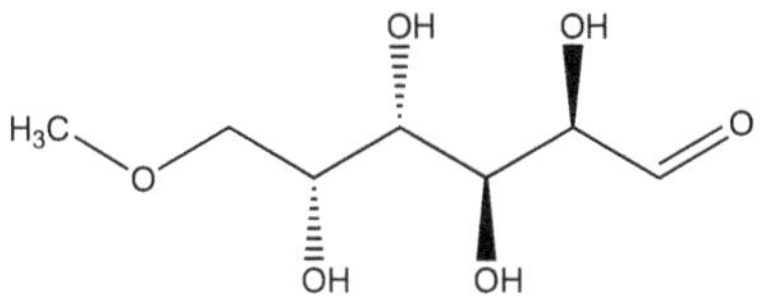

6-O-methyl-D-galactose

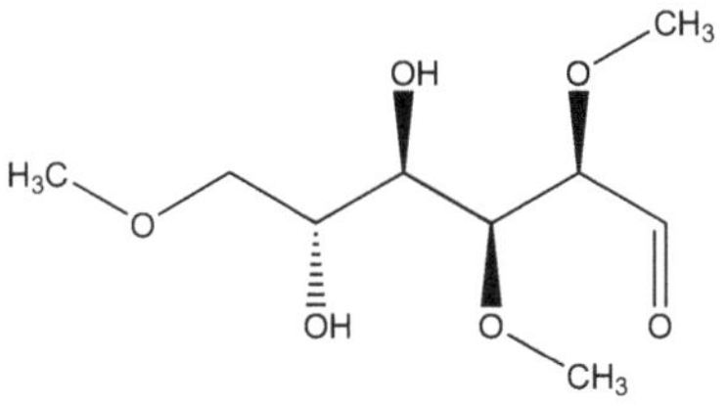

2,3,6-Tri-O-methyl-D-glucose

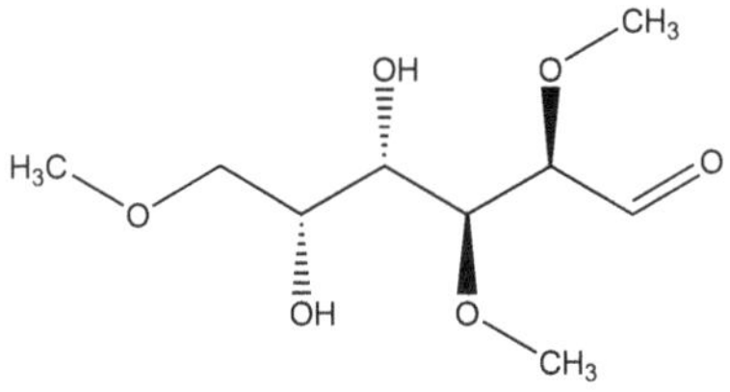

2,3,6-Tri-O-methyl-D-galactose

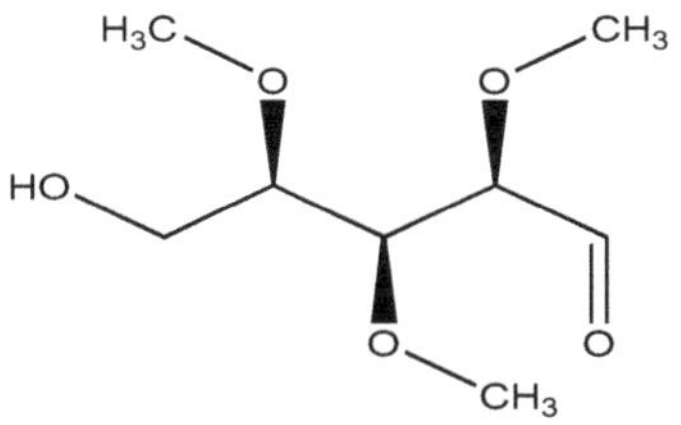

2,3,4-Tri-O-methyl-D-xylose

Fig. 46.5 2,5-Di-O-methyl-L-arabinose, 2,3,4-Tri-O-methyl-L-arabinose, Arabinose, 2,3,5-Tri-O-methyl-L-arabinose, 2,3,4,6-Tetra-O-methyl-D-glucose, 3,5-Di-O-methyl-L-arabinose, 2,4-Di-O-methyl-L-arabinose, 6-O-methyl-D-galactose, 2,3,6-Tri-O-methyl-D-glucose, 2,3,6-Tri-O-methyl-D-galactose, 2,3,4-Tri-O-methyl-D-xylose, 2-O-methyl-L-arabinose, 2,3,4,6-Tetra-O-methyl-D-galactose, etc. compounds isolated by ammonium oxalate method

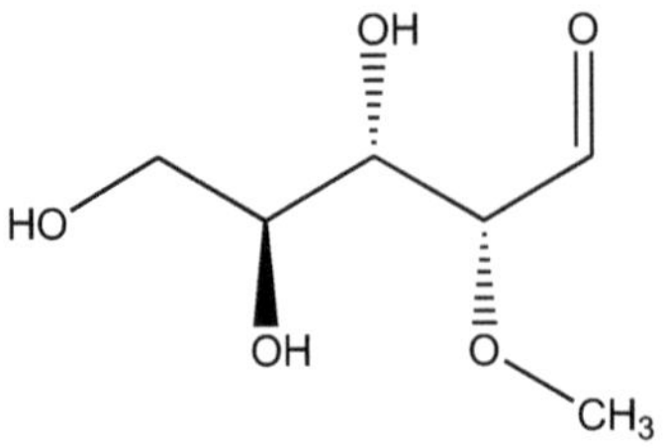

2-O-methyl-L-arabinose

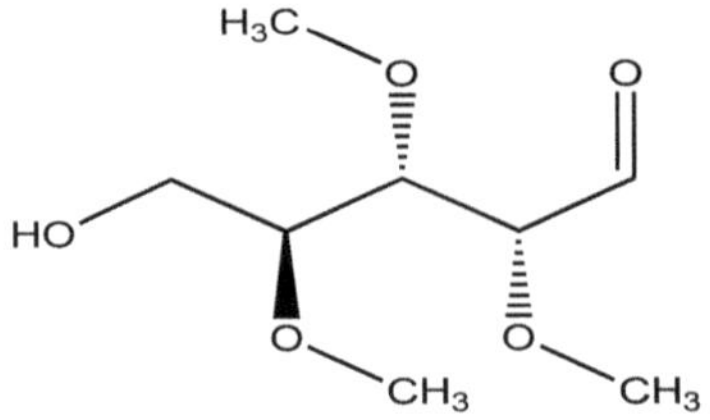

2,3,4-Tri-O-methyl-L-arabinose

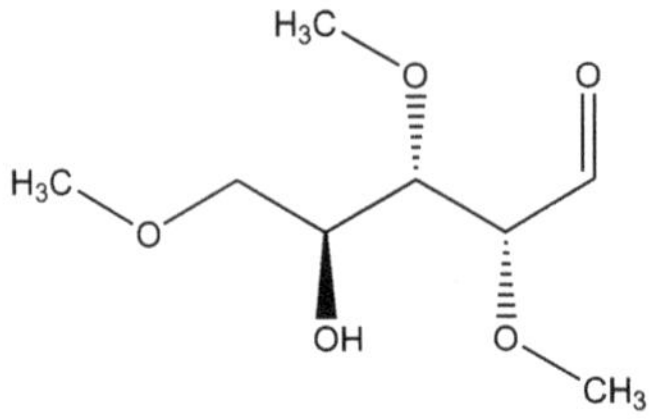

2,3,5-Tri-O-methyl-L-arabinose

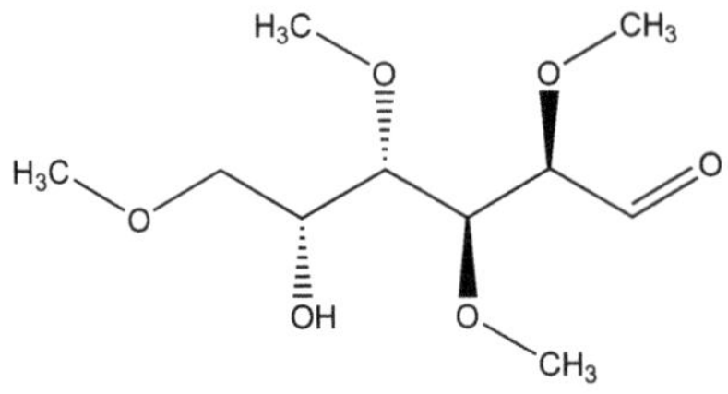

2,3,4,6-Tetra-O-methyl-D-galactose

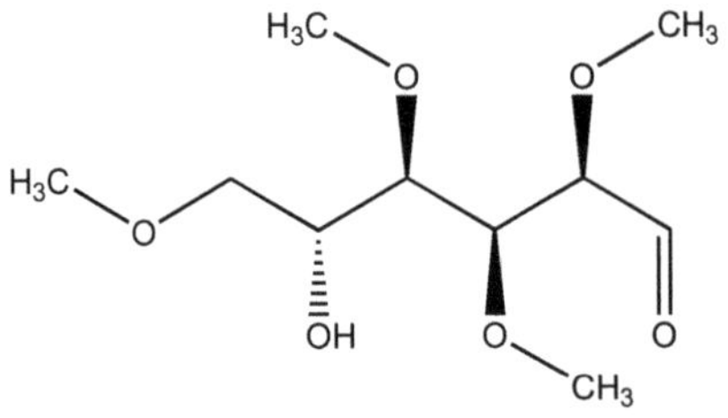

2,3,4,6-Tetra-O-methyl-D-glucose

Fig. 46.5 (continued)

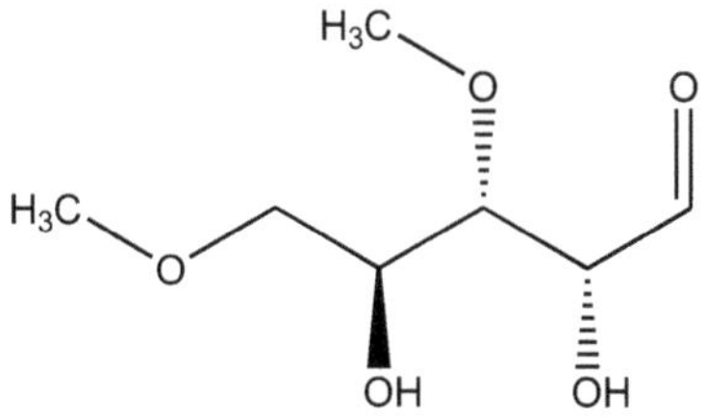

3,5-Di-O-methyl-L-arabinose

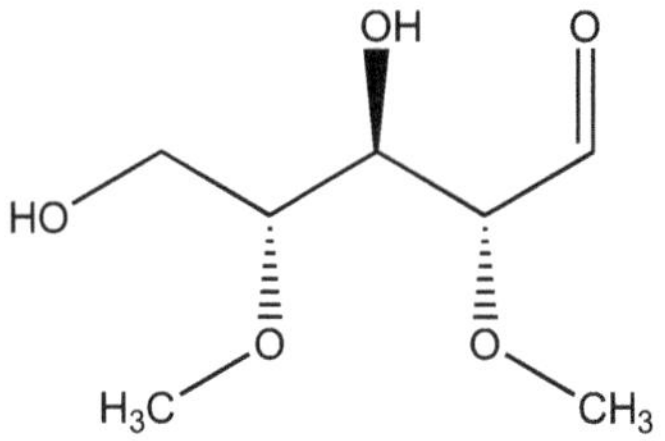

2,4-Di-O-methyl-L-arabinose

Fig. 46.5 (continued)

manner in food and chemical compounds. Thus, it became necessary to check the elements in the samples (Rajubabu et al. 2017) (Tables 46.5–46.7).

46.8 Biological Activities

46.8.1 Antifungal

Antifungal activity was performed using agar cup plate diffusion method with different extracts of three dose levels on various fungal organisms. All the crude extracts of *Spongomorpha indica* L. showed maximum activity against *Aspergillus niger* and minimum against *Saccharomyces cerevisiae.* When compared to other extracts, ethanolic extract showed the highest activity on *Aspergillus niger* at a dose level of 500 mg/ml, and water extract showed the lowest activity on *Saccharomyces cerevisiae* at a dose level of 100 mg/ml. Nystatin (100 mg/ml) was used as a standard drug that showed maximum activity on *Aspergillus niger* and minimum activity on *Mucor racemosus* (Rangaiah et al. 2010).

46.8.2 Antibacterial

Antibacterial activity was performed using agar cup plate diffusion method with different extracts of three dose levels on various bacterial organisms. Among gram-positive organisms, all the crude extracts of *Spongomorpha indica* L. showed maximum activity against *Bacillus subtilis* and minimum against *Streptococcus*

Table 46.5 Major elements present in ppm

Sample name	Variety	Na	P	K	Ca	Mg
Spongomorpha indica L.	Green	5.7	203.0	802.5	4112.1	1.3

Table 46.6 Essential trace elements present in ppm

Sample name	Variety	I	Zn	Se	Cu	Mo	Cr	Mn	Fe	Co
Spongomorpha indica L.	Green	215.5	145.3	2.4	33.5	ND	11.1	246.5	5950.7	2.9

Table 46.7 Potentially toxic elements present in ppm

Sample name	Variety	Pb	Cd	Hg	As
Spongomorpha indica L.	Green	28.0	0.3	2.4	16.6

mutans. When compared to other extracts, ethanolic extract showed the highest activity on *Bacillus subtilis* at a dose level of 500 mg/ml, and water, chloroform extracts showed the lowest activity on *Streptococcus mutans* at three dose levels. Among gram-negative organisms, all the crude extracts of *Spongomorpha indica* L. showed maximum activity against *Pseudomonas aeuruginosa* and minimum against *Lactobacillus acidophilus*. When compared to other extracts, ethanolic extract showed the highest activity on *Pseudomonas aeuruginosa* at a dose level of 500 mg/ml, and water extract showed the lowest activity on *Lactobacillus acidophilus* at three dose levels. Chloramphenicol (100 mg/ml) was used as a standard drug that showed maximum activity on *Staphylococcus aureus* and minimum activity on *Streptococcus anginosus* in gram-positive organisms, whereas in gram-negative organisms, maximum activity showed on *Proteus vulgaris* and minimum activity on *E. coli* (Rangaiah et al. 2010).

46.8.3 Anti-Inflammatory

In vitro anti-inflammatory activity was performed using erythrocyte stabilization method and egg albumin method with methanolic and chloroform extracts. Percentage inhibitions of extracts were compared with diclofenac sodium at 1000 mg/ml. In erythrocyte stabilization method, percentage inhibition obtained for diclofenac sodium was 91.41% and 17% for the methanolic extract of *Spongomorpha indica* L. at 1000 mg/ml. In egg albumin method, percentage inhibition obtained for diclofenac sodium was 87% and 48.66% for the methanolic extract of *Spongomorpha indica* L. at 1000 mg/ml. Percentage inhibition of methanolic extract showed maximum activity when compared to chloroform extract (Satyalakshmi et al. 2018).

46.8.4 Antioxidant

In vitro antioxidant activity was performed using the reducing power method and radical scavenging method (DPPH). In reducing power method, methanolic extract of *Spongomorpha indica* L. showed significant activity at absorbance 0.08 under 700 nm and zone of inhibition was observed. In radical scavenging method (DPPH), methanolic extract of *Spongomorpha indica* L. showed 48.76% zone of inhibition, whereas L-ascorbic acid showed 88.7% zone of inhibition at 200 mg/ml concentration under 517 nm (Satyalakshmi et al. 2018).

46.8.5 Antimicrobial

Antimicrobial activity was performed using agar cup plate diffusion method with different extracts of three dose levels on various bacterial organisms. Among gram-positive organisms, ethanolic extract of *Spongomorpha indica* L. showed the highest activity on *Bacillus subtilis* at a dose level of 500 mg/ml, and water, chloroform extracts showed the lowest activity on *Streptococcus mutans* at three dose levels. Among gram-negative organisms, ethanolic extract of *Spongomorpha indica* L. showed the highest activity on *Pseudomonas aeuruginosa* at a dose level of 500 mg/ml, and water extract showed the lowest activity on *Lactobacillus acidophilus* at three dose levels (Rangaiah et al. 2010).

46.9 Summary

The morphology of seaweed was based on the previous literature about the plant. It is dark seaweed green in color with or without weeds and epiphytes attached to it, having a fishy odor. It consists of thallus that is attached to rocks and unbranched or uniseriate-branched filaments. The seaweed is having a rough texture and irregular clustered rosette shape. The phytochemical reports showed the isolation of sulfated heteropolysaccharide $[\alpha]_D + 59°$ by ammonium oxalate method. Reported pharmacological or biological activities were antibacterial, antimicrobial, antifungal, anti-inflammatory, and antioxidant activities. According to an article on vitamins of marine algae, the seaweed was reported for having high vitamin C content.

The seaweed *Spongomorpha indica* L. also has major elements like Na, P, K, Ca, and Mg; essential trace elements like I, Zn, Se, Cu, Mo, Cr, Mn, Fe, and Co; and potentially toxic elements like Pb, Cd, Hg, and As.

Previous studies showed that ethanolic extract of seaweed *Spongomorpha indica* L. had the highest activity of antibacterial, antifungal, and antimicrobial actions when compared to chloroform, water, and methanolic extracts.

In vitro activities such as anti-inflammatory activity, methanolic extract of seaweed *Spongomorpha indica* L. showed potent zone of inhibition in egg albumin method than erythrocyte stabilization method. In the antioxidant activity, methanolic extract of *Spongomorpha indica* L. showed significant activity in radical scavenging method (DPPH) than reducing power method.

46.10 Conclusion

The above literature survey was reviewed to show that marine sources such as seaweeds are also having an important role in isolating compounds and studying biological activities other than terrestrial plants and animals. It is used for different uses like industrial, ethnobotanical, traditional, and medicinal uses among which traditional uses are more preferred as water waste treatments, soil conditioners, livestock as fodder, organic fertilizers and foliar feeding for plants, etc. The seaweed *Spongomorpha indica* L. also has major elements, essential trace elements, and potentially toxic elements. The seaweed has been reported to possess anti-inflammatory, antioxidant, antibacterial, antifungal, and antimicrobial activities. Hence the review on seaweed *Spongomorpha indica* L. concludes that the seaweed is useful for compounds discovery and have biological aspects other than traditional uses. However it can be considered as a medicinal plant with greater benefits in the future.

Acknowledgments The authors express sincere thanks to Dr. G. Mohan Narasimha Rao, Phycologist, Department of Botany, Andhra University, Visakhapatnam for his help throughout. Thanks to UGC for providing all the necessities and one of the authors Dr. Devarakonda Ramadevi (File No.: PDF- SS- 2015 and 2017-10498).

Conflicts of Interest: None.

References

Alves C, Silva J, Pinteus S et al (2018) From marine origins to therapeutics: the antitumor potential of marine algae-derived compounds. Front Pharmacol 9:3–51. https://doi.org/10.3389/fphar.2018.00777

Bartle WR, Madorin P, Ferland G (2001) Seaweed, vitamin K, and warfarin. Am J Health Syst Pharm 58(23):2300. https://doi.org/10.1093/ajhp/58.23.2300

Bhakuni DS, Rawat DS (2005) Bioactive marine natural products. Springer, Netherland

Flewelling AJ, Ellsworth KT, Sanford J et al (2013) Macroalgae Endophytes from the Atlantic Coast of Canada: a potential source of antibiotic natural products? Microorganisms 1 (1):175–187. https://doi.org/10.3390/microorganisms1010175

Gudbjarnason S (1999) Bioactive marine natural products. Rit Fiskideildar 16:107–110

Guiry MD, Guiry GM (2019) *Spongomorpha indica* Thivy & V. Visalakshmi, 1963. World Register of Marine Species, Galway. http://www.marinespecies.org/aphia.php?p=taxdetails&id=660655. Accessed on 20 Jun 2019

Guiry MD, Guiry GM (2020) *Spongomorpha* Kützing, 1843. World Register of Marine Species, Galway. http://www.marinespecies.org/aphia.php?p=taxdetails&id=143804. Accessed on 2 Jan 2020

Hoek CVD (1963) Revision of the European species of Cladophora. EJ Brill, Netherland

Hudson JB, Kim JH, Lee MK et al (1999) Multiple antiviral activities in extracts of seaweeds from British Columbia. Pharm Biol 37(4):300–306. https://doi.org/10.1076/phbi.37.4.300.5804

Indergaad M (1983) The aquatic resource. The wild marine plants: a global bio-resource. In: Cote WA (ed) Biomass utilization, vol I. Plenum Publishing Corporation, New York, pp 137–168
Kato A, Aruga H, Motomura T (2001) Identification of a cDNA homologous to the cell-cycle-controlling cdc2 gene in Acrosiphonia duriuscula (Acrosiphoniales, Chlorophyta). Phycol Res 49(3):201–205. https://doi.org/10.1046/j.1440-1835.2001.00240.x
Keshini B, Arvind G, Deepeeka K et al (2016) Challenges and opportunities in the present era marine algal applications. In: Thajuddin N, Dhanasekaran D (eds) Algae organisms for imminent biotechnology. IntechOpen, Croatia, pp 238–273
Khalid S, Abbas M, Saeed F et al (2018) Therapeutic potentials of seaweed bioactive compounds. In: Maiti S (ed) Seaweed biomaterials. IntechOpen, Croatia, pp 7–25
Manjula E (2015) Studies on ecological, phytochemical and antimicrobial, antioxidant activities on some seaweeds of Visakhapatnam. Ph.D. Thesis, Andhra University, Visakhapatnam
McHugh DJ (2003) A guide to the seaweed industry. In: FAO fisheries technical paper 441. Food and Agricultural Organization of the United Nations, Rome. DIALOG. http://www.fao.org/docrep/006/y4765e/y4765e00.HTM. Accessed on 15 Jun 2019
McKinnell JP, Perecival E (1962) 606. Structural investigation on the water-soluble polysaccharides of the green seaweed Enteromorpha compressa. J Chem Soc 1:3141–3148. https://doi.org/10.1039/JR9620003141
Newman DJ, Cragg GM (2004) Marine natural products and related compounds I clinical and advanced preclinical trials. J Nat Prod 67(8):1216–1238. https://doi.org/10.1021/np040031y
Nishimura Y, Ishii N, Sugita Y et al (1998) A case of carotenodermia caused by a diet of the dried seaweed called Nori. J Dermatol 25(10):685–687. https://doi.org/10.1111/j.1346-8138.1998.tb02482.x
PRI (2019) Sea plants. Paleontological Organization of Paleontological Research Institute, New York. https://www.priweb.org/index.php/education/education-projects-programs/under-siege-marine-life-vs-the-gulf-oil-spill/species-marine-life-of-the-gulf-coast/sea-plants#top. Accessed on 27 Jun 2019
Purshotam K, Abhishek C (2009) Cyanobacteria: antibacterial activity. New India Publishing Agency, New Delhi
Rajubabu DCH, Srinivasulu A, Aruna B et al (2017) Nutritional and toxicological importance of selected seaweed from Visakhapatnam coast. IJSRSET 3(6):840–850
Rangaiah GS, Lakshmi P, Sruthikeerthi K (2010) Antimicrobial activity of the crude extracts of Chlorophyceae seaweed Ulva and Spongomorpha sps. Against clinical and Phytopathogens. Drug Invent Today 2(6):311–314
Rao EV, Rao NV, Ramana KS (1991) Structural features of the sulphated polysaccharide from a green seaweed, Spongomorpha indica. Phytochemistry 30(4):1183–1186. https://doi.org/10.1016/s0031-9422(00)95199-9
Russell TH (2009) Marine natural products biotechnology. In: Doelle HW, Rokem JS, Berovic M (eds) Biotechnology-volume IX: fundamental in biotechnology. Encyclopedia of life support system, vol 9. EOLSS Publisher, Oxford, pp 120–137
Salem WM, Galal H, Eldeen FN (2011) Screening for antibacterial activities in some marine algae from the red sea (Hurghada, Egypt). Afr J Microbiol Res 5(15):2160–2167. https://doi.org/10.5897/ajmr11.390
Sarojini Y, Sharma NS (1999) Vitamin c contents of some macroalgae of Visakhapatnam, east coast of India. Indian J Mar Sci 28:408–412
Satyalakshmi S, Rekha SJ, Sravani T et al (2018) Biological activities of red and green algae from Visakhapatnam. J Glob Pharm Technol 10(07):07–16
Sekar D (2015) Isolation and Identification of Bioactive Compounds from Certain Marine Macroalgae. Ph.D. Thesis, Annamalai University, Tamil Nadu
Setchell WA, Gardner NL (1920) The marine algae of the Pacific coast of North America, part II Chlorophyceae. In: University of California Publications in botany 8.AlgaeBase. DIALOG. https://img.algaebase.org/pdf/8CCB0C050c93f13B75pV985E4D35/17511.pdf. Accessed 17 June 2019

Stein JR, Borden CA (1984) Causative and beneficial algae in human disease conditions: a review. Phycologia 23:485–501

Taylor WR (1957) Marine algae of the Northeastern coast of North America. University of Michigan Press, Ann Arbor, MI

Taylor WR (1960) Marine algae of the eastern tropical and subtropical coasts of the America. University of Michigan Press, Ann Arbor, MI

Teas J, Baldeon ME, Chiriboga DE et al (2009) Could dietary seaweed reverse the metabolic syndrome? Asia Pac J Clin Nutr 18(2):145–154

Thomas NV, Kim SK (2013) Beneficial effects of marine algal compounds in Cosmeceuticals. Mar Drugs 11(12):146–164. https://doi.org/10.3390/md11010146

Veeresham C (2012) Natural products derived from plants as a source of drugs. J Adv Pharm Technol Res 3(4):200–201. https://doi.org/10.4103/2231-4040.104709

Wu Q, Nay B, Yang M et al (2018) Marine sponges of the genus Stelletta as promising drug source: chemical and biological aspects. Acta Pharm Sin 9(2):237–257. https://doi.org/10.1016/j.apsb.2018.10.003

Validation of Traditional Claim of *Oxalis debilis* Kunth: An Ethnomedicinal Plant from Northeastern Region of India

47

Julfikar Ali Junejo, Mithun Rudrapal, and Kamaruz Zaman

Abstract

Ethnomedicinal survey documents the traditional uses of *Oxalis debilis* Kunth leaves in the management of diarrhea and helminthiasis. In our study, the antidiarrheal and anthelmintic activities were evaluated for ethyl acetate and hydro-alcoholic leaf extracts of *O. debilis*. Both the leaf extracts exhibited antidiarrheal and anthelmintic activities. The hydro-alcoholic extract exhibited better activity profile than that of ethyl acetate extract. Moreover, the anthelmintic activity was found to be more significant in comparison to the antidiarreal activity for the hydro-alcoholic extract. From phytochemical and TLC studies, the presence of phenolic compounds and flavonoids has been demonstrated in the extracts indicating their possible role in the bioactivity of *O. debilis* leaves. However, the potent antidiarrheal and anthelmintic activities are claimed with the hydro-alcoholic extract of *O. debilis* leaves. Our present work justifies the ethnomedicinal claims of *O. debilis* leaves in the treatment of diarrhea and helminthiasis, particularly in the northeastern region of India.

Keywords

Traditional claim · *Oxalis debilis* · Hydro-alcoholic extract · Ethnomedicinal · Phytochemical · Antidiarrheal · Anthelmintic

J. A. Junejo · M. Rudrapal · K. Zaman (✉)
Department of Pharmaceutical Sciences, Dibrugarh University, Dibrugarh, Assam, India
e-mail: kzaman71@dibru.ac.in

S. C. Mandal et al. (eds.), *Evidence Based Validation of Traditional Medicines*,
https://doi.org/10.1007/978-981-15-8127-4_47

Abbreviations

CPCSEA	Committee for the Purpose of Control and Supervision of Experiments on Animals
IAEC	Institutional Animal Ethics Committee
LOD	Loss on drying
OECD	Organisation for Economic Co-operation and Development
SEM	Standard error of mean
TLC	Thin-layer chromatography
UV	Ultraviolet
WHO	World Health Organization

47.1 Introduction

Medicines of plant origin have been used in traditional health practices for the prevention, treatment, and management of a variety of human ailments since ancient time (Rahman et al. 2013). According to the World health Organization (WHO), about 60% of population depends on remedies based on traditional practices worldwide, and 80% of the population living in developing world relies on traditional herbal medicines for their primary healthcare needs (Ekor 2014; Mukherjee et al. 2012).

Oxalis debilis Kunth is a perennial herb belonging to the family of Oxalidaceae (Fig. 47.1). It has three-part, clover-like leaves and deep pink flowers found during July to September (Patiri and Borah 2012). This edible plant is grown as wild species used in ethnomedicines for skin infections. It also has potent antipyretic, antidiarrheal, antiasthma, and anthelmintic activities (Junejo et al. 2014b; Singh and Dubey 2012). In traditional practice, the leaf extract of *O. debilis* is taken orally for curing fever (Lourteig 2000; Luo et al. 2006). Literature review reveals that no scientific work has been carried out so far on antidiarrheal and anthelmintic activities

Fig. 47.1 *O. debilis* plant

of this plant. Our present investigation, therefore, aims at the evaluation of certain leaf extracts of *O. debilis* for antidiarrheal and anthelmintic activities.

Diarrhea is increasingly responsible for mortalities in developing countries. Children under 5 years old are more susceptible to this disease (Amole et al. 2010; Saralaya et al. 2010). To circumvent this disease, the WHO has begun a program for the control and prevention program of diarrhea disease using traditional medicine practices, particularly herbal remedies (Damiki and Siva 2011). Helminth infections (worm infestations or helminthiases) are the most common parasitic illness in humans. This chronic and debilitating illness affects about half of the population of the world. The morbidity rate of helminthiasis is increasing worldwide particularly in subtropical regions. Herbal remedies have potential in the treatment and management of helminthiases (Sirama et al. 2015).

47.2 Materials and Methods

47.2.1 Drugs and Chemicals

Loperamide hydrochloride and albendazole were purchased from Sigma-Aldrich, Mumbai. These are used as standard drugs. Crude extracts of ethyl acetate and hydro-alcoholic solvent of *O. debilis* leaves were used as test drugs. The extracts were formulated at the concentrations of 250 and 500 mg/kg body weight for antidiarrheal study and at the concentrations of 15, 60, and 100 mg/kg body weight for anthelmintic study.

47.2.2 Collection of Plant

Fresh *O. debilis* Kunth leaves were collected from Dibrugarh University Campus, Dibrugarh, Assam, India, during the month of April to May, 2013. Leaves were properly washed, dried under the shade, and coarsely powdered. The plant species was authenticated by the Botanical Survey of India, Eastern Regional Centre, Shillong. A voucher specimen (Specimen Number DU/KM/2013/07, Reference Number BSI/ERC/2013/Tech/PlantIdentification/636) was submitted for future references.

47.2.3 Preparation of Extracts

The extraction of powdered leaves (approximately 300 g) of *O. debilis* was done by cold maceration for about 15–20 days. Ethyl acetate and hydro-alcohol (70:30) were used as solvents of extraction. After extraction is over, the solvent was allowed to reduce by evaporation at 45 °C under negative pressure to yield the concentrated extract.

47.2.4 Physicochemical and TLC Analyses

Various physicochemical properties such as %LOD, ash value, and extractive value of the powdered leaves were evaluated in accordance with the standard methods (Kaneria and Chanda 2011; Evans 2002). The extractive values of solvent extracts were also determined.

Silica gel 60 F254 (0.25 mm thick, precoated on aluminum sheet) and TLC plates (Merck, Germany) were used for TLC fingerprinting using the mixture of toluene/ethyl acetate/formic acid (4:5:1) as a mobile phase. The visualization of spots was carried out under UV light at 254 and 365 nm (Indian Pharmacopoeia 2014). The spots obtained with their relative R_f values were recorded.

47.2.5 Phytochemical Screening

Preliminary screening of both the leaf extracts of *O. debilis* for the detection of various phytoconstituents was systematically performed (Kokate 1994; Harborne 1998).

47.2.6 Experimental Animals

Healthy Wistar albino rats (200–250 g) were used for the antidiarrheal activity. Animals were procured from the Laboratory Animal House, Department of Pharmaceutical Sciences, Dibrugarh University, with the approval of the IAEC (Approval no. IAEC/DU/50 Dated 24.9.13) under guidelines set by the CPCSEA, New Delhi (India). Standard laboratory conditions (temperature 25 ± 2 °C, relative humidity 50 ± 5%, 12 h light/dark cycle) were maintained for the animals. Prior to experiment, acclimatization of animals was carried out for 1 week. During this period, animals were provided with standard diet and drinking water.

47.2.7 Acute Oral Toxicity Study

Animals were randomly grouped with six animals in each group. One group was kept as normal control. Increasing doses (250, 500, 1000, 2000, and 5000 mg/kg bw) of extracts were given to the test groups. The observation for acute toxicity for 14 days was done as per OECD guidelines (Junejo et al. 2014a).

47.2.8 Antidiarrheal Activity

For antidiarrheal activity, the in vivo model of castor oil-induced diarrhea in rats was used (Shoba and Thomas 2001; Uddin et al. 2005). Four groups of animals were prepared, allocating five rats each: group I, control received normal saline (10 ml/

kg); group II, standard treated with loperamide (50 mg/kg); and groups III and IV, test groups treated with ethyl acetate and hydro-alcoholic extract (250 mg/kg and 500 mg/kg, respectively). The percent inhibition of defecation was calculated as follows:

$$\%\text{Inhibition of defecation} = [(A - B)/A] \times 100$$

where *A* and *B* are mean value of number of defecation in control and drug/extract, respectively. The results are presented as the mean ± standard error of mean (SEM) of three replicate observations.

47.2.9 Anthelmintic Activity

Adult Indian earthworms (*Pheretima posthuma*) and tapeworms (*Raillietina spiralis*) were used for the evaluation of in vitro anthelmintic activity (Panda et al. 2011). The earthworms and tapeworms were obtained from moist soil and intestines of chicken, respectively. The worms were prepared before performing the experiment. The average size of earthworms and tapeworms used were 7.3–8.5 and 6.4–7.6 cm, respectively (Pillai and Nair 2011; Thorn et al. 1977; Zaman 1984). A total of 60 worms were divided into 10 groups, each group containing 6 worms. Different concentrations (15, 60, 100 mg/ml) of extracts (ethyl acetate or hydro-alcoholic) and albendazole (standard) were prepared in distilled water and used as test and standard solutions, respectively. The time required for paralysis and thereby death of experimental worms were calculated to assess the anthelmintic activity. The results are presented as the mean ± standard error of mean (SEM) of three replicate observations.

47.3 Results and Discussion

47.3.1 Physicochemical and TLC Analyses

The results of physicochemical analyses are given in Table 47.1. Physicochemical parameters such as moisture content, ash values, and extractive values were within limit. Physicochemical analyses standardize the crude leaf powder and the leaf extracts of *O. debilis* considering their purity and quality.

Table 47.1 also depicts the TLC profiling of *O. debilis* leaf extracts. Characteristic spots with their relative R_f values were obtained. The chromatograms are displayed in Fig. 47.2. TLC is a useful analytical tool for chemical fingerprinting as well as chemotaxonomic identification of plants (Junejo et al. 2016). TLC analysis exhibited the detection of certain phytoconstituents or marker components such as phenolic substances, tannins, and flavonoids in both the leaf extracts (ethyl acetate and hydro-alcoholic) of *O. debilis*.

Table 47.1 Physicochemical and TLC analyses of leaf and leaf extracts of *O. debilis*

Physicochemical[a]		TLC[b]			
Parameter	% w/w ± SEM	Ethyl acetate		Hydro-alcoholic	
		No. of spots	R_f value	No. of spots	R_f value
Moisture content (% LOD)	11.09 ± 0.17	02	0.924 0.886	04	0.943 0.905 0.792 0.584
Ash values					
Total ash	34.56 ± 0.29				
Acid insoluble	7.18 ± 0.47				
Water soluble	12.56 ± 0.33				
Extractive values (leaf)					
Water soluble	15.73 ± 0.26				
Alcohol soluble	8.22 ± 0.18				
Extractive values (extract)					
Ethyl acetate	1.59 ± 0.23				
Hydro-alcoholic	5.46 ± 0.37				

[a]Values are expressed as mean ± SEM of triplicate observations
[b]Solvent system: toluene/ethyl acetate/formic acid = 5:4:1

47.3.2 Phytochemical Screening

The results of phytochemical screening depicted in Table 47.2 revealed the presence of certain phytoconstituents such as phenolics and tannins, flavonoids, saponins, and alkaloids in the leaf extracts of *O. debilis* (Junejo et al. 2016).

47.3.3 Acute Toxicity

No signs of acute toxicity and/or mortality were observed in animals up to the maximum dose of 2000 mg/kg bw during the experimental period. Significant changes in behavior, food intake, water consumption, hair loss, postural abnormalities, etc. were not observed. Acute toxicity study, therefore, validates the folloric claim on the plant as a safe herbal medicine without having any adverse effects (Junejo et al. 2016).

47.3.4 Antidiarrheal Activity

Both the extracts exhibited antidiarrheal activity in rats. Hydro-alcoholic extract exhibited higher activity profile than that of the ethyl acetate extract. The antidiarrheal effects of extracts are presented in Table 47.3. Upon administration of extracts, the production of wet feces was decreased causing prevention of defecation to an appreciable extent. At 250 mg/kg bw dose, the extract with ethyl acetate produced 24.83% inhibition of defecation. At a higher dose, i.e., 500 mg/kg

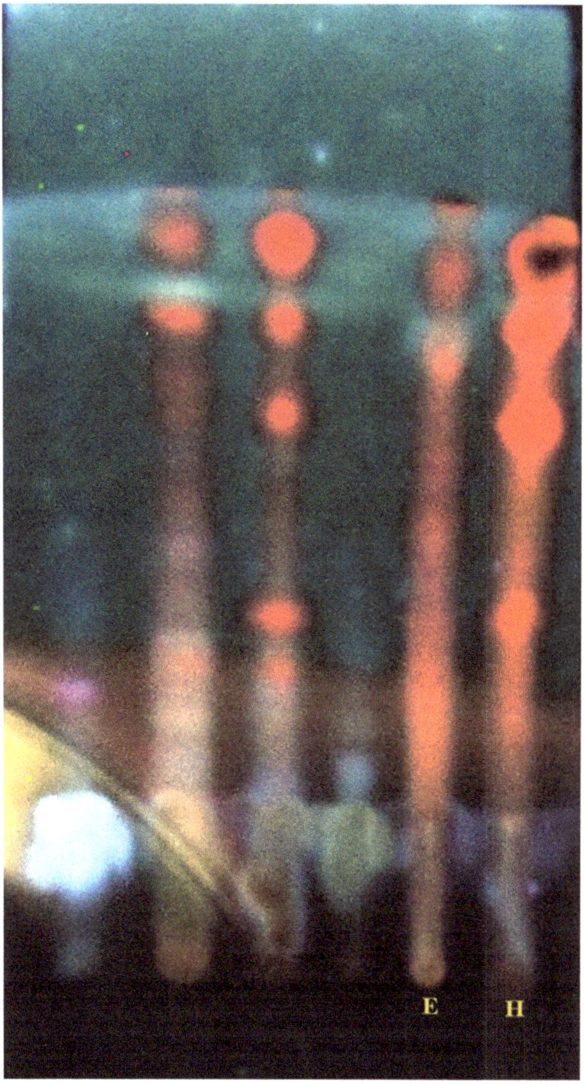

Fig. 47.2 TLC of *O. debilis* leaf extracts under 254 nm (E is ethyl acetate extract; H is hydro-alcoholic extract)

bw, the activity was improved up to 44.15% inhibition of defecation. Similarly, the extract with hydro-alcohol showed 40.01% of inhibition at 250 mg/kg bw dose, while 73.37% defecation inhibition was observed at 500 mg/kg bw dose. Administration of loperamide (standard) at 50 mg/kg dose produced 3.12 $\pm$ 1.32 numbers of diarrheal feces in 4 h causing 74.67% inhibition of defecation. It was also seen that with increasing dose of the extracts, antidiarrheal activity was also increased. The decrease in defecation frequency was significantly inhibited by the extract with hydro-alcoholic solvent as compared with untreated group. The hydro-alcoholic

Table 47.2 Phytochemical screening of *O. debilis* leaf extracts

Phytoconstituents	Ethyl acetate extract	Hydro-alcoholic extract
Alkaloids	–	+
Glycosides	–	–
Phenolic compounds and tannins	+	+
Flavonoids	+	+
Saponins	–	+
Terpenoids	–	–
Steroids	–	
Carbohydrates	–	+
Amino acids and proteins	–	+
Fats and oils	–	–

+, presence; –, absence

Table 47.3 Antidiarrheal effect of leaf extracts of *O. debilis*

Group	Treatment	Dose (mg/kg)	Number of diarrheal feces in 4 h	% Inhibition of defecation
I	Normal control	10 (ml/kg)	12.32	–
II	Loperamide (standard)	50	3.12 ± 1.32	74.67
III	Ethyl acetate extract	250	9.26 ± 1.67	24.83
		500	6.88 ± 2.11	44.15
IV	Hydro-alcoholic extract	250	7.39 ± 1.82	40.01
		500	3.28 ± 1.38	73.37

Data are presented as the means ± SEM of three replicate studies

extract produced similar extent of antidiarrheal effect as that of loperamide but at considerably higher dose.

47.3.5 Anthelmintic Activity

Table 47.4 depicts details of anthelmintic activity produced by extracts of *O. debilis* leaves. In earthworms, the extract with ethyl acetate solvent exhibited highest anthelmintic activity at the dose of 100 mg/ml causing paralytic effect in 43.21 ± 1.57 min and death in 69.24 ± 0.68 min. In a similar way, the extract with hydro-alcoholic solvent showed the highest anthelmintic activity in earthworms at 100 mg/ml dose producing paralytic effect in 23.23 ± 0.33 min and death in 34.48 ± 0.39 min. In tapeworms, ethyl acetate extract produced maximum paralytic response causing paralysis in 41.51 ± 1.14 min and death in 63.71 ± 0.68 min at 100 mg/ml dose. Similarly, the highest activity was demonstrated with hydro-alcoholic extract against tapeworms at 100 mg/ml causing paralysis in 21.87 ± 0.88 min and death in 39.83 ± 0.69 min. Albendazole (standard drug) showed paralytic effect in 21.27 ± 0.74 min causing death in 33.12 ± 1.47 and

Table 47.4 Anthelmintic activity of leaf extracts of *O. debilis*

Group	Treatment	Conc. (mg/ml)	*Pheretima posthuma* (earthworm) Time taken for paralysis (min)	*Pheretima posthuma* (earthworm) Time taken for death (min)	*Raillietina spiralis* (Tapeworm) Time taken for paralysis (min)	*Raillietina spiralis* (Tapeworm) Time taken for death (min)
I	Normal control	–	–	–	–	–
II	Experimental control	–	–	–	–	–
III	Albendazole (standard)	15	35.28 ± 0.68	52.32 ± 1.12	37.73 ± 0.76	53.14 ± 0.69
		60	29.23 ± 1.53	47.35 ± 1.33	27.14 ± 1.86	47.21 ± 1.42
		100	21.27 ± 0.74	33.12 ± 1.47	20.31 ± 1.57	40.56 ± 1.57
IV	Ethyl acetate extract	15	68.44 ± 1.35	97.62 ± 0.23	62.46 ± 0.58	95.73 ± 0.97
		60	51.31 ± 0.46	77.46 ± 0.42	49.19 ± 0.92	71.74 ± 0.35
		100	43.21 ± 1.57	69.24 ± 0.68	41.51 ± 1.14	63.71 ± 0.68
V	Hydro-alcoholic extract	15	49.68 ± 0.95	71.39 ± 1.13	45.86 ± 1.45	67.97 ± 1.39
		60	32.96 ± 1.68	51.12 ± 0.58	29.23 ± 0.98	46.86 ± 0.73
		100	23.23 ± 0.33	34.48 ± 0.39	21.87 ± 0.88	39.83 ± 0.69

Data are presented as the mean ± SEM of three replicate studies

20.31 ± 1.57 min causing death in 40.56 ± 1.57 min against earthworms and tapeworms, respectively, at 100 mg/ml. It was apparent from our observation that the anthelmintic activity was significantly improved with increasing the concentration of extract. The maximum activity was recorded with 100 mg/ml dose, while 60 and 15 mg/ml doses exhibited considerably reduced activity. The extract with hydro-alcohol exhibited better profile of anthelmintic activity as compared to the extract with ethyl acetate. The activity of hydro-alcoholic extract could be compared with that of the standard albendazole.

To summarize, the hydro-alcoholic extract was found superior to possess antidiarrheal and anthelmintic activities over ethyl acetate extract. Both the extracts possess bioactivity at the highest tested dose. The activities of hydro-alcoholic extract are comparable with that of the standard drugs. The hydro-alcoholic leaf extract possesses significant level of anthelmintic efficacy, whereas its antidiarrheal efficacy was found comparatively less significant. From phytochemical and TLC studies, the presence of phenolic substances and flavonoids has been demonstrated in the extracts, indicating their possible role in the bioactivity of *O. debilis* leaves. However, antidiarrheal and anthelmintic potencies lie with the hydro-alcoholic leaf extract of *O. debilis*. Finally, our present work justifies the ethnomedicinal uses of *O. debilis* leaves in the treatment of diarrhea and helminthiasis in the northeastern region of India.

47.4 Conclusion

Our work reports the significant role of ethyl acetate and/or hydro-alcoholic leaf extracts of *O. debilis* in the prevention as well as treatment of diarrhea and helminthiasis. This study confirms the traditional claim of *O. debilis* leaves as its use in ethnomedicine for the management of diarrhea and worm infestation in the northeastern parts of India. Since it is a preliminary study, further phytochemical, toxicological, and pharmacological studies are necessary for further exploration of phytoconstituents responsible for the bioactivity of *Oxalis debilis* leaves.

References

Amole OO, Salahdeen HM, Onyeahialam AE (2010) Evaluation of the antidiarrhoeal effect of *Lanneawel witschii* Hiern (Anacardiaceae) bark extract. Afr J Pharm Pharmacol 4(4):165–169

Damiki L, Siva H (2011) Ethnomedicinal plants used for diarrhea by tribals of Meghalaya, Northeast India. Pharmacogn Rev 5(10):147–154

Ekor M (2014) The growing use of herbal medicines: issues relating to adverse reactions and challenges in monitoring safety. Front Pharmacol 4:1–10

Evans WC (2002) Trease and Evans' Pharmacognosy. WB Sounders Company Ltd., London

Harborne JB (1998) Phytochemical methods: a guide to modern techniques of plant analysis. Chapman & Hall, London

Indian Pharmacopoeia (2014) (vol. IV), Published by the Indian Pharmacopoeia Commission (IPC), Ministry of Health and Family Welfare, Government of India, New Delhi

Junejo JA, Gogoi G, Zaman K, Islam J, Hazarika G (2016) Phytochemical profiling and toxicological studies of *Oxalis debilis* Kunth leaves. Int J Green Pharm 10(3):165–171

Junejo JA, Zaman K, Rudrapal M, Mondal P (2014a) Antidiabetic assessment of the hydro-alcoholic leaf extracts of the plant *Tetrastigma angustifolia* (Roxb.), a traditionally used north-eastern Indian vegetable. Biomed Pharmacol J 7(2):635–644

Junejo JA, Zaman K, Rudrapal M, Mondal P, Singh KD, Verma VK (2014b) Preliminary phytochemical and physicochemical evaluation of *Carallia brachiata* (Lour.) Merr. Leaves. J Appl Pharm Sci 4:123–127

Kaneria M, Chanda S (2011) Phytochemical and pharmacognostic evaluation of leaves of *Psidium guajava* L. (Myrtaceae). Phcog J 3:41–45

Kokate CK (1994) Practical Pharmacognosy. Vallabh Prakashan, New Delhi

Lourteig A (2000) Oxalis L. Subgenero Monoxalis (Small) Lourt., Oxalis y Trifidus Lourt. Bradea 7:201–629

Luo S, Zhang D, Renner S (2006) *Oxalis debilis* in China: distribution of flower morphs, sterile pollen and polyploidy. Ann Bot 8:459–464

Mukherjee PK, Nema NK, Venkatesh P, Debnath PK (2012) Changing scenario for promotion and development of Ayurveda—way forward. J Ethnopharmacol 24:654–342

Panda SK, Das D, Tripathy NK (2011) Evaluation of anthelmintic activity of *Chlorophytum borivilianum* santapau & fernandes. Int J Res Pharm Biomed Sci 2(2):676–679

Patiri B, Borah A (2012) Wild edible plants of Assam, NE India. Geetakhi Printers and Publishers, Guwahati

Pillai LS, Nair BR (2011) A comparative study of the anthelmintic potential of *Cleome viscosa* L. and *Cleome burmanni* W. Indian J Pharm Sci 73(1):98–100

Rahman MK, Barua S, Islam MF, Islam MR, Sayeed MA, Parvin MS et al (2013) Studies on the anti-diarrheal properties of leaf extract of *Desmodium puchellum*. Asian Pac J Trop Biomed 3 (8):639–664

Saralaya MG, Patel P, Patel M, Roy SP, Patel AN (2010) Anti-diarrheal activity of methanolic extract of *Moringa oleifera* lam roots in experimental animal models. Int J Pharm Res 2 (2):35–39

Shoba FG, Thomas M (2001) Study of antidiarrheal activity of four medicinal plants in castor-oil induced diarrhea. J Ethnopharmacol 76(1):73–76

Singh A, Dubey NK (2012) An ethnobotanical study of medicinal plants in Sonebhadra District of Uttar, Pradesh, India with reference to their infection by foliar fungi. J Med Plants Res 6:2727–2746

Sirama V, Kokwaro J, Owuor B, Yusuf A, Kodhiambo M (2015) *In-vitro* anthelmintic activity of *Vernonia amygdalina* Del. (asteraceae) roots using adult *Haemonchus contortus* worms. Int J Pharmacol Res 5(1):1–7

Thorn GW, Adams RD, Braunwald E, Isselbacher KJ, Petersdrof RG (1977) Harrison's principles of internal medicine. McGraw-Hill Book Co., New York

Uddin SJ, Sjilpi JA, Alam SM, Alamgir M, Rahman MT, Sarker SD (2005) Antidiarrhoeal activity of the methanol extract of the barks of *Xylocarpus moluccensis* in castor oil and magnesium sulphate induced diarrhea models in mice. J Ethnopharmacol 101:139–143

Zaman V (1984) Atlas of medical parasitology. ADIS Health Science Press, Sydney

Traditional Herbal Medicine Practiced in Plateau-Fringe and Rarh Districts of West Bengal, India

48

Debarati Jana, Kartik Chandra Guchhait, Saroj Ballav, Amiya Kumar Panda, and Chandradipa Ghosh

Abstract

Traditional medicine comprises the healthcare support indigenously developed over generations among diverse cultural groups inhabiting different geographical locations all over the globe since time immemorial. According to several reports, around 80% of world populations especially from the developing countries rely on traditional medical practices of which usage of plant products remains in highest position due to their proven medicinal values. India also has a very rich tradition of practicing herbal medicine particularly in the rural and tribal communities for prevention and cure of diseases. The reasons behind popularity and widespread uses of herbal products are their low cost, easy availability and lesser side effects in addition to poor access of the common people with socio-economic vulnerability to primary healthcare system and conventional medicine. The great biodiversity of medicinal plants of India also remained pivotal for developing such a very rich tradition of medical practices since ancient time, mainly practiced at individual and familial levels, and usually transferred orally from one generation to other. West Bengal is not an exception to this therapeutic culture. Presently research is being encouraged on herbal medicine from different stakeholders aiming development of more life-saving modern drugs at low cost as there is an increasing trend of interest in traditional medicine worldwide. A systematic approach for documentation of traditional knowledge of herbal medicine is required in this view. In this review such an attempt is made to document the plants and their products with potential therapeutic uses in the plateau-fringe

D. Jana · K. C. Guchhait · S. Ballav · C. Ghosh (✉)
Department of Human Physiology, Vidyasagar University, Midnapore, West Bengal, India
e-mail: ch_ghosh@mail.vidyasagar.ac.in

A. K. Panda
Department of Chemistry, Vidyasagar University, Midnapore, West Bengal, India
e-mail: akpanda@mail.vidyasagar.ac.in

S. C. Mandal et al. (eds.), *Evidence Based Validation of Traditional Medicines*,
https://doi.org/10.1007/978-981-15-8127-4_48

and rarh districts of state of West Bengal, India, that is not only distinct from geomorphological and climatic points of view but also in terms of natural and human resources.

Keywords

Traditional medicine · Herbal · Plateau-Fringe · Rarh · Tribal

Abbreviations

AYUSH	Department of Ayurveda Unani Siddha and Homeopathy
CM	Complementary Medicine
IDRC	International Development Research Centre
T&CM	Traditional & Complementary Medicine
TM	Traditional Medicine
WHO	World Health Organization

48.1 Introduction

In the era of globalization, medical sciences has made remarkable progresses resulting in an increase in the life expectancy and decrease in overall rate of mortality and morbidity through the formulation of newer drugs against different infectious and life-threatening diseases. In spite of such successes, quality healthcare facilities are unreachable to all, as people residing in rural and remote corners of the globe cannot access all conventional or modern medical facilities (Sen and Chakraborty 2017). Hence a huge fraction of the world population dwelling in the underdeveloped and developing countries largely depend on indigenously developed traditional system of medicine in their respective societies for primeval healthcare needs since human civilization came into being. As a result traditional medicine plays significant role among the diverse communities and cultural groups in those rural and far-flung areas where the magical touches of conventional medicine have largely failed to reach. According to the report of World Health Organization (WHO), around 80% of world population especially from the developing countries relies on traditional medical practices to meet their basic healthcare needs (Akerele 1993).

According to WHO Traditional Medicine Strategy 2014–2023, the definition of traditional medicine (TM) is “the sum total of the knowledge, skill, and practices based on the theories, beliefs, and experiences indigenous to different cultures, whether explicable or not, used in the maintenance of health as well as in the prevention, diagnosis, improvement or treatment of physical and mental illness. Traditional medicine has got a long history of its use in maintenance of health and prevention and treatment of diseases” (World Health Organization 2013). Alternatively, traditional medicine or nonconventional medicine may be termed complementary medicine (CM). Complementary medicine (CM) has been defined as “a broad set of health care practices that are neither part of a country's own tradition nor part of conventional medicine, and are not fully integrated into the dominant health care system. CM practices are used interchangeably with TM practices in some countries" (World Health Organization 2013). The term T&CM encompasses both TM and CM products, practices and practitioners. T&CM can further be described as an important but mostly underestimated health resource that includes diverse applications, especially in dealing with lifestyle-related chronic diseases, and the health issues of ageing populations (World Health Organization 2019).

This also includes ancient indigenous experiences used to maintain health and comfort as well as to prevent illnesses, assist diagnosis and help cure. WHO has also defined "indigenous traditional medicine" as the "sum total of knowledge and practices, whether explicable or not, used in diagnosing, preventing or eliminating physical, mental and social diseases. This knowledge or practice may rely exclusively on past experience and observation handed down orally or in writing from generation to generation. These practices are native to the country in which they are practiced. The majority of indigenous traditional medicine has been practiced at the primary health care level. The terms for these medicines may vary from country to country and region to region, so specific definitions are used" (World Health Organization 2019).

Due to ready availability and close associations of human beings with plants in their surroundings since ancient time, plant-based herbal medicine has gained predominance in traditional healthcare system. In most Asian and African countries where folk medicines are mostly preferred, the search for herbal cures remains to be very common practice (Akpanabiatu et al. 2005). In developing countries the basic approach of the traditional healers and their plant-based herbal medicines is curative rather than preventive for common ailments (Gabriel et al. 2007). According to WHO, "herbal medicines include herbs, herbal materials, herbal preparations and finished herbal products that contain, as active ingredients, parts of plants, other plant materials or combinations thereof. In some countries, herbal medicines may contain, by tradition, natural organic or inorganic active ingredients that are not of plant origin (e.g. animal and mineral materials)" (World Health Organization 2019). In traditional system the use of medicinal plants, their parts, doses and mode of applications vary with the herbal practitioners and/or healers among diverse ethnic groups and communities. The geographical and climatic conditions are also found to be major regulatory factors in the variations of the herbal products, their medicinal values and efficacies and also their indigenous applications. But research on medicinal plants and herbal medicines targets to identify and isolate the active principles and attempts to standardize their adequate applications (Das and Ghosh 2017).

The effective uses, easy availability, low cost and lesser side effects of the herbal medicines besides remote access of the common people to primary healthcare system and conventional medicine might be the underlying reasons for their enormous popularity and extensive uses especially in poor and vulnerable rural communities residing in developing countries (Pattanayak et al. 2015). At the International Conference on Traditional Medicine for Southeast Asian Countries in February 2013, the WHO Director-General Dr. Margaret Chan stated that "traditional medicines, of proven quality, safety, and efficacy, contribute to the goal of ensuring that all people have access to care. For many millions of people, herbal medicines, traditional treatments, and traditional practitioners are the main source of health care, and sometimes the only source of care. This is care that is close to homes, accessible and affordable. It is also culturally acceptable and trusted by large numbers of people. The affordability of most traditional medicines makes them all the more attractive at a time of soaring health-care costs and nearly universal austerity. Traditional medicine also stands out as a way of coping with the relentless rise of chronic non-communicable diseases" (World Health Organization 2013).

Three major categories of traditional medicine have been suggested: the codified medical systems, folk medicine and allied forms of health knowledge. The great traditions of healthcare practices in different parts of the world that have been evolved through time immemorial are codified medical systems. This includes Ayurveda, Siddha and Unani in the Indian subcontinent and traditional Chinese medicine and acupuncture in China. This system has unique and systematic approach towards understanding of physiology, pathogenesis, pharmacology and pharmaceuticals and thus has been professionalized. Folk medicine involves the diverse groups of non-formalized medicines that can be adaptable under changing conditions. Folk medicines show similarities over diverse cultural groups and geographical locations. On the other hand, allied forms of health knowledge include techniques attributed for well-being, such as yoga, qi-gong, tai chi and different meditations and breathing techniques (Payyappallimana 2010).

At present, environmental pollution, sedentary and unhealthy lifestyle, increased consumption of junk foods and use and overuse of conventional drugs are responsible for putting human lives in danger by increasing risk of life-threatening diseases that are sometimes difficult to treat by conventional medicines (Imam and Ismail 2017). Moreover, healthcare costs and side effects of these medicines depict them less popular. Nowadays most of the developing countries including India are also facing the threat of a wide range of non-communicable and communicable diseases due to lack of literacy and awareness, poor living conditions, lack of hygiene, high population density and changing lifestyles intensifying the risks. In addition to this, there are growing numbers of incidences of microbial resistance against different commonly used antibiotics leading to increased healthcare cost, longer days of hospital stay and increased mortality rate (Li and Webster 2018). These are the main reasons that have directed to increasing trend of interests and demand for traditional medicine all over the world in recent times as traditional medicine has a very long history of its usefulness in maintenance of health and prevention and cure of diseases. Formulation of quite a good number of conventional medicines has been done taking into account the traditional knowledge (Manna and Mishra 2018). The initiative of World Health Organization through *WHO Traditional Medicine Strategy 2002–2005* has generated increased interests among several countries, governments and consumers in adopting traditional medicine for health and well-being at individual level and also comprehensiveness of the healthcare systems. On the basis of that, *WHO Traditional Medicine Strategy 2014–2023* has been proposed with the strategic objectives in (1) building the knowledge base and formulating national policies; (2) strengthening safety, quality and effectiveness through regulation; and (3) promoting universal health coverage by integrating T&CM services and self-healthcare into national health systems (World Health Organization 2013). The overall objective is that traditional medicine of proven quality, safety and efficacy to be extended to every part of the world and it has to be ensured that all people have access to care.

48.2 Traditional Medicine: *The Global Scenario*

Traditional medicines are used by about 60% of the world's population. These are not only used in the rural areas in developing countries but also in the developed countries as well where the modern medicines are predominantly used. According to WHO, about three-quarters of the world population depends upon traditional remedies (mainly herbs) for the healthcare of its people. Plant or herbs have not only been providing food and shelter but also serving the humanity to cure different dysfunctions. Almost 80% of the population of African and Asian countries believes on traditional medicine for their primordial healthcare needs. Traditional medicines include traditional Chinese medicine, traditional Korean medicine, traditional European medicine, traditional African medicine, Ayurveda, Siddha, Unani, Iranian Medicine, etc. The traditional medicines also sometimes are coined as herbal or natural medicine existed in one way or another in different cultures/civilizations, such as Egyptians, Western, Chinese, Kampo (Japan) and Greco-Arab or Unani/Tibb (South Asia). According to the historians, ancient people used plants as the remedial sources. Before the discovery of malarial disorder, quinine from *Cinchona* bark was generally used. By the middle of the nineteenth century, at least 80% of all medicines were derived from plants (Ansari and Inamdar 2010).

Since 2600 BCE, the uses of herbal medicine are evident through record of 1000 plant-based products in Mesopotamia. That demonstrated utilization of extracted oils of *Cupressus arizonica* Greene, *Commiphora acuminata* Mattick, *Cedrus libani* A. Rich, *Glycyrrhiza glabra* L. and *Papaver somniferum* L. These are still used for the treatment of common colds, coughs, swelling and parasitic diseases (Cragg and Newman 2013). The Chinese traditional medicine has been extensively known throughout the centuries (Huang 1998), with the records obtained in *Wu Shi Er Bing Fang* (sometimes pronounced as *Wushi'er Bingfang*), translated as *recipes for 52 ailments*, written in ancient Chinese seal script (between 1100 and 800 BCE), the Shennong herbal, a manual that covers the use of medicinal herbs (100 BCE; 365 medicines) and the Tang herbal (659 CE; 850 medicines). The Romans and Greeks also had great contributions to the intelligible development of practices in traditional medicine in the ancient Western world. Dioscorides, a Greek physician ~100 CE, documented the collection, storage and proper use of traditional medicine in the then "known world" (Adhikari and Paul 2018).

Traditional medicine, being used by the African populations, includes indigenous herbalism and African spirituality. This type of practice ails several diseases, viz. high blood pressure, cancer, psychological disorders, diarrhoea, venereal diseases, epilepsy, asthma, eczema, fever, anxiety, depression, urinary tract infections, etc. Large numbers of African populations still remain far away from the modern pharmaceuticals and sophisticated ailment procedures. As a consequence, they are forced to rely on the traditional practices. This makes the herbalists popular among the Africans. In Durban herb trading market, 700,000 to 900,000 traders come per year from South Africa, Zimbabwe and Mozambique (Halberstein and Saunders 1978). According to the International Development Research Centre (IDRC), an estimate puts the number of Africans who routinely use these services for primary healthcare as high as 85% in sub-Saharan Africa (Antwi-Baffour et al. 2014). In Kwahu district of Ghana, for every traditional practitioner, there are 224 people, against one university trained doctor there are nearly 21,000.

In recent time there has been an immense upsurge in the interest in traditional medicine all over the globe that includes also a large number of developed countries. WHO has reported a considerable amount of rise in interest since the publication of the first global strategy in 2002 with the targets of linking TM contribution to health, wellness and people-centred healthcare and also promotion of safe and effective use of TM by regulating, researching and integrating TM products, practitioners and practice into health systems, where appropriate. Gradually more countries have been observed to accept the contribution of T&CM in maintenance of health and well-being of any individual. Studies have revealed that besides widespread uses of T&CM in developing countries, i.e. India (65–70%), Rwanda (~75%), Tanzania (50–60%), Uganda (55–60%), Benin (80%) and Ethiopia (90%), more than 40% population in the United States, 48–50% population in Australia, 65–70% population in Canada and more than 20–60% population in Europe (for instance, 30% in Belgium, 50% in France) use some form of T&CM (Bell et al. 2004; Ekor 2014; Fabricant and Farnsworth 2001; Katz and Baltz 2016; Lim et al. 2005). On the basis of second WHO global survey respondents only, 2012 among 133 member states the 9 indigenous TM and other T&CM mostly practiced on the basis of preferences all over the world being acupuncture, Ayurvedic medicine, chiropractic, herbal medicine, homeopathy, naturopathy, osteopathy, traditional Chinese medicine and Unani medicine (World Health Organization 2019). Besides interest in herbal medicines, more countries and consumers have been found with interests in aspects of T&CM practices and practitioners, and there are gradually increasing demand in integration of these into effective health service delivery (World Health Organization 2013). As per WHO report, at present 88% (i.e. total 170) of member states are using T&CM by developing policies, laws, regulations, programmes and offices for T&CM, and the actual number is even higher. Greater numbers of countries at present are offering high-level T&CM education programmes including Bachelor's, Master's and Doctoral degrees at university level. Some countries have also developed university curricula for health professionals and also started training programmes to develop the knowledge of traditional health practitioners (World Health Organization 2019). Systematic research under regulated actions includes finding out effective bioactive compounds in herbal products with therapeutic actions. In most developed countries, the increasing trend of using TM is more due to confidence in the treatment, ease of access and convenience in the choice of a traditional healer rather than the low cost. Another reason for preference is lesser side effects (Debas et al. 2006). It is apprehended that integrating at the areas of convergence of T&CM and conventional medicine would act best to help tackle the twenty-first century unique health challenges rather than looking into the differences of the two. In an ideal world, traditional medicine would be an option in a sound, people-centred health system that will function through balancing curative services with preventive care (World Health Organization 2019).

48.3 Traditional Medicine in India

India is the second largest country in the world in terms of population after China – with population density of 460 per km^2. The 2019 population was 1,366,417,754 at mid-year according to United Nations data which is equivalent to about 17.71% of total world population (Worldmeters United Nation 2019). According to Registrar General, and Census Commissioner, India (2011), around 70% of Indian population lives in rural areas, while 34.5% of the population is in urban (471,828,295 people) as per data available in 2019 (Worldmeters United Nations 2019).

India not only has a rich cultural heritage but also has been bearing a very strong tradition of indigenous healthcare practices among its people. The practice of traditional medicine is observed mostly in the rural parts of the country especially among diverse tribal groups and also others. As these people have very close associations with nature, forests and plants in their daily living, they have relied predominantly on plant-based herbal medicines for healing as well as preventing different ailments since their existence came into being (Tripathi et al. 2013). The easy availability of the herbal products from their surroundings at low and affordable costs, lesser side effects and non-narcotic nature of the products have made them as one of the major resorts for the disease-free living and well-being for these poverty-prone rural population of India to whom the modern medicine remains mostly inaccessible. Additionally, easy availability of plants with medicinal values in close proximity has made it possible for the inhabiting population in India to develop such rich age-old practices of herbal medicine. The rural people mainly use leaf, root, bark, rhizome, stem, fruit, seed and latex of these plants as medicine for various ailments. The plant parts are mostly administered as decoction, extracts, paste, juice, etc. India has been identified as a leading centre for biodiversity with more than 450,000 plant species, of which more than 6000 are used in traditional herbal medicine and folklore. India contains over 5% of the worlds' diversity though it covers only 2% of the earth's surface area. India has two major biodiversity hotspots, the Western Ghats and Eastern Himalayas, the rich, unique-natured and highly endangered ecoregions of the world. Out of the 20,000 medicinal plants listed globally by WHO, 15–20% contribution comes from India (Chakraborty and Paul 2014). 1500 medicinal plants have been recognized by Indian traditional medicine of which 500 are of common use (Pattanayak et al. 2015).

The tribal populations of India constitute the ethnic communities, known as the 'adivasis' who are indigenous social groups in their respective inhabiting places.

Total population of tribes in India is 84,326,240 as per the Census 2001 which is 8.2% of the total population and is the largest tribal population in the world. Majority of the tribes inhabit in rural areas, i.e. in a total number of 593,615 villages in India, and their population is 10.4% of the total rural population of the country (Registrar General, and Census Commissioner, India 2001). Report reveals that there are around 427 tribal communities in India (Tripathi et al. 2013). All these ethnic groups have their own culture, tradition, language and lifestyle. The tribal populations because of their inherent dependence on nature use a large number of plants in running their daily livelihood. Plants have become integral part of their life and play vital roles in their culture, customs, rituals, and most importantly, traditional healthcare system. This interrelationship generally has evolved over generations through experiences and practices. The tribal people possess substantial traditional knowledge on the use of various biotic resources including plants. Most healers/ practitioners prepare their own formulations and dispense to the patients. Due to lack of own scripts and written language in most of the tribal or ethnic communities, the knowledge about formulations, prescription and approach towards diseases, diagnosis, etc. of the indigenously developed age-old practices of tribal traditional medicine is remaining almost untaken (Chakraborty and Paul 2014). Such knowledge, mostly verbal, is passed from one generation to other and thus has a chance of getting eroded due to gradual changes in the lifestyle of these communities in modern era. Hence systematic documentation of this invaluable knowledge regarding the plants and their products with exclusive therapeutic uses are essential for future development of life-saving drugs.

The history of traditional medicine in India dates back to 5000 BCE and that may be acquired from statements in terms of healthcare needs and diseases in ancient literatures, viz. *Rigveda* (1700–1100 BCE, therapeutic uses of 67 plants enlisted), *Yajurveda* (1400–1000 BCE, therapeutic uses of 81 plants enlisted) and *Atharvaveda* (1200 BCE, utilities of 290 plants enlisted). Later, the scripts, "Charaka Samhita" (990 BCE, uses of 341 plants described), "Sushruta Samhita" (660 BCE, 395 plants mentioned) and "Dhanwantari Nighantu" (1800 CE, mentioned many plants), play momentous role wherefrom the information regarding the uses of plants and poly-herbal preparations can be attained (Kumar et al. 2007). Herbs are widely mentioned in *Vedic* hymns, the most famous hymn is the *The Healing Plants* hymn in *Rigveda*. The hymns found in *Vedas* about herbs specify the philosophies predominated in *Vedic* society that the herbs and the amulets made from them can do miracles. According to *Vedic* belief, through their indigenous knowledge about their surroundings, herbs can cure all the ailments, remove and clean poison from the blood (Swaminathan 2013). It is believed that the aforementioned practices of medicine are of Indian origin or some of these practices have come to India from outside and with time accepted into Indian culture to develop the Indian traditional medicine. India has the exclusive diversity of its own renowned traditional medicine, such as Ayurveda, Siddha, Unani, Yoga and naturopathy, and homoeopathy. Homoeopathy came to India in the eighteenth century and thereafter is totally accepted and enriched by Indian culture and has become most integral part of Indian traditional medicine (Kumar et al. 2007).

48.3.1 Ayurveda

Ayurveda is one of the oldest, probably the oldest medical systems in the world. Literally, Ayurveda means "the science of life". It came from two Sanskrit words "*ayur*" (life) and "*veda*" (science or knowledge) (Prasad 2002). Beginning of Ayurveda can be idealized from different ancient scriptures including *Rigveda* and *Atharvaveda* in between 2500 and 500 BCE in India (Mukherjee 2001). The two basic fundamental rules of Ayurvedic treatments are (a) to keep the reason for illness and (b) to make the patient more cautious about the reason for the sickness. The crucial objective of Ayurveda is "Ayurveda deals with happy and unhappy life. It explains what is appropriate and what is inappropriate in relation to the life, as well as it measures the life expectancy and the quality of life" (Mukherjee et al. 2017; Singh 2008). According to this holistic arrangement of medical services, seven fundamental tissues, i.e. *Rasa* (fluid), *Rakta* (blood), *Mansa* (flesh) *Meda* (fat), *Asthi* (bone), *Majja* (bone marrow) and *Shukra*, (germ cells) make a network within human body and the waste outcomes of the body, such as urine, excretion and sweat are derived by the five essential components, i.e. fire, water, air, ether and earth. It also encompasses three dynamic functional philosophies, "vatha, pitta and kapha"—the Tridosha, meaning the fundamental principles of energy or biological humour forming life. Any inequality or disturbing influence in these basic standards of the body causes disease (Lad 2002; Mukherjee and Wahile 2006). Ayurveda treats a patient in general, not the sickness alone. Ayurvedic preparations are frequently poly-herbal mixtures of plant and/or animal-derived products, metals and minerals. However, ancient manuscript shows the dominance of natural products than other derived products (Adhikari and Paul 2018; Joshi et al. 2017). Due to easy acceptability, affordability, traditional acceptance and socio-economic benefit, Ayurvedic medicines are increasingly gaining popularity all over the globe. Quality, safety, stability and efficiency of Ayurvedic medicines are now being assured through extensive research as bioactive compounds present in those medicinal plants are taking a major role in the development and management of healthcare issues in India.

48.3.2 Siddha

Siddha was established in India, around 10,000–4000 BCE. It was developed through everyday skills of applying natural resources for sustaining good health. Like Ayurveda, it is an oldest medicinal practice in South India. This system is based on Saiva philosophy of Hindu religion, which is a major tradition that worships Lord Shiva, the Hindu god as the Supreme Being. The word "Siddha" specifies, "attaining excellence" or "holy harmony" or "recognized fact". The "Siddhars" were supernatural beings with intellectual powers by continuous practice of such type of medicine. Siddha system of medicine is supposed to be established by 18 *Siddhars*, i.e. Agathiyar, Thirumoolar, Bogar, Konganar, Therayar, Korakkar, Karuvurar, Edaikkadar, Chattamuni, Sundaranar, Ramadevar, Pambatti, Machamuni,

Kudambai, Azhukanni Siddhar, Agapai siddhar, Nandhidevar and Kakapusundar. This medicinal practice believes the philosophical idea that "food is medicine, medicine is food" and "sound mind makes a sound body" (Sathasivampillai et al. 2017). The philosophy and the ideologies of this system are similar to Ayurveda. In this system of medicine, the human body is an accumulation of seven basic materials, three humours and the unwanted products. A balance among humours reflects a healthy well-being, and its unevenness leads to disease or sickness (Piḷḷai 1998). Siddha medicines are less known in the Western world as large portion of the literature is not well understood from the Tamil language (Thas 2008) but rather well accustomed as an alternative biomedicine among the Tamil communities (Adhikari and Paul 2018).

48.3.3 Unani

The Unani system of medicine was introduced in Greece, and it was urbanized in Arabs into graceful medical science by the Greek philosopher and physician Hippocrates (460–377 BCE). Greek and Arab scholars such as Galen (131–212 CE), Raazes (850–925 CE) and Avicenna (980–1037 CE) heightened the system significantly and it is recognized as Greco-Arab Medicine. According to Hippocratic hypothesis, Unani system is built on the four conditions of living, such as sodden, hot, frosty and dry, and four humours such as blood, dark bile, yellow bile and mucus. Human body is made up of seven standards: *Mizaj* (temperaments), *Anza* (organs), *Quo* (resources), *Arawh* (spirits), *Arkan* (components), *Afal* (capacities) and *Aklath* (humours) (Chopra 2006; Kalim et al. 2010). In the Unani system of medicine, three types of medications are recommended: diet therapy, pharmaco-treatment and regimental treatment. Diet treatment is extended by appropriate diet plans or by modifying the nature and quantity of diet in a routine manner for treating certain illnesses; pharmaco-treatment is done by the application of natural products; and regimental treatment includes diuresis, diaphoresis, Turkish shower, knead cleansing, etc. Unani system of medicine is a wide-ranging medication where crude form is preferred in single or in the poly-herbal formulation that miraculously deals with numerous diseases such as, gastrointestinal, nervous disorders and cardiovascular diseases.

The use of plants in the different Indian systems of medicine is as follows: Ayurveda 2000, Siddha 1300, Unani 1000, homeopathy 800, Tibetan 500, modern 200 and folk 4500. In Indian traditional and folk medicine, approximately 25,000 effective plant-based formulations are used, and more than 1.5 million traditional practitioners are there. It is estimated that more than 7800 manufacturing units are involved in the production of natural health products and traditional plant-based formulations in India (Pandey et al. 2008). In 1995 the Government of India developed the Department of Indian Systems of Medicine and Homeopathy as a separate department; later it was recognized as the Department of Ayurveda, Yoga, Unani, Siddha and Homeopathy (AYUSH), to serve as T&CM national office. AYUSH has four separate councils for research: the Central Council for Research in Ayurveda and Siddha, the Central Council for Research in Unani Medicine, the Central Council for Research in Yoga and Naturopathy and the Central Council for Research in Homeopathy. A national plan began in 2014 for integrating T&CM into national health delivery. Numbers of T&CM practitioners registered (as at 1 January 2016) under each practice are Ayurveda, 419,217; Unani, 48,196; Siddha, 8528; naturopathy, 2220; and homeopathy, 293,307. The total number is 771,468 (6.4 per 10,000 people) (World Health Organization 2019).

48.3.4 Some Evidences of Bioactive Components of Indian Herbal Medicine

The widespread pharmacological impacts of medicinal plants depend fundamentally on their phytochemical constituents. Commonly the phytochemical constituents of plant products are of two categories, i.e. primary and secondary metabolites, depending on their involvement in the metabolic processes. Primary plant metabolites are almost similar in all living cells as they exert their effects on basic life processes, whereas secondary plant metabolites originate from subsidiary pathways, namely, the shikimic acid pathway (Hussein and El-Anssary 2018). In traditional and folk uses of plant products, the secondary plant metabolites have been identified to be highly significant in alleviation of several ailments, and also in modern medicine, they have gained importance for providing the principal compound for preparing medicines to combat several diseases ranging from migraine up to cancer (Bourgaud et al. 2001).

It has been found that many communities use unripe green banana for the folk treatment of different intestinal disorders that also includes diarrhoea. Green banana, rich in amylase-resistant starch (ARS), has been reported to protect against chemically induced damage of gastrointestinal mucosa in animals. It has been postulated that high content of ARS that remains undigested in human small intestine is responsible for exerting this protective action. The resident bacteria in colon ferment it into the short-chain fatty acids (SCFA) i.e. butyrate, propionate and acetate as it reaches colon. SCFA in turn promote salt and water absorption in the colon, provide energy and drive a trophic effect on the mucosa of colon as well as small intestine (Rabbani et al. 2001).

Indigenous people from diverse parts of India use leaves of *Azadirachta indica* A. Juss (Meliaceae) (locally known as neem) for treating gastrointestinal disorder such as diarrhoea and cholera (Thakurta et al. 2007). Azadirachtin has been found as the foremost active component in these leaves and the other compounds being nimbin, nimbidin, nimbolinin, nimbidol, sodium nimbinate, quercetin, gedunin and salannin. In addition to these, other ingredients those are also found are nimbanene, 6-desacetylnimbinene, n-hexacosanol, nimbiol, nimbandiol, nimbolide, 7-desacetyl-7-benzoylazadiradione, 7-desacetyl-7-benzoylgedunin, 17-hydroxyazadiradione, ascorbic acid and amino acid (Ali 1998; Hossain et al. 2013). Azadirachtin and nimbolide have showed concentration-dependent antiradical scavenging activity (Alzohairy 2016). *Azadirachta indica* aqueous leaf extract has been found to normalize the levels of bilirubin, protein, alanine aminotransferase, aspartate aminotransferase and alkaline phosphatase in vitro under hepatotoxic condition (Kale et al. 2003).

The unripe fruit of *Aegle marmelos* (bael) is said to be an excellent remedy for diarrhoea and is especially useful in chronic diarrhoeas (Chopra and Chopra 1994). This plant has been described as "Rasayana" in Charaka Samhita. Root, trunk, seed and fruit are also found effective against several ailments. Besides, in intermittent fever, hypochondriasis, melancholia and palpitation of the heart, the root bark has been found useful (Nadkarni 1954). The leaves and bark have been found beneficial as laxative, and the leaves are also used in diabetes mellitus (Chopra and Chopra 1994). Presence of mucilage, pectin, coumarins such as marmelosin and marmelide and tannins in *Aegle marmelos* fruits play central role in its biological actions (Das and Das 1995; Morton and Dowling 1987). Tannins and flavonoids in general have been described to possess antidiarrhoeal property because of their potential to act as antimicrobial and antisecretory agents and also inhibitor of intestinal motility (Di Carlo et al. 1993; Lutterodt et al. 1999).

Moreover reports reveal that in chemically induced hepatotoxicity rat model, the hepatoprotective impacts of *Epaltes divaricata* was assessed on the basis of record of traditional use of the herb in Ayurveda for treatment of acute dyspepsia, jaundice and urethral discharge. The plant extract decreased the hepatic injury, as evident by serum liver enzyme profile and histopathological evaluation (Hewawasam et al. 2004). Medicinal plants from Nilgiris were screened for antiviral activity, which have potential to treat herpes simplex virus (HSV)-1 infection in vitro. Eighteen plants with record of traditional use as anti-infective agents were identified. The study was done to detect the active plant and its part for further elucidation of active constituents. Three plants were found to have potent antiviral activity that could be evaluated comprehensively for therapeutic potential (Vijayan et al. 2004).

48.4 Traditional Herbal Medicine in Plateau-Fringe and Rarh Regions of West Bengal

The state of West Bengal present in eastern part of India is the 13th largest Indian state (area of 88,752 km^2) and fourth most populous state in India, the population being 101,438,931 as of January, 2019; 91,276,115 as per Census 2011 constituting 7.54% of Indian population in 2011 (Registrar General, and Census Commissioner, India 2011). West Bengal extends from the Himalayas in the north and happens to be the only state in India to have such feature in its geographic location. This state of India is surrounded by the countries Bangladesh in the east, Bhutan in the north-east, and Nepal in the north-west, the Indian states Assam and Sikkim in the north-east, Bihar and Jharkhand in the west and Odisha in the south-west. West Bengal has great diversity in terms of geomorphological and climatic points of view but also is distinct in relation to natural and human resources. In the course of its extension of 483 km from the Himalayas in the north through the Gangetic plains to the Bay of Bengal in the south, it exhibits almost all geophysical features and associated biodiversity. A large section of the population of West Bengal also embraces various tribes mostly residing in the rural areas. The rich culture, ethnicity and biodiversity in this state remained crucial for century-old practices of traditional medicine. In the present article, an attempt has been made to document in detail the herbal products with their traditional therapeutic uses (Tables 48.1 and 48.2, Fig. 48.1) among people residing in two geomorphologically distinct locations of West Bengal, (a) the plateau-fringe region and (b) Rarh region. These regions have several variations, like natural forests, mainly dense dry deciduous forests dominated by Sal tree (*Shorea robusta*) in association with *Pterocarpus marsupium* (Vijayasar), *Madhuca latifolia* (mahua or mohul), *Schleichera oleosa* (kusum), *Terminalia arjuna* (arjun), *Terminalia bellirica* (bahera), *Bombax ceiba* (red cotton or shimul), *Butea monosperma* (palash or flame of the forest), *Ficus religiosa* (pial), *Terminalia chebula* (haritaki), *Artocarpus hirsutus* (wild jackfruit), etc. (mostly found in Purulia, Bankura, Paschim Medinipur and Jhargram districts) (Ghosh 2003); lateritic wastelands, afforested lands with eucalyptus (*Eucalyptus sp.*), akashmoni (*Acacia auriculiformis*) etc.; and water bodies like rivers, tidal creeks, canals and ponds of various sizes (Bandyopadhyay et al. 2014; Das and Ghosh 2017). Rural people including the tribals in these regions hold a great deal of ethno-medicinal knowledge. The deep faith in religion, culture, economic vulnerability and lack of access to conventional medicine are the major reasons for poor rural people to depend on herbal products for healthcare. Despite great inclination of common people towards the traditional herbal medicine, the preparation and mode of applications remain confined to the local healers. In the face of rapid socio-economic and cultural changes in the tribal populations and others with increased literacy among young generations and migration of the youths to the urban areas, this traditional healthcare system is facing substantive threat of abolishment and so do the natural resources and plant species due to increased anthropogenic interventions (Ghosh 2008). For this reason documentation of this endogenous knowledge is essential to preserve the opulent traditional knowledge and practices and also aiming for conservation of crucial and treasure plant species that may pave the way of formulation of precious drugs.

The **plateau-fringe** region is characterized by rolling upland with isolated small hills in a very irregular pattern. This feature is observed in western part of West Bengal (Fig. 48.2), which is the continuation of Chhota Nagpur Plateau region. This region is made of gneiss and granite of Archean era along with old igneous rock and quartz- and coal-bearing mudstone of Carboniferous period. The total area under plateau-fringe has undergone massive erosional phase, and therefore the morphology of this region has been transformed into a rolling peneplain comprising of small monadnocks (locally named “tila”). Some of the important hills are Ajodhya hills (677 m) in Purulia district and Susunia (442 m) in Bankura district. The western part of Bankura, Birbhum, Purulia, Paschim Medinipur and Bardhaman districts exhibits the plateau-fringe geomorphologic features (Table 48.3). The altitude of this region ranges from 100 to 500 m, and the higher slope towards eastern alluvial plain can be seen in this region. The tributaries of Kangsabati-Haldi, Subarnarekha and Dwarakeswar are the important rivers here (Bandyopadhyay et al. 2014; Das

Table 48.1 Details of medicinal plants in common uses in the plateau-fringe and Rarh districts of West Bengal, India

Plants	Local name	Family	Habit	Parts used	Phytochemicals or Bioactive compounds present	Therapeutic uses	References
Acacia nilotica	Babla	Fabaceae	Tree	Leaf	Leaf: Catechin, calycanthidine, D galactose, malic acid, linoleic acid, erythritol, D-glucitol, pipecolic acid, 2,5 dihydroxyacetophenone, oxalic acid, butanoic acid, ribitol and palmitic acid	Immature leaf buds are used with common salt as antiscorbutic Also used for stimulating digestive system, protect against anorexia and in reducing acidity	Bai et al. (2014); Gupta and Bhat (2016)
Achyranthes aspera L.	Apang	Amaranthaceae	Herb/ subshrub/ shrub	Whole plants, seeds	Root: Triterpenoid saponins, betaine, achyranthine, hentriacontane, ecdysterone (polypodineA), ecdysone Fruit: Two oleanolic acid base saponins. Two new saponins C and D isolated from fruits Seed oil: Linoleic, oleic, palmitic, stearic, behenic, arachidic, myristic and lauric acids	Plant is used in cough, asthma, bronchitis, dyspepsia, flatulence, colic, painful inflammation, dropsy, helminthiasis, renal calculi, cardiac disorders Seeds are used as anti-emetic and in treatment of hydrophobia Unripe fruits are used in treatment	ENVIS (n.d.) Centre, Govt. of India

						of respiratory diseases	
Aegle marmelos	Bael	Rutaceae	Tree	Fruit, leaf	Leaf: Marmenol, a new 7-geranyloxycoumarin [7-(2, 6-dihydroxy-7- methoxy-7-methyl-3-octaenyloxy) coumarin], rutacine, ɣ-sitosterol, aeglenine, aegeline, marmeline, fragrine, dictamine, cinnamide, lupeol, rutin, marmesin, β-sitosterol, flavone, glycoside, O-isopentenyl halfordiol, cuminaldehyde, eugenol, cineol, citral, citronellal and phenylethyl cinnamides Fruit: Umbelliferone, dictamine, xanthotoxol, xanthotoxin, scaparone, isopimpinellin, isoimperatorin, N-2 methoxy-2-[4 methoxyphenyl] ethylenamide	Unripe fruit pulp is used against dysentery and diarrhoea Ripe fruit pulp is used in constipation Leaf juice with honey is used for fever Leaf paste acts as antibacterial agent	Sharma and Dubey (2016)
Allium cepa	Peyaj	Amaryllidaceae	Bulb	Bulb	Bulb: Quercetin 3,4′-diglucoside, quercetin 4′-monoglucoside, myricetin, quercetin aglycone, isorhamnetin, peonidin 3′-glucoside, petunidin 3′-glucoside acetate, delphinidin 3′-glucoside, malvidin 3′-glucoside	Onion paste is used locally in case of insect bite Onion juice is applied on scalp to prevent hair fall and promote hair growth Onion juice is also used with	Fredotović et al. (2017)

(continued)

Table 48.1 (continued)

Plants	Local name	Family	Habit	Parts used	Phytochemicals or Bioactive compounds present	Therapeutic uses	References
						warm water for constipation	
Allium sativum L.	Rasun	Liliaceae	Bulb	Bulb	Bulb: Diallyl thiosulfonate (allicin), diallyl sulphide (DAS), diallyl disulphide (DADS), diallyl trisulphide (DATS), E/Z-ajoene, S-allylcysteine (SAC), S-allylcysteine sulphoxide (alliin), desgalactotigonin-rhamnose, proto-desgalactotigonin, proto-desgalactotigonin-rhamnose, voghieroside dl, sativoside Bl-rhamnose, and sativoside R1, β-resorcylic acid, pyrogallol, gallic acid, rutin, protocatechuic acid, quercetin	Seed vessels are fried in mustard oil, and the oil is massaged in case of rheumatic/ arthritic pains It reduces high blood pressure A single seed vessel with hot boiled rice is consumed in the ailment of heart	Shang et al. (2019)
Aloe vera	Ghritokumari	Asphodelaceae	Succulent	Leaf	Leaf: Aloe emodin, aloectic acid, anthranol, aloin A and B, isobarbaloin, emodin, ester of cinnamic acid, ure mannan, acetylated mannan, acetylated glucomannan, glucogalactomannan, galatan, galactogalacturan, arabinogalactan, galactoglucoarabinomannan, pectic substance, xylan, cellulose,	Fresh leaf extract is used for wound healing and skin problems Leaf juice is used to control wound caused by burn It is also used to support digestive system	Raksha et al. (2014)

					chromones, isoaloeresin-D, isoarabaichromone, neoaloesin A. Arachidonic acid, linolenic acid, triglycerides, triterpenoid, gibberellins, lignin, potassium sorbate, salicylic acid, uric acid, campesterol, cholesterol and β-sitosterol		
Alstonia scholaris (L.)	Chhatim	Apocynaceae	Tree	Bark, leaf, latex, wood	Stem bark: α-amyrine acetate, lupeol acetate, echitamine, picrinine, venoterpine glucoside, akuammidine, strictamine, tetrahydroalstonine, Δ3-carene, citral, citronellol, geraniol, linalool, α-pinene, terpinolene, angustilobine B, losbanine, alschomine, isoalschomine, alstonamine, (20S)-19, 20-dihydrocondylocarpine, 19-hydroxytubotaiwine, (+)-lochneridine, (−)-scholarine, picralinal, rhazimanine, scholaricine, Nb-methyl-scholaricine, 19, 20-E-vallesamine, 19,20-Z-vallesamine, betulin, ditamine, Nb-demethylechitamine, echitenine, 17-O-acetyl-echitamine, 6,7-angustilobine B, akuammigine, N-demethyl-6-7-seco alstonamine, 6,7-seco alstonamine and its N-demethyl	Leaf decoction is used in beriberi Bark decoction or powdered bark employed in malaria, diarrhoea, fevers, dysentery, skin diseases, pruritus, asthma, bronchitis, worms helminthiasis, toxaemia, cures gastrointestinal troubles Latex applied in rheumatic pains, sores, toothache Latex mixed with oil is used as ear drop Wood paste applied in	ENVIS (n.d.) Centre, Govt. of India

(continued)

Table 48.1 (continued)

Plants	Local name	Family	Habit	Parts used	Phytochemicals or Bioactive compounds present	Therapeutic uses	References
					derivative, etc. Leaf: Picrinine, picralinal, strictamine, tetrahydroalstonine, pscudoakuammigine, betulin, ursolic acid, β-sitosterol, (−)-scholarine, (+)-lochneridine, 20 (S)-19, 20-dihydrocondylocarpine, 19,20-vallesamines, alstonamine, rhazimanine, alschomine along with its C-5 isomer, isoalschomine, neraline Roots: Akuammigine, tubotaiwine, akuammicine, akuammicine-Nb-oxide, hydroxy-19,20-dihydroakuammicine, α-amyrine, lupeol acetate, stigmasterol, α-sitosterol, scholaricine, isookanin-7-O-rhamnoside (8,3′,4′-trihydroxyflavanone-7-O-α-L-rhamnopyranoside)	rheumatism and wounds	
Ananas comosus	Anaras	Bromeliaceae	Succulent	Leaf	Leaf: Dodecanoic acid, tetradecanoic acid, malic acid, pentadecanoic acid, cis-9-hexadecenoic acid, palmitic acid, cis-10-heptadecenoic acid, margaric acid, 9,12-octadecadienoic acid, stearic acid,	Soft part of leaf is used against helminthic infection	Guedes et al. (2018)

					cis-10-nonadecenoic acid, nonadecanoic acid, citric acid, cis-ll-eicosenoic acid, behenic acid, tricosanoic acid, tetracosanoic acid, vitamin E, α-tocopherolquinone, campesterol, ergostanol, stigmasterol, γ-sitosterol, stigmastanol, sitostenone, stigmasterol, δ-tocopherol		
Andrographis paniculata	Kalmegh	Acanthaceae	Herb	Leaf and stem	Leaf: Andrographolide, deoxyandrographolide, neoandrographolide, 14-deoxy-11,12-didehydroandrographide, andrograpanin, isoandrographolide, 5-hydroxy-7,8-dimethoxyflavone, 5-hydroxy-7,8,2′,5′-tetramethoxyflavone, 5-hydroxy-7,8,2′,3′ tetramethoxyflavone, 5-hydroxy-7,8,2′-trimethoxyflavone, 7-O-methylwogonin, 2′-methyl ether, 19-O-acetylanhydroandrographolide	Leaf and stem juices are taken at empty stomach to protect liver and helminthic infection	Chao and Lin (2010)
Asparagus racemosus	Satavari/ satamuli	Asparagaceae	Climber	Root	Root: Essential oils, asparagine, arginine, tyrosine, kaempferol, quercetin, rutin, resin, tanninsteroidal, asparagosides, bitter glycosides, diosgenin, shatavarins I and IV	Root juice mixed with milk is used to cure blood dysentery and hematemesis	Mishra and Verma (2017)

(continued)

Table 48.1 (continued)

Plants	Local name	Family	Habit	Parts used	Phytochemicals or Bioactive compounds present	Therapeutic uses	References
						Used to control blood pressure	
Azadirachta indica	Neem	Meliaceae	Tree	Leaf	Leaf: Azadirachtin, nimbolinin, nimbin, nimbidin, nimbidol, sodium nimbinate, gedunin, salannin, quercetin, nimbanene, 6-desacetylnimbinene, nimbandiol, nimbolide, ascorbic acid, n-hexacosanol and amino acid, 7-desacetyl-7-benzoylazadiradione, 7-desacetyl-7-benzoylgedunin, 17-hydroxyazadiradione, ß-sitosterol and nimbiol	Neem leaf is used to control high blood pressure and diabetes Soft newborn leaf is used against pox and measles and used as blood purifying agent Neem oil is applied against lice and to cure ulcer/wounds Leaf paste or decoction of leaves is used in different skin diseases or skin allergy	Alzohairy (2016)
Bacopa monnieri	Brahmi	Plantaginaceae	Herb	Leaf	Leaf: Alkaloids (brahmine, nicotine and herpestine); saponins (monnierin, hersaponin); sterols (b-sitosterol, stigmasterol); d-mannitol, acid A, betulinic acid,	Leaf fried in ghee is then eaten to promote memory	Tembe-Fokunang et al. (2019)

					bacosides and bacopasaponins; jujubogenin; pseudojujubogenin; phenylethanoid glycosides, namely, monnierasides I–III; phenylethanoid glycosides, namely, monnierasides I–III; plantainoside B		
Barleria lupulina Lindl.	Kanta bishalyakarani	Acanthaceae	Shrub/ small tree	Leaf	Leaf: Benzofuranone, hexadecanoic acid, ethyl 9,12, 15-octadecatrienoate and 3,7,11,15-tetramethyl-2-hexadecanoic acid, barlerin, acetylbarlerin, shanzhiside methyl ester, acetylshanzhiside methyl ester, ipolamiidoside and iridoid glucosides	Leaf juice is used to stop bleeding Leaf paste is used for alleviating pain and acne	Kumari and Dubey (2016)
Blumea lacera	Kukshima	Asteraceae	Herb	Root, leaf	Aerial part: (Z)-lachnophyllum ester, (Z)-lachnophyllic acid, germacrene D, (E)-P-famesene, bicyclogermacrene, (E)-caryophyllene, (E)-nerolidol, (E)-lachnophyllic acid, (E)-lachnophyllum ester, (Z)-lachnophylum ester Root: Specific compounds are not known	Root paste with sugar and water is used to cure digestive ailment Leaf juice is used as antibacterial agent	Satyal et al. (2015)
Boerhavia diffusa	Punarnaba	Nyctaginaceae	Herb	Leaf	Leaf: Bezyloxy-1, 2(4H)-benzofuranone, 5,6,7,7-a-tetrahydro-6-hy, 3,7,11,15-tetramethyl-2-hexadecen-1-ol, hexadecanoic acid, phytol,	Leaf is boiled in water, and then the water is taken to reduce oedema	Bhardwaj and Sharma (2019)

(continued)

Table 48.1 (continued)

Plants	Local name	Family	Habit	Parts used	Phytochemicals or Bioactive compounds present	Therapeutic uses	References
					cyclododecene, ethyl 9-hexadecenoate, trans-2-decen-1-ol, cholest-5-en-3-ol (3.Beta.)-, propanoate, cholest-5-en-3-ol (3 beta.)-, tetradecanoate, stigmasterol		
Bryophyllum pinnatum	Patharkuchi	Crassulaceae	Succulent herb	Leaf	Leaf: Bryophillin A, bersaldegenin-3-acetate, and bryophillin C, butyrolactone, 3,4-epoxytetrahydrothiophene-1,1-dioxide, l-octen-3-ol, 3,5-dihydroxy-6-methyl-2,3-dihydro-4H-pyran-4-one, benzaldehyde, alpha-D-glucopyranoside, methyl, n-hexadecanoic acid, oleic acid, octadecanoic acid	Leaf juice mixed with common salt is consumed in bloody enteritis Leaf juice is taken for flatulence Freshly crushed leaves are taken to treat tumour	{Uchegbu, #76} Uchegbu et al. (2017)
Calotropis gigantea (L.)	Akanda	Apocynaceae	Shrub	Root bark, leaf, latex	Root bark: Isovalerates of a kundrol, mudarol, giganticine (non-protein amino acid), β-amyrin, β-amyrin acetate, tetracyclic resinols, sterols, acetic, isovaleric acids, isogiganteol, taraxasterol and its ψ-isomer, triterpenes identified as lup-13 (18), 19(29)-dien-9α-yl acetate, lupeol acetate, urs-18β-H-12, 20 (30)-dien-3β-yl-acetate and	Root bark is used in intestinal worms, dysentery, piles and elephantiasis Roasted leaf (hot poultice of leaf) is applied to painful joints on swellings and rheumatic	ENVIS (n.d.) Centre, Govt. of India.

					17β-hydroxy-28-normethyl urs-18α-H-12, 20(30)-dien-3β-yl-acetate, etc. Leaf: Holarrhetine, cyanidin-3-rhamnoglucoside, tarasasterol isovalerate, taraxasteryl acetate, β-sitosterol, amino acids Latex:α- and β-calotropeol, β-amyrin, gigantin, (toxic principle), proteases: Calotropins DI and DII; calotropins FI and FII, ester of acetic and isovaleric acids, glutathione, 3′-methylbutanoates of α-amyrin and ψ-taraxasterol, uscharin, etc.	arthritis Leaf powder boiled in oil is useful in curing eczema, skin eruptions, toothache and wounds Latex is applied locally to cut wounds	
Carica papaya	Penpe	Caricaceae	Tree	Leaf, fruits	Leaf: Tocopherol, ascorbic acid, carpaine, deoxykaempferol, kaempferol, deoxyquercetin, quercetin, dicoumarol, coumaroylquinic acid, coumarin, folic acid, cystine, homocysteine, cysteine sulphoxide, L-glutamic acid, p-coumaroyl alcohol, dimethoxyphenol, umbelliferone, phenylalanine, caffeoyl alcohol, methyl nonyl ketone	Leaf juice is fed to "dengue" patients to increase platelet count Boiled fruit or fruit curry is used in curing constipation and liver disorders Fruit latex is taken orally with "batasa" (sugar or jiggery candy) in fever or to increase appetite	Akhila and Vijayalakshmi (2015)

(continued)

Table 48.1 (continued)

Plants	Local name	Family	Habit	Parts used	Phytochemicals or Bioactive compounds present	Therapeutic uses	References
Catharanthus roseus	Nayantara	Apocynaceae	Small subshrub	Leaf	Leaf: Mono-inositol, hexadecanoic acid, 9-octadecanoic acid(z), heptadecen(8)-carbonic acid, octadecanoic acid, eicosanoic acid, methyl ester, icosanoic acid, 1,2-benzenedicarboxylic acid, squalene, desmethoxy-vindoline, tetracontane	Fresh soft leaves or leaf juice are used to control diabetes and hypertension. It also acts against intestinal worm infection	Thanwar et al. (2017)
Centella asiatica L.	Thankuni	Apiaceae.	Herb	Leaf	Leaf: 3-glucosylquercetin, 3-glucosyl-and 7-glucosyl-kaemfrol, polyacetylenes I–V and nine other acetylenes, amino acids, asiatic, centic, centellic, centoic, pectic, madasiatic acids, carotene, centellose (oligosaccharide), hydrocotylin (alkaloid), lipid, protein, pectin, saponins, vallerine, asiaticoside (2,3,23-trihydroxy-urs-12-en-28-oic-acid-o-6-deoxy-α-L-mannopyranoyl-(1 → 4)-O-β-D-glucopyrono-syl-(1 → 6)-O-β-D-gluco-pyranosyl ester), aslaticosides A [O-α-L-rhamnopyranosyl-(1 → 4)-O-β-D-glucopyranosyl(1 → 6)]-O-β-D-glucopyranose ester of 2α, 3β, 6β, 23α-tetra hydroxyurs-12-	Leaf juice is rubbed on the forehead to treat severe headache Leaves are consumed by chewing or in the form of juice in empty stomach in dysentery, diarrhoea and amoebiasis Curry of leaves is used to cure chronic dysentery	ENVIS (n.d.) Centre, Govt. of India.

					en-28-oic acid; asiaticoside B ([O-α-L-rhamnopyranosyl-(1 → 4)-O-β-D-glucopyranosyl (1 → 6)]-O-β-D-glucopyranose ester of 2α, 3β, 6β, 23α-tetrahydroxyolean-12-en-28-oic acid, indocentelloside, brahmoside, brahminoside, isothankuniside, madecassoside, thankcuriside, brahmic (madecassic or made gascaric acid, 2α, 3β, 6β, 23- tetrahydroxy-urs-12-en-28-oic acid), and thankunic acids; β-caryophyllene, trans-β-farnesene, germacrene D, mesoinositol, polyphenols, α-terpinene, thymol methyl ether, betulic acid, β-sitosterol, campesterol, stigmasterol, tannins, vitamins B1, B2 and C		
Citrus aurantiifolia	Kagji lebu/ pati lebu	Rutaceae	Shrub	Fruit, leaf	Fruit: 5,7-dimethoxycoumarin, 3-methyl-1,2-cyclopentanedione, 1-methoxy-cyclohexene, corylone, palmitic acid, 5,8-dimethoxypsoralen, α-terpineol and umbelliferone Leaf: Sabinene, myrcene, limonene, trans-β-ocimene, γ-terpinene, linalcol, limonene oxide, isopulegol, citronellal, α-terpineol, citronellol, geraniol, geranial, citronellyl acetate,	Fruit juice with warm water is used to boost the digestive system Leaves are used as anti-vomiting agent	Lemes et al. (2018); Sandoval-Montemayor et al. (2012)

(continued)

Table 48.1 (continued)

Plants	Local name	Family	Habit	Parts used	Phytochemicals or Bioactive compounds present	Therapeutic uses	References
					trans-β-caryophyllene, α-trans-bergamotene, germacrene D, bicyclogermacrene P-bisabolene, caryophyllene oxide		
Coccinia grandis	Telakucha	Cucurbitaceae	Vine	Leaf, fruits	Aerial part: Heptacosane, cephalandrol, β-sitosterol, alkaloids cephalandrins A and B Fruits: β-amyrin acetate, lupeol, cucurbitacin b, taraxerone, taraxerol, β-carotene, lycopene, cryptoxanthin, xyloglucan, carotenoids, β-sitosterol, stigma-7-en-3one	Leaf juice is used in empty stomach to control diabetes and dysentery Fruits are used as vegetable for hepatoprotective activity	Mathews and Sunny (2019)
Curcuma longa L.	Holud	Zingiberaceae	Herb	Rhizome	Rhizome: Bisacurone, α-turmerone, ß-turmerone, ar-curcumyl alcohol, curcuminoids, coronadiene, cyclocurcumin, 8,12-epoxygermacra-1(10),4,7,11-tetraen-6-one, cyclohexanecarboxylic acid methyl ester, isopulegol, 2-menthen-1-ol, menth-1-en-9-ol, octahydrocurcumin and labda-8(17)-12-diene-15,16-dial	Tender rhizome or paste of both rhizome and neem leaves, dried and stored as tablets, is taken daily at empty stomach in liver disorders, diabetes, skin diseases and helminthic infection Dried rhizome paste is eaten	Rajkumari and Sanatombi (2017)

						with hot boiled rice in chronic amoebiasis Lukewarm paste of rhizome powder and lime is applied locally in traumatic swelling	
Cynodon dactylon	Durba	Gramineae	Grass	Leaf	Leaf: Glycerin, 4H-pyran-4-one, 2,3-dihydro-3,5-dihydroxy-6-methyl, thymol, conhydrin, 1,2-cyclopentanediol, 3-methyl, benzenepropanol, 4-hydroxy-à-methyl, ethyl à-d-glucopyranoside, 3,7,11,15-tetramethyl-2-hexadecen-1-ol, n-hexadecanoic acid, hexadecanoic acid, ethyl ester, phytol, linoleic acid ethyl ester, 9,12-octadecadienoylchloride, octadecanoic acid, ethyl ester, pentanal, 2-methyl, 1-(cyclopropyl-nitromethyl)-cyclopentanol, 2-propenamide, N-[2-(dimethylamino)ethyl], hexadecanoic acid, 2-hydroxy-1-(hydroxymethyl) ethyl ester, didodecyl phthalate, 13-tetradece-11-yn-1-ol, 10-undecyn-1-ol, squalene, 9,12-octadecadienoic acid (Z,Z), phenylmethyl ester,	Leaf juice is used in dysentery Leaf paste mixed with sugar is applied to treat cut injury	Al-Snafi (2016)

(continued)

Table 48.1 (continued)

Plants	Local name	Family	Habit	Parts used	Phytochemicals or Bioactive compounds present	Therapeutic uses	References
					diazoprogesterone, apigenin, luteolin, 6-C-pentosyl-8-C-hexosyl apigenin and 6-C-hexosyl-8-C-pentosyl luteolin		
Cyperus rotundus	Mutha ghas	Cyperaceae	Herb	Rhizome	Rhizome: 12-methyl cyprot-3-en-2-one-13-oic acid, n-dotriacontan-16-one, n-pentadecanyl linoleate, n-hexadecanyl linoleate, n-hexadecanyl oleate, stigmasteryl laurate, stigmasteryl myristate, n-tetracontan-7-one, n-pentacos-13′-enyl oleate, β-sitosterol-3-β-o-glucoside, lupenyl 3β-o-arabinpyranosyl 2′-oleate	Rhizome paste, mixed with common salt, is used to cure enteritis	Sultana et al. (2017)
Datura metel L./ *Datura stramonium*	Dhutra	Solanaceae	Shrub	Leaf, flower, seed	Whole plant: 3-phenylacetoxy-6, 7-epoxynortropane,7-hydroxyapoatropine, scopoline, 3-(hydroxyacetoxy) tropane, 3-hydroxy-6-(2-methylbutyryloxy) tropane, 3a-tigloyloxy-6-hydroxytropane, 3,7- dihydroxy-6-tigloyloxytropane, 3-tigloyloxy-6-propionyloxytropane, 3-phenylacetoxy-6,7-epoxytropane, 3-phenylacetoxy-	Leaf paste or juice mixed with mustard oil is used in skin diseases and rheumatic pain Flower petal juice is used to treat ear pain Seed powder and/or dust is/are used as laxative	Aqib and Mohib (2014)

					6-hydroxytropane, aponorscopolamine, 3a,6a-ditigloyloxytropane, 7-hydroxyhyoscyamine Essential oil: Sterols and their derivatives, 5-alpha-ergosta-7,22-dien-3-beta-ol, 3-hydroxycholestan-5-yl, acetate 26, 26-dimethyl-5, 24(28)-ergostadien-3-beta-ol		
Eclipta prostrata	Keshut	Asteraceae	Herb	Leaf	Leaf: Propanedinitrile dimethyl, pentadecane, heptadecane, neophytadiene, 1, 3-propanediol-2-hydroxymethyl-citr-1-llyl butyrate, citr-1-llyl propionate, heptadecanoic acid, methyl ester, 1-allyloxyl-octa-2,7-diene, 6(E), 9(Z), 13(E)-pendectriene, phytol, pentanoic acid, 4-methyl-ethanamine, 2, 2′-oxybsin [N, N-dimethyl, 2 methoxysulpholane, cyclopropane, methoxymethylen, ethyl-2-oxo-4-methyl-3-pentenoate, squalene, butanoic acid, 4-(ethoxyhydroxyphosphinyl) and dl-α- tocopherol	Fresh leaf juice is applied over scalp to promote hair growth	Priya et al. (2018)
Emblica officinalis	Amlaki	Phyllanthaceae	Tree	Fruit	Fruit: Apigenin, gallic acid, ellagic acid, chebulinic acid, quercetin, chebulagic acid,	Fruit juice is used in urinary tract infection	Hasan et al. (2016)

(continued)

Table 48.1 (continued)

Plants	Local name	Family	Habit	Parts used	Phytochemicals or Bioactive compounds present	Therapeutic uses	References
					corilagin, isostrictiniin, methyl gallate, luteolin, emblicanin A, emblicanin B, phyllaemblicin B, punigluconin, pedunculagin, glutamic acid, proline, aspartic acid, alanine and lysine	Fruit juice boiled in coconut oil to promote hair growth Used as a supplement of vitamin C Fruits are dried with common salt, and used to reduce acidity and digestive problems	
Enydra fluctuans	Hingcha/ Helencha	Asteraceae	Herb	Leaf, Stem	Leaf and stem: Enhydrin, fluctuanin; fluctuadin, tiglate, 1-hydroxy-methacrylate, 4-hydroxytiglate, 2,3-epoxyisobutyrate of 8-desacylenhydrin (I´-IV´), 8β-methacryloyloxy-9α-acetoxy-14-oxo-acanthospermolide (V´); (−) kauran-16α-ol, 16α-hydroxy-(−)kauran-19-oic acid (I), (−) kaur-15-en-19-oic acid, 15α-hydroxy (−) kaur-15-en-19-oic acid, 17-hydroxy (−) kaur-15-en-19-oic acid, (II) kaur-16-en-19-oic acid, 15α-angeloyloxy (IV) and 15α-isovaleroyloxy-(III)	Leaf extract is used as antibilious and antidyspeptic It is also used in torpidity of liver, in nervous ailments and skin diseases It is consumed as vegetable to reduce digestive ailment and constipation	ENVIS (n.d.) Centre, Govt. of India

					(−)kaur-16-en-19-oic-acid; 4-hydroxy farnesyl acetate, myricylalcohol; β-amyrin, (24S)-24-ethyl-5, 22, 25-cholestatrien-3β-ol (24S)-24-ethyl-5,22,25-cholestatrienyl stearate, (24S)-24-ethyl-5,22,25-cholestatrien-3β, (24S)-24-etyl-5,22,25-cholestatrienyl stearate, (24S)-24-ethyl-4, 22, 25-cholestatrien-3-one, stigmasterol, stigmasta-5, 22, 25-trien-3β-ol, 4,22-stigmastadien-3-one, gibberellins A9 and A13		
Evolvulus nummularius	Mushakani	Convolvulaceae	Herb	Leaf	Leaf: β-sitosterol and its glucoside, stigmasterol, d-mannitol, ursolic acid, oleanolic acid, 3-O-(4-stearoyl-Z-coumaroyl)-stigmast-5, E-22-dien-3β-ol; 16-(E-coumaroyloxy)-palmitic acid, 3p-hydroxy-urs-12-en-29β-oic acid, plectranthoic acid (3a-hydroxy-urs-12-en-29β-oic acid)	Fresh leaf juice is used at empty stomach to treat dysentery	Dinda et al. (2007)
Ficus bengalensis Linn.	Bot	Moraceae	Tree	Leaf, root	Leaf: Catechin and genistein, quercetin-3-galactoside, rutin, leucocyanidin, friedelin and β-sitosterol Root: Bengalensinone, benganoic acid	Bot leaves are immersed in water and the extract is used to regulate blood sugar	Khaliq (2017)

(continued)

Table 48.1 (continued)

Plants	Local name	Family	Habit	Parts used	Phytochemicals or Bioactive compounds present	Therapeutic uses	References
						Root cap of bot fried with ghee is used in dysentery One drop of fresh latex mixed with a pinch of camphor is applied externally to the eye in case of cataract	
Ficus racemosa	Dumur	Moraceae	Tree	Fruits	Fruits: Gluanol, glucanol, tiglic acid, taraxasterol, lupeol acetate, friedelin and hydrocarbons	Fruits are taken as vegetable to reduce menorrhagia, leucorrhoea, different blood disorders and anaemia Fruits also act as laxative	Yadav et al. (2015)
Glinus oppositifolius	Gimeshak	Molluginaceae	Herb	Leaf, stem	Leaf and stem: L-(−)-(N-trans-cinnamoyl)-arginine, kaempferol 3-O-galactopyranoside, isorhamnetin 3-O-β-D-xylopyranosyl-(1 → 2)-β-D-galactopyranoside, vitexin, vicenin-2, adenosine,	Leaves and stems are used in daily meal as a stimulant of digestive system	Ragasa et al. (2015)

					L-phenylalanine, oppositifolone, squalene, spinasterol, oleanolic acid, phytol, lutein, spergulagenin A		
Hemidesmus indicus	Anantamul	Asclepiadaceae	Shrub	Root	Root: 2-hydroxy-4-methoxy benzaldehyde, 2-hydroxy-4-methoxy benzoic acid, vanillin, (E,Z)-nonadienal, lupeol octacosanoate, hexatriacontane, lupeol (III), lupeol acetate, β-amyrin (I), β-amyrin (II), β-amyrin acetate, β-sitosterol, hemidesminine, hemidesmine-1 and hemidesmine-2	Root juice is used to treat the infection caused by *Miliaria rubra* that causes prickly heat	Banerji et al. (2017)
Hibiscus rosa-sinensis	Joba	Malvaceae	Shrub	Leaf, flower	Leaf: n-tetracosanyl cyclopentylcarboxylate, taraxerol acetate, 26β-cyclopentanyl n-hexacosan-5β-ol Flower: Ethanimidic acid, ethyl ester, propanal, 2,3dihydroxy, propanamide, methyl-ethylenediamine, O-methylisourea hydrogen sulphate, hexadecanoic acid, methyl ester, 7-formylbicyclo heptanes, 2-butanamine, (S)-, 1,3,5-triazine-2,4,6-triamine, n-formylβ-alanine, (z)6, (z)9-pentadecadien-1-ol, butanedial, 1-propanol, 2-methyl-, and methane carbothiolic acid	Fresh leaf extract with sugar is used to cure digestive ailments Leaves and flowers are used in alopecia and blackening of hair	Missoum (2018)

(continued)

Table 48.1 (continued)

Plants	Local name	Family	Habit	Parts used	Phytochemicals or Bioactive compounds present	Therapeutic uses	References
Hygrophila spinosa T./ *Hygrophila auriculata*	Kulekhara	Acanthaceae	Herb	Leaf	Leaf: Butane, 1, 1-diethoxy-3-methyl-pentane, 1,1-diethoxy-3, 3-diethoxy-2-butanone, propane, 1,1,3-triethoxy-1,1,3-triethoxybutane, benzene, nonane, 3,7-dimethyl-, diethyl phthalate, isopropyl myristate, 2-hexadecen-1-ol, 3,7,11,15-tetram,5-undecene, 3-methyl-,(E), 2,6,10-trimethyl, 14-ethylene-14-pe, 7-octadecyne, 2-methyl-, heptadecanoic acid, ethyl ester, phytol isomer, phytol acetate, tridecanol, 2-ethyl-2-methyl-, squalene, 1,2-benzenedicarboxylic acid	Raw leaf juice with cow milk or boiled leaf juice with small amount of salt is given to patients having anaemia to increase the amount of haemoglobin in blood It is also used as vegetable	Ravisankar and Ester (2017)
Jatropha gossypiifolia L.	Varenda	Euphorbiaceae	Shrub	Stem, latex, Fruit & Seed	Whole plant parts: Propacin, venkatasin, citlalitrione, jatrophenone Stem: 4-O-demethyl retrochinensin, cleomiscosin A, gossypidien, isogadain, jatrodien, prasanthaline Latex: Mainly cyclogossine A, cyclogossine B Seeds: Arachidic acid, caprylic acid, lauric acid, lignoceric acid, linoleic acid, myristic acid oleic acid, palmitic acid, palmitoleic	Latex is used in cuts, wounds and skin burn Stem is used for toothache Fruit powder is used as a laxative	Félix-Silva et al. (2014)

					acid, ricinoleic acid, stearic acid, vernolic acid, 12-deoxy-16-hydroxylphorbol		
Justicia adhatoda L.	Basak	Acanthaceae	Shrub	Leaf, Stem Bark	Leaf and stem: Vasicine (1,2,3,9-tetrahydropyrrolo [2,1-b] quinazolin-3-ol), vasicinone (3-hydroxy-2,3-dihydropyrrolo [2,1-b]quinazolin 9(1H)-one), adhatodine, anisotine, vasicoline, vasicolinone, β-sitosterol, tritriacontane, vasicinine, vasicinol, 1,2,3,9-tetrahydro-5-methoxypyrrolo[2,1-b] quinazolin-3-ol, adhavasinone, 7-hydroxyvasicine, vasicinolone, 3-deoxyvasicine, vasicol, triterpenes, anisotine, steroids carbohydrate and alkanes	Fresh leaf juice mixed with honey is used to treat chronic bronchitis, cold and cough Leaf paste with black pepper is used in leukoderma Dry leaves are used to smoke in asthma Dry bark powder is taken in acidity and indigestion	Haq et al. (1967); Singh et al. (2017)
Lantana camara	Bhut-bhoirab/ chotra	Verbenaceae	Shrub	Leaf, Fruit	Leaf: Gautin, acacetin-7-O-β-D-rutinoside 2, tricin 3, hispidulin 4,3,5,7,8-tetra hydroxyl-6,3′-dimethoxy flavones 5 and pectolinarigenin 10, (25S)-spirostan-5-ene-3β-21-diol-3-O-α-L-rhamnopyr-anosyl-(1,2)-[α-L-rhamnopyranosyl-(1,4)]-β-D-glucopyranoside 6, ursolic acid 7, lantanilic acid 8, icterogenin 9, betulonic acid 11, betulinic acid 12	Leaf juice acts as antiseptic in cut injury Leaves give relief from pain of piles and tumour Fruit paste is used in fever and wounds	Patil et al. (2015)

(continued)

Table 48.1 (continued)

Plants	Local name	Family	Habit	Parts used	Phytochemicals or Bioactive compounds present	Therapeutic uses	References
Litsea glutinosa (Lour)	Piplas	Lauraceae	Tree	Leaf	Leaf: (E)-β-ocimene, α-pinene, β-pinene, (Z)-β-ocimene, myrcene, limonene, (E)-anethol, caryophyllene, bicyclogermacrene, α-humulene, nerolidol, caryophyllene oxide, germacrene D, linalool, nonanal, menthone, iso-menthone, (Z)-, (E)- anethol, decanal, octyl acetate, linalyl acetate, 2-undecanone, bornyl acetate, undecanal, neryl acetate, dodecanal, (E)-nerolidol, bourboneol, germacrene-D-4-ol, spathulenol, caryophyllene oxide, cerdrol, ledol, ι-muurolol, nerolidol, β-eudesmol, α-cadinol, (Z)-β-asarone, benzyl benzoate, tetradecanal, 6,10,14-trimethyl 2-pentadecanone, phytol, n-paraffin: n-eicosane, n-heneicosane, n-docosane and n-heptacosane	After smashing the leaves, a thick agar-like extract is formed. It is used as a curative agent for chronic digestive ailment and anorexia	Hien et al. (2010)
Mangifera indica	Aam	Anacardiaceae	Tree	Fruit, leaf	Fruit: Cycloartenol, α-amyrin, β-amyrin, ocotillol, 3b-hydroxycycloart-24-en-26-al, 24-methylene-cycloartan-3b,26-diol, dammarenediol II, and psi-taraxastane-3b. Polyphenols	Boiled or roasted green mango is used to prepare syrup and taken during summer for avoiding sun	Ediriweera et al. (2017)

					and phenolic acids include ascorbic acid, quercetin, mangiferin, quercetin 3-ara, quercetin 3-rha, isomangiferin gallate, mangiferin gallate, methyl mangiferonate, methyl mangiferolate, tetra-O-galloylglucose, hexa-O-galloylglucose, methyl isomangiferolate, caffeic acid, ferulic acid, gallic acid, cinnamic acid, vanillin, rhamnetin-3-O-galactoside, kaempferol, and kaempferol-hexose. Resorcinolic lipids include 5-(11′Z-heptadecenyl)-resorcinol and 5-(8′Z,11′Z-heptadecadienyl)-resorcinol. Carotenoids include β-carotene, cis-violaxanthin, neochrome, cis-neoxanthin, luteoxanthin, zeaxanthin and 9- or 9′-cis-lutein. Long-chain fatty acids include oleic acid, linoleic acid, linolenic acid and n-pentacosanol	stroke Mustard oil taken in aam leaf is heated, and then the oil is used to massage the throat region, chest and back to reduce common cold and cough	
Marsilea minuta	Susni	Marsileaceae	Herb	Leaf	Leaf: n-hexadecanoic, (Z)6,(Z)9-pentadecadien-1-ol, phytol, 2-cyclohexane-1-one,4-hydroxy-3,5,6-trimethyl-4 (3-oxo-1-butenyl), 9,12,15-Octadecatrienoic acid (Z,Z,Z),	Boiled leaf juice mixed with cow milk is taken in sleep disorder mainly in insomnia	Sabithira and Udayakumar (2017)

(continued)

Table 48.1 (continued)

Plants	Local name	Family	Habit	Parts used	Phytochemicals or Bioactive compounds present	Therapeutic uses	References
					3, 7, 11, 15-tetramethyl-2-hexadecan-1-ol and benzofuran	It is also used as vegetable	
Mentha piperita L.	Pudina	Lamiaceae	Herb	Leaf	Leaf: 3-octanol, eucalyptol, 3-octanol acetate, borneol, dihydrocarveol, pulegone, carvone, caryophyllene, β-cuvebene, hexadecylene oxide, N-hexadecylene oxide, phytol, α-linolenic acid, 2-monopalmitin, α-amyrin, squalene, β-sitosterol, vitamin E	Curry prepared from leaves is taken in nausea, flatulence, vomiting Leaf juice acts as carminative and gastric stimulant It is also used as antiseptic	Das and Ghosh (2017); Hossain et al. (2014)
Mikania micrantha	Bonlata	Asteraceae	Vine/ climber	Leaf	Leaf: 2,2-dimethoxybutane, nonane, Undecane, cyclohexyl (dimethoxy) methylsilane, α-cubebene, α-longipinene, copaene, germacrene d, caryophyllene, cedrene, α-caryophyllene, humulene, acoradiene, 4,6-dimethyldodecane, 3,5-bis (1,1-dimethylethyl)phenol, β-himachalene, δ-cadinene, hexadecane, 2, 6, 11-trimethyldodecane, eicosane	Leaves are smashed and then juice is used in cuts to stop bleeding	Ishak et al. (2018)
Momordica charantia	Karala	Cucurbitaceae	Vine	Fruit	Fruit: β-sitosterol, daucosterol, campesterol, stigmasterol,	Fruit juice is taken in empty	de Oliveira et al. (2018)

					β-sitosterol, 25,26-dihydroelasterol, charantagenins D and charantagenins E, 5β,19-epoxy-25- methoxy-cucurbita-6,23-diene-3b,19-diol (EMCD), charantin A, charantin B, momordicines I and II, momordicoside K, palmitic, stearic, myristic, pentadecanoic, arachidic, α-linolenic, linoleic, oleic and palmitoleic, capric, lauric acids, caffeic, p-coumaric, ferulic, o-coumaric, chlorogenic, m-coumaric, p-hydroxybenzoic, gallic, protocatechuic, βresorcylic, vanillic, syringic, gentisic, salicylic, veratric, tcinnamic and homogentisic acids	stomach especially for diabetes	
Moringa oleifera L.	Sajina	Moringaceae	Tree	Leaf, fruits, root bark, flower	Leaf: n-hexadecanoic acid, tetradecanoic acid, cis-vaccenic acid, octadecanoic acid, palmitoyl chloride, beta-l-rhamnofuranoside, 5-O-acetyl-thio-octyl, gamma-sitosterol, and pregna-7- diene-3-ol-20-one and E-lutein Fruit: Pods contain isothiocyanate, thiocarbamates, nitrile, O-[2′-hydroxy-3′-(2′′-heptenyloxy)]-propyl undecanoate, methyl-p-hydroxybenzoate and O-ethyl-4-	Tender leaf, flower and fruits are used as vegetables to prevent chickenpox Root bark paste is applied locally in traumatic swelling and rheumatism Leaf juice or fried leaf is taken	Bhattacharya et al. (2018)

(continued)

Table 48.1 (continued)

Plants	Local name	Family	Habit	Parts used	Phytochemicals or Bioactive compounds present	Therapeutic uses	References
					[(α-l-rhamnosyloxy)-benzyl] carbamate. Fruits contain cytokines, whereas seeds contain high concentrations of benzylglucosinolate, 4-(α-l-rhamnopyranosyloxy)-benzylglucosinolate, 4-(α-l-rhamnosyloxy) benzylisothiocyanate, 4-(α-l-rhamnosyloxy) phenylacetonitrile, and O-ethyl-4-(α-l-rhamnosyloxy) benzyl carbamate Flower: Flowers contain sucrose, amino acids, alkaloids and flavonoids, such as rhamnetin, isoquercitrin and kaempferitrin Stem: Stem contains alkaloids (moringine and moringinine), 4-hydroxymellein, octacosanoic acid and β-sitosterol Root: Spirochin, anthonine, beta-sitosterone, vanillin, 4-hydroxymellein, β-sitosterol and octacosanoic acid	to control blood pressure	
Nyctanthes arbor-tristis	Shewli	Oleaceae	Shrub/ small tree	Leaf	Leaf: β-sitosterol, triterpenes-β-amyrin, oleanolic acid, friedeline, lupeol, nyctanthine, flavonol glycosides-	Leaf juice mixed with honey is given to children in case of fever	Gulshan et al. (2015)

					astragaline, nicotiflorine, iridoid glycosides-arborsides A,B,C, 6β-hydroxyloganin, desrhamnosylverbacoside, 6,7-di-obenzoylnycthanoside, 6-o-transcinnamoyl-6β-hydroxyloganin, 7-Otrans-cinnamoyl-6β hydroxyloganin, nicotiflorin, mannitol, tannic acid, ascorbic acid, methyl salicylate, traces of volatile oil, an amorphous resin, carotene, glucose, fructose, hexatriacontane, benzoic acid and benzoic ester of loganin	Leaf juice is also useful in worm infection and liver disorders	
Ocimum sanctum	Tulsi	Lamiaceae	Subshrub	Leaf	Leaf: Oleanolic acid, ursolic acid, rosmarinic acid, eugenol, carvacrol, linalool, β-caryophyllene, eugenol, β-elemene, β-caryophyllene and germacrene	Fresh leaf or leaf juice with honey is used in common cold, cough and fever Leaf juice is rubbed at the infected site of insect bite Leaf juice mixed with lime is used in ringworm Leaves are smashed and then extract is used in headache	Padalia and Verma (2011); Sundaram et al. (2012); Tumpa et al. (2014)

(continued)

Table 48.1 (continued)

Plants	Local name	Family	Habit	Parts used	Phytochemicals or Bioactive compounds present	Therapeutic uses	References
Oxalis corniculata	Amrul	Oxalidaceae	Shrub	Leaf	Leaf: Tartaric acid, citric acids, calcium oxalate, flavones (acacetin and 7, 4′-di-OMe apigenin), glycoflavones (4’-OMe vitexin, 4’-OMeiso-vitexin and 3′,4’-diOMe orientin), flavonols (3′,4’-diOMe quercetin) and phenolic acids such as p-hydroxybenzoic, vanillic, syringic acids, 6-C-glucosyl luteolin (isoorientin), 6-C-glucosylapigenin (isovitexin) and isovitexin 7-methylether (sertisin)	Leaf is used as vegetables in chronic amoebiasis It is also used to cure digestive ailment and for quick recovery after suffering from any disease	Srikanth et al. (2012)
Paederia foetida	Gandal	Rubiaceae	Vine	Leaf	Leaf: Kaempferol 3-O-glucoside, kaempferol 3-O-rutinoside, kaempferol 3-O-rutinoside-7-O-glucoside, kaempferol 7-O-glucoside quercetin 3-O-glucoside, quercetin 3-O-rutinoside, quercetin 3-O-rutinoside-7-O-glucoside, quercetin 7-O-glucoside, delphinidin, pelargonidin, peonidin malvidin, quercetin 3-O-rutinoside-7-O-xylosylglucoside, linarin, daidzein, pentan-2-one, s-methyl thioacetate, dimethyl disulphide, (E)-pent-3-en-2-one, pent-1-en-3-one, limonere,	Leaf is used as vegetable in diarrhoea, amoebiasis and dysentery It is also used to increase taste and appetite during fever	Wang et al. (2014)

					3-methylbutan-1-ol, pentan-1-ol, 3-methylbut-2-en-1-ol, hexanol, (E)-hex-3-en-1-ol, dimethyl trisulphide, (Z)-hex-3-en-1-ol, phenylacetaldehyde, 2-furanmethanol, benzofuran, α-terpineol, methyl salicylate, nerol, geraniol, 2,3-dihydrobenzofuran		
Pergularia daemia	Chagal-Bati	Asclepiadoideae	Vine	Whole plant, leaf, root bark, fruits	Plant: Coroglaucigenin, uzarigenin, hentriacontane, β-amyrin, betaine, 5β-stigmast-7(8)-en-3α-ol, 5β-stigmast-8(14)-en-3α-ol; sugar residues of the hydrolysates of cardiac glycosides are D-cymarose, D-glucose, L-oleandrose, D-sermentose Leaf: 3β-hydroxyfriedelan-7-one, lupeol and its acetate, oleanolic acid, putranjivadione, β-sitosterol Root: Lupeol, α-amyrin and their acetate, β-sitosterol and its glucoside, calactin, calotropin Seed: Coroglaucigenin, uzarigenin, calactin, calotropin, calotropigenin, corotoxigenin, dihydrocalotropigenin, protouscharin, uscharidin, uscharin, etc.	Plant is used in cough, asthma, amenorrhoea, dysmenorrhoea, intermittent fever, leucoderma; extract given in uterine and menstrual disorders and to facilitate parturition Leaf juice is useful in cough, helminthiasis, asthma, haemorrhoids dyspepsia Leaf juice is given in catarrhal affections, infantile	ENVIS (n.d.) Centre, Govt. of India

(continued)

Table 48.1 (continued)

Plants	Local name	Family	Habit	Parts used	Phytochemicals or Bioactive compounds present	Therapeutic uses	References
						diarrhoea Leaf paste combined with lime is applied to rheumatic swellings Poultice prepared from fresh leaf (soft mass of crushed leaves tied in a small piece of cloth) is applied to carbuncles, piles Root bark mixed with cow's milk used as purgative Fruits are useful in cough, asthma, bronchitis, dyspepsia	
Piper betle	Pan	Piperaceae	Vine	Leaf	Leaf: 1,3,5-triazine-2,4,6-triamine, Tetradecanoic acid, 2,5-dimethoxybenzoic acid, phenol, 2-methoxy-3-(2-propenyl)-, 3,7,11,15-	Leaf of pan with "supari" (*Areca catechu*) and "chuna" (calcium hydroxide) is	Zahira and Thamilmani (2016)

					tetramethyl-2-hexadecen-1-ol, phytol, piperine	used as digestive stimulator Extract of pan leaf is used to control lice of head	
Piper nigrum	Gol-marich	Piperaceae	Vine	Seed	Seed: Piperine, piperanine, piperettine, piperylin A, piperolein B, pipericine, δ-cadinol, δ-guaiene, (Z) (E)-farnesol, (E)-β-ocimene and guaiol	Seed is chewed with common salt and then swallowed in sore throat, cough, tonsillitis and pharyngitis	Abdallah and Abdalla (2018)
Pluchea indica (L.)	Bontulsi	Asteraceae	Shrub	Leaf	Leaf: Eudermane derivative: 3-(2′,3′-diacetoxy-2′-methyl-butyryl)-cuauhtemone, protein, 4-allyl-2, 6-dimethoxyphenyl, benzyl eugenyl glucosides, hedyotisols A and B, (Z)-3-hexenylglucosides, 1-(4-hydroxy-3-phenyl)-2-{2-methoxy-4-[1-(E)-propene-3-ol]-phenoxyl}-propane-1, 3-diol (erythro) and its threo form, 1,2-bis (4-hydroxy-3-methoxyphenyl)-propane-1, 3-diol (erythro) and its threo form; 9-hydroxylinaloyl, linaloyl-, linaloylapisoyl-, methyl salicylate and phenylethyl-glucosides, pinoresinol monoglucoside, plucheosides A and B;	Leaf juice is consumed in fever and dysentery Leaf juice or paste is applied in insect's bites Decoction (plant parts are boiled to dissolve the phytochemicals) of leaves is used to stop bleeding, to gastric disorder of children and cutaneous diseases	ENVIS (n.d.) Centre, Govt. of India

(continued)

Table 48.1 (continued)

Plants	Local name	Family	Habit	Parts used	Phytochemicals or Bioactive compounds present	Therapeutic uses	References
					syringaresinol monoglucoside, stigmasteryl glucoside; chlorogenic acid		
Psidium guajava	Peyara	Myrtaceae	Shrub/ small tree	Fruit, leaf	Leaf: Caryophyllene oxide, caryophyllene, α-cubebene, calamenene, α-humulene, α-bulnesene, humulene epoxide II, neryl acetate, γ-bisabolene, alpha-famesene, γ-bisabolene, Δ-cadinene, γ-cadinene, longifolene, α-bergamotene, γ-gurjunene, geranyl acetate, α-cadinene, sabinene, quercetin, vanillic acid, syringic acid, m-coumaric acid, cinnamic acid Fruit: Eehane, 1,1-dietoxi, n-hexanal, 2-hexanol, 2-hexenal, 2(5H)-furanone, 5-ethyl-, eucalyptol, 1,3,6-octatriene, 3, 7-dimethyl-, (E)-, 1,4-cyclohexadiene-1-methanol, 4-(1-methylethyl)-,4-(1-Hidroxyethyl) benzaldehyde, 3,4-dimethylacetophenone, 2,5-dimethylacetophenone, 2-hexanone	Extract prepared with overnight immersion of mature fruit in warm water is used to control blood sugar Fresh soft leaves are used to protect teeth and gum	Afzal et al. (2019); Bodini et al. (2019)
Ricinus communis L.	Reri	Euphorbiaceae.	Shrub	Branch, oil	Oil: Palmitic, stearic, arachidic, hexadecenoic, oleic, linoleic,	Luke warm oil is massaged over	Kumar (2017)

					linolenic, ricinoleic and dihydroxy stearic acids Stem: Stem contains ricinine, ergost-5-en-3-ol, stigmasterol, Y-sitosterolfucosterol	abdomen in abdominal flatulence with pain especially in children Branches are used as tooth brush in bleeding gum Oil is used to protect from hair fall	
Rauvolfia serpentina (L.)	Sarpagandha	Apocynaceae	Shrub	Roots, leaf	Root: Indole and indoline alkaloids, viz. ajmalicidine (1-carbomethoxy-17-α-hydroxy-16-decarbomethoxy-16-17-dihydroajmali-cine), ajmalimine, (21-trimethoxybenzoyl ajmaline), ajmaline, ajmalinimine, ajmalinine, ajmalicine, isoajmaline, ajmalidine, alstonine, aricine, corynanthine, canembine, deserpidine, eosarpegine, mitroridine, obscuridine, obscurine, picrinine, purpeline, perakine, isopsuedoresperpine, isoraunesicine, isoreserpine indobine, indobinine, isoreserpiline, isoreserpinine, neoreserpiline, rescinnamidine, rescinnamine, rescinnaminol, renoxidine, rescinnamine,	Roots are useful in heart disorders, worms, fever, wounds, colic, insomnia, hypertension, epilepsy, giddiness The decoction of the root is used to increase uterine contractions Leaf juice is used to remove corneal opacities Reserpine reduces noradrenaline and serotonin in	ENVIS (n.d.) Centre, Govt. of India

(continued)

Table 48.1 (continued)

Plants	Local name	Family	Habit	Parts used	Phytochemicals or Bioactive compounds present	Therapeutic uses	References
					rescinnaminol, renoxidine, reserpiline, reserpine, raujemedine, rauvomitine, rauwolfinine, rauwoxine, raucaffridine, raudaffricine raunamineraucaffriline, raunamine, rauwolfine, serpentine, serpentinine, seredine, sarpagine, rauniticine, raunitidine, raumitorine, raunatine, rauvanine, resajmaline, tetraphylline, tetraphyllicine, yohimbine, α-yohimbine (rauwolscine), 3-epi-α-yohimbine, yohimbinine, sandwicensine, samatine, semperflorine, sandwicolidine, sandwicoline, etc.	the brain and thus possesses neurotropic properties and is used in anxiety, fear, tension, aggressiveness and chronic schizophrenia	
Syzygium cumini	Jam	Myrtaceae	Tree	Seed, leaf	Leaf contains acylated flavonol glycosides, quercetin, myricetin, myricitin, myricetin 3-O-4-acetyl-L-rhamnopyranoside, triterpenoids, esterase, galloyl carboxylase and tannin Fruits: The fruits are rich in raffinose, glucose, fructose, citric acid, mallic acid, gallic acid, anthocyanins, delphinidin-3-gentiobioside, malvidin-3-laminaribioside, petunidin-3-	Seed powder is taken for diabetes Leaf juice is also beneficial for diabetes patients	Ayyanar and Subash-Babu (2012)

					gentiobioside, cyanidin diglycoside, petunidin and malvidin		
Tagetes erecta	Ganda	Asteraceae	Herb	Leaf	Leaf: 2,6,10-trimethyl,14-ethylene-14-pentadecane, tetradecanoic acid, butanoic acid, 3,7-dimethyl-6-octenyl ester, citronellyl isobutyrate, heptadecanoic acid, methyl ester, 9-mexadecenoic acid, N-hexadecanoic acid, phosphorothioic acid, 0,0-diethyl-0-(3,5,6-trichloro-2-pridinyl) ester, tetradecanoic acid, ethyl ester, 15-hydroxy pentadecanoic acid, hexadecadienoic acid, methyl ester, phytol, 9,12,15-octadecatrienoic acid, ethyl ester (2,2,2), cis-9-hexadecenal, 15-hydroxypentadecanoic acid, 2-hydroxy-3-[(9E)-9-octadecenoyloxy] propyl (9E)-9-octadecenoate, celidoniol, deoxy, alpha-tocopherol-beta-D-mannoside, stigmasterol	Leaf juice is applied over cut injury to stop bleeding.	Devika and Justin (2014)
Tamarindus indica	Tentul	Fabaceae	Tree	Leaf, fruit pulp	Leaf: l-methyl-4-propylbenzene (p-cymene), limonene, diphenyl-ether longifolene, caryophyllene, 2,6-di-ter/–butyl-4-methylphenol (BHT), methyl 3,5- di-tert-butyl-4-hydroxybenzoate, methyl	Fresh new leaves and buds are used for treatment of chronic anorexia Fruit pulp is used	Escalona-Arranz et al. (2010)

(continued)

Table 48.1 (continued)

Plants	Local name	Family	Habit	Parts used	Phytochemicals or Bioactive compounds present	Therapeutic uses	References
					hexadecanoate, 6,10,14-trimethylpentadeca-5,9,13-trien-2-one, linalool anthranilate, 3-eicosyne, methyl-9, 12,15-octadecatrienoate, phytol, methyl-7,10-octadecadienoate Fruit pulp: 3-hexen-2-one, 2-furaldehyde, 4-hydroxyl-4-methyl pentan-2-one, 7-hydroxyl-oct-3-ene-2-carboxylic acid, 2-furaldehyde, 2-furanmethanol, malonic ester, 4-hydroxylmethyl-2-fliraldehyde	to make curry (Ambol or Tak with fish) to cure digestive upset and acidity	
Terminalia arjuna	Arjun	Combretaceae	Tree	Stem bark	Stem: Triterpenoids (arjunin, arjunic acid, arjunenin, terminic acid, terminoltin, arjunolic acid), glycosides (arjunetin, arjunoside I and II, arjunolone, arjunolitin, arjunaphthanoloside, arjunglucoside IV and V, arjunasides A-E, Olean-3b, 22b-diol-12-en-28 b-D-glucopyranosie-oic acid, terminarjunoside I and II, terminoside A, thermionic acid), flavonoids and phenolics (arjunone, luteolin, baicalein, ethyl gallate, gallic acid, kaempferol, oligomeric	Stem bark juice or powder is used in heart disease Dried stem bark with common salt is used as stimulator of digestive system and also reduces acidity	Gupta et al. (2018)

					proanthocyanidins, pelargonidin, quercetin, (þ)-catechin, (þ)-gallocatechin, epigallocatechin, gallic acid, ellagic acid), tannin (pyrocatechols, punicallin, castalagin, casuariin, casuarinin, punicalagin, terchebulin, terflavin C)		
Terminalia bellirica	Bahera	Combretaceae	Tree	Fruit	Fruit: Methyl-7(2,4,6-trimethylphenyl)-5H furo(2,3c) thiopyran-4-corboxylate, D-fructose, 1,3,4,5,6-pentakis-O-(trimethylsilyl), quinic acid, glucopyranose, 3,4,5-trimethoxybenzoic acid ethyl ester, 3,4,5 tris (trimethylsiloxyl) benzoic acid, inositol-trimethylsilyl	*Terminalia bellirica* is one ingredient of triphala (amlaki, haritaki and bahera). It acts as laxative and is used in the purgation therapy It promotes hair growth	Chipiti et al. (2015)
Terminalia chebula	Haritaki	Combretaceae	Tree	Fruit	Fruit: Punicalagin [2, 3-(S)-hexahydroxydiphenoyl-4,6-(S,S)-gallagyl-D-glucose], terflavin A, terchebulin, terchebin (1, 3, 6-trigalloyl glucose), terflavins B, terflavin C, terflavin D, punicalin, neo-chebulic acid, 1, 6-di-O-galloyl-D-glucose, gallic acid (3, 4, 5-trihydroxybenzoic acid), casuarinin, chebulanin, corilagin, ellagic acid (2, 3,	Fruit powder is taken at bedtime in constipation. Fruits are soaked in water overnight and is drunk next morning for treating acidity, constipation	Riaz et al. (2017)

(continued)

Table 48.1 (continued)

Plants	Local name	Family	Habit	Parts used	Phytochemicals or Bioactive compounds present	Therapeutic uses	References
					7, 8-tetrahydroxy-chromeno [5, 4, 3-cde]chromene-5, 10-dione), chebulagic acid, chebulinic acid (1, 3, 6-tri-O-galloyl-2, 4-chebuloyl-β-D-glucopyranoside), 1, 2, 3, 4, 6-penta-O-galloyl-D-glucose,2,3,4,6-tetra-O-galloyl-B-D-glucose, ethyl gallate (ethyl 3,4,5-trihydroxybenzoate), methyl gallate(methyl 3,4,5-trihydroxybenzoate), chebulagic acid, 4-O-methylgallic acid, methyl(S)-flavogallonate, methyl neochebulagate, eugenol, ascorbic acid, triethyl chebulate, shikimic acid, ferulic acid, vanillic acid, p-coumaric acid, caffeic acids, melilotic acid, phloroglucinol [benzene-1,3,5-triol], pyrogallol [1,2,3-trihydroxybenzene], rutin, quercetin, luteolin [2-(3,4-dihydroxyphenyl)-5,7-dihydroxy-4-chromenone], β-sitosterol, daucosterol, arjungenin, arjunolic acid, arjunic acid, terminolic acid, arjunglucoside I, arjunglucoside		

					II, arjunetin, chebuloside II, bellericoside, etc.		
Tinospora cordifolia	Guduchi	Menispermaceae	Vine	Stem	Stem: Propanoic acid, 2-[3-acetoxy-4,4, 14-trimethylandrost-8-en-17-yl], (22S)-21-acetoxy-6a, 11a-dihydroxy-16a, 17apropylmethylenedioxypregna-1, 4-diene-3, 20-dione, 2-butenoic acid, pregnane-3,20-dione	Tender stem juice is taken for diabetes	Albinjose et al. (2015)
Vitex negundo	Swet nishinda	Lamiaceae	Shrub	Root	Root: Vitexoside, agnuside, R-dalbergiphenol, negundin A, negundin B, 6-hydroxy-4-(4-hydroxy-3-methoxy)-3-hydroxymethyl-7-methoxy-3,4-dihydro-2-naphthaledehyde, vitrofolal E, (+)-lyoniresinol, (+)-lyoniresinol-3α-O-βd-glucoside, (+)-(−)-pinoresinol, (+)-diasyringaresinol, 2β,3α-diacetoxyoleana-5,12-dien-28-oic acid, 2α,3α-dihydroxyoleana-5,12-dien-28-oic acid, 2α,3β-diacetoxy-18-hydroxyoleana-5,12-dien-28-oic acid, vitexin and isovitexin, acetyl oleanolic acid, sitosterol, 3-formyl-4.5-dimethyl-8-oxo-5H-6,7-dihydronaphtho (2,3-b) furan (a new furanoeremophilane)	Crushed root bark with cold water is given to snake bite patients	Suganthi and Dubey (2016)

(continued)

Table 48.1 (continued)

Plants	Local name	Family	Habit	Parts used	Phytochemicals or Bioactive compounds present	Therapeutic uses	References
Withania somnifera	Ashwagandha	Solanaceae	Shrub	Leaf	Leaf: Withanolide IV, withanoside V, VI, withaferin A, withastromonaloide, withanolide A, withanone, withanolide B, ergostane	Leaf juice is boiled and taken with milk as energy drink or health drink	Dwivedi et al. (2015); Kumar et al. (2015)
Zingiber officinale	Ada	Zingiberaceae	Herb	Rhizome	Rhizome: 6-gingerol, 6-shogaol, zingerone with phenolics and flavonoids. 4-, 6-, 8-, and 10-gingerdiols, 6- and 10-gingerdiones, 6-methylgingerdiol, 6-hydroxyshogaol, 6-, 8-, 10-dehydroshogaols, diarylheptanoids and zingerone	Tender rhizome paste is boiled in water with common salt and gurgled in sore throat, cough and tonsillitis	Mele (2019)

Table 48.2 Herbal medicines used in some common ailments in plateau-fringe and rarh districts of state of West Bengal, India

Disease	Plants used	Scientific name	Plant parts used	Mode of uses
Anaemia	Dumur	*Ficus racemosa*	Fruits	• Fruits are taken as vegetable to reduce anaemia
	Kulekhara	*Hygrophila spinosa* T.	Leaf	• Raw leaf juice with cow milk or boiled leaf juice with small amount of salt is given to anaemia patients to increase the amount of haemoglobin in blood
Common cold and cough	Basak	*Justicia adhatoda* L.	Leaf	• Fresh leaf juice mixed with honey is used to treat chronic bronchitis, cold and cough
	Ada	*Zingiber officinale*	Rhizome	• Tender rhizome paste boiled in water with common salt and then gurgled in sore throat, cough and tonsillitis
	Marich	*Piper nigrum*	Seed	• Seeds are chewed with common salt and then swallowed in sore throat, cough, tonsillitis and pharyngitis
	Tulsi	*Ocimum sanctum*	Leaf	• Fresh leaves or leaf juice with honey is used in common cold and cough
	Aam	*Mangifera indica*	Leaf	• Mustard oil taken in aam leaf is heated, and then the oil is massaged at throat region, chest and back to reduce common cold and cough
Constipation	Bael	*Aegle marmelos*	Fruit	• Ripe fruit pulp with cow milk or curd is used as syrup
	Haritaki	*Terminalia chebula*	Fruit	• Fruit powder is taken at bedtime in case of constipation
	Varenda	*Jatropha gossypifolia* L.	Fruit	• Dried fruit powder is used as a laxative
	Bahera	*Terminalia bellirica*	Fruit	• It is one ingredient of triphala (combination of amlaki, haritaki and bahera). It acts as laxative

(continued)

Table 48.2 (continued)

Disease	Plants used	Scientific name	Plant parts used	Mode of uses
	Peyaj	*Allium cepa*	Bulb	• Onion juice with warm water is used for constipation
	Penpe	*Carica papaya*	Fruit	• Boiled fruit or fruit curry is taken for constipation
	Hingcha	*Enydra fluctuans*	Leaf, stem	• It is consumed as vegetable to reduce constipation
Cuts and wounds	Varenda	*Jatropha gossypiifolia* L.	Latex	• Latex is used locally in cuts, wounds and burns on skin
	Bhut bhairab	*Lantana camara*	Leaf, fruit	• Leaf juice acts as antiseptic in cut injury • Fruit paste is used to cure wounds
	Neem	*Azadirachta indica*	Leaf	• Neem oil is applied to cure wounds
	Kanta bishalyakarani	*Barleria lupulina* Lindl.	Leaf	• Fresh leaf juice is used at cut injury to stop bleeding
	Akanda	*Calotropis gigantea* L.	Latex, leaf	• Latex is applied locally to cut and wounds • Leaf powder boiled in oil is useful to treat wounds
	Durba	*Cynodon dactylon*	Leaf	• Leaf paste mixed with sugar is applied over cut injury to stop bleeding
	Ganda	*Tagetes erecta*	Leaf	• Leaf juice is applied over cut injury to stop bleeding
	Bonlata	*Mikania micrantha*	Leaf	• Leaves are smashed, and then juice is used in cuts to stop bleeding
Diabetes	Nayantara	*Catharanthus roseus*	Leaf	• Fresh four to five leaves can be taken by chewing or as leaf juice in empty stomach in the morning
	Karala	*Momordica charantia*	Fruit	• Fruit juice is drunk in empty stomach especially for diabetes
	Peyara	*Psidium guajava*	Fruit	• Extract prepared with overnight immersion of mature fruit in warm

(continued)

Table 48.2 (continued)

Disease	Plants used	Scientific name	Plant parts used	Mode of uses
				water is used to control blood sugar
	Bot	*Ficus benghalensis* Linn.	Leaf	• Bot leaves are immersed in water, and the extract is used to regulate blood sugar
	Jam	*Syzygium cumini*	Leaf, seed	• Seed powder is taken for diabetes • Leaf juice is also beneficial for diabetic patients
	Neem	*Azadirachta indica*	Leaf	• Fresh soft leaves are taken as such by chewing or by making juice at empty stomach
Digestive ailments	Babla	*Acacia nilotica*	Leaf buds	• Immature leaf buds are used for stimulating digestive system and also reducing acidity
	Kukshima	*Blumea lacera*	Root	• Root paste with sugar and water is used to cure indigestion
	Kagjilebu	*Citrus aurantiifolia*	Fruit	• Fruit juice with warm water is used to boost the digestive system
	Amlaki	*Emblica officinalis*	Fruit	• Fruits are dried with common salt and are taken after meal to reduce acidity and digestive problems
	Hingcha	*Enydra fluctuans*	Leaf, stem	• It is consumed as vegetable to reduce digestive ailment
	Gimeshak	*Glinus oppositifolius*	Leaf, stem	• Leaves and stems are used in daily meal as a stimulant of digestive system
	Piplas	*Litsea glutinosa* (lour)	Leaf	• After smashing the leaves, a thick agar-like extract is formed. It is used as a curative agent for chronic digestive ailment
	Pan	*Piper betle*	Leaf	• Leaf of pan with "supari" (*Areca catechu*) and "chuna" (calcium hydroxide) is used as digestive stimulator

(continued)

Table 48.2 (continued)

Disease	Plants used	Scientific name	Plant parts used	Mode of uses
	Tentul	*Tamarindus indica*	Fruit pulp	• Fruit pulp is used to make sour curry with fishes to cure digestive ailment and acidity
	Arjun	*Terminalia arjuna*	Stem bark	• Dried stem bark with common salt is used as stimulator of digestive system and also reduces acidity
Dysentery/ diarrhoea/ enteritis	Telakucha	*Coccinia grandis*	Leaf	• Leaf juice is used in empty stomach to control dysentery
	Bael	*Aegle marmelos*	Fruit	• Unripe fruit pulp is used against dysentery and diarrhoea
	Gandal	*Paederia foetida*	Leaf	• Leaf is used as vegetable in diarrhoea, amoebiasis and dysentery
	Satamuli	*Asparagus racemosus*	Root	• Root juice mixed with milk is taken in blood dysentery
	Durba	*Cynodon dactylon*	Leaf	• Leaf juice is drunk in dysentery at empty stomach
	Mushakani	*Evolvulus nummularius*	Leaf	• Fresh leaf juice is consumed in empty stomach to treat dysentery
	Patharkuchi	*Bryophyllum pinnatum*	Leaf	• Leaf juice mixed with common salt is consumed in bloody enteritis
	Thankuni	*Centella asiatica* L.	Leaf	• Leaves are consumed in empty stomach by chewing or by preparing juice in dysentery, diarrhoea and amoebiasis • Curry of leaves is also useful to cure chronic dysentery
	Muthaghas	*Cyperus rotundus*	Rhizome	• Rhizome paste mixed with common salt is used to cure enteritis
	Bot	*Ficus bengalensis* Linn.	Root	• Root cap of bot fried with ghee is used for dysentery

(continued)

Table 48.2 (continued)

Disease	Plants used	Scientific name	Plant parts used	Mode of uses
Fever	Bontulsi	*Pluchea indica* (L.)	Leaf	• Leaf juice is consumed in fever
	Shewli	*Nyctanthes arbor-tristis*	Leaf	• Leaf juice mixed with honey is given to children during fever
	Bhut bhairab	*Lantana camera*	Fruit	• Fruit paste or juice is useful in fever
	Tulsi	*Ocimum sanctum*	Leaf	• Fresh leaf or leaf juice with honey is used in fever
	Penpe	*Carica papaya*	Latex	• Fruit latex is taken orally with sugar/jaggery candy known as "batasha" in Bengali in fever
Hypertension/ high blood pressure	Neem	*Azadirachta indica*	Leaf	• Leaf juice or leaf paste, dried as a tablet, is taken at empty stomach
	Nayantara	*Catharanthus roseus*	Leaf	• Leaf juice is taken to combat hypertension
	Sajina	*Moringa oleifera* L.	Leaf	• Leaf juice or fried leaf is taken to control blood pressure
Skin diseases	Ghritokumari	*Aloe vera*	Leaf	• Fresh leaf extract or jelly-like pulp is rubbed in the affected area
	Holud	*Curcuma longa* L.	Rhizome	• Tender rhizome is taken at empty stomach • Paste of rhizome and neem leaves together is applied in skin diseases
	Neem	*Azadirachta indica*	Leaf	• Leaf paste with "kancha (raw) holud (*Curcuma longa* L.)" is used in skin diseases and skin allergy. • Leaves are boiled in water first, and then the water is used in affected area of the skin
	Akanda	*Calotropis gigantea* L.	Leaf	• Leaf powder boiled in oil is useful in eczema, skin eruptions
	Tulsi	*Ocimum sanctum*	Leaf	• Leaf juice mixed with lime is used in ringworm
	Dhutra	*Datura metel* L.	Leaf	• Leaf paste or juice mixed with mustard oil is used in skin diseases

Fig. 48.1 (**a**) Some important medicinal plants of plateau-fringe and Rarh regions of West Bengal, India. (a) *Centella asiatica* L. (Thankuni), (b) *Coccinia grandis* (telakucha), (c) *Andrographis paniculata* (kalmegh), (d) *Oxalis corniculata* (amrul), (e) *Azadirachta indica* (neem), (f) *Ananas comosus* (anaras), (g) *Emblica officinalis* (amlaki), (h) *Aloe vera* (ghritokumari). (**b**) Some important medicinal plants of plateau-fringe and Rarh regions of West Bengal, India. (i) *Marsilea minuta* (susni), (j) *Zingiber officinale* (ada), (k) *Psidium guajava* (Peyara), (l) *Acacia nilotica* (babla), (m) *Mikania micrantha* (bonlata), (n) *Curcuma longa* L. (holud), (o) *Blumea lacera* (kukshima), (p) *Jatropha gossypiifolia* L. (varenda), (q) *Cynodon dactylon* (durba). (**c**) Some important medicinal plants of plateau-fringe and Rarh regions of West Bengal, India. (r) *Bacopa monnieri* (brahmi), (s) *Hibiscus rosa-sinensis* (joba), (t) *Catharanthus roseus* (nayantara), (u) *Nyctanthes arbortristis* (shewli), (v) *Tagetes erecta* (ganda), (w) *Datura metel* L. (dhutra), (x) *Bryophyllum pinnatum* (patharkuchi), (y) *Calotropis gigantea* L. (akanda), (z) *Eclipta prostrata* (keshut). (**d**) Some important medicinal plants of plateau-fringe and Rarh regions of West Bengal, India. (aa) *Ocimum sanctum* (tulsi), (bb) *Carica papaya* (penpe), (cc) *Citrus aurantiifolia* (kagjilebu), (dd) *Aegle marmelos* (bael), (ee) *Mangifera indica* (aam), (ff) *Barleria lupulina* Lindl. (kantabishalyakarani), (gg) *Rauvolfia serpentina* L. (sarpagandha), (hh) *Evolvulus nummularius* (mushakani), (ii) *Cyperus rotundus* (muthaghas), (jj) *Hygrophila spinosa* T. (kulekhara)

Fig. 48.1 (continued)

1983). Some regions, i.e. the places in close vicinities of Ajodhya hills and Dalma reserve forest area, contain dense to moderate forest area, mainly dominated by Sal tree (*Shorea robusta*). Santal, Munda, Savara, Bhumij, Saoria and Oraon are the predominant tribal communities found in this region.

A symbolic north to south lateritic region of West Bengal is recognized as **Rarh Bengal** (Bagchi and Mukherjee 1983) (Fig. 48.2). This region lies between the south-flowing (e.g. Bhagirathi-Hooghly) and east-flowing river (Kangsabati, Dwarkeswar, etc.) fan systems of West Bengal (Bandyopadhyay et al. 2014). It is mainly covered by older alluvium and laterite soil with kaolinite and ferruginous gravels deposits. Moreover distinct depositional pattern in-between Archean-Gondwana time at west and recent alluvial deposits of quaternary period at east of

Fig. 48.1 (continued)

Bengal basin defines the rarh region (Biswas 1987; Niyogi 1975). Elevation of this geomorphic pattern ranges from 75 to 150 m (Mukherjee 1995). The important rivers of Rarh Bengal are Ajay, Damodar, Dwarkeswar, Shilabati, Mayurakshi and Kangasabati (Ghosh and Guchhait 2015). Santal, Munda, Oraon, Mahali, Kheria/ Lodha, Kora, Chakma and Bhumij are mostly found tribal communities in this region.

Fig. 48.1 (continued)

48.5 Conclusions

In the face of rapid urbanization, there is an incredible shift in socio-cultural and socio-economic domains of life of an individual, overall in lifestyle. However, majority of the rural population including the tribes in India still depend on traditional herbal medicine for their overall well-being and treatment of ailments. Similar faith and dependence on traditional medicine is also observed in other developing countries. Besides, a growing popularity for several T&CM practices from different corners of the globe is observed in developed countries in recent times. Here the

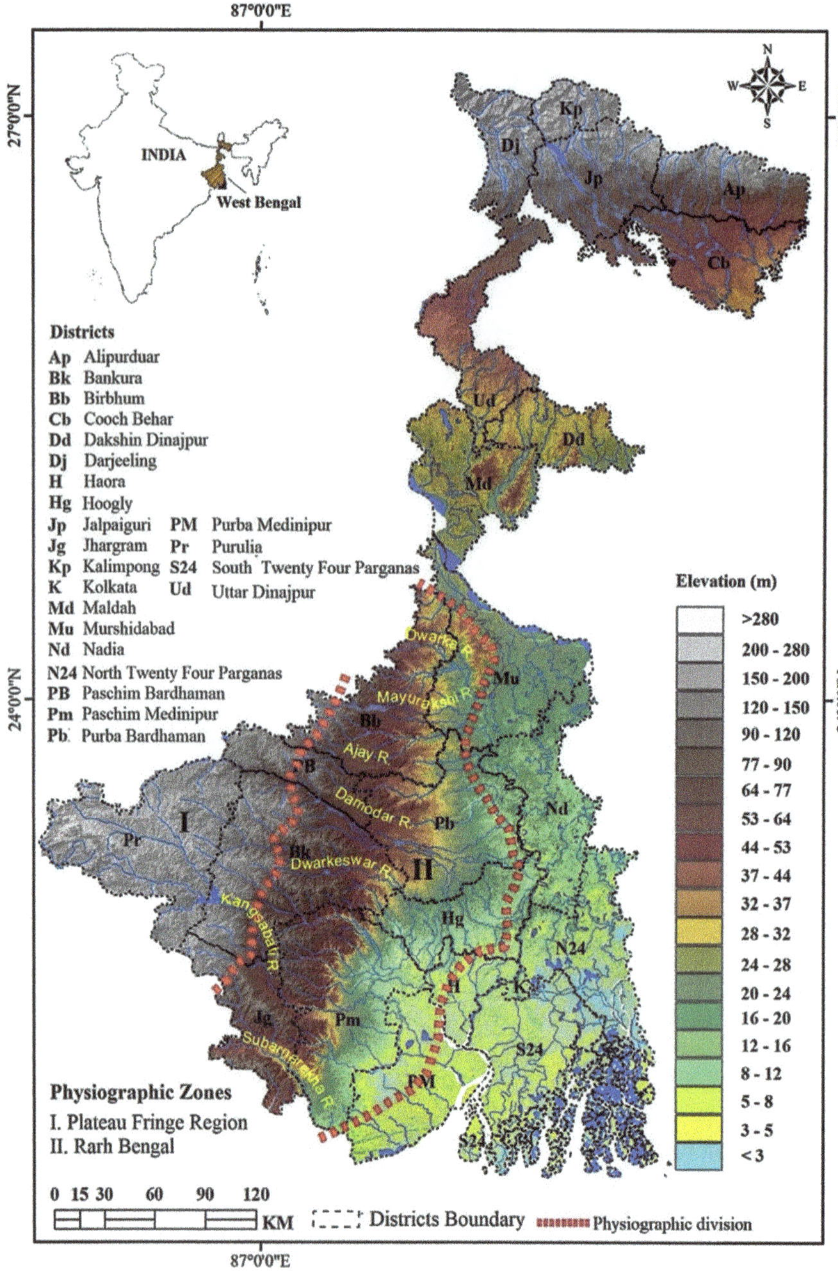

Fig. 48.2 The localizations of plateau-fringe and Rarh districts of state of West Bengal, India

Table 48.3 Population in the districts (fully or partly) falling under Plateau-fringe and Rarh regions of West Bengal, India (Registrar General, and Census Commissioner, India 2011)

Districts	Total area (km^2)	Total population	Number of villages	Total rural population	Total urban population	Tribal population		
						Total	Rural	Urban
Bardhaman[a]	7024	7,717,563	2502	4,639,264	3,078,299	489,447	379,262	110,185
Birbhum	4545	3,502,404	2455	3,052,956	449,448	242,484	232,666	9818
Bankura	6882	3,596,674	3823	3,296,901	299,773	368,690	365,380	3310
Purba Medinipur	4736	5,095,875	2994	4,503,161	592,714	27,952	24,028	3924
Hooghly	3149	5,619,145	1866	3,390,646	2,128,499	229,243	211,620	17,623
Purulia	6259	2,930,115	2667	2,556,801	373,314	540,652	531,822	8830
Murshidabad	5324	7,103,807	2166	5,703,115	1,400,692	91,035	86,004	5031
Paschim Medinipur[b]	9345	5,913,457	8694	5,190,771	722,686	880,015	853,031	26,984

[a]The district Bardhaman has been divided into Purba (east) and Paschim (west) Bardhaman districts on April 7, 2017
[b]The district Paschim Medinipur has been divided into Jhargram and Paschim Medinipur districts on April 4, 2017

major reasons for reliance being the effectiveness of the drugs, lesser side effects and confidence on the practitioners, not the low cost. As a whole the present scenario is that millions of people all over the world are trusting on these practices. As a result more and more countries are coming forward with the intentions and establishment of formal education and meticulous research in this area. One of the major targets of the global healthcare leaders is to frame a universal healthcare system *for all* by integrating prospective sides of several T&CM practices and conventional medicine in future. In this view proper documentation of traditional knowledge is essential as systematic record of these age-old practices especially from tribal societies is unavailable due to being mostly confined with the practitioners/healers and transferred orally from one generation to others. The present approach of documentation of precious herbal medicines from a distinct biodiversity zone with very rich ethnic background and tradition of practicing herbal medicine might be important input to this approach.

Acknowledgement The authors are indebted to Mr. Kousik Das, Research Scholar, Department of Geography, Vidyasagar University in preparing the map (Fig. 48.2). Authors also express their sincere gratitude to Ms. Krishna Jana for providing photographs of medicinal plants.

References

Abdallah EM, Abdalla WE (2018) Black pepper fruit (*Piper nigrum* L.) as antibacterial agent: a mini-review. J Bacteriol Mycol 6(2):141–145

Adhikari PP, Paul SB (2018) History of Indian traditional medicine: a medical inheritance. Asian J Pharma Clin Res 11(1):421–426

Afzal M, Iqbal R, Mahmood Z et al (2019) Study of GC-MS and HPLC characterized metabolic compounds in guava (*Psidium guajava* L.) leaves. Pak J Agric Sci 56(3):709–713

Akerele O (1993) Summary of WHO guideline for the assessment of herbal medicines. Herbalgram 28:13–17

Akhila S, Vijayalakshmi NG (2015) Phytochemical studies on *Carica papaya* leaf juice. Int J Pharma Sci Res 6(2):880–885

Akpanabiatu MI, Umoh IB, Udosen EO et al (2005) Rat serum electrolytes, lipid profile and cardiovascular activity on *Nauclea latifolia* leaf extract administration. Ind J Clin Biochem 20 (2):29–34

Albinjose J, Jasmine E, Selvankumar T et al (2015) Bioactive compounds of *Tinospora cordifolia* by gas chromatography-mass spectrometry (GC-MS). Int J Multi Disc Res Dev 2(1):88–97

Ali M (1998) Textbook of pharmacognosy. CBS Publishers & Distributors, New Delhi

Al-Snafi AE (2016) Chemical constituents and pharmacological effects of *Cynodon dactylon*-a review. IOSR J Pharm 6(7):17–31

Alzohairy MA (2016) Therapeutics role of *Azadirachta indica* (Neem) and their active constituents in diseases prevention and treatment. Evid Based Complement Alt Med 2016:1–11

Ansari JA, Inamdar NN (2010) The promise of traditional medicines. Int J Pharmacol 6(6):808–812

Antwi-Baffour SS, Bello AI, Adjei DN et al (2014) The place of traditional medicine in the African society: the science, acceptance and support. Am J Public Health Res 2(2):49–54

Aqib S, Mohib S (2014) Phytochemistry, pharmacological and traditional uses of *Datura stramonium* L. review. J Pharmacogn Phytochem 2(5):123–125

Ayyanar M, Subash-Babu P (2012) *Syzygium cumini* (L.) Skeels: a review of its phytochemical constituents and traditional uses. Asian Pac J Trop Biomed 2(3):240–246

Bagchi K, Mukherjee KN (1983) Diagnostic survey of Rarh Bengal. In: Pantg Delta & Rarh Bengal. Kolkata, The Dept. of Geography, University of Calcutta, pp 17–19

Bai S, Seasotiya L, Malik A et al (2014) GC-MS analysis of chloroform extract of *Acacia nilotica* L. leaves. J Pharmacogn Phytochem 2(6):79–82

Bandyopadhyay S, Kar NS, Das S et al (2014) River systems and water resources of West Bengal: a review. Geo Soc Ind Spec Publ 3(2014):63–84

Banerji A, Banerji J, Das M et al (2017) Some aspects of investigation of the Indian medicinal plant *Hemidesmus indicus* R. Br.: chemical constituents and anti-diabetic activity. J Chem Pharm Res 9(4):50–64

Bell IR, Lewis DA, Brooks AJ et al (2004) Improved clinical status in fibromyalgia patients treated with individualized homeopathic remedies versus placebo. Rheumatology 43(5):577–582

Bhardwaj R, Sharma RA (2019) Boerhavia diffusa-GC-MS analysis of alkaloids and their inhibitory activity against pathogenic microorganisms. J Pharmacogn Phytochem 8(2):756–760

Bhattacharya A, Tiwari P, Sahu PK et al (2018) A review of the phytochemical and pharmacological characteristics of *Moringa oleifera*. J Pharm Bioallied Sci 10(4):181–191

Biswas A (1987) Laterites and lateritoids of Rarh Bengal. In: Datye VS et al (eds) Explorations in the tropics. Prof. KD Dikshit Felicitation Volume Committee, pp 48–54

Bodini RB, Montini EM, de Carvalho CC et al (2019) Sensory and composition analyses of the aqueous phases from the concentration of guava (L.) and mango (L.) juices and the process-induced losses of vitamin C. Open Food Sci J 11(1):44–55

Bourgaud F, Gravot A, Milesi S et al (2001) Production of plant secondary metabolites: a historical perspective. Plant Sci 161(5):839–851

Chakraborty NR, Paul A (2014) Traditional knowledge on medicinal plants used by the tribal people of birbhum district of West Bengal in India. Intl J Agric Environ Biotech 7(3):547–554

Chao WW, Lin BF (2010) Isolation and identification of bioactive compounds in *Andrographis paniculata* (Chuanxinlian). Chin Med 5(1):17

Chipiti T, Ibrahim MA, Koorbanally NA et al (2015) In vitro antioxidant activity and GC-MS analysis of the ethanol and aqueous extracts of *Cissus cornifolia* (baker) Splanch (Vitaceae) parts. Acta Pol Pharm 72:119–127

Chopra D (2006) In the seven spiritual laws of yoga. John Wiley & Sons, Hoboken, NJ

Chopra RN, Chopra IC (1994) Indigenous drugs of India. Academic publishers, Kolkata

Cragg GM, Newman DJ (2013) Natural products: a continuing source of novel drug leads. Biochim Biophys Acta 1830(6):3670–3695

Das B, Das R (1995) Medicinal properties and chemical constituents of *Aegle marmelos* correa. Indian Drugs 32:93–99

Das D, Ghosh P (2017) Some important medicinal plants used widely in Southwest Bengal, India. Int J Eng Sci Invent 6(6):28–50

Das RP (1983) Some remarks on the bengali deity dharma: its cult and study. Anthropos 78:661–700

de Oliveira MS, da Costa WA, Bezerra FWF (2018) Phytochemical profile and biological activities of *Momordica charantia* L. (Cucurbitaceae): a review. Afr J Biotechnol 17(27):829–846

Debas TH, Laxminarayan R, Straus SE (2006) Complementary and alternative medicine. In: Jamison DT, Breman JG, Measham AR et al (eds) Disease control priorities in developing countries. Oxford University Press, New York, pp 1281–1291

Devika R, Justin K (2014) Screening and evaluation of bioactive components of Tagetes erecta L. by GC–MS analysis. Asian J Pharm Clin Res 7(2):58–60

Di Carlo G, Autore G, Izzo AA et al (1993) Inhibition of intestinal motility and secretion by flavonoids in mice and rats: structure-activity relationships. J Pharmacol 45(12):1054–1059

Dinda B, Ghosh B, Arima S et al (2007) Chemical constituents of *Evolvulus nummularius*. Chem Ind J 46B:492–498

Dwivedi D, Thanwar M, Gharia AK (2015) Study of phytochemical active compounds in extract of Withania somnifera. RASAYAN J Chem 8:522–526

Ediriweera MK, Tennekoon KH, Samarakoon SR (2017) A review on ethnopharmacological applications, pharmacological activities, and bioactive compounds of *Mangifera indica* (mango). Evid Based Complement Alt Med 2017:1–24

Ekor M (2014) The growing use of herbal medicines: issues relating to adverse reactions and challenges in monitoring safety. Front Pharmacol 4:177

ENVIS Resource Partner on Biodiversity (n.d.). Important plants with chemical constituents. Hosted by Botanical Survey of India, Kolkata, West Bengal. Ministry of Environment, Forest and Climate Change, Government of India. Available at:http://www.bsienvis.nic.in/Database/ImportantPlantswithChemicalConstituents_3946.aspx

Escalona-Arranz JC, Pérez-Rosés R, Jiménez IL et al (2010) Chemical constituents of *Tamarindus indica* L. leaves. Revista Cubana De Química 22(3):65–71

Fabricant DS, Farnsworth NR (2001) The value of plants used in traditional medicine for drug discovery. Environ Health Perspect 109(1):69–75

Félix-Silva J, Giordani RB, Silva-Jr AAD et al (2014) *Jatropha gossypiifolia* L. (Euphorbiaceae): a review of traditional uses, phytochemistry, pharmacology, and toxicology of this medicinal plant. Evid Based Complement Alt Med 2014:1–32

Fredotović Ž, Šprung M, Soldo B et al (2017) Chemical composition and biological activity of *Allium cepa* L. and *Allium cornutum* (Clementi ex Visiani 1842) methanolic extracts. Molecules 22(3):448–454

Gabriel A, Agbor DK, Julius EO (2007) Medicinal plants can be good sources of antioxidants: case study in Cameroon. Pak J Biol Sci 10(4):537–544

Ghosh A (2003) Herbal folk remedies of Bankura and Medinipur districts, West Bengal. Ind J Trad Knowl 2(4):393–396

Ghosh A (2008) Ethnomedicinal plants used in West Rarh region of West Bengal. Nat Prod Rad 7 (5):461–465

Ghosh S, Guchhait SK (2015) Characterization and evolution of laterites in West Bengal: implication on the geology of Northwest Bengal Basin. Transactions 37(1):93–119

Guedes JAC, de Godoy AFE, Rodrigues THS et al (2018) Metabolic profile and cytotoxicity of non-polar extracts of pineapple leaves and chemometric analysis of different pineapple cultivars. Ind Crop Prod 124:466–474

Gulshan B, Suri KA, Parul G (2015) A comprehensive review on Nyctanthes arbortristis. Int J Drug Dev Res 7(1):183–193

Gupta AK, Bhat JL (2016) GC-MS analysis of methanol extract of *Acacia nilotica* (L.) leaves. Int J Pharm Chem 6(2):50–53

Gupta S, Bishnoi JP, Kumar N et al (2018) Terminalia arjuna (Roxb.) Wight & Arn: competent source of bioactive components in functional food and drugs. Pharma Innov 7(3):223–231

Halberstein RA, Saunders AB (1978) Traditional medical practices and medicinal plant usage on a Bahamian island. Cult Med Psychiatry 2(2):177–203

Haq ME, Ikram M, Warsi SA (1967) Chemical composition of *Adhatoda vasica* (L.) II. Pak J Sci Ind Res 10:224–225

Hasan MR, Islam MN, Islam MR (2016) Phytochemistry, pharmacological activities and traditional uses of *Emblica officinalis*: a review. Int Curr Pharmacol J 5(2):14–21

Hewawasam RP, Jayatilaka KAPW, Pathirana C et al (2004) Hepatoprotective effect of Epaltes divaricata extract on carbon tetrachloride induced hepatotoxicity in mice. Indian J Med Res 120:30–34

Hien NT, Thang TD, Dai DN, Thai TH (2010) Chemicao composition of the leaf oil of *Litsea glutinosa* (lour.) CB rob. From ha Tinhprovinve. VNU J Sci Nat Sci Technol 26:161–164

Hossain MA, Al-Hdhrami SS, Weli AM et al (2014) Isolation, fractionation and identification of chemical constituents from the leaves crude extracts of *Mentha piperita* L. grown in Sultanate of Oman. Asian Pac J Trop Biomed 4:S368–S372

Hossain MA, Al-Toubi WAS, Weli AM et al (2013) Identification and characterization of chemical compounds in different crude extracts from leaves of Omani neem. J Taibah Univ Sci 7 (4):181–188

Huang KC (1998) The pharmacology of Chinese herbs, 2nd edn. Taylor & Francis, Routledge, New York
Hussein RA, El-Anssary AA (2018) Plants secondary metabolites: the key drivers of the pharmacological actions of medicinal plants. In: Builders PF (ed) Herbal medicine. IntechOpen, London, pp 11–30
Imam MU, Ismail M (2017) The impact of traditional food and lifestyle behavior on epigenetic burden of chronic disease. Global Chall 1(8):1700043
Ishak AH, Shafie NH, Esa NM et al (2018) From weed to medicinal plant: antioxidant capacities and phytochemicals of various extracts of Mikania micrantha. Int J Agric Biol 20(3):561–568
Joshi VK, Joshi A, Dhiman KS (2017) The ayurvedic pharmacopoeia of India, development and perspectives. J Ethnopharmacol 197:32–38
Kale BP, Kothekar MA, Tayade HP et al (2003) Effect of aqueous extract of Azadirachta indica leaves on hepatotoxicity induced by antitubercular drugs in rats. Ind J Pharmacol 35(3):177–180
Kalim MD, Bhattacharyya D, Banerjee A et al (2010) Oxidative DNA damage preventive activity and antioxidant potential of plants used in Unani system of medicine. BMC Complement Altern Med 10(1):77
Katz L, Baltz RH (2016) Natural product discovery: past, present, and future. J Ind Microbiol Biotechnol 43(2–3):155–176
Khaliq HA (2017) A review of pharmacognostic, physicochemical, phytochemical and pharmacological studies on Ficus benghalensis L. J Sci Innov Res 6(4):151–163
Kumar B, Vijayakumar M, Govindarajan R et al (2007) Ethnopharmacological approaches to wound healing—exploring medicinal plants of India. J Ethnopharmol 114(2):103–113
Kumar M (2017) A review on phytochemical constituents and pharmacological activities of Ricinus communis L. plant. Int J Pharma Phytochem Res 9(4):466–472
Kumar V, Dey A, Hadimani MB et al (2015) Chemistry and pharmacology of Withania somnifera: an update. TANG [Humanitas Medicine] 5(1):1–13
Kumari R, Dubey RC (2016) Phytochemical analysis and antibacterial and cytotoxic properties of *Barleria lupulina* Lindl. Extracts. J Plant Pathol Microbiol 7(380):2
Lad V (2002) Textbook of Ayurveda: fundamental principles, vol 1. Ayurvedic Press, New Mexico
Lemes RS, Alves CC, Estevam EB et al (2018) Chemical composition and antibacterial activity of essential oils from Citrus aurantifolia leaves and fruit peel against oral pathogenic bacteria. An Acad Bras Cienc 90(2):1285–1292
Li B, Webster TJ (2018) Bacteria antibiotic resistance: new challenges and opportunities for implant-associated orthopedic infections. J Orthop Res 36(1):22–32
Lim MK, Sadarangani P, Chan HL et al (2005) Complementary and alternative medicine use in multiracial Singapore. Complement Ther Med 13:16–24
Lutterodt GD, Ismail A, Basheer RH et al (1999) Antimicrobial effects of *Psidium guajava* extract as one mechanism of its antidiarrhoeal action. Malays J Med Sci 6(2):17–20
Manna SS, Mishra SP (2018) Ethnomedicinal survey of plants used by tribal in Lalgarh forest range, WB, India. J Phytopharmacol 7(2):199–202
Mathews MM, Sunny B (2019) A compendious write-up on Coccinia grandis. Int J Pharm Sci Rev Res 54(2):29–36
Mele MA (2019) Bioactive compounds and biological activity of ginger. J Multidiscip Sci 1(1):1–7
Mishra J, Verma NK (2017) Asparagus racemosus: chemical constituents and pharmacological activities - a review. Eur J Biomed Pharma Sci 4(6):207–213
Missoum A (2018) An update review on *Hibiscus rosasinensis* phytochemistry and medicinal uses. J Ayur Herb Med 4(3):135–146
Morton JF, Dowling CF (1987) Fruits of warm climates. JF Morton, Miami, FL
Mukherjee AR (1995) Forest resources conservation and regeneration: a study of West Bengal plateau. Concept Publishing Company, New Delhi
Mukherjee PK (2001) Evaluation of Indian traditional medicine. Drug Inf J 35(2):623–632
Mukherjee PK, Harwansh RK, Bahadur S et al (2017) Development of Ayurveda–tradition to trend. J Ethnopharmacol 197:10–24

Mukherjee PK, Wahile A (2006) Integrated approaches towards drug development from Ayurveda and other Indian system of medicines. J Ethnopharmacol 103(1):25–35
Nadkarni KM (1954) Indian Materia Medica. Karnataka Printing Press, Bombay
Niyogi D (1975) Quaternary geology of the coastal plain of West Bengal and Orissa. Ind J Earth Sci 2(1):51–61
Padalia RC, Verma RS (2011) Comparative volatile oil composition of four Ocimum species from northern India. Nat Prod Res 25(6):569–575
Pandey MM, Rastogi S, Rawat AKS (2008) Indian herbal drug for general healthcare: an overview. Internet J Altern Med 6(1):3
Patil G, Khare AB, Huang KF et al (2015) Bioactive chemical constituents from the leaves *of Lantana camara* L. Ind J Chem 54B:691–697
Pattanayak S, Mandal TK, Bandyopadhyay SK (2015) A study on use of plants to cure enteritis and dysentery in three southern districts of West Bengal, India. J Med Plants Stud 3(5):177–183
Payyappallimana U (2010) Role of traditional medicine in primary health care: an overview of perspectives and challenges. Yokohama J Soc Sci 14(6):723–743
Piḷḷai NK (1998) History of siddha medicine. Department of Indian Medicine & Homoeopathy, Government of Tamil Nadu
Prasad LV (2002) Indian system of medicine and homoeopathy traditional medicine in Asia. WHO-Regional Office South East Asia, New Delhi, pp 283–286
Priya K, John P, Usha PTA et al (2018) Phytochemical analysis of *Eclipta prostrata* L. leaves. Int J Curr Microbiol App Sci 7(8):1069–1075
Rabbani GH, Teka T, Zaman B et al (2001) Clinical studies in persistent diarrhea: dietary management with green banana or pectin in Bangladeshi children. Gastroenterology 121 (3):554–560
Ragasa CY, Cabrera EC, Torres OB et al (2015) Chemical constituents and bioactivities of *Glinus oppositifolius*. Pharm Res 7(2):138
Rajkumari S, Sanatombi K (2017) Nutritional value, phytochemical composition, and biological activities of edible curcuma species: a review. Int J Food Prop 20(3):S2668–S2687
Raksha B, Pooja S, Babu S (2014) Bioactive compounds and medicinal properties of Aloe vera L.: an update. J Plant Sci 2(3):102–107
Ravisankar M, Ester VCJ (2017) Determination of bioactive compounds in *Hydrophila auriculata* leaf extract using GC MS. IJCS 5(3):729–733
Registrar General & Census Commissioner, India (2001) Provisional population totals. In: Census of India 2001. Government of India, New Delhi
Registrar General & Census Commissioner, India (2011) Provisional population totals. In: Census of India 2011. Government of India, New Delhi
Riaz M, Khan O, Sherkheli MA et al (2017) Chemical constituents of *Terminalia chebula*. Nat Prod Indian J 13(2):112–127
Sabithira G, Udayakumar R (2017) GC-MS analysis of methanolic extracts of leaf and stem of *Marsilea minuta* (Linn.). J Complement Altern Med Res 3(2):1–13
Sandoval-Montemayor NE, García A, Elizondo-Treviño E et al (2012) Chemical composition of hexane extract of *Citrus aurantifolia* and anti-*Mycobacterium tuberculosis* activity of some of its constituents. Molecules 17(9):11173–11184
Sathasivampillai SV, Rajamanoharan PRS, Munday M et al (2017) Plants used to treat diabetes in Sri Lankan Siddha Medicine–an ethnopharmacological review of historical and modern sources. J Ethnopharmacol 198:531–599
Satyal P, Chhetri BK, Dosoky NS et al (2015) Chemical composition of *Blumea lacera* essential oil from Nepal. Biological activities of the essential oil and (Z)-lachnophyllum ester. Nat Prod Commun 10(10):1749–1750
Sen S, Chakraborty R (2017) Revival, modernization and integration of Indian traditional herbal medicine in clinical practice: importance, challenges and future. J Tradit Complement Med 7 (2):234–244

Shang A, Cao SY, Xu XY et al (2019) Bioactive compounds and biological functions of garlic (*Allium sativum* L.). Foods 8(7):246
Sharma N, Dubey W (2016) Bioactive compounds present in *Aegle marmelos* and their role in medicinal properties: a review. Int J Pharm Bio Sci 7(3):170–176
Singh LK (2008) Indian cultural heritage perspective for tourism. Gyan Publishing House, New Delhi
Singh SK, Patel JR, Dangi A et al (2017) A complete over review on *Adhatoda vasica* a traditional medicinal plants. J Med Plants 5(1):175–180
Srikanth M, Swetha T, Veeresh B (2012) Phytochemistry and pharmacology of *Oxalis corniculata* Linn.: a review. Int J Pharm Sci Res 3(11):4077–4085
Suganthi N, Dubey S (2016) Phytochemical constituents and pharmacological activities of *Vitex negundo* Linn. J Chem Pharm Res 8(2):800–807
Sultana S, Ali M, Mir SR (2017) Chemical constituents from the rhizomes of *Cyperus rotundus* L. Open Plant Sci J 10(1):82–91
Sundaram RS, Ramanathan M, Rajesh R et al (2012) LC-MS quantification of rosmarinic acid and ursolic acid in the Ocimum sanctum Linn. Leaf extract (holy basil, Tulsi). J Liq Chromatogr Relat Technol 35(5):634–650
Swaminathan S (2013) 107 Miracle herbs in the Hindu Vedas. In: Tamil Vedas
Tembe-Fokunang EA, Charles F, Kaba N et al (2019) The potential pharmacological and medicinal properties of neem (*Azadirachta indica* a. Juss) in the drug development of phytomedicine. J Complement Altern Med Res:1–18
Thakurta P, Bhowmik P, Mukherjee S et al (2007) Antibacterial, antisecretory and antihemorrhagic activity of *Azadirachta indica* used to treat cholera and diarrhea in India. J Ethnopharmacol 111 (3):607–612
Thanwar M, Dwivedi D, Gharia AK (2017) GC-MS study of methanolic extract of leaves of *Catharanthus roseus*. Int J Chem Stud 5(2):517–518
Thas JJ (2008) Siddha medicine—background and principles and the application for skin diseases. Clin Dermatol 26(1):62–78
Tripathi S, Ray S, Mondal AK, Verma NK (2013) Rare ethno medicinal plants of South West Bengal, India with their different medicinal uses: needs conservation. Int J Life Sci Biotech Pharm Res 2:2250–3137
Tumpa SI, Hossain MI, Ishika T (2014) Ethnomedicinal uses of herbs by indigenous medicine practitioners of Jhenaidah district, Bangladesh. J Pharmacogn Phytochem 3(2):23–33
Uchegbu RI, Ahuchaogu AA, Amanze KO et al (2017) Chemical constituents analysis of the leaves of *Bryophyllum pinnatum* by GC-MS. AASCIT J Chem 3(3):19–22
Vijayan P, Raghu C, Ashok G et al (2004) Antiviral activity of medicinal plants of Nilgiris. Indian J Med Res 120:24–29
Wang L, Jiang Y, Han T et al (2014) A phytochemical, pharmacological and clinical profile of *Paederia foetida* and *P. scandens*. Nat Prod Commun 9(6):879–886
World Health Organization (2013) WHO traditional medicine strategy 2014–2023. World Health Organization, Geneva. https://www.who.int/medicines/publications/traditional/trm_strategy14_23/en/, accessed 27 January 2019
World Health Organization (2019) WHO global report on traditional and complementary medicine 2019. World Health Organization, Geneva
Worldmeters United Nation (2019) India population. https://www.worldometers.info/world-population/india-population
Yadav RK, Nandy BC, Maity S et al (2015) Phytochemistry, pharmacology, toxicology, and clinical trial of *Ficus racemosa*. Pharmacogn Rev 9(17):73–80
Zahira A, Thamilmani K (2016) Evaluation of bioactive compounds present in Piper betle Linn. By elution chromatography coupling technique. WIPPS 5(5):1405–1413

Pharmacology and Mechanisms of Natural Medicine in Treatment of Type 2 Diabetes Mellitus

49

Tanushree Karmakar, Rituparna Chaki, and Nilanjan Ghosh

Abstract

Diabetes mellitus (DM) is a chronic metabolic disorder that is affecting the worldwide population with high rates of morbidity and mortality. The most prevalent case of DM is type 2 diabetes mellitus (T2DM) which is characterized by insulin resistance and insulin secretion defect due to impaired β-cell functioning. Epidemiological studies implied that the number of diabetic cases has been doubled during the past two decades and has turned out to be a global epidemic accompanied by severe metabolic and endocrine complications over time. The rising emergence of T2DM in children, adolescents, and young adults signifies it to be one of the most dominant perturbing features of its kind. Proper diet control, moderate exercise, and hypoglycemic and lipid-lowering agents are some evident strategies that have been employed in the management of T2DM so far. However, a distinct possibility of severe diabetic complications still exists despite the therapeutic benefits achieved by these drugs. Various in vitro and in vivo models have suggested that phytochemicals exert several pharmacological effects on metabolic disorders such as in hyperglycemia, hypertension, and hyperlipidemia by modulating oxidative stress, inflammatory response, autophagy, and anti-apoptosis effects. Additionally, they have positive modulatory actions on molecular targets of T2DM like insulin signaling, IRS, glucose transporters, α-glucosidase, PPARγ, DPP-IV, PTP1B, NF-κB, etc. Regulation of these maladaptive pathophysiological mechanisms can improve insulin-resistant state, lower blood glucose levels, and protect against various macrovascular and microvascular complications. Thus, natural products have garnered significant interest as bioactive agents in the management of T2DM. This chapter aims to overview the activities and underlying mechanisms of natural medicine in the management of T2DM.

T. Karmakar · R. Chaki · N. Ghosh (✉)
Dr. B.C. Roy College of Pharmacy and Allied Health Sciences, Durgapur, India
e-mail: nils.bhu@gmail.com

S. C. Mandal et al. (eds.), *Evidence Based Validation of Traditional Medicines*,
https://doi.org/10.1007/978-981-15-8127-4_49

Keywords

Antidiabetic drug · Herbal medicine · Type 2 diabetes mellitus · Insulin resistance · Hyperlipidemia · Obesity · Inflammation

Abbreviations

AGE	Advanced glycation end products
AGT II	Angiotensin II
Akt	Also known as protein kinase B (PKB)
AMPK	Adenosine monophosphate-activated protein kinase
Bcl-2	β-cell lymphoma protein 2
BMP 4	Bone morphogenic protein 4
cAMP	Cyclic AMP
CB1	Cannabinoid receptor type 1
CCL 5	Chemokine ligand 5
CCR 5	Chemokine receptor type 5
CTGF	CCN2 or connective tissue growth factor
DM	Diabetes mellitus
DPP	Dipeptidyl peptidase
eNOS	Endothelial nitric oxide synthase
ERK	Extracellular signal-regulated kinases
FFA	Free fatty acids
Foxp3	Forkhead box
G6PD	Glucose-6-phosphate dehydrogenase
GIP	Gastric inhibitory polypeptide
GLP	Glucagon-like peptide
GLUT	Glucose transporter
GSK 3	Glycogen synthase kinase
HbA1C	Glycosylated hemoglobin
HDL	High-density lipoproteins
Hep G2	Human hepatocyte carcinoma cell line
HFD	High-fat diet
HGP	Hepatic glucose production
HMG-coA	3-Hydroxy-3-methyl-glutaryl-Coa reductase
ICAM	Intercellular adhesion molecule
IDF	International Diabetes Federation
IKK	IκB kinase
IL-6	Interleukin 6
IRS-1	Insulin receptor substrate-1
JAK	Janus kinase
JNK-c	Jun N-terminal kinases
LDH	L*actate dehydrogenase*

LDL	Low-density lipoproteins
MAPK	Mitogen-activated protein kinases
MCP-1	Monocyte chemoattractant protein-1
MMP	Matrix metalloproteinase
NF-κB	Nuclear factor kappa-light-chain-enhancer of activated B cells
NLRP	Nod-like receptor protein
NO	Nitric oxide
Nrf2	Nuclear factor erythroid 2-related factor 2
PCSK9	Pro-protein convertase subtilisin-kexin type 9
PI3K	*Phosphoinositide* 3-kinases
PKC	Protein kinase C
PPAR	Peroxisome proliferated-activated receptor
PTP-IB	Protein tyrosine phosphatase 1B
RAAS	Renin angiotensin aldosterone system
RAGE	Receptor for advanced glycation end products
ROS	Reactive oxygen species
S1P	*Sphingosine-1-phosphate*
SIRT1-NAD	dependent protein deacetylase sirtuin-1
SphK1	*Sphingosine kinase 1*
SREBP	*Sterol regulatory element* binding proteins
STAT	Signal transducers and activators of transcription
STZ	Streptozotocin
T2DM	Type 2 diabetes mellitus
TBARS	Thiobarbituric acid-reactive substances
TG	Triglycerides
TGF	Tumor growth factor
TNF-α	Tumor necrosis factor-α
VEGF	Vascular endothelial growth factor
VLDL	Very-low-density lipoprotein
α-SMA	Alpha-smooth muscle actin

49.1 Introduction

The co-occurrences of type 2 diabetes mellitus (T2DM) and a nexus of associated detrimental events designate itself as a complex metabolic disorder to manage therapeutically. It originates with multiple pathophysiological abnormalities and pathway anomalies, thus affecting the worldwide population due to its high rate of mortality which is emphasized by several epidemiological studies. The natural history of T2DM represents the core defect in β-cell functioning, which ultimately leads to an inherent decrease in the production as well as decreased secretion of insulin over time, thereby leading to severe progressiveness of the disease. The molecular and metabolic milieus associated with DM elicit a series of malfunctioned stimuli of hyperinsulinemia, increased circulating FFA and TG, cardiac dysfunction, and increased levels of inflammatory cytokines. Numerous mechanisms of enhanced free radical generation, lipid accumulation and elevated lipid peroxides, mitochondrial and endothelial dysfunction, overexpression of PPARs, and overproduction of advanced glycation end products act in concordance with the systemic progression and complications developed in DM.

β-cell mass deterioration determines the onset and rate of progression of dietary failure in T2DM (Hadden et al. 1975; Wilson et al. 1980). In light of this dysfunctional behavior of the endocrine disorder and the hepatic system, β-cells, and the muscle comes into play which produces some of the significant characteristic pathophysiological defects. Among these, the predominant hallmark pathophysiological defects in T2DM are mainly due to insulin resistance, decreased glucose uptake from the peripheral tissues, increased hepatic glucose production, and impaired insulin secretion due to long-term pancreatic β-cell destruction resulting in absolute insulin deficiency. These collective effects thereby hamper normal glucose homeostasis in individuals (DeFronzo 1988).

In addition to this, several other associated factors like age, sedentary lifestyle, obesity, stress, and uncontrolled dietary intake intensify the advancement of T2DM in due course of time. The co-existence of obesity and diabetes mellitus following interminable changes in the environmental conditions and lifestyle has projected the extent of complications in an alarming situation. Various acute and chronic comorbidities have raised due to uncontrolled glucose levels, insulin insensitivity, dysregulated incretin levels (modulator of β-cell survival and function), and glucose intolerance which are potentiated by obesity (Golay et al. 1988). The complications, however, intensify along with chronic hyperglycemic state leading to disturbances in carbohydrate, fat, and protein metabolism. This, in turn, gives rise to the negative metabolic disturbances like glucotoxicity and lipotoxicity which further worsen the situation along with β-cell failure by increasing disturbances in substrate levels in the physiological system. Dysfunction of insulin sensitivity with an imbalance in substrate metabolism leads to a state of glucose and lipid overloading, thereby depicting the role of obesity-linked predisposition which further converges to produce cellular oxidative stress, inflammatory cytokine release, and lipid-mediated complex impairments, thus leading to cellular dysfunction (Rossetti et al. 1990; Unger 1995).

On the other hand, excess nutrient content, rise in the levels of systemic free fatty acids, ectopic fat deposition, and inflammation of the peripheral tissues, as well as disturbances in the regulation of different metabolic hormones such as leptin, adiponectin, and glucagon, reinforce instances of the insulin-resistant state (Boden 1997; Goodpaster et al. 2000; McGarry 2002; Maciel et al. 2013; Seppälä-Lindroos et al. 2002). In addition to this, different pathophysiological mechanisms have also attributed the role of inflammatory parameters like generation of free radicals and its abrupt signaling pathways which elevate consequences of oxidative stress, apoptosis, and significant destruction of β-cells. Inflammatory mediators and free radicals trigger different negative regulations at the molecular level, thereby causing a cascade of downstream signaling pathways involved in insulin receptor activity. These detrimental effects arise due to a significant cluster of abnormalities like hyperglycemia, hyperinsulinemia, hypertriglyceridemia, activation of the inflammatory markers, and substrate imbalance which lead to several macrovascular and microvascular complications like diabetic retinopathy, neuropathy, nephropathy, myocardial infarction, etc. It also leads to the development of various potential cardiovascular events causing subsequent heart failures in the later stages, peripheral vascular disease, ischemic heart disease, and several cerebrovascular diseases seen in diabetic patients. Also, due to severe hyperglycemic conditions, the role of various oxidative derivatives such as advanced glycation end products comes into play which is further categorized as the possible risk factor in inducing β-cell injury (Baynes and Thorpe 1999; Dandona et al. 1996).

Taking this into account, the additional theory of prediabetes as a prior and existing phenomenon with T2DM has provided some insight into the prognosis theory. Prediabetes is a state of diverse entity involving impaired glucose tolerance and impaired fasting glucose levels, characterized by insulin insensitivity. It develops into a transitional state of hyperglycemia with glycemic parameters above normal levels but below diabetes-producing threshold. It is found to be associated with dysfunction of cardiac autonomic activity which is reflected by decreased heart rate variability and altered parasympathetic modulation of the heart (Tabák et al. 2012). According to several epidemiological studies conducted by the International Diabetes Federation (IDF), 91% of 415 million people have type 2 diabetes mellitus which comprise 8.8% of the total population worldwide (Abraham et al. 2015). This estimate is expected to rise to 642 million by 2040. In India, the prevalence has increased by 64% according to a report by the Indian Council for Medical Research, in November 2017 (Abraham et al. 2015). The current diabetic strain in India is about 49% of the world's total diabetic cases and which is about to leap to 134 million by 2025.

These epidemiological data have produced some significant information on the ongoing prevalence and incidence of diabetes complications in different populations which also predict the association of different risk factors determining susceptibility to diabetes complications. The management of diabetes, as well as its numerous related complications, implies for the use of different treatment modalities like oral hypoglycemic agents to control glycemic levels, exogenous insulin administration to regulate glycemic homeostasis and for additional physiological benefits, traditional herbal agents, and change of diet with distinguished and intense lifestyle modifications.

However, with the rising advent of disgruntled outcomes of higher treatment costs and potential side effects, existing conventional therapies and modern medicines complex the prevailing situation of treatment. These situations have directed towards the approaches and use of traditional natural products by making it more patient-compliant along with potential therapeutic benefit as compared to existing allopathic formulations and limited side effects as compared to synthetic medications. The mentioned phytoactive agents provide an effective and rational alternative in the treatment regimen of diabetes. The abundant content, potential treatment modalities, and less adverse effects have given rise to the demand for use of these agents in the present scenario. The compounds have shown increased therapeutic benefits as antidiabetics either as a single agent or in combination with multiple extracts which in turn reduces the extent of drug loading in patients. This chapter summarizes these different types of herbal and natural phytochemicals that are being used and proved to be beneficial as the line of treatment of diabetes. It focuses on the potential mechanisms of these agents.

49.2 An Emerging Detrimental Outlook of Key Factors in Play

The multiple etiologies of diabetes mellitus have designated it to be a distinct and complex metabolic-endocrine clinical manifestation. The convergence of several malfunctions arising due to persistent hyperglycemia, altered metabolic pathways, and dysregulation in the levels of lipids, carbohydrates, and proteins thereby causes a significant increase in the chances of vascular and cerebrovascular complications.

The deranged state of hyperglycemia causes more complications due to prolonged exposure of tissues to elevated concentrations of glucose and reduction in the conversion of glucose from glycogen in the liver which is developed in due course of time. The associated effects due to insulin resistance in the physiological structures lead to the development of the associated complexities such as increased hepatic glucose production, decreased peripheral glucose utilization, and impaired glucose tolerance which is induced by reduction in glucose-responsive insulin secretion, in the early phase, and a decrease in additional insulin secretion after meals augmenting postprandial hyperglycemia (Gastaldelli et al. 2004).

Additional pathogenic mechanisms have also come into play in recent years which involve the triumvirate, disharmonious quartet, and the ominous octet structures which primarily comprise the malfunctioning of the liver, muscle, and β-cell. The liver is the main target organ of insulin sensitivity and plays a key role in maintaining normal glucose levels by dynamically regulating balances between glycogenolysis and gluconeogenesis (Gastaldelli et al. 2004). However, these hepatic effects of insulin are greatly opposed due to rising insulin resistance and systemic FFA levels (Kashyap et al. 2003). This includes significant adipocyte insulin resistance leading to accelerated lipolysis due to the antilipolytic effect of insulin, reduced incretin secretion/sensitivity as a result of incretin hormone deficiency like GLP-1 and GIP, increased glucagon secretion, and enhanced glucose reabsorption.

Alterations in the levels of insulin and glucagon secretions remarkably affect the lipid, ketone, and protein metabolism. GLP-1 and GIP are the hormones that augment insulin secretion and inhibit glucagon secretion which in turn produce beneficial effects in regulating insulin functions and normal homeostasis (Fujioka 2007). In T2DM, increased glucagon sensitivity plays a key role in the maintenance of the levels of basal HGP. The elevated concentration of glucagon due to increased pancreatic α-cell secretion enhances HGP and predominantly aggravates hepatic insulin resistance. The response of the diabetic kidney to the predisposed hyperglycemic condition further enhances glucose reabsorption and thereby contributes to the consequences of glucose intolerance. A strong correlation thus exists between the increase in hepatic glucose production and the increase in plasma glucose concentration in T2DM (Baron et al. 1987; McGarry 2002).

The dysfunctional structures produce consequences by altering fat topography due to elevated plasma FFA levels and ectopic fat. This rise in the levels of plasma FFA further stimulates gluconeogenesis, induces hepatic and muscle insulin resistance, and impairs insulin secretion in genetically susceptible individuals. These FFA-induced disturbances are designated as the state of lipotoxicity. Dysfunctional fat cells produce excessive amounts of insulin insensitivity, which in turn leads to a rise in the content of inflammatory and atherosclerotic cytokines, and fail to secrete normal amounts of insulin-sensitizing adipocytokines. Insulin resistance causes the diminished capacity of the enlarged fat cells to store fat. This malfunction leads to overflowing of the lipids into the muscle and liver when adipocyte storage capacity is exceeded, thereby augmenting potential actions of muscle and hepatic insulin resistance and hence impairing insulin secretion (Gastaldelli et al. 2004; Kashyap et al. 2003).

Neurotransmitter dysfunction in CNS plays a significant role in the etiology of metabolic disorders. The pathophysiologic concept of ominous octet describes the converging and detrimental effects under long-term hyperglycemic state and suggests neurotransmitter dysfunction in the brain contributing to the progression of the disease (Matsuda et al. 1999). The compensatory mechanism causes an increase in insulin secretion in response to β-cell insulin resistance. The pathogenesis of obesity-linked insulin resistance and T2DM also involves various chronic and low-grade inflammation as well as immune reactions. Persistent hyperglycemia significantly stimulates oxidative stress by the autoxidation of glucose in the presence of several transition metals and hence causes the generation of reactive oxygen species (ROS) during the process of glycation (Sakurai and Tsuchiya 1988). It induces vascular injury through activation of complex overlapping pathways leading to the formation of advanced glycation end products following activation of protein kinase C and generation of ROS. The activated inflammatory markers are designated as some of the major factors contributing to developing macro- and microvascular complications. These activated markers and elevation in the levels of pro-inflammatory cytokines interfere with blood glucose level and pancreatic β-cell function. It has already been suggested that a rise in oxidative stress is further attenuated in systemic inflammation, impaired glucose utilization in peripheral tissues, endothelial dysfunction, and impaired pancreatic β-cell function (Brownlee 2001; Ihara et al. 1999; Prentki and Nolan 2006). Outlook of multiple metabolic anomalies involved in progression of type 2 diabetes mellitus is given in Fig. 49.1.

49.3 Type 2 DM and Outline Reviews of Major Inimical Pathways Involved

49.3.1 Insulin-Resistant State

The key pathophysiologic defect of insulin resistance under the prevailing condition of hyperglycemia gives rise to various macrovascular and microvascular complications. It causes an imbalance in the levels of metabolic as well as growth

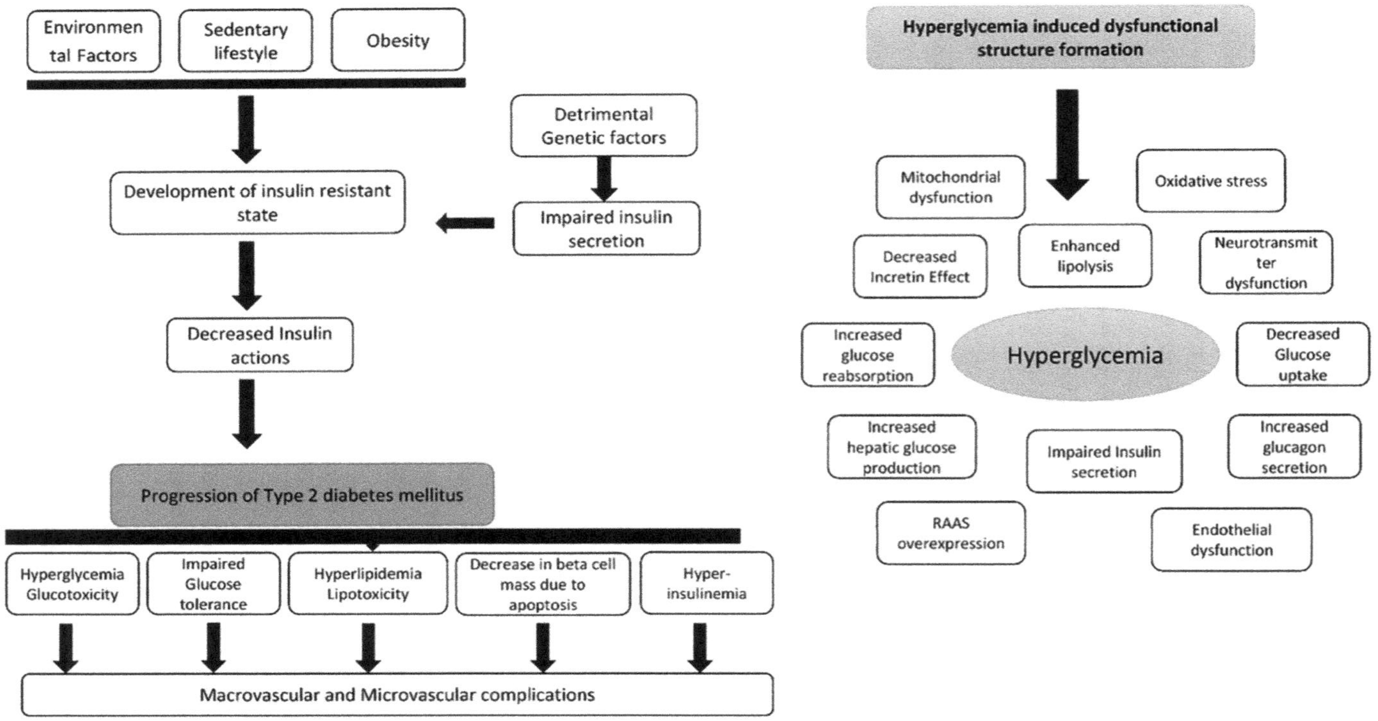

Fig. 49.1 Outlook of multiple metabolic anomalies involved in progression of type 2 diabetes mellitus

effects of insulin. This also gives rise to modulation of the insulin signaling pathway by causing impairment in the PI3K signaling pathway and stimulation of Akt by increased phosphorylation of IRS-1 serine residue (Jia et al. 2016). Excess nutrient content as in insulin-resistant state enhances phosphorylation in mTOR/S6 kinase 1 pathway which in turn promotes insulin resistance in the heart as well, thus inducing myocardial fibrosis and hypertrophy. The impairment of the PI3K signaling pathway causes a reduction in the levels of glucose uptake and increases intracellular Ca^{2+} levels due to reduced Ca^{2+} ATPase activity which then leads to abnormalities of Ca^{2+} handling and subcellular remodeling (Kim et al. 2012a, b). Persistent hyperglycemia also gives rise to the generation of superoxide radicals due to the induction of oxidative stress which then leads to cellular DNA damage and initiation of inflammation at the cellular level. In diabetic conditions, there is an abrupt change in the utilization of substrates such as glucose and fatty acids. It leads to increased metabolism of free fatty acids and downregulation of GLUT-4 transporter expression causing a reduction in glucose uptake mechanism and decreased glucose utilization which alters the rate of glycolysis and glucose oxidation. The progressive hyperglycemia and induced glucotoxicity activate the process of protein glycation reaction and lead to rising levels of AGEs/RAGEs which contribute to the detrimental inflammatory expression of genes involved in MAPK and JAK pathways and hence promote inflammation and fibrosis of the vascular tissues. It also subsequently activates aldose-reductase pathway, hexosamine pathway, protein kinase C, and mitogen-activated protein kinases with an elevation in expressions levels of various growth factors such as platelet-derived growth factor, insulin-like growth factor, and vascular endothelial growth factor which causes a cascade of activated transcription events followed by intracellular signal transduction which leads to increased oxidation of lipids, proteins, and nucleic acids (Jia et al. 2016; Kim et al. 2012a, b).

49.3.2 Lipotoxicity

Change in substrate utilization in diabetes also involves the metabolic condition of hyperlipidemia. It takes place under the metabolism shift of free fatty acids, and hence lipid accumulation predominates which is also known as the state of lipotoxicity. It is due to the elevated levels of toxic lipid metabolites such as fatty acyl-coenzyme A, diacylglycerol, and ceramide which produces a negative feed back loop of metabolic decompensation and further contributes to β-cell failure and worsening of the insulin-resistant state (Atkinson et al. 2003). Ceramide is a potent activator of inflammation which can thereby activate the JNK pathway and NF-κB/IKK, producing favorable conditions to insulin resistance. Increased oxidation of free fatty acids causes elevation of ROS levels and promotes myocardial inefficiency and damage. It also modulates multiple gene transcription involved in PPAR-α expression. High levels of TG are a potential predictor of hyperglycemia-induced cardiac events. It produces a compensatory relationship between insulin resistance, glucose intolerance, and central adiposity. It causes transcription of DNA and production of several inflammatory and atherosclerosis-provoking cytokines such as TNF-α, IL-6, and monocyte chemotactic protein-1 by NF-κB activation and hence further increases AGE/RAGE signaling causing subsequent cardiac fibrosis (Baynes and Thorpe 1999; Boden 1997; Kashyap et al. 2003; Unger 1995).

49.3.3 Dysfunctional RAAS

In diabetes, there exists a predominant role of the renin-angiotensin-aldosterone system. Increased expression of RAAS and overproduction of angiotensin II (AGT II) produce different anomalies such as increased vascular resistance and arterial pressure due to elevated inflammation and oxidative stress under the insulin-resistant state. Impairment in insulin signaling is potentiated due to the activation of RAAS and increased levels of mineralocorticoids through activation of the mTOR/S6 kinase 1 signaling pathway. These overexpressions produce maladaptive immune responses by triggering leucocyte adhesion, cytokine expression, and macrophage infiltration (Miller et al. 1996; Miller 1999).

49.3.4 Maladaptive Immune and Inflammatory Responses

Over time, hyperglycemia has been recognized as the predominant oxidative stress-inducing clinical condition which causes an imbalance between the cellular reactive oxygen species generation and action of natural counteracting antioxidant mechanisms. The abnormalities in substrate utilization owing to severe glucotoxicity and lipotoxicity alter the associated expression of various pro-inflammatory cytokines, TNF-α, and interleukins. These overexpressions further interfere with endothelial relaxing factor production which in turn causes significant modulation of cardiovascular homeostasis under substrate imbalances. Cytokine-mediated inflammation possesses the potential to reduce the activity of peroxisome proliferator-activated receptors (PPARs) and hence accelerate fat cell death and inflammation on account of central adiposity (Jia et al. 2016; Ma et al. 2009; Sell et al. 2012).

49.4 Current Natural Therapies in the Management of Diabetes Mellitus

With the rising advent of persistent limitations and adverse effects offered by the ongoing conventional treatment strategies, the need of the hour is directed towards therapeutic benefits offered by phytoconstituents. Diabetes, a multidimensional metabolic syndrome with a principal defect in glucose and lipid profiles in individuals, has become very difficult to manage. Moreover, in addition to the series of undesirable adverse effects and lack of efficacy offered by the various conventional therapies, it is evident that all current antidiabetic medications elevate systemic complications and glycemic parameters rather than control and cure it (Zhang and Moller 2000). Thus, the treatment regimen should emphasize acting on multiple metabolic pathways which can reduce the chances of side effects.

Insulin secretagogues are often associated with weight gain, hypoglycemia, hepatotoxicity, and subsequent β-cell injury (Mozaffari et al. 1989). Biguanides also lead to weight gain, kidney toxicity, and an impaired cardiac functioning recovery over time causing potential induction of myocardial fibrosis (Forcheron et al. 2009; Verma and McNeill 1994). The risk correlation between PPAR-α agonists and heart failure is greatly dependent upon persistent plasma lipid reduction (Giles et al. 2008; Hemmeryckx et al. 2013; Komajda et al. 2010). Significant gastrointestinal problems such as diarrhea, flatulence, sour stomach, belching, nausea, vomiting, and indigestion result due to persistent use of α-glucosidase inhibitors and incretin-based drugs over time. DPP-4 inhibitors and GLP-1R agonists are also significantly associated in producing higher chances of cardiac dysfunction, heart failures as a matter of inconsistent cardiac function recovery, and vascular remodeling (Hemmeryckx et al. 2014). In addition to these theories, several other predominant side effects also result in the virtue of hypoglycemia causing additional serious cardiovascular events. In the context of this present scenario, several complementary and alternative natural-based treatment regimens and constituents have come into play which holds the potential to provide better glycemic control as well as reduce chances of metabolic complications and extent of drug loading. These bioactive agents serve as potential therapy in managing diabetes and its related complications possessing antidiabetic, anti-inflammatory, and metabolic pathway agonists, inhibitors, and modulators (Lu et al. 2016; Mirza and Siddiqui 2014). This chapter aims to summarize and highlight the mechanisms of such natural products along with the bioactive phytochemical constituents implicated in diabetes treatment.

49.4.1 Flavonoids

This group of plant metabolites is comprised mostly of polyphenolic compounds. These are widely found in fruits, vegetables, herbs, etc. They contain hydroxylated phenolic structure having structure-dependent activities (Heim et al. 2002). Chemically, flavonoids contain a 15- carbon skeleton with two phenyl rings and a heterocyclic ring. These are designated as one of the beneficial classes of compounds having significant hypoglycemia potential and promote antioxidant activity as well as modulate several cell signaling pathways. These are widely classified as anthocyanidins, catechins, flavonols, flavones, flavanones, flavan-3-ols, isoflavonoids, etc. (Heim et al. 2002; Mohan and Nandha Kumar 2014). These are typically found in the form of glycosides or acyl glycosides, which can be water-soluble and can get accumulated in cell vacuoles. Various approaches have been made so far in utilizing these natural sources in diabetes treatment which help in regenerating pancreatic cells, enhancing insulin release, and increasing intracellular calcium levels, thus attenuating complications of diabetes mellitus (Mohan and

Nandha Kumar 2014). The various flavonoid compounds, their modes of actions, and biological diversities are discussed below.

Rutin: Chemically, it belongs to a polyphenolic bioflavonoid group. It is found in many fruits, plants, etc. It improves insulin secretion, helps to restore glycogen content by reducing chances of conversion to glucose, causes a reduction in the levels of free radicals, inhibits the formation of AGE/RAGE production, decreases pro-inflammatory cytokine levels, reduces glycosylated hemoglobin level, and restores normal hepatic antioxidant content (Sharma et al. 2013). It has already been evident that rutin tends to reduce fasting and non-fasting glucose levels with a prominent reduction in serum TG, VLDL, and LDL levels and increased HDL levels in diabetic models (Ghorbani 2017). It has shown to improve glucose insensitivity in high-glucose content states. As given by several mechanistic approaches, it causes activation of GLUT-4 glucose translocation pathway, hence augmenting glucose uptake in muscles (Hunyadi et al. 2012). It also participates in the upregulation of the hexokinase enzyme and hence modulates the glycolysis pathway. It stimulates PKC and MAPK pathways and further acts on tissue glucose uptake. In addition to this, it is found that rutin helps in glomerulosclerosis of diabetic nephropathy significantly by decreasing cell content of G1 phase and inhibition of the expression of Smad 2/3, laminins, type IV collagen, TGF-β, and mRNA level, thus augmenting renal protective actions (Fernandes et al. 2010; Hsu et al. 2014; Stanely Mainzen Prince and Kannan 2006).

Naringin: It is a major flavanone glycoside that is mainly found in citrus fruits and vegetables, providing antidiabetic, antioxidant, and anti-inflammatory properties in abundance. It acts as a promoter for carbohydrate metabolism and an immune system modulator. It attenuates anti-inflammatory and anti-fibrotic activities providing renal protective actions in diabetic nephropathy (Bharti et al. 2014). It has also shown to reduce plasma glucose levels and blood urea nitrogen levels with a predominant increase in insulin levels. It causes inhibition in the α-glucosidase effect and activates AMPK. It is a significant DPP-4 inhibitor and can elevate hepatic glucose utilization to prevent persistent hyperglycemic state. It has shown to decrease glucose levels, HbA1C, NO, TNF-α, IL-6, mRNA expression, and protein production in the kidney, thus providing a renal protective action as in the case of diabetic nephropathy (Parmar et al. 2012; Jung et al. 2004). It causes an increase in the levels of serum insulin, glutathione, and vitamins C and E which suggests that naringin provides protective antioxidant effects and also suppresses pro-inflammatory cytokine productions in high-fat-diet- and STZ-induced diabetic models. It also diminishes insulin resistance and pancreatic β-cell dysfunction, through upregulation of PPAR pathways in diabetic rats. In addition to these treatment theories, it also tends to show predominant protection and improvement in diabetes-induced memory dysfunction by enhancing cognitive behavior and certain biochemical changes. Naringin also improved cholinergic function by inhibition of elevated cholinesterase activity (Mahmoud et al. 2012).

Mangiferin: The extensive studies of various in vivo and in vitro diabetic models have emphasized that mangiferin which is a type of C-glucosyl xanthone (chemically as 1,3,6,7-tetrahydroxyxanthone-C2-β-D-glucoside) holds some bioactive potential by protecting against various complications related to diabetes mellitus. It is predominantly isolated from mango tree and exerts its milieu of pharmacotherapeutic effects as an antidiabetic, anti-inflammatory, antioxidant, and regulator of lipid metabolism, immune-modulation and cardio-protection (Dar et al. 2005; Duang et al. 2011; Guha et al. 1996). Administration of mangiferin in high-fat and STZ-induced diabetic rats have significantly shown improved insulin resistance, enhanced β-cell function, and also reduced serum lipid profiles such as TG, LDL, VLDL, etc., thus decreasing the atherogenic index (Saleh et al. 2014). It has also been demonstrated that modulation of cell cycle genes by mangiferin can augment β-cell proliferation and pancreatic islet regeneration (Wang et al. 2014). It also provides protective mechanism against hyperglycemia-induced oxidative stress and inflammation by inhibiting activation of AGE/RAGE and MAPK, c-Jun N-terminal kinase (JNK), and p38 molecular pathways (Suchal et al. 2017). It reduces chronic inflammatory cytokine levels such as TNF-α and IL-18 and also decreases expressions of myocardial enzymes such as LDH and creatine kinase MB which in turn reduces left ventricular myocardium stress helping in diabetic cardiomyopathy (Suchal et al. 2017). In addition to these predominant effects, it also provides a wide range of beneficial effects, thus designating it to be a potential drug candidate in the management of T2DM complications.

Quercetin: It is a flavonol which is predominantly found in many vegetables, fruits, grains, leaves, etc. It produces renal protective effects by providing action against glomerulosclerosis of diabetic nephropathy. It has shown to reduce ubiquitin and NF-κB p65 expressions and decrease the cell percentages in G0/G1 phase, Smad 2/3 expression, laminins and type IV collagen, and TGF-β1 mRNA level in the high-fat-diet STZ-induced rat model. It also improved cognitive deficits in diabetes-induced memory dysfunction state and ameliorated neuronal damages. It also tends to improve hyperglycemia-induced cellular injury and activated Akt/cAMP responses. It exhibits its antidiabetic potential by augmenting several significant antioxidant properties, decreasing lipid peroxide levels, and reducing renal glucose reabsorption by inhibiting GLUT-2. It also has shown to recover cell proliferation by blocking tyrosine kinases (Chen et al. 2012a; Chen et al. 2016; Eid et al. 2015; Maciel et al. 2013; Wang et al. 2013b).

Baicalein: A type of flavonoid which is mainly found in the stem bark of *Oroxylum indicum* and roots of *Scutellaria baicalensis*. It was shown to improve glucose intolerances, hyperglycemic states, and increased insulin levels by improving pancreatic β-cell survival as given by several diabetic models. Researchers found that baicalein also improved the related complications of this metabolic-endocrine disorder by activating or upregulating AMPKα2 selectively and reduced inflammation and insulin resistance by increasing phosphorylation of IRS-1 and Akt. It also causes a reduction in fatty acid synthesis and increased mitochondrial β-oxidation by suppressing and dephosphorylating extracellular signal-regulated kinase (ERK), JNK, and NF-κB. It potentially augmented its antidiabetic potential by modulating PI3K/Akt/GSK3 signaling pathway, inhibited levels of oxidative stress factors, and reduced inflammatory markers levels by scavenging free radicals (Ahad et al. 2014a;

Fu et al. 2014; Pu et al. 2012; Qi et al. 2015; Stavniichuk et al. 2011; Yang et al. 2009).

Silymarin: It is a type of flavonoid which is mainly derived from the milk thistle plant *Silybum marianum*. Chemically it is a flavonolignan containing a C-25 backbone. It is generally a complex mixture of silybin, silydianin, silychristin, silybin A, silybin B, isosilybin A, and isosilybin B. It causes a significant reduction in blood glucose levels, glycosylated hemoglobin, serum creatinine, and uric acid levels which further supports it to be a nephroprotective agent and proves its effectiveness in various T2DM models (Sheela et al. 2013). It acts as a potent free radical scavenger, as a hepatoprotective agent, and also as a modulator of elevated lipid levels by inhibiting lipid peroxidation processes. It mediates its antioxidant properties by protecting against increased oxidative stress by reducing the levels of glutathione reduction process (Zhao et al. 2015). It is also evident that silymarin inhibits TNF-α, NF-κB transcription, expression of IL-6, and iNOS (Dehmlow et al. 1996; Zhao et al. 2015). In addition to this, isosilybin, a constituent of silymarin, is reported to possess activity as a PPAR-γ agonist (Pferschy-Wenzig et al. 2014).

Chrysin (5,7-dihydroxyflavone): It is a type of naturally occurring flavone which is usually found in flowers of *Passiflora caerulea*, *Passiflora incarnata*, and *Oroxylum indicum*. Chemically it consists of ubiquitous 15-C flavone backbone with fused phenyl rings. It possesses multiple therapeutic benefits such as antioxidant, immunomodulatory, and pro-apoptotic characteristics. Molecular mechanism of chrysin underlies its diverse pleotropic effects involving modulation of several cell signaling pathways. It has also shown to induce nephroprotective effects in the high-fat-diet STZ-induced T2DM rat model. It produces anti-inflammatory effects by targeting TNF-α pathway, reduced pro-inflammatory cytokine and interleukin levels, and suppressed TGF-β factor, fibronectin, and type IV collagen expressions in the kidney all of which are found to be helpful in nephropathy cases (Ahad et al. 2014b). It also have shown some promising antihyperlipidemic and hepatoprotective effects due to an evident decrease in serum lipid profiles and its ability to modulate TNF-α expression and block TNF-α-converting enzyme activity (Hermenean et al. 2017).

Hesperidin and neohesperidin: These are generally isolated from several species of genus *Citrus* like *Citrus aurantium*. It has shown antidiabetic potential with significant antioxidant and anti-inflammatory effects and hence can attenuate hyperglycemic complications. It plays an important role in activating AMPK pathways, thus modulating hypoglycemic and hyperlipidemic effects under lipid normalization. It downregulates free radical generation and pro-inflammatory cytokine release. It alters glucose-metabolizing enzyme activities, causes a reduction in TBARS levels, and elevates LDH activities (Jia et al. 2015; Mahmoud et al. 2012).

Genistein: It is a type of isoflavone flavonoid, found in species of *Sophora subprostrala* and *Genista tinctoria*. It improves glucose intolerance states with a significant increase in blood insulin levels. It significantly reduces high-glucose-induced monocyte adhesion, monocytechemotactic protein-1 (MCP-1), and IL-8 production, thereby reducing instances of inflammation. Treatment with genistein has proved to be helpful in diabetic retinopathy. Genistein is a tyrosine kinase inhibitor that suppresses the release of TNF-α and inhibits ERK and p38 MAPK pathways which are involved in inflammatory mechanisms of diabetic retinopathy (Gupta et al. 2015; Liu et al. 2006; Vinayagam and Xu 2015).

Diosmin: It is a type of flavonoid which is typically found in citrus plant origins, isolated from *Scrophularia nodosa.* It is also obtained from the dehydrogenation of hesperidin. It holds potential antidiabetic activities which are predominantly mediated by a significant decrease in HbA1C levels and increases the efficiency of glutathione peroxidase, hepatic hexokinase, and G6PD enzymes causing a reduction in plasma glucose levels and increases in insulin levels. Treatment with diosmin showed improved lipid profiles in diabetic rat models (Jain et al. 2014).

Icariin: It is a flavonoid glycoside that is predominantly found in various species of *Epimedium.* It provides a protective action against high-glucose-induced type IV collagen and fibronectin accumulation by inhibiting transforming growth factor-β downstream cascades, G-protein estrogen receptor, and extracellular signal-regulated kinase (ERK)1/2 signaling pathway which in turn reduces Smad 2/3 and ERK1/2 phosphorylation (Li et al. 2011).

Isoangustone A: It is an isoflavone that is present in licorice. It possesses the ability to inhibit renal fibrosis and inflammation induced by high glucose content. It suppresses high-glucose-induced overexpression of MMP-1, MMP-2, and TGF-β1-SMAD-responsive signaling pathway. It also reduces hyperglycemia-induced inflammation in response to suppressing intracellular cell adhesion molecule-1 (ICAM-1) level and monocyte chemoattractant protein-1 (MCP-1) and mRNA expression (Li et al. 2011).

Isoliquiritigenin: It is obtained from *Glycyrrhiza glabra*. It is a type of chalcone flavonoid. It has been evident that it holds the ability to inhibit hyperglycemia-induced fibrosis by reducing the formation of type IV collagen. It reverses collagen secretion and connective tissue growth factor expression. It augments the degradation of the mesangial matrix and reduces its accumulation by diminishing the TGF-β1-SMAD signaling transduction pathway (Li et al. 2010).

Kaempferol: It is a type of flavonol which is present in *Gingko biloba*, grapefruits, and berries. It holds potential for inhibition of programmed cell death such as in apoptosis and necrosis with a reduction in the activity of caspase-3 in β-cells, improves cAMP signaling pathway, and elevates insulin synthesis and secretion with an improvement of the insulin-resistant state. In addition to this, its activity is also associated with the regulation of antioxidant properties by decreasing pro-inflammatory cytokine levels through the mediation of AMPK pathway activation (Häkkinen et al. 1999; Nirmala and Ramanathan 2011).

Fisetin: This type of flavonoid induces antidiabetic activity by inhibiting gluconeogenesis by decreasing cytosolic NADH/NAD levels and mitochondrial pyruvate. It causes a reduction in glycogen breakdown, plasma glucose, and glycosylated hemoglobin levels. It suppresses the expression of gluconeogenic genes like phosphoenolpyruvate carboxykinases. It induces insulin release and provides protective action against inflammation by modulating nuclear factor kappa B p65 unit, IL-1β, and NO expressions (Constantin et al. 2010; Prasath et al. 2014).

Morin: It is found in many medicinal herbs such as *Prunus dulcis*, *Cholorophora tinctoria*, and *Psidium guajava*. It provides antidiabetic activities with a prominent reduction of insulin resistance and oxidative stress factors. It normalizes levels of lipids and lipoproteins and reduces TNF-α and inflammatory cytokine levels. It also induces recovery from insulin-resistant states and leptin insensitivity and reduces instances of hyperlipidemia and lipid accumulation, thereby proving it helpful in diabetic complications (Abuohashish et al. 2013; Vanitha et al. 2014).

Luteolin: It is primarily found in carrots, peppers, cabbage, apple, and many other fruits and vegetables. It acts by potentiating insulin actions and secretion; causes transcriptional activation of PPARs; reduces circulating levels of pro-inflammatory cytokines, MCP-1, and resistin; and increases adiponectin expression (Ding et al. 2014b; Liu et al. 2014b; Miean and Mohamed 2001).

Apigenin: It is a flavone flavonoid that is ubiquitously found in fruits of citrus origins and vegetables. It provides antioxidant properties, upregulates GLUT-4 translocation, and inhibits detrimental lipid synthesis and β-cell preservation over time. It improves blood glucose, HbA1C, and insulin-resistant states (Ohno et al. 2013; Zhang et al. 2014).

Eriodictyol: It is abundantly extracted from lemon fruits. It helps to suppress oxidative stress and causes upregulation of mRNA expression of PPAR-γ. It also leads to the reduction in the levels of TNF-α, ICAM-1, VEGF, and eNOS levels, thus proving its anti-inflammatory actions (Zhang et al. 2012).

Anthocyanin (cyanidin-3-O-β-glucoside): It is a type of anthocyanin which is found in various types of fruits, grains, and vegetables. It possesses an anti-inflammatory potential with a protective action on adiponectin expression, thus proving it beneficial in hyperglycemia-induced endothelial dysfunction (Liu et al. 2014a, b).

Catechin: A type of flavonol which is found in high concentrations in grapes, berries, dark chocolate, tea, cocoa, etc. Catechin is effective in suppressing NF-κB system activation by inhibiting pro-inflammatory cytokine secretion and hence reduces instances of chronic inflammation which are potentiated in the case of T2DM, obesity, and atherosclerosis (Stofkova 2009).

Tangeretin: It is abundantly found in citrus fruits, it reduces total cholesterol, resistin, leptin, IL-6, and MCP-1 levels. It also tends to decrease glycosylated hemoglobin levels and elevated insulin levels and enhance glucose metabolism pathways in the hepatic system. In addition to this, it reduces insulin-resistant factors in adipocytes, thus helping in minimizing associated complications (Kim et al. 2012b).

Daidzein: It is a type of isoflavone flavonoid which is predominantly present in soybeans, fruits, nuts, etc. It tends to improve glucose and lipid metabolism with a significant improvement in insulin insensitivity and also elevates AMPK phosphorylation in skeletal muscles (Cheong et al. 2014).

49.4.2 Polyphenols

These are also known as tannins which contain polyhydroxy phenolic molecule and are predominantly found in many natural products such as berries, legumes, spices, herbs, etc. It can be further categorized into three major classes such as hydrolysable tannins, non-hydrolysable or condensed tannins, and phlorotannins. The therapeutic benefits of polyphenols include antidiabetic, anti-inflammatory, and anti-fibrotic effects along with modulations of metabolic pathways. These agents are also capable of mediating cell cycle arrest and elevating plasma antioxidant capacity which takes place by its capability of accepting reactive oxygen species, thus forming stable phenoxyl complexes. These associated chemical modifications thereby lead to decreased lipid peroxidation and protein oxidations as well as reduce instances of programmed cell death mechanisms. This reduction in oxidative stress and inflammation ultimately helps in minimizing complications related to T2DM. Phenol-rich compounds, their mode of action, and different biological species of phenolic origin are discussed below.

Epigallocatechin gallate (EGCG): A very potent and bioactive polyphenol that is abundantly present in green tea. It holds the potential to regulate glucose uptake mechanism in skeletal muscle cells by inhibiting gluconeogenesis via the PI3K pathway and also suppresses hepatic gluconeogenesis taking place in isolated hepatocyte cells. It also inhibits insulin signaling in HepG2 cells via the AMPK pathway. In addition to antidiabetic activity, it also caused the induction of lipid-reducing capabilities in various high-fat-diet-induced T2DM models. The anti-inflammatory actions of ECCG include reduction of the free radical generation with its improvement of glucose intolerance state, enhance insulin secretion, and also upregulate pancreatic cell regeneration. It promotes GLUT-4 translocation and also reduces JNK phosphorylation which further improve insulin resistance in adipocytes. It increases PPARγ expression in high-fat-diet-induced diabetic animals (Collins et al. 2007; Jung et al. 2008). The rich polyphenolic compounds obtained from EGCG thus exert anti-obesogenic properties by decreasing body weight and reduce content of adipose mass and plasma lipid profiles. It also regulates numerous mechanisms involving improvement in fat oxidation process in adipose tissues and skeletal muscles and suppresses lipogenesis and dietary fat absorption (Bose et al. 2008; Chen et al. 2009; Klaus et al. 2005; Lu et al. 2012; Wolfram et al. 2006).

Resveratrol: Resveratrol has been extensively used in regulating multiple metabolic pathways in various animal models owing to its pleitropic effects on the metabolic system. This is extensively found in grapes, peanuts, etc. It has shown to improve the insulin-resistant state in high-fat-diet-induced rats. Long-term administration of resveratrol causes improvement in visceral insulin sensitivity in diet-induced animal models. It also ameliorates mitochondrial β-oxidation and also promotes structural biogenesis to mitochondria to elevate fatty acid oxidation and further decrease lipid content of the skeletal muscle. It activates the SIRT1 pathway to increase phosphorylation of the insulin receptor and hence adjust protein levels of insulin receptor substrate-1. Resveratrol leads to reduction of hyperinsulinemic state and improved pancreatic cell content level to normalize the function of the pancreatic islets which helps in maintaining blood glucose and lipid levels. As a potent free radical scavenger, it reduces lipid peroxidation which is induced by oxidative stress. It also attenuates the release of pro-inflammatory cytokines by causing downregulation of the NF-κB pathway (Chen et al. 2012b; Kim et al. 2011; Zheng et al. 2013).

Curcumin: It is isolated from the species of *Curcuma longa*. It exhibits several therapeutic benefits such as antioxidant, anti-tumorigenic, anti-inflammatory, etc. It significantly reduces blood glucose level and glycosylated hemoglobin and also improves insulin insensitivity. It inhibits lipid peroxidation and lysosomal enzyme activity and hence reduces the instances of pro-inflammatory cytokine content thus helpful in decreasing inflammation. It also activates nuclear factor erythroid 2-related factor (Nrf2), PPARγ, and lipoprotein lipase. In addition to these important effects, another potent analog, namely, bisdemethoxycurcumin, is also used in treating complications of T2DM due to its ability to inhibit pancreatic α-amylase. Hyperglycemia-induced free radical generation stimulates NF-κB activation and causes an increase in the expression of ICAM-1 and induces vascular inflammation. However, this stimulation further develops the consequences of diabetes-associated atherosclerosis. In light of this, it has been seen that curcumin supplementation reduces diabetes-induced vascular inflammation by decreasing ROS overproduction and ICAM-1 overexpression (El-Azab et al. 2011; He et al. 2012; Soetikno et al. 2011).

Oligonol: It is a low molecular weight polyphenol that is isolated from lychee fruit and also green tea. It acts to reduce renal glucose levels and the generation of superoxide radicals. It also leads to the improvement of expression of anti-apoptotic β-cell lymphoma protein 2 (Bcl-2) and pro-apoptotic proteins such as Bcl-2-associated X protein, cytochrome c, and caspase-3. It also involves the reduction of ROS generation, lipid peroxidation, and serum TG levels and thus attenuates the formation and expression of AGEs (Park et al. 2014).

Salvianolic acid A (SalA): It happens to be a potent polyphenolic antioxidant and ROS scavenger. It shows the reversal of hyperlipidemic conditions and the reduction of hepatic TG levels with a decrease of type I and III collagens. It also improved hepatic mitochondrial function in a high-fat-diet- and STZ-induced diabetic rats. These series of effects also proposed the theories of oxidative stress which gets reduced under the SalA application by suppressing α-SMA and TGF-β1 expression and producing mitochondrial protective effects (Qiang et al. 2014).

Eugenol: It has shown inhibition of α-glucosidase and AGE formation in HepG2 cells and HFD-induced C57BL/6 J mice. It also reduces blood glucose levels through inhibition of hepatic gluconeogenesis and modulation of the CAMKK-AMPK-CREB signaling pathway (Jeong et al. 2014).

Hydroxytyrosol: It decreases the instances of central adiposity and improves impaired glucose sensitivity and insulin intolerances in high-fat-diet-induced animal models. It also functionalizes to inhibit 3 T3-L1 cell differentiation by modulating CB1 receptor gene transcription in 3 T3-L1 cells (Poudyal et al. 2017; Tutino et al. 2016).

Polydatin: It regulates lipid and glucose metabolism by downregulating pro-protein convertase subtilisin/kexin type 9 (PCSK9) in palmitic acid-induced HepG2 cells and db/db mice as well as through activation of Akt signaling pathway in high-fat-diet- and STZ-induced rats. db/db mice models also show evident downregulation of SphK1-S1P signaling pathway upon administration of polydatin (Wang et al. 2019).

p-Coumaric acid: It ameliorates glucose tolerance, improves antioxidant status, and hence declines inflammation and inhibits apoptosis. It also causes a reduction in blood glucose level as well as gluconeogenic enzymes, which in turn modulates glucose and lipid metabolism via GLUT 2 activation in STZ-induced diabetic rats (Amalan et al. 2016).

Caffeic acid: db/db mice models have shown that caffeic acid holds the potential to attenuate hepatic glucose output which results in enhancement of adipocyte glucose uptake and improves insulin secretion and subsequent antioxidant activity (Jung et al. 2006).

49.4.3 Saponins

These phytoactive compounds generally constitute the naturally occurring primary and secondary plant metabolites. Their primary mode of action involves a defense mechanism which acts as a chemical barrier. Chemically they usually contain sugar moieties which are linked with a hydrophobic aglycone part (sapogenin). These are also called as steroid or triterpenoid glycosides. The structural diversities and potent biological profiles have designated saponins as a beneficial treatment regimen that can be used as antidiabetic, anti-inflammatory, antibacterial, antifungal, etc. Saponins exert their antidiabetic effects by potentially inducing insulin production, activate calcium-dependent AMPK pathways, modulate gluconeogenesis, and hence also regulate glucose uptake mechanisms by upregulating GLUT translocation. It has been reported that saponins can regulate PPAR expression as well as lipid metabolism in the diabetic population (Seo et al. 2015). It also possesses significant antioxidant activity owing to its amelioration of oxidative stress and inhibition of certain AGE formation. Several groups of sapogenin compounds, their respective plant species, and their mode of action are discussed below.

Panax notoginseng: This species contains geinoside saponin which possesses the ability to inhibit gluconeogenesis. It also improves PPAR expression and also regulates lipid metabolism. Another compound, namely, ginsenoside compound K, achieves hypoglycemic and insulin-sensitizing activities by causing downregulation of phosphoenolpyruvate carboxykinases. In addition to these metabolites, ginsenoside Rg3, ginsenoside Rg1, and ginsenoside Re show antihyperglycemic effect by stimulating GLP-1 and promote glucose uptake through activation of the AMPK pathway in insulin-resistant muscles. This also tends to decrease serum TG levels and increase insulin and HDL levels by stimulating p38 MAPK, ERK1/2, and JNK signaling pathways in several diabetic animal models (Lee et al. 2011).

Astragaloside IV: It is isolated from the species of *Astragalus membranaceus*. Chemically it is composed of 3-O-β-Dxylopyranosyl-6-O-β-D-glucose-pyranose cycloastragenol. Administration of astragaloside IV significantly has shown a decrease in blood glucose and TG levels and also has the ability to inhibit glycogen phosphorylase in diabetic rats. It has been also found that astragaloside also inhibited TNF-α-associated accelerated lipolysis, downregulated key enzymes involved in lipogenesis, and also caused an improvement of TNF-α-induced insulin resistance in 3 T3-L1 adipocytes (Jiang et al. 2008).

Platycodi radix: It produces platyconic acid which is a saponin and is involved in insulin-mediated glucose uptake mechanisms in 3 T3-L1 adipocytes, thereby decreasing serum glucose levels. It also enhances glycogen accumulation and reduces TG levels in the liver by inhibiting JAK phosphorylation and modulation of STAT-3 signaling cascades and causing upregulation of Foxp3 expression. In addition to this, it also holds the ability to stimulate adiponectin and PPARγ expression in adipose tissues (Kwon et al. 2012).

Diosgenin: It is isolated from the species of *Dioscorea rotundata*. It significantly increases the activity of glucose-6-phosphate and thereby modulates the insulin signaling pathway. It also decreases blood glucose levels and instances of insulin resistance with modulation of lipid profiles. In addition to this, it can also attenuate ER stress and free radical-generated oxidative damage in the pancreas (Uemura et al. 2010).

Helicteres isora: Saponins from this species tend to reduce serum lipid content and plasma glucose levels and also lead to the upregulation of fatty acid-binding protein expression and glucose-6-phosphatase enzyme with additional potentiated expression of GLUT-4 translocation (Bhavsar et al. 2009).

Polygonatum kingianum: Isolated saponins possess antihyperglycemic potential causing upregulation of GLUT-4 expression and downregulation of glucose-6-phosphate expression involved in the insulin signaling pathway (Deng et al. 2012).

Polygonatum odoratum: Its mode of action generally comprises of improvement of hyperglycemia-induced complications. It enhances glucose uptake mechanisms. It also increases the activity of superoxide dismutase enzyme with a reduction in the levels of malondialdehyde (Deng et al. 2012).

49.4.4 Alkaloids

A group of naturally occurring chemical compounds which are generally secondary plant metabolites and contain basic nitrogen atoms in their chemical structures. These are usually classified as true alkaloids, protoalkaloids, polyamine alkaloids, and cyclopeptide alkaloids. Structurally these are designated as amphipathic glycosides having one or more hydrophilic glycoside moieties which are further combined with some lipophilic triterpene groups.

Berberine: A type of benzyl tetra-isoquinoline alkaloid which holds some significant therapeutic potential against type 2 diabetic cases. It can reduce blood glucose levels, enhance insulin secretion, and decrease instances of lipid profiles which in turn can modulate insulin-resistant and glucose-intolerant states. It attenuates insulin resistance by causing activation of the AMPK pathway, elevates GLP-1 content, reduces glycosylated hemoglobin content, decreases free radical generation, suppresses inflammatory mediators levels, and reverses endothelial as well as mitochondrial dysfunction under decreased oxidative stress. In addition to this, it also showed a reduction in the activation of gluconeogenic enzymes as given in STZ nicotinamide-induced diabetic animals (Punitha and Shirwaikar 2005).

Boldine: It is a type of aporphine-benzylisoquinoline alkaloid which is isolated from the species of *Peumus boldus. Several experimentations on T2DM models such as in db/db mice have shown its potential natural antioxidant property by reducing overproduction of superoxide radicals by inhibiting expression of angiotensin II-mediated BMP4 (*Lau et al. 2013).

Neferine: It belongs to the class of bisbenzyl isoquinoline which is obtained from the species of *Nelumbo nucifera*. Treatment with neferine has shown reduced CCR5 and CCL 5 mRNA expressions. It also decreases fasting blood glucose levels and lipid profiles such as triglycerides and cholesterols but elevates HDL levels (Li et al. 2013).

Vindogentianine: It is an indole alkaloid that possesses hypoglycemic activity by causing upregulation of glucose uptake mechanisms which results in inhibition of PTP-1B, thus proving its therapeutic benefits inT2DM (Tiong et al. 2015).

Tetrandrine: It decreases the expression of miRNA-155, reduces instances of oxidative stress, and also regulates NF-κB signaling pathways which in turn shows converging actions in reducing pancreatic β-cell injury (Song et al. 2015).

Matrine: It has shown a significant reduction of glucose intolerances and plasma insulin levels in high-fat-diet-induced diabetic animals. It also causes a reduction in hepatic triglyceride content, suppresses lipogenesis, and increases fatty acid oxidation (Zeng et al. 2015).

Evodiamine: It has been found that evodiamine causes a reduction in blood glucose levels, improves glucose intolerances, and also prevents insulin resistance by inhibiting mTOR-S6K signaling and IRS-1 serine phosphorylation pathways in KK-Ay mice models (Wang et al. 2013a, b, c).

49.4.5 Terpenoids

These natural compounds constitute a structurally related diverse group which is basically secondary metabolites derived from different plant species. These are also known as isoprenoids and are modified terpenes with a series of different functional groups. Terpenoids are classified as monoterpenes, diterpenes, sesquiterpenes, and triterpenes depending on the number of carbon units present in their molecular structure which greatly modulates their variety of biological effects. Apart from their hypoglycemic abilities, they also exert hypolipidemic, anti-obesity, and antioxidant effects. They modulate insulin-resistant state by inhibiting a series of enzymes involved in glucose metabolism and hence normalize glucose homeostasis and insulin levels. They also have proven to be helpful in preventing diabetic complications of nephropathy and neuropathy due to strong antioxidant activity and inhibiting generation of AGEs.

Andrographolide: This active constituent is found in species of *Andrographis paniculata*. It possesses activity against hyperglycemia and hyperlipidemia. AL-1, a novel andrographolide derivative, has shown cytoprotective effects by augmenting antioxidant and anti-inflammatory activities. It causes a reduction of sterol regulatory element-binding protein (SREBP) expressions which in turn leads to decrease lipid accumulation in high-fat-diet-induced mice by reducing serum lipid levels (Ding et al. 2014a, b). It also has potential to improve insulin-resistant states. It attenuates extent of oxidative stress by inhibiting activation of the NF-κB signaling pathway (Chen et al. 2013). Recent advances also prove that AL-1 suppresses phosphorylation of both p65 and IκBα proteins and hence protects from high-glucose-induced oxidative damage to islet cells (Li et al. 2015).

Ursolic acid: It is a pentacyclic triterpenoid that has reduced fasting blood glucose level and plasma TG, FFA, and other lipid profiles in high-fat-diet-induced diabetic models. It also significantly reduced hepatic G6P activity and predominantly increased glucokinase activity. The modulation of GLUT-2 mRNA levels and attenuation of diabetes-induced renal fibrosis are also achieved by ursolic acid administration (Lee et al. 2014).

Abscisic acid: It can stimulate β-pancreatic cells to release insulin as well as adjust to GLUT-4-mediated glucose uptake mechanisms in addition to its rejuvenating capabilities in glucose intolerance (Magnone et al. 2015).

Genipin: It can attenuate hepatic insulin resistance with its associated effects to decrease the extent of hyperinsulinemia, high TG levels, and hepatic steatosis to reduce hepatic oxidative stress and insulin-resistant-mediated mitochondrial dysfunction (Guan et al. 2013).

Oleanolic acid: It causes a reduction in blood glucose levels and improves glucose and insulin insensitivity. It also enhances insulin signaling cascades and inhibits gluconeogenesis (Wang et al. 2013c).

49.4.6 Quinones

This group of compounds contains many diverse therapeutic properties. Their mechanism of action greatly depends on their structural variations. There are different derivatives of quinones depending on the number of carbon skeleton present such as the benzoquinones, naphthoquinones, and anthraquinones.

Emodin: Recent experimentation and advances have shown that emodin possesses the ability to reverse hyperglycemic parameters by decreasing insulin resistance and other associated metabolic complications.It reduces serum glucose levels, augments insulin sensitivity, and enhances hepatic and skeletal muscle glycogen content in KK-Ay diabetic mice models. Administration of emodin also caused reduction in pro-inflammatory cytokine levels such as IL-6 and TNF-α. It modulated PI3K/Akt pathway in hepatic tissues as well as in skeletal muscle in diabetic KK-Ay mice models (Yuping Song et al. 2017).

Rheum palmatum: Purified anthraquinone glycosides which are generally obtained from this species have shown significant antidiabetic potential with anti-inflammatory and detoxification effects in STZ and high-fat-diet-induced type 2 diabetes mellitus animal models (Fang-Rong et al. 2019). *Rhein* (4,5-dihydroxy anthraquinone-2-carboxylic acid) which is predominantly found in *Rheum* sp. ameliorates glucose tolerance by subsequent preservation of β-cell mass by attenuating inflammation-mediated β-cell injury and apoptosis. It also leads to a reduction of blood glucose levels, enhances insulin level, attenuates oxidative stress, and decreases hepatic steatosis (Du et al. 2012). It has shown beneficial effects in renal hypertrophy preventing development of glomerulosclerosis. It also helps in several chronic inflammatory conditions due to its ability of inhibiting interleukin (IL)-1β and further NF-κB activation (Du et al. 2012).

Pyrroloquinoline quinone (PQQ): It is an ubiquitous, water-soluble anionic compound which primarily acts as a cofactor for various bacterial enzymes. However recent advances suggest that these phytoconstituents also hold multiple therapeutic benefits of preventing oxidative stress- induced cellular injury by free radical scavenging mechanisms and reducing extent of lipid peroxidation. These antioxidant properties help in restoring diabetes-mediated cellular complications and cardiac histopathological features. Experimentation on STZ-induced diabetic animals has indicated that POQ potentially regulates cardiac oxidative damages and related cardiovascular events with a marked decrease in serum TG, LDL, and VLDL levels. It also has shown to increase insulin levels with a reduction in glucose concentration and reverse activity of elevated lactate dehydrogenase being a predominant hallmark enzyme in inducing myocardial injury due to its overexpression (Misra et al. 2004, 2012; Narendhirakannan et al. 2006; Smidt et al. 1991).

49.4.7 Miscellaneous Agents

Momordica charantia: Chemical constituents such as momordicoside S, momordicoside T, karaviloside XI, etc. are generally isolated from the fruits of this species. These compounds collectively consist of saponins and triterpenes which exert some potent antidiabetic benefits. Isolated triterpenes compounds act by inhibiting disaccharidase activity and have the ability to activate AMPK pathway (Chang et al. 2011; Tan et al. 2008), whereas saponin constituents have shown evident upregulation of GLUT-4 translocation and also augment insulin secretion from pancreatic β-cells in L6 muscle cells, 3 T3-L1 adipocytes, and MIN6 β-cells (Keller et al. 2011). Momordicoside S and momordicoside T have shown increased glucose tolerance and fatty acid oxidation in insulin-sensitive and insulin-resistant mice models (Tan et al. 2008).

Oleuropein: It is a phenolic compound which specifically belongs to the class of phenyl-propanoids. It is generally isolated from olive leaves and predominantly possesses potential antidiabetic potentials by reducing blood glucose levels. Recent advances have also shown that administration of oleuropein causes decrease in infarct size and coronary effluent creatine kinase MB and hence reduces the chances of developing coronary resistance in type 2 diabetic models. It also reduces rate of ventricular pressure by decreasing left ventricular pressure (Nekooeian et al. 2014).

Poria cocos: Triterpenes and saponins such as lanostane, dehydrotumulosic acid, dehydrotrametenolic acid, and pachymic acid decrease postprandial blood glucose level and enhance insulin sensitivity in STZ-induced db/db mice diabetic models (Sato et al. 2002). Huang et al. have also demonstrated that pachymic acid upregulates GLUT-4 translocation and elevates phosphorylation of insulin receptor substrate-1 (IRS-1) and also enhances activity of Akt and AMPK (Huang et al. 2010).

Swertiamarin: It belongs to class of seco-iridoid glycosides and is generally isolated from *Enicostemma littorale* species. Experiments on T2DM models have shown that swertiamarin reduces serum TG, cholesterol, and LDL levels. It also predominantly reduces fasting blood glucose levels and elevates insulin sensitivity (Vaidya et al. 2012).

Allium sativum: It stimulates insulin secretion from pancreatic β-cells, hence increasing glucose utilization and uptake by GLUT upregulation. It also serves as a potent HMG co-A reductase inhibitor and as antioxidant and anti-inflammatory agent (Kumari et al. 1995).

Aloe barbadensis: Its methanolic extract has shown potential antidiabetic actions in streptokinin-induced diabetic rats. It causes increase in insulin secretion from pancreatic β-cells and also inhibits pancreatic α-amylase activity. It also increases insulin sensitivity and also exerts anti-inflammatory and antioxidant potential in treatment therapies in streptokinin-induced diabetic rats (Grindlay and Reynolds 1986).

Coptis chinensis: Ethyl acetate extract of its rhizome has significantly shown regeneration of pancreatic β-cells, stimulates fatty acid oxidation process, attenuates lipogenesis, and also enhances glucose uptake by upregulating GLUT-4 translocation in high-fat-diet-induced diabetic mice (Cui et al. 2016).

Catharanthus roseus: It is also known as Madagascar periwinkle. It shows significant antidiabetic modulation in streptokinin-induced diabetic models with an enhancement of glucose uptake mechanisms and glucose utilization. It also possesses antioxidant activity and alpha glucosidase inhibition and exerts insulin sensitivity upregulation (Tiong et al. 2015).

Loganin: It is also a type of iridoid glycoside isolated from *Corni fructus*. Administration of loganin results in amelioration of hyperglycemia and dyslipidemic condition by reduction of TG levels following decreased oxidative stress. It also inhibits advanced glycation end-product formation and expression (Yamabe et al. 2010).

Catalpol: A type of iridoid glycoside which is usually isolated from *Catalpa* species. It reduces diabetes-mediated endothelial dysfunction by decreasing ROS generation (Liu 2014).

In addition to the abovementioned phytoconstituents, several other lead molecules belonging to the aforesaid class of compounds have also been explored so far which hold potential in treatment of T2DM and its related complications. Schematic overview of phytoconstituents and their therapeutic benefits in management of diabetic complications is presented in Fig. 49.2. Those phytochemicals and their respective mechanisms of actions are mentioned in Table 49.1.

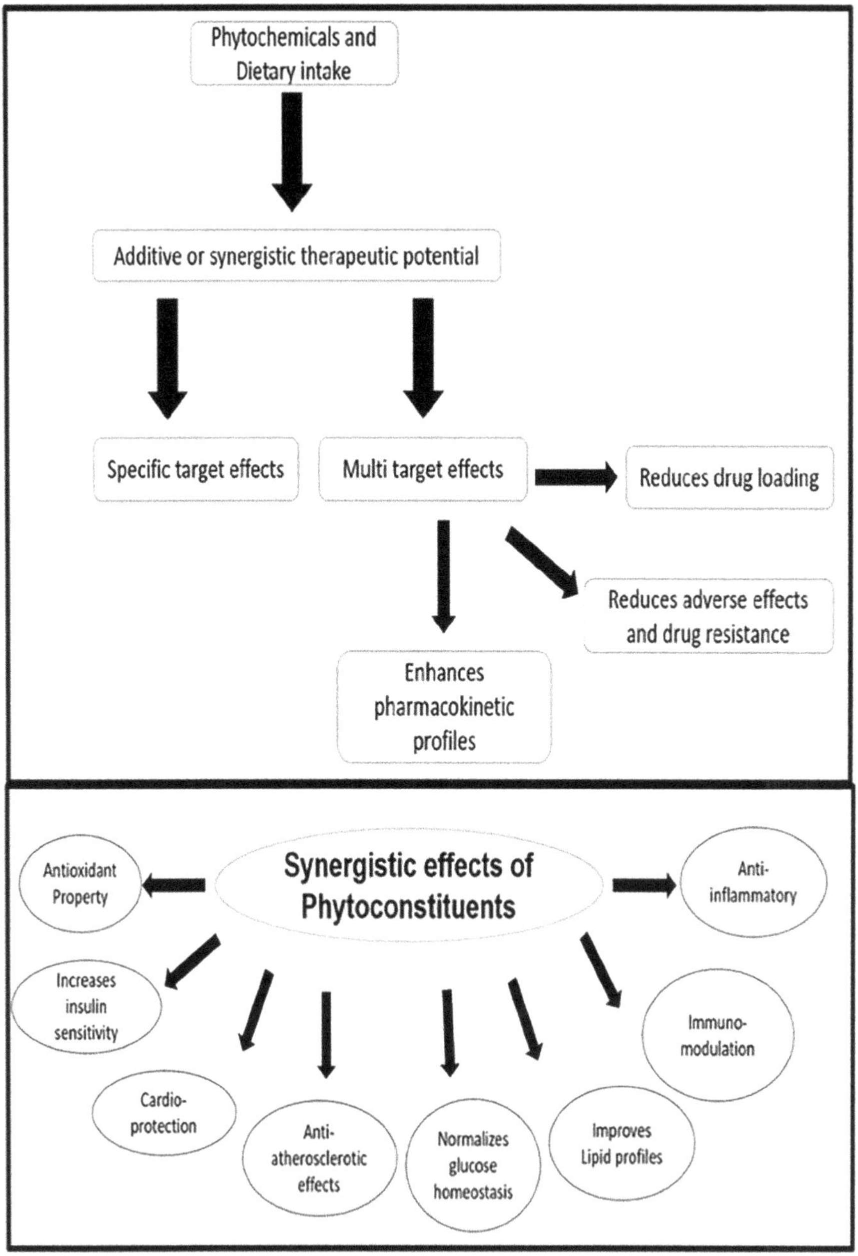

Fig. 49.2 Schematic overview of different effects of phytoconstituents obtained from various medicinal plants and dietary supplementations and their therapeutic benefits in the management of diabetic complications

Table 49.1 Additional phytoconstituents and their mechanisms of action in the management of type 2 diabetes mellitus and its associated complications

Phytoconstituents or. plant species	Mechanism of action	References
Flavonoids		
Delphinidin	Exerts its antidiabetic action by augmenting associated antioxidant activities	Gharib et al. (2013)
Pelargonidin	Administration of pelargonidin causes a reduction in oxidative stress and also tends to stimulate insulin secretion	Roy et al. (2008)
Myricetin	Elevates insulin sensitivity with a positive modulation of the insulin signaling pathway and also upregulates glycogen synthase activity, thereby attenuating gluconeogenesis and glycogenolysis	Kandasamy and Ashok Kumar (2014)
Bavachin	Upregulates glucose uptake mechanism through modulation of the GLUT-4 translocation pathway by activating Akt and AMPK signaling cascades	Lee et al. (2016)
Wogonin	It is isolated from the roots of *Scutellaria baicalensis*. It attenuates insulin insensitivity and lipid metabolism by modulating AMPK and PPAR signaling pathways	Bak et al. (2014)
Isorhamnetin	It is isolated from *Gingko biloba* and *Oenanthe javanica*. It decreases oxidative stress and interferes with lipid metabolism	Yokozawa et al. (2002)
Polyphenols		
Paeonol	It decreases blood glucose, glycosylated serum proteins, and AGE levels via modulating AGEs/RAGE and NF-κB pathway in STZ-induced diabetic rats	Liu et al. (2013)
Ferulic acid	It decreases blood glucose level, increases serum adiponectin levels, and reduces oxidative stress and hence ROS-mediated inflammation in OLETF rats	Choi et al. (2011)
Ellagic acid	It exerts its antidiabetic potential by improving hepatic steatosis and serum lipid composition by reducing serum resistin levels and activating hepatic PPAR-α expression in KK-ay mice models	Yoshimura et al. (2013)
Chlorogenic acid	It decreases plasma glucose and glycosylated hemoglobin levels via modulation of adiponectin receptor signaling pathway in STZ-induced diabetic rats	Mubarak et al. (2013)
Saponins		
Solanum anguivi	Saponins isolated from this species have the potential to induce antioxidant activity and also decrease serum glucose levels in diabetic rats	Elekofehinti et al. (2013)
Garcinia kola	Saponins from this species significantly reduced plasma glucose levels with a predominant increase in insulin level as given by alloxan-induced diabetic models	Smith and Adanlawo (2012)

(continued)

Table 49.1 (continued)

Phytoconstituents or. plant species	Mechanism of action	References
Trigonella foenum-graecum	Saponins originated from this species cause significant inhibition of TG accumulation, reduce TG levels, and also decrease expression of lipogenic genes which in turn collectively lead to reduced lipid levels	Uemura et al. (2010)
Terminalia arjuna	Arjunolic acid, a type of triterpene saponin, exerts its antidiabetic benefits by scavenging free radicals and also causes inhibition of excessive reactive oxygen species generation. It also can inhibit α-amylase and α-glucosidase enzymes	Hemalatha et al. (2010)
Entada phaseoloides	These plant species generally contain saponins which show an elevation in serum insulin levels, thereby attenuating chances of hyperglycemia, and also tend to decrease lipid levels	Zheng et al. (2012)
Alkaloids		
Jatrorrhizine	It holds the potential to reduce hyperlipidemic conditions by suppressing lipogenesis and lipid oxidation in high-fat-diet-induced animals	Yang et al. (2016)
Colchicine	It augments its antidiabetic potential by reducing activation of NLRP-3-mediated inflammation and also improves the metabolic dysregulated state	Demidowich et al. (2016)
Galantamine	It has shown significant improvement of insulin-resistant states and also alleviated inflammation in STZ-induced rats	Consolim-Colombo et al. (2017)
Huperzine A	It tends to modulate several inflammatory modulators, thereby attenuating further inflammation which are subjected due to a rise in oxidative stress and apoptosis	Mao et al. (2014)
Koenidine	It is isolated from the plant species of *Murraya koenigii* It causes an increase in insulin sensitivity which in turn alleviates glucose uptake mechanisms	Yadav et al. (2002)
Terpenoids		
Pinitol	It modulates glucose uptake mechanisms mediated by insulin signaling pathway, hence regulating the PI3K/Akt pathway in T2DM rats	Gao et al. (2015)
Benzofuran-2-carboxaldehyde	A diterpene in nature which is isolated from the species of *Globba pendula* and which exerts its hypoglycemic activity by glucose-lowering activities	Maulidiani et al. (2009)
Betulinic acid	It proved to improve the hyperglycemic state by inhibiting hepatic gluconeogenesis through modulation of the CAMKK-AMPK-CREB signaling pathway in HepG2 cells and HFD-fed mice	Kim et al. (2014)

(continued)

Table 49.1 (continued)

Phytoconstituents or. plant species	Mechanism of action	References
Dehydroabietic acid	Suppresses expression of pro-inflammatory cytokines such as TNF-α whereas causes upregulation of anti-inflammatory cytokine levels as in adiponectin and hence improves plasma parameters	Kang et al. (2009)
Tormentic acid	It possesses antihyperglycemic and antihyperlipidemic potential via the downregulation of hepatic SREBP-1c and apolipoprotein C-III and subsequently increasing PPAR-α pathway expression	Wu et al. (2014)
Quinones		
Shikonin	It reduces central adiposity and also enhances glucose tolerance by upregulating tyrosyl phosphorylation of the insulin receptor in high-fat-diet-induced diabetic mice	Bettaieb et al. (2015)
Tanshinone I	It alleviates insulin resistance by modulating the IRS-1 pathway and hence regulates blood glucose and TG levels	Wei et al. (2017)
Cryptotanshinone	It possesses potential antidiabetic effect with AMPK pathway activation	Kim et al. (2007)
Plumbagin	It causes upregulation of GLUT-4 translocation and hence contributes to normal glucose homeostasis	Sunil et al. (2012)
Thymoquinone	It can reduce oxidative stress by decreasing the extent of pro-inflammatory cytokine generation	Safhi et al. (2019)

49.5 Conclusion

Diabetes mellitus, being a multifactorial disorder, engages a number of metabolic and molecular pathways which get detrimental in due course of time producing major complications. It encompasses a series of key players which augments fatal consequences, thus producing many challenges in therapeutic regimens. However, it has already been suggested that this metabolic-endocrine disorder can be managed by several forms of medications, lifestyle modification, and diet. Over the years, the use of synthetic and conventional medicines had been on the brink due to their chain of adverse cascades and their unavoidability as a matter of cost in a great population of diabetic patients. The growing mortalities and consequences arising due to diabetes mellitus explain and demand more promising therapeutic regimens which can be subjected to any need.

Recent advances in phytochemicals have ameliorated such complications owing to their greater availability and abundance in nature. Several plant species comprising of flavonoids, quinones, saponins, polyphenols, alkaloids, etc. serve as some potential agents in minimizing diabetes state and its related metabolic complications.

These agents exert a circle of beneficial effects of hypoglycemic action by increasing insulin secretion, sensitivity, glucose uptake, and utilization along with potent antioxidant, anti-inflammatory, and free radical scavenging effects. Some phytochemicals are also beneficial in managing hyperlipidemic and hyperinsulinemia states by decreasing TG and lipid levels, inhibiting lipolysis, and reducing FFA contents. The chapter summarizes such phytoconstituents, their effects, and natural species which are novel and effective and have gained enormous momentum in reducing various ailments relating to diabetes, hyperlipidemia, and atherosclerosis.

References

Abraham TM, Pencina KM, Pencina MJ et al (2015) Trends in diabetes incidence: the Framingham heart study. Diabetes Care 38:482–487

Abuohashish HM, Al-Rejaie SS, Al-Hosaini KA et al (2013) Alleviating effects of morin against experimentally-induced diabetic osteopenia. Diabetol Metab Syndr 5(1):5

Ahad A, Ganai AA, Mujeeb M et al (2014b) Chrysin, an anti-inflammatory molecule, abrogates renal dysfunction in type 2 diabetic rats. Toxicol Appl Pharmacol 279(1):1–7

Ahad A, Mujeeb M, Ahsan H et al (2014a) Prophylactic effect of baicalein against renal dysfunction in type 2 diabetic rats. Biochimie 106:101–110

Amalan V, Vijayakumar N, Indumathi D et al (2016) Antidiabetic and antihyperlipidemic activity of p-coumaric acid in diabetic rats, role of pancreatic GLUT 2: in vivo approach. Biomed Pharmacother 84:230–236

Atkinson LL, Kozak R, Kelly SE et al (2003) Potential mechanisms and consequences of cardiac triacylglycerol accumulation in insulin-resistant rats. Am J Physiol Endocrinol Metab 284: E923–E930

Bak EJ, Kim J, Choi YH et al (2014) Wogonin ameliorates hyperglycemia and dyslipidemia viaPPARα activation in db/db mice. Clin Nutr 33:156–163

Baron AD, Schaeffer L, Shragg P et al (1987) Role of hyperglucagonemia in maintenance of increased rates of hepatic glucose output in type II diabetics. Diabetes 36:274–283

Baynes JW, Thorpe SR (1999) Role of oxidative stress in diabetic complications: a new perspective on an old paradigm. Diabetes 48(1):1–9

Bettaieb A, Hosein E, Chahed S et al (2015) Decreased adiposity and enhanced glucose tolerance in Shikonin treated mice. Obesity (Silver Spring) 23(11):2269–2277

Bharti S, Rani N, Krishnamurthy B et al (2014) Preclinical evidence for the pharmacological actions of Naringin: a review. Planta Med 80:437–451

Bhavsar SK, Föller M, Gu S et al (2009) Involvement of the PI3K/AKT pathway in the hypoglycemic effects of saponins from Helicteres isora. J Ethnopharmacol 126:386–396

Boden G (1997) Role of fatty acids in the pathogenesis of insulin resistance and NIDDM. Diabetes 46:3–10

Bose M, Lambert JD, Ju J et al (2008) The major green tea polyphenol, (−)-Epigallocatechin-3-Gallate, inhibits obesity, metabolic syndrome, and fatty liver disease in high-fat–fed mice. J Nutr 138:1677–1683

Brownlee M (2001) Biochemistry and molecular cell biology of diabetic complications. Nature 414 (6865):813–820

Chang CI, Tseng HI, Liao YW et al (2011) In vivo and in vitro studies to identify the hypoglycaemic constituents of Momordica charantia wild variant WB24. Food Chem 125:521–528

Chen N, Bezzina R, Hinch E et al (2009) Green tea, black tea, and epigallocatechin modify body composition, improve glucose tolerance, and differentially alter metabolic gene expression in rats fed a high-fat diet. Nutr Res 29:784–793

Chen P, Chen JB, Chen WY et al (2012a) Effects of quercetin on nuclear factor-κB p65 expression in renal ubiquitin-proteasome system of diabetic rats. Zhonghua Nei Ke Za Zhi 51(6):460–465

Chen S, Jiang H, Wu X et al (2016) Therapeutic effects of quercetin on inflammation, obesity, and type 2 diabetes. Mediat Inflamm:9340637

Chen S, Li J, Zhang Z et al (2012b) Effects of resveratrol on the amelioration of insulin resistance in KKAy mice. Can J Physiol Pharmacol 90:237–242

Chen YY, Hsu MJ, Sheu JR et al (2013) Andrographolide, a novel NF- κ B inhibitor, induces vascular smooth muscle cell apoptosis via a Ceramide-p47phox-ROS Signaling Cascade. Evid Based Complement Alternat Med 2013:821813

Cheong SH, Furuhashi K, Ito K et al (2014) Daidzein promotes glucose uptake through glucose transporter 4 translocation to plasma membrane in L6myocytes and improves glucose homeostasis in type II diabetic model mice. J Nutr Biochem 25:136–143

Choi R, Kim BH, Naowaboot J et al (2011) Effects of ferulic acid on diabetic nephropathy in a rat model of type 2 diabetes. Exp Mol Med 43:676–683

Collins QF, Liu HY, Pi J et al (2007) Epigallocatechin-3-gallate (EGCG), a green tea polyphenol, suppresses hepatic gluconeogenesis through 5'-AMP-activated protein kinase. J Biol Chem 282:30143–30149

Consolim-Colombo FM, Sangaleti CT, Costa FO et al (2017) Galantamine alleviates inflammation and insulin resistance in patients with metabolic syndrome in a randomized trial. JCI Insight 2 (14):pii: 93340

Constantin RP, Constantin J, Pagadigorria CL et al (2010) The actions of fisetin on glucose metabolism in the rat liver. Cell Biochem Funct 281:49–58

Cui L, Liu M, Chang X et al (2016) The inhibiting effect of the Coptis chinensis polysaccharide on the type II diabetic mice. Biomed Pharmacother 81:111e119

Dandona P, Thusu K, Cook S et al (1996) Oxidative damage to DNA in diabetes mellitus. Lancet 347(8999):444–445

Dar A, Faizi S, Naqvi S et al (2005) Analgesic and antioxidant activity of Mangiferin and its derivatives: the structure activity relationship. Biol Pharm Bull 28:596–600

DeFronzo RA (1988) Lilly Lecture: the triumvirate: -cell, muscle, liver: a collusion responsible for NIDDM. Diabetes 37:667–687

Dehmlow C, Erhard J, de Groot H (1996) Inhibition of kupffer cell functions as an explanation for the hepatoprotective properties of silibinin. Hepatology 23:749–754

Demidowich AP, Davis AI, Dedhia N et al (2016) Colchicine to decrease NLRP3-activated inflammation and improve obesity-related metabolic dysregulation. Med Hypotheses 92:67–73

Deng Y, He K, Ye X et al (2012) Saponin rich fractions from Polygonatum odoratum (mill.) druce with more potential hypoglycemic effects. J Ethnopharmacol 141:228–233

Ding L, Li J, Song B et al (2014a) Andrographolide prevents high-fat diet-induced obesity in C57BL/6 mice by suppressing the sterol regulatory element-binding protein pathway. J Pharmacol Exp Ther 351:474–483

Ding Y, Shi X, Shuai X et al (2014b) Luteolin prevents uric acid-induced pancreatic β-cell dysfunction. J Biomed Res 28:292–298

Du H, Shao J, Gu P et al (2012) Improvement of glucose tolerance by Rhein with restored early-phase insulin secretion in db/db mice. J Endocrinol Investig 35:607–612

Duang XY, Wang Q, Zhou XD et al (2011) Mangiferin: a possible strategy for periodontal disease to therapy. Med Hypotheses 76:486–488

Eid HM, Nachar A, Thong F et al (2015) The molecular basis of the anti-diabetic action of quercetin in cultured skeletal muscle cells and hepatocytes. Pharmacogn Mag 11:74–81

El-Azab MF, Attia FM, El-Mowafy AM (2011) Novel role of curcumin combined with bone marrow transplantation in reversing experimental diabetes: effects on pancreatic islet regeneration, oxidative stress, and inflammatory cytokines. Eur J Pharmacol 658:41–48

Elekofehinti OO, Kamdem JP, Kade IJT et al (2013) Hypoglycemic anti-per-oxidative and antihyperlipidemic effects of saponins from Solanum anguivi lam. Fruits in alloxan-induced diabetic rats. S Afr J Bot 88:56–61

Fang-Rong C, Hong-Xin C, Ji-Li F et al (2019) Ameliorative effect and mechanism of the purified Anthraquinone-glycoside preparation from Rheum Palmatum L. on type 2 diabetes mellitus. Molecules 24(8):1454

Fernandes AA, Novelli EL, Okoshi K et al (2010) Influence of Rutin treatment on biochemical alterations in experimental diabetes. Biomed Pharmacother 64:214–219

Forcheron F, Basset A, Abdallah P et al (2009) Diabetic cardiomyopathy: effects of fenofibrate and metformin in an experimental model—the Zucker diabetic rat. Cardiovasc Diabetol 8:16

Fu Y, Luo J, Jia Z et al (2014) Baicalein protects against type 2 diabetes via promoting islet beta-cell function in obese diabetic mice. Int J Endocrinol 2014:846742

Fujioka K (2007) Pathophysiology of type 2 diabetes and the role of incretin hormones and beta-cell dysfunction. JAAPA Suppl:3–8

Gao Y, Zhang M, Wu T et al (2015) Effects of D-pinitol on insulin resistance through the PI3K/Akt signaling pathway in type 2 diabetes mellitus rats. J Agric Food Chem 63:6019–6026

Gastaldelli A, Miyazaki Y, Pettiti M et al (2004) Separate contribution of diabetes, total fat mass, and fat topography to glucose production, gluconeogenesis, and glycogenolysis. J Clin Endocrinol Metab 89:3914–3921

Gharib A, Faezizadeh Z, Godarzee M (2013) Treatment of diabetes in the mouse model by delphinidin and cyanidin hydrochloride in free and liposomal forms. Planta Med 79:1599–1604

Ghorbani A (2017) Mechanisms of antidiabetic effects of flavonoid Rutin. Biomed Pharmacother 96:305–312

Giles TD, Miller AB, Elkayam U et al (2008) Pioglitazone and heart failure: results from a controlled study in patients with type 2 diabetes mellitus and systolic dysfunction. J Card Fail 14:445–452

Golay A, Felber JP, Jequier E et al (1988) Metabolic basis of obesity and noninsulin-dependent diabetes mellitus. Diabetes Metab Rev 4(8):727–747

Goodpaster BH, Thaete FL, Kelley DE (2000) Thigh adipose tissue distribution is associated with insulin resistance in obesity and in type 2 diabetes mellitus. Am J Clin Nutr 71:885–892

Grindlay D, Reynolds T (1986) The Aloe vera phenomenon: a review of the properties and modern uses of the leaf parenchyma gel. J Ethnopharmacol 16(2e3):117e151

Guan L, Feng H, Gong D et al (2013) Genipin ameliorates age-related insulin resistance through inhibiting hepatic oxidative stress and mitochondrial dysfunction. Exp Gerontol 48:1387–1394

Guha S, Ghosal S, Chattopadhyay U (1996) Antitumor, immunomodulatory and anti-HIV effect of Mangiferin, a naturally occurring glucosyl xanthone. Chemotherapy 42:443–451

Gupta SK, Dongare S, Mathur R et al (2015) Genistein ameliorates cardiac inflammation and oxidative stress in Streptozotocin-induced diabetic cardiomyopathy in rats. Mol Cell Biochem 408(1-2):63–72

Hadden DR, Montgomery DA, Skelly RJ et al (1975) Maturity onset diabetes mellitus: response to intensive dietary therapy. Br Med J 2:276–278

Häkkinen SH, Kärenlampi SO, Heinonen IM et al (1999) Content of the flavonols quercetin, myricetin, and kaempferolin 25 edible berries. J Agric Food Chem 47(6):2274–2279

He HJ, Wang GY, Gao Y et al (2012) Curcumin attenuates Nrf2 signaling defect, oxidative stress in muscle and glucose intolerance in high fat diet-fed mice. World J Diabetes 3:94–104

Heim KE, Tagliaferro AR, Bobilya DJ (2002) Flavonoid antioxidants: chemistry, metabolism and structure-activity relationships. J Nutr Biochem 13(10):572–584

Hemalatha T, Pulavendran S, Balachandran C et al (2010) Arjunolic acid: a novel phytomedicine with multifunctional therapeutic applications. Indian J Exp Biol 48:238–247

Hemmeryckx B, Hoylaerts MF, Gallacher DJ et al (2013) Does rosiglitazone affect adiposity and cardiac function in genetic diabetic mice. Eur J Pharmacol 700:23–31

Hemmeryckx B, Swinnen M, Gallacher DJ et al (2014) Effect of Sitagliptin treatment on metabolism and cardiac function in genetic diabetic mice. Eur J Pharmacol 723:175–180

Hermenean A, Mariasiu T, Navarro-González I et al (2017) Hepatoprotective activity of Chrysin is mediated through TNF-a in chemically-induced acute liver damage: an in vivo study and molecular modeling. Exp Ther Med 13:1671–1680
Hsu CY, Shih HY, Chia YC et al (2014) Rutin potentiates insulin receptor kinase to enhance insulin-dependent glucose transporter 4 translocation. Mol. Nutr. Food Res 58:1168–1176
Huang YC, Chang WL, Huang SF et al (2010) Pachymic acid stimulates glucose uptake through enhanced GLUT4 expression and translocation. Eur J Pharmacol 648:39–49
Hunyadi A, Martins A, Hsieh TJ et al (2012) Chlorogenic acid and Rutin play a major role in the in vivo anti-diabetic activity of Morus alba leaf extract on type II diabetic rats. PLoS One 7(11): e50619
Ihara Y, Toyokuni S, Uchida K et al (1999) Hyperglycemia causes oxidative stress in pancreatic β-cells of GK rats, a model of type 2 diabetes. Diabetes 48(4):927–932
Jain D, Bansal MK, Dalvi R et al (2014) Protective effect of Diosmin against diabetic neuropathy in experimental rats. J Integr Med 12:35–41
Jeong KJ, Kim DY, Quan HY et al (2014) Effects of eugenol on hepatic glucose production and AMPK signaling pathway in hepatocytes and C57BL/6J mice. Fitoterapia 93:150–162
Jia G, DeMarco VG, Sowers JR (2016) Insulin resistance and hyperinsulinaemia in diabetic cardiomyopathy. Nat Rev Endocrinol 12:144–153
Jia S, Hu Y, Zhang W et al (2015) Hypoglycemic and hypolipidemic effects of Neohesperidin derived from Citrus aurantium L. in diabetic KK-A(y) mice. Food Funct 6:878–886
Jiang B, Yang Y, Jin H et al (2008) Astragaloside IV attenuates lipolysis and improves insulin resistance induced by TNF-alpha in 3T3-L1 adipocytes. Phytother Res 22(11):1434–1439
Jung KH, Choi HS, Kim DH et al (2008) Epigallocatechin gallate stimulates glucose uptake through the phosphatidylinositol 3-kinase-mediated pathway in L6 rat skeletal muscle cells. J Med Food 11:429–434
Jung UJ, Lee MK, Jeong KS et al (2004) The hypoglycemic effects of hesperidin and Naringin are partly mediated by hepatic glucose-regulating enzymes in C57BL/KsJ-db/db mice. J Nutr 134:2499–2503
Jung UJ, Lee MK, Park YB et al (2006) Antihyperglycemic and antioxidant properties of caffeic acid in db/db mice. J Pharmacol Exp Ther 318:476–483
Kandasamy N, Ashok Kumar N (2014) Protective effect of bioflavonoid Myricetin enhances carbohydrate metabolic enzymes and insulin signaling molecules in Streptozotocin-cadmium induced diabetic nephrotoxic rats. Toxicol Appl Pharmacol 279:173–185
Kang MS, Hirai S, Goto T et al (2009) Dehydroabietic acid, a diterpene improves diabetes and hyperlipidemia in obese diabetic KK-ay mice. Biofactors 35(5):442–448
Kashyap S, Belfort R, Gastaldelli A et al (2003) A sustained increase in plasma free fatty acids impairs insulin secretion in non-diabetic subjects genetically predisposed to develop type 2 diabetes. Diabetes 52:2461–2474
Keller AC, Ma J, Kavalier A et al (2011) Saponins from the traditional medicinal plant *Momordica charantia* stimulate insulin secretion in vitro. Phytomedicine 19:32–37
Kim EJ, Jung SN, Son KH et al (2007) Anti-diabetes and anti-obesity effect of Cryptotanshinone via activation of AMP-activated protein kinas. Mol Pharmacol 72:62–72
Kim JA, Jang HJ, Martinez-Lemus LA et al (2012a) Activation of mTOR/p70S6 kinase by ANG II inhibits insulin stimulated endothelial nitric oxide synthase and vasodilation. Am J Physiol Endocrinol Metab 302:E201–E208
Kim MS, Hur HJ, Kwon DY et al (2012b) Tangeretin stimulates glucose uptake via regulation of AMPK signaling pathways in C2C12 myotubes and improves glucose tolerance in high-fat diet-induced obese mice. Mol Cell Endocrinol 358:127–134
Kim S, Jin Y, Choi Y et al (2011) Resveratrol exerts anti-obesity effects via mechanisms involving down-regulation of adipogenic and inflammatory processes in mice. Biochem Pharmacol 81:1343–1351
Kim SJ, Quan HY, Jeong KJ et al (2014) Beneficial effect of betulinic acid on hyperglycemia via suppression of hepatic glucose production. J Agric Food Chem 62:434–442

Klaus S, Pültz S, Thöne-Reineke C et al (2005) Epigallocatechin gallate attenuates diet-induced obesity in mice by decreasing energy absorption and increasing fat oxidation. Int J Obes Relat Metab Disord 29:615–623

Komajda M, McMurray JJ, Beck-Nielsen H et al (2010) Heart failure events with rosiglitazone in type 2 diabetes: data from the RECORD clinical trial. Eur Heart J 31:824–831

Kumari K, Mathew BC, Augusti KT (1995) Antidiabetic and hypolipidemic effects of S-methyl cysteine sulfoxide isolated from Allium cepa Linn. Indian J Biochem Biophys 32:49–54

Kwon DY, Kim YS, Ryu SY et al (2012) Platyconic acid, a saponin from platycodi radix, improves glucose homeostasis by enhancing insulin sensitivity in vitro and in vivo. Eur J Nutr 51:529–540

Lau YS, Tian XY, Mustafa MR et al (2013) Boldine improves endothelial function in diabetic db/db mice through inhibition of angiotensin II-mediated BMP4-oxidative stress cascade. Br J Pharmacol 170(6):1190–1198

Lee H, Li H, Noh M et al (2016) Bavachin from Psoralea corylifolia improves insulin-dependent glucose uptake through insulin signaling and AMPK activation in 3T3-L1 adipocytes. Int J Mol Sci 17:527

Lee J, Lee HI, Seo KI et al (2014) Effects of ursolic acid on glucose metabolism, the polyol pathway and dyslipidemia in non-obese type 2 diabetic mice. Indian J Exp Biol 52:683–691

Lee KT, Jung TW, Lee HJ et al (2011) The antidiabetic effect of genosenoside Rb2 via activation of AMPK. Arch Pharm Res 34:1201–1208

Li G, Xu H, Zhu S et al (2013) Effects of Neferine on CCL5 and CCR5 expression in SCG of type 2 diabetic rats. Brain Res Bull 90:79–87

Li J, Kang SW, Kim JL et al (2010) Isoliquiritigenin entails blockade of TGF-beta1-SMAD signaling for retarding high glucose-induced mesangial matrix accumulation. J Agric Food Chem 58(5):3205–3212

Li J, Lim SS, Lee ES et al (2011) Isoangustone a suppresses mesangial fibrosis and inflammation in human renal mesangial cells. Exp Biol Med (Maywood) 236(4):435–444

Li Y, Yan H, Zhang Z et al (2015) Andrographolide derivative AL-1 improves insulin resistance through down-regulation of NF-κB signaling pathway. Br J Pharmacol 172(12):3151–3158

Liu D, Zhen W, Yang Z et al (2006) Genistein acutely stimulates insulin secretion in pancreatic beta-cells through a CAMP-dependent protein kinase pathway. Diabetes 55:1043–1050

Liu J, Wang S, Feng L et al (2013) Hypoglycemic and antioxidant activities of Paeonol and its beneficial effect on diabetic encephalopathy in Streptozotocin-induced diabetic rats. J Med Food 16:577–586

Liu JY (2014) Catalpol protect diabetic vascular endothelial function by inhibiting NADPH oxidase. Zhongguo Zhong Yao Za Zhi 39(15):2936–2941

Liu Y, Fu X, Lan N et al (2014b) Luteolin protects against high fat diet-induced cognitive deficits in obesity mice. Behav Brain Res 267:178–188

Liu Y, Li D, Zhang Y et al (2014a) Anthocyanin increases adiponectin secretion and protects against diabetes-related endothelial dysfunction. Am J Physiol Endocrinol Metab 306(8):E975–E988

Lu C, Zhu W, Shen CL et al (2012) Green tea polyphenols reduce body weight in rats by modulating obesity-related genes. PLoS One 7(6):e38332

Lu H, Wang J, Wang T et al (2016) Recent progress on nanostructures for drug delivery applications. J Nanomater 2016:20

Ma H, Li SY, Xu P et al (2009) Advanced glycation end product(AGE) accumulation and AGE receptor (RAGE) up-regulation contribute to the onset of diabetic cardiomyopathy. J Cell Mol Med 13:1751–1764

Maciel RM, Costa MM, Martins DB et al (2013) Antioxidant and anti-inflammatory effects of quercetin in functional and morphological alterations in Streptozotocin-induced diabetic rats. Res Vet Sci 95:389–397

Magnone M, Ameri P, Salis A et al (2015) Microgram amounts of abscisic acid in fruit extracts improve glucose tolerance and reduce insulinemia in rats and in humans. FASEB J 29:4783–4793

Mahmoud AM, Ashour MB, Abdel-Moneim A et al (2012) Hesperidin and Naringin attenuate hyperglycemia-mediated oxidative stress and proinflammatory cytokine production in high fat fed/Streptozotocin-induced type 2 diabetic rats. J Diabet Complications 26:483–490

Mao XY, Cao DF, Li X et al (2014) Huperzine a ameliorates cognitive deficits in Streptozotocin-induced diabetic rats. Int J Mol Sci 15:7667–7683

Matsuda M, Liu Y, Mahankali S et al (1999) Altered hypothalamic function in response to glucose ingestion in obese humans. Diabetes 48:1801–1806

Maulidiani, Shaari K, Paetz C, Stanslas J et al (2009) Naturally occurring labdane diterpene and benzofuran from Globba pendula. Nat Prod Commun 4(8):1031–1036

McGarry JD (2002) Banting lecture 2001: dysregulation of fatty acid metabolism in the etiology of type 2 diabetes. Diabetes 51:7–18

Miean KH, Mohamed S (2001) Flavonoid (myricetin, quercetin, kaempferol, luteolin, and apigenin) content of edible tropical plants. J Agric Food Chem 49:3106–3112

Miller JA (1999) Impact of hyperglycemia on the renin angiotensin system in early human type 1 diabetes mellitus. J Am Soc Nephrol 10:1778–1785

Miller JA, Floras JS, Zinman B et al (1996) Effect of hyperglycaemia on arterial pressure, plasma renin activity and renal function in early diabetes. Clin Sci (Lond) 90:189–195

Mirza AZ, Siddiqui FA (2014) Nano medicine and drug delivery: a mini review. Int Nano Lett 4:94

Misra HS, Khairnar NP, Barik A et al (2004) Pyrroloquinoline–quinone: a reactive oxygen species scavenger in bacteria. FEBS Lett 578(1–2):26–30

Misra HS, Rajpurohit YS, Khairnar NP (2012) Pyrroloquinoline–quinone and its versatile roles in biological processes. J Biosci 37(2):313–325

Mohan S, Nandha Kumar L (2014) Role of various flavonoids: hypotheses on novel approach to treat diabetes. Iran J Med Hypotheses Ideas 8(1):1–6

Mozaffari MS, Allo S, Schaffer SW (1989) The effect of sulfonylurea therapy on defective calcium movement associated with diabetic cardiomyopathy. Can J Physiol Pharmacol 67:1431–1436

Mubarak A, Hodgson JM, Considine MJ et al (2013) Supplementation of a high-fat diet with chlorogenic acid is associated with insulin resistance and hepatic lipid accumulation in mice. J Agric Food Chem 61:4371–4378

Narendhirakannan RT, Subramanian S, Kandaswamy M (2006) Biochemical evaluation of anti diabetogenic properties of some commonly used Indian plants on Streptozotocin–induced diabetes in experimental rats. Clin Exp Pharmacol Physiol 33(12):1150–1157

Nekooeian AA, Khalili A, Khosravi MB (2014) Oleuropein offers cardio protection in rats with simultaneous type 2 diabetes and renal hypertension. Indian J Pharmacol 46(4):398–403

Nirmala P, Ramanathan M (2011) Effect of kaempferol on lipid peroxidation and antioxidant status in 1,2-dimethyl hydrazine induced colorectal carcinoma in rats. Eur J Pharm 654:75–79

Ohno M, Shibata C, Kishikawa T et al (2013) The flavonoid Apigenin improves glucose tolerance through inhibition of microRNA maturation in miRNA103 transgenic mice. Sci Rep 3:2553

Park CH, Yokozawa T, Noh JS (2014) Oligonol, a low-molecular-weight polyphenol derived from lychee fruit, attenuates diabetes-induced renal damage through the advanced glycation end product-related pathway in db/db mice. J Nutr 144(8):1150–1157

Parmar HS, Jain P, Chauhan DS et al (2012) DPP-IV inhibitory potential of Naringin: an in silico, in vitro and in vivo study. Diabetes Res Clin Pract 97:105–111

Pferschy-Wenzig EM, Atanasov AG, Malainer C et al (2014) Identification of Isosilybin a from milk thistle seeds as an agonist of peroxisome proliferator-activated receptor gamma. J Nat Prod 77:842–847

Poudyal H, Lemonakis N, Efentakis P et al (2017) Hydroxytyrosol ameliorates metabolic, cardiovascular and liver changes in a rat model of diet-induced metabolic syndrome: pharmacological and metabolism-based investigation. Pharmacol Res 117:32–45

Prasath GS, Pillai SI, Subramanian SP (2014) Fisetin improves glucose homeostasis through the inhibition of gluconeogenic enzymes in hepatic tissues of streptozotocin induced diabetic rats. Eur J Pharmacol 740:248–254

Prentki M, Nolan CJ (2006) Islet β cell failure in type 2 diabetes. J Clin Invest 116(7):1802–1812

Pu P, Wang XA, Salim M et al (2012) Baicalein, a natural product, selectively activating AMPK alpha (2) and ameliorates metabolic disorder in diet-induced mice. Mol. Cell Endocrinol 362:128–138

Punitha ISR, Shirwaikar A (2005) Antidiabetic activity of benzyl tetra isoquinoline alkaloid Berberine in Streptozotocin-nicotinamide induced type 2 diabetic rats. Diabetol Croat 34:117–128

Qi Z, Xu Y, Liang Z et al (2015) Baicalein alters PI3K/Akt/GSK3 beta signaling pathway in rats with diabetes-associated cognitive deficits. Int J Clin Exp Med 8:1993–2000

Qiang G, Yang X, Xuan Q et al (2014) Salvianolic acid a prevents the pathological progression of hepatic fibrosis in high-fat diet-fed and Streptozotocin induced diabetic rats. Am J Chin Med 42 (5):1183–1198

Rossetti L, Giaccari A, DeFronzo RA (1990) Glucose toxicity. Diabetes Care 13:610–630

Roy M, Sen S, Chakraborti AS (2008) Action of pelargonidin on hyperglycemia and oxidative damage in diabetic rats: implication for glycation-induced hemoglobin modification. Life Sci 82:1102–1110

Safhi MM, Qumayri HM, Masmali AUM et al (2019) Thymoquinone and fluoxetine alleviate depression via attenuating oxidative damage and inflammatory markers in type-2 diabetic rats. Arch Physiol Biochem 125(2):150–155

Sakurai T, Tsuchiya S (1988) Superoxide production from non-enzymatically glycated protein. FEBS Lett 236(2):406–410

Saleh S, El-Maraghy N, Reda E et al (2014) Modulation of diabetes and dyslipidemia in diabetic insulin-resistant rats by Mangiferin: role of adiponectin and TNF-α. An Acad Bras Cienc 86:1935–1948

Sato M, Tai T, Nunoura Y et al (2002) Dehydrotrametenolic acid induces pre adipocyte differentiation and sensitizes animal models of noninsulin-dependent diabetes mellitus to insulin. Biol Pharm Bull 25:81–86

Sell H, Habich C, Eckel J (2012) Adaptive immunity in obesity and insulin resistance. Nat Rev Endocrinol 8:709–716

Seo WD, Lee JH, Jia Y et al (2015) Saponarin activates AMPK in a calcium-dependent manner and suppresses gluconeogenesis and increases glucose uptake via phosphorylation of CRTC2 and HDAC5. Bioorg Med Chem Lett 25(22):5237

Seppälä-Lindroos A, Vehkavaara S, Häkkinen AM et al (2002) Fat accumulation in the liver is associated with defects in insulin suppression of glucose production and serum free fatty acids independent of obesity in normal men. J Clin Endocrinol Metab 87:3023–3028

Sharma S, Ali A, Ali J et al (2013) Rutin: therapeutic potential and recent advances in drug delivery. Expert Opin Investig Drugs 22(8):1063–1079

Sheela N, Jose MA, Sathyamurthy D et al (2013) Effect of Silymarin on Streptozotocin-nicotinamide-induced type 2 diabetic nephropathy in rats. Iran J Kidney Dis 7(2):117–123

Smidt CR, Steinberg FM, Rucker RB (1991) Physiologic importance of pyrroloquinoline quinine. Proc Soc Exp Biol Med 197(1):19–26

Smith YR, Adanlawo IG (2012) Hypoglycaemic effect of saponin from the root of Garcinia kola on alloxan–induced diabetic rats. J Drug Delivery Ther 2:9–12

Soetikno V, Sari FR, Veeraveedu PT et al (2011) Curcumin ameliorates macrophage infiltration by inhibiting NF-kappaB activation and proinflammatory cytokines in Streptozotocin induced-diabetic nephropathy. Nutr Metab 8:35

Song C, Ji Y, Zou G et al (2015) Tetrandrine down-regulates expression ofmiRNA-155 to inhibit signal-induced NF-kappa B activation in a rat model of diabetes mellitus. Int J Clin Exp Med 8:4024–4030

Song Y, Fan X, Guo Z et al (2017) Therapeutic effects of emodin in type 2 diabetes mellitus in KKAy mouse model. Int J Clin Exp Med 10(10):14408–14413

Stanely Mainzen Prince P, Kannan NK (2006) Protective effect of Rutin on lipids, lipoproteins, lipid metabolizing enzymes and glycoproteins in Streptozotocin-induced diabetic rats. J Pharm Pharmacol 58(10):1373–1383

Stavniichuk R, Drel VR, Shevalye H et al (2011) Baicalein alleviates diabetic peripheral neuropathy through inhibition of oxidative-nitrosative stress and p38 MAPK activation. Exp Neurol 230 (1):106–113

Stofkova A (2009) Leptin and adiponectin: from energy and metabolic dysbalance to inflammation and autoimmunity. Endocr Regul 43:157–168

Suchal K, Malik S, Khan SI et al (2017) Protective effect of Mangiferin on myocardial ischemia-reperfusion injury in Streptozotocin-induced diabetic rats: role of AGE-RAGE/MAPK pathways. Sci Rep 7:42027

Sunil C, Duraipandiyan V, Agastian P et al (2012) Antidiabetic effect of plumbagin isolated from Plumbago zeylanica L. root and its effect on GLUT4 translocation in Streptozotocin-induced diabetic rats. Food Chem Toxicol 50:4356–4363

Tabák AG, Herder C, Rathmann W et al (2012) Prediabetes: a high-risk state for diabetes development. Lancet 379(9833):2279–2290

Tan MJ, Ye JM, Turner N et al (2008) Antidiabetic activities of triterpenoids isolated from bitter melon associated with activation of the AMPK pathway. Chem Biol 15:263–273

Tiong SH, Looi CY, Arya A et al (2015) Vindogentianine, a hypoglycemic alkaloid from Catharanthus roseus (L.) G. Don (Apocynaceae). Fitoterapia 102:182–188

Tutino V, Orlando A, Russo F et al (2016) Hydroxytyrosol inhibits cannabinoid CB1 receptor gene expression in 3T3-L1 pre adipocyte cell line. J. Cell Physiol 231:483–489

Uemura T, Goto T, Kang MS et al (2010) Diosgenin, the main aglycon of fenugreek, inhibits LXRa activity in HepG2 cells and decreases plasma and hepatictriglycerides in obese diabetic mice. J Nutr 141:17–23

Unger RH (1995) Lipotoxicity in the pathogenesis of obesity-dependent NIDDM: genetic and clinical implications. Diabetes 44:863–870

Vaidya H, Prajapati A, Rajani M et al (2012) Beneficial effects of swertiamarin on dyslipidaemia in streptozotocin-induced type 2 diabetic rats. Phytother Res 26(8):1259–1261

Vanitha P, Uma C, Suganya N et al (2014) Modulatory effects of morin on hyperglycemia by attenuating the hepatic key enzymes of carbohydrate metabolism and β-cell function in Streptozotocin-induced diabetic rats. Environ Toxicol Pharmacol 37:326–335

Verma S, McNeill JH (1994) Metformin improves cardiac function in isolated Streptozotocin-diabetic rat hearts. Am J Phys 266(2 Pt 2):H714–H719

Vinayagam R, Xu B (2015) Antidiabetic properties of dietary flavonoids: a cellular mechanism review. Nutr Metab (Lond) 12:60

Wang HL, Li CY, Zhang B et al (2014) Mangiferin facilitates islet regeneration and β-cell proliferation through upregulation of cell cycle and β-cell regeneration regulators. Int J Mol Sci 15:9016–9035

Wang T, Kusudo T, Takeuchi T et al (2013a) Evodiamine inhibits insulin-stimulated mTOR-S6K activation and IRS-1 serine phosphorylation in adipocytes and improves glucose tolerance in obese/diabetic mice. PLoS One:8e83264

Wang W, Wang C, Ding XQ et al (2013b) Quercetin and allopurinol reduce liver thioredoxin-interacting protein to alleviate inflammation and lipid accumulation in diabetic rats. Br J Pharmacol 169:1352–1371

Wang X, Liu R, Zhang W et al (2013c) Oleanolic acid improves hepatic insulin resistance via antioxidant, hypolipidemic and anti-inflammatory effects. Mol Cell Endocrinol 376:70–80

Wang Y, Ye J, Li J et al (2019) Polydatin ameliorates lipid and glucose metabolism in type 2 diabetes mellitus by downregulating pro-protein convertase subtilisin/kexin type 9 (PCSK9). Cardiovasc Diabetol 15:19
Wei Y, Gao J, Qin L et al (2017) Tanshinone I alleviates insulin resistance in type 2 diabetes mellitus rats through IRS-1pathway.Biomed. Pharmacotherapy 93:352–358
Wilson EA, Hadden DR, Merrett JD et al (1980) Dietary management of maturity onset diabetes. Br Med J 3:1367–1369
Wolfram S, Wang Y, Thielecke F (2006) Anti-obesity effects of green tea: from bedside to bench. Mol Nutr Food Res 50:176–187
Wu JB, Kuo YH, Lin CH et al (2014) Tormentic acid, a major component of suspension cells of Eriobotrya japonica, suppresses high-fat diet-induced diabetes and hyperlipidemia by glucose transporter 4 and AMP-activated protein kinase phosphorylation. J Agric Food Chem 62:10717–10726
Yadav S, Vats V, Dhunnoo Y et al (2002) Hypoglycemic and anti-hyperglycemic activity of Murraya koenigii leaves in diabetic rats. J Ethnopharmacol 82(2-3):111–116
Yamabe N, Noh JS, Park CH et al (2010) Evaluation of loganin, iridoid glycoside from Corni Fructus, on hepatic and renal glucolipotoxicity and inflammation in type 2 diabetic db/db mice. Eur J Pharmacol 648(1-3):179–187
Yang LP, Sun HL, Wu LM et al (2009) Baicalein reduces inflammatory process in a rodent model of diabetic retinopathy. Invest Ophthalmol Vis Sci 50:2319–2327
Yang W, She L, Yu K et al (2016) Jatrorrhizine hydrochloride attenuates hyperlipidemia in a high-fat diet-induced obesity mouse model. Mol Med Rep 14:3277–3284
Yokozawa T, Kim HY, Cho EJ et al (2002) Antioxidant effects of isorhamnetin 3, 7-di-O-beta-Dglucopyranoside isolated from mustard leaf (Brassica juncea) in rats with Streptozotocin-induced diabetes. J Agric Food Chem 50:5490–5495
Yoshimura Y, Nishii S, Zaima N et al (2013) Ellagic acid improves hepatic steatosis and serum lipid composition through reduction of serum resistin levels and transcriptional activation of hepatic PPAR-a in obese, diabetic KK-A(y) mice. Biochem Biophys Res Commun 434:486–491
Zeng XY, Wang H, Bai F et al (2015) Identification of Matrine as a promising novel drug for hepatic steatosis and glucose intolerance with HSP72 as an upstream target. Br J Pharmacol 172:4303–4318
Zhang BB, Moller DE (2000) New approaches in the treatment of type 2 diabetes. Curr Opin Chem Biol 4(4):461–467
Zhang WY, Lee JJ, Kim Y et al (2012) Effect of Eriodictyol on glucose uptake and insulin resistance in vitro. J Agric Food Chem 60:7652–7658
Zhang X, Wang G, Gurley EC et al (2014) Flavonoid apigenin inhibits lipo-polysaccharide-induced inflammatory response through multiple mechanisms in macrophages. PLoS One 9(9):e107072
Zhao F, Shi D, Li T et al (2015) Silymarin attenuates paraquat-induced lung injury via Nrf2-mediated pathway in vivo and in vitro. Clin Exp Pharmacol Physiol 42(9):988–998
Zheng T, Shu G, Yang Z et al (2012) Antidiabetic effect of total saponins from Entada phaseoloides (L.)Merr. in type 2 diabetic rats. J Ethnopharmacol 139:814–821
Zheng X, Zhu S, Chang S et al (2013) Protective effects of chronic resveratrol treatment on vascular inflammatory injury in Streptozotocin-induced type 2 diabetic rats: role of NF-kappa B signaling. Eur J Pharmacol 720:147–157

Indian Traditional Herbs and Alzheimer's Disease: Integrating Ethnobotany and Phytotherapy

50

Jai Malik and Sunayna Choudhary

Abstract

The use of plants for therapeutic reasons is perhaps as old as history of man itself. Different traditional systems of medicine that developed across the world depended largely on the usage of plants for treatment of numerous diseases. Indian system of medicine, i.e., "Ayurveda," is one such system, which is one of the oldest systems of medicine known to mankind. According to Ayurveda, the wellness of our body depends upon the delicate balance between the three physiological indicators, known as doshas, i.e., vata, pitta, and kapha. Their imbalance causes various physiological, metabolic, and also degenerative disorders. Among these, neurodegenerative disorders, such as Alzheimer's disease (AD), Huntington's disease, Parkinson's disease, etc., have attracted ample attention from the scientific community in the recent past. AD is one of the most extensively studied NDs throughout the globe. Besides modern therapeutic interventions, various herbal drugs have also been found beneficial in AD. Ayurveda has categorized certain plant drugs for such disorders under a separate class known as "Medhya Rasayana" (brain tonics). Traditional plants like Shankhpushpi, Brahmi, Gotu kola, Jyotishmati, Bach, Ashwagandha, etc. and their phytoconstituents have shown beneficial effects against AD. This chapter will highlight the protective role of these Indian traditional plants and their phytoconstituents, along with their mechanisms, against AD.

Keywords

Ayurveda · Alzheimer's disease · Ashwagandha · Brahmi · Shankhpushpi · Jyotishmati

J. Malik (✉) · S. Choudhary
Department of Pharmacognosy, University Institute of Pharmaceutical Sciences, Panjab University, Chandigarh, India
e-mail: jmalik_pu@hotmail.com; jmalik@pu.ac.in

S. C. Mandal et al. (eds.), *Evidence Based Validation of Traditional Medicines*,
https://doi.org/10.1007/978-981-15-8127-4_50

Abbreviations

AC	*Acorus calamus*
AChE	Acetylcholinesterase
AChEI	Acetylcholinesterase inhibitor
AD	Alzheimer's disease
APP	Amyloid precursor protein
Aβ	Amyloid beta
BM	*Bacopa monnieri*
C99	C-terminal fragment
CA	*Centella asiatica*
CL	*Curcuma longa*
CP	*Convolvulus pluricaulis*
CPa	*Celastrus paniculatus*
CREB	c-AMP response element binding protein
CT	*Clitoria ternatea*
DLB	Dementia with Lewy bodies
GG	*Glycyrrhiza glabra*
HD	Huntington's disease
IL-1 β	Interleukin-1 β
NDs	Neurodegenerative disorders
NFT	Neurofibrillary tangles
PD	Parkinson's disease
PSEN1	Presenilin 1
PSEN2	Presenilin 2
sAPPα	Soluble fragment of APP
sAPPβ	Soluble peptide APPβ
SDH	Succinate dehydrogenase
SLNs	Solid lipid nanoparticles
STZ	Streptozotocin
TC	*Tinospora cordifolia*
TNF-α	Tumor necrosis factor-α
WA	Withanamide A
WC	Withanamide C
WHO	World Health Organization
WS	*Withania somnifera*

50.1 Introduction

Neurodegeneration is a progressive loss or atrophy of neuronal function and structure leading to different disorders known as neurodegenerative disorders (NDs). Commonly encountered NDs are dementia, Alzheimer's disease, Parkinson's disease, Huntington's disease, stroke, etc. (Rubinsztein 2006). Among these, dementia is one of the most common NDs illustrated by cognitive loss which interferes mainly

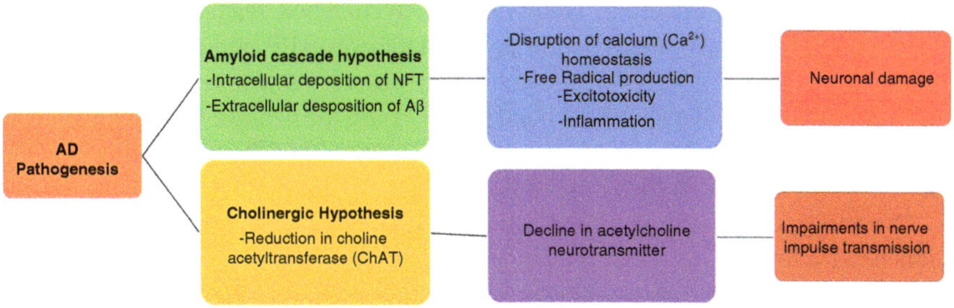

Fig. 50.1 Major hypothesis involved in AD pathogenesis

with the person's ability to perform daily chores (Ulep et al. 2018). Dementia has severely impacted the lives of not only the demented patients but also people associated with them (caregivers, family members, colleagues, friends, etc.). The global societal cost of the disease is US$1 trillion, and it is projected to get doubled by 2030. In addition to this, the annual global number of informal care is estimated to consume about 82 billion working hours (Patterson 2018). The impact of dementia is so sinister that the World Health Organization (WHO) in 2008 declared it is a priority condition in both developed and developing countries, like China, India, etc., and is also the fifth leading cause of death (Duthey 2013).

According to an estimate, in 2018, 50 million people suffered from dementia, and the figure is expected to reach 82 million in 2030. Every year nearly ten million new cases [1 new case every 3 s] of dementia are added across the globe (Patterson 2018). Besides aging, Alzheimer's disease (AD), Huntington's disease (HD), Parkinson's disease (PD), dementia with Lewy bodies (DLB), stroke, vascular dementia, and frontotemporal dementia have been enlisted as the common causes of dementia. Among these, AD has been identified as the most prominent cause as 60 to 70% of demented people suffer from AD, and the percentage increases to 80–90% when a person reaches to an age of more than 75 years (Gaugler et al. 2014). Therefore, most of the research for finding the newer therapeutic strategies for dementia is focused around AD.

The major pathological causes of AD include (a) plaques of amyloid β peptide (Aβ) deposition due to abnormal amyloid precursor protein (APP) processing, (b) neurofibrillary tangles (NFT) formed due to hyperphosphorylation of the tau proteins, and (c) cholinergic deficits which cause significant loss of neurons and synaptic changes in the brain areas responsible for memory functions like the hippocampus, cerebral cortex, etc. (Bhushan et al. 2018) (Fig. 50.1). APP is processed by the amyloidogenic pathway and the non-amyloidogenic pathway. In the amyloidogenic pathway, β-secretase cleaves APP into soluble peptide APPβ (sAPPβ) and a C-terminal fragment (C99). Further, γ-secretase acts on C99 to generate the Aβ and an intracellular APP domain. Non-amyloidogenic pathway involves APP cleavage by α-secretase, thereby producing a soluble fragment of APP (sAPPα) and C-terminal fragment. The latter is further acted upon by γ-secretase to produce APP domain and p3 peptide. AD is also caused by mitochondrial dysfunction, neurovascular dysfunction, calcium dysregulation, oxidative stress, neuro-inflammation and mutation in APP and presenilin (PSEN) genes (PSEN1 and PSEN2). APP gene has been associated with neural plasticity and synapse formation, whereas PSEN1 and PSEN2 form the heart of γ-secretase and are responsible for the stability and activity of the complex. Mutation in these genes increases Aβ42 levels in the brain (Bekris et al. 2010).

The only approved drugs for the management of AD are cholinesterase inhibitors [donepezil (Aricept®), rivastigmine (Exelon®), galantamine (Razadyne®)] and NMDA receptor antagonists [memantine (Namenda®)] (Birks 2006). The modern therapeutics only provide symptomatic relief and do not stop the underlying neuronal degeneration; therefore, alternative therapeutic strategies, which can prevent this neuronal loss, are always warranted. Over the last two to three decades, extensive interest has been developed in the therapeutic agents from natural sources as ample plants effective in NDs have been provided by "Mother Nature." Galantamine, an approved AChE inhibitor for the symptomatic treatment of AD, isolated from *Galanthus nivalis* (Amaryllidaceae), is one such example. Some of the lead compounds and plant extracts, viz., huperzine A, EGb761 (standardized extract of *Ginkgo biloba*), essential oil of *Salvia lavandulifolia*, etc., have shown promising effects in both preclinical and clinical studies (Hussain et al. 2018; Natarajan et al. 2013). Various traditional systems of medicine have mentioned different plants for such disorders. Ayurveda, Indian traditional system of medicine, has categorized many plants helpful in NDs under "Medhya Rasayana" (nervine tonic). *Medhya* means brain or intellect, and *Rasayana* means "tonic" or preparation that improves overall body functioning, including physical, immunological, and cognitive, leading to longevity (Kulkarni et al. 2012). Different plants that are mentioned under this category are Brahmi (*Bacopa monnieri*), Shankhpushpi (*Convolvulus pluricaulis*), Ashwagandha (*Withania somnifera*), Gotu kola (*Centella asiatica*), Jyotishmati (*Celastrus paniculatus*), Jatamansi (*Nardostachys jatamansi*), Guduchi (*Tinospora cordifolia*), and Yashtimadhu (*Glycyrrhiza glabra*) (Chunekar and Pandey 2002). The present chapter puts together the phytochemical and pharmacological aspects of different plant drugs mentioned for their beneficial effects against AD in Ayurveda.

50.2 Plants Used in Ayurveda to Treat Alzheimer's Disease

50.2.1 *Bacopa monnieri* Linn. (Brahmi, Family: Scrophulariaceae)

Bacopa monnieri (BM) or *Herpestis monniera* also referred to as water hyssop or Jalanimba has been used for centuries in the Ayurveda as *Medhya Rasayana*. The name *brahmi* was derived after the name of "Brahma," the mythical creator in the Hindu pantheon (Gohil and Patel 2010). Traditionally, it has been used in cognitive disorders and as anticonvulsant, anxiolytic, and sedative (Shinomol and Muralidhara 2011). The plant contains dammarane-type triterpenoid saponins known as bacosides as the major constituents. Bacopa also contains other saponins including bacopasaponins A–G, pseudojujubogenin, and bacopasides I–V, X, N1, and N2;

flavonoids, apigenin and luteonin; alkaloids, brahmine, herpestine, and hydrocotyline; glycosides, asiaticoside and thanakunicide; and other constituents like brahmic acid, brahamoside, brahminoside, and isobrahmic acid have also been isolated from BM (Chaudhari et al. 2017).

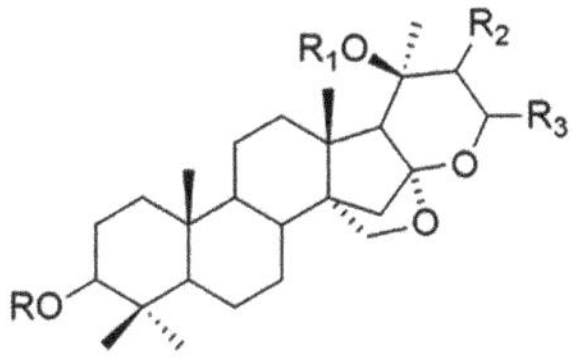

	R	R_1	R_2	R_3
Bacoside A_3	[β-D-glucopyranosyl-(1→3)]-O-{α-L-arabinofuranosyl-(1→2)}-O-(β-D-glucopyranosyl]	H	-CH=C(CH3)-CH3	H
Bacopaside II	[α-L-arabinofuranosyl-(1→2)-{β-D-glucopyranosyl-(1→3)}-β-D-glucopyranosyl]	H	H	-CH=C(CH3)-CH3
Jujubogenin	H	H	-CH=C(CH3)-CH3	H
Bacopasaponin C	[β-D-glucopyranosyl-(1→3)]-{α-L-arabinofuranosyl-(1→2)}-α-L-arabinopyranosyl]	H	H	-CH=C(CH3)-CH3

Among various bacosides, bacosides A and B have been found responsible for the neuroprotective and memory-enhancing activity of the plant (Vollala et al. 2011). Furthermore, bacoside A is a mixture of four different saponins, viz., bacoside A_3, bacopaside II, jujubogenin, and bacopasaponin C (Deepak and Amit 2004). Strong antioxidant activity, by metal chelation, free radical scavenging, and augmentation of antioxidant enzymatic activity, has been the mainstay in its memory-enhancing and neuroprotective activity (Russo et al. 2003). BM also exerts a protective effect against DNA damage in astrocytes and human fibroblasts (Chaudhari et al. 2017). Bacoside A protects the neurons by boosting the kinase activity, synthesizing neurons, and also improving synaptic activity, thereby improving the nerve impulse transmission. It also ameliorated the cognitive and memory impairment induced by scopolamine, colchicine, phenytoin, sodium nitrite, diazepam, and BN52021 (a platelet-activating factor antagonist) possibly by improving acetylcholine levels in the brain (Chaudhari et al. 2017). Moreover, methanol extract of BM decreased the formation of amyloid fibrils in vitro and also reduces Aβ40 and Aβ42 levels in brains of PSAPP mice (Mathew and Subramanian 2012). BM improves the cognitive skills by increasing the levels of serotonin and activating 5-HT3A receptors and CREB in the hippocampus of rodents. BM has also shown to mitigate the ethylcholine aziridinium ion-induced memory impairment and reduction in density of neurons and cholinergic neurons (Uabundit et al. 2010). Apart from alleviating memory, Brahmi also plays a vital role in treating other NDs like PD and 3-NP-induced HD-like symptoms (Mathur et al. 2016).

Furthermore, BM has remarkably improved verbal learning, memory acquisition, and logical memory in many double-blind, randomized, placebo-controlled trials (Calabrese et al. 2008; Chaudhari et al. 2017). Different nano-drug delivery systems, like polymeric nanoparticles and solid lipid nanoparticles (SLNs), have been employed for the successful delivery of bacosides to the brain. These delivery systems have considerably improved the solubility, bioavailability, and stability of bacosides (Bonifácio et al. 2014; Josea et al. 2014).

50.2.2 *Convolvulus pluricaulis* Chois. (Shankhpushpi, Family: Convolvulaceae)

Convolvulus pluricaulis (CP), also known as *C. prostratus* Forsk. and *C. microphyllus* Sieb., is a perennial herb, mentioned as a *Medhya Rasayana* in Ayurveda for its brain-stimulating and brain-rejuvenating properties (Bhowmik et al. 2012). In Ayurveda, its aerial parts along with small part of the roots are used in prophylaxis of various disorders (Chunekar and Pandey 2002). An alkaloid shankhpushpine ($C_{17}H_{23}NO_3$); coumarins-scopoletin, scopolin, ayapanin, and esculetin; a phytosterol-*β*-sitosterol; aliphatic compounds, viz., glycosidic acid, microphyllic acid, *n*-triacontane, and *n*-tetracontane; primary alcohols, viz., *n*-triacontanol, *n*-octacosanol, and *n*-hexacosanol; and flavonoids-kaempferol and its glycosides have been reported from the plant (Malik et al. 2016). Various fatty acids, viz., myristic, palmitic, and linoleic acid, have also been extracted from the plant (Agarwal et al. 2014).

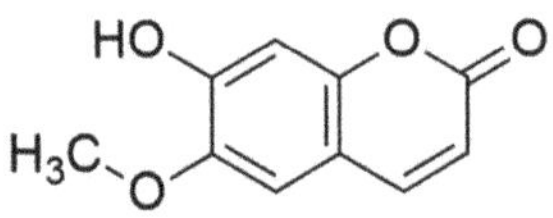

Scopoletin

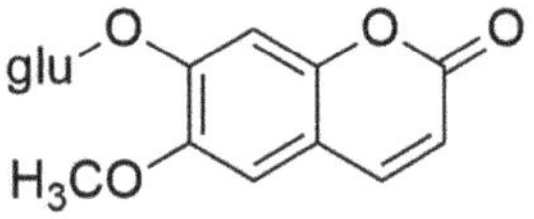

Scopolin

The plant possesses antioxidant, antifungal, antiulcerogenic, anti-cholesterolemic, anti-anxiety, and antidepressant activity. The memory-enhancing activity of CP has been established by numerous studies on various animal models (Malik et al. 2011; Nahata et al. 2008). CP inhibits AChE activity and attenuated the increase in levels of tau protein, Aβ levels and their aggregation, and oxidative stress in rat brains (Kizhakke et al. 2019). In a comparative study, CP showed maximum memory-enhancing activity among *Evolvulus alsinoides* and *Clitoria ternatea*, which are also commonly known as Shankhpushpi (Malik et al. 2011). In a bioactivity-guided fractionation and isolation study, scopoletin and scopolin were found to be responsible for the memory-enhancing activity as they attenuated the scopolamine-induced amnesia and also inhibited acetylcholinesterase enzyme (Malik et al. 2016). Clinically, when patients were treated with Shankhpushpi tablets (made of powder and juice of *C. pluricaulis*), their auditory and visual recognition and long-term memory improved (Chen et al. 2017). The plant, by virtue of its strong antioxidant action, has also showed protective effect against 3-NP-induced HD-like symptoms in rats (Malik et al. 2015).

50.2.3 *Withania somnifera* Dunal (Ashwagandha, Family: Solanaceae)

Withania somnifera (WS) is a well-known and widely used *Rasayana* drug in Ayurveda for increasing longevity and vitality (Winters 2006). The plant has reported antioxidant, anti-inflammatory, immunomodulating, anti-stress, memory-enhancing, and antiepileptic properties. Chemical analysis of Ashwagandha shows the presence of steroidal lactones, known as withanolides (withaferin A, withanolides A–P, withanone, sitoindosides VII–X) and alkaloids (withanine, somniferine, somnine, somniferinine, etc.) (Mirjalili et al. 2009).

The plant increases cortisol circulation and physical performance, decreases fatigue and depression in animals subjected to stress, and influences various neurotransmitter receptor systems (Naidu et al. 2006). WS has also shown neuroprotective activity in AD by enhancing the peripheral clearance of Aβ and increasing its degradation (Sehgal et al. 2012). WS and its constituents also possess significant antioxidant activity that further helps in improving the memory. Sitoindosides VII–X and withaferin A exhibited strong antioxidant effect by increasing the levels of endogenous antioxidant enzymes and decreasing lipid peroxidation (Bhatnagar et al. 2005; Gupta et al. 2003). These compounds have also protected neurons against the ibotenic acid-induced neurotoxicity (Bhattacharya et al. 2006). Withanolide A and withanosides IV and VI have protected cortical neurons against Aβ peptide-induced neurotoxicity (Kuboyama et al. 2006). Withanone showed preventive effect against scopolamine-induced DNA damage and oxidative stress in C6 cells (Kuboyama et al. 2014). Withanamides A (WA) and C (WC) protect PC-12 cells from beta-amyloid-induced cell damage by binding to the active motif of β-amyloid and preventing the formation of fibrils (Jayaprakasam et al. 2010). Moreover, WS and its active constituents modulate various targets associated with APP processing and also enhance Aβ clearance from the brain, a property which is valuable in designing newer and potent therapeutics for AD (Patil et al. 2010).

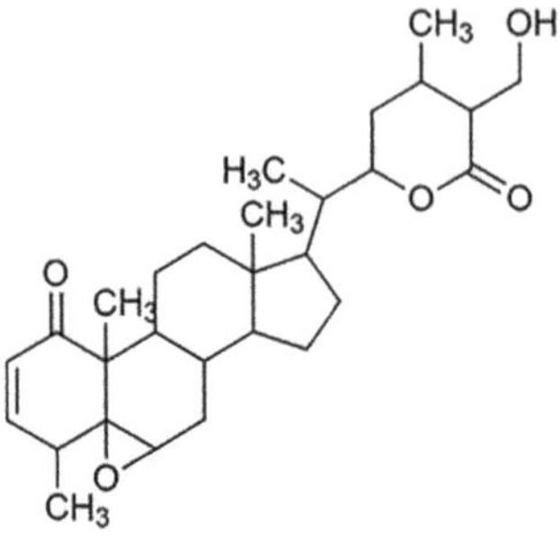

Withaferin A

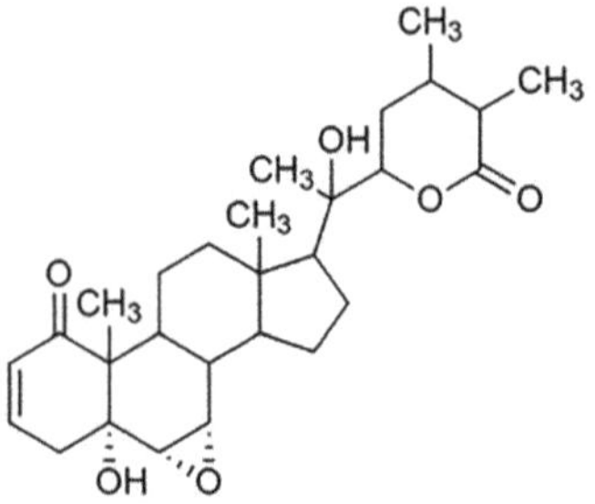

Withanolide A

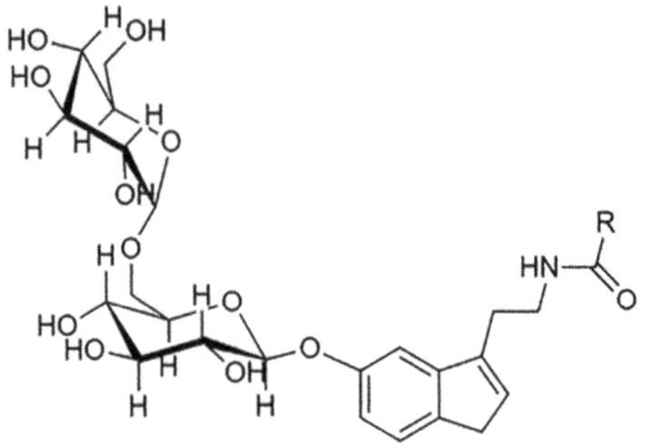

R 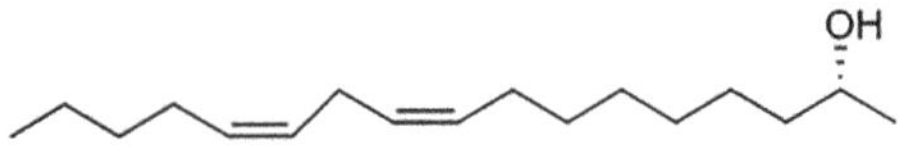

WA

R 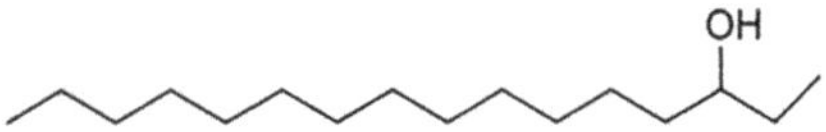

WC

R consisting of Withanamide A (WA) and Withanamide C (WC)

50.2.4 *Centella asiatica* Linn (Jal Brahmi, Family: Apiaceae)

Centella asiatica L. (CA) (*Syn. Centella coriacea* Nannfd., *Hydrocotyle asiatica* L., *Hydrocotyle lunata* Lam., and *Trisanthus cochinchinensis* Lour.) is commonly known as Gotu kola (miracle elixirs of life), Mandukaparni, Indian pennywort, Asiatic pennywort, Indian water navelwort, wild violet, and tiger herb (Gohil et al.

2010). It is also used as a *Medhya Rasayana* drug by virtue of its effectiveness in cognitive disorders and other NDs. CA has exhibited various neuropharmacological effects comprising of antiepileptic, anxiolytic (Gohil et al. 2010), memory-enhancing, neurite elongation, and nerve regeneration activities (Lokanathan et al. 2016; Soumyanath et al. 2005). Triterpenoidal saponins, namely, asiaticoside, madecassoside, asiatic acid, and madecassic acid, are the major constituents of CA. Centellasaponins A–D, brahmoside, and brahminoside are the saponins present in minor quantities (Siddiqui et al. 2007). Various terpenes, scentellin, asiaticin and centellicin, triterpene acids, betullic acid, brahmic acid, isobrahmic acid, thankunic acid, and isothankunic acid have also been reported from the plant (James and Dubery 2009).

The plant has shown protective effect against colchicine- and STZ-induced memory impairment and oxidative damage in rat brain (Kumar et al. 2009; Kumar and Gupta 2003). It increases the phosphorylation of CREB both in neuroblastoma cell line expressing Aβ (1–42) and in rat embryonic cortical primary cell culture, thereby illustrating the mechanism of its memory-enhancing property (Xu et al. 2008). Asiatic acid protected neurons from the glutamate-induced oxidative damage (Lee et al. 2000) and aluminum chloride-induced amyloid burden (Rather et al. 2019) in cultured neurons and rats, respectively. CA ethanol extract inhibits acetylcholinesterase, butyrylcholinesterase, and tyrosinase enzymes, thereby restoring the levels of important neurotransmitters in the brain (Orhan et al. 2008). Furthermore, CA extract improves the neuronal morphology by increasing the dendritic arborization in CA3 neurons located in the hippocampus, thereby improving the memory (Rao et al. 2005). In addition to this, CA extract also reduces the levels of hippocampal Aβ plaques in PSAPP mice (Dhanasekaran et al. 2009) and mitigates the Aβ-induced oxidative burden and mitochondrial dysfunction in neuroblastoma cells (Gray et al. 2015). CA aqueous extract and its triterpene compounds attenuate the amyloid-β-induced neurodegenerative spine loss and dendritic simplification, indicating their broader therapeutic, beyond AD (Gray et al. 2017).

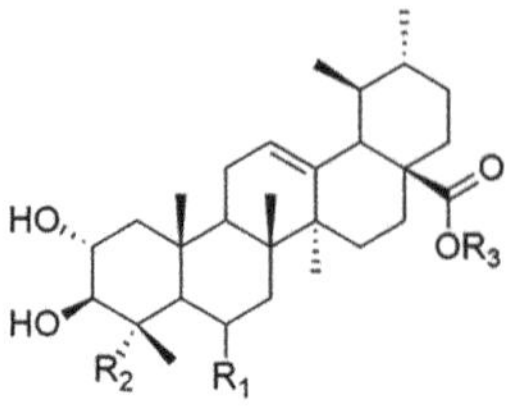

	R1	R2	R3
Asiatic Acid	H	CH_2OH	H
Madecassic Acid	OH	CH_2OH	H
Madasiatic acid	OH	CH_3	H
Asiaticoside	H	CH_2OH	Glucose-glucose-rhamnose
Madecassoside	H	CH_2OH	Glucose-glucose-rhamnose

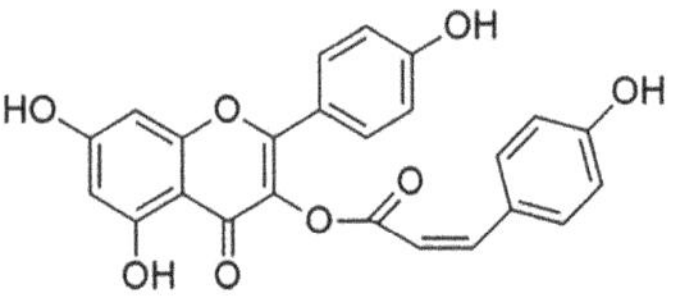

Castilliferol

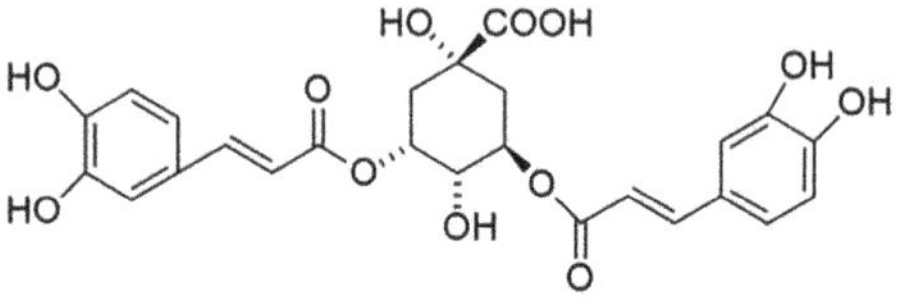

Isochlorogenic acid

50.2.5 *Curcuma longa* (Haldi, Family: Zingiberaceae)

Curcuma longa (CL), also known as turmeric, is a rhizomatous perennial herb with a long history of use as a condiment and coloring agent. It is used in household remedies for the healing wounds, inflammation, and different skin problems. CL contains various curcuminiods responsible for its yellow color, sesquiterpenes, volatile oil, and starch. Most of the curcuminiods are diarylheptanoid derivatives, and among them curcumin is the major bioactive component (Choudhary et al. 2013). Curcumin is known for its wide range of activities like neuroprotective, anticancer, wound healing, antioxidant, anti-inflammatory, antimicrobial, antiseptic, antiparasitic, choleretic, analgesic, hepatoprotective, and anti-mutagenic activity.

CL has shown its neuroprotective action through different mechanisms of action such as antioxidant, anti-amyloid, anti-tau, anti-inflammatory, etc. The antioxidant potential of CL is due to free radical scavenging, metal chelation, and upregulation of endogenous antioxidant enzymes (Choudhary et al. 2013; Kumar and Sakhya 2013). Curcumin, due to its potent antioxidant property, has also been found effective against vascular dementia and in the treatment of age-related cognitive dysfunctions. It also antagonizes the amyloid pathogenesis by reducing the Aβ burden in AD transgenic mice (Lim et al. 2001; Yang et al. 2005). Curcumin, being lipophilic, crosses the BBB and binds to Aβ plaques. It destabilizes the Aβ polymer and inhibited β-amyloid-40 aggregation (Mishra and Palanivelu 2008). It also inhibited the APP metabolism by binding to the amyloid beta peptide (Narlawar et al. 2008). Curcumin also reduces the levels of Aβ40/42, PSEN1 protein, and mRNA in neuroblastoma cells with overexpressing APP and also inhibits A-β-induced tau phosphorylation via PTEN/Akt/GSK-3bm pathway (Huang et al. 2014). Very recently, curcumin has abrogated scopolamine-induced $A\beta_{40/42}$ and tau hyperphosphorylation by downregulating GSK3 and Cdk5 (Das et al. 2019). Curcumin reduced the levels of $A\beta_{40}/42$ and phosphorylated tau at Ser396 (PHF13) and Ser202/Thr205 in the brain and in the plasma. It also increases the clearance of Aβ by increasing its phagocytosis in AD patients (Fiala et al. 2007).

Apart from curcumin, other compounds like calebin A, demethoxycurcumin, bisdemethoxycurcumin, and 1,7-bis(4-hydroxyphenyl)-1-heptene-3,5-dione isolated from CL protected PC12 cells against *Aβ*-induced toxicity (Park and Kim 2002). Lipid-core nanocapsules of curcumin attenuate Aβ-induced increase in TNF-α and IL-1 β levels in rodents and mitigate the cognitive impairment. It also protected brain cells against stroke by stimulating Sirt1 and Bcl-2 expression. Though curcumin did not exhibit any toxic effect during short-term use, studies on long-term safety and efficacy on human subjects are warranted.

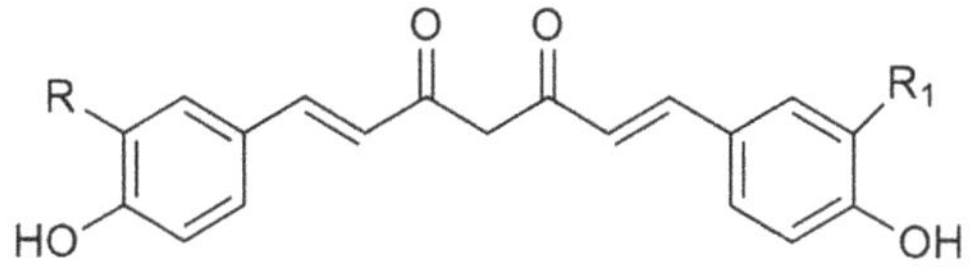

$R, R_1 = OCH_3$ **(Curcumin)**
$R = H, R_1 = OCH_3$ **(Demethoxycurcumin)**
$R, R_1 = H$ **(Bisdemethoxycurcumin)**

50.2.6 *Acorus calamus* (Bach, Family: Araceae)

Acorus calamus Linn. (AC), also known as Vacha or sweet flag, is a valued Ayurvedic brain rejuvenator. The words "Acoron (eye pupil) and Calamos (a reed)" led to the genesis of the name of genus and species, i.e., *Acorus* and *Calamus*, respectively. Traditionally, the plant was used to treat anorexia, GIT disorders, bronchitis, chest pain, and nervous disorders. AC rhizomes are one of the most common ingredients in Chinese herbal remedy for age-related dementia (May et al. 2016). Phytoconstituents like tannins, glycosides, polyphenolic compounds, flavonoids, sesquiterpenes, saponins, mucilage, essential oil, and bitter principle have been reported in AC rhizomes. It contains calamen, clamenol, calameon, calamendiol, β-asarone, α-asarone, acorine, acoradin, acorone, and galangin along with eugenol, pinene, and camphene (Chandra and Prasad 2017).

The hydro-alcohol extract and essential oil of AC rhizomes showed in vitro acetylcholinesterase inhibitory activity with oil exhibiting better activity than the extract (Mukherjee et al. 2007b). Oral administration of AC mitigated hyoscine-induced cognitive loss in rats (Howes et al. 2017). Among different chemical constituents, β-asarone and α-asarone showed in vitro acetylcholinesterase inhibitory activity which was more prominent in case of β-asarone than α-asarone (Mukherjee et al. 2007a). β-Asarone also exhibited neuroprotective action and curb neuronal apoptosis by downregulating Bcl2, Bcl-w, and caspase-3 and preventing JNK phosphorylation (Nair et al. 2019). α-Asarone also exhibited memory improvement, through cholinergic and antioxidant effects, in rodents injected with β-amyloid (Howes et al. 2017; Kumar et al. 2012). AC also exhibited neuroprotective effect by preventing oxidative stress induced by the histomorphological changes in the CA1 and CA3 regions of rat hippocampus (Subamalani et al. 2019).

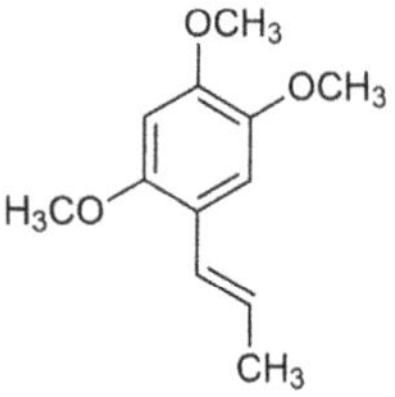

α- Asarone

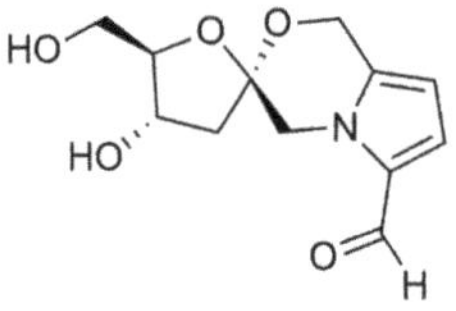

Acortatarin A

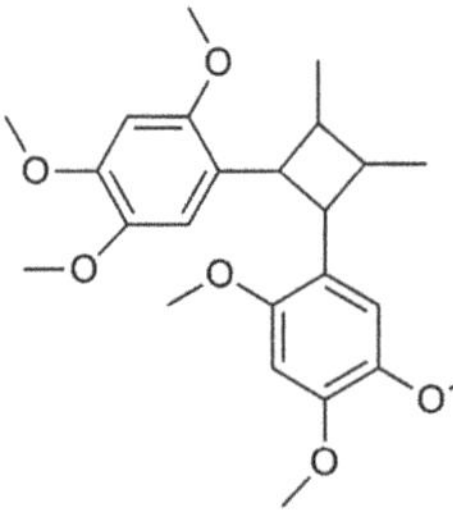

Acoradin

50.2.7 *Glycyrrhiza glabra L.* (Mulhatti, Family: Fabaceae)

Glycyrrhiza glabra Linn. (GG), also known as licorice, Jashtimadhu, and Yashtimadhu, is one of the most widely used herbs in Ayurveda for its medicinal and flavoring properties. Hippocrates (400 BC) mentioned it as an antiulcer and thirst-quenching agent. Licorice is also mentioned in Siddha system of medicine for its cough suppressant, expectorant, demulcent, laxative, and sweetening properties (Kaur et al. 2013). The roots and rhizomes of GG contain saponin (glycyrrhizin), flavonoids (liquirtin, isoliquertin, liquiritigenin, rhamnoliquirilin, glucoliquiritin apioside, prenyllicoflavone A, shinflavanone, shinpterocarpin, 1-methoxyphaseolin, and glabridin), and carbenoxolone and volatile components (pentanol, hexanol, linalool oxides A and B, isoangustone A, and licoriphenon). The essential oil contains various constituents such as methyl ethyl ketone, 2,3-butanediol, furfuraldehyde, furfuryl formate, 1-methyl-2-formylpyrrole, trimethylpyrazine, and maltol (Pastorino et al. 2018).

The aqueous extract of GG significantly improved memory and learning in diazepam- and scopolamine-induced amnesia (Dhingra et al. 2004). Licorice also augments succinate dehydrogenase (SDH) activity and energy supply in brain parts (Oganisyan et al. 2005). Glycyrrhiza is one of the major components of the traditional "Japanese Kampo" medicine, "Yokukansan," that antagonizes α2A adrenoceptors. In addition, isoliquiritigenin inhibited NMDA receptors, suppressed caspase-3, and showed protection against cerebral ischemia-reperfusion injury in rats (Zhan and Yang 2006). Carbenoxolone protected skeletal muscles and hippocampal neurons against acute ischemic-reperfusion by suppressing superoxide anions and hydrogen peroxide generation in macrophages in rats (Hosseinzadeh et al. 2005). Glabridin showed cognitive improvements, due to its antioxidant, neuroprotective, and anticholinesterase activities, in diabetic rats (Hasanein 2011).

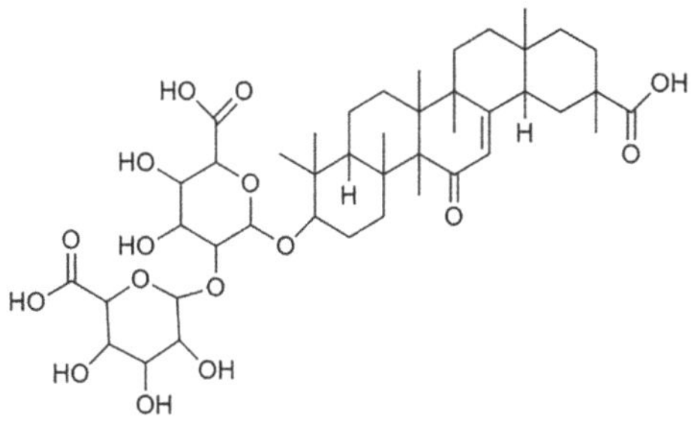

Glycyrrhizic acid

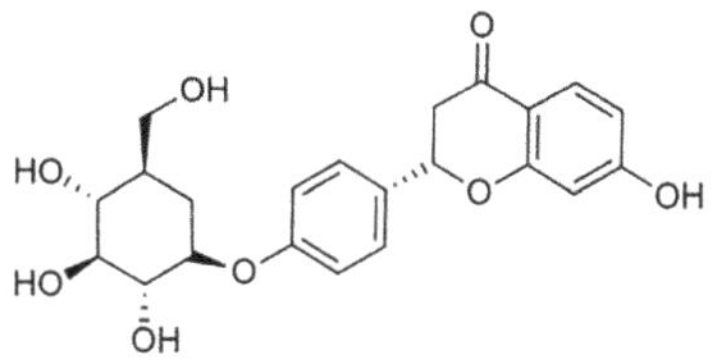

Liquiritin

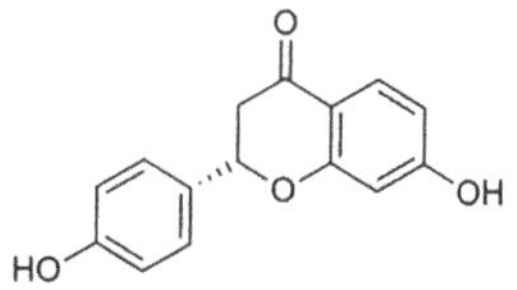

Liquiritigenin

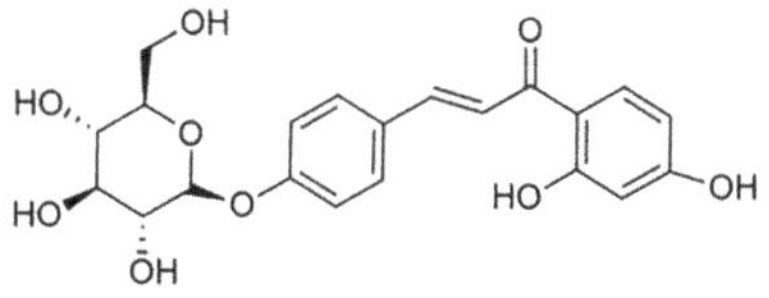

Isoliquiritin

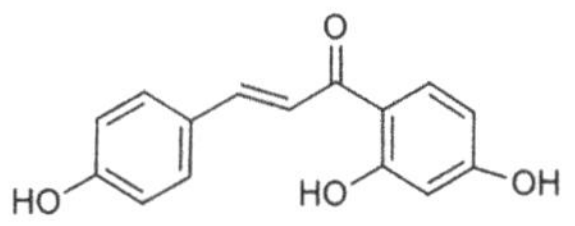

Isoliquiritigenin

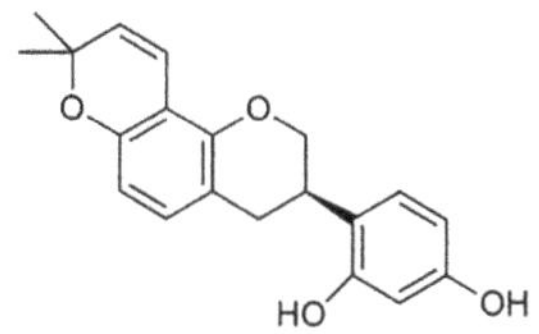

Glabridin

50.2.8 *Tinospora cordifolia* Willd. (Guduchi, Family: Menispermaceae)

Tinospora cordifolia (TC) or Giloya, a large, glabrous, deciduous climbing shrub, is categorized as "Rasayana" drug in Ayurveda due to its immunomodulating properties (Singh et al. 2003). According to Hindu mythology, "Giloya or Giloe" denotes to the blissful elixir that has saved celestial beings from old age and kept them forever young. The other common names are Guduchi (means one which protects the entire body), Amrita (means ability to impart youthfulness, vitality, and longevity), Amritavalli, Madhuparni, Guduchika, Gilo, and heartleaf moonseed. Traditionally, it has been widely used as general tonic, antiperiodic, anti-spasmodic, anti-inflammatory, antiarthritic, anti-allergic, anti-stress, anti-leprotic, anti-malarial, and anti-diabetic agent. Constituents belonging to different phytochemical groups like polyphenolics, sesquiterpenoids (tinocordifolin), diterpenoid lactones (tinosporin, tinosporide, columbin), alkaloids (tinosporine, tinosporidine, berberine, tembertarine, jatrorrhizine, magniflorine, palmatine), glycosides (tinosporaside, tinocordiside, cordifolide), steroids (β-sitosterol, gilosterol), aliphatic compounds (heptacosanol), and carbohydrates (syringing) have been isolated from TC (Joshi and Kaur 2016).

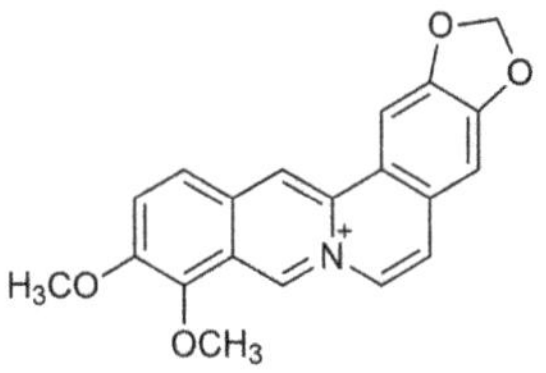

Berberine

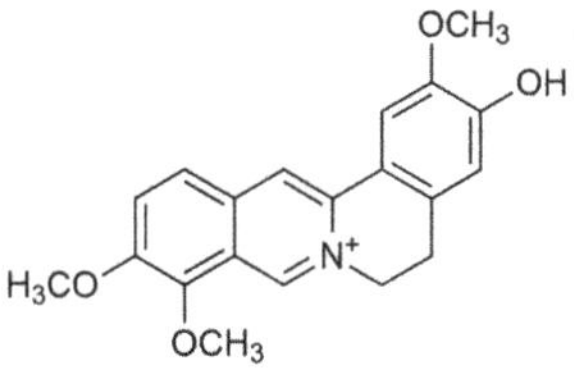

Jatrorrhizine

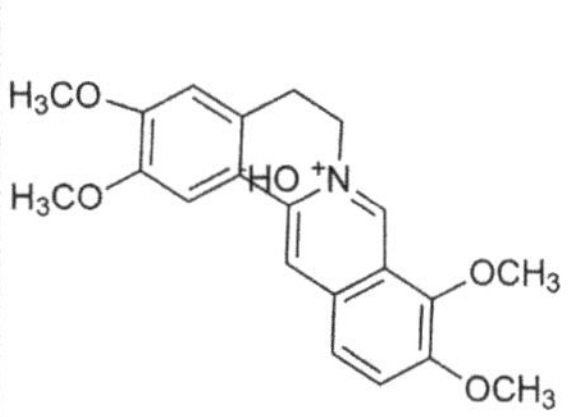

Palmatine

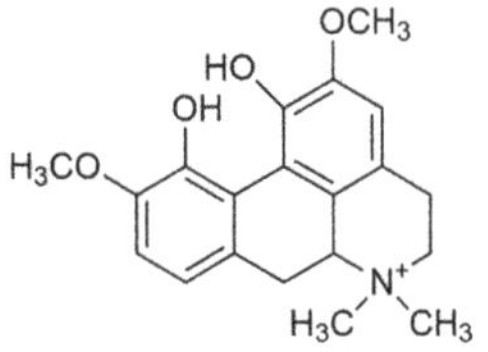

Magnoflorine

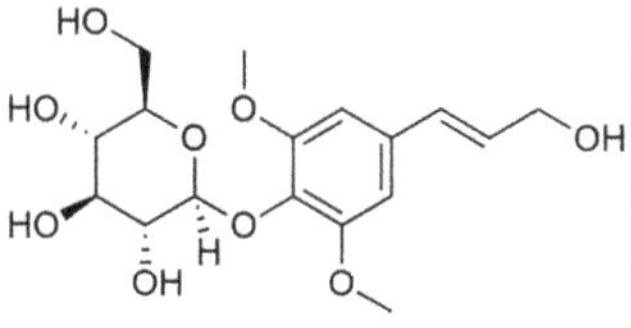

Syringin

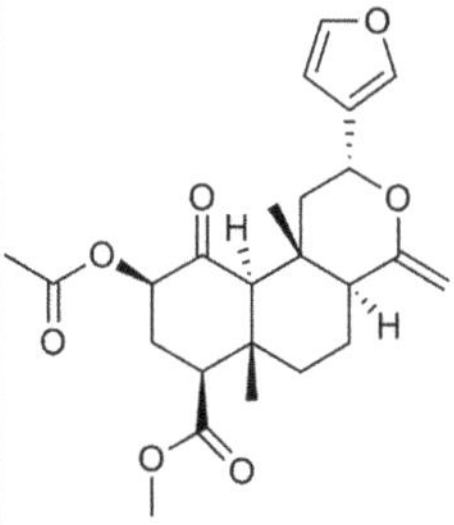

Furanolactone

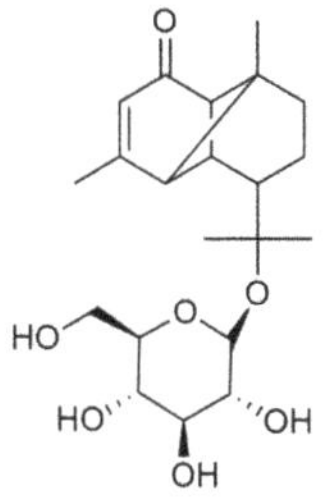

Tinocordiside

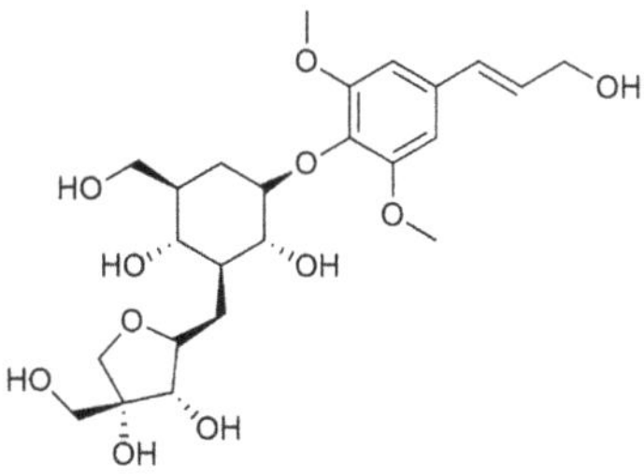

Cordifolioside A

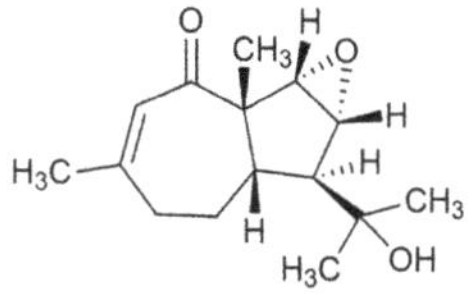

Tinocordifolioside

T. cordifolia along with *Phyllanthus emblica* and *Ocimum sanctum* showed nootropic activity against scopolamine-, diazepam-, and cyclosporine-induced amnesia in Wistar rats (Malve et al. 2014). Its combination with *Bacopa monnieri* and *Evolvulus alsinoides* showed significant cognitive improvement against scopolamine-induced amnesia (Gupta et al. 2013). In a 21-day randomized, double-blind, placebo-controlled study, pure aqueous extract of TC root improved the verbal learning and logical memory of the volunteers (Bairy et al. 2004). The alcoholic and aqueous extracts of TC have shown improved cognition against cyclosporine-induced memory deficit in rats. The memory enhancing effect was a evident from the histopathological examination of hippocampal cells and improved scores in Hebb-Williams maze test (Upadhyay et al. 2010). The plant has shown protective effect against neuronal degeneration, and exploring it further may enlighten the exact mechanism responsible for the action.

50.2.9 *Clitoria ternatea* (Aparajita, Family: Fabaceae)

Clitoria ternatea (CT) also known as Koyal, butterfly pea, blue bell vine, and Darwin pea is a tropical perennial climber with white or blue flowers. An Ayurvedic preparation "Medhya Rasayana" for different cognitive and CNS disorders has CT as a major ingredient. CT roots are also used to relieve inflammation, constipation, arthritis, indigestion, bronchitis, fever, eye and throat disorders, and asthma. Different plant parts of CT contain a wide array of constituents like pentacyclic triterpenoids (taraxerol, taraxerone), sterols (β-sitosterol, γ-sitosterol), and flavonoids and their glycosides in roots; anthocyanins (ternatins A1, A2, B1, B2, D1, and D2) in flowers; lactones (aparajitin and clitorin) in leaves; and fatty acids and proteins in seeds (Mukherjee et al. 2008).

Clitoria ternatea is also used under the common name "shankhpushpi," along with *Convolvulus pluricaulis* and *Evolvulus alsinoides*, in different parts of our country. The memory-enhancing, anxiolytic, and antidepressant activities of these three plants were evaluated and compared (Malik et al. 2011). In a comparative study between the memory-enhancing and cholinergic activities of roots and aerial parts, roots were found to be more biologically active than aerial parts (Taranalli and Cheeramkuzhy 2000). CT roots have also shown to augment Ach in hippocampus region (a region responsible for learning and memory) of rat brain (Rai et al. 2002). They also improved dendritic arborization in the neurons of amygdala region and increased the proliferation and growth of neurospheres leading to improved cognition in rats (Rai et al. 2005). Roots of CT have also attenuated the streptozotocin (STZ)-induced Alzheimer-like conditions via AChE inhibitory and antioxidant activity and also by downregulating the ROCK II expression in the brain (Mehla et al. 2013). Additionally, "Medhya Rasayana" formulation of CT improved cognition and protected neurons against kainic acid-induced neurotoxicity (Raghu et al. 2017). CT roots also boost the synthesis of neurotransmitter such as acetylcholine similar to synthetic drugs such as nefiracetam, dehydroepiandrosterone sulfate, and FG7142 (Esmail Al-Snafi 2016; Muhammad Ezzudin and Rabeta 2018). Various studies on CT have indicated its potential that could be exploited in AD treatment (Gollen et al. 2018).

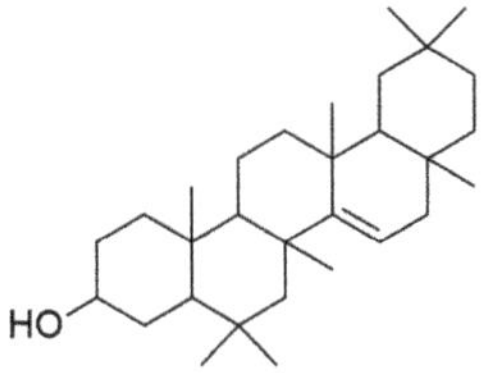

Taraxerol

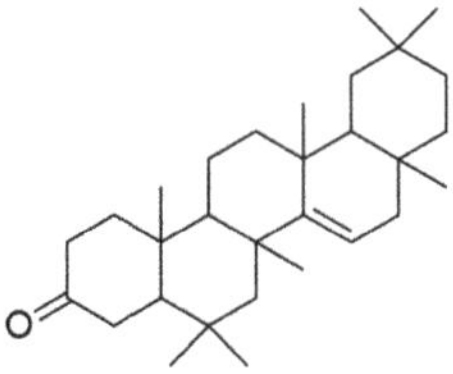

Taraxerone

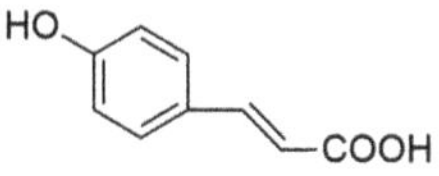

p-Hydroxycinnamic acid

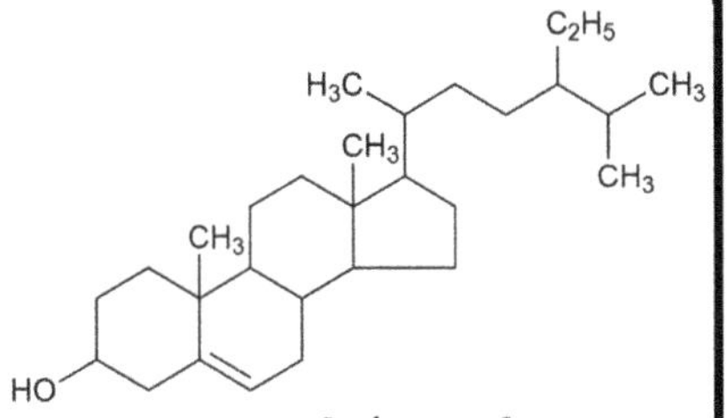

β-sitosterol

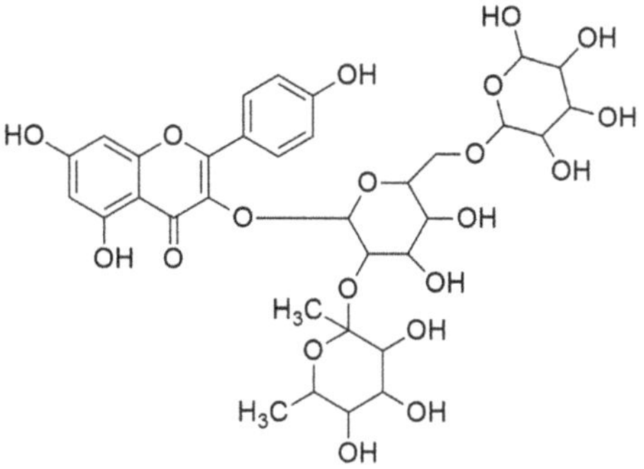

Clitorin

50.2.10 *Celastrus paniculatus* (Malkangani, Family: Celastraceae)

C. paniculatus (CPa), also known as "Jyotishmati" or "tree of life," is a valued "Medhya Rasayana" herb used in Ayurveda for centuries to improve learning and memory (Bhanumathy et al. 2010). Traditionally it has been used for its benefits against rheumatism, gout, leprosy, leukoderma, paralysis, and asthma. The seed oil of CPa has been used to improve memory and increase mental awareness (Lekha et al. 2010). Different parts of CPa have been used for various disorders like roots against malaria, wood for tuberculosis, stems for urinary disorders, fruits as antiflatulent, and leaves in dysentery (Shen et al. 2019). The plant contains alkaloids such as wifornine F and paniculatines A and B in the stem and celastrine, celapagine, celapanigine, and celapanine in the seeds; polyhydric alcohol such as malkanguniol, malkangunin, paniculatadiol, and malkanginnol; triterpenoids like pristimerin; triterpenes such as β-amyrin and phytosterol, β-sitosterol (Shashank et al. 2017).

Seeds of CPa protected neurons against H_2O_2-induced and glutamate-induced toxicity by their free radical scavenging properties, reducing lipid peroxidation, stimulating endogenous antioxidant enzymes, and modulating glutamate receptor activity, respectively (Godkar et al. 2006). CPa seeds have shown to improve cognitive function by exhibiting antioxidant (Kumar and Gupta 2002), cholinergic (da Rocha et al. 2011), and cholinesterase inhibitory (Alama and Haque 2011) activities. The seeds also increase myelination and brain phospholipid content (Bidwai et al. 1987).

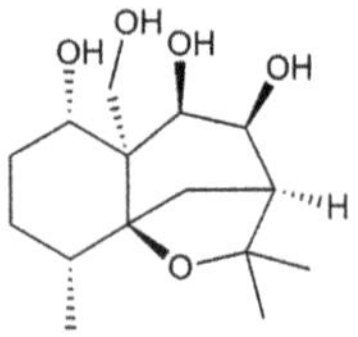

Malkanguniol

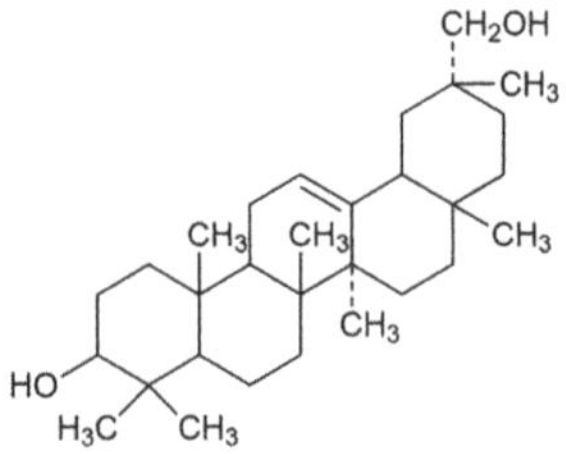

Paniculatadiol

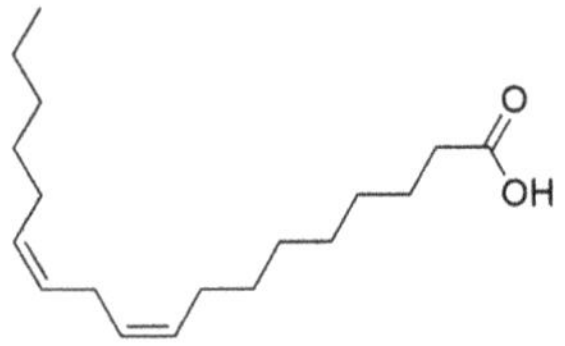

Linoleic acid

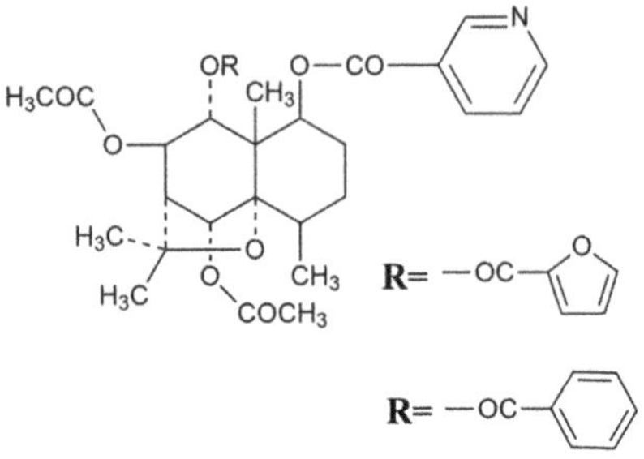

Celapanin

Celapanigin

50.3 Conclusion

Plant drugs form the central core of all the traditional systems of medicines. In recent years, a remarkable amplification in the use of herbal products/medicines, in both developing and developed countries, has been observed. Besides ample data and knowledge of AD pathophysiology, the presently available treatments (approved by US FDA) provide only symptomatic relief with various side effects. In addition to this, these drugs do not affect or slow down the progression of the underlying disease. It is now well anticipated that alternative systems, especially plant-based therapeutics, can be helpful in delaying and altering the course of this dreadful disease. Therefore, thorough and target-oriented investigations of various plants and/or their constituents, which can prove beneficial in the treatment of AD, are being done. Many plants and their constituents have been found effective in various clinical trials. But still, there are lot many plants that have not been explored for their potentiality in combating this disease. Very less scientific data authenticating the role of such plants or their active constituents and their mechanism of action is available. Thus, meticulous scientific exploration of these plants, which will help in developing newer therapeutics for the treatment of dementia, is highly warranted.

References

Agarwal P, Sharma B, Fatima A, Jain SK (2014) An update on Ayurvedic herb Convolvulus pluricaulis Choisy. Asian Pac J Trop Biomed 4:245–252

Alama B, Haque E (2011) Anti-alzheimer and antioxidant activity of Celastrus paniculatus seed. Iran J Pharm Res 7:49–56

Bairy KL, Rao Y, Kumar KB (2004) Efficacy of Tinospora cordifolia on learning and memory in healthy volunteers: a double-blind, randomized, placebo controlled study. Iran J Pharmacol Ther 3:57–60

Bekris LM, Yu CE, Bird TD, Tsuang DW (2010) Genetics of Alzheimer disease. J Geriatr Psychiatry Neurol 23:213–227

Bhanumathy M, Harish MS, Shivaprasad HN, Sushma G (2010) Nootropic activity of Celastrus paniculatus seed. Pharm Biol 48:324–327

Bhatnagar M, Sisodia SS, Bhatnagar R (2005) Antiulcer and antioxidant activity of Asparagus racemosus Willd and Withania somnifera Dunal in rats. Ann N Y Acad Sci 1056:261–278

Bhattacharya SK, Kumar A, Ghosal S (2006) Effects of glycowithanolides from Withania somnifera on an animal model of Alzheimer's disease and perturbed central cholinergic markers of cognition in rats. Phytother Res 9:110–113

Bhowmik D, Kumar KPS, Paswan S, Srivatava S, Yadav AP, Dutta A (2012) Traditional Indian herbs Convolvulus pluricaulis and its medicinal importance. J Pharmacogn Phytochem 1:44–51

Bhushan I, Kour M, Kour G, Gupta S, Sharma S, Yadav AP (2018) Alzheimer's disease: causes & treatment—a review. Ann Biotechnol 1:1002–1010

Bidwai PP, Wangoo D, Bhullar NK (1987) Effect of Celastrus paniculatus seed extract on the brain of albino rats. J Ethnopharmacol 21:307–314

Birks J (2006) Cholinesterase inhibitors for Alzheimer's disease. Cochrane Database Syst Rev 25: CD005593

Bonifácio BV, Silva PB, Ramos MAS, Negri KMS, Bauab MT, Chorilli M (2014) Nanotechnology-based drug delivery systems and herbal medicines: a review. Int J Nanomedicine 9:1–15

Calabrese C, Gregory WL, Leo M, Kraemer D, Bone K, Oken B (2008) Effects of a standardized Bacopa monnieri extract on cognitive performance, anxiety, and depression in the elderly: a randomized, double-blind, placebo-controlled trial. J Altern Complement Med 14:707–713

Chandra D, Prasad K (2017) Phytochemicals of Acorus calamus (sweet flag). J Med Plants Stud 5:277–281

Chaudhari KS, Tiwari NR, Tiwari RR, Sharma RS (2017) Neurocognitive effect of nootropic drug Brahmi (Bacopa monnieri) in Alzheimer's disease. Ann Neurosci 24:111–122

Chen GT, Lu Y, Yang M, Li JL, Fan BY (2017) Medicinal uses, pharmacology, and phytochemistry of convolvulaceae plants with central nervous system efficacies: a systematic review. Phytother Res 32:823–864

Choudhary S, Kumar P, Malik J (2013) Plants and phytochemicals for Huntington's disease. Pharmacogn Rev 7:81–91

Chunekar KC, Pandey GS (2002) Bhavaprakasa Nighantu. Indian Materia Medica of Sri Bhavamisra Chaukhambha Bharati Academy, Varanasi

da Rocha MD et al (2011) The role of natural products in the discovery of new drug candidates for the treatment of neurodegenerative disorders II: Alzheimer's disease. CNS Neurol Disord Drug Targets 10:251–270

Das TK, Jana P, Chakrabarti SK, Abdul Hamid MRW (2019) Curcumin downregulates GSK3 and Cdk5 in scopolamine-induced Alzheimer's disease rats abrogating Aβ40/42 and tau hyperphosphorylation. J Alzheimers Dis Rep 3:257–267

Deepak M, Amit A (2004) The need for establishing the identities of bacoside a and B, the putative major bioactive saponins of Indian medicinal plant Bacopa monnieri. Phytomedicine 11:264–268

Dhanasekaran M, Holcomb LA, Hitt AR, Tharakan B, Porter JW, Young KA, Manyam BV (2009) Centella asiatica extract selectively decreases amyloid beta levels in hippocampus of Alzheimer's disease animal model. Phytother Res 23:14–19

Dhingra D, Parle M, Kulkarni SK (2004) Memory enhancing activity of Glycyrrhiza glabra in mice. J Ethnopharmacol 91:361–365

Duthey B (2013) Background paper 6.11: Alzheimer disease and other dementias. A public health approach to innovation, pp 1–74. https://www.who.int/medicines/areas/priority_medicines/BP6_11Alzheimer.pdf. Accessed 1 June 2020

Esmail Al-Snafi A (2016) Pharmacological importance of Clitoria ternatea—a review. IOSR J Pharm 6:68–83

Fiala M et al (2007) Innate immunity and transcription of MGAT-III and toll-like receptors in Alzheimer's disease patients are improved by bisdemethoxycurcumin. Proc Natl Acad Sci 104:12849–12854

Gaugler J, James B, Johnson T, Scholz K, Weuve J (2014) 2014 Alzheimer's disease facts and figures. Alzheimers Dement 10:e47–e92

Godkar PB, Gordon RK, Ravindran A, Doctor BP (2006) Celastrus paniculatus seed oil and organic extracts attenuate hydrogen peroxide- and glutamate-induced injury in embryonic rat forebrain neuronal cells. Phytomedicine 13:29–36

Gohil KJ, Patel JA (2010) A review on Bacopa monniera: current research and future prospects. Int J Green Pharm 4:1–9

Gohil KJ, Patel JA, Gajjar AK (2010) Pharmacological review on Centella asiatica: a potential herbal cure-all. Indian J Pharm Sci 72:546–556

Gollen B, Mehla J, Gupta P (2018) Clitoria ternatea Linn: a herb with potential pharmacological activities: future prospects as therapeutic herbal medicine. J Pharmacol Rep 3:141–149

Gray NE, Sampath H, Zweig JA, Quinn JF, Soumyanath A (2015) Centella asiatica attenuates amyloid-β-induced oxidative stress and mitochondrial dysfunction. J Alzheimers Dis 45:933–946

Gray NE et al (2017) Centella asiatica: Phytochemistry and mechanism of neuroprotection and cognitive enhancement. Phytochem Rev 17:161–194

Gupta A, HemRaj, Singh KM, Upmanyu N (2013) Comparative evaluation of ethanolic extracts of Bacopa monnieri, Evolvulus alsinoides, Tinospora cordifolia and their combinations on cognitive functions in rats. Curr Aging Sci 6:239–243

Gupta SK, Dua A, Vohra BP (2003) Withania somnifera (ashwagandha) attenuates antioxidant defense in aged spinal cord and inhibits copper induced lipid peroxidation and protein oxidative modifications. Drug Metabol Drug Interact 19:211–222

Hasanein P (2011) Glabridin as a major active isoflavan from Glycyrrhiza glabra (licorice) reverses learning and memory deficits in diabetic rats. Acta Physiol Hung 98:221–230

Hosseinzadeh H, Nassiri Asl M, Parvardeh S (2005) The effects of carbenoxolone, a semisynthetic derivative of glycyrrhizinic acid, on peripheral and central ischemia-reperfusion injuries in the skeletal muscle and hippocampus of rats. Phytomedicine 12:632–637

Howes MJ, Fang R, Houghton PJ (2017) Effect of Chinese herbal medicine on Alzheimer's disease. Int J Neurobiol 135:29–56

Huang HC, Tang D, Xu K, Jiang ZF (2014) Curcumin attenuates amyloid-β-induced tau hyperphosphorylation in human neuroblastoma SH-SY5Y cells involving PTEN/Akt/GSK-3β signaling pathway. J Recept Signal Transduct Res 34:26–37

Hussain G et al (2018) Role of plant derived alkaloids and their mechanism in neurodegenerative disorders. Int J Biol Sci 14:341–357

James JT, Dubery IA (2009) Pentacyclic triterpenoids from the medicinal herb, Centella asiatica (L.) urban. Molecules 14:3922–3941

Jayaprakasam B, Padmanabhan K, Nair MG (2010) Withanamides in Withania somnifera fruit protect PC-12 cells from beta-amyloid responsible for Alzheimer's disease. Phytother Res 24:859–863

Josea S, Sowmyaa S, Cinuab TA, Aleykuttya NA, Thomasc S, Souto EB (2014) Surface modified PLGA nanoparticles for brain targeting of bacoside-a. Eur J Pharm Sci 63:29–35
Joshi G, Kaur R (2016) Tinospora cordifolia: a phytopharmacological review. Int J Pharm Sci Res 7:890–897
Kaur R, Kaur H, Dhindsa AS (2013) Glycyrrhiza glabra: a phytopharmacological review. Int J Pharm Sci Res 4:2470–2477
Kizhakke PA, Olakkaran S, Antony A, Tilagul KS, Hunasanahally PG (2019) Convolvulus pluricaulis (Shankhapushpi) ameliorates human microtubule-associated protein tau (hMAPT) induced neurotoxicity in Alzheimer's disease Drosophila model. J Chem Neuroanat 95:115–122
Kuboyama T, Tohda C, Komatsu K (2006) Withanoside IV and its active metabolite, sominone, attenuate Aβ (25–35)-induced neurodegeneration. Eur J Neurosci 23:1417–1426
Kuboyama T, Tohda C, Komatsu K (2014) Effects of ashwagandha (roots of Withania somnifera) on neurodegenerative diseases. Biol Pharm Bull 37:892–897
Kulkarni R, Girish KJ, Kumar A (2012) Nootropic herbs (Medhya Rasayana) in Ayurveda: an update. Pharmacogn Rev 6:147–153
Kumar A, Dogra S, Prakash A (2009) Neuroprotective effects of Centella asiatica against intracerebroventricular colchicine-induced cognitive impairment and oxidative stress. Int J Alzheimers Dis. https://doi.org/10.4061/2009/972178
Kumar H et al (2012) Cognitive enhancing effects of alpha asarone in amnesic mice by influencing cholinergic and antioxidant defense mechanisms. Biosci Biotechnol Biochem 76:1518–1522
Kumar MH, Gupta YK (2002) Antioxidant property of Celastrus paniculatus Willd.: a possible mechanism in enhancing cognition. Phytomedicine 9:302–311
Kumar MHV, Gupta YK (2003) Effect of Centella asiatica on cognition and oxidative stress in an intracerebroventricular streptozotocin model of Alzheimer's disease in rats. Clin Exp Pharmacol Physiol 30:336–342
Kumar N, Sakhya SK (2013) Ethnopharmacological properties of Curcuma longa: a review. Intl J Pharm Sci Res 4:103–112
Lee MK et al (2000) Asiatic acid derivatives protect cultured cortical neurons from glutamate-induced excitotoxicity. Res Commun Mol Pathol Pharmacol 108:75–86
Lekha G, Kumar BP, Rao SN, Arockiasamy I, Mohan K (2010) Cognitive enhancement and neuroprotective effect of Celastrus paniculatus Willd. Seed oil (Jyothismati oil) on male wistar rats. J Pharm Sci Technol 2:130–138
Lim GP, Chu T, Yang F, Beech W, Frautschy SA, Cole GM (2001) The curry spice curcumin reduces oxidative damage and amyloid pathology in an Alzheimer transgenic mouse. J Neurosci 21:8370
Lokanathan Y, Omar N, Puzi NNA, Saim A, Idrus RH (2016) Recent updates in neuroprotective and neuroregenerative potential of Centella asiatica. Malays J Med Sci 23:4–14
Malik J, Choudhary S, Kumar P (2015) Protective effect of Convolvulus pluricaulis standardized extract and its fractions against 3-nitropropionic acid-induced neurotoxicity in rats. Pharm Biol 53:1448–1457
Malik J, Karan M, Vasisht K (2011) Nootropic, anxiolytic and CNS depressant studies on different plant sources of Shankhpushpi. Pharm Biol 49:1234–1242
Malik J, Karan M, Vasisht K (2016) Attenuating effect of bioactive coumarins from Convolvulus pluricaulis on scopolamine-induced amnesia in mice. Nat Prod Res 30:578–582
Malve HO, Raut SB, Marathe PA, Rege NN (2014) Effect of combination of Phyllanthus emblica, Tinospora cordifolia, and Ocimum sanctum on spatial learning and memory in rats. J Ayurveda Integr Med 5:209–215
Mathew M, Subramanian S (2012) Evaluation of the anti-amyloidogenic potential of nootropic herbal extracts in vitro. Int J Pharm Sci Res 3:4276–4280
Mathur D, Goyal K, Koul V, Anand A (2016) The molecular links of re-emerging therapy: a review of evidence of Brahmi (Bacopa monniera). Front Pharmacol 7:1–15
May BH et al (2016) Memory impairment, dementia, and Alzheimer's disease in classical and contemporary traditional Chinese medicine. J Altern Complement Med 22:695–705

Mehla J, Pahuja M, Gupta P, Dethe S, Agarwal A, Gupta YK (2013) Clitoria ternatea ameliorated the intracerebroventricularly injected streptozotocin induced cognitive impairment in rats: behavioral and biochemical evidence. Psychopharmacology 230:589–605

Mirjalili MH, Moyano E, Bonfill M, Cusido RM, Palazon J (2009) Steroidal lactones from Withania somnifera, an ancient plant for novel medicine. Molecules 14:2373–2393

Mishra S, Palanivelu K (2008) The effect of curcumin (turmeric) on Alzheimer's disease: an overview. Ann Indian Acad Neurol 11:13–19

Muhammad Ezzudin R, Rabeta MS (2018) A potential of Telang tree (Clitoria ternatea) in human health. Food Res 2:415–420

Mukherjee PK, Kumar V, Kumar NS, Heinrich M (2008) The Ayurvedic medicine Clitoria ternatea—from traditional use to scientific assessment. J Ethnopharmacol 120:291–301

Mukherjee PK, Kumar V, Mal M, Houghton P (2007a) Acorus calamus.: scientific validation of ayurvedic tradition from natural resources. Pharm Biol 45:651–666

Mukherjee PK, Kumar V, Mal M, Houghton PJ (2007b) In vitro acetylcholinesterase inhibitory activity of the essential oil from Acorus calamus and its main constituents. Planta Med 73:283–285

Nahata A, Patil UK, Dixit VK (2008) Effect of Convolvulus pluricaulis Chois. On learning behaviour and memory enhancement activity in rodents. Nat Prod Res 22:1472–1482

Naidu PS, Singh A, Kulkarni SK (2006) Effect of Withania somnifera root extract on reserpine-induced orofacial dyskinesia and cognitive dysfunction. Phytother Res 20:140–146

Nair A, Chattopadhyay D, Saha B (2019) Plant-derived immunomodulators. In: Khan MSA, IAaDC (eds) New look to phytomedicine, advancements in herbal products as novel drug leads. Academic Press, pp 435–499

Narlawar R et al (2008) Curcumin-derived pyrazoles and isoxazoles: Swiss army knives or blunt tools for Alzheimer's disease? ChemMedChem 3:165–172

Natarajan S, Shunmugiah KP, Kasi PD (2013) Plants traditionally used in age-related brain disorders (dementia): an ethanopharmacological survey. Pharm Biol 51:492–593

Oganisyan AO, Oganesyan KR, Minasyan SM (2005) Changes in succinate dehydrogenase activity in various parts of the brain during combined exposure to vibration and licorice root. Neurosci Behav Physiol 35:545–548

Orhan I, Aslan S, Kartal M, Şener B, Hüsnü Can Başer K (2008) Inhibitory effect of Turkish Rosmarinus officinalis L. on acetylcholinesterase and butyrylcholinesterase enzymes. Food Chem 108:663–668

Park SY, Kim DS (2002) Discovery of natural products from Curcuma longa that protect cells from beta-amyloid insult: a drug discovery effort against Alzheimer's disease. J Nat Prod 65:1227–1231

Pastorino G, Cornara L, Soares S, Rodrigues F, Oliveira MBPP (2018) Liquorice (Glycyrrhiza glabra): a phytochemical and pharmacological review. Phytother Res 32:2323–2339

Patil SP, Maki S, Khedkar SA, Rigby AC, Chan C (2010) Withanolide a and asiatic acid modulate multiple targets associated with amyloid-β precursor protein processing and amyloid-β protein clearance. J Nat Prod 73:1196–1202

Patterson C (2018) World Alzheimer report 2018 the state of the art of dementia research: new frontiers. Alzheimer's Disease International (ADI), London

Raghu KS et al (2017) Age dependent neuroprotective effects of medhya rasayana prepared from Clitoria ternatea Linn. In stress induced rat brain. J Ethnopharmacol 197:173–183

Rai KS, Murthy KD, Karanth KS, Nalini K, Rao MS, Srinivasan KK (2002) Clitoria ternatea root extract enhances acetylcholine content in rat hippocampus. Fitoterapia 73:685–689

Rai KS, Murthy KD, Rao MS, Karanth KS (2005) Altered dendritic arborization of amygdala neurons in young adult rats orally intubated with Clitorea ternatea aqueous root extract. Phytother Res 19:592–598

Rao SB, Chetana M, Devi PU (2005) Centella asiatica treatment during postnatal period enhances learning and memory in mice. Physiol Behav 86:449–457

Rather MA, Thenmozhi AJ, Manivasagam T, Saravanababu C, Guillemin GJ, Essa MM (2019) Asiatic acid attenuated aluminum chloride-induced tau pathology, oxidative stress and apoptosis via AKT/GSK-3B signaling pathway in wistar rats. Neurotox Res 35:955–968

Rubinsztein DC (2006) The roles of intracellular protein-degradation pathways in neurodegeneration. Nature 443:780–786

Russo A, Borrellib F, Campisia A, Acquavivaa R, Racitia G, Vanellaa A (2003) Nitric oxide-related toxicity in cultured astrocytes: effect of Bacopa monniera. Life Sci 73:1517–1526

Sehgal N et al (2012) Withania somnifera reverses Alzheimer's disease pathology by enhancing low-density lipoprotein receptor-related protein in liver. Proc Natl Acad Sci U S A 109:3510–3515

Shashank D, Rajendra SV, Mistry A (2017) An overview of phytoconstituents and pharmacological activities of Celastrus paniculatus Willd. J Pharm Res 16:307–313

Shen Y, Chen BL, Zhang QX, Zheng YZ, Fu Q (2019) Traditional uses, secondary metabolites, and pharmacology of Celastrus species—a review. J Ethnopharmacol 241. https://doi.org/10.1016/j.jep.2019.111934

Shinomol GK, Muralidhara (2011) Bacopa monnieri modulates endogenous cytoplasmic and mitochondrial oxidative markers in prepubertal mice brain. Phytomedicine 18:317–326

Siddiqui BS, Aslam H, Ali ST, Khan S, Begum S (2007) Chemical constituents of Centella asiatica. J Asian Nat Prod Res 9:407–414

Singh SS, Pandey SC, Srivastava S, Gupta VS, Patro B, Ghosh AC (2003) Chemistry and medicinal properties of Tinospora cordifolia (guduchi). Indian J Pharmacol 35:83–91

Soumyanath A, Zhong YP, Gold SA, Yu X, Koop DR, Bourdette D, Gold BG (2005) Centella asiatica accelerates nerve regeneration upon oral administration and contains multiple active fractions increasing neurite elongation in-vitro. J Pharm Pharmacol 57:1221–1229

Subamalani S et al (2019) Effect of Acorus calamus Linn on histomorphometric changes in the CA1 and CA3 regions of hippocampus in wistar albino rats. Res J Pharm Technol 12:3531–3536

Taranalli AD, Cheeramkuzhy TC (2000) Influence of Clitoria ternatea extracts on memory and central cholinergic activity in rats. Pharm Biol 38:51–56

Uabundit N, Wattanathorn J, Mucimapura S, Ingkaninan K (2010) Cognitive enhancement and neuroprotective effects of Bacopa monnieri in Alzheimer's disease model. J Ethnopharmacol 127:26–31

Ulep MG, Saraon SK, McLea S (2018) Alzheimer disease. J Nurse Pract 14:129–135

Upadhyay AK, Kumar K, Kumar A, Mishra HS (2010) Tinospora cordifolia (Willd.) Hook. f. and Thoms. (Guduchi)—validation of the Ayurvedic pharmacology through experimental and clinical studies. Int J Ayurveda Res 1:112–121

Vollala VR, Upadhya S, Nayak S (2011) Learning and memory-enhancing effect of Bacopa monniera in neonatal rats. Bratisl Lek Listy 112:663–669

Winters M (2006) Ancient medicine, modern use: Withania somnifera and its potential role in integrative oncology. Altern Med Rev 11:269–277

Xu Y, Cao Z, Khan I, Luo Y (2008) Gotu Kola (Centella asiatica) extract enhances phosphorylation of cyclic AMP response element binding protein in neuroblastoma cells expressing amyloid beta peptide. J Alzheimers Dis 13:341–349

Yang F et al (2005) Curcumin inhibits formation of amyloid beta oligomers and fibrils, binds plaques, and reduces amyloid in vivo. J Biol Chem 280:5892–5901

Zhan C, Yang J (2006) Protective effects of isoliquiritigenin in transient middle cerebral artery occlusion-induced focal cerebral ischemia in rats. Pharmacol Res 53:303–309

Correction to: Natural Excipients in Pharmaceutical Formulations

Pradeep Singh, Garima Mishra, and Subas Chandra Dinda

Correction to:
Chapter 40 in: S. C. Mandal et al. (eds.), ***Evidence Based Validation of Traditional Medicines,***
https://doi.org/10.1007/978-981-15-8127-4_40

The original version of this chapter was inadvertently published with dual affiliations for Dr. Dinda which has been corrected now as follows:

S. C. Dinda
Department of Pharmaceutics, College of Pharmacy, Teerthanker Mahaveer University, Moradabad, UP, India

The updated online version of this chapter can be found at
https://doi.org/10.1007/978-981-15-8127-4_40

S. C. Mandal et al. (eds.), *Evidence Based Validation of Traditional Medicines*,
https://doi.org/10.1007/978-981-15-8127-4_51

GPSR Compliance
The European Union's (EU) General Product Safety Regulation (GPSR) is a set of rules that requires consumer products to be safe and our obligations to ensure this.

If you have any concerns about our products, you can contact us on

ProductSafety@springernature.com

In case Publisher is established outside the EU, the EU authorized representative is:

Springer Nature Customer Service Center GmbH
Europaplatz 3
69115 Heidelberg, Germany

www.ingramcontent.com/pod-product-compliance
Ingram Content Group UK Ltd.
Pitfield, Milton Keynes, MK11 3LW, UK
UKHW020943200826
474603UK00003B/51
* 9 7 8 9 8 1 1 5 8 1 2 6 7 *